AF565056

HYDRAULIC MACHINERY AND CAVITATION

VOLUME I

HYDRAULIC MACHINERY AND CAVITATION

Proceedings of the XVIII IAHR Symposium on Hydraulic Machinery and Cavitation

VOLUME I

Edited by

E. CABRERA, V. ESPERT and F. MARTÍNEZ
Polytechnic University of Valencia, Spain

Polytechnic University of Valencia
Hydraulic and Environmental Eng. Dept.
FLUID MECHANICS GROUP

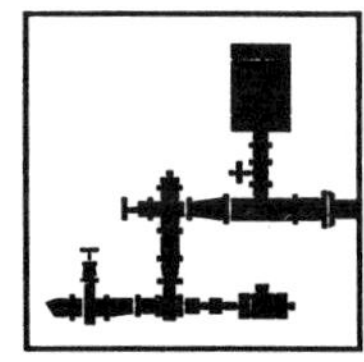

Co-sponsors:

KLUWER ACADEMIC PUBLISHERS
DORDRECHT / BOSTON / LONDON

A C.I.P. Catalogue record for this book is available from the Library of Congress.

ISBN 0-7923-4208-9 (Volume I)
ISBN 0-7923-4209-7 (Volume II)
ISBN 0-7923-4210-0 (Set of 2 Volumes)

Published by Kluwer Academic Publishers,
P.O. Box 17, 3300 AA Dordrecht, The Netherlands.

Kluwer Academic Publishers incorporates
the publishing programmes of
D. Reidel, Martinus Nijhoff, Dr W. Junk and MTP Press.

Sold and distributed in the U.S.A. and Canada
by Kluwer Academic Publishers,
101 Philip Drive, Norwell, MA 02061, U.S.A.

In all other countries, sold and distributed
by Kluwer Academic Publishers Group,
P.O. Box 322, 3300 AH Dordrecht, The Netherlands.

Printed on acid-free paper

Printed in the Netherlands

TABLE OF CONTENTS

Acknowledgements xvii
Preface xix
Foreword xxi

VOLUME I

INVITED LECTURES

The hydroelectricity in the world-Present and future.
Brekke, H. 3

Analysis of transients caused by hydraulic machinery.
Chaudhry, M.H. 17

Some present trends in hydraulic machinery research.
Fanelli, M. 23

Rapid prototyping of hydraulic machinery.
Schilling R., Riedel, N., Bader, R., Ascherbrenner, T., Weber, Ch., Fernandez, A. 40

Fluid transients in flexible piping systems.
A perspective on recent developments
Wiggert, D.C. 58

SYMPOSIUM CONTRIBUTIONS

GROUP 1 HYDRAULIC TURBINES. ANALYSIS AND DESIGN

1. 1 NUMERICAL ANALYSIS OF COMPONENTS

A decision aid system for hydraulic power station refurbishment procedure.
Bellet, L., Parkinson, E. Avellan, F., Cousot, T., Laperrousaz, E. 71

A three dimensional spiral casing Navier-Stokes flow simulation.
Bruttin, C.H., Kueny, J.L., Boyer, B., Héon, K,. Vu, T.C. and Parkinson, E. 81

Tip clearance flow in turbomachines- experimental flow analysis.
Ciocan, G.D. and Kueny, J.L. 91

Numerical prediction of hydraulic losses in the spiral casing of a Francis turbine.
Drtina, P. and Sebestyen, A. 101

Improvements of a graphical method for calculation of flow on a Pelton bucket.
Hana, M. 111

Flow behavior and performance of draft tubes for bulb turbines.
Kanemoto, T.,Uno, M., Nemoto, M., and Kashiwabara, T. 120

Performance analysis of draft tube for GAMM francis turbine.
Kubota, T., Han, F., and Avellan, F. 130

Modelling complex draft-tube flows using near-wall turbulence closures.
Ventikos, Y., Sotiropoulos, F. and Patel, V.C. 140

1.2 NUMERICAL ANALYSIS AND DESIGN OF HYDRAULIC MACHINERY

Fluid flow interactions in hydraulic machinery.
Aschenbrenner, T., Riedel, N. and Schilling, R. 150

Numerical optimization of high head pump-turbines
Buchmaicher, H,. Quaschnowitz, B., Moser, W. and Klemm, D. 160

Numerical hill chart prediction by means of CFD stage simulation for a complete Francis turbine.
Keck, H. Drtina, P. and Sick, M. 170

Development of a new generation of high head pump-turbines, Guangzhou II.
Klemm, D., Jaeger, E.U., and Hauff, C. 180

Development of integrated CAE tools for design assessment and analysis of hydraulic turbines.
Massé, B,. Pastorel, H. and Magnan, R. 190

Design and analysis of a two stage pump turbine
Mazzouji, F. , Francois, M. Hebrard, F., Houdeline, B. and Bazin, D. 200

Study of high speed and high head reversible pump-turbine
Nakamura, T., Nishizawa, H., Yasuda, M., Suzuki, T. and Tanaka, H. 210

Analysis of the performance of a bulb turbine using 3-D viscous numerical techniques
Qian, Y., Suzuki, R. and Arakawa, C. 220

Analysis of the inlet reverse flows in a pump turbine using 3-D viscous numerical techniques
Qian, Y., Suzuki, R. and Arakawa, C. 230

Importance of interaction between turbine components in flow field simulation
Riedelbauch, S,. Klemm, D. and Hauff, C. 238

From components to complete turbine numerical simulation
Sabourin, M., Labrecque, Y. and de Henau, V. 248

Validation of a stage calculation in a Francis turbine
Sick M., Casey, M.V. and Galpin, P.F. 257

Simulation of flow through Francis turbine by LES method
Song, C.C.S., Xiangying, Ch., Ikohagi,T., Sato, J., Shinmei, K. and Tani, K. 267

Simulation of flow through pump-turbine
Song, C.C.S., Changsi, C., Ikohagi,T., Sato, J., Shinmei, K. and Tani, K. 277

1.3 LOSS ANALYSIS AND SCALE EFFECTS

The scale effect in Kaplan turbines. New relationships for the calculation of scalable and non-scalable hydraulic losses and of the coefficient V.
Anton, I.M. 284

Analysis of losses in hydraulic turbines.
Brekke, H. 294

Scaling-up head discharge characteristics from model to prototype.
Couston, M. and Philibert, R. 304

Recent development of studies on scale effect
Ida, T., Kubota,T. , Kurokawa, J. and Tanaka, H. 313

Prediction of scalable loss in Francis runners of different specific speed
Kitahora, T., Jurokawa, J., Matumoto, M. and Suzuki, R. 323

Scale effect of jet interference in multinozzle Pelton turbines
Nakanishi, Y. and Kubota, T. 333

Further development of step-up formula considering surface roughness
Nichtawitz, A. 342

Numerical simulations of jet in a Pelton turbine.
Nonoshita, T. , Matsumoto, Y., Kubota,T. and Ohashi, H. 352

An assessment of the loss distribution in Francis turbines
Suzuki, R., Qian, Y., Kitahora, T. and Kurokawa, J. 361

GROUP 2. HYDRAULIC PUMPS

Unsteady flow calculation in a centrifugal pump using a finite element method.
Bert, P.F., Combes, J.F. and Kueny, J.L. 371

Steady and unsteady flow pattern between stay and guide vanes in a pump-turbine
Ciocan, G., Kueny, J. L. and Mesquita, A.L.A. 381

Self-sustained oscillation of gas-liquid flow in a centrifugal pump with semi-open impeller
Kurokawa, J., Matsui, J., Takada, H. and Hirayama, T. 391

Measurements in the dynamic pressure field of the volute of a centrifugal pump
Parrondo, J.L., Fernández, J., Santolaria, C. and González, J. 401

Functional modelling of pump volute geometry
Thackray, P.R. and James, R.D. 411

Analysis of flow measurements in the impeller and vaned diffuser of a centrifugal pump operating at part load.
Toussaint, M. and Hureau, F. 419

Influence of the blade roughness on the hydraulic performance of a mixed-flow pump. A viscous analysis
Undreiner, S. and Dueymes, E. 428

Improvement of performance of centrifugal pumps based on computational and theoretical methods and experimental design
Vinokurov, A.F., Volkov, A.V., Morgunov, G.M. and Pankratov, S.N. 438

Liquid-particulate two-phase flow in centrifugal impeller by turbulent simulation
Wu, Y., Dai, J., Mei, Z., Oba, S. and Ikoagi, T. 445

GROUP 3. HYDRAULIC ELEMENTS. DYNAMIC CHARACTERIZATION AND HYDRAULIC BEHAVIOUR.

Application of the method of kinetic balance for flow passages forming.
Benisek, M., Cantrak, S., Ignatovic, B. and Pokrajac, D. 455

Dynamics of large hydrogenerators.
Brito, G.C., Weber, H.I. and Fuerst, A.G.A. 464

Instabilities in a flow-control valve.
Cigada, A., Guadagnini, A. and Orsi, E. 474

Study of stayvane vibration by hydroelastic model.
Deniau, J.L. 484

Study of dynamic behaviour of non-return valves.
François, P. 494

Optimum hydraulic design of two-way inlet conduit of Wangyuhe pumping station
Linguang, L., Jiren, Z. and Rentian, Z. 504

Flow analysis for the intake of low-head hydro power plants
Ruprecht, A., Maihofer, M. and Gode, E. 514

GROUP 4. CAVITATION AND SAND EROSION

Efficiency alteration of Francis turbines by travelling bubble cavitation.
Arn, Ch., Dupont, Ph. and Avellan, F. 524

Cavitation erosion prediction on Francis turbines - Part 1. Measurements on the prototype.
Bourdon, P., Pfarhat, M., Simoneau, R., Pereira, F., Dupont, P., Avellan, F. and Dorey, J.M. 534

Determination of critical cavitation limit in the pressure control devices.
Castorani, A., De Martino, G. and Fratino, U. 544

Stability of air cavities in tip vortices.
Crespo, A., Castro, F., Manuel, F. and Fruman, D.H. 554

Cavitation erosion prediction on Francis turbines - Part 3 Methodologies of prediction.
Dorey, J.M., Laperrousaz, E., Avellan, F., Dupont, P., Simoneau, R. and Bourdon, P. 564

Cavitation erosion prediction on Francis turbines - Part 2. Model test and flow analysis.
Dupont, P.H., Caron, J.F., Avellan, F., Bourdon, P., Lavigne, M., Farhat, M., Simoneau, R., Dorey, J.M., Archer, A., Laperrousaz, E. and Couston, M. 574

Impact of vapour production and cavity dynamics on the estimation of thermal effects in cavitation.
Fruman, D.H., Reboud, J.L. and Stutz, B. 584

Aireation versus cavitation in dam spillways: self-aeration and artificial aeration (aerators).
Gutiérrez , R. 594

Leading edge cavitation in a centrifugal pump: Numerical predictions compared with model tests
Hirschi, R., Dupont, P.H., Avellan, F., Favre, J.N., Guelich, F., and Handloser, W. 604

The relation between erosion ripples on the wetted surface of hydraulic turbine and instability waves in the turbulent boundary layer
Huang, S. and Cheng, L. 614

Acoustic method and its applications on measuring and judging cavitation of hydraulic turbine.
Kehuang, L. and Chun, Y. 622

Numerical simulation for dilute sandy water flow in plane cascade
Liu, X.B., Zeng, Q.C. and Zhang, L.D. 632

Review of research on abrasion and cavitation of silt-laden flows through hydraulic turbines in China
Mei, Z. and Wu, Y. 641

Effect of the leading edge design on sheet cavitation around a blade section
Reboud, J.L., Rebattet, C. and Morel, P. 651

VOLUME II

GROUP 5. HYDRAULIC TRANSIENTS AND CONTROL SYSTEMS RELATED WITH HYDRAULIC MACHINERY AND PLANTS

Optimal closure of a valve for minimizing waterhammer.
Abreu, J., Cabrera, E., García-Serra, J. and López, P.A. 661

Qualitative flow visualizations during fast start-up of centrifugal pumps.
Barrand, J.P. and Picavet, A. 671

Analysis of a numerical model for the oscillatory properties of a Francis turbine group.
Cattanei, A., Capozza, A., and Molinaro, P. 681

Transients analysis and dynamic criteria for HPP exploatation
Gajic, A., Pejovic, S., Krsmanovic, L.J. and Stojanovic, Z. 691

Modelling a protection device in a low pressure lifting system.
Giustolisi, O. and Mastrorilli, M. 701

Dynamic compression of entrapped air pockets by elastic water columns.
Guarga, R., Acosta, A. and Lorenzo, E. 710

Generalization of pump station boundary condition in hydraulic transient simulation.
Izquierdo, J., Iglesias, P., Espert, V. and Fuertes, V. 720

Analysis of unsteady characteristics of flows through a centrifugal-pump impeller by an advanced vortex method.
Kamemoto, K., Kurasawa, H., Matsumoto, H. and Yokoi, Y. 729

Expert system for analysis of pumped storage schemes.
Koelle, E., Andrade, J.G.P. and Luvizotto Jr., E. 739

Model-based analysis of active PID-control of transient flow in hydraulic networks.
Lauria, J.C. and Koelle, E. 749

Simulation of transients in pressurized hydraulic systems with visual tools.
Martínez, F., Izquierdo, J., Pérez, R. and Vela, A. 759

Dynamic behaviour of governing turbines sharing the same electrical grid
Nielsen, T.K. 769

Prediction of natural frequencies in a hydro power plant supplying an electric network by itself having a known load type
Raabe, J. 779

Modelling and practical analysis of the transient overspeed effect of small Francis turbines.
Ramos, H. and Betamio, A. 789

Simulation of turbine governing in time domain.
Stuksrud, D.B. 799

Parametrical modelling of power characteristics of the Francis and Kaplan hydraulic turbines.
Tolea, M.F. and Kueny, J.L. 809

Unsteady frictions in pipelines.
Vennatro, R. 819

GROUP 6. OSCILLATORY AND VIBRATION PROBLEMS IN HYDRAULIC MACHINERY AND POWER STATIONS

Swirl flow in conical diffusers.
Dahlhaug, O.G. 827

Experimental investigation of vortex core in reverse swirl flow from Francis runner.
Furuie, Y., Mita, H. and Hosoi, Y. 835

Hydraulic oscillation analysis using the fluid-structure interaction model.
Gajic, A., Pejovic, S. and Stojanovic, Z. 845

Francis turbine surge: discussion and data base.
Jacob, T. and Prenat, J.E. 855

Non-stationary flow in reversible Francis turbine runner due to wakes trailing the guide vanes.
Jernsletten, J. 865

The swirling inlet flow effects on the pressure recovery of a low head water turbine draft tube.
Kikuyama, K., Hasegawa, Y., Augusto, G., Nishibori, K. and Nakamura, S. 875

Two kinds of whirl on fixed-blade propeller type turbine.
Léonard, F. 885

Self-excited hydraulic oscillations dued of unstable valve behaviour. A case study.
Mateos, C., Pérez-Andújar, T., Andreu, M. and Cabrera, E. 895

An experimental study on fins. Their role in control of the draft tube surging
Nishi, M., Wang, X.M., Yoshida, K., Takahashi, T. and Tsukamoto, T. 905

Model for vortex rope dynamics in Francis turbine outlet.
Pedrizzetti, G. and Angelico, G. 915

Vortices rotating in the vaneless space of a Kaplan turbine operating under off-cam high swirl flow conditions.
Pulpitel, L., Skotak, A. and Kontnik, J. 925

Experimental investigation of frequency characteristics of draft tube pressure pulsations for Francis turbines.
Qinghua, S. 935

On the suppression of coupled liquid/pipe vibrations.
Tijsseling, A.S. and Vardy, A.E. 945

Unsteady hydraulic force on an impeller due to rotor-stator interaction in a diffuser pump.
Tsukamoto, H., Uno, M., Qian, W., Teshima, T., Sakamoto, K., and Okamura, T. 955

Swirling flow with helical vortex core in a draft tube predicted by a vortex method.
Wang, X.M. and Nishi, M. 965

GROUP 7. EXPERIMENTAL INVESTIGATIONS RELATED WITH HYDRAULIC MACHINARY AND ITS APPLICATIONS

Friction loss of rough wall passage in a turbomachinery
Akaike, S. 975

Redesign of sharp heel draft tube- Results from tests in model and prototype.
Dahlbäck, N. 985

Model and prototype draft tube pressure pulsations
Kerkan V., Bajid, M. Djelic, V. Lipej, A. and Jost, D. 994

Fish bypass system impact upon turbine runner performance at Rocky Reach dam.
Lang, A. and Christman, B. 1004

LDV measurements in an impeller-generated turbulent jet developing in a new coflow.
Peterson, P., Larson, M and Jönsson, L. 1014

Inline radial force measurement of turbine runners
Riener, J., Egger, A. and Schnur, G. 1024

Different types and locations of part-load recirculations in centrifugal pumps found from LDV measurements.
Stoffel, B. and Weiss, K. 1034

Turbulent 3D flows near the impeller of a mixed-flow pump
Wang, B. and Hellman, D.H. 1044

GROUP 8 PRACTICAL APPLICATIONS OF THE HYDRAULIC MACHINERY

Overload in Kaplan turbines of Salto Grande hydropower complex.
Baccino, M. and Bonecarrere, E. 1053

Full sized tests on a french main coolant pump under two-phase flow.
Huchard, J.C., Bore, C. and Dueymes, E. 1063

Computer simulations of dynamic performance of 400 Mw adjustable speed pumped storage units.
Kita, E., Nakagawa, H., Kuwabara, T. and Harada, M. 1073

Performance comparison of nuclear reactor recirculation pumps tested under large reynolds number difference
Saiki, K., Ikura, T., Matsumoto, K., Komita, H., Kobayashi, M., Saito, T. and Tanaka, H. 1083

Performance of Candu heat transport pumps under two-phase flow conditions.
Samarasekera, H. and Kumar, A.N. 1093

Interdependence of draft tube and tailwater flow in bulb turbine power plants.
Schneider, C.H., Knapp, W. and Schilling, R. 1103

Study of hydraulic transients using the bond graphs method.
Tiago Filho, G.L. 1113

GROUP 9. MONITORING, PREDICTIVE MAINTENANCE AND REFURBISHMENT.

Numerical flow analysis of a Kaplan turbine.
Jost, D., Lipej, A., Oberdank, K., Jamnik, M. and Velensek, B. 1123

18-Paths acoustic flowmeter and transducer protrusion.
Lévesque, J.-M., Néron, J. and Tran, C.M. 1133

Step-up in rehabilitation: not a myth, a science
Mahé, B., de Henau, V. and Sabourin, M. 1142

Super system: a hydroelectric unit condition monitoring system in operation at Hydro-Quebec.
Mossoba, Y. 1152

Application of rotor response analysis to fault detection in hydro powerplants.
Nascimento, L.P. and Egusquiza, E. 1162

The placement exploration of diagnostic measuring points on operating unit equipment.
Liu, X.T. 1172

ACKNOWLEDGEMENTS

There are many and happy reasons for the XVIII Symposium on Hydraulic Machinery to be held in Valencia (Spain). But, without any kind of doubt the most important one can be given a proper noun: Michele Fanelli. The start, November 1990. Professor Fanelli is looking for a place in Spain for the next meeting of the Work Group on "Hydraulic Transients with Water Column Separation" to be held in September 1991. Somebody suggests Valencia. And Professor Fanelli, who has heard about our U.D.M.F, makes a bid for it. It is the start of this XVIII Symposium. The Unidad Docente Mecánica de Fluidos will be, for ever, greatly indebted to Professor Fanelli, Miguel for his Spanish friends.

Secondly, we have to be very grateful to Prof. Brekke, to Prof. Schilling and to Dr. Dueymes for the impressive work they have done for the Symposium and for their wise advice given as members of the Executive Technical Committee. We have shared long but fruitful and pleasant meetings to prepare everything.

We cannot forget to mention, -and thank for their support-, our sponsors: Cedex, Iberdrola and Red Eléctrica de España, that from the very beginning envisioned the importance of hosting such an event would have for our country. We also have to give thanks to the Universidad Politécnica de Valencia for its unconditional support, basically embodied by our Rector and friend, Justo Nieto.

We must also acknowledge Kluwer Academic Publishers and especially their Publisher Petra van Steenbergen for her assistance, careful presentation and production of the book.

There is no doubt that the organization of an event like this Symposium is not possible without the contribution of a very important working group. So, last but not least, it is fair to underline the great dedication of the members of our U.D.M.F. Their work, -naturally inconspicuous but highly dedicated-, has been indispensable. Amongst them, those working full-time within the Group are:

Angeles Alvarez
Miguel Andreu
Vicent Espert
Vicente Fuertes
Antonio Gallardo
Marta García
Jorge García-Serra
Francisco García
Pedro Iglesias
Joaquín Izquierdo
P. Amparo López
Gonzalo López
Javier Martínez
Fernando Martínez
Rafael Pérez
Jose V. Ribelles
Antonio F. Vela

On behalf of the Fluid Mechanics Group,
Enrique Cabrera.
Chairman of the XVIII IAHR Symposium
on Hydraulic Machinery and Cavitation.

PREFACE

These proceedings include the papers presented at the 18th Symposium on Hydraulic Machinery and Cavitation held in Valencia - Spain from September 16 - 19, 1996. The scope of the symposium covers a wide range of important hydraulic machinery problems. Following the first announcement more than 140 abstracts were submitted to the symposium and screened by the Technical Committee. 128 abstracts were accepted and 120 final papers were received and reviewed by the committee. In the end, 114 technical papers were accepted for inclusion in the proceedings. Together with 5 invited papers they were grouped in the following 9 sections:

1 Hydraulic Turbines, Analysis and Design.
2 Hydraulic Pumps
3 Hydraulic Elements. Dynamic Characterization and Hydraulic Behavior.
4 Cavitation and Sand Erosion.
5 Hydraulic Transients and Control Systems
6 Oscillation and Vibration Problems in Hydraulic Machinery and Power Stations.
7 Experimental Investigations Related with Hydraulic Machinery and its Applications.
8 Practical Applications of the Hydraulic Machinery.
9 Monitoring, Predictive Maintenance and Refurbishment.

Compared with the last symposium held in Beijing, China 1994, the contributions in the field of numerical flow analysis and performance predictions have increased while the number of papers dealing with experimental investigations have decreased. The high level reached in this field is demonstrated by various papers in which not only the real flow within an isolated component but also the interaction of the flow between rotor and stator of hydraulic machinery is considered thus yielding a more realistic performance prediction. Some technical papers deal with the simulation of the flow through the whole hydraulic machinery and one presents the first application of a Large-Eddy Simulation in a Francis turbine. Despite of this progress in CFD applications and turbulence modelling, highly accurate flow measurements are still indispensible; they are still needed for a better understanding of complex flow phenomena and testing of numerical models and hence for CFD software validation. Thus the experimental flow analysis remains an important section in the scope of this symposium. The 18. symposium was held at the Polytechnic University of Valencia and hosted by the Fluid Mechanics Group. I would like to thank the Local Organization Committee especially its Chairman, Prof. E. Cabrera, for having taken over the task to organize this symposium, the Technical Committees for their help making this conference a success and the sponsors: CEDEX, IBERDROLA, RED ELECTRICA de ESPAÑA for their kind support and cooperation. Finally, I express my sincere appreciation for the cooperation with Kluwer Publishers, Dordrecht - The Netherlands, in the preparation of the proceedings.

R. Schilling
Chairman of the IAHR,
Section on Hydraulic Machinery and Cavitation.

FOREWORD

Water is for the future like food is for the human being. And so, even leaving out that food and humans are essentially water. Water is linked to the future like tears (that are sacred water) are linked to our eyes. For this reason, a civilized society devotes science, technology, resources and thinking to optimize and cherish, to rationalize and democratize water use, possibilities and circumstances. Thus, when a SYMPOSIUM ON HYDRAULIC MACHINERY AND CAVITATION gathers around three hundred experts, scientists and professionals all around the world to debate and offer to the society the results of their research, experiences and endeavors, we have to be proud and grateful for the event.

It goes without saying that only the work of many allows this kind of events. But, let me personalize my congratulations and merit the Symposium success to the man-tree-of-water, Professor Cabrera. With steady, controlled and, at the same time, inconspicuous work, he achieves, like good sportsmen, the highest goals.

Justo Nieto
Chancellor
Valencia Polytechnic University

INVITED LECTURES

THE HYDROELECTRICITY IN THE WORLD - PRESENT AND FUTURE

PROFESSOR HERMOD BREKKE
Norwegian University of Science and Technology
Division of Thermal Energy and Hydro power

1. Introduction

Only about 15% of the exploitable hydro power potential in the world has so far been utilized. However, in the industrialized countries in Europe and North America the majority of the hydro power has been developed for electricity production.

In some western countries however, more hydro power is available for production of clean renewable energy, but so far the collaboration with the environmentalists has not been organized to develop the potential even in not protected rivers.

The history of hydropower development for electricity production started in the 19th, century. As an example I may use the development of hydroelectric power in Norway where 99.8% of the electricity production comes from hydro power today. The first known hydro electric power station was built at a farm in 1877 producing approximately 200 Watts at Lisleby Brug in southern Norway. Later in 1890 the world's most northern town, Hammerfest, at 70° north in Northern Norway got electricity supply for street lightening driven by a hydro electric power plant. The street lightening was important because of 24 hours nights in winter time.

Up to the 20th century the electricity production for industry and private consumption in Norway increased to approximately 100 MW in total. The largest power station was Hammeren in Oslo with 1,800 kW and Hafslund power station located south east of Oslo with 5,200 kW. It is also of interest to know that 4 vertical Francis 1,600 kW turbines in another power station built in 1905, are still in operation today during the flood season.

However, the first boom in developing hydro power in Norway came in 1907-1916 when the development of electrochemical and metallurgical industry started. The two Norwegian engineers, Birkeland and Eyde, had in 1908 started the nitrogen fertilizer production by an electro chemical process producing 25,000 tons of "Norges salpeter" containing 13% nitrogen. These men planned to produce 300,000 tons per year and the demand for hydro electric power was estimated to be 360 MW. With a total installation of 200 MW in Norway at that time, this industrial demand was a great challenge. To meet this challenge Vemork power plant was built and put in operation in 1910 and then Saaheim power plant was built and put in operation in 1916.

In addition to the electrochemical industry, the mechanical industry grew up to meet the demand for turbo machinery for hydro power production in general.

E. Cabrera et al. (eds.), Hydraulic Machinery and Cavitation, 3–16.

In 1955 the total capacity of Norwegian hydro electric power production was 6,000 MW. Up to 1990 the electro hydraulic power production increased to nearly 27,000 MW and 105 TWh annual energy production.

The development of hydropower in Norway is only an example of the world wide development. In table I are listed the hydroelectric power production of the largest hydropower producing countries in 1992.

COUNTRY	Hydropower production		Installed capacity	
	TWh	Ann. Change %	MW	Ann. Change %
Canada	316.5	+1.9	61,668	+2.1
USA	248.9	-2.2	96,038	+2.0
Brazil	223.4	+4.7	47,709	+3.8
Russia	172.6	-	43,136	-
China	132.5	+5.9	42,000	+6.3
Norway	117.1	+2.4	27,030	+2.0
Japan	89.6	+0.6	39,535	+1.7
Sweden	74.9	+3.1	16,380	+0.7
France	72.5	+0.4	24,857	+1.7
India	69.8	+3.7	19,573	+4.1

Table 1. Hydroelectric power production and installed capacity, 1992.

In the table the growth in hydro is also shown and it is quite clear that the increase of hydropower is moderate in western countries and in USA even a decrease in production was recorded in 1992. On the contrary an increased growth is recorded in the far east and in particular in China where huge resources are available.

It is the main task for the hydraulic engineers to propose hydro power projects with an acceptable environmental profile. With a collaboration between environmental experts and engineers the huge remaining potentials worldwide can be utilized and the growth of CO_2 production can be reduced..

2. Development of mechanical equipment for hydropower production in the past

Water power has been utilized by different types of water wheels back to the ancient Rome. In 398 A.D. the first public water wheel was erected in the ancient Rome. However, the first report of the existence of water wheels is given by Philon of Byzantium, who was a technician of the 3rd century B.C. [Ref. 1]. Later water wheels have been used for different purposes, but were most commonly used in flour mills.

Water wheels were also used in flour mills in Scandinavia from about 1150 A.D. i.e. after the Viking period. The water wheels did not utilize the reaction force of the water flow and the efficiency was low.

2.1. REACTION TURBINES

The first documented description of a reaction water wheel has been found in Segner's water wheel and the Barker reaction wheel which was developed in Germany and England around 1750 (see fig. 1).

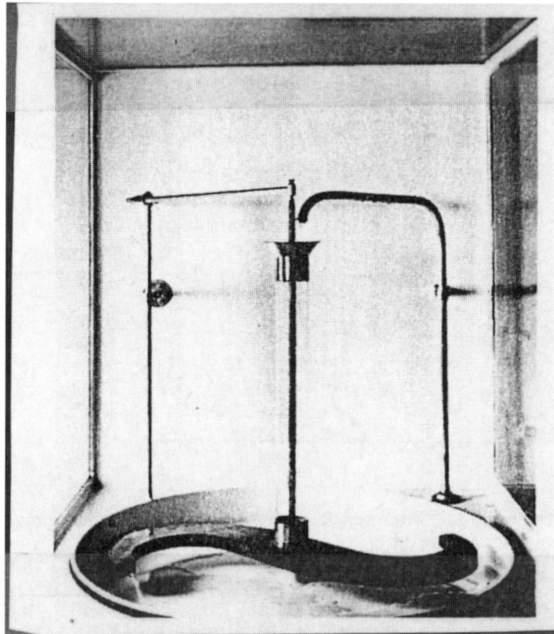

Figure 1. Segner's reaction water wheel of 1770.

In 1754 Leonard Euler from Basel presented his turbine theory in Berlin after a stay of 14 years in St. Petersburg until 1744. Based on the improved theory an improvement of the Segner's water wheel was made by introducing a swirl flow at the inlet (see fig. 2).

The first effective turbine for industrial use was developed by Benoit Fourneyron from France in 1826. The essential feature of this turbine type was a stationary cascade in a movable cylinder gate at the inlet of a radial runner (see fig. 3).

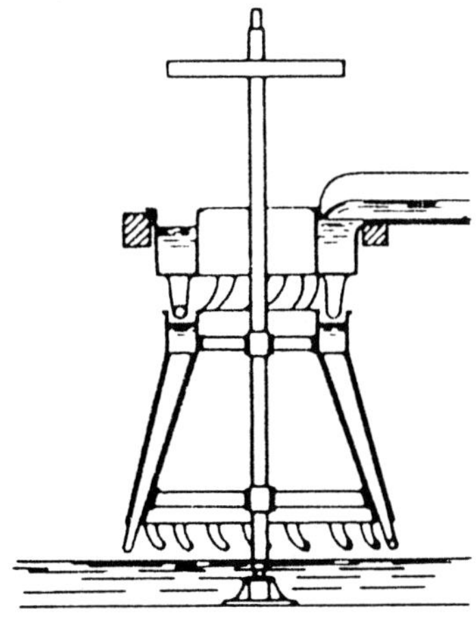

Figure 2. Euler's turbine from 1754. From Raabe: *Geschichte und derzeitiger Stand der Wasserkraftmaschine.*

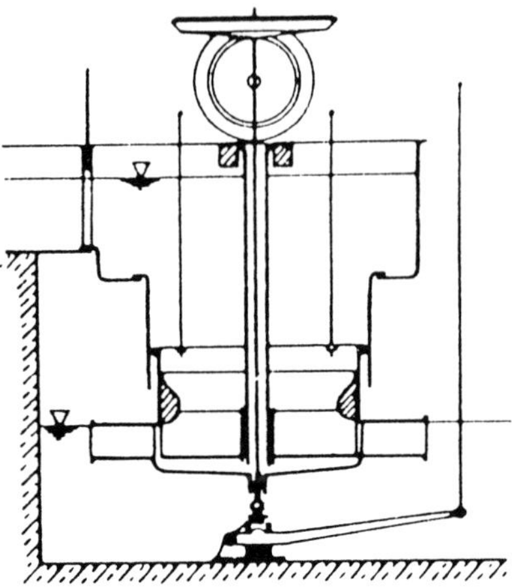

Figure 3. Fourneyron turbine in 1827. From Raabe: *Geschichte und derzeitiger Stand der Wasserkraftmaschinen.*

By means of the movable cascade cylinder the turbine output could be controlled, but the inlet losses at part load were very high due to separation. A further development of reaction turbines was patented in USA by S. Hovd in 1836 with the development of a centripetal reaction turbine. The British engineer, James B. Francis, worked on a further development of the patented Hovd turbine by introducing movable guide vanes which improved the efficiency and reduced the noise. The name of Francis has been linked to this turbine later during its development up to present time. At present time high head Francis turbines have been built for 700 m heads and output up to 750 MW.

An other main type of reaction turbines was patented by the Czechoslovakian-Austrian professor, Victor Kaplan as late as in 1913. This well known Kaplan turbine with axial propeller like runner with movable blades was for the first time successfully installed in the Austrian power plant Velm. Later problems with cavitation caused problems until the Swedish company Karlstad Mekaniska Werkstad built a cavitation test rig for model turbine testing.

Based on tests from this laboratory a Kaplan turbine with runner diameter 5.8 m for Lilla Edet power plant was successfully commissioned in 1925 (see fig. 4). From experience of these tests modern Kaplan turbines have been developed over the whole world. However, the development of the oil hydraulic servomotor and leverage system located in the runner hub for the turbine Lilla Edet still stands as a basic design for large modern Kaplan turbines.

Today Kaplan turbines are used up to 70 m head and is chosen mainly because of a flat efficiency curve compared with a Francis turbine.

Later the horizontal Kaplan bulb turbines with conical arrangement of the guide vanes were developed. This type of turbines are often located in the dam for heads up to approximately 15 m. The reason for this is that a Bulb turbine requires less space and less depth than a vertical Kaplan turbine.

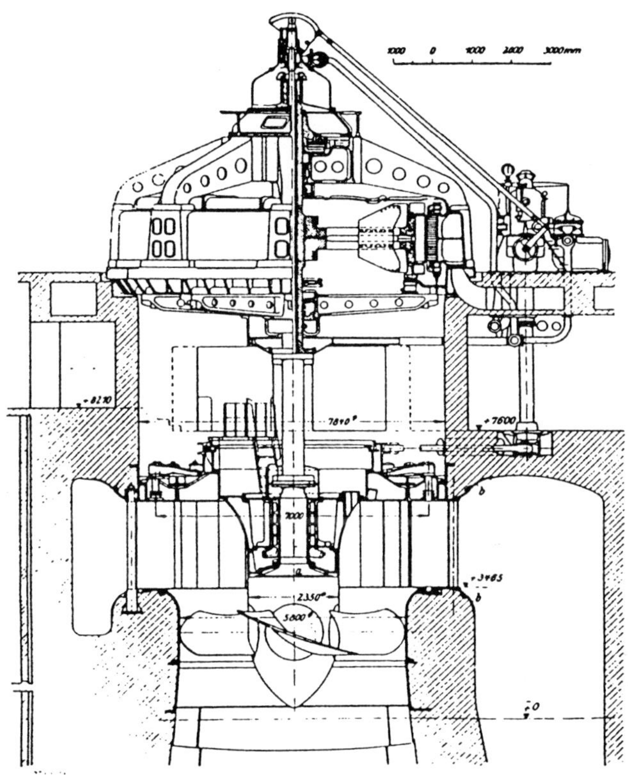

Figure 4. Cross section of the Kaplan turbine for Lilla Edet power station in Sweden.

2.2. IMPULSE TURBINES

The dominating impulse turbine used world wide is the Pelton turbine. This turbine type was patented by Lester Allen Pelton in 1880 in USA. The development of the bucket shape is illustrated in fig. 5.

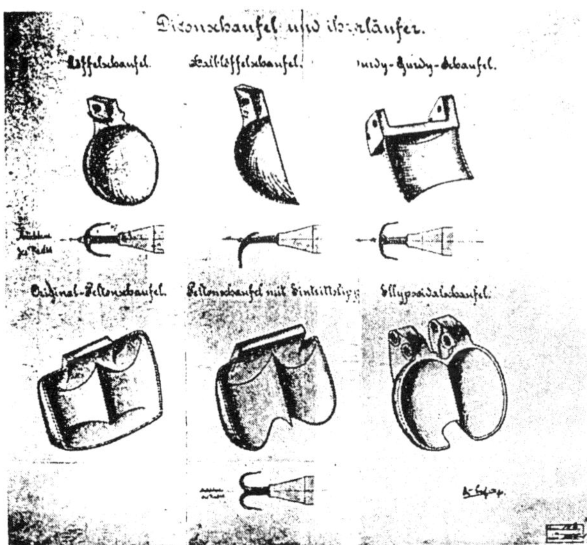

Figure 5. Schematic drawings of the Pelton bucket (Raabe).

The development of this turbine type included in the first time single and twin jet horizontal shaft turbines. Later the vertical types were developed with a number of jets up to 6 and output up to 315 MW. Within 2 years turbines with 400 MW for 1882 m net head will be in operation in Switzerland.

2.3. REVERSIBLE PUMP TURBINES

The last turbine type developed after world war II is the reversible pump turbines. This type of turbine was developed in the fifties and has been designed for increasing head especially in the industrialized part of the world for peak load operation. A diagram of the development of commissioned and planned reversible pump turbine projects up to 1985 is shown in fig. 6. This diagram was presented during the IAHR Symposium in Tokyo 1980, by Dr. Hitoshi Muray [Ref. 2].

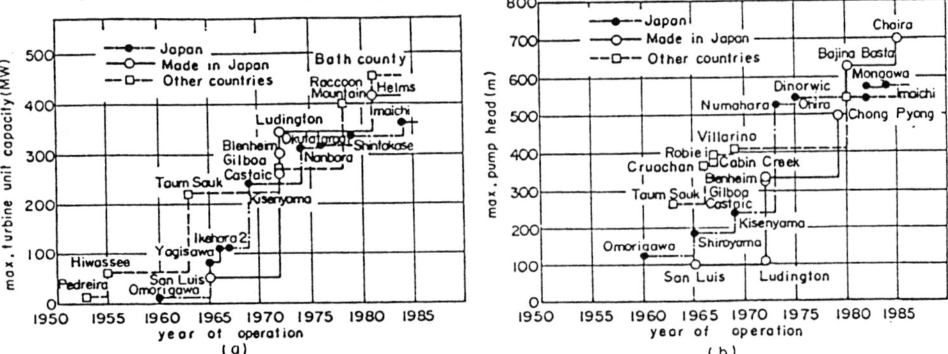

Figure 6. Trends in (a) unit capacity and (b) total head of pump turbines.

2.4. STATE OF THE ART IN DEVELOPMENT

As the present state of the art of Francis turbines, Itaipu in Brazil/Uruguay with 740 MW for 118.4 m, could be mentioned as the largest Francis turbines. For Pelton turbines Sy Sima in Norway with 315 MW for 885 m head is the largest. For Kaplan turbines the turbine for Ligga III in Sweden with 182 MW for 39 m head could be mentioned as one of the largest turbines of this type.

3. Development of hydraulic machines for electricity production in the future

In general all turbine types will be developed for increased output and head for a given specific speed. An increase in the efficiency by 0.5-1.0% may be expected. Simplified design of small hydro turbines for utilization of non-regulated small rivers will also be a challenge in order to give a contribution to the production of clean renewable energy for electricity production.

In the following a brief description of the design challenges for future turbines will be given. The initial data for a turbine design will be the flow (Q) and the range of operating head (H). From this a decision will be made, based on experience from the manufacturer and consultants. In the future, border lines will be broken concerning specific speed versus head and range of operation.

In order to do so it is important to establish an analytical approach to the problem besides the available CFD tool which is improving very fast. I believe that it is important to be able to make an initial design for a complete new runner by means of an analytical study of the different geometry parameters. This is because the requirements for new turbines may be outside the available data base for runner design based on the experience of the turbine manufacturer.

Such initial analytical parameter design must of course be analysed by modern CFD codes and tested in laboratory for the fine tuning of blades.

The main problem at present time is that young engineers may have been "too computerized" and have too little geometry feeling and decreased physical thinking. The engineer's education does not include creative drawing board training. This training in three-dimensional shaping without a computerized geometri shaping aid will always be valuable for a turbine runner designer.

In order to clarify and explain the meaning of an analytical preliminary design a brief description is given for the use of some parameters.

3.1. THE INFLUENCE FROM THE SPEED NUMBER $^*\Omega$ ON CAVITATION

The speed number is defined as $^*\Omega = {^*\omega}\ {^*Q}^{1/2}/(2g\ {^*H})^{3/4}$ based on the best efficiency point (BEP, denoted *). The relation between the defined speed number and the specific speed is:

$$n_s = {^*\Omega} \cdot K^{1/2} (2g)^{3/4} 30/\pi, \qquad \text{where } K = Q_n/{^*Q}.$$

The speed number $^*\Omega$ is a dimensionless parameter which is more convenient to be used instead of the specific speed in this lecturer. The speed number may also be expressed as a function of the circumferential speed of the blade outlet and the blade angle:

$$^*\Omega = \sqrt{\pi}\ \underline{u}_2^{3/2}\sqrt{\tan(\pi-\beta_2)} \tag{1}$$

where $\underline{u}_2 = u_2/\sqrt{2gH}$ which is the dimensionless outlet velocity and β_2 = blade angle in the stream way direction (see fig. 7).

From eq. (1) we obtain

$$\underline{u}_2 = \ {}^*\Omega^{3/2}/(\pi\tan(\pi-\beta_2))^{1/3}$$

From this equation we find that within the same speed number an increased diameter and circumferential speed may be compensated by a decreased outlet blade angle within a limited range of angles.

3.2. THE REACTION RATIO

Besides the specific speed the reaction ratio, which is a measure of the pressure drop through the runner, is a very important parameter for the blade loading and cavitation performance. The reaction ratio can be increased by changing the blade shape for a given speed number and thus inlet cavitation problems may be solved. The reaction ratio is defined by the pressure drop from runner inlet to outlet divided by the total available net head at best efficiency flow for $c_{u2} = 0$. By introducing the dimensionless expression for pressure $\underline{h} = h/H$ the reaction ratio yields:

$$\underline{h}_1 - \underline{h}_2 = \frac{h_1 - h_2}{H} \tag{3}$$

f friction losses are ignored we can establish the following equation by combining the Euler turbine equation, the absolute specific energy (E) and the relative specific stagnation energy (I) (*rothalpy*):

$$gH\eta_h = u_1 c_{u1} - u_2 c_{u2} = (E_1 - I_1) - (E_2 - I_2) \tag{4}$$

Because the relative stagnation energy is constant along a streamline through the runner if friction is ignored, we get $I_1 = I_2$ and then

$$gH\eta_h = u_1 c_{ul} - u_2 c_{u2} = E_1 - E_2 = (gh_1 + \frac{c_1^2}{2}) - (gh_2 + \frac{c_2^2}{2}) \quad (5)$$

Here the difference in geodetic height from inlet to outlet is ignored. When introducing reduced or dimensionless variables $\underline{c} = c/\sqrt{2gH} = \underline{c}_u^2 + \underline{c}_m^2$, remembering that $c_{u2} = 0$ at best efficiency point (BEP) and assuming $\underline{c}_{m1} = \underline{c}_{m2}$ i.e. constant meridional cross section through the runner we can establish eq. (6) by substituting for η_h by eq. (4) in eq. (5):

$$\frac{h_1 - h_2}{H} = \underline{h}_1 - \underline{h}_2 = 2\underline{u}_1 \underline{c}_{ul} - \underline{c}_{ul}^2 = \underline{u}_1^2 [2 \frac{\underline{c}_{ul}}{\underline{u}_1} - (\frac{\underline{c}_{ul}}{\underline{u}_1})^2] \quad (6)$$

By studying the vector diagrams for runners with increasing speed number and constant hydraulic efficiency $\eta_h = 2\underline{u}_1\underline{c}_{u1}$ as shown in fig. 7, we find increasing pressure difference from inlet to outlet when $\underline{u}_1$ and $\underline{u}_2$ are increased.

However, it must be emphasized that an increased speed number increases $\underline{u}_2$. Thus, decreased pressure at the outlet = $\underline{h}_2$ may also lead to a reduction of the inlet pressure $\underline{h}_1$ even if $\underline{h}_1$-$\underline{h}_2$ is increased, if the reduction in $\underline{h}_2$ is not compensated by an increased submergence.

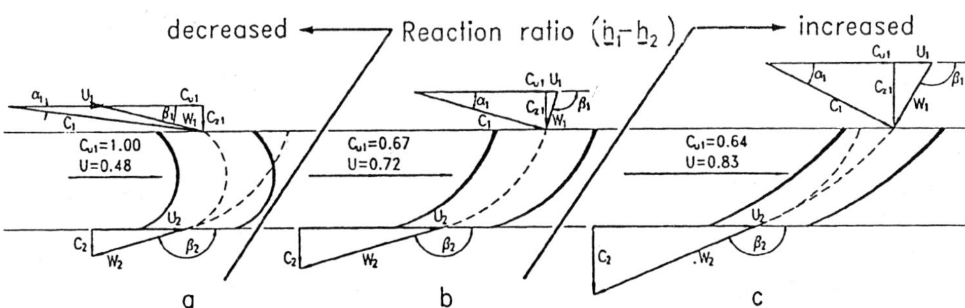

Figure 7. Blade cascades for different reaction ratios for c_{u2}=0.

Two examples illustrate the increase in the reaction ratio for $\eta_h = 2\underline{u}_1\underline{c}_{u1}$ = const. For a high head turbine $\underline{c}_{u1}$= 0.69 and $\underline{u}_1$ = 0.72: $\underline{h}_1 - \underline{h}_2 = 0.518$. For a low head turbine $\underline{c}_{u1}$ = 0.6 and $\underline{u}_1$ = 0.83: $\underline{h}_1 - \underline{h}_2 = 0.634$.

However, for a given reduced circumferential speed $\underline{u}_1$ a maximum pressure ratio is obtained for $\underline{c}_{u1}/\underline{u}_1$=1 i.e. β_1 = 90°. From this we find that the inlet pressure may be increased locally by bending the inlet of the blade towards β_1= 90°. The efficiency will also increase with decreasing β_1 if no rotation occurs at the outlet of the runner, but this is impossible with the given boundary conditions. However, local bending of the blade on the shroud side may also lead to a local negative blade lean which in turn leads to a low pressure zone near the blade inlet with cavitation problems.

3.3. THE INFLUENCE FROM THE MERIDIONAL CROSS SECTION OF CROWN AND SHROUD

During the initial stage of design work on a runner the choice of inlet and outlet angles have been proven to be of great importance. The meridional cross section of the runner i.e. crown and shroud shape will also be influenced by the blade angles.

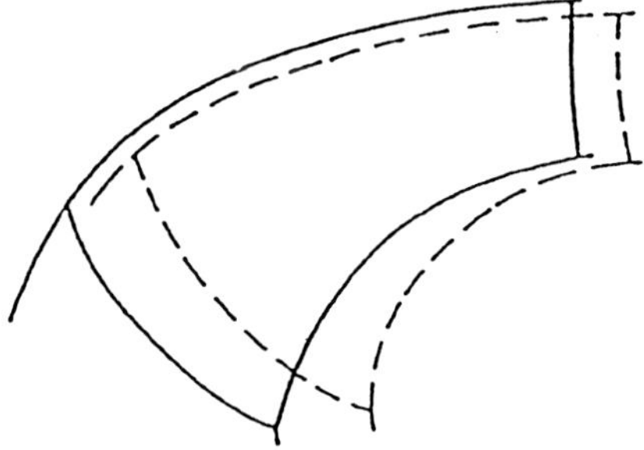

Figure 8. Different shapes of Francis runners for the same specific speed i.e. same speed, output and head.

For example, the blade outlet angle [= (π - β_2), see fig. 7] may have a variation range from 13° to 18° and the inlet angle [= (π - β_1)] may have a variation 50° to 70° for a medium head turbine. The meridional cross section may have a variation in shape and outlet diameters as shown schematically in fig. 8, for the same speed number of a high head turbine.

At best efficiency point of operation both a large outlet diameter and a small outlet diameter may operate stably. However, at off-design, with swirl flow in the draft tube, it will be more difficult to avoid large pressure surges for runners with large outlet diameters.

The advantage of a large outlet diameter will be that the required submergence to avoid cavitation of the turbine, is smaller according to eq. (2), but the disadvantage may be a more unstable operation at off-design head or at off-design flow.

3.4. THE INFLUENCE FROM BLADE LEAN

The blade lean (also called blade raking) is defined by the blade angle with the meridional plane normal to the streamlines (Θ) and other geometry parameters as illustrated in fig. 9. The dimensionless pressure gradient $d\underline{h}/dy$ = $d(h/H)/dy$ in a stream surface can be established by eq. (7) expressed by the dimensionless meridional velocity $\underline{c}_m = c_m/(2gH)^{0.5}$ and angular speed $\underline{\omega} = \omega/(2gH)^{0.5}$ (m^{-1}).

$$\frac{d\underline{h}}{dy}=2\left\{\left[\left(\frac{1}{R}-\frac{\cos\delta\cos^{3}\beta}{r}\right)\frac{\underline{c}_{m}^{2}}{\sin^{3}\beta}-\frac{\underline{c}_{m}\left(\partial\underline{c}_{m}/\partial z\right)}{\tan\beta}+2\underline{\omega}\cos\delta\,\underline{c}_{m}\right]\tan\theta\right.$$
$$\left.+\left(\frac{\sin\delta}{r\tan^{2}\beta}+\frac{1}{\rho}\right)\underline{c}_{m}^{2}+\left(2\underline{\omega}\frac{\sin\delta}{\tan\beta}\right)\underline{c}_{m}+\underline{\omega}^{2}r\sin\delta\right\} \qquad (7)$$

Eq. (7) is based on the equilibrium of forces and is valid for a runner with infinite number of blades i.e. potential flow. In addition to eq. (7), the rothalpy equation (8), based on a given specific inlet energy, must be used together with the equation of continuity. The theory is 2-dimensional only, but is still useful in order to obtain a physical understanding of the quantitative influence from the geometric parameters for improvement of the blade shape in the basic design followed by a full 3D viscous analysis.

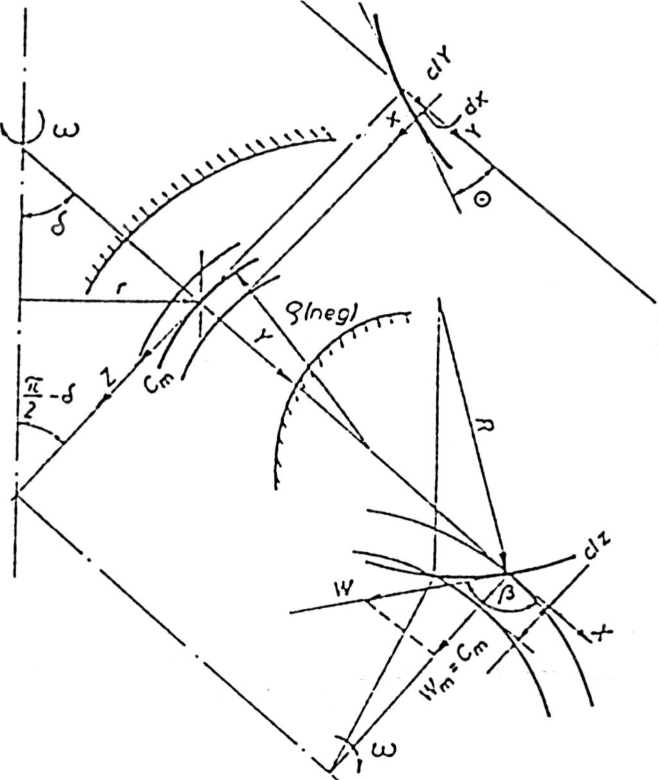

Figure 9. Definition of blade lean and the blade geometry.

The *rothalpy equation* yields:

$$\underline{h} = \underline{\omega}^2 r^2 - \underline{c}_m^2 / \sin^2\beta + (1 - 2\underline{u}_1 \underline{c}_{u1}) - J \tag{8}$$

J is the estimated losses along a streamline from inlet to the regarded point. In addition to eq. (7) and eq. (8), the equation of continuity must be used, as mentioned earlier (see fig.9). The advantage using eq. (7) is that the designer can calculate the influence of the blade lean i.e. it will be possible to calculate the change in the pressure gradient from hub to shroud by a change in the blade lean.

Even if a low pressure at the shroud occurs because of its curvature, this low pressure may be compensated by changing the blade lean angle Θ.

The first stage of a new blade design will be to chose main dimensions of shroud and crown. Then the curvature, the inlet and outlet angles and finally adjusting the blade lean by the use of eq. (7). The second stage will be a CFD analysis in order to study the details.

4. The second step of hydraulic analysis

When the geometry of a new runner is made in the first stage, a CFD analysis must be made as the second stage in the analysis. At present time various CFD programs with different turbulence models are available. An example of such analysis is illustrated in fig. 10 where two alternatives of runner blades with different blade lean and outlet curvature are shown.

In fig. 10a an original runner is shown and in fig.10b an improved alternative with a 10° difference in blade lean angle at the inlet is shown for comparison. No changes in shroud and crown have been made in this case. (For a final optimization, the pressure distribution will be balanced by adjusting both the shroud and crown curvature as well as the blade lean angle Θ at the outlet). This example clearly shows that the pressure gradient calculated during the initial analysis is a valuable tool, together with the CFD analysis.

5. Model test

The model tests in laboratory will not be described in this lecture. However, it must be emphasized that dynamic test of a model will not give a correct picture of the pressure oscillations identical to the oscillations in the prototype at site unless the Froude number and the stiffness and lengths of the conduits and model turbine are homologous to the prototype.

A closed test loop will also give transients created by the main supply pumps unless free water surface tanks are located upstream and downstream of the model with intake and draft tube outlet modelled correctly. The water hammer reflection time versus the model turbine speed must also be adjusted by the pipe wall flexibility and pipe lengths in the test loop.

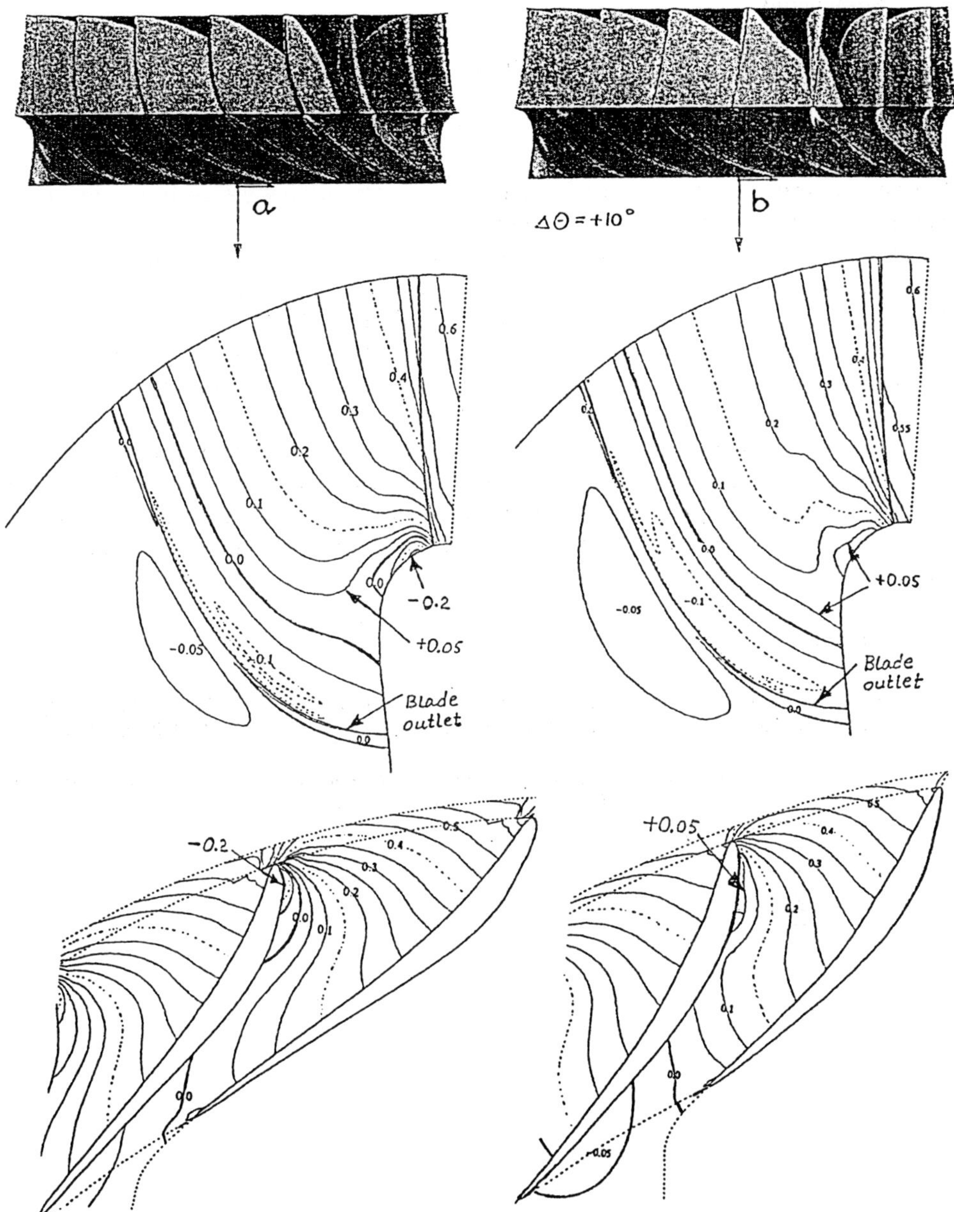

Figure 10. Comparison of the reduced pressure $\underline{h}$ = h/H for a runner (a) compared with modified inlet of the blades by a change in the blade lean by **ΔΘ**=10° (b).

6. Conclusion

An example of design work for Francis turbines for the future electricity production has been given.

Other examples are the introduction of splitter blades which for the majority of turbine manufacturers are in the initial stage of design. For Pelton turbines the task will be increased performance of high specific speed multi nozzle turbines and turbines operating at heads up to 2,000 m without fatigue and cavitation pitting problems.

Development of hard surface coating and surface treatment in order to reduce sand erosion is an ongoing research work. Improved methods for efficient coating at site will also be a main task in the future. On the structural design side the main problems are to fulfil the requirement of *leakage before rupture* on the thick plates for large turbines. The toughness needed to fulfil this requirement for high tensile strength steel for large turbines will be a challenge for the steel mills in the future.

The last challenge I will emphasize in my lecture, is to encourage experienced manufacturers of large turbines to be engaged in small hydro. This can be done if the quantity and quality requirements are increased in order to reduce the number of poorly performing small hydro turbines delivered by unskilled manufacturers under the green umbrella of environment.

References

1. Raabe, Joachim. *Great names and development of hydraulic machinery.* IAHR 50 years, 1935-1985. Hydraulics and Hydraulic Research - A historical review, Jubilee volume. A.A. Balkema, 1987.

2. Muray, Hitoshi. *Hydrodynamic topics in development of high head pump turbines and investigation on related problems in Japan.* IAHR Symposium, Tokyo, Japan 1980.

Analysis of Transients Caused by Hydraulic Machinery

M. Hanif Chaudhry
Professor, Dept. of Civil and Environ. Engg.
Washington State University
Pullman, WA, USA 99164-2910

1. Introduction

The analysis of transients in systems having hydraulic machinery is necessary during the design of these systems. It may be needed to upgrade, modify, or retrofit an existing system or if the mode of operation is changed. Sometimes, such analyses are done to determine the cause of incidents, accidents, and failures; and to develop procedures or system modifications to prevent their re-occurrence.

Normal and emergency operation of control devices produce transients in piping systems. For the analysis of these transients, mathematical models are commonly used. These models have been developed by solving the governing equations by numerical methods. The solution may be in the time or in the frequency domain. The solution in the time domain are more common and are discussed in this paper. The frequency domain analysis, done to determine resonating characteristics and the stability of systems, are not as common and thus will not be covered herein.

In this paper, the available methods of analysis and their advantages and disadvantages are first discussed. This is followed by a discussion of the inclusion of the hydraulic machinery in these models. A number of suggestions are outlined which the author has gained through his experience during the solution of many real-life projects.

2. Governing Equations

Based on the assumptions of one-dimensional flow, linear elasticity of the conduit wall material, slightly deformable walls, and steady-state friction formula, the following equations may be written (Chaudhry, 1987) to describe transient flows in closed conduits

E. Cabrera et al. (eds.), Hydraulic Machinery and Cavitation, 17–22.

$$\frac{\partial Q}{\partial t}+\frac{gA}{a}\frac{\partial Q}{\partial x}+RQ|Q|=0 \qquad (1)$$

$$\frac{\partial H}{\partial t}+\frac{a^2}{gA}\frac{\partial Q}{\partial x}=0 \qquad (2)$$

in which Q = rate of discharge, in m^3/s; H = piezometric head, in m; g = acceleration due to gravity, in m/s^2; A= cross-sectional area, in m^2; R = $f/(2DA)$; f = Darcy-Weisbach friction factor; D = pipe diameter, in m; and a = wave speed, in m/s. In these equations, the convective acceleration terms and the pipe-slope terms have been neglected. The convective acceleration terms are usually small and may be neglected. Similarly, the exclusion of the slope term is a commonly used procedure since it simplifies the computation of the steady-state conditions that are compatible with the transient equations.

Equations 1 and 2 are hyperbolic partial differential equations. The presence of the friction-loss term makes these equations nonlinear for which closed form solutions are not available except for very simplified cases. Therefore, numerical methods are used to integrate them. Several different numerical methods are available for the solution of hyperbolic partial differential equations, such as the method of characteristics, finite-difference methods, finite element method, etc. (Chaudhry, 1987, Almeida and Koelle, 1992). Among these methods, the method of characteristics has become very popular and is discussed in this paper.

3. Method of Characteristics

By multiplying Eq. 2 by an unknown multiplier λ, adding it to Eq. 1, making a number of algebraic manipulations, and integrating along the positive characteristic AP and the negative characteristics BP, the following equations are obtained

$$\int_A^P dQ+\frac{gA}{a}\int_A^P dH+R\int_A^P Q|Q|dt=0 \qquad (3)$$

$$\int_B^P dQ-\frac{gA}{a}\int_B^P dH+R\int_B^P Q|Q|dt=0 \qquad (4)$$

Transient flow and head are known at points A and B and their values are to be determined at point P.

If a first-order approximation is used for the friction-loss term, then these equations may be written as

$$Q_P=C_p+C_aH_P \qquad (5)$$

$$Q_P = C_n - C_a H_P \qquad (6)$$

where $C_a = gA / a$; $C_p = Q_A + C_a H_A - R\,\Delta t Q_A\,|Q_A|$ and $C_n = Q_B - C_a H_B - R\Delta\,t\,Q_B\,|Q_B|$. A simultaneous solution of these equations gives transient flows and heads at the interior points. At the boundaries, however, one of these equation is available and additional information relating H and Q is needed to determine a unique solution.

4. Boundary Conditions

The additional information needed to develop the boundary conditions for hydraulic machinery defines the relationship between flow and head for the machine under consideration. While defining this relationship, other variables upon which the flow and head depend are introduced, thereby making the inclusion of the hydraulic machine in the mathematical model more difficult and cumbersome.

Usually limited information relating the head and flow for hydraulic machines is available. This information is mostly obtained from steady-state tests conducted on scale models in the laboratory. The validity of this information during transient state has not been confirmed. However, this information is the best we have (and this is also not available most of the times) and thus is a commonly utilized in transient analysis.

The head and discharge through centrifugal pumps depend upon its rotational speed. For turbines, the wicket gate opening and the blade angle, if it can be varied, are also important. The relationships among these variables are presented in the form of machine characteristics. Each manufacturer presents this data in different forms and there is no standard accepted procedure for this presentation. Thus, depending upon the model being utilized, the information provided by the manufacturer has to be reduced to a particular form needed by the mathematical model.

5. Machine Characteristics

Several different formulations (Knapp, 1937; Kittredge, 1956; Donsky, 1961; Parmakian, 1963; Marchal, et al., 1965; Martin, 1982; Boldy and Walmsley, 1983; Chaudhry, 1987; Almeida and Koelle, 1992; Andrade and Martin, 1992; Streeter and Wylie, 1993) are available for including the machine characteristics in numerical models. Each of them has some advantages and disadvantages and preference for a particular one is more of a personal choice. If properly included in the analysis, each should give acceptable results.

Depending upon the stage of a particular project during planning or design, the machine characteristics may not be available. In such a case, the characteristics of a machine with similar specific speed may be used, as a preliminary estimate. However, once the machine has been selected, then these calculations should be repeated with the proper characteristics.

5.1 CENTRIFUGAL PUMPS

If the pump may reverse during transient conditions, then the characteristics in all four quadrants will be needed. The form presented by Marchal et al. (1965) is concise and easy to use and is recommended as compared to several others reported in the literature. However, note that different expressions are used by different authors for θ: some define it as $\tan^{-1}(v/\alpha)$ and others use $\tan^{-1}(\alpha/v)$, etc. In these expressions, v, is the non-dimensional pump discharge and α is the non-dimensional pump speed; in both cases the rated values being used as reference conditions. Similarly, different expressions are used for the ordinates of the characteristics, F_h and F_b. Each formulation should give acceptable results if properly included in the model.

5.2 TURBINES

Turbine manufacturers usually present the characteristic data in terms of unit speed, n_{11}, unit discharge, q_{11}, and unit torque, T_{11}, or unit power, p_{11}. Expressions for these unit variable vary from one manufacturer to another. Some authors transform these characteristics to a form similar to Suter form for pumps. In addition to additional effort in the transformation of data, little is gained as far as computational efficiency is concerned. In addition, the characteristics in the form of unit values have some physical meaning and should be preferred, if possible. It is possible to extend the characteristic curves by simple extrapolation. Normally, little or no data is available beyond the run-away conditions. During transient conditions, it is possible for the machine to enter these regions and the missing data in this region may be determined by extrapolation. Similarly, unit torque at gate openings less than the speed-no-load gate may be extrapolated from the available information.

Some authors have tried to use in transient analysis a valve boundary for Francis and Kaplan turbines. This could give totally unacceptable results since the turbine flow may decrease due to choking or increase during overspeed. Thus the computed transient pressures could be significantly in error if these effects are not properly included. As a preliminary estimate, the data for flow reduction or increase at runaway conditions may be used to include the effects of overspeed on flow changes. For this purpose, the procedure developed by Ramos and Almeida (1996) to include the effects of overspeed may be utilized.

5.3 PUMP TURBINES

For the analysis of transients caused by pump turbines, it is necessary to include the machine characteristics in all four zones of operation. Unfortunately, the characteristics cross each other in certain zones or they may have multiple values for the same unit speed. In order to avoid the possibility of the iterative solution to diverge, several procedures have been reported. Of these formulations, the procedure presented by Boldy (1983) in which s-curves are imposed on the characteristics curves instead of using a rectangular grid or the one proposed by Martin (1982) in which the characteristic curves are opened up appear to have some advantages as compared to the other formulations. However, the problem of multi-valued functions remain in the former and a special treatment for the zero gate is necessary for the latter.

6. Turbine Governors

The addition of a turbine governor in the mathematical model adds another complexity. This is further compounded by several different types of governors, e.g., temporary droop (proportional-integral, PI), proportional-derivative-integral (PID), etc. Since there are many different designs in each type of governor depending upon the feedback, it is not possible to write a general formulations that will handle all different types. Since each manufacturer usually keeps the feedback diagram for their governors as a proprietary information, difficulties may be encountered during planning and design stages since the governor manufacturer may not have been selected at the time the transient studies are conducted. In such a case, a PID governor may be used since this type of governor is becoming more and more common.

7. Draft Tubes

As discussed earlier, Eqs. 1 and 2 describing transient flow are based on the assumption of one-dimensional flow. In the draft tube, however, the flow is rarely one-dimensional, especially at partial wicket gate opening. Thus, although these equations are extensively used to compute transients in hydroelectric power plants with long tailrace conduits, the results from such analyses may not be accurate. In addition, since the upstream and the downstream conduits are interconnected through the turbo-machinery, the results for the upstream conduit system may be affected as well due to inaccuracies introduced in the analysis of the downstream conduit system. For example, while studying the formation of an unusual pressure spike in the penstock of a power plant with long draft tubes, the author was unsuccessful in reproducing the spike in transient analyses by trying different reduced areas for the draft tubes.

8. Transient Studies

In the past, rapid transients (e.g. the waterhammer) and slow transients (e.g., surge oscillations) have been analyzed separately. This goes back to the days of hand calculations. Nowadays, however, sophisticated computer programs are available for both rapid and slow transients. However, both types of analyses may be done simultaneously utilizing computer models for rapid transients because presently fast personal computers are readily available. However, care must be exercised in such analyses for determining the stability of oscillations. Although, the effects of most of the parameters are properly included, thereby excluding their uncertain effects on the final results, it may be somewhat difficult to determine the cause of instability in case one occurs in a particular system. The instability may be due to smaller tank areas, improper governor settings, multiple-valued characteristics, or due to instability in the computational procedure. In addition, some simplified analyses are needed in order to roughly select system dimensions and other parameters for a detailed analysis.

9. References

Almeida, A. B. and Koelle, E. (1992) *Fluid Transients in Pipe Networks,* Elsevier, Southampton, UK.

Andrade, J. G. P. and Martin, S. (1992) "Representation of Pump-Turbine Characteristics using Fourier Series," *Proc., Inter. Conf. on Unsteady Flow and Fluid Transients*, British Hydromech. Research Assoc., UK.

Boldy, A. P. and Walmsley, N., (1983) "Representation of the Characteristics of Reversible Pump Turbines for Use in Waterhammer Simulations," *Proc., Fourth International Conf. on Pressure Surges, British Hydromech. Research Assoc.*, Sept., pp. 287-296.

Chaudhry, M. H. (1987) *Applied Hydraulic Transients*, 2nd ed., Van Nostrand Reinhold, New York, NY.

Chaudhry, M. H. and Yevjevich, V., eds., (1981), *Closed Conduit Flow*, Water Resources Publications, Littleton, CO.

Donsky, B. (1961) "Complete Pump Characteristics and the Effect of Specific Speed on Hydraulic Transients," *Jour. Basic Engineering, Amer. Soc. Mech. Engrs.*, Dec., pp. 685-699.

Knapp, R. T., "Complete Characteristics of Centrifugal Pumps and their Use in Prediction of Transient Behavior," *Trans. Amer. Soc. Mech. Engrs.*, vol. 59, pp. 683-689.

Kittredge, C. P. (1956) Hydraulic Transients in Centrifugal Pump Systems," *Trans., Amer. Soc. Mech. Engrs.,* vol. 78, pp. 1307-1322.

Marchal, M., Flesh, G., and Suter, P. (1965) "The Calculations of Waterhammer Problems by Means of Digital Computer," pp. 168-188.

Martin, C. S. (1982) "Transformation of Pump-Turbine Characteristics for Hydraulic Transient Analysis," *Proc., Symp. on Operating Problems of Pump Stations and Power Plants*, International Assoc. for Hydraulic Research, vol. 2, Sept.

Parmakian, J. (1963) *Waterhammer Analysis*, Dover Publications.

Ramos, H. and Almeida, A. B. (1996) "Hydrotransients Induced By Dynamic Behavior of Turbo-Generators at Hydroelectric Power Plants," *Proc., Conf. on Pressure Surges and Fluid Transients in Pipeline sand Open Channels,* BHR Group., pp. 417-429.

Stepanoff, A. J. (1957) *Centrifugal and Axial Flow Pumps*, John Wiley & Sons, New York, NY.

Wiley, E. B. and Streeter, V. L. (1993) *Fluid Transients in Systems*, Prentice Hall, Englewood Cliffs, NJ.

SOME PRESENT TRENDS IN HYDRAULIC MACHINERY RESEARCH

MICHELLE FANELLI
Consultant- V.L.B. Alberti n° 5 - 20149 MILANO (Italy); Former Director of the Centre for Hydraulic & Structural Research (CRIS) of ENEL

Summary

An overview is given of the recent developments in the field of research about the dynamic behaviour of hydraulic machines, both as concerns their reaction to external disturbances (passive behaviour) and their role as a source of excitation for the rest of the installation (active behaviour). The interest of such investigations for manufacturers and users of hydraulic machines is outlined. The case of the draft tube surge caused by the interaction with the elbow of the cavitated vortex rope at the outlet of a Francis turbine operating at partial load is treated in some detail. Hints are offered as to the likely future trends of research about these topics.

Resume

On esquisse l'histoire des développements récents de la recherche sur le comportement dynamique des machines hydrauliques, en entendant par là soit leur réaction aux perturbations engendrées ailleurs dans l'installation (comportement passif), soit leur rôle en tant que source d'excitation pour le reste de l'installation (comportement actif). On souligne l'intérêt de telles recherches pour les applications industrielles. Le cas des oscillations dans l'aspirateur causées par l'interaction avec le coude de la torche à la sortie d'une turbine Francis fonctionnant sous charge partielle est traité en particulier. En conclusion on offre quelques considérations sur les tendances probables de la recherche future dans ce domaine.

1. Introduction

Hydraulic machinery is, more than ever, an essential component of modern technology. Its versatility, coupled to very high efficiency and minimum environmental impact, make for an irreplaceable role as a high-quality source of peak power (and energy storage) in electricity generation systems, as well as a versatile power link in hydraulic pressure circuits for the most diverse technical applications.

In the field of hydropower, the technological advances in material and fabrication techniques have allowed manufacturers to build machines of extremely high unit power, making it possible to match the exponentially increasing energy demand without a correspondingly unacceptable increase in the number of generating units.

E. Cabrera et al. (eds.), Hydraulic Machinery and Cavitation, 23–45.

Also, reversible machines for extremely high specific energies have been developed under the pressure for overall economical and technical optimization, and special machines (bulb or similar types) of increasing capacity have been developed for very low-head sites.

As the performance of these modern machines has attained very satisfactory levels in terms of efficiency at the rated power, it is only natural that the focus of requirements has progressively shifted toward such qualities as reliability, durability, smoothness of operation at reduced loads, last but not least limited noise and vibration generation, both at régime and during transients. In fact, while in former times the outage of one small-power machine did not seriously affect the overall operation of the typical power plant based on a large number of such small units, nowadays the outage of a group rating some hundreds of Megawatts, or even its inability to perform satisfactorily at reduced loads, can severely downgrade the service features of a major power plant (not to mention the very high costs of service interruption for interventions). In a modern large power plant, indeed, possibly only a few such giant units are installed, and load variations caused by network power regulation inevitably imply that one of these huge groups be operated far from its rated parameter values during substantial periods of time. It is also appropriate to consider that the ever more prevailing trend toward use of hydropower for peak production and for pumped energy storage entails a much higher frequency of transients than in the formerly common run-of-the-river, or pre-programmed, mode of operation.

Within this changing framework, modern research in the field of hydraulic machinery, while on one hand serving the need for more advanced high (and low) specific energy reversible machines, is on the other hand more and more concerned with the following aspects:

a) - cavitation and fatigue resistance;

b) - vibration mitigation;

c) - transient behaviour;

d) - smoothness of operation at régime under reduced loads, both as concerns the "passive" behaviour, i. e. response to periodic disturbances generated elsewhere in the installation, and the "active" behaviour, i. e. as a source of periodic excitation for the installation as a whole or for parts thereof;

e) - reliability of the machine proper and of the ancillary components;

etc.etc.

It is not without good cause that the purchasers of hydraulic machinery require more and more frequently, besides traditional guarantees on rated power, efficiency etc., also the respect of well defined limits on the above aspects of behaviour; and that the international codes of practice (see e.g. the I.E.C. recommendations) also set - or propose to do so in the near future - limits and guidelines about the acceptability of the intensity of the above-mentioned phenomena.

In the following, some considerations will be developed in particular about the last but one (d) of the above points, in which the Author and many of his colleagues have been personally engaged for many years. It will be seen that comprehensive, updated consideration of all the implications leads to a revision of the experimental and analytical tools necessary to investigate the relevant problems in a correct way.

2. The passive dynamic behaviour of hydraulic machines

At the beginning of this line of investigation (the sixties), it was thought - naively on hindsight - that, under *externally generated* sinusoidal variations, the machine could be adequately represented as a *linear* system characterized by its complex "hydraulic impedance" Z. In complex notation:

$$Z = \frac{\Delta H_1 - \Delta H_2}{\Delta Q} \quad , \quad Z = Z(\omega) \tag{2.1}$$

where $\omega = 2\pi.f$ is the circular frequency of the variations, ΔH_1 is the complex amplitude of the variations of piezometric head on the "high-pressure" side, ΔH_2 the analogous amplitude on the "low-pressure" side, ΔQ being the complex amplitude of the discharge oscillations through the machine. It was implicitly assumed that no capacitive effects were present (incompressible fluid, undeformable casing) and that variations of other important machine parameters, such as angular speed, efficiency etc. could be neglected without inconvenience.

The problem was then reduced to the determination of the function $Z(\omega)$ - its real as well as its imaginary parts - in principle for all values of ω in correspondence of every "average" steady-state of interest around which the oscillations were assumed to occur. Such a determination could be attempted either experimentally or analytically, the latter avenue of approach entailing some kind of representation of the flow field inside the runner (as well as in the adjoining waterways, notably on the high-pressure side the spiral case and the distributor and on the low-pressure side the draft tube).A straightforward extension of the classical Euler theorem to non-permanent, periodic flow constitutes the simplest possible version of such a theoretical approach; it yields a function

$$Z(\omega) \cong R + i.\omega.I \quad , \tag{2.2}$$

with constant R and I. In this simplistic model, R is the slope of the machine H/Q characteristic (at rated angular speed), while $i.\omega.I$ is the *inertance* of the waterways. It was soon apparent, however, that on one hand the experimental approach called for very sophisticated setups and for very delicate measurements, especially as concerned the difficulty of accurately evaluating the minute (and fast) discharge variations. On the other hand, the capacitive effects were shown quite often to be non-negligible; besides, the experimental conditions were difficult to define in a reproductible way. Last but not least, the variations of angular speed, wicket gate opening, efficiency etc., in particular through the agency of the rotating masses inertia, the electrical and hydraulic torque variations, the regulation loops, made it next to impossible to define accurately the conditions under which the "impedance" was determined, thus rendering the measurements virtually useless. In other words, according to the conditions of testing, one could well obtain widely different values of the "impedance" under apparently quite equivalent circumstances.

It was thus learned "the hard way" that the correct manner of posing the problem - albeit still in the drastically simplified framework of the linear system model - consisted in representing the machine as a *"multi-port" system*, including explicitly the influence of the regulating loops, the electrical system, etc. Instead of a simple "impedance", one should define the machine behaviour through a *"transfer matrix"*, or better through the combination of the transfer matrices of each component of the multi-port block-diagram (Fig. 1) .

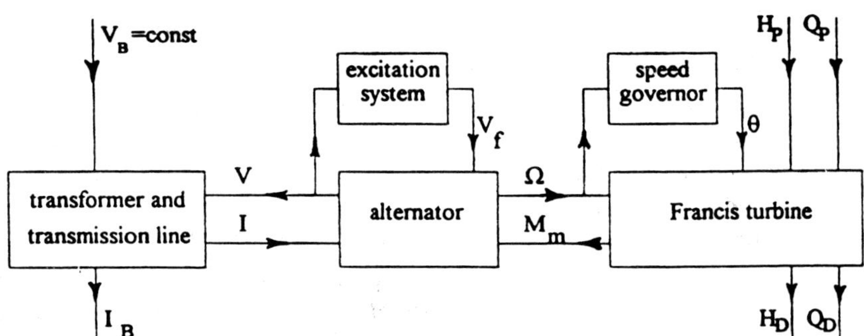

Figure1. Multiport block diagram of a hydraulic machine

In this scheme, the hydraulic machine proper is to be thought of as a system of channels, some fixed (e.g. scroll case and guide vane passages) and some rotating (e.g. runner blade-to-blade passages). In case of analytical modelling, such passages could be represented by ad-hoc continuity and equilibrium equations, together with the appropriate boundary conditions (particularly delicate at the interface between fixed and rotating parts): a daunting task, that only in recent times has been carried out with some measure of success (Figure2).

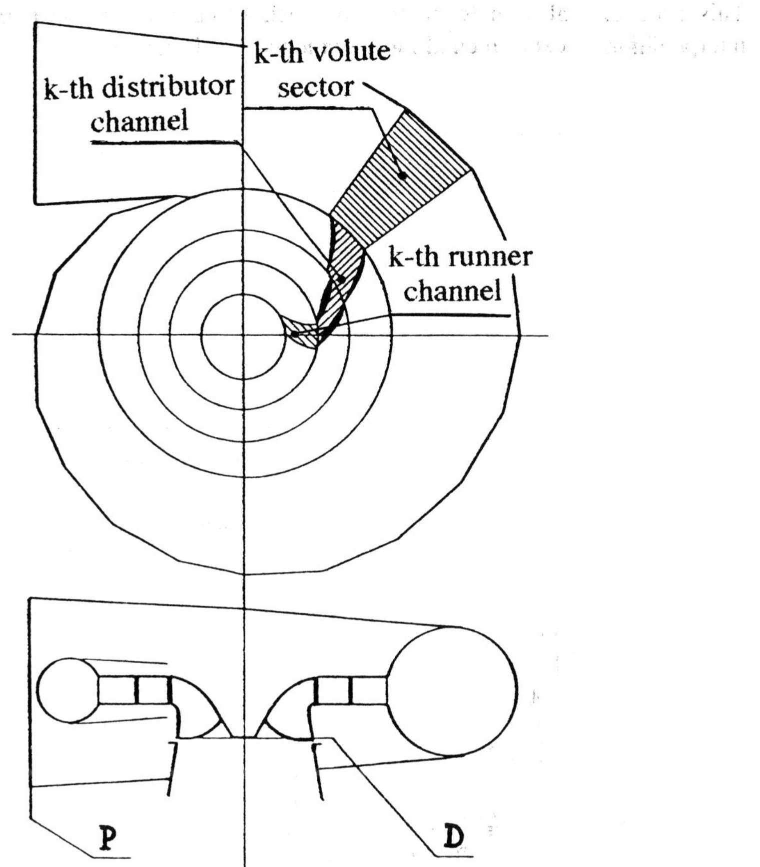

Figure2. A hydraulic machine as an aggregate of fixed and rotating waterways

(It is worth noting that, under restrictive conditions making the "hydraulic impedance" concept applicable, these recent studies, as well as the experimental campaigns- see below-, substantially confirmed the Euler-like result (2.2), at least for $\omega \to 0$).

This more rational transfer-matrix approach also allows a better planning and an easier interpretation of experimental measurements (see Figure 3).

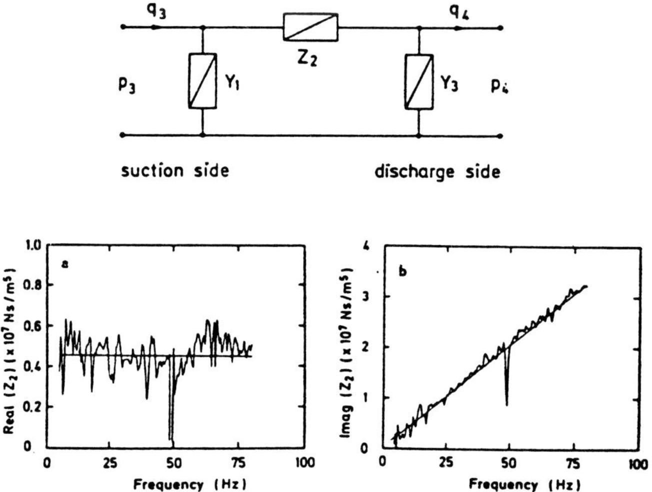

Figure 3. A few experimental results for the transfer matrix of a hydraulic machine

3. The active dynamic behaviour of hydraulic machines

In their role as a source of disturbances, hydraulic machines can generate both "high-frequency" and "low-frequency" excitations for the rest of the installation.

High-frequency oscillations (from some tens to some hundreds of Hertz) are typically connected with fixed-to-rotating blades interaction. This case will not be dealt with in the present note for brevity sake.

Low-frequency oscillations (around one Hertz) are typically linked to the *"vortex rope"* phenomenon at the low-pressure side of a **Francis turbine** running at partial load (i.e. at discharge $n.Q_0$, $n<1$, Q_0 being the "*zero-swirl*" discharge at which runner exit is radial), and to the ensuing "*draft tube surge*", especially common in the case of a *cavitated* vortex rope (low σ installations) in presence of an *elbow* along the draft tube axis (as is the case in practically all the vertical-axis units).

This phenomenon can lead, under special circumstances, to local amplification of pressure and discharge fluctuations. Sometimes also the regulation loops and the electrical components of the installation can be affected, to the point of causing intense power oscillations (*"power swings"*). In particularly severe instances, the pressure and power fluctuations can grow so large (up to 60% of the nominal values) as to make it impossible to operate the unit in the range of reduced loads in which the swings occur.

The phenomenon is a very complex one, and it is fair to say that a detailed quantitative comprehension of all the interacting mechanisms has not yet been completely attained (although fairly representative analytical models are becoming more and more available). Besides, it has recently been shown that the *non-linearities* inherent in the physics of the system may be essential to its effective analytical representation. Indeed, the linear approximation is in all likelihood insufficient to gain a true insight into the richness and diversity of the observed manifestations.

In the following, some considerations will be developed about the current state of research in this particular field.

3.1. Elementary models of the draft tube surge: the linear, mass-oscillation model

This model, first introduced by FRITSCH and MARIA (1987) , considers the draft tube pressure and discharge fluctuations as dynamic variables of a one-degree-of-freedom linear oscillator (Fig.4) in which:

- the "*stiffness*" is provided by an isotherm gas cavity, identified with the cavitated vortex rope volume;

- the "*mass*" is provided by the inertance of the liquid column in the draft tube, considered as formed by incompressible fluid ("*rigid water column*" assumption);

- the "*resistance*" is provided by the hydraulic friction.

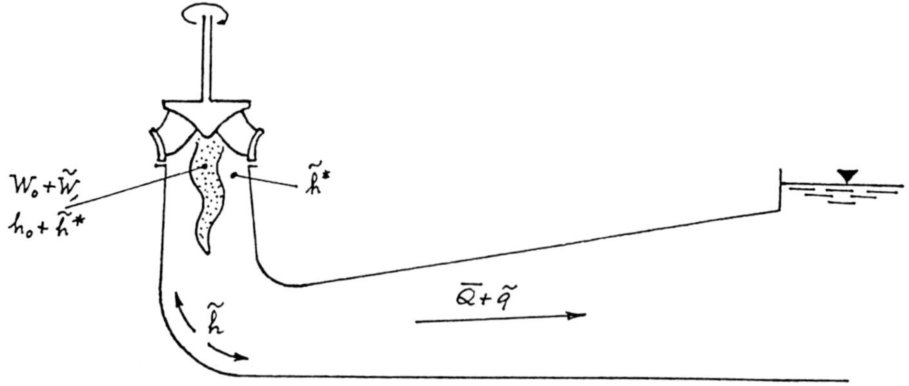

Figure.4. Overall scheme of mass oscillations in the draft tube of a francis turbine at partial load, with cavitated vortex rope

On these basis, it is straightforward to establish the following equations:

a) - *Continuity equation*:

$$\frac{dw}{dt} = q \ , \qquad (3.1.1)$$

where $w(t)$= volume variations of the gas cavity, $q(t)$=oscillating discharge in the draft tube; discharge variations at machine exit are assumed to be nil;

b) - *Dynamic equilibrium equation*:

$$I.\frac{dq}{dt} + k.q = h \ , \qquad (3.1.2)$$

where $h(t)$= variations of piezometric head at the initial section of the draft tube, $I = \frac{l}{g.S_m}$ = inertance of the rigid water column (l being the draft tube length, and S_m its average cross-section surface; in case of linear increase of the transverse dimensions of cross-section with abscissa, it is $S_m = \sqrt{S_0.S_1}$, S_0, S_1 being respectively the initial and final cross-section surfaces); k is a linearized friction factor;

c) - *Equation of state of the gas cavity:*

$$p_0.W_0 = (p_0 + \rho.g.h).(W_0 + w), \tag{3.1.3}$$

ρ being the liquid density and p_0, W_0 the average values of the pressure, respectively the volume, of the gas cavity.

By *linearizing* (3.1.3) and eliminating h , W between (3.1.1)...(3.1.3), the final equation is obtained for the *free oscillations* of the liquid column:

$$I.\frac{d^2w}{dt^2} + k.\frac{dw}{dt} + \frac{p_0.w}{\rho.g.W_0} = 0 \tag{3.1.4}$$

From this, the **natural circular frequency** ω_D of mass draft tube surges is obtained:

$$\omega_D^2 \cong \frac{p_0.S_m}{\rho.I.W_0} \quad \text{for} \quad k \to 0 \tag{3.1.5}$$

The interaction of the vortex rope with the elbow produces a periodic piezometric-head *excitation* h_e of the above system which at a first approximation can be represented by a sinusoidal function of time:

$$h_e \cong F_e.\sin\omega_T t \ , \tag{3.1.6}$$

where ω_T =circular frequency of the precessionary rotation of the vortex rope: usually

$$\omega_T \cong 0{,}25 \div 0{,}35.\omega_0 \ , \tag{3.1.7}$$

ω_0 being the rotational speed of the runner (rad/s).
Taking into account the presence of this excitation, the dynamic equilibrium equation becomes:

$$\frac{d^2w}{dt^2} + \frac{k}{I}.\frac{dw}{dt} + \omega_D^2.w = \frac{F_e}{I}\sin\omega_T t \ , \tag{3.1.8}$$

i.e. the classical equation of the forced harmonic motion of a linear oscillator. It is evident that in the case of *resonance*, i.e. for:

$$\omega_T \cong \omega_D \ , \tag{3.1.9}$$

important *amplifications of response* will occur.

Many critical remarks can be advanced in connection with this simple model; also, it has important positive aspects.

Let us consider first the main ***quality*** of this model.

One can reasonably assume that both the frequency ($\frac{\omega_T}{2\pi}$) and the intensity (F_e) of the excitation do not vary (at a given constant value n of the reduced load) when the downstream conditions are changed, i.e. the σ of the installation is varied. Moreover, a linear relationship can be assumed between p_0 and σ :

$$p_0 = C.(\sigma + \sigma_0) \quad , \tag{3.1.10}$$

and an approximate inverse relationship can be assumed between W_0 and p_0 :

$$W_0 \cong \frac{p_{ref}.W_{ref}}{p_0} \quad ; \tag{3.1.11}$$

indeed, $W_0 \to 0$ (no cavitation) for high values of p_0 , and $\frac{dW_0}{dp_0} < 0$.

With these assumptions, the natural frequency of draft tube surges is given by:

$$\omega_D^2 \cong \frac{S_m}{\rho.l}.\frac{p_0}{W_0} \cong \frac{S_m}{\rho.l}.\frac{p_0^2}{p_{ref}.W_{ref}} \quad , \tag{3.1.12}$$

so that ω_D is approximately proportional to p_0 , i.e. to $\sigma + \sigma_0$, see (3.1.10); on the basis of (3.1.8) and of classical results of linear oscillator theory, then, the response of our oscillator, i. e. the amplitude A of draft tube surge, will vary with σ as follows ($\nu =$ damping factor$= \frac{k}{2.I.\omega_D}$):

$$\frac{A(\sigma)}{A(\sigma \to \infty)} =$$

$$= \frac{1}{\sqrt{\left[1 - \left(\frac{\omega_T}{\omega_D}\right)^2\right]^2 + 4\nu^2.\left(\frac{\omega_T}{\omega_D}\right)^2}} \cong \frac{1}{\sqrt{\left[1 - \left(\frac{\sigma_{max} + \sigma_0}{\sigma + \sigma_0}\right)^2\right]^2 + 4\nu^2.\left(\frac{\sigma_{max} + \sigma_0}{\sigma + \sigma_0}\right)^2}}$$

(3.1.13)

where σ_{max} is evidently the value of σ at which the maximum response amplitude is observed.

Now relationship (3.1.13) is quite well in agreement with experimental results, *provided the values of* $\sigma_{max}, \sigma_0, \nu$ *are suitably calibrated*; see e.g. Figure 5.

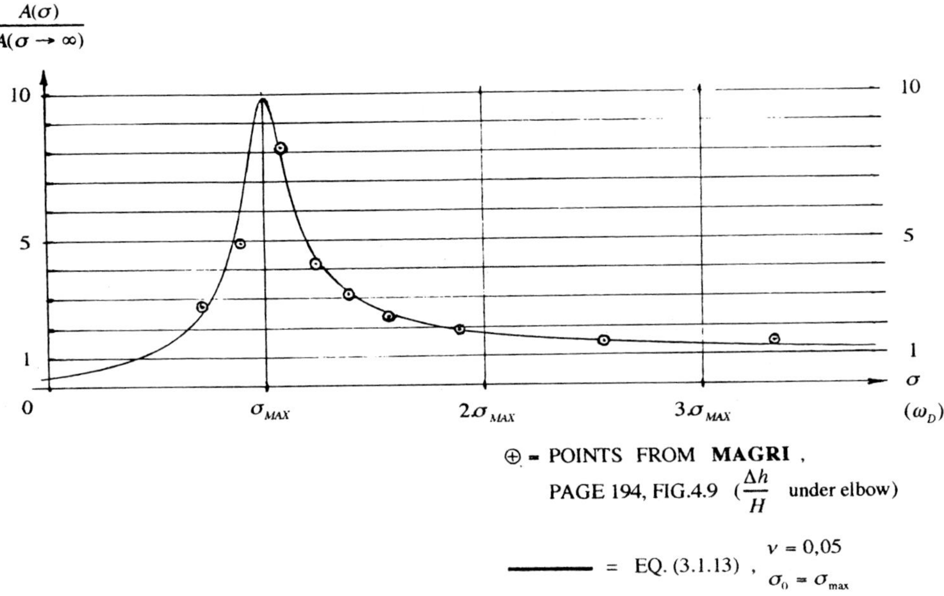

FIG. 5 - LINEAR MASS-OSCILLATION MODEL: THEORETICAL VS. EXPERIMENTAL DRAFT TUBE SURGE RESPONSE AMPLITUDE AT FIXED PARTIAL LOAD AND VARIABLE σ

From this comparison one can gain some important insights, e.g. an evaluation of the order of magnitude of the damping factor ν : it is found that usually

$$\nu \cong 0{,}05 \quad \text{to } 0{,}07 \qquad , \tag{3.1.14}$$

and from this the linearized friction factor $k = 2.I.\nu.\omega_D$ can be "identified", which may be interesting in order to investigate the dissipation mechanisms of the fluctuations.

It may be observed that the above interpretation covers only the *amplitude* variations with σ ; the *phase* information, which could confirm or disprove this interpretation, is unfortunately not always available. In this connection, it seems logical to suggest that in future tests also the *phase* data of synchronous pressure response should *always* be recorded (with reference of course to the *excitation* phase). This could be done quite easily, since the excitation phase is related to the position of the "corkscrew" vortex rope at elbow inlet, and this could be gained either by optical imaging or by recording the phase of the local rotating component of pressure field .

Let us now turn our attention to the ***drawbacks*** of the simple model under examination.

A) The dynamics of draft tube surges has been modeled *in isolation from other components* of the installation.The machine reactions ensuing from wicket gate opening, torque and angular speed variations have been disregarded; discharge variations coming from the machine exit section have been assumed to be nil.

B) The *intensity and frequency of the excitation* appear in the model as given, external data. No explanation of the *genesis of excitation*, nor of its dependence on the degree of partial load *n,* is attempted. The excitation is assumed as sinusoidal in $\omega_T.t$; its amplitude must be "identified" a posteriori from observational data.

C) Also the *model parameters* (e.g. the ratio $\dfrac{p_0}{W_0}$, on which ω_D depends) are to be "*identified*" by calibration using experimental data, since they are not accessible to direct measurement in the present state of the art, nor can they be reliably computed from theoretical considerations.

D) The *"mass oscillation"* model implies *infinite celerity* in the draft tube. In reality, *the celerity is finite* and it can even be quite low (a few hundreds of m/s) since it is not infrequent to have a certain amount of diffused air bubbles in the stream (either as a consequence of low pressure or by artificial air introduction as a mitigation measure). The actual system, thus, is not a "*concentrated parameters*" one, with a single eigenfrequency, but a "*distributed parameters*" one, with a discrete spectrum of eigenfrequencies.

E) Most important, the *several physical non-linearities* inherent in the nature of the system are not modeled. Probably the more important non-linear effect comes from the *stiffness of the isothermal gas cavity*. From (3.1.3) we get indeed:

$$\frac{\rho.g.h}{p_0} = \frac{W_0}{W_0 + w} - 1 = -\frac{w}{W_0 + w} = -\frac{Y}{1+Y} \quad , \tag{3.1.15}$$

where $Y = \dfrac{w}{W_0} =$ relative volume variation of the cavity; in the above linear model it has been assumed instead :

$$\frac{\rho.g.h}{p_0} = -Y \quad , \tag{3.1.16}$$

only valid for $Y << 1$ (very small fluctuations).

Figure 6 compares the "real" stiffness characteristic (3.1.15) with the linear approximation (3.1.16); it is seen that for finite values of Y the latter formula can lead to quite gross errors.

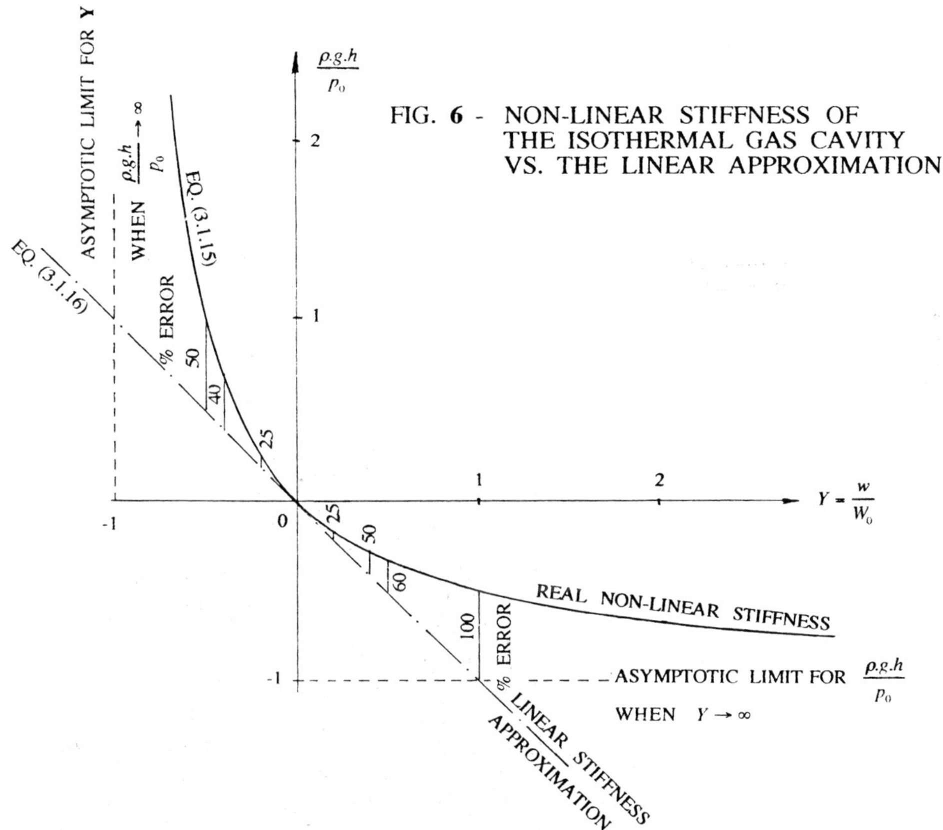

FIG. **6** - NON-LINEAR STIFFNESS OF THE ISOTHERMAL GAS CAVITY VS. THE LINEAR APPROXIMATION

Some of the above drawbacks could be eliminated, albeit still in a *linear model* framework.

As concerns heading A), e.g., models have been developed in which the *interaction between the components of the installation* in the pulsating régime are correctly represented (see Figure 7). To be noted that in this global formulation the *transfer matrix* of the machine (see § 2) has to be explicitly considered.

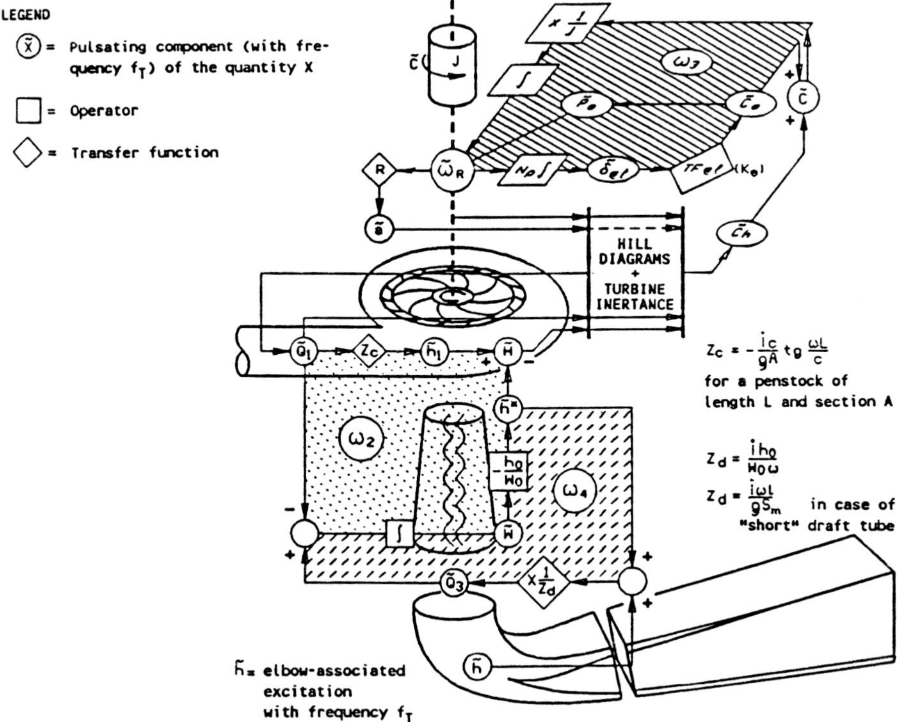

Figure 7. Possible oscillation loops in the penstock and in the electric system under cavitated vortex rope conditions

It is thus seen that *oscillations in penstock pressure and discharge* could be excited when, besides satisfying condition (3.1.9) for draft tube surge amplification, the rope frequency is near one of the penstock eigenfrequencies. Likewise, these extended linear models show that *electrical power fluctuations* could be excited if, again besides satisfying condition (3.1.9), the rope frequency is near one of the eigenfrequencies of the electric system. Thus a more comprehensive point of view is attained; however, some of the defects of the present theory are not easily removed. Among these, it is worth recalling on one hand the failure to model the excitation mechanism, on the other hand the gross approximations introduced by linearization.

In this last connection, it can be of interest to report some of the results of investigations carried out on *simplified non-linear models.*

3.2. Elementary non-linear models of draft-tube surge

The first step toward introducing non-linear effects in the draft tube surge model consisted in using eq. (3.1.15) for the stiffness of the gas cavity instead of its linearized form (3.1.16). After suitable *non-dimensionalization*, the following *non-linear dynamic equation* was obtained for the non- dimensional volume variation, Y :

$$\ddot{Y} + \frac{2\zeta}{\eta}.\dot{Y} + \frac{Y}{\eta^2.(1+Y)} = \frac{sin\tau}{K.\eta^2} \quad , \qquad (3.2.1)$$

where $\eta = \frac{\omega_T}{\omega_D}$, $\tau = \omega_T.t$, K is an amplitude parameter of the excitation, and dots denote differentiation with respect to τ .

The character of the solutions of (3.2.1) was extensively investigated by numerical integration for a wide spectrum of variability of "*control parameters*" $\eta, K, Y_0, \dot{Y}_0$ (the last two representing the *initial conditions*); the damping parameter ζ was kept constant at $\zeta = 0,1$. A number of important differences in behaviour emerged with respect to the *linearized* system ($Y << 1$; $\frac{Y}{1+Y} \to Y$),whose dynamic equation would be, instead of (3.2.1), the following:

$$\ddot{Y} + \frac{2\zeta}{\eta}.\dot{Y} + \frac{Y}{\eta^2} = \frac{sin\tau}{K.\eta^2} \quad . \qquad (3.2.2)$$

Only a few important aspects of said differences will be discussed here for brevity sake. It is assumed that the reader is familiar with the basic tools and terminology of standard non-linear dynamics.

For *large amplitude* fluctuations, i.e. beyond certain *thresholds* of the excitation, it is possible to get, according to different initial conditions, *multiple régime solutions* for the same excitation; one can also find for $\eta > 1$, besides the so-called "*synchronous*" (order 1) régime solutions, *subharmonics* of order 2, 3,... until the obtention, in particular conditions, of so-called "*chaotic*" responses.

If one attempts to draw a curve of the peak-to-peak synchronous régime response amplitude as a function of η , analogously to what is customary for linear systems, one finds for instance the results depicted in Figure 8 : secundary peaks and, near resonance, *multiple response régimes* with different amplitudes, which can lead to "*dynamic hysteresis*" (Figure 9: sudden jumps from one régime to the other), if the eigenfrequency (ω_D) of the system is not perfectly stable.

LINEAR OSCILLATOR
FREQUENCY RESPONSE AT REGIME

$$\ddot{Y} + \frac{2\zeta}{\eta}.\dot{Y} + \frac{Y}{\eta^2} = \frac{\sin\tau}{K.\eta^2}$$

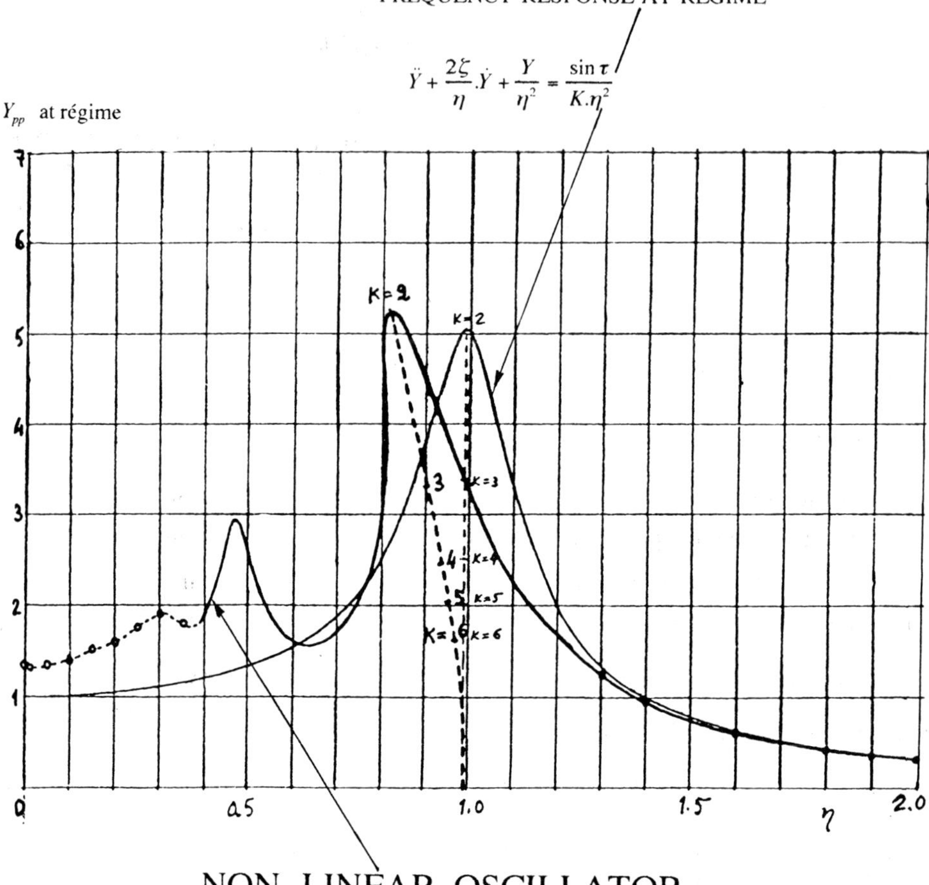

NON- LINEAR OSCILLATOR
FREQUENCY RESPONSE (PEAK-TO-PEAK) AT REGIME

$$\ddot{Y} + \frac{2\zeta}{\eta}.\dot{Y} + \frac{Y}{\eta^2.(1+Y)} = \frac{\sin\tau}{K.\eta^2}$$

Figure 8. frequency response curve of a non-linear oscillator representing the draft tube surge

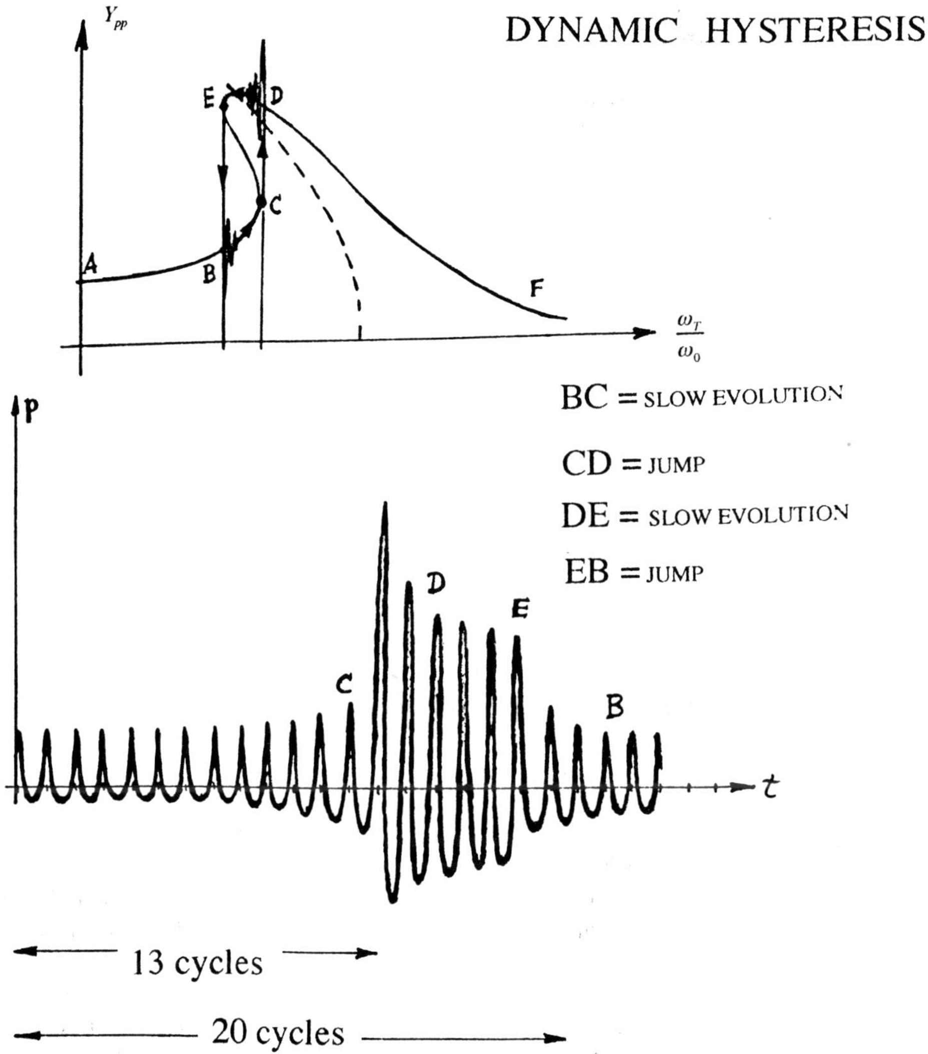

Figure 9. Dynamic hysteresis for a non-linear oscillator with softening stiffness

The system is *very sensitive to small variations in initial conditions*; near the "chaotic" régimes, any long-term prediction of the system evolution becomes impossible.

It is thus evident that the usual linearized treatment, and even the ordinary analytical tools used for investigation of linear systems, could lead to serious misgivings in the appraisal of the system *large-fluctuation* behaviour . The specialized methods and concepts of non linear dynamics (*space phase portraits*, *dynamic attractors* and their *attraction basins* etc.) must necessarily be used.

The above considerations, however, are *theoretical* ones, based on the nature of the equation assumed to represent the physical phenomenon. It might be asked with reason (especially considering such *prima facie* successes of simple linear models as evidenced in Figure 5) whether there is any *experimental evidence* in support of the necessity to switch over to the much more complicated non-linear models. In fact (besides the irregular, never purely periodic, appearance of time histories), sophisticated techniques of data processing, applied to recordings e.g. of pressure swings, not only give quite conclusive proof of the presence in the system of appreciable non-linearities, but also allow to affirm that the number of "degrees of freedom" of the system is very low (of the order of 2), thus lending hope to the search for simple (global) analytical models .

Also, image processing techniques applied to cavity volume variations have shown evidence of chaotic-like behaviour.

In conclusion, it can be said that there is a strong case for the necessity to investigate the draft tube surges with the tools of non-linear dynamics.

3.3. Current research and open questions for partial load "active" behaviour of Francis turbines

Some preliminary numerical investigations have been carried out on slightly more advanced *non-linear models having more than 1 d.o.f.*

In particular, the supplementary degrees of freedom thus far considered have been on one hand the *temperature* of the gas inside the cavity, on the other hand the *gaseous mass* contained in it.

It seems that *temperature variations* are not important; as for the *mass variations*, which physically could be caused by "*rectified diffusion*" effects and by *turbulent ablation* exerted by the draft tube flow on the cavitated vortex rope terminal part, no conclusive results have thus far been obtained, and it must be recognized that it is difficult to set up credible mathematical models of this complex phenomenon. It is thought that this aspect is potentially important, because it could provide a plausible mechanism for the onset of "*dynamic hysteresis*", which could help interpret certain *sudden amplitude transitions* found in some experimental records.

Let us now review briefly some of the main *open questions* calling for **future research efforts.**

A satisfactory model for the *excitation mechanism* is still largely lacking. It is necessary to investigate the detailed structure of the helicoidal flow field inside the draft tube "cone" and its interaction with the draft tube "elbow". Some partial analytical and numerical models of the flow field generated by a *precessing helicoidal vortex filament* inside a *cylindrical boundary* have been set up (this problem can be approximated by a potential flow, inviscid incompressible fluid scheme); however, the modeling of a precessing helicoidal vortex filament inside an *elbow* is immensely more complicated and calls essentially for integration of the full Navier-Stokes equations under non-permanent conditions.

The final result of such models would ideally provide an estimate of the *intensity and frequency of the excitation term* as well as indications about its *dependence on the index n*
of partial load.

At present, two main mechanisms are being considered for the excitation: the "*elbow pressure gradient sweep*" and the "*parametric head loss excitation*".

Elbow pressure gradient sweep : it is considered that the cavitated vortex rope, in its portion near the elbow, sweeps periodically zones of higher pressure (near the outside of the elbow) and zones of lower pressure (near the inside of the elbow). By setting up the equation for the volume variations of the cavity under the joint action of the synchronous pressure variations and of the elbow gradient sweep, together with the equation for dynamic equilibrium of the draft tube water column under the synchronous pressure variations, the excitation is made to depend explicitly on the ratio between the radius of the cone and the radius of curvature of the elbow axis:

$$h_e \cong \frac{n^2.Q_0^2}{2g.S_0^2}.\left[1-\frac{1}{(1-\varepsilon.\kappa.\sin\omega_T.t)^2}\right], \qquad (3.3.1)$$

where S_0=cone exit section, ε =excentricity of helicoidal vortex rope at cone exit, $\kappa = \frac{r}{R}$ = ratio between radius of cone exit section and radius of curvature of elbow axis (ε = function of n).

It is seen, in contrast with (3.1.6), that under this hypothesis *the excitation would not be exactly sinusoidal* in time (it would be nearly so only for $\varepsilon.\kappa << 1$); it is also seen that the excitation would be all the more intense as the radius of curvature *R* grows smaller, which is in accordance with known facts.

Parametric head loss excitation : it is assumed that the head losses in the draft tube, and particularly those concentrated immediately downstream of the elbow, are dependent on the position of the helicoidal vortex rope. Indeed, the zone of low pressure surrounding the helicoidal vortex filament interacts more or less with the zone of low pressure near the inside of the elbow during the precessionary motion of the vortex rope; thus more or less pronounced flow separation will occur, with quasi-periodic variations in the head loss factor. This makes for a *quasi-periodic blocking effect*; the conjecture is in accordance with experience, in the sense that numerical models based on a periodic variation of the head loss factor in the draft tube produce (by suitable *calibration* of said variations) effects in general agreement with experimental data (see DOERFLER, 1982).

The two above-mentioned effects could well coexist in the physical reality. More detailed experimental data (including *phase* informations, see § 3.1) will be necessary to confirm or disprove either model, as well as to "identify" the values to be assigned to the relevant parameters.

A non-linear numerical model based on the above considerations and in which the finite value of celerity in the draft tube is introduced (*"distributed parameters model"*) is under development by the Author; in it, besides the "control parameters" already seen for the elementary non-linear model of § 3.2, two new control parameters appear, related to propagation times along the draft tube axis. More accurate identification tools could thus be derived.

4. Conclusions

From the foregoing synthetic overview it is easy to see what will be the likely development of *future research* in the field.

As concerns the *analytical-numerical models*, an effort of *synthesis* is necessary to fuse together the partial models so far developed: dynamics of the cavitated vortex rope, its interactions with the machine on one hand, with the draft tube elbow (excitation mechanism) on the other hand, interplay of the different components of the installation in fluctuating régimes, effects of non-linearities.

It is also worth mentioning that, in alternative to the search for an "excitation mechanism", it might be more correct to reformulate the problem in terms of "*instability*" of the non-linear system, e.g. through the presence in the representative equations of terms having the nature of "*negative stiffness*" or of "*negative damping*".

The final aim would be, needless to say, in line with the twofold goals of every theoretical model:

- on one hand, *a posteriori* interpretation of observations and correct guidelines for *identification* of non-measurable parameters. An obvious example is the identification of the excitation parameters keeping into account the effect of mutual interactions of the system components; indeed, what is measured is always a *local response* and, in particular, what is measured on a reduced scale model cannot be transposed directly to reality because the mutual interactions will in general produce different amplifications in the model and in the prototype;
- on the other hand, *a priori* forecasting of pulsating phenomena: correct transposition from the scale model to the prototype, i.e. forecasting (after the above said identification of excitation) of the amplitude of local responses in the real system; the provision of reliable guidelines for setting up experiments and for locating measuring instruments in the "right" positions; the forecasting of the effects of mitigating measures (such as e.g. air introduction or "*active control*" devices); etc.

Important progress is under way, of course, and more is to be expected, in the perfection of *experimental techniques*. Velocity fields can now be investigated in detail with non-intrusive techniques such as LDV; pulsating discharges can be measured directly by US or EM sensors, optical image processing allows shape and volume of cavitated vortex rope to be evaluated....

In contrast to the experimental stands so far used for "industrial" reduced-scale models, mainly aimed at determination of efficiency, steady-state characteristics etc., *test loops of new conception*, particularly suitable for pulsating régime investigations, are under development as well. These new hydraulic circuits will have a large degree of *flexibility*: the length of penstock, of draft tube, of rotating shaft between hydraulic and electric machine will be adjustable to ensure a more satisfactory dynamic similarity between test installation and industrial system (it is quite possible to provide even for adjustable celerity in the penstock); devices for "active control" of pulsations will be a standard component of the test loop, and will also be used to simulate, through an on-line, real-time computer link, components not physically modeled in the circuit ("*hybrid model*" technique). Powerful new ways of on-line or off-line signal processing (together with more effective identification strategies, like e.g. *acoustic intensimetry*) will be used to extract all relevant information from measurements.

Some prototypical test loops of this kind have already been built, suffice it to mention the HESPERIA circuit set up in co-operation by CRIS/ENEL and CISE in Italy. (Figure 10)

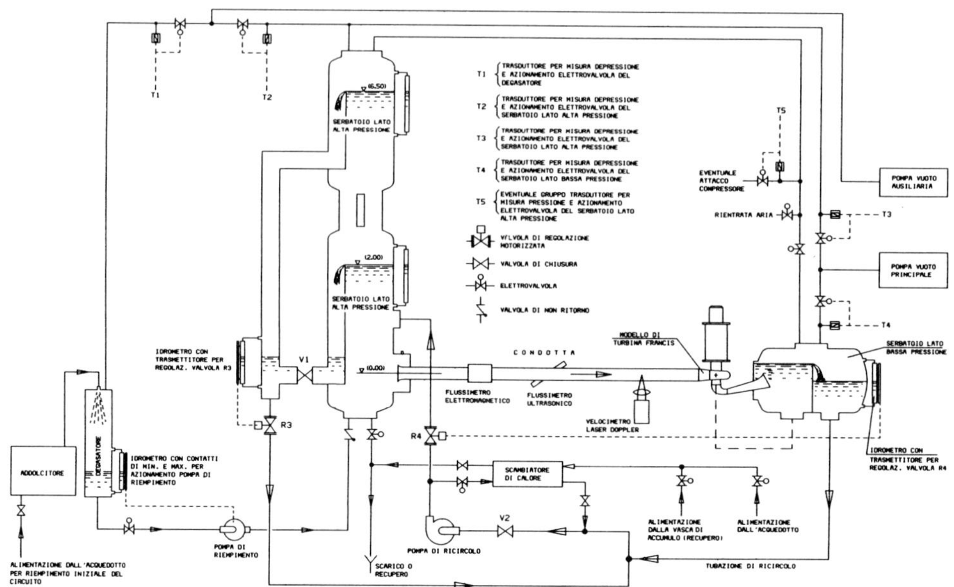

Figure 10. Functional scheme of the ìhesperiaî test loop for experimental investigations on the oscillatory behaviour of hydraulic machinery

As in the past, the progresses in analytical and experimental tools will interact synergically with each other, fostering a deeper understanding and more effective mastery of the phenomena herein considered.

A very important role in this long-range effort has been, and will be, played by the IAHR, through specialized working groups such as the **W.G. for the study of oscillating régimes in hydraulic machinery and systems**, with its biennal International Round Tables, as well as through the more general discussion forum provided by the Symposia of the Section for Hydraulic Machinery, Equipment and Cavitation (new W.G.s, such as one on *complex dynamics and chaotic phenomena* in hydraulics, could even be established in the future under the impulse provided by the investigations in question). This effort is also seconded by some national W.G.s, e.g. the **Italian W.G.** bearing the same title, established since 1971 and meeting at least once a year.

Everyone who has witnessed the first uncertain steps taken along the arduous road cannot but appreciate the great strides already made toward attainment of the important objectives in sight, and feel confident that the final goal is within our grasp.

5. Aknowledgments

I am indebted to my good friend and colleague Ing. Giuseppe Angelico for helping me in the preparation of the present lecture.

6. Bibliographic List

W.J. Rheingans: Power Swings in Hydroelectric Power Plants , Transactions ASME, Vol.62,1940

P. Doerfler: Vortex rope excited oscillations in the draft tube of Francis turbines working at partial load (in German) - Thesis, 1982

A. Stirnemann, J. Eberl: Experimental determination of the Dynamic Transfer Matrix for a Pump, ASME Winter Annual Meeting, Miami Beach, Nov. 1995

A.Fritsch, D.Maria: Dynamic Behaviour of a Francis turbine at partial load , Third Round Table of IAHR W. G. on behaviour of hydraulic machines under steady oscillatory conditions, Lille, Sept. 1987

L.Magri: Problematiche delle turbomacchine idrauliche per produzione di energia elettrica (Ed. Pitagora, 1987)

M.Fanelli: The vortex rope in the draft tube of Francis turbines operating at partial load: a proposal for a mathematical model (internal report Enel/CRIS,1988)

J. Raabe: Hydraulische Maschinen und Anlagen, VDI Verlag, 1989

J.M.T. Thompson, H.B. Stewart: Nonlinear Dynamics and Chaos, Mc Graw-Hill,1989

H. Ohashi: Vibration and Oscillation of Hydraulic Machinery, Avebury, 1991

Proceedings of the International Round Tables of the IAHR W.G. "Behaviour of Hydraulic Machinery and Systems under Steady Oscillatory Conditions" from 1981 to 1995

Proceedings of the International Symposia of the Section for Hydraulic Machines, Equipment and Cavitation of IAHR from 1982 to 1996

RAPID PROTOTYPING OF HYDRAULIC MACHINERY

R. SCHILLING, N. RIEDEL, R. BADER , T. ASCHENBRENNER, CH. WEBER AND A. FERNANDEZ

Institute for Hydraulic Machinery and Plants
Technical University of Munich, Germany

Abstract.

The paper presents a philosophy of "Rapid Prototyping of Hydraulic Machinery" and describes the main tools for the efficient tailor-made design and optimization.

1. Introduction

Due to the widely spread Computational Fluid Dynamics (CFD) methods the hydraulic performances of almost all types of hydraulic machinery, i.e. pumps, fans, compressors and turbines, could be improved considerably. Despite of this progress, still further improvements may be expected. However, with increasing efficiencies the amount of time and money for the tailor-made development of big hydraulic machinery, especially turbines having a power output of several hundred mega watt, is augmenting progressively. Opposite to this trend the prices which may be realized for hydraulic turbines on the worldwide market are going down due to a strong competition. To save time and expenses numerical systems have been established in the past for the efficient and specification controlled layout and development of hydraulic machinery. However, these systems are not so efficient as they could be if the processes of geometrical designing and flow analysis would be integrated into a powerful design and optimization system. A philosophy of rapid prototyping is presented. Its main modules for the efficient tailor-made design and optimization being necessary to speed up the daily R&D work are described in detail.

E. Cabrera et al. (eds.), Hydraulic Machinery and Cavitation, 46–57.

2. The rapid prototyping system (RPTS)

The philosophy of rapid prototyping (RPT) applied to hydraulic machinery is shown in Fig. 1. The first phase of RPT is to establish the hydraulic design of the various components, especially the impeller or the runner respectively, by means of a CFD-Aided Design System on a very high level within a few days. Then, the geometrical data are provided for the numerically controlled milling of the blades. At the end of phase 1 the model tests have to be carried out yielding a comparison between predicted and measured performances of the model machine. Based on the existing experience and by means of a mature CFD tool only a very few iterations or adoptions should be needed to satisfy the requirements of the specification. The second phase consists of the calibration of the CFD-code, generally a Navier-Stokes code, used in the design and optimization process by comparing the predicted efficiency behaviour with the experimental data measured on a highly accurate test rig. In the third phase the prototyping itself has to be performed in three steps. After having designed the prototype components a numerical loss analysis and performance prediction has to be carried out by means of the calibrated CFD-code to check whether the prototype performances guaranteed may be reached. Additionally, a structural analysis by means of a commercial FEM-software package should be performed to compare the occuring maximum stresses and strains with the maximum allowable values of critical components, such as guide vanes and runner blades due to the hydrodynamic and mechanical forces. Finally, in the fourth phase it has to be checked whether the installation of the prototype at site i.e. the interaction of the turbine with the plant, could lead to missing efficiency and/or power output.

2.1. MODEL DESIGN AND PERFORMACE PREDICTION

The strategy of designing the geometry of the model machine and predicting its performances is shown in Fig. 2. Based on a preliminary design corresponding to an existing data basis the CFD-Aided Design of the bladings may be performed using a simplified but very fast CFD-code. Therefore, this design has to be recalculated first by a calibrated full 3D Euler code with respect to the cavitation behaviour at turbine inlet and outlet, the velocity distribution at the outlet to obtain the maximum pressure recovery in the draft tube and the specified power output. Finally, the performance prediction has to be carried out by a viscous code over the whole range of operation defined by the specification. To realize an optimum design of impellers or runners within a few days with a high reliability the geometrical design tool and a suitable fast working CFD-code have to be integrated into a CFD-Aided Design System, s. /1/. However, such a system working

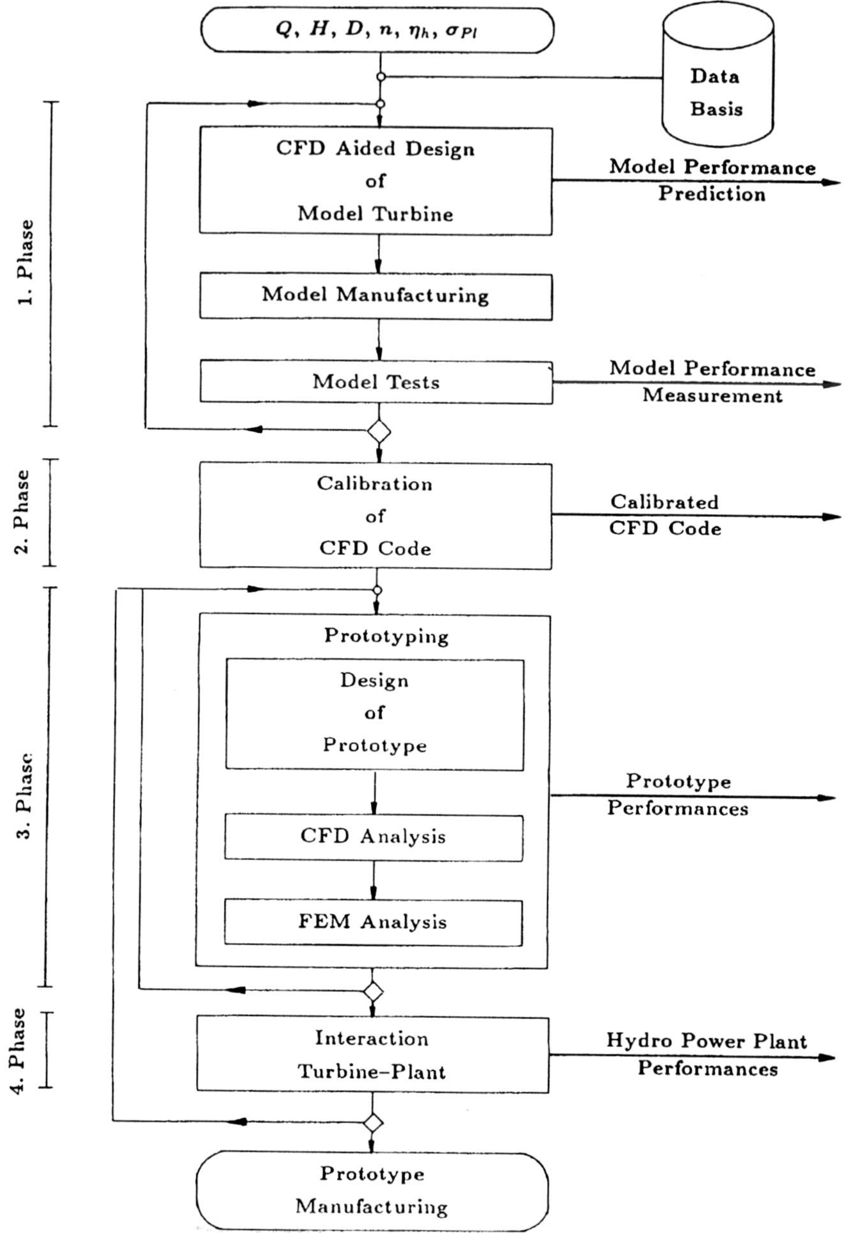

Fig. 1: Rapid prototyping system for hydraulic machinery

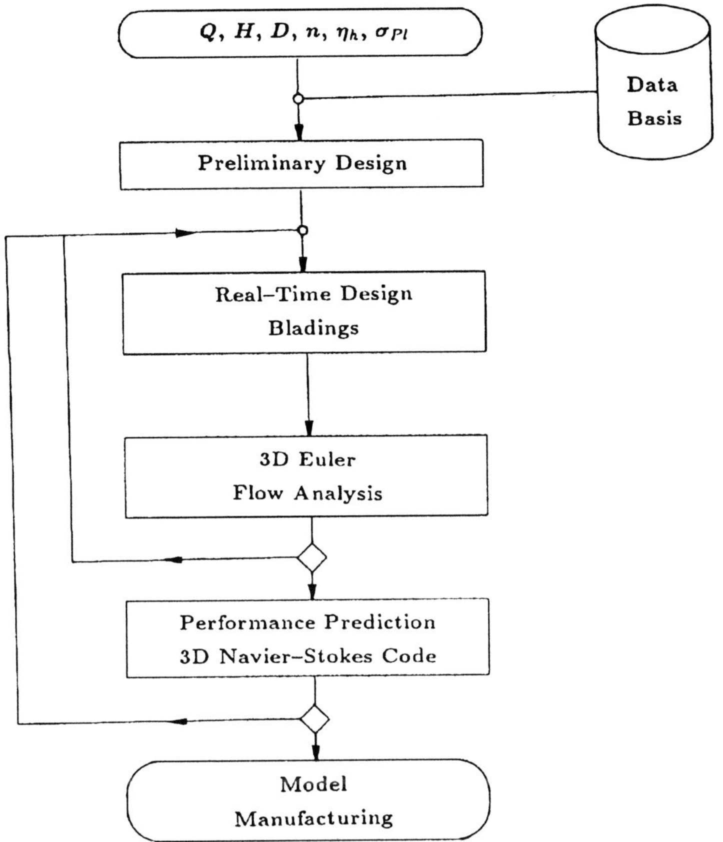

Fig. 2: Strategy of model design and performance prediction

sequentially has proven to be not efficient enough. A much better performance of designing has been reached by the Real-Time Design (RTD) System developed, s. /2,3/, which consists of an efficient design tool and a very fast quasi three dimensional (Q3D) Euler code. The basic idea of the RTD-System is to modify the whole blade geometry on three representative stream surfaces interactively and to obtain the flow analysis answer within less than one second. The real-time behaviour may be reached by using either a multi-processor system or a workstation cluster controlled by the PVM-Software, s. /4/. Applying the developed RTD-System a high specific speed Francis runner FT 115 has been designed. Fig. 3 shows one step of the RTD of the runner near the band. In Fig. 4 a comparison of pressure distributions near the band computed by the 3D Euler code developed, s. /5/, and by the Navier-Stokes code TASCflow, s. /6/, is shown for the optimum point of operation.

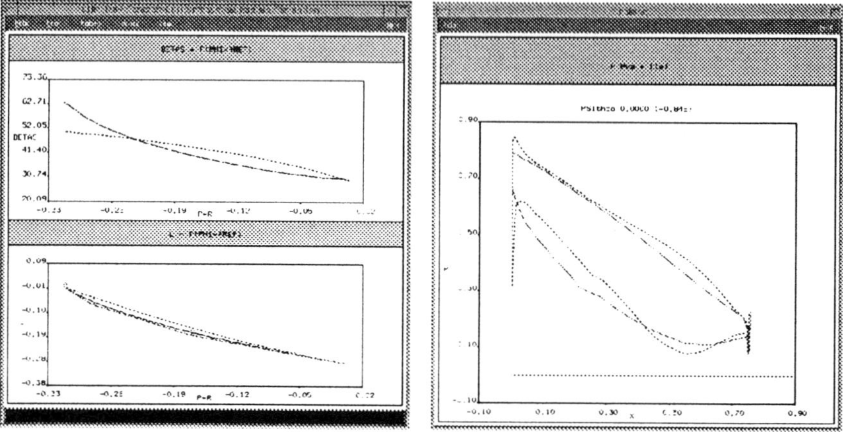

Fig. 3: Real-time design of a Francis runner blade FT 115 near the band

The 3D Euler computations have been carried out for the full stage consisting of the stay vanes, the guide vanes and the runner bladings taking into account the rotor-stator interaction. Former investigations have proven that the prediction of exact inflow angles can only be performed on the basis of coupled guide vane-runner computations, s. /7/. The results of this coupled 3D Euler flow analysis have been used as input data for the Navier-Stokes computations. Following the strategy proposed the 3D Navier-Stokes code

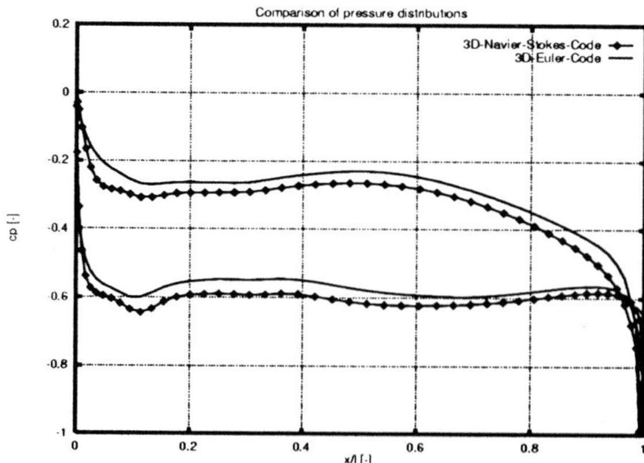

Fig. 4: Comparison of pressure distributions near the band of FT 115 computed by 3D Euler and Navier-Stokes code

is applied only within the last iteration cycle since the CPU-time needed for one 3D viscous recalculation of the real flow is about 20 times bigger than that of an 3D inviscid computation. For economical reasons only the robust and fast 3D Euler code is used to predict the relative inflow angles β_1 depending on the geometry of the guide vanes and runner blades. In the phase of preliminary design the theoretical Euler head of the Francis runner ψ_t may be computed economically as a function of the guide vane opening $\Delta\gamma$ in comparison with the Euler head required, s. Fig. 5. In a similar way the combination of guide vane and runner blade opening $\varphi(\Delta\gamma)$ can be determined for a bulb Turbine, s. /8/.

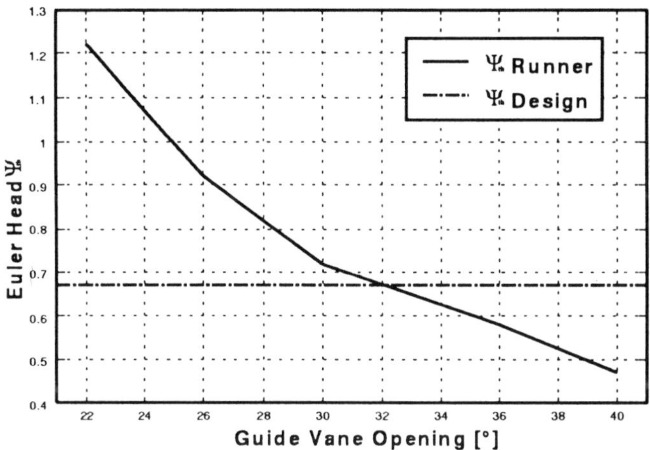

Fig. 5: Determination of the optimum guide vane opening for a Francis turbine FT 115

2.2. CALIBRATION OF CFD-CODES

Since the available Euler and Navier-Stokes codes are based on physical models simplifying the complex structure of the real flow through hydraulic machinery, they need a priori to be calibrated, s. /9/. Calibration means an adoption to the flow field data measured either by pressure probes and/ or by the more accurate Laser-Velocimetry. Assuming a second order accurate discretization of the governing partial differential equations the adoption to the reality may only be performed first by the mesh refinement, second by the improvement of the turbulence modelling and third by the improvement of the coupling algorithm to simulate the flow interaction between two components of hydraulic machinery, such as the rotor-stator interaction. Applying Euler methods principally only the numerical mesh can be modified to check the accuracy of the code. Basic investigations have shown that the number of grid points has no significant influence on the computed energy conversion. In the case of Navier-Stokes codes using the standard k-ϵ turbulence model with wall functions the influence of the mesh refinement is also relatively small. Table 1 shows the influence of the number of grid points on the total pressure increase and hydraulic efficiency of the BRITE-Impeller NQ 50 in comparison with the measured data, s. /10/, and with the CPU-time needed by the 3D Euler and Navier-Stokes code. To be able to detect not only surface friction losses but also the effects of

Nodes	21120	35190	62500	102400	19200
$\psi_{t,3DNS}$	0.815	0.840	0.822	0.831	0.809
$\psi_{th,3DNS}$	0.840	0.889	0.873	0.881	0.850
$\eta_{h,3DNS}$	97.12	94.42	94.21	94.38	95.20
$\psi_{th,E3D}$	0.853	0.849	0.844	-	-
$\psi_{t,Meas}$	0.857	-	-	-	-
$\psi_{th,Meas}$	0.922	-	-	-	-
$\eta_{h,Meas}$	93.00	-	-	-	-
CPU_{3DNS}	11.25	20.75	35.50	60.00	112.00
CPU_{E3D}	2.30	6.26	11.62	-	-

TABLE 1. Influence of number of grid points on the total pressure increase ψ_t, hydraulic efficiency η_h and the CPU-time in hours needed.

3D boundary flows without or with separation with higher accuracy the more sophisticated and much more time-consuming two layer turbulence model has to be used. Finally, it should be mentioned that Euler codes can be upgraded remarkably by implementing body forces which simulate the influence of friction relatively close to the reality without encreasing the CPU-time remarkably. The results of calibration are shown in Fig. 6 and 7. First, a comparison between predicted and measured inflow angles of the GAMM-Turbine NQ80 is shown in Fig. 6. Obviously, the developed coupling algorithm is predicting the relative inflow accurately, especially near the band being very important for an optimum design of Francis runners, s. /11/. Second, the influence of body forces on the adoption of the Euler computations to measured data is to be seen in Fig.7. The prediction of circumferential velocity distribution including body forces leads to a much better coincidence with the experimental data of /12/.

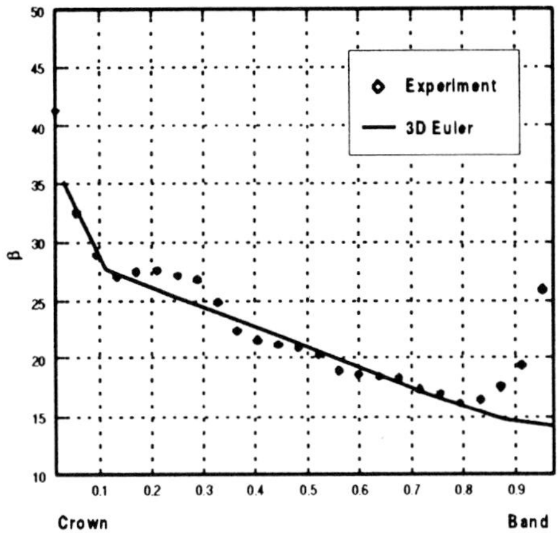

Fig. 6: Comparison of predicted and measured relativ flow

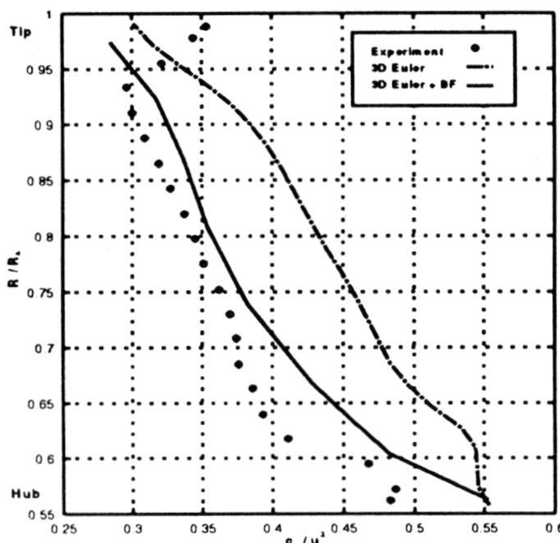

Fig: 7: Upgrade of a 3D Euler code by introducing body forces

2.3. PROTOTYPING

After having reached the specified model performances of the hydraulic machinery and calibrated the CFD-code to predict the prototype performances first the prototype drawings have to be established in the design department. Based on these drawings both a CFD- and a FEM-analysis have to be performed to guarantee the prototype performances with respect to the efficiency and cavitation behaviour as well as to the maximum occuring stresses and strains.

2.3.1. *CFD-analysis*

Applying the calibrated Navier-Stokes code the losses of all components of a hydraulic machinery may be evaluated on one side for the model and on the other side for the prototype characterized by a much higher Reynolds number defined by a reference velocity u_{ref}, the rotor diameter D and the kinematic viscosity ν $Re = u_{ref} * D/\nu$. Thus, the model efficiency η_M may be scaled up to that of the prototype η_P

$$\Delta\eta = \eta_P - \eta_M = f\{Geometry, Re\}$$

A scale-up has been performed for the Francis runner FT 115 assuming an ideal geometrical and kinematic similarity but a Reynolds number ratio $Re_P/Re_M = 10$, first using Ackeret's scale-up formula and second applying the 3D Navier-Stokes code, s. Table 2.

Furthermore, the leakage flow between the pressure and suction side of the rotor may be computed by a 3D Navier-Stoke code. The inlet and

		η in %	$\Delta\eta$ in %	
	Re	NS3D	NS3D	Ackeret
Model	$1.5 * 10^7$	93.47	-	-
Prototype	$1.5 * 10^8$	94.06	0.59	1.18

TABLE 2. Scale-up of model runner efficiency from $Re_M = 1.5 * 10^7$ to $Re_P = 1.5 * 10^8$

outlet flow conditions are to be taken of the rotor flow analysis. Thus, the coefficients of friction torque C_T and axial forces C_{Fax} as well as the relative leakage flow rate $q_L = Q_L/Q$ can be determined as a function of the geometry of the rotor and the side spaces as well as of the point of operation (φ, ψ) and the Reynolds number Re, s. /13/.

$$C_T, C_{Fax}, q_L = f\{Geometry, \varphi, \psi, Re\}$$

Since the design of the rotor is usually carried out neglecting the a priori unknown leakage flow this influence on the inflow to the rotor has to be checked after having fixed the geometrical properties of the rotor and its side spaces. A numerical study for a pump turbine impeller PT35 has been carried out to show the effect of leakage flow and swirl defined by the operating point on the pressure distribution at the shroud with respect to cavitation. Fig. 8 shows a comparison of pressure distributions near the shroud computed for leakage free inflow and for inflow distorted by the leakage flow and swirl.

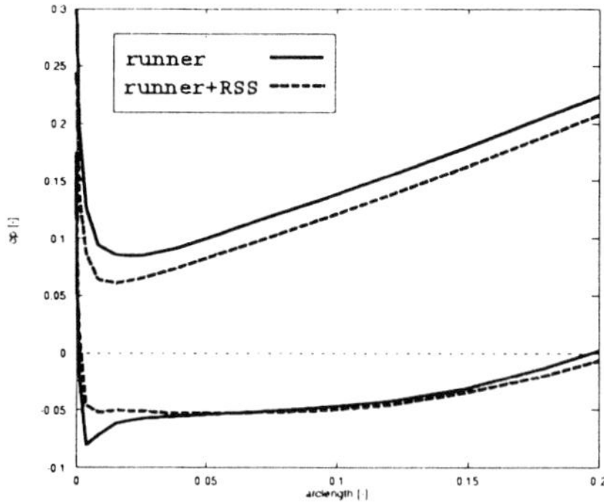

Fig. 8: Influence of leakage flow and swirl on the pressure distribution at the shroud of the pump turbine impeller PT35

2.3.2. *FEM-analysis*

Based on the final drawings a Finite Element Method (FEM) analysis may be carried out to evaluate the maximum stresses and strains in critical components. The hydrodynamic loading, i.e. the static pressure, is taken out of the flow analysis. Usually, the FEM-analysis is performed only one time without taking into account the effect of blade bending on the flow. Considering large axial turbine runners the bending of the blades due to the pressure forces may lead to a smaller turning of the flow and therefore to a lower power output. The numerical simulation of the fluid-structure interaction of a 4-bladed Kaplan turbine runner with a diameter $D = 4.2m$ has shown a maximum displacement of $\delta = 7.7mm$ yielding a maximum increase of blade angles at runner outlet of $\Delta\beta_{s2} = 3°$ and consequently a reduction of the power output of about $\Delta P/P \cong 2.5\%$, s. /14,15/. Fig. 9 shows the bending of the Kaplan turbine runner considered due to hydrodynamic forces. These basic numerical investigations indicate that the

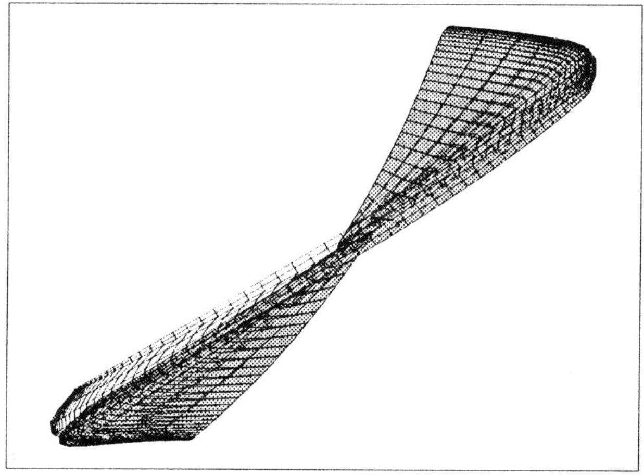

Fig. 9: Fluid-structure interaction in a 4-bladed Kaplan turbine runner

effect of fluid-structure interaction should be taken into account primarily for bigger low head turbines with small hub to tip ratio $\nu \leq 0.4$ and a small number of blades $Z \leq 4$.

3. Extrapolation to the plant performances

The performances of the prototype, i.e. discharge, head, power output, are usually scaled up assuming the similarity of geometrical dimensions and kinematic relations, whereas the efficiency is to be scaled up by a prescribed scale-up formula as a function only of the similarity parameter Reynolds number. In the case of low head machines, especially bulb turbines, an additional similarity parameter, the Froude number $Fr = 2 * H/D$, defined by the head H and runner diameter D is becoming of interest. With increasing specific speed of installed bulb turbines the prototype Froude number has reached the magnitude $Fr^P \approx 4$ while the model tests have mostly been carried out with a model Froude number $Fr^M \cong 40$. Contrary to the Froude number relation the Reynolds number of the Prototype is about 20 times bigger than that of the model. Thus, the Froude similarity is also infringed but about a factor 2 less than the Reynolds similarity law. Despite of these known facts the scale-up formula used today are focused only on the major effect of friction and inertia forces and neglect the influence of gravity on the performances. The basic physics of bulb turbine performance tests in an open flume test rig and of the interdependence of the draft tube and tail water flow is being studied in detail, s. /16/. The first results are indicating that the submersion of the draft tube outlet may have a significant influence on the performances of the power plant. The experimental results gained up to now show that the efficiency and the power output may drop down about 1% for higher values of submersion. However, much more experimental experience and theoretical knowledge is needed to scale up the performances of low head turbines with the required accuracy and reliability.

4. Conclusion

The philosophy of rapid prototyping presented might be a suitable approch to overcome existing problems and speed up the daily work in the R&D departments. However, more sophisticated methods have to be developed and integrated in a powerful system of rapid prototyping. Thus, highly efficient 3D design tools and improved CFD-codes taking into account the effects of cavitation and/ or erosion are needed for a more reliable prediction of the preformances and working life of hydraulic machinery. Nevertheless, the goal of fully numerical treatment of all hydraulic machinery problems seems to become much more realistic than before.

5. References

1. R. Schilling: CFD-Aided Design of Hydraulic Machinery Bladings CFD'91 Intensive Course on Computational Fluid Dynamics, Ed.: Velensek, B., Ljubljana, June 1991.
2. R. Schilling, Ch. Watzelt, H. Haas, L. Sporer, A. Fernandez: Abschlußbericht zum AIF-Forschungsvorhaben Nr. 8897 Echtzeit - Entwurfssystem, FKM Nr. 620075, TU München, 1993.
3. Ch. Watzelt: Entwicklung eines Echtzeit-Entwurfssystems für radiale Beschaufelungen auf einem Mehr-Prozessor-System, Dissertation, Lehrstuhl für Hydraulische Maschinen und Anlagen, TU München, 1995.
4. Ch. Watzelt, N. Riedel, A. Fernandez, R. Schilling: Real-Time Design of Hydraulic Machinery Bladings on a Parallel Environment System. ASME International Symposium on Numerical Simulations in Turbomachinery, Hilton Head, South Carolina, August 13-16, 1995.
5. N. Riedel: Entwicklung eines 3D Euler Codes mit Riemann-Solver, Int. Bericht des Lehrstuhls für Hydraulische Maschinen und Anlagen, 1995.
6. TASCflow User Documentation, Theory Documentation, 1995.
7. N. Riedel, R. Schilling: Water Turbine Runner Design Including Rotor-Stator Interaction. 6-ISCFD, Lake Tahoe, Nevada, USA, Sept. 04 - 08, 1995
8. N. Riedel, Ch. Weber: Bestimmung der optimalen Leitradöffnung für Francis- und Kaplanturbinen mit Hilfe eines 3D Euler Codes, Int. Bericht des Lehrstuhls für Hydraulische Maschinen und Anlagen, 1996.
9. R. Schilling, N. Riedel, S. Ritzinger: A Critical Review of Numerical Models Predicting the Flow through Hydraulic Machinery Bladings, 17th IAHR Symposium, Section on Hydraulic Machinery and Cavitation, Beijing, pp. 15-31, September 1994.
10. B. Stoffel, K. Weiss: Study of Internal Recirculation in Rotordynamic Pumps Operating at Partial Capacity. BRITE-Project No. RIB-0238-C.
11. N. Riedel, T. Aschenbrenner, R. Schilling: Rotor-Stator Interaction in Hydraulic Machinery. ISROMAC-6, Honolulu, Hawaii, USA, Febr. 25 -28 1996.
12. G. Kosyna: Ergebnisse der Strömungsuntersuchungen an Axialventilatoren, Private Communication, 1996.
13. Th. Aschenbrenner, N. Riedel, R. Schilling: Fluid flow interactions in hydraulic machinery, 18th IAHR Symposium, Section on Hydraulic Machinery and Cavitation,Valencia, 1996.
14. L. Sporer: Numerische Untersuchung der Fluid-Struktur-Wechselwirkung an axialen Beschaufelungen, Dissertation, Lehrstuhl für Hydraulische Maschinen und Anlagen, TU München, 1995.
15. L. Sporer, R. Schilling: Fluid-Struktur Wechselwirkung in axialen Beschaufelungen, Pumpen und Kompressoren, Vulkan-Verlag Essen, 2, 1996.
16. Ch. Schneider, W. Knapp, R. Schilling: Interdependence of draft tube and tailwater flow in bulb turbine powerplants, 18th IAHR Symposium, Section on Hydraulic Machinery and Cavitation, Valencia, 1996

FLUID TRANSIENTS IN FLEXIBLE PIPING SYSTEMS

A Perspective on Recent Developments

D.C. WIGGERT
Dept. of Civil and Environmental Engineering
Michigan State University
East Lansing, MI 48824 USA

1. Introduction

Between 1970 and 1990 a substantial amount of research activity focused on understanding the mechanical interaction between unsteady flow in piping and the resulting vibration of the pipe elements and support structures. As a result, several analytical and numerical techniques were developed and reported. A primary motivation for this work emanated from the nuclear power industry, where a number of water hammer incidents and resulting pipe motion occurred. In the 1980's, analytical methods were advanced to the point where dynamically coupled numerical predictions proved to be attainable without extensive complication.

What remains elusive today is the aspect of bringing the recent advances into practice, especially from a design viewpoint. One encounters a number of uncertainties when predicting pipe motion resulting from fluid transients, and vice versa. One uncertainty in power plant operation is the nature of the fluid transient that initiates the pipe motion; a significant example is condensation-induced water hammer, a highly stochastic event. Another uncertainty is knowing when a water hammer event is to occur, and whether that event will cause a piping failure. Presently, the accepted procedure is either to perform an analysis after the failure occurs, attempting to determine the root cause, or to anticipate a worst case scenario and offer an acceptable design alteration to the system.

The objective of this paper is to present some of the recent advances and understanding of fluid transients in flexible piping systems, and to suggest some ways in which they can be used in industrial situations. The topics to be covered include coupling mechanisms, methods of solution, and single- and two-phase fluid transients that cause pipe motion.

E. Cabrera et al. (eds.), Hydraulic Machinery and Cavitation, 58–67.

2. Background

An excellent and thorough review of fluid transients in liquid-filled piping systems has been presented by Tijsseling (1996). His paper relates the historical contributions that lead to the present physical and mathematical understanding of unsteady liquid-pipe interaction. In addition, the paper describes various interactive fluid-piping numerical algorithms that provide predictions in the time domain and discusses their validation by physical experiments. Tijsseling's paper is recommended reading for those who desire an historical perspective on fluid-structure interaction in liquid filled piping, as well as an accounting of the significant contributions made by many investigators. The brief historical review that follows is excerpted from Tijsseling (1996).

Some type of pipe motion has always been considered when formulating the water-hammer problem; for example, Korteweg in 1878 accounted for the elasticity of pipe walls in the acoustic wave speed in a pipe. Lamb, in his treatment of acoustic velocities in tubes in 1898, recognized that an acoustic wave propagating in the liquid is modified by the tube wall motion, and that both axial and radial vibrations occur in the tube. Several researchers extended Lamb's work to include bending and rotation of the pipe element. Frequency-domain analyses that incorporated dynamic coupling between contained liquid and junction-like structural components were developed for simple systems such as a short single-elbow pipe.

With the advent of high-speed computing in the 1960's, water hammer analysis followed a traditional approach, progressing with the basic assumption that the piping and support structures are rigid and do not affect the predicted wave forms. The one exception was the adjustment of the acoustic wave speed based on static-elastic interaction between the liquid pressure and the pipe circumferential and axial strains. The possible vibration of a pipe component due to a water hammer pressure loading has been recognized in textbooks, and early attempts at coupled time-domain analysis incorporated, for example, a lumped mass-spring to approximate a portion of the piping system.

The nuclear power industry in particular has been required to analyze large-scale piping systems to maintain their integrity while being subjected to thermal stresses and seismic loads. Typically, standard structural finite element and modal analysis procedures have been used to predict piping vibrations without regard to acoustic waves propagating in the contained liquid. Water hammer loads had been considered secondary and often were not the major focus of a dynamic design analysis of the piping. From 1970 to 1980 in both Europe and North America, water hammer incidents in nuclear power plants kindled an interest in water hammer induced piping vibration; as a result, fluid transient and structural dynamics disciplines began to merge. Work was carried out simultaneously by a number of institutions and agencies, resulting in several computational methodologies, and more significantly, an improved understanding of the intricate and subtle interaction between fluid transient and pipe motion. The characteristics of systems studied and analytical methodologies employed to date are summarized in Table 1.

TABLE 1. Various approaches to fluid-structure interaction in piping systems.

System component characteristics:
- Piping: elastic or inelastic
- Fluid: continuous or discontinuous (liquid slug or column separation)

Analytical approach:
- Coupled versus uncoupled
- Time domain versus frequency domain

Uncoupled analysis consists of first obtaining the dynamic pressures from a water hammer calculation without regard to any piping motion, and then using them as loads to predict the piping response in a separate structural dynamics analysis. By contrast, coupled analysis requires that the water hammer and structural motion be solved simultaneously. In the sections that follow, we will focus on elastic piping, and on both continuous and discontinuous fluid components. Time-domain solutions will be emphasized, and the validity of coupled versus uncoupled analysis will be evaluated.

3. Essentials of Liquid-Piping Interaction

Three coupling mechanisms can be identified in liquid-piping interaction. *Poisson coupling* is associated with the radial stress perturbations produced by liquid pressure transients that translate to axial stress perturbations by virtue of the Poisson ratio coefficient. The axial stress and accompanying axial strain perturbations travel as waves in the pipe wall at approximately the speed of sound in the pipe. Typically its magnitude is three to five times greater than the acoustic velocity in the contained liquid in the pipe. *Friction coupling* is created by the transient liquid shear stresses acting on the pipe wall; usually it is insignificant when compared to the other coupling mechanisms. Both Poisson and friction coupling are distributed along the axis of a pipe element.

The third, and often the most significant, coupling mechanism is *junction coupling*, which results from the reactions set up by changes in liquid momentum at discrete locations in the piping such as bends, tees, valves, and orifices. In flexible piping, water hammer waves impacting at junctions may set up a vibration which in turn may translate to a variety of structural responses (bending, rotation, shear, axial stresses) at locations away from the junction. In addition, the vibrating junction will induce fluid transients in the contained liquid column, with acoustic waves traveling upstream and downstream from the junction. Clearly, the net effect is a complex motion in both the piping and the contained liquid, with resulting waveforms being highly dependent upon the geometry of the pipe system. Sources of excitation include not only those associated with liquid motion, but may also come from the structural side. Figure 1 relates some of the various excitation mechanisms.

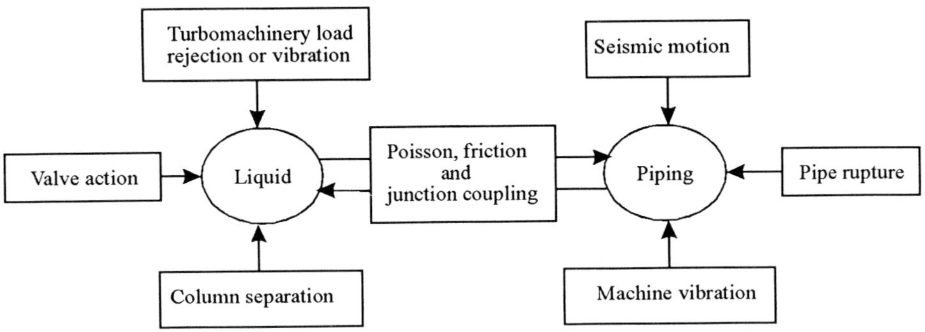

FIGURE 1. Sources of excitation in fluid-piping interaction.

Most large-scale industrial piping systems are comprised of liquid-filled, thin-walled piping elements, so that the transient behavior of the liquid can be described as one-dimensional wave phenomena. Consider a fundamental liquid-filled prismatic pipe element, with the z-coordinate positioned along the pipe axis, and the x- and y-coordinates normal to the z-axis. The pipe can transmit torsional motion θ_z in the axial direction, lateral motion u_x and u_y normal to the axis, and longitudinal motion u_z and stress σ_z along the axis. In addition, the contained liquid can exhibit transient velocity V and pressure p. Neglecting radial inertia of the pipe wall, a set of six partial differential equations can be derived that relate the liquid pressure and velocity, the axial displacement and stress of the pipe, the lateral pipe displacement, and the tortional rotation of the pipe (Heinsbroek and Kruisbrink, 1993):

$$\frac{\partial V}{\partial t} + \frac{1}{\rho_f}\frac{\partial p}{\partial z} + \frac{f V_r |V_r|}{2D} = 0 \tag{1}$$

$$\frac{1}{\rho_f c_f^2}\frac{\partial p}{\partial t} + \frac{\partial V}{\partial z} - \frac{2v}{E}\frac{\partial \sigma_z}{\partial t} = 0 \tag{2}$$

$$\rho_t A_t \frac{\partial^2 u_z}{\partial t^2} - EA_t \frac{\partial^2 u_z}{\partial z^2} = \frac{vDA_t}{2e}\frac{\partial p}{\partial z} + \frac{\rho_f A_f f V_r |V_r|}{2D} + \rho_t A_t g \sin\gamma \tag{3}$$

$$\left(\rho_t A_t + \rho_f A_f\right)\frac{\partial^2 u_x}{\partial t^2} + EI\frac{\partial^4 u_x}{\partial z^4} = -\left(\rho_t A_t + \rho_f A_f\right) g \cos\gamma \tag{4}$$

$$\left(\rho_t A_t + \rho_f A_f\right)\frac{\partial^2 u_y}{\partial t^2} + EI\frac{\partial^4 u_y}{\partial z^4} = 0 \tag{5}$$

$$\rho_t J \frac{\partial^2 \theta_z}{\partial t^2} - GJ\frac{\partial^2 \theta_z}{\partial z^2} = 0 \tag{6}$$

Additional parameters in the equations are: $c_f = 1/\sqrt{\rho_f(1/K + D/Ee)}$, the liquid wave speed; f = Darcy-Weisbach friction factor; V_r = liquid velocity relative to the axial pipe motion; D = pipe diameter; E = Young's modulus of the pipe material; ν = Poisson's ratio; K = bulk modulus of liquid; ρ_f = fluid density; A_f = cross-sectional flow area; ρ_t = density of pipe material; A_t = cross-sectional pipe area; γ = pipe angle; I = moment of inertia of pipe; G = pipe material shear modulus, and J = polar moment of inertia of pipe. The extended liquid and axial displacement equations are coupled by friction and Poisson coupling along the axis of the pipe element. At pipe junctions, flow, liquid pressure forces, moments, displacements and torsion are translated from one pipe to another. Standard boundary conditions for fluid transients are employed; these include valve motion, turbomachinery transients such as load rejection or pump trip, and column separation. An example of a structural boundary condition is seismic excitation (Hatfield and Wiggert, 1990).

4. Methods of Solution

Solutions can be formulated in either the time domain or in the frequency domain. In the time domain, the most successful technique is to solve the water hammer equations by the method of characteristics and the structural equations are either by finite element representation (Lavooij and Tijsseling, 1991) or by modal component synthesis (Hatfield and Wiggert, 1991). Under special circumstances the method of characteristics has been used to solve the Poisson-coupled axial pipe motion (Budny, *et al.*, 1991, Heinsbroek, *et al.*, 1991) as well as lateral pipe motion (Wiggert, *et al.*, 1985a; Tijsseling, Vardy, and Fan, 1996). Frequently it is advantageous to simplify the structural motion to account for only those modes that interact significantly with the liquid transient. Under such circumstances, for small degree-of-freedom structural motion the method of lumped parameters (Wiggert, *et al.*, 1985b) can be employed, and for larger degrees of freedom, modal component synthesis is useful.

5. Water Hammer Interaction with Piping Structure

A simple example of how a water hammer load interacts with a vibrating pipe (Budny, 1988) is demonstrated in Fig. 2. The pipe is horizontally suspended so that it can move unrestrained in the axial direction. At the downstream end a valve was rapidly closed, and the resulting pressure waveform immediately upstream of the valve is shown. For the numerical analysis, the method of characteristics was utilized for both the liquid column and the axial mode of piping vibration. Two fundamental frequencies are superposed in the wave form: 1) water hammer $c_f/4L \cong 6.6$ Hz, and pipe wall $c_t/4L \cong$ 19 Hz, where $c_t = \sqrt{E/\rho_t}$, the acoustic speed in the pipe wall. Even though both Poisson coupling and axial pipe vibration are taking place, the axial vibration

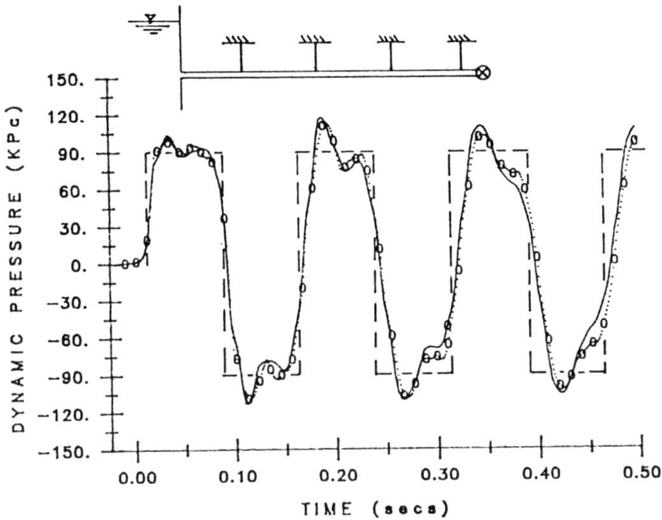

FIGURE 2. Axially-coupled motion in the liquid and structural modes (Budny, 1988).
Legend: o·····o experimental; —— numerical; – – – Joukowsky

dominates the feedback into the pressure wave form. One noteworthy observation is that the Joukowsky pressure rise of 90 kPa is exceeded because of the dynamic interaction.

A more complicated interaction takes place when the piping system has an intricate geometry (Kruisbrink and Heinsbroek, 1992; Heinsbroek and Kruisbrink, 1993). For example, consider the system shown in Fig. 3. From a water hammer viewpoint, the system is identical to that shown in Fig. 2: a single liquid column with a reservoir at the upstream end and a rapid valve closure excitation at the downstream end, resulting in one primary mode of fluid motion. However, because of the complexity of the piping layout, there may be three or more significant modes of vibration in the piping excited by the valve closure. The pressure waveform at the valve, Fig. 4, reflects the feedback due to bending and translation of the piping. The pressure rise predicted by the Joukowsky approximation is approximately 3.8 bar, whereas the recorded peak dynamic pressure is about 5 bar. In the computation, the authors used a combined method of characteristics/finite element solution.

Both Figs. 2 and 4 demonstrate that in many similar industrial systems, when the piping is relatively compliant and not rigidly supported, the water hammer waveforms will deviate from the ideal Joukowsky predictions. Quite often, with higher frequency components the peak pressures in the waveforms will exceed the Joukowsky prediction, but normally not by a significant amount. What may be more important from an integrity viewpoint are the dynamic pipe strains and stresses resulting from the water hammer activity. It has been found that measured pipe displacements and strains are accurately predicted by coupled analysis, but are not accurately predicted using uncoupled analysis (Heinsbroek, 1993; Heinsbroek and Tijsseling, 1994).

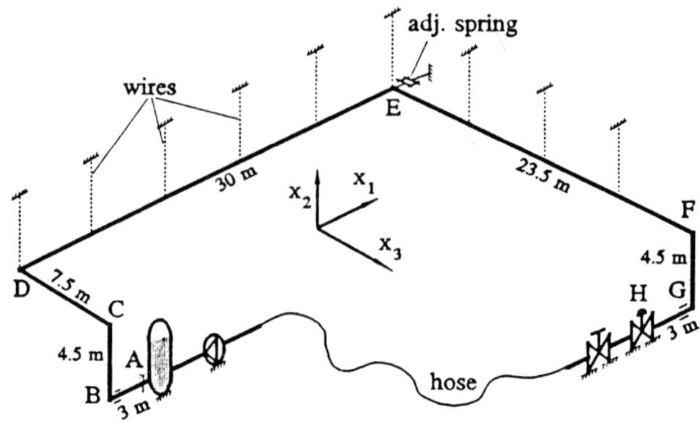

FIGURE 3. Experimental compliant pipeline system (Kruisbrink and Heinsbroek, 1992).

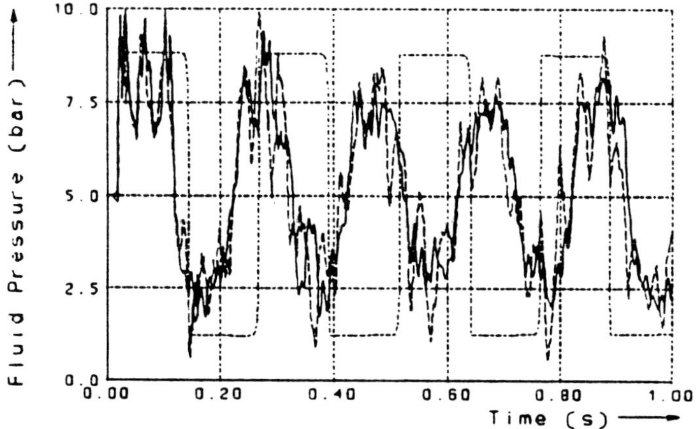

FIGURE 4. Dynamic pressure at the shutoff valve (Kruisbrink and Heinsbroek, 1992).
Legend: —— measured; – – – computed; ········Joukowsky

The significance of piping restraints has received little attention. The interaction of a pipe rack with the axial translation of a pipe was recently investigated by Tijsseling and Vardy (1996). They linked the interaction between the moving pipe and the rack by means of dry Coulomb friction, and successfully predicted the pipe axial velocity and the water hammer in the contained liquid.

6. Two-Component Flows

A series of studies related to fluid structure interaction and cavitation have been conducted in a novel manner by impacting a solid metal rod at the end of a freely suspended liquid-filled horizontal pipe (Tijsseling and Fan, 1991a,b; Fan and Tijsseling, 1992). In addition, Tijsseling, Vardy and Fan (1996) applied the same experimental procedure to a single-elbow pipe arrangement. The authors obtained experimental verification of the interaction between vapor cavities, the liquid, and the axial pipe wall motion. They utilized a discrete cavity model combined with fluid-structure interaction models to obtain predictions that compared well with experiments. These investigations show that some of the differences reported in earlier water hammer literature between measured and predicted cavitation can be attributed to fluid structure interaction.

A significant cavitation phenomenon that can lead to pronounced fluid-structure interaction is condensation-induced water hammer (Chou and Griffith, 1989). It is known to occur in piping when subcooled water comes in to contact with steam, and it has also been identified in refrigeration piping systems. Because of the stochastic nature of the void collapse, it is difficult to isolate an individual event and accurately estimate the loads that occur on the piping structure.

Liquid slug motion in voided lines leading to pipe support failure has been documented to occur in power plant piping. A slug, driven by a pressure gradient, can accelerate to high velocities and transfer significant loads when it impacts at a pipe junction, Fig. 5. There have been very few studies conducted that focus on this phenomenon (Fenton and Griffith, 1990; Bozkus and Wiggert, 1992). Experiments have shown that the impact waveforms and resulting peak pressures are highly stochastic, thereby increasing the uncertainty of a numerical prediction. Because of the

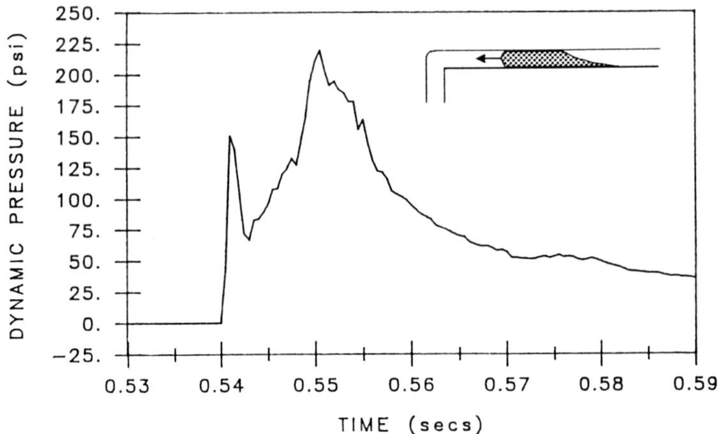

FIGURE 5. Pressure waveform at a pipe elbow due to liquid slug impact (Bozkus and Wiggert, 1992). Slug velocity = 17.3 m/s, slug length = 2.1 m, pipe diameter = 50 mm.

short duration of the impact pulse, when estimating piping stresses it is reasonable to uncouple the fluid transient from the structural motion. Additional experimental and analytical research is recommended, both for condensation-induced water hammer and liquid slug impact in full scale piping systems.

7. Discussion

Frequency domain studies have not been highlighted herein. Component synthesis techniques have been successfully applied to oscillatory flows that induce vibrations in small slender piping systems (Lesmez, Wiggert, and Hatfield, 1990). Recently, there has been interest in adapting these to large piping systems (Charley and Caignaert, 1993; Svingen, 1994, 1995; Gajic, *et al.*, 1995). In hydropower and piping systems, vibrations in the flow may be induced by turbomachinery, leading to possible resonance and significant fluid-structure interaction. Additional research and development in this area is likely to occur.

A fundamental question often raised is: Can uncoupled analysis be used to determinine whether a piping system can withstand a water hammer load without substantial vibration or possible support failure? One must deal with an amount of uncertainty when estimating water hammer loads on piping. Dynamic pressures larger than those predicted by the Joukowsky approximation are known to exist when the piping is compliant, and coupled analysis has been shown to accurately predict such situations. For most commercial piping situations, the deviations in dynamic pressure are not significant, so that uncoupled analysis may be adequate, especially when safety factors are incorporated. However, with the present understanding, coupled analysis is possible without undue additional effort. Component synthesis enables one to combine conventional structural dynamic analysis techniques with extended water hammer algorithms, producing solutions that accurately predict the true dynamic interaction.

Acknowledgment

Appreciation is extended to Arris Tijsseling and Bjørnar Svingen for their assistance during preparation of the manuscript.

References

Bozkus, Z., and Wiggert, D.C. (1992) Hydromechanics of slug motion in a voided line, in R. Bettess and J. Watts (eds), *Unsteady Flow and Fluid Transients*, A.A. Balkema, Rotterdam, pp. 77-86.

Budny, D.D. (1988) The influence of structural damping on the internal fluid pressure during a fluid transient pipe flow, Ph.D. dissertation, Michigan State University, Department of Civil and Environmental Engineering, E. Lansing.

Budny, D., Wiggert, D.C., and Hatfield, F.J. (1991) The influence of structural damping on internal pressure during a transient flow, *ASME Journal of Fluids Engineering*, **113**, 424-429.

Charley, J., and Caignaert, G. (1993) Vibroacoustical analysis of flow in pipes by transfer matrix with fluid-structure interaction, *Proc. Work Group on The Behavior of Hydraulic Machinery Under Steady Oscillatory Conditions, 6th International Meeting*, IAHR, Lausanne.

Chou, Y., and Griffith, P. (1989) Avoiding steam-bubble collapse-induced water hammers in piping systems, EPRI Research Project NP-6647.

Fan, D., and Tijsseling, A. (1992) Fluid-structure interaction with cavitation in pipe flows, *ASME Journal of Fluids Engineering*, **114**, 268-274.

Fenton, R.M., and Griffith, P. (1990) The forces at a pipe bend due to the clearing of water trapped upstream, *Transient Thermal Hydraulics and Resulting Loads on Vessel and Piping Systems-1990*, American Society of Mechanical Engineers PVP Vol. 190, New York, pp. 59-67.

Gajic, A., Pejovic, S., Stojanovic, Z., and Josef, K. (1995), *Proc. Work Group on The Behavior of Hydraulic Machinery Under Steady Oscillatory Conditions, 7th International Meeting*, IAHR, Ljubljana.

Hatfield, F.J., and Wiggert, D.C. (1990) Seismic pressure surges in liquid-filled pipelines, *ASME Journal of Pressure Vessel Technology*, **112**, 279-283.

Hatfield, F.J., and Wiggert, D.C. (1991) Water hammer response of flexible piping by component synthesis, *ASME Journal of Pressure Vessel Technology*, **113**, 115-119.

Heinsbroek, A.G.T.J., Lavooij, C.S.W., and Tijsseling, A.S. (1991) Fluid-structure interaction in non-rigid piping: a numerical investigation, *SMiRT 11 Transactions*, B12/1, Tokyo, pp. 309-314.

Heinsbroek, A.G.T.J. (1993) Fluid-structure interaction in non-rigid pipeline systems: comparative analysis, *Proc. ASME/TWI 12th International Conference on Offshore Mechanics and Arctic Engineering*, Glasgow, pp. 405-410.

Heinsbroek, A.G.T.J., and Kruisbrink, A.C.H. (1993) Fluid-structure interaction in non-rigid pipeline systems: large scale validation experiments, *SMiRT 12 Transactions*, J08/1, Stuttgart, pp. 205-210.

Heinsbroek, A.G.T.J., and Tijsseling, A.S. (1994) The influence of support rigidity on waterhammer pressures and pipe stresses, *Proc. Second International Conference on Water Pipeline Systems*, BHR Group, Edinburgh, pp. 17-29.

Kruisbrink, A.C.H., and Heinsbroek, A.G.T.J. (1992) Fluid-structure interaction in non-rigid pipeline systems: large scale validation tests, *Pipeline Systems*, Kluwer Academic Publishers, Dordrecht, pp. 151-164.

Lavooij, C.S.W., and Tijsseling, A.S. (1991) Fluid-structure interaction in liquid-filled piping systems, *Journal of Fluids and Structures*, **5**, 573-595.

Lesmez, M.W., Wiggert, D.C., and Hatfield, F.J. (1990) Modal analysis of vibrations in liquid-filled pipe systems, *ASME Journal of Fluids Engineering*, **112**, 311-318.

Svingen, B. (1994) A frequency domain solution of the coupled hydromechanical vibrations in piping systems by the finite element method, *Proc. IAHR International Symposium on Hydraulic Machinery and Cavitation*, Beijing.

Svingen, B. (1996) Fluid structure interaction in slender pipes, *Proc. Pressure Surges and Fluid Transients in Pipelines and Open Channels*, Mechanical Engineering Publications Ltd., Suffolk.

Tijsseling, A.S., and Fan, D. (1991a) The response of liquid-filled pipes to vapour cavity collapse, *SMiRT 11 Transactions*, J10/2, Tokyo, pp. 183-188.

Tijsseling, A.S., and Fan, D. (1991b) The concentrated cavity model validated by experiments in a closed tube, *Proc. International Meeting on Hydraulic Transients with Water Column Separation: 9th Round Table of the IAHR Group*, Session A-3, Paper 2, Valencia.

Tijsseling, A.S. (1996) Fluid-structure interaction in liquid-filled pipe systems: a review, *Journal of Fluids and Structures*, **10**, 109-146.

Tijsseling, A.S., and Vardy, A.E. (1996) Axial modeling and testing of a pipe rack, *Proc. Pressure Surges and Fluid Transients in Pipelines and Open Channels*, Mechanical Engineering Publications Ltd., Suffolk, UK.

Tijsseling, A.S., Vardy, A.E., and Fan, D. (1996) Fluid-structure interaction and cavitation in a single-elbow pipe system, *Journal of Fluids and Structures*, to be published.

Wiggert, D.C., Hatfield, F.J., and Stuckenbruck, S. (1985a) Analysis of liquid and structural transients in piping by the method of characteristics, *Fluid transients in fluid-structure interaction–1985*, American Society of Mechanical Engineers, New York, pp. 97-102.

Wiggert, D.C., Otwell, R.S., and Hatfield, F.J. (1985b) The effect of elbow restraint on pressure transients, *ASME Journal of Fluids Engineering*, **107**, 402-406.

SYMPOSIUM CONTRIBUTORS

A DECISION AID SYSTEM FOR HYDRAULIC POWER STATION REFURBISHMENT PROCEDURE

Francis and Kaplan Turbine

BELLET L., PARKINSON E.* AND AVELLAN F.
IMHEF/LMH/EPFL - Av. de Cour,33
1007 LAUSANNE - Switzerland

AND

COUSOT T. AND LAPERROUSAZ E.
EDF-CNEH - Savoie Technolac
73373 LE BOURGET DU LAC - France

1. Project

Maintenance policies for hydrogenerating equipment vary considerably from one operator to another, from a least cost strategy to the greatest care and tightest scheduling. However, whatever policy is adopted, sooner or later maintenance is no longer enough to uphold acceptable levels of performance or safety. When that happens, renovation or refurbishment takes over.

In general, the decision to refurbish a scheme is inspired by a combination of causes, which can be grouped in three types: age, downgraded performance, and unsuitable or costly operating techniques.

It is fairly difficult to evaluate the effect of age, although it is well known that, as time passes, the risk of a serious accident increases. Obviously, refurbishment consecutive to an accident cannot be scheduled and can therefore not take place at the same cost nor with the same results as a scheduled refurbishment that is correctly integrated into the generating program.

Even the most painstaking maintenance works are not enough to preserve initial performance levels. Energy losses increase inexorably everywhere in the system: hydraulic, mechanic and electric losses.

Finally, operation becomes increasingly difficult. The original design requirements, such as operating limits imposed by cavitation or instability, are compounded by constraints due to wear and permanent deformations (watertightness of guide vanes or gates and valves) and the safety of the plant in general. Operating losses due to the frequency of maintenance must also be considered, and even, within a cascade of plants, the unsuitability of one scheme versus those upstream and downstream from it.

E. Cabrera et al. (eds.), Hydraulic Machinery and Cavitation, 71–80.

Furthermore, although the *raison d'être* of power plants remains the generation of energy at least cost, other needs or requirements have grown up over the years such as:

- the safety of facilities;
- the match between performance levels and network requirements;
- compliance with new legislation;
- new operating requirements (irrigation, low stage support, recreational uses).

A well-managed refurbishment must take all these new data into account, as well as technological and scientific advances. Depending on those new requirements, a refurbishment operation can take many forms, from replacement of elements with identical equipment to the construction of a whole new plant. Compromises also exist, such as replacing a runner with one of modern design. In general the objective is to minimise the cost of the operation or guarantee that it is cost-effective because of improved generating performances.

Improving performance by replacing the runner alone, and when necessary modifying the generator, means extra cost and risk in comparison to replacement with identical equipment, extra cost because more study is required (and even model tests), and risk because it can never be absolutely guaranteed that performance will actually improve.

The IMHEF and EDF have been working together for several years on various subjects relating to hydraulic machinery. Since 1995, the two have joined forces to develop a procedure for the evaluation of older turbines. This joint work was inspired by the realisation of two facts:

1- EDF is confronted with an ageing stock of hydro facilities and the renew al of many of its concession agreements. EDF's engineers working in hydro generation are reflecting on how to enhance the value of French facilities, as part of a project called "Hydraulique Demain" ("The Hydraulics of Tomorrow"). This project is aimed at implementing the technico-economic tools and methods required to rationalise expenditures and to analyse the potential of hydro facilities.

2- The IMHEF and EDF have long worked for industrial groups, operators or water resources managers as consulting engineers. With the resulting awareness of those entities' needs in refurbishment projects, they have decided to adapt modern techniques to that field.

Our work is part of the global process required to prepare for a refurbishment, which among other things includes technical, operating and economic studies. Our objective is to ascertain how suitable a machine is for its site and to evaluate the energy potential of the stationary elements of an existing turbine. We therefore first focus on those stationary parts (spiral casing, distributor and where necessary headrace pipes) and then on the runner if necessary.

Such a study can be scheduled in various stages of a project, and can have various goals. When it is scheduled before design, it is a decision-making tool that helps in establishing the technical and economic balance sheet on a facility and in orienting the choice of what type of refurbishment to plan. If the decision is made to change the runner, it will serve in setting up the technical specifications for the new runner on solid bases and when necessary in predicting whether changes will be needed to stationary parts.

2. Refurbishment procedure

There are a multitude of components in a hydroelectric scheme, and therefore the engineer in charge of study of a refurbishment is faced with a wide and complex range of options. To guide his choices, he must take into account both technical and economic aspects, without neglecting legal aspects (the concession agreement) and environmental aspects, which may also play a decisive role. The procedure described herein concerns only technical aspects and is intended to help in establishing a rapid and accurate balance sheet on the site under study in order to be able to conduct an economic study. This procedure is divided into three stages, progressing from a global approach to a specific approach.

2.1. PHASE 1:RENOVA

The first stage in a refurbishment study consists in gathering all the available information on the project at hand concerning the hydraulic elements in the scheme, with the purpose of drawing up a balance sheet on the following main elements: water intakes, penstocks, gates and valves, turbine, headrace and tailrace channels.The information collected is used to identify critical zones in the scheme that the study will focus on. A complete and systematic summary can be transcribed in computer files of characteristics to create a data base, called *Renova*. This data base was created as part of the present project in order to:

- inventory existing machines described in the literature, whether refurbished or not;
- run a statistical study based on those machines;
- establish a data base for refurbishment projects.

In the statistical study, the machine at hand is compared to other machines, whether modern or older, and where appropriate a new type of turbine can be defined that is better suited to the site's present characteristics. This data base is for the moment essentially concentrated on the turbine, but can easily be extended to the other elements in a scheme.

2.2. PHASE 2:RENOVATURB

The second phase is more specific, as it concerns the high pressure parts of the turbine, i.e., the spiral casing and distributor. Its objective is to precisely determine conditions upstream from the runner. The *Renovaturb* computer program, which was partially developed during this project, allows geometrical analysis of those components:

- verification and computation of geometrical quantities (sections laws, skeletons lines, distributor opening law, opening, etc.);
- prediction of flow angles at the spiral casing outlet;
- compatibility between the estimated angles and actual angles of stay vanes;
- compatibility of geometrical angles between stay vanes and guide vanes;
- automatic generation of a structured mesh of the distributor and semi-automatic of the spiral casing in preparation for phase 3.

2.3. PHASE 3:COMPUTER MODELLING

Knowledge of hydraulic conditions in the high pressure parts of the turbine (spiral casing and distributor) is very important, and a computer model allows a more in-depth analysis of each component. Phase 2 of this procedure gives an estimation of the average flow angle, defined as $\alpha = atan[\frac{C_m}{C_u}]$ where C_m and C_u are respectively the meridional and tangential velocity, at the spiral casing outlet. If a more precise idea of the phenomena in the spiral casing is desired, a numerical calculation can be done, although it must be borne in mind that considerable time is needed for calculation and generation of the mesh (as the tong is specific to each spiral casing, generation of it must be modified in the *Renovaturb* program).

Calculation of the distributor gives the changes in flow angle as well as those in kinetic moment, so that the maximum energy available at the runner inlet and the flow angle are known. The runner computation will indicate whether it is capable of transforming this available energy. The geometry of the blades and the meridian channel are defined on the basis of drawings or from site measurements (articulated arm or theodolite) if no drawings are available or if the runner's geometry has been changed.

A study of the draft tube has not been envisaged for the moment, although it is a major element in the turbine, because of the complexity of the unsteady phenomena that take place in it. Calculation of this element with a Navier-Stokes program has still not been validated experimentally, and so cannot be systematically employed. Two types of computation have been envisaged for this procedure:

- Calculation using a Euler type code: the EULER-IMHEF [5]. The CALECHE © code (Metraflu), a finite element code based on a structured mesh, will be used for future calculation of the runner;
- Calculation with a Navier-Stokes type code: the N3S © [1] code (EDF-Simulog) in its turbomachine version.

3. Application to a real-life case

The Pinet hydropower plant on the Tarn river in France was chosen to validate this procedure. It was commissioned in 1929, with five vertical axis Francis turbines, each outputting maximum mechanical power of 8 MW at 250 rpm^{-1} for a net head of 32 m and a discharge of $Q = 30.4\ m^3/s$. The Thoma number σ, or cavitation factor, defined as being the ratio between net positive suction specific energy and specific energy is close to 0.3.

3.1. PHASE 1

Data gathered on the site, including some of the original drawings, were used in this study. A statistical comparison of the existing turbine with the Francis turbines inventoried in the data base is presented in Figures 1 and 2. The dimensionless discharge and energy coefficients are defined as $\varphi = \frac{Q}{\pi.\omega.R^3}$ and $\psi = \frac{2.E}{\omega^2.R^2}$, where ω is the rotating speed and R the raduis. The following observations can be made:

- Specific speed, defined as $\nu = \frac{\omega.\sqrt{\frac{Q}{\pi}}}{(2.E)^{3/4}}$, which is representative of the type of turbine used, is well suited to the available head.
- The reference runner diameter, representative of the turbine's characteristic dimensions, is well suited to the discharge through the turbine.
- The reference diameter, on the other hand, is not suited to the mechanical power output by the runner. It is therefore possible, with a better runner, to supply more mechanical power to the shaft.
- The coefficient φ^2/ψ, that is representative of the specific kinetic energy at the runner outlet versus head, is within the statistical average. This coefficient is also proportional to energy losses in the draft tube, and therefore would seem to be suited to the runner being used.
- The Thoma number σ, or cavitation factor, is not isolated from the other values for Francis turbines (the number of values is reduced, since σ is not routinely given in the literature). However, this factor alone cannot give information on turbine elevation setting, which is why the coefficient κ, defined as $\kappa = \frac{\psi_{\overline{1}}}{tg\beta_{\overline{1}}}.\sigma_{min}$ where $\beta_{\overline{1}}$ is the relative flow angle at the runner trailing edge, has been calculated.
- The lower turbine setting acceptable for this machine requires that coefficient $\kappa \simeq 1$. A considerable margin of safety was taken by using a value of 3.5.

On the basis of these observations, it was proposed that a Francis runner with a smaller diameter would be better suited to the present operating conditions. A statistical comparison with the Kaplan turbines in the inventory (figure 1) shows that the use of a Kaplan (or Impeller) turbine with this reference diameter could also be envisaged.

It should be noted that almost none of the original equipment has been replaced (generators, transformers, runner, etc.), and, as there is no reg-

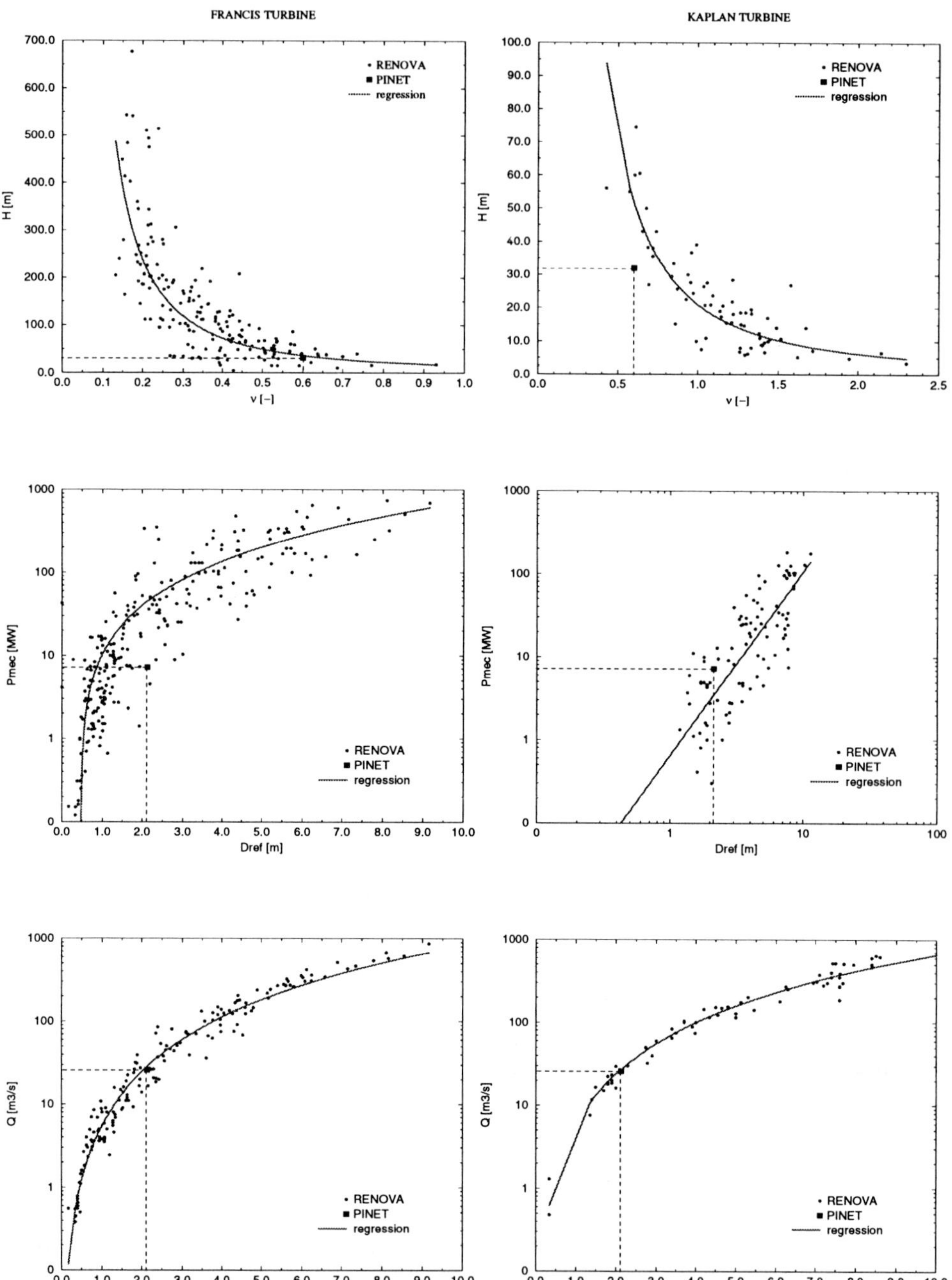

Figure 1. Head/Specific speed - Mechanical power/Diameter - Discharge/Diameter

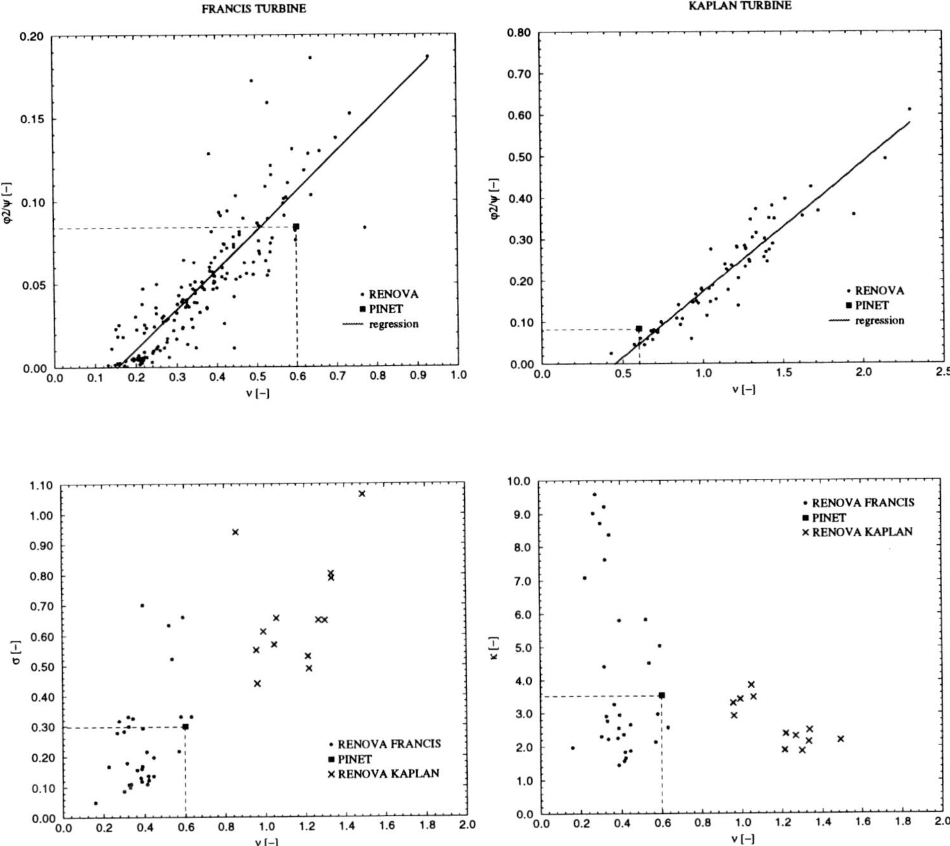

Figure 2. Coefficient $\frac{\varphi^2}{\psi}$/Specific speed - Thoma number/Specific speed - Coefficient κ/Specific speed

ulating device, the turbines always run at the same operating point. The essential information gathered on the turbine was the following:

- Major losses were measured in the inlet valves, as well as in the five-branch manifold upstream from those valves.
- The double curve spiral casing has two by-passes that divert flow directly downstream from the turbine in the event of any problem, and leakage has been observed at the outlet from those elements.
- The runner is suffering from relatively major erosion due to cavitation at the pressure-side inlet of the blades near the runner band.

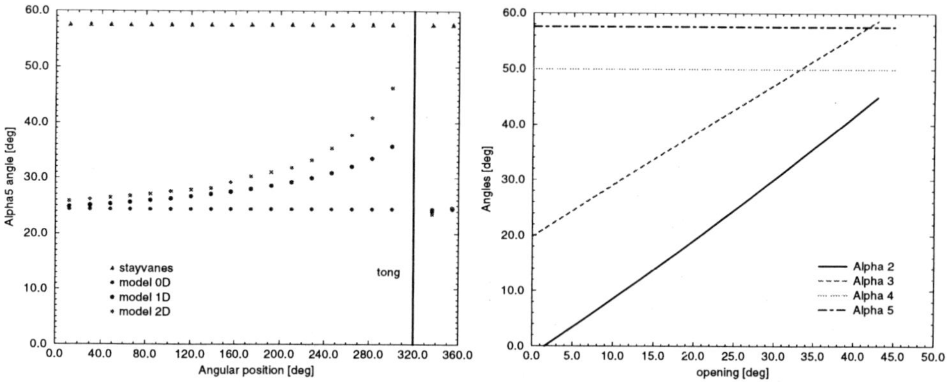

Figure 3. Estimated angles at the spiral casing outlet

Figure 4. Geometrical angles in the distributor

3.2. PHASE 2

Index 5 & 4 and 3 & 2 are corresponding respectively to the stay vane leading & trailing edge and guide vane leading & trailing edge. Three models for evaluation of average hydraulic flow angle at the spiral casing outlet have been programmed into *Renovaturb* and are shown in figure 3. The assumptions applied are the following:

- model 0D: $C_u = cste$ at the inlet and $C_r = cste$ at the outlet.
- model 1D: discharge is absorbed evenly around the machine.
- model 2D: the spiral casing is considered as a coiled pipe where headloss is assumed constant per unit of length.

It can be noted that the skeleton angle at the leading edge of the stay vanes is 57.6 [deg] while the models used give an average angle close to 28 [deg]. We may therefore expect major incidence losses at this location. On Figure 4, we note an intersection of the curves representing angles α_3 and α_4 for an opening of 33.18 [deg], where losses would be at a minimum in the distributor. As the opening used was 33.5 [deg], the setting of the stay vane is well suited to the setting of the guide vane.

3.3. PHASE 3

The distributor is of the radial type with adjustable guide vanes. It is composed of symmetrical stay vanes and guide vanes. For the Euler calculation, a structured mesh of a channel between blades, shown in Figure 5, was created with 29,600 nodes: $I_{max} = 74 \quad J_{max} = 20 \quad K_{max} = 20$, where direction i corresponds to the main direction of flow, and k to the height of the distributor. 10,000 iterations, i.e. one hour of computation time on an IBM Risc 6000-3AT work station (196 Mb) were required. The mesh for the

N3S computation was obtained from a structured hexahedral mesh measuring 70x16x15 divided into tetrahedrons to give an unstructured mesh of 125,941 nodes. The computation was done on EDF's Research and Development Department's Cray X-MP, with 1000 iterations and 3.5 hours of computation time.

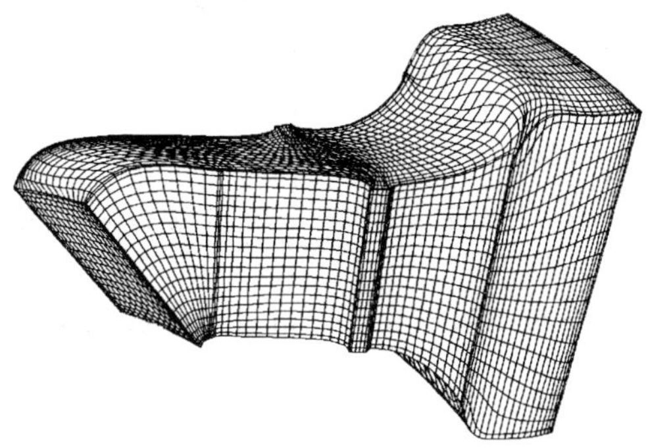

Figure 5. Structured mesh of the computational domain

3.3.1. *Results of the Euler computation*

For each side of the mesh i the average value of kinetic moment and hydraulic flow angle α was calculated (Figure 6), using the following formula:

$$\overline{r.c_u} = \frac{\int_{A_i} r c_u \vec{c}.\vec{n} dA}{\int_{A_i} \vec{c}.\vec{n} dA} \qquad \overline{tg\alpha_i} = \frac{\overline{r.c_m}}{\overline{r.c_u}} = \frac{\int_{A_i} r c_m \vec{c}.\vec{n} dA}{\int_{A_i} r c_u \vec{c}.\vec{n} dA} \qquad (1)$$

Two calculations, one with an initial angle of 28 [deg] and the other with 24.37 [deg], were run. Despite the difference in the flow angle at the distributor inlet, the average kinetic moment and the average flow angle at the outlet remained unchanged. It was observed that the average flow angle at the distributor outlet α_2 is about 37.25 [deg], while the geometric angle of the camber line α_{2geom} is 34 [deg] (see Figure 4). Maximum dimensionless energy available at the runner inlet, assuming that we are close to the top point and that $U.C_u \simeq 0$, is $\psi_{disp} = 0.628$ $(2.r.c_u = 0.7)$.

3.3.2. *Results of the Navier-Stokes computation*

For the same flow inlet conditions, the average flow angle at the distributor outlet is about 33.0 [deg], so 4 [deg] lesser than the Euler computation. The viscosity effect on the outlet flow angle is significant. Maximum dimensionless energy available at the runner inlet is $\psi_{disp} = 0.73$. Since the net dimensionless energy available for the turbine is $\psi_{\overline{1e}} = 0.806$ and the efficiency (measured) is close to 87%, the dimensionless transformed energy is about $\psi_{mec} = 0.701$, which is in accordance with the computational ψ_{disp} value.

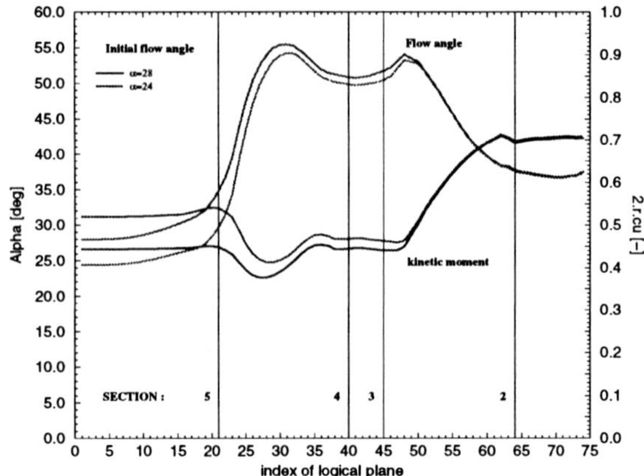

Figure 6. Flow angle and kinetic moment

4. conclusion

This first study was the opportunity to set up an overall balance sheet on the hydraulic elements of the scheme and to emphasis sensitive points. The next stages in the project will be to measure runner geometry on site, in order to analyse it and to have all the elements needed to define, from a technical standpoint, the modifications required.

5. Acknowledgments

The authors wish to thank J-F Combes and E. Dueymes of EDF-DER for their help in achieving the N3S computations. They wish, also, thank Hydro-Vevey S.A and Sulzer Hydro for their contribution of the development of *Renovaturb.*

References

1. J.-P. Chabard, B. Metivet, G. Pot, and B. Thomas. An efficient finite element method for the computation of 3d turbulent incompressible flows. *Finite Elements in Fluids*, vol. 8, 1992.
2. EPRI. *Hydropower Plant Modernization Guide-Hydroplant modernization*, volume 1. Electric Power Research Institute, 1989.
3. EPRI. *Hydropower Plant Modernization Guide-Turbine Runner Upgrading*, volume 2. Electric Power Research Institute, 1989.
4. P. Henry. *Turbomachines hydrauliques.* Presses polytechniques et universitaires romandes, 1992.
5. E. Parkinson, P. Dupont, R. Hirschi, J. Huang, and F. Avellan. Comparison of flow computation results with experimental flow surveys in a Francis turbine. In *XVII IAHR International Symposium*, Beijing, China, September 1994.

(*)Now at Hydro-Vevey

A THREE DIMENSIONAL SPIRAL CASING NAVIER-STOKES FLOW SIMULATION

A comparative study

Ch. Bruttin
J.-L. Kueny

IMHEF-LMH / EPFL
Av. de Cour, 33
1007 Lausanne
Switzerland
christophe.bruttin@imhef.dgm.epfl.ch

B. Boyer
K. Héon
T. C. Vu

General Electric Hydro
795 First Avenue
Lachine, Québec H88 258
Canada
boyerb@hydro.ge.com

E. Parkinson

Hydro-Vevey S.A
Av. des deux gares, 6
1800 Vevey
Switzerland
hydro@hydro.ch

Abstract

Navier-Stokes flow simulations are now commonly used in the process of hydraulic turbomachinery design. Comparison with experiments and careful understanding of the behavior of such tools is however very important to the correct interpretation of the results. Following this trend, Navier-Stokes flow simulations, using three solvers, applied to a pump turbine spiral casing in turbine mode, are presented and discussed in this paper. The physics of the flow and guidelines for the interpretation of the results are emphasized

Figure 1. Pump-turbine model on test-rig (Hydro-Vevey SA)

1. Introduction

The understanding and the prediction of the flow behavior in a turbine spiral casing can be of crucial importance in a refurbishment project where the slightest efficiency

E. Cabrera et al. (eds.), Hydraulic Machinery and Cavitation, 81–90.

increase in any component of the turbine can make the difference. Even more, non classical flow patterns can occur when dealing with old designs. Therefore, sole local analysis provide accurate information to design and/or optimize hydraulic components [1, 2, 3, 4]. Three dimensional flow simulations appear then as the optimum solution, not talking yet of experimental analysis. Navier-Stokes simulations, when interested in energetic loss predictions and local vane optimization, short cutting all mesh generation problems, suffer from the amount of required mesh nodes and boundary definitions in such components. Indeed, the spiral casing outlet flow is strongly coupled with the stay vanes' hydraulic definition. Therefore, either the numerical domain of the simulation includes the stay vanes, implying large mesh sizes and mesh generation complexity [5], either the outlet boundary condition simulates the stay vane ring influence. This solution requires reasonable mesh sizes and is therefore affordable on standard workstations. Such an option requires extensive validations.

This paper examines a pump turbine spiral casing in turbine mode and various codes are applied. Following a brief description of the geometry, different outlet boundary conditions are detailed and tested with three codes. The influence of the turbulence modeling, either mixing length or k-epsilon types, is also analyzed. Both global and local analysis outline the possibilities and the limits of such flow simulations in spiral casings, with regards to reasonable CPU time and memory sizes.

2. Spiral casing description

2.1 GEOMETRY

The geometry used for this study a pump turbine spiral casing, illustrated when operated on the test rig at Figure 1. The geometry of the spiral casing, see Figure 2, is defined with 21 sections along the azimuth direction.

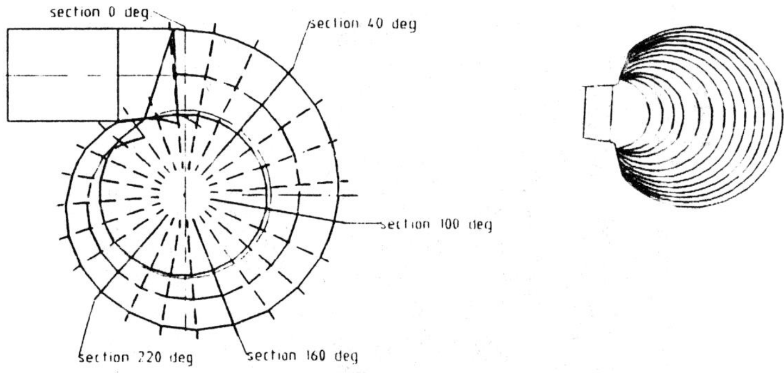

Figure 2. Geometry of the spiral casing

2.2 MESH GENERATION

Two different meshes are generated. In the first case, a block structured grid was created with FIMESH© of FDI. The blocks are structured with a butterfly description of the computing domain, see Figure 3, for the inlet and spiral pipes. A single block is

added for the inlet section of the runner. This method ensures a correct grid distribution near the wall. The main hypothesis made for this grid generation is that the stay vane connected to the spiral tongue is not taken into account in the mesh generation, i.e. no solid walls boundaries exist in this part. Consequently, discharge can cross over from the end of spiral casing pipe to the inlet sections. Mesh dimensions vary, according to the considered solvers, from 70'000 to 90'000 nodes.
As a second case, a 3-D structured single block body fitted grid system of the casing is defined, see Figure 3, using GE Hydro's CFD-based computer aided engineering system. On the opposite to the previous mesh definition, no fluid connectivity exists in the tongue domain. Therefore, no discharge exists in the tongue domain between the inlet pipe and the spiral end. The structured mesh dimensions are approximately of 60'000 nodes.

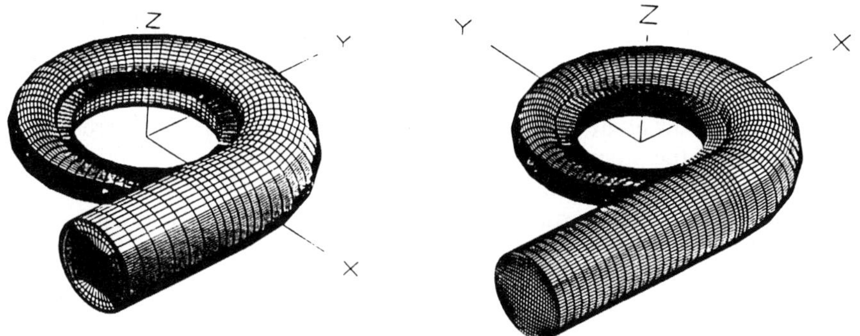

Figure 3. Butterfly and single bloc spiral casing meshes

3. Flow simulation strategy

3.1 TWO DIFFERENT APPROACHES

Three turbulent Navier and Stokes packages were used to perform the simulations: FIDAP© of FDI, TASCflow© of ASC, both commercial packages and GE Hydro's proprietary software. A short description of their main features is provided below. According to these different features, two different approaches are applied to calculate the spiral casing's flow. The spiral casing flow is calculated as an isolated turbine component with FIDAP and TASCflow, whereas with GE Hydro's solver, applied as a two stage approach for the casing / distributor assembly.

3.2 ISOLATED CALCULATIONS WITH FIDAP AND TASCFLOW

FIDAP is a finite element Navier-Stokes software, with k-ε and mixing length turbulence modeling capabilities. TASCflow is a multiblock finite volume code using a standard k-ε turbulence modeling. In all cases treated with FIDAP and TASCflow, the boundary condition at the inlet is given as an uniform normal velocity profile at the inlet section. At the outlet, different options are used to simulate the stay vanes. The first option is to impose a constant pressure at the outlet of the domain. These cases are referenced as F1, F3 and F4 in table 1, where all tested configurations are listed. « F » stands for FIDAP. A second option is to impose an open boundary condition, cases T1 and T2, where « T » stands for TASCflow, and last an open boundary with a drag coefficient (1), case F2.

$$f_n = 0{,}5\rho C_l u_n^2 \qquad (1)$$

As the simulated flow is turbulent, a zero velocity including logarithmic laws is imposed for all walls. A constant intensity of 5% and an eddy length of one tenth of the inlet pipe diameter are imposed at the inlet section for cases F3, F4, T1 and T2, for which k-ε turbulence modeling is applied.

3.3 CASING / DISTRIBUTOR CALCULATIONS WITH GE HYDRO'S SOLVER

As stated previously, a casing/distributor assembly is analyzed in a two-level flow analysis approach with GE Hydro's solver. A global flow analysis is performed for the combined geometry where the distributor region is treated as a porous medium in order to simulate the flow resistance or head loss produced by the distributor. On the other hand, to provide input to the porous medium treatment, a series of 2D viscous flow analysis is conducted to determine the individual distributor head loss where the incoming attack flow angle is obtained by the global flow analysis. Therefore, near the outlet portion of the casing flow domain, where the presence of the distributor is simulated by a porous medium, Darcy's coefficient has to be specified. For this specific application, the Reynolds-averaged Navier-Stokes equations with an extra inclusion of the porous medium treatment based on the Darcy's law are adopted. Validation of the computational model against experimental data and its application as a hydraulic design tool are explained in [6, 7, 8, 9]. The casing / distributor flow analysis is performed with the GE Hydro CFD-based computer aided engineering system developed specifically for hydraulic turbine application [10, 11]. The solution, referenced as G1, where « G » stands for GE Hydro. This information is obtained with a series of 2D flow analysis for the distributor at 20° of wicket gate opening.

TABLE 1. Characteristics of considered flow simulations cases

Case	Solver	Nodes	Turbulence	Outlet Condition	CPU time
F1	Fidap	70'000	mixing-length	Pressure	8 hours
F2	Fidap	70'000	mixing-length	Drag coefficient	8 hours
F3	Fidap	70'000	k - ε	Pressure	20 hours
F4	Fidap	90'000	k - ε	Pressure	35 hours
T1	TASCflow	70'000	k - ε	Pressure	6 hours
T2	TASCflow	90'000	k - ε	Pressure	8 hours
G1	GE Solver	60'000	k - ε	Free Outlet Porous media	5 hours

4. Global flow analysis

4.1 DISCHARGE DISTRIBUTION

The first analysis to be performed when considering the amount of data generated by the flow simulations is the discharge distribution along the main flow direction. It is indeed a key factor for the runner inflow conditions, in turbine mode. It is illustrated at

Figure 4, considering a single case per solver, where the discharge in radial cross sections of the spiral is presented versus the azimuth direction. As observed, the discharge remains constant in the inlet pipe. The chart shows the influence of the assumption made for the spiral tongue in the FIDAP and TASCflow computations. In these two cases, the discharge decreases faster than in GE's computations due to the lack of the stay vane. The opposite appears at the casing end.

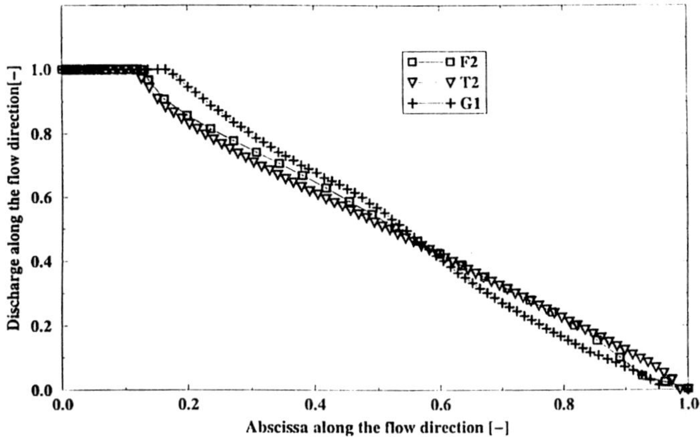

Figure 4. Discharge distribution along the main flow direction

4.2 AVERAGED FLOW ANGLES

The knowledge of flow angles in the spiral casing is of a great importance for stay vane optimization. Two flow angles can be extracted from the flow computations: a mean flow angle based upon the velocity integration along the stay vane height and an energetic flow angle. This mean flow angle is defined as

$$\overline{\alpha} = \frac{\int \alpha C_r dz}{\int C_r dz} \tag{2}$$

Its distribution along the azimuth abscissa for all calculated cases is illustrated at Figure 5. This chart is directly related with Figure 4 and calls for similar comments. Indeed, F2 and T2 profiles cross over G1's profile at the same position, i.e. roughly 140°. The comparison between F and T computed angles shows two different patterns:

• TASCflow and FIDAP with mixing length turbulence modeling where flow angles remain within a margin of 4°.

• FIDAP computations with k-ε.

The energetic transfer between C_r and C_u components is analyzed using an energetic flow angle (3).

$$\overline{tg\alpha_e} = \frac{\overline{C_r U}}{\overline{UC_u}} \tag{3}$$

It is also used to defined an optimum stay vane angular setting, with regards to incidence losses. In such a case, the boundary layer domain should be eliminated from the integration domain, as shown later, within the local flow analysis. Its azimuthal distribution is presented at Figure 6.

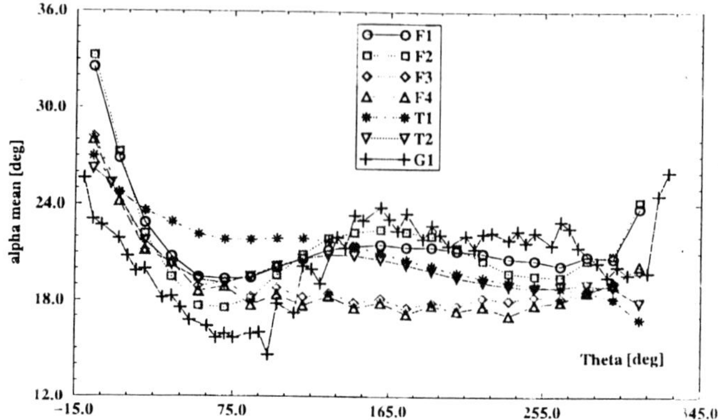

Figure 5. Mean flow angle distribution at the inlet of the stay vanes

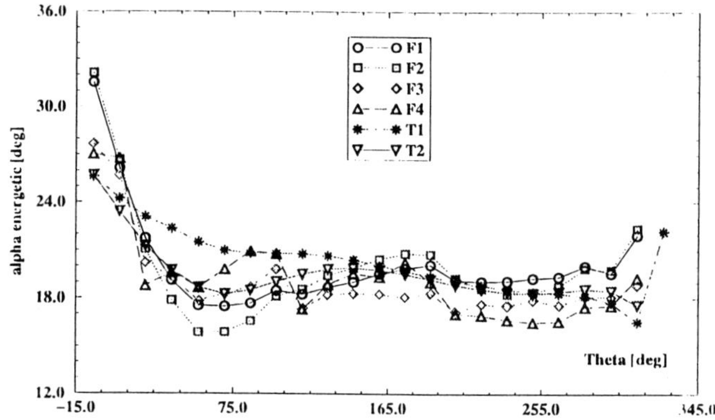

Figure 6. Energetic flow angle distribution at the inlet of the stay vanes

5. Local flow analysis

To analyze the different results given by the three solvers, 4 vertical survey sections are chosen at the stay vanes' inlet section. Their respective azimuth positions are 40, 100, 160 and 220 degrees, see Figure 2.

5.1 TANGENTIAL AND RADIAL VELOCITIES

Both tangential C_u and radial C_r velocity components for all four survey sections are plotted versus the vertical z axis in Figures 7 and 8 respectively. All solutions present similar quantitative values, forgetting the boundary layers. Sole G1's results in sections 160 and 220 degrees show a clear difference. It is related to the discharge distribution, itself function of the tongue treatment, considered as a solid wall for G1. The thinnest boundary layers are calculated by both finite volume solves (T1, T2, G1), independently of FIDAP's turbulence model. It will thus directly influence the predicted energetic losses.

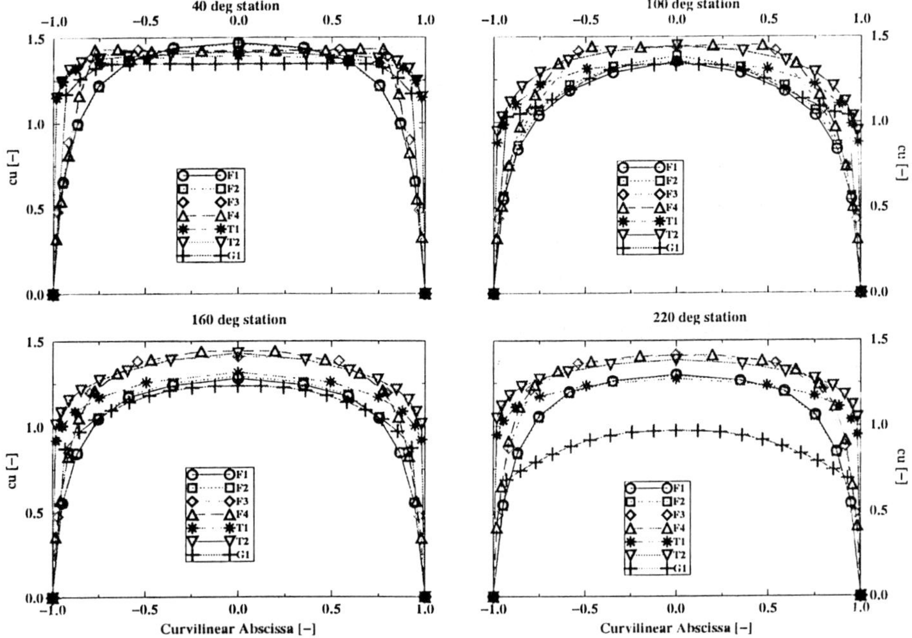

Figure 7. Cu distribution for 4 reference stations

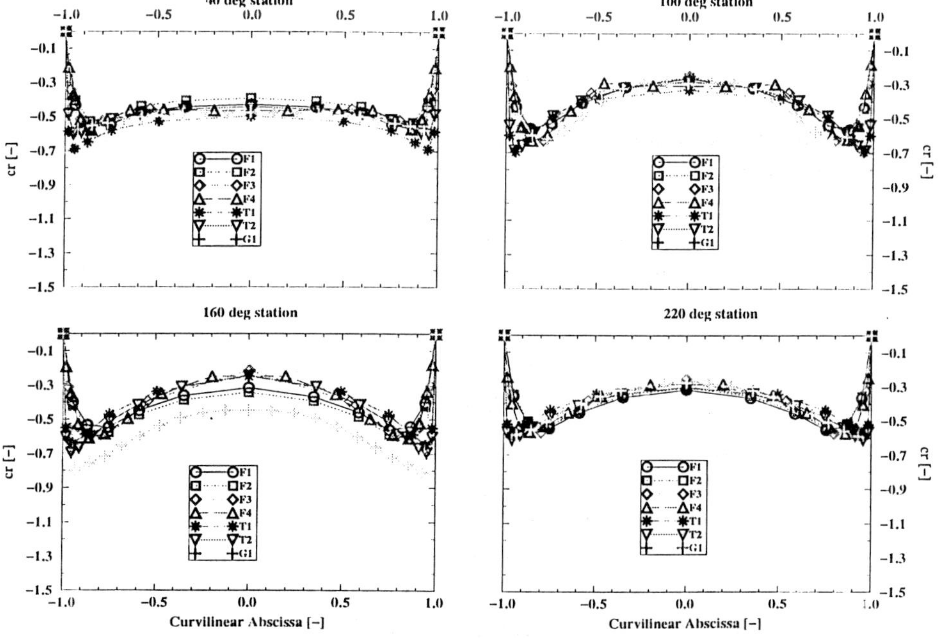

Figure 8. Cr distribution for 4 reference stations

5.2 LOCAL PRESSURE TRANSFER

To illustrate the pressure to kinetic energy transfer, considering for example case T2, the momentum equation in the radial direction is considered in (4).

$$c_z \frac{\partial c_u}{\partial z} + c_r \frac{\partial c_r}{\partial r} + \frac{c_u}{r}\frac{\partial c_r}{\partial \theta} + \frac{c_u^2}{r} = -\frac{1}{\rho}\frac{\partial p}{\partial r} + \upsilon(\nabla^2 c_r - \frac{c_r}{r^2} - \frac{2}{r^2}\frac{\partial c_u}{\partial \theta}) \qquad (4)$$

In a first approximation, it can be simplified as (5) :

$$c_r \frac{\partial c_r}{\partial r} + \frac{c_u^2}{r} \approx -\frac{1}{\rho}\frac{\partial p}{\partial r} \qquad (5)$$

Considering in the following the azimuth position θ= 270°, vertical iso-pressure values are observed in the meridian section, see Figure 9.

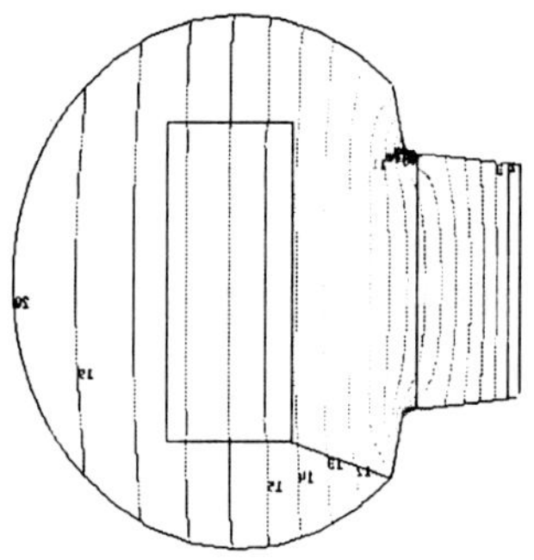

Figure 9. Iso-pressure in meridian section θ= 270°

To analyze this pressure gradient source terms, the momentum equation terms are estimated along a line in the middle plane of the spiral casing. Referring to Figure 10,

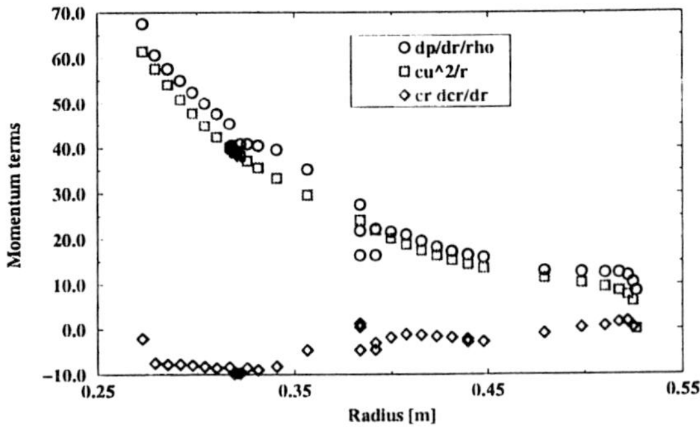

Figure 10. Momentum equation terms along a mid-line in the meridian plane

the pressure transfer is completely due to the centrifugal term along this line, as pressure and C_u terms profiles are equivalent. On the opposite, the bottom near wall pressure gradient is mainly due to the radial component of the velocity. It induces, as seen in Figure 8, higher values of C_r near the wall than at the middle of the section

6. Conclusion

Following the analysis of all cases from the various codes in use for this study, conclusions arise on the geometry handling, turbulence modelling and two stage calculations. Correct modelling of the tongue geometry is a key factor for such flow analysis. Indeed, it will directly influence the discharge distribution and therefore all main flow parameters. The blockage effect of the distributor, which creates a non uniform azimuth pressure distribution, as a second order influence, is also unavoidable and justifies all calibrated porous medium type analysis. Indeed, when comparing these simplified approaches to a complete geometric modeling of casing / stay vanes / wicket gate, limited CPU time and memory sizes remain very important factors if such analysis is to be used in the design process. As observed with the butterfly type meshes, used for FIDAP and TASCflow, a high sensitivity of the flow angles to nodes distribution is illustrated. Care is thus needed when using such meshes. Moving to turbulence modeling, different behaviors are observed between k-ε and mixing length models with FIDAP. Further study is needed to complete this analysis.
These first numerical steps of an on-going development work on high pressure turbine components, as discussed in the paper, raise many points. Future work should include comparisons with a complete geometric model (fully modelling casing, stay vanes and wicket gate) and of course, experimental results.

7. Acknowledgments

The authors wish to thank Hydro-Vevey SA for allowing the publication of this study and GE Hydro for its contribution to this work.

8. List of symbols

Symbol	Name	Unit
x, y, z	Cartesian coordinates	m
r, θ, z	Cylindrical coordinates	m, °, m
A_{ref}	Inlet reference area	m^2
Q_{ref}	Inlet reference discharge	$m^3.s^{-1}$
U_x, U_y, U_z	Cartesian velocity components	$m.s^{-1}$
C_r, C_u, C_z	Cylindrical velocity components	$m.s^{-1}$
C_{ref}	Reference velocity (Q_{ref}/A_{ref})	$m.s^{-1}$
p	Pressure	Pa
k	Turbulent kinetic energy	$m^2.s^{-2}$
ε	Turbulent dissipation rate	$m^2.s^{-3}$
α	Flow angle	°
α_e	Energetic flow angle	°

9. Bibliography

1. Vu, T.C., Héon, K., Coulson, S., Neury, C., Winkler, S. - *A comparative study of computational methods for a high specific speed Francis runner flow analysis* - XVII IAHR Symposium, Beijing, China, 1994.
2. Parkinson, E., Neury, C., Vuilloud, G., Walther, W. - *An optimum combination of numerical and experimental tools for pump-turbine developments* - Modelling, testing & Monitoring for Hydro-Powerplants - II, Lausanne, Switzerland, 1996.
3. Avellan, F., Dupont, Ph., Fahrat, M., Gindroz, B., Henry, P., Parkinson, E., Santal, O. - *Numerical and experimental analysis of the flow in a Francis turbine* - XV IAHR Symposium, Belgrade, 1990.
4. Parkinson, E., Dupont, Ph., Hirschi, R., Huang, J., Avellan, F. - *Comparison of flow computational results with experimental flow surveys in a Francis turbine model* - XVII IAHR Symposium, Beijing, China, 1994.
5. Soares Gomez, F., Mesquita, A., Ciocan, G., Kuény, J.L. - *Numerical and experimental analysis of the flow in a pump-turbine spiral casing in pump operation* - XVII IAHR Symposium, Beijing, China, 1994.
6. Vu, T.C., Héon, K., Shyy, W. - *A comparative study of three dimensional viscous flows in semi and full spiral casings* - XV IAHR Symposium, Belgrade, 1990.
7. Vu, T.C., Héon, K., Desbiens, E. - *Development of a new hydraulic design tool for spiral casing / distributor geometry* - Water Power Conference, Denver, Colorado, USA, 1991.
8. Shyy, W., Vu, T.C. - *Modeling and computation of flow in a passage with 360° turning and multiple airfoils* - Journal of Fluid Engineering, Vol. 115, pp. 103-108, 1993.
9. Vu T.C., Shyy, W. - *Viscous Flow Analysis as a Design Tool for Hydraulic Turbine Components* - Journal of Fluid Engineering, Vol.112, March 1990.
10. Vu, T.C., Héon, K., Shyy, W. - *An integrated CFD tool for hydraulic turbine efficiency prediction* - 5th International Symposium on Refined Flow Modeling and Turbulence Measurements, Paris, France, 1993.
11. Vu, T.C., Héon, K., Shyy, W. - *A CFD based computer aided engineering system for hydraulic turbines* - XVII IAHR Symposium, Beijing, China, 1994.

TIP CLEARANCE FLOW IN TURBOMACHINES - EXPERIMENTAL FLOW ANALYSES

Gabriel Dan CIOCAN
LEGI - ENSHMG - INPG
Laboratoire des Ecoulements Géophysique et Industriels
B.P. 53 - 38041 Grenoble Cèdex 9, FRANCE

Jean Louis KUENY
LEGI - ENSHMG - INPG
Laboratoire des Ecoulements Géophysique et Industriels
B.P. 53 - 38041 Grenoble Cèdex 9, FRANCE

1. Abstract

To analyse the tip leakage flow, it is considered as necessary the set up of an experimental database. For that, a test's water tunnel, without rotation, that represents the blade-to-blade canal of a rocket inducer, has been constructed and qualified at Centre de Recherches et d'Essais de Machines Hydrauliques de Grenoble (CREMHyG - France). For these configurations the flow evolution in the water tunnel has been analysed, and specially the vortex of the leading edge by the 3D velocity measurements, performed by LDV. The tip clearance flow analysis has been carried out by velocity measurements: to the inlet and the exit of the tip clearance by LDV and into the tip clearance by PIV. The different levels of the turbulence kinetic energy has been measured at the different positions. The analysis of the phenomenon has been thus achieved.

2. Introduction

Among the rotating machines, a large number operates with an open runner and a more or less important tip clearance, between the extremity of blades and the annulus wall. Many pumps, inducers, Kaplan turbines, bulb turbines and compressors operate on this principle, similarly as the propeller's boats. At the level of this tip clearance a leakage flow develops from the pressure side to the suction side and this flow generates efficiency losses that can be very important, up to 15-20 % of total power. These phenomena of the tip clearance flow are badly known and modelled, and, if it want to advance from a practical viewpoint this machines type, it is important to minutely analyse the mechanisms and to research influence parameters, to find construction solutions which also avoid to the maximum their pernicious effects.

E. Cabrera et al. (eds.), Hydraulic Machinery and Cavitation, 91–100.

Owing to the development of the experimental investigation means and the numerical modelling, the analysis and the understanding of the tip clearance flows have largely progressed.

From the phenomenology viewpoint, the pressure gradient from the pressure surface to suction surface generates the transverse flow that crosses the endwall. The geometry and dimensions of the tip clearance can determine a complex tip clearance flow. Thus one can find three mechanisms: a separation, a reattachment and a zone of the flow mixing due to a strong leakage vortex.

The separation is due to the flow acceleration in the pressure side with linestreams that are convergent to the endwall, and therefore linked to the form the blade extremity. This behaviour induces a locally increased velocity by passage section decrease (Venturi effect).

According to Bindon [2], for the forms well-adapted forms of the blade extremity profile this phenomenon can be avoided. An other solution consists in the utilisation of the blades supply tablets coming to prolong the pressure surface (Morphis and al. [13]), but this system puts problems of the mechanical resistance viewpoint. This mechanism is important because it is a furnace of cavitation for liquids and it is the place of the acoustic emission and the erosion of the blade extremity.

The reattachment in the endwall is an other very sensitive phenomenon of the geometry. In this case, besides the form of the blade extremity, an important parameter is the proportion between the thickness of the endwall τ and its length e. Thus according to Wadia and al. [15] for a proportion $\tau / e < 0.25$ the reattachment takes place on the blade. This behaviour type is often the case of turbines and more rarely of pumps and compressors. Moore and al. [12] has measured the pressure distribution on the blade extremity of a subsonic turbine. One notices a rapid decrease of the pressure on the pressure surface, with a minimum from the edge. In the separation zone the pressure is nearly constant and increases abruptly in the reattachment zone, while being constant for the rest of the endwall. This separation zone favours thermal changes which can damage the blade. It can be eliminated by the adjustment of the profile form.

A flow mixture due to a strong leakage vortex at the exit of the endwall is generated in the interaction zone between the tip clearance flow and the main flow. According to several authors the leakage vortex is separate from the blade in a well precise position. This position and the trajectory of this vortex, the centre of a strong turbulent dissipation, and the separation point are a function of the geometry and of the blade loading, as well as the pressure gradient sense, direct (turbine) or reverse (pump) - see Bindon [2], Dishart [6].

At the tip clearance exit, the tip clearance flow is in interaction with the main flow and with the other secondary flows: the annulus wall boundary layers and the blade boundary layers. The annulus wall boundary layers represent the 3D boundary layers which develop on the annulus and that induces a supplementary viscous perturbation according to the machine type. In the turbine one can even suppress the tip clearance flow for weak value of the tip clearance. In the pump or in the compressor the effect of this phenomenon comes to increase the leakage flow.

Lakshminarayana and al. [9] has shown both experimentally and numerically the existence of the blades boundary layers deviation from the hub to the annulus. Thus, into an inducer with a wedging angle of 80°, radial velocities in the boundary layers reach 20% the upstream main velocity, they cannot therefore be neglected for a correct representation of the phenomenon in interaction. In the pressure surface, this blade boundary layers feeds the tip clearance flow and in the suction surface, this centrifugal boundary layer will roll with the leakage flow in the leakage vortex.

All these mechanisms cause important losses, and for this reason they have become the object of the numerous studies. The most recent ones Dishart and al. [6], Yaras and al. [18], Storer and al. [14] the different mechanisms of the losses generation for compressors have been identified and separated: the reattachment viscous effects in the tip clearance, the losses by mixture in the tip clearance, the viscous effects due to the interaction of the tip clearance flow with blade boundary layers, the viscous effects due to the interaction of the tip clearance flow with annulus wall boundary layers and the losses by mixture of the tip clearance vortex with the main flow. The authors are generally agreed from a qualitative point of view. There remain however important differences to the level of the qualitative estimation of these phenomena. According to Dishart and al. [6], for a tip clearance value of 2.5% of the chord, 40% of losses are generated in the tip clearance and mainly in the separation bulb. Yaras and al. [18], finds 17% of losses to the exit of the tip clearance for a turbine and with a tip clearance of 2.1% of the chord. Others authors estimate to 10-15%.

These phenomena have been studied both experimentally that numerically. The complete calculation of the flow in a turbomachine, taking into account the tridimensional effect and interaction runner-stator is not again possible with the modern computers. Knowing the importance of the tip clearance flow for a correct phenomena's representation on the whole calculation, at this phase, the introduction of the simplified patterns of the tip clearance appears as a necessity. To this end an experimental database has been constituted. This database will serve to validate tip clearance flow calculation. From the numerical and experimental results, a physical pattern of the tip clearance flow will be introduced in the complete calculations of the turbomachines.

Further on we present the constitution of the database and the first results obtained.

3. Experimental facility and methods

A test tunnel simulating, without rotation, the blade-to-blade channel of a rocket inducer has been conceived to CREMHyG laboratory. This tunnel has the runner blades reduced to two fixed profiles allowing to analyse the tip clearance effects. The final form of the tunnel is presented in Le Devehat [5] - see the figure 1.

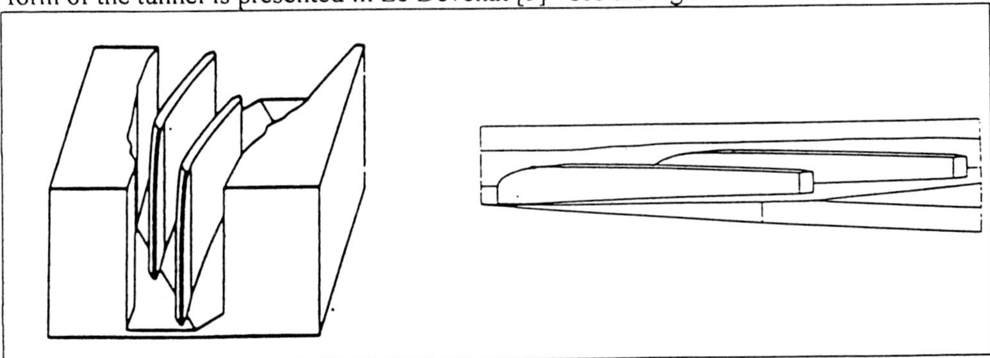

Figure 1. Experimental facility

The definition of the section and profiles has been obtained by a meridian calculation in accordance with Katsanis and al [7]. This has been performed for the nominal flow rate of the inducer. To obtain a similitude of pressure coefficients on the blade and the profiles, a blade-to-blade calculation, quasi 3D, S1-S2 type, in finished element, has been performed -see Kueny et al. [8].

The Reynolds number based on the chord is $4x10^6$ and the tip clearance can vary from 1 to 5 mm, for a constant thickness of the profile of 10 mm.

For the tunnel validation and to determine the pressure evolution on the tip of the profiles and on the profiles the followings have been provided: 60 pressure tapes on the suction surface, 48 on the pressure surface and 48 in the tip clearance and on the wall. A water gate cupboard with electrically driven gates was used for the connection between the static pressure taps and two Desgranges and Huot balances. The electro-gates' cupboard was driven by a micro-computer. The precision of the pressure measurements is of +/- 1mb.

The test tunnel has been performed in plexiglas to allow the carrying out of the velocity measurements by LDV and PIV. The quality performance of walls has allowed the carrying out of the measurements.

The LDV system used is a DANTEC system whit two components, in the back scatted light and transmission by optical fibre, with a laser of 5W argon - ion source. Optical characteristics of the system are found in the table 1.

Laser wave lengths	488 / 514.5 nm
Probe diameter	60 mm
Beam expander	80 mm
Beam spacing probe expander	38 mm 70.5 mm
Focal length	310 mm
Fringe spacing	2.16 /2.277 nm
Volume of measurements $\sigma_x = \sigma_y \sim$ $\sigma_z \sim$	0.1 mm 1 mm

Table 1 Characteristics of the LDA system

The signal processing for the Doppler frequency was accomplished by a spectrum analysis. The measurements precision has been estimated following Mofat and al. [11] of 4%. The displacement of the laser probe has been made by a carriage. The precision of displacement is 0.1 mm. The reference position for the decreasing of the measurement volume has been obtained by the crossing of beams on the internal part of the wall.

For the PIV part a cooperation with DANTEC company has been performed. The stroboscopic lightsheet (thickness of ~1 mm) has been performed with a double-cavity Nd:YAG laser that performed 200 mJ during 9 ns. An optical high power light guide has allowed the performance of measurements on all the length of the chord. The images have been obtained with a CCD camera developed by DANTEC, in a succession of a pair of images at 100 μs, that constitutes a measure. The interval between two measures has been 100 ms. The camera resolution has been 768x480 pixels for a spacial resolution of 30x40 mm. The synchronisation of the shot segment with luminous flashes, the acquisition and the image processing have been performed with the

The two LASER measurements systems have been performed with the spherical particles, in silver covered glass, empty inside to have the water density. The size of these particles is 20 μm.

4. Results of measurements

After the qualification of the test tunnel from the pressure similitude viewpoint, a control of the inlet section has been performed. Thus the secondary velocity in this section does not exceed 3% of the main velocity.

4.1. PRESSURE MEASUREMENTS

The static pressure measurements in the profiles have allowed to validate the tunnel in similitude viewpoint - see Le Devehat et al.[4]. The pressure distribution on the profile are shown in figure 2.

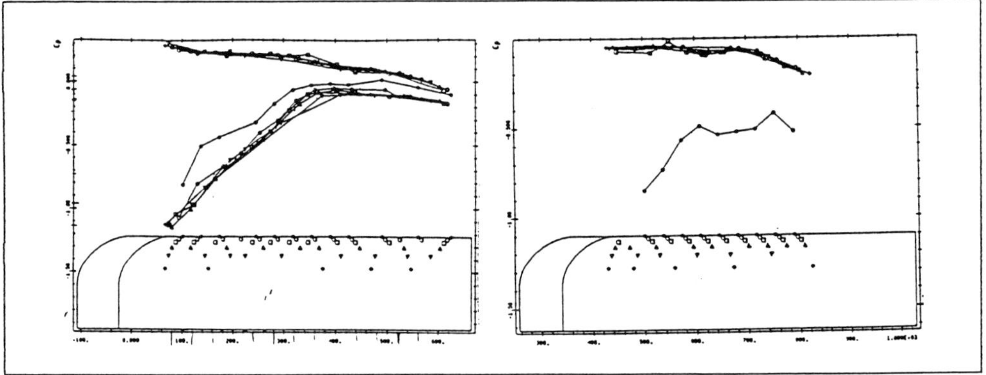

Figure 2. Pressure distribution

4.2. VELOCITY MEASUREMENTS BY LDV

To perform the 3D measurements, one carries out two orthogonal measurements. The calculation of the optical arrangements is presented in Mesquita and al [1]. For the flow description 6 sections considered representative have been chosen - see figure 3 and 4.

The measurements have been carried out for three different tip clearance values: 1 mm; 3 mm and 5 mm. In figures 3, 4, 5 and 6 have been represented the comparisons between the tip clearance flow of 1 mm and 5 mm.

The velocities have a-dimensioning by the average velocity in the inlet section and the turbulence rate by the square of this velocity. The length has been divided whit the characteristics dimensions of the inlet section: height - ho, width - lo and the length of the chord - lch.

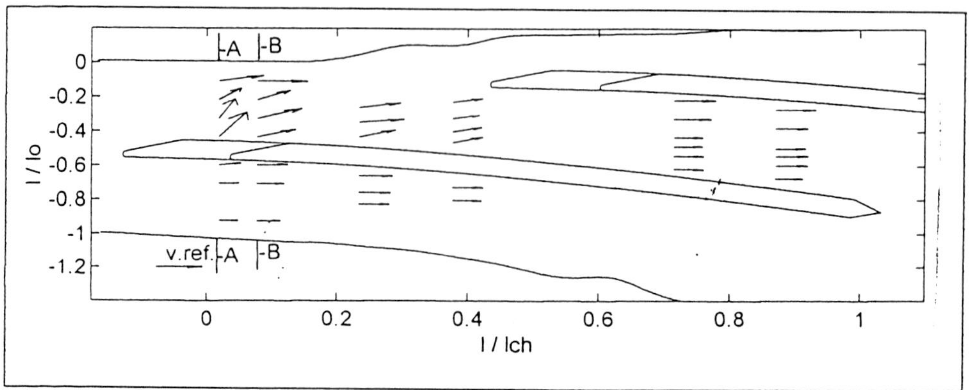

Figure 3. Longitudinal section to 3 mm of walls to a gap clearance of 1 mm

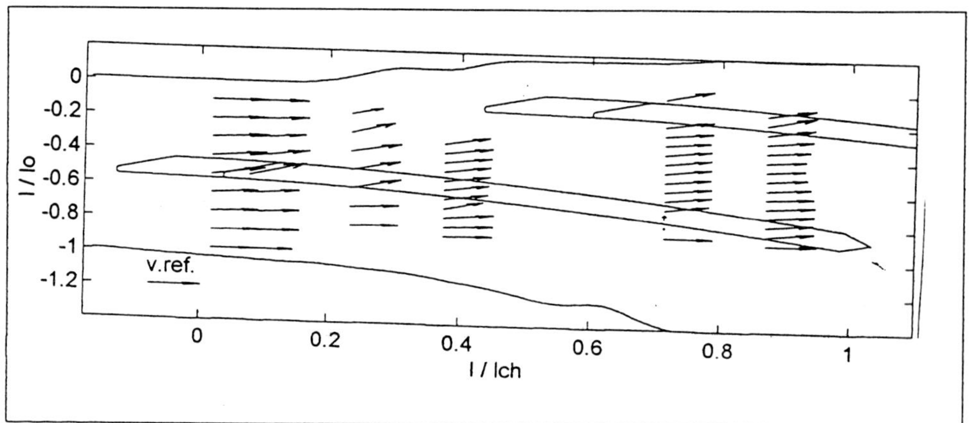

Figure 4. Longitudinal section to 3 mm of walls to a gap clearance of 5 mm

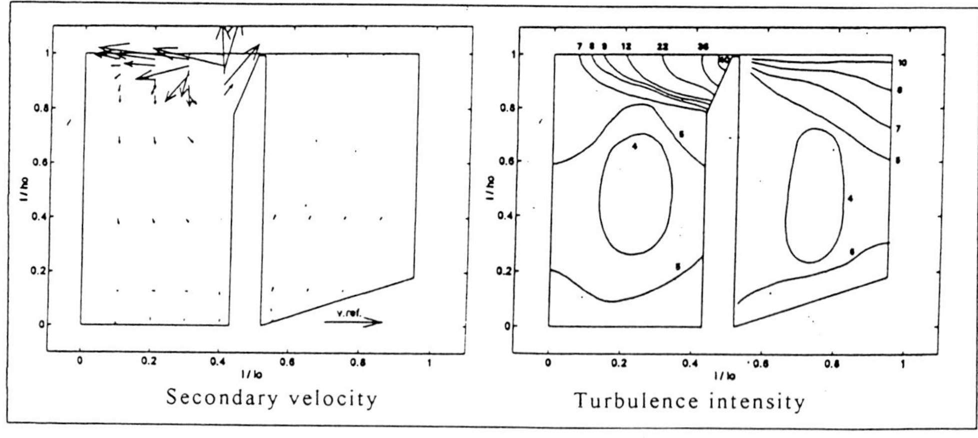

Figure 5. Cross section B-B to a gap clearance of 1 mm

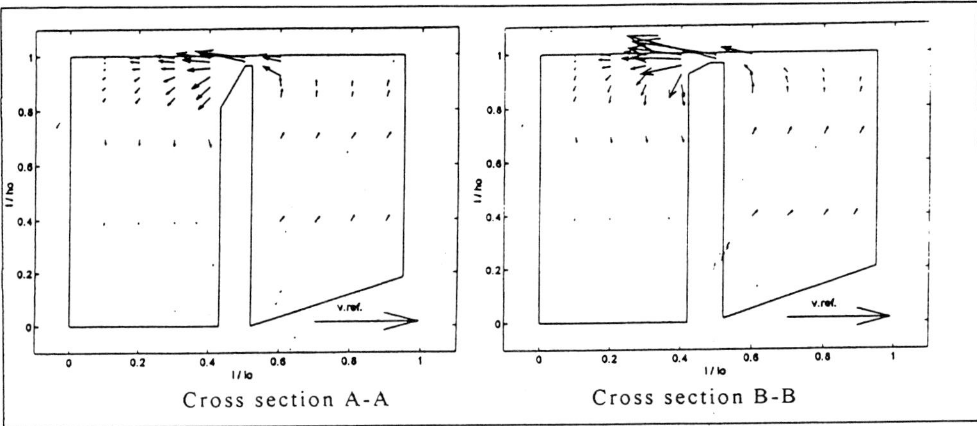

Figure 6. Secondary velocity to a gap clearance of 5 mm

4.3. VELOCITY MEASUREMENTS BY PIV

The measurements presented in the figure 7 and 8 are accomplished at half of the tip clearance of 5 mm. The figure 7 corresponds of the leading edge and 8 to the half of the chord. The lack of the data in the figure 8 is due to the reflections produced by taps of pressure.

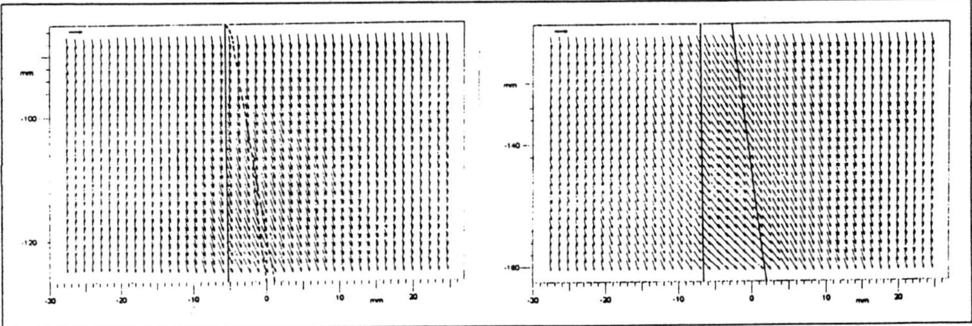

Figure 7. Velocity toleadingedge to 3 mm of walls to a gap clearance of 5 mm

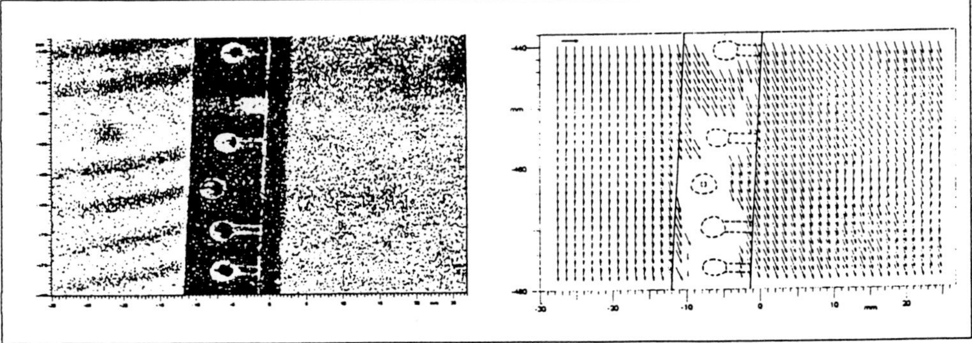

Figure 7. Velocity to mi-chord to 3 mm of walls to a gap clearance of 5 mm

In the figure 9 one presents a comparison between a measurements' profile performed by LDV and by PIV. A direct quantitative comparison of the different LDA and PIV measurements gives excellent agreement - less 4% .

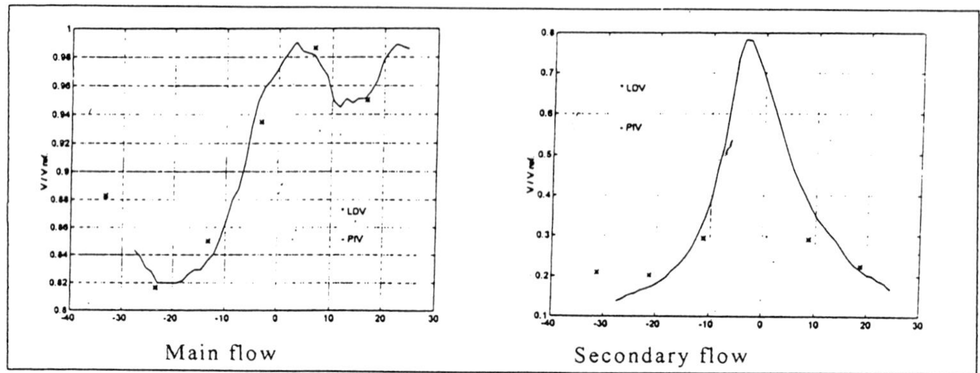

Figure 9.LDV - PIV comparison

5. Discussion

For a tip clearance of 1 mm a flow mixing due to a strong leakage vortex develops to the superior limit of the leading edge. The reasons for the formation of this vortex are both the effect of theleadingedge and the leakage flow generated by the strong pressure gradient in this area, as well as the interaction of this flow with main flow. Vortex develops on 25% of the section (in depth) and is deviated to the second profile, without penetrating between profiles. For the rest of the cord, the tip clearance flow is less important, even if it is always present. The lacks of the kinetic energy prevents the development of a new vortex flow to the exit of the tip clearance and it is convected by the main flow. This is linked also to the decrease of the pressure gradient along the chord.

For the tip clearance of 5 mm one does not observe the same behaviour. The vortex disappears and it is replaced by a tip clearance flow that develops in the blade-to-blade channel but that no longer rolls in the vortex. The increase of tip clearance (5% of the height of the profile) has reduced the velocity values and the height of the tip clearance has allowed the development of the flow which crosses the tip clearance (the same gradient of pressure by for the tip of 1 mm). The tip clearance flow intensity decreases along the cord in the direction of the decrease of the blade loading.

This behaviour - disappearing of the leading edge vortex for a increasing in the gap, was found out also in the experiments performed on a prototype inducer at the KSB-GUINARD Company. This is favorable for the off-design flow rate, with a view to decreasing the NPSH - see Ciocan and al. [3].

The turbulence intensity, calculated in the isotope hypothesis, has increased values upon coming out of the tip clearance and in the vortex action area (up to 40%) for lowers values than 5% in the rest of the section. This distribution is not very different for several cross sections and tip clearance.

6. Conclusion and Perspectives

The study performed on the tunnel permitted to obtain the evolution of the leakage phenomenon and the estimation of the leakage flow rate for several values of the tip clearance. A database has been carried out on this pattern. The behaviour observed has been confirmed by the tests performed with inducer prototype.

To be able to understand the phenomena interaction, a more complete running of the possibilities offered by this pattern is therefore necessary: the study of the the interaction of the boundary layers limit with the tip clearance flow, the influence and the appearance of the cavitation. The instrumentation of the profile with 8-10 unsteady miniature pressure sensors, placed on the profile, will allow to quantify the level of depression on the suction side corresponding to the starting point of the leakage vortex. The utilisation of an unsteady pressure probe for the turbulent losses determination is envisaged.

Using these experimental data as boundary conditions and the validation data bank, a numerical simulation, using a uses 3D Navier-Stokes numerical code with a k-ε model of the turbulence, will be performed. A simplified geometry allowing to preserve the pressure gradients from the pressure to the suction surface will be used to represent the studied phenomena.

The experimental and numerical approaches will allow to describe the tip clearance flow and the mixing area formed by this flow and the main flow. Finally we shall try to specify the influence of each parameter in the losses' generation process.

7. Acknowledgements

The authors take this opportunity to thank both DANTEC Company for their participation to PIV measurements and the Région Rhônes-Alpes, KSB-GUINARD Company and SEIM-MGI COUTIER for their financial support and participation in this project.

8. References

1. Amarante Mesquita, A. L., (1993) Experimental Techniques for the Flow Characterisation in Hydraulic Turbomachines Models, *Proceedings of the 2nd Meeting of the Latin-American Division of the International Association of the Hydraulic Research* - IAHR, Vol. 1, pp. 193-209, Ilha, Solteira Brazil
2. Bindon, (juin, 1988) Mesurement and formation of tip clearance loss, *Gas turbine and Aeroengine Congress*, Amsterdam,
3. Ciocan, GD, Kueny, JL, Analyse experimentale de l'ecoulement de jeu dans les inducteurs, *PEH 96 Conference*, Bordeaux, France, 1996.
4. Le Devehat, E., Kueny, J.L., Dellby, J., (decembre, 1992) Tip leakage flow in blade tip clearance of a 3D test-facility, *Seminar on 3D Turbomachinery Flow Prediction*, Val d'Isere, France,
5. Le Devehat, (1993) Ecoulement de fuite dans l'entrefer d'une turbomachine. Analyse expérimentale et numérique - *Thèse de doctorat à l'INPG*
6. Dishart, Moore, (1989) Tip leakage losses in a linear turbine cascade - *ASME paper*, Brussels, Belgium,
7. Katsanis, « Katsanis, revised FORTRAN program for calculating velocities and streamlines on the hub-shroud midchannel stream surface of axial, radial or mixed flow turbomachines of annular duct »,

8. Kueny, J.L., Le Devehat, E., (january, 1994) Tip clearance flow calculation - 3D channel with tip clearance flow, *Seminar and workshop on 3D turbomachinery flow prediction II*, Vol. III, pp. 200-223, Val d'Isere, France,
9. Lakshminarayana, Sitaram, Zhang, (janvier, 1986) End-wall and profil losses in a low-speed axial flow compressor rotor - *Journal of Engineering for Gas turbines and Power*, vol. 108,
10. McCluskey, D.R., (july 1995) Obtaining high-resolution PIV vector maps in real time - *International Workshop on PIV*, Fukui, Japan,
11. Mofat, R. J., (1985) Using Uncertainty Analysis in Planning of an Experiment, *Journal of Fluid Engineering*, Vol. 107, pp. 173-178,
12. Moore, Le Fur, (juillet, 1990) Computational study of 3D turbulent air-flow in a helical rocket pomp-inducer - 26th *Propulsion Conference*, Orlando,
13. Morphis, Bindon, (juin, 1988) The effects of relative motion, blade edge radius and gap size on the blade tip pressure distribution in annular turbine cascade with clearance - *Gas turbine and Aeroengine Congress* - Amsterdam,
14. Storer, Cumpsty, (1991) Tip Leakage flow in axial compressors - *Journal of Turbomachinery*, vol. 113
15. Wadia, Booth, (janvier, 1982) Rotor-typ leakage. Part II: Design optimization through viscous analysis and experiment - *Journal of Engineering for Power* - vol.104,
16. Yaras, Sjolander, (1989) Losses in the tip leakage flow of a planar cascade of turbine blades - Secondary Flows in Turbomachines, Luxembourg

NUMERICAL PREDICTION OF HYDRAULIC LOSSES IN THE SPIRAL CASING OF A FRANCIS TURBINE

PETER DRTINA
Sulzer Innotec AG
CH-8401 Winterthur, Switzerland

AND

ANDREAS SEBESTYEN
Sulzer Hydro AG
CH-8023 Zürich, Switzerland

1. Introduction

Power plants that have been in operation for several decades show a considerable potential for improvement in hydraulic performance by bringing the design up to modern standards. This situation has led to numerous rehabilitation projects for existing hydraulic power plants.

Up to now most of the modernisation of Francis turbines has been aimed at replacing the runner while retaining all of the existing stationary parts of the machines, e.g. spiral casing, stay vanes, guide vanes and draft tube. In order to improve the hydraulic performance of a Francis turbine as a whole, attention should also be paid to the non-rotating elements of the machine.

For a long time the design of stay vane profiles has been influenced mainly by the considerations of safety, stress analysis and manufacturing technology. Indeed, the high pressure part of the Francis-machine is one of the most critical elements from the point of view of material loading. The blade profiles of stay vane rings of many existing machines mainly reflect the result of mechanical and not hydraulic optimization: they are robust and of very simple geometrical shape. Modern safety engineering routinely uses three-dimensional finite element stress calculus for this part of the machine. This allows arbitrary shaped stay vane profiles to be applied with sufficient safety margin. Hence it is now possible to design stay vane profiles of high performance which optimize both hydraulic and stress considerations.

E. Cabrera et al. (eds.), Hydraulic Machinery and Cavitation, 101–110.

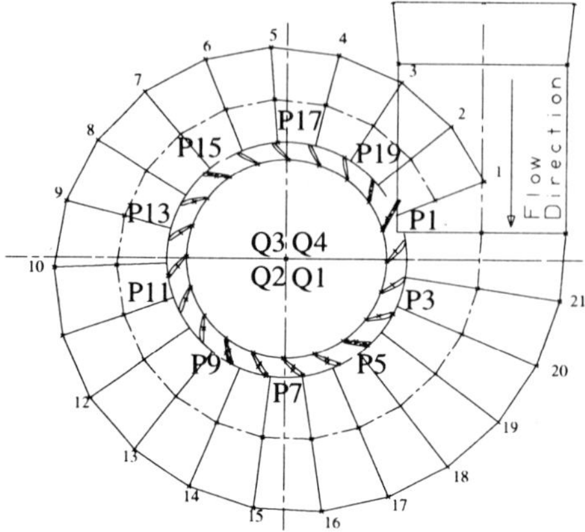

Figure 1. Draft of spiral casing in the symmetry plane.

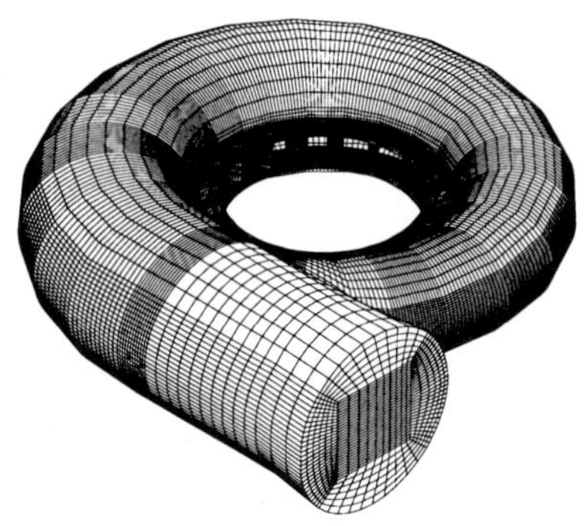

Figure 2. Entire computational grid for the spiral casing.

The spiral casing transforms the rectilinear penstock-flow into a uniformly distributed swirling inflow over the circumferential inlet surface of the stay vane ring. The stay vane passages turn the flow to the guide vanes under a given flow angle. The requirement from the hydraulic point of view is that the flow turning in the annulus cascade of the stay vanes produces a very low level of hydraulic loss. The use of an appropriate 3D viscous calculation procedure enables the hydraulic loss to be minimized by the choice of the best possible hydraulic profiles for the stay vane ring.

The present paper shows that the application of a sophisticated calculation method is capable of evaluating the difference in hydraulic loss between two differently shaped stay vane rings. Consequently, it is possible to estimate theoretically the gain in energy terms which can be reached by the modification and improvement of an existing stay vane ring.

Shyy and Vu (1993) presented a simplified two-level approach to simulate the non-rotating components upstream of a hydraulic turbine runner. In a first step a flow calculation for the spiral casing has been performed with the details of the distributor smeared out by applying a porous medium treatment instead of modelling each stay vane. In a second step an individual cascade is investigated with the inlet conditions extracted from the global spiral casing analysis. Lipej et al. (1994) describe numerical results for all components of a Francis turbine obtained by applying FIDAP and TASCflow. For the spiral casing less than 7000 nodes were used and neither the stay vane ring nor the guide vanes were taken into account, which certainly leads to a weak prediction of the flow. Parkinson et al. (1994) point out that the circumferential discharge distribution obtained from their spiral casing computation does not correspond correctly to the experimental data due to the absence of the stay vanes in their calculation. Instead of including them into the computational domain, which they claim to be very time consuming and node increasing, they tend to prefer the application of a loss model to include the effects of the distributor. Ruprecht et al. (1994) also performed separate computations for the spiral casing and the the distributor. The inlet condition for their single passage computations are extracted from the spiral casing results. They propose an iterative procedure to account for the coupling between spiral casing and distributor. DeHenau and Markovich (1995) and DeHenau (1995) use a combination of coarse grid flow simulations for the complete spiral casing including the distributor and a refined calculation on a selected stay vane/wicket gate passage with inlet boundary conditions taken from the coarse grid simulation. This approach yields a reliable loss prediction enabling the authors to compare the original stay vane design with an optimised design.

In the present case a flow analysis for a spiral casing including 20 stay vanes is performed by solving the incompressible Reynolds-averaged Navier-Stokes equations closed with a standard k-ϵ turbulence model and logarithmic wall functions. The results obtained for an original stay vane design are used to optimise a new design with respect to losses. The second key issue of the present investigation is to prepare the inlet velocity profile for the downstream components (wicket gate, runner, draft tube) of the Francis turbine. Stage calculations carried out for all components from the stay vane inlet to the draft tube outlet leading to a numerically predicted hill chart are described in detail by Keck et al. (1996). A description of the basic numerical procedure as well as an extensive comparison between stage calculation results and experimental data is published by Sick et al. (1996).

2. Computational method, grid generation and boundary conditions

Early potential flow models (Athanassiadis, 1984; Kubota *et al.*, 1984) and 3D finite element models (Ruprecht *et al.*, 1994; Soares-Gomes *et al.*, 1994) have been developed for studying the flow in spiral casing assuming the flow to be inviscid. An accurate study requires today the solution of the laws of conservation for viscous flow by using the finite element method or the finite volume method in the discretization scheme.

The flow in the spiral casing has been considered as incompressible, steady and viscous, leading to the need of applying a Navier-Stokes solver. The commercially available CFD code TASCflow (ASC Canada) which solves the conservation equation for mass and the Reynolds-averaged Navier-Stokes equations has been used. This code is nowadays very often used in the industrial environment (Casey *et al.*, 1995). Validation of the code (Sick *et al.*, 1996) promises high reliability and therefore its use deserves to win confidence as a design tool.

For the generation of the computational grid the ICEM-CFD mesh generator P^3 (PCUBE) has been applied. In combination with its advanced CAD capabilities the model setup time could be reduced remarkably compared to the previously applied grid generation technique. The translator interface of the mesh generator allows the transformation of the constructed mesh to most of the well-known commercial CFD solvers. For the case under investigation a block-structured grid has been generated which consists of 190 blocks. In order to avoid highly skewed grid cells the main body of the mesh which fills the casing region has been constructed using a butterfly arrangement of subblocks. In order to capture all significant flow phenomena the flow field resolution had to be sufficiently high. On the other hand the available computer resources restrict the mesh size. In the past spiral casing grids have been set up with different refinement levels ranging from 7000 nodes (Lipej *et al.*, 1994), 31185 nodes (Shyy and Vu, 1993) up to 240000 nodes (DeHenau and Markovich, 1995). Even unstructured grids have already been applied (Soares-Gomes *et al.*, 1994). With the grid shown in figure 2 the above mentioned constraints are sufficiently met. It consists of about 280,000 nodes for one half of the spiral casing and all stay vane passages.

At the spiral casing inlet the mass flow and appropriate turbulence conditions for the k,ϵ-equations are defined, while at the outflow boundary only the average pressure is set. This allows the pressure distribution to be a result of the computations even in the outlet area. Due to geometrical symmetry only half of the spiral casing and the stay vane ring had to be modelled. At all solid surfaces the logarithmic law of the wall was applied.

3. Simulations

Numerical simulations have been performed in a three-step design procedure. First, flow calculations for a spiral casing including a set of original stay vanes have been carried out. Five different operating points were considered ranging from $0.75\,\dot{Q}_1$ to $1.25\,\dot{Q}_1$ with $\dot{Q}_1$ being the design volume flow rate. Overall and local flow phenomena as well as loss sources have been identified and studied. In the second step an improved set of stay vanes has been developed based on the computational results and on the engineering knowledge available. In the third step, the new design was transfered to the grid generator via the IGES data interface. After performing moderate

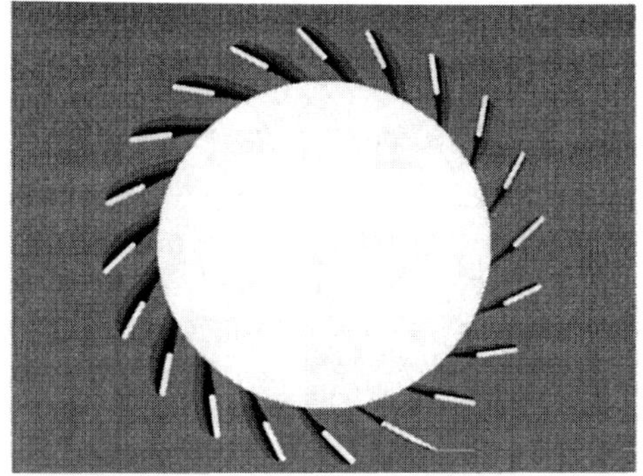

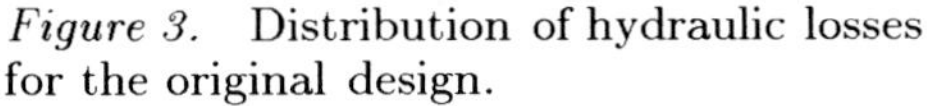

Figure 3. Distribution of hydraulic losses for the original design.

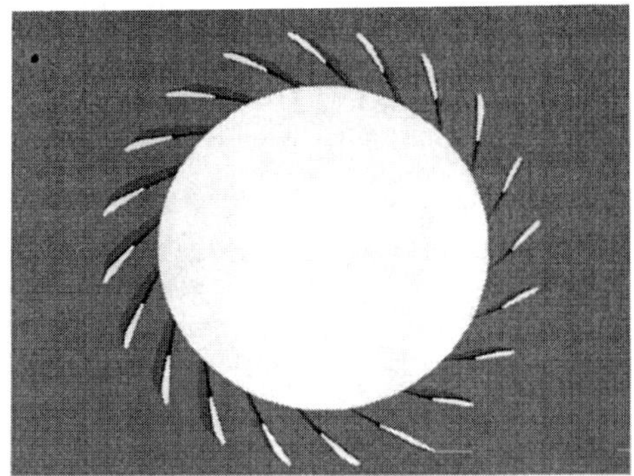

Figure 4. Distribution of hydraulic losses for the improved design.

modifications, the new grid has been generated and the calculations were repeated for three different operating points. To achieve a satisfactorily converged solution (maximum residuals $\leq 10^{-4}$) the CPU time required for a single run is in the range of 20h - 40h on a SGI Power Challenge. A restart run from a previous solution may converge within 5h - 10h.

4. Computational results and analysis

For an industrial engineer involved with the design and optimisation of hydraulic turbine components the most important criteria to judge a design is the overall component efficiency. In order to see the effect of modifying the stay vane profiles, the overall component loss defined by

$$L_{global} = \frac{\int\limits_{A_{inlet}} \dot{m} p_{tot} dA - \int\limits_{A_{interface}} \dot{m} p_{tot} dA}{\int_{A_{inlet}} \dot{m} p_{tot} dA} \cdot 100 \qquad (1)$$

has been plotted in figure 5 as a function of the volume flow rate. A parabola which fits to the original stay vane data has been added. The numerical results show the expected behaviour. With increasing flow rate (increasing Reynolds number) the loss increases quadratically. Comparing the original and the optimised design shows clearly the improvement which has been achieved by the geometrical modification of the vanes. In the design point the loss is reduced from 3.72% to 3.20% that is a reduction of 14%. Thus the global goal of the optimisation procedure has been reached. The next question which arises automatically is: How is the loss distributed in the spiral casing and the stay vane ring and what are the reasons for the improvement?

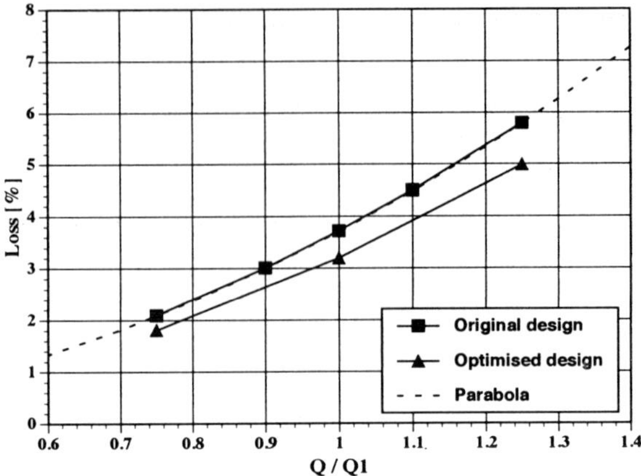

Figure 5. Global loss of spiral casing and stay vane ring as a function of volume flow rate.

Figure 3 (original design) and figure 4 (optimised design) indicate the regions of high losses which are located, as expected, on the suction sides of the stay vanes. The grey scale patterns range from 0% loss (light grey) to 10% loss (black) based on:

$$L_{local} = \frac{p_{tot,local} - \overline{p}_{tot,inlet}}{\rho g H} \cdot 100 \tag{2}$$

Comparing both loss distributions reveals some interesting aspects.

- At each of the 20 stay vanes the losses have been reduced.
- For both designs losses are high within the first and the second quarter (Q1 and Q2, see figure 1). A loss minimum occurs in the third quarter (passages P15-P17).
- The location of the major loss changes from the suction to the pressure side for the last 3 vanes.

Figure 6 shows quantitatively the circumferential variation of the normalised total pressure $p_{tot}(\Theta)/\overline{p}_{tot,inlet}$ (averaged over the span) up- and downstream of the stay vanes. $\Theta = 0^o$ is located at the 'leading edge' of the tongue and increases in main flow direction. Note: While the R1-lines start with the first passage (P1), the R2-lines (downstream of the stay vanes) start within passage P16 (same circumferential position Θ!). The peaks indicate the circumferential position of the stay vanes wakes. For the optimised design - marked with symbols - the size of the wakes is clearly reduced. Even upstream of the stay vanes the losses decrease slightly. Some

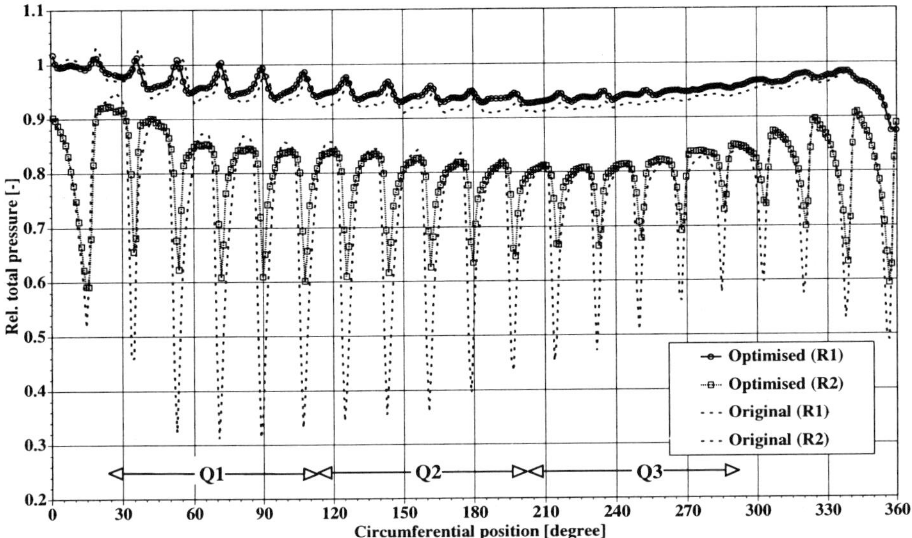

Figure 6. Circumferential variation of normalised total pressure.

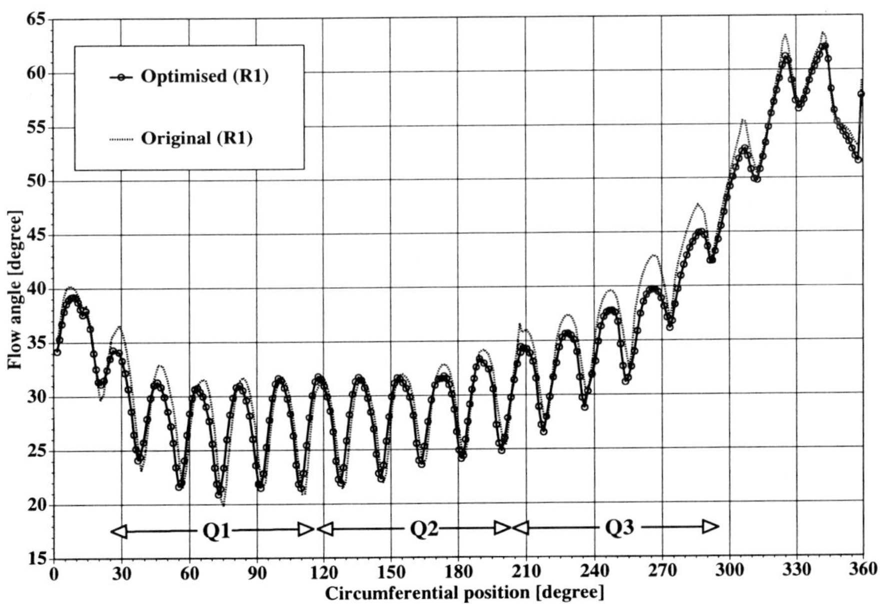

Figure 7. Circumferential variation of flow angle.

more insight can be gained by looking onto the vector plots in figures 8 to 11.

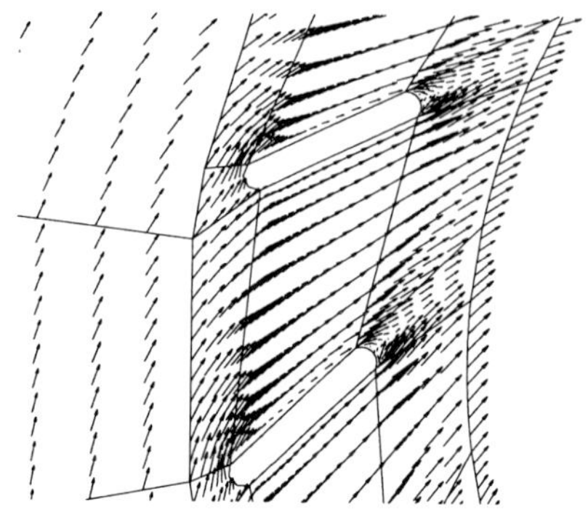

Figure 8. Velocity vectors in passage P7 at $z = 0$ for the original design.

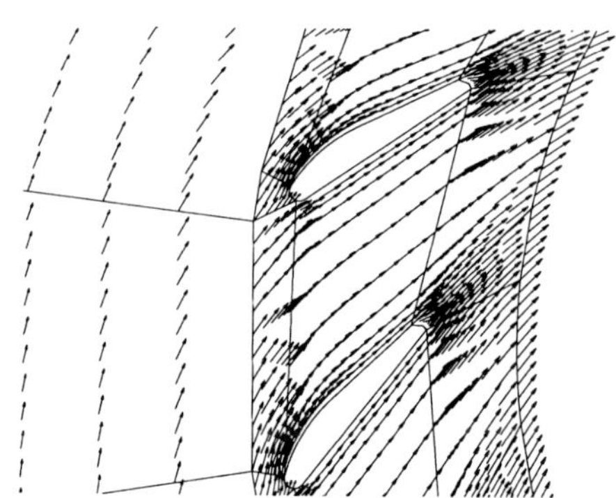

Figure 10. Velocity vectors in passage P7 at $z = 0$ for the improved design.

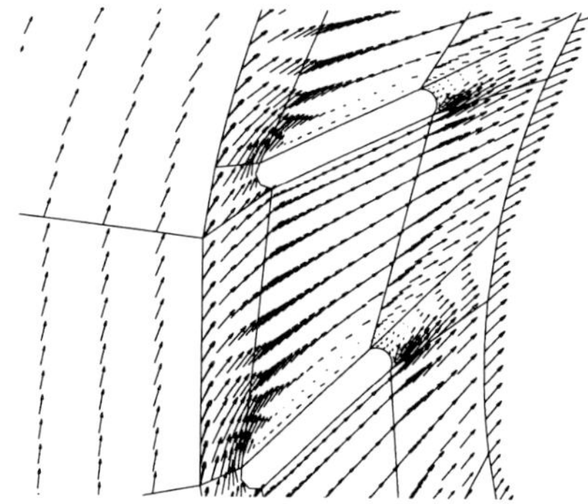

Figure 9. Velocity vectors in passage P7 at $z = 0.8 * H$ for the original design.

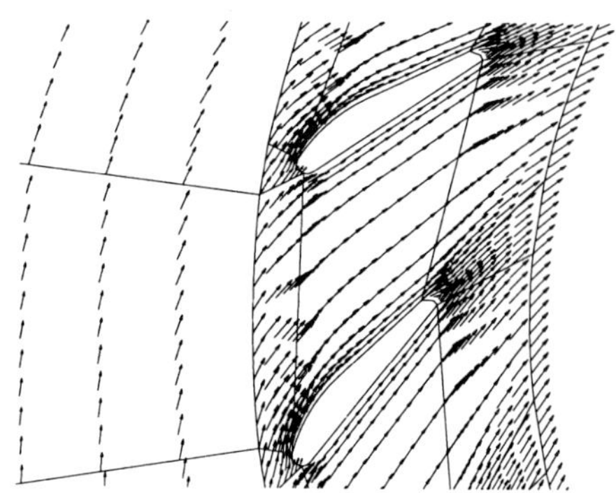

Figure 11. Velocity vectors in passage P7 at $z = 0.8 * H$ for the improved design.

The original design shows a separation zone close to the head cover, which extends far downstream. For the optimised design this separation zone is removed by the improved alignment of flow and stay vanes. From figure 12 the same conclusion can be drawn. While the velocity and flow angle profiles upstream of the stay vanes only change very slightly from the original to the optimised design, significant differences occur in the downstream profiles. The scatter (difference of minimum and maximum values) is reduced remarkably with the mean profiles being only marginally affected. Velocities and flow angles are non-dimensionalised with the average values at the corresponding radius. Figure 7 shows that in the circumferential sectors Q3/Q4 the mean flow angle is in the range of about $40° - 50°$ which corresponds very well to the inclination of the stay vanes which is $\approx 53°$

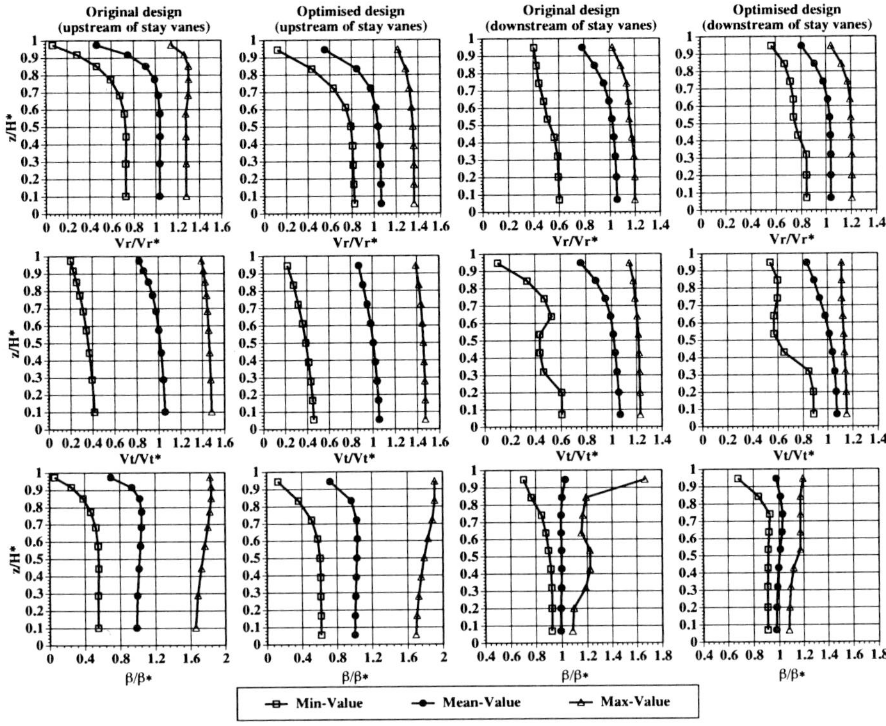

Figure 12. Velocity and flow angle profiles. First and second columns show velocity and flow angle profiles upstream of the stay vanes for the original and optimised design, respectively. Third and fouth columns show downstream conditions for the original and optimised design, respectively. Rows: Non-dimensional radial velocity component (top), non-dimensional tangential velocity component (middle) and non-dimensional flow angle (bottom).

(original) and $\approx 48°$ (optimised) respectively, and accounts for the low loss in passages P15 to P17. Note the weak design of this particular spiral casing which does not lead to constant flow angle around the circumference.

5. Conclusion

The results presented in this paper show that numerical flow simulations can be integrated successfully into a design procedure for hydraulic turbine components which can dramatically shorten the time needed for a new development. The visualisation of the flow (pressure, velocity vectors, flow angles) and derived quantities (losses) helps the designer to analyse the flow and to decide which modifications may be most appropriate for improvements to important local flow phenomena. In the present case the position and strength of the separation zones, and the detailed loss distribution, made it possible to adjust the stay vane profiles very quickly leading to the demonstrated improvements.

The described flow problem could only be solved with the help of a sophisticated CAD/grid generation software and an efficient robust flow solver. Both are available as commercial software packages offering a wide range of capabilities to tackle complex industrial problems.

In particular, the spiral casing calculations presented here show the important effect of flow angle distribution into the stay vanes and the potential for improvements with stay vane redesign.

References

Athanassiadis, N. (1960) Potential Flow through Spiral Casings, Diss. ETH Zürich, Mitteilungen aus dem Institut für Aerodynamik Nr.30

Casey, M.V., Borth, J., Drtina, P., Hirt, F., Lang, E., Metzen, G., Wiss, D. (1995) The Application of Computational Modeling to the Simulation of Environmental, Medical and Engineering Flows, *SPEEDUP Journal*

DeHenau, V. Markovich, M.S. (1995) Optimization of the Sir Adam Beck II Turbine Distributor using Computational Fluid Dynamics, *Updating and Refurbishing Hydro Power Plants*, International Water Power and Dam Construction and National Hydropower Association, Nice

DeHenau, V. (1995) Turbine Rehabilitation: CFD Analysis of Distributors, *Waterpower'95*, Proc. Intl. Conf. Hydropower

Sick, M., Casey, M.V., Galpin, P. (1996) Validation of a Stage Calculation in a Francis Turbine, *Proc. XVIII IAHR Symposium*

Keck, H., Drtina, P., Sick, M. (1996) Numerical Hill Chart Prediction by means of CFD Stage Simulation for a Complete Francis Turbine, *Proc. XVIII IAHR Symposium*

Kubota, T., Takimoto, S., Kawashima, M. (1984) Effect of Stay Vane Shape on the Performance of Hydraulic Turbine, *Proc. 12th IAHR Symposium*

Lipej, A., Jost, D., Oberdank, K., Velensek, B., Jamnik, M. (1994) Numerical and Experimantal Flow Analysis in a Francis Turbine, *Proc. XVII IAHR Symposium*

Parkinson, E., Dupont, P.,Hirschi, R.,Huang, J., Avellan, F. (1994) Comparison of Flow Computation Results with Experimental Flow Surveys in a Francis Turbine, *Proc. XVII IAHR Symposium*

Ruprecht, A., Bauer, C., Riedelbauch, S. (1994) Numerical Analysis of Three-Dimensional Flow Through Turbine Spiral Casing, Stay Vanes and Wicket Gate, *Proc. XVII IAHR Symposium*

Shyy, W., Vu, T.C. (1993) Modeling and Computation of Flow in a Passage with 360-Degree Turning and Multiple Airfoils, *J.Fluids Engineering*, **115**

Soares-Gomes, F., Mesquita, A.A., Ciocan, G., Kueny, J.L. (1994) Numerical and Experimental Analysis of the flow in a Pump-Turbine Spiral Casing in Pump Operation, *Proc. XVII IAHR Symposium*

IMPROVEMENTS OF A GRAPHICAL METHOD FOR CALCULATION OF FLOW ON A PELTON BUCKET

MORTEN HANA
Ph.D. student
Norwegian University of Sience and Technology
Trondheim, Norway

Abstract

A graphical method for calculation of particle paths over Pelton buckets is presented. This method involves a laborious graphical derivation. This paper presents a new method for deriving particle velocities and accelerations. This method is based on cubic splines, and is calculated on a computer. A comparison with a graphical method is performed.

Resume

Nous présentons une méthode graphique de calcul de la trajectorie des particules liquides dans une aube de turbine Pelton qui implique une dérivation graphique trés laborieuse. Une nouvelle méthode de calcul par ordinateur d'obtention des vitesses et des accélérations est présentée, utilisant des fonctions splines cubiques. Nous comparons ses résultats avec ceux de la méthode graphique.

E. Cabrera et al. (eds.), Hydraulic Machinery and Cavitation, 111–119.

1. Introduction

At the Hydropower laboratories at the Norwegian University of Science and Technology, there has been a tradition for graphical calculations of flow on a Pelton bucket. This tradition dates back to 1918, when Henrik Christie wrote the first diploma thesis at the Laboratory. In this thesis he introduced the method, which is based on the simple formulation that a particle on the surface always has a acceleration that is perpendicular to the surface.

Later Hermod Brekke continued this tradition at Kværner, where Christie was his superior. Brekke modernized the method by introducing reduced quantities, which made the method more universal. In 1984 he presented his work in a paper called "A general study on the design of vertical Pelton turbines" [1].

As part of a Ph.D. thesis the graphical method has been investigated. Improvements done to the method will be presented in this paper. The scope of this work has been to make part of the graphical analysis into a numerical analysis.

2. Theoretical Background

The flow over Pelton buckets is non-stationary. Thus making it very difficult to establish a thorough theoretical analysis of the flow over Pelton buckets.

By using a graphical method that traces particles, a analysis of the flow is possible. This graphical method is based on the following facts [1]:

1. The acceleration on a particle on the jet surface is zero before entering the buckets. And in the same way is the acceleration zero on the particle after the outlet of the buckets.
2. The acceleration on a particle must always be normal to the water surface.
3. The water volume on a bucket included the outlet water behind the buckets is equal to the water volume cut out of the jet.

These three simple facts makes a graphical analysis possible.

2.1. REDUCED VALUES

All the values used for velocity, acceleration and time in this paper is reduced values. The use of reduced quantities in the graphical method was introduced by Brekke [1].

Reduced values are calculated the following way:

$$\underline{V} = \frac{V}{\sqrt{2gH}}$$

Where $\underline{V}$ is the reduced dimensionless velocity, V is the velocity, g gravity and H is the net head.

The reduced angular velocity is defined as:

$$\underline{\omega} = \frac{\omega}{\sqrt{2gH}}$$

Which gives the redusec angular velocity the dimension $[m^{-1}]$.

The reduced acceleration is found by examining equation (1), which is presented later, which gives that the acceleration should have the same dimension as $2\omega V$ which is $[m^{-1}]$:

$$\underline{a} = \frac{a}{2gH}$$

If we consider that $a = \frac{dV}{dt}$, the reduced time become:

$$\underline{t} = t \cdot \sqrt{2gH}$$

In the remaining of this paper all quantities will be reduced as described above

3. Procedure for the graphical analysis

The first step in analyzing flow over a Pelton bucket is to establish the geometrically description of the water surface as it moves over the bucket. One way to do this is by obtaining stroboscopic photos of the surface.

Modern visualization tools like CCD-cameras and multiflash stroboscopic light together with computer tools for photographic analysis will in the future make this part of the process faster and more precise.

It is assumed that a geometrical description of the water surface is established for a Pelton bucket for the following calculations.

The next step is to trace particle paths and derive both velocities and acceleration for the particles. The particle path is at first guessed, but experience soon makes it good guesses. Acceleration is calculated by these basic equations:

$$\begin{aligned}
a_x &= \frac{d^2x}{dt^2} - \omega^2 R \cos\varphi - 2\omega V_y \\
a_y &= \frac{d^2y}{dt^2} + \omega^2 R \sin\varphi + 2\omega V_x \\
a_z &= \frac{d^2z}{dt^2}
\end{aligned} \tag{1}$$

These equations are for a relative coordinate axis on the Pelton wheel, as shown in figure 1. They demonstrates that when moving to a reference system fixed on the Pelton wheel one has to take into account the centrifugal and Coriolis accelerations.

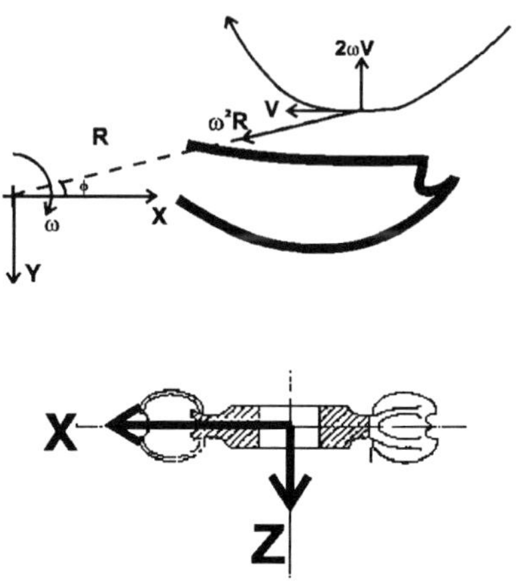

Figure 1. Coordinate axis on a Pelton bucket

The different signs on the Coriolis and centrifugal forces is deducted when considering the coordinate axis, velocities and spin axis, as shown in figure 1. The different terms in the equations are defined in the nomenclature list at the end of this paper.

When the acceleration has been calculated it is checked according to the second statement that said that it must always be perpendicular to the surface. If this is not fulfilled the particle path is adjusted and acceleration calculated again. This iteration process is repeated until the acceleration is perpendicular to the surface.

In figure 2 two shapes of water is showed along with some of the particle paths. The particle paths is named a, b, c and so on. Figure 2 also shows the graphical calculations and derivations of first and second derivatives. Figure 2 are from a calculation done by Hermod Brekke[1].

4. Cubic Splines

When deriving the velocities and accelerations for the particle moving along a path, the laborious and time consuming method of graphical derivation has been used in the past, as shown in figure 2.

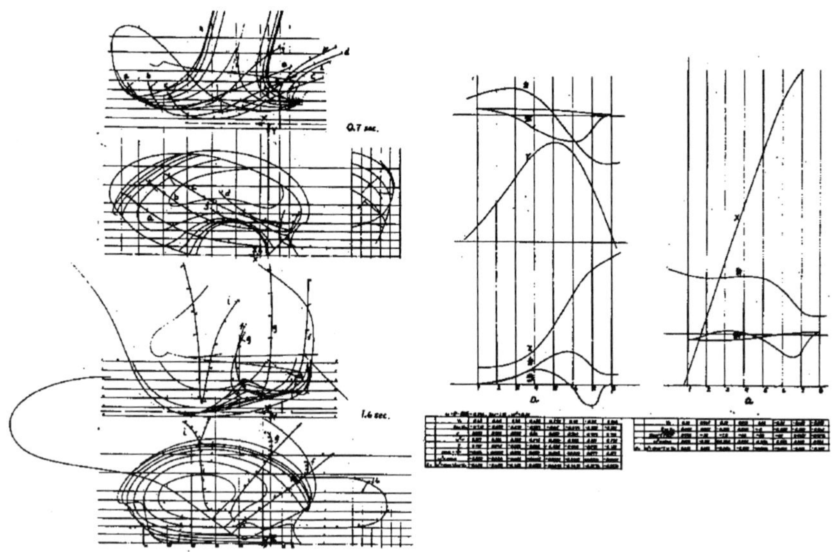

Figure 2. Example of water surfaces and graphical derivation

By using cubic splines to approximate the path, one gets the velocities and acceleration directly from the splines calculations. The cubic spline method is described in numerous numerical books, for instance the book by Cheney and Kinciad [2], on which this method is based.

Cubic Spline fitting is based upon Taylor series expansion. Taylor series expansion with regard to the time t may be written [2]:

$$S_i(t_i) = A_i + B_i t_i + C_i t_i^2 + D_i t_i^3 \tag{2}$$

Where the index i is a counting variable that indicates the internal splines between the particle points. Thus: $i = 1, 2, \ldots.., (n-1)$. Where $n =$ number of particle points. This gives $(n-1)$ equations with $4(n-1)$ unknown constants. These unknowns can be found by using boundary conditions. These boundary conditions are for x-direction:

$$S_i(t_i) = x_i \tag{3}$$

$$\lim_{t \to t_i^-} S^{(k)}(t_i) = \lim_{t \to t_i^+} S^{(k)}(t_i), \text{ where } k = 0, 1, 2 \tag{4}$$

$$S''(t_1) = S''(t_n) = 0 \tag{5}$$

The last condition (5) makes this a Natural Cubic Spline (NCS) approach. The NCS condition is reasonable if we consider that the second derivative of S, in either x, y or z direction, is the acceleration in that direction. According to the first statement made earlier the acceleration is zero on a particle on the jet surface before entering and after leaving the buckets, which makes (5) a reasonable boundary condition.

The two first conditions (3) and (4) are internal boundary condition ensuring continuity in the spline fitting. The boundary conditions gives:

Boundary condition	No. of conitions
1	$2\,(n-1)$
2	$2\,(n-1)$
3	2

Totally this gives $4\,(n-1)$ boundary conditions, and makes the system of equations solvable. This leads to a simple and fast method that substitute the laborious method of manual derivatives.

5. Computer Implementation

When the cubic splines system of equations is established, there are several ways to implement the solution procedure. Solving can simply be done by solving each spline separately, matrix solving or commercial available numerical mathematics programs. Since the cubic spline computer algorithms as described in text books like Cheney and Kinciad [2] is quite simple, the solution algorithm was implemented by programming a C/C++ code.

By programing the algorithm it was possible to have control of the adding of Coriolis and centrifugal acceleration according to equation (1).

The code has a simple input/output format. The input reads the x, y and z-coordinates, the code then calculates the respective first and second derivatives, and then writes the result to a file. The result file can be plotted.

The input file can contain one, several or all particle paths that is to be calculated. It also reads start time, time increment and how many time steps it shall calculate.

6. Test case: Skjåk Pelton turbine

In 1963 Hermod Brekke graphically calculated particle paths on the Skjåk Pelton turbine. These calculations has been used for comparing the cubic spline method with the graphical method.

The calculation began by reading the different paths, velocities and accelerations, from drawings as showed in figure 2. The coordinates was send as input to the program, and the result was compared to the read velocities and accelerations.

Figure 3 shows the results from the calculation of particle path B. This is a particle path which enters the bucket early and hence leaves at the back of the bucket. Other more complicated paths, which enters later, was also calculated.

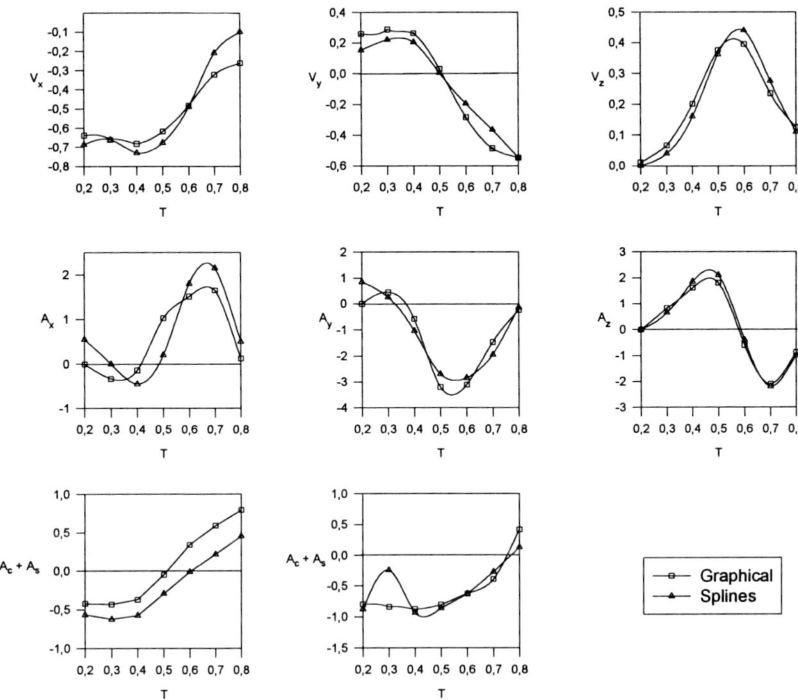

Figure 3. Example of results from Spline calculations

The results shows very good agrement between the graphical and spline calculations. The spline calculations show a tendency of oscillation around the graphical calculations.

By examining the spline approach one can see that a small change in the coordinate gives a relative big change in acceleration. When the acceleration was used as a guideline to make small changes in the coordinate, the particle path, velocity and acceleration became smoother. The changes in the particle path was made within the reading error.

7. Further improvements

As mentioned adjustment of the particle path were done manually, but it would be possible to implement a smoothing function on the acceleration path and integrate numerically the spline function to obtain a new particle path. And by allowing a magnitude of displacement of the particle path, smoother results will be obtained. A smoothing scheme that could be used could be it Gauss-Seidel iteration. Which is similar to relaxation schemes known from multigrid methods.

For future calculations of new paths, velocities and accelerations it is necessary to check that the sum of the acceleration always is perpendicular to the surface. If the surface can be described by surface vectors, a numerical comparison can be obtained. By developing certain rules for changing the particle paths depending the difference in acceleration and surface vector a almost complete numerical calculation can be achieved.

The surface description can, as mentioned earlier, be obtained by a CCD-camera. Numerical manipulation of these results can give a files for surface vectors.

8. Concluding remarks

The Natural Cubic Spline algorithm has been successfully implemented and tested. The splines show a oscillating nature, which could be smoothed manually. Future development will include numerical smoothing schemes.

The development of modern visualization tools will make a closer link between experiments and numerical analysis of flow over Pelton buckets.

Nomenclature

a_x, a_y, a_z	$[m^{-1}]$	Reduced acceleration in x, y and z directions, respectively
$V_x = \frac{dx}{dt}$		Reduced velocity in x direction, respectively
$V_y = \frac{dy}{dt}$		Reduced velocity in y direction, respectively
$R = \sqrt{x^2 + y^2}$	$[m]$	Radius
ω		Reduced angular velocity
$\varphi = \tan^{-1}\left(\frac{x}{y}\right)$		Angle
n		Number of particles in one path
g	$\left[\frac{m}{s^2}\right]$	Gravity $= 9.81 m/s^2$
H	$[m]$	Net head

References

1. Brekke, Hermod: "A general study on the design of vertical Pelton turbines", 1984, presented at 25th anniversary Turbuinstitut, Lubljana.
2. Kinciad, Ward and Cheney, David: "Numerical mathematics and computing", 2nd edition 1985, Brooks/Cole Publishing Company.

FLOW BEHAVIOR AND PERFORMANCE OF DRAFT TUBES FOR BULB TURBINES

T. KANEMOTO
M. UNO
Kyushu Institute of technology
Sensui 1-1, Tobata, Kitakyushu 804, Japan
M. NEMOTO
Kanagawa Institute of Technology
Shimo-ogino 1030, Atsugi 243-02, Japan
T. KASHIWABARA
Koochi Technical College
Shinkai-otsu 200-1, Monobe, Nankoku 783, Japan

1. Introduction

To cope with the warming global environment, the development of the ultra-low head energy is active by welcoming as the clean and cool energy resources. Bulb turbines, which are suitable to the ultra-low head, have the very high velocity head at the runner outlet, relative to the net head. Since the role of the draft tube on the hydraulic performance is extremely high, the bulb turbine normally uses the straight draft tube which is composed of the inlet annular diffuser with the hub followed by the diffuser having the transitional cross section from conical to rectangular shape. The performance of such a draft tube can not be predicted successfully by usual data for two-dimensional, conical and/or annular diffusers. Thus, the pressure recovery of the straight draft tube for the bulb turbine and the internal flow in the S-shaped draft tube for the tubular turbine have been investigated [1]-[3].

The internal flow and performance of the straight draft tube equipped with the conically shaped hub with the cusp end were also investigated by us, under the free vortex type swirling flow at the draft tube inlet [4]. The flow condition at the draft tube inlet is determined from the turbining operation point, and the above free vortex type swirling flow corresponds to the flow condition discharged from the runner at the partial load operation point. As making the load increase, the swirling direction on the tip side of the runner blade becomes to differ from that on the hub side (for instance, clockwise on the tip side and counter-clockwise on the hub side). Besides, the profile of the prototype hub is not the conical shape with the cusp end but the trapezoid shape, in general.

Accordingly, this paper discusses the effects of the inlet swirling flow type and the hub shape on the internal flow structure and the performance of the straight draft tube for the bulb turbine, through the experimental results.

E. Cabrera et al. (eds.), Hydraulic Machinery and Cavitation, 120–129.

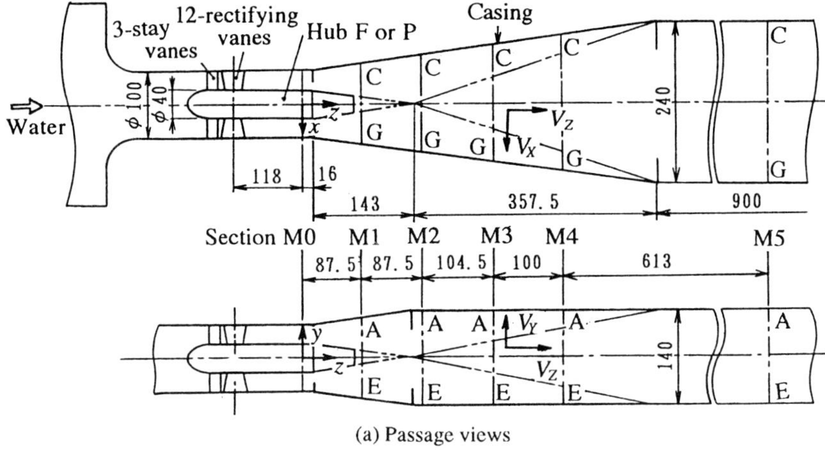

(a) Passage views

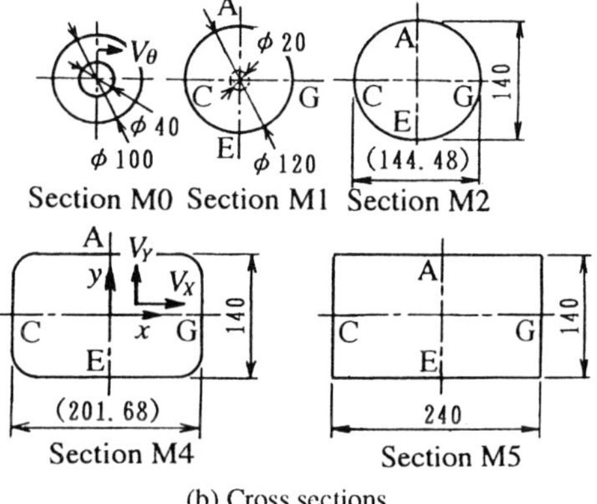

(b) Cross sections

Figure 1. Model draft tube

2. Model Draft Tube and Inlet Flow Condition

2.1. MODEL DRAFT TUBE

Figure 1(a) shows the tested model passage of the draft tube with the inlet approach and the downstream diffuser. Figure 1(b) illustrates the cross sections on which the traverse lines for the flow measurements are displayed with dot and dash lines. The twelve adjustable rectifying vanes can generate the necessary swirling flow in place of the rotational runner, and Section M0 which corresponds to the runner outlet was considered as the inlet datum section for the model.

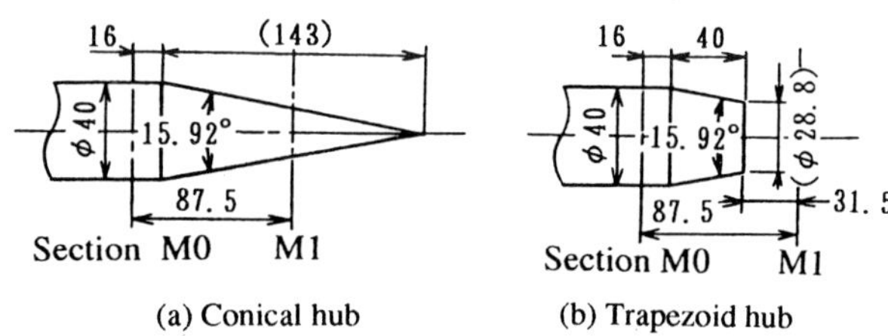

(a) Conical hub (b) Trapezoid hub

Figure 2. Model hubs

Figure 2 is the tail shape of the hub. The tail of Hub F is the conical shape with the cusp end (a), and the tail of Hub P is the trapezoid shape with the flat end (b). The distance from the tail end of Hub P to Section M1 is 31.5mm.

2.2. INLET FLOW CONDITION

Figure 3 shows the flow conditions at Section M0, under which the effects of the inlet swirling flow type on the internal flow and the performance were investigated with Hub

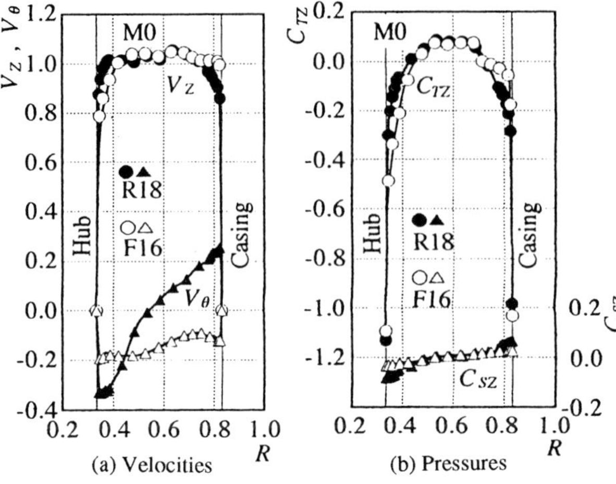

TABLE 1. Swirling intensity at Section M0 (R18, F16)

Flow type	R18	F16
|m|	0.183	0.159
m	0.0952	-0.159

Figure 3. Inlet Flow condition at Section M0 (R18, F16)

F. The abbreviations V_Z and V_θ give the dimensionless velocities in the z and θ directions (Fig.1), divided by the sectional avaraged velocity v_{ZO} at Section M0. The abbreviations C_{TZ} and C_{SZ} are the total and static pressure coefficients estimated with the difference from the mass flow avaraged pressures and v_Z at Section M0, and R is the dimensionless radius divided by double the inlet width B_0. These flow conditions are axissymmetrical and the Reynolds numbers are 5.6×10^5 estimated with the mean abso- lute velocity and the equivalent pipe diameter at Section M0.

The inlet swirling intensities $m = \Sigma r V_\theta V_Z dA \ / \ B_0 \Sigma V_Z{}^2 dA$ and $|m| = \Sigma r \ |V_\theta| V_Z dA \ / \ B_0 \Sigma V_Z{}^2 dA$ are presented in Table 1, where r is the radius and dA is the segmental area. Above inlet flow conditions are hereafter described as R18 and F16. The inlet swirling flow type is distinguished with alphabet R or F and the numerical value alludes to the inlet swirling intensity. The swirling intensity of R18 on the casing side is slightly larger than that on the hub side, as recognized from value of m in Table 1. The intensities $|m|$ estimated from the absolute swirling velocity component $|V_\theta|$, however, are nearly the same irrespective of the inlet swirling flow type.

Figures 4, 5 and Table 2 show the flow conditions at Section M0, under which the effects of the hub shape are investigated. The flow type with alphabet P means the case equipped with Hub P, the flow type with alphabet F means the case equipped with Hub F, and numerical value alludes to the inlet swirling intensity. These flows are also axissymmet-

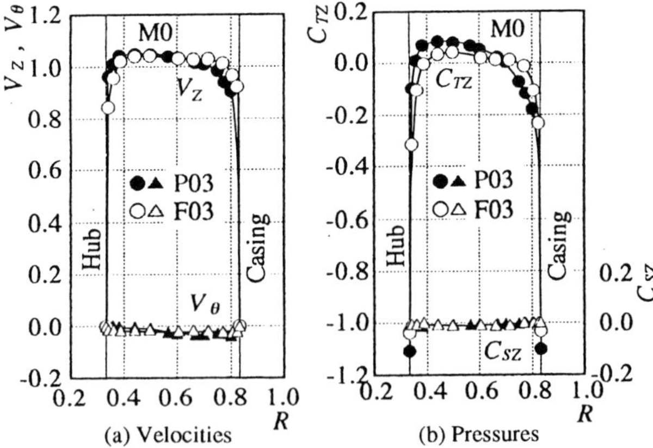

Figure 4. Inlet flow condition at Section M0 (P03, F03)

rical and the Reynolds numbers are the same as those of R18 and F16 mentioned above.

3. Effect of Inlet Swirling Flow Type on Flow Behavior

In this chapter, the effects of the inlet swirling flow type on the flow behavior in the draft tube equipped with Hub F are discussed, under the inlet swirling flow types of R18 and F16 given in Fig.3 and Table 1.

3.1. FLOW IN ANNULAR DIFFUSER

The velocity distributions at Section M2, just behind the cusp end of Hub F, are shown in Fig.6. The abbreviations V_X, V_Z are the dimensionless velocities divided by the sectional avaraged inlet velocity v_{ZO}, and Y is the dimensionless vertical distance divided by $2B_0$ (see Fig.1). The axissymmetrical flow condition of R18 at Section M0 already collapses in the annular diffuser composed of the axissymmetrical cross section, because the flow condition near the boundary where the swirling direction changes from clockwise to counter-clockwise (near Y= -0.6 and Y=0.7) is very unstable. On the contrary, the axissymmetrical flow condi-

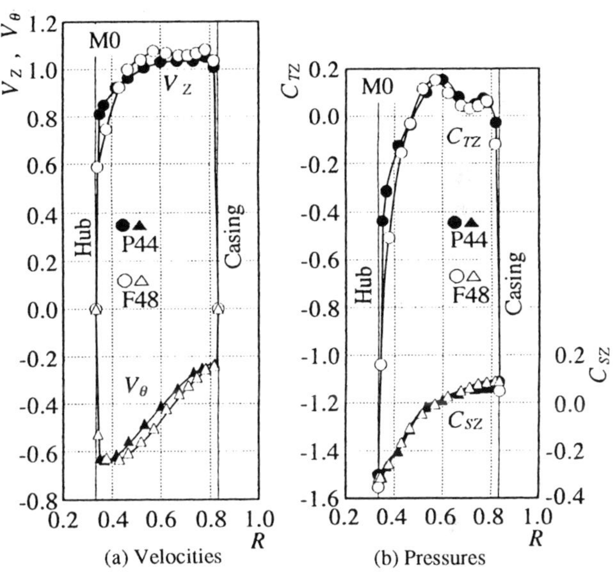

(a) Velocities (b) Pressures

Figure 5. Inlet flow condition at Section M0 (P44, F48)

TABLE 2. Swirling intensity at Section M0 (P03, F03, P44, F48)

Hub	P		F	
Flow type	P03	P44	F03	F48
| m |	0. 0301	0. 442	0. 0284	0. 478

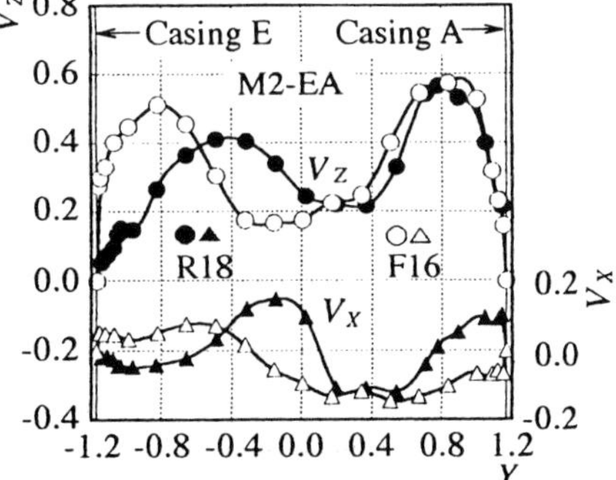

Figure 6. Velocity distributions on E-A at Section M2 (R18,F16)

tion of F16 at Section M0 is maintained moderately, but the velocity defect due to the hub is large as compared with one of R 18. Such a velocity defect is directly induced from the boundary layer along the hub wall [4]. By means of the boundary layer analysis [5] in the main flow predicted with the steramline curvature method [6], it was clear that the boundary layer flow of R18 does not separates but the flow of F16 separates at the first half of the hub wall.

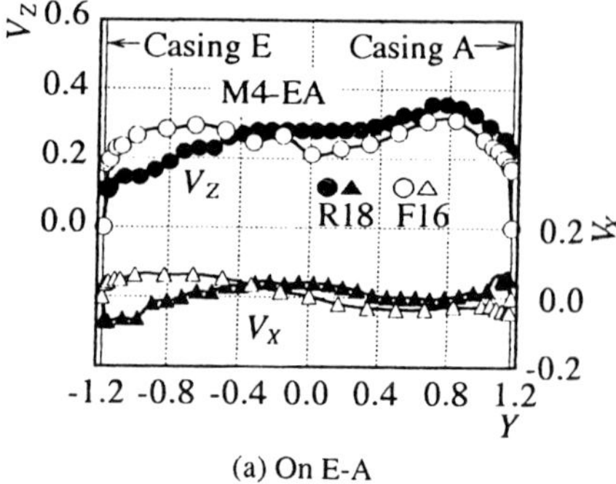

(a) On E-A

3.2. FLOW IN DOWNSTREAM DIFFUSER

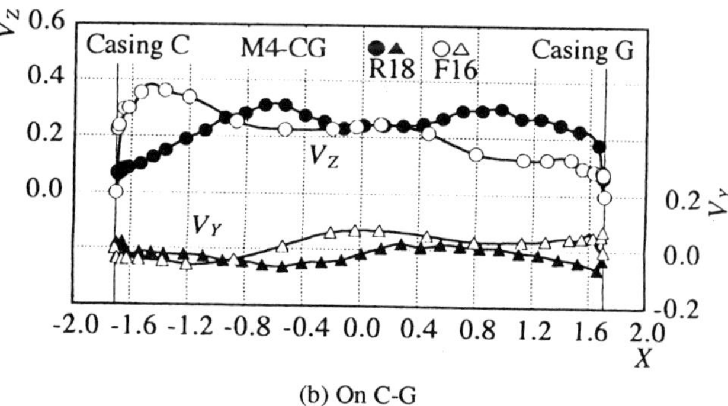

(b) On C-G

Figure 7. Velocity distributions at Section M4 (R18, F16)

The flows at Section M4 are shown in Fig. 7, where V_Y and X are the dimensionless velocity and horizontal distance (see Fig.1). The velocity deffects due to the hub come to disappesr considerably, and the flow of F16 tends to incline toward the one of the casing wall. Besides, it is noticeable that the swirling component V_X or V_Y, of R18 decreases conspicuously from Section M2 to M4 owing to the shearing stress. Such a flow mixing induced from the clockwise and counter-clockwise swirling components may make the hydraulic loss increase and the total pressure drop from Sections M2 to M4 of R18 is also larger than that of F16, as recognized in Fig.8.

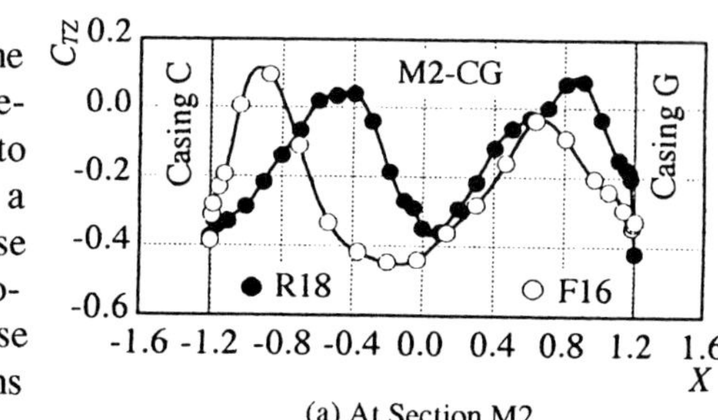

(a) At Section M2

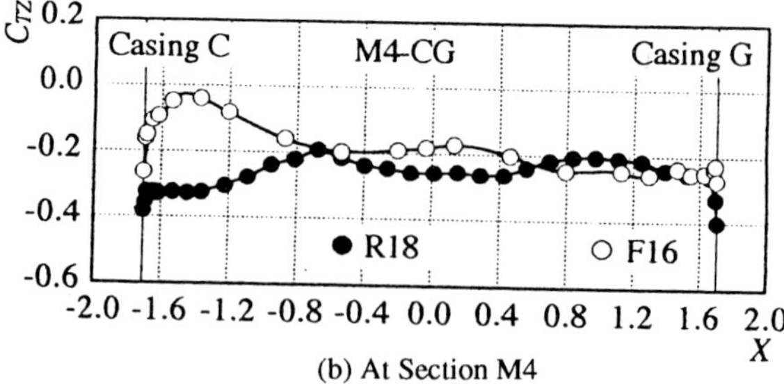

(b) At Section M4

Figure 8. Total pressure distributions on C-G (R18, F16)

At Section M5 in the straight passage composed of the rectangular cross section connected to the model draft tube outlet, the velocity V_Z distributions become uniform except for the thin boundary layer region on the casing wall, and the swirling velocity components V_Y are nearly the same value, irrespective of the inlet swirling flow type (Fig.9).

4. Effect of Hub Shape on Flow Behavior

In this chapter, the effects of the hub shape on the flow behavior in the draft tube is discussed, under the inlet flow conditions given in Figs.4, 5 and Table 2.

4.1. SWIRLLESS FLOW

The inlet flow conditions of P03 and F03 given in Fig.4 were here regarded as the swirlless flow, because of tiny swirling velocity components. The velocity distributions at Section M1, just behind the tail end of Hub P with the trapezoid shape, are shown in Fig.10, under the inlet flow conditions given in Fig.4. When Hub P is installed, the dead water region whose wide is nearly equal to the tail end diameter is observed obviously as marked with P03, and the axissymmetrical flow condition at the inlet already collapses in spite of the axissymmetrical passage. The flow of F03, however, is axissymmetrica and stable because Hub F makes the annular and narrow cross section.

The dead water region due to Hub P disappears at Section M2, as confirmed in Fig.11 (P03). Such a flow mixing makes the jet flow along the casing, surrounding the dead water region, diffuse suddenly into the whole passage and causes the unexpected pressure rise in the meridional direction.

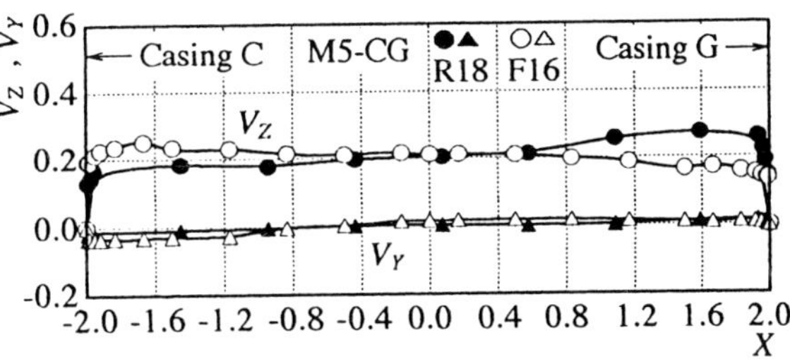

Figure 9. Velocity distributions on C-G at Section M5 (R18,F16)

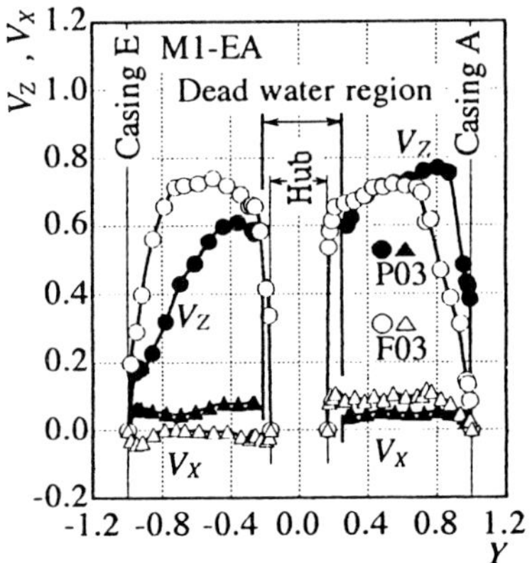

Figure 10. Velocity distributions on E-A at Section M1 (P03, F03)

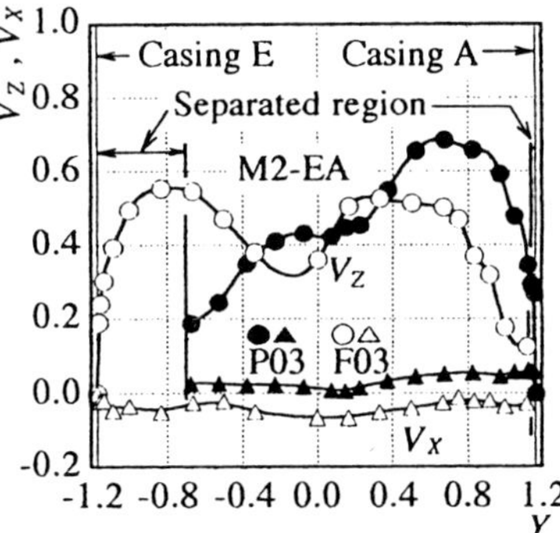

Figure 11. Velocity distributions on E-A at Section M2 (P03, F03)

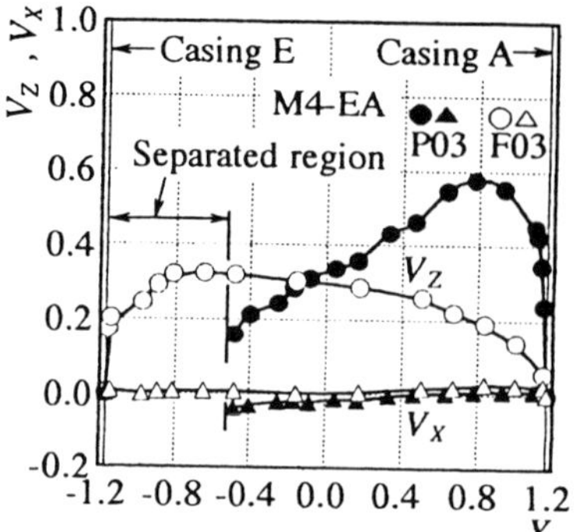

Figure 12. Velocity distributions on E-A at Section M4 (P03, F03)

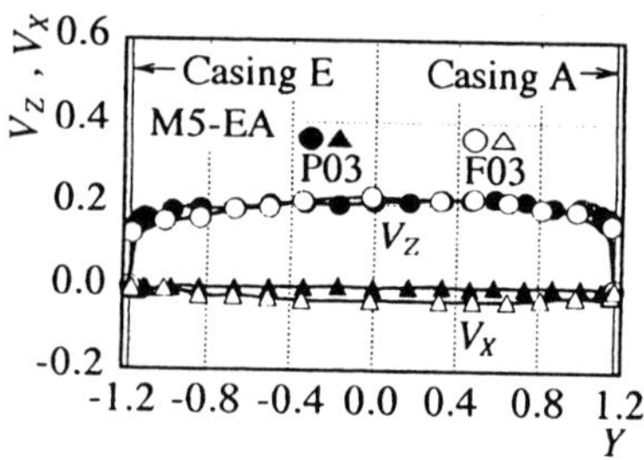

Figure 13. Velocity distributions on E-A at Section M5 (P03, F03)

This pressure rise promotes remarkably the undesirable boundary layer separation on the casing wall. The separation is naturally observed in the flow of F03, but its region is very narrow and the separated flow reattaches soon. On the contrary, the separation of P03 never reattaches and the dead water region spreads undoubtedly at downstream Section 4 close to the draft tube outlet, as shown in Fig.12. Such a separation affects markedly the pressure recovery of the draft tube. Accordingly, this result should be taken into account, in the design of the prototype draft tube whose outlet is directly opened in the dischrging water way. As our model draft tube outlet is connected to the straight passage composed of the rectangular cross section, however, the separation reattaches finally and velocity distributions become uniform irrespective of the hub shape (Fig. 13).

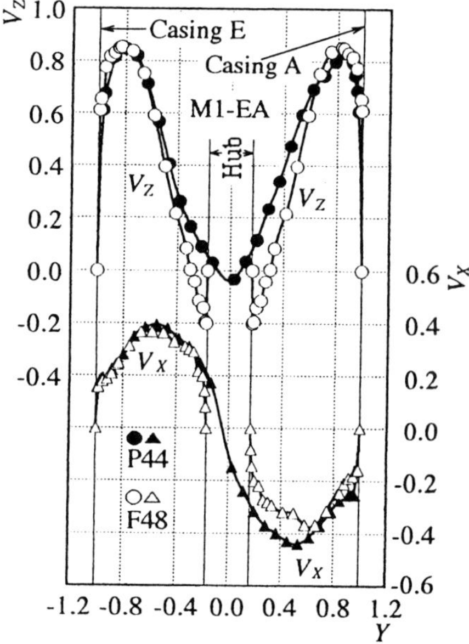

Figure 14. Velocity distributions on E-A at Section M1 (P44, F48)

4.2. SWIRLING FLOW

The flow conditions at Section M1, under the inlet flow with strong swirling velocity compnent, are shown in Fig.14. Contrary to the flow behavior without the swirling velocity component (P03), the flow of P44 never stagnates and runs somewhat backward accompanying with the forced vortex type swirling velocity component, in the center of this section just behind Hub P. This reason may be that the flow mixing is promoted well by the strong pressure gradient in the radius direction due to the swirling velocity component. Moreover, the flow condition of P44 is axissymmetrical and its axissym-

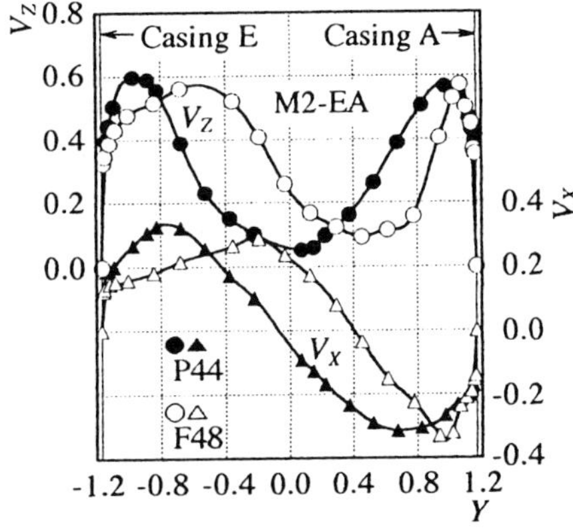

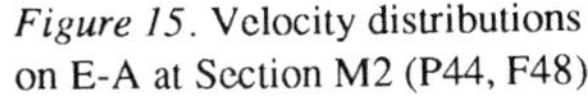
Figure 15. Velocity distributions on E-A at Section M2 (P44, F48)

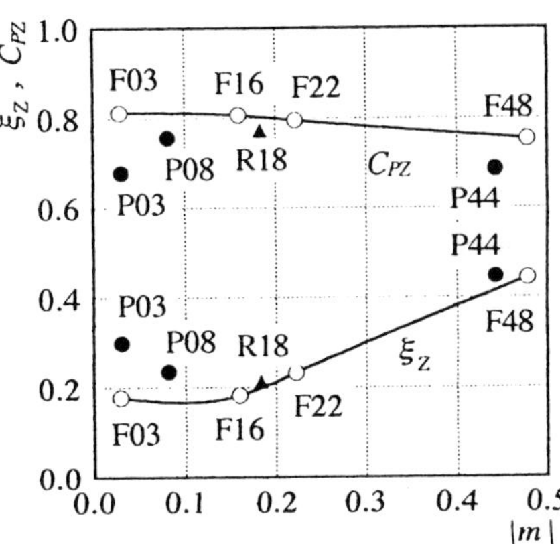

Figure 16. Model draft tube performances

metry is maintained as far as Section M2 (Fig. 15). On the other hand, the axissymmetrical flow condition of F48 at Section M1 becomes to collapse at Section M2 owing to the comparatively large reverse flow occured on the hub wall (see Fig. 14).

Except for the difference of the axissymmetrical flow conditon, the velocity profiles of P44 are almost the same as those of F48 at Section M2. Thus, the more downstream velocity profiles are also scarcely affected by the hub shape.

5. Hydraulic Loss and Pressure Recovery

Figure 16 shows the hydraulic loss coefficient ξ_Z and the pressure recovery C_{PZ} for the inlet swirling intensity $|m|$, whose values are estimated with the difference of the mass flow avaraged pressures between Sections M0 and M5 and the dynamic pressure with the sectional avaraged velocity v_{ZO} at Section M0. The abbreviation F with the numerical value on the full line is the data for Hub F, under the free vortex type swirling flow at Section M0 [4] . As compared with these values, the performances of R18, P03, P08, and P44 deteriorate at the same inlet swirling intensity $|m|$. These interesting results are discussed below.

5.1.EFFECT OF INLET SWIRLING FLOW TYPE

For R18 and F16 with Hub F, the hydraulic loss coefficient ξ_Z and the pressure recovery C_{PZ} in the segmental passage are presented in Table 3. In the annular passage from Sections M0 to M2, ξ_Z of F16 is larger than one of R18, owing to the difference of the boundary layer condition on the hub wall as discussed in the heading 3.1. On the contrary, ξ_Z of R18 becomes conspicuous in the downstream passage from Sections M2 to M4, as compared with one of F16. That is, the flow mixing accompanying with the decline of the swirling velocity component with the differnt direction may contribute

TABLE 3. Effect of swirling type

(a) Loss coefficient ξ_Z

Flow type	R18	F16
M0～M2	0.080	0.115
M2～M4	0.110	0.058
M4～M5	0.023	0.010

(b) Pressure recovery C_{PZ}

Flow type	R18	F16
M0～M2	0.665	0.646
M2～M4	0.053	0.110
M4～M5	0.065	0.051

TABLE 4. Effect of hub shape

(a) Loss coefficient ξ_Z

Flow type	P03	F03	P44	F48
M0～M1	0.090	0.012	0.271	0.100
M1～M2	0.134	0.072	0.074	0.256
M2～M5	0.076	0.094	0.102	0.087

(b) Pressure recovery C_{PZ}

Flow type	P03	F03	P44	F48
M0～M1	0.466	0.553	0.315	0.348
M1～M2	0.041	0.114	0.173	0.155
M2～M5	0.173	0.145	0.201	0.256

considerably to the increase of the hydraulic loss.

The static pressure recoveres mostly in the annular passage from Sections M0 to M2 and the quantity of the pressure recovery C_{PZ} in each segmental passage corresponds conversely to the hydraulic loss, irrespective of the inlet swirling flow type.

The effect of the inlet swirling flow type on the draft tube performance is small in comparison with the same $|m|$, but the performance of R18 cannot be helped estimating low in comparison with another inlet swirling intensity m defined in the heading 2.2, as confirmed in Fig.16.

5.2.EFFECT OF HUB SHAPE

Table 4 shows the effect of the hub shape on the hydraulic loss coefficient ξ_Z and the pressure recovery C_{PZ} in each segmantal passage. Under the swirlless flow condition, ξ_Z of P03 is larger than one of F03, in the segmental passage from Sections M0 to M2. It may be the reason why the mixing loss hehind Hub P with the trapezoid end is larger than the friction loss on Hub F with the cusp end. Such a flow mixing undoubtedly contributes to the increase of the hydraulic loss in the segmental passage from Sections M0 to M1, too.

Under the swirling flow condition, the hydraulic loss coefficient ξ_Z of P44 is large in the segmental passage from Sections M0 to M1 owing to the mixing loss just behind Hub P, while ξ_Z of F48 is large in the segmental passage from Sections M1 to M2 owing to the friction loss on the hub wall and the mixing loss induced from the reverse flow discussed in Fig.14. The increase of the hydraulic loss of P44, however, is almost the same as that of F48. This result means that the effect of the hub shape on the hydraulic loss in the segemntal passage from Sections M0 to M2 becomes comparatively small as increase of the inlet swirling intensity.

The pressure recovery C_{PZ} is affected by not only the hydraulic loss but also the displacement effect of the velocity defect in the wake flow and the boundary layer separation. That is, the pressure recovery C_{PZ} of P03 can not be expected from Sections M1 to M2 but expected from Sections M2 to M5 where the boundary layer separation reattaches.

6. Conclusion Remark

The effects of the swirling flow type at the inlet section and the hub shape on the flow behavior and performance of the model draft tube are summarized as follows.
(1) The swirling flow with the clockwise and counter-clockwise velocity components make the internal flow unstable, and the axissymmetrical inlet flow condition collapses at the first half of the annular duffuser.
(2) The clockwise and counter-clockwise swirling velocity components decrease obviously behind the annular passage owing to the shearing stress, and that makes the hydraulic loss increase.
(3) The trapezoid shaped hub accompanies with the dead water region behind the tail end, under the swirlless inlet flow condition. The diffusion of the jet flow surrounding this dead water region makes not only the mixing loss increase but also the flow along the casing separate. On the contrary, the water behind the trapezoid shaped hub hardly stagnates, under the strong swirling inlet flow condition.

Aknowledgement

The authors would like to express them thanks to Professor T. Kubota of Kanagawa University, Professors M. Nishi and H. Tsukamoto of Kyushu Institute of Technology for kindly giving valuable guidance and advice, and thanks to Messrs. H. Akamatsu, Y. Nagai, S. Kurahashi, Y. Yamane, A. Fujishima and K. Horikawa for devotedly helping these experiments

References

1. Kikuyama,K. et al., The Swirling flow effect on low head water turbine draft tube, *Proc.JSME Symp.*(1994),411=413(in Japanese).
2. Shimizu,Y. et al., Study on S-shaped draft tube for Tubular turbines, *Trans.JSME*,**52**-474,B(1986), 585-592, and **53**-492,B(1987),2500-2506(in Japanese).
3. Loeffler,A. Jr., A navier-Stokes code for S-shaped diffusers-review, *Int. J. for Numerical Methods in Fluids,* **8**(1988),463-474.
4. Kanemoto,T. Kubota,T. and Ishibashi,H, Swirling flow in straight draft tube with annuala and rectangular cross section, *11th Australasian Fluid Mechanics Conf.* **2**(1992),845-848.
5. Schlichting,H.,*Boundary Layer Theory*, 4th Ed.(1960),574.
6. Kanemot,T. and Toyokura,T., Flow in annular diffusers(2nd Report), *Bull. JSME*,**26**-218(1983), 1323-1329.

PERFORMANCE ANALYSIS OF DRAFT TUBE FOR GAMM FRANCIS TURBINE

T. KUBOTA
Kanagawa University
Yokohama, Japan

F. HAN
Huazhong University
Wuhan, China

F. AVELLAN
IMHEF, EPFL
Lausanne, Switzerland

1. Introduction

To precisely analyze the hydraulic losses in the various components of a Francis turbine, it is necessary to know the internal flow velocity distributions in the respective components. Since, however, the hydraulic energy loss $\Delta\psi$ and the performance characteristics such as energy coefficient ψ and discharge coefficient ϕ etc. are all the one-dimensional information for a whole turbine, it is sufficient to know the sectional mean velocities along a single representative mid-streamline through the turbine. Thus, the one-dimensional flow theory is decisively useful for the loss analysis corresponding to the performance diagrams. Performance diagrams acquired by the precise model test are so reliable that utilizing the diagrams to the loss analysis is highly recommendable. Two of the authors have presented, applying the one-dimensional flow theory, a new algorithm of extracting the various component losses in bulb turbines from the performance diagrams measured with the model tests [1].

Hydraulic loss in a draft tube is strongly affected by the intensity of swirl flow at the runner outlet. Since the swirl intensity of runner outflow varies with the operating conditions of ψ and ϕ, the draft tube loss also depends on ψ and ϕ. So far, the effect of swirl intensity on the flow behavior in the draft tube is well-investigated relating to the pressure pulsations [2]. Little is reported, however, concerning its effect on the hydraulic energy loss in the draft tube. To analyze the draft tube losses versus ψ and ϕ is essential to improve the hydraulic performance of Francis turbines, and to convert the model performance to its prototype considering the reliable scale effect.

GAMM Francis model turbine was designed and manufactured by IMHEF in EPFL to provide the experimental data on the hydraulic performances and flow distributions in the turbine for the 1989 GAMM Workshop on 3D-computation of incompressible internal flows [3]. The model test results revealed that the efficiency hill-diagram has the unusual two optimum peaks with the unsatisfactory efficiencies of 0.905, respectively. To investigate the unusual performance of the turbine, a *special* test was added by relocating

E. Cabrera et al. (eds.), Hydraulic Machinery and Cavitation, 130–139.

the low pressure measuring section for the determination of the specific hydraulic energy of turbine, SHET from the draft tube outlet to its *inlet.* The hill-diagram of the special test showed a single optimum peak with the improved efficiency of 0.920 [4].

The aim of this investigation is to extract the hydraulic energy loss in the bend draft tube for the GAMM Francis turbine from the above two performance diagrams, the one acquired by the *normal* model test with the draft tube *outlet* section, and the other by the *special* test with the draft tube *inlet* section.

2. Specific Hydraulic Energy of Turbine SHET

The specific hydraulic energy of turbine SHET shall, by nature, be defined as the difference of the true specific hydraulic energies between the inlet and outlet of the turbine. The true flow energy in a cross section can be determined by integrating the distributions of mass-averaged pressures and kinetic energies including the swirl velocity under the assumption of the same potential energies between inlet and outlet. Practically, however, the IEC code stipulates to calculate SHET E_{nor} with the following equation (cf. Fig. 1);

$$E_{nor} = E_{Ci} - E_{Do} = \left(\frac{P_{Ci}}{\rho} + \frac{V_{Ci}^2}{2} \right) - \left(\frac{P_{Do}}{\rho} + \frac{V_{Do}^2}{2} \right) \tag{1}$$

where P_{Ci} and V_{Ci} is the *wall* pressure and the *sectional mean* velocity at the spiral case inlet, and P_{Do} and V_{Do} at the draft tube outlet, respectively. Equation (1) would be approximately true only when the flows at both the inlet and outlet sections are uniform without swirl. In general, when the flow has a swirl component, it is necessary to correct the swirl energy to the normal SHET E_{nor} of Eq. (1).

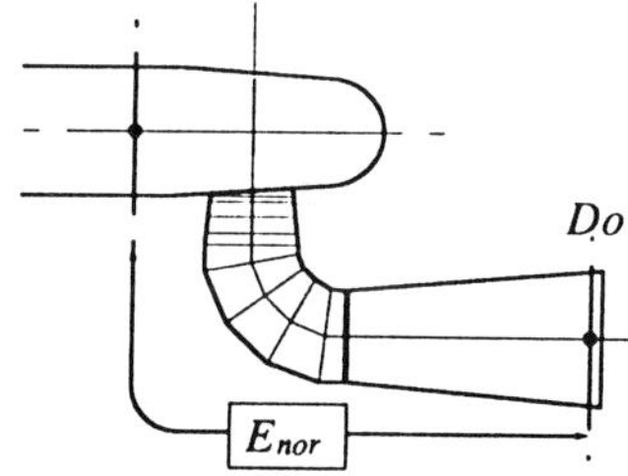

Fig. 1 *Enor* at normal test

Generally, a model test stand supplies rather uniform flow to the model turbine from the upper tank with rectifier via short straight inlet pipe. The flow in the high pressure measuring section is, therefore, nearly uniform without swirl irrespective of the operating conditions. The SHE (specific hydraulic energy) E_{Ci} at turbine inlet calculated with the *wall* pressure and the *sectional mean* velocity is approximately close to the true energy. On the other hand, the flow at runner outlet greatly varies with the operating conditions, and the bend draft tube distorts the flow toward the outlet. The flow at draft tube outlet is not uniform with strong distortion, swirl and/or reverse flows. As a result, the SHE E_{Do} at the draft tube *outlet* calculated with the wall pressure and the sectional mean velocity differs from the true energy.

Under the operating condition of swirl-free flow at the runner outlet, the SHE E_{Di} at draft tube *inlet* obtained from the wall pressure and the sectional mean velocity is rather close to the true energy. If we measure the *special* SHET E_{spc} by relocating the low pressure measuring section to the draft tube *inlet* as shown in Fig. 2 that differs from the IEC code specified, then, the SHET is close to the true energy at the swirl-free condition.

$$E_{spc} = E_{Ci} - E_{Di} = \left(\frac{P_{Ci}}{\rho} + \frac{V_{Ci}^{\ 2}}{2} \right) - \left(\frac{P_{Di}}{\rho} + \frac{V_{Di}^{\ 2}}{2} \right) \tag{2}$$

This E_{spc} reaches maximum at an operating point where the E_{Di} becomes minimum under the given E_{Ci}, and where the Euler energy of runner becomes maximum. Since E_{Di} reaches minimum at the point where the runner outflow has no swirl, the energy efficiency obtained by E_{spc} reaches maximum at this point. The special SHET E_{spc} does not include the hydraulic loss in the draft tube, so the efficiency based on E_{spc} must be higher than the one based on E_{nor}. At the point where the runner outflow has swirl velocity, the swirl flow energy shall be corrected to both E_{nor} and E_{spc} for getting the true energy.

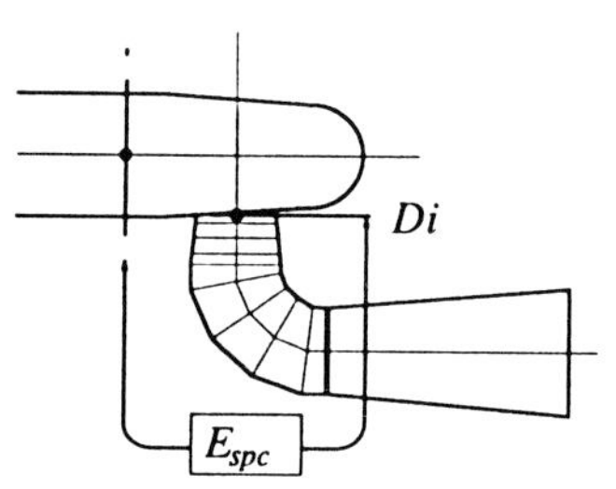

Fig. 2 *Espc* at special test

3. Model Test Results

3.1. MODEL TURBINE AND TEST PROCEDURE

The tested model Francis turbine has the discharge specific speed n_{sQ} of 76, the number of runner vanes Z_R 14, the runner outlet radius R_{ref} 0.200 m and the runner vane outlet radius R_2 on the representative mid-streamline of 0.1394 m. The adopted draft tube has an inlet conical diffuser with half cone angle θ_{Di} of 6.6 deg., a 90 deg.-bend of circular cross section with the constant area, an outlet conical diffuser with half cone angle θ_{Do} of 5.0 deg and the area ratio of outlet to inlet AR of 3.23.

The *normal* model test was executed by the experts of IMHEF by selecting the draft tube *outlet* to the low pressure measuring section according to the IEC code for the ten guide vane opening angles α_G from 15 deg. to 35 deg. The tested energy E_{nor} was set to be a constant value of 98 J/kg according to Eq. (1).

After that, the *special* test followed by relocating the low pressure measuring section to the draft tube *inlet* with the same range of guide vane angles. The tested energy E_{spc} was also set to be a constant value of 98 J/kg according to Eq. (2).

3.2. DIMENSIONLESS REPRESENTATION OF PERFORMANCES

To find the corresponding operating points between the two tests, we need the dimensionless performances that are normalized to the identical reference conditions according to the similarity law. Since both the E_{nor} and E_{spc} are adjusted to be the same value of 98 J/kg during the tests, the actual energy for the special test is higher than the normal test, even under the same guide vane angle. Since the speed- and discharge-factor n_{DE} and Q_{DE} etc. based on the SHET E can not be used, therefore, discharge-, torque- and energy- coefficient ϕ, τ and ψ must be adopted based on the peripheral speed at runner outlet U_{ref} ($=\omega R_{ref}$) as follows;

$$\phi = \frac{Q}{\pi\omega R_{ref}{}^3} \tag{3}$$

$$\tau = \frac{2T}{\rho\pi\omega^2 R_{ref}{}^5} \tag{4}$$

$$\psi = \frac{2E}{\omega^2 R_{ref}{}^2} \tag{5}$$

The above ϕ and τ do not include E, so if the complete similarity to the internal flow through the model turbine is kept between the normal and the special test, the relation of τ versus ϕ shall coincide between the two tests irrespective of the tested energy E. As an example, Fig. 3 shows the correlation of τ and ψ versus ϕ under the guide vane angle of 30 deg. The blank marks illustrate the normal test results and the filled marks correspond to the special test results. The torque coefficients for the both tests well coincide against the discharge coefficients and demonstrate the dimensionless performances of Eqs. (3) to (5) are applicable.

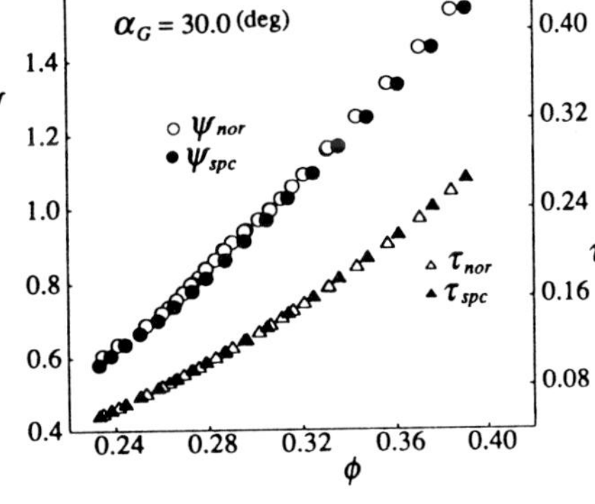

Fig. 3 Dimensionless Performances (α_G=30 deg.)

When the slight deviation is detected in the setting of α_G between the two tests that are executed separately, the value of ϕ is fine tuned to hold the consistency of ϕ versus α_G for both tests. The corresponding τ and ψ are corrected with the square of ϕ by confirming the coincidence of τ and ϕ. From the difference in ψ thus corrected, now we can get the specific hydraulic energy of draft tube SHED E_{spc}-E_{nor}.

3.3. NORMAL TEST RESULTS

The diagram of normal performances measured with E_{nor} are shown in Fig. 4. Strangely, the diagram has two efficiency peaks with the insufficient values of 0.905 under the lower discharge (ϕ_{opt1} = 0.26, α_{Gopt1} = 22.5 deg.) and the higher discharge (ϕ_{opt2} = 0.31, α_{Gopt2} = 25 deg.).

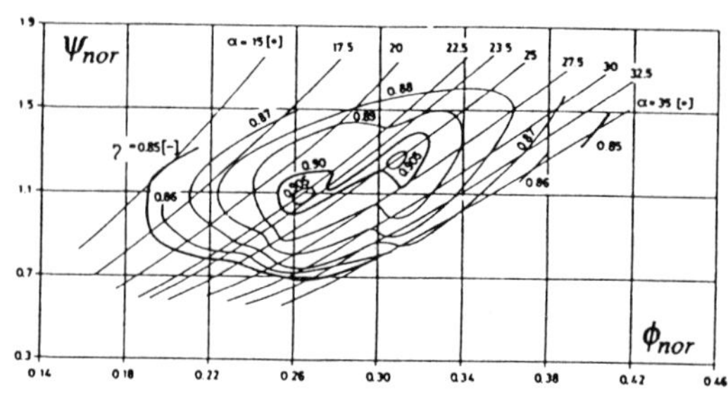

Fig. 4 Normal Performances based on E_{nor}

3.4. SPECIAL TEST RESULTS

The diagram of special performances measured with E_{spc} are shown in Fig. 5. The diagram has an improved single efficiency peak of 0.92 at the discharge coefficient of 0.29. As mentioned above, the efficiency based on E_{spc} reaches optimum at the operating condition of swirl-free runner outflow. We can designate the optimum discharge in the special performances as the swirl-free discharge ϕ_{cf} (=0.29).

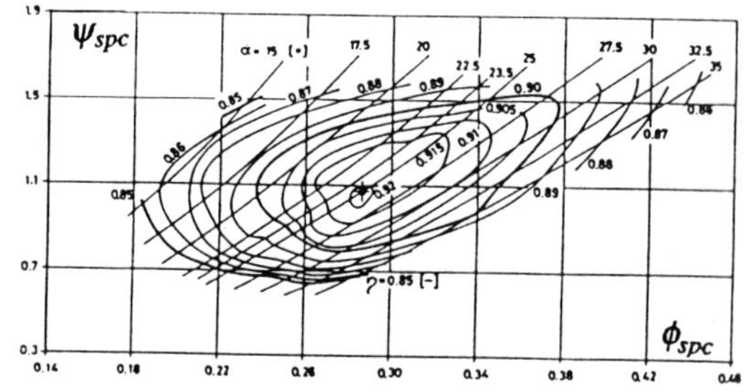

Fig. 5 Special Performances based on E_{spc}

4. Extraction of Draft Tube Energy

As confirmed with Fig. 3, the difference in ψ between the two performances is very small. To increase the sensitivity of detecting the small difference $\Delta\psi$ on the diagram, it is better to represent the diagram so as that the variation of ordinate is minimum against the variation of abscissa. For this reason, the energy ψ and the torque τ of the two performances are transformed by multiplying the square and the cube of the ratio of the swirl-free discharge ϕ_{cf} to the respective discharge ϕ, respectively. As the typical examples, Fig. 6 illustrates the magnified ordinates $\psi(\phi_{cf}/\phi)^2$ and $\tau(\phi_{cf}/\phi)^3$ versus ϕ for the guide vane angles of 25, 30 and 35 deg., respectively. Blank and filled mark corresponds to the normal and special performance, respectively. Both the τ_{nor} and τ_{spc} are coincide well within a wide range of ϕ irrespective of α_G. The magnified ψ_{nor} varies much against ϕ with increasing α_G, whereas the variation of ψ_{spc} is moderate.

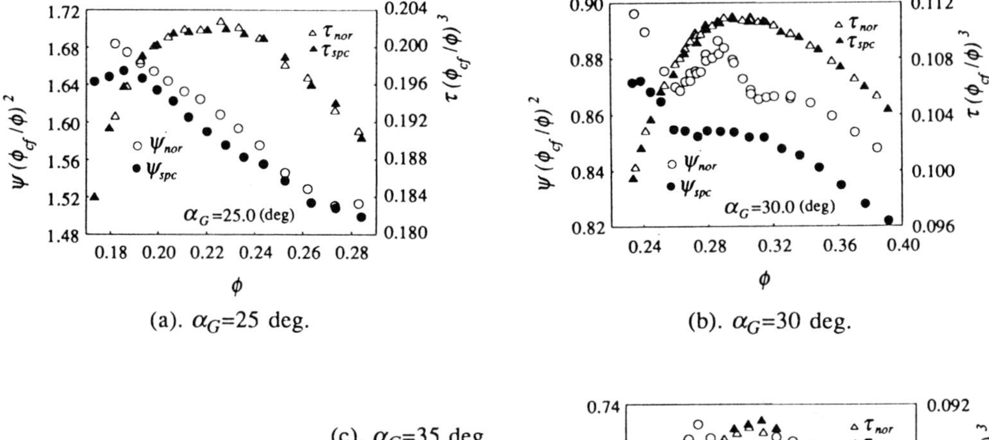

(a). α_G=25 deg.

(b). α_G=30 deg.

(c). α_G=35 deg.

Fig. 6 Magnified performances

The differential energy coefficient ψ_{nor}-ψ_{spc} under the identical discharge coefficient ϕ implies the specific hydraulic energy of draft tube SHED based on the IEC definition (wall pressures and sectional mean velocities) as follows;

$$\Delta\psi_{IEC} \equiv \psi_{nor} - \psi_{spc} = \frac{2(E_{Di} - E_{Do})}{\omega^2 R_{ref}^2} \tag{6}$$

The SHED thus extracted from the both performance diagrams are illustrated in Fig. 7 with the entire guide vane angles. The curves of $\Delta\psi_{IEC}$ repeat up and down against ϕ irrespective of α_G except the minimum angle of 15 deg. The local maxima on the middle of curves appear at the swirl-free discharge ϕ_{cf} of 0.29 irrespective of α_G. According to the typical Fig. 6(b), this results from ψ_{nor} including the draft tube loss and not from ψ_{spc}. So, the local maxima is the evidence of sudden increase in draft tube energy necessary for the swirl-free flow at runner outlet.

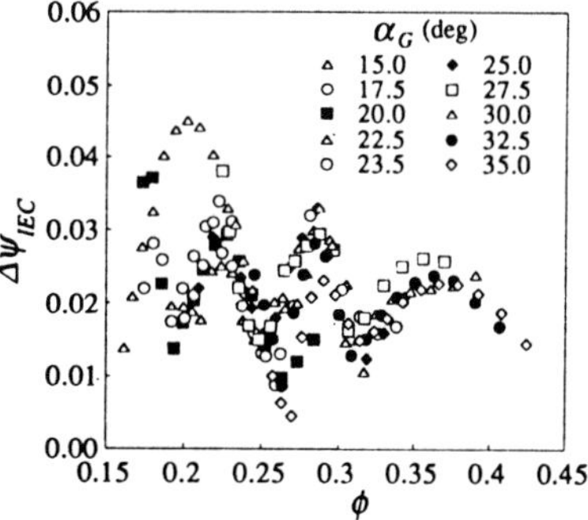

Fig. 7 Specific hydraulic energy of draft tube $\Delta\psi_{IEC}$

5. Correction of Swirl Flow at Draft Tube Inlet

As understood by Eq. (6), $\Delta\psi_{IEC}$ does not include the swirl energy at draft tube inlet. Actually, a Francis runner generates the swirl flow at runner outlet except the swirl-free operating conditions. The total kinetic energy of the swirl flow shall be the axial energy based on the sectional mean velocity plus the swirl energy. Also, the measured wall pressure includes the pressure rise due to the centrifugal force of swirl flow in addition to the static pressure. To obtain the actual energy at draft tube inlet, we have to correct the above effects of swirl flow as follows:

5.1. SWIRL VELOCITY AT RUNNER OUTLET

The velocity triangle at the intersection of a representative mid-streamline with the trailing edge line of runner vanes (subscript 2) forms the right triangle at the swirl-free discharge ϕ_{cf}. The relative angle β_{2cf} of runner outflow can directly be obtained from ϕ_{cf} as follows;

$$\tan\beta_{2cf} \equiv \frac{V_{m2cf}}{U_{2cf}} = \phi_{cf}\left(\frac{\pi R_{ref}^{3}}{A_2 R_2}\right) \tag{7}$$

where A_2 is the cross sectional area at runner vane trailing edge. The fact that the swirl-free discharge ϕ_{cf} does not depend on the guide vane angle as shown in Fig. 7, tells us that also the runner outflow angle β_{2cf} is not dependent on the guide vane angle. This is understandable because the solidity of Francis runner vanes is high and, in general, the runner outflow angle is approximately constant within the normal operating range.

Assuming the runner outflow angle β_2 is constant ($=\beta_{2cf}$) irrespective of ϕ, the swirl velocity V_{u2} of the runner outflow can be calculated for the arbitrary ϕ as follows;

$$V_{u2} = U_2 - \frac{V_{m2}}{\tan\beta_{2cf}} = U_2\left(1 - \frac{\phi}{\phi_{cf}}\right) \tag{8}$$

5.2. SWIRL ENERGY AT DRAFT TUBE INLET

The swirl energy coefficient $\Delta\psi_{vu}$ of the runner outflow can be obtained with the swirl velocity of Eq. (8) for the correction of the draft tube inlet energy as follows;

$$\Delta\psi_{vu} \equiv \frac{V_{u2}^2/2}{U_{ref}^2/2} = \left(1 - \frac{\phi}{\phi_{cf}}\right)^2 \left(\frac{R_2}{R_{ref}}\right)^2 \tag{9}$$

5.3. WALL PRESSURE RISE DUT TO SWIRL FLOW

Assuming the flow field between the runner vane trailing edge (position 2) and the low pressure measuring section at draft tube inlet (for P_{Di}) is of free-vortex, the wall pressure rise coefficient $\Delta\psi_p$ due to the swirl velocity of Eq. (8) is calculated with the following equation for the low pressure measuring section;

$$\Delta\psi_p \equiv \frac{\Delta p/\rho}{U_{ref}^2/2} = \Delta\psi_{vu}\left(\frac{R_2}{R_{ref}}\right)^2 \tag{10}$$

6. Extraction of Draft Tube Loss

By correcting the swirl energy $\Delta\psi_{vu}$ of Eq. (9) and the wall pressure rise $\Delta\psi_p$ of Eq. (10) to the IEC draft tube energy $\Delta\psi_{IEC}$ of Eq. (6), the effect of swirl flow on the energy at draft tube inlet has been corrected. Theoretically, the similar correction would be necessary for the swirl flow at draft tube outlet. As mentioned in Chap. 2, however, the flow at draft tube outlet is distorted greatly with the complicated swirl and reverse flow. The maximum velocity in the outlet section with swirl component would be much higher than the sectional mean velocity. Nevertheless, the kinetic energy of the tube outflow is negligibly small than the runner peripheral speed energy, except the case of high specific speed machines. The wall pressure at tube outlet is close to the pressure in the tail water tank, and less affected by the pressure rise because the swirl flow is not dominant. According to the above consideration, the actual energy loss $\Delta\psi_D$ in the draft tube flow can be to calculated with the following formula;

$$\begin{aligned}\Delta\psi_D &\equiv \frac{2\Delta E_D}{\omega^2 R_{ref}^2} = \Delta\psi_{IEC} + \Delta\psi_{vu} - \Delta\psi_p \\ &= \Delta\psi_{IEC} + \left(1-\frac{\phi}{\phi_{cf}}\right)^2\left(\frac{R_2}{R_{ref}}\right)^2\left(1-\frac{R_2^2}{R_{ref}^2}\right)\end{aligned} \tag{11}$$

Figure 8 shows the actual draft tube loss $\Delta\psi_D$ obtained from SHED $\Delta\psi_{IEC}$ in Fig. 7 by using Eq. (11). Roughly speaking, the extracted draft tube loss coefficients are aligned on a single line of W-shape except the minimum guide vane angle. The tube loss reaches maximum at the swirl-free discharge, and decreases once when the discharge leaves ϕ_{cf}. As understood by Eq. (8), when the discharge decreases from the swirl-free discharge, the positive swirl velocity increases, and vice versa. There are two optimum discharges ϕ_{opt} of 0.26 and 0.31

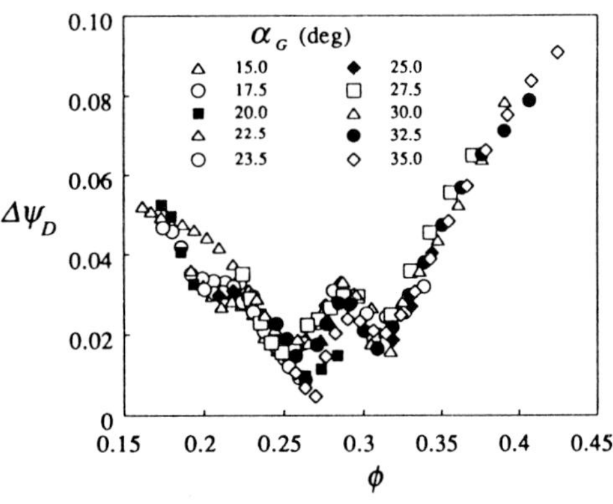

Fig. 8 Actual draft tube energy loss $\Delta\psi_D$

where the tube loss reaches minimum. The discharges can be designated as the optimum swirl discharge. The tube loss at the lower side of ϕ_{opt} is smaller than the loss at the higher side. At the lower side, the swirl energy at runner outflow is positively optimum, and vice versa. When the discharge decreases or increases apart from the optimum swirl discharges, the tube loss increases especially at high discharge range.

In general, if there is no-swirl flow at the inlet of a bend draft tube, the flow is used to separate from the inner bend wall, and the energy loss increases. The draft tube applied for this study has a circular bend of constant sectional area. The aspect ratio of the bend (the ratio of the radius of curvature at outer bend wall to that at inner bend wall) is high, and the secondary flow is apt to develop in the bend. This is the reason why the draft tube loss at the swirl-free discharge is large compared with the normal bend draft tube. The swirl energy at tube inlet increases when the discharge leaves the swirl-free discharge and suppresses the flow separation at the bend, resulting in the decrease of tube loss irrespective of the direction of swirl. Between the two minima of tube loss at the optimum swirl discharges, it is natural that the loss at the lower side of discharge is smaller. Since the tube loss at the swirl-free discharge is large, the contribution of swirl to the loss reduction is also large. Apart from the two optimum swirl discharges, with increasing positive or negative swirl energy, the tube loss becomes large.

Now we can understand the reason why the normal performance diagram has two efficiency peaks as follows: At the swirl-free discharge, the separation loss is large in the bend draft tube. At the two optimum swirl discharges smaller or larger than the swirl-free discharge, the separation loss is drastically suppressed by the optimum swirl. As a result, the two efficiency peaks appear at the two optimum swirl discharges. On the other hands, in the case of an ordinary Francis turbine with the well designed bend draft tube, since the separation loss in the bend is not so large, only a single efficiency peak will appear at the lower side of the two optimum swirl discharges.

The draft tube loss coefficient ζ_D based on the kinetic energy of the sectional mean velocity V_{mDi} at draft tube inlet can be deduced from $\Delta\psi_D$ in Fig. 8 as follows;

$$\zeta_D \equiv \frac{\Delta E_D}{V_{mDi}^2/2} = \frac{\Delta\psi_D}{\left(V_{mDi}/U_{ref}\right)^2} \tag{12}$$

The obtained results are illustrated in Fig. 9. The extracted draft tube loss coefficients are also aligned on a W-shaped curve. The increase of coefficient is sharp at the lower discharge range, whereas the slope of curve is small at the higher discharge region with approaching to a constant value asymptotically.

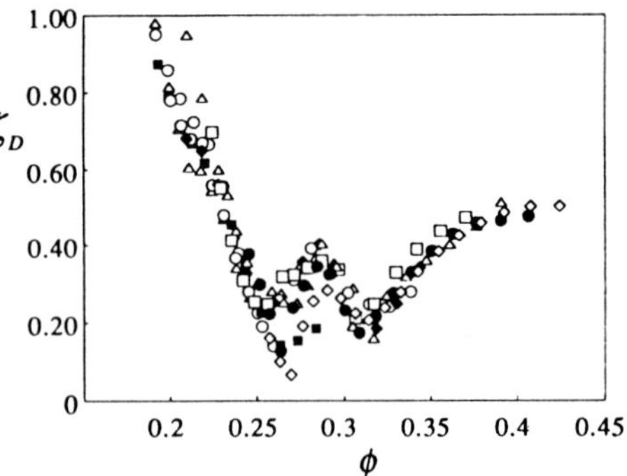

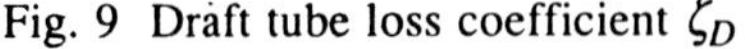
Fig. 9 Draft tube loss coefficient ζ_D

7. Conclusion

Applying the GAMM Francis model turbine, the normal model test based on the draft tube outlet for the low pressure measuring section and the special test based on the draft tube inlet were executed separately. The SHED (specific hydraulic energy of draft tube) $\Delta\psi_{IEC}$ is extracted from SHET (specific hydraulic energy of turbine) for both performances under the identical flow conditions of the model turbines. By correcting the swirl energy at draft tube inlet to $\Delta\psi_{IEC}$, tne draft tube loss coefficient ζ_D can be empirically extracted from the both performance diagrams

8. References

1. Han, F., Ida, T. and Kubota, T.: Analysis of Scalable Loss in Bulb Turbine in Wide Operating Range, 17tn IAHR Symposium - Beijing (1994), Vol.2, G5, 853 - 864.
2. Kubota, T. and Yamada, S.: Effect of Cone Angle at Draft Tube Inlet on Hydraulic Characteristics of Francis Turbine, 11th IAHR Symposium - Amsterdam (1982), 53, 1-14.
3. Sottas, G. and Ryhming, I.L.; "3D-Computation of Incompressible Internal Flows", Notes on Numerical Fluid Mechanics, Vol.39, (1993) Vieweg.
4. Avellan, F., Dupont, P., Farhat, M., Gindroz, B., Henry, P., Hussain, M., Parkison, E. and Santal, O.; "Flow Survey and Blade Pressure Measurements in A Francis Turbine Model", 15th IAHR Symposium - Belgrade (1990), Vol.2, I5.

MODELLING COMPLEX DRAFT-TUBE FLOWS USING NEAR-WALL TURBULENCE CLOSURES

Y. VENTIKOS
Postdoctoral Associate

F. SOTIROPOULOS
Assistant Professor

School of Civil and Environmental Engineering
Georgia Institute of Technology
Atlanta, Georgia 30332, U.S.A.

and

V.C. PATEL
Professor and Director

Iowa Institute of Hydraulic Research
The University of Iowa
Iowa City, Iowa 52242, U.S.A.

Abstract

This paper presents a finite-volume method for simulating flows through complex hydroturbine draft-tube configurations using near-wall turbulence closures. The method employs the artificial-compressibility pressure-velocity coupling approach in conjunction with multigrid acceleration for fast convergence on very fine grids. Calculations are carried out for a draft tube with two downstream piers on a computational mesh consisting of $1.2\text{x}10^6$ nodes. Comparisons of the computed results with measurements demonstrate the ability of the method to capture most experimental trends with reasonable accuracy. Calculated three-dimensional particle traces reveal very complex flow features in the vicinity of the piers, including horse-shoe and longitudinal vortices and regions of flow reversal.

1. Introduction

Understanding hydroturbine draft-tube flows is a crucial prerequisite for addressing numerous operational and environmental challenges facing the hydropower industry today. From the operational standpoint, the draft-tube is of paramount importance for the overall efficiency and smooth operation of a hydraulic turbine, particularly at off-design conditions. Its importance is best demonstrated by the fact that the primary

E. Cabrera et al. (eds.), Hydraulic Machinery and Cavitation, 140–149.

consideration when designing turbine blades is to ensure that they deliver well-conditioned flow at the draft-tube entrance (Fisher, 1995). Poor inflow conditions are associated with several undesirable flow phenomena--flow reversal downstream of the runner, formation of rope vortices, and cavitation--which could induce large efficiency losses, devastating pressure pulsations in the entire system, and even failure. From the environmental standpoint, draft tube flows are important for understanding the causes of injury and/or mortality of passing fish as well as for developing effective strategies for improving the tailrace water quality. Regions of large flow gradients, intense streamwise vortices, cavitation, areas of flow recirculation, and formation of the so-called "back-roll" vortices at the draft-tube/tailrace interface may be responsible for injuring and/or disorienting passing fish. On the other hand, tailrace water quality, which is affected by the depletion of Dissolved Oxygen (DO) in the lower levels of the reservoir during warm months of the year (Bohac and Ruane, 1991), may be substantially improved using autoventing hydroturbines (AVT). AVT technology relies on turbulence mixing within the draft tube to ensure transfer of oxygen from the air-bubbles--injected into the water at strategic locations downstream of the runner--to the water at a rate sufficient to increase the tailwater DO concentration at environmentally acceptable levels (Carter, 1995).

A typical draft tube consists of a short conical diffuser followed by a strongly curved 90° elbow of varying cross-section and then a rectangular diffuser section. Its cross-sectional shape changes continuously from circular at the inlet, to elliptical within the elbow, and finally to rectangular at the exit. Additional geometrical complexities include the presence of one or more piers, downstream of the elbow, splitter blades, guide vanes, slots, etc. The flow that enters the draft tube--the wake of the turbine runner--is turbulent and three dimensional, with high swirl levels. This already complex inlet flow undergoes additional straining as it passes through the elbow, induced by the rapid area changes, the very strong longitudinal curvature, and the presence of various obstacles. The resulting flow is extremely complicated with regions of strong induced pressure gradients, intense longitudinal and horse-shoe vortices, regions of flow reversal, etc. These complexities make the numerical simulation of draft tube flows particularly challenging for even the most advanced numerical methods available today. Yet modern computational fluid dynamics (CFD) methods offer the most promising alternative for elucidating the physics of draft tube flows at a level of detail necessary for addressing the operational and environmental issues noted above.

Numerical simulations of draft tube flows have been reported, among others, by Vu and Shyy (1988), Agouzoul et al. (1990), Sotiropoulos and Patel (1993), and Reidelbauch et al. (1995). With the exception of Sotiropoulos and Patel (1993)--who employed a two-layer near-wall k-ε model and a moderately fine computational mesh (approximately 200,000 nodes)--all these studies adopted the standard, high Reynolds number, k-ε model with wall functions, and reported results on rather coarse meshes (40,000 to 100,000 nodes). Despite reproducing general physical trends, regarding the effect of inflow swirl on the flow development, none of these studies demonstrated their ability to quantitatively predict the flow details.

The objective of this work is to develop the computational framework that would enable accurate quantitative predictions of turbulent flows through complex draft-tube geometries over a range of operating conditions. An efficient, finite-volume numerical method is presented for solving the three-dimensional Reynolds-averaged

Navier-Stokes equations in conjunction with near-wall turbulence closures on very fine, highly stretched and skewed computational meshes. Numerical solutions are obtained for one of the Norris Power Plant draft tubes (Tennessee Valley Authority) at model-scale Reynolds numbers. The two-layer k-ε model of Chen and Patel (1988) is employed for turbulence closure. The computed solutions are compared with available mean velocity measurements at several locations downstream of the elbow (Hopping, 1992) and analyzed in terms of three-dimensional particle traces.

2. The numerical method

The numerical method of Sotiropoulos and Lin (1996) is modified and used in the present study. This method solves the three-dimensional Reynolds-averaged Navier-Stokes (RANS) equations, in conjunction with two-equation, near-wall, turbulence closures, formulated in generalized curvilinear coordinates in strong conservation form. Pressure-velocity coupling is achieved using the artificial compressibility approach. The governing equations are discretized on a non-staggered computational mesh using finite-volume discretization schemes. Three-point central differencing is employed for the viscous fluxes and source terms in the turbulence closure equations. The method features a number of options for approximating the spatial derivatives of the convective flux-vectors. These include second-order, central--with scalar and matrix valued fourth-difference artificial dissipation terms--and flux-difference splitting upwind (ranging from first to fifth-order accuracy) differencing schemes. The spatial resolution of these schemes has been carefully evaluated in both laminar (Lin and Sotiropoulos, 1996a) and turbulent flow simulations (Sotiropoulos and Lin, 1996b).

The discrete mean flow and turbulence closure equations are integrated in time using a four-stage explicit Runge-Kutta algorithm (Jameson, 1983) enhanced with local time-stepping, implicit residual smoothing, and multigrid acceleration. A three-grid level V-cycle algorithm with semi-coarsening in the transverse plane (that is, coarse grids are constructed by doubling the grid spacing only in the transverse directions) is employed in the present study. One, two, and three iterations are performed on the first, second, and third grid level, respectively. The present multigrid method is capable of solving the turbulence closure equations in both loosely and strongly-coupled fashion. In the first approach, multigrid is applied only to the mean-flow equations while the turbulence closure equations are solved only on the finest mesh (the eddy-viscosity values are injected to the coarser meshes and held constant during the cycling process). In a strongly coupled strategy, on the other hand, multigrid is applied simultaneously to both the mean and turbulence closure equations and the eddy-viscosity values are updated at each grid level (see Sotiropoulos and Lin (1996) for a detailed discussion and comparison of the various methods). All subsequently presented calculations have been obtained using the loosely coupled algorithm with three iterations performed on the turbulence closure equations per multigrid cycle.

The present method features a number of isotropic and non-isotropic (non-linear) two-equation turbulence models (see Sotiropoulos and Ventikos (1996) for details). In the present study, however, only the isotropic two-layer k-ε model of Chen and Patel (1988) is employed. Work is currently underway to implement and validate the various non-linear models for draft-tube geometries.

3. Test case and computational details

The draft tube configuration, used for the present computations, is one of the TVA Norris Autoventing Power Plant (Norris, Tennessee) draft tubes designed to operate with 66,000 HP hydroturbines. The area expansion ratio for this draft tube (ratio of the exit to inlet cross-sectional area) is approximately 4.4:1 while the radius of curvature of the elbow is 1.34 diameters of the inlet circular cross-section. Two vertical piers, symmetrically placed about the centerline, support the downstream rectangular diffuser.

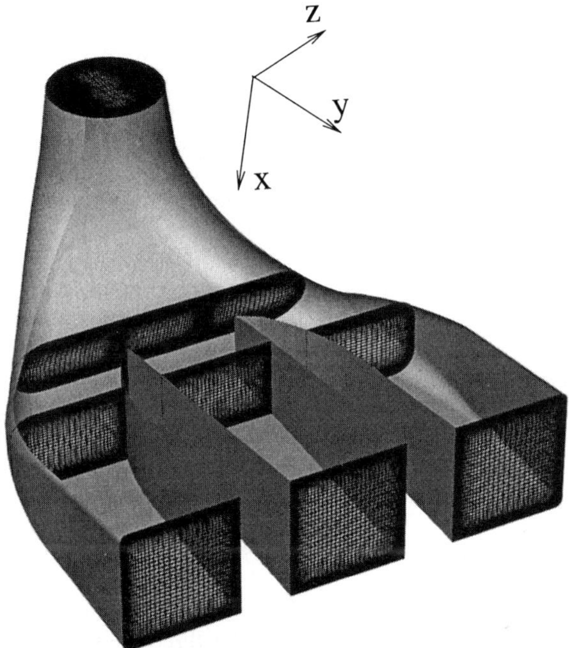

Figure 1. Cross-sectional views of the computational grid

The computational grid for every cross-section is generated using an efficient algebraic grid generation method which employs linear and third-order spline interpolation. The grid lines are concentrated near the walls using the hyperbolic tangent stretching function. The cross-sectional grids are then stacked along the centerline of the tube to complete the three-dimensional grid. To accurately resolve the flow in the vicinity of the piers, the streamwise planes are clustered around the pier leading edges also using hyperbolic tangent stretching. Typical cross-sectional views of the computational mesh and the relevant coordinates are shown in figure 1. All the subsequently reported calculations were carried out on a grid with 85 x 73 x 193 nodes (a total of approximately 1.2×10^6 nodes), in the streamwise, vertical, and horizontal (ξ, η, and ζ) directions, respectively, which is the finest mesh to be used so far for draft-tube calculations. The near-wall coordinate surfaces are located everywhere such that $1 < n^+ < 5$, where $n^+ = u_\tau n/\nu$ (u_τ is the shear velocity, n denotes the distance from the wall, and ν is the molecular kinematic viscosity).

Inlet conditions are specified using the experimental measurements (Hopping, 1992). These include both the axial and transverse mean velocity components at a plane downstream of the runner (see Figure 2). The measurements were carried out using a laser velocimeter for a range of operating conditions with and without air injection. The present simulations correspond to experimental run No. 1 (see Hopping, 1992) which was performed with the air off, runner speed 898 rpm, net head 24.8m, and water discharge 0.44cm^3/sec. These conditions correspond to a Reynolds number Re=$1.1x10^6$, based on the diameter D, and bulk velocity U_b at the inlet of the draft tube. It should be noted that inlet measurements were obtained along two mutually perpendicular radii (see Figs. 2a, and c), which suggest that the flow is not circumferentially symmetric. Due to lack of more detailed data, however, the calculations were carried out by arbitrarily choosing one of the two profiles and assuming that the inlet flow is axisymmetric. To facilitate the application of outflow boundary conditions, an artificial straight extension (of total length 10D) was added downstream the end of the draft tube. The flow quantities at the downstream end of this extension were obtained by assuming zero streamwise diffusion. On the solid walls the velocity components and turbulence kinetic energy are set equal to zero. The pressure at all boundaries is calculated by using linear extrapolation form the interior nodes.

The computational domain is treated as a single block with the piers accounted for by using a blanking technique. This treatment necessitates the use of several two-dimensional arrays to store the Jacobian and metrics of the geometric transformation and the pressure field on each pier wall. Converged solutions (four orders of magnitude reduction in residuals) are obtained after approximately 800 multigrid cycles. The computational time per grid node per cycle is $2.2x10^{-4}$ secs on a single-processor Silicon Graphics, 90MHZ, R8000, Power Challenge workstation.

4. Results and Discussion

In this section we present comparisons of calculated mean streamwise velocity profiles with measurements at several locations within the three bays. The numerical solutions are also interrogated using particles traces to elucidate the structure of the three-dimensional flow separation and vortex formation phenomena within the elbow and the downstream diffuser.

Figure 2 shows comparisons of measured (Hopping, 1992) and calculated streamwise mean velocity profiles at two streamwise locations, downstream the start of the piers, in all three bays. The velocity profiles, are plotted at two y = constant planes (see Figures 2a and 2c for axes definition) along the horizontal (Fig. 2a), and vertical (Figs. 2b, c, and d) centerlines of each cross-section. Figures 2a and 2c also include the measured streamwise and swirl velocity components at the inlet section, which, as discussed above, were used to provide inlet conditions for the calculations. All velocities in these figures have been scaled by the bulk velocity at the inlet of the draft tube.

The measurements in Fig. 2 suggest that most of the flow passes through the left (with respect to an observer standing at the draft-tube inlet looking downstream) bay. This is evident by the overall larger velocities through that bay and is obviously associated with the clockwise direction of the inflow swirl. The calculations reproduce this flow feature and appear to capture reasonably well most experimental trends. Some discrepancies are observed at the downstream location in the right bay (Fig. 2a), where the calculated streamwise velocity profile indicates the presence of a small reversed flow

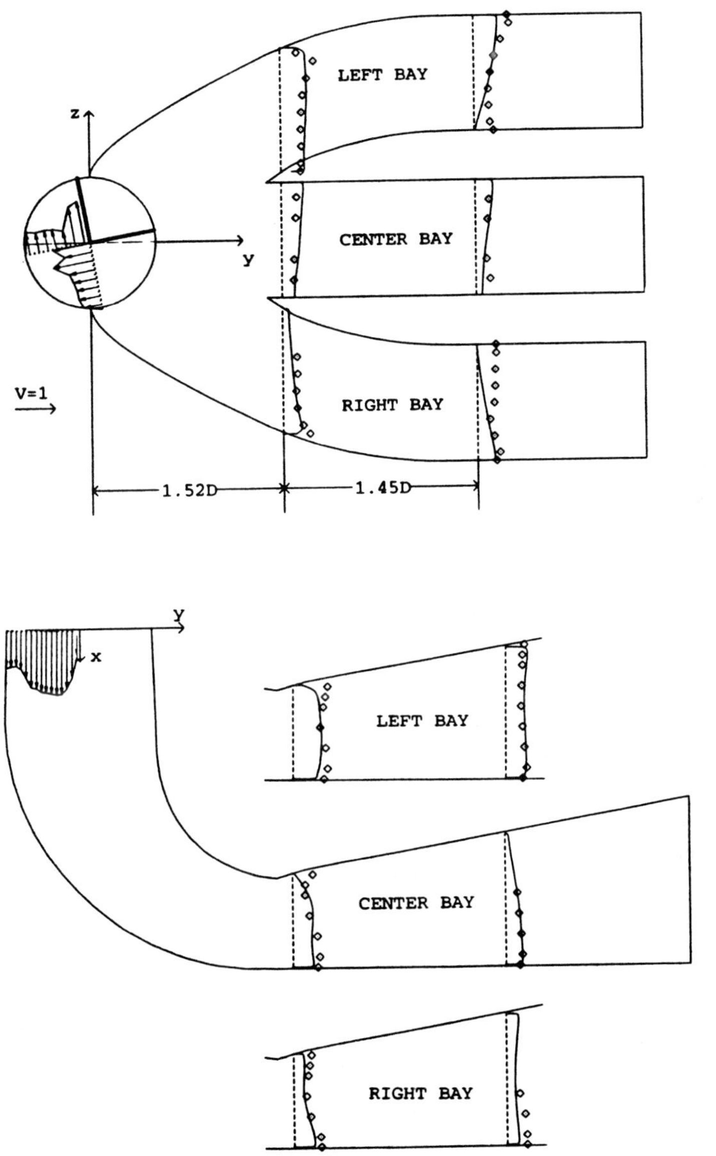

Figure 2. Comparisons of measured and computed streamwise mean velocity profiles

region near the inner wall. Contour plots of the calculated streamwise velocity component, not shown here due to space considerations, reveal a recirculating flow region starting upstream of that section and ending immediately downstream. The measurements, on the other hand, suggest a fuller and almost uniform velocity profile there which appears to have recovered very rapidly from its upstream distorted shape. Similar discrepancies, albeit not as pronounced, are observed at the downstream section in the left bay as well. It should be noted, however, that the experimental measurements are not detailed enough to allow a comprehensive assessment of the accuracy of the numerical solutions. Given the continuous area expansion downstream of the elbow, it is very likely that reversed flow does exist in the experiment, although may be not at the same locations indicated by the calculations, but could not be resolved by the few available velocity measurements. Yet another source of uncertainty is the lack of detailed velocity measurements at the inlet. As discussed in the previous section, the inlet flow was assumed axisymmetric, although the limited available measurements do not support such an assumption (see inlet swirl profile in Fig. 2a). Given the complexity of the draft-tube geometry, even small differences in inlet conditions could account for the observed discrepancies. Obviously, the present calculations can not offer positive answers to all these questions. They do, however, underscore the need for carefully designed, very detailed laboratory experiments.

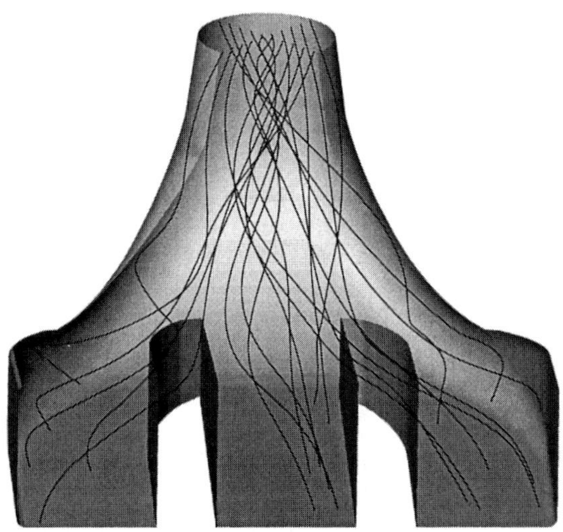

Figure 3. Three-dimensional particle traces: General view

Figures 3, 4 and 5 depict particle traces released at strategically selected locations to clarify various three-dimensional flow features. A global view of the flowfield is given in Fig. 3, which shows the paths of particles originating along two mutually perpendicular diameters at the inlet plane. It is seen that most of the flow passes through the left bay and the left half of the center bay, which is consistent with the trends exhibited by the velocity profiles discussed above. Particles released near the center of the inlet section are seen to form a coherent, rope-like, vortical structure which appears to pass through the left half of center bay. Significant secondary motion is also

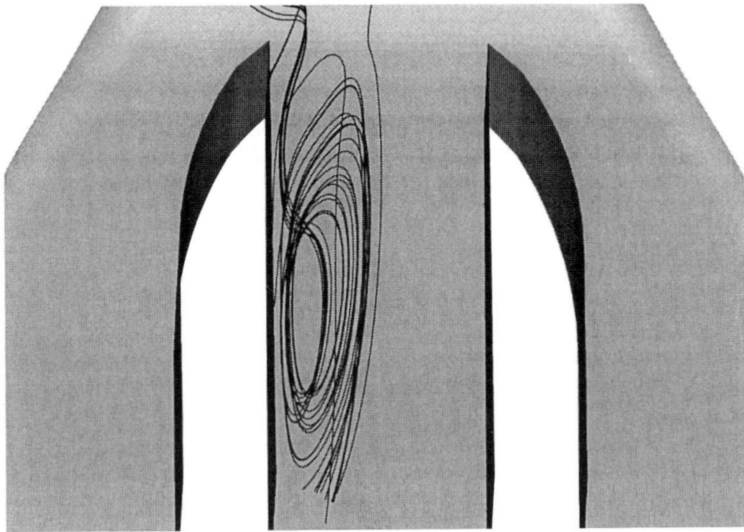

Figure 4. Three-dimensional particle traces: Reversed flow region

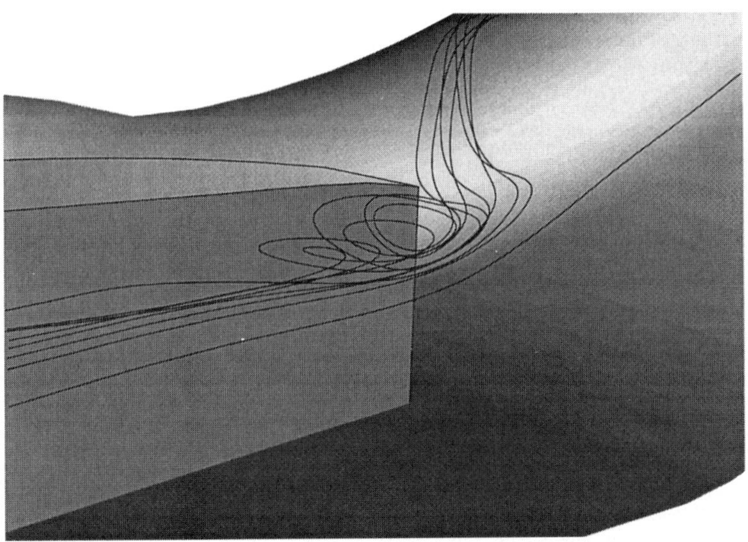

Figure 5. Three-dimensional particle traces: Horse-shoe vortex

present in the right bay as indicated by the twisting particle trajectories there. Figures 4 and 5 reveal some very complex three-dimensional flow patterns along the flat wall of the right pier. Figure 4 indicates the existence of a recirculation region which is located near the top (diverging) wall of the draft tube--although not shown herein due to space limitations, the particles that are trapped in this area originate from the near-wall region at the left side of the inlet section. Underneath this recirculating flow region there is a very intense longitudinal vortical structure which is shown in Figure 5. This structure is similar to horse-shoe like vortices known to form at wing-body junctions and is produced by lateral skewing of the vorticity vector. These flow patterns--which to the best of our knowledge have not been reproduced before numerically for draft-tube geometries--serve to demonstrate the enormous complexities of such flows, underscore the challenges for advanced CFD methods, and point, once again, to the need for very detailed laboratory experiments to provide data for numerical validation.

5. Summary and Conclusions

An efficient finite-volume method was presented for carrying out fine-grid calculations with near-wall turbulence models for complex draft-tube geometries. The computed solutions are compared with mean velocity measurements downstream of the elbow. The calculations reproduce most experimental trends with reasonable accuracy. Three-dimensional particle traces reveal, for the first time, the presence of very complex three-dimensional flow patterns around the piers. These include longitudinal and horse-shoe vortex formation, and regions of reversed flow. The present study underscores that detailed three-dimensional flow measurements are of crucial importance for further advancements in numerical modelling of real-life draft-tube geometries. Current work focuses on further improving the efficiency of the multigrid method, by implementing grid sequencing techniques, as well as implementation and testing of advanced turbulence models that account for turbulence anisotropy.

6. Acknowledgments

The initial phase of this work, while the second author was at the University of Iowa, was supported by a grant from the Tennessee Valley Authority. The continuation of this work at the Georgia Institute of Technology is funded by Voith Hydro, Inc. and the U.S. Department of Energy. The authors are most grateful to Richard K. Fisher, Jr., of Voith Hydro, Inc., and Patrick March and Paul Hopping, of TVA, for their advice and support.

7. References

Agouzoul, M., Reggio, M., and Camarero, R. (1990), "Calculation of Turbulent Flows in a Hydraulic Turbine Draft Tube," *ASME J. Fluids Eng.*, 112, pp. 257-263.

Bohac, C. E., and Ruane, RT. J. (1991) "Tailwater Concerns and the History of Turbine Aeration," ASCE National Hydraulic Engineering Conference.

Carter, J., Jr. (1995) "Recent Experience with Turbine Venting at TVA," *Water-Power 95*, ASCE Int. Conf. & Exposition on Hydropower, San Francisco, California, Vol. 2, pp. 1396-1405.

Chen, H. C. and Patel, V. C. (1988), "Near-Wall Turbulence Models for Complex Flows Including ," *AIAA J.*, Vol. 26, pp. 641-648.

Fisher, Jr., R. K. (1995), private communication.

Hopping, P. N. (1992), "Draft Tube Measurements of Water Velocity and Air Concentration in the 1:11.71 Scale Model of the Hydroturbines for Norris Dam," *Tennessee Valley Authority Engineering Laboratory*, Report No. WR28-2-2-116, Norris, Tennessee.

Jameson, A. (1983), "Solution of the Euler Equations by a Multigrid Method," *Applied Mathematics and Computation*, Vol. 13, pp. 327-356.

Lin, F., and Sotiropoulos, F. (1996), "Assessment of Artificial Dissipation Models for Three-Dimensional, Incompressible Flow Solutions," to appear in the *ASME J. of Fluids Engineering* (April 1995).

Riedelbauch, S., Fisher, Jr., R. K., Faigle, P., and Franke, G. (1995), "The Numerical Laboratory Gets Better," *Water-Power 95*, ASCE Int. Conf. & Exposition on Hydropower, San Francisco, California, pp. 1386-1395.

Sotiropoulos, F., Lin. F. (1996), "Strongly-Coupled Multigrid method for 3-D Incompressible Flows Using Near-Wall Turbulence Closures," to appear in the *ASME J. of Fluids Engineering.*

Sotiropoulos, F., and Patel, V. C. (1992), "Flow In Curved Ducts Of Varying Cross-Section," Iowa Institute of Hydraulics Research, IIHR report No. 358, The University of Iowa, Iowa City, Iowa.

Sotiropoulos, F., and Ventikos, Y. (1996), "Assessment of Some Non-Linear Two-Equation Turbulence Models in a Complex, 3-D, Shear Flow," to be presented at and appear in the proceedings of the *6th International Symposium on Flow Modelling and Turbulence Measurements*, September 8-10, Tallahassee, Florida.

Shyy, W. and Braaten, M. E., (1986), "Three-dimensional Analysis of the Flow in a Curved Hydraulic Turbine Draft Tube," *Int. J. Numerical Methods in Fluids*, Vol. 6, pp. 861-882.

Vu, T. C. and Shyy, W. (1990), "Viscous Flow Analysis as a Design Tool for Hydraulic Turbine Components," *ASME J. Fluids Engineering*, Vol. 112, pp. 5-11.

FLUID FLOW INTERACTIONS IN HYDRAULIC MACHINERY

T. ASCHENBRENNER, N. RIEDEL AND R. SCHILLING
Institute for Hydraulic Machinery and Plants
Technical University of Munich, Germany

Abstract.
The paper deals with the fluid flow interactions between rotor and stator as well as rotor and side spaces of hydraulic machinery. It is shown that the flow through the rotor strongly influences the leakage flow through the side space, the friction torque and the axial forces acting on the rotor.

1. Introduction

In the past years the development of hydraulic machinery has made much progress due to very sophisticated numerical tools. After having defined the point of operation a recalculation of each component is started to find out the potential of improvement. Based on these flow analysis results each component may be optimized seperately so that they reach a high hydraulic quality. However, this strategy does not necessarily yield the optimum performance of the hydraulic machinery since the fluid flow interactions between different components is not yet considered.
Due to the increasing performance of modern workstations and parallel computation techniques the optimization of the whole hydraulic machinery becomes more realistic.
A potential to improve the efficiency of hydraulic machinery still exists by taking into account the fluid flow interactions between the components of the machinery in the design and optimization process. The present paper describes two main fluid flow interactions in hydraulic machinery.

2. Definition of problems

In order to take into account the fluid flow interactions in hydraulic machinery during the design process a setup of numerical tools has to be

E. Cabrera et al. (eds.), Hydraulic Machinery and Cavitation, 150–159.

developed. These tools have to be fast and accurate enough to give reliable results within some hours.
At the institute a 3D-Euler Code which is used for rotor-stator-interaction (RSI) -calculations has been developed, see /3/. Also a Navier-Stokes code for the recalculation of the flow in the side space between the rotor and the housing of hydraulic machinery has been developed, see /4/. Various turbulence models are beeing tested. Combining these codes both the interaction between rotor-stator (RSI) and rotor-side space (RSSI) may be considered as a coupled fluid flow problem. A research programme has been defined to investigate the effects in hydraulic machinery induced by these fluid flow interactions. The aim is to find out the best computational model for RSI and RSSI interaction, to calibrate the codes with experimental data and to study the interactions in different types of hydraulic machinery.

3. Rotor-Stator-Interaction (RSI)

3.1. THE 3D EULER CODE

Within the research programme a fast block structured 3D-Euler code has been developed. The governing equations for the incompressible fluid flow are modified using Chorins artificial compressibility concept yielding a hyperbolic set of equations. Since the rotor calculations are performed in the rotating frame of reference Coriolis and centrifugal forces are added to the right hand side. The governing equations are discretized by means of an implicit cell-centered finite volume scheme. The numerical fluxes in the cell faces are constructed using a second order upwind scheme similar to Roe's method, and an advanced limiter function is used. The system of algebraic equations is solved by means of a point-Gauss-Seidel scheme.

3.2. NUMERICAL MODELS

Four different approaches to perform a rotor-stator calculation were checked. First, an isolated runner calculation has been carried out taking the exit angle of the guide vane camberline as inflow boundary condition. This approach is used for a standard recalculation. Second, only the flow through the guide vanes has been analyzed without rotor using a computational domain from the guide vane inlet down to the draft tube. This calculation is performed using the massflux and the approximate outflow angle of the spiral casing as inflow boundary conditions. The stator solution then defines new boundary conditions for a rotor calculation. Using this approach the evaluated rotor inflow angle can be expected to be more accurate then in the first case. The third test case is defined to check whether the flow field downstream of the stator is influenced by the rotor. Using overlapping grids, the computational domain of the stator ends at the leading edge of the rotor and the rotor domain begins at the trailing edge of the stator. The basic idea is to start a stator calculation then to take this solution as rotor inflow boundary condition to recalculate the runner flow and to feed back the pressure field as boundary condition for the stator outflow. Thus, convergence may be ecpected after some iterations.

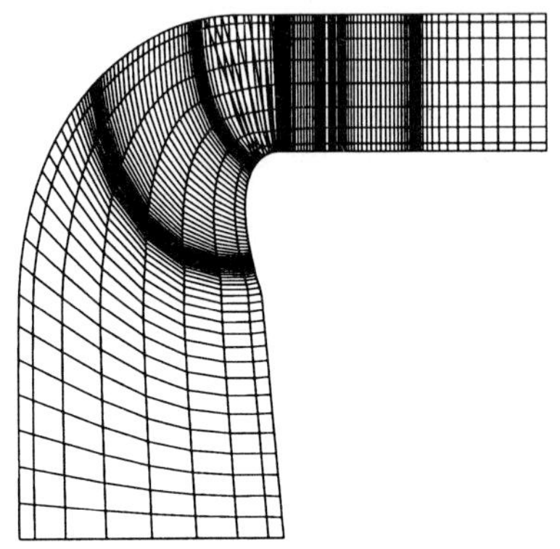

Fig. 1: Overlapping grid, Test Case 3

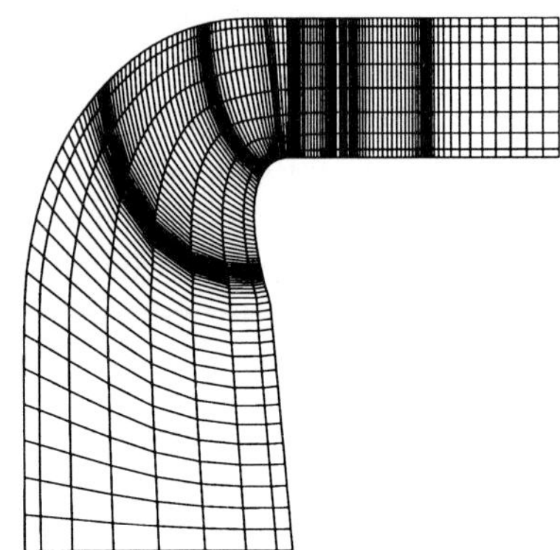

Fig. 2: Connected grid, Test Case 4

The last test case is a coupled rotor-stator analysis where rotor and stator are calculated in parallel on two workstations. The two computational domains are linked together at an intersection plane, exchanging the boundary conditions after each time step. Since the flow within each component does react directly on changes at the intersection plane a faster convergence may be optained.

3.3. BOUNDARY CONDITIONS

For all test cases periodic boundary conditions are prescribed in circumferential direction except at blade surfaces where the normal velocity is set to zero. At in-and outflow a special approach is used. Since the boundaries are very close to leading and trailing edges, a non-uniform flow field has to be assumed and only momentum averaged values can be specified. The distribution of a characteristic quantity f is momentum averaged in circumferential direction on a torus surface for line correction and over the whole plane for areal correction. The averaged value $\bar{f}$ is then compared with the boundary condition f_{bc} yielding a difference Δf.

$$\Delta f = f_{bc} - \bar{f}$$

This difference is then added to the discrete value of f in each boundary cell face so that the new momentum average exactly yields f_{bc} and the circumferential distribution of f remains unchanged. For stability reasons this correction can be reduced by an relaxation parameter α

$$f'_k = f_k - \alpha \Delta f \,, \qquad k = k_1 \cdots k_2$$

In cases where f is an integral value as the massflux, $\bar{f}$ has to be transformed to a velocity correction.

Boundary conditions for test case 1

Since the inflow cells are located in the radial part of the computational domain, the axial velocity is set to zero. Massflux and inflow angle β are specified by the velocities c_r and c_ϕ . The velocities are corrected using the integral method described above.

Boundary conditions for test case 2

Here the boundary conditions of the rotor inflow are evaluated from the stator calculation. The conditions specified are $\bar{c}_r, \bar{c}_\phi, \bar{c}_z$ and the corrections are added using the prescribed line method. For the rotor and stator calculation the radial equilibrium for the pressure is used at the outlet.

Boundary conditions for test case 3

In this test case a stator calculation is performed using an estimated pressure field at the outlet. Then a rotor calculation is started with the boundary conditions on the torus surface computed at the trailing edge of the present stator solution as prescribed for test case 2. After convergence the rotor solution yields a new pressure distibution for the next stator calculation. This iterative procedure is continued until the changes of the boundary values reach a convergence criterion.

Boundary conditions for test case 4

For this test case the rotor and stator meshes are linked together at an intersection plane. After each time step the boundary conditions are updated. The circumferentially averaged velocity components $\bar{c}_r, \bar{c}_\phi, \bar{c}_z$ on each torus surface are passed from the stator to the rotor and the static pressure $\bar{p}$ is transferred from the rotor to the stator. The boundary conditions are then compared and updated using the line method.

3.4. RESULTS

To calibrate the numerical models used, the GAMM-Turbine stage consisting of stay vanes, wicket gates and runner has been analyzed. At the EPFL in Lausanne velocity measurements up-and downstream of the runner have been carried out /6/.

Fig. 3 shows the distribution between stator and rotor of c_r for the four test cases in comparison with the measurements. The radial velocity component shows up the absolute value and the distribution of the massflux in the measuring plane. The c_r-distributions considering testcases 1 and 2 are approximately uniform due to the nearly two dimesional boundary conditions. Test case 3 and 4 show a much more accurate distribution of c_r especially near the shroud. This improvement results from the pressure field at the leading edges of the rotor which has a strong influence on the meridional flow distribution at the trailing edges of the guide vanes. As a direct consequence similar effects can be found looking at the absolute inflow angle,see Fig. 4. There is a bigger difference between test case 1 and 2 due to the fact that the boundary condition in test case 1 was only a geometrically defined angle. Nevertheless in test cases 1 and 2 the inflow angel α is strongly underpredicted at the shroud by 15 degree. Test cases 3 and 4 coincide much bettter with the experimental data. The results clearly show that the rotor-stator-interaction has to be taken into account during the design process to optimize a rotor blading. The rotor strongly influences the flow within the stator and this effect can be predicted well applying the numerical models used in test case 3 and 4.

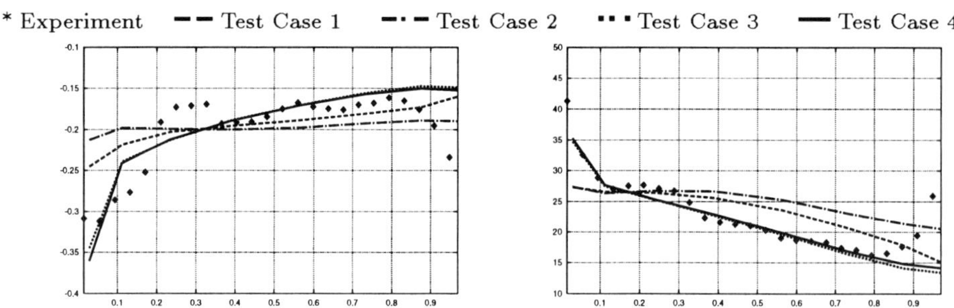

Fig. 3: Radial velocity c_r *Fig.* 4: Absolute inflow angle α

3.5. COMPUTATIONAL EFFICENCY

To analyze the computational effiency the effort to set up the test cases and the CPU time needed for a converged solution has to be considered. Since the meshes are generated using the Integated Flow Analysis System developed at the Institute here the time needed for the mesh generation can be neglected.

A remarkable difference between the four methods can be found looking at the CPU time needed. Since the test case 1 represents the common approach to a 3D runner flow analysis, all CPU times are related to this case, which requires about 70 minutes on a RS6000 workstation having a performance of 50 mflops using a grid of 95 * 30 * 9 mesh point. Table 1 shows a comparison of the CPU times needed for the four test cases.

Due to the slightly faster convergence of the stator computation the effort for test case 2 is a bit less then twice the effort for test case 1. Since the accuracy is approximately the same the approach 2 is not very useful. Test case 3 needs the largest amount of CPU time because only one calculation can be performed at the same time, so the rotor computation has to wait for the stator result and vice versa. Test case 4 shows almost identical results as test case 3 but it needs only 60% of the computational effort. Using two CPUs the numerical effort can be halved for this test case due to the minimum communication effort between the processes compared with the time needed for one time step. Here the most important point is that two domains can be calculated in parallel wheras the calculations in test case 3 have to be performed one after the other, and the CPU time directly depends on the number of iterations needed to satisfy the boundary conditions.

Test Case	1	2	3	4
1 CPU	1	1.8	5.4	3.2
2 CPUs				1.6

TABLE 1. Comparison of the Computational Efficiency

4. Rotor-Side-Space-Interaction (RSSI)

4.1. THE NAVIER-STOKES CODE

Based on the Euler code described a Navier-Stokes code has been developed. A special derivate is in use for rotor side space calculations. The flow field in the rotor side space is assumed to be constant in circumferential direction. Applying this simplification, the three dimensional flow field can be computed in an two and a half dimensional approach. A 2D mesh is set up in the meridional plane of the side space. On this mesh the governing equations for p, c_r, c_z and a circumferentially averaged value $\bar{c}_\phi$ are solved. This approach yields a short CPU-time and a fast convergence as the number of mesh points is relatively small. As the turbulence model of the Navier-Stokes code is still beeing tested the turbulent flow has been computed by the commercial code TASCflow /5/, using the standard $k-\epsilon$ model with log-law or two-layer approach for solid walls.

4.2. ROTOR SIDE SPACE GEOMETRY

As a first step a simplified rotor side space geometry has been defined. Based on this simplified geometry the influence of Reynolds number Re, leakage flow rate q_L and circumferential momentum c_ϕ of the leakage flow on the coefficients of friction torque C_T and axial force C_{Fax} acting on the rotor is studied.

$$C_T = \frac{2T_f}{\rho\omega^2 r_a^5}; \quad F_{ax} = \frac{2F_{ax}}{\rho\omega^2 r_a^4}; \quad q_L = \frac{Q_L}{Q}; \quad Re = \frac{\omega r_a^2}{\nu}$$

The computational domain is characterized by the relative axial gap width of $\lambda_a = 0.0046$ and relative radial gap width $\lambda_r = 0.0232$. The mesh consist of 48300 points so that a sufficiently accurate resolution of the boundary layer and recirculating flow is guaranteed. At the inflow the mass flux and direction of the velocity vector are specified as boundary conditions, at the outflow the averaged static pressure is prescribed.

Fig. 5 and 6 show the influence of Reynolds number on the friction torque and the axial force acting on the rotating disk.

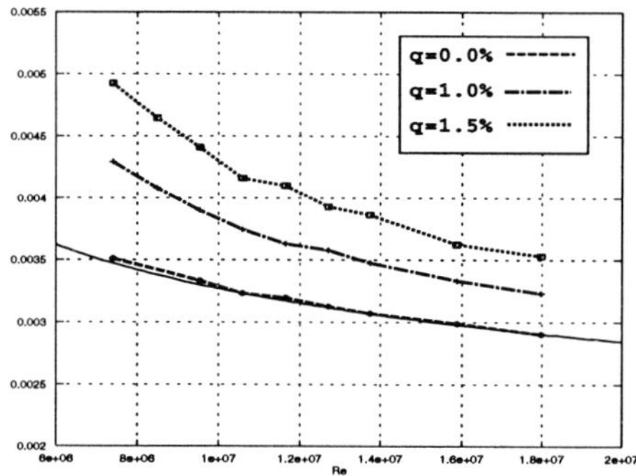

Fig. 5: Torque coefficient C_T

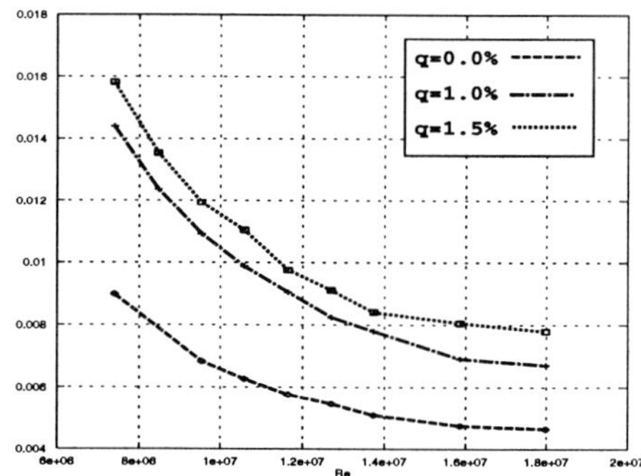

Fig. 6: Axial force coefficient C_{Fax}

Fig. 7 shows the variation of friction torque and axial force for different leakage flow rates q_L. Due to the higher pressure difference between inflow and outflow at increasing flow rates the axial forces are growing linearly with the flow rate. The momentum coefficient also grows linear with increasing flow rates. Similar effects were presented in /7/.

Aditional calculations were made to show the influence of inflow swirl. The circumferetial velocity profile is changed by the increasing inflow swirl leading to a higher average circumferential velocity and so the friction torque is reduced. The axial force is increased due to a higher pressure gradient.

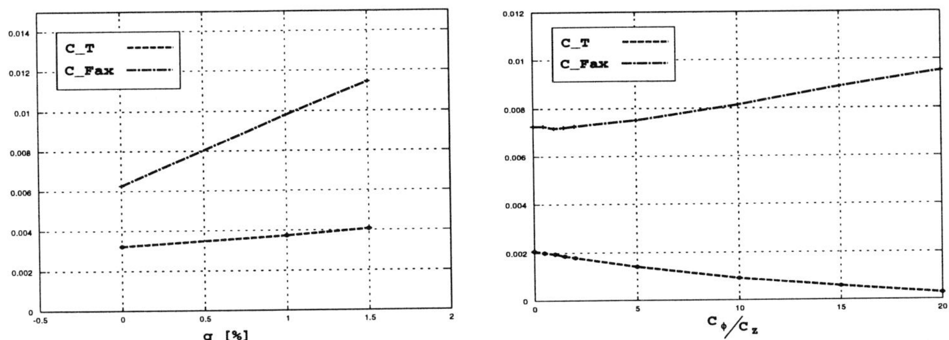

Fig. 7: C_T, C_{Fax} for different flow rates

Fig. 8: C_T, C_{Fax} for different c_ϕ at inlet

4.3. COUPLED IMPELLER SIDE SPACE COMPUTATIONS

A pump turbine having a specific speed $n_q = 35$ was chosen for a coupled calculation of the flow within the rotor and side space. Fig. 9 shows a meridional section of the pump turbine. The radial gap width has been assumed to be 0.1% of the rotor diameter. A pump turbine configuration was chosen to be able to study the flow in both operating modes without changing the geometry. Thus comparison of the occuring effects and of the interaction between runner and side space may be studied.

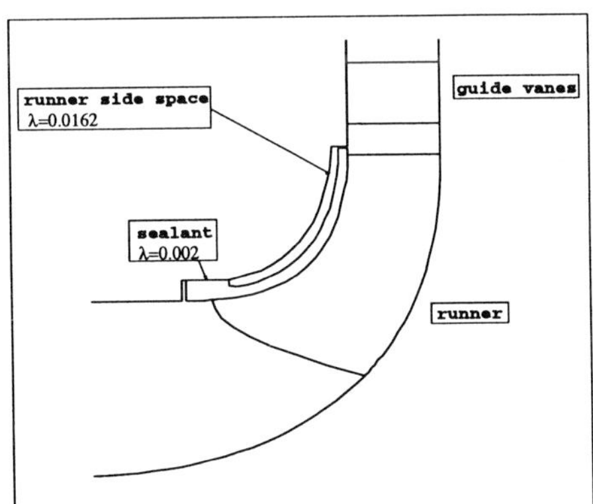

Fig. 9: Meridional section of pump trubine PT35

The flow in the runner is calculated by means of the 3D Euler code described above. As the Reynolds number in the rotor side space is $Re = 1.1 * 10^7$ the flow in the rotor side space is assumed to be tubulent. The coupling of the runner and side space calculation was realized in a similar way as the rotor stator coupling shown in test case 3. The velocities at the inlet and the static pressure at the outlet of the side space are circumferentially averaged and fed into the rotor side space calculation as boundary conditions. The following side space calculation yields new boundary coditions for the rotor at the interfaces. The correction of the boundary values is carried out, as explained above. This iterative process is continued until the exchanged boundary conditions satisfy a prescribed convergence criterion.

4.4. RESULTS

In turbine mode only one calculation has been performed considering the optimum point. As the leakage flow leaves the main flow passage between stator and rotor and is fed back on the suction side of the runner the flow rate in the runner is smaller then the design flow rate. Fig. 10 shows the pressure distributions of the runner near the shroud comparing the flow with and without the RSSI. The reduced flow rate leads to a smaller relative flow angle β and consequently to a reduced loading at the leading edge, see Fig. 10. Beside this local effect on the pressure distribution at the runner inlet near the shroud the rotor side space interaction has no significant influence on the runner flow. The leakage flow rate has been computed to be $q_L = 0.0225$, i.e. 2.25% of the discharge.

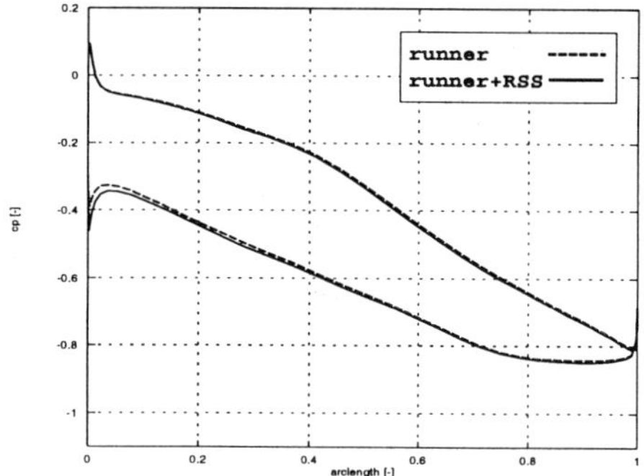

Fig. 10: Turbine pressure distribution

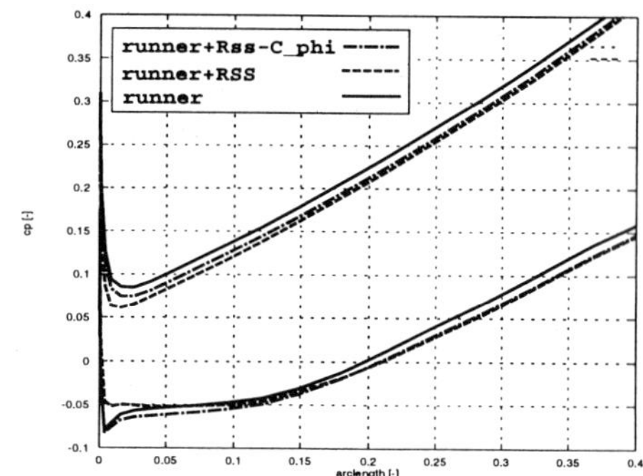

Fig. 11: Pump pressure distribution

In pumping mode three different operating points have been considered, 100%, 85% and 75% of the optimum flow rate. Fig. 12 shows ψ_t versus the flow rate φ_r for the three operating points. The Euler calculation of the runner yields a linear increase of ψ_t with decresing flow rates. The leakage flow which is pumped through the rotor side space, is added to the overall mass flow in the runner. Fig. 11 shows the pressure distribution on the blade near the shroud. Similar to the effects in turbine mode the loading at the leading edge is reduced due to a higher relative flow angle β. An aditional runner calculation was performed setting the inflow swirl of the the leakage flow to zero. As shown in Fig. 11 the effect of the swirl brought into the runner flow is only minor. The leakage flow rate has been calculated to be $q_L = 0.0249$ at optimum point and increasing with reduced discharge. In Fig. 13 Torque and axial Force are shown. They both show a similar behaviour as in the simplyfied geometry, s. Fig. 8. The torque is reduced due to the increasing circumferential velocity component entering the rotor side space.

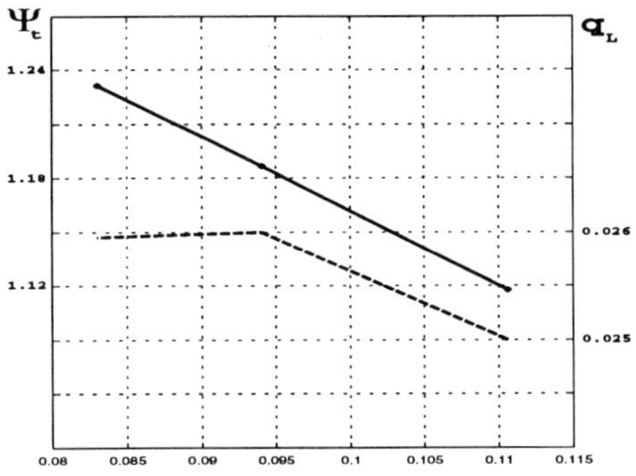

Fig. 12: Ψ_t, q_L versus φ_r

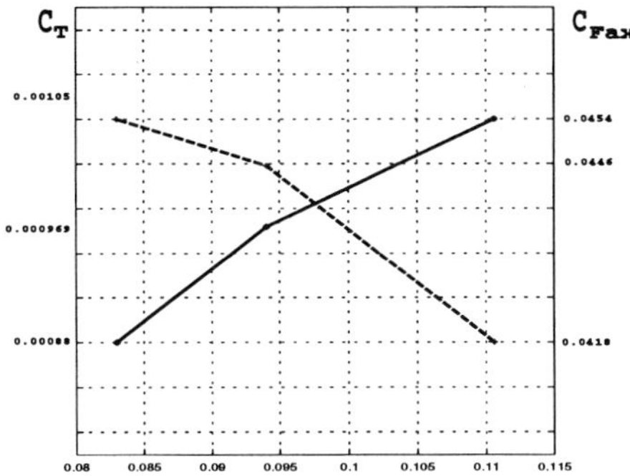

Fig. 13: C_T, C_{Fax} versus φ_r

5. Conclusions

In this study different approaches to predict the fluid flow interactions in hydraulic machiney are presented. Numerical tools for calculation of RSI and RSSI have been developed. It could be shown that the fluid flow interactions have sigificant influence on performance and efficiency of hydraulic machinery.

6. References

1. W.N. Dawes: Torward Improved Throughflow Capability: The Use of 3D Viscous Flow Solvers in Multistage Enviroment, ASME Paper 90-GT-18, 1990.

2. H. Haas, N. Riedel, A. Fernandez, ,Ch Watzelt, R. Schilling: An Integrated Flow Analysis System for the Recalculation of Hydraulic Bladings, Preprints; 5th International Symposium of Transport Phenomena and Dynamic of Rotating Machinery ISROMAC-5 Maui USA 1994.

3. N. Riedel: Entwicklung eines 3D Euler Codes mit Riemann-Solver, Int. Bericht des Lehrstuhls für Hydraulische Maschinen und Anlagen, 1995.

4. J. Fritz: Numerische Berechnung der Strömung im Radseitenraum hydraulischer Maschinen, Diplomarbeit Lehrstuhls für Hydraulische Maschinen und Anlagen, 1995.

5. TASCflow User Documentation, Theory Documentation, 1995.

6. E. Parkinson: Turbomachinery Workshop ERCOFTAC II, Test Case 8: Francis Turbine, EVA94-PAE, IMHEF Ecole Polytechnique Federal de Lausanne

7. R. Schilling: Strömungen in Radseitenräumen von Kreiselpumpen, Habilitationsschrift Universität Karlsruhe, 1980

NUMERICAL OPTIMIZATION OF HIGH HEAD PUMP-TURBINES

A Challange for Mechanical and Hydraulic Design

H. BUCHMAIER, B. QUASCHNOWITZ,
W. MOSER, D. KLEMM
Voith Hydro GmbH
D-89509 Heidenheim, Germany

1. Summery

In order to cope with the technical challenges of high head pump-turbines, new methods like simultaneous engineering and numerical studies are required. In this paper the application of simultaneous engineering is shown with an example, as well as typical pump-turbine concepts and design features.

2. Introduction

The design process of the High Head Pump-Turbine is an increasing challenge to the development teams of today:
Operational behavior and efficiency trends have to be predicted exactly, requiring an optimized hydraulic layout. At the same time, the development time between the placement of the order and the start of manufacturing is decreasing drastically. This forces a trend toward numerical studies. In addition, the strong competition among the hydro turbine companies can only be met with cost optimized designs.

The process of numerical design and optimization will be presented from both a hydraulic and a mechanical standpoint, as they are realized using simultaneous engineering.

3. Pump-Turbine Design Optimization (typical example: Spiral Case)

Because of increased evaluation of efficiency, more importance has to be attached to the optimization of (pump-)turbines. The optimization should not only be conducive to the reduction of losses, but should improve the relation of costs to benefit as well.

E. Cabrera et al. (eds.), Hydraulic Machinery and Cavitation, 160–169.

Optimizing an assembly group as complex as a pump-turbine spiral case means that single parameters are varied while the degree of target-fulfillment has to be checked with every step.

Figure 1 shows the results of the optimization steps of one parameter for a high head pump-turbine. Civil construction limits are not considered.

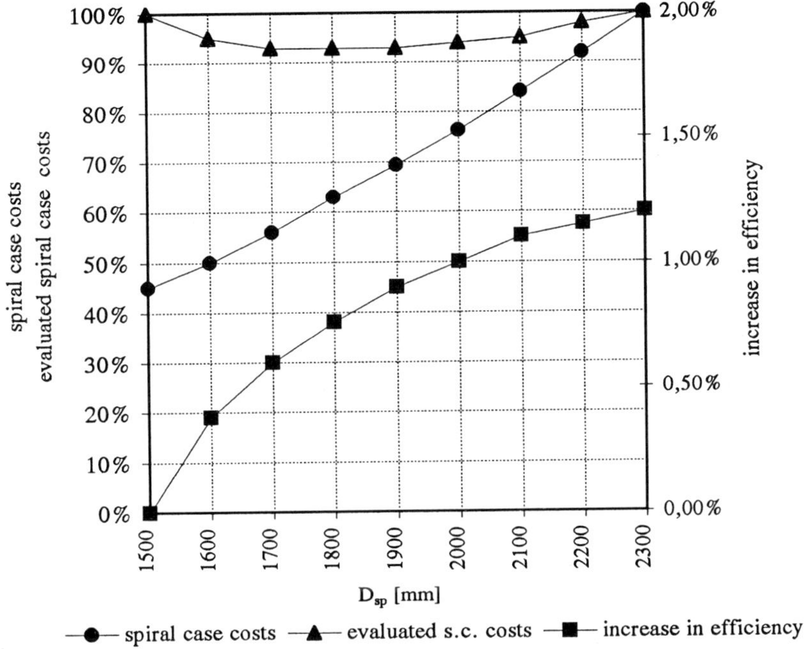

Figure 1. Sample of design optimization of a high head pump-turbine: throat diameter D_{sp} versus increase in efficiency and costs.

The diagram shows clearly that the evaluated costs of the spiral case do have their minimum between minimum throat diameter D_{sp} on the one hand and maximum efficiency on the other hand.

To reliably define all effects of the design variations within a short period of time, and to estimate the quality in addition, a fast tool with sufficient accuracy is needed.

The use of simultaneous engineering proved to be very efficient for this task (see figure 2). The backbone is formed by a program system called KE that permits the generation of the most important hydro-mechanical (pump-)turbine assembly groups as 3-D computer internal models in the shortest time possible.

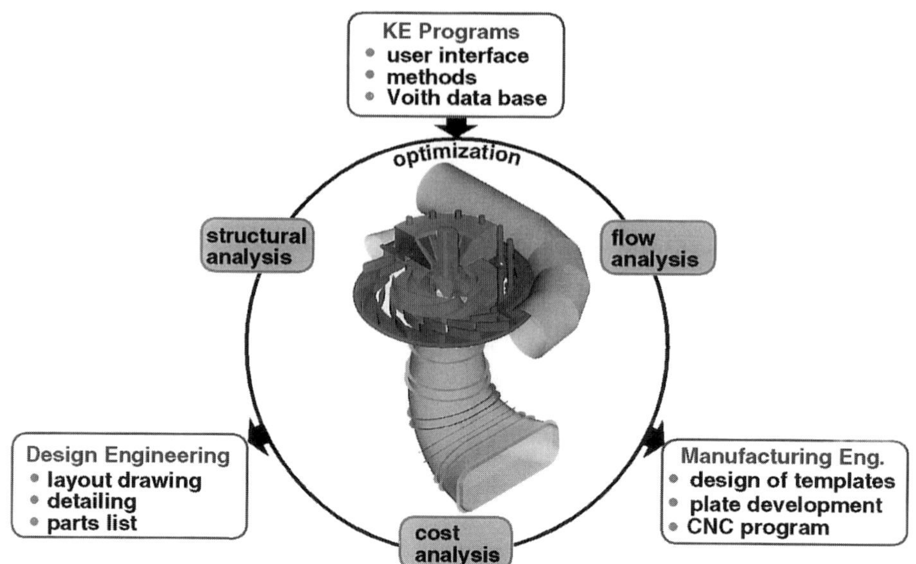

Figure 2. The concept of Simultaneous Engineering at VOITH.

With the aid of attached programs for flow-, structural- and cost analysis, the design variants can be discussed with regard to losses, stability and costs. This allows the run through of several optimization cycles in a fairly short time period and is therefore a guarantor for a good layout.

Besides design optimization, the 3-D model is used as a base for detailing and to derive manufacturing aids in the next project phase.

With our example pump-turbine spiral case, the enlargement of the throat diameter causes some changes in the conditions: e.g. the pressure area is increased due to the larger spiral case sections, while the velocity of the flow is decreased due to the larger cross section. All changes have to be examined carefully.

Since the loads on the stay vanes are increased, a repeated structural analysis is definitely necessary. A large number of load cycles with a considerable portion of dynamic stress have to be expected during the life cycle of pump-turbines. Therefore a load universe as input for the fatigue analysis is generated. In doing so, the flow-induced vibrations of the stay vanes must not be ignored. They are determined through experiments and calculations.

The resulting loads are input into the Finite Element Method calculation. Stress and strain at the most critical spots can be extracted from the results of the computation (see Figure 3).

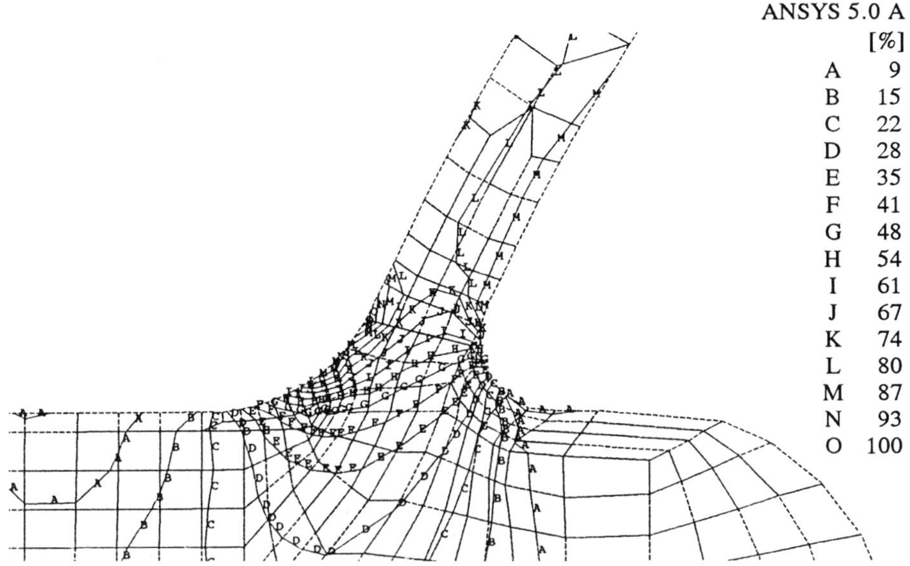

Figure 3. Equivalence-stress distribution, shown as iso stress lines at the transition of spiral case section to upper stay ring deck. Sample of a design variant of GUANGZHOU II at load rejection (H=775 m).

Locations of max. stress (see fig. 3 e.g.), are examined with methods of fracture mechanics. To prove that certain defects which are smaller than the detectible size do not become critical during the life cycle, crack-propagation is applied. It is desirable to select materials and working stresses such that a defect will grow through the wall of the component before it will fracture due to brittle failure.

The spiral case losses are relatively small in many cases. Nevertheless, it is important to find the best individual design, and for that the losses have to be determined.

During the simultaneous engineering, different flow calculation methods are used. In order to compare different designs or to optimize some dimensions regarding losses, fast and rather simple algorithms are necessary. For a deeper understanding of the loss mechanism, more time consuming numerical tools which account for the three dimensional and turbulent behaviour of the flow are applied. Due to the basic equations which describe this kind of flow, they are called Navier-Stokes-methods. These sophisticated methods are also used to calibrate the above mentioned simple approaches. For more detailed information see [1].

Fig. 4 exemplarily shows the distribution of the total pressure in the symmetry plane of the spiral case of a pump turbine (turbine mode).

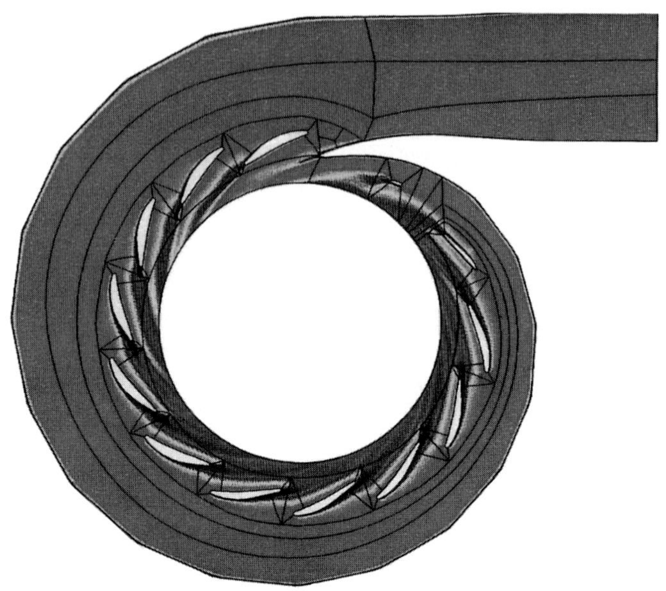

Figure 4. Distribution of the total pressure in a pump-turbine spiral case.

With such detailed results and the total component losses, the designer has the appropriate hydraulic information. Together with the stress analysis results and the manufacturing costs, an optimization of the spiral case according to the special wishes of the customer is easily possible. With similar methods all important components can be designed/optimized separately or together with neighboring components.

4. General Pump-Turbine Concepts and Design Features [2]

During the last decades the application of reversible pump-turbines has been extended to high heads above 300 .. 400 m. Due to the lack of knowledge about the dynamic forces occurring primarily in transient operating conditions, some premature component failures occurred in the beginning, and reduced the availability of reversible pump-turbines significantly compared to separate pump and turbine units as used for many such high head applications.

In the meantime, the know-how progressed due to extensive theoretical and practical studies. The combination of experience gained on prototypes together with simultaneous engineering including FE-computations enables the building of pump-turbines with heads of more than 600 m, with an expected availability similar to that of other types of turbines. In order to meet this high demand, special design concepts have been developed. Figure 5 shows a typical concept for a high head pump-turbine.

This paper highlights typical design features of these concepts and gives some recommendations based on VOITH's experience for the optimized design of high head pump-turbines. These are in detail:

- Shaft systems and dismantling methods.
- Embedment of stay ring/bottom ring assembly.

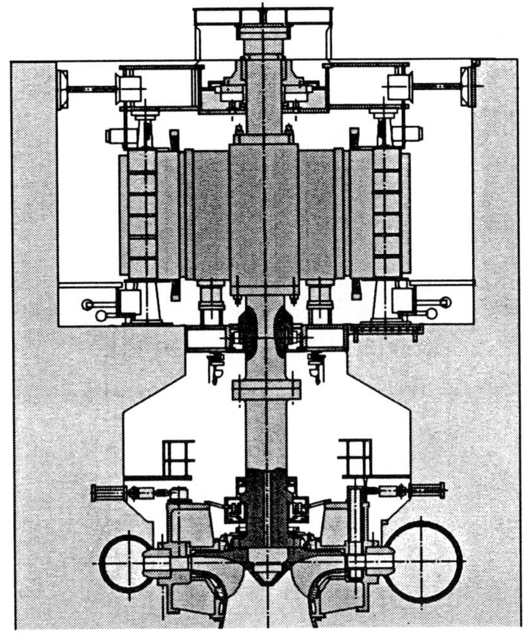

Figure 5.
Typical concept for a high head pump-turbine.

4.1. SHAFT SYSTEMS AND DISMANTLING

The following three different concepts for the arrangement of the shaft system of pump-turbines are discussed (figure 6):

1- Intermediate shaft with bracket supported thrust bearing on top of the motor-generator.
2- Separate turbine/generator shafts with bracket supported thrust bearing below the motor-generator.
3- Single shaft with head cover supported thrust bearing.

For the first concept, the components are to be removed laterally (side removal). For the concepts no. 2 and 3, they are to be removed vertically (top removal).

4.1.1. *Top Removal of the Pump-Turbine Parts*

Up to approx. 300 m head, the motor-generator stator is larger than the pump-turbine head cover, therefore it is possible to remove a one-piece head cover vertically. This is an advantage, if the transportation limits do not prevent it. With regard to powerhouse volume considerations, VOITH's recommendation for top removal is for runner diameters larger than 4 m and heads smaller than 600 m.

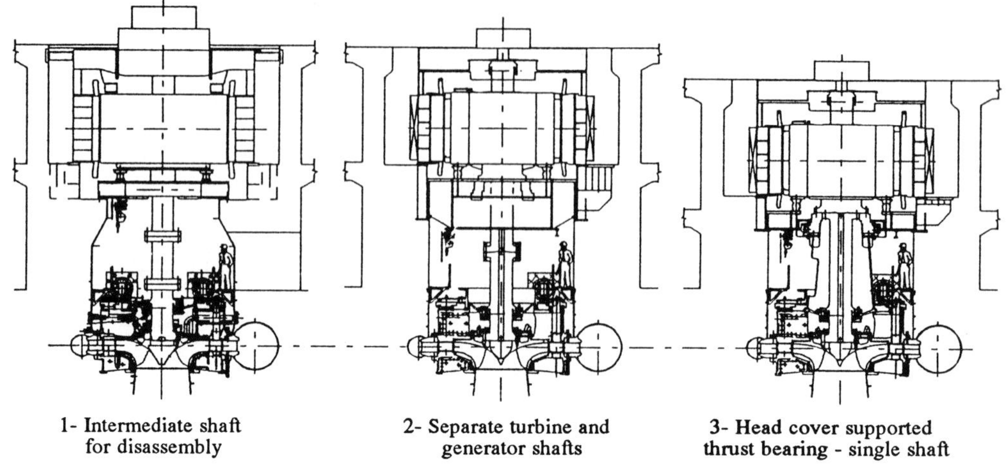

Figure 6. Different concepts for shaft systems.

4.1.2. *Side Removal of the Pump-Turbine Parts (Figure 7)*

From approx. 300 m head, the motor-generator stator diameter becomes smaller than the pump-turbine head cover diameter; thus for vertical removal it is necessary to split the head cover in 2 or 4 sections. This is a disadvantage, if the diameter of a one-piece head cover is below the transportation limits. For plants with a runner diameter smaller than 4 m, all components are generally shipped in one piece.

Plants with lateral removal normally need an intermediate shaft, thus an extension of the shafting is required. For a surface power plant with the spherical valve directly ahead of the spiral case, lateral removal can be performed in the upstream direction without extending the powerhouse dimensions.

For plants with a runner diameter larger than 3 m the vertical clearance between the motor-generator and the pump-turbine due to the required vertical height in the turbine pit, is such that for lateral removal no significant increase of the shafting length is required.

One of the most important arguments for lateral removal however is the reduction of time for the dismantling of a pump-turbine by approx. 2-3 weeks, which is one reason why this solution was finally adopted for the GUANGZHOU II power plant.

In general, the VOITH recommendations for lateral removal consist of the following:

- Lateral removal is advisable for plants up to runner diameter of approx. 4 m.
- Lateral removal is possibly advisable for heads higher than 600 m.

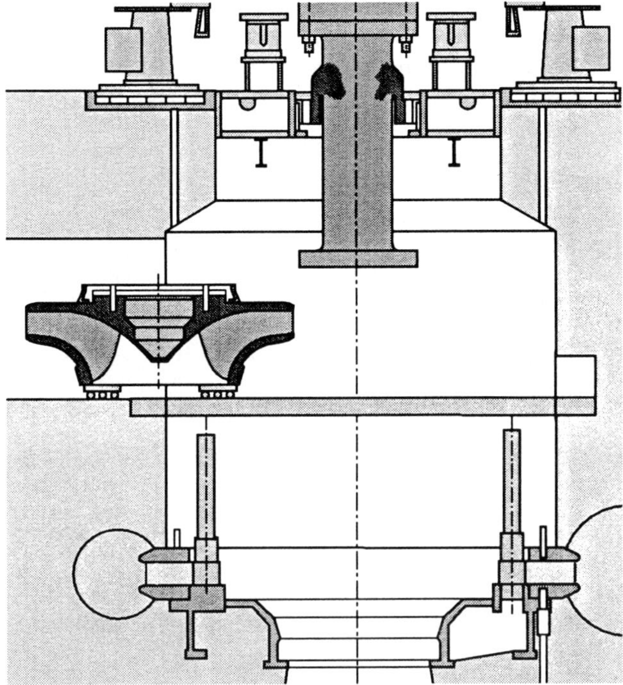

Figure 7. Side removal of the runner

4.1.3. *Bottom Removal of the Runner*

Bottom removal of the runner, while possible, has been adopted very rarely, since it requires an expensive design of the foundations and the removable draft tube section. Furthermore this design can lead to detrimental vibrations and a high level of noise.

4.2. EMBEDDING OF THE STAY RING/BOTTOM RING ASSEMBLY

The design criteria for the stay ring/bottom ring assembly of a pump-turbine are of much more importance than for a Francis turbine.

A comparison of many pump-turbines shows that, in general, the following two design concepts are applied:

- Non-embedded removable bottom ring, freely accessible from below.
- Embedded stay ring/bottom ring assembly without access to the lower wicket gate bearings.

Non-Embedded Bottom Ring (Fig. 8). The stay ring is partially embedded and fixed in the foundation. The bottom ring is not embedded but, like the head cover, bolted to the stay ring and accessible and removable. The static vertical hydraulic forces acting on the head cover (upward) and on the bottom ring (downward) are balanced and result in small foundation loads only. For the absorbtion of the dynamic forces of the bottom ring and discharge ring however, the stay ring requires heavy anchorage in the foundation.

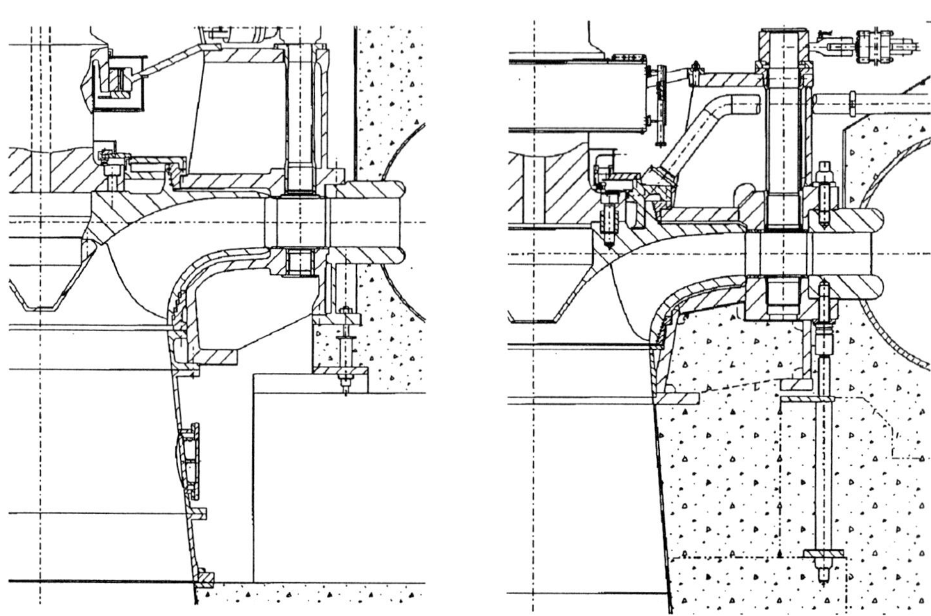

Figure 8.
Non-embedded stay ring/ bottom ring assembly.

Figure 9.
Embedded stay ring/ bottom ring assembly.

Embedded Stay Ring/Bottom Ring (Fig. 9). The stay ring/bottom ring assembly is completly embedded and fixed in the foundation. Due to a special embedment procedure, the vertical hydraulic forces from the head cover and bottom ring are nearly balanced and result in reduced foundation forces. The different forces acting on the stay ring/bottom ring assembly are determined by finite element computation (fig. 10). The results are given to the civil engineer (fig. 11).

VOITH´s recommendation. If sand erosion and early wear of the wicket gate bearings are expected, an accessible and removable bottom ring is more favourable. In pumped storage plants, however, water normally does not carry abrasive materials. Therefore erosion does not play an important part.

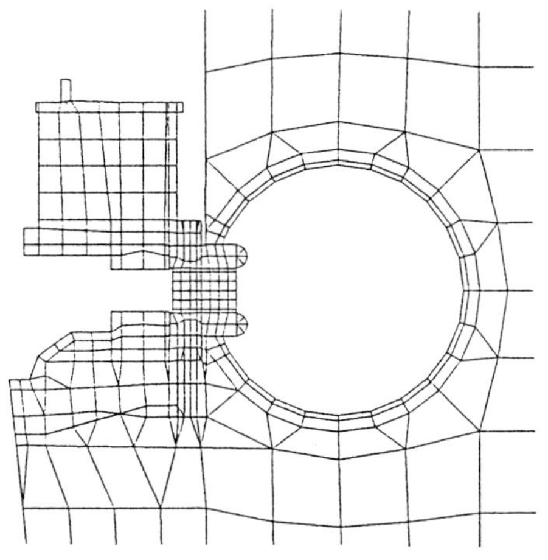
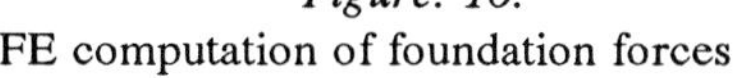

Figure. 10.
FE computation of foundation forces

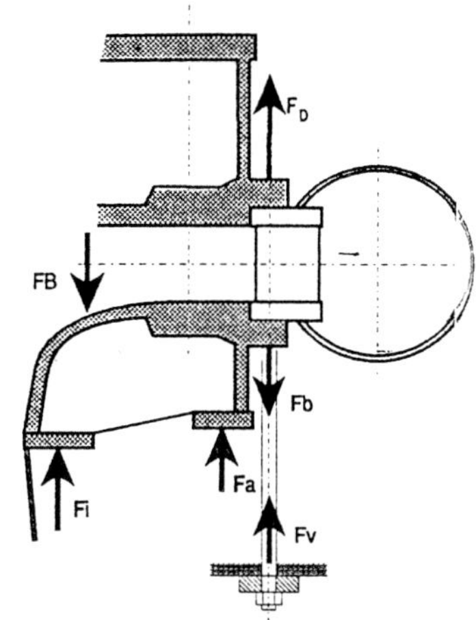

Figure. 11.
Forces on stay ring/bottom ring assembly

Generally a number of important considerations speak in favour of embedding the bottom ring:

- The dynamic forces are transmitted to the foundation directly.
- No axial movement of the discharge ring and draft tube occurs in operation.
- The rigid embedding reduces deflection and vibrations of the bottom ring, thus also reducing the wear of the lower wicket gate bushings.
- No vibrations and therefore low noise level in the lower distributor.
- The smaller deflection reduces the clearance and therefore minimizes the cross flow between the wicket gate ends and the bottom ring. This has a positive effect on part load efficiency and reduces the risk of leakage erosion.

Cosidering all these aspects, this solution was finally adopted for the GUANGZHOU II power plant.

References

1. Riedelbauch, S., Klemm, D. and Hauff, C. (in press) Importance of interaction between turbine components in flow field simulation, IAHR 1996, Valencia.

2. Heine, W., Meazza, G. and Wüst, M. (1995) Comparison of mechanical design concepts for high head pump-turbines.

NUMERICAL HILL CHART PREDICTION BY MEANS OF CFD STAGE SIMULATION FOR A COMPLETE FRANCIS TURBINE

HELMUT KECK
Sulzer Hydro AG
CH-8023 Zürich, Switzerland

AND

PETER DRTINA AND MIRJAM SICK
Sulzer Innotec AG
CH-8401 Winterthur, Switzerland

Abstract. A full stage simulation method is used to calculate the first numerically predicted hill chart of a high specific speed Francis turbine. The numerical method simulates rotating and non-rotating components and their mutual interactions within a single computation. An efficiency hill chart is determined numerically and compared to experimental results from a model test whereby good agreement is obtained.

1. Introduction

Three-dimensional Computational Fluid Dynamics (CFD) is a key technology without which a modern design of a hydraulic machine could not be imagined. Sulzer Hydro has now applied a 3D-Euler code for advanced runner design for 10 years. This was developed together with the EPF-Lausanne (Goede and Rhyming, 1987; Keck *et al.*, 1990) and has been used for over 70 different contracts, allowing wide experience to be gained.

Parallel to the 3D-Euler code, Sulzer Hydro also applies 3D-Navier-Stokes codes for the design of components with adverse pressure gradients, i.e. impellers in pump mode (Goede *et al.*, 1992), for the analysis of the losses in turbine components (Drtina *et al.*, 1992), and for the prediction of turbine hill charts by calculating the flow in a complete turbine from the spiral casing inlet to the draft tube outlet.

E. Cabrera et al. (eds.), Hydraulic Machinery and Cavitation, 170–179.

The calculation of a complete turbine has been made possible by the development and validation of stage capability in a Navier-Stokes code (Sick *et al.*, 1996). The Reynolds-averaged Navier Stokes equations are solved using the TASCflow-code applying the k-ε-turbulence model and with a new interface condition to extend the code to include stage capability (see (Galpin *et al.*, 1995)). The new stage capability is a major innovation compared to conventional CFD-calculations of individual components linked together in a stacking technique ((Shyy and Vu, 1994)).

With the ability to carry out a CFD calculation of a complete turbine an old dream has come true. But what do we really gain from it? Simply adding the calculations of individual components together is not very challenging except for the memory and CPU requirements of the computer. The real challenge is the prediction of the hill chart and for this not only the interactions between the components are important, but also the increase of losses due to off-design conditions of adjacent components.

Experiences will be presented from a CFD-simulation of a Francis turbine of very high specific speed ($n_q = 116$). We have selected a high specific speed turbine for this work because:

- For high specific speeds the streamline curvature at wicket gate outlet (runner inlet) is more pronounced than for low specific speeds. Hence, specifying boundary conditions in this inter-cascade passage is more difficult and the advantage of the stage capability can be fully exploited.
- For high specific speeds the losses of the draft tube have a higher influence on the total turbine efficiency than for low specific speeds. Therefore, a stage simulation where runner and draft tube are calculated simultaneously is of high significance.
- High specific speed turbines have strongly "distorted" hill charts which represent a larger challenge for numerical predictions than the more regular hill charts of low specific speed machines.

The numerical analysis has been performed for a representative number of operating points within the hill chart. Of special interest is the question whether the shape of the efficiency curves $\eta = f(ku)$ for different guide vane openings can be predicted correctly.

2. Computational Method

2.1. DISCRETISATION METHOD AND TURBULENCE MODEL

The calculations were carried out using TASCflow. This CFD code solves the 3D Reynolds-averaged Navier-Stokes equations in strong conservative form for structured multi-block grids. The system of transport equations is discretized using a conservative finite element based finite volume method and is solved for the primitive flow variables (pressure and cartesian velocity components) using a coupled algebraic multigrid method. The discretization scheme is second order accurate. Turbulence effects are modeled using the standard k-ε model.

2.2. STAGE CAPABILITY

The steady-state interaction between stationary and rotating components in a turbomachine is simulated by a mixing plane between the components. Each component is calculated in its own frame of reference and the blade rows can be reduced to single blade channels with periodic boundaries. The method is based on stage simulation ideas of Denton (1992) which were extended by Galpin (1995) and installed into TASCflow under the partnership of ASC, Sulzer Hydro and Sulzer Innotec. Details of the validation of the stage capability are described in the companion paper (Sick *et al.*, 1996) and by (Galpin *et al.*, 1995).

2.3. GEOMETRY AND GRID GENERATION

The geometry data were directly transfered from the CAD system to the CAD based grid generating software ICEM-CFD. This software enables complex multi-block grids in complicated geometries to be generated with a high degree of freedom.

The computational grids are generated for each component of the machine separately. The spiral casing grid also includes the stay vanes. The grid of the distributor consists of stay vanes (2 passages) and wicket gate (3 passages). The grids were overlapped so that the inlet conditions for the distributor could be obtained from the calculation of the spiral casing flow.

The spiral casing and the draft tube are discretised by a butterfly grid. The computational grids for the blade rows (stay vanes, wicket gate and runner) consist of a combination of O-grids around the blades and H-grids. This grid topology resolves steep flow gradients near the blades accurately and is flexible enough for different blade rows. The combination of all grids for the entire Francis turbine is shown in figure 1.

2.4. BOUNDARY CONDITIONS

The flow in the spiral casing is a function of the Reynolds number only, just as the flow in the penstock or the branch pipe. To determine the losses in the spiral casing over the whole hill chart it is only necessary to calculate the viscous flow in the spiral casing at one operating point. The losses are then a parabolic function of the volume flow (Q^2), see Drtina and Sebestyen (1996). For the stage simulation it is therefore sensible to calculate the flow

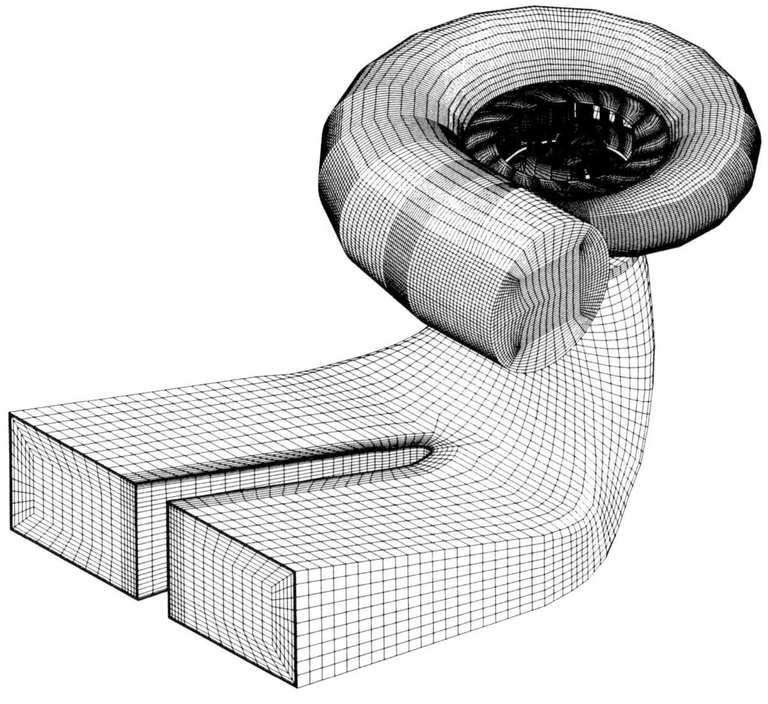

Figure 1. Grid for entire Francis turbine.

in the spiral casing separately (and in one half of the symmetric casing only) to avoid unnecessary large grids for each operating point.

The flow field depends on the operating point between the inlet of the distributor and the outlet of the draft tube. This whole flow regime, including two stationary cascades (stay vanes and guide vanes), one rotating cascade (runner) and one stationary outlet section (draft tube) was computed simultaneously in a single stage simulation. No intermediate boundary conditions are needed and this minimizes the amount of user intervention during the whole analysis. The boundary conditions for the spiral casing are given as follows:

Inlet: mass flow and turbulence intensity

Outlet: constant pressure

The outlet of the spiral casing simulation is chosen to be downstream of the stay vanes such that the flow at the inlet to the stay vane cascade is correctly modelled. The boundary conditions for the stage including distributor, runner and draft tube are:

Inlet: swirl in a reference plane upstream of the distributor as calculated in the spiral casing analysis

Outlet: constant pressure in an outlet plane downstream of the draft tube exit (this allows for the typical non-uniformities of the flow at draft tube exit)

2.5. COMPUTATIONAL DATA

The calculations presented were carried out on a SGI workstation (Power Challenge XL with six R8000 processors) on a computational grid consisting of 279,300 nodes for half of the spiral casing including the stay vanes and 265,795 nodes for distributor, runner and draft tube.

The solutions were estimated to be sufficiently accurate when the maximum residual was below 10^{-4}. The flow simulation in the spiral casing took 170 iteration steps and required 22 hours CPU, that is 1.7 msec cpu per node. The first stage simulation of the combined distributor, runner and draft tube needed 600 iterations whereby each iteration took 700 cpu sec, giving a total time of 5 days and a computational time per node of 2.6 msec. The solution of the first simulation is taken as initial condition for other operating points. This reduces the number of subsequent iteration steps to 100-300 (1 to 3 days) depending on the operating point.

3. Hill chart prediction

3.1. EXPERIMENTAL DATA BASE

The hill chart for a high specific speed Francis turbine has been determined experimentally on a model test rig. The turbine efficiency has been evaluated for more than 200 operating points which yields a good approximation of the turbine performance for the operating range of interest. Figure 2 shows the resulting non-dimensionalized η/η_{opt} hill chart.

3.2. EVALUATION OF EFFICIENCY

To calculate the turbine efficiency from the numerical simulation data the spiral casing (casing and stay vane ring) and stage (distributor, runner and draft tube) calculation results are examined together in order to evaluate the total pressure loss over the entire machine.

Efficiencies are determined for each component (i) by applying

$$\eta_i = \frac{\overline{p}_{tot,i,in} - \overline{p}_{tot,i,out}}{\Delta \overline{p}_{tot,em}} \tag{1}$$

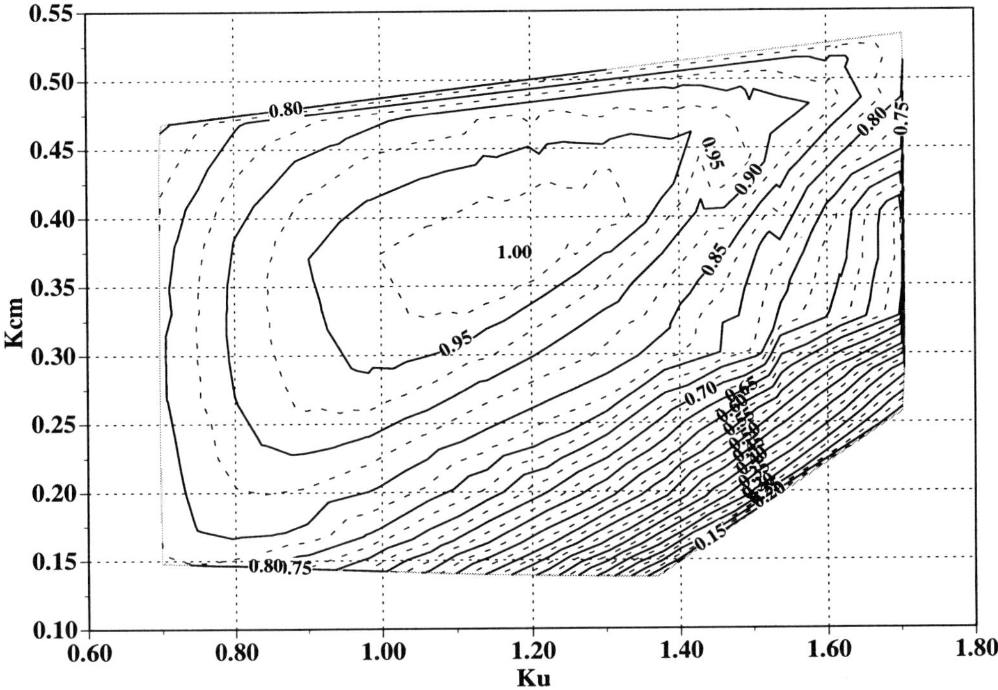

Figure 2. Hill chart based on all experimental data points.

with $\Delta\overline{p}_{tot,em}$ being the difference in total pressure over the entire machine. Note that for the runner one has to take into account the work delivered by the runner.

3.3. NUMERICAL HILL CHART

Due to the high demand in computer memory and CPU-time the number of operating points which could be simulated was restricted. In the present case 14 operating points have been chosen, whereby three different guide vane openings (volume flow rates) and six different heads are investigated.

A comparison between experimentally and numerically evaluated hill charts only makes sense if the number and distribution of data points is identical. For this purpose, the experimental hill chart shown in figure 3 is based only on the experimental data corresponding to the operating points that were also calculated. Figure 4 shows the resulting hill chart based on the 14 operating points from the stage calculations.

The qualitative agreement of both hill charts is impressive. All general features are captured by the simulations. The best efficiency point is iden-

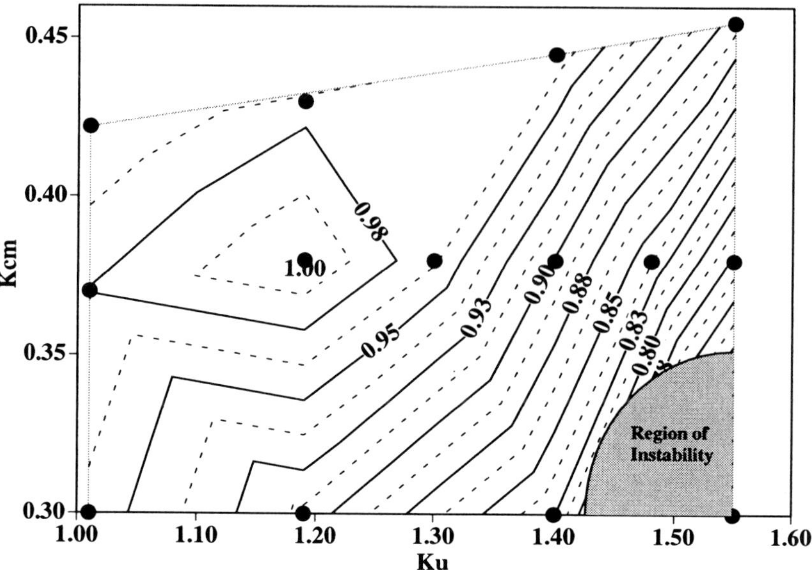

Figure 3. Hill chart based on 14 experimentally obtained efficiency values. Data points are marked by dots.

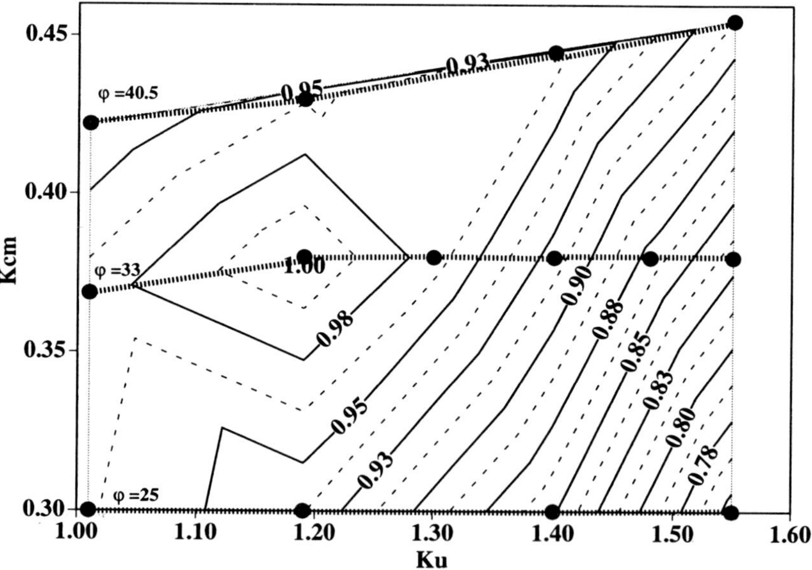

Figure 4. Hill chart based on 14 numerically obtained efficiency values. Data points are marked by dots. Constant guide vane opening is indicated by dashed lines.

tical in both hill charts and all gradients show a very similar behaviour. Please note that the two scales Ku and Kcm are in absolute, not in relative terms, to be able to check the correct location of the best efficiency point. Quantitatively there is some discrepancy in the steepness of the gradients, in particular for severe off-design points with high values of Ku.

A detailed comparison for the three different guide vane openings under investigation can be obtained from figures 5 - 7. For each guide vane opening the normalized efficiency is plotted as a function of Ku.

The shape of the efficiency curves for all three guide vane openings is predicted correctly. For small openings the best efficiency point occurs at low Ku-values (high head). At large openings the high efficiency region occurs at high Ku-values (low head). This typical characteristic of a high specific speed turbine is perfectly represented in the analysis.

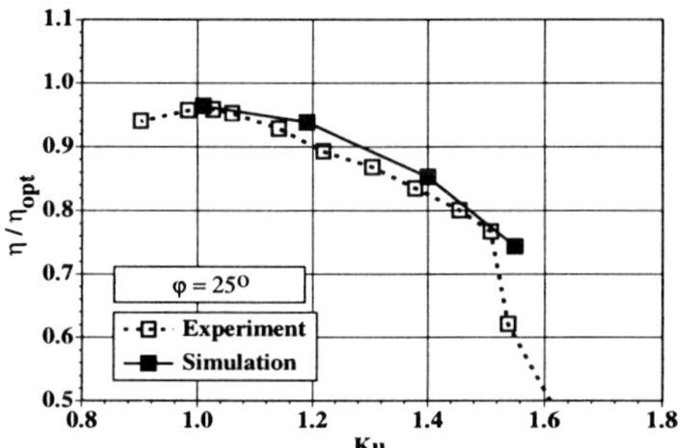

Figure 5. Turbine efficiency for $\varphi = 25°$.

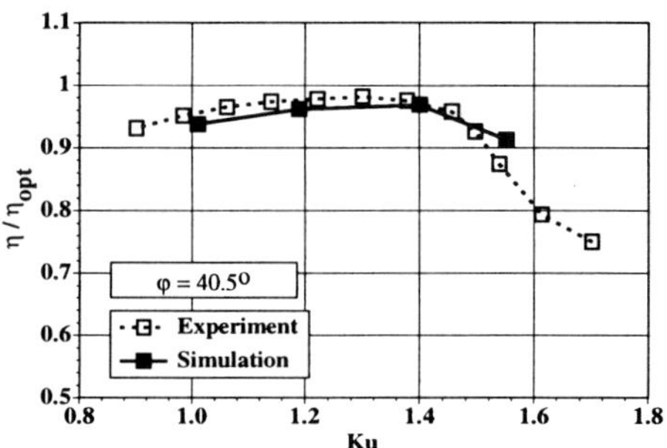

Figure 6. Turbine efficiency for $\varphi = 40.5°$.

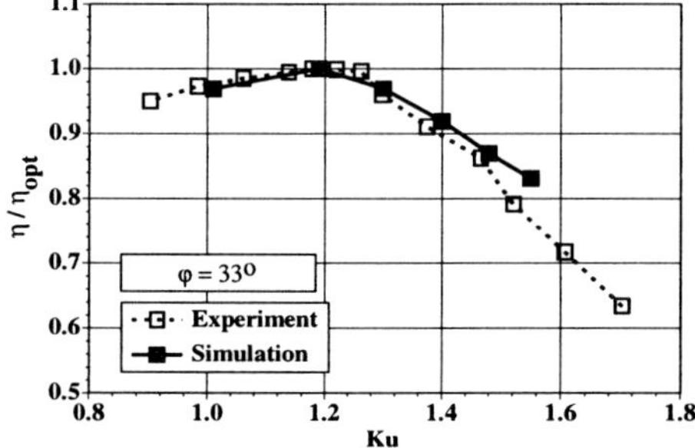

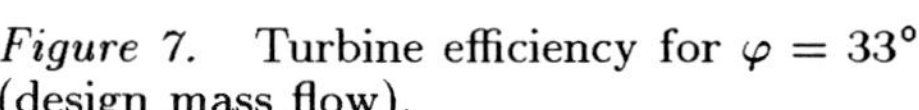

Figure 7. Turbine efficiency for $\varphi = 33°$ (design mass flow).

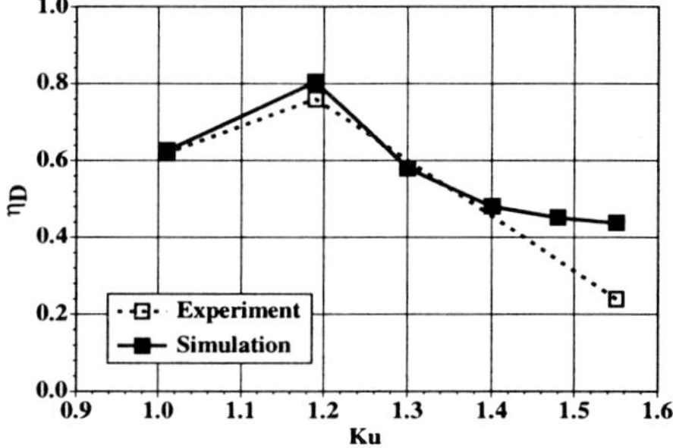

Figure 8. Draft tube efficiency.

Two weaknesses of the present analysis should be mentioned. Firstly, whereas the relative shape of the hill chart is perfectly simulated, the absolute level of the calculated efficiency is about 3% lower than that measured.

This is attributed to the use of a coarse grid. A check of one operating point calculated with a finer mesh produces a reduction of friction losses.

Further improvements are expected if the near wall regions are treated by advanced turbulence formulations, such as two-layer models. Calculations in individual components have shown that the application of two-layer models increase the accuracy of loss prediction, but also increase the computational time.

Secondly, the rate of convergence is rather poor for high Ku-values at small guide vane opening. Large separation zones occur in the draft tube in this case. This is probably the reason why the steep gradients of the measured efficiency curves could not be reproduced in the numerical analysis. However, it is known from the model test that the turbine exhibits a very rough and unstable operational behaviour at Ku $\geq$ 1.5 especially for small openings. The unstable conditions make it difficult to compare the time averaged data of strongly unsteady experimental signals with numerical data of a basically steady analysis.

This is confirmed by the comparison of measured and predicted diffusor efficiency of the draft tube, see figure 8. At high values of Ku, where the flow separates, the numerical simulation currently overestimates the off-design performance of the draft tube.

4. Conclusion

The 3D-turbulent flow analysis in a complete Francis turbine for 14 operating points over a full hill chart has been presented. Of special interest is the fact that the whole assembly of distributor, runner and draft tube has been calculated in a single computation using the newly developed stage capability of the TASCflow code. The advantages of this approach are:

- no need for the user to specify interface boundary conditions between stationary and rotating parts of the machine, thereby eliminating possible errors and time-consuming interventions of the CFD-engineer.
- more accurate simulation of interaction phenomena between adjacent turbine components, especially at off-design conditions. This enables the optimisation of the turbine as a whole and not only of individual components.
- prediction of the shape of a complete hill chart. This is of high practical interest if by modification of certain turbine components special shapes of the hill chart can be generated or if for upgrading projects the impact of existing poorly designed components on the performance of the unit equipped with a new runner has to be predicted.

Future work will concentrate on improvements of the absolute accuracy without excessively increasing the CPU-time and on further automation of the chain of calculations from pre- to post-processing.

5. Acknowledgements

The authors gratefully acknowledge the helpful contributions of Dr. M.V. Casey and D. Wiss, Sulzer Innotec, to the preparation of this paper.

6. List of symbols

$Ku = \omega R(2E)^{-\frac{1}{2}}$	$[-]$	speed coefficient
$Kcm = Q(R^2\pi)^{-1}(2E)^{-\frac{1}{2}}$	$[-]$	flow coefficient
φ	$[-]$	guide vane angle
R	$[m]$	radius at runner blade outlet
ω	$[s^{-1}]$	rotational frequency
E	$[Jkg^{-1}]$	specific hydraulic energy
Q	$[m^3s^{-1}]$	flow rate
$n_q = 157.8 \cdot Ku \cdot Kcm^{\frac{1}{2}}$	$[-]$	specific speed

References

Drtina,P., Goede,E. and Schachenmann,A. (1992) Three-dimensional turbulent flow simulation for two different hydraulic turbine draft tubes, Proceedings of the First European Computational Fluid Dynamics Conference, September 7-11, Brussels, Belgium

Drtina, P., Sebestyen, A. (1996) Numerical Prediction of Hydraulic Losses in the Spiral Casing of a Francis Turbine, *Proc. XVIII IAHR Symposium*

Galpin,P.F., Broberg,R.B. and Hutchinson,B.R.(1995) Three-dimensional Navier-Stokes predictions of steady state rotor/stator interaction with pitch change, Third Annual Conference of the CFD Society of Canada, June 25-27, 1995, Banff, Alberta, Canada

Goede,E. and Rhyming,I.L.(1987) 3-D computation of the flow in a Francis runner, Sulzer Technical Review 4/87

Goede,E., Sebestyen,A. and Schachenmann,A. (1992) Navier-Stokes Flow Analysis for a Pump Impeller, 16th Symposium of the IAHR, september 14-19 1992, Sao Paulo, Brazil

Keck,H., Goede,E. and Pestalozzi,J. (1990) Experience with 3D Euler flow analysis as a practical design tool, IAHR Symposium, Belgrade, September 1990

Shyy, W., Vu, T.C. (1994) A CFD-Based Computer Aided Engineering System for Hydraulic Turbines, *Proc. XVII IAHR Symposium*

Sick, M., Casey, M.V., Galpin, P. (1996) Validation of a Stage Calculation in a Francis Turbine, *Proc. XVIII IAHR Symposium*

DEVELOPMENT OF A NEW GENERATION OF HIGH HEAD PUMP-TURBINES, GUANGZHOU II

D. KLEMM, E.-U. JAEGER, C. HAUFF
Voith Hydro GmbH
D-89509 Heidenheim, Germany

1. Introduction

For the Guangzhou Pumped Storage Plant in the People's Republic of China we had to develop the Pump-Turbines of the second phase considering the very tight delivery and erection schedule. The hydraulic, as well as the mechanical design, the manufacture of a new complete homologous model machine and the detailed model tests had to be performed within a very short time.
In order to achieve all the goals and requirements the application of most advanced methods of numerical flow calculation was necessary (CFD = Computational Fluid Dynamics). Furthermore new ideas for the mechanical design of the model were necessary for a time saving, low cost and very accurate manufacture of the model machine. A powerful test rig with automatic data acquisition and processing of the static and dynamic measurements made possible a very fast and accurate evaluation of the test results.
According to the customer's requirements after the preliminary model tests in our own laboratory in Germany model acceptance tests in an independent laboratory had to be performed. These tests confirmed all our results. The extremely high number of the individual guarantees was met for all performance characteristics and also for the extremely low pressure fluctuations.

2. Guangzhou II

2.1 DESCRIPTION OF THE PLANT

The Guangzhou Pumped Storage Power Station is located at Lutian Town,

E. Cabrera et al. (eds.), Hydraulic Machinery and Cavitation, 180–189.

Conghua County, north-east of Guangzhou city in the Peoples Republic of China. The plant is constructed in two phases as an underground powerhouse equipped with 4 sets of reversible pump-turbines for each phase. Phase one was completed in 1994. Phase two, similar to the project data of phase one, is planned to have the first unit in commercial operation in March 1999, and is scheduled to be completed at the end of 1999.

2.1.1 *Main Data*

The main data of the plant are given in the following table 1.

TABLE 1. Guangzhou II main data

	Gross head range	509,6 - 541,8	m
	Rated rotational speed	500	1/min
	Rated frequency	50	Hz
	Number of sets (vertical)	4	
Turbine	Rated gross head	522	m
	Rated output	306	MW
	Max. output	352	MW
Pump	Head range	514,5-550	m
	Min. discharge	50,6	m^3/s
	Max.input	333	MW

A special challenge for pump operation was the requirement for an absolutely stable Q-H-curve between 49 to 51 Hz grid frequency.

2.2 TIME SCHEDULE

The time for the completion of the preliminary model tests was 210 days after the date of commencement of the contract. Within this time frame the hydraulic design of the machine and its optimization was done. Also the mechanical design of the new model and the manufacturing of all parts as well as the model tests with an extended measuring program was done within this time.

2.3 GUARANTEES

We have got the order for GUANGZHOU II based on very high efficiency guarantees. The more operation points are guaranteed the more difficult is the hydraulic design and its optimisation. In turbine operation the guaranteed weighted efficiency was calculated based on 36 operating points, but furthermore some other individual operation points were guaranteed. In pump operation the number of points for the calculation of the guaranteed weighted efficiency is much less, but also some individual points with respect to the efficiency were guaranteed as well as the minimum and maximum discharge. This means that the steepness of the Q-H-curve was guaranteed too.
For the hydraulic design it was important that pressure pulsations at a very low level were guaranteed for different operating conditions and at different locations of the water passage way.

3. Hydraulic Development

3.1 K - VALUES

Fig. 1 shows the relation between the K - factor and n_q in the pump cycle for Pump-Turbines developed by the VOITH - Group.

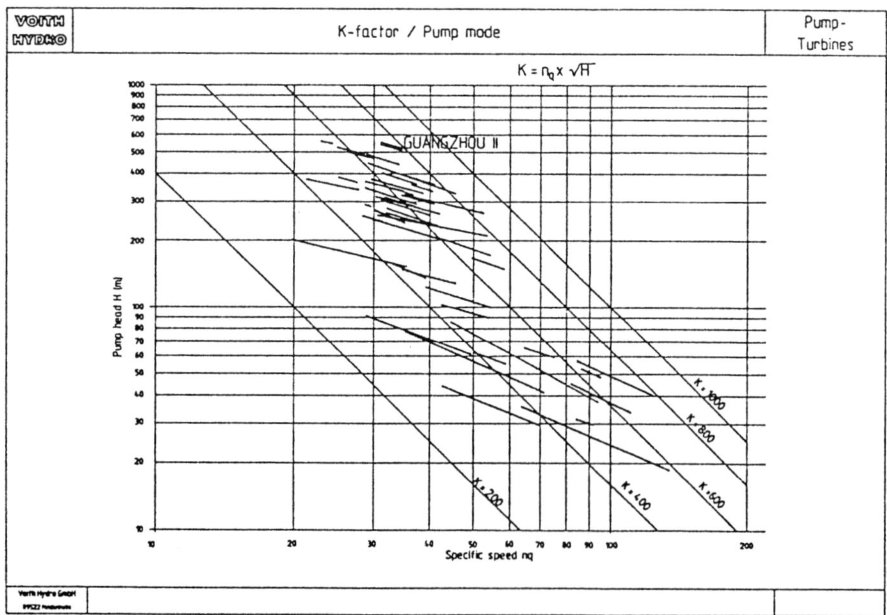

Figure 1. Pump-Turbine K - Factors over n_q for pump cycle

Definition: $K = n_q * H^{1/2}$

Specific speed: $n_q = n * Q^{1/2} / H^{3/4}$

In order to get an economical solution for the hydraulic design the K - factor was chosen as high as possible. For the given head and the K - factor close to 800 the specific speed results as about 34. It is obvious that K = 800 is a marginal value but not too high for a mechanical design preventing excellent hydraulic performance characteristics.

3.2 SIMULTANEOUS ENGINEERING

The hydraulic development was done in a simultaneous engineering mode, combining numerical flow field simulations, stress analysis, model manufacturing and model testing [1].
Computer designed runner blades, guide vanes and stay vanes were numerically preoptimized with a very fast Q3D-Code and finally optimised with the help of a 3D-Navier-Stokes Code.
One example of a hydraulic optimization is shown in the figures 2 and 3. A different leading edge configuration of the runner blade profiles for the turbine operation resulted in an improvement of the flow behaviour without local flow separation.
Special attention was paid to the interaction of the individual machine components [2]. This was important to get high efficiencies, stable characteristics and low pressure fluctuations.

3.3 COMBINATION OF THE NUMBER OF BLADES

With respect to a smooth operational behaviour and low pressure fluctuations a blade number of 9 blades for the runner was selected in combination with 20 guide vanes and 20 stay vanes. This combination of blade numbers will prevent the forced excitation of oscillations with the basic frequency of the runner blade number times runner rotational frequency. Higher frequencies such as harmonics of the basic frequency are less dominating. They are in no case representative at the mechanical resonances of the machine parts and at the hydraulic resonance of the penstock for high head machines, because with increasing length of the penstock its natural frequency decreases.

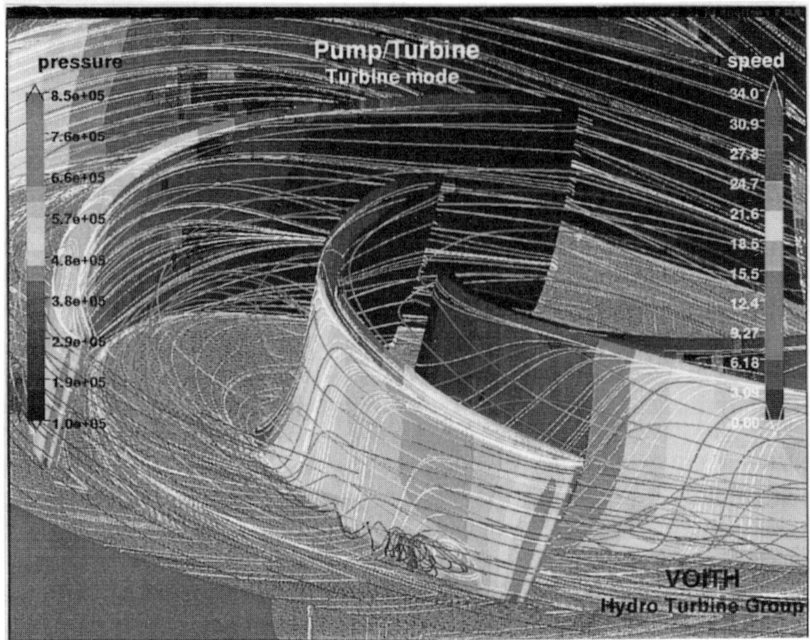

Figure 2. Local flow separation close to the crown in the turbine mode

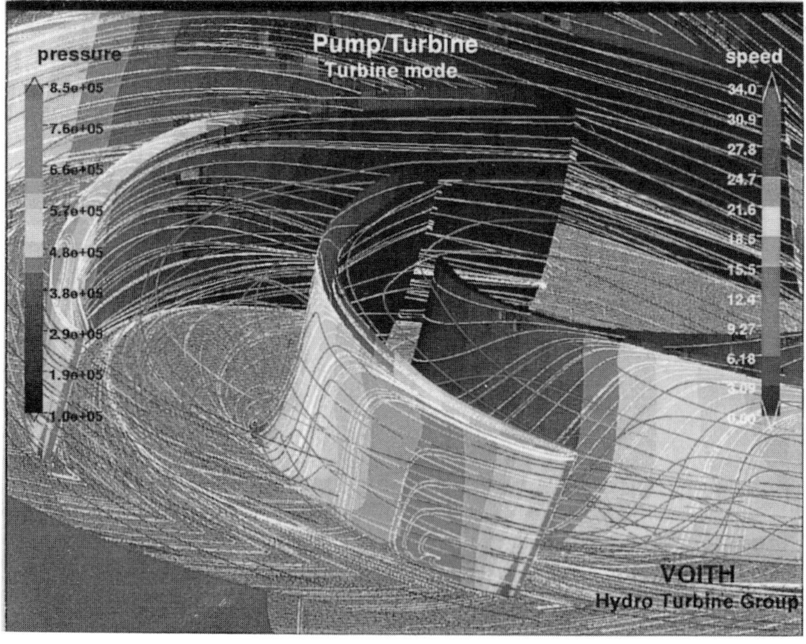

Figure 3. Improved flow behaviour without local flow separation in the turbine mode

4. Model Machine

Based on the optimised hydraulic design a new fully homologous model machine was designed with a scale factor to the prototype of 7.655. The design of the model machine, needless to say using CAD, was done with respect to time saving and low cost manufacturing, but high quality for all hydraulic and mechanical components.
For example the wetted contour of the model spiral case was CNC-milled in two halves from two square aluminium blocks (Fig. 4) using the original data of the hydraulic computer design. The manufacturing time was only half of the time needed normally for a cast spiral case. The dimensions of the hydraulic shape could be improved significantly.

Figure 4. One half of the Model Spiral Case after CNC-milling

The stay ring was manufactured from a cast ring in a similar way.
The runner blades as well as the guide vanes were CNC-milled as individual blades.

5. Model Test Results

5.1 MODEL TESTS AT VOITH LABORATORY

According to contract requirements preliminary model tests were performed in our own laboratory followed by model acceptance tests on a third party's test rig.
We used our vertical shaft test rig. The test speed was 1500 rpm, the test head was about 80 m.

For the final tuning of the model two modifications were prepared for the stay vanes and guide vanes. The test results confirmed the calculated characteristics. All the guarantees of the contract could be met without any further modifications. The test report was handed over to the customer according to the schedule of the contract.

5.1.1 *Performances*

Efficiencies. The measured efficiencies confirmed that we have found a new generation of high head Pump-Turbines. Fig. 5 for pump operation and fig. 6 for turbine operation show the best efficiency of GUANHZHOU II compared with Pump-Turbines of the VOITH group within the relevant specific speed range. The best efficiency of the pump as well as that of the turbine was measured about 0.8 %-points higher than the previous level for this n_q.

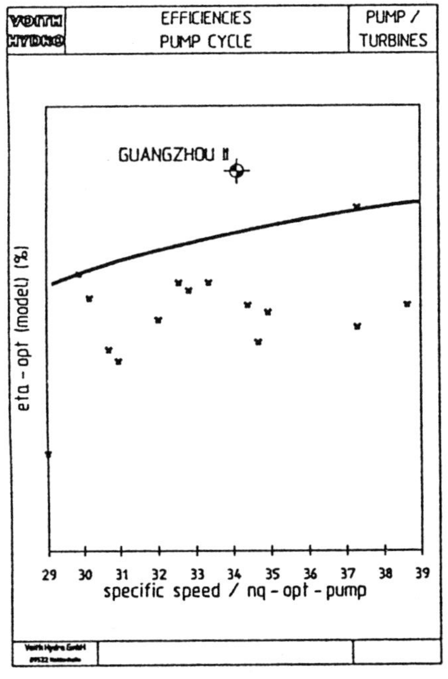

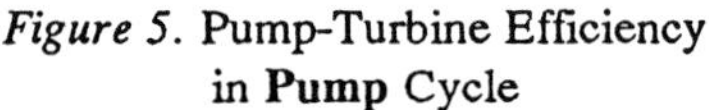
Figure 5. Pump-Turbine Efficiency in **Pump** Cycle

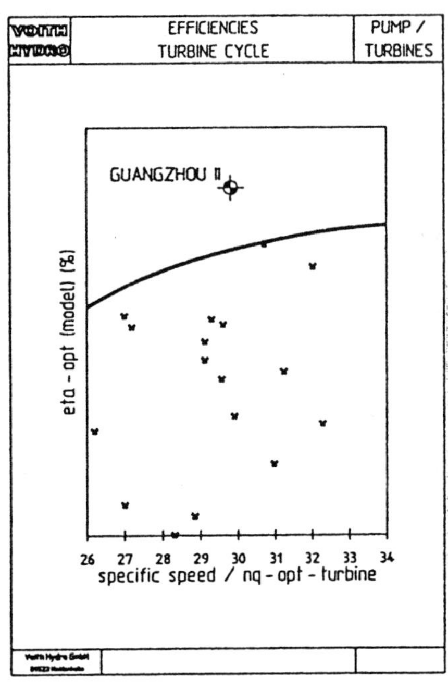

Figure 6. Pump-Turbine Efficiency in **Turbine** Cycle

All other efficiencies, first of all the mean averaged efficiencies for pump and turbine were equal or higher than the very high guaranteed efficiencies.

Characteristics. All other guarantees with respect to discharge and power output or input were met. Even the very extreme pump operating point with reduced grid frequency of 49 Hz at highest pump head can be operated without any problems due to the very stable Q-H characteristic curve.

5.1.2 *Cavitation*

The model test visual observations of the cavitation phenomena at the leading edges of the blades confirmed the CFD-results. In the normal operating ranges the blades are absolutely free of cavitation.
The range of the suction heads at site is between -70 m and -82.4 m.

5.1.3 *Pressure Fluctuations*

According to the contract for different operating conditions maximum pressure fluctuations at several locations of the water passage way were guaranteed. For the measurement of these fluctuations pressure transducers were used in the spiral, in the vaneless space between runner and guide vanes and in the draft elbow at three different positions, one at the outer side, one at the inner side and one at the draft cone.
The fluctuations at the different locations were measured simultaneously. The analysis regarding characteristic amplitudes of the fluctuations in the time domain as well as dominating amplitudes and their frequencies in the frequency domain was established with the help of the computerised data acquisition and processing system. The following tables No. 2 and 3 show the most interesting results.

TABLE 2. Amplitudes of pressure fluctuations in the draft elbow

Draft Tube	Measured amplitudes in % of H		
	Cone	Outer side	Inner side
Turbine rated load	0.3	0.3	0.3
Turbine bestpoint	0.18	0.16	0.15
Turbine partial load	1.7	1.5	2.1
Turbine without load	1.1	1.4	1.2
Pump mode	0.25	0.21	0.42
Runaway	5.5	4.2	4.3
Pump no discharge	2.1	2.1	2.1

TABLE 3. Amplitudes of Pressure Fluctuations between Runner and Guide Vanes

Between runner and and guide vanes	Measured amplitudes in % of H
Pump best point	1.68
Normal pump worst case	1.73
Turbine 50 % load	3.9
Turbine rated load	2.9
Pump no discharge	17.8
Runaway	33

These pressure fluctuations are very low. They represent the high hydraulic quality of this Pump-Turbine and they meet the guarantees. However it should be remarked that older Pump-Turbines are operated with higher amplitudes very smoothly and without any damages caused by pressure fluctuations.
The three locations at the draft elbow show nearly the same amplitudes of the pressure fluctuations. Normally it seems to be sufficient to make measurements at only one position of the draft elbow..

5.1 MODEL TESTS AT A THIRD PARTY'S TEST RIG

The results of the internal development tests and the official model acceptance tests in an independent laboratory in the presence of the customer and his consultant showed in all cases a very good agreement. All guarantees for the model as well as for the prototype could be met.
For example the best efficiency of the turbine mode was measured 0.071 %-points higher at the acceptance test than at VOITH test; the weighted average turbine efficiency was only 0.007 %-points lower. In pump mode the differences of the weighted average efficiency was 0.173 %-points and 0.183 %-points at the best point. All these differences are within the systematic uncertainty of the measurements which is +/- 0.218 % for both test rigs. The pump Q-H-curve was shifted by about 1 %. This may be caused by different inflow conditions to the draft tube at the different test rigs.
With respect to the measurement of the pressure fluctuations at the acceptance tests, most of the piezoelectric transducers were those of the Voith tests. However some transducers and the total measuring technique such as amplifiers, the data acquisition and processing as well as the evaluation method were the normal standards of the other company.
Amplitudes very similar to our preliminary tests have been measured. In the normal operating modes the differences of amplitudes between the two

measurements were in the range of +/- 0.5 %-points. Only at off-design operating conditions, such as runaway, higher differences occurred. Here for comparison the dynamic behaviour of the whole test rig should be taken into account.

6. Conclusion

- ☐ Modern design methods such as CFD and CAD are necessary to get high efficiencies and specially defined characteristics precisely within a relative short time and with low costs.
- ☐ The number of modifications of a model configuration and the number of model tests can be minimised essentially applying CFD and CAD.
- ☐ Model tests at two different test rigs, with an independent staff of the third party, confirmed the calculated results.
- ☐ Model tests still remain essentially for the final verification of a hydraulic design.

References

[1] Buchmaier, H., Quaschnowitz, B., Moser, W., Klemm, D. , (1996) Numerical Design and Optimization of High Head Pump-Turbines, *Proceedings of the XVIII IAHR Symposium, Valencia., Spain*, Kluwer Academic Publisher, to appear

[2] Riedelbauch, S., Klemm, D., Hauff, C. (1996) Importance of Interaction between Turbine Components in Flow Field Simulation, *Proceedings of the XVIII IAHR Symposium, Valencia., Spain*, Kluwer Academic Publisher, to appear

DEVELOPMENT OF INTEGRATED CAE TOOLS FOR DESIGN ASSESSMENT AND ANALYSIS OF HYDRAULIC TURBINES

BERNARD MASSÉ, HENRI PASTOREL AND ROBERT MAGNAN
Hydro-Québec (IREQ)
1800 boul. Lionel-Boulet
Varennes, Québec, Canada, J3X 1S1.

1. Abstract

Recent advances in numerical analysis can lead to substantial improvements in turbine design. In light of the present tendency to increase the unit power and operating speed of these machines at better cost, electric utilities have to develop their expertise in the field of numerical simulation with a view to improve technical decisions, validate conceptions and, in the long term, enhance plant reliability. Applications of numerical simulation range from analysis of problems specific to hydraulic machines such as manoeuvre margin estimation of turbines in operation, correction of design flaws, design validation and technical assistance for planners, designers and operators. This paper describes an integrated system of Computer Aided Engineering (CAE) tools developed to assess designs, optimize retrofits and provide insight for problem analysis.

This system is described using examples to illustrate the use and limitations of the different tools. Selected applications are presented such as the flow simulation in a water intake, the geometric measurements of Francis runners, computed and measured pressure and stress distributions on runner blades, vibration analysis of a Francis turbine, draft tubes flow validation, spiral casing flow and friction losses due to the roughness in a hydraulic tunnel.

2. Introduction

Manufacturers have improved their design methods to use numerical simulations more intensively. Public utilities have several advantages to gain from this technology to improve the reliability, safety and life expectancy of their equipment, especially in view of today's high capital costs and mounting environmental pressures. Numerous challenges are involved in improving plant design and in rehabilitating or making better use of existing plants.

With almost 96% of its production based on hydroelectric resources, Hydro-Québec has a vested interest in increasing its hydraulic turbine expertise in order to design and maintain generating systems in top operating condition. An effort is ongoing within the framework of a research project called MATH, directed toward simulating the fluid flow in hydraulic turbines and estimating the stress level of the mechanical structure for various operating conditions. The MATH project makes use of several numerical techniques to analyze the

E. Cabrera et al. (eds.), Hydraulic Machinery and Cavitation, 190–199.

behavior of water turbines components. Various software tools were developed and integrated in a CAE environment focused on the needs of electric utilities.

Utilities do not always share the same viewpoint as manufacturers. Performance and high-quality products are important for both, of course, but the owner of a hydraulic machine is particularly interested in its good dynamic behavior, high life expectancy, and low cavitation damage. Numerical simulation tools developed so far do not allow easy computation of unsteady conditions, forced vibration analysis and life prediction. With the rapid growth of computational capabilities, such computations will be available in the near future. The CAE system is designed to allow the use of this numerical simulation technology.

3. CAE tools for turbine analysis.

The CAE system is a flexible network of commercial software, custom tools and data translators set up to insure the flow of information from the geometric and operational parameters to the required engineering results. This environment integrates geometric modeling and meshing tools, fluid-flow computation codes, stress and vibration software, performance analysis codes, all linked to a common data base.

The tools are designed to allow the analysis from the water intake to the draft tube exit. For this purpose, various simulations are required in the different hydraulic components, and interpolation of the boundary conditions insures flow continuity from one component to the next. This approach allows specific modelling of each hydraulic passage taking advantage of possible optimization. Numerical tools capable of simulating the whole range of turbine operation do not yet exist and the development of relevant algorithms can be a very difficult task.

Despite all the progress made, numerical simulation methods approximate the physical phenomena more or less accurately. The modelling of turbulence, unsteady flows, cavitation and hydroelastic effects, as encountered in hydraulic turbines, are good examples of such limitations. Consequently, the confirmation of numerical models by experimental testing is very important and takes time and resources. The CAE system is still under development to include unsteady fluid flow, hydroelastic effects, forced response and fatigue analysis.

Figure 1 shows the overall structure of the various tools in use. Geometries of the hydraulic turbine components, collected from drawings and measurements, are stored in a database which is the input for geometric modeling and meshing. The numerical solvers use the data produced in the preceding steps to perform different numerical simulations. The last step consists of viewing and analyzing the solutions.

To exchange data between the different software tools, a common data format and program translators are needed. Although this approach is not general, it works well and requires a minimum of conversion programs to maintain. Conversion of geometric data is among the most involved and a specific program is endowed with this responsibility: VEDGE. It handles a variety of geometric format among which IGES is perhaps the most important since I/O with CAD packages is done using that standard exchange format.

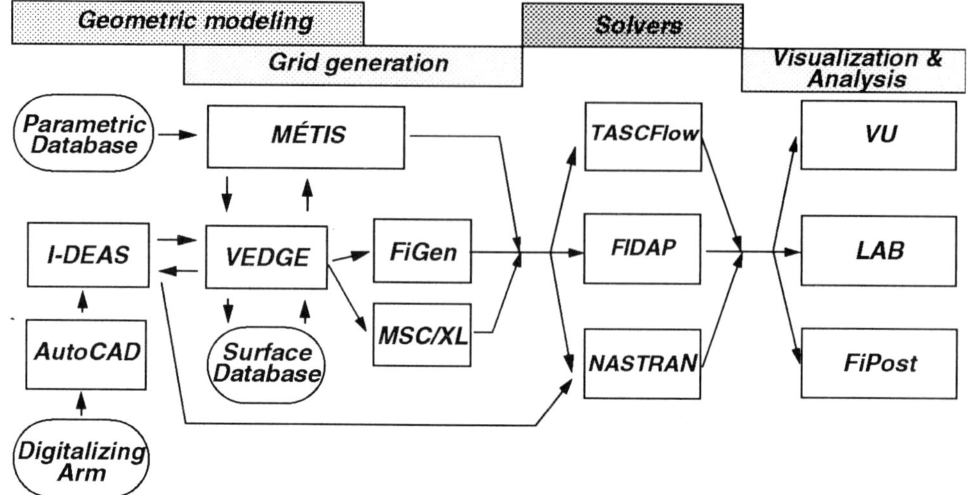

Fig. 1. Flow of information between the software tools

4. Modeling and Meshing of Geometries

Important efforts were spent on geometric modelling and mesh generation. A custom made tool (MÉTIS) can automatically construct a geometric model and meshes for spiral casings, runners and draft tubes. MÉTIS starts with parametric data describing the main features of the components. For other geometries or to model fine details that cannot be described through parametric data, commercial CAD packages are used to hand-build a geometric model.

MÉTIS can automatically produce meshes for most geometries. The geometric model computed by MÉTIS can also be used by itself as an input to FiGen (a mesh generator included in FIDAP) to allow a better control of the meshing. The geometry constructed with CAD packages follows a similar path. In particular, FiGen makes it possible to use unstructured meshes in part of the domain. This can prove useful in many instances by greatly reducing the effort required to mesh complex geometries and sometimes considerably reducing the number of nodes. The flexibility of the tool network allows us to combine coarse geometry coming from MÉTIS with precise surfaces measured in the field and modeled with CAD software. This can greatly simplify the task of analyzing complex, measured or detailed geometries.

Figure 2, produced using VU, shows a spiral casing modelled from drawing informations and on-site measurement for details near the baffle vane region. This example illustrates the use of several tools to obtain the complete mesh. Geometric sections were generated using MÉTIS. FiGen was used to create the mesh and measured curves were used where needed.

Figure 3 shows an example of complex modelling and meshing for structural analysis of a Kaplan blade. In this case, most of the work has been done using I-DEAS, the blade profiles coming from the database.

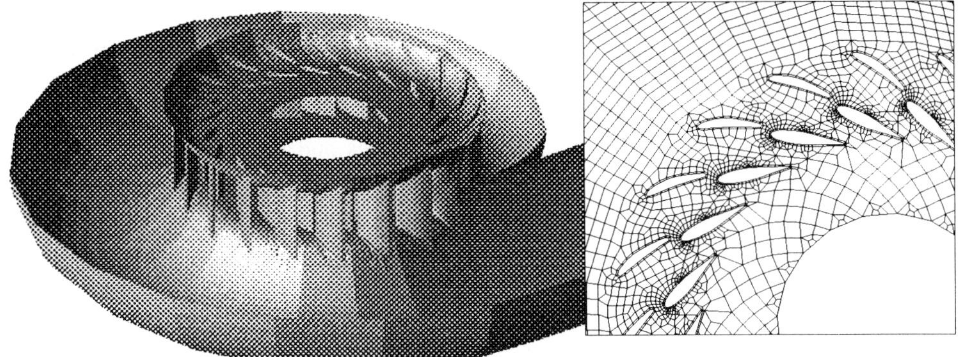

Fig. 2. Geometric model and meshing of a spiral casing

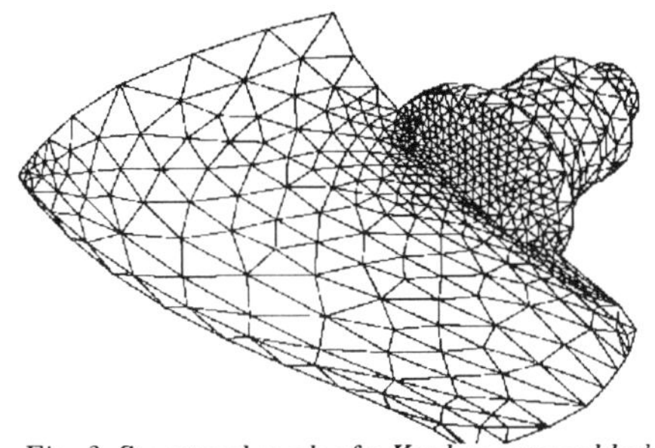

Fig. 3. Structural mesh of a Kaplan runner blade

Fig. 4. Measure of two hydraulic passages at LG3

4.1. GEOMETRIC MEASUREMENT

On-site experimental measurements are required for several parts such as runners, stay and guide vanes and spiral casing baffle vane. Fine fluid-flow analysis asks for improved measurement methods and geometric modelling.

Measurement can be used as a quality control method and as a method to obtain surface data of critical parts such as fillets, leading and trailing edges and complex surfaces. These methods can be useful for retrofitting turbine units, allowing comparison between old and new designs. Measured data are stored in the geometric database for fluid-flow and structural analysis and could be used for rebuilding cavitated areas.

In the past, a laser interferometer and a mechanical digitizing arm were used. The former is expensive, accurate (in the order of 0.01 mm) but not well coupled to CAD systems. The latter is cheaper, less accurate (0.1mm) but has a direct interface with a CAD system, which allows on-site generation and validation of measured surfaces.

Figure 4 shows the reconstructed geometry of a runner at LG-3. Measurement were taken in the field using the mechanical digitizing arm.

5. Solving the equations

The equations to calculate fluid flow, stress distributions, frequencies and modes of the structural components are well known. Commercial tools such as FIDAP and TASCFlow for fluid flow and NASTRAN for structures are generally used. In fluid-flow simulations, the convergence of the numerical solution depends on several parameters such as the upwinding scheme, relaxation factors, and boundary and initial conditions. An inappropriate choice of solution parameters can lead to divergence of the solution. A cyclic symmetry formulation has been applied for calculating the modes and frequencies of the turbine runner. Solving the various equations using numerical tools calls for optimized parameters. However, the process demands a significant amount of experience.

5.1. FLUID-FLOW

In fluid-flow simulations, the 3D Reynolds-averaged Navier-Stokes equations are solved. The flow is assumed steady and incompressible. However, the first assumption is not always valid at partial load and full power, where a rope forms in the draft tube and oscillations occur, but it is imposed by the available computer resources. The Reynolds number is very high ($>10^6$) and the flow can be considered fully turbulent. The momentum equations to be solved with these assumptions are the Reynolds-averaged Navier-Stokes equations including Coriolis, centrifugal and gravitational components. The Reynolds stress tensor is calculated using the k-ε model of turbulence. In this model, the characteristic turbulent velocity and length scales are related to the kinetic energy of the turbulence and its rate of viscous dissipation. It adds two semi-empirical transport equations to be solved with the momentum equations. The k-ε model has been calibrated for several turbulent fluid flow problems but validity of this model in hydraulic turbine is still under investigation. The standard k-ε turbulence model is based on the Boussinesq eddy-viscosity model which assumes a linear function between the Reynolds stress and the strain rate tensor. Anisotropic eddy-viscosity models which renders Reynolds stress tensor a higher order function of the strain rate tensor are now being considered.

5.2. STRESSES AND VIBRATIONS

For mechanical structures, in addition to the well known stress equations from a pressure loading, the development effort is oriented toward solving for different modes and frequencies of rotating runners. Boundary conditions are then applied, including cyclic symmetry, together with loading forces such as centripetal acceleration and flow pressure.

The whole runner (not only a sector), the presence of the fluid and rotating inertia terms must be considered to compute vibration characteristics of turbines. To take the whole runner into account, classical theory of cyclic symmetric structures is applied [7], computing what are called rotating mode shapes. Each family is characterized by a phase angle relating the deflection of one blade to its neighbors. Table 1 shows typical results for a Francis runner[5]. Addition of Coriolis terms is straightforward but work is ongoing for the integration of added mass terms. Forced response will have to include all these effects and compute the dynamic stresses due to unsteady hydrodynamic loads.

Table 1. Natural frequencies (Hz) of a Francis runner

Mode shape	Characteristic displacement phase angle					
	0	2 π / 11	4 π / 11	6 π / 11	8 π / 11	10 π/11
1	25.44	54.28	103.34	144.92	165.37	171.95
2	54.81	162.48	192.73	207.52	211.45	211.81
3	179.39	213.39	235.10	254.61	273.86	281.27
4	232.63	235.40	259.21	286.39	307.34	324.50
5	243.24	259.86	341.38	347.54	365.55	354.05

6. Post-processing

To extract engineering data from numerical solutions several post-processing tools are needed. Validation must be done to insure the quality of the solutions. This can imply visualization of results and computation of subsequent quantities for analysis. For this task we can use directly the built-in tools in the case of commercial packages or the custom software VU[3]. The later option gives more flexibility and control over the computation of relevant quantities. We are also developing a complementary tool called LAB to extract engineering quantities from the numerical solutions.

7. Validation

Validation test cases are of key importance to evaluate the mathematical and numerical model parameters used to simulate a physical phenomenon. For example, the computation of a turbulent swirling flow in a conical diffuser is presented. This test case has been chosen from the ERCOFTAC database [6] for its similarity with the first diffusing part of a hydraulic draft tube. It is well-documented and experimental data are available in several sections along the diffuser. For this test case, inlet condition includes measured velocities and turbulent kinetic energy suitable for fluid flow simulations. Figure 5 shows comparison between computed and measured axial flow velocity at different positions in the diffuser. Details of this analysis can be found in reference 4.

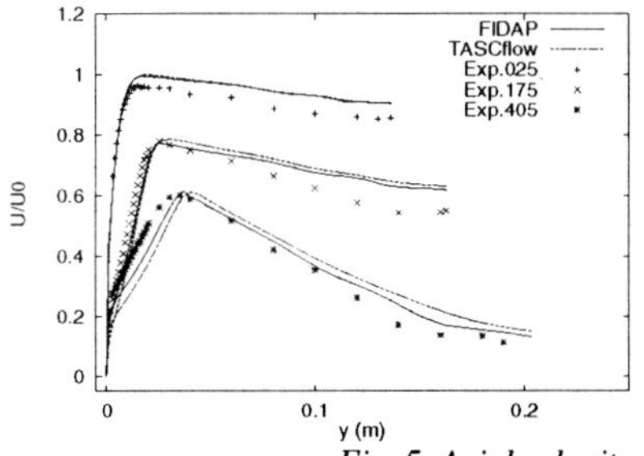

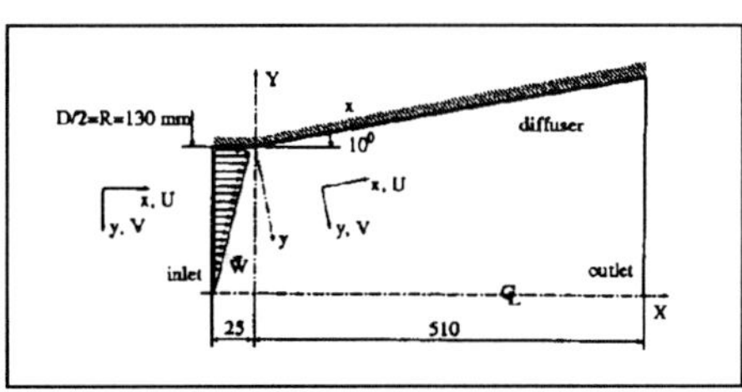

Fig. 5. Axial velocity at three position along the diffuser

Among validation test cases examined to date, let us mention: velocity measurement at the inlet of a Francis prototype runner and pressure and stresses on that runner (see figure 9); comparison of computed losses in a straight pipe with rough walls to losses predicted from the Moody diagram[1]; modal analysis on a scaled propeller runner to validate computed frequencies and mode shapes.

8. Selected results

8.1. WATER INTAKE

Fluid flow in the water intake and in the penstock is responsible for the flow profile entering the spiral casing. At the LG3 power station, questions were raised about the velocity profile at the entrance of the spiral casing due to the presence of an elbow just upstream and also because the flow at the water intake itself arrives at various angles depending on the position of the actual intake.

Experimental measurements were available on a scaled model of the water intake where the shape of the reservoir has been reproduced and the inlet angle could be changed. Recently, a water intake analysis using fluid flow computation has been applied to this case. Figure 6 shows typical computed velocity profiles compared to the model measurements and good agreement was obtained, showing the usefulness of CFD.

If the free surface interacts with the flow field and vortices form near the intake, fluid flow analysis becomes more difficult. We are starting to develop new capabilities in this area.

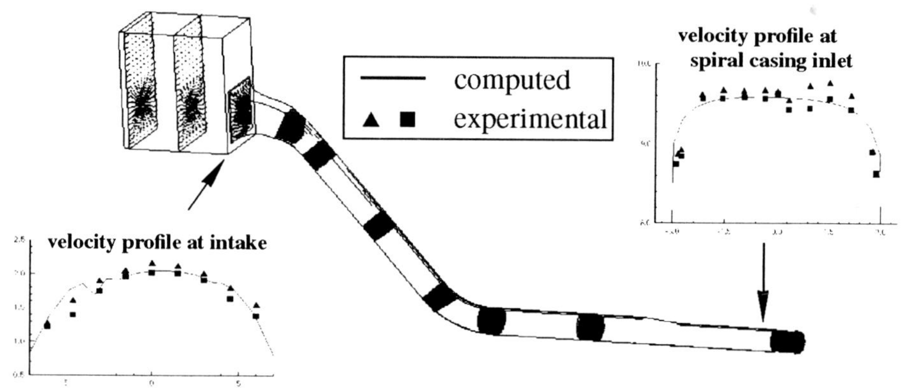

Fig. 6. Water intake analysis

8.2. TUNNEL

Friction loss in the Bersimis-1 tunnel was studied using fluid-flow simulations[1]. This tunnel is 12 km long with an average diameter of 9.6 m. A black substance of varying thickness gradually formed on the walls increasing the roughness and reducing the efficiency of the tunnel. As the tunnel is used to bring water into the Bersimis power plant, head loss implies power loss at the plant. On the perimeter, the black substance does not always have the same thickness so the roughness varies. The purpose of the study was to calculate the loss associated with specific portions of the tunnel perimeter to indicate where repairs are more profitable. Using FIDAP's capability to model surface roughness, simulations were made to estimate losses in the tunnel.

Several numerical simulations were performed to compare the computed head losses to available measurements on-site in 1958, 1978 and 1993. The flow field for the 1958 roughness conditions is shown in figure 7 for a mass flow of 425 m^3/s, corresponding to a Reynolds number of 4.28 x 10^7. Note that the velocity profile is not symmetric from bottom to top of the tunnel. Roughness is more important at bottom, which tends to round the profile in that region, since the losses are greater.

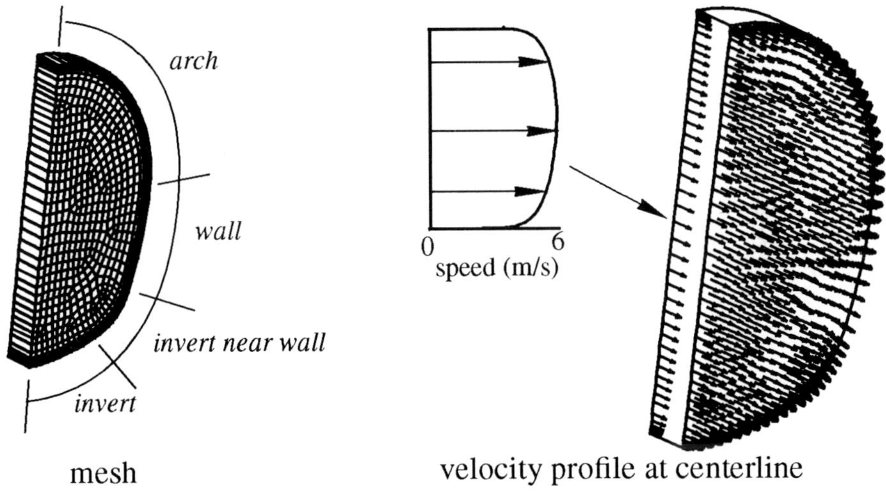

Fig. 7. Velocity field computed in the tunnel for 1958 roughness

Although friction losses due to surface roughness are modeled in the present work, cracks, construction joints, undulations and other large-scale defaults existing in a real tunnel cannot be considered. Comparing computed and measured data, large-scale defaults such as these account for 8.5% of the total head loss. In this analysis, the tunnel section is divided in three parts: the arch and walls section, the invert near the wall and the invert central strip. Different values of roughness were used and the tunnel loss compared to measured values (see Table 2).

Table 2. Bersimis-1 tunnel

	Roughness (mm)			Losses (m)		
year	*arch-walls*	*invert-near walls*	*invert-central strip*	*computed values*	*computed +8.5%*	*measured values*
1958	0.381	1.080	2.540	21.48	**23.30**	**23.2**
1978	2.032	2.794	2.794	26.93	**29.22**	**29.2**
1993	1.700	2.500	2.500	26.15	**28.38**	**30.0**

8.3. SPIRAL CASING

In most numerical investigations of flow in turbine runners, the flow field is assumed to be time-independent. For Francis turbines, where the runner is located close to the wicket gates, the wakes produced by the vanes introduce oscillations in the flow at the inlet. Numerical methods that allows unsteady flow conditions are being developed and their use will increase in the near future. Figure 8 shows velocities in a spiral casing computed to obtain the flow distribution at a runner inlet.

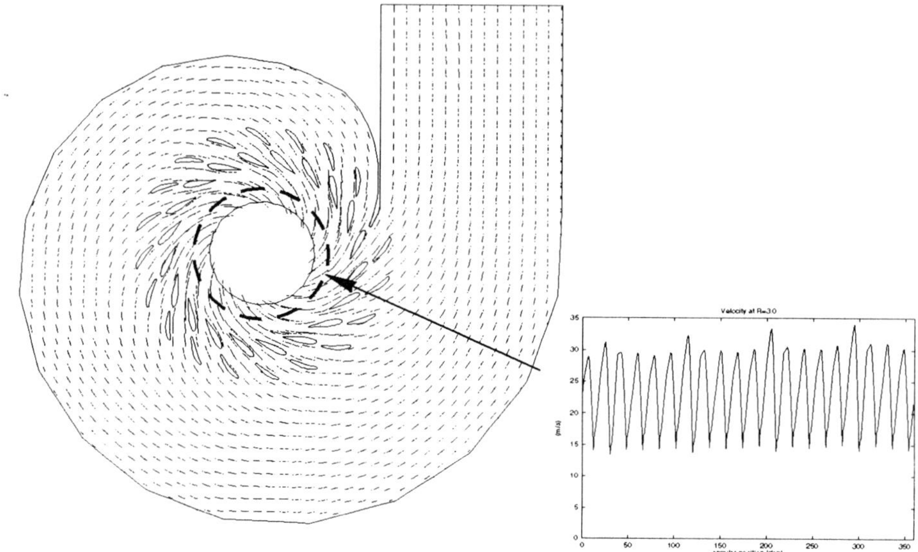

Fig. 8. Velocities in a spiral casing and its distributor

8.4. TURBINE RUNNER

An experimental investigation was performed on a 35-MW Francis turbine at

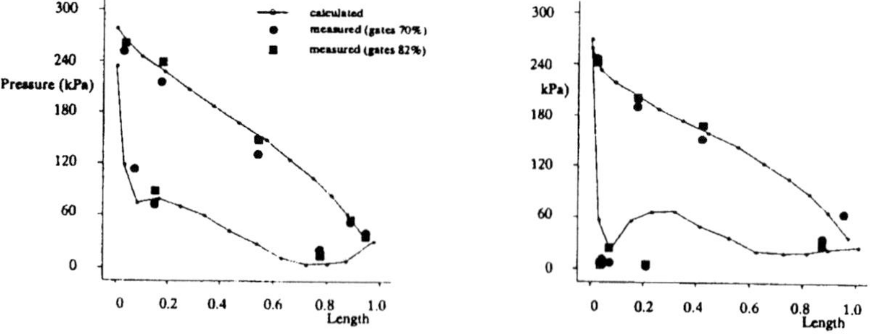

Fig. 9. Comparison of measured and computed pressure distributions on two mesh lines of a Francis runner

Hydro-Québec Rapide Blanc power plant using flush-mounted pressure sensors on the runner. Comparison of the numerical and experimental analyses shows generally good agreement, especially for the line near the middle of the runner as seen on the left side of figure 9 [2]. Figure 10 gives the stress distribution obtained from the computed pressure on the blades. It compares well with measured values in most cases.

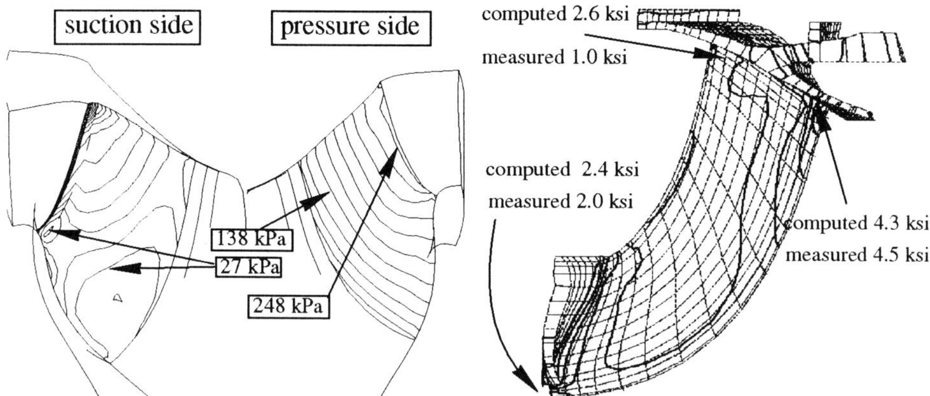

Fig. 10. Pressure and stress distribution on Francis runner

9. Conclusions

Water turbine technology is evolving toward refined design to optimize performance, reliability and life span. Numerical tools contribute to a better knowledge of the hydraulic parameters needed to improve stability, reduce noise and vibrations and ensure a greater life expectancy. It is important to integrate these tools into a flexible CAE environment to insure efficient use of these numerical techniques. This environment is a key factor to produce numerical simulations of complex phenomena. Results obtained so far and ongoing validation activities bring more confidence in numerical simulation technologies.

10. References

1. Massé, B., *"Turbulent-flow head loss calculation in hydraulic tunnels"*, The 6th FIDAP Users Conference, Chicago, USA, April 30 - May 2 1995.
2. Massé, B. and Pastorel, H., *"Numerical simulation and experimental verification of flow in a Francis turbine"*, ASME paper FED-Vol. 171, 1993 ASME Winter Annual Meeting, New Orleans, USA, December 1993.
3. Ozell, B., *"VU, A Configurable Visualization Program"*, User's Manual, CERCA, Montréal, 1995.
4. Page, M., Giroux, A.-M., Massé B., *"Turbulent swirling flow computation in a conical diffuser with two commercial codes"*, 4th annual conference of the CFD Society of Canada, Ottawa, June 2-6, 1996.
5. Pastorel, H., *"Calcul des fréquences naturelles et des modes normaux d'une turbine hydraulique de type Francis"*, Colloque de conception vibratoires et acoustiques, 62 eme Congrès de l'ACFAS, Mai 1994, UQAM, Montréal.
6. Rodi, W., Bonnin, J.-C. and Buchal, T., editors. *"ERCOFTAC Workshop on Data Bases and Testing of Calculation Methods for Turbulent Flows"*, 4th ERCOFTAC/IAHR Workshop on Refined Flow Modelling, University of Karlsruhe, Barlsruhe, Germany, April 3-7 1995. Preliminary proceeding.
7. Thomas, D.L., *"Dynamics of rotationnaly periodic structures"*, International Journal for Numerical Methods in Engineering, 14, pp. 81-102, 1979.

DESIGN AND ANALYSIS OF A TWO STAGE PUMP TURBINE

FARID MAZZOUJI - MARYSE FRANCOIS - FRANK HEBRARD - JEAN BERNARD HOUDELINE - DANIELE BAZIN
GEC ALSTHOM NEYRPIC
82, Rue Léon Blum BP 75, 38041 Grenoble Cedex, FRANCE

In this paper we present the Computation Fluid Dynamics and design tools used in the hydraulic conception at GANP applied to a two-stage pump-turbine with nq37. Each device of the machine is generated and calculated using the most appropriated C.F.D. tool: direct or inverse methods for shape generating as well as inviscid and viscous flow calculation.

1. INTRODUCTION:

Nowadays the design of all turbomachinary components can hardly be conceived without taking numerical flow modelisation into account. This is due to the advances in numerical methods and in computer technology .

Thus, the design process has become a two-way procedure between numerical conception and analysis software. The use of the appropriate numerical tool at the right moment allows to reduce the development duration of new machines with a better behaviour in all areas such as cavitation or fluid velocities. Model tests are still indispensable, but their number are significatively reduced and their goal are more cibled. They allow to investigate the machine more thoroughly .

Since the seventies, GANP began to develop softwares based on quasi-3D approximation

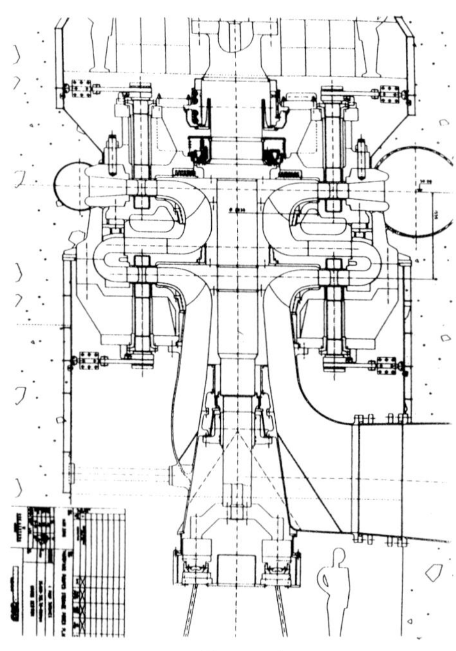

figure 1

E. Cabrera et al. (eds.), Hydraulic Machinery and Cavitation, 200–209.

for runners (1) and spiral cases (2); and they are still excellent tools for first analysis.
But GANP has acquired a new generation of design and analysis numerical tools and uses the two CFD concepts: direct and inverse approach with different levels of approximation, from the inviscid fluid computations to fully 3D Navier-Stokes models (N3S,Calèche) (3) (4).

In this paper we present a summary of the theoretical basis, and give examples on an 800 m net head and nq37 two stage pump turbine.
The design of the return channel second blade was performed using an inverse method, and the flow in the draft tube in turbine mode, and in the spiral case, stay vane and guide vane and in the runner in pump mode was performed using a Navier Stokes software.

2 THEORETICAL BASIS OF THE INVERSE DESIGN AND VISCOUS ANALYSIS SOFTWARE

2-1 3D INVERSE METHOD USING STREAM FUNCTION

Local equations
We model an incompressible, inviscid flow in a turbomachine channel rotating with a constant angular velocity $\omega_0 \mathbf{k}$. We consider as steady the relative flow wich assume the upstream conditions to be steady and axisymetrical. The relative velocity **W** and the absolute velocity **V** are introduced.
The mass conservation is given by :

$$\nabla.\mathbf{W}=0$$

and the conservation of momentum is given by :

$$(\nabla \times \mathbf{V}) \times \mathbf{W} = -\nabla\left(h+\frac{W^2}{2}-\frac{(\omega_0 r)^2}{2}\right)$$

The S1S2 method is used to separate the 3D problem into two sets of 2D problems being solved each one on stream surfaces S1 and S2. Equation 1 allows to introduce two stream functions $\psi 1$ and $\psi 2$ to describe the relative velocity as :

$$\rho W=\nabla\psi_1 \, x \, \nabla\psi_2$$

The stream function are fixed up to arbitrary functions which define the source lines of stream surfaces : S1 for $\psi 1$ constant and S2 for $\psi 2$ constant.
Thus under the axisymetrical alimentation hypothesis the equation of momentum allows to obtain the following system :

$$\nabla.(W \, x \, \nabla\psi 2)+\rho\frac{dI}{d\psi_1}+2\omega_0\frac{\partial\psi 2}{\partial z}=0$$

$$\nabla.(W \, x \, \nabla\psi 1)+2\omega_0\frac{\partial\psi 1}{\partial z}=0$$

It can be proved that the $\psi 1$ and $\psi 2$ operators in the preceding system are a 2D elliptic operators .In order to use this property, it leads to use a system of co-ordinate and mesh depending on the flow .Thus the mesh is one of the unknown parameters of the problem.

Boundary conditions

We impose sliding conditions on hub, shroud, pressure and suction side. It comes to impose $\psi 1$ on hub and shroud and $\psi 2$ on pressure and suction side. Upstream we impose an absolute swirl Γ. At the outlet the quantities corresponding to the velocity direction are unknown and determined by informations given by pressure and vorticity fields. For the pressure we impose a variation pressure model along $\xi 1$ ($\xi 1$ is a meridian abscissa). We then use the fact that the absolute helicity can be calculated along a stream line with the transport equation obtained by Hawthorne and Horlock

$$W.\nabla\tau = \frac{1}{\rho W^4}\left(\nabla\left(W^2\right).(\nabla I\, x\, W) - 2\omega_0 W^2 \frac{\partial I}{\partial z}\right)$$

where the helicity is defined as :

$$\tau = \frac{(\nabla\, x\, V).W}{\rho W^2}.$$

We then solve the problem using an iterative process to linearise equations .

2-2 3D VISCOUS MODELISATION (N3S)

We recall the Reynolds-averaged Navier Stokes equation for a relative incompressible flow in a rotating frame of reference

$$\frac{\partial W}{\partial t} + W.\nabla W = -\frac{1}{\rho}\nabla p + \nabla\left((v + v_t)\vec{\nabla}\vec{W}\right) - 2\Omega\, x\, W + \Omega^2\, r$$

$$\nabla.W = 0$$

where ν_t is the eddy viscosity linked to the kinetic energy of turbulence k and to its dissipation rate ε by :

$$\nu_t = C_\mu \frac{k^2}{\varepsilon}$$

k and ε are solution of the transport-diffusion equation.

Boundary conditions

At the inlet we generally give Dirichlet condition on all the velocity components in the rotating frame as well as for k and ε.

At the outlet, constrain conditions are introduced naturally through the variational formulation. For our calculation we generally use a constant normal constrain or a Neuman condition on normal and azimutal velocities .

For walls, even rotating ones, we use wall function according to the Reichard law , to avoid mesh refinements .

If the calculation domain is periodic (inter blade channel for example) we first construct a periodic meshing and apply the periodicity on all the variables of the problem

Numerical method

Standard Galerkin finite element method is used for space discretisation .P1isoP2 tetrahedrons with 10 nodes are also used. Linear interpolation is used for the pressure and linear interpolation on sub-elements for the velocity ,k and ε.

The characteristics method with a Runge Kutta integration is used to solve the advection step.

The diffusion step is solved using an Uzawa or a Chorin algorithm

3-DESIGN AN ANALYSIS APPLICATION TO THE TWO STAGE PUMP TURBINE

3-1 DESIGN OF A RETURN CHANNEL BLADE USING THE INVERSE APPROACH

The aim of the return channel is to supply the second runner in pump and turbine mode. This return channel included two kinds of blades: one is on the elbow and the second on the flat part. We present here the design of the second one. With the usual direct method, we could not find a shape satisfying at no flow separation and a good velocity profile for the second runner inlet in pump mode or for the other blade inlet in turbine mode. The inverse method permitted to avoid these problems.

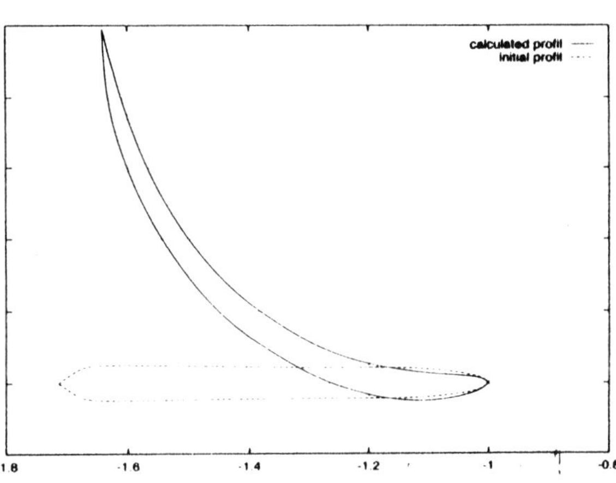

figure 2 : initial and calculated blade profil

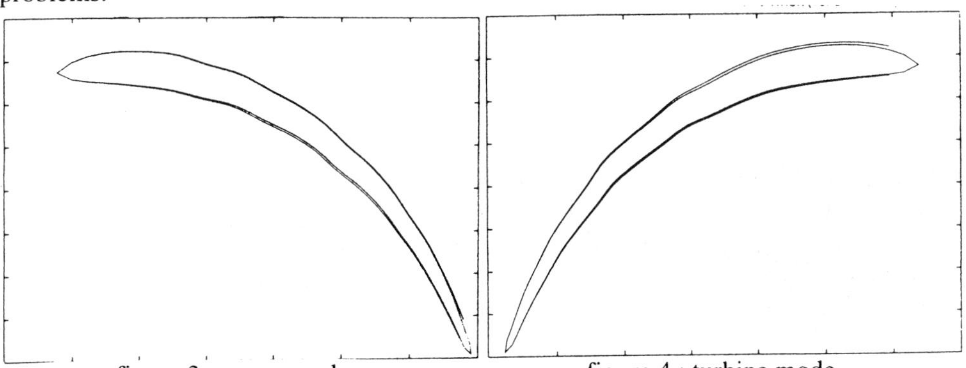

figure 3: pump mode figure 4 : turbine mode

Return channel : boundary layer comutation

The data were the velocities at the inlet and outlet, and the pressure difference between the pressure and the suction side of the profile .The profile azimuth thickness was also given.The obtained blade satisfied all the design criteria such as the boundary layer no separation and the velocity field at the outlet.
A boundary layer computation in pump mode and in turbine mode is shown on figure 3 and 4.

3-2 ANALYSIS OF THE FLOW IN THE GUIDE-VANES, STAY-VANES AND THE SPIRAL CASE IN PUMP RUNNING USING N3S

In this paragraph we present the results of an N3S computation in the domain constituted by the guide vanes , stay vanes and the spiral case of the two stage pump turbine.
The calculation was performed at the best delivery point Q=168l/s on a mesh of about 110 000 nodes .At the inlet (of guide vanes) we have imposed the axisymetrical velocity field delivered by a meridian calculation in an S1S2 computation of the runner (figure 5). The k-ε values were approximated to $V^2/100$ for k and $V^3/1000$ for ε .we have noted that the calculation was not sensible to these values . At the outlet we have imposed a zero normal stresses condition .

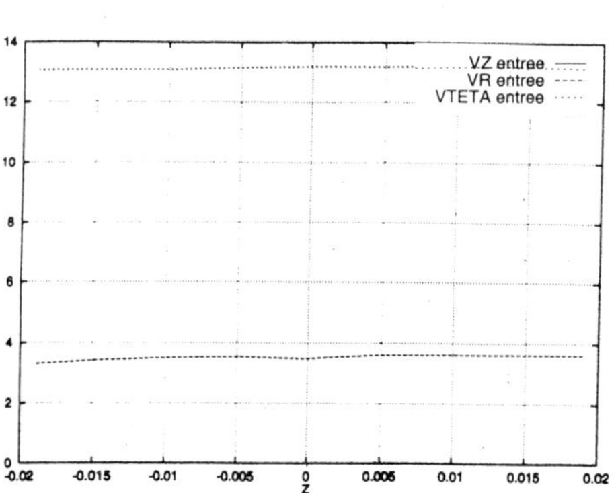

figure 5 : inlet boundary conditions

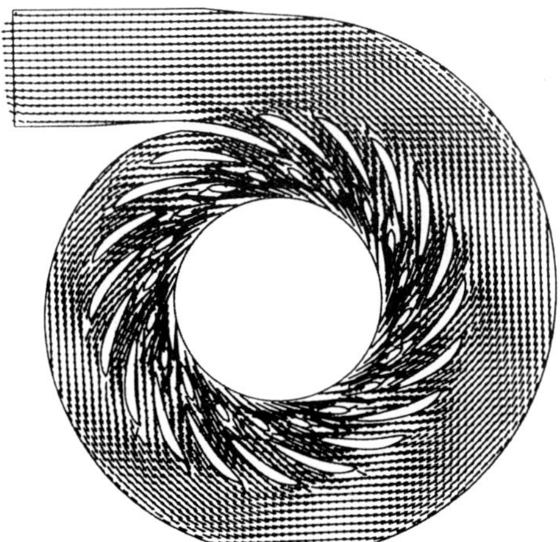
figure 6

On figure 6 we present the velocity field in the median plan. The computation shows that the delivery and the pressure are equi-distributed in the all the inter stay and guide vanes channels . And no recirculation have been detected .
The radial velocity field, presented on two plans (figure 7 and 8) (versus the azimuth θ= 180°, 360°) shows that :

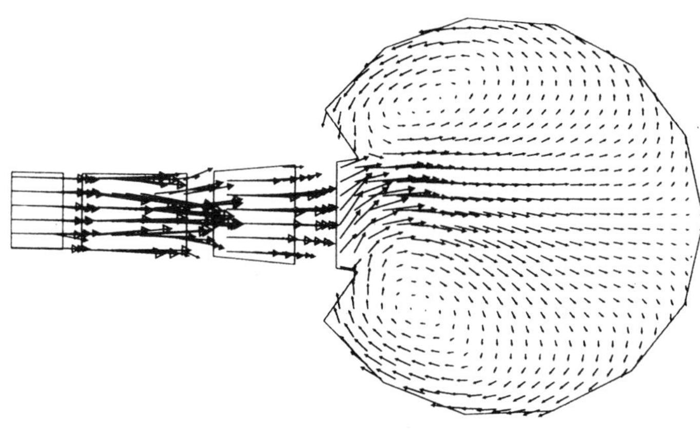

figure 7: θ = 180°

- issued from a quasi uniform radial velocity at the inlet , the radial flow in the stay vanes and in the guide vanes present an acceleration in the median region and a deceleration near the walls .
- the flow issued from the stay vanes generates two contra-rotative eddies in the spiral case

Besides the calculation of the total pressure at the inlet and outlet of the guide vanes stay vanes and the hole domain let us guess that the major part of the losses have been between the domain inlet and the stay vanes inlet .

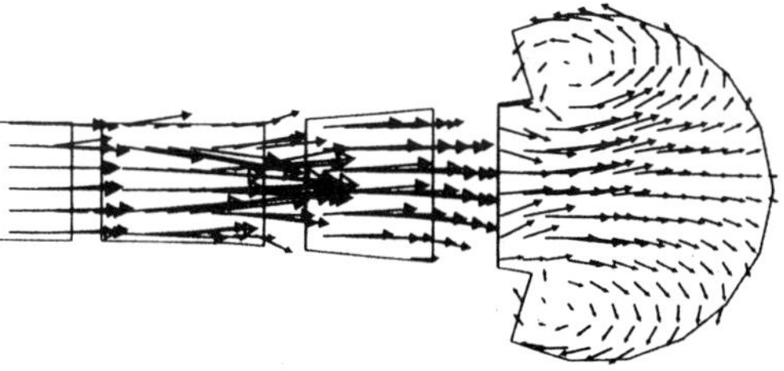

figure 8: θ = 360°

To confirm this analysis we have performed a computation with the same code in only one inter stay and guide vanes channel. The mesh was refined(60 000 nodes) and the same boundary condition were taken .

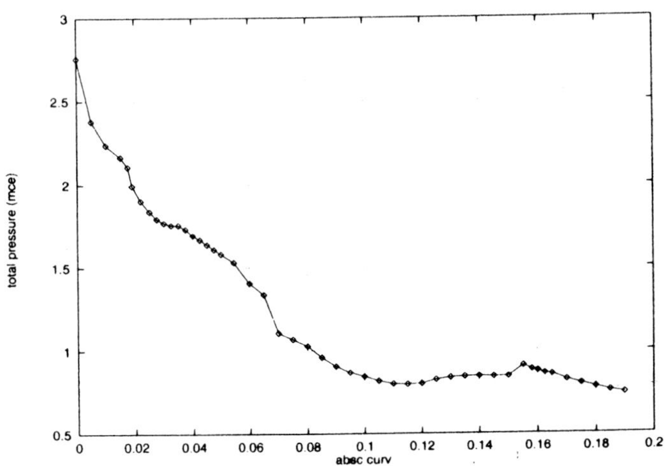

figure 9 :evolution of total pressure along stay vane and wicket gate

Figure 9 shows the total pressure evolution in the channel and confirm the assumption we have made while the hole domain computation. Besides, this run, with a refined mesh, shows not only a flow deceleration near the walls but also a recirculation zone in the channel .

3-3 ANALYSIS OF A DRAFT TUBE IN TURBINE RUNNING USING N3S

In a two-stages Pump Turbine, the draft tube is characterised by the presence of the machine shaft which crosses it, due to the necessary bearing under the lower runner.

The calculation of this draft tube is a Navier-Stokes calculation in turbine running. The geometry and the mesh are shown on figures 1 and 10.

The running conditions for calculation are corresponding at a partial load:

Q= 119.1 l/s H= 30 mwc (reduced model)

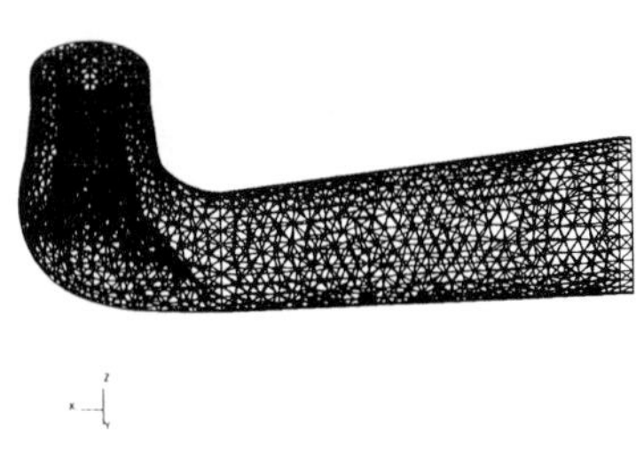

figure 10

Two calculations have been realised with two sets of boundary conditions at the inlet (figure 11), to study the influence of the velocities distribution under the runner.

- the first distribution is the same as that measured in the laboratory on an old reduced model (figure 12)
- the second is a theoretical ideal velocities distribution, which is aimed during the design of the runner (figure 13)

The boundary conditions at the outlet are a zero normal stresses conditions.

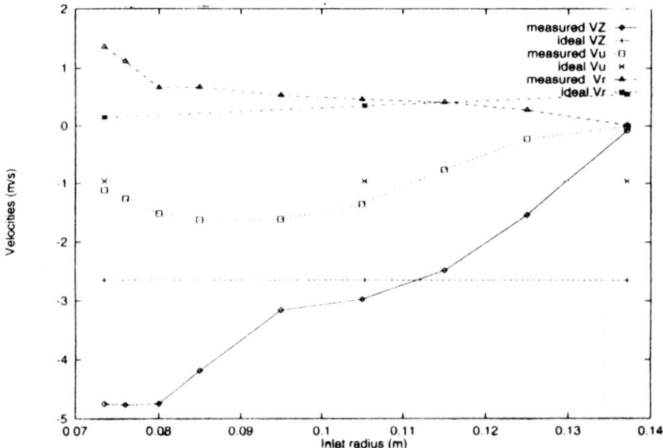

figure 11 : ideal and measured inlet velocity

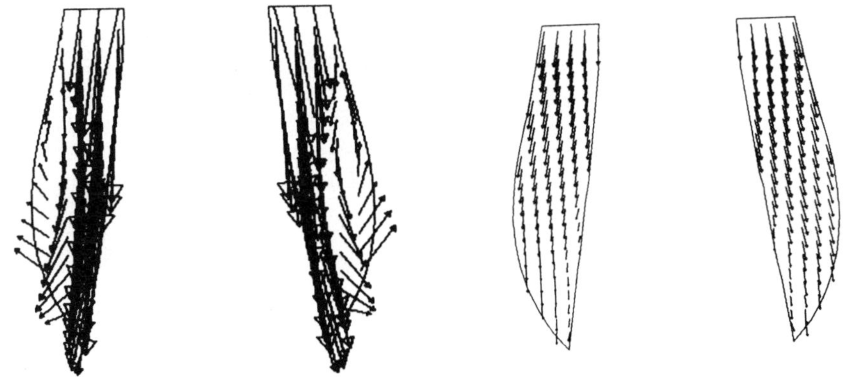

figure 12: Measured boundary conditions　　figure 13 : Ideal boundary conditions

Figures 12 and 13 are a comparison of velocities distribution in a vertical cut of the draft tube for the two inlet boundary conditions.

Figure 15 is a comparison of the losses curves for the two inlet boundary conditions.

When the velocities distribution at the inlet is that measured, the calculation head losses in the draft tube are $\Delta H = 2\%$ .The same measured head losses are $\Delta H = 2.21\%$. When the velocities distribution at the inlet is ideal, the head losses are $\Delta H = 0.57\%$.

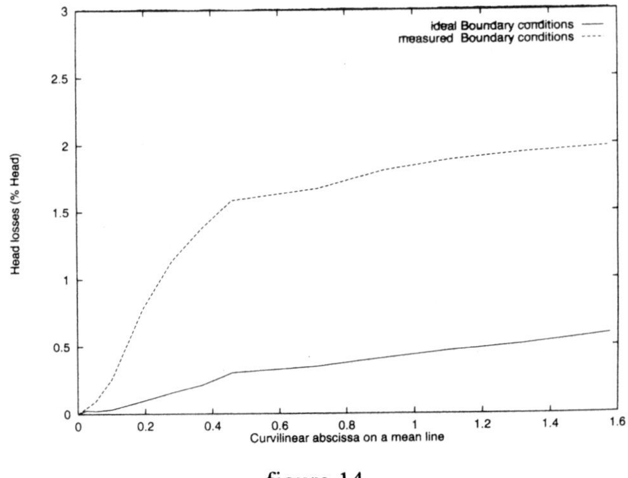

figure 14

We observe that the losses are concentrated in the elbow of the draft tube, due to the recirculation (figure 13), and calculation shows clearly the relation between the reduction of the losses and the velocities distribution along the elbow.
Thus, the runner design will respect at best the ideal velocity distribution at the inlet of the draft tube.

3.4 ANALYSIS OF RUNNER CALCULATION IN PUMP MODE

This calculation was realised with an unstructured mesh of about 40 000 nodes of speed.

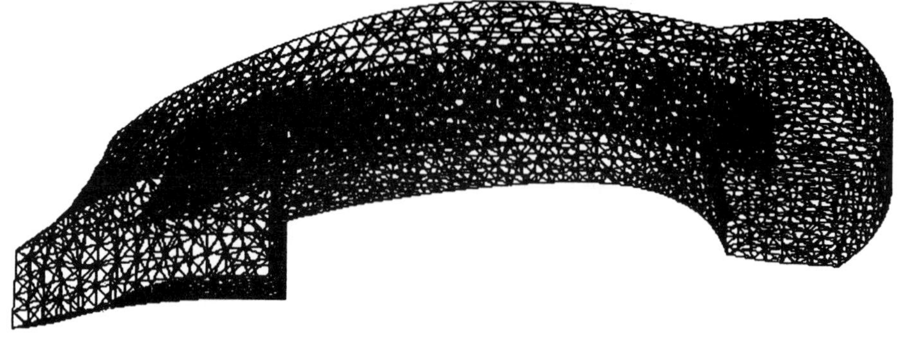

figure 15

The used boundary conditions are:
* At the inlet , an homogenous velocity field without rotational speed.
* At the oulet , a Neuman condition has been used
* In the rotating frame, an wall condition is imposed on blade, shadow and crown.
* Out of the runner , a wall condition is imposed in the fixed frame.

All the calculation is done with a speed of 1200 rpm for a runner diameter of 400 mm. Two discharges are calculated: Q= 168 l/s (nearly the maximum of the hill chart) and Q= 154 l/s.The figure 17 shows the comparison between the calculated values and the experimental' ones.

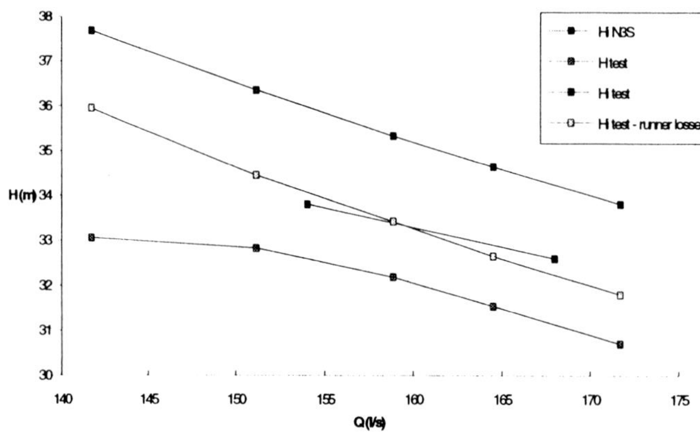

Hi test represents a measured value:

$$H_i test = \frac{N_a - N_{fr}}{\omega(Q+q)}$$

Na output measured on shaft
Nfr losses by friction
q : clearance discharge
Q : discharge of runner
ω : rotational speed

Hi N3S represents the intern head calculated of the runner:

$$H_i = \omega \frac{\Delta r V u}{g},$$

The value of Hi test - runner losses (measured) are to be compared with the calculated value of Hi . We find a good correlation between the two values.

The figure 18 shows the distribution of meridian velocity at the runner outlet which allow to better the flow in guide vane.

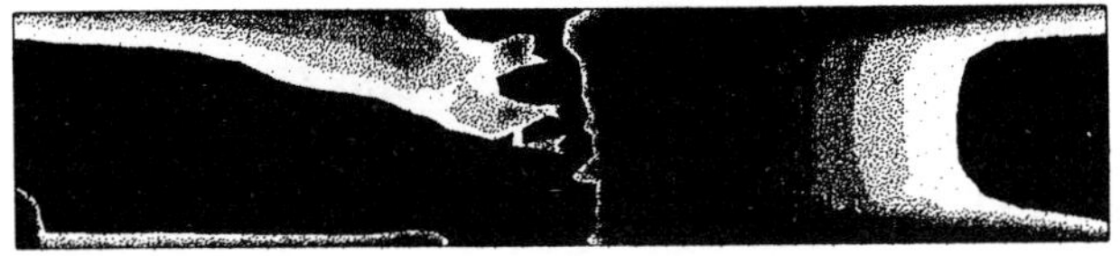

figure 18 : velocity at the runner outlet

CONCLUSION:

This paper deals with a design methodology in the specific case of a two-stage pump-turbine. The type of calculations used has been described for each component of the machine. Among the series of codes available, it is essential to choose the best suited tool for each shape to be generated. The model tests are still required as final confirmation of the design and the calculation / measure comparisons permitted to validate our approach.

REFERENCES

(1) Eremeef L.R, Philibert R :(1985) :*Modélisation quasi-tridimensionnelle des écoulements dans les turbomachines hydrauliques* ,La Houille Blanche n°7/8

(2) Francois M, Philibert R (1988) :*Ecoulements dans les baches spirales*, Symposium AIRH, TRONDHEIM

(3) Soares Gomes F, Kueny JL, Reynaud G, Combes JF (1992) :*Numerical analysis of three-dimensional flow in a pump turbine spiral casing*, Symposium AIRH, SAO POLO

(4) Caudiu E, Grimbert I, ElGhazzani EM, Verry A, Philibert R (1988) : *3D flow computation in turbomachinery* ,

STUDY ON HIGH SPEED AND HIGH HEAD REVERSIBLE PUMP –TURBINE

T.NAKAMURA *Toshiba Corporation 20-1,Kansei-cho,Tsurumi-ku, Yokohama 230* Japan

H.NISHIZAWA *Electric Power Development Co.,Ltd.* Japan

M.YASUDA *Electric Power Development Co.,Ltd.* Japan

T.SUZUKI *Toshiba Corporation* Japan

H.TANAKA *Toshiba Corporation* Japan

ABSTRACT

In this paper, improvement of hydraulic performance, particularly pressure fluctuation and cavitation, in order to speed up the runner revolution for a pump-turbine of 500m head class, is numerically and experimentally studied. High speed design of a pump-turbine offers an economical merit, because the size of the pump-turbine and generator-motor can be made smaller and the efficiency is improved. However, there are some problemes to be solved in order to develop a high speed pump-turbine. The first one is the increase of pressure fluctuation of runner outlet at partial load turbine operation. The second one is the deterioration of cavitation performance at pump operation. The third one is the increase of stress and vibration for runner. In order to overcome these problemes, the optimization of runner shape is carried out at design stage by using numerical simulation and its results are experimentally verified.

1 Introduction

Recently, the demand for larger capacity of pumped-storage power station is increasing because of expansion of power system and for inprovement of the plant economy. In the planning of a pumped-storage power station, higher head and larger unit capacity are selected usually to reduce the cost per capacity (cost/kW). But these involve some problems, for example, size limitation of

E. Cabrera et al. (eds.), Hydraulic Machinery and Cavitation, 210–219.

transportation, efficiency decline due to using low specific speed machines, etc. In Japan, several large scale pumped-storage projects are now being planned. Most of the projects are over 500m head and unit capacity is more than 300MW to reduce the size of reservoirs and machine cost. Advanced type high specific speed pump-turbines (Nsq= 40~45 m-m^3/s) are envisaged for a pumped-strage project, now Electric Power Development Co. is planning to make the machine size smaller and to achieve higher efficiency(Figure 1). There are some problems to be solved in order to adopt high specific speed machines. The first one is the increase of pressure fluctuation of runner outlet at partial load of turbine operation. The second one is the deterioration of cavitation performance in pump operation. These are attributed to the faster peripheral velocity of runner outlet and the larger absolute swirl velocity at runner outlet thereby in turbine operation, and to the lower static pressure of runner outlet in pump operation. The third one is the increase of stress and vibration of runner due to higher speed.

In order to overcome the above problems and to adopt higher rated speed, the optimization of runner shape is carried out at design stage by using numerical simulation and its results are experimentally verified. Further, for the strength of runner, FEM analysis and real head model test are carried out, then the stress level and resonance characteristic of the runner are verified.

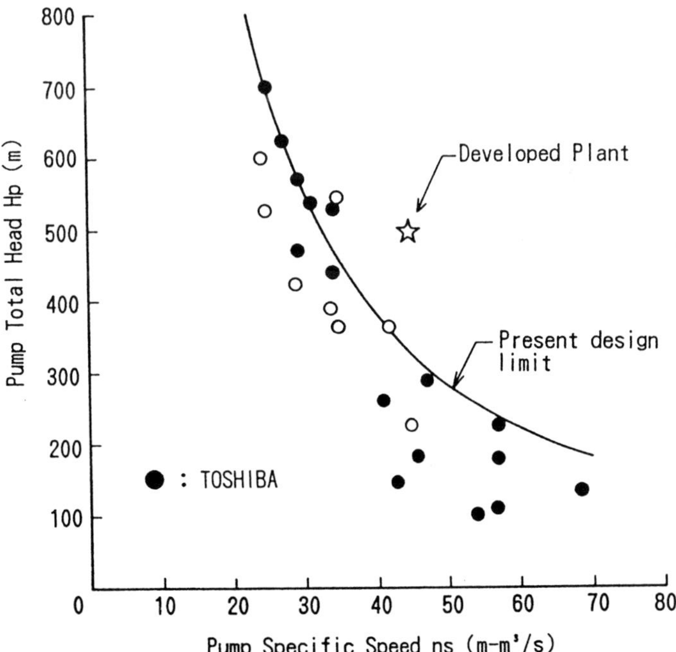

Fig. 1 Pump head versus spacific speed

2 Computational Method

2.1 Computational Grid

The three-dimensional Euler code with the pseudo-compressibility[1] is applied to flow simulation in pump-turbine runner. Figure 2 shows the solid-model and the grid in one runner blade channel of pump-turbine. In this analysis, in order to obtain flow distribution at runner outlet related to pressure fluctuation in the draft tube at partial load turbine operation, the simulation is carried out for the flow domain between runner inlet and conical part of draft tube. For the boundary surfaces other than runner blade surfaces, velocity and pressure are added periodically. The number of grid points is about 44,000.

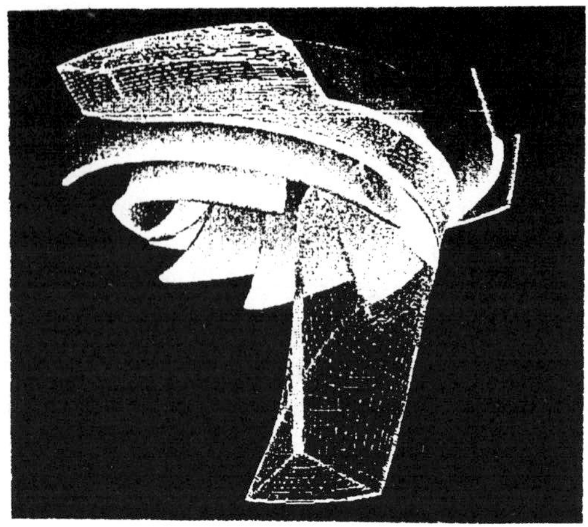

Fig. 2 Solid-model and Grid

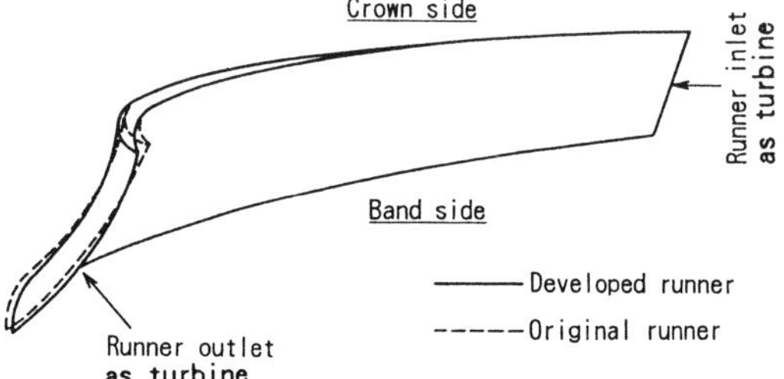

Fig. 3 Comparison of runner profiles

2.2 Calculated Condition

Flow analysis is carried out for model pump-turbine. In the case of turbine flow, calculated conditions are three cases about design point, maximum power point and partial load point. In the case of pump flow, calculated conditions are two cases about maximum and minimum head.

2.3 Geometry

Figure 3 shows the comparison of the profiles for the original runner(A) and the newly developed one(B). The modifications are limited to the area near the outlet of blades because the head-discharge curve and the optimum point must be kept as those of the original runner as possible. The purpose of the development of the new runner is improvement of cavitation characteristics in pump operation and that of pressure fluctuation in partial load turbine operation. In the course of this development, attention is paid so that no significant change in principal turbine and pump performance is involved to retain excellent efficiency characteristics of the original runner.

3 Computational Results

3.1 Cavitation Performance

Figures 4 and 5 show the pressure contours on the surface of blade at runner inlet for pump flow. The static pressure on blade surfaces which is obtained from the flow analysis is normalized by the pressure coefficient(Cp). Since it is thought that the cavitation inception occurs when the minimum static pressure (Hst) on blade surfaces becomes lower than the vapour pressure(Hv), the following relation is led :

$$\text{Cpmin} = \frac{(\text{Hv} + \text{Hsi} - \text{Ha})}{\text{H}} = -\sigma_i \qquad (1)$$

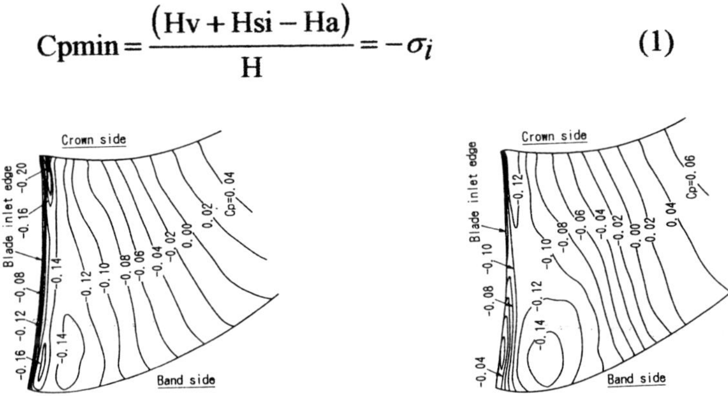

(a) Original runner A (b) Developed runner B

Fig. 4 Pressure coefficient contours on suction surface of blade at maximum head operation

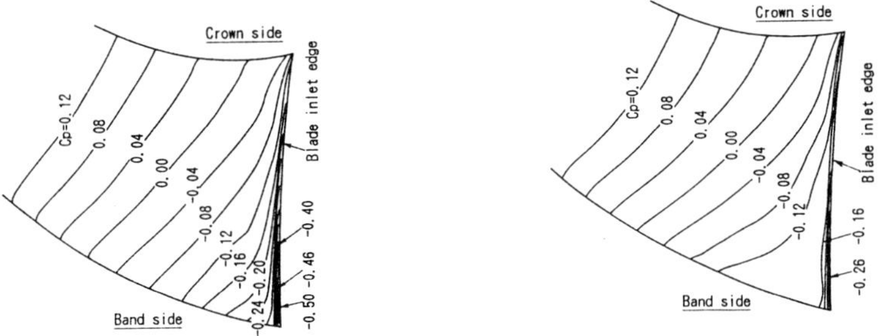

(a) Original runner A (b) Developed runner B

Fig. 5 Pressure coefficient contours on pressure surface of blade at minimum head operation

Therefore, the minimum pressure coefficient which is obtained from the flow analysis is compared by using eq.(1) with the inception sigma of cavitation obtained from the experiment.

For the maximum head operation, as shown in Figure 4, the pressure on the suction surface at runnner inlet edge becomes lower. As observed in these figures, the low pressure region of the developed runner(B) becomes smaller than that of the original runner(A). It can be seen that the developed runner(B) shows remarkable improvement in the pressure distribution at the crown side by the optimization of the runner profile. Also, in the minimum head operation, the improvement on pressure side surface is remarkable. In this case(Figure 5), the extre-mly low pressure region exists on pressure side surface of blades in the vicinity of the inlet near band, especialy for runner(A). It is recognized that the minimum pressure region of the modified runner(B) is reduced to narrow limited area. In addition, the minimum value of pressure coefficient Cp is considerably improved. As shown in Figure 6, the numerical results show very good agreement with experiments. It indicates that the numerical simulation technique presented in this paper can predict the cavitation characteristics of pump-turbines satisfactorily.

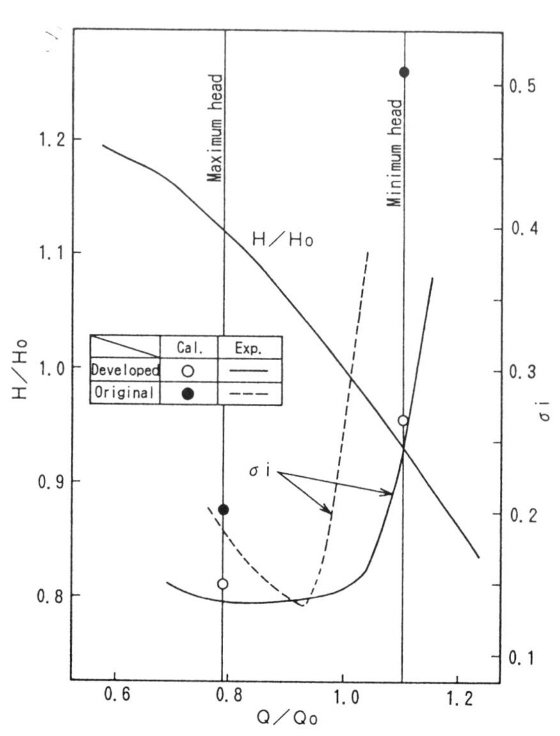

Fig.6 Comparison of cavitation performance

3.2 Flow Patterns and Velocity Distributions

When the runner revolution speed for a pump-turbine is increased, the increase of pressure fluctuation due to whirl in the draft tube at the partial load is concerned because of the faster peripheral velocity at runner outlet. Since this problem is attributed to the runner outlet flow and blade-to-blade flow, numerical simulation to investigate them is conducted. As an example, Figure 7 and Figure 8 show the numerical results of the flow patterns and velocity distributions at the partial load for the original runner(A) and the newly developed one(B) respectively. It is very different from each other. The flow in runner(A) is distorted, particularly the partial flow from crown side to band side is observed in the result. Then the streamlines near band is twisted. On the other hand, in case of the runner(B), the profile of runner outlet passage is rectified properly and the flow in it becomes much smoother than runner(A). The effect of this rectification is also recognized clearly in the velocity distributions at the runner outlet shown in Figure 8. Here, Ca and Cu mean the axial and absolute swirl velocity normalized by the mean axial velocity of runner outlet V0. This figure indicates that the swirl velocity coefficient Cu becomes smaller in case of the newly developed runner. It indicates that the intensity of the swirl is reduced. The decrease of swirl velocity is related to the decrease of rotating energy. Namely it suggests that the pressure fluctuation becomes smaller.

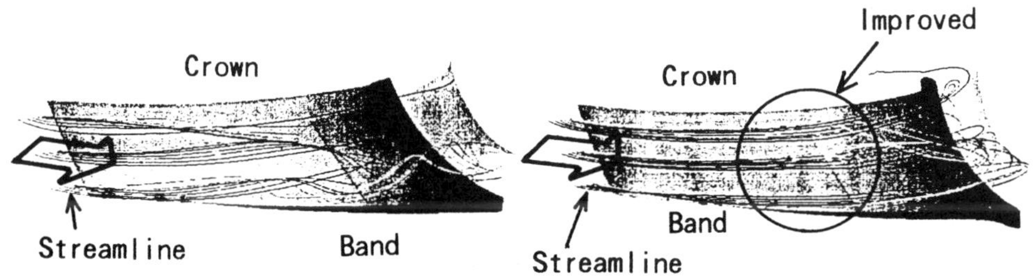

(a) Original runner A (b) Developed runner B

Fig. 7 Flow patterns in runner at 65% partial load

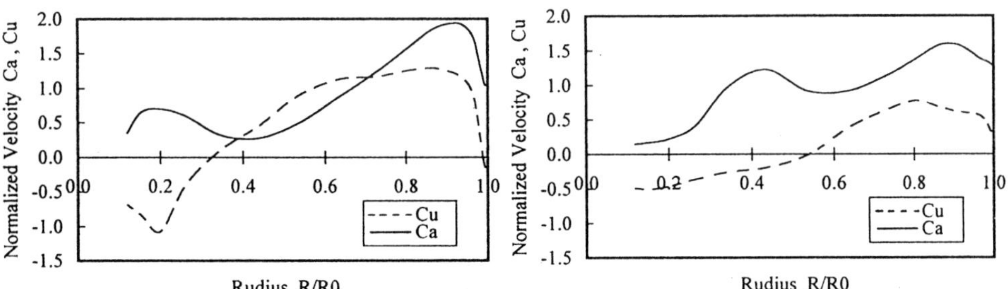

(a) Original runner A (b) Developed runner B

Fig. 8 Velocity distributions of runner outlet at 65% partial load

4 Pressure Fluctuation of Runner Outlet

Pressure fluctuation of the newly developed runner was measured by the model test. Figure 9 shows the comparison of the experimental results of pressure fluctuation for the pump-turbine of conventional design (specific speed Ns=34 min^{-1},m^3/s,m) and for the newly developed one (Ns=41). It is demonstrated from the figure that the pressure fluctuation of the developed runner is less than that of the conventional one.

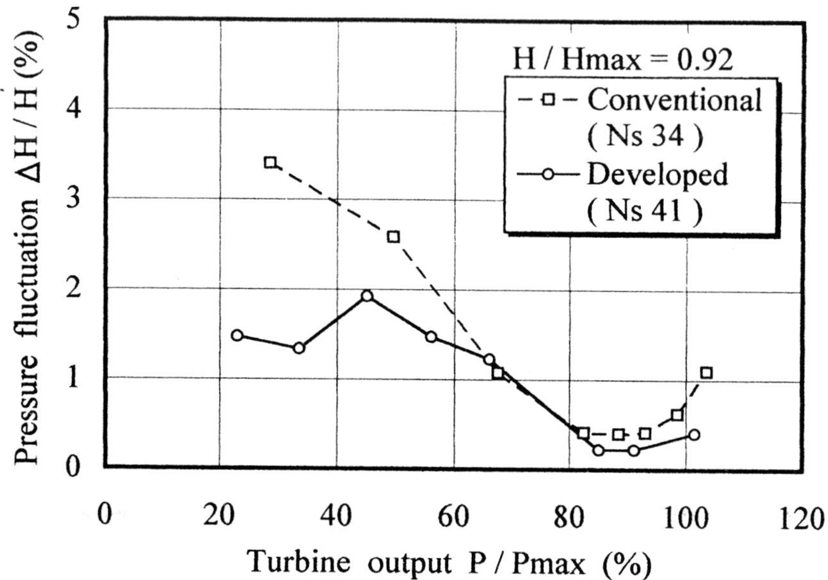

Fig. 9 Pressure fluctuation of draft tube in turbine operation

5 Stress and Vibration Analysis

When high specific speed design is adopted to increase the rotational speed of the runner, the width of a channel becomes relatively larger, so that the rigidity of runner and guide vanes may deteriorate. For static strength, FEM analysis was carried out and it was verified that the stress of the runner can be retained below the conventional stress level.

Regarding the dynamic strength, there has been no experimental result available for high head and high specific speed runner with larger height of inlet channel. Therefore, the experiments by means of real head model testing which conformed to hydro-elastic similitude were conducted to simulate the dynamic behaviour of the runner and to make clear the magnitude of the dynamic stress fluctuation.

5.1 Measurements

The real head model test arrangement consists of two model pump-turbines. The effective head for the model pump-turbine to be tested as turbine is provided by the other model pump-turbine operated as pump which is coupled to the end of a common shaft. The model has homologous passages and a runner made of the same material with prototype. Verious quantities(for example,stress and vibration,pressure fluctuation etc.)in model are related to those of the prototype by the hydro-elastic similitude law. The dynamic stress is measured by strain gauges stuck on the runner. The model speed is changed from 3,000min^{-1} to 6,600min^{-1}.

5.2 Results

During the 1980s, experimetal studies, as well as field measurements of the dynamic stress of runners, were conducted by Toshiba to explore the vibrational behaviour of high head pump-turbine runners. Detailed results of the studies were published in 1990 [2]. Similar to the past studies, investigation on vibrational behaviour of the high specific speed runner developed this time was conducted. An example of the mesurement results is shown in Figure 10 and Figure 11. Figure 10 shows the trend diagrams of frequency spectra with the change of test speed. It shows a dominant frequency corresponding to (NZg)Hz. Figure 11 shows a resonance curve obtained by the model test. From this figure, no significant resonance peak is observed in the characteristics of the developed runner. This means the runner vibration stress is sufficiently low, as long as it is used within the tested speed range. In order to confirm the strength against fatigue failure, the modified Goodman's diagram is applied to the assessment of the fatigue strength (Figure 12). It is indicated that the stress amplitude is sufficiently low, which reveals that the safety factor against fatigue strength is 9.2 .

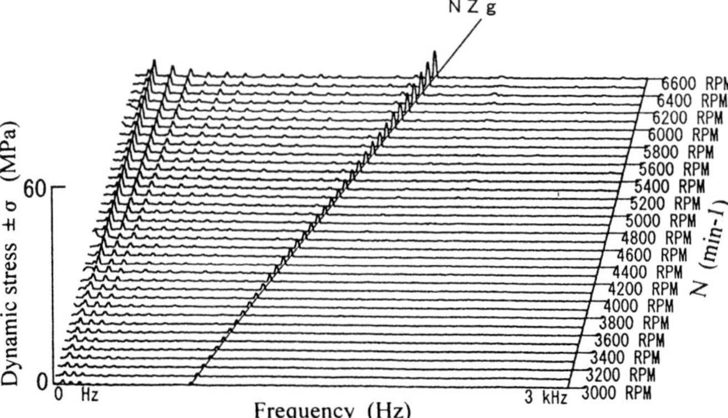

Fig. 10 Diagrams of frequency spectra

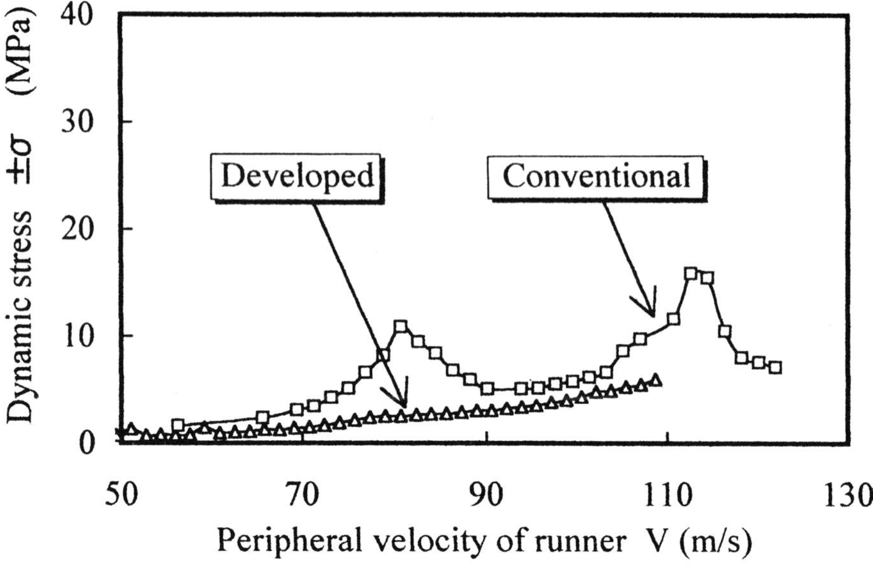

Fig. 11 Resonance characteristics of the runner

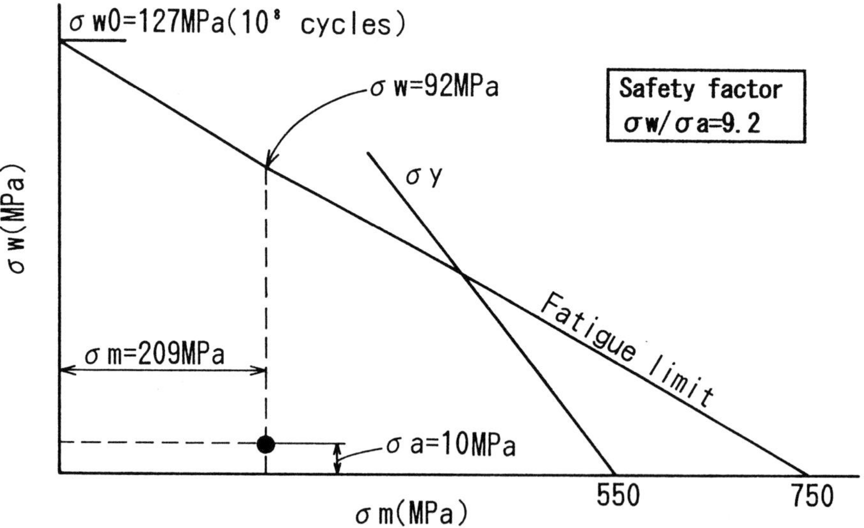

Fig. 12 Assessment of the dynamic stress of the runner using the modified Goodman's diagram

6 Conclusion

In order to accomplish the speeding up the runner revolution and reducing the size of pump-turbines of 500m head class, there are three subjects to be explored. The first one is the decraese of pressure fluctuation of runner outlet at partial load turbine operation. The second one is the improvement of cavitation performance at pump operation. The third one is the decrease of stress and vibration for runner.

This paper describes the summary of the studies conducted on the above subjects. The optimization of the design of the blade profile was carried out at the design stage by using flow analysis and FEM analysis and its effect was verified by model test. Regarding the strength of runner, a real head model test was conducted to verify the stress level. Finally, it was confirmed that the newly developed high speed pump-turbine of specific speed 41 can be applied to 500m class high head pumped storage project.

References

[1] Nagafuji.T., Uchida.K., Nakamura.T. 1995, "Numerical and experimental studies of three-demensional flow fields in hydraulic turbine runners", proc. of the 1995 ASME/JSME Fluids Engng. Conf., FED-Vol. 227, pp.37-44.

[2] Tanaka.H., "Vibration Behaviour and Dynamic Stress of Runners of Very High Head Reversible Pump-turbines", *Proceedings,* IAHR Belgrade, Yugoslavia; 1990.

ANALYSIS OF THE PERFORMANCE OF A BULB TURBINE USING 3-D VISCOUS NUMERICAL TECHNIQUES

Y. QIAN *Fuji Electric Co., Ltd.* *Japan*
R. SUZUKI *Fuji Electric Co., Ltd.* *Japan*
C. ARAKAWA *The University of Tokyo* *Japan*

Abstract

In this paper, a high-accuracy 3-D viscous numerical technique, which has been developed previously by the authors, is introduced to analyze the flows through a bulb turbine passing through the guide vanes, runner and draft tube. The techniques of predicting the interaction between the guide vanes(stationary) and runner vanes(rotating), the tip clearance, and a new turbulent model which was previously developed by the authors, are introduced to this analysis. In order to check the shortcomings of the k- ε two equations turbulent model, the limitation of the k- ε turbulent model in the application to vortex flows are discussed by various swirling rates in this paper. The purpose of this paper is to attempt to show that 3-D numerical techniques have been developed to analyze the flows through the all flow channels, guide vanes, runner, and draft tube. The flow patterns, separation phenomenon and performance of bulb turbine can be simulated depending on the characteristics of discharge rate and head only. The simulations are carried out under a series of discharge rate and rotating speed conditions. That is, when the runner opening is fixed and when the guide vane openings are changed to several opening angles to determine the On-cam operating conditions following the discharge rate and rotational rate.

1. Introduction

The authors have previously published several papers in analyzing the flow phenomenon through hydraulic turbine. Arakawa and Qian(1991) successfully developed a 3-D numerical code to simulate the flow through a hydraulic turbine. Qian(1994) proposed an interface method to deal with the interaction between the rotating block(runner) and stationary block (stay and guide vanes) in Francis type turbines. This code considers the influence of viscosity in some turbulent models and can be successfully applied to simulate the phenomenon of channel vortexes in Francis turbines. This interface method provides very good agreement in comparison with the experimental results. This interface technique is introduced in this paper to solve the flow associated with the rotor-stator vane configurations. Furthermore, Qian(1995) developed the code to analyze the flow in a bulb turbine runner. An embedded H-type mesh topology was successfully utilized to resolve the gap at the blade tip region. In order to simulate the turbulent flows in the gap, a new wall condition were proposed

E. Cabrera et al. (eds.), Hydraulic Machinery and Cavitation, 220–229.

to provide the turbulent parameters near the wall.

It is well known that the k- ε turbulent model is not suitable for predicting strong vortex flows as for example, discussed for basic confined swirling flows by Hogg and Leschziner(1989). The swirling flow in the draft tube is always one of the most formidable problems in hydraulic turbine design. Clearly, rationally controlling the swirling flow will greatly improve the loss creation and pressure recovery in the draft tube. Several research papers concerning draft tubes have been reported. One of the researches of this topic is Vu (1990), who comparatively provide complete numerical researches of hydraulic turbine draft tubes with viscous numerical techniques. However, these previous researches were limited in individual parts. In this paper, a highly-accurate 3-D numerical code is introduced to analyze the flow patterns in bulb turbines including all the flow channels of the guide vane, runner, and draft tube. The boundary flow conditions are obtained by exchanging the information substituting the results from each flow parts into the iterations of the others. The authors check the limitations of the k- ε turbulent model for draft tubes under several uniform swirling rates. The shortcomings of k- ε the turbulent model are found under strong swirling flows. This phenomena shows that the k- ε turbulent model has a limitations in the case of strong vortex flows.

2. Major symbols of the bulb turbine model

The major parameters of bulb turbines are specified by the following symbols.

n_{11} : Unit rotating speed
Q_{11} : Unit discharge rate
H : Effective head
D_1 : Runner diameter
U_1 : Runner peripheral speed
V : Flow velocity
C_i : Flow velocity coefficient
P : Pressure
Pref : Reference pressure
Cp : Pressure coefficient
Ci : Velocity coefficient
m : Swirling rate

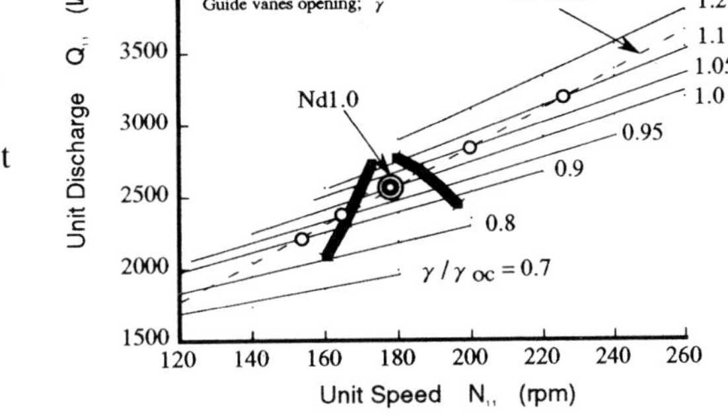

Figure 1. Discharge characteristics

3 . The geometry of the bulb turbine

The specifications of this bulb turbine under design operating conditions is presented in table 1. The computations are carried out under several operating condition shown in figure 1. The outline of this model turbine is shown in figure 2. The computational domain consists of the guide vanes, runner, and draft tube.

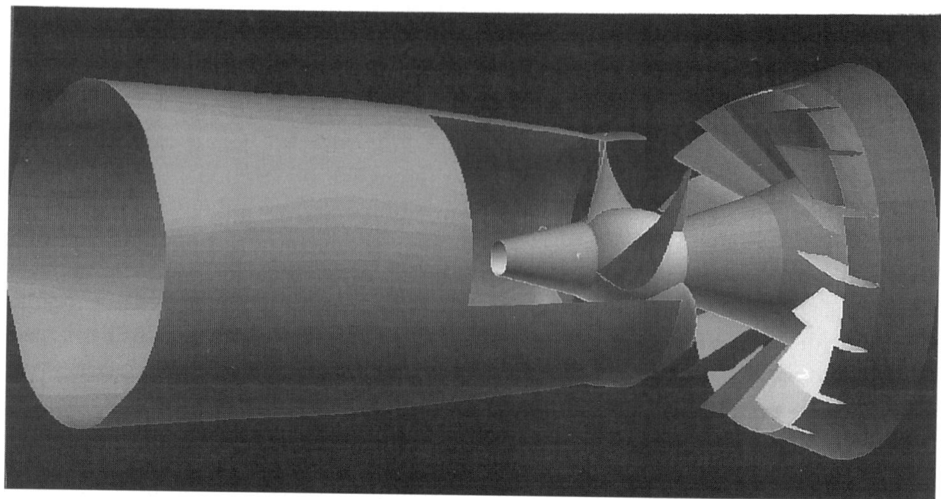

Figure 2. Outline of the bulb turbine

TABLE 1. Specification of the bulb turbine under design operating conditions

H(m)	D1(mm)	Z_R(blade)	Z_g(blade)	n_{11}(l/s)	Q_{11}(l/s)
12	500	4	16	150	1800

4. Block structured grid geometry and boundary conditions

The guide vanes consist of 16 blades and the runner consists of 4 blades. The guide vanes are set in an absolute coordinate system and the runner is set in a relative coordinate system under the constant rotational speed. The boundary conditions between the runner inlet and stator outlet are in the unsteady flow conditions. However, to overcome these difficulties, a mesh is used which is divided into three blocks, one that includes the runner, one that includes the guide vanes and another that includes the draft tube. The flow region associated with the runner and guide vanes can be described by patched grid blocks as proposed by Qian(1994). The inlet boundary conditions of the draft tube are obtained from the exit of the runner. The meshes chosen for the calculation are described with and H-type grid. The size of guide vanes grid is 70x21x21, runner is 70x28x26, and the draft tube is 41x21x21. All of the computations are carried out on an EWS computer.

5. Numerical method and turbulent model

The pseudo-compressibility idea is introduced into this research, which employs the numerical technique of compressible flows to solve hyperbolic-type partial differential equations. This means that the scheme conveniently covers both in incompressible flows and air turbines using the same algorithm. We take the so-called Beam-Warming scheme(1976) proposed for compressible flows, which is based on an implicit algorithm, such as an approximately factored method, to obtain a large time step. The upwind based highly-accurate TVD scheme(1985) is used without any artificial dissipation terms. The diagonalized ADI method is used to improve computational efficiency, and the operator is inverted by the forward and backward sweep techniques.

The standard k- ε turbulent model is employed to simulate the flow of a high Reynolds number. The wall function is used to decrease the number of grid points near the wall. A modified wall function is introduced to describe the turbulent boundary conditions on the points very close to the wall, as in the gap region of the runner.

6. Results

6.1 Guide vanes

To confirm the numerical techniques, the flows through the guide vanes (eliminate the runner) are computed independently. Figure 3 shows the velocity and pressure coefficients on the exit of the guide vanes. These results are in very good agreement with the experimental results. After this confirmation, the computations are carried out including the influence of the runner. Figure 4 shows the tangential velocity (V_θ)under two types of channel shapes from the guide vanes outlet to the runner inlet under the same operating parameters and same guide vanes and runner. The discrepancy between the two types of channels can be found by analysis of the results in figure 4. It is well known that the V_θ volumes in the span direction are one of the important conditions for the runner design. Hence, optimally designing the V_θ at the leading edge will not only approve the maximum efficiency, but will also extend the operating regions by suppressing cavitation. After optimally considering the attack angle of the runner, operating region, and cavitation, the best channel form is selected in this paper.

Figure 5 shows the guide vane pressure coefficients on the three sections from the boss side to the discharge ring side under one of the On-cam operating conditions, where sections 2, 5, 9 represent the section near the discharge, middle, and boss ring respectively. Figure 6 shows one of the On-cam velocity and pressure coefficients on the exit (as the inlet boundary condition of runner).

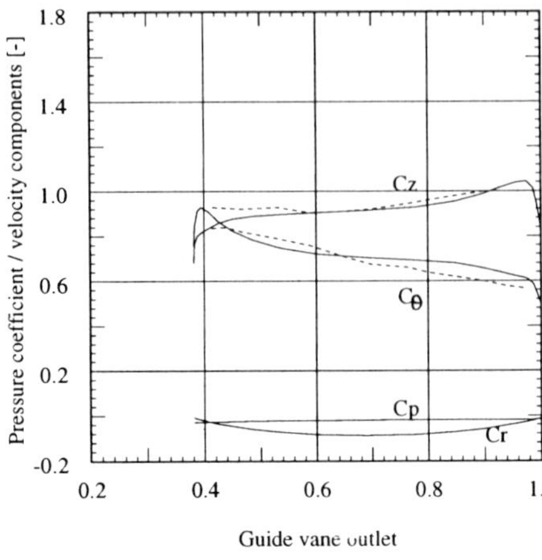

Figure 3. Velocity and pressure coefficient on the exit guide vanes
——: Computation, ······· : Experiment

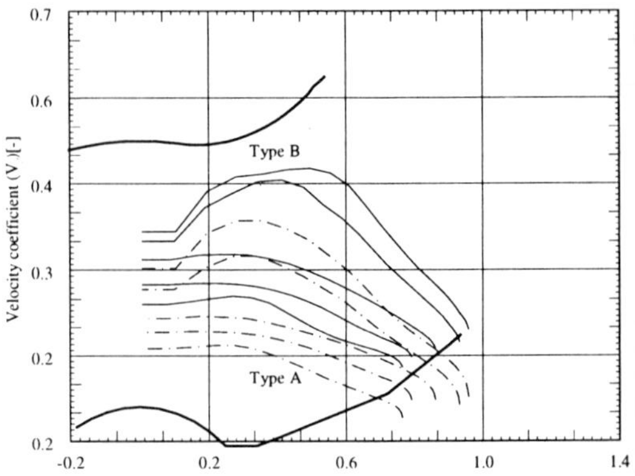

Figure 4. V_θ along the two types shape of channels

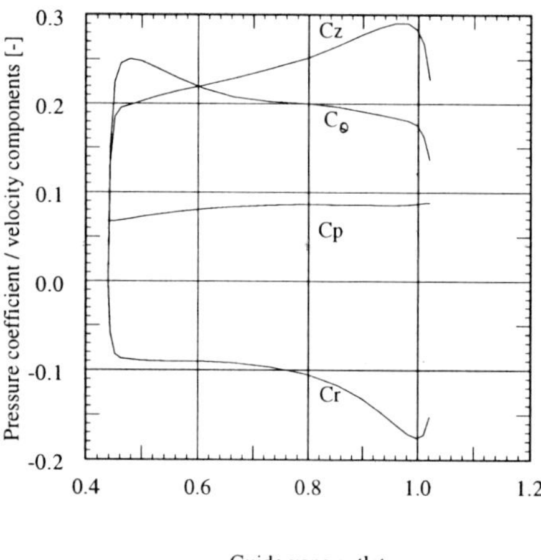

Figure 6. Velocity and pressure coefficient on the exit guide vanes

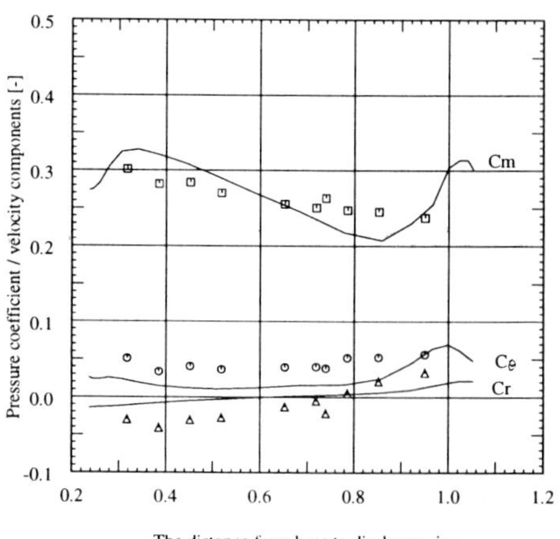

Figure 8. Velocity and pressure coefficient on the the runner exit. Symbols: Experiment

6.2 Runner

In the computation of the runner, the boundary conditions are obtained from the exit of the guide vanes at the initial iteration. After performing a back iteration and exchange on the overlapping region, final boundary conditions are computed in figure 6. A series of computations are carried out under various rotating speeds and discharge rates for each guide vane opening. The operating condition at the maximum efficiency is defined as the On-cam operating condition. Figure7 shows the pressure coefficients on

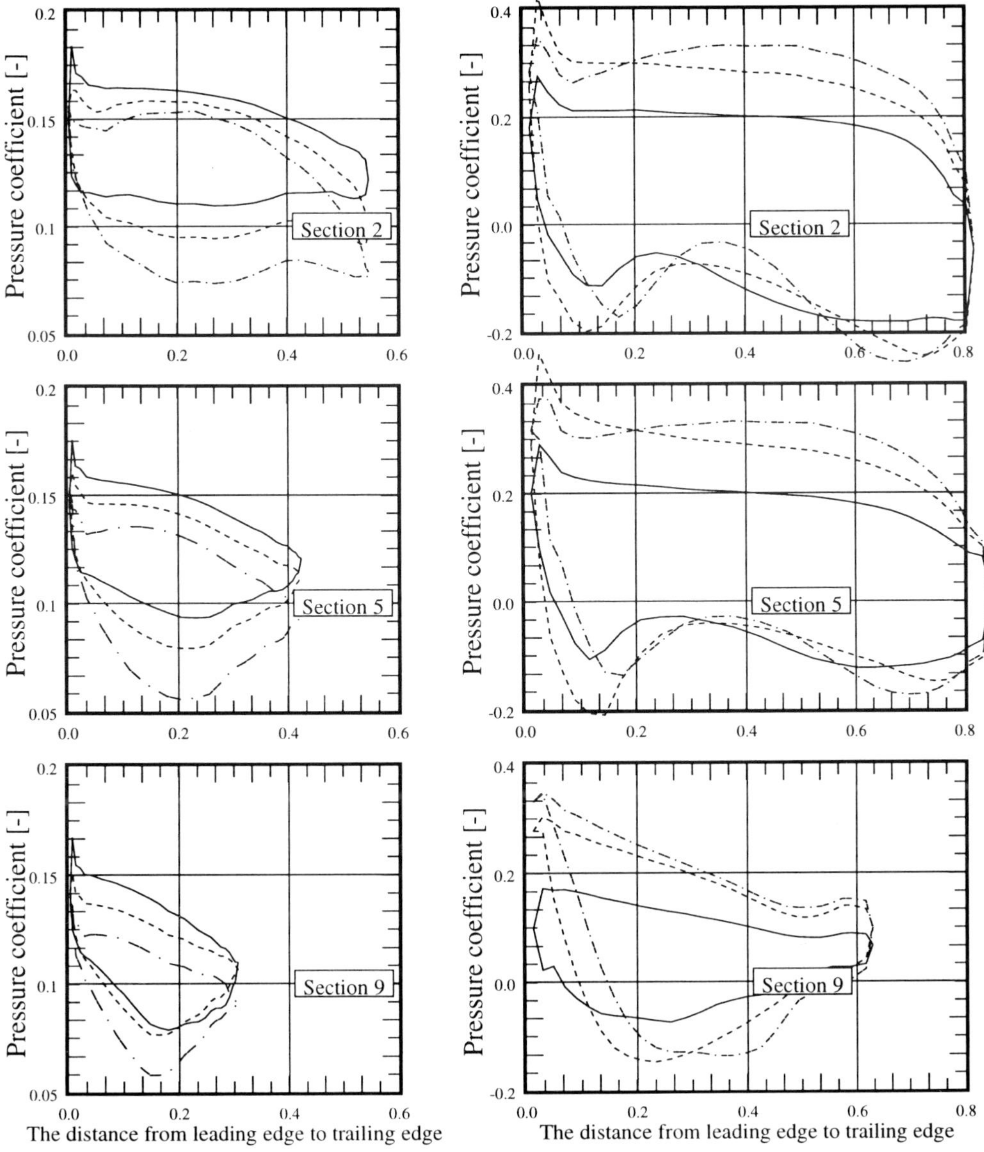

Figure 5 Pressure coefficient on the guide vanes ——— γ / γ_{opt}=1 -·-·-·- γ / γ_{opt}=1.1 ······ γ / γ_{opt}=1.2

Figure 7 Pressure coefficient on the runner ——— γ / γ_{opt}=1 -·-·-·- γ / γ_{opt}=1.1 ······ γ / γ_{opt}=1.2

the sections along the streamwise direction under three On-cam operating conditions. The velocity and pressure coefficients on the exit of the runner are shown in figure 8.

These exit results are selected as the inlet boundary conditions for the computation of the draft tube. A peak of C_m can be found near the discharge ring, This phenomena may be considered that the boundary layer becomes thicker along the blade tip from the leading edge to the trailing edge, so that the flow can easily pass through the gap near the exit of the runner as a wall jet, as observed in the pressure dumping near the trailing edge in the figure 7.

6.3 Draft tube

In order to check the limitations of k- ε turbulent model application for swirling flows, a series of uniform rotation velocities(ω x R) are introduced to the test. The swirling rates(= $\int \rho VzVudA$ / R $\int \rho Vz^2 dA$) are displayed in figure 9 under these 5 kinds of flow

patterns. The pressure recovery characteristics Cp_q and pressure coefficients P_q are shown in figure 10 and figure 11. Cp_q and P_q are defined as follows;

$$Cp_q = \frac{Po - Pq}{Ps}, \qquad P_q = \frac{Pqi - Po}{\rho \bar{u} / 2} \qquad (1)$$

where P_o is the static pressure of the exit, u is the axial velocity, and P_d $\left(= \frac{\int (\rho V^2) V_z dA}{Q}\right)$ is the dynamic pressure coefficient. The parameter of Cp_q is the average pressure-rise and P_q represents the relation of the transport energy from draft tube inlet to the outlet. When the swirling rate is larger than m=0.5, the computational iteration processing appears to not converge in the turbulent dissipation equation. However, the swirling rate under the On-cam operating conditions are almost always less than 0.3 as shown by the dotted line on the figure 9. Figure 12 shows the velocity vectors from inlet to outlet under one of the On-cam inlet boundary conditions obtained from the computation of the runner.

Following the successfully development of the above techniques for the bulb turbine design, the total performance of the bulb turbine can be analyzed the using the numerical tools. The flow patterns and performance of all features of the bulb turbine, such as the blade attack angles, loss phenomena, and cavitation, can be provided by computation before the experiment. One set of the analysis results is provided in this paper. Figure 13 shows the pressure contours and velocity vectors from the guide vanes to the runner under the design operating conditions, and figure 14 displays the cavitation coefficient at the inlet of the bulb turbine runner. A more detailed report will be presented in later research paper.

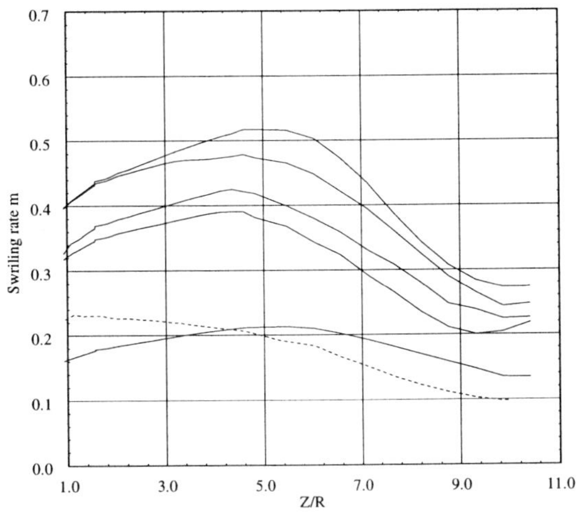

Figure 9 . Swirling rate in the draft tube

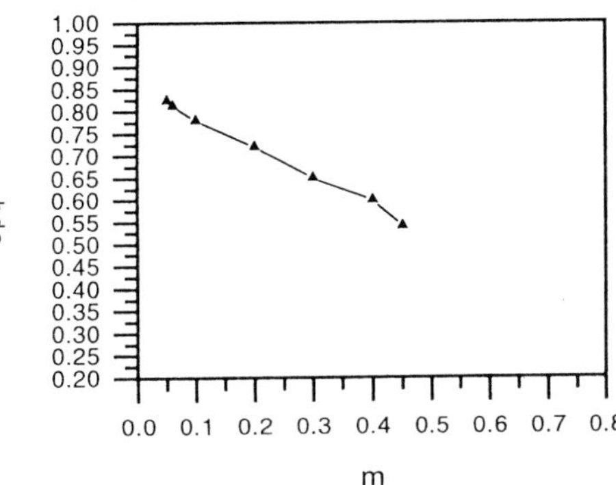

Figure 10. Pressure recovery under various swirling rate

7. Conclusion

High-accuracy 3-D viscous numerical techniques have been successfully introduced to analyze the performance of bulb turbines including the guide vanes, runner, and draft tube. In the simulation of guide vanes and runner, the authors select a series of different channel shapes to the optimum shape. The computations provide very good cavitation results at the leading edge of the runner because the influence of tip clearance is accurately considered in the numerical tools. These results show the operating regions of the bulb turbine and provide guidance to improve the cavitation phenomenon and extend the operating regions. The computations also present several flow patterns in the draft tube. The pressure recovery and pressure coefficient will be important data for draft tube design.

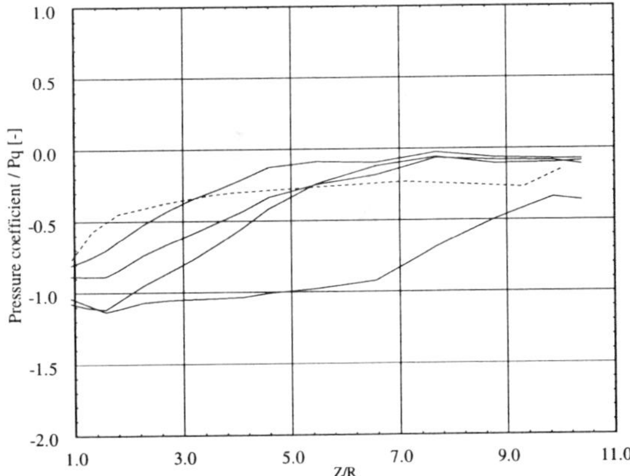

Figure 11. P_q under various swirling rate

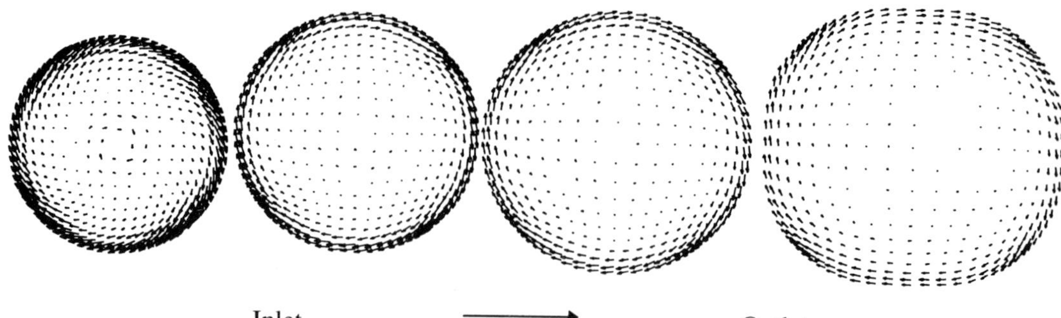

Figure 12. Velocity vectors on the sections from the inlet to the outlet of draft tube

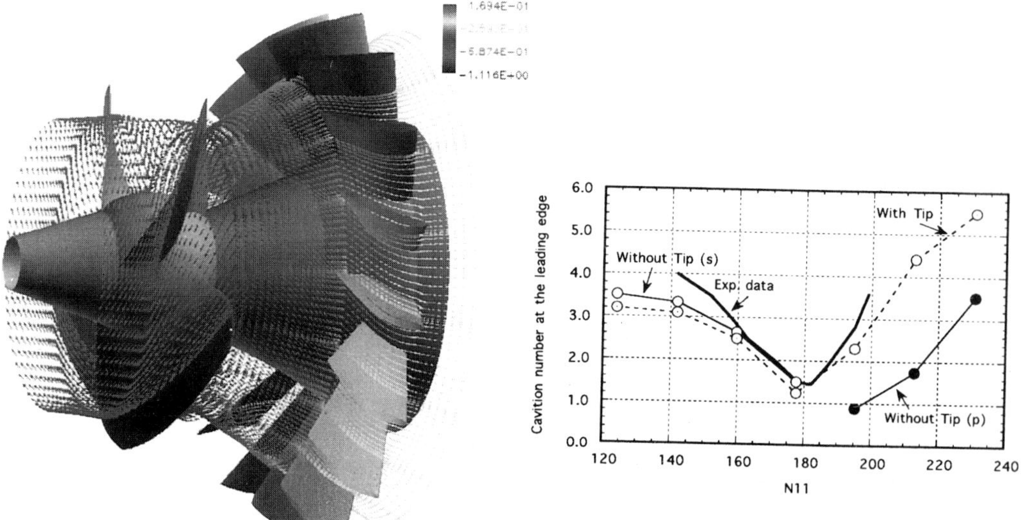

Figure 13. Pressure and velocity vectors

Figure 14 Cavitation parameters at the runner leading edge

As a consequence, the computational results can be obtained depending only on the characteristics of the head and discharge rate. The computational results provide a lot of important information to correct the initial design. These results have already been adopted as a part of our design routine. However, in order to verify the applicable regions of the computations, extended studies will be required. to improve the shortcomings of the turbulent model and the develop highly-accurate turbulent models. The rotor-stator interaction under Off-cam operating conditions will also have to be addressed in future work.

Reference

Arakawa, C., Qian, Y., and, Samejima, M.(1991) Turbulent flow Simulation of Francis Water Runner With Pseudo-Compressibility, *GAMM-Conference on Numerical Methods in Fluid Mechanics, 259-268*

Beam,R.F. and Warming, R.M.,(1976), An Implicit Finite-Difference Algorithm for Implicit Finite-Difference Hyperbolic System in Conservation Laws, Hyperbolic System in Conservation Laws, *Journal of Computational Physics, Vol.22,*

Chakravarthy,S.R. and Oshers,S.,(1985), An new Class of High Accuracy TVD Schemes for Hyperbolic Equations for Steady-State and Time-Dependent Program, *AIAA Paper 89-0463.*

Chorin,A.J., (1967) A Numerical Method for Solving Incompressible Viscous Flow Problems, *Journal of Computational Physics, Vol.2,pp.12-26*

Hogg, S. and Leschziner,M.(1989), Computation of Highly Swirling confined Performance Prediction of Mixed-Flow with a Reynolds Stress Turbulence Model, *Vol.27 No.1, pp.57-63*

Obayashi,S. and Fujii,K.,(1985), Computation of Three-Dimensional Viscous Transonic Flows with the LU Factored Scheme, *AIAA Paper 85-1510*

Qian,Y.,Arakawa,C.,and Kubota,T.,(1994), Numerical Flow Simulation on Channel Vortex in Francis Runner, *IAHR Symposium ,1994*

Qian, Y.,Arakawa,C.(1995), 3-D Numerical Analysis of Bulb Turbine Runner Performance with and Without Tip Clearance, FED-Vol.227, Numerical Simulation in Turbomachinery ASME 1995.

Vu, T.C., Shyy, W. (1990), Navier-Stokes Flow Analysis for Hydraulic Turbine Draft Tube, *ASME Journal of Fluid Engineering, pp. 199-204*

ANALYSIS OF THE INLET REVERSE FLOWS IN A PUMP TURBINE USING 3-D VISCOUS NUMERICAL TECHNIQUES

Y. QIAN	*Fuji Electric Co., Ltd.*	*Japan*
R. SUZUKI	*Fuji Electric Co., Ltd.*	*Japan*
C. ARAKAWA	*The University of Tokyo*	*Japan*

Abstract

The purpose of this paper is to apply advanced technology to predict the phenomenon of reverse flows created at the inlet during pump operation of Francis type pump-turbines. The authors introduce 3-D viscous numerical techniques, which were developed for hydro-turbine runners several years previous, to predict and analyze the recirculation flows at the inlet during pump operation. The motions of the separation flows are predicted by reducing the discharge rate from design to off-design operation conditions. In order to accurately capture the strong vortex, a new boundary condition of turbulent parameters on the wall is introduced in this simulation. The computational results are found to be in reasonable agreement with the experimental results. The limitations of the k- ε two equations turbulent model in the application of vortex flows are discussed in this paper. Finally the authors present a basic scale for use with Navier-Stokes simulation techniques with k- ε two equations model when using the model as design tool in the case of reverse flows.

1. Introduction

Many researchers have published numerous papers on the reverse flow phenomena using measurement techniques. In a series of combined analytical and experimental researches, Kurokawa(1994) provided experimental results of reverse flow patterns over a wide discharge range. These experimental results provided a detailed description and became well established as research and design tools in industry. On the other hand, following the development of computational hardware and Computational Fluid Dynamics(CFD), CFD techniques have been widely used to simulate the flow patterns and predict the performance of hydraulic turbines and are known as "numerical experiment". Qian and Arakawa(1991) successfully developed a 3-D numerical code to simulate the flow through a hydraulic turbine. The code considers the influence of viscosity with some turbulent models and can be successfully applied to simulate the phenomenon of the channel vortex in Francis turbines(1994). This work is a attempt to apply this code to predict the inlet reverse flows of pump turbines during pump operation.

E. Cabrera et al. (eds.), Hydraulic Machinery and Cavitation, 230–237.

It has been made known that the k- ε turbulent model is not suitable for predicting strong vortex flows, for example, discuss for basic confined swirling flows by Hogg and Leschziner(1989). Qian(1996) checked the difference vortex flows through the draft tube against the k- ε turbulent model simulation. The shortcomings of the k- ε turbulent model are found in strong swirling flow. When the swirl rate 'm' is increased to greater than 0.5, the computational iterations can be diverged to relational results comparable with the experiments. This phenomena confirms the fact that the k- ε turbulent model have a limitation in the case of strong vortex flows. However, it is well known that if higher level turbulent models are selected, such as Rynolds stress model and LES model, longer computational time are needed. At present, it is very difficult for industrial design to simulate the flow patterns, due to the huge CPU time required. Therefore, in this paper, the k- ε model is still selected to model the eddy viscosity in the Reynolds average Navier-Stokes equations, even though its shortcomings have been made known. According to the above reason, Qian(1995) proposed a new ideal model to modify the turbulent parameters near the wall in the k- ε model. This new method is also introduced in this paper. The simulation results can be used with confidence as a design tool, and the process of the iterations converge smoothly despite the instability of the vortex progress during the iteration.

2. Major symbols of the pump turbine model

The major parameters of pump-turbines are specified by the following symbols.

n	: Rotating speed
Q	: Discharge rate
H	: Effective head
D_1	: Runner diameter
U_1	: Runner peripheral speed = $\pi D_1 n/60$
C_i	: Flow velocity coefficient = V_i/U_1
P	: Pressure
Pref	: Reference pressure
Cp	: Pressure coefficient = $2(P\text{-}Pref)/\rho U_1^2$

3 . The geometry of the pump turbine

The pump-turbine model runner analyzed is one of the test models being researched and developed by the authors. The specifications of this pump-turbine are presented in table 1, and the outline of this model runner is shown in figure 1. Several new hydraulic design ideas are introduced in this series pump turbines, such as have 9 blades. As one of the most important tools, numerical analysis techniques predict the performance before the model test and support the new model design.

TABLE 1 Specifications of the pump-turbine

H(m)	D1(mm)	Z_R(blade)	n(r/min)	Qd(l/s)
60	500	9	1200	400

Figure 1 . Outline of pump-turbine runner

4. Numerical method and boundary conditions

Pseudo-compressibility was adopted as in previous papers (Arakawa, 1991,1994). The Navier-Stokes equation is transformed into the linear Delta form using the Beam-Warming method(1976), and moreover the TVD scheme (Chakravarth, 1985) based on the upwind and approximately factored method (Obayashi, 1985) was used. The turbulent model that the authors use for the simulation is the standard k- ε 2 equations turbulent model with wall function condition and a modified new wall boundary condition proposed by Qian(1995).

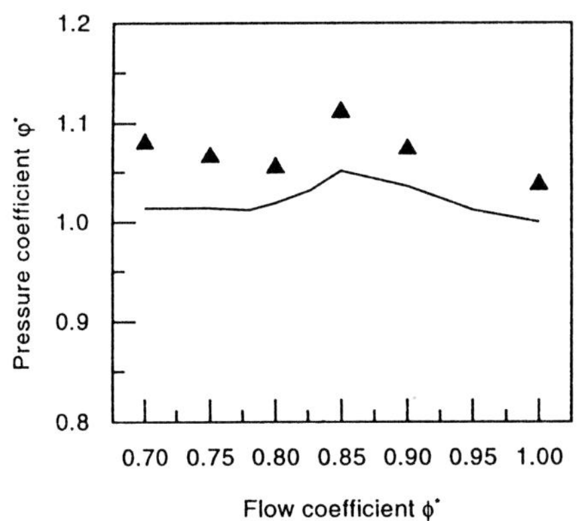

Figure 2. Characteristics of pump

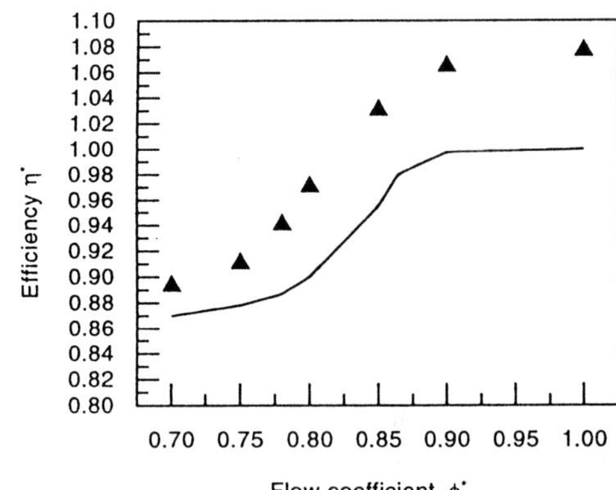

Figure 3 . The efficiency of pump

The computational domain includes the runner and the prolonged parts at the inlet and outlet of the runner. The inlet boundary conditions are given by the discharge rates. For the boundary at the exit, the differentials of velocity components in the streamwise direction are set at 0 for the initial iteration, and then a non-reflective boundary condition proposed by Dulikravich(1993) is introduced and the velocities are corrected in order to conserve the mass flow rate to be the same as the inlet discharge. On the crown, band and blade surfaces, impermeable conditions are specifying. The tangency condition is enforced by specified the contravariant velocity W=0 on the crown and band surfaces, and V=0 on the blade surfaces. The contravariant velocities and their pressures are obtained by linear extrapolation from the interior points when required.

5. Analysis results and performance characteristics of the pump turbine

The analyses are carried out under the design discharge(Qd) and partial discharge at 90%(0.9Qd), 85%(0.85Qd), 80%(0.8Qd), 75%(0.75Qd) and 70%(0.7Qd) of the design discharge. The simulation results are presented in the related forms of the flow coefficient ($\phi = \frac{Q}{nD_1}$) and pressure coefficient ($\psi = \frac{P - \mathrm{Pr}ef}{\frac{1}{2}\rho U_1^2}$) as shown in equation 1.

$$\phi^* = \phi / \phi_{opt}$$
$$\psi^* = \psi / \psi_{opt} \qquad (1)$$

ϕ_{opt}, ψ_{opt} are the experimental flow coefficient and pressure coefficient under the highest efficiency. The numerical results(symbols) of ϕ^*, ψ^* are shown comparison with experimental results in figure 2. When ψ^* rises to the peak of ϕ^*, ψ^* distribution, the ψ^* tends to become less against the decrease of ϕ^*. The decrease in experimental ψ^* starts after $\phi^* < 0.85$ and that of the computational ψ^* starts after $\phi^* < 0.80$. According to the visualization of the experiment, the separation flows near the inlet of the pump are first appear close to this operation condition. The efficiency presented in this paper is defined by the authors as equation 2.

$$\eta_c^* = \eta_c / \eta_{opt}$$
$$\eta_e^* = \eta_e / \eta_{opt} \qquad (2)$$

Where, η_c, η_e, η_{opt} are the computational efficiency, experimental efficiency, and highest experimental efficiency in the experiment respectively. Figure 3 shows the results of the efficiency. The efficiency reduction is observed against ϕ^*. The numerical solutions predict a larger efficiency when compared with the experiments. The computational efficiency is higher than the experimental results, because the loss given by computation are only taken from the runner, whereas the measurements not only cover the rotating part but also include the losses created in the stator such as guide vanes and casing. However, the discrepancy between the two results are almost constant following the different ϕ^*.

Figure 4 shows the results of the absolute velocity and pressure distributions at the inlet of the pump under design operation conditions. C_θ, C_r, C_z are respectively the tangential, centrifugal, and axial velocities averaged between the rotating blades in the circumferential direction, and C_p is the pressure coefficient, all of which are made non-dimensional using the spouting velocity corresponding to the net head $\sqrt{2gH}$. Figure 5 shows the results of the velocity and pressure coefficients, which are defined in figure 4 under the partial discharge Q=0.7Qd. The axial velocity created a negative volume near the band side and the tangential velocity rise at the same region. The results show that the reverse flows are created near the band side of the pump inlet when the discharge is reduced to less than 75% of the design discharge. The reverse flow separation point can be accurately obtained by data analysis. On the other hand, the reverse flow patterns may be observed by computational graphic visualization on the computer display. Some of the results are shown in figures 6,7. Figure 6 shows the pressure contours of Q=Qd, where figure 7 displays the results under Q=0.7Qd. Figures 8, and 9 show the velocity volume along the streamline from the inlet to the outlet of the runner at the Qd and 0.7Qd operation conditions, and the curve numbers represent the sections from the crown side to the band side. This result clearly shows that a vortex exists at the inlet near the band and extends a long way in the downstream direction along the band surface.

In order to observation the reverse flow pattern in more detail, several velocity vectors on side sections are shown in figure 10. When the discharge is reduced to less than 75%Qd, a reverse flow is created at the inlet and the reverse region enlarges from the suction surface to the pressure surface of runner blades.

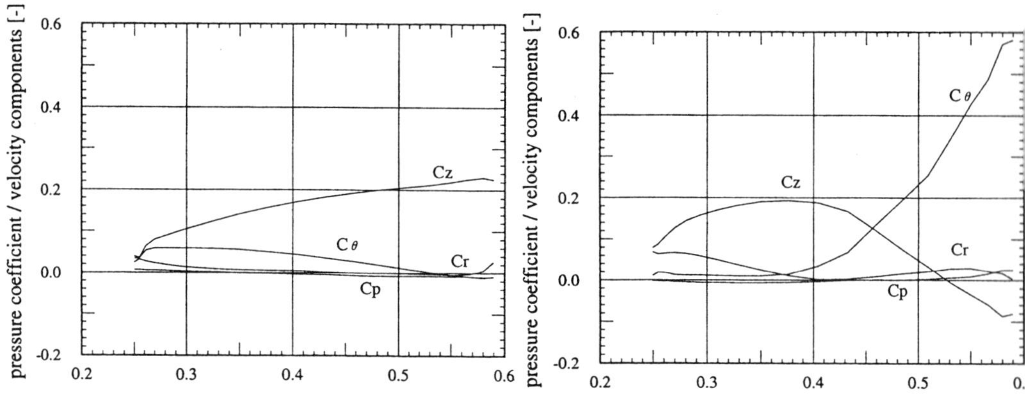

Figure 4. Velocity and pressure coefficient under Q=Qd

Figure 5. Velocity and pressure coefficient under Q=0.7Qd

Figure 6. Pressure contours under Q=Qd

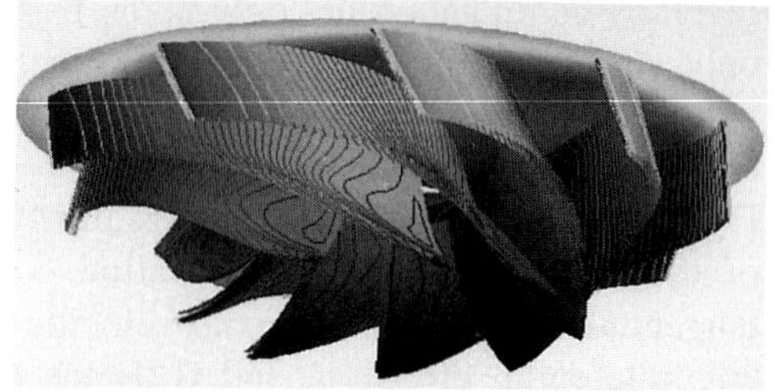

Figure 7. Pressure contours under Q=0.7Qd

Figure 11 shows the pressure coefficient distributions along three chord-wise sections of a runner at several discharge rates, while the 2,5,9 sections present the sections near the crown, in the middle and the band respectively. A negative pressure peak can be found at the leading edge on the suction surface near the band side when the discharge rate is reduced to less than 0.75Qd.

6. Conclusion

This paper reports an attempt to analyze and observe the pump reverse flow patterns using numerical techniques. The recriculation flow phenomena, reverse flow starting conditions, and the characteristics of the pump-turbine have been predicted in the paper. These results have already been adopted as a part of the design routine for daily use after introduction connection scales.

The authors believe this paper is only a first report for the analysis of pump-turbines. In the near future, a further series of analyses will be reported. Several points will be studied, for example, introducing a conservative algorithm to handle the unstable flows, computing the flow including the guide vanes, and developing highly-accurate turbulent models which can be computed on small scale computers such as desk-top workstations.

Reference

Arakawa, C., Qian, Y., and, Samejima, M.(1991) Turbulent flow Simulation of Francis Water Runner With Pseudo-Compressibility, *GAMM-Conference on Numerical Methods in Fluid Mechanics, 259-268*

Beam,R.F. and Warming, R.M.,(1976), An Implicit Finite-Difference Algorithm for Implicit Finite-Difference Hyperbolic System in Conservation Laws, Hyperbolic System in Conservation Laws, *Journal of Computational Physics, Vol.22,*

Chakravarthy,S.R. and Oshers,S.,(1985), An new Class of High Accuracy TVD Schemes for Hyperbolic Equations for Steady-State and Time-Dependent Program, *AIAA Paper 89-0463.*

Chorin,A.J., (1967) A Numerical Method for Solving Incompressible Viscous Flow Problems, *Journal of Computational Physics, Vol.2,pp.12-26*

Dulikravich, G. Ahuja, V. and Lee, S. (1993), Three-Dimensional Solidfication With Magnetic Fields and Reduced Gravity, *31st Aerospace Sciences Meeting & Exhibt, AIAA 93-0912.*

Hogg, S. and Leschziner,M.(1989), Computation of Highly Swirling confined Performance Prediction of Mixed-Flow with a Reynolds Stress Turbulence Model, *Vol.27 No.1, pp.57-63*

Kurokawa, J., Kitahora,T. and Jiang, J.(1994), Pumps Using Inlet Reverse Flow Model, *IAHR Symposium, No. A13,1994*

Obayashi,S. and Fujii,K.,(1985), Computation of Three-Dimensional Viscous Transonic Flows with the LU Factored Scheme, *AIAA Paper 85-1510*

Qian,Y.,Arakawa,C.,and Kubota,T.,(1994), Numerical Flow Simulation on Channel Vortex in Francis Runner, IAHR Symposium ,1994

Qian, Y.,Arakawa,C.(1995), 3-D Numerical Analysis of Bulb Turbine Runner Performance with and Without Tip Clearance, FED-Vol.227, Numerical Simulation in Turbomachinery ASME 1995.

Qian, Y., Suzuki, R., Arakawa, C., (1996) Analysis of the Performance of Bulb Turbine Using 3-D Viscous Numerical Techniques, *IAHR, Symposium, 1996.*

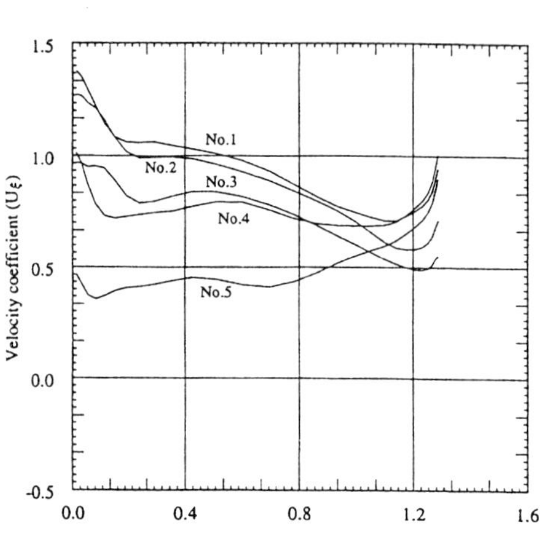

Figure 8. Axial velocity from inlet to outlet under Q=Qd

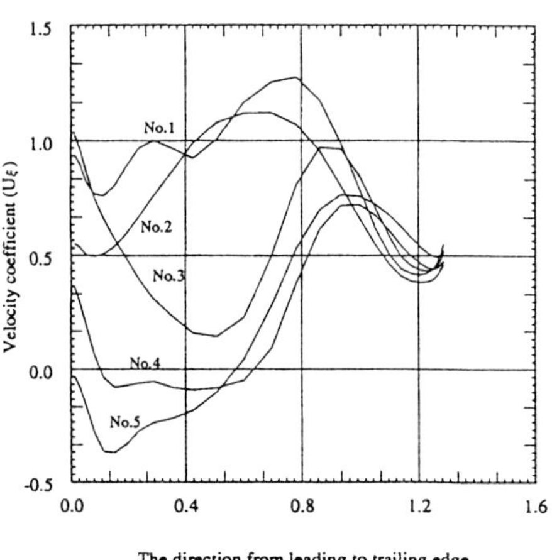

Figure 9. Axial velocity from inlet to outlet under Q=0.7Qd

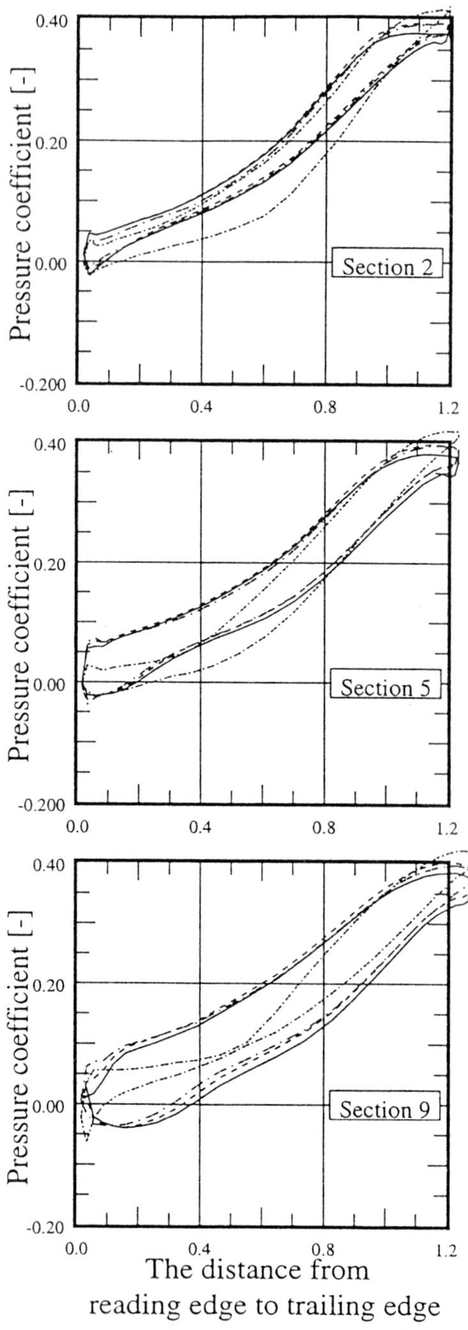

Figure 11. Pressure coefficient along the streamwise direction. ——— :Qd, ········ :0.9Qd, –·–·– :0.8Qd, –··–··: 0.7Qd

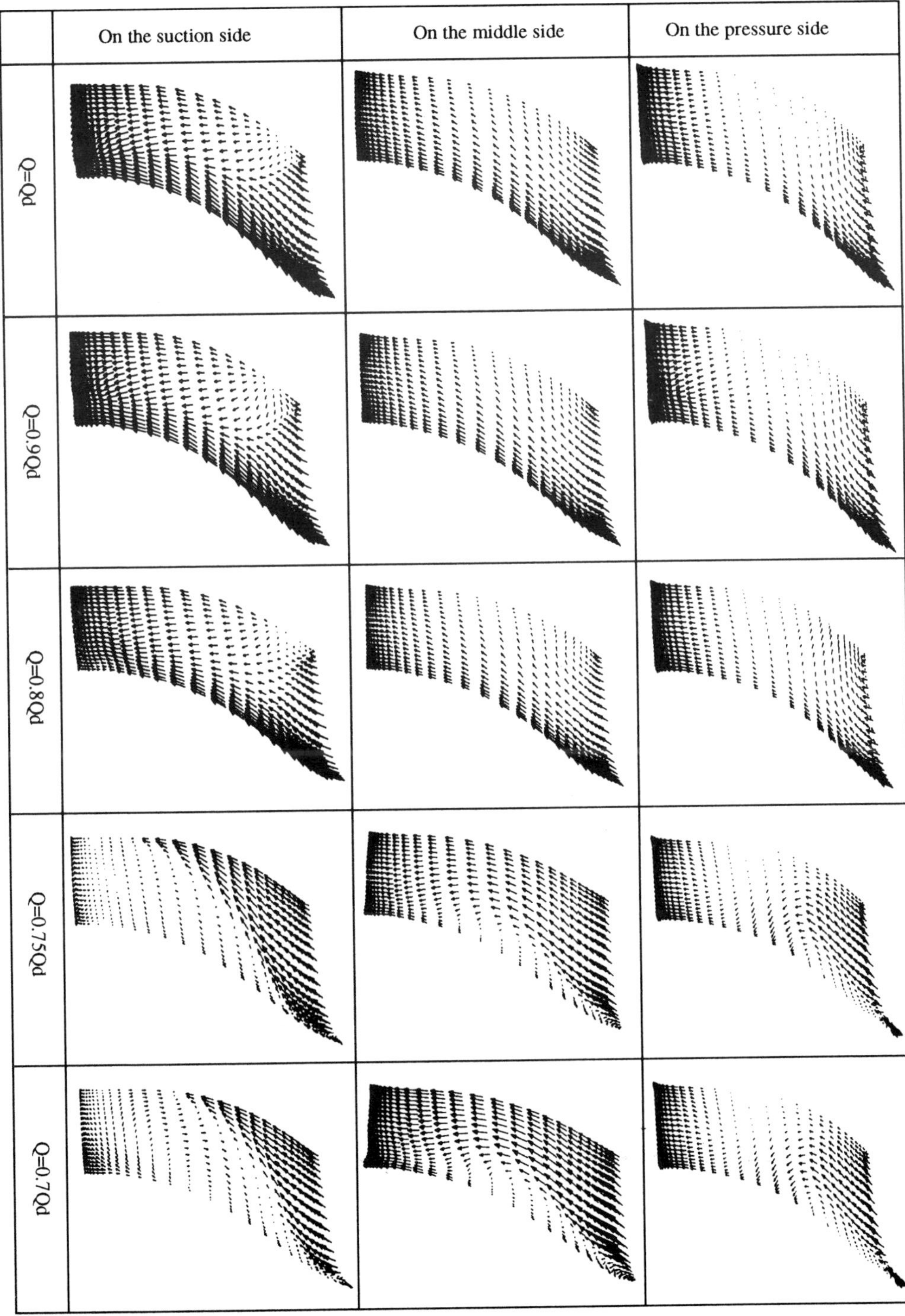

Figure 10. velocity vectors on the side from suction to pressure side

IMPORTANCE OF INTERACTION BETWEEN TURBINE COMPONENTS IN FLOW FIELD SIMULATION

S. RIEDELBAUCH, D. KLEMM, C. HAUFF
Voith Hydro GmbH
D-89509 Heidenheim, Germany

1. Introduction

The application of viscous flow prediction methods, i.e. Navier - Stokes methods, has drastically increased during the development of hydraulic machines in the course of the last several years. This proceeding supports the design process through a detailed knowledge of the viscous flow behaviour in all components of the hydraulic machine. Usually, a broad operation range is investigated, covering the optimum conditions, too.

Another reason for the large increase of theoretical flow prediction applications is the increased reliability of the corresponding results. In most cases, these predictions show a flow behaviour being comparable to that observed during model test investigations.

The main reason for this agreement is the following: The numerical flow field simulation also takes into consideration the influence of the neighbouring hydraulic components. Thus, the flow is approximated in a much more realistic way.

This paper will focus on the used numerical method including coupling procedures. Some selected results will demonstrate the influence of component interaction and the related success during the product development phase.

2. Numerical method and component coupling

The numerical method used here is based on the incompressible Reynolds averaged Navier - Stokes equations and employes the well-known finite volume concept. The influence of turbulence is considered with the two - equation k-ε turbulence model.

E. Cabrera et al. (eds.), Hydraulic Machinery and Cavitation, 238–247.

Owing to restrictions in computer capacity, a complete hydraulic machine is usually subdivided into several components:

- semispiral or full spiral including stay vanes
- tandem cascade of stay vane and wicket gates
- runner
- draft tube

The modelling of individual turbine components is more simple than that of a complete turbine in a product development environment, even if the turbine model is available in a CAD system (Buchmaier *et al.*, 1996).

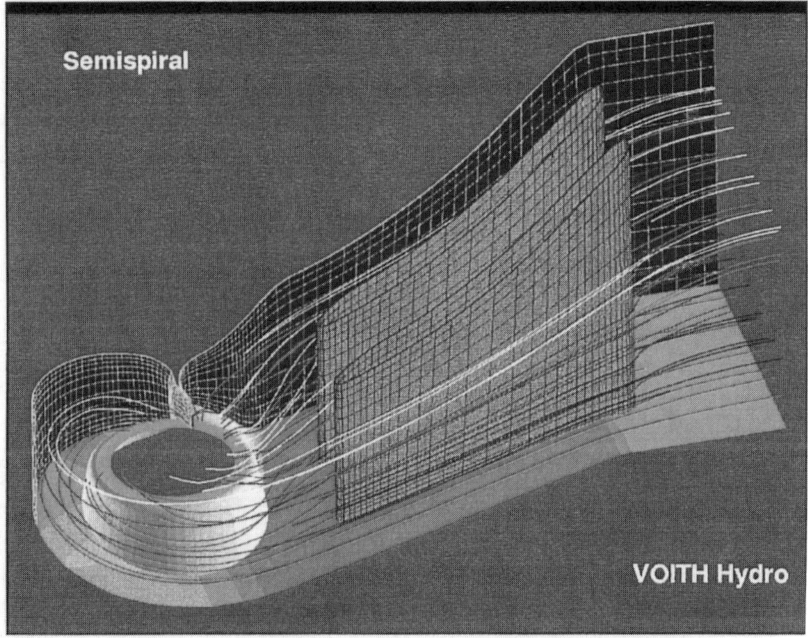

Figure 1. Semispiral with two piers and selected streamlines.

The viscous flow through these components has been analyzed for many years (Riedelbauch *et al.*, 1994 and Ruprecht *et al.*, 1994). The simulation of the flow through a semispiral is one example (fig. 1). This picture presents a semispiral with two piers and a set of selected streamlines coloured with the height level at the entrance. The nose vane is also well seen. The streamlines show a very smooth flow behaviour. Flow separation is only present in the wakes of the piers. Here, constant inflow conditions were chosen. However, the result of this flow case is strongly dependent on the boundary conditions - formulated overrefined: Any boundary condition will give any result possibly being far away from the real flow situation. This is especially true for the tandem cascade, the runner and the draft tube.

Of course, the main goal is to predict the flow in each of these components as realisticly as possible for steady state flow conditions. Here, the word prediction is emphasized in contrast to analysis. E.g. for turbine operation, necessary input parameters for prediction are only discharge, wicket gate opening, angular velocity of the runner and, of course, the entire geometry. All other magnitudes are results of the flow prediction, e.g. operation point and efficiency. This minimum set of parameters is sufficient for the simulation of the complete machine. However, additional assumptions have to be introduced, if individual components are considered alone. These assumptions are basically boundary conditions obtained through simple physical considerations. In most cases, however, these are constant conditions and stiff in this sense. Appropriate coupling procedures permit the adjustment of those boundary conditions in a non-stiff automatic way resulting in more realistic flow approximations. This is particularly important for the stay vane / wicket gate - runner interaction and the runner - draft tube interaction.

In principle, three different types of coupling are possible:

- three dimensional unsteady simulation
- frozen rotor, i.e. fixed relative location of runner and wicket gate and/or runner and draft tube
- circumferential flow averaging

Of course, each option may vary due to implementation details.

The three dimensional unsteady simulation is supposed to be the most realistic approach, provided that the corresponding flow code is designed for time accurate flow simulation. Real machines possess a non-integer ratio of wicket gates and runner blades. As a consequence, all wicket gates combined with the entire runner have to be considered. The number of nodes of these computational grids is huge - and so is the needed simulation time. Today, three dimensional unsteady simulation of viscous flow is considered to be of academic interest only. Effective support of product development needs shorter turn-around times.

The second possible approach is the so-called frozen rotor approach. This terminology means that the relative location between rotating parts and non-rotating parts remains fixed. At a certain surface of revolution, the flow components have to be transformed from a fixed frame of reference to a rotating frame of reference. Due to the non-integer ratio of wicket gates and runner blades, the complete geometry has to be considered being very similar to the unsteady approach. Consequently, no problem reduction is possible through the use of cyclic periodic boundary conditions.

Another coupling approach is to average the flow variables in circumferential direction at the boundary between rotating and non-rotating parts (sliding boundary). This proceeding allows to reduce the size of the problem to one periodic flow channel for axisymmetric geometries, i.e. runner and tandem cascade. Of course, the size of the draft tube grid and the spiral grid is not influenced. The choice of the location of the sliding boundary is dictated by the flow physics. The runner - draft tube sliding boundary has to be defined within the axisymmetric part of the geometry. The wicket gate - runner sliding boundary has to be positioned between the trailing edge of the wicket gate and the leading edge of the runner. From a physical point of view, it should be as far upstream as possible for the runner to account for the upstream influence of the runner blades.

For the tandem cascade, however, it should be as far downstream as possible. For this reason, we are working with overlapping grids and not only with a sliding boundary, i.e. the runner grid extends very close to the trailing edge of the wicket gate, while the tandem cascade grid extends to the leading edge of the runner. The runner - draft tube boundary is treated in a similar way.

3. Selected results

This chapter is devoted to present some selected results, considering the flow influence of neighbouring turbine components. Here, the flow averaging in circumferential direction has been employed.

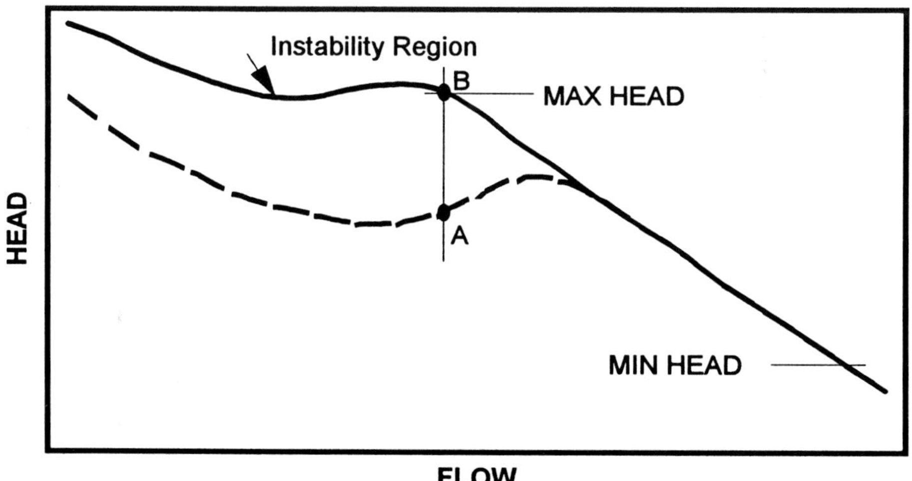

Figure 2. **Sketch of the pump characteristic of a Pump-Turbine.**

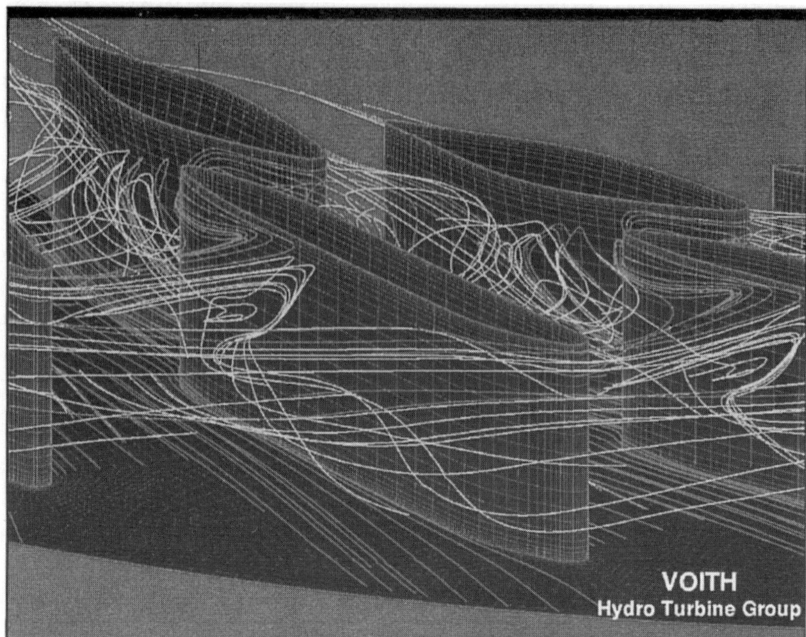

Figure 3. Selected streamlines through tandem cascade with runner interaction.

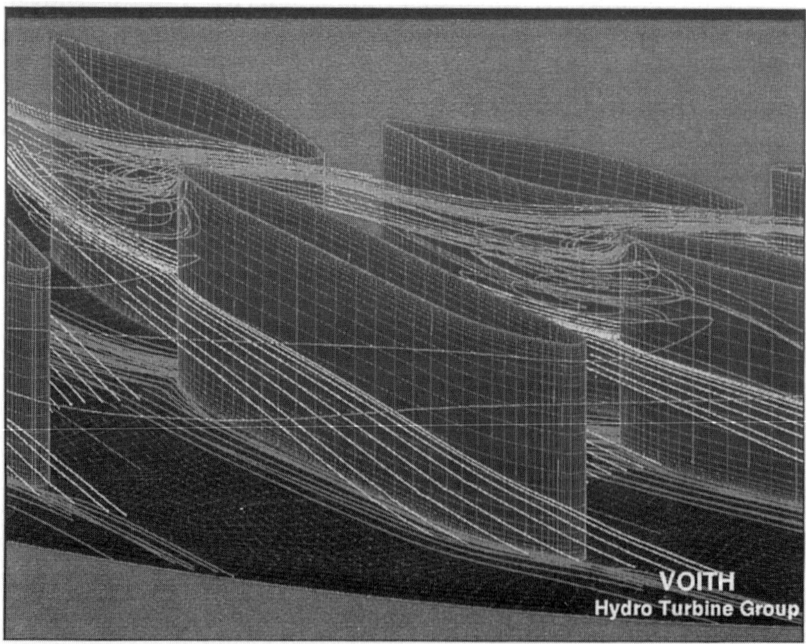

Figure 4. Selected streamlines through tandem cascade without runner interaction.

3.1 TANDEM CASCADE - RUNNER FLOW INTERACTION

Pump storage plants are often operated by pump-turbines. This type of machine is capable of pumping fluid into the upper reservoir as well as running in turbine mode generating energy. One runner handles both, pump and turbine operation. Thus, the development of high performance pump-turbines is a challenging task. The efficiency level has to be very high for pump as well as for turbine mode. See Klemm *et al.*, (1996) for a recent development.

Typically, the pump operation is limited by an instability region at high heads (Fig. 2). This instability region is related to massive flow separation. It may occur within the stay vane/wicket gate component and/or the runner (Fisher *et al.*, 1978). The separation of flow and the formation of appreciable regions of unsteady recirculating flow is characterized as stall. Rotating stall may occur if stall appears only in one part of the flow channels, while it periodically moves around in a certain rotating sense. Pump-turbines should not operate within this instability region because considerable pressure fluctuations are induced which, finally, lead to component damage.

The qualitative characteristic of a Pump-Turbine at pump operation is displayed in figure 2. The plant head range is indicated as well as the instability region. Each of the pump characteristics belongs to a specific geometry, denoted with A and B at a considered flow rate.

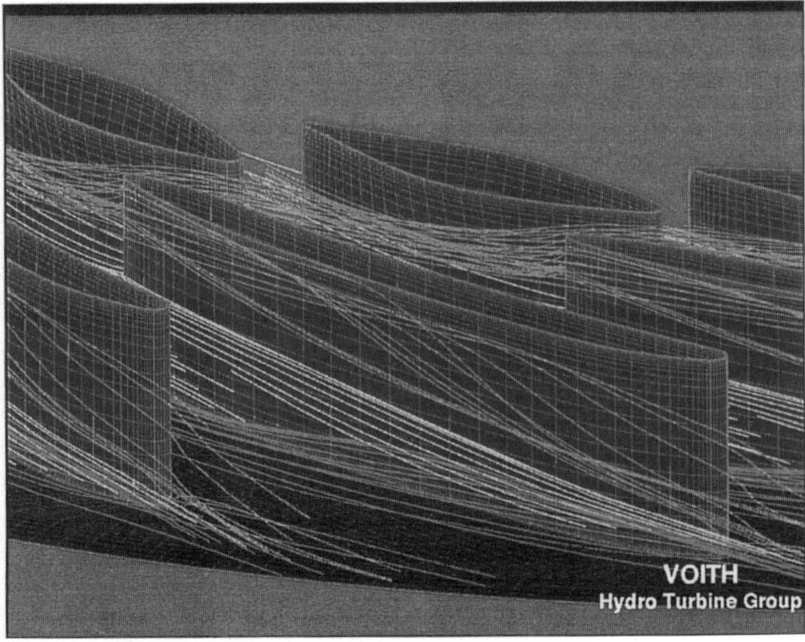

Figure 5. Selected streamlines through modified tandem cascade with runner interaction.

At first, consider operation point A indicating that severe flow separation phenomena occur. The visualization of the flow through the runner shows a very smooth behaviour (not shown here). The flow through the tandem cascade is presented in figure 3. The streamlines close to the crown meridional contour clearly emphasize a large region of vortical flow between the stay vanes creating high flow losses. Close to the band meridional contour, the flow moves smoothly towards the spiral. Note, the runner influence on the stay vane/wicket gate component has been considered for this prediction.

If runner interaction is not taken into account, Euler's angular momentum equation has to be applied to determine the inflow conditions at the tandem cascade for pump operation. The resulting streamlines are presented in figure 4. Flow separation is also observed close to the crown meridional contour, while everywhere else smooth flow prevails. The comparison of figure 3 with figure 4 reveals a considerable difference of the flow field.

Owing to the project requirements, the pump-turbine has to operate at conditions B, too (cf. fig. 2). Thus, no large regions with flow separation are permitted. A modification of the tandem cascade geometry meets this requirement (fig. 5). The flow moves fairly smooth towards the spiral over the entire pad height. Note, this result has been obtained with consideration of runner interaction. In this case, the runner flow also behaves very smooth.

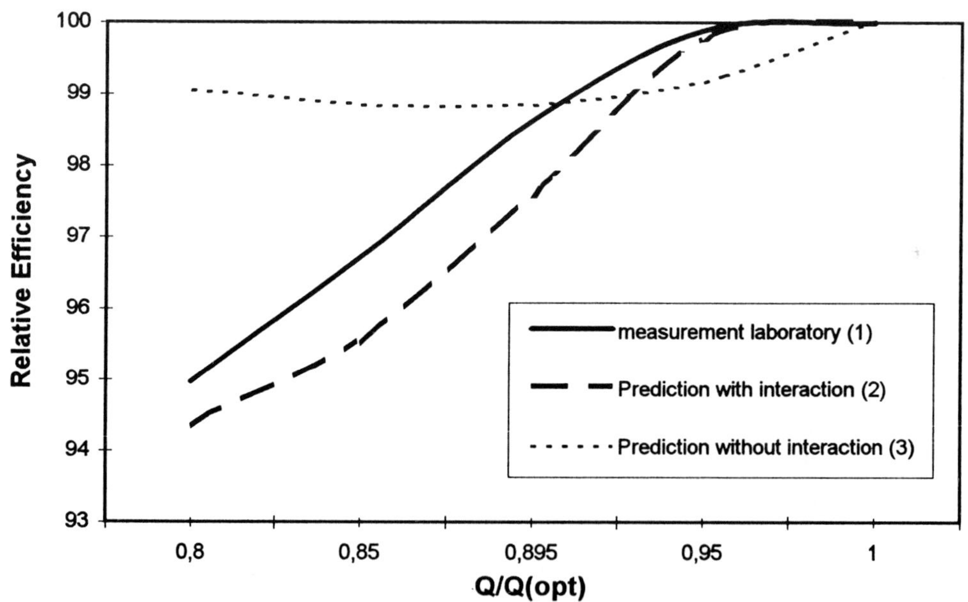

Figure 6. Relative efficiency versus non-dimensional flow rate.

These described results are summarized in figure 6. It presents the relative efficiency, i.e. η/η_{max}, versus a non-dimensional flow rate, i.e. Q $/Q_{opt}$. Curve one through curve three belong to geometry 1. The first curve represents measured data from the laboratory, while curve two presents the flow prediction with flow interaction of runner and tandem cascade (cf. fig.3). Curve three shows the flow result without the consideration of runner and tandem cascade interaction (cf. fig.4). There is a good agreement between the measurements in the laboratory and the flow prediction, if component interaction is considered. The neglect of flow interaction between neighbouring turbine components shows a wrong flow behaviour under high head operation conditions as well as near optimum operation. The flow characteristics of the modified geometry (geometry 2) are fairly different. It produces nearly constant flow losses over a large portion of the operation range (cf. fig. 5).

3.2 RUNNER - DRAFT TUBE FLOW INTERACTION

A high pressure recovery in the draft tube is a necessary prerequisite for high turbine efficiencies. The kinetic energy remaining in the flow downstream of a runner has to be transformed into static pressure. This process should occur with minimum losses to maintain high efficiencies.

The flow through the draft tube is strongly dependent on the flow distribution and the flow properties downstream of the runner. This automatically implies the necessity to take the runner - draft tube flow interaction into consideration. A corresponding flow field simulation result of a Francis turbine draft tube is presented in figures 7 and 8. Both flow fields differ in their operation point. Figure 7 corresponds to near machine optimum conditions (OP1), while figure 8 represents a larger flow rate and gate opening at the same head (OP2).

Operation point OP1 shows a counter-clockwise inlet swirl distribution (runner rotation is counter-clockwise) across the entire draft tube entrance. The streamlines reveal a vortex originating at the rotation axis. Their colour represent the absolute speed. This vortex moves into the right half of the draft tube. Near the end of the draft tube, the transport velocity (main flow component) displays a low mass flow rate through the right half of the draft tube, while it is considerable higher on the left half. The wake downstream of the pier is clearly visible, too.

Operation point OP2 produces a different inlet swirl distribution. It is counter-clockwise at large radii, but clockwise near the rotation axis. The center vortex is less marked. This flow field exhibits a very similar transport velocity on both sides of the pier. The flow losses of the draft tube at operation

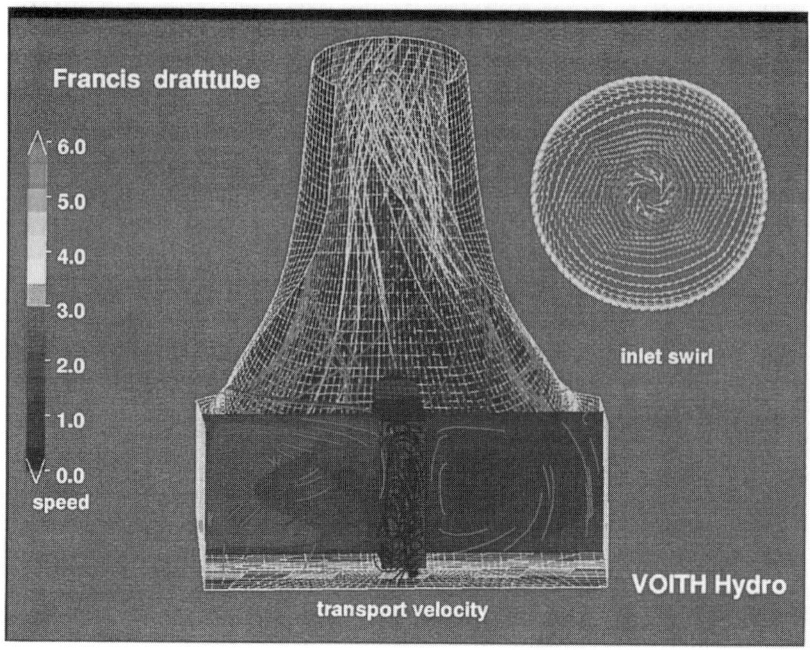

Figure 7. Inlet swirl, streamlines and transport velocity of a Francis turbine draft tube (OP1).

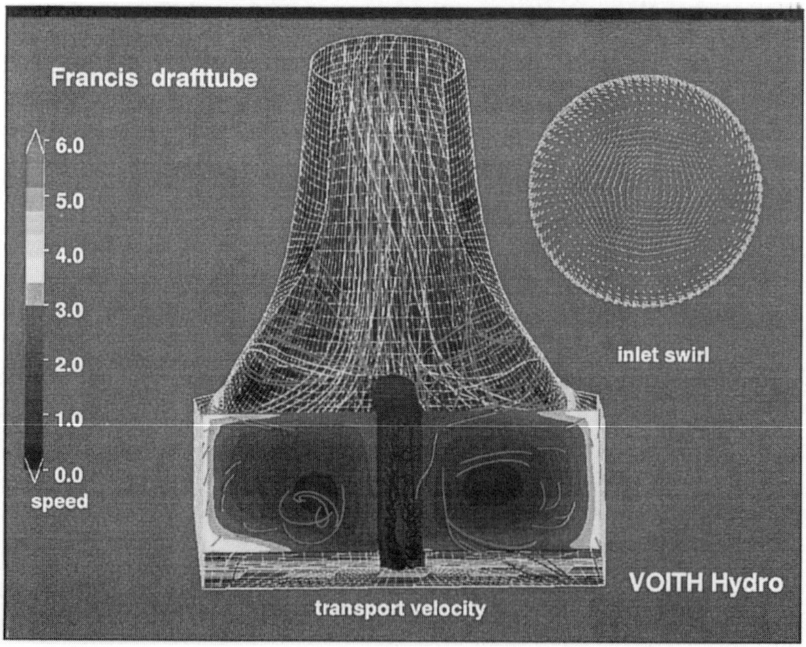

Figure 8. Inlet swirl, streamlines and transport velocity of a Francis turbine draft tube (OP2).

point 1 are less than one percent, while those of operation point 2 amount to about two percent. This difference can not be neglected for turbines with high efficiencies.

4. Concluding remarks

Numerical simulation of viscous flow fields has proven to be a very powerful tool for the design and development of hydraulic machinery. It gives the designer detailed insight into the flow field and leads to an improved understanding of the flow physics.

Different methods of considering flow interaction between turbine components are shortly described. For the present investigation, a method has been chosen which allows an effective support during the turbine development phase. Results being obtained through the flow simulation without component interaction will be very likely subject to misinterpretation of the flow behaviour and, consequently, of the turbine design.

The presented results strongly emphasize the necessity to consider flow interaction between turbine components. During the development of our hydraulic machines, these advanced methods are applied for the benefit of power generation.

5. References

Buchmaier, H., Quaschnowitz, B., Moser, W. and Klemm, D.(1996) Numerical Design and Optimization of High Head Pump-Turbines, *Proceedings of the XVIII IAHR Symposium, Valencia, Spain,* Kluwer Academic Publishers, to appear.

Fisher, R.K. and Webb, D.R. (1978) Effect of Cavitation on the Discontinuity Point and On Alternating Pressures and Gate Torques on a Pump/Turbine Model in the Pump Cycle, ASCE-IAHR/AIHR-ASME Joint Symposium, Colorado Springs, Colorado, USA.

Klemm, D., Jaeger, E.-U. and Hauff, C. (1996) Development of a new Generation of High Head Pump-Turbines, GUANGZHOU II, *Proceedings of the XVIII IAHR Symposium, Valencia, Spain,* Kluwer Academic Publishers, to appear.

Riedelbauch, S., Fisher, R.K. and Riva, P. (1994) Utilization of Three Dimensional Viscous Flow Simulation for the Design and Optimization of Hydraulic Machinery, *Proceedings of the XVII IAHR Symposium, Beijing, China 1994* **1**, 353-364.

Ruprecht, A., Bauer, C. and Riedelbauch, S. (1994) Numerical Analysis of Three-Dimensional Flow through Turbine Spiral Case, Stay Vanes and Wicket Gates, *Proceedings of the XVII IAHR Symposium, Beijing, China 1994* **1**, 71-82.

FROM COMPONENTS TO COMPLETE TURBINE NUMERICAL SIMULATION

M. SABOURIN
Gec Alsthom
Tracy, Qc, Canada

Y. LABRECQUE
Gec Alsthom
Tracy, Qc, Canada

V. DE HENAU
Gec Alsthom
Tracy, Qc, Canada

1. Introduction

In the quest for turbine improvement, computational fluid dynamics (CFD) is the emerging tool. The fast development of computers combined with the new capabilities of turbulent flow codes make possible the calculation of an entire turbine unit. Until recently, fluid simulations for turbines were done considering the various components separately. This convenient way of analysis can however present some major drawbacks. As components can be strongly coupled, the unknown introduced by assumed boundary conditions needs to be clarified. One consequence of a complete turbine numerical simulation is to avoid assumptions of internal boundary conditions.

In this paper, the implementation of a strategy to simulate the interactions between rotating and non-rotating components is presented. The capacity of the method to predict beginning of cavitation, power output, head losses and other hydraulic behaviour makes this tool very useful for the turbine designer.

2. Historic

Since the beginning, the design of hydraulic turbines has been an experimental science. However, step by step CFD has emerged as a cost effective complement to model testing. Commercial codes for three dimensional turbulent flow simulation, which are now available, yield significant improvements in accuracy of simulation due to their capacity to solve real fluid flowing in real geometry.

E. Cabrera et al. (eds.), Hydraulic Machinery and Cavitation, 248–256.

The calculation of losses in turbine components treated as separate entities has been the subject of numerous papers [1,5,6]. This approach brings to the designer interesting information on the behaviour of the turbine. It presents however some difficulties:

- the specification of boundary conditions stands on experimental data or assumptions based on designer experience,
- the analysis of results must segregate local improvement from overall benefit,
- comparison with experimental results is partial as only the performance of the complete turbine is measured with accuracy.

Additional information concerning the strength of the hydraulic coupling between component can be evaluated.

A publication from De Henau [1] demonstrated the importance of interaction between static components like the casing and the distributor. The full understanding of this interaction can be accomplished only through a complete 3D model including the casing and the complete distributor. The pressure condition at the distributor outlet coming from the runner modifies the flow distribution with more or less influence, depending on the turbine type and the hydraulic conditions.

This interaction between runner and distributor has been integrated for 20 years in our flow simulation software used in the design of GEC ALSTHOM's turbines [4]. Although this software uses ideal fluid concept the prediction of inlet cavitation is more accurate with this method. The interaction stator-rotor has a major influence in turbine behaviour.

Similarly, the draft tube, responding to the flow field delivered by the runner provides a back pressure resulting from its kinetic energy recuperation capacity. During the optimization process of the draft tube, a modification of the geometry can have significant influence on the runner back pressure and therefore, on the velocity field at draft tube inlet. This feed back loop escapes to the simulation of the isolated draft tube. A precise evaluation of losses can therefore be achieved only through precise boundary conditions.

For all these reasons, the adequate simulation of the coupling between turbine components is highly desirable. Moreover, for turbulent flow codes, the passage from the status of analysis tool to the one of design tool necessitates the simulation of interaction between components. The stator-rotor interaction takes more importance in this perspective.

3. Implementation of the stator-rotor interaction

3.1 NUMERICAL CONCEPTS

The basic constraint to the implementation of coupling between components is the limit of computer resources. Experience with calculation of isolated components demonstrated that in many cases the solution is mesh dependent. This makes the implementation strategy more delicate as it must embed two aspects in the numerical process: the numerical zoom and the circumferential averaging.

The numerical zoom is based on the assumption that velocity field is more accurate than the pressure field for a given grid, creating a lever effect that permits to transfer velocity field from a coarse mesh to a fine mesh in order to obtain a more accurate pressure field solution. Thus, the solution is found after two sequences of calculations. The first one has the goal to determine velocity boundary conditions and the second step must refine the losses calculation. Figure 1 shows a local concentration of nodes in the distributor in order to calculate the losses.

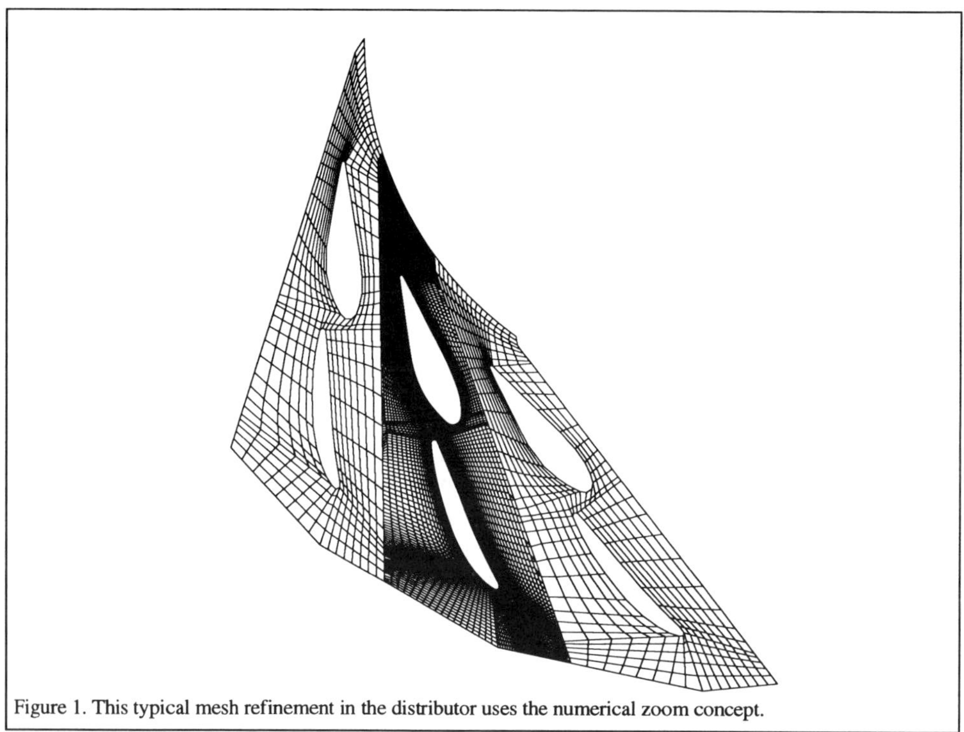

Figure 1. This typical mesh refinement in the distributor uses the numerical zoom concept.

The circumferential averaging is the method chosen to simulate the stator-rotor interaction. Galpin et al. [2] reviewed various methods. He concluded that a practical way of simulation is to use a steady state method using a conservative averaging in the circumferential direction at the sliding interface. A sliding interface is a surface of revolution and most commonly a conical surface. There are two sliding interfaces. A first one is located at the runner inlet between the trailing edge of guide vane and the blade leading edge of the runner blades. A second sliding interface is below the trailing edge of the runner blades. These interfaces constitute internal boundaries where node variables are treated and communicated upstream and downstream.

Two approaches implement circumferential averaging.

The first one, the *simulation by linked components*, manages three numerical models: the casing and distributor model, the runner in a rotational frame of reference and the draft tube. This method, described in a recent paper by Labrecque et al. [3], permits to use natural boundary conditions at the ends of the complete turbine, see Figure 2. At inlet, the total pressure level is specified. Static pressure is specified at the outlet of the draft tube. These are the only boundary conditions needed for the entire simulation. Between the components, at interfaces, boundary conditions are initially estimated and after few iterations, calculated boundary conditions can be obtained. On Figure 2, the descending arrows indicate the communication of velocity and turbulence fields from

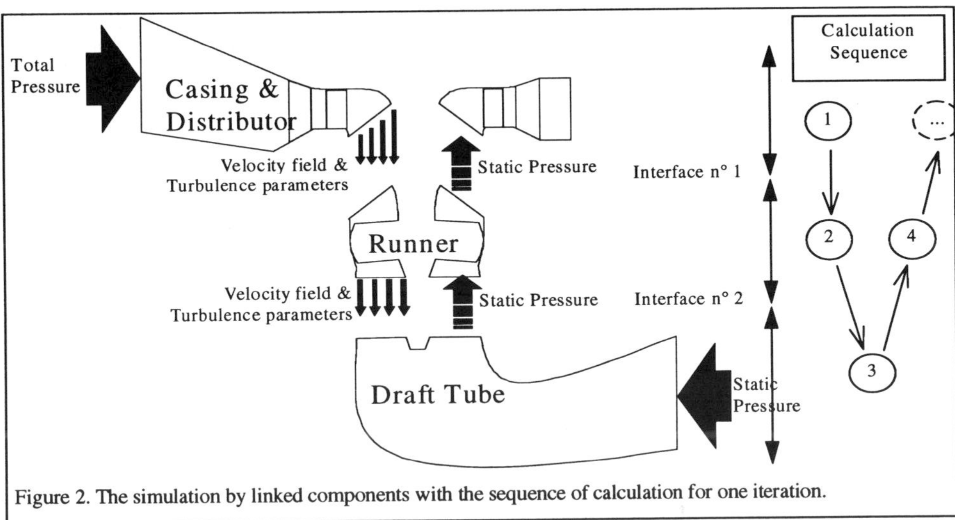

Figure 2. The simulation by linked components with the sequence of calculation for one iteration.

the upstream components to the downstream ones. The ascending arrows communicate the static pressure from the downstream components to the upstream ones. One iteration over the complete turbine necessitates 4 calculations of isolated models. Depending on the initial conditions and on the level of convergence desired, a solution ca be reached within about 10 iterations.

The second approach needs the new software package *stage* now available within TASCflow [2,7]. This package handles multiple frame of reference (MFR) giving the possibility to calculate the entire turbine in one single calculation automatically. However, this type of calculation over an entire turbine may require millions of nodes to obtain reasonable precision.

3.2 THE COMPUTING PROCESS

The calculation of an entire turbine unit is made in an industrial environment: a result must be obtained on a low price computer within a reasonable time. The software

TASCflow can run on various computers. A rule of thumb for memory requirement is 1 Meg for each thousand nodes Although the speed and memory capacity of today's computers have increased significantly in recent years, there is a limit to the largest calculation possible and this limit is a lot smaller than required.

The *linked components* method permits to work with different mesh sizes and topology for each model as no mesh connection is necessary at sliding interface. The sequence of

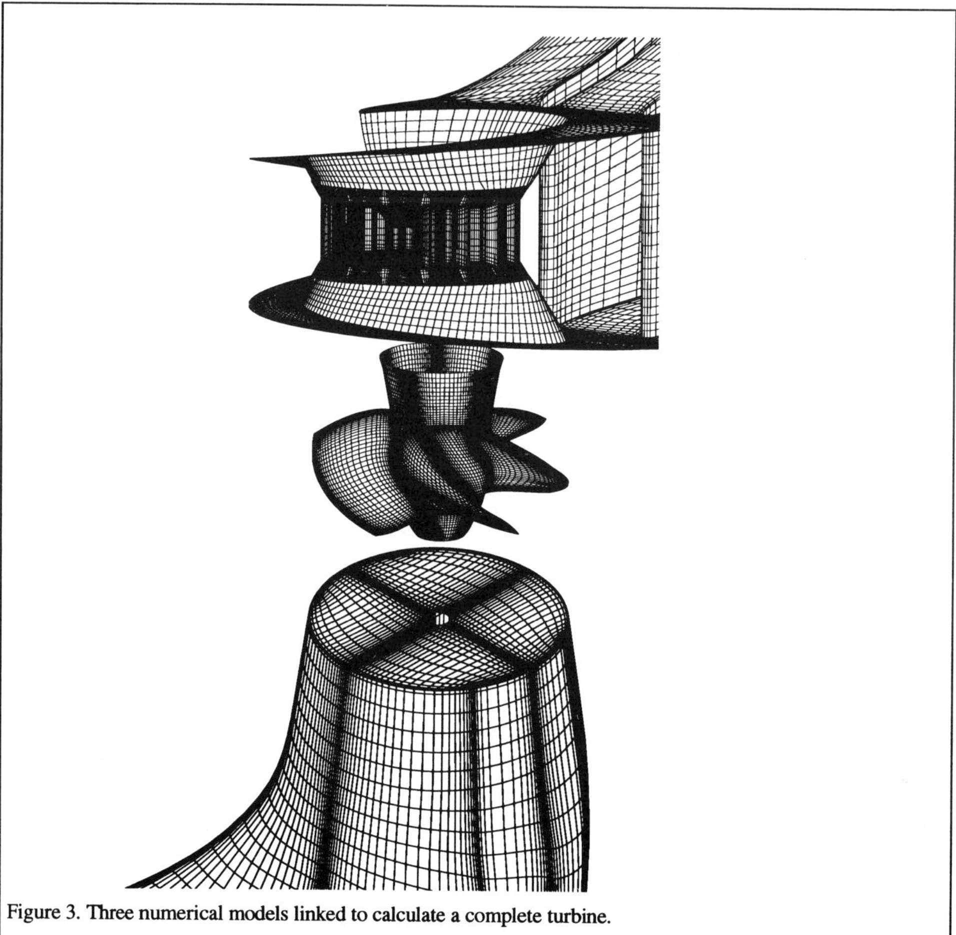

Figure 3. Three numerical models linked to calculate a complete turbine.

calculation is not automated. Depending on the hydraulic behaviour to evaluate, the mesh of each component can be refined at any step, concentrating all the computer resources over the area of interest. Also, at the beginning of the problem, the convergence criteria do not need to be stringent as boundary conditions are approximate at this time. During processing, the draft tube calculation may be skipped a while, as the coupling between it and the runner is weaker than the one between the distributor and the runner. This strategy can contribute to reduce the machine calculation time.

The *stage* method is theoretically able to solve the entire turbine in a single calculation. However, the computer resources limit the accuracy of the solution and a marriage with the linked components method is necessary. The calculation of the casing-distributor model gives the boundary condition to a distributor model limited to one or two guide vane passages. This distributor model is connected through the stage interface with the runner. Therefore, the distributor and the runner are calculated in a single calculation, see Figure 4. The draft tube is calculated separately and the pressure condition at the runner outlet is adjusted when necessary.

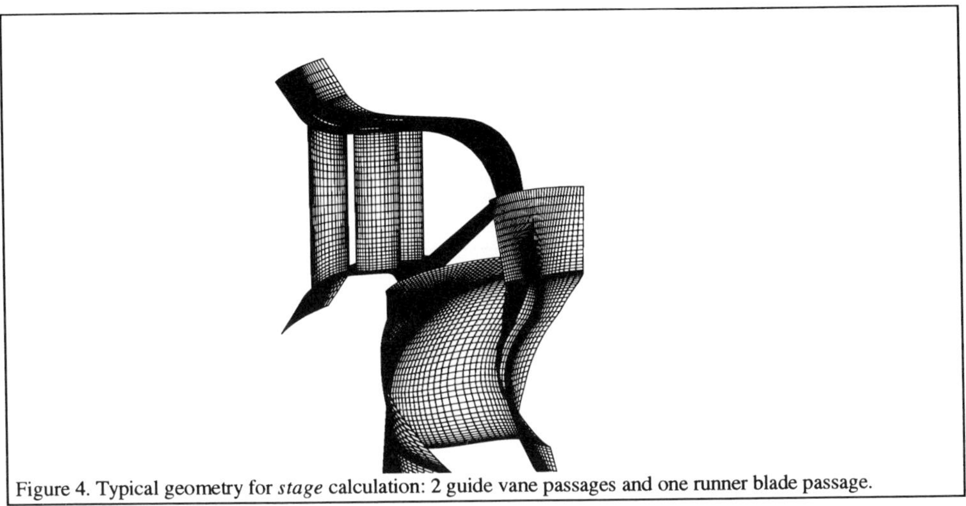

Figure 4. Typical geometry for *stage* calculation: 2 guide vane passages and one runner blade passage.

The optimal strategy of the computing process is to embed the *stage* method in the *linked components* method benefiting the efficiency of *stage* to simulate coupling and the accuracy of *linked components* to obtain more reliable results.

4. Application

In particular, two aspects of the turbine design and analysis are improved with the calculation of a complete turbine. They are cavitation and losses evaluation.

4.1 CAVITATION ANALYSIS

The cavitation in the runner can be found at inlet close to the leading edge or on the profile.

The inlet cavitation originates from the mis-adaptation of the blade angle. Therefore, the coupling between distributor and runner demands the simulation of the stator-rotor interaction. On Figure 5, the result of two simulations is presented. The first one involves the distributor alone where the outlet pressure condition comes from an ideal fluid simulation of stator-rotor interaction. The second one is a calculation of the complete turbine with a turbulent flow code. At the sliding interface between the runner and the distributor corresponding to the outlet condition on the distributor, a comparison is made on velocity and pressure. It is observed that the two methods yield different incidence angles near the shroud. The complete turbine calculation indicates a greater risk of inlet cavitation as observed during model test.

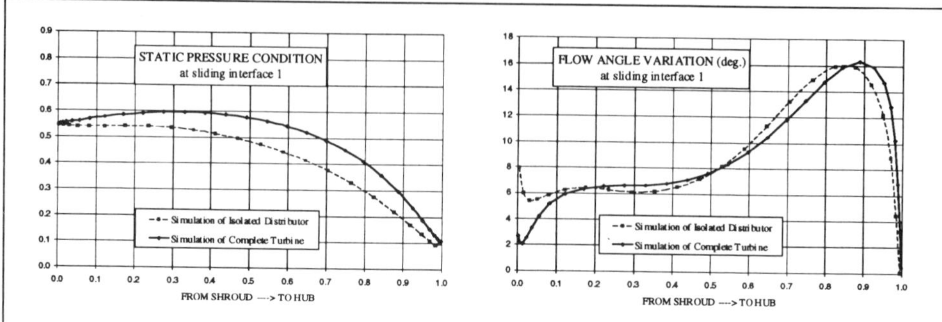

Figure 5. Comparison of calculation over isolated distributor and complete turbine showing a better detection of inlet cavitation for the complete simulation.

In the case of the profile cavitation, the tailwater level and losses between it and the runner must be calculated to obtain the absolute static pressure. On Figure 6, the absolute static pressure field on suction side of blade is compared to the observation. It can be observed that the area where the pressure is below the vapour pressure is in agreement with the location of bubbles conveyed by the flow.

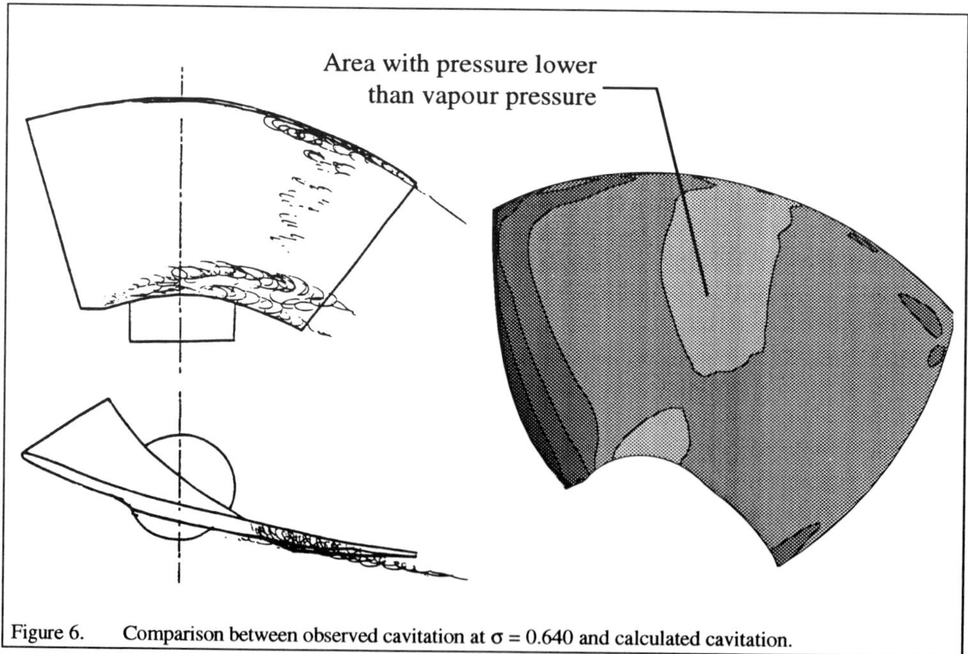

Figure 6. Comparison between observed cavitation at $\sigma = 0.640$ and calculated cavitation.

4.2 LOSSES ANALYSIS

As seen, the other aspect improved with entire turbine simulation is the evaluation of losses. Losses calculations are linked to the other hydraulic parameters. They are speed, flow and torque coefficient.

Experience shows that torque and power are calculated precisely either by the pressure field or the kinetic momentum. However, the numerical solution is mesh dependent regarding losses calculation. The fact that the mesh is coarse introduces a drift in the static pressure. This drift increases from the outlet to the inlet in the flow direction. As the drift of the static pressure is similar on both sides of the blade, the pressure differential on the blade is precise. The evolution of losses with mesh concentration indicates a convergence to the experimental result as presented on Figure 7.

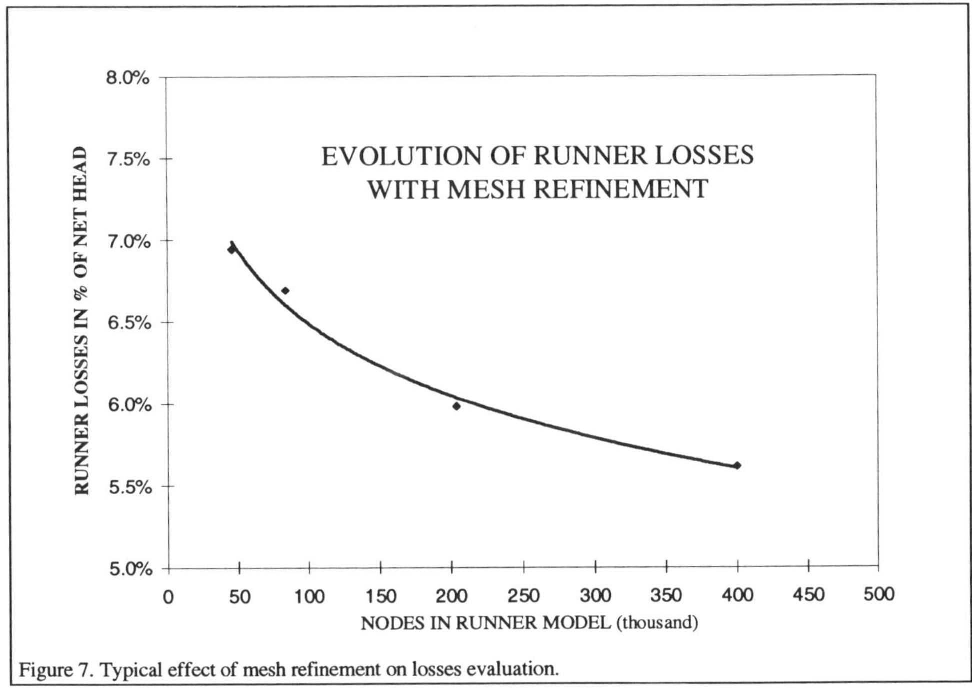

Figure 7. Typical effect of mesh refinement on losses evaluation.

Using similar meshes on various geometries leads to a precise evaluation of losses difference. The calculation of an entire turbine eliminates the need for assumptions on boundary conditions and therefore gives a more precise evaluation of losses. Moreover, this method can evaluate if a local improvement will produce the expected effect on the complete turbine. This makes the calculation of the entire turbine an excellent design tool.

5. Conclusion

The calculation of an entire turbine needs numerical simulation of the interaction between components. This simulation involves numerical zoom and circumferential averaging. The combination of the *linked component* with the *stage* methods permits to optimize the calculation process.

The complete turbine numerical simulation eliminates the need for experimental data to specify boundary conditions. Although this method is less accurate than model testing, it is faster and more economical. However, with continuous development of computers, the actual level of precision is improving.

The results, compared to separated components simulation, provide better information for the turbine designer and permit to see the effect of a local modification on the entire turbine behaviour.

The CFD involving three dimensional turbulent flow codes and stator-rotor interaction can now be considered as the best tool for turbine design.

References.

[1] De Henau V., 1995, "Turbine Rehabilitation: CFD analysis of distributors", Water Power Symposium, San Francisco, USA

[2] Galpin P.F., Broberg R.B., Hutchinson B.R., 1995, "Three-Dimensional Navier Stokes Predictions of Steady State Rotor / Stator Interaction with Pitch Change", CFD Conference, Banf, Alberta, Canada.

[3] Labrecque Y., Sabourin M., Deschênes C., 1996, "Numerical simulation of a complete turbine and interaction between components", Modelling, Testing & Monitoring for Hydro Powerplants, Lausanne, Switzerland.

[4] Moulin C., Wegner M., Eremeef R., Vinh P., 1977, "Méthode de tracé des turbomachines hydrauliques", La houille blanche/n° 7/8.

[5] Sabourin M., Eremeef R., De Henau V., 1995, "Extensive Use of Computational Fluid Dynamics in the Upgrading of Hydraulic Turbine", Canadian Electrical Association, Vancouver, British Colombia, Canada.

[6] Sabourin M., Couston M., 1994, "Turbine Rehabilitation: Experience Gained through Systematic Draft Tube Evaluation", IAHR Symposium, Beijing, China.

[7] TASCflow Users Manual, 1995, Version 2.4, Advanced Scientific Computing Ltd., Waterloo, Ontario, Canada.

VALIDATION OF A STAGE CALCULATION IN A FRANCIS TURBINE

MIRJAM SICK AND MICHAEL V. CASEY
Sulzer Innotec Ltd
CH-8401 Winterthur, Switzerland

AND

PAUL F. GALPIN
Advanced Scientific Computing Ltd.
Waterloo, Ontario, Canada

Abstract. This paper describes the verification of a simulation method for complete hydraulic turbine systems, from spiral casing through distributor and runner to the outlet draft tube. The method solves the steady 3D Reynolds-averaged Navier Stokes equations with a mixing plane at the interfaces between the components, so that the flow in each frame of reference is steady. The steady-state interactions between the components are taken into account but unsteady interactions are neglected. The paper verificates the method by comparison of computations with detailed flow measurements from a high specific speed Francis turbine model. Computations in which each component is computed in isolation are compared to computations using the mixing interface between components to demonstrate the advantages of the new technique.

1. Introduction

The calculation of flows in hydraulic turbines has undergone a rapid transformation during the last ten years. The first major breakthrough in the use of CFD methods for hydraulic turbine design allowed Euler methods to be used for design of water turbine components and, combined with suitable design rules, drastically reduced the number of model tests needed for the achievement of a satisfactory design (Keck *et al.*, 1990). The relatively simple Euler methods predict the important features of the flow, such as

E. Cabrera et al. (eds.), Hydraulic Machinery and Cavitation, 257–266.

incidence levels at runner inlet, pressure levels at runner outlet (cavitation) and swirl in the draft tube inlet. They are, however, limited in their ability to predict the losses, as viscous forces are neglected.

Recent developments have aimed to solve the Reynolds-averaged Navier-Stokes equations to obtain detail of the loss sources in the individual components and to improve component performance prediction. Mature CFD methods for predicting the flow and losses in the individual components of the turbine stage are now well established features of most hydraulic turbine design systems, e.g.(Keck *et al.*, 1994), and these have made substantial improvements in the cost and speed with which new designs can be assessed.

This paper is concerned with the next major improvement in turbine design methods, that is the ability to calculate the performance of components taking into account their interaction with the upstream and downstream elements. Some steps in this direction have already been reported, for example the Kaplan turbine simulations reported by Goede *et al.* (1990).

The method examined in this paper includes a mixing plane interface between the rotating and stationary components so that a steady solution can be calculated in each domain. The mixing plane interface has been incorporated into a commercially available CFD code (TASCflow from Advanced Scientific Computing Ltd., Waterloo, Canada). The development has been carried out under partnership between Sulzer Hydro, Sulzer Innotec and ASC. This paper describes aspects of the verification of the method by comparison with detailed laser anemometry data from a high specific speed Francis turbine (Sulzer Innotec and Sulzer Hydro).

A companion paper to this describes the use of this newly developed tool for the hill-chart performance prediction of a Francis turbine (Keck, Drtina and Sick, 1996).

2. Theoretical method

2.1. NAVIER-STOKES CODE

The TASCflow commercial software package was used for this work. This software solves the 3D Reynolds-averaged Navier-Stokes equations in strong conservation form. A colocated variable arrangement is employed to solve for the primitive variables (pressure, Cartesian velocity components) in either rotating or stationary coordinate systems. The transport equations are discretized using a conservative, finite element based finite-volume method. Turbulence effects are modeled using the standard k-ε model. A second order accurate skew upwind differencing scheme with physical advection correction is employed. A coupled algebraic multigrid method solves the system of equations (coupled solution of mass and momentum). Liquids, subsonic, transonic and supersonic gas flows can be analyzed. Details regarding the theoretical basis of the software are reported by Raw (1994) and in the ASC Theory Documentation (1995). The software has been previously applied to turbomachinery flows and for flow calculations in a wide variety of other components (Casey, Borth *et al.*, 1995).

2.2. TURBOMACHINERY STAGE CALCULATIONS WITH AN INTERFACE MIXING PLANE

The first three-dimensional flowfield calculations to predict the performance of a whole turbomachinery stage were made by Denton using the steady Euler equations, Denton (1983). The flow in adjacent rotating and stationary blade rows is unsteady, so to carry out steady Euler calculations in each blade row some modelling of the flow processes has to be carried out to remove the unsteadiness. Denton achieved this by simple circumferential averaging of the flow at an intermediate calculation plane (the mixing plane) between the adjacent blade rows. The upstream blade row experienced a circumferentially uniform downstream boundary condition, and the downstream row saw a circumferentially uniform flow approaching it.

Following the first Euler stage computations by Denton, several years elapsed for the development of practical viscous flow solvers for isolated blade rows based on the three-dimensional Navier-Stokes equations. As soon as these techniques were available, it was natural that turbomachinery stage capability with Navier-Stokes codes would be developed, see Denton (1992), Dawes (1992) and more recently, Galpin *et al.* (1995).

A problem with the original circumferential mixing model of Denton was that a circumferentially uniform flow may be forced to exist too close to the leading edge of the downstream blade row. This does not allow the flow to adjust circumferentially to the presence of the blade. As a result the leading edge loading on the blade row may be wrong. The solution of this problem (Denton, 1992) was to allow a circumferential variation of fluxes at the mixing plane (by extrapolation from the upstream and downstream planes) while adjusting the level of the fluxes to satisfy overall conservation.

2.3. GALPIN MODEL FOR SLIDING STAGE INTERFACE PLANE

The frame change and pitch change is accomplished at a sliding grid interface using a control surface approach. A Stage sliding interface is defined when the control surface grid forms bands parallel to the machine motion, forming a conservative "mixing plane".

The interface fluxes are assembled into the volume and surface equations in a fully conservative and implicit manner. The pressure forces at the Stage interface are computed such that the average interface pressure equals the

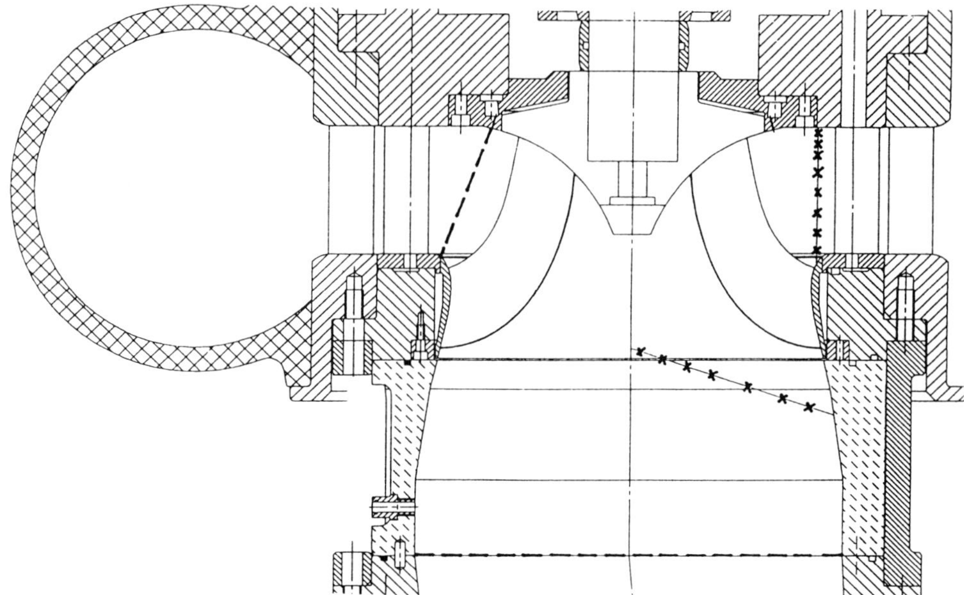

Figure 1. Francis Turbine with measurement positions (x) and interfaces (dashed lines)

control surface pressure, while supporting local nodal pressure variations due to elliptic effects, as shown in fig. 2.

For convenience the control surface equations are placed in the stationary absolute frame of reference, with the pitch-wise extent equal to the pitch of the larger side of the interface. Once all fluxes have been evaluated at all surfaces exposed to the sliding interface, the equation set consists of control surface equations and control volume equations. An algebraic multigrid solver is used to solve this equation set.

3. Test case

3.1. MODEL FRANCIS TURBINE

The test case considered in this paper for the verification of the stage interface method is a model Francis turbine of high specific speed, see fig. 1. The full-scale turbine has a runner diameter of 3.4 m and was designed for a head of 32.7 m at a nominal flow rate of 90 $\frac{m^3}{s}$ with a rotational speed of 166 rpm and a specific speed of 422. The model turbine is a scale model of the original machine with a runner diameter of 0.3 m. The experiments were carried out in the hydraulic turbine test stands of Sulzer Hydro in Zürich. Standard measurement techniques for modern hydraulic practice were used for the derivation of all performance parameters.

The velocity near the draft tube outlet was measured by means of a

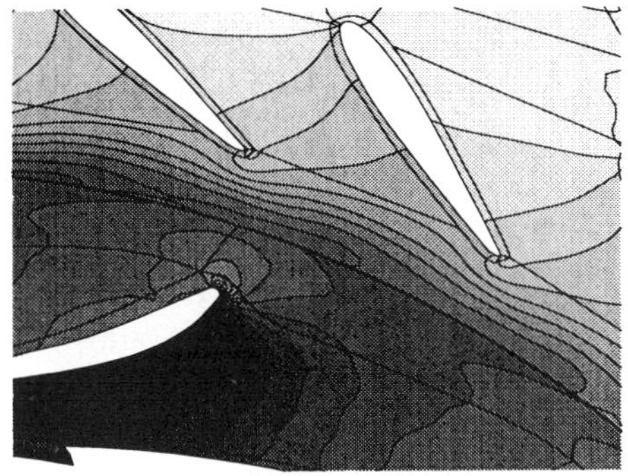

Figure 2. Pressure contours towards the interface between wicket gate and runner

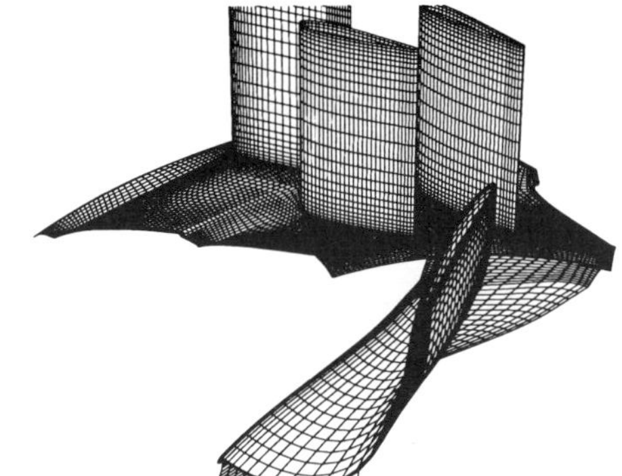

Figure 3. Computational grid: distributor and runner

propeller anemometer. Meausurements of the detailed flow velocities were also carried out in the hydraulic turbine test stands of Sulzer Hydro using L2F laser velocimetry. Flow traverse planes were established at the runner inlet and at the runner outlet by insertion of suitable windows in the casing, see fig. 1. Flow velocities at each of these measurement planes were obtained at five operating points of the turbine. These operating points were selected to provide a good overview of the flow at the best operating point and at a variety of different off-design conditions.

3.2. THE COMPUTATIONAL GRID

Fig. 1 shows a part of the spiral casing, the stay vanes, the wicket gate, the runner and the beginning of the draft tube and also the locations of the measurement planes and the position of the grid interfaces between wicket gate and runner and between runner and draft tube.

Grids were generated with the ICEM-CFD software for the wicket gate, the runner and the draft tube separately and for combinations of these three components with interface planes. For simulations of the whole machine the distributor, i.e. stay vanes and wicket gate, was modelled as a single unit consisting of two stay vanes and three guide vanes for which different grids were needed for each wicket gate opening. Two different types of grid topology were used for the wicket gate and the runner grid:

- a combination of block structured H-grids only
- a combination of O-grids around the blades inserted within H-grids

The latter more sophisticated grid topology (see fig. 3) was developed for better accuracy near the blades and to improve the convergence rate with less skewed grid cells, especially near the trailing edge. The grid for the whole distributor also consists of a combination of O-grids around the blades with H-grids. The draft tube is discretised by a butterfly grid.

All grids are refined towards the walls in order to get a good prediction of the boundary layer with reasonable values of y^+. In all calculations the values of y^+ were between 10 and 300, and in most cases between 20 and 100. The effect of grid refinement on the computed velocity field has been examined for the runner and the wicket gate and has been found to be small. The results for a runner grid consisting of 27,000 grid points are not largely different from those for a grid of 128,000. The calculated efficiency depends slightly on the grid size within a range of 1 percent.

4. Results

This section presents the results of the CFD simulations and compares these with measurement data. Where possible, useful details of the simulations are highlighted to explain features of the flow pattern of relevance to the simulation and to the performance of the turbine. The simulations are used to highlight the advantages of the full stage simulation over simulations in which the individual components are calculated separately.

4.1. SIMULATION OF INDIVIDUAL COMPONENTS

4.1.1. *Wicket Gate alone up to the Location of the Interface Plane*

The simulation of the flow through the wicket gate alone at design point used boundary conditions of fixed mass flow and flow angle at the inlet and average pressure at the outlet. The grid consisted of 55680 nodes.

The upper part of fig. 4 shows the comparison between experimental data and numerical prediction at the measurement plane close to the wicket gate outlet. The circumferential component of velocity is well predicted. The measurement of the radial velocity component shows a strong increase towards the band because of the strong curvature of the band in the runner region. The numerical simulation predicts a stronger increase of the radial velocity towards the band because the simple outlet boundary condition does not adequately represent the interaction of the flow with the runner. Note that a small part of the measurement plane between the stator and the rotor is not part of the stator grid. Therefore the comparison between experimental data and prediction ends shortly before the band.

This comparison demonstrates the need to account properly for the interaction between the gate and the runner. Without the stage capability the computational domain of the wicket gate needs to be extended artificially into the runner region with "source terms" to model the effect of the runner, or a known pressure profile must be prescribed at the outlet of the gate.

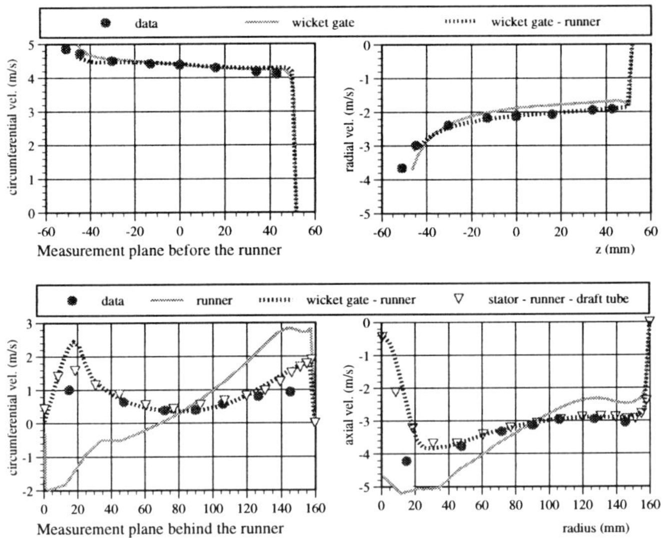

Figure 4. Comparison measurement data and predictions before and behind the runner

4.1.2. *Simulation of the Runner alone*

The runner flow is simulated with a constant energy distribution and a fixed flow angle as inlet condition and average pressure at the outlet. The inlet flow angle results from the wicket gate simulation. The grid consisted of a total of 90390 nodes.

The comparison between experimental data and numerical prediction made at the measurement plane behind the runner can be seen in the lower part of fig. 4. For both the circumferential and the axial component the prediction produces a trend which differs clearly from the measured data. The axial velocity distribution is biased towards the axis of the machine with a corresponding swirl distribution. It can be inferred from the simulations shown below that this is due to the strong influence of the velocity and pressure distribution at the runner inlet (which is very closely spaced to the runner blades) on the flow at the runner outlet.

4.2. SIMULATION OF COMBINED WICKET GATE AND RUNNER

The wicket gate and the runner are joined as one computational domain whereby each part is calculated in its own frame of reference, separated by a sliding interface. The inlet condition at the wicket gate inlet is the same as in Section 4.1.1 (mass flow and flow angle are fixed). The outlet condition below the runner is the same as in Section 4.1.2 (average static pressure). The conditions at the sliding interface are an implicit result of the calculation. The grid consisted of a total of 146,070 nodes.

The comparison of the prediction with experimental data at the measurement plane between wicket gate and runner can be seen in the upper part of fig. 4. The differences between the calculations of single components and the calculation including the interaction of the wicket gate and the runner can be clearly seen. The agreement of both circumferential and radial velocity components with the experimental data is very good. The interaction between wicket gate and runner affects the velocity distribution within the domain of the wicket gate leading to a more accurate prediction.

Fig. 4 also demonstrates the good agreement between experiment and numerical prediction in the measurement plane behind the runner for both the axial and the circumferential component of the velocity. The improvement at the runner outlet results from the better prediction of the flow at the runner inlet by the sliding interface.

4.3. SIMULATION OF THE WHOLE MACHINE

The successful simulation of the combined runner and wicket gate with the new stage capability encouraged us to take a further step, that is the simulation of the whole machine including the interaction between the stay vanes, wicket gate, runner and draft tube.

The inlet conditions to the stay vanes for this simulation result from a circumferential average of a separate spiral casing calculation. Behind the draft tube outlet, the computational domain is extended in order to remove any disturbance of the outlet boundary condition on the draft tube flow. At the draft tube outlet the static pressure is set constant. The grid consisted of a grand total of 265, 795 nodes.

The comparison between test data and prediction behind the runner outlet shows a further slight improvement in the predictions of the circumferential component of the velocity compared to wicket gate and runner computation, see lower part of fig. 4. This demonstrates that there is a small but not insignificant influence of the interaction with the draft tube on the flow in the runner. Fig. 5 shows a comparison of the simulation of the draft tube alone and the simulation of the whole machine with measured data near the draft tube outlet. Both computations show a good agreement between measurement and prediction. The flow through the draft tube alone was simulated by prescribing the circumferentially averaged velocity distribution resulting from the wicket gate - runner simulation at the draft tube inlet. The outlet condition (constant pressure) was the same as for the simulation of the whole machine.

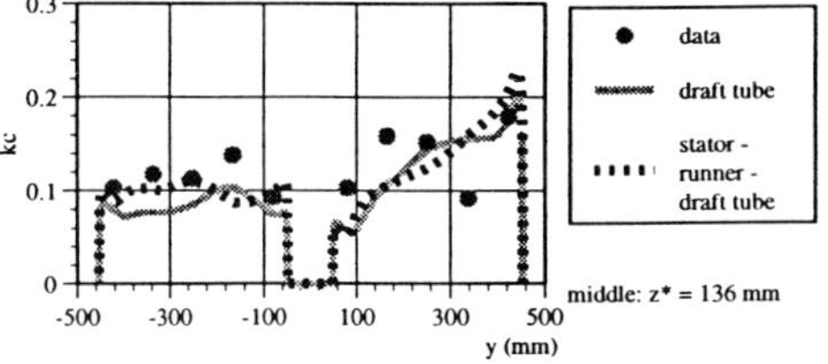

Figure 5. Comparison measurement data and predictions near the draft tube outlet: nondimensionalized velocity

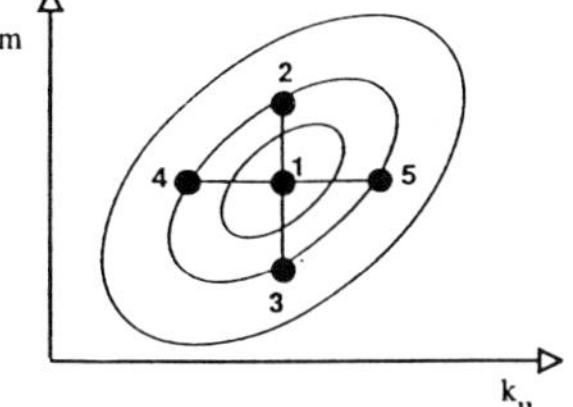

Figure 6. Hill Chart: Situation of the five operating points examined in this paper

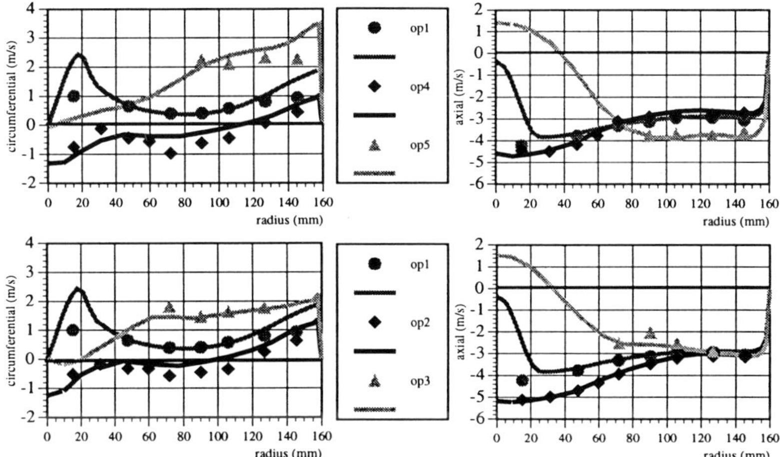

Figure 7. Off-Design: Measurement plane behind the runner. Data shown by symbols, calculated results shown by curves.

4.4. OFF-DESIGN POINTS

Calculations were carried out for the four off-design points shown in the simplified hill chart in fig. 6. The comparison of the measured and calculated circumferential and axial velocity components at the runner outlet (see fig. 7) demonstrates the reliability of the computational method in detail. The good agreement shows that the component interactions are well predicted at both design and off-design points. Further details of the simulations of the whole machine including hill chart prediction are presented in a companion paper (Keck, Drtina and Sick, 1996).

5. Conclusions

The main conclusions of this work are, as follows:

- The stage calculation procedure as incorporated in TASCflow allows for the simultaneous simulation of stationary and rotating frame components by use of a mixing plane at sliding interfaces.
- The mixing plane interface supports circumferential pressure variations due to elliptic effects.
- The stage calculation procedure provides better predictions of the flowfield both for design and off-design operating points as it takes into account the steady-state interactions between all components.
- The most important interaction is seen to be the effect of the runner on the wicket gate and vice versa. Calculations which do not include this interaction may be substantially in error at the runner outlet.

The results show with encouraging clarity that the stage calculation method presented here is a viable method for the hydraulic analysis of a complete turbine. Further work, described in the companion paper (Keck *et al.*, 1996), demonstrates the ability of the method to predict losses and hill charts for a complete turbine.

References

Casey,M.V., Borth,J., Drtina,P., Hirt,F., Lang,E., Metzen,G. and Wiss,D. (1995) The Application of Computational Modeling to the Simulation of Environmental, Medical and Engineering Flows, Speedup Journal, vol 9/2, pages 62-69

Dawes,W.N.(1992) Towards improved throughflow capability: the use of three dimensional viscous solvers in a multistage environment, Trans. ASME Journal of Turbomachinery, vol 114, pages 8-17

Denton,J.D. (1983) An improved time-marching method for turbomachinery flow computation, Trans. ASME Journal of Engineering for Power, vol 105, pages 514-523

Denton,J.D. (1992) The calculation of three-dimensional viscous flow through multistage turbomachines, Trans. ASME Journal of Turbomachinery, vol 114 pages 18-26

Galpin,P.F., Broberg,R.B. and Hutchinson,B.R.(1995) Three-dimensional Navier-Stokes predictions of steady state rotor/stator interaction with pitch change, Third Annual Conference of the CFD Society of Canada, June 25-27, 1995, Banff, Alberta, Canada

Goede,E., Cuenod,R. and Grunder,R.(1990) An advanced flow simulation technique for hydraulic turbomachinery, IAHR Symposium, Belgrade, September 1990

Keck,H., Goede,E. and Pestalozzi,J. (1990) Experience with 3D Euler flow analysis as a practical design tool, IAHR Symposium, Belgrade, September 1990

Keck,H. Goede,E. and Sebestyen,A. (1994) 3-Dimensionale Stroemungsberechnungen in Pumpturbinen, VDI-Kolloquium Wasserkraftanlagen, Dresden, June 28, 1994

Keck,H., Drtina,P. and Sick,M. Numerical Hill Chart Prediction by means of CFD Stage Simulation for a Complete Francis Turbine, 18th IAHR Symposium Valencia, Spain

Raw,M.J. (1994) Coupled Algebraic Multigrid for the Solution of the Discretized 3D Navier-Stokes Equations, CFD 94, CFD Society of Canada, Toronto

TASCflow Theory Documentation, Version 2.4., Advanced Scientific Computing Ltd., Waterloo, Ontario, Canada (1995)

SIMULATION OF FLOW THROUGH FRANCIS TURBINE BY LES METHOD

Charles C. S. Song & Xiangying Chen
Univ. of Minnesota, Mpls, MN, USA
Toshiaki Ikohagi
Tohoku Univ., Sendai, Japan
Johshiro Sato, Katsumasa Shinmei & Kiyohito Tani
Hitachi Ltd., Hitachi-shi, Japan

I. INTRODUCTION

The traditional approach of Francis turbine design which is based on the steady potential flow theory and heavily dependent on model testing and engineering experience, has come a long way in producing efficient and relatively cavitation free turbines. But further improvement of performance for design and off design operating conditions will be extremely difficult with the traditional method because it will depend more on those phenomena such as, boundary layer separation, vortex dynamics, interactions between different components, vibrations, etc., which are not predictable with conventional approach and difficult to measure in physical models.

In recent years the Euler Equation Model has became quite popular, but it still can't deal with the problems listed above. The Reynolds time averaged method such as the k-e model is also being widely tried, but aside from the time averaged boundary layer flow separation effect, most of the problems listed above still remain intractable.

This paper describes recent development and application of the compressible hydrodynamics theory based Large Eddy Simulation model on simulation of flow through spiral case, runner, and draft tube of a Francis turbine system. The results described herein are mostly based on numerical simulations through each of the three basic units separately because the development of a totally unified model has not been completed. But the importance of mutual interactions is suggested in this paper. Some of mutual interaction phenomena are described in a companion paper entitled SIMULATION OF A PUMP-TURBINE.

II. GOVERNING EQUATIONS AND METHOD OF SOLUTION

The governing equations for compressible hydrodynamic flow or weakly compressible flow derived by Song and Yuan (1988), also see Chen (1995), can be written in a conservative form as

$$\frac{\partial G}{\partial t} + \nabla \bullet F = 0 \qquad (1)$$

where

E. Cabrera et al. (eds.), Hydraulic Machinery and Cavitation, 267–276.

$$G = \begin{bmatrix} p \\ u \\ v \\ w \end{bmatrix} \tag{2}$$

$$F = \begin{bmatrix} Ku & Kv & Kw \\ uu + \frac{p}{\rho_o} - \frac{\tau_{xx}}{\rho_o} & uv - \frac{\tau_{xy}}{\rho_o} & uw - \frac{\tau_{xz}}{\rho_o} \\ uv - \frac{\tau_{yx}}{\rho_o} & vv + \frac{p}{\rho_o} - \frac{\tau_{yy}}{\rho_o} & vw - \frac{\tau_{yz}}{\rho_o} \\ uw - \frac{\tau_{zx}}{\rho_o} & vw - \frac{\tau_{zy}}{\rho_o} & ww + \frac{p}{\rho_o} - \frac{\tau_{zz}}{\rho_o} \end{bmatrix} \tag{3}$$

Song and Chen (unpublished) has shown that this set of equations are the Navier-Stokes equations for small Mach number flows and contains the effect of compressibility inside the time boundary layer (inner solution). For an example, the flow due to impulsively started circular cylinder is highly transient and compressibility dependent within a small time interval, $\delta_t = 5D/a_o$, where D is the diameter and a_o is the speed of sound. Its outer solution is the incompressible flow solution regardless of the value chosen for a_o or the Mach number, M. This set of equations is hyperbolic and can be efficiently solved with a simple explicit method using artificially large value of M.

The idea of LES method is to directly calculate the large scale motions that can be resolved with the finite grid volume averaged quantity but use a subgrid scale turbulence model to approximate the effect of unresolvable small scale eddies. The finite volume averaged equations are identical to Eqs. 1, 2, and 3 except that the shear stress terms should be regarded as the sum of the viscous shear stress and the subgrid scale turbulent shear stress. The subgrid scale turbulent stress can be written as,

$$\tau = \tau_l + \tau_t = \rho(\nu + \nu_t)\left\{\frac{\partial u_i}{\partial x_j} + \frac{\partial u_j}{\partial x_i}\right\} \tag{4}$$

where ν_t is the subgrid scale eddy viscosity. We use the simple model due to Smagorinsky (1963) as follows.

$$\nu_t = (C_s \Delta)^2 (2\overline{S}_{ij}\overline{S}_{ij})^{0.5} \tag{5}$$

$$\overline{S}_{ij} = \frac{1}{2}\left(\frac{\partial u_i}{\partial x_j} + \frac{\partial u_j}{\partial x_i}\right) \tag{6}$$

In Eq.5 Δ represents the grid size and C_s is an unknown parameter to be determined. For homogeneous isotropic turbulence, Lilly (1966) analytically determined that $C_s \approx 0.23$. Deardorff (1970) found that $C_s = 0.1$ is optimal for turbulent channel flow. For present study we found that using $C_s = 0$ on the wall and increase it to 0.12 outside of the boundary layer makes the numerical results agree with experimental data well.

For the flow through a runner, it is more convenient to use the equations transferred into a rotating coordinate system. Denoting Ω the rotational speed about the z-axis, the governing equations can be transferred to

$$\frac{\partial G}{\partial t}+\nabla \bullet F=T \tag{7}$$

where,

$$T=\begin{bmatrix} 0 \\ 2\Omega v+\Omega^2 x \\ -2\Omega u+\Omega^2 y \\ 0 \end{bmatrix} \tag{8}$$

and G and F are given by Eqs. 2 and 3, respectively. All velocity components now represent the relative velocity with respect to the rotating coordinates.

The governing equations, Eqs.1 & 7, are first averaged over a finite volume $\Delta\forall$ before discritization. During the process of integration, the volume interaction of $\nabla \bullet F$ is converted into a surface integral using the divergence theorem, and the following equation is produced.

$$\frac{\partial \overline{G}}{\partial t}=-\frac{1}{\Delta\forall}\int_s n \bullet F dS+\overline{T} \tag{9}$$

In the above equation bar represents the volume averaged value and n is the unit outward normal vector on the bounding surface S.

MacCormack's (1969) predictor corrector method is used for numerical integration. For detailed description of the method, reference is made to Chen's thesis (1995).

III. GRID SYSTEMS

In earlier days when computers were not so powerful, only very limited space such as one or two flow passages of a runner was isolated from the total system and modeled. In a complex and compact system, such as the spiral case-stay vanes-wicket gates-runner-draft tube of a Francis turbine facility, a small localized model cannot be expected to simulate all the complex flow phenomena. The power of the computer and the computational method can now justify development of a large model containing many related components.

Fig.1 is a typical view of boundary surface grid system of a spiral case containing 20 stay vanes and 20 wicket gates. Note that only small part of the grids are shown in this figure. The flow field is divided into 44 zones, having a total of 78,381 nodes and 58,752 elements. Grids in any of 44 zones can be subdivided into finer grids if so desired. One of the most difficult problem is how to determine the boundary condition on the artificial boundary so that the influence of the unmodeled part can be accurately accounted for. For an example, the most logical boundary condition at the downstream end of the spiral case model is

$$\frac{\partial f}{\partial r} = 0 \tag{10}$$

where f represents p, u, v, and w and r is the radial coordinate. Experience shows, however, that there is a sharp turn of the flow toward the z direction and the boundary condition causes significant amount of error on the velocity and pressure distribution near the spiral case exit or the runner entrance. To improve on this deficiency, an enlarged region as sketched in Fig. 2 was used to model the flow through the spiral case. The crown and shroud of the runner, excluding the runner, plus a fictitious draft tube are attached to the spiral case. The downstream boundary of the model is shifted to the end of the fictitious draft tube. As will be seen later, the velocity profile at the entrance to the runner is significantly modified.

At this stage of the modeling effort, only partial flow passage, has been used to model the runner. It was found by Song and Chen (1991), two flow passage model is actually more efficient than the more commonly used single flow passage model. Two types of draft tubes, one with and one without a divider wall namely a draft tube pier, were modeled.

IV. BOUNDARY CONDITIONS

Expecting upstream effect to be more important than the downstream effect, the simulation process starts with the spiral case, followed with the runner and end with the draft tube. At the entrance to the spiral case, the flow is assumed to be the fully established turbulent flow and the time averaged velocity profile is specified. The pressure is assumed to be constant on the cross-section. No fluctuating components of velocity and pressure were included at the upstream end although it is possible to do so.

As mentioned in the last section, the downstream boundary condition of the spiral case model is applied at the downstream end of the fictitious draft tube. Here the zero gradient condition, Eq. 10 with r replaced with z is imposed. The partial slip condition as described by Chen (1995) is applied to all solid boundary.

The velocity and pressure distribution at the junction between the spiral case and the runner calculated for the spiral case model is used as the upstream boundary condition for the runner model. The zero gradient condition is applied at the downstream end of the runner model.

Because only a partial model is used for the runner, it is not possible to apply a realistic boundary condition at the inlet of the draft tube. For this reason only uniform velocity distributions with and without swirl is applied at the upstream end of the draft tube. As usual, the zero gradient condition is applied at the down stream end of the draft tube.

V. RESULTS

5.1 Calculation Conditions

Fig.3 shows the hill chart determined by the physical model test. The model test was conducted in accordance with International Electro-technical Commission codes. The computed characteristics are compared with measured values at several wicket gate openings such as point A, B, C, D and E in Fig.3.

5.2 Typical Calculated Flow Pattern and Pressure Distribution

The computed time averaged velocity vector field on the middle cross section of the spiral case near some of the stay vanes and wicket gates for 80 % opening condition is shown in Fig.4. The flow appears to behave well and no significant flow separation exists. The corresponding dimensionless pressure distribution is shown in Fig.5. Here, the velocity at the entrance and the pressure at the exit were used as the reference quantities. The time averaged pressure distribution on the solid boundaries of the spiral case at 80 % opening is shown in Fig.6. It appears that the flow in the spiral case is subjected to a slight adverse pressure gradient suggesting that there is a room for some improvement. The computed time averaged velocity distribution at the exit of the wicket gates at three different cross sections are plotted in Fig.7. Note that the angle is measured from the first stay vane and the subscripts t, m, and z, represent circumferential, meridian, and vertical components of the velocity. It is also impotent to note that, without adding the runner casing and the fictitious draft tube, there would be no vertical velocity component. The fact that the velocity on the lower cross section is higher than that of the upper cross section is also due to the flow turning effect which can't be modeled with the zero gradient boundary condition.

Even though only two flow passage model was used to simulate the runner, very fine grids were used and as much computational time was used for the runner as for the spiral case to insure good accuracy. It appears that there is a small flow separation on the suction side of the blade near the leading edge and the hub. Assuming complete symmetry, the pressure distribution on the entire runner can be assembled from the results of the partial model as shown in Fig.8.

Typical time averaged velocity distributions in a draft tube with central divider using the time averaged out flow from the runner for the case of 80% wicket gate opening are shown in Fig.9. The corresponding figure for 60 % wicket gate opening case is shown in Fig.10. Large eddies and flow separation caused by the increased swirl is noteworthy. Although time averaged inflow condition is used, the computed flow in the draft tube is very unsteady. There are unsteady flow separations from the outer walls and horse shoe vortex on the upper part of the central pier. These eddies interact each other and exhibit preferred frequencies of oscillations. Fig.11 shows an instantaneous pressure distribution on the walls of the draft tube with center wall. When the center wall is removed the draft tube becomes a diffuser of large angle and the flow separation becomes much more severe. Flow separates alternatively from two side walls and the jet like flow swings from one wall to the other wall. Fig.12. shows an instantaneous pressure distribution on the wall of the draft tube without divider wall.

All calculated data are normalized by the value of the best efficiency point of the model test. The calculated power outputs of a model turbine at five gate opening conditions, 60%, 70%, 80%, 90%, and 100%, are compared with the model test data in Fig.13. No analysis was made to determine why the agreement is best at the 80% opening case which happens to be the design point. The calculated head loss through each of the three components as functions of the wicket gate opening are shown in Fig.14. The solid line Clearly indicate that most of energy loss in spiral case occurs at wicket gates.

The over all efficiency predicted numerically is compared with the measured value in Fig.15. Here an estimated friction loss and the leakage loss of 1.5% has been added to the computed loss because the numerical simulation doesn't account for this kind of losses.

V. CONCLUSIONS

Although a Large Eddy Simulation approach was used there may still be a significant amount of numerical error because rather coarse grids were used and the coefficient, C_s, need to be selected. As demonstrated in Fig.7 there is a strong downstream effect on the exit velocity distribution of the spiral case. This non uniform exit velocity should produce strong unsteadiness effect on the runner. The calculated flow in the draft tube is not expected to be very realistic because the real outflow from the runner is expected to be much deferent from what was used as the inflow to the draft tube model. The out flow from the runner should be unsteady and much more complex.

The current computation is largely based on three individual models performed in series from spiral case to draft tube. Therefore, much of the detailed interactions between units, especially the unsteady vortex phenomena, are not obtained. The next objective of the development should be an integrated model.

VI. ACKNOWLEDGMENTS

The entire computations were carried out with the super computing facility of the Minnesota Super computer Institute of the University of Minnesota through its Super computer Research Grant program. The authors are very grateful for this support.

VII. REFERENCES

Chen, X.Y. (1995) Multi-Dimensional Finite Volume Simulation of Fluid Flows on Fixed, Moving and Deforming Mesh Systems, Univ. of Minnesota, Ph.D. Thesis.

Deardorff,J.W. (1970) A Numerical Study of Three-Dimensional Turbulent Channel Flow at Large Reynolds Number, *Journal of Fluid Mechanics,* Vol. 41, 492-480.

Lilly, D.K. (1966) On the Application of the Eddy Viscosity Concept in the Inertial Subrenge of Turbulence, *NCAR Manuscript,* No. 123.

MacCormack, R.W. (1969) The Effect of Viscosity in Hyper Velocity Impact Crating, *AIAA Paper,* No. 69-354.

Song, C.C.S. and Yuan, M. (1988) A Weakly Compressible Flow Model and Rapid Convergence Methods, *Journal of Fluid Engineering,* Vol. 110, 441-445.

Song, C.C.S. and Chen, X.Y. (1991) Calculation of Three-Dimensional Turbulent Flow in Francis Turbine Runner, *Waterpower'91,* Denver, July 24-26.

Song, C.C.S. and Chen, X.Y. (unpublished) Compressibility Boundary Layer Theory and Its Significance in Computational Hydrodynamics.

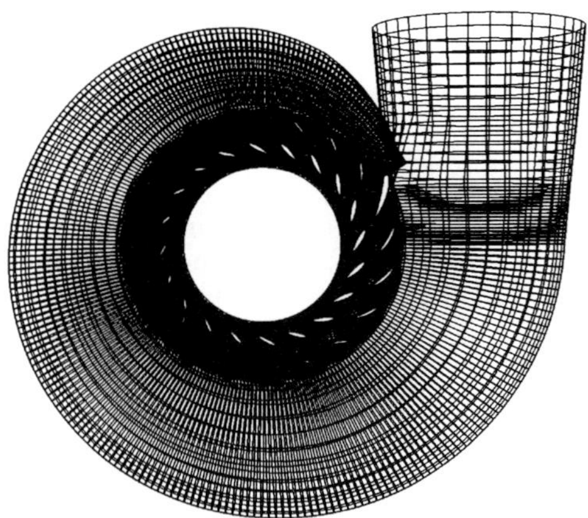

Figure 1 A top view of the mesh system on selected surfaces for spiral case computation.

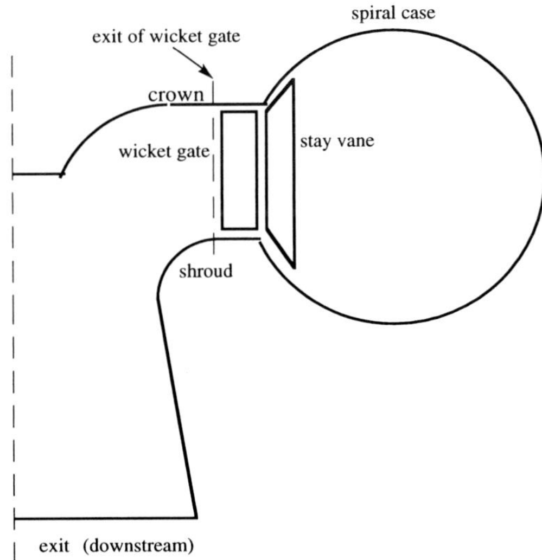

Figure 2 A sketch of the configuration for the computation of spiral case.

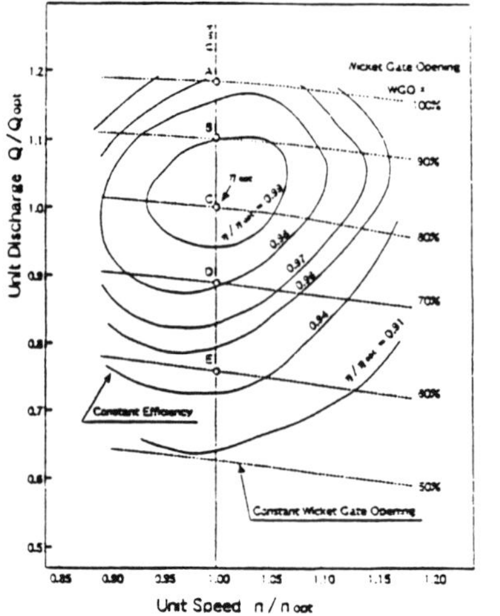

Figure 3 Measured model hill chart.

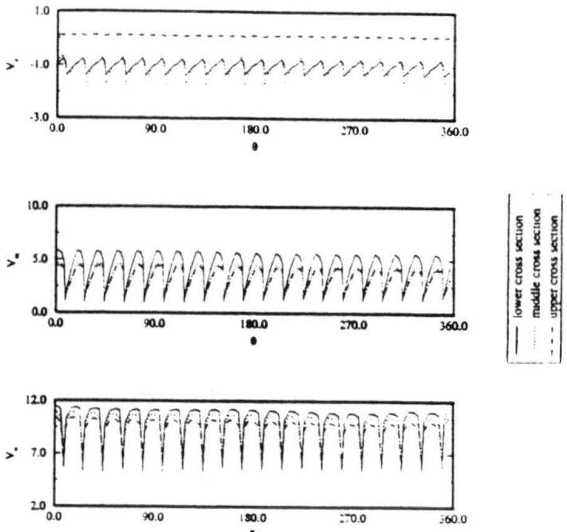

Figure 7 Computed time-averaged velocity components at the exit of wicket gates for WGO = 80%.

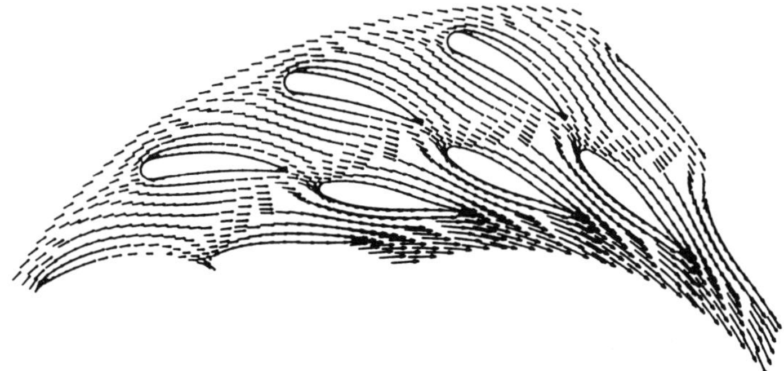

Figure 4 Computed time-averaged velocity field around several stay vanes and wicket gates for WGO = 80 %.

Figure 5 Computed time-averaged pressure field around several stay vanes and wicket gates for WGO = 80 %.

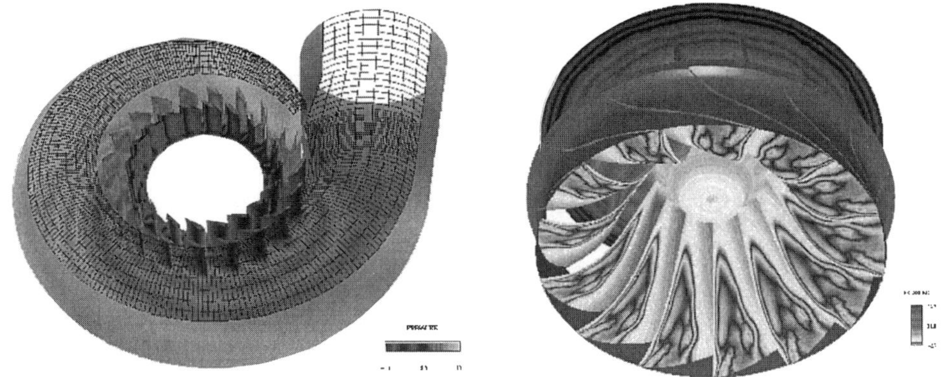

Figure 6 computed time-averaged pressure distribution on the solid surfaces of the entire spiral case for WGO = 80 %.

Figure 8 Computed time-averaged pressure distribution on the solid surfaces of the turbine runner for WGO = 90 %.

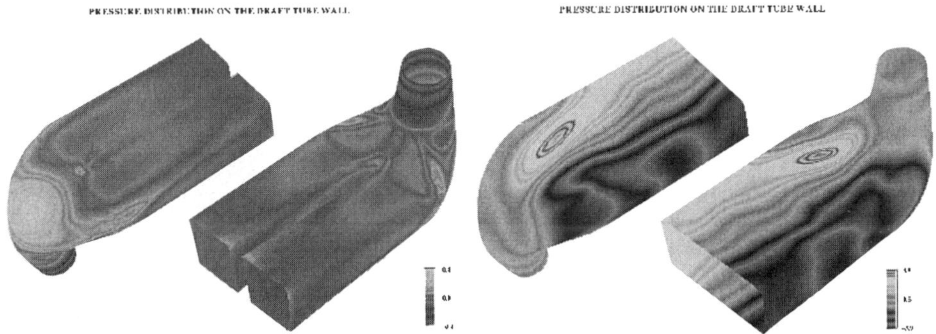

Figure 11 Instantaneous pressure distribution on the solid surfaces of the draft tube with a center pier.

Figure 12 Instantaneous pressure distribution on the solid surfaces of the draft tube without a center pier.

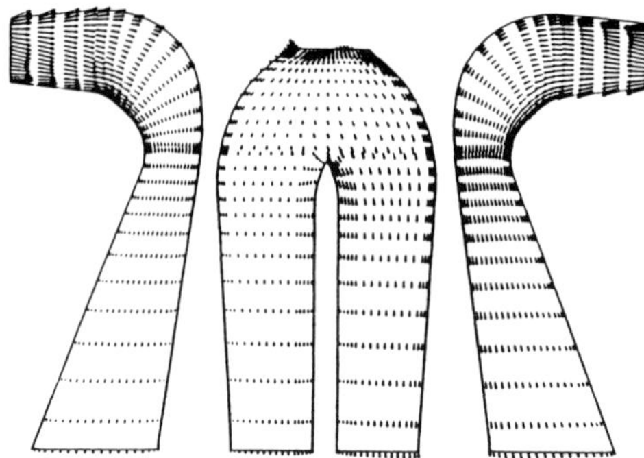

Figure 9 Time-averaged velocity field with inlet swirling flow for WGO = 80 %.

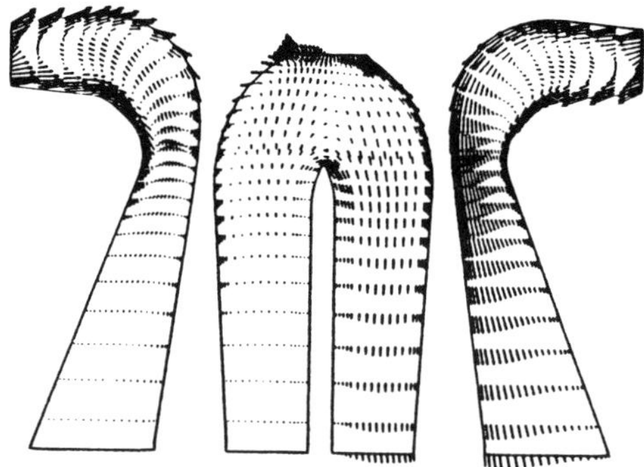

Figure 10 Time-averaged velocity field with inlet swirling flow for WGO = 60 %.

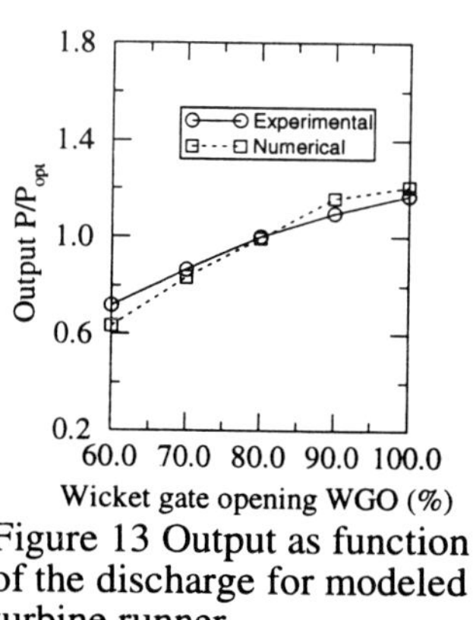

Figure 13 Output as function of the discharge for modeled turbine runner.

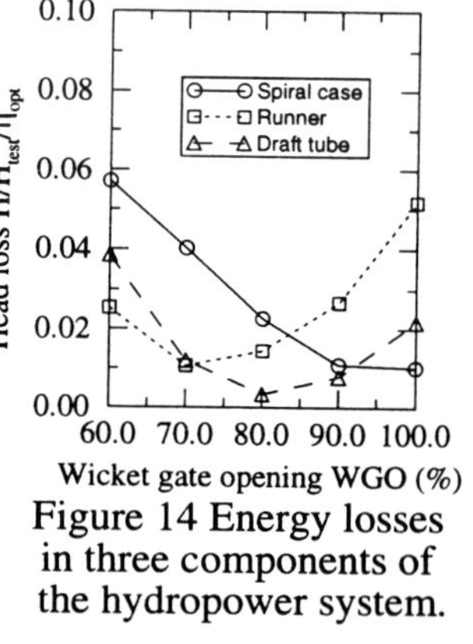

Figure 14 Energy losses in three components of the hydropower system.

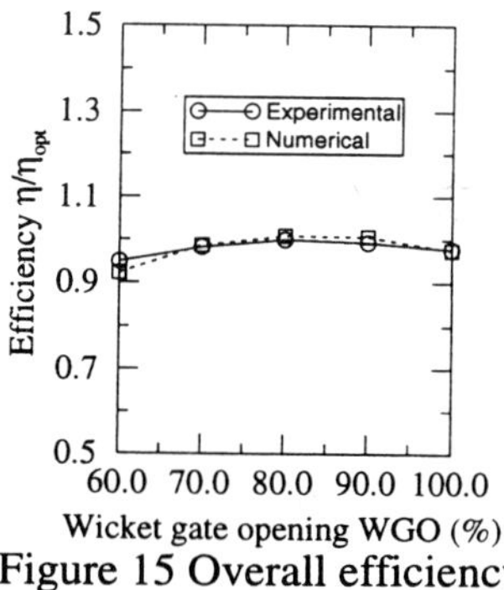

Figure 15 Overall efficiency of entire hydropower system.

SIMULATION OF FLOW THROUGH PUMP-TURBINE

Charles C. S. Song & Changsi Chen
Univ. of Minnesota, Mpls, MN, USA
Toshiaki Ikohagi
Tohoku Univ., Sendai, Japan
Johshiro Sato, Katsumasa Shinmei & Kiyohito Tani
Hitachi Ltd., Hitachi-shi, Japan

1. INTRODUCTION

For years Francis - type reversible pump - turbines have been applied to many pumped storage power plants over the world. So far, they have been used under peak shaving generating operation with AFC (automatic frequency control) function during daytime, and plain pumping - energy consuming - operation at night time. Recently, an adjustable speed pump - turbine was realized in Japan. It begins to add a remarkable AFC function in pumping mode to the conventional pump-turbines.

Figure 1 schematically shows the range of pump input variation of a typical adjustable - speed pump - turbine. The line of rated speed is located usually in the middle of the whole operating area. There is a line of maximum speed on the upper - right side. A line of minimum speed is on the lower - left side. These two lines are determined mainly by the specifications of the frequency converters. On the other hand, there is a line of runner inlet cavitation on the pressure side on the upper - left area of the range. The line of reverse flow at the inlet of the runner is located on the lower - right side. These two lines are determined by the hydraulic performance of the pump - turbine.

In order to make the range of pump input variation larger (to make AFC operation more effective), it is necessary to get those two lines (cavitation and reverse flow) farther from the range. Hitachi has been using various CFD techniques including the industry standard steady - state turbulent flow analysis using $k-\varepsilon$ model to improve those hydraulic performances of the pump - turbine. It was shown that the cavitation inception is predicted very well by industrial standard technique. However, the reverse flow at the runner inlet as well as other flow related to low pump discharge were not simulated very well by the conventional steady - state simulation techniques. In this report the flow in the pump - turbine in pumping mode is simulated by Large Eddy Simulation technique, and it is shown that unsteady flows play very important roles in the hydraulic performance of the pump - turbine.

In turbine mode it is of great importance to further improve the efficiencies not only at an optimum point but at off-design points for both the conventional pump - turbines and the adjustable pump - turbines. A lot of analytical works have been made by using industry standard techniques, however, as we analyze the flow in the turbomachineries more and more in details, we begin to feel that the unsteadiness of the flow should be taken into account. In this report the flow in the pump - turbine in turbine mode is simulated by LES technique, and it is shown that vortex shedding exists even in turbine mode.

E. Cabrera et al. (eds.), Hydraulic Machinery and Cavitation, 277–283.

2. FLOW IN PUMPING MODE

The flow through runner and spiral case in pumping mode is basically an expending flow with strong downstream effect which is intrinsically unstable. For this reason, it is extremely difficult to model the runner and the spiral case separately. Because there is a strong downstream effect, a proper downstream boundary condition of a runner model cannot be determined a priory. Many such trials failed to produce a convergent solution. Therefore, it was decided to develop a complete model containing the runner and the spiral case with all the runner blades, wicket gates, and stay vanes. An overall view of this model is shown in Fig. 2. The inflow and the outflow boundary conditions are relatively simple to specify. The model turned out to be quite stable but requires large amount of memory and computer time. Some of the more interesting results are described below.

2.1 FLOW SEPARATION ON SHROUD SURFACE

The designer of this machine has been bothered by the existence of reverse flow near the leading edge of the runner blades and sudden drop in efficiency when flow rate is reduced below a critical point. This phenomenon was observed in a physical model but the flow detail and the reason for its occurrence was not clear. The output from the mathematical model with the help of numerical flow visualization technique enabled detailed study of the flow near the junction between the runner blades and the shroud surface. Boundary layer flow separation is observed on the shroud surface slightly upstream of the runner blade when the model flow rate is equal to or slightly less than 0.182 cms. The separation bubble is rather thin but extends beyond the leading edge of the runner blades and affects all blades. As shown in Fig. 3, reverse flow can be observed near part of a runner blade submerged under the separation bubble. Because this separation bubble is not very thick, there is no reverse flow in the mid span part and near the crown of the flow passage. This type of flow separation does not occur for larger flow rate. For an example, no separation can be observed when the discharge is equal to 0.200 cms. The most likely reason for the flow separation on the shroud surface is the existence of large adverse pressure gradient due to small wicket gate opening which constrict the flow downstream of the runner.

2.2 MOVING SEPARATION IN STAY VANE CASCADE

Another unique problem only a comprehensive model can solve is that of unsteady flow separation on stay vanes that was found to occur for relatively small discharge cases. If observation is concentrated on one vane, then the separation bubble will appear to periodically change its size and intensity. But when all vanes are observed simultaneously, there appears to be four wave pattern that rotates slowly, much slower than the rotational speed of the runner. There appears to be four waves of varying separation in the stay vane cascade. An instantaneous velocity field around a few neighboring vanes is shown in Fig. 4. No detailed analysis of the wave length and wave speed as related to other flow conditions has been carried out.

2.3 PRESSURE OSCILLATION DUE TO RUNNER - WICKET GATE INTERACTION

Very strong pressure oscillation generated by the runner blades and the wicket gates interaction is observed every where in the computational domain. Let the rotational frequency of the runner be F_r, the number of the runner blades be N_r, and the number of wicket gates be N_w. Then the pressure on a runner blade will have a strong wicket gates induced pressure oscillation at frequency equal to F_rN_w. This is the frequency at which a runner blade passes by wicket gates. A typical time history of pressure at a point on a runner blade is shown in Fig. 5. On the other hand the pressure at a point on a wicket gate will have a strong oscillation at the frequency equal to F_rN_r which is the frequency of a wicket gate passing by the runner blades. A typical time history of pressure at a point on a wicket gate is shown in Fig. 6. These two figures also show a beating phenomenon at the frequency roughly equal to F_rN_w / N_r.

3. FLOW IN TURBINE MODE

3.1 STARTING PROBLEM

The numerical model is intended to simulate the actual physical process closely. Therefore, the initial condition as well the boundary conditions should also closely reflect the physical process. A lesson learned is that the runner cannot be given full rotational speed before the flow is nearly established. If the runner is given its full rotational speed too soon, then it will act like a pump generating reverse flow resulting in computational instability. This problem doesn't occur in pumping mode because the runner is an active element in that case, but it is a passive element in turbine mode.

3.2 VORTEX SHEDDING FROM RUNNER BLADES

This particular pump-turbine has six runner blades. Each blade produces a vortex sheet that will detach from the blade and carried by the main stream into the draft tube. Therefore, there are six vortex sheets trailing the runner blade, in this case. However, it is interesting to note that the six vortex sheets quickly merge into three vortex sheets as they move downstream and migrate toward the center line. The detached vortex sheets appear to rotate at slower speed than runner's rotational speed. Fig. 7 shows an instantaneous pressure distribution on a cross - section a short distance downstream of the trailing edges of the runner blades. Fig. 8 shows the corresponding distribution of the vertical velocity component at the same instant and the same cross - section. There are also only three sets of spiral shaped ribbons extending out from the center of cross - section. Three spiral ribbons of relatively high pressure zones and equal number of low pressure zones indicate that there are only three vortex sheets. Notice that the areas of positive velocity correspond to high pressure areas and the areas of negative velocity (reverse flow) correspond to low pressure zones. It would be interesting to extend the present model to include the draft tube and study further evolution of vortex sheets in the draft tube.

4. CONCLUSIONS

In order to meet recent requirement of the customers worldwide, HITACHI has been focusing its effort to develop CFD technique for designing reliable pump-turbines. Among various CFD techniques developed so far LES is the most accurate one and will be the industrial standard in the near future. Present study indicates that flow in each component of a pump-turbine system is so much dependent on other components, a comprehensive model is essential to a good over all design.

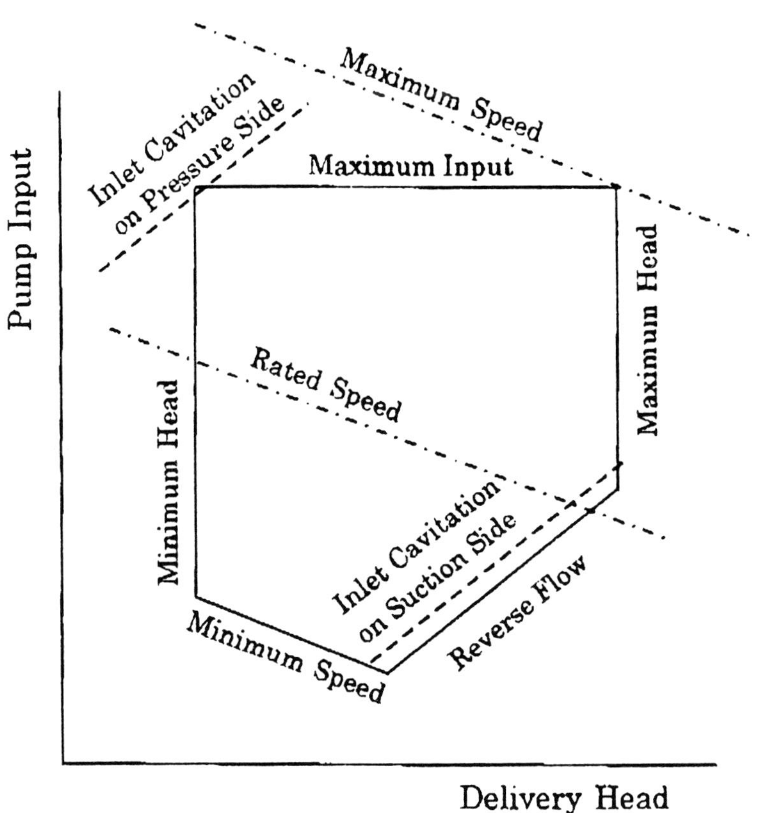

Figure 1 Schematic diagram of pump input variation.

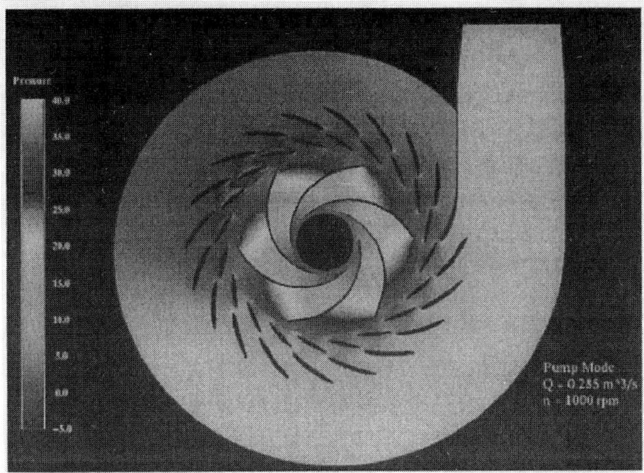

Figure 2 Instantaneous pressure distribution on the middle cross section of pump-turbine system for pump mode (Q=0.285 m^3/s; n=1000 rpm).

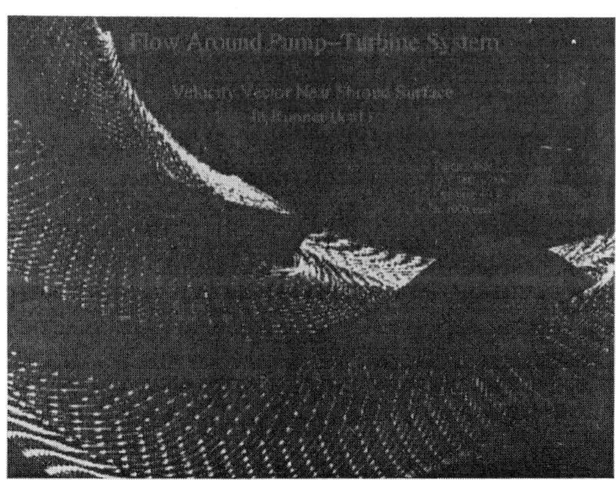

Figure 3 Instantaneous velocity field near shroud surface in runner for pump mode (Q=0.285 m^3/s; n=1000 rpm).

Figure 4 Illustration of instantaneous velocity pattern around vanes for pump mode (Q=0.285 m^3/s; n=1000 rpm).

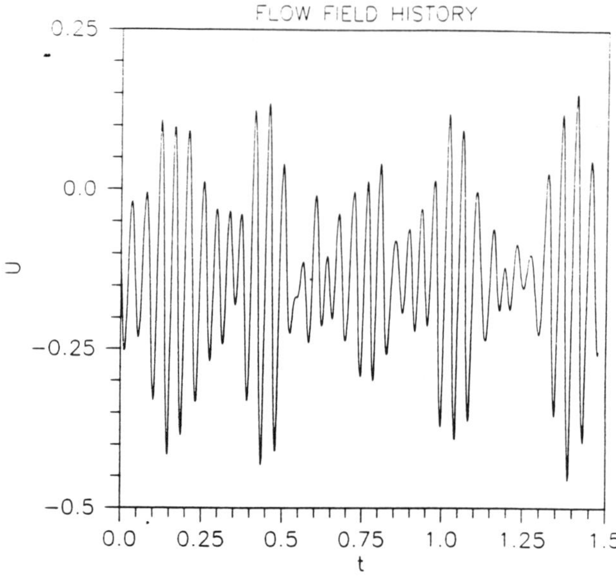

Figure 5 Time history of velocity at a point on a runner blade.

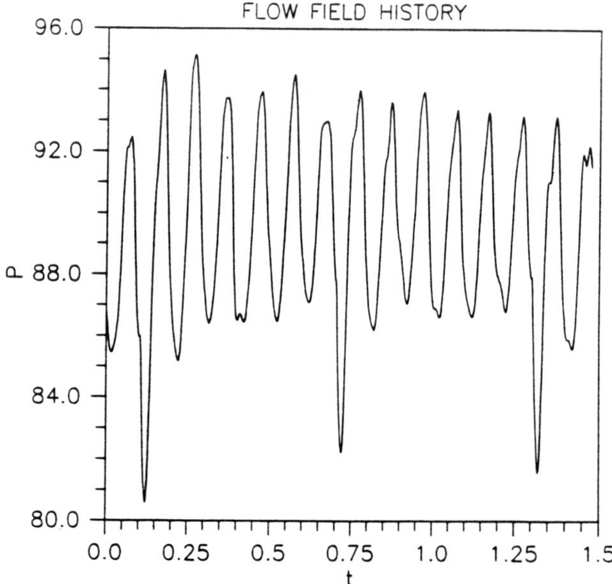

Figure 6 Time history of pressure at a point on a wicket gate.

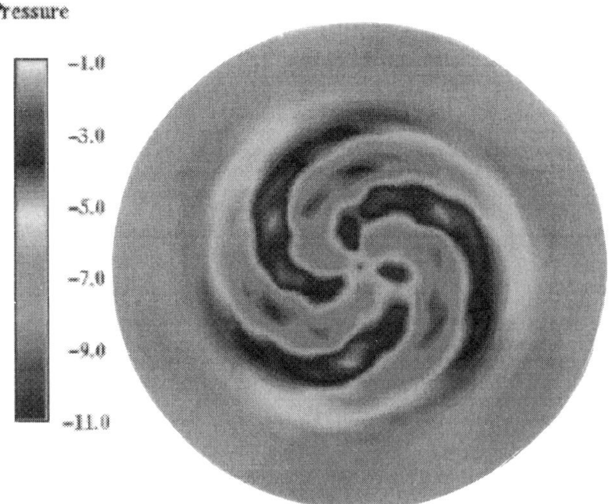

Figure 7 Instantaneous pressure distribution on a cross - section a short distance downstream of the trailing edges of the runner for turbine mode (Q=0.0712 m^3 / s ; n=169.3 rpm).

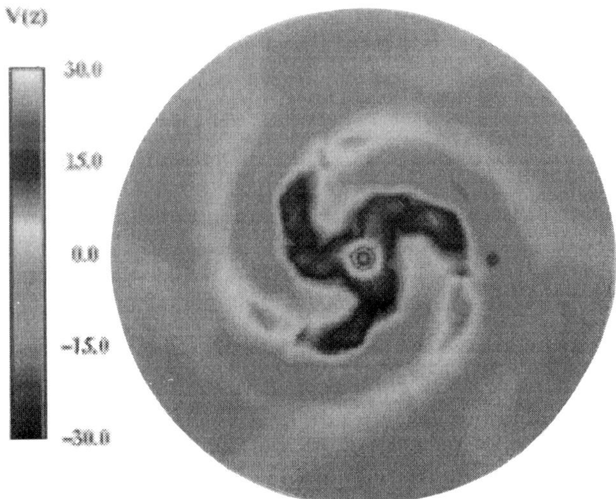

Figure 8 Instantaneous vertical velocity distribution on a cross - section a short distance downstream of the trailing edges of the runner for turbine mode (Q=0.0712 m^3 / s ; n=169.3 rpm).

THE SCALE EFFECT IN KAPLAN TURBINES. NEW RELATIONSHIPS FOR THE CALCULATION OF SCALABLE AND NON-SCALABLE HYDRAULIC LOSSES AND OF THE COEFFICIENT V

I.M. ANTON
Polytechnical University of TIMISOARA
Bd. Mihai Viteazu Nr. 1 TIMISOARA 1900
ROMANIA

Abstract

In this paper new directions are opened concerning the scale effect at Kaplan turbines. The original relationships are developed for the calculation of coefficients δ_M, δ_{ns} and V in the whole domain of operating of Kaplan turbines. Their use needs only the hill diagram of the turbine. Applying a new method there were obtained new relations, for the coefficients δ_{Mo}, δ_{nso} and V_o corresponding to the optimum operating regime of a Kaplan turbine.

1. Introduction

The first paper on this subject was published in 1909 by Cammerer [1]. Since then many researchers worked on the scale effect in Kaplan turbines. S.P. Hutton [2] made a substantial step toward the solution of this problem. The idea of Ackeret [3] to separate hydraulic losses from the turbine in friction and kinetical losses proved to be very useful.

The formula developed by S.P. Hutton [2] was accepted by IEC [4] in 1965 and recommended to predict the efficiency of prototype turbine (P) when one knows the efficiency of the model (M).

J. Osterwalder [5,6] defined the coefficient V of the distribution of hydraulic losses in the turbine and using the results of laboratory measurements established correlations of the type $V=f(Q_x/Q_o)$ for k_u=const. Using such correlations he tried to find the value of V for regimes different from the optimum one.

In order to find out the correct parameters to be used in the formula developed by S.P. Hutton, a lot of research work was necessary. Important contributions are due, among others, to Hutton [7,8], Osterwalder [5,6], Chevalier [9], Fáy [10,11] and Ida [12].

In spite of these the IEC cod from 1992 [13] recommended a unique value for the coefficient V, for all the types of Kaplan turbines and every operating regime.

E. Cabrera et al. (eds.), Hydraulic Machinery and Cavitation, 284–293.

In this paper new relationships are presented, which describe the dependence on the efficiency η_M of the friction losses, δ_M=f(η_M), kinetic losses, δ_{ns}=f(η_M) and coefficient V, V=f(η_M), for the whole operating domain of the model turbine.

A. THEORETICAL BASIS

2. Scalable and nonscalable hydraulic losses in Kaplan turbines

2.1. GENERALITIES

Related to the scale effect formula for turbines, Ackeret [3] suggests the idea to separate the hydraulic losses δ_h into two categories: friction losses, δ and kinetic losses, δ_{ns}. Thus we have

$$\delta_h = \delta + \delta_{ns} \tag{1}$$

$$\eta_h = 1 - \delta_h \tag{2}$$

The hydraulic losses through friction δ depend on the Reynolds number (Re) and on the relative surface roughness (e/D) of the solid flow surfaces, which confer them the property of being scalable losses and thus δ=f(Re, e/D). On the contrary, the kinetic losses δ_{ns}, which are independent of Re and depend only on the geometry of the flow channel, are nonscalable.

The coefficient V of distribution of hydraulic losses in a turbine, which appears explicitly in the scale effect formula of Hutton and Osterwalder, it is defined in the following way:

$$V = \delta/\delta_h = \delta/(1 - \eta_h) \tag{3}$$

The correct use of the scale effect formula for Kaplan turbines needs to know the values δ_M, δ_{ns}, and V for the model for its whole operating domain.

2.2. THE SCALABLE, δ_{Mo} AND NONSCALABLE, δ_{nso} HYDRAULIC LOSSES AT THE OPTIMUM WORKING REGIME

The solution of this problem, i.e. to establish relations of the form δ_{Mo}=f(η_{qo}) and δ_{nso}=f(η_{qo}), according to the method presented in what follows, it is possible when one has the hill diagram of various Kaplan turbine types, to be found in the computer program libraries of the great turbine manufacturing companies.

For example, in Fig. 1 it is represented the diagram of Kaplan turbine types developed by NOHAB Co. [14]. The closed curves are equal efficiency curves, η_M=0.91, for various Kaplan turbines.

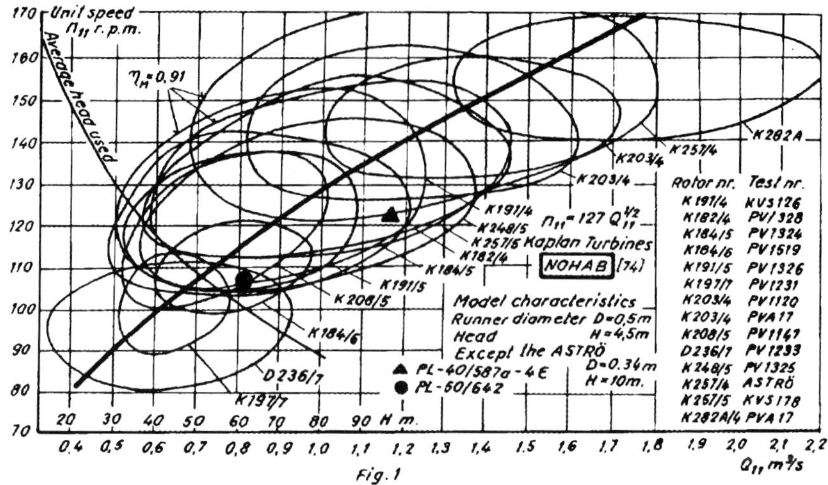

Fig. 1

The locus of the central points of the surfaces delimited by these curves represents the curve of optimum regimes, which has the form [15-17]

$$n_{11ok} = a_1 \cdot (Q_{11ok})^{p/q} \tag{4}$$

According to NOHAB this curve may expressed as [15,16]

$$n_{11ok} = 127.5 \cdot (Q_{11ok})^{1/2} \tag{4a}$$

Here n_{11o} and Q_{11o} are the unitary rotation speed and unitary flow rate, respectively, corresponding to the optimal working regime, were k is the rank of the turbine from the series considered. Taking into account the specific speed $n_q = n_{11}\sqrt{Q_{11}}$, relation (4a) becomes

$$n_{11ok} = 11.284 \cdot n_{qok}^{1/2} \tag{5}$$

When one of the Kaplan turbines developed by the company (in this case NOHAB) is considered as reference turbine, defined by n_{11oref} and Q_{11oref}, then it follows from (3) that

$$\frac{V_{ok}}{V_{oref}} = \frac{\delta_{Mok}/(1-\eta_{Mok})}{\delta_{Moref}/(1-\eta_{Moref})} \tag{6}$$

Without introducing significant errors, it may be admitted that $\eta_{hMok} \cong \eta_{hMoref}$. In this case (6) becomes

$$\frac{V_{ok}}{V_{oref}} = \frac{\delta_{Mok}}{\delta_{Moref}} \tag{7}$$

However, the current in Kaplan turbine models at the optimum operating regime, may be considered hydraulically smooth [7, 9, 12], therefore it may be written

$$\lambda_{ok} \equiv \delta_{Mok} = C/(Re_{Mok})^{1/n} \tag{8}$$

and

$$\frac{V_{ok}}{V_{oref}} = \frac{\delta_{Mok}}{\delta_{Moref}} = \left(\frac{Re_{Moref}}{Re_{Mok}}\right)^{1/n} \tag{9}$$

If $Re = u_M \cdot D_M / \nu_M$ [13], n=6.2 [5, 13], and n_{11} [15] is expressed as below

$$n_{11} = n_M D_M / \sqrt{H_M} \left(\frac{\eta_{hM}}{\eta_{h11}}\right)^{1/2} \tag{10}$$

then it follows that

$$\frac{V_{ok}}{V_{oref}} = \frac{\delta_{Mok}}{\delta_{Moref}} = \left(\frac{n_{11oref}}{n_{11ok}}\right)^{1/n} \tag{11}$$

and

$$\delta_{Mok} = \delta_{Moref}\left(n_{11oref}/n_{11ok}\right)^{1/n} \tag{12}$$

The new relation (12) gives the dependence of hydraulic losses δ_{Mok} on the unitary speed n_{11ok} of a Kaplan turbine working at the optimum regime.

In particular, in the case of HOHAB turbines the above relation becomes

$$\delta_{Mok} = \delta_{Moref}\left(\frac{n_{qoref}}{n_{qok}}\right)^{\frac{1}{2}\times\frac{1}{n}} \tag{13}$$

Once the scalable losses δ_{Mo} are known from (1), (2), (12) and (13), it follows the expression of nonscalable hydraulic losses:

$$\delta_{nsok} = 1 - \eta_{hMok} - \delta_{Moref} \cdot \left(n_{11oref}/n_{11ok}\right)^{1/n} \tag{14}$$

$$\delta_{nsok} = 1 - \eta_{hMok} - \delta_{Moref} \cdot \left(n_{qoref}/n_{qok}\right)^{\frac{1}{2}\times\frac{1}{n}} \tag{15}$$

In conclusion, the relations (12) and (14) are new and permit the calculus of scalable δ_{Mo}, and nonscalable δ_{nso} hydraulic losses in Kaplan turbine models, at the aptimum operating characteristics, to be considered as reference machine.

2.3. THE SCALABLE δ_M AND NONSCALABLE δ_{ns} LOSSES IN KAPLAN TURBINES OPERATING IN AN ARBITRARY REGIME.

The operating characteristics of Kaplan turbines are known from the hill-diagram Fig. 2, whose analytic expression is of the form

$$F\left(n_{11}, Q_{11}, a_0, \varphi^o, \eta, \sigma\right) = 0 \tag{16}$$

The analysis of a great number of Kaplan turbines developed by great companies like Escher-Wyss, J.M. Voith, LMZ etc., conducted us to the idea that a direct correlation must exists between the friction losses and hydraulic efficiency of the turbine, valid over the whole operating domain.

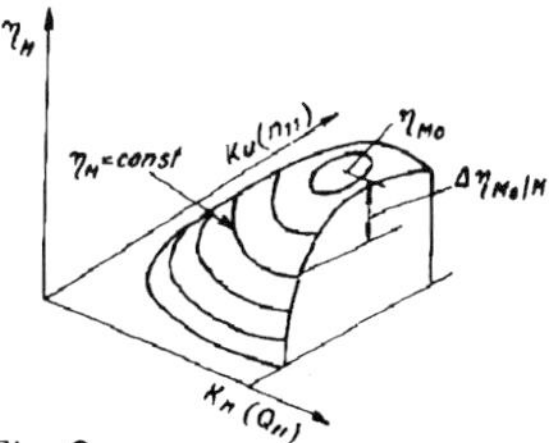

Fig. 2.

Thus, offer many trials, the following relation was obtained

$$\delta_{M0} - \delta_M = K \cdot V_o \cdot (\eta_{hMo} - \eta_{hM}) \tag{17a}$$

$$\frac{\delta_{Mo} - \delta_M}{\delta_{Mo}} = K \frac{\eta_{hMo} - \eta_{hM}}{1 - \eta_{hMo}} \tag{17}$$

The relation (17) is identical to that developed by us for Francis turbines [18]. The value of constant K for Kaplan turbines is K=1/5, which is different from that obtained for Francis turbines.

The constant K depends on the design method of the turbine, its degree of novelty, manufacturing technology applied and on the precision of laboratory measurements.

The problem of nonscalable hydraulic losses is simple and from (1), (2) and (17) it follows that

$$\delta_{ns} - \delta_{nso} = (K \cdot V_o + 1)(\eta_{hMo} - \eta_{hM}) \tag{18}$$

In conclusion, relations (17) and (18) allow the calculation of scalable δ_M and nonscalable δ_{ns} hydraulic losses for Kaplan turbines in the whole operating domain, provided that the hill-diagram of the turbine is known.

3. The coefficient V of hydraulic losses.

3.1. GENERALITIES.

The coefficient V of distribution of hydraulic losses in Kaplan turbines it was introduced by Hutton [2] and Osterwalder [5, 6] in the scale effect formula:

$$\frac{1 - \eta_{hP}}{1 - \eta_{hM}} = (1 - V) + V \cdot \left(\frac{Re_M}{Re_P} \right)^{1/n} \tag{19}$$

Ida [12] underlines the fact that in the case when the geometrically similar Kaplan turbines operate in hydraulically similar regimes ($\phi_P = \phi_M$), the scale effect formula becomes

$$\frac{1 - \eta_{hP}}{1 - \eta_{hM}} \cdot \frac{\eta_{hM}}{\eta_{hP}} = (1 - V) + V \cdot \left(\frac{Re_M}{Re_P} \right)^{1/n} \tag{20}$$

Here $\phi = v_m/u$ defines the flow rate coefficient of the turbine. More recently, Ida [19] proposed for axial bulb turbines the following scale effect formula:

$$\eta_{hP} = \frac{\eta_{hM}}{1 - (1 - \lambda_P / \lambda_M)\delta_M} \tag{21}$$

where λ represents the friction losses in the turbine (or runner). It seems that the introduction of δ_M in the scale effect formula ensures a higher precision in the prediction of operating characteristics of the prototype turbine, when the hill-diagram of the model is known. The correct application of scale effect relations recommended by IEC [13] needs to know the right value of the coefficient V in the whole operating domain of the turbine.

3.2. THE COEFFICIENT V_o FOR KAPLAN TURBINES.

Several year ago Hutton [8] demonstrated that the value V_o depends on the type of the Kaplan turbine, that is on n_{qo}. In this scope he developed an analytical relation of the following form:

$$1 - V_0 = \frac{\eta_{dP}}{1 - \eta_{dM} + \dfrac{C_{D_M}}{\sin^3 \beta_{eM}} (1/t)_e} \qquad (22)$$

Taking into account the results published by Anton [15] concerning the dependencies $(l/t)_e = f(n_s)$ and $\beta_e = f(n_s)$, with $\eta_{dP} = \eta_{dM} = 0.85$ and drag coefficient $C_{DM} = 0.012$, Hutton obtained the values of coefficient V_o for various specific speed values. These allowed us to draw the curve $V_o = f(n_{qo})$ gives in Fig. 3. On the same figure was drawn also the curve $V_o = f(n_{qok})$ calculated using relation (11), which in the case of NOHAB turbines becomes

$$V_{ok} = 1.65 \cdot \left(n_{11ok}\right)^{-0.16} \quad \text{or} \quad V_{ok} = 1.107 \cdot \left(n_{qok}\right)^{-0.08} \qquad (23)$$

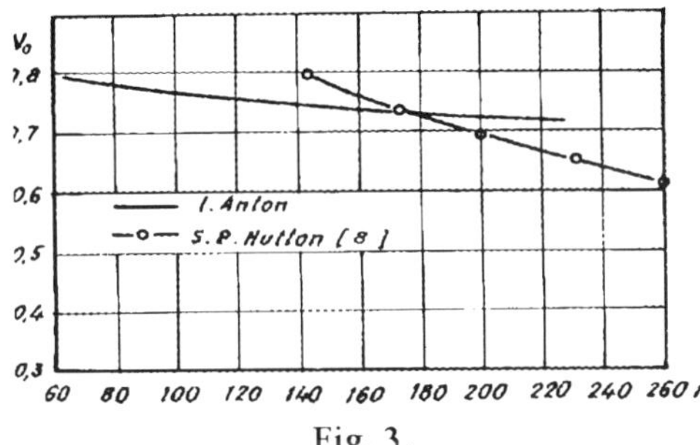

Fig. 3.

The reference turbine has $n_{11oref} = 128.65$ rpm or $n_{qoref} = 130$ and $V_{oref} = 0.75$.

The differences between the two curves of Fig. 3 are due to the fact that Hutton calculated V_o using the values of $(l/t)_e$ and β_e corresponding to the peripheral zone (e) of the runner blade, while relation (23) gives the global values.

3.3. THE DISTRIBUTION COEFFICIENT V OF KAPLAN TURBINES

Using the relation which defined V and the relation (17), it follows a new relation for this coefficient. Thus

$$\frac{V_o - V}{V_o} = (K+1) \frac{\eta_{hMo} - \eta_{hM}}{1 - \eta_{hM}} \qquad (24)$$

Relation (24) permits the calculation of coefficient V of the distribution of hydraulic losses in Kaplan turbines in the whole operating domain when the hill diagram of the model is known from laboratory measurements.

B. APPLICATION

4. The coefficient δ_M, δ_{ns} and V for a Kaplan turbine type PL 40/587-4ε

The Kaplan turbine model PL 40/587-4ε [20] was at the origin of the Kaplan turbines manufactured for the Iron Gate Hydroelectric Power Plant (HEPP) [21]. The hill diagram corresponding to this turbine is given in Fig. 4.

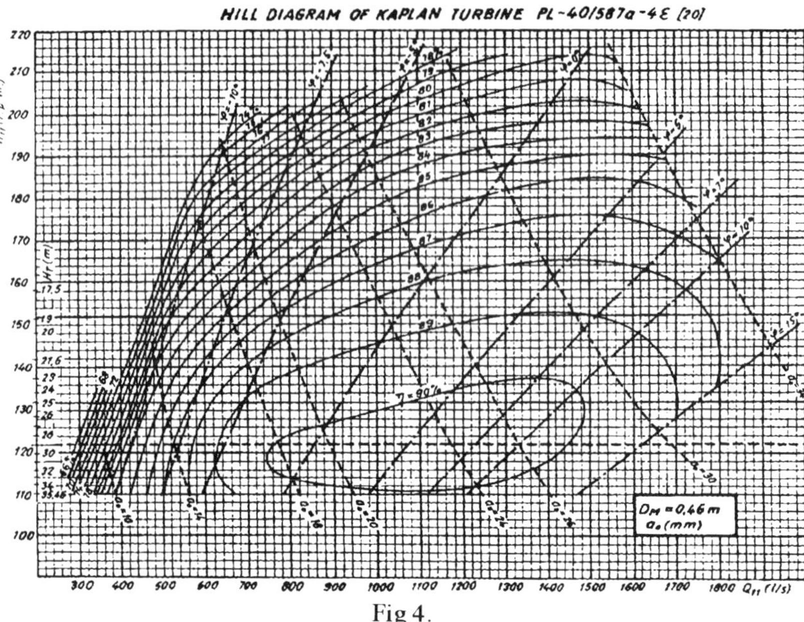

Fig 4.

The parameters of the prototype turbine are: P=178MW, H=27.16m, n=71.43 rpm, η=0.94 and D=9.5 m. If one admits the hypothesis that η_{vM}=0.989 and η_{mM}=0.989, i.e. the volumic and mechanical efficiencies are constant in the whole domain of the turbine, it follows that $\eta_h=1.022\eta_M$.

From the hill diagram we have the values: n_{11o}=122 rpm, Q_{11o}=1.16 m^3/s, n_{qo}=131.99 and η_{Mo}=0.908. From relation (23) it follows V_o=0.755. Using the above data and relations (3) and (2), it follows δ_{Mo}=0.054 and δ_{nso}=0.018. Taking into account these values, relations (17), (18) and (24) particularized for the turbine PL40/587-4ε becomes:

$$\delta_M = 0.054 - 0.1543 \cdot (\eta_{Mo} - \eta_M) \tag{25}$$

$$\delta_{ns} = 0.018 + 1.1763 \cdot (\eta_{Mo} - \eta_M) \tag{26}$$

$$V = 0.755 - 0.906 \cdot \left(\frac{\eta_{Mo} - \eta_M}{0.978 - \eta_M} \right) \tag{27}$$

With the relations (25-27) and the hill diagram of Fig. 4 may be calculated and represented the curves $\delta_M=f(n_{11},a_0)$, $\delta_{ns}=f(n_{11},a_0)$ and $V=f(n_{11},a_0)$ in Fig. 5 and $\delta_M=f(n_{11},\varphi°)$, $\delta_{ns}=f(n_{11},\varphi°)$, $V=f(n_{11},\varphi°)$ in Fig. 6.

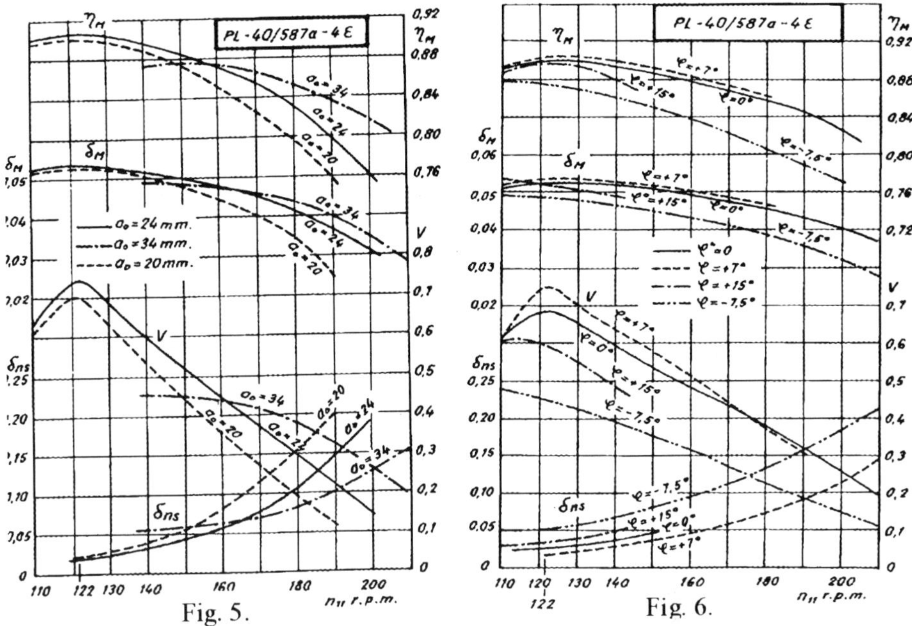

Fig. 5. Fig. 6.

From these figures it follows that δ_M and V have maximum values at η_h maximum, while δ_{ns} indicates a minimum value, results which were obtained also for Francis turbines [18].

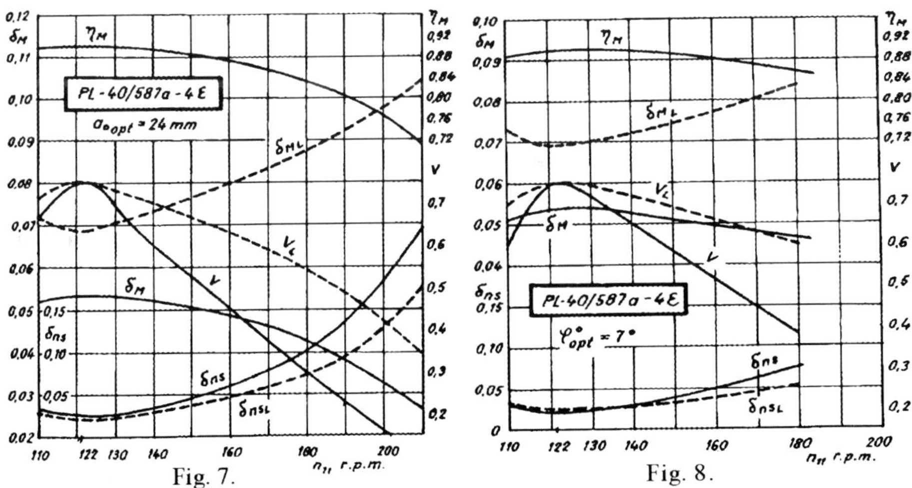

Fig. 7. Fig. 8.

In Figs. 7-9 there are represented the variations of these coefficients with unitary speed n_{11} or unitary flow rate Q_{11}/Q_{11o}, for the optimum opening angle a_{0opt} of the guide vane and for the runner opening $\varphi^{\circ}_{opt}=7°$.

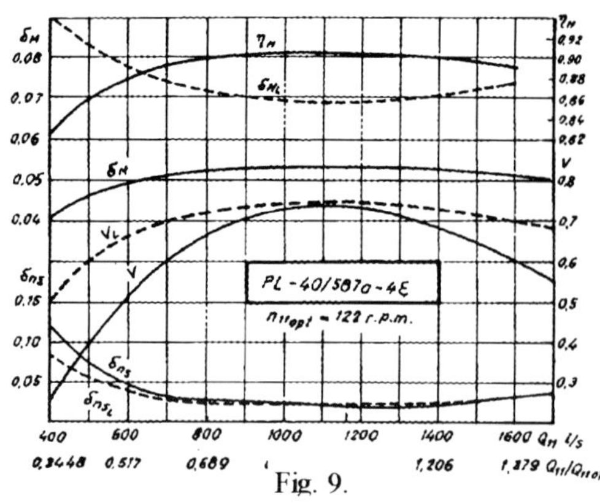

Fig. 9.

On the same diagrams are represented also the curves corresponding to the distribution coefficient V_L of hydraulic losses in Kaplan turbines, given by the relation of Guschin et al. [22]:

$$V_L = V_0 \frac{2 \cdot (1 - \eta_{Mo})}{2 - \eta_{Mo} - \eta_M} \tag{28}$$

Taking into account the relation (3) of definition of V, there were obtained the following expressions:

$$\delta_{ML} = \delta_{MoL} \frac{2 \cdot (1 - \eta_M)}{2 - \eta_{Mo} - \eta_M} \tag{29}$$

$$\delta_{nsL} = \left[2 \cdot \delta_{nsoL} + (\eta_{Mo} - \eta_M)\right] \frac{(1 - \eta_M)}{2 - \eta_{Mo} - \eta_M} \tag{30}$$

Using the values V_o=0.755, δ_{Mo}=0.069 and δ_{nso}=0.023, as well as the relations (29) and (30), there were obtained the curves $\delta_{ML}=f(n_{11})$, $\delta_{nsL}=f(n_{11})$. As it may be observed from Figs. 7-9, the differences appear between the coefficients V and V_L. The greater the departure from the optimum operating regime the greater are the differences between V and V_L.

The greatest diferences is observed between the values of δ_M and δ_{ML}, as well as between the curves of their variation with flow rate. The curves $\delta_M=f(n_{11})$ or $\delta_M=f(Q_{11}/Q_o)$ have a maximum at the efficiency η_{Mo}, situation corresponding to the greatest friction losses, for inlet edge of runner blades without shock ($i^o_o=i_{so}$). At this regime the flow covers completely, on both sides the blade profile. At regimes corresponding to inlet flow with shock on a side or other of the blade detachament of the flow occurs. Thus the friction losses reduce, while the kinematical losses increase. It follows that the relation developed by Guschin et al [22] for the calculation of V_L does not correspond to the process from turbines. Moreover, the use of coefficient V_L in the scale effect formula of Hutton, implies the over-estimate of the efficiency η_{PL}, i.e. $\eta_{PL}>\eta_P$.

In conclusion, the relations (17), (18) and (24) developed in the paper for the coefficients δ_M, δ_{ns} and V correspond to the process from the turbine and generalizes diparate results published in various journals.

5. Conclusions

In this paper there are opened new perspectives for the use of scale effect relations of S.P. Hutton or T. Ida in the whole operating regime of a Kaplan turbine. New relations were developed to calculate the following coefficients:

(i) the coefficient δ_M of scalable hydraulic losses;

(ii) the coefficient δ_{ns} of kinetic losses;

(iii) the coefficient V of distribution of hydraulic losses in Kaplan turbines.

These new relations are valid for all the operating regimes of Kaplan turbines. Their use needs only the hill diagram of the turbine. Applying a new method there were obtained new relations for the coefficients δ_{Mo}, δ_{nso} and V_o corresponding to the opening regime of a model Kaplan turbine.

The relations given in this paper are simple and enough precise, having a high level of generality, a good reason to standardize them.

6. References

1. Cammerer, R.. (1909) Die Abhängigkeit das wirkungsgrades bei Wasser turbinen von Gefälle, Wasserwärme Turbinengrosse and Rauheit der Kanale, *Z. VDI* **53,** 1541.
2. Hutton, S.P. (1954) Component losses in Kaplan turbines and the prediction of efficiency from model tests, *Proc. Inst. Mec. Eng.* **168**.
3. Ackeret, J. (1931) Theoretische Betrachtunghen Zur Kaplanturbine, *E. W. Mitt* **4**.
4. *** (1965) IEC International Code for model acceptance tests of hydraulic turbines scale effect, *Publication* **993**.
5. Osterwalder, J. (1967) Model testing for Kaplan turbine design, including studies on efficiency scale effects, *Water Power* **19**.
6. Osterwalder, J. (1972) Considerations on the revaluarization behaviour of hydraulic turbomachines, *AIRH Symposium*, Rome.
7. Hutton, S.P.and FÁY Á (1974) Scaling up head flow and power curves for water turbines, *AIHR Symposium*, Vienna.
8. Hutton, S.P. (1983) An alternative to the Fay pipe flow analogy for scaling up water turbine efficiency, *Proc. of the Seventh Conf. of Fluid Machinery* **1**, Akad. Kiado., Budapest,.
9. Chevalier, J. (1965) Contribution à l'étude de l'effect d'échelle dans les turbines Kaplan, *Houille Blanche* **20**.
10. Fáy, Á. (1976) Admissible roughness values of model hydraulic turbines, *Conf. Hydroturbo*, Brno.
11. Fáy, Á. (1994) On the accuracy of hydro-turbine perfomance prediction based an model tests, *Conf. Modelling Testing and Monitoring for Hydro Power Plants IAHR*, Budapest.
12. Ida, T. (1989) Analysis of scale effects on performance characteristics of hydraulic turbines (Part 1,2), *Journal of Hydraulic Research* **27and 28**, No. 6,7.
13. *** (1992) IEC/TC or SC:TC4, Model acceptance tests to determine the hydraulic performance of hydraulic turbines storage pumps and pump turbines.
14. *** 1970 Test stand for Francis turbine models, *A.B. NOHAB Trollhättan*, Sweden.
15. Anton, I. (1979) *Turbine hidraulice*, Editura Facla, Timisoara.
16. Anton, I.M. (1993) New relations for the design of Kaplan and bulb turbines, *16th IAHR Symposium*.
17. Anton, I.M. (1993) Basic parameters and characteristics coeficients of hydraulic turbine operation, *Conf. Hydraulic Machinery and Hydrodinamics*, Timisoara.
18. Anton, I.M., (1995) The scale effect and the coefficient V of hydraulic losses for FRANCIS turbines, *Rev. Roumaine Sci., Tech. Mec. Appl.* **4,** No 44, Bucharest.
19. Ida, T. (1994) A new proposal for efficiency scale-up formulae for hydraulic turbines, *XVII IAHR Symposium*, Beijing, China.
20. *** (1962) Nomenclaturî krupnîh verticalnîh povorotnolopastîh i radialnoasevîh ghidroturbin, *Leningradski Metalliceskii Zavod*.
21. Costre, F., Bitang, A. and Anghel, A. (1974) Aspects referred to the hydromechanical equipment manufacturing and characteristics from Hydroenergetic and Navigation System "Portile de Fier" Danube, *IAHR Symposium*, Vienna.
22. Guschin, M.U., Pîljev J.M. and Sidorenko, V.F. (1980) Naturnîje ispitanija gidroagregata GES Djenepeg Energomasinostroienie, No. **9**.

ANALYSIS OF LOSSES IN HYDRAULIC TURBINES

HERMOD BREKKE
The Norwegian University of Science and Technology
Division of Thermal Energy and Waterpower, Waterpower Laboratory,
Alfred Getz vei 4, N-7034 Trondheim, NORWAY

Abstract

The paper gives a brief discussion on the influence of the different losses in a hydraulic turbine. Of special interest are the friction losses which decreases with Re. This leads to a scale up of the turbine efficiency from model to prototype as expressed in the IEC code according to the theory by prof. Spurk [1].

The importance of homology in geometry and flow vector fields for the validity of the efficiency scale effect will be discussed.

The shift of the best efficiency point (BEP) from model to prototype caused by flow friction loss and disk friction loss will be discussed.

Losses from guide vane end leakages and non-scalable losses will be described.

A brief discussion on the problemes with reliable scale up formulas for Pelton turbines will be included

Résumé

Nous étudions l'influence des différentes pertes dans une turbine hydraulique, et en particulier celles qui décroissent avec le nombre de Reynolds. Ceci nous conduit à un effet d'échelle modèle-prototype tel qu'il est exprimé dans le code de la CEI en conformité avec la théorie du professeur Spurk [1].

L'importance de la similitude géométrique et de l'écoulement est discutée.

Le décalage du point de meilleur rendement du prototype par rapport au modèle est examinée.

Les pertes par fuites du distributeur ainsi que les pertes non transposables sont également examinées.

E. Cabrera et al. (eds.), Hydraulic Machinery and Cavitation, 294–303.

Nous incluons aussi une brève discussion des problèmes liés à l'effet d'échelle des turbines Pelton.

1. Introduction

For high head turbines the efficiency of prototype turbines can be measured by the thermodynamic method with sufficient accuracy.. Model tests are only required as a preliminary indication of the turbine performance especially on cavitation behavior.. For low head turbines the result of the efficiency on the models are used to calculate the efficiency of prototype turbines by scale effect codes according to the IEC norm or other national normes.

Because of this it is important to make further work for improvement of the existing scale effect codes.

The ongoing work in Norway on this field is the back ground for this paper. The given information is based on theoretical studies, laboratory tests and field measurements. Of special interest is the shift in the theoretical best efficiency point i.e. at the point of non rotating outlet flow from runner which is discussed in the presented work

2. Similarity requirements of model and prototype for step up efficiency of reaction turbines

When assuming non rotating outlet flow for both models and prototypes we get following relationship form the Euler equation where η_{hP} and η_{hM} are the hydraulic efficiency of the model and prototype respectively:

$$\eta_{hP} = \left[\frac{u_1 c_{u1} - u_2 c_{u2}}{gH}\right]_P = \left[\frac{u_1 c_{u1}}{gH}\right]_P$$

$$\eta_{hM} = \left[\frac{u_1 c_{u1} - u_2 c_{u2}}{gH}\right]_M = \left[\frac{u_1 c_{u1}}{gH}\right]_M$$

Here c_u is the absolute velocity component in circumferential direction and u is the circumferential blade velocity. The index 1 indicates inlet and the index 2 indicates outlet. For non rotating outlet flow $c_{u2} = 0$.

When introducing dimensionless velocities $\underline{c} = \frac{c}{\sqrt{2gH}}$ and $\underline{u} = \frac{u}{\sqrt{2gH}}$ the following equation yields

$$[\underline{u}_1 \underline{c}_{u1}]_M = \frac{\eta_{hM}}{\eta_{hP}} [\underline{u}_1 \underline{c}_{u1}]_P \tag{1}$$

If the inlet and outlet flow angles and vector velocity diagrams shall be homologous for prototype and model, we must fulfil the following relationship:

(c_m is the meridional velocity component)

$$\left[\frac{\underline{c}_{m1}}{\underline{u}_1}\right]_M = \left[\frac{\underline{c}_{m1}}{\underline{u}_1}\right]_P \qquad \text{and} \qquad \left[\frac{\underline{c}_{u1}}{\underline{u}_1}\right]_M = \left[\frac{\underline{c}_{u1}}{\underline{u}_1}\right]_P$$

Then

$$\frac{\underline{u}_{1M}}{\underline{u}_{1P}} = \frac{\underline{c}_{1M}}{\underline{c}_{1P}} = \sqrt{\frac{\eta_{hM}}{\eta_{hP}}} \tag{2}$$

In order to obtain similarity in flow and guide openings for model and prototype the specific speed or speed number must then have a small difference between model and prototype expressed by the ratio $\frac{\eta_{hM}}{\eta_{hP}}$. This relationship can be proven by the equation of the speed number where ω = angular velocity and Q = flow :

$$\Omega = \frac{\omega\sqrt{Q}}{(2gH)^{3/4}} = \sqrt{\pi} \cdot \underline{u}_2 \cdot \sqrt{\underline{c}_{m2}} \tag{3}$$

The relationship between the specific speed of the model and prototype based on eq. 2 yields:

$$\frac{n_{sM}}{n_{sP}} = \frac{\Omega_M}{\Omega_P} = \frac{\underline{u}_{2M} \cdot \sqrt{\underline{c}_{m2M}}}{\underline{u}_{2P} \cdot \sqrt{\underline{c}_{m2P}}} \cdot \left(\frac{\eta_{hM}}{\eta_{hP}}\right)^{\frac{3}{4}} \tag{4}$$

The physical reason for this is that the scalable losses i.e. the friction loss is higher in the model than in the prototype. The available specific energy must then be increased in the model versus the prototype in order to obtain similar velocity vector diagrams for both model and prototype.

Further it is important to refer to the guide vane opening which gives non rotating flow in the draft tube i.e. $c_{u2} = 0$ as reference point for both prototype and model turbine. Often the best efficiency point is used for the model turbine as reference point even if there normally is a shift in best efficiency from non rotation outlet flow and best efficiency. This is in the authors opinion not correct.

3. Shift in best efficiency point (BEP) of prototype compared with models

A shift in BEP will always be found caused by reduced flow friction. On the other hand a different shift of BEP is found caused by reduced disk friction losses as explained in the following.

3.1. REDUCED FLOW FRICTION LOSS IN PROTOTYPE

The loss of specific energy caused by flow friction loss from spiral casing inlet to draft tube outlet may be expressed by

$$\Delta(gH) = \xi Q^2 \tag{5}$$

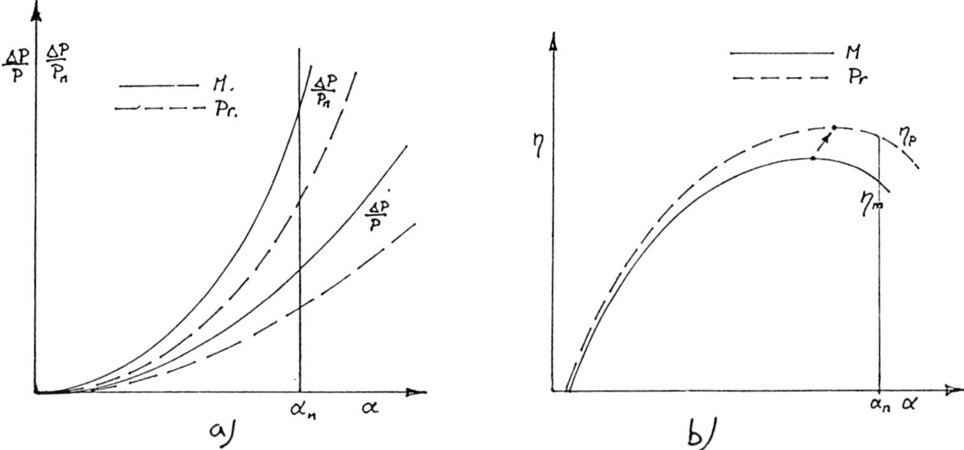

Figure 1. a) Schematic illustration of the loss relative to the nominal max. output P_A and the regarded output P. b) Schematic illustration of the shift in efficiency from model (full line) and prototype (dotted line).

Here $\xi = f(\mathrm{Re})$ and $Q = f(\alpha)$ where α is the guide vane opening angle. The loss in power $\Delta P = \rho Q \Delta(gH) = \rho \xi Q^3$ and then the relative power loss will be:

$$\frac{\Delta P}{P} = \frac{\rho \xi Q^3}{\rho g Q H} = \text{const} \cdot Q^2 \tag{6}$$

The loss relative to the nominal max. output will be:

$$\frac{\Delta P}{P_n} = \frac{\rho \xi Q^3}{\rho g Q_n H_n} = \text{const} \cdot Q^3 \tag{7}$$

Because the relative loss is increasing with the square of the flow the BEP of the efficiency will shift to a larger guide vane opening for the prototype than for the model caused by the reduced friction factor $\xi = f(\mathrm{Re})$. The relative loss and shift in efficiency is illustrated schematically in fig 1a and fig 1b respectively.

3.2. REDUCED DISK FRICTION LOSS IN PROTOTYPE

The disk friction loss will be reduced in the prototype compared with the model turbine if both are homologous in shape and hydraulically smooth. It should be emphasized that the requirement of homologous geometry of prototype and model not always is fulfilled for the surfaces of the covers

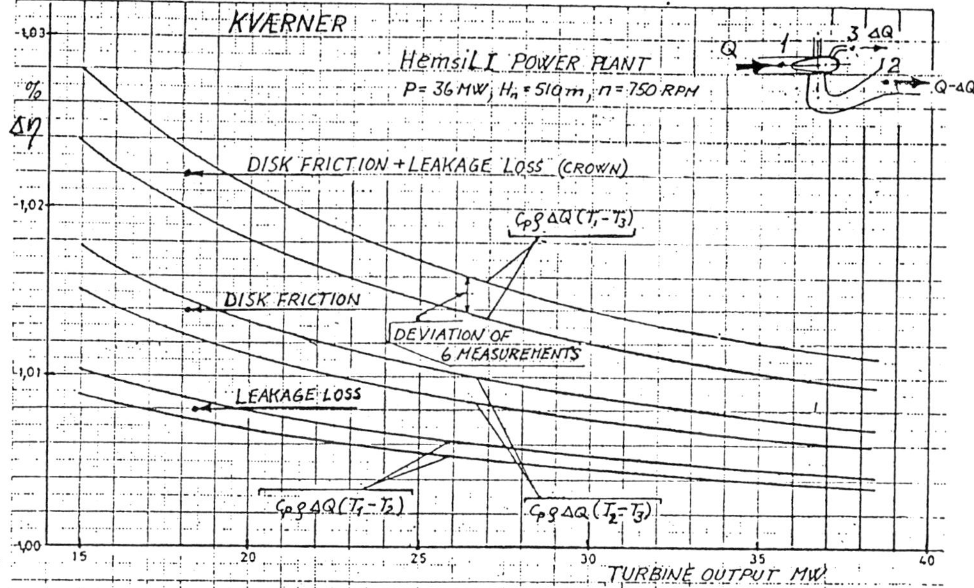

Figure 2. Relative disk friction losses as function of turbine output for a high head Francis turbine. (Courtesy KVRNER.)

enveloping the runner and the outside surface of the runner. [Accept criteria for deviations are also not given with sufficient accuracy in the IEC code.]

The scalable disk friction loss has been measured during thermodynamic measurement in several high head turbines in Norway. The result shows that the powerloss is almost constant and not a pure function of the guide vane openings. In fig 2 is shown the result of 6 different measurements for a high head Francis turbine by means of the thermodynamic method from 1960 to 1963 during development of two different runners with homologous outside of crown and band.

When assuming the disk friction loss as a loss independent of the guide vane opening the relative loss will decrease with increasing output as shown in fig 2 and as illustrated schematically in fig 3a.

Because the disk friction loss decreases in the prototype due to increased Re, the BEP of the efficiency for the prototype will shift to a smaller guide vane opening than for the model . This is illustrated in fig 3b.

At the Norwegian University of Science and Technology research work

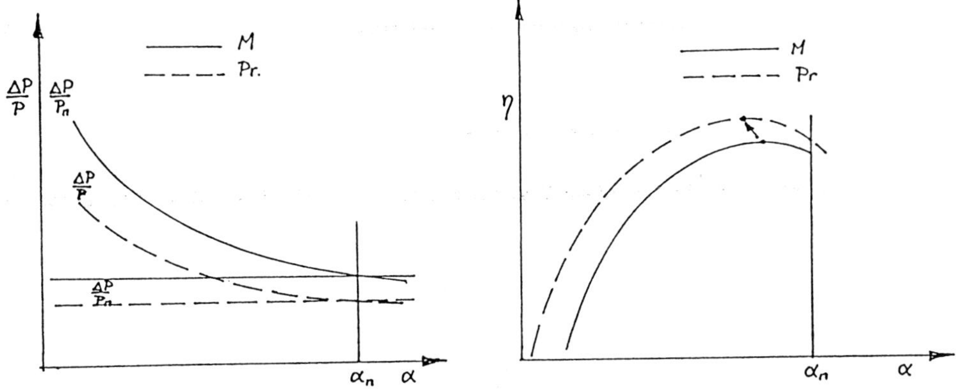

Figure 3. Relative loss of model and prototype (a), and increased efficiency and shift of BEP from model to prototype (b). Model full lines, prototype dotted lines.

is going on on disk friction losses. Of special interest is the influence of the geometry and leakage flow on the loss. Some data may be given during presentation of this paper.

4. The total losses in reaction turbines

In fig 4 is given an illustration of the different losses in a high head turbine. It can be found that the disk friction losses cover for approximately 1.5% of the total loss for a typical high head Francis turbine while 2.0% – 2.5% cover for the flow friction losses. Because the larger flow friction loss which gives a shift of BEP towards increased output for the prototype it should be expected a total shift towards higher output. However, another complex loss caused by the guide vane end clearance leakage has been measured to be as large as 1.5% on a prototype and approximately 0.5% for a model at BEP. See fig 5. The reason for the higher loss in the prototype may be found in the stronger guide vane end clearance leakage in the turbulent domain versus the laminar leakage in the model. An illustration of the end clearance losses in a model is given in fig 6 and for a prototype compared with models in fig 7.

The guide vane end clearance loss is increasing inverse proportional to the guide vane opening angle due to higher increasing leakage and stronger cross flow effect. For a prototype with homologous geometry to the model the relative clearance should be the same. Then the contribution from the

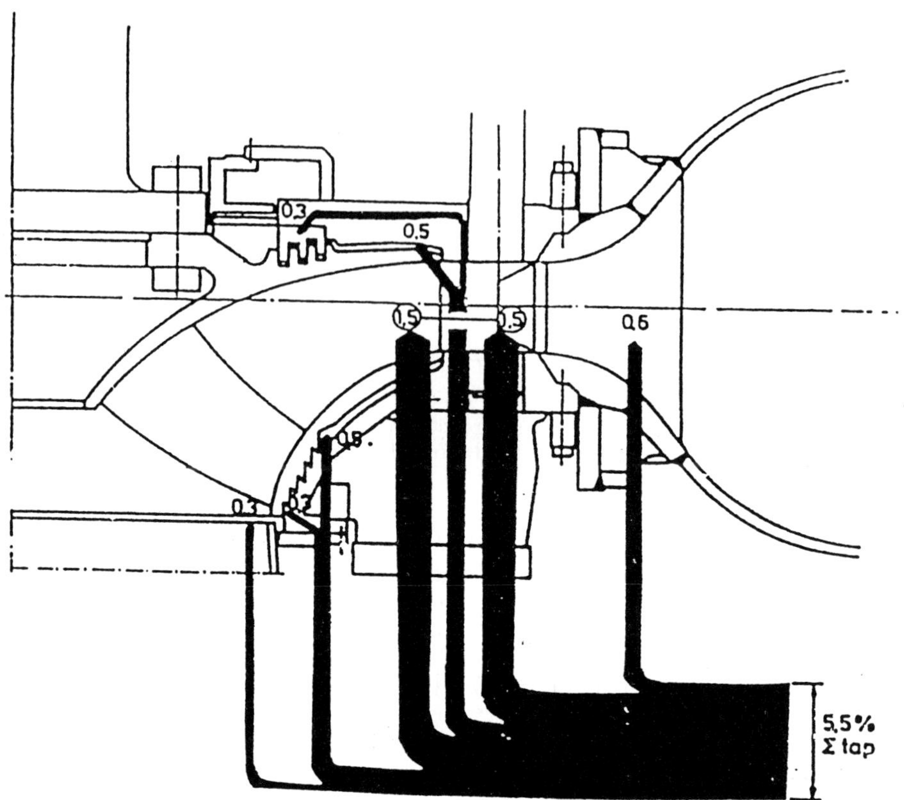

Figure 4. The different losses in a high head Francis turbine. Note the loss created by the guide vanes occures mainly in the runner and not 1.5% + 1.5% distributed as shown in the figure.

end clearance loss pushes the best efficiency point BEP towards higher guide vane openings for the prototype. In addition a "negarive scale effect" occures from the higher end clearance loss of the prototype.

Besides the described losses separation losses at the runner blade inlets, guide vanes and stay vanes occures. These losses gives normally not a contribution to the scale effect, but are losses which may be reduced to a minimum by an optimum design. An important parameter for an optimum design will be the blade lean angle θ, which is the blade angle normal to the stream line direction (Ref [2]). It should be emphasized on the fact that an analytical analysis must be made of the blade design prior to and in collaboration with modern CFD analysis. Without an analytical study of the influence of the geometry parameters prior to the CFD analysis for a direct solution it will be very difficult to obtain a sucesful design without an expensive "cut and try" period. However, CFD analyses are very useful

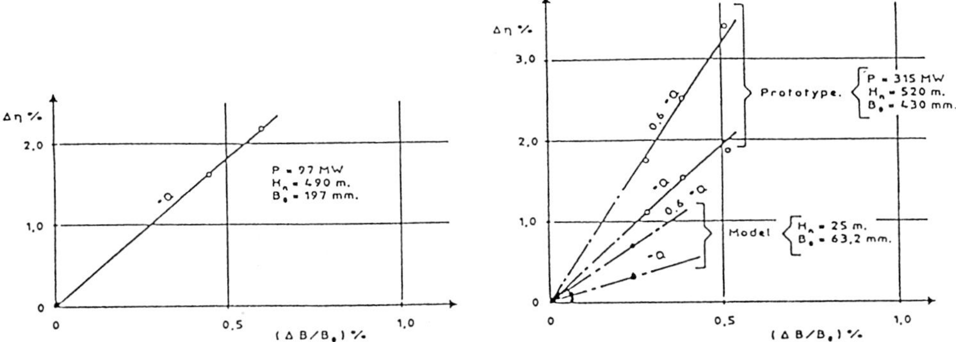

Figure 5. a) Efficiency loss versus guide vane clearance gap for prototype turbine as given by Mc.Hamish BOVING in IAHR Symp. Fort Collins 1978. b) Efficiency end clearance loss versus clearance gap for model and prototype at Kvilldal Power Station.

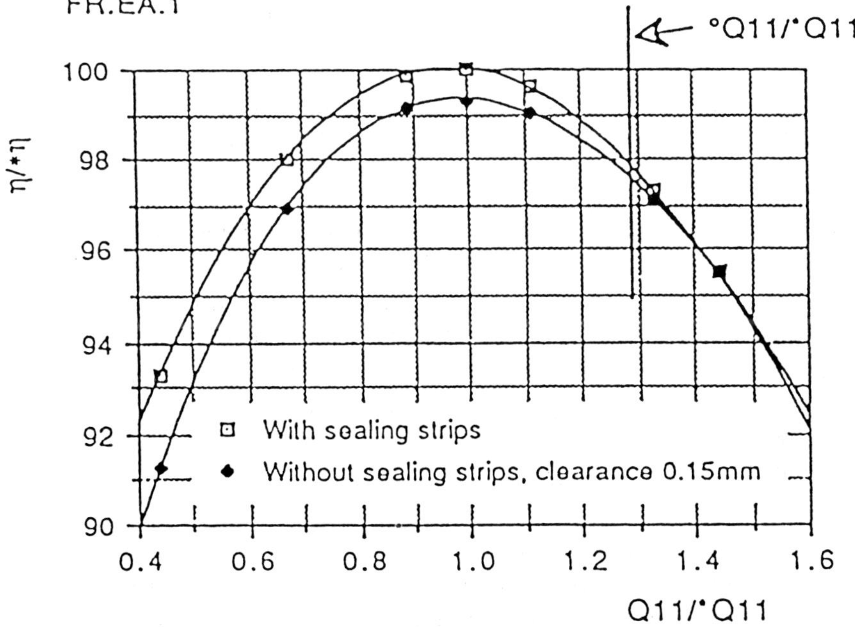

Figure 6. The measuring result of model turbine with sealing strips on the guide vane faces and compared with $0.15mm$ clearance gaps measured at $25m$ net head.

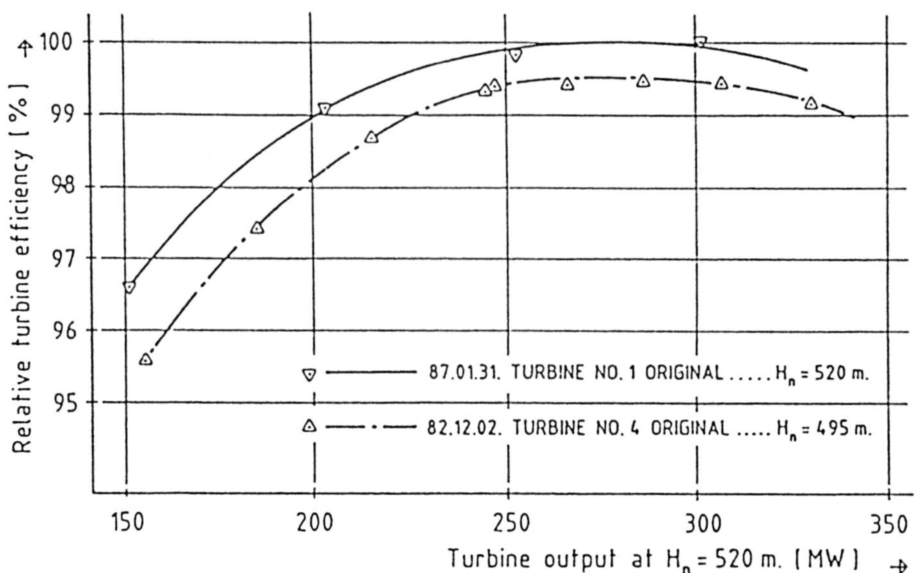

Figure 7. NVE Kvilldal Power Plant. Difference in efficiency of turbine No. 1 and turbine No. 4 with a difference in the guide vane clearance gap of $0.3mm$.

in order to optimize a given blade structures within a turbine manufactorers geometry variation in the runner collection. Because a complete CFD analysis of the draft tube flow has so far not been available, there are still a need for laboratory work to be made on turbine models.

For low head turbines the lack of recovery of the runner outlet energy in the draft tube is the largest contribution to the losses.

5. Losses in impulse turbines

For impulse turbines i.e. Pelton turbines loss analysis and scale effect formulas have been discussed for years. However, up to now scale effect formula seems to be perfect even if the work by Dr. Grein and prof. Spurk [3] seems to be a promissing approach.

The reason for failour to calculate the losses are the complex outlet flow. A typical example is outlet water in the bucket which may escape through the buckets entrance cut out when the last water is leaving the bucket. Secondary effect of such high velocity outlet water for a 6 jet turbine is fatal for the efficiency and the scale effect formula is not valid.

Following conclusion may be drawn. There is a possibility to make a useful loss analysis for a well designed low specific speed Pelton turbine, but a useful loss analysis to be used in a scale effect formula in general will fail.

6. Conclusion

For reaction turbines the understanding of the losses is improving by use of modern analyses and testing technology. However, computation of the draft tube flow is still not solved.

For impulse turbine the non stationary free surface flow is difficult to handle theoretically, and no CFD analysis has so far been succesfully presented.

However, high speed video technique and analytical flow analysis has improved the performance of Pelton turbines.

Scale effect formulas cannot be regarded to be valid for all types of Pelton turbines in general.

Aknowledge

The author is grateful for admission to measuring results from Norwegian Power Plant owners and Kvœrner.

References

1. Spurk, Joseph H. ", Dimensionsanalyse in der strohmungslehre", Springer Verlag, 1992, Germany. (German language.)
2. Brekke, H. ",A parameter study on cavitation performance of Francis turbines", ASCE Water Power Conf., 25-28 July 95, San Francisco.
3. Grein, H., Meier, J., and Klicov D. ", Efficiency scale effects in Pelton turbines", Proceedings IAHR Symposium, Montreal, 1986.

Symbols

u = circumferential speed
c = absolute velocity
H = Head in mWC (gH specific energy (J/kg))
Q = Flow (m^3/s)
n_s = Specific speed
g = gravity accelaration (m/s^2)
η = efficiency
Ω = Speed number
ω = angular velocity (rad/s)

SCALING-UP HEAD-DISCHARGE CHARACTERISTICS FROM MODEL TO PROTOTYPE

Michel COUSTON,Robert PHILIBERT
GEC-ALSTHOM NEYRPIC
82, Rue Léon Blum BP 75, 38041 Grenoble Cedex, FRANCE

We present here a method to calculate the scale effect on discharge and output that is correlated with the scale effect on efficiency . A theoretical analysis is proposed and general formulas are induced . Different comparisons betwen model and prototype measurements confirm the proposed analysis.

1-INTRODUCTION

Often the acceptance tests of an hydraulic turbine are based on model tests . So prototype characteristics have to be determined from model results . The scaling-up for head , discharge , output and efficiency is made according to IEC standards :output is corrected by efficiency , head and discharge are not corrected. But on-site measurements for different types of hydraulic turbines are not in good agreement with this evaluation [1,2,3].

So it seems necessary to foresee the prototype characteristics with a better accuracy from model tests .

In this paper we present a method that has been validated by comparisons between model results and on-site measurements .

E. Cabrera et al. (eds.), Hydraulic Machinery and Cavitation, 304–312.

The difference of characteristics between model and prototype can be explained by viscous effects, by geometrical differences and by differences in the general lay-out of a test rig and a power plant.

One of the big difficulties comes from the inacurracy of on-site measurements,thus it is not easy to separate what is due to scale effect and what is due to inaccurate measurements.

Nevertheless we have made an attempt to evaluate the predictable differences between model and prototype and we have applied this approach to different cases. The agreement between calculation and measurement is reasonably good and we can propose to use our method in order to improve the prediction of prototype characteristics from data obtained on model.

2-BASIC ANALYSIS

2-1- EXISTING APPROACH

In IEC Publications the main concern is the evaluation of efficiency increase. For scaling-up head , discharge , output no specific analysis is proposed .

In customer's requirements the weighted operating points are based on discharge or output . According to the scale-up of these parameters different results can be obtained.

In japanese standards for pump-turbines an evaluation of the modification of head/discharge curve is given and it is a reasonable approach .

We can say that there is no internationally recognized method to define the global scale effect including
discharge and output.

2-2 PHYSICAL APPROACH

The basic difference existing between a model and a prototype is the Reynolds number so that the viscous phenomenons are not in similitude .The viscous effects on the flow can be localised everywhere in the flow or in immediate proximity of the boundaries ; in this case the boundary layer analysis can be used to describe the differences .

According to the components and the operating point ,one of these hypothesis can be retained .For example the draft tube is the component where viscosity has an effect in all the flow ,and for other components the operating point and the quality of hydraulic design are the parameters that determine the type of viscous effects.

In our paper we do not consider the quantification of these viscous effects on the efficiency but we try to correlate shift of discharge and output with the increase in efficiency .

2-3-PROPOSED APPROACH

The proposed analysis consists in separating the global scale effect in two aspects :

-the **"efficiency scale-up"**

-the **"shift of head-discharge characteristic"**

and deals separately with the "shift " effect .

This shift analysis is based upon the splitting of the flow in two parts :

-a basic flow of a perfect fluid;this flow will be independent of turbine size and REYNOLDS number

-an adjustment of flow due to viscous effects

The purpose is therefore to evaluate the "shift of head-discharge characteristic" in function of the efficiency scale-up .

3- SHIFT OF HEAD - DISCHARGE CHARACTERISTIC

3-1 SHIFT OF HEAD-DISCHARGE CHARACTERISTIC -GENERAL FORMULAS

3-1-1 TURBINE MODE

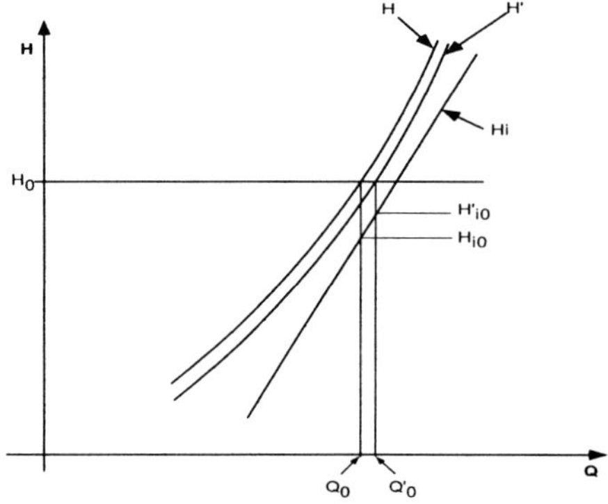

On this diagram we can see :

-head/discharge characteristic H(Q) obtained by scaling-up model characteristic h(q) in n^2D^2 and nD^3

-head/discharge characteristic H'(Q) for prototype

-head/discharge characteristic Hi(Q) supposed to be an invariant for turbine (same characteristic for model and prototype) .There is no shift on $H_i(Q)$.[hypothesis]

We can note that all these characteristics correspond to geometries in perfect similitude ; particularly same wicket gate opening and blade angle .

Basic relations are :

$$P = \eta \rho g Q H$$
$$H_i = \eta_h H$$
$$\eta = \eta_{frot} \eta_{vol} \eta_h$$

with the classical definitions
$$P_{11} = \frac{P}{D^2 H \sqrt{H}}$$
$$Q_{11} = \frac{P}{D^2 \sqrt{H}}$$
$$n_{11} = \frac{nD}{\sqrt{H}}$$

We assume that the curve Hi(Q) is a straigth line of equation :

$$H_i = \lambda Q + \mu$$

We are going to study evolution of output and discharge under a head H_0

Dévelopment

with the following definitions

$$\Delta Q = Q'_0 - Q_0$$
$$\Delta H_i = H'_{i0} - H_{i0}$$
$$\Delta P = P' - P$$
$$\Delta \eta = \eta'(Q'_0) - \eta(Q_0)$$

we can easily obtain :(assuming $\Delta \eta_h = \Delta \eta$)

$$\frac{\Delta P}{P} = \frac{\Delta P_{11}}{P_{11}} = (1 + \frac{\eta H_0}{\lambda Q_0}) \frac{\Delta \eta}{\eta}$$
$$\frac{\Delta Q}{Q} = \frac{\Delta Q_{11}}{Q_{11}} = \frac{H_0}{\lambda Q_0} \Delta \eta$$

Evaluation of λ

We assume that the curves $Q_{11}(n_{11})$ are linear near the considered operating point :

$$Q_{11} = \alpha n_{11} + \beta$$

From the definition of Q11 and n11 we have :

$$\frac{dQ_{11}}{Q_{11}} = \frac{dQ}{Q} - \frac{1}{2}\frac{dH}{H} \quad \Rightarrow \quad dQ_{11} = \frac{Q_{11}}{Q} dQ - \frac{1}{2}\frac{Q_{11}}{H} dH$$
$$\frac{dn_{11}}{n_{11}} = -\frac{1}{2}\frac{dH}{H} \quad \Rightarrow \quad dn_{11} = -\frac{1}{2}\frac{n_{11}}{H} dH$$

So we have the following relation :

$$\frac{Q_{11}}{Q}dQ - \frac{1}{2}\frac{Q_{11}}{H}dH = \alpha(-\frac{1}{2}\frac{n_{11}}{H}dH)$$

$$\Rightarrow \quad \lambda = \frac{dH}{dQ} = 2\frac{H}{Q}\frac{Q_{11}}{Q_{11} - \alpha n_{11}}$$

assuming that $\frac{dH}{dQ} \approx \frac{dH_i}{dQ}$

In this case the following formulas will be obtained :

$$\frac{\Delta Q}{Q} = \frac{1}{2}\frac{\Delta\eta}{\eta}\frac{Q_{11} - \alpha n_{11}}{Q_{11}} \qquad \text{with} \quad \alpha = \frac{dQ_{11}}{dn_{11}}$$

$$\frac{\Delta P}{P} \approx \frac{\Delta\eta}{\eta}\left\{1 + \frac{1}{2}\frac{Q_{11} - \alpha n_{11}}{Q_{11}}\right\}$$

3-1-2 PUMP MODE

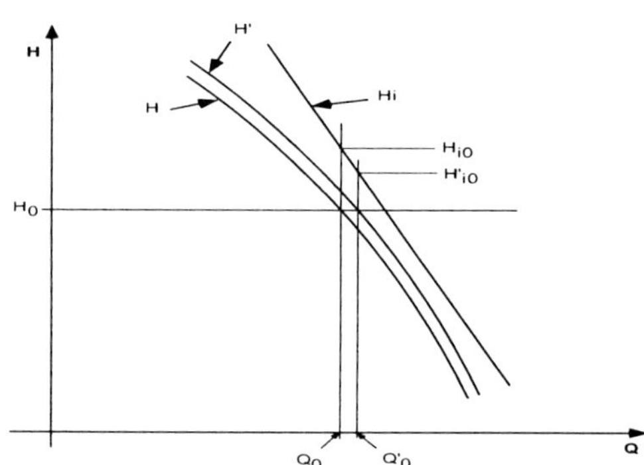

In this case the following formulas will be obtained

$$\Rightarrow \quad \Delta Q = -\frac{\Delta\eta}{\eta^2}\frac{H_0}{\lambda}$$

$$et \quad \frac{\Delta P}{P} = -\frac{\Delta\eta}{\eta}(1 + \frac{H_0}{\eta\lambda Q_0})$$

$$\lambda = -\frac{n}{60 g b_2 tg\beta'_2}$$

where :

n	rotational speed
b2	runner width at pump outlet
g	9,81 ms^{-2}
β'2	relative flow angle at pump outlet

3-2 SHIFT OF CHARACTERISTICS -SIMPLIFIED FORMULAS

3-2-1 TURBINE MODE

The formulas of §3-1-1,while not really complicated, are not so convenient to handle as α coefficient depends on both the specific speed and the operating point .However or FRANCIS type turbines and pump-turbines operating as turbine the coefficient $\frac{1}{2}\frac{Q_{11}-\alpha n_{11}}{Q_{11}}$ is quite close to 0,5 .

So in case of FRANCIS turbine the simplified formulas are :

$$\frac{\Delta Q}{Q}=\frac{\Delta Q_{11}}{Q_{11}}=0,5\frac{\Delta\eta}{\eta}$$

$$\frac{\Delta P}{P}=\frac{\Delta P_{11}}{P_{11}}\approx 1,5\frac{\Delta\eta}{\eta}$$

However for low head turbine , such BULB units the α value becomes important . Thus the above simplification is not valid , for example at large blade angle (as can be seen on § 4) we get :

$$\frac{\Delta Q}{Q}=\frac{\Delta Q_{11}}{Q_{11}}=0$$

$$\frac{\Delta P}{P}=\frac{\Delta P_{11}}{P_{11}}\approx\frac{\Delta\eta}{\eta}$$

3-2-2 PUMP MODE

As for FRANCIS turbines ,There is **an important case for pump-turbine in pump mode** where the operating point is close to :

$$H_{i0}=\frac{1}{2}\frac{U_2^{\,2}}{g}$$

and if the curve Hi(Q) has the value 0,85 U^2/g (statistical value) for Q=0 we obtain the simplified formulas:

$$\frac{\Delta Q}{Q}=1,4\frac{\Delta\eta}{\eta}$$

and $$\frac{\Delta P}{P}\approx 0,4\frac{\Delta\eta}{\eta}$$

4-SOME PROTOTYPE RESULTS

Several cases of BULBS , FRANCIS , PUMP-TURBINES in pump and turbine mode are presented and demonstrate the validity of our analysis .

4-1 FRANCIS

To demonstrate the applicability of the method we present hereafter two comparisons for generating operation . The first case corresponds to the medium head ITAIPU turbine .Achieved results are conforting our assumptions .

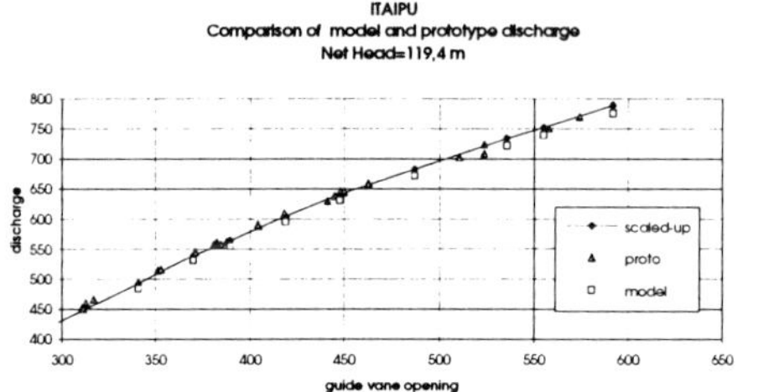

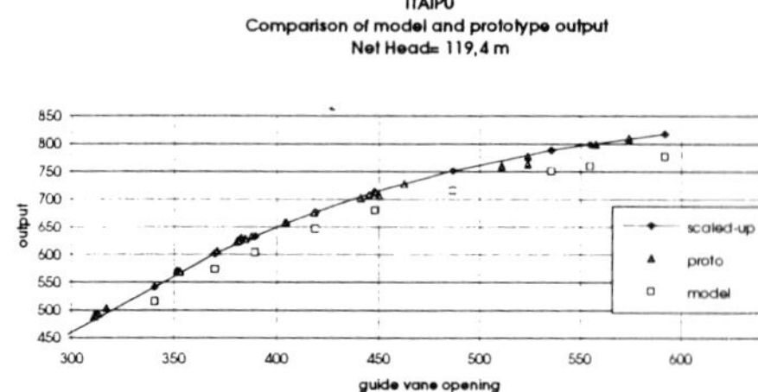

The second case corresponds to the high head MUJU pump-turbine in generating mode . While not as satisfactory as ITAIPU, is comparison is still good and full load operations are well reproduced .

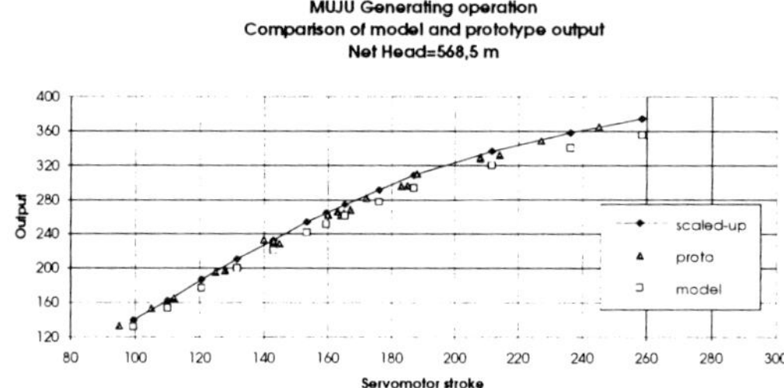

4-2 BULBS

For low head machine it is very difficult to have prototype results with enough accuracy .However results obtained on a model with different testing conditions give indications on the scale-up of discharge and output .On the following diagrams we can see two different results that were obtained on our test rig.

We can observe that the application of the formulas give a better evaluation than the usual way in which discharge is unchanged .

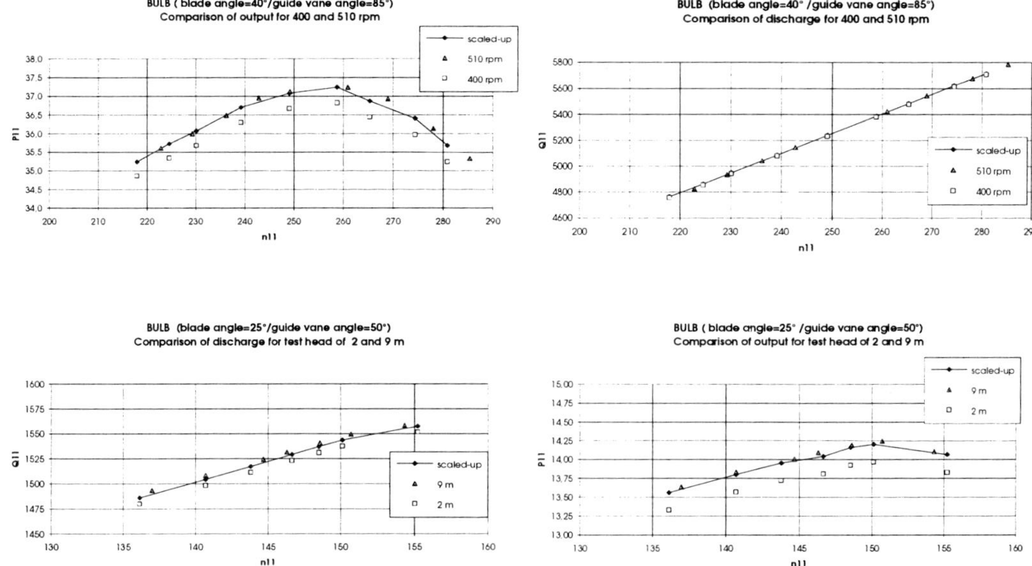

4-3 PUMP-TURBINE in PUMP MODE

High-head pump-turbines correspond to the straighforward application of the method presented in § 3-1-2 as such machines are almost non regulated .

Hereafter we present the achieved results for two recent single stage pump-turbines : GUANGZHOU in CHINA and MUJU in KOREA .

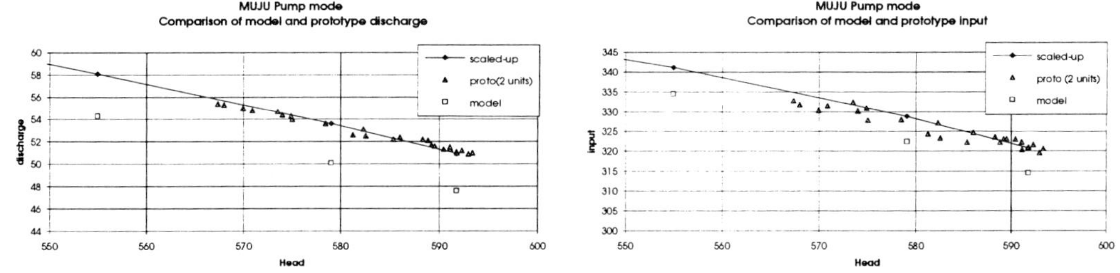

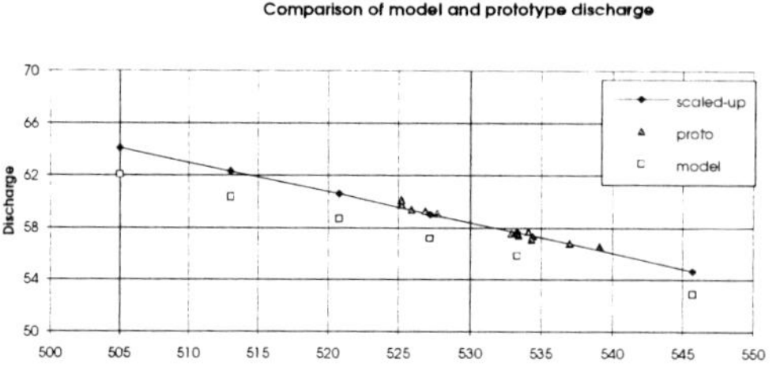

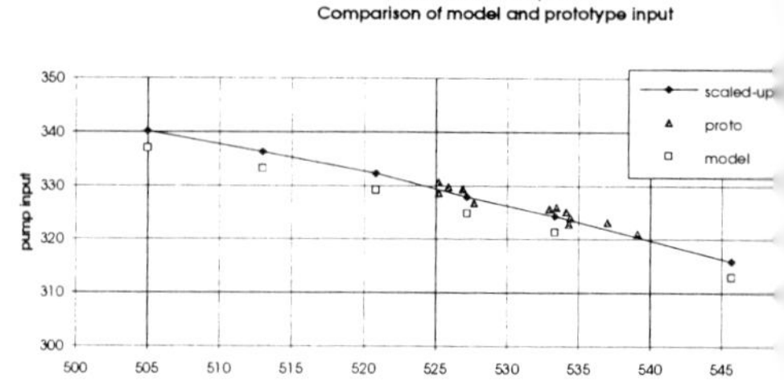

In both cases the shift of characteristic at a given head is very accurately predicted by the proposed simplified formula . However additional results for lower head pump-turbines have to be investigated to confirm the range of validity of the proposed approach .

5-CONCLUSIONS

The main conclusions of this paper are :

-Whatever the type of turbine the "efficiency scale-effect" induces a "shift of the head-discharge characteristic "and by a "shift of the head-output characteristic "

-The proposed formulas give an accurate evaluation of this shift for single regulated turbines as well as for double regulated turbines ; in pump mode as well as in turbine mode , for most of the cases available . Additional results have to be analysed in order to confirm the proposed approach.

6-REFERENCES

1-IDA,T. New formula for efficiency step-up of hydraulic turbines
XVII AIRH Symposium (1994) BEIJING CHINA
2-SPURK,J.H. and GREIN,H. Performances prediction of hydraulic machines by dimensionnal considerations
Water Power & Dam Construction (1993-11) pp 42-49
3-COUSTON,M. ,MACHADO FERNANDES FILHO A. and TEMEZ RUIZ DIAZ P.P. Hydraulic performances of ITAIPU Francis turbines .
XVII AIRH Symposium (1994) BEIJING CHINA

RECENT DEVELOPMENT OF STUDIES ON SCALE EFFECT

T.IDA
Kanagawa University
Yokohama, Japan

J. KUROKAWA
Yokohama National University
Yokohama, Japan

T. KUBOTA
Kanagawa University
Yokohama, Japan

H. TANAKA
Toshiba Corporation
Yokohama, Japan

1. Introduction

In 1986, a sub-committee was organized in Japan Society of Mechanical Engineers (JSME) to standardize a new scale effect formula by studying the outstanding problems of scale effect such as dependence of V (loss distribution coefficient) on specific speed, roughness effect and so on. Then the sub-committee established a new performance conversion method and published it as JSME S 008 in 1989 [1]. After its publication, the second sub-committee has been organized to revise the conversion method by collecting the results of further study.

The most difficult problem in the study of scale effect is that the accuracy of the measurement of prototype efficiency is incompetent to verify the scale effect. Therefore, most of the study was conducted by using both theoretical analyses on numerical models of hydraulic machines and its verification by laboratory model tests. For this study, various numerical models which may simulate the performance of a whole machine were developed independently at institutes and industries in Japan. Most of them employed 3D flow analysis for major parts of the machines. In order to assess the wall friction, they used boundary layer calculation or wall friction loss calculation using flat plate analogy. These numerical models were refined and improved by comparing their simulation results with performance characteristics obtained by laboratory model tests. By using these models, the hydraulic loss in the machines were separated into scalable loss and non-scalable loss and the scale effect on the scalable loss was assessed quantitatively.

This paper describes the summary of the results of these studies and the essential part of the newly proposed performance conversion method which will be published soon as a new version of JSME S 008.

E. Cabrera et al. (eds.), Hydraulic Machinery and Cavitation, 313–322.

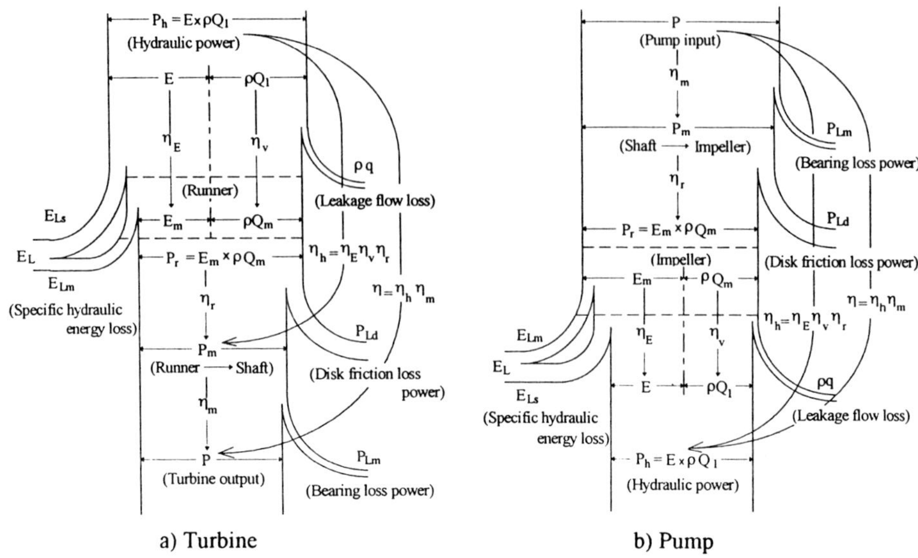

Fig. 1 Flux diagrams of hydraulic machines

2. Structure of the New Scale Effect Formula

The energy flux through hydraulic machines is conceptually represented by the diagrams set out in Fig. 1 a) and b). As indicated in these diagrams, the hydraulic efficiency of hydraulic machine η_h is given by the product of the following three internal efficiencies.

$$\eta_h = \eta_E \eta_v \eta_r \tag{1}$$

where, η_E is specific energy efficiency concerning all specific hydraulic energy loss in main water passages, η_v is volumetric efficiency concerning leakage loss and η_r is power efficiency of runner/impeller concerning disk friction.

Scale effect observed in performance characteristics of geometrically homologous hydraulic machines is caused by the change of wall friction loss due to the difference of Reynolds number and the relative roughness of the wall surface. Since the wall friction mostly causes dissipation of the specific hydraulic energy of the main flow, scale effect appears primarily in the specific hydraulic energy of the machine E. Scale effect also appears in leakage flow loss and in disk friction loss of the runner/impeller due to the difference of Reynolds number.

In the new performance conversion method, conversion factors for each internal efficiencies proposed by T. Ida are newly introduced. [5] They are specific energy efficiency scale factor $F_E = \eta_{EP}/\eta_{EM}$, volumetric efficiency scale factor $F_v = \eta_{vP}/\eta_{vM}$ and power efficiency scale factor $F_r = \eta_{rP}/\eta_{rM}$, where subscript P indicates the values of the prototype and M for those of the model. By using these scale factors, the hydraulic efficiency of a model can be converted to the prototype efficiency as set out below.

$$\eta_{hP} = F_E F_v F_r \eta_{hM} \tag{2}$$

These scale factors are standardized in the new conversion method as explained hereinafter.
Scale effect causes also shifting of performance. These scale factors are also applicable to the conversion of E, Q and P, respectively. [2]

Turbine	Pump	
$E_P = F_E^{-1}\left(\frac{n_P}{n_M}\right)^2\left(\frac{D_P}{D_M}\right)^2 E_M$	$E_P = F_E\left(\frac{n_P}{n_M}\right)^2\left(\frac{D_P}{D_M}\right)^2 E_M$	(3)
$Q_P = F_v^{-1}\left(\frac{n_P}{n_M}\right)\left(\frac{D_P}{D_M}\right)^3 Q_M$	$Q_P = F_v\left(\frac{n_P}{n_M}\right)\left(\frac{D_P}{D_M}\right)^3 Q_M$	(4)
$P_P = F_r\left(\frac{n_P}{n_M}\right)^3\left(\frac{D_P}{D_M}\right)^5 P_M$	$P_P = F_r^{-1}\left(\frac{n_P}{n_M}\right)^3\left(\frac{D_P}{D_M}\right)^5 P_M$	(5)

3. New Formula for Conversion of Specific Energy Efficiency

One of the major changes from the existing method to the new method is the use of relative scalable loss δ_E instead of loss distribution coefficient V. The definition of δ_E is given by the following equation.

$$\delta_E = E_{Lf}/E \tag{6}$$

where, E_{Lf} is the scalable specific energy loss due to wall friction.
On the other hand, the loss distribution coefficient V used in IEC Publication 995, which is denoted as V_{IEC} or that in the existing version of JSME S 008, which is denoted as V_{JSME}, is defined as follows;

$$V_{IEC} = \frac{(\text{Total scalable losses including disk friction, etc.})}{[(\text{Total scalable loss}) + (\text{Total non-scalable loss})]} \tag{7a}$$

$$V_{JSME} = E_{Lf}/E_L = E_{Lf}/(E_{Lf} + E_{Lns}) \tag{7b}$$

Whichever V is used, the problem to use these V values is that they include non-scalable loss E_{Lns} and others in their denominators which could be significantly affected by the difference of specific speed or individual design and, consequently, they could vary considerably from their standardized values depending on individual machine design. Also, it has been pointed out that, since these V values were standardized for average machines, the prototype efficiency stepped up from poor efficiency model by using these values would be unreasonably too high.
It is expected that relative scalable loss δ_E is less affected than V by the difference of specific speed or individual design since it does not include non-scalable loss like E_{Lns}. The values of δ_E for Francis turbines and pump turbines, which are obtained

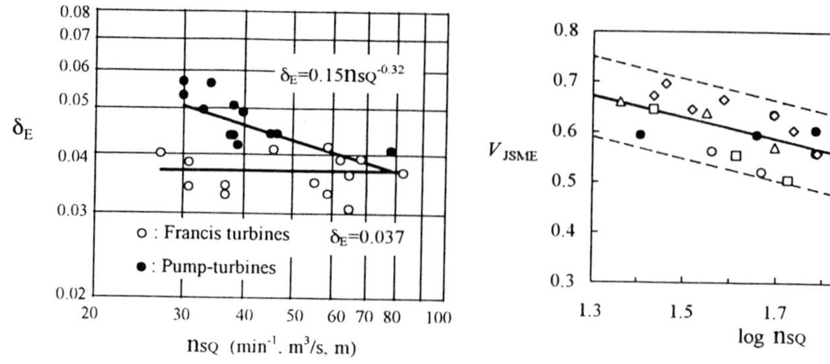

Fig. 2 Scalable loss δ_E versus specific speed

Fig. 3 V_{JSME} versus specific speed

from the investigation using various numerical models are shown in Fig. 2. Apparently, their scattering is less than that of V_{JSME} [2], which is shown in Fig. 3, for example.

Hence, δ_E is selected instead of V as the standardized reference of the scalable loss.

When the flow condition in both prototype and model is homologous, the ratio of their relative scalable loss denoted as Λ is given as follows;

$$\Lambda = (\delta_{EP} / \delta_{EM}) = (C_{fP} / C_{fM}) \tag{8}$$

where, C_f is the friction loss coefficient for main water passages.

Then new scale effect formulae for the conversion of the specific energy efficiency are proposed by T. Ida.[3] They are written as follows;

$$\text{For turbine:} \quad F_E = \frac{1}{1-(1-\Lambda)\delta_E} \quad \text{or} \quad \Delta\eta_E = \eta_{EM}\left[\frac{1}{1-(1-\Lambda)\delta_E} - 1\right] \tag{9}$$

$$\text{For pump:} \quad F_E = 1+(1-\Lambda)\delta_E \quad \text{or} \quad \Delta\eta_E = \eta_{EM}(1-\Lambda)\delta_E \tag{10}$$

For simplification, all the model values in this paper are assumed to be those measured at the reference Reynolds number.

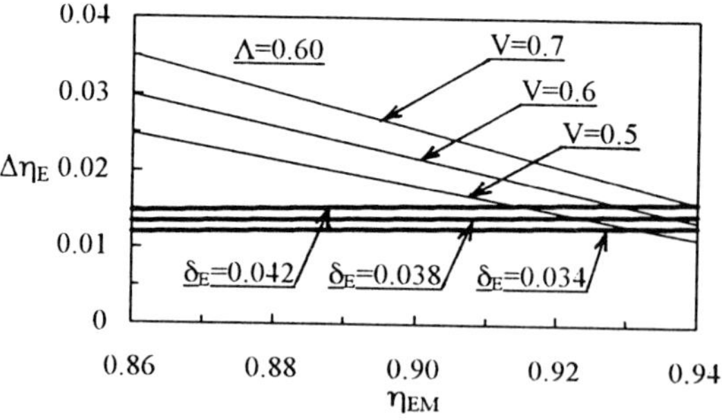

Fig. 4 Efficiency step up $\Delta\eta_E$ versus model efficiency η_{EM}

The values of δ_E and V', which were collected from the results of the numerical model analysis, show some scattering as shown in Figs. 2 and 3. The uncertainty of $\Delta\eta_E$ caused by such scattering is examined and shown in Fig. 4. It is observed that the scattering of $\Delta\eta_E$ caused by the scattering of δ_E is less than that caused by V'. Thus, by using the new formulae based on δ_E, the accuracy of $\Delta\eta_E$ can be improved considerably and the unreasonably large efficiency step up for poor efficiency turbines can be avoided.

4. Friction Coefficient Ratio Considering Roughness Effect

The effects of surface roughness on the wall friction loss are studied by T. Ida using the analogy of the friction loss of flat plate. Basic equations of the fricion loss coefficient for flat plate which are used as a basis for the discussions hereinafter are shown below.

For smooth surface: $C_{fS} = 0.455(\log_{10} Re_L)^{-2.58}$ (11)

For rough surface : $C_{fR} = \left[1.89 - 1.62 \log_{10}(k_S / L)\right]^{-2.5}$ (12)

where, k_S is the equivalent sand roughness and L is the length of the plate.
The admissible sand roughness can be obtained as the solution of $C_{fS} = C_{fR}$. The result is;

$$(k_{Sadm} / L) = 10^{\left\{(1.89/1.62)-(1/1.62)\left[(\log_{10} Re)^{2.58}/0.455\right]^{0.4}\right\}} \approx 24.0 Re_L^{-0.931} \quad (13)$$

The extreme right is the approximation for the middle equation. The admissible roughness thus obtained is shown in Fig. 5. The admissible roughness given by the Schlichting's equation is also plotted for comparison.

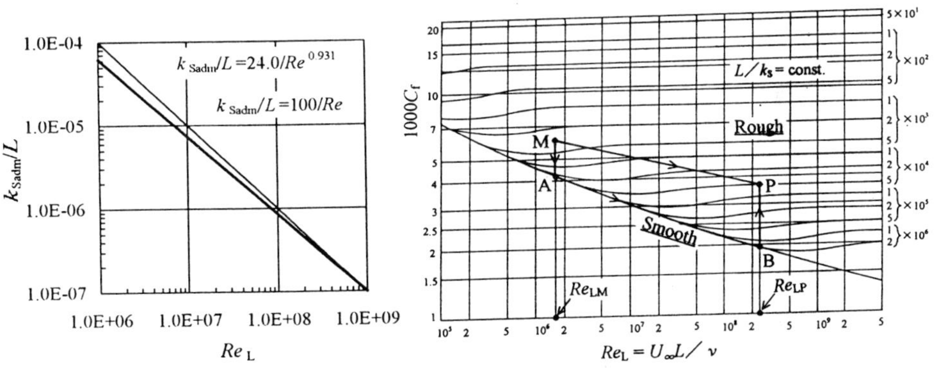

Fig. 5 Admissible sand roughness versus Reynolds number

Fig. 6 Friction loss coefficient of sand roughened flat plate

In general, actual surface roughness e, which is usually expressed in terms of Ra, is not the same as k_S. Therefore, when using the admissible sand roughness obtained above, the following conversion is required.

$$e_{\mathrm{adm}} = k_{\mathrm{Sadm}} / C_{\mathrm{k}} \tag{14}$$

where, C_{k} is a coefficent to be obtained experimentally. Unfortunately, although a few results of the experimental studies on C_{k} were published, they did not sufficiently coincide with each other. Further study is required in this regard.

The variation of friction loss coefficient C_{f} for different Reynolds number and surface roughness is shown in Fig. 6. In this diagram, point M represents C_{f} for a rough model (strictly speaking, a model not perfectly smooth) and point P for a rough prototype (similarly, a prototype not perfectly smooth). Here, we consider a smooth model A which has the same Reynolds number with M and a smooth prototype B with the same Reynolds number with P. Then, the conversion of the friction loss coefficient from the rough model M to the rough prototype P is divided into three steps; namely, M→A→B→P. Thus, the ratio of friction loss coefficient, $\Lambda_{\mathrm{M}\to\mathrm{P}}$, may be expressed by the following equation.

$$\Lambda_{\mathrm{M}\to\mathrm{P}} = \Lambda_{\mathrm{M}\to\mathrm{A}} \times \Lambda_{\mathrm{A}\to\mathrm{B}} \times \Lambda_{\mathrm{B}\to\mathrm{P}} = \Lambda_{\mathrm{A}\to\mathrm{B}} \times \Lambda_{\mathrm{B}\to\mathrm{P}} / \Lambda_{\mathrm{A}\to\mathrm{M}} \tag{15}$$

Firstly, by Eq. (11), $\Lambda_{\mathrm{A}\to\mathrm{B}}$ (smooth model to smooth prototype) is given as the first line below and it can be approximated with good accuracy by the equation at the second line.

$$\begin{aligned} \Lambda_{\mathrm{A}\to\mathrm{B}} &= \left(C_{\mathrm{fSB}} / C_{\mathrm{fSA}}\right) = \left[\left(\log_{10} Re_{\mathrm{LP}}\right) / \left(\log_{10} Re_{\mathrm{LM}}\right)\right]^{-2.58} \\ &\approx \left(Re_{\mathrm{LP}} / Re_{\mathrm{LM}}\right)^{-0.155} \end{aligned} \tag{16}$$

Secondly, $\Lambda_{\mathrm{B}\to\mathrm{P}}$ (smooth machine to rough machine) is obtained. By combining Eqs. (11) and (12), it is given as the first line below and it can be closely approximated by the equation at the second line.

$$\begin{aligned} \Lambda_{\mathrm{B}\to\mathrm{P}} &= \left(C_{\mathrm{fRP}} / C_{\mathrm{fSB}}\right) = \left[1.89 - 1.62 \log_{10}\left(k_{\mathrm{SP}} / L_{\mathrm{P}}\right)\right]^{-2.5} / 0.455\left(\log_{10} Re_{\mathrm{LP}}\right)^{-2.58} \\ &\approx 0.678\left(k_{\mathrm{SP}} / L_{\mathrm{P}}\right)^{0.18} Re_{\mathrm{LP}}{}^{0.155} \end{aligned} \tag{17}$$

Similarly $\Lambda_{\mathrm{A}\to\mathrm{M}}$ is obtained as follows;

$$\Lambda_{\mathrm{A}\to\mathrm{M}} = 0.678\left(k_{\mathrm{SM}} / L_{\mathrm{M}}\right)^{0.18} Re_{\mathrm{LM}}{}^{0.155} \tag{18}$$

By substituting Eqs. (16), (17) and (18) into Eq. (15), we obtain;

$$\begin{aligned} \Lambda_{\mathrm{M}\to\mathrm{P}} &= \left(Re_{\mathrm{LP}} / Re_{\mathrm{LM}}\right)^{-0.155}\left[\left(k_{\mathrm{SP}} / L_{\mathrm{P}}\right) / \left(k_{\mathrm{SM}} / L_{\mathrm{M}}\right)\right]^{0.18}\left(Re_{\mathrm{LP}} / Re_{\mathrm{LM}}\right)^{0.155} \\ &= \left[\left(k_{\mathrm{SP}} / L_{\mathrm{P}}\right) / \left(k_{\mathrm{SM}} / L_{\mathrm{M}}\right)\right]^{0.18} \end{aligned} \tag{19}$$

In the above equation, L represents the length of vanes or passages. Therefore, in case of homologous machines, $\left(L_{\mathrm{P}} / L_{\mathrm{M}}\right) = \left(D_{\mathrm{P}} / D_{\mathrm{M}}\right)$. Also, using actual roughness e, $\left(k_{\mathrm{SP}} / k_{\mathrm{SM}}\right)$ can be substituted by $\left(e_{\mathrm{P}} / e_{\mathrm{M}}\right)$. After all, $\Lambda_{\mathrm{M}\to\mathrm{P}}$ is obtained as follows;

$$\Lambda_{M\to P} = (D_P / D_M)^{-0.18} (e_P / e_M)^{0.18} \tag{20}$$

If the surface roughness of the model is smooth i.e. $e_M \le e_{admM}$, the following equation can be used for the conversion from a smooth model to a rough prototype.

$$\Lambda_{M\to P} = (D_P / D_M)^{-0.18} (e_P / e_{admM})^{0.18} \tag{21}$$

Since a prototype has generally larger Reynolds number and smaller admissible roughness, its surface roughness is mostly hydraulically rough. Especially, surfaces of runner/impeller and distributor are exposed to high velocity flow and subjected to significant roughness effect. Therefore, it is generally necessary to apply Eq. (20) or (21) to the wall friction loss on these areas taking into account the difference of the roughness of both model and prototype; namely, $(e_P / e_M) \ne 1$. Regarding the other areas, most of them are painted and their roughness is almost identical for both model and prototype. Therefore, for these areas, Eq. (20) or (21) assuming $(e_P / e_M) = 1$ can be applied.

Then, similar to the existing code JSME S 008, a coefficient β which is defined by the following equation is introduced.

$$\beta = \frac{(\text{Friction loss in runner / impeller and distributor})}{(\text{Total friction loss in main water passage})} \tag{22}$$

By using β, the friction loss coefficient ratio $\Lambda_{M\to P}$ applicable to actual machine is obtained as follows;

$$\begin{aligned}\Lambda_{M\to P} &= (1-\beta)(D_P / D_M)^{-0.18} (1/1)^{0.18} + \beta (D_P / D_M)^{-0.18} (e_P / e_M)^{0.18} \\ &= (D_P / D_M)^{-0.18} \left[(1-\beta) + \beta (e_P / e_M)^{0.18} \right]\end{aligned} \tag{23}$$

The value of β can be estimated by the simulation analysis using numerical models. Some examples of its value at optimum condition are shown in Fig. 7. It becomes larger for higher specific speed machines, however, mostly it remains within the range of 0.6~0.85.

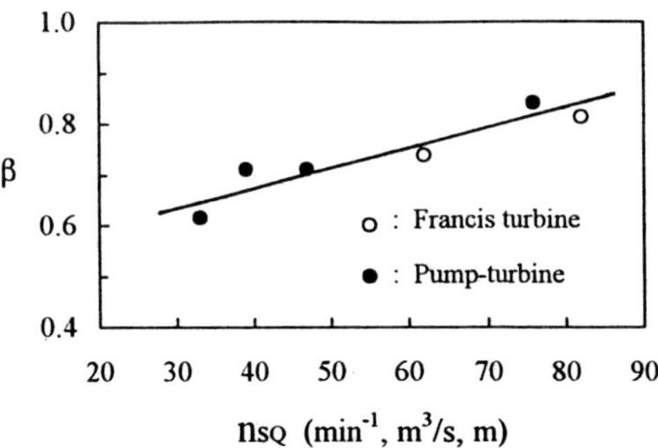

Fig. 7 β versus specific speed

5. Influence of the Approximation of Principal Coefficients

5.1 INFLUENCE OF RELATIVE SCALABLE LOSS δ_E

As shown in Fig. 2, the scattering of δ_E ranges over 0.037 ± 0.004. The result of an example to examine the influence of this scattering is shown in Fig. 8. As observed in this diagram, the difference of $\Delta\eta_E$ caused by the above scattering of δ_E is only $\pm 0.13\%$. It suggests that a single value of $\delta_E = 0.037$ can be used practically as a standardized value for all Francis turbines of various specific speed.

5.2 INFLUENCE OF DEPENDENCE COEFFICIENT OF ROUGHNESS RATIO β

As shown in Fig. 7, the value of β varies over a range of 0.60 to 0.85. Another examination to see the influence of the different values of β was conducted. An example of the results of the conversion using different values of β is shown in Fig. 9. It indicates that, if β is selected within a range of 0.8 ± 0.2, the dispersion of the resultant prototype efficiency does not exceed ± 0.1 %. Thus the influence of the deviation of β on $\Delta\eta_E$ is small. It suggests that a constant value $\beta = 0.7$ may be used for all Francis turbines and pump-turbines in practical application.

In this figure, K_m and K_p are the dimensionless flow coefficient and output power coefficient, respectively, which are defined as follows:

$$K_m = (Q_1 / A_{R1}) / (2E)^{1/2} \tag{24}$$

$$K_p = P / \left[\rho A_{R1} (2E)^{1/2} E \right] \tag{25}$$

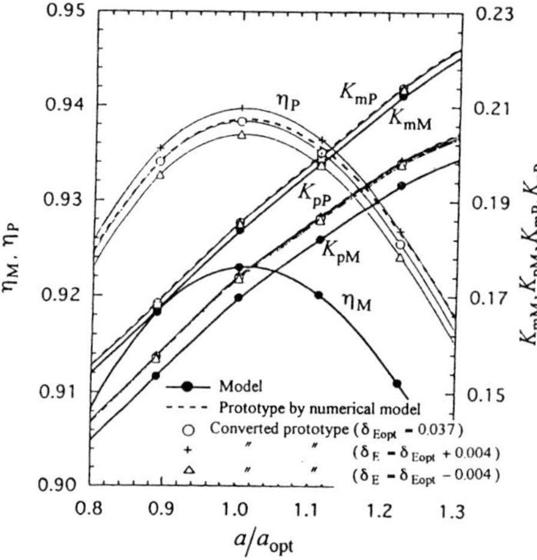

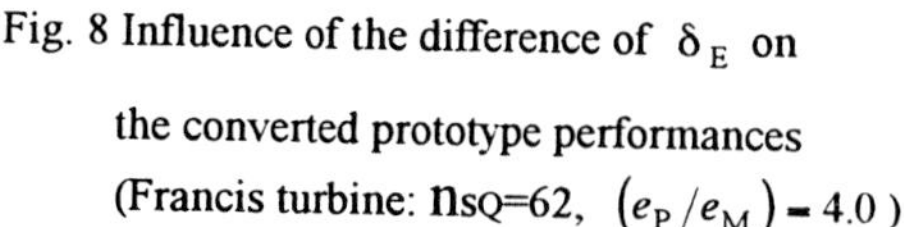

Fig. 8 Influence of the difference of δ_E on the converted prototype performances (Francis turbine: n_{SQ}=62, $(e_P/e_M) = 4.0$)

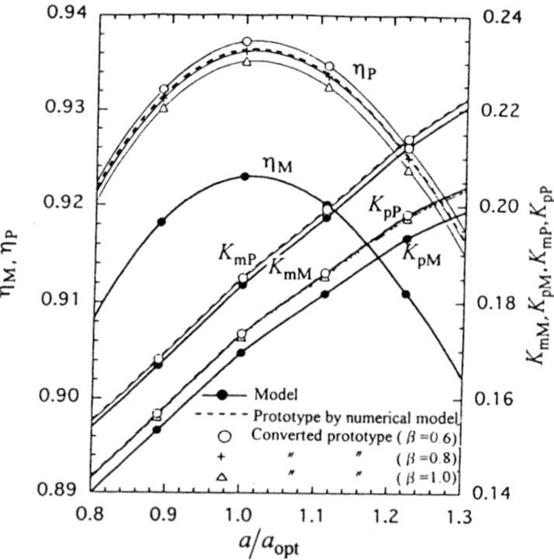

Fig. 9 Influence of the difference of β on the converted prototype performances (Francis turbine: n_{SQ}=62, $(e_P/e_M) = 4.0$)

It should be noted that K_m and K_p of the prototype are significantly influenced by scale effect since they include E, which is most likely to be affected by scale effect, in their denominators.

6. Conversion of Volumetric Efficiency

Volumetric efficiency η_v is determined by leakage flow loss and it is subjected to scale effect. In the new conversion method, it is converted to the prototype value by the following equation.

$$\eta_{vP} = F_v \eta_{vM} \tag{26}$$

The value of η_{vM} obtained by the numerical analysis is plotted against specific speed and shown in Fig. 10. Those for Francis turbines and pump-turbines are represented by a common single curve. [4]

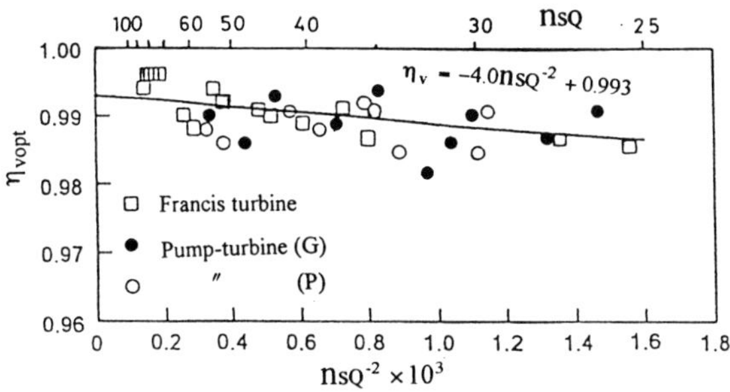

Fig. 10 Volumetric efficiency η_{vM} versus specific speed

Detailed study to determine Λ_v and F_v is conducted by T. Ida. [5] The scale factor F_v can be expressed as follows;

$$F_v = 1/\left[\eta_{vM} + \Lambda_v\left(1-\eta_{vM}\right)\right] \qquad \text{(for turbine)} \tag{27}$$

$$F_v = \left[\left(1-\Lambda_v\right)/\eta_{vM}\right] + \Lambda_v \qquad \text{(for pump)} \tag{28}$$

As a result of the study, it is made clear that F_v may be regarded as 0.997～0.998 within the practical range of $\left(Re_P/Re_M\right) = 5 \sim 25$ and $n_{sQ} = 25 \sim 100$, when the geometry of the seal clearances of both model and prototype is homologous.

7. Conversion of Power Efficiency of Runner/impeller

Similar to volumetric efficiency, power efficiency η_r concerning disk friction of runner/impeller can be converted by the following formula.

$$\eta_{rP} = F_r \eta_{rM} \tag{29}$$

The value of η_{rM} is investigated by using the numerical models and the results are shown in Fig. 11.

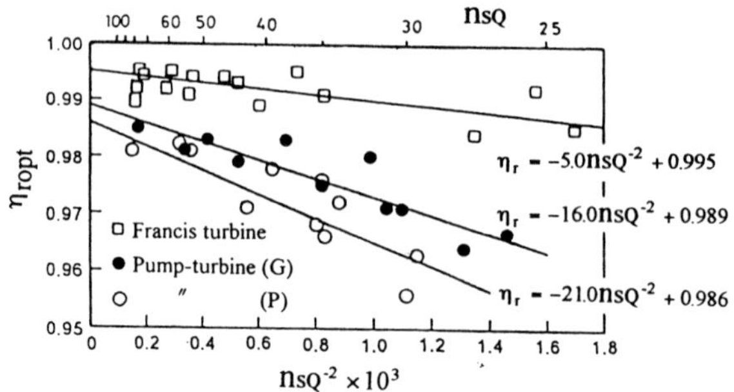

Fig. 11 Power efficiency of runner/impeller η_{rM} versus specific speed

The scale factor F_r is given by the following equation. [5]

$$F_r = \left[(1-\Lambda_r)/\eta_{rM}\right] + \Lambda_r \qquad \text{(for turbine)} \tag{30}$$

$$F_r = 1/\left[(1-\Lambda_r)\eta_{rM} + \Lambda_r\right] \qquad \text{(for pump)} \tag{31}$$

where, Λ_r is the ratio of the disk friction loss coefficient. If it is assumed that the surface roughness relevant to the disk friction is hydraulically rough, the following approximate equation is obtained.

$$\Lambda_r = (D_P/D_M)^{-0.27}(e_P/e_M)^{0.12} \tag{32}$$

Finally, the following equations for F_r are derived.

$$F_r = (0.388/\eta_{rM}) + 0.612 \qquad \text{(for turbine)} \tag{33}$$

$$F_r = 1/\left(0.995 - 8.2n_{SQ}^{-2}\right) \qquad \text{(for pump)} \tag{34}$$

8. Conclusion

New performance conversion method with regard to scale effect is proposed for the new version of JSME S 008. Compared with the existing version, it is expected that the new method gives more reasonable and accurate conversion. In addition, introduction of scale factors makes the calculation of performance shifting due to scale effect much easier.

Reference

1 JSME S 008 (1989) Performance conversion method for hydraulic turbines and pumps

2 Ida, T. et al. (1990) Advanced scale-up formulae for performances of hydro-turbines and pump-turbines considering surface roughness, *Proc. IAHR Symp. in Belgrade* **1**, B4

3 Ida, T. (1994) New formula for efficiency step-up of hydraulic turbines, *Proc. IAHR Symp. in Beijing* **2**, 827-840

4 Kurokawa, J. et al. (1994) Accurate determination of volumetric and mechanical efficiencies and leakage behaviour of Francis turbine and Francis pump-turbine, *Proc. IAHR SYMP. in Beijing* **2**, 889-900

5 Ida, T. (1995) New formulae for scaling-up hydraulic efficiency of hydraulic turbines, *Journal of Hydraulic Research* **33**, 2, 147-162

PREDICTION OF SCALABLE LOSS IN FRANCIS RUNNERS OF DIFFERENT SPECIFIC SPEED

T.KITAHORA, J.KUROKAWA, M.MATUMOTO
Yokohama National University
156 Tokiwadai, Hodogaya-ku, Yokohama 240, Japan
R.SUZUKI
Fuji Electric Co.,Ltd.
1-1 Tanabeshinden, Kawasaki-ku, Kawasaki 210, Japan

1. Introduction

Hydraulic losses are composed of scalable loss and non-scalable loss. For the determination of scale-up formulae in hydraulic turbines, it is important to predict the scalable loss, that is the friction losses in a runner and fixed channels, as precisely as possible. However, it is still difficult to determine them both theoretically and experimentally, partly because the boundary layer is too thin to prepare sufficient mesh size for CFD, and partly because the scalable loss cannot be separated experimentally from the measured total hydraulic loss.

Scale-up formulae have long been discussed in IEC-WG18[2] and IAHR-WG3, and also in JSME[3]. In the JSME Standard S-008 "Performance Conversion Method for Hydraulic Turbines and Pumps", the new method of theoretically derived scale-up formulae are presented[5]. In order to further improve the accuracy of scale-up, it is necessary to develop a method of more precisely predicting scalable loss.

The present study presents a prediction of scalable loss in a runner channel of arbitrary three-dimensional shape. The prediction is composed of a 3-D Euler calculation for the main flow in a runner channel and the 3-D boundary layer calculation. The latter is based on the momentum integral equation by use of an orthogonal curvilinear coordinates.

The present method is applied for five Francis runners of different specific speed, and the calculated results of scalable losses are compared with the results by conventional two-dimensional loss analysis.

2. Boundary layer equation

A three-dimensional boundary layer along a Francis turbine runner is analyzed taking

E. Cabrera et al. (eds.), Hydraulic Machinery and Cavitation, 323–332.

the surface curvature into account. Z-axis of the absolute Cartesian coordinate (X,Y,Z) is set on the rotating shaft and the rotating local relative curvilinear orthogonal coordinates (x,y,z) is set on the runner wall as shown in Fig.1. Each intersection of calculated grid lines on the walls is selected as the origin of this local coordinates. For simplicity of theory, the directions of the grid lines on the wall are neglected, and the main flow direction at the outside of boundary layer is taken as the x direction in each origin. Moreover, the axis perpendicular to this direction on the wall is defined as z-axis and the normal direction to the wall is taken as y- axis.

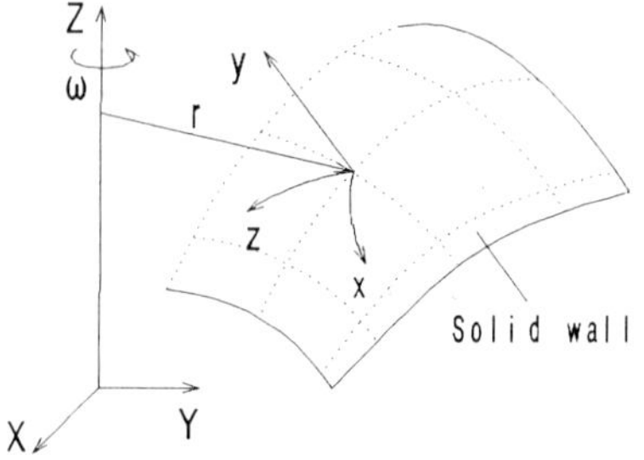

Figure 1. Coordinates system

The Navier-Stokes equations and the continuity equation applied to this coordinate system are given as follows.

$$\frac{\partial u}{\partial t}+u\frac{\partial u}{\partial x}+v\frac{\partial u}{\partial y}+w\frac{\partial u}{\partial z}-\kappa_z uw+\kappa_z w^2=-\frac{1}{\rho}\frac{\partial p}{\partial x}+\nu\frac{\partial^2 u}{\partial y^2}+f_{cex}+f_{coy} \tag{1}$$

$$\frac{\partial w}{\partial t}+u\frac{\partial w}{\partial x}+v\frac{\partial w}{\partial y}+w\frac{\partial w}{\partial z}-\kappa_x uw+\kappa_z u^2=-\frac{1}{\rho}\frac{\partial p}{\partial z}+\nu\frac{\partial^2 w}{\partial z^2}+f_{cez}+f_{coz} \tag{2}$$

$$\frac{\partial u}{\partial x}+\frac{\partial v}{\partial y}+\frac{\partial w}{\partial z}-\kappa_x u-\kappa_z w=0 \tag{3}$$

To stabilize the calculation, the method of calculating unsteady flow is adopted. In this method, the real length is used for the length of each axis, therefore the scale factors are fixed to unity, because the grid interval distance and the boundary layer thickness are much smaller than the radius of curvature of the axis. The velocity v in y-direction is determined by the following equation derived from Eq.(3) neglecting the last two terms.

$$v=-\int_0^\delta \frac{\partial u}{\partial x}dy-\int_0^\delta \frac{\partial w}{\partial z}dy$$

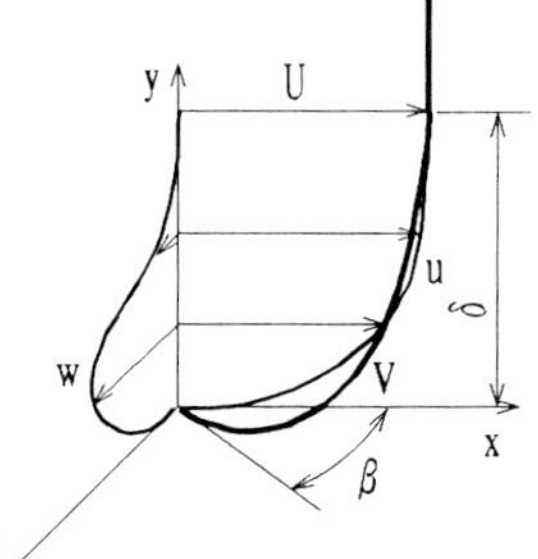

Figure 2. Velocity distribution in boundary layer

As shown in Fig.2, $1/n$ th-power law model is adopted for the velocity profile u in the x-direction. The Marger's, shown as the followings, model is adopted for secondary flow in z-direction.

$$u=U\left(\frac{y}{\delta}\right)^{\frac{1}{n}} \tag{4}$$

$$w=u\left(1-\frac{y}{\delta}\right)^2 \tan\beta \tag{5}$$

Here, δ is the boundary layer thickness, and β is the angle between the direction of the main flow U at the outside of boundary layer and the direction of the wall stream line.

Boundary layer integral equations in x and z directions are obtained by substituting the above velocity distributions into Eqs. (1) and (2), and by integrating them from the wall to the edge of boundary layer under the assumption that the main flow pressure prevails into the wall surface. The pressure terms can be deleted by combining the equation of the main flow with that of boundary layer. The following equations are finally obtained for the boundary layer.

$$U\frac{1}{n+1}\frac{\partial\delta}{\partial t}+\frac{\partial}{\partial x}C_1-U\frac{\partial}{\partial x}C_2+U\frac{\partial}{\partial z}C_3-\frac{\partial}{\partial z}C_4+\kappa_z C_4$$
$$-\kappa_x C_5=\frac{\tau_x}{\rho}-2\Omega\left(\frac{\partial x}{\partial X}\frac{\partial Y}{\partial z}-\frac{\partial x}{\partial Y}\frac{\partial X}{\partial z}\right)C_3 \tag{6}$$

$$-\frac{\partial}{\partial t}C_3-\frac{\partial}{\partial x}C_4-\frac{\partial}{\partial z}C_5+\kappa_x C_4+\kappa_z C_1$$
$$=\frac{\tau_z}{\rho}+2\Omega\left(\frac{\partial z}{\partial X}\frac{\partial Y}{\partial x}-\frac{\partial z}{\partial Y}\frac{\partial X}{\partial x}\right)C_2 \tag{7}$$

Here, the coefficients $C_1 \sim C_5$ in the above equations are expressed as follows.

$$C_1=\int_0^\delta (U^2-u^2)\,dy=\frac{2U^2\delta}{n+2}$$
$$C_2=\int_0^\delta (U-u)\,dy=\frac{U\delta}{n+1}$$
$$C_3=\int_0^\delta w\,dy=\frac{2U\delta n^3\tan\beta}{6n^3+11n^2+6n+1}$$
$$C_4=\int_0^\delta uw\,dy=\frac{U^2\delta n^3\tan\beta}{3n^3+11n^2+12n+4}$$
$$C_5=\int_0^\delta w^2\,dy=\frac{6U^2\delta n^5\tan^2\beta}{30n^5+137n^4+225n^3+170n^2+60n+8} \tag{8}$$

and κ_x and κ_z in Eqs. (6) and (7) are the curvature of the z-axis and x-axis respectively. Ω is rotating angular velocity of runner, and the last terms in Eqs. (6) and (7) show the decrease of the Corioli's force in the boundary layer.

The following experimental formula by Ludwieg-Tillman is adopted for wall shear stress τ_x in the main flow direction.

$$\tau_x=\rho U^2\times 0.123\times 10^{-0.678\left(\frac{n+2}{n}\right)}\left(\frac{U\theta}{\nu}\right)^{-0.268} \tag{9}$$

$$\theta=\frac{\delta n}{(n+1)(n+2)}$$

$$\tau_z=\tau_x\tan\beta \tag{10}$$

Equation (3) of continuity is also integrated, yielding the following equation.

$$\frac{\partial}{\partial x}C_2 - U\frac{\partial \delta}{\partial x} - \frac{\partial}{\partial z}C_3 - \kappa_x C_2 + \kappa_z C_3 + 0.0306U(He-3)^{-0.653} = 0 \tag{11}$$

Here, He is the entrainment of the main flow and the following equation by Standen is assumed.

$$H_e = \frac{n\delta}{(n+1)\theta} \tag{12}$$

After all, the variables to be determined by use of the numerical method are δ , n and $tan\ \beta$ in Eqs. (6), (7) and (11).

3. Finite difference method

The main flow velocity U at each intersection of curvilinear grid lines, shown in Fig.3, on the wall is given by the main flow calculation. In the present study, the main flow velocity was determined by solving the Euler's equation by use of a finite difference method.

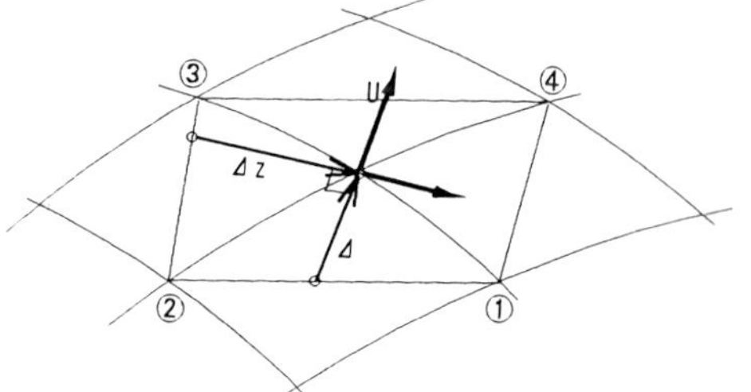

Fig.3 Local upstream FDM

As for the boundary layer calculation, Eqs.(6),(7) and (11) are converted into the finite difference equations under the local coordinates system. The upstream difference scheme is used to stabilize the calculation. The upstream reference point used for FDM is set on the intersection between the lines connecting four grid points ①~④ and the main flow line passing the origin. Each variable at the upstream reference point is determined by interpolation from the values of two grid points ① and ② . And also, the reference point in the direction perpendicular to the main flow is similarly determined for the secondary flow. The implicit method of Crank-Nicolson is used for unsteady flow calculation.

For the inlet boundary condition, very small values of δ ,n=5 and $tan\ \beta$ =0 are given. The same values are also given to the whole area as the initial condition of the unsteady flow calculation. It is generally difficult to determine the flow characteristics near the corner. In the present study the equations on each wall surface is solved separately under the assumption of $tan\ \beta$ =0 at the corner boundary.

The solutions at each time step are determined by applying an iterative method to these equations, and the final steady solutions can be determined when the unsteady solution does not change any more.

4. Application to Francis Turbine Runner and Main Flow Characteristics

The preceding calculation is applied to five runners of model Francis turbines. The design parameters of these models are shown in Table 1. The turbine E is well known

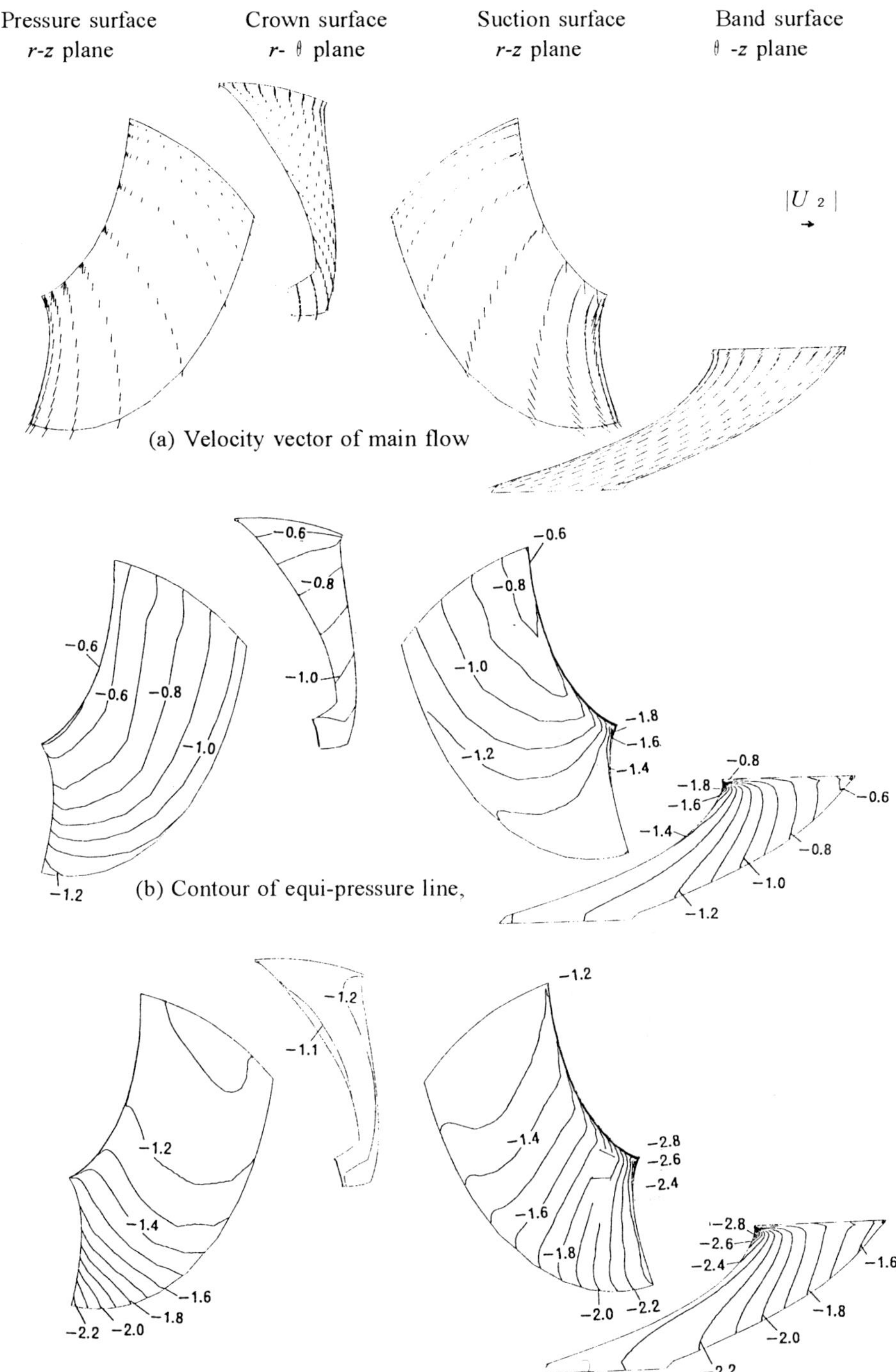

(a) Velocity vector of main flow

(b) Contour of equi-pressure line,

(c) Equi pressure line when pressure rize due to cetrifugal force is neglected

Figure 4. Main flow characteristics (Runner E)

Table 1. Parameters of model Francis turbines

Runner	$Re = \frac{D_2 U_2}{\nu}$	D_2 (mm)	n_{SQ} (rpm,m, m^3/s)	No. of runner vanes	Head	Main Flow Calc.
A	5.2×10^6	350	31	17	45	FEM-P
B	7.1×10^6	319	46	18	50	FDM-E
C	8.2×10^6	410	55	14	35	FDM-E
D	6.3×10^6	350	75	11	20	FDM-E
E(GAMM)	4.2×10^6	400	80	13	6	FDM-E

P : Potential Flow, E : Euler

as GAMM Runner.

The boundary layer characteristics are analyzed only at the maximum efficiency point.

In order to show the main flow characteristics, the results of the runner E, as an example, are illustrated in Fig.4. These results are obtained by numerical calculation of a three-dimensional Euler's equation under the assumption of wall slip condition. The calculated velocity vector of the main flow is shown in Fig.4(a) and the contour of equi-pressure lines are shown in Fig.4(b). Here, U_2 is the peripheral velocity of the runner at reference diameter. Four surfaces of a runner channel are developed and are projected from the inside of the channel. The main flow velocity U at the edge of boundary layer is used as the input data of the following boundary calculation.

The calculated results of the main flow in the runner exit section is compared with the measured data in Fig.5 for the case of runner E. The horizontal axis is the distance from the outer wall of draft tube. The numerical results are averaged over the vane-to-vane channel. The calculated results agree well with the experimental results.

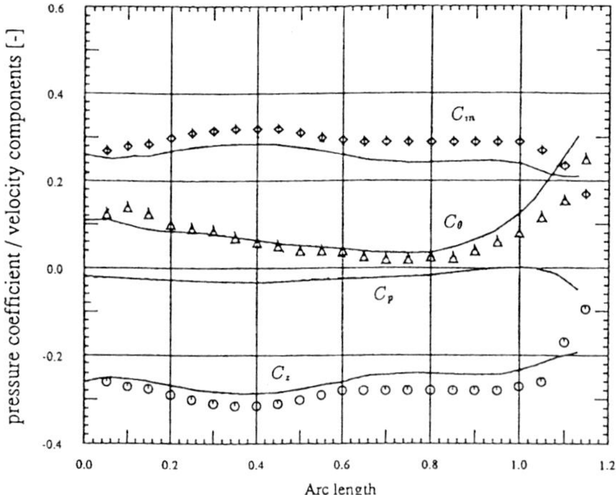

(C_p :Pressure,C_z :axial velocity,C_θ :Tangential velocity,C_m : meridional velocity)

Figure 5. Outlet Flow characteristice in runner E and comparison with measurments

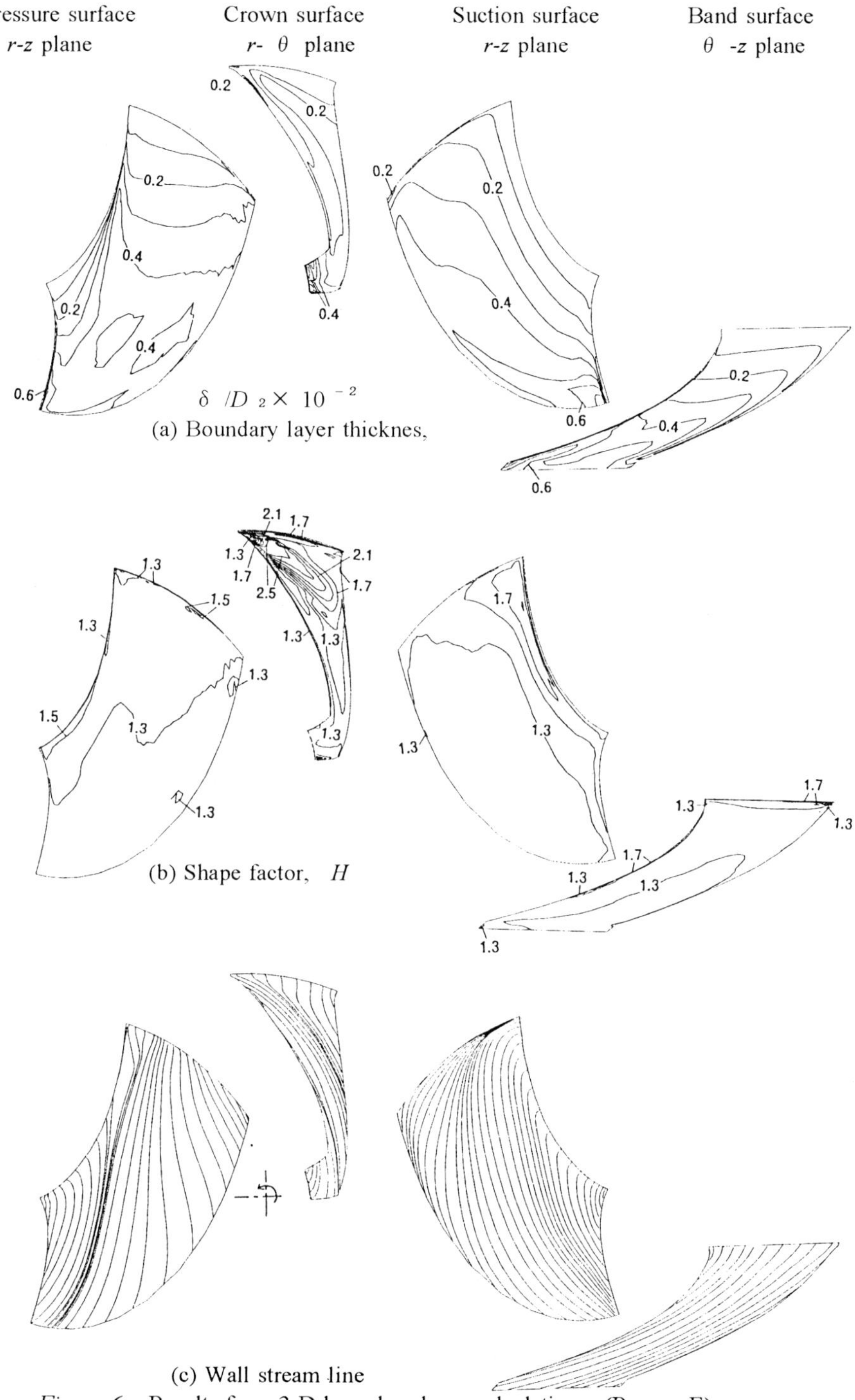

(a) Boundary layer thicknes,

(b) Shape factor, H

(c) Wall stream line

Figure 6. Results from 3-D boundary layer calculation (Runner E)

5. Results and Discussions of Boundary Layer Calculation

To discuss the boundary layer characteristics, the contours of the equal boundary layer thickness δ , of the equal shape factors H, and the wall stream lines are illustrated in Fig.6 for the case of runner E. The boundary layer calculation were performed without any trouble, although reverse flow occurs near the crown wall of the pressure side in the main flow.

On the vane suction surface and on the band surface of the boundary layer grows from the leading edge, but on the vane pressure surface it grows from the entrance of the crown surface, since the main flow enters into the pressure surface through the crown surface as shown in Fig.6(a).

The Corioli's force works strongly in the main flow where the velocity is large, and balances with both the pressure gradient and the inertia force. Because the Corioli's force works weakly near the wall where flow is slow, the wall stream line turns to the opposite direction of Corioli's force, or turns to the direction in which the real pressure decreases if the pressure rise due to centrifugal force is removed as shown in Fig.4(c).

On the crown surface shown in Fig.6(b), the shape factor becomes very large near the corner between the inlet and the pressure surface where flow velocity is slow, and the boundary layer fluid is transferred along the wall stream line which deviated from the pressure side to the suction side between vanes. This flow behavior is mainly caused by the pressure gradient from the pressure side to the suction side rather than the Corioli's force.

In the suction surface of the vane, the boundary layer gradually grows toward the downstream near the leading edge, but the stream lines turns to the band side with an increase in the boundary layer thickness. This is because the pressure decreases from the crown side to the band side due to the meridian curvature of channel.

Using the present method, the hydraulic energy loss due to friction can be predicted by use of energy thickness of the boundary layer. The friction loss on the wall can be estimated as a energy defect at the channel exit. The energy defect is calculated from the energy thickness θ^* at the exit surface for the case of three dimensional boundary layer.

$$\theta^* = \frac{1}{U^3} \int_0^{\upsilon} V' \left(U^2 - V^2 \right) dy \tag{13}$$

Here, V' is the velocity component perpendicular to V, which is velocity in the boundary layer(Fig.2), on the exit section.

On the other hand, in the performance conversion, the ratio δ_{Er} of friction head to effective head is very convenient for standardization, as adopted in IEC Code or JSME Code. The ratio δ_{Er} is obtained by the following equation.

$$\delta_{Er} = \oint (U^3 \theta^*) \, dl \; / \; gQH \tag{14}$$

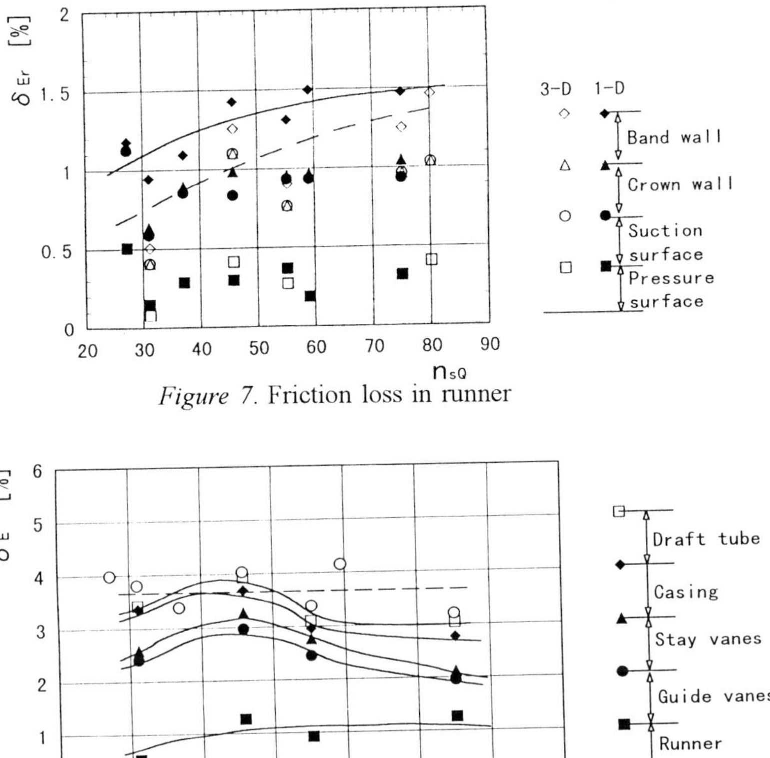

Figure 7. Friction loss in runner

Figure 8. Friction loss of turbine

Here, l is the distance along the contour of the exit wall. The curvilinear integration concerning l is performed along the exit wall edge of all channels of the runner.

The present calculation was applied to all turbine runners A-E and the calculated friction loss ratio δ_{Er} are plotted against specific speed n_{sQ} Fig. 7. The notations ○◇□△ are the results from the present calculation, and the notation ●◆■▲ are those of the conventional two dimensional analysis, in which the flow is assumed to be along the grid line, and the energy equation and Rotta's equation are used.

The total friction loss in a runner becomes larger with an increase in the specific speed , but the scatter of the data is large because of the difference in the design philosophy and manufacturing tolerance of each turbine. The solid line indicates total friction loss obtained by two dimensional analysis.

Figure.7 shows that the total friction loss from two dimensional analysis becomes larger than that from the present analysis. Due to the secondary flow, the boundary layer thickness must be thinner in the thee-dimensional analysis than in the two dimensional analysis, because the fluid in the boundary layer is removed by the secondary flow in the 3Danalysis.

Figure.7 also reveals that the friction loss on the vane occupies large part of the total friction loss of a runner. On the other hand, on the crown surface, the main flow velocity is slow and the wall stream line leans. Therefore the boundary layer thickness in the exit section is thin and the friction loss becomes thus very small.

The friction loss ratio δ_E of the casing, stay vanes, guide vanes, and draft tube of turbines from A to D are shown in Fig.8. All the results except for loss of the runner are obtained by two dimensional calculation. Figure.8 reveals that the friction loss on guide vanes becomes smaller with an increase in specific speed. The notation ○ shows the sum of friction losses on all surfaces obtained by using two dimensional analysis. The dotted line is the averaged total scalable loss and is seen to be almost independent of the specific speed, and δ_E takes the value around 0.037.

6. Conclusions

The calculation method of three dimensional boundary layer developed by use of upstream FDM with local curvilinear coordinate system, is applied to five types of Francis turbines. The following conclusions are obtained.

(1) The present method can be solved without trouble, even if complicated flow including reverse flow is induced such as the flow in the pressure surface.

(2) The present method is able to estimate the influence of the secondary flow, which is perpendicular to the main flow. The friction loss obtained by this method becomes smaller than that obtained by two dimensional method.

(3) The averaged total friction loss ratio on all surfaces obtained by using two dimensional analysis is about 0.037 in Francis turbine, and it is almost independent of the specific speed.

Acknowledgements

Calculation was performed as a part of the activity of JSME WG "Performance Conversion Method for Hydraulic Turbines and Pumps". The main flow data for 3D Euler, 3D potential flow and 2D boundary layer data were offered by the committee members, Dr. Nagafuji and Mr. Suzuki in Toshiba Co.Ltd., Mr. Miura in Hitachi Co.Ltd. and Mr. Miyagawa in Mitsubishi Heavy Industry Co.Ltd.. The authors would like to express sincere gratitude to the WG members.

References

1. Arakawa, C., Samejima, M., et al., A 3D Euler solution of Francis runner using pseudo-compressibility, 3D-Computation of Incompressible Internal Flows (NNFM39), (1989),65-69
2. IEC draft code 4(CO)47-1987,Determination of the prototype performance from model tests of hydraulic machines with consideration of scale effects
3. JSME S008(1989), Performance conversion method for hydraulic turbines and pumps
4. Suzuki, R., An assessment of the loss distribution in Francis Turbines, Proc. of 18th IAHR Sympo., (1996)
5. Ida T., et al., Recent development of studies on scale effect in Japan, Proc. of 18th IAHR Sympo.,

SCALE EFFECT OF JET INTERFERENCE IN MULTINOZZLE PELTON TURBINES

YUJI NAKANISHI
TAKASHI KUBOTA
Kanagawa University
3-27-1, Kanagawa-ku, Yokohama 221, Japan

1. Introduction

The IEC model acceptance test code stipulates that there is no scale effect in Pelton turbines[1]. One of the primary reasons why the scale effect in Pelton turbines is ignored, so far, is that the complicated unsteady free water-sheet flow on Pelton buckets prevents to numerically analyze the loss mechanism. Another reason is that the positive scale effect in the closed conduit flow from the turbine inlet to the nozzle outlet tends to compensate for the negative scale effect in free water-sheet flow on the buckets.

In 1986, Grein and Spurk proposed a formula for efficiency conversion of Pelton turbines[2]. The formula directly converts the hydraulic efficiency considering the scale effects of Froude, Reynolds and Weber numbers. The origin of scale effect on the hydraulic efficiency of Pelton turbines, however, results primarily from the scale effect of energy coefficient ψ, and secondarily of the discharge coefficient ϕ and power coefficient Π. Therefore, the conversion of specific hydraulic energy, discharge and power shall theoretically precedes the conversion of the hydraulic efficiency. The Grein/Spurk formula can not convert the performance shifts neglecting the background of fluid dynamics, and the most detrimental scale effect of the jet interference.

The jet interference is strongly related to the spilt flow from the cutout. The free water-sheet flow on a bucket is apt to spill out of the cutout (CutFlow) under the operating condition of the higher speed factor with the larger needle stroke. This spilt flow may collide with the approaching jet out of the succeeding nozzle in the case of multi-nozzle units, namely, the jet interference. The collision of CutFlow with the approaching jet amplifies CutFlow with domino effect. Eventually, CutFlow is a trigger for the jet interference. Since the prototype CutFlow is larger than the model due to the scale effect of the jet, the jet interference shows very strong negative scale effect. Seriously, therefore, the jet interference can not be predicted by the model test.

The CutFlow increases with weakening of free jet in the course of nozzle outlet

E. Cabrera et al. (eds.), Hydraulic Machinery and Cavitation, 333–341.

to the entrainment on the bucket. The weakened free jet leads to the decrease of jet velocity and the increase of jet diameter at the bucket landing. In the present study, the intensity of CutFlow from a bucket is tried to predict using the relative jet paths considering the weakened free jet velocity based on the scale effects of the conduit flow and of the free water-sheet flow on the buckets. The proposed CutFlow Intensity correlates well with the negative scale effect of the jet interference.

2. Closed Conduit Flow and Contraction Jet Diameter

2.1 POSITIVE SCALE EFFECT IN CLOSED CONDUIT FLOW

The mono-phase water flow is steady in the closed conduit of Pelton turbines from its inlet to the nozzle exit. All the hydraulic energy losses appear as a pressure loss in the closed conduit. While the kinetic energy deficiency $\delta_{C\text{-}ns}$ due to the branch loss and bend loss in the distributing pipe etc. is non-scalable, the wall friction deficiency δ_C is scalable depending on the Reynolds number R_e and relative roughness R_a/L as follows:

$$\delta_c = \Delta E_c / E = f(R_e, R_a / L) \tag{1}$$

where, ΔE_c and E denote the friction loss energy in the conduit and the specific hydraulic energy of the turbine. The prototype has smaller δ_c than the geometrically homologous model due to the difference in Reynolds number and relative roughness, in other words, the conduit flow has the 'positive' scale effect.

2.2. CONTRACTION DIAMETER OF FREE JET

The boundary layers developed in the closed conduit flow are confined close to nozzle and needle walls due to strong acceleration through the nozzle. The acceleration leads to the core flow structure at the exit of the nozzle as shown in Fig. 1 [3]. Since the most specific energy in the closed conduit flow is transformed to the kinetic energy, the core velocity becomes approximately $(2E)^{1/2}$. After that, the flow leaves the nozzle as a free jet. Since the jet becomes free from the solid wall, the low velocity region is rapidly accelerated up to the core velocity toward downstream. On the other

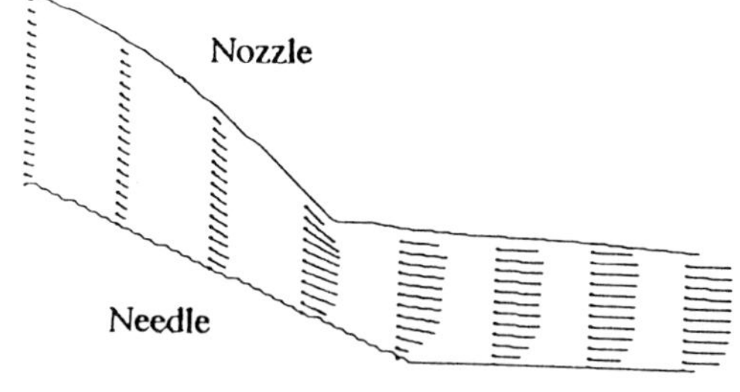

Figure 1. Velocity profiles in jet

hand, the boundary layer on the needle surface continues to develop in the jet, because the needle tip protrudes out of the nozzle, and finally the boundary layer is concentrated at the needle tip. Downstream of the needle tip, the low velocity region is also accelerated because of becoming free from the solid wall. Eventually, the uniform velocity distribution appears downstream of the needle tip at the circular cross-section of the jet. The uniform velocity is nearly equal to the throat core velocity, because the loss in the flow downstream the throat is negligibly small.

In the free jet, the increase of the jet velocity results in the decrease of the jet diameter from the mass continuity. This is the reason why the contraction of jet is generated at a little downstream of the needle tip. The boundary layer thickness at the nozzle tip is thinner in the prototype than in the model, because the thickness or the friction loss is dependent on the Reynolds number R_e, so that the prototype jet is not so contracted as the model jet under the same dimension less needle stroke $s_n(=S_n/D_T)$.

3. Scale Effect of Free Jet and Water-Sheet Flows

3.1. INCREASE OF FREE JET DIAMETER

Since the free jet travels through the air in the housing, the surface velocity of the jet is decelerated by the shear stress due to contact with the air flow. Since the free jet trys to retain its uniform velocity distribution across its cross section by dissipating its energy, the air resistance causes the jet velocity to weaken.

In addition, the water droplets splashed with the Pelton buckets exists around the free jet. Entraining the water droplets into the surface of the jet, also causes the kinetic energy of the jet to weaken. Since the amount of the water droplets rotating in the housing is proportional to the nozzle discharge, the kinetic energy loss in the jet depends on the needle stroke. After all, the jet velocity is gradually decreased toward downstream and the diameter is simultaneously increased to fulfill the mass continuity.

3.2. OPTIMUM JET DIAMETER AND INHERENT SCALE EFFECT OF JET

In general, a Pelton bucket has an optimum jet diameter d_{opt}, relating to its inner width B at optimum needle stroke $s_{n\text{-}opt}$. Although the prototype and its model have the same optimum diameter d_{opt}, the $s_{n\text{-}optP}$ for the prototype is smaller than $s_{n\text{-}optM}$ for the model because the jet diameter of the prototype is increased due to the scale effect in the closed conduit flow. Under the operating condition with smaller s_n than $s_{n\text{-}optP}$, the increased jet diameter of the prototype affects positively by approaching close to the optimum one. On the other hand, for large s_n, the increased jet diameter causes the spilt flow from the cutout CutFlow. Since the blade angle at the cutout is quite large than the angle at the brim, the amount of angular momentum $u_{cut}v_{u\text{-}cut}$ increases with increase of CutFlow. This leads to the decrease in specific energy of runner, or the appearance of the negative scale effect.

3.3. JET INTERFERENCE AND DOMINO EFFECT

Under the operating condition at the higher speed factor with the larger needle stroke, the free water-sheet flow in the bucket is apt to spill from the cutout. This CutFlow may collide with the approaching jet out of the succeeding nozzle in the case of a multinozzle unit more than 4 nozzles. The collision of the spilt flow with the approaching jet is called as the jet interference. The jet interference weakens the approaching jet, leads to the decrease in the velocity or the increase in the diameter. The weakened jet causes the delayed landing on the bucket, and that at the position near the cutout. This results in the more spilt flow from the cutout. Therefore, once the jet interference is occurred, the amount of the CutFlow is strikingly amplified with this chain reaction, namely, the domino effect.

Eventually, CutFlow is a trigger for the jet interference. Since the prototype CutFlow is larger than the model, the jet interference shows very strong negative scale effect. Seriously, therefore, the jet interference can not be predicted by the model test.

4. Modeling the Diameter of the Weakened Free Jet

In order to predict the intensity of CutFlow from a bucket using relative jet paths considering the scale effects of the conduit flow and of the free water-sheet flow on the buckets, the weakened jet diameter model is introduced as follows:

The free jet spouted from a nozzle, loses its kinetic energy or velocity gradually toward downstream. Then, the jet increases its diameter d_j ($\equiv D_j/D_T$, D_T denotes the nozzle tip diameter) to hold the continuity law. In this study, the increase in the jet diameter is represented by the following equation;

$$d_j = \left[1 + 0.001 \frac{\left(x_0 - x_j \right)}{d_0} + 0.001 \left(\frac{d_0}{B/D_T} \right)^2 + 2.2 \times 10^3 \left(\frac{V_{res}}{B^3} \right)^2 \right] d_0 \tag{2}$$

where, d_0 and x_0 denote the contraction diameter and its position respectively. The jet velocity c_j ($\equiv C_j/(2E)^{1/2}$) is calculated from the continuity equation using the jet diameter given by Eq. (2).

4.1. SCALE EFFECT ON CONTRACTION DIAMETER

As described in 2.2, the contraction diameter of the prototype d_{0P} is larger than the model due to the positive scale effect in the conduit flow. To take into account the effect of Reynolds number, the prototype contraction diameter is given by following equation;

$$d_{0P} = \left(\frac{R_{eP}}{R_{eM}} \right)^{\frac{1}{7}} d_{0M} \tag{3}$$

The weakened jet diameter for the prototype is calculated from Eq. (2) by substituting the obtained d_{0P} into the d_0.

4.2. EFFECT OF CONTACT WITH AIR ON FREE JET VELOCITY

Since the free jet travels through the air in the housing, the surface velocity of the jet is decelerated by shear stress due to contact with the air and the jet diameter is bloated toward downstream. Assuming the contribution of the travel to the increase in the jet diameter to be linearly proportional to the distance from the contraction position, the effect of resistance of the air on the jet velocity is considered as the second term in the right side of Eq. (2)

4.3. INCREASE IN JET DIAMETER DEPENDENT ON THE NEEDLE STROKE

The water droplets splashed with the Pelton buckets exists around the free jet. Entraining the water droplets into the surface of the jet, causes the jet diameter to bloat. Since the amount of the water droplets rotating in the housing is proportional to the nozzle discharge or the square of the jet diameter, the increase in the jet diameter dependent on the needle stroke is taken into account by the third term of Eq. (2).

4.4. MODELING OF THE DOMINO EFFECT CAUSED BY CutFlow

In a multinozzle turbine, once the spilt flow from the cutout CutFlow is occurred, CutFlow interferes with the succeeding jet and causes the jet diameter to bloat. The bloated jet diameter leads to the much more spilt flow from the cutout due to the domino effect. In order to consider the domino mechanism triggered by CutFlow, the contribution of CutFlow to the increase in the jet diameter is modeled by the fourth term of Eq.(2) so as to be proportional to the square of the residual jet volume V_{res} will be defined below.

4.5. DEDUCTION OF EMPIRICAL EQUATION FOR MODELING THE DIAMETER OF THE WEAKENED JET

As described in (4.1) through (4.4), the increase in jet diameter is modeled by considering the effects of Reynolds number of the conduit flow, of contact with the air flow, of entrainment of the rotating water droplets and of the domino mechanism. Each coefficient of the terms in the Eq.(2) was determined so that the variation of the

CutFlow Intensity I_{cut} versus the speed factor and the needle stroke could represent the tendency of the deterioration of the model efficiency at the test head of 115m as shown in Fig. 2. To realize the remarkable jet inference, 17 buckets was selected. The specific speed of the model turbine B/D_{ref} was 0.35, where D_{ref} denotes the jet pitch diameter.

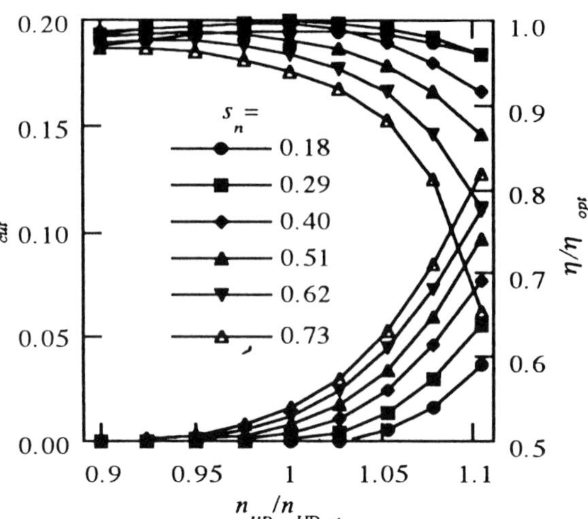

Figure 2. I_{cut} and model efficiency (H=115)

5. Relative Jet Path Considering the Weakened Jet Velocity

Suppose a splitter tip of a rotating bucket exits at an arbitrary position **M** in a jet as shown in Fig. 3. The position $\mathbf{I_r}$ of the jet particle of the impinging jet (ImpJet) which should catch up the splitter tip at the position **I**, can be calculated by integrating backward the jet velocity given from Eqs. (2) and (3) in the time dt=MI/u, where u denotes the peripheral velocity of the splitter tip. Similarly, The position $\mathbf{D_r}$ of the jet particle which passed the splitter tip at the position **D** just dt ago, can be also obtained. Since the passed jet is a little deviated due to contact with the bucket, we call this the deviated jet (DevJet). $\mathbf{MI_r}$ and $\mathbf{MD_r}$ both represent the parts of the relative jet path for the given bucket. While $\mathbf{MI_r}$ means the head of ImpJet which has not entered by this instant, **MDr** the head of ImpJet which has already entered by this instant. As for DevJet, while $\mathbf{MI_r}$ means the tail of the DevJet which will pass the bucket, $\mathbf{MD_r}$ the tail of DevJet which has already passed.

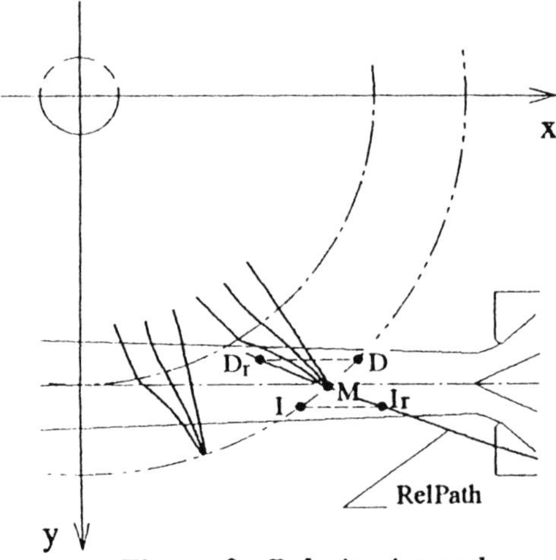

Figure 3. Relative jet path

Let us consider the two adjacent buckets to investigate the relationship between the jet and the two buckets. If the following bucket of the two adjacent buckets is abbreviated to FolBuc, the preceding bucket to PreBuc. We can get FolPath for FolBuc and PrePath for PreBuc as shown in Fig. 4. Considering these two relative paths, the ImpJet for a bucket can be obtained as $\mathbf{A_r'C_r'C_rA\,A_r'}$. The jet portion is designated as the responsible jet for a bucket V_{duty}. The portion $\mathbf{A_r'C_r'Q_bP_bA_r'}$ of the responsible jet represents the entrained jet on the PreBuc at this instant. On the other hand, the remaining portion $\mathbf{P_bQ_bC_rAP_b}$ represents the residual jet V_{res} that has not been entrained yet.

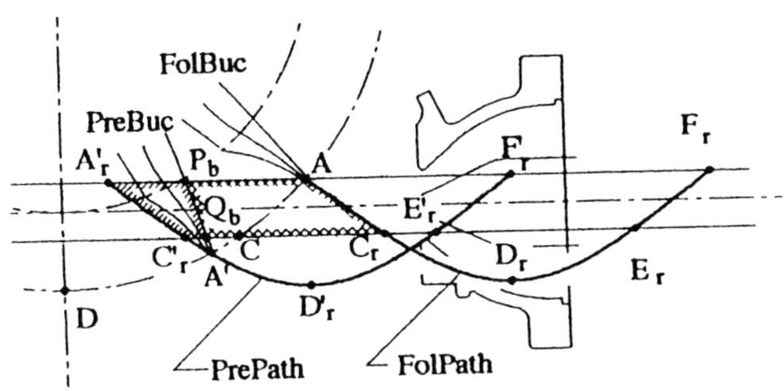

Figure 4. Fol- and Pre- relative paths

In order to predict the intensity of the jet interference, it is important to predict the amount of CutFlow quantitatively. However, the prediction of CutFlow is difficult and time-consuming. In this study, the residual jet related to the jet interference is to predict in lieu of the CutFlow. To predict the amount of the residual jet, the relative jet paths for PreBuc and FolBuc are drawn at a flash when the PreBuc first touches the second jet as shown in Fig. 5. Since the second jet starts to enter into the PreBuc after the flash, the residual portion $\mathbf{B_rQ_bC_rB_r}$ of the first jet collides with the second jet in the PreBuc, and may be discharged as CutFlow. Therefore, A new term of CutFlow Intensity I_{cut} that is defined as the ratio of the residual jet volume V_{res} to the responsible jet volume V_{duty} is introduced to assess the jet interference.

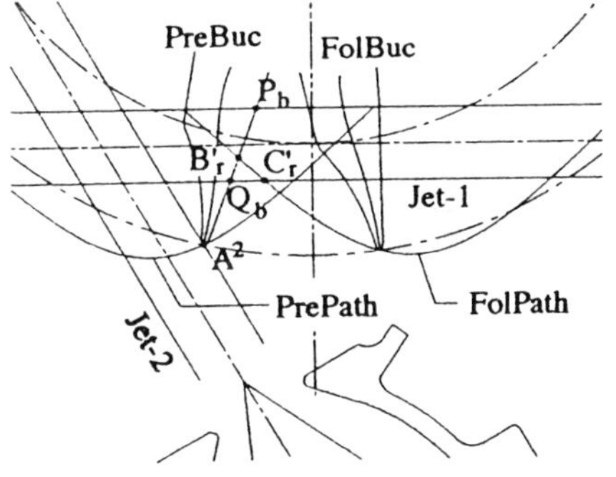

Figure 5. Residual jet volume

6. Comparisons of CutFlow Intensity with the Model Efficiency

The CutFlow Intensity I_{cut} was calculated using the relative jet paths considering the weakened jet diameter by Eqs. (2) and (3) under the operating conditions of the needle strokes s_n of 0.4 and 0.73, the heads of 40, 60, 90 and 115m, and 9 speed factors n_{HD}/n_{HDopt} from 0.9 through 1.1. The variations of obtained CutFlow Intensity I_{cut} versus n_{HD}/n_{HDopt} are shown in the lower parts of Fig. 6 for $s_n = 0.73$ and Fig. 7 for $s_n = 0.4$.

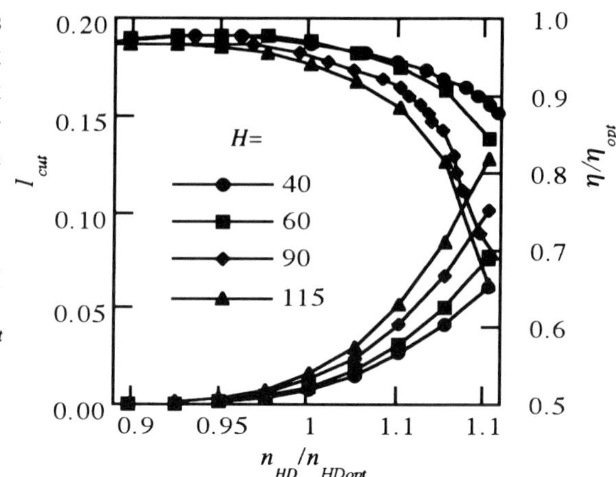

Figure 6. I_{cut} *and model efficiency* (s_n*=0.73)*

From the both figures, it is followed that the larger the n_{HD}/n_{HDopt} is, the larger Cutflow intensity I_{cut} becomes. The reason is that the large n_{HD}/n_{HDopt} caused the residual jet V_{res} to increase and the contribution of fourth term in the right side of Eq. (2) to the increase of the jet diameter was strongly intensified. In other words, the fast rotation of the runner made the jet catching up the bucket difficult. Since the higher heads causes the larger contraction diameter d_0 of Eq. (2), this tendency became outstanding at the larger heads. The CutFlow Intensity I_{cut} is larger at s_n=0.73 than at s_n=0.4 due to the effect of entrainment of water droplets considered by the third term of Eq. (2).

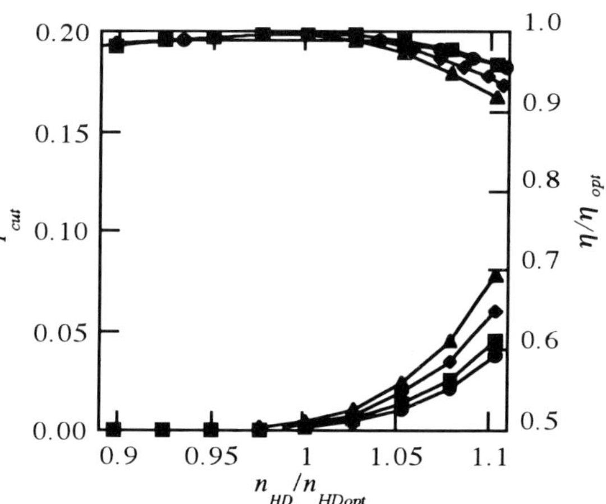

Figure 7. I_{cut} *and model efficiency* (s_n*=0.40)*

In order to ascertain the relationship between I_{cut} and the intensity of the jet interference, the CutFlow Intensity I_{cut} was compared with the model efficiency deteriorated by the jet interference. The jet pitch diameter D_{ref} of the model runner is 0.325m, the bucket inner width B 0.114m and the number of the bucket 17. The uncertainty of the efficiency measurements is +-0.2% with the 95% confidence level.

In the upper parts of Figs. 6 and 7, the variations of measured efficiency in the model tests were shown, where n_{HDopt} and η_{opt} denote the speed factor and the efficiency at the optimum operating condition, respectively. The jet interference was not recognized only at the operating condition of H=40m and s_n=0.40. In that case, the efficiency curve is roughly symmetric to the vertical line through the best efficiency point.

The model efficiency is deteriorated at the operating condition of high speed factor with high test head due to the jet interference. The higher the test head is, the stronger the deterioration of the efficiency is. This reveal that the jet interference has the strong negative scale effect. This tendency is remarkable at the operating condition of s_n=0.73. The variation of the CutFlow Intensity I_{cut} well demonstrates the tendency of the model efficiency deteriorated due to the scale effect.

7. Conclusions

The scale effect in a Pelton turbine has not been investigated so far. Since the jet interference has very strong negative scale effect, the intensity of it in the prototype can not be predicted by the model tests. The jet interference is strongly related to the spilt flow from the cutout CutFlow, and so the intensity of CutFlow from a bucket was tried to predict using the relative jet paths considering the weakened free jet velocity based on the scale effects of the conduit flow and of the free water-sheet flow on the buckets. The proposed CutFlow Intensity correlated well with the negative scale effect of the jet interference in model tests.

References

1. IEC (1994), 4/111/CDV .
2. Grein,H. et al. (1986), Efficiency scale effects in Pelton turbines, IAHR-Montreal, 76.
3. Nonoshita T.,Matsumoto Y., Haneda Y. and Kubota T. (1993), Behavior of the Jet from a Pelton Turbine Nozzle, Proc. 4th Asian Int. Conf. Turbo Machinery, Suzhou, 499-503.

FURTHER DEVELOPMENT OF STEP-UP FORMULA CONSIDERING SURFACE ROUGHNESS

Alois Nichtawitz
Voest Alpine M.C.E.
Linz, Austria

Abstract: The paper is based on the proposal for a new step-up formula made by the author at the IAHR Symposium 1994 in Beijing. In its first part an upgrading of proposed formula to also consider influence of surface roughness is derived. In a second step it is demonstrated that the scalable loss S_0 which is originally a constant can be extended to a function of flow and specific hydraulic energy. Efficiency step-up at a constant flow coefficient and geometry automatically involves a change of specific hydraulic energy, so a formula to describe this effect is also given in the paper, which is concluded by a set-up for the power shift which can be derived from the above. Summarizing the paper offers a complete set of formulae describing a method for the transposition of characteristics from model to prototype.

1. Introduction

A large variety of step-up procedures has been developed in the past. With the code IEC 995 a procedure for efficiency step-up in reaction machines is effective since 1991. It seems that it is not accepted on a worldwide basis mainly because of its complicated structure.

IEC 995

$$\Delta\eta = \delta_{ref} \cdot \left[\left(\frac{R_{ref}}{Re_M} \right)^{0.16} - \left(\frac{R_{ref}}{Re_P} \right)^{0.16} \right]$$

$$\delta_{ref} = \frac{1 - \eta_M}{\left(\frac{Re_{ref}}{Re_M} \right)^{0.16} + \frac{1 - V_{ref}}{V_{ref}}}$$

During the IAHR Symposium 1994 in Beijing various existing step-up formulae and new approaches were discussed by the author. In the course of the paper a new step-up formula was presented and discussed:

NEW PROPOSAL $$\Delta\eta = S_0 \cdot \left[\left(\frac{Re_0}{Re_M} \right)^n - \left(\frac{Re_0}{Re_P} \right)^n \right]$$

E. Cabrera et al. (eds.), Hydraulic Machinery and Cavitation, 342–351.

Based on the set-up as presented attempts were made by the author to further develop the step-up procedure in four directions.

- Implementation of different surface roughness at model and prototype.
- Discussion on the scalable loss S_0 and its upgrading to a function dependent on φ and ψ.
- Development of a step-up formula for the energy coefficient ψ.
- From the above a step-up formula for the power output / consumption can be derived.

2. Surface Roughness

As a starting point to extend the formula presented in Beijing the following equation by Colebrook and White was taken which is describing the friction coefficient also in the transition zone between smooth and fully rough pipe flow.

$$\frac{1}{\sqrt{\lambda}} = -2 \cdot \log\left(0.27 \cdot \frac{K_S}{D} + \frac{2.51}{Re \cdot \sqrt{\lambda}}\right)$$

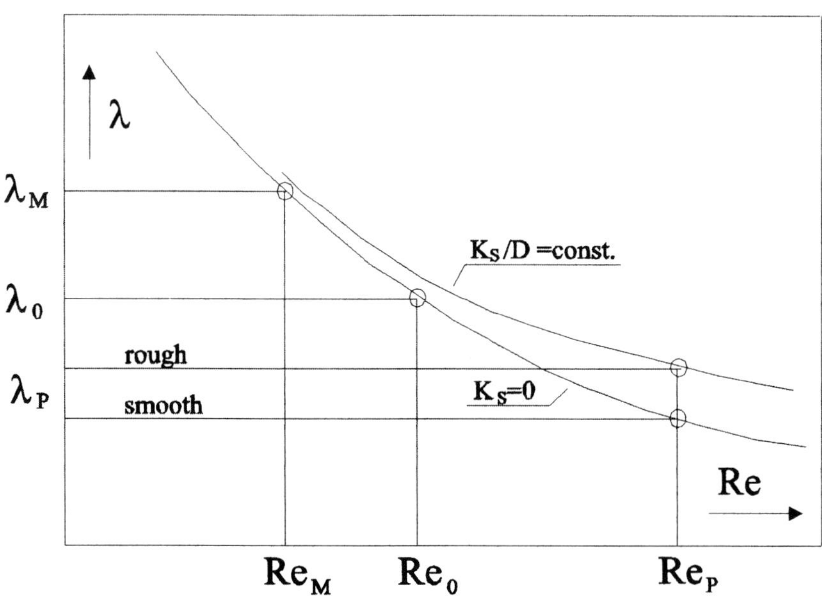

Fig. 1 Friction coefficient λ depending on Re and K_S/D

This is an implicit formulation for λ which can be converted by some tricky transformations and small simplifications to the following explicit shape as demonstrated in the following.

$$\frac{1}{2\cdot\sqrt{\lambda_0}}-\frac{1}{2\cdot\sqrt{\lambda}}=\log\left(0.18\cdot\frac{K_S}{D}\cdot Re_0\cdot\sqrt{\lambda_0}+\frac{Re_0}{Re}\cdot\frac{\sqrt{\lambda_0}}{\lambda}\right)$$

$$\frac{10^{\left(\frac{1}{2\cdot\sqrt{\lambda_0}}-\frac{1}{2\cdot\sqrt{\lambda}}\right)}}{\sqrt{\frac{\lambda_0}{\lambda}}}=0.108\cdot\frac{K_S}{D}\,Re_0\cdot\sqrt{\lambda}+\frac{Re_0}{Re}\cong\left(\frac{\lambda-\lambda_\infty}{\lambda_0-\lambda_\infty}\right)^n$$

a best fit is given by $\lambda_\infty = 0.00222$ and n = 5.0

$$\frac{\lambda}{\lambda_0}=0.74\cdot\left(0.108\cdot\frac{K_s}{D}\cdot Re_0\cdot\sqrt{\lambda}+\frac{Re_0}{Re}\right)^{0.2}+0.26$$

$$\lambda=0.00632\cdot\left(C_R+\frac{Re_0}{Re}\right)^{0.2}+0.00222 \qquad \text{with} \quad C_R=0.108\cdot\frac{K_S}{D}\cdot Re_0\cdot\sqrt{\lambda}$$

Friction losses in a hydraulic machine cannot be modeled just by frictional losses in a pipe but also friction at a flat plate and disc friction contribute to the total scalable losses due to friction. In a quite similar way a formulation by A. Fay describing friction coefficients for a flat plate can be simplified and brought to an explicit structure too.

$$\frac{1}{C_f^{0.4}}=-1.62\log\left(\frac{1}{14.7}\cdot\frac{K_S}{L}+A\cdot\frac{0.241}{Re\cdot\sqrt{C_f}}\right)$$

$$\frac{C_f}{C_{f0}}\cong 0.81\cdot\left(C_R+\frac{Re_0}{Re}\right)^{0.2}+0.19$$

$$C_f\cong 0.0025\cdot\left(C_R+\frac{Re_0}{Re}\right)^{0.2}+0.0006 \qquad \text{with} \quad C_R=0.28\cdot\frac{K_S}{L}\cdot Re_0\cdot\sqrt{C_f}$$

To the knowledge of the author there is no similar formula for the disc friction but it is reasonable to assume that disc friction also can be described with sufficient accuracy in an explicit shape in quite a similar way.

$$\frac{C_m}{C_{m0}}=B\cdot\left(C_R+\frac{Re_0}{Re}\right)^{0.2}+(1-B)$$

Taking the same explicit structure for all types of friction (pipe, plate, disc) and with $\Delta\eta \hat{=} (\lambda_M - \lambda_P)$ but also $\Delta\eta \hat{=} (C_{fM} - C_{fP})$ and $\Delta\eta \hat{=} (C_{mM} - C_{mP})$ there follows

$$\Delta\eta = S_0 \cdot \left[\left(C_{R,M} + \frac{Re_0}{Re_M} \right)^{0.2} - \left(C_{R,P} + \frac{Re_0}{Re_P} \right)^{0.2} \right]$$

In case of hydraulic smooth surfaces at model and prototype the terms C_{RM} and C_{RP} become zero and the formula is reduced to the one as presented in Beijing. C_R=1 corresponds to a roughness where no step-up will occur even at an infinite Reynolds number of prototype. It has to be mentioned that there is a difficult question how to determine an arithmetic mean roughness equivalent to a sand roughness K_S.

The quality of surface roughness of model machines normally is excellent so C_{RM} can be set zero. For the prototype machines it is proposed to introduce classes of roughness similar to the ones established for machining / grinding. It is advisable to follow a geometric sequence starting with C_{RP} =1, 0.5, 0.25, 0.125, 0.063, 0.032, 0.016 and finally zero.

Particularly at runner replacement projects the surface roughness of existing machine components may be far away from hydraulically smooth conditions which makes it absolutely necessary to also consider surface roughness.

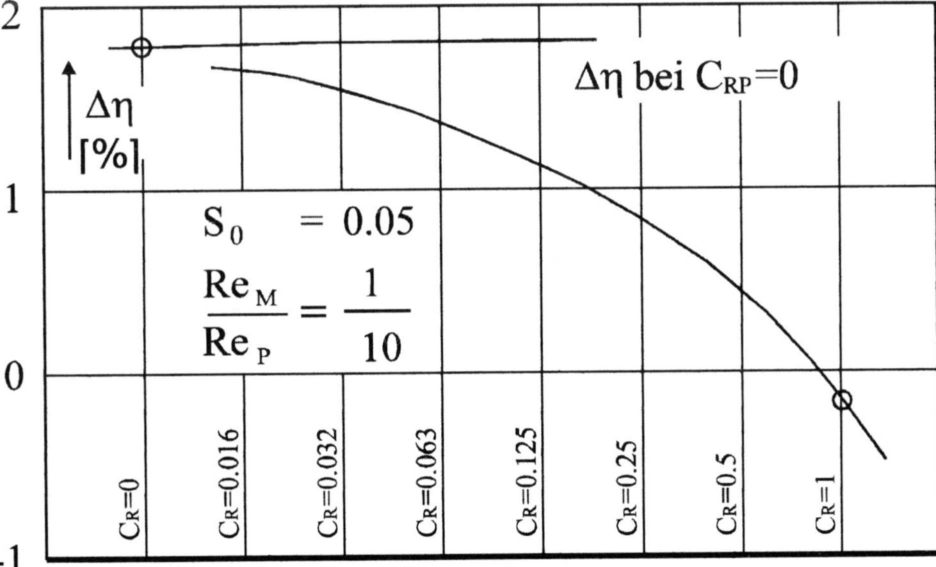

Fig. 2 Efficiency step-up for different surface roughness

Also this chart is a helpful tool for judgment if an improvement of roughness of remaining components should be made due to economic reasons. In many cases it will lead to the conclusion that it is economic to improve surface roughness of old machines. The increase in efficiency reached then could be even higher than by a newly designed runner however at lower costs.

The new formula also opens the possibility to calculate a step up for the frequent case that just the runner is replaced but nothing is done at surrounding components.

$$\Delta\eta = \left(S_0 - S_{0,RU}\right)\cdot\left[1-\left(C_{R,P} + \frac{Re_M}{Re_P}\right)^{0.2}\right] + S_{0,RU}\left[1-\left(\frac{Re_M}{Re_P}\right)^{0.2}\right]$$

Numerical example:

$S_0 = 0.05$ $\quad \frac{Re_M}{Re_P} = \frac{1}{10}$ $\quad C_{RP} = 0.5$ in general, just the runner with $C_{RP} = 0$

$S_{0,RU} = 0.02$

$$\Delta\eta = 0.0029 + 0.0074 = 0.0103 \Rightarrow \Delta\eta = 1.03\ \%$$

compared to

$\Delta\eta = 1.85\ \%$ at smooth surfaces at all components including runner
$\Delta\eta = 0.49\ \%$ at surfaces with class 0.5 at all components including runner

3. Scalable loss S_0

The scalable loss So is related to flow and specific hydraulic energy at peak efficiency. This value is a constant in the original approach depending on specific speed of machine which means that the step up derived therefrom is a constant adder independent of point of operation. Obviously it is very unlikely that this is the case in reality. In fact S_0 depends on coefficients of flow and specific hydraulic energy as well.

The losses due to friction in a hydraulic machine occur in all of the components and can be classified after the place where they are originated.

These main components and losses are:

- spiral case
- stay vanes
- wicket gates
- runner
- draft tube
- leakage
- disc friction

Losses due to main flow in the various components contribute to a loss in specific hydraulic energy. A different class of losses is given by disc friction and seal leakage resulting in a power loss.

For sake of simplicity the procedure is outlined in the following by example of scalable losses in a spiral case.

$$\Delta\eta_{SP} = (\lambda_M - \lambda_P)\cdot\left(\frac{L}{D}\right)_{SP}\cdot\left(\frac{D}{D_{SP}}\right)^4\cdot\frac{\varphi}{\psi_0}^2 \qquad \left(\frac{L}{D}\right)_{SP} \text{ equivalent relative length of spiral case}$$

The friction coefficient λ has to be calculated according to the Reynolds number in the spiral case

$$Re_{SP} = \frac{c_{SP}\cdot D_{SP}}{\nu} \qquad Re = \frac{u\cdot D}{\upsilon}$$

$$\frac{\lambda}{\lambda_0} = 0.74\cdot\left(0.108\cdot\frac{K_S}{D_{SP}}\cdot Re_0\cdot\sqrt{\lambda} + \frac{Re_0}{Re_{SP}}\right)^{0.2} + 0.26$$

$$\frac{\lambda}{\lambda_0} = 0.74\cdot\left(\frac{D_{SP}}{D}\cdot\frac{1}{\varphi}\right)^{0,2}\cdot\left(\underbrace{0.108\frac{K_S}{D_{SP}}\cdot\frac{D}{D_{SP}}\cdot Re_0\cdot\sqrt{\lambda}\cdot\varphi}_{C_R} + \frac{Re_0}{Re}\right)^{0.2} + 0.26$$

The term $0.108\cdot\frac{K_S}{D}\cdot\left(\frac{D}{D_{SP}}\right)^2\cdot Re_0\cdot\sqrt{\lambda}\cdot\varphi$ corresponds to a class of roughness C_R.

For the example of a spiral case the step-up can be written in the following way:

$$\Delta\eta_{SP} = S_{0,SP}(\varphi,\Psi)\cdot\left[\left(C_{RM} + \frac{Re_0}{Re_M}\right)^{0.2} - \left(C_{RP} + \frac{Re_0}{Re_P}\right)^{0.2}\right]$$

The term $S_{0,SP}$ in front of the brackets represent the scalable losses of a spiral case at the reference Reynolds number Re_0 and with hydraulically smooth surfaces. Assuming that various components show equivalent quality of roughness the term C_R is a constant. Equivalent means that with smaller hydraulic diameters of a component the surface roughness must be smaller which is in agreement with the practice. Therefore the roughness has to be chosen carefully respecting the individual Reynolds numbers and hydraulic diameters within the water passage. In this case the individual partial step-ups can be summed up to

$$\Delta\eta = S_0(\varphi,\Psi)\cdot\left[\left(C_{RM} + \frac{Re_0}{Re_M}\right)^{0.2} - \left(C_{RP} + \frac{Re_0}{Re_P}\right)^{0.2}\right]$$

4. Energy Coefficient

In the next step an attempt is made to develop a formula describing the step-up of energy coefficients which is strictly linked to the efficiency step-up at a given flow coefficient φ and a given geometry.

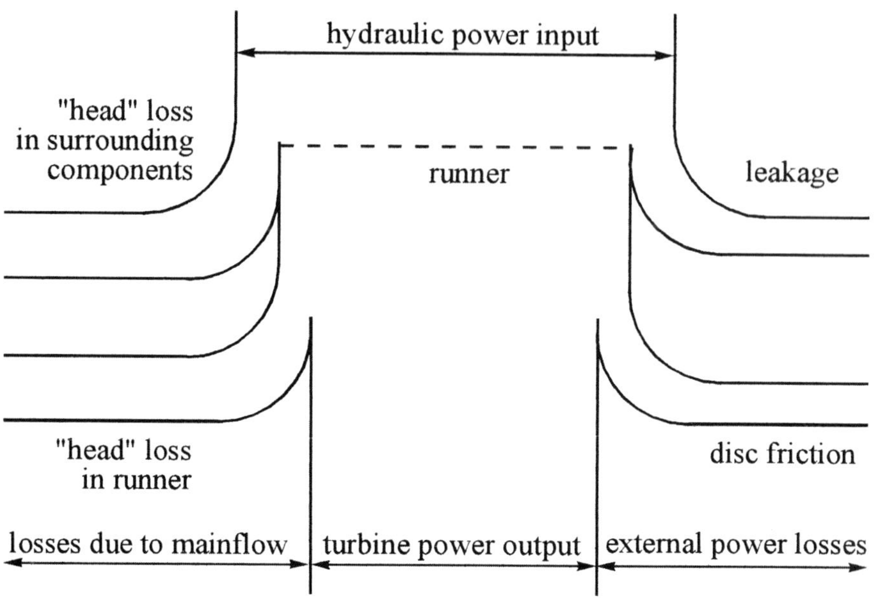

Fig. 3 Flow chart for losses in a turbine

$$g \cdot H = \Sigma\, g \cdot \Delta H_{HL} + g \cdot H_{RU} \qquad \Delta u \cdot c_u = \frac{M_{RU} \cdot \omega}{\rho \cdot Q_{RU}} = g \cdot H_{RU} \cdot \eta_{RU}$$

$$M = M_{RU} - M_{DF} = M_{RU} \cdot \eta_{DF} \qquad Q = Q_{RU} + Q_{LF} = \frac{Q_{RU}}{\eta_{LF}}$$

$$1 = \frac{\Sigma g \cdot \Delta H_{HL}}{g \cdot H} + \frac{\eta}{\eta_{RU} \cdot \eta_{DF} \cdot \eta_{LF}} \qquad \Rightarrow \eta = \underbrace{(\eta_{HL} \cdot \eta_{RU})}_{\text{main flow}} \cdot \underbrace{(\eta_{DF} \cdot \eta_{LF})}_{\text{external}}$$

HL... head losses in surrounding components
RU... head loss in runner
DF... disc friction
LF ... leakage flow

Starting from Euler's equation and introducing φ and ψ one can derive from simple velocity triangles at runner inlet and outlet the following equations.

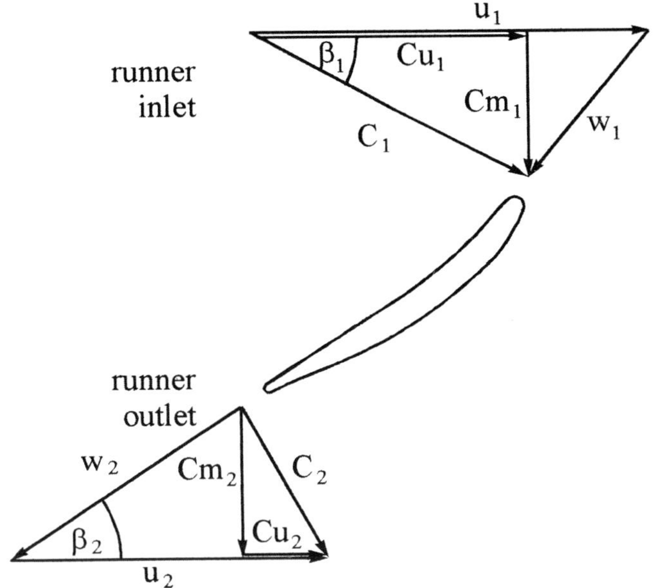

Fig. 4 Velocity triangles at runner inlet and outlet

$$u_1 \cdot c_{u1} - u_2 \cdot c_{u2} = g \cdot H_{RU} \cdot \eta_{RU} = g \cdot H \cdot \eta_{HL} \cdot \eta_{RU} = g \cdot H \cdot \frac{\eta}{\eta_{DF} \cdot \eta_{LF}}$$

$$tg\beta_1 = \frac{c_{m1}}{u_1 - c_{u1}} \qquad tg\beta_2 = \frac{c_{m2}}{u_2 - c_{u2}}$$

$$ctg\beta_2 \cdot \frac{c_{m2}}{u} \cdot \frac{u_2}{u} - ctg\beta_1 \cdot \frac{c_{m1}}{u} \cdot \frac{u_1}{u} + \frac{u_1^2 - u_2^2}{u^2} = \frac{g \cdot H}{u^2} \cdot \frac{\eta}{\eta_{DF} \cdot \eta_{LF}}$$

$$\varphi \cdot \left(a \cdot ctg\beta_2 - b \cdot ctg\beta_1\right) + c = \frac{\psi}{2} \cdot \frac{\eta}{\eta_{DF} \cdot \eta_{LF}}$$

$$\varphi \cdot f\left(C_R + \frac{Re_0}{Re}\right) + c = \frac{\psi}{2} \cdot \frac{\eta}{\eta_{DF} \cdot \eta_{LF}}$$

The constants a, b and c depend on the geometry of a hydraulic machine. On the other hand f (β_1, β_2) is a function of flow angles in front of and behind a rotating runner and is assumed to depend slightly on Reynolds number Re and surface roughness C_R. This function changes slightly only when varying surface roughness and Reynolds number therefore a Taylor progression can be carried out.

$$\varphi \cdot \left[\frac{f(1)}{0!} + \frac{f'(1)}{1!} \cdot \left(C_R + \frac{Re_0}{Re} - 1 \right) + \frac{f''(1)}{2!} \cdot \left(C_R + \frac{Re_0}{Re} - 1 \right)^2 + .. \right] + c = \frac{\psi}{2} \cdot \frac{\eta}{\eta_{DF} \cdot \eta_{LF}}$$

The Taylor progression can be stopped after the second term.

$$\varphi \cdot \left[\frac{f(1)}{0!} + \frac{f'(1)}{1!} \cdot \left(C_R + \frac{Re_0}{Re} - 1 \right) \right] + c = \frac{\psi}{2} \cdot \frac{\eta}{\eta_{DF} \cdot \eta_{LF}}$$

Applying above equations to model and prototype conditions one gets

$$\frac{1 + h \cdot \left(C_{R,P} + \frac{Re_0}{Re_P} \right)}{1 + h \cdot \left(C_{R,M} + \frac{Re_0}{Re_M} \right)} = \frac{\psi_P}{\psi_M} \cdot \frac{\eta_P}{\eta_M} \cdot \frac{(\eta_{DF} \cdot \eta_{LF})_M}{(\eta_{DF} \cdot \eta_{LF})_P}$$

and consequently

$$\frac{\psi_P}{\psi_M} = \frac{\eta_M}{\eta_P} \cdot \frac{(\eta_{DF} \cdot \eta_{LF})_P}{(\eta_{DF} \cdot \eta_{LF})_M} \cdot \frac{1 + h \cdot \left(C_{RP} + \frac{Re_0}{Re_P} \right)}{1 + h \cdot \left(C_{RM} + \frac{Re_0}{Re_M} \right)}$$

in case of $C_{RM} = 0$, $Re_M = Re_0$ and $\eta_{LF,P} = \eta_{LF,M}$ there follows

$$\frac{\psi_P}{\psi_M} = \frac{\eta_M}{\eta_P} \cdot \frac{\eta_{DF,P}}{\eta_{DF,M}} \cdot \frac{1 + h_{TU} \cdot \left(C_{R,P} + \frac{Re_M}{Re_P} \right)}{1 + h_{TU}}$$

for generating mode

$$\frac{\psi_P}{\psi_M} = \underbrace{\frac{\eta_P}{\eta_M} \cdot \frac{\eta_{DF,M}}{\eta_{DF,P}}}_{\text{head loss}} \cdot \underbrace{\frac{1 + h_{PU} \cdot \left(C_{R,P} + \frac{Re_M}{Re_P} \right)}{1 + h_{PU}}}_{\text{change of runner/impeller head}}$$

for pumping mode

It is assumed that the coefficient h strongly depends on the specific hydraulic design e.g. number and geometry of wicket gates and of runner vanes/blades as well.

It is also suggested to analyze and compare corresponding points from model and field tests in order to determine the value of h. Excellent homology between model and prototype is necessary to get reliable results.

5. Power

Knowing the step-up of efficiency together with the shift of specific hydraulic energy between model and prototype conditions one can calculate now the power at a given value of phi. The power to be achieved operating a turbine at corresponding points can be written in the following way

$$P_P \hat{=} \varphi \cdot \Psi_P \cdot \eta_P \cdot n_P{}^3 \cdot D_P{}^5 \qquad P_M \hat{=} \varphi \cdot \Psi_M \cdot \eta_M \cdot n_M{}^3 \cdot D_M{}^5$$

Using formula for step-up of specific hydraulic energy coefficient one gets

$$P_{P,TU} = P_{M,TU} \cdot \left(\frac{n_P}{n_M}\right)^3 \cdot \left(\frac{D_P}{D_M}\right)^5 \cdot \frac{1 + h_{TU}\left(C_{RP} + \frac{Re_M}{Re_P}\right)}{1 + h_{TU}} \cdot \frac{(\eta_{DF} \cdot \eta_{LF})_P}{(\eta_{DF} \cdot \eta_{LF})_M}$$

For many types of hydraulic reaction machines differences in disc friction and in leakage losses can be neglected. In this case power step-up can be well estimated by testing homologous machines at a given φ-value just at two different Reynolds numbers whereby efficiencies do not appear in above equation (!)

6. Conclusion

The paper offers a complete set of formulae for the problem of transposition from model to prototype conditions. The procedures cover efficiency step-up considering the influence of surface roughness and demonstrate how the proposed formula can be extended in order to calculate efficiency step-up also in regions off the peak condition. It also outlines a new approach how to predict the shift in specific hydraulic energy between model and prototype machines together with ist effect to the power output respectively power consumption.

The presented paper is considered also to stimulate the work in WG 5 of IAHR. Based on the proposals worldwide experiences from various manufacturers should be collected in order to determine the parameters for an optimum fit with the experience.

The main goal of all step-up activities is to better predict the prototype performance, however, dealing with this subject certainly also will improve knowledge in loss distribution of hydraulic machines, which again will stimulate the future progress in technology of hydro machinery.

7. References

- IEC Publication 995: Determination of the prototype performance from model acceptance tests of hydraulic machines with consideration of scale effects, 1991
- Nichtawitz, A.: Discussion on step-up procedures in hydraulic machines, IAHR Symposium, Beijing 1994

NUMERICAL SIMULATION OF JET IN A PELTON TURBINE

TOMOYASU NONOSHITA
Dept. of Mechanical Engineering, Sophia University
7-1, Kioicho, Chiyoda-ku, Tokyo 102, Japan
YOICHIRO MATSUMOTO
Dept. of Mechanical Engineering, The University of Tokyo
7-3-1, Hongo, Bunkyo-ku, Tokyo 113, Japan
TAKASHI KUBOTA
Dept. of Mechanical Engineering, Kanagawa University
3-27-1, Rokkakubashi, Kanagawa-ku, Yokohama 221, Japan
HIDEO OHASHI
Dept. of Mechanical System Engineering, Kogakuin University
1-24-2, Nishi-shinjuku, Shinjuku-ku, Tokyo, 163-91, Japan

1. Abstract

In case of predicting the performance of prototype Pelton turbine from a model test, a scale effect has been ignored so far. However, it is getting clear that there is an outstanding deviation from a classical similarity laws. The direction of the deviation is completely inverse against the case of reaction turbines, namely, the performance of prototype is worse than the predicted one. In order to analyze this negative scale effect, the behavior of jets should be considered more precisely. In this study, the unsteady jet issuing from a Pelton turbine nozzle was calculated numerically using HSMAC method under the assumption of axisymmetric flow. The calculation shows that the needle considerably affects the velocity distribution in the jet and the velocity distribution in the jet changes by the head.

E. Cabrera et al. (eds.), Hydraulic Machinery and Cavitation, 352–360.

2. Nomenclature

D	: diameter of nozzle exit		ρ	: density
h	: free surface height from axis		σ	: surface tension
p	: pressure			
Q	: volumetric flow rate		subscript	
r, z	: radial and axial coordinates		a	: air
t	: time		d	: droplet
v	: velocity		s	: free surface
θ	: incline angle of nozzle exit		j	: jet
μ	: viscosity		nz	: nozzle
ν	: kinetic viscosity		r, z	: radial and axial component

3. Introduction

In Pelton turbines, the mechanism of motion has been simply explained by using the angular momentum theory for time averaged flow. However, the actual flow is very complicated due to the periodic change of relative position between a jet and a bucket(Bachmann et al., 1990) and the interaction among jets(Kubota, 1989). It is found that there is an outstanding deviation from the classical similarity law between the model test data and the prototype performance. Various reasons can be considered with respect to the scale effect which is caused by the complicated flow. Grein et al.(1986) made it clear that the specific flow rate, Reynolds number, Froude number and Weber number have to be considered for the scale effects in Pelton turbines by dimensional analysis. Kubota and Nakanishi(1994) classified the flow in Pelton turbines for the sake of studying the scale effect.

In this study, we focus on the jet from a Pelton turbine nozzle as a basic research in order to obtain a hint about the scale effect. The diameter of the jet is getting larger after the contraction because of the change of velocity distribution. Additionally, there are lots of water droplets in the casing which are generated by the reflection of jet at the buckets. Some of them are entrained into the jet, and it causes the momentum loss and the enlargement of jet diameter.

Behavior of jets has been investigated concerning with not only Pelton turbines, but also water jet cutting. The length of jet which maintains its continuity (so-called breakup length) was investigated by Phinney(1973). He extended the laminar jet stability theory, which is derived from Weber's theory, into the turbulent jet. Hoyt et al.(1977) observed the complicated jet surface using high-speed photography. They showed that the instabilities occurs in high Reynolds number water jet. These instabilities include the axisymmetric mode accompanying the transition from laminar to turbulent flow at the nozzle exit, spray formation as a culmination of the axisymmetric disturbances, and further down stream, helical disturbances which result in the entire jet assuming a helical form. In regard to a Pelton turbine jet, Guilbaud et al.(1992) measured the head distribution inside the jet under two kinds of conditions. One is the jet from a secondary branch of the wye-piece and another is the jet from a straight nozzle. The former one has a curvature effect imposed to the flow (deviated flow) and the latter is axisymmetric flow (non-deviated flow). The measurements disclosed that the jet from a deviated flow leads to a divergence located on the curvature inner side and it induces a smaller head pressure than that measured on the curvature outer side.

4. Numerical Analysis Method

4.1. GOVERNING EQUATIONS

The jet issuing from a Pelton turbine nozzle which is shown in Fig.1 is numerically analyzed in this study. The shape of the nozzle is as same as that of the former report (Nonoshita et al., 1994). The diameter of the nozzle exit is 52mm and that of the needle at the nozzle exit is 20mm when the needle is in its normal position. The conical angle of the needle tip is 50 degrees, and the incline angle of the nozzle

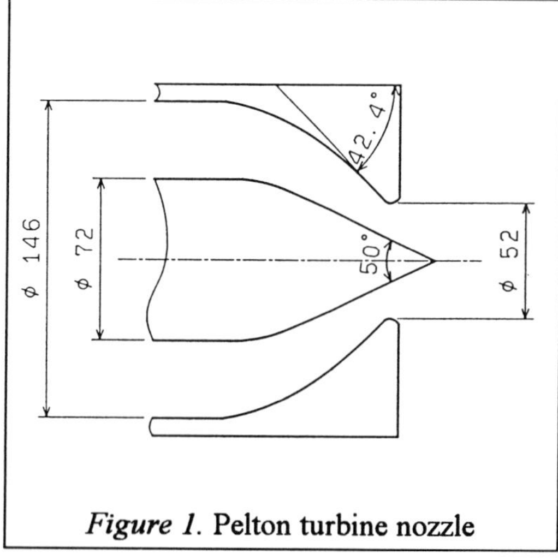

Figure 1. Pelton turbine nozzle

exit is 42.4 degrees.

The flow is assumed to be axisymmetric and without swirl flow, then it is governed by the following equations, namely the equation of continuity and NS equations.

$$\frac{1}{r}\frac{\partial}{\partial r}(rv_r)+\frac{\partial v_z}{\partial z}=0 \tag{1}$$

$$\frac{\partial v_z}{\partial t}+\frac{\partial v_z^2}{\partial z}+\frac{1}{r}\frac{\partial}{\partial r}(rv_z v_r)=-\frac{1}{\rho}\frac{\partial p}{\partial z}+\nu\frac{1}{r}\frac{\partial}{\partial r}\left\{r\left(\frac{\partial v_z}{\partial r}-\frac{\partial v_r}{\partial z}\right)\right\} \tag{2}$$

$$\frac{\partial v_r}{\partial t}+\frac{1}{r}\frac{\partial}{\partial r}\left(rv_r^2\right)+\frac{\partial}{\partial z}(v_z v_r)=-\frac{1}{\rho}\frac{\partial p}{\partial r}+\nu\frac{\partial}{\partial z}\left(\frac{\partial v_r}{\partial z}-\frac{\partial v_z}{\partial r}\right) \tag{3}$$

These equations are solved by finite differential method using rectangular mesh system for convenience of handling the free surface. The number of effective grid, which varies by the position of the free surface, is approximately 50 thousands. The calculation area is up to 4D from the nozzle exit, where D is the nozzle exit diameter. The motion of the free surface is governed by the following equation.

$$\frac{\partial h}{\partial t}+v_{zs}\frac{\partial h}{\partial z}=v_{rs} \tag{4}$$

The pressure just below the free surface is calculated by the following equation.

$$p_s=\sigma\frac{1}{r}-\sigma\frac{\partial^2 h}{\partial z^2}+2\mu\frac{\partial v_r}{\partial r}+p_a \tag{5}$$

Any turbulent flow model is not introduced into the calculation yet. Therefore, the turbulence whose scale is smaller than the mesh size is ignored.

4.2. BOUNDARY CONDITIONS

The following boundary conditions are introduced.

At the inlet,

$$v_z=\text{const.},\qquad v_r=0 \tag{6}$$

At the outlet,

$$\frac{\partial v_z}{\partial z} = 0, \qquad v_r = 0, \qquad \frac{\partial p}{\partial z} = 0, \qquad \frac{\partial h}{\partial z} = 0 \tag{7}$$

On the axis,

$$\frac{\partial v_z}{\partial r} = 0, \qquad v_r = 0 \tag{8}$$

On the solid wall of the nozzle and the needle,

$$v_z = 0, \qquad v_r = 0 \tag{9}$$

At the exit of the nozzle, only for the outermost cell,

$$\frac{v_r}{v_z} = \tan\theta \tag{10}$$

where, θ is the incline angle of the nozzle exit.
On the free surface,

$$\frac{\partial v_z}{\partial r} = 0, \qquad \frac{\partial v_r}{\partial r} = 0, \qquad p_a = 0 \tag{11}$$

4.3. ALGORITHM

The basic solution algorithm is HSMAC(Hirt et al., 1972) method. The equations are represented in a finite difference form by forward differencing in time and centered differencing in space except for convective terms. For the convective terms, the third-order up-wind differencing method is used.

5. Results and Discussions

5.1. INFLUENCE OF NEEDLE STROKE

The mean axial velocity component at the nozzle exit is 7.4m/s, the Reynolds number using this mean velocity and the nozzle outlet diameter (D=52mm) is 3.8×10^5. The needle is in its normal position. Time of each figure is (a)t=0.391s, (b)t=0.693s, (c)t=0.992s.

These instantaneous jets are averaged for a certain duration to obtain a stable

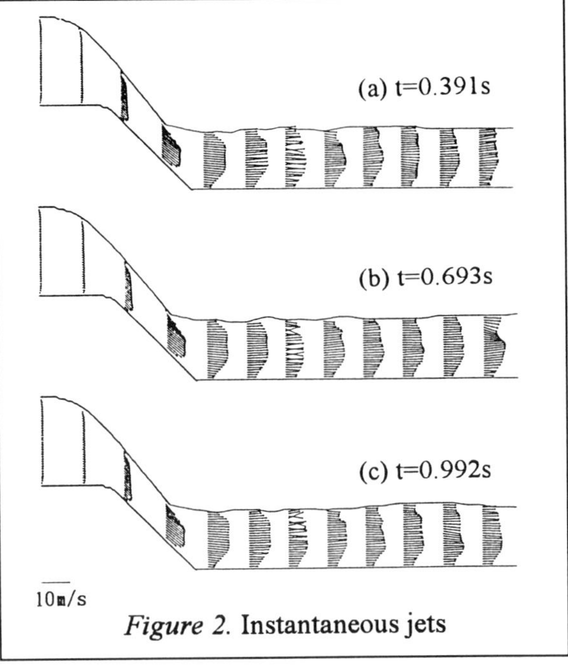

Figure 2. Instantaneous jets

surface and velocity vectors. Figure 3 shows the averaged jets (Nonoshita et al., 1994). The time period being averaged and the number of calculation time steps are listed in Table 1. The number of time steps for averaging is so decided in order that a stable jet surface is obtained. Each time step in calculation is determined by the Courant number and the cell Reynolds number, therefore, the duration of the same time step number (i.e., (a),(b) and (e)) differ each other. The positive needle position corresponds to closing direction. In every case, the mean axial velocity component at the nozzle exit is set equal.

From Fig. 3, it is revealed that the profile of the velocity distribution in the jet changes remarkably by the needle opening. The profile is relatively flat when the opening is large, i.e., the flow rate is large. On the other hand, when the opening is small, i.e., the flow rate is small, the profile is no more flat and has a

TABLE 1. Data for averaging of Fig.3

Figure Number	Needle Position [mm]	Time Period		Duration of Averaging [ms]	Corresponding Time Steps
		From [s]	To [s]		
Fig. 3 (a)	-8	0.931	0.972	41	2000
Fig. 3 (b)	0	0.992	1.036	44	2000
Fig. 3 (c)	+8	1.020	1.381	361	15000
Fig. 3 (d)	+12	1.423	1.904	481	20000
Fig. 3 (e)	+16	0.883	0.935	52	2000

clear peak value. In any case, there is low velocity region which is caused by the boundary layer of the needle around the axis. There is also another low velocity region near the free surface. It is caused by the boundary layer of the nozzle, and this region is accelerated easier than that around the axis. The position of the peak value moves toward outer direction as the needle opening becomes smaller, and the velocity gradient around the peak becomes steeper. The flow pattern in the bucket will also change by such variation of the velocity profile of the jet. Therefore, it can be concluded that the change of the velocity distribution by the flow rate may be one of the reasons of the inverse scale effect.

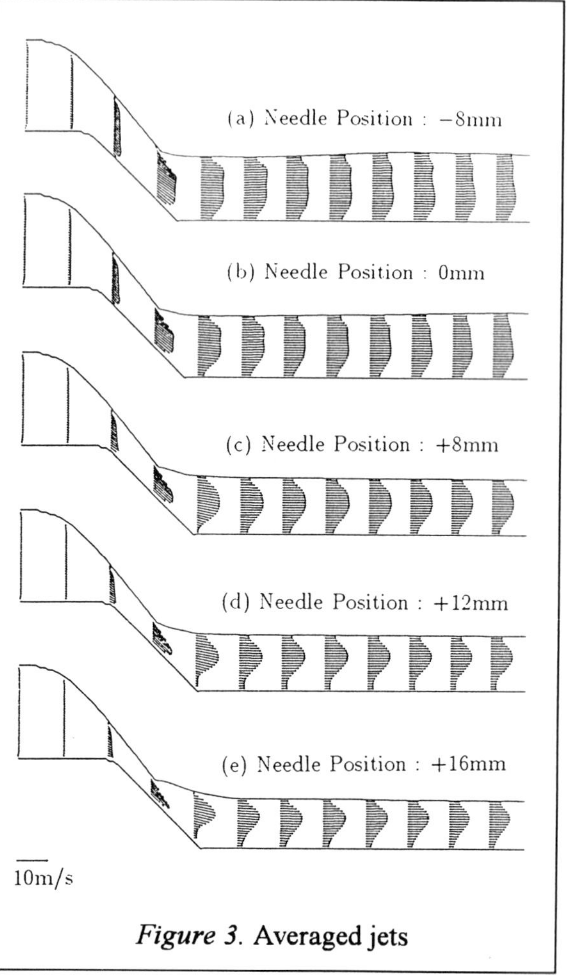

Figure 3. Averaged jets

5.2. INFLUENCE OF HEAD

In order to examine the influence of head change, the mean velocity at the inlet is set twice as large as the former cases. Figure 4 shows the comparison of velocity distribution in the jet at the distance of (a) z=1D and (b) z=2D from the nozzle exit. The needle is in its normal position, and 2000 instantaneous data are averaged. The solid line shows the case of lower head (H=2.8m), which corresponds to that the axial velocity component at the nozzle exit $\overline{v_{nz,z}}$ is 7.4m/s, and the

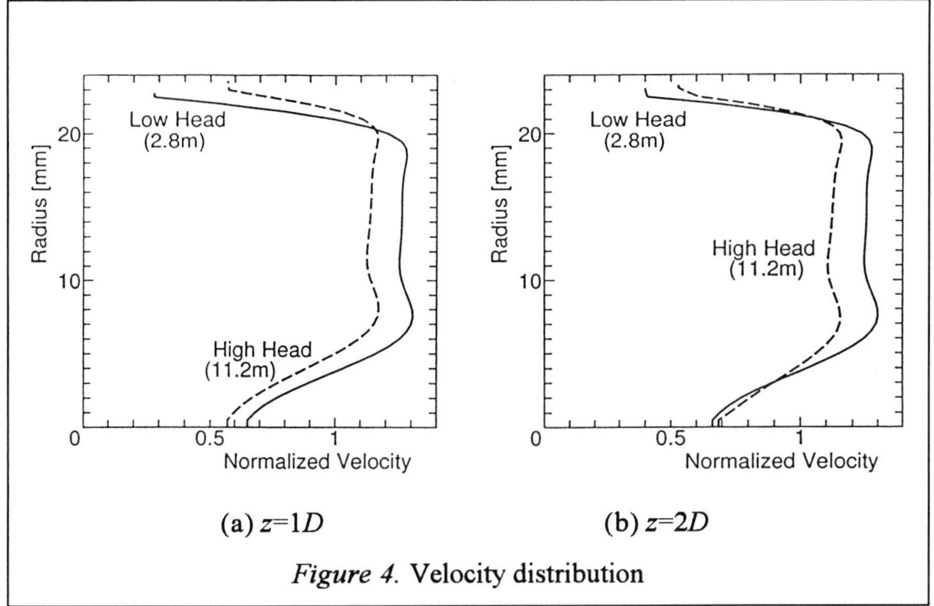

(a) z=1D (b) z=2D

Figure 4. Velocity distribution

broken line shows the case of higher head (H=11.2m) when $\overline{v_{nz,z}}$ is 14.8m/s , where the head is calculated by the following equation.

$$H = \frac{\overline{v_{nz,z}}^2}{2g} \tag{12}$$

The jet velocity $v_{j,z}$ is normalized by $\overline{v_{nz,z}}$. The main jet flow, which is so-called potential core, has two peaks at both edges. The peak of free surface side is very gentle, but that of axis side is clear. If we draw a straight line which passes the both two peaks, it is almost flat especially in case of higher head. The low velocity region around the axis is much wider than that near the surface as already seen in Fig. 3. At the axis side, the higher head flow is more accelerated than the lower head flow. At the surface side, the velocity gradient is unexpectedly large especially in the lower head case. We cannot find out the reason up to the present, anyhow, the steeper velocity gradient at the both sides induces the higher velocity of potential core. Therefore, in case of higher head, the normalized velocity of potential core becomes lower relatively comparing to the lower head case. It would cause the negative scale effect.

The jet diameter at $z=1D$ is 45.6mm when the head is low, and 47.2mm when the head is high. That at $z=2D$ is 46.1mm and 46.6mm respectively. In either case, the higher head causes the larger diameter. The larger jet may have a possibility of decreasing the efficiency of momentum transmission at a bucket.

6. Concluding Remarks

The jets from a Pelton turbine nozzle are calculated numerically concerning to the negative scale effect. In order to simplify the calculation, the flow is assumed to be axisymmetric. The calculated results reveal the followings.

There is a clear low velocity region in jet near the axis, and it spreads with the decrease of needle opening. Hence the velocity distribution changes by the flow rate. Moreover, the velocity distribution is also affected by the head. These should be some of the reasons of the negative scale effect in Pelton turbines.

References

Bachmann, P., Schärer, Ch., Staubli, T. and Vullioud, G. (1990) Experimental Flow Studies on a 1-jet Model Pelton Turbine, *Proc. 15th IAHR Symp.* **2**, Paper R3.

Grein, H., Meier, J. and Klicov, D. (1986) Efficiency Scale Effects in Pelton Turbines, *Proc. 13th IAHR Symp.* **2**, Paper 76.

Guilbaud, M., Houdeline, J. B. and Philibert, R. (1992) Study of the Flow in the Various Sections of a Pelton Turbine, *Proc. 16th IAHR Symp.* **2**, 819-831.

Hirt, C. W. and Cook, J. L. (1972) Calculating Three-dimensional Flows around Structures and over Rough Terrain, *J. Computational Physics* **10**, 324-340.

Hoyt, J. W. and Taylor, J. J. (1977) Waves on Water Jets, *J. Fluid Mechanics* **83** part 1, 119-127.

Kubota, T., (1989) Observation of Jet Interference in 6-nozzle Pelton Turbine, *J. Hydraulic Research* **27**-6, 753-767.

Kubota, T. and Nakanishi, Y. (1994) Classification of Flow in Pelton Turbines for Study of Scale Effect, *Proc. 17th IAHR Symp.* **2**, 865-876.

Nonoshita, T. et al. (1993) Behavior of the Jet from a Pelton Turbine Nozzle, *Proc. 4th Asian Int. Conf. Fluid Machinery* II, 499-503.

Nonoshita, T. et al. (1994) Effect of Water Droplets on the Jet from a Pelton Turbine Nozzle, *Proc. 17th IAHR Symp.* **1**, 165-176.

AN ASSESSMENT OF THE LOSS DISTRIBUTION IN FRANCIS TURBINES

R. SUZUKI, Y. QIAN
Fuji Electric Co., Ltd.
1-1, Tanabeshinden, Kawasaki-ku, Kawasaki, 210 JAPAN

T. KITAHORA, J. KUROKAWA
Yokohama National University
156 Tokiwadai, Hodogaya-ku, Yokohama, 240 JAPAN

Abstract

Scalable loss in Francis model turbine is analyzed for four runners with various number of blades which are tested in the same model turbine. The flow in the stay vane, the guide vane and the runners are analyzed by a 3D Euler flow analysis code, and then three methods of boundary layer calculations are carried out in order to obtain scalable loss. As for spiral casing and draft tube, scalable loss is obtained by the analogy of pipe flow.

From the comparison between runners, the difference in peak efficiency seems to be mainly caused by non-scalable losses. Therefore scalable loss itself is most suitable to be prescribed in the scale-up formula than loss distribution coefficient.

Problems on the effect of number of blades might be solved by using scalable loss in scale-up formulae.

1. Introduction

The efficiency performance of prototype turbines is predicted by applying so-called scale-up formulae to the performance of model turbines. Some famous formulae, for example Moody, Ackeret and Hutton formulae, had been used for many years for this purpose. Recently the study on the scale effect has been activated, and some new methods of scale-

E. Cabrera et al. (eds.), Hydraulic Machinery and Cavitation, 361–370.

up were proposed. Japan Society of Mechanical Engineers (JSME) established a standard based on numerical loss analyses and including the effect of wall roughness. IEC established a new standard based on the strict consideration on the step-up to a hydraulically smooth prototype emphasizing the effect of test Reynolds number of model turbines. Both in JSME and IEC the loss distribution coefficient V was used. Spurk/Grein, starting from dimensional analysis, introduced a formula using the efficiency at infinite Reynolds number. Nichtawitz claimed it was not reasonable that a model turbine with lower peak efficiency would be rewarded with a higher step-up. Ida and Nichtawitz proposed new formulae in order to avoid this absurdity. They introduced δ_{ref} in their formulae.

From the viewpoint of the value to be prescribed, there are three standpoints;

(1) To prescribe loss distribution coefficient

Ackeret, Hutton, JSME and IEC formulae belong to this group, and they are based on the idea that the ratio of scalable (frictional) loss to the total hydraulic loss would be almost constant for various turbines.

(2) To prescribe non-scalable loss

The proposal of Spurk/Grein corresponds to this standpoint if η_∞ is to be accumulated experimentally. It is based on the idea that non-scalable loss is almost constant for various turbines. In other words, the difference in peak efficiency is mainly caused by scalable loss.

(3) To prescribe scalable loss

Ida and Nichtawitz formulae belong to this group, and they are based on the idea that the difference in peak efficiency is mainly caused by non-scalable loss.

All the ideas about the cause of difference in peak efficiency mentioned above are, of course, idealized ones. But it is necessary to clarify which is most realistic.

On the other hand, there was a discussion in Europe that turbines with large number of runner blades would gain high efficiency step-up because of high ratio of scalable loss. This claim has never been adopted in any standard and V or other values are prescribed constant regardless of the number of runner blades. Even which formula is to be adopted, it is better to be discussed how the number of blade affect to efficiency step-up.

The purpose of this paper is to present some data for judging the key factor of the difference in peak efficiencies, and also the effect of the number of runner blades.

2. Specified Model Turbine

The specifications of Francis turbine used for the investigation is listed in Table 2. Spe-

TABLE 1. Scale-up Formulae

Proposer	Formula	
JSME S008 (1989)	$\frac{1-\eta_{E\,\mathrm{P}}}{1-\eta_{E\,\mathrm{M}}}=\frac{\eta_{E\,\mathrm{P}}}{\eta_{E\,\mathrm{M}}}\left\{(1-V)+V\Lambda\right\}$	(1)
	$\Lambda=(D_{\mathrm{P}}/D_{\mathrm{M}})^{0.18}\left\{(1-\beta)+\beta\,(e_{\mathrm{P}}/e_{\mathrm{M}})^{0.16}\right\}$	(2)
IEC 995 (1991)	$\Delta\eta_h=\delta_{ref}\left[\left(\frac{Re_{u\,ref}}{Re_{u\,\mathrm{M}}}\right)^{0.16}-\left(\frac{Re_{u\,ref}}{Re_{u\,\mathrm{P}}}\right)^{0.16}\right]$	(3)
	$\delta_{ref}=\frac{1-\eta_{h\,opt\,\mathrm{M}}}{\left(\frac{Re_{u\,ref}}{Re_{u\,opt\,\mathrm{M}}}\right)^{0.16}-\frac{1-V_{ref}}{V_{ref}}}$	(4)
Spurk/Grein (1992)	$\Delta\eta_h=(\eta_{h\,\infty}-\eta_{h\,\mathrm{M}})\left[1-\left(\frac{Re_{\mathrm{M}}}{Re_{\mathrm{P}}}\right)^{\alpha}\right]$	(5)
Ida (1993)	$\frac{\eta_{E\,\mathrm{P}}}{\eta_{E\,\mathrm{M}}}=\frac{1}{1-\delta_E\,(1-\Lambda')}$	(6)
	$\Lambda=(D_{\mathrm{P}}/D_{\mathrm{M}})^{-0.18}\left\{(1-\beta)+\beta\,(e_{\mathrm{P}}/e_{\mathrm{M}})^{0.18}\right\}$	(7)
Nichtawitz (1994)	$\Delta\eta_h=S_0\left(\frac{Re_0}{Re_{\mathrm{M}}}\right)^{n}-\left(\frac{Re_0}{Re_{\mathrm{P}}}\right)^{n}$	(8)

Note; Some formulae include the conversion of operating condition as well as the efficiency step-up procedure for pumping operation. But only the efficiency step-up formulae for turbine are contrasted here.
η_{E} in the formulae by JSME and Ida is the specific energy efficiency, the ratio of the specific hydraulic energy available for the runner to the specific hydraulic energy of the machine (see equations (9) and (10) below).

cific speed $n_{sQ}=n\,Q^{0.5}/H^{0.75}$ is around 40 (min^{-1}, m^3/s, m). Runners A and B have the same number of blades but Runner A shows higher efficiency. Runners C and D have smaller number of blades, and shows lower efficiency than Runners A and B. Here, η_{E} denotes specific energy efficiency;

$$\eta_E=E_m\,/\,E=1-E_{L12}\,/\,E \tag{9}$$

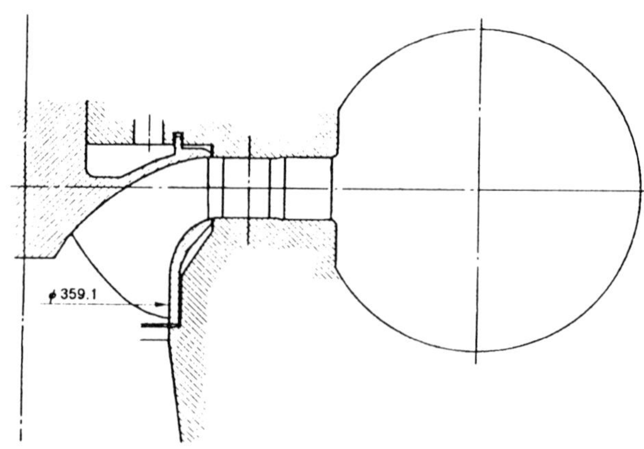

FIGURE 1. Sectional View of the Specified Model Turbine

TABLE 2. Specifications of Runners

Runner	A	B	C	D
D (mm)	359.1	359.1	359.1	359.1
Re	6.3×10^6	6.3×10^6	6.3×10^6	6.3×10^6
n_{SQopt}	40.4	38.3	37.0	35.6
$n_{ED}/n_{ED\,(A)}$	1.00	1.00	1.00	1.00
$Q_{ED}/Q_{ED\,(A)}$	1.00	1.00	0.963	0.969
n_{ED}/n_{EDopt}	1.00	1.03	1.00	1.05
Q_{ED}/Q_{EDopt}	1.00	1.04	1.15	1.12
η_E (%)	95.0	93.6	92.0	91.3
η_{Eopt} (%)	95.0	93.7	92.4	91.9
Z_R	13	13	9	7

where E_m, E_{L12} and E indicate specific hydraulic energy of the runner, specific energy loss of the machine and specific hydraulic energy of the machine respectively. Therefore;

$$\eta_E = \eta_h \,/\, \eta_Q \,\eta_R \qquad (10)$$

where η_h, η_Q and η_R are hydraulic (internal) efficiency, discharge (volumetric) efficiency and power efficiency (related to disk friction) respectively.

The values n_{ED} and Q_{ED} listed in Table 2 without subscript are the speed factors and discharge factors of the operating conditions analyzed. Those with subscript (A) indicate the values for Runner A, and subscripts opt denotes the best efficiency point of each runner. Runner A is analyzed at the best efficiency point. Analyses for other runners are carried out at the same guide vane opening and the same speed factor as Runner A. The discrepancy between the analyzed point and the best efficiency point for each runner is not so large, and the maximum difference in efficiency is 0.6% in Runner D.

3. Numerical Analysis of Scalable Loss

3.1 INTERNAL FLOW ANALYSES IN CASCADES

The flow in stay vanes, guide vanes and runner blades are analyzed with 3D Euler code with pseudo-compressibility and implicit formulation of finite difference. This Euler code has the same numerical schemes as the Navier-Stokes code [6], but the effect of viscosity is neglected and the boundary condition on the wall surfaces is different. Figure 2 shows the flow field for Runner A obtained from the analysis.

FIGURE 2. Flow Velocity Vectors in Cascades

3.2 CALCULATION OF SCALABLE LOSS

3.2.1 *Scalable Loss in Runner*

Three Dimensional Boundary Layer Analysis. Three dimensional boundary layer is calculated on the wall surface in runner using the velocity distribution obtained by 3D Euler flow analysis. Coriolis force and centrifugal force are considered. The numerical scheme of calculation is described in [7]. Figure 3 illustrates the result of calculation for Runner A.

The scalable loss is obtained from the energy thickness of boundary layer at the outlet of blade surfaces and of crown and band surfaces.

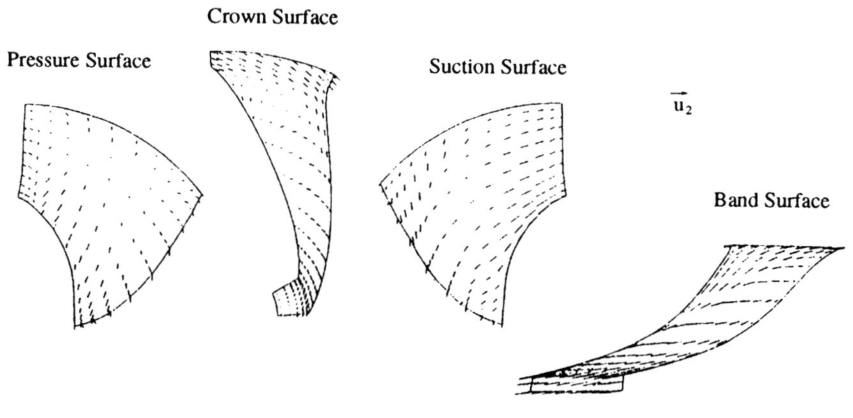

(a) Main Flow Velocity on the Walls

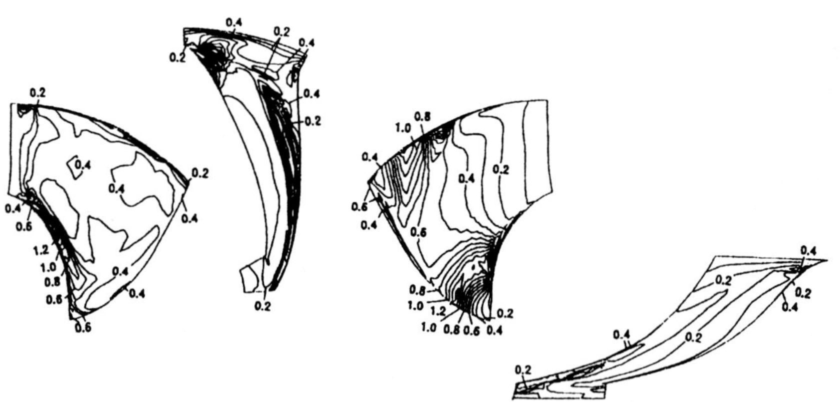

(b) Boundary Layer Thicness δ / D_2 (%)

FIGURE 3. Three Dimensional Boundary Layer on Runner Surface

One Dimensional Boundary Layer Calculation. One dimensional approximate boundary layer calculation is made along the grid lines using the velocity distribution obtained by 3D Euler flow analysis. Coriolis force and centrifugal force are not considered. The discrepancy in flow direction from the grid lines are also neglected.
The scalable loss is obtained from the energy thickness of boundary layer at the outlet of blade surfaces and of crown and band surfaces.

Integration of Shearing Stress. The scalable loss is approximately estimated integrating the local shearing stress of a flat plate, which is obtained from the local velocity and the distance from the inlet. The velocity distribution on the wall surfaces obtained from 3D Euler flow analysis is used.

3.2.2 *Scalable Loss in Stay Vanes and Guide Vanes*

The scalable losses in stay vanes and guide vanes are obtained only by the integration of shearing stress mentioned above, because the flow distortion is not so large.

3.2.3 Scalable Loss in Spiral Casing and Draft Tube

The scalable losses in spiral casing and draft tube are calculated with Colebrook's equation by analogy of pipe friction.

4. Results and Discussions

Table 3 shows the results of scalable loss analyses. δ_E denotes the total scalable deficiency (relative scalable loss of specific hydraulic energy). δ_{ES}, δ_{ER} and δ_{ED} are the scalable deficiencies in the upstream parts of runner (spiral casing to guide vanes), in the runner and in the draft tube respectively. δ_{Ens} denotes the non-scalable deficiency which is obtained subtracting δ_E from the total specific hydraulic loss $E_{L12} = 100 - \eta_E$. The loss distribution coefficients are also tabulated. The subscripts S, R and D correspond to those of scalable deficiencies. Three values in array divided by slash marks correspond to the methods of scalable loss calculation.
Figure 4 illustrates the variation of the total scalable deficiency. Three kinds of bars correspond to the loss calculation methods. In spite of large difference in efficiency among four runners, the scalable deficiency is almost constant.
Figure 5 shows the variation of the non-scalable deficiency. δ_{Ens} steeply increases from Runner A to D, indicating that the discrepancy in efficiency is mainly caused by non-

TABLE 3. Scalable and Non-scalable Deficiencies

Runner	A	B	C	D
η_E	95.04	93.60	91.97	91.25
δ_{ES} (%)	1.75	1.75	1.65	1.64
δ_{ER} (%)	0.71 / 0.96 / 1.00	1.01 / 1.49 / 1.20	1.24 / 1.72 / 1.40	1.26 / 1.67 / 1.35
δ_{ED} (%)	0.04	0.04	0.04	0.04
δ_E (%)	2.50 / 2.76 / 2.79	2.80 / 3.28 / 2.99	2.91 / 3.39 / 3.07	2.95 / 3.35 / 3.04
δ_{Ens} (%)	2.46 / 2.20 / 2.17	3.60 / 3.12 / 3.41	5.12 / 4.64 / 4.96	5.80 / 5.40 / 5.71
V_S	0.35	0.27	0.20	0.19
V_R	0.14 / 0.19 / 0.20	0.16 / 0.23 / 0.19	0.16 / 0.21 / 0.17	0.14 / 0.19 / 0.16
V_D	0.01	0.01	0.01	0.01
V	0.50 / 0.56 / 0.56	0.44 / 0.51 / 0.47	0.36 / 0.42 / 0.38	0.34 / 0.38 / 0.35

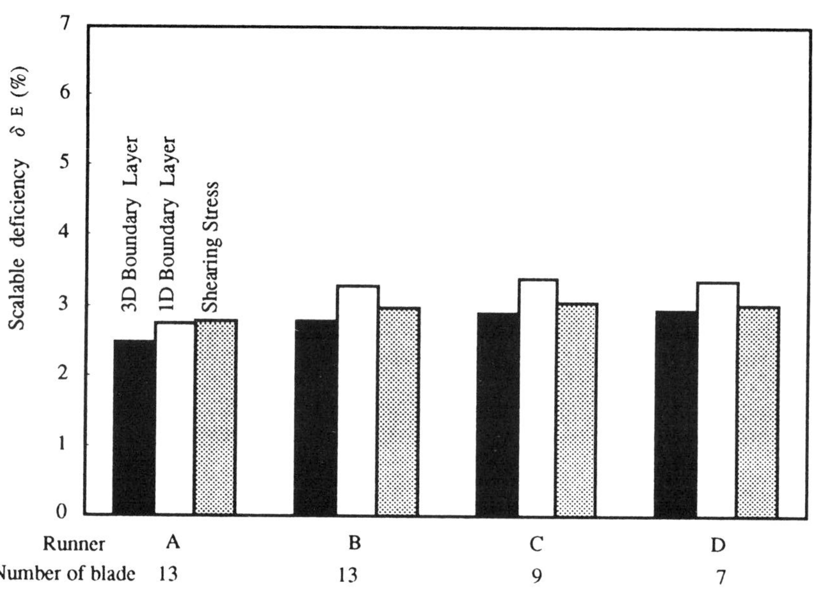

FIGURE 4. Scalable Deficiency

scalable loss. From Figure 6 illustrating the variation of the loss distribution coefficient, the same conclusion can be drawn.

Changing the point of view to the effect of the number of blades, the loss distribution coefficient decreases apparently with decreasing number of blades. If this result is extrapolated toward larger number of blades, it would be concluded that the increase in the number of blades would increase the value of V.

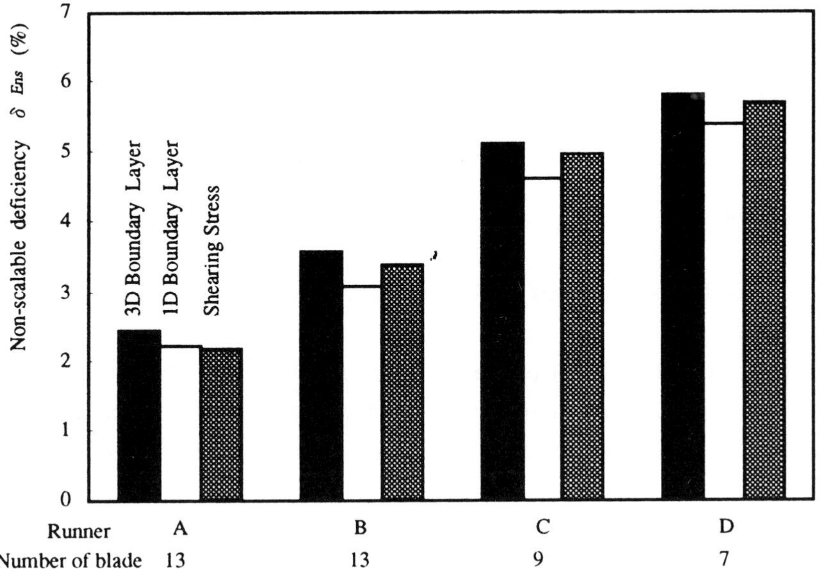

FIGURE 5. Non-scalable Deficiency

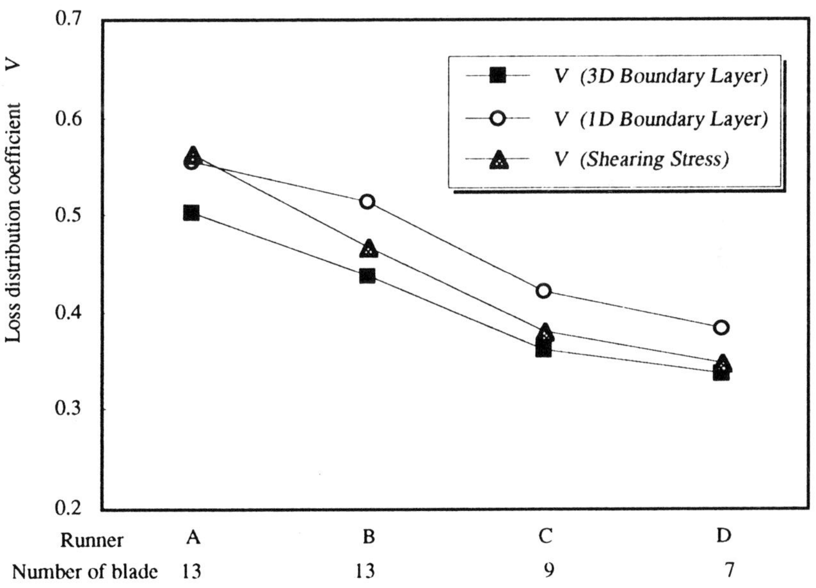

FIGURE 6. Loss Distribution Coefficient

Irrespective of the above mentioned tendency in loss distribution coefficient, however, the discussion about the effect of the number of blades might be disposed of when scalable loss is prescribed in scale-up formulae.

5. Conclusion

Results of scalable loss analyses for four Francis runners tested in the same model turbine are presented.

Analyzed results showed best regularity in scalable loss. Non-scalable loss steeply increased and loss distribution coefficient considerably decreased in proportion as efficiency dropped. Then lower peak efficiency seems to be mainly caused by non-scalable loss. This means that scalable loss is most suitable for prescription in scale-up procedures. Although the possibility could be seen on the increase in loss distribution coefficient with increasing number of blades, yet the problem would be solved by prescribing scalable loss instead of loss distribution coefficient.

6. Acknowledgement

This study was strongly motivated by the activities of the committee on preparing the revised JSME S008. The 3D boundary layer analyses are performed by Mr. Matsumoto, the graduate student in Yokohama National University.

7. References

1. JSME Standard S008, Performance Conversion Method for Hydraulic Turbines and Pumps (1989).
2. IEC Publication 995, Determination of the Prototype Performance from Model Acceptance Tests of Hydraulic Machines with Consideration of Scale Effects (1991).
3. Spurk, J.H. and Grein, H., Performance Predictions of Hydraulic Machines by Dimensional Considerations, Water Power & Dam Construction (1993-11), 42-49.
4. Ida, T., New Formula for Efficiency Step-up of Hydraulic Turbines, Proc. 17th IAHR Symposium (Beijing, 1994), 827-840.
5. Nichtawitz, A., Discussion on Step-up Procedures in Hydraulic Machines, Proc. 17th IAHR Symposium (Beijing, 1994), 841-852.
6. Qian, Y., Arakawa, C., Kubota, T. and Suzuki, R., Numerical Flow Simulation on Channel Vortex in Francis Runner, Proc. 17th IAHR Symposium (Beijing, 1994), 237-247.
7. Kurokawa, J., Kitahora, T., et al., Prediction of Scalable Loss in Francis Runners of Different Specific Speed, Proc. 18th IAHR Symposium (Valencia, 1996).

UNSTEADY FLOW CALCULATION IN A CENTRIFUGAL PUMP USING A FINITE ELEMENT METHOD

P.F. BERT, J.F. COMBES
EDF-DER, 6 quai Watier
78400 Chatou, FRANCE

J.L. KUENY
INPG-ENSHMG, St Martin D'Hères
38400 Grenoble, FRANCE

1. Abstract

In order to predict rotor-stator interactions in hydraulic turbomachinery, a new multidomain method was implemented in a finite element code developed in the Research Division of Electricité de France. This code deals with 2D or 3D laminar or turbulent flows in complex geometries. This method was used to study the unsteady flow in a simplified centrifugal pump model. The model consists in a 420 mm diameter unshrouded centrifugal impeller with seven untwisted constant thickness backswept blades and a radial vaned diffuser with twelve vanes and a six percent vaneless radial gap in which important unsteady interactions are expected. Unsteady numerical results are compared with the experimental results published by M. UBALDI et al. and also with a multistage calculation on the same geometry using a mixing plane to communicate circumferential average of flow properties.

2. Introduction

Centrifugal turbomachines with vaned diffuser or scroll have always a vaneless radial gap lower than the length scale of the blade row interactions. These interactions generate very important unsteady flows that extend both upstream and downstream of the gap. The two major sources of unsteadiness are potential interaction and blade/wake interaction (Dring et al. 1982). In centrifugal turbomachines, the effects of these sources of unsteadiness become comparable (Arndt et al. 1990). This remark is very important for numerical analysis of unsteady flows. Simplified approach used by Yu et al. (1995) taking into account bladed row effects as an unsteady boundary condition cannot be applied. A full modelling of the flow in the machine must be undertaken, that suppose a computational code able to solve rotor-stator problems with sliding meshes.

To study the unsteady flow in a centrifugal pump with vaned diffuser, experimented at the University of Genova (Italy), a multidomain method was implemented in a finite element industrial code. A full bidimensional discretization of the machine is used to calculate the flow and to analyse the basic fluid phenomena. Major objectives of this

E. Cabrera et al. (eds.), Hydraulic Machinery and Cavitation, 371–380.

paper are to complement the unsteady flow analysis performed by Ubaldi et al. (1994) and to show that the computational method developed is able to apprehend unsteady flow effects.

3. Simulation Test Case

The simplified model of centrifugal turbomachine studied by Ubaldi et al. (1994) has been utilised firstly to validate unsteady calculations and secondly to improve the understanding of basic fluid dynamic phenomena. The model consists of a 420 mm diameter unshrouded centrifugal impeller and a 664 mm diameter radial vaned diffuser. The geometry of the machine is shown in figure 1.a/ and the co-ordinates of the impeller blade and diffuser vane profiles are given in Ubaldi et al. (1994). The impeller has seven untwisted constant thickness backswept blades with rounded off leading and trailing edges. The numerical model cannot describe the manufacturing rounded off : the leading edge is pointed and the trailing edge is squared. The diffuser used has 12 vanes and a 6 percent vaneless radial gap. The tip clearance is set at a value of 0.4 mm, 1 percent of the blade span. Even if the geometry of the blades is bidimensional, the tip leakage effects and the fixed casing lead to tridimensional secondary flows that create difficulties for comparisons with bidimensional computations.

The measuring techniques used were hot wire anemometer and fast response pressure transducers. The hot wire probe was used to measure the unsteady tridimensional flow in the vaneless gap at a radial distance of 4 mm from the blade trailing edge and 8 mm from the vane leading edge. The unsteady static pressure was measured at the front cover facing the unshrouded impeller passages.

The model operates in an open circuit with air directly discharged into the atmosphere from the radial diffuser. The inlet air temperature was 298 K and the air density 1.2 kg/m^3. The results were obtained at the nominal operating conditions, at a constant rotational speed of 2000 rpm. The main geometric data of the model and the operating conditions are summarised in table 1.

Table 1 : Geometric data and operating conditions

Impeller				
inlet blade diameter	D_1	=	240	mm
outlet diameter	D_2	=	420	mm
blade span	b	=	40	mm
number of blades	z_i	=	7	
Diffuser				
inlet vane diameter	D_3	=	444	mm
outlet vane diameter	D_4	=	664	mm
vane span	b	=	40	mm
number of vanes	z_d	=	12	
Operating conditions				
rotational speed	n	=	2000	rpm
flow rate coefficient	φ_e	=	0.048	
total pressure rise coefficient	ψ_e	=	0.65	
Reynolds number	Re	=	6.5×10^5	

4. Numerical Approach

In order to solve industrial flow problems in complex geometries, a finite element code, N3S, has been developed at Electricité de France. It allows the computation of a wide variety of 2D or 3D unsteady incompressible flows, by solving the Reynolds-averaged Navier-Stokes equations together with a k-ε turbulence model.

The space discretization uses a standard Galerkin finite element method. In the present calculations we used isoP2 6-noded triangles verifying the LBB condition. The time discretization is based on a fractional step method ; at each time step, one has to solve successively :

- an advection step for the non-linear convection terms of the Navier-Stokes and k and ε equations. It is treated by a characteristics method.
- a diffusion step for the remaining part of the k and ε equations : the finite element discretization leads to linear systems solved by a preconditioned conjugate gradient algorithm.
- a generalised Stokes problem for the velocity and the pressure. It is solved by a projected gradient Uzawa or Chorin algorithm preconditioned in order to ensure a fast convergence.

The numerical method is described in details by Chabard et al (1991).

The most recent development of this code concerns the extension to multidomain problems. The method has to solve the N3S system equation on several domains with possibility of translation/rotation with or without periodic boundaries. The resolution of the problem in each mesh reference frame with sliding boundary was considered as the better solution. The fractional step method, used in N3S code, compels us to implement two different coupling methods. The first one assumes the coupling of the convection step and the second assumes the coupling of the diffusion and Stokes step.

The continuity of the non-linear convection terms is achieved through the integration of characteristics through subdomain boundaries. The second method is the most important step in solving the incompressible Reynolds-averaged Navier-Stokes and k-ε equations on multidomain meshes. A mortar element was used to assume the continuity of aerodynamics quantities on sliding meshes. It assumes also the local and global divergence of velocity to stretch to zero. The mortar element method is a non-conforming discretization based on the explicit construction of an optimal approximation space. This method preserves element-based locality, which distinguishes it from other more global techniques.

5. Steady computation

The 2D computational domain correspond to a meridian plane with a radial inlet. The total domain was treated by two parallel computations which exchange boundary informations. At the impeller outlet, an azimutal average of the pressure calculated at the previous step at the diffuser inlet was used. At the diffuser inlet, we use an azimutal average of the velocity and the turbulent quantities estimated at the previous step at the

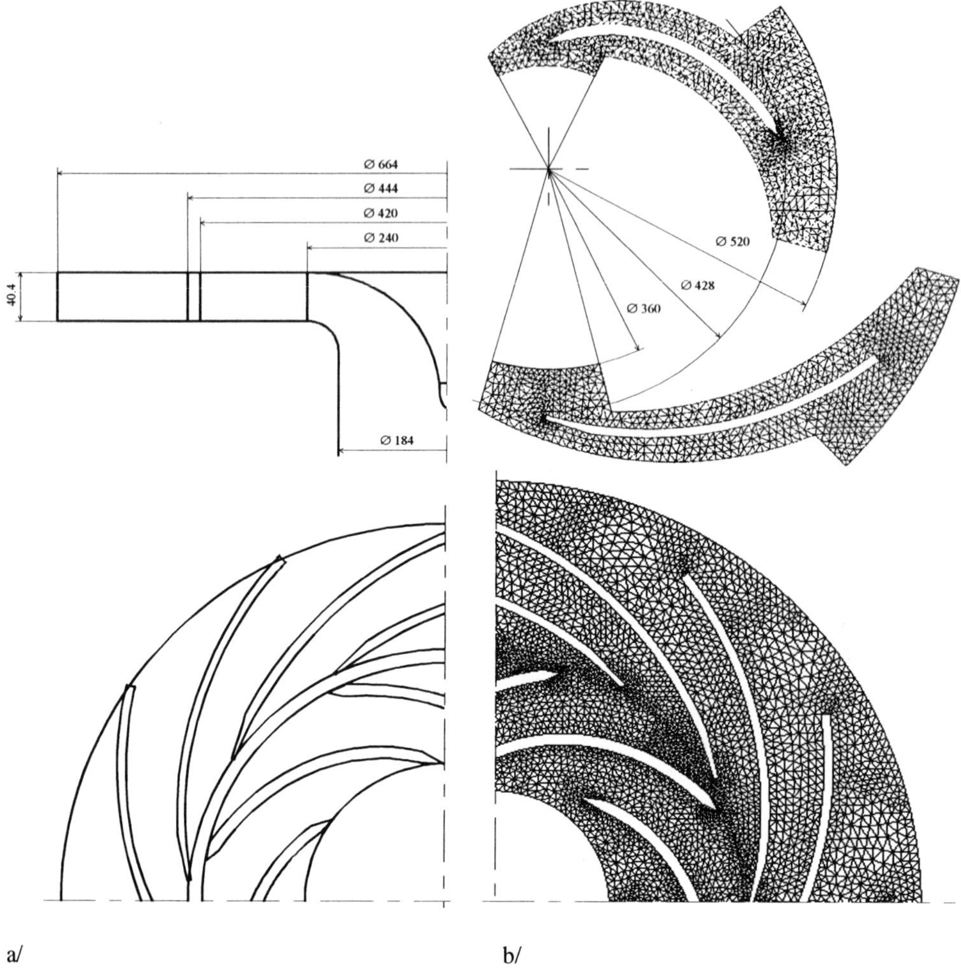

Figure 1 : a/ Experimental radial geometry pump ; b/ Navier-Stokes simulation meshes (each element is divided in 4 sub-elements), upper figure : steady calculation mesh, lower figure : unsteady calculation mesh.

impeller outlet. These dynamic type boundary conditions are naturally imposed for an incompressible computation. For a compressible computation these conditions are not suited, energetic conditions must be implemented. The average process and the exchange of information is consistent with the conservative equations and they integrate the change of frame.

The impeller finite element mesh consists in 1610 triangular elements and 3478 nodes, for the diffuser a 1398 triangular elements and 3066 nodes mesh is used (figure 1.b/). The impeller mesh was extended downstream about 46 mm. It was estimated that a constant pressure on this new outlet boundary does not modify the flow around the blade and the gradient constraint calculated on the real frontier was physically better. The same line of argument can be used for the diffuser with the velocity at the inlet where we have extended the mesh upstream about 34 mm.

The other boundary conditions are :

- at the inlet of the impeller, a uniform velocity was prescribed, according to the flowrate measured. Also k and ε were prescribed, corresponding to a turbulence level of 2%,
- at the outlet of the diffuser, a null constraint was prescribed. It correspond to a constant static pressure if the shear stress is negligible (atmospheric pressure),
- on the walls Reichardt's function was used,
- periodic conditions were also used to limit the calculation at one inter-blade channel.

6. Unsteady computation

The unsteady computation was achieved on the same meridian plane used in the steady case. The whole domain, impeller and diffuser, was treated on a global computation using a sliding mesh interface. In this global modelling the geometric periodicity does not exists, periodic boundary conditions can not be applied and the seven inter-blade channels of the impeller and the twelve inter-blade channels of the diffuser must be discretized. The total 2D finite element mesh is formed of 23000 triangular elements and 49800 nodes (figure 1.b/). The boundary conditions applied at the impeller inlet and at the diffuser outlet are the same as in the steady calculation.

7. Comparison results

We can make two kinds of comparisons between the steady, the unsteady calculations, and the experimental results : comparison of the temporal average values and comparisons of instantaneous values. Naturally, comparisons of instantaneous values have been only performed between unsteady computational results and experimental ones. Time averaged unsteady results are not equal to the steady calculation results because non-linear unsteady quantities are not evaluated by the second approach. As it is shown in figure 2, the differencies are not uniformly distributed in the azimutal direction.

In figure 2, we represent the time average relative velocity at the impeller outlet, normalized with the rotor tip speed U_2, in function of the normalized rotor circumferential co-ordinate. The pressure side peaks and the wake deficit are correctly represented around the blade trailing edge ; both calculations are able to estimate the maximum of the azimutal fluctuations generated by the blade passage. The large smooth maximum measured in the mid-passage, at mid-span ($z/b=0.5$), is underestimated by the calculations, but a detailed analysis of the experimental results

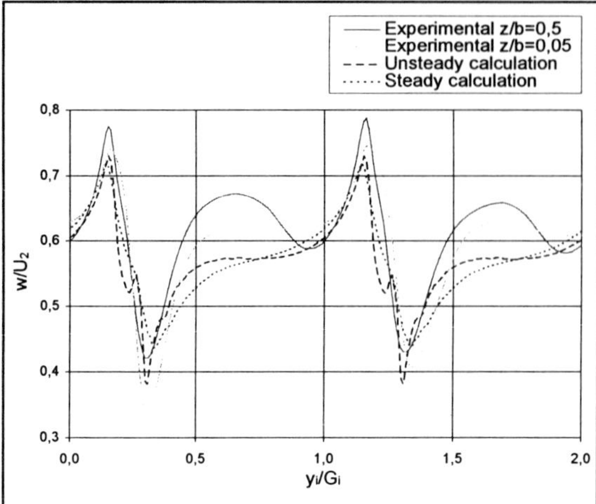

Figure 2 : Measured and calculated time average relative velocity at the impeller outlet.

show that this maximum disappears when z decreases. The relative velocity, near the hub (z/b=0.05), increases continuously from the wake suction side toward the pressure side peak as it was calculated. In the mid-passage, the velocity gradients seem to be better predicted by the unsteady calculation.

The instantaneous distributions, at the impeller outlet, of the radial velocity, measured (on the top) and calculated, are plotted with dark line in figure 3. The dotted line represent the temporal average of the velocity. The conservation of the flowrate prescribes in the calculation the global level of the radial velocity. This level is a little less than that measured because we consider a uniform distribution in the axial pump direction. In reality the effect of the hub and casing boundary layers is to

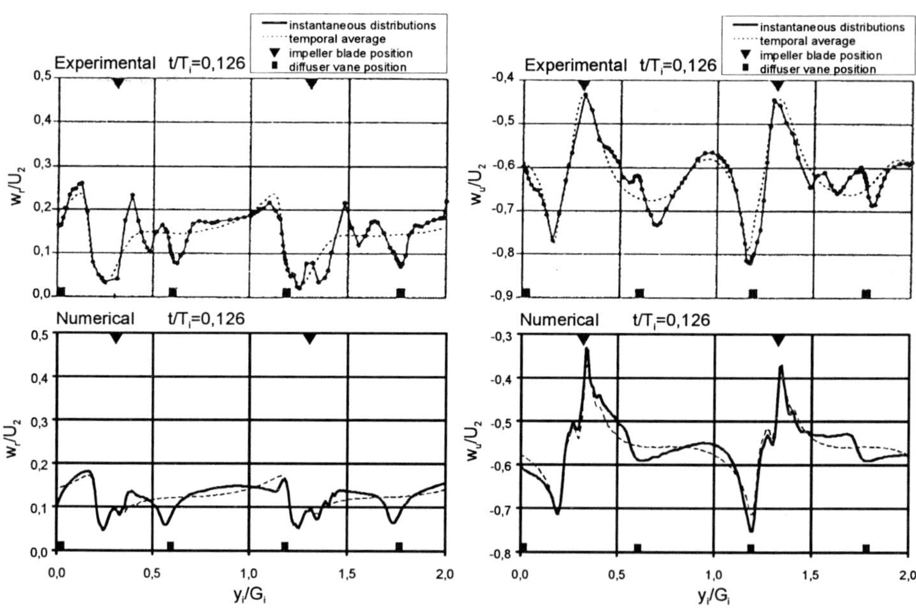

Figure 3 : Instantaneous distributions of the radial velocity at the impeller outlet.

Figure 4 : Instantaneous distributions of the relative tangential velocity at the impeller outlet.

increase the flowrate in the mid-passage. The amplitude of the fluctuations calculated is underestimated about thirty percent without any explication except a bad representation of the velocity on the walls by the Reichardt's wall function. The most important is that the peaks corresponding to the wake velocity deficit and the leading edge vaned diffuser interaction are well represented and positioned. Figure 4 shows the instantaneous distributions, plotted as in figure 3, of the tangential relative velocity. The weak mean level calculated is directly a consequence of the weak radial velocities. The amplitude of the fluctuations calculated is well estimated as the position of the peaks. The absence of a large smooth maximum in the mid-passage on the calculated results can be explained by the same line of arguments used in figure 2 for the temporal average of the relative velocity.

8. Flow Analysis

For incompressible fluid, the unsteady total pressure is representative of the local instantaneous fluid energy. Figure 5 show that impeller wakes are high energy region because the tip speed $U_2=\omega R_2$ is higher than the outlet absolute velocity.

The wakes decay very rapidly in the gap, but after their chopping by the vanes leading edge, they are convected in the vanes passage and we can clearly notice their presence at a radial distance of 1.4 R_2. In the impeller, the total pressure is dominated by the pressure gradient except in the blade boundary layers where friction losses are important.

In figure 6.a/, we represent the temporal average relative velocity in the impeller (unsteady calculation). The maximum of velocity is located on the leading edge suction side, the minimum is just on the other side of the blade. The adverse pressure gradients decelerate the flow from the leading edge to the trailing edge on the suction side and

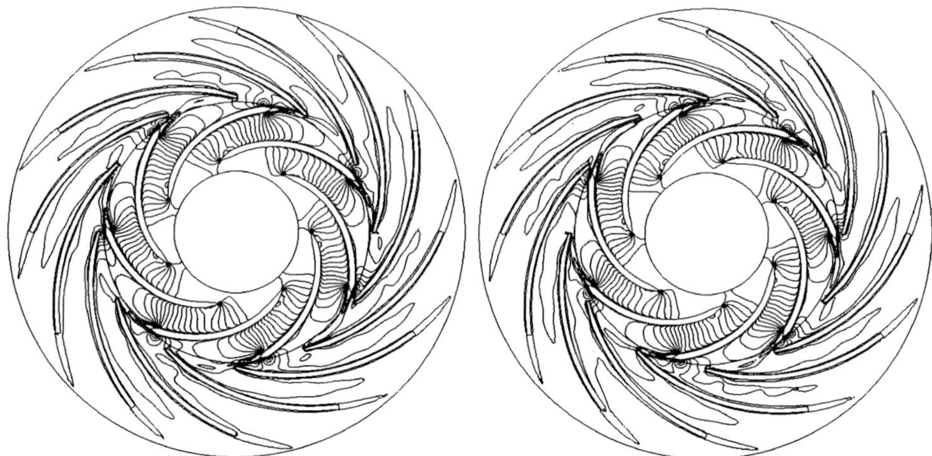

Figure 5 : Isovalues of instantaneous total pressure at $t/T_i=0$ and $t/T_i=0.5$.

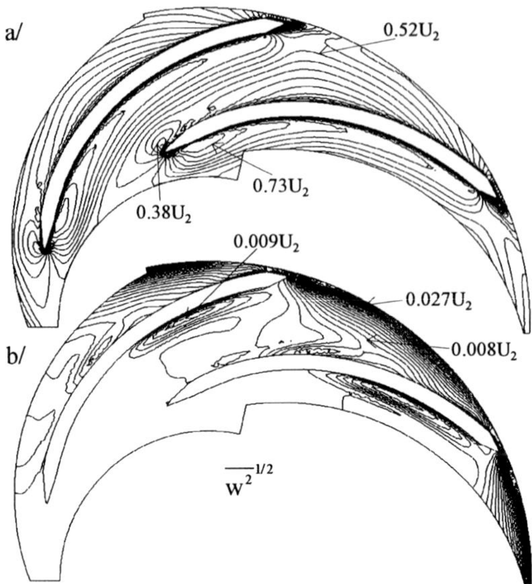

Figure 6 : a/ Isovalues of the temporal average relative velocity in the impeller ; b/ Isovalues of stator induced velocity fluctuations.

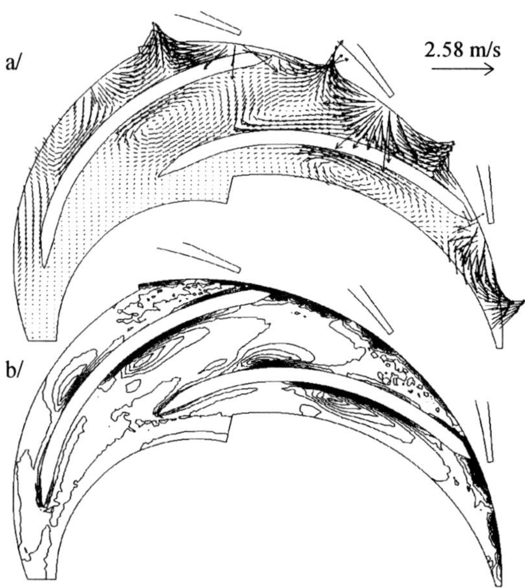

Figure 7 : a/ Unsteady secondary relative velocity vectors $w(t)-\overline{w}$; Unsteady vorticity contours.

between 10% and 50% of the chord on the pressure side. In these decelerate flow regions, the vanes passage induce a high level of fluctuations, highest than in the other regions (figure 6.b/). Apart from these two regions, the level of fluctuation decreases with the distance from the vanes, but the azimutal distribution is not uniform. On the trailing edge suction side gradients fluctuations are very concentrated, on the pressure side the fluctuations seem to decrease rather slowly.

The most important secondary instantaneous velocities are located in the regions of highest fluctuations (figure 7.a/). The secondary velocity is calculated as the difference between the instantaneous velocity and the time average one. We have plotted the velocity scale, it must be compared to the tip speed U_2=43.98 m/s and to the discharge velocity $v_{r2}=Q/(2\pi R_2 b)=$ 5.47 m/s. The structure of the secondary flow induced by the stator is very complex ; the vortex, located near the suction surface is always present but its centre position change with time. The second vortex structure noticeable in figure 7.a/ disappears periodically with the vane passage. The leading edge acts as a source point and any vortex structure is associated to the potential interaction. We can remark that the vorticity is not the best parameter to analyse this mode of interaction (figure 7.b/).

In figure 8, the temporal average relative velocity and rotor induced rms velocity fluctuations are plotted

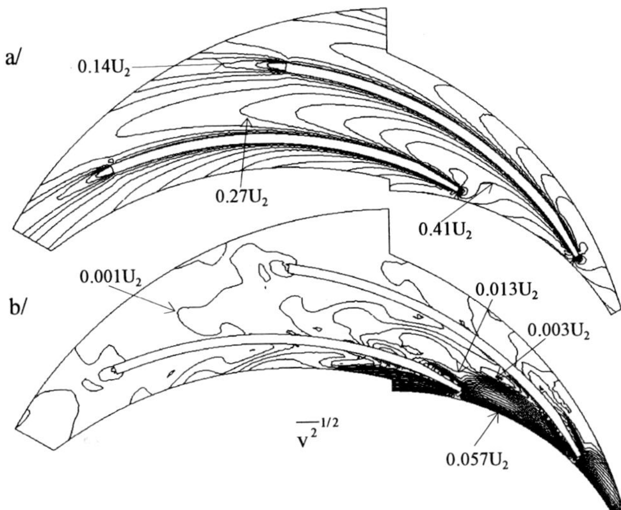

Figure 8 : a/ Isovalues of the temporal average absolute velocity in the vaned diffuser ; b/ Isovalues of stator induced velocity fluctuations.

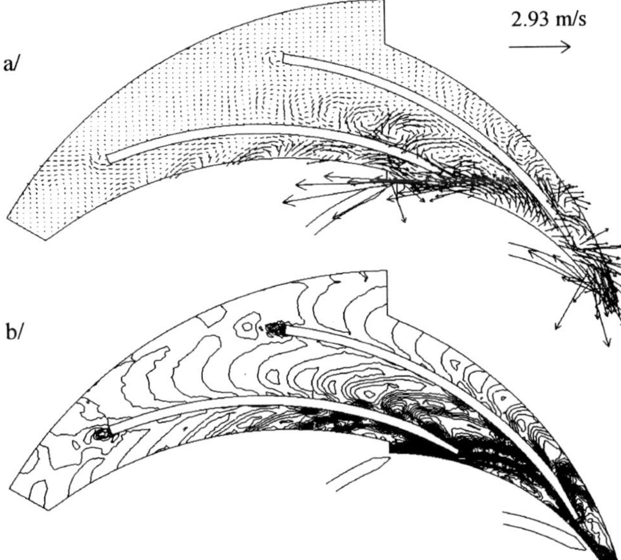

Figure 9 : a/ Unsteady secondary absolute velocity vectors $v(t) - \overline{v}$; b/ Unsteady vorticity contours.

in the vaned diffuser. The gradients of fluctuation are concentrated in the zone of maximum mixing process, before the wakes were chopped by the vanes. At the inlet, the fluctuations are maximum on the blades walls, in the inter-blade channel this observation seems to be inverted and the maximum is located on the centre passage. This centre passage maximum is associated with the secondary vortex velocity structure convected in the channel (figure 9.a/). The presence of wakes in the diffuser is clearly noticeable and it results in two counter rotating vortices on either side as it was indicated by several earlier investigators in axial turbomachines. Each wake is separated by large positive and negative vorticity regions (figure 9.b/), the isovalue zero representative of the wake structure is always present at the outlet.

On the contrary of in the axial machines (Yu et al. 1995), the velocity incidence angles tends to increase before the wake passage, it decreases when the wake impiges the leading edge and it increases again after its passage. An other difference with the kinematics of the wake in an axial machine (Hodson 1984) is that the wake is bowed only at the inlet in the maximum fluctuation region. In the diffuser channel, the wake becomes progressively perpendicular to the pressure side wall and we observed a concentration of the main structure vortex (line of iso-maximum velocity in the vortex).

9. Conclusion

To study the rotor-stator interactions in a centrifugal pump with vaned diffuser, a steady and an unsteady simulation were carried out using a finite element Navier-Stokes code. The results were compared with the experimental investigation of Ubaldi et al. (1994), good agreements were found and the main differencies can be explained by the 3D secondary flows generated by the unshrouded impeller. The stator induced flow in the impeller and the rotor induced flow in the diffuser are studied : rms fluctuations, secondary instantaneous velocity and unsteady vorticity are plotted. In the impeller, it was shown that high fluctuation regions exist in the centre of the channel induced by the periodic passage of the leading edge vane which acts as a source point. In the diffuser, it was observed that the wake mechanisms are different in an axial and in a radial machine.

10. References

Anagnoustou, G., Maday, Y., Patera, A.T., 1991, "A Sliding Mesh Method for Partial Differential Equations in Nonstationary Geometries : Application to the Incompressible Navier-Stokes Equations", Publ. of Numeric Lab., Université Pierre et Marie Curie

Arndt, N., Acosta, A.J., Brennen, C.E., Caughey, T.K., 1990, "Experimental Investigation of Rotor-Stator Interaction in a Centrifugal Pump with Several Diffusers", A.S.M.E. Transactions, Vol. 112, pp 98-108

Chabard, J.P., Metivet, B., Pot ,G., Thomas, B., 1991, "An efficient finite element method for the computation of 3D turbulent incompressible flows", Finite Element in Fluids, vol.8

Chen, Y.S., 1988, "3-D Stator-Rotor Interactions of the SSME", A.I.A.A. Paper 88-3095

Dring, R.P., Joslyn, H.D., Hardin, L.W., Wagner, J.H., "Turbine Rotor-Stator Interaction", Journal of Engineering for Power, Vol. 104, pp 729-742

Gallus, H.E., Grollius, H., Lambertz, J., 1982, "The Influence of Blade Number Ratio and Blade Row Spacing on Axial-Flow Compressor Stator Blade Dynamic Load and Stage Sound Pressure Level", Journal of En gineering for Power, Vol. 104, pp 633-641

Giles, M.B., 1988, "Stator/Rotor Interaction in a Transonic Turbine", A.I.A.A. Paper 88-3093

Hodson, H.P., 1984, "Measurements of Wake-Generated Unsteadiness in the Rotor Passages of Axial-Flow Turbines", A.S.M.E. Paper 84-GT-189

Korakianitis, T., 1991, "On the Propagation of.Viscous Wake and Potential Flow in Axial-Turbine Cascades", A.S.M.E. Paper 91-GT-373

Krain, H., 1981, "A Study on Centrifugal Impeller and Diffuser Flow", Journal of Engineering for Power, Vol. 103, pp 688-697

Rai, M.M., 1987, "Unsteady Three Dimensional Navier-Stokes Simulations of Turbine Rotor-Stator Interaction", A.I.A.A. Paper 87-2058

Ubadi, M., Zunino, P., Barigozzi, G., Cattanei, A., 1994, "An Experimental Investigation of Stator Induced Unsteadiness on Centrifugal Impeller Outflow", A.S.M.E. Paper 94-GT-105

Yu W.S., Lakshminarayana, 1995, "Numerical Simulation of the Effects of Rotor-Stator Spacing and Wake/Blade Count Ratio on Turbomachinery Unsteady Flows", Journal of Fluid Engineering, Vol. 117, pp 639-646

STEADY AND UNSTEADY FLOW PATTERN BETWEEN STAY AND GUIDE VANES IN A PUMP-TURBINE

Gabriel Dan CIOCAN
LEGI - ENSHMG - INPG
Laboratoire des Ecoulements Géophysique et Industriels
B.P. 53 - 38041 Grenoble Cèdex 9, FRANCE

Jean Louis KUENY
LEGI - ENSHMG - INPG
Laboratoire des Ecoulements Géophysique et Industriels
B.P. 53 - 38041 Grenoble Cèdex 9, FRANCE

André Luiz Amarante MESQUITA
Department of Mechanical Engineering
Federal University of Pará
66075-900 Belém, PA, BRAZIL

1. Abstract

Due to a collaboration INPG, EDF and NEYRPIC an experimental database concerning the flow in stay and guide vanes, on an industrial test pattern of a turbine - pump has been achieved. The structure of the flow, be it steady or unsteady, has also been determined. An analysis of these flows as well as their evolution according to the functioning point have been determined. These data have been used as boundary layers and validation data for a steady numerical calculation and are available for the unsteady calculations.

2. Introduction

The turbines and pumps conception still contains an important empirical feature. The three-dimensional complex flow understanding which develops in these machines is indispensable to improve their tracing. Similarly, the progress achieved in the numerical techniques also allow to envisage the flow calculations in these configurations, but it is necessary to have the experimental data to provide their boundary layers validation.

Despite the large diversity of the turbines and pumps and the interest presented by the flow description for the improvement performances, the measurements in the fixed parts are badly represented in the speciality literature.

E. Cabrera et al. (eds.), Hydraulic Machinery and Cavitation, 381–390.

A first experimental research part refers to the multi-hole pressure probes performed in the spiral casing. These measurements have first shown the complex character and 3D of the flow, even if they do not contain the information on the unsteady part of the flow. Thus Shuliag and al [8] has realised the measures by the five holes pressure probe in a turbine - pump in pump regime. The measurements have been carried out at the runner outlet and in the spiral casing. To facilitate the access, the stay and guide vanes have been removed, which causes the elimination of the wake effect, therefore the speed profiles are uniform following the angular position. The secondary flow in the spiral casing has been set forth in the three explored sections.

Even if they are not frequent in the literature, the detailed measurements by the LDV technique have recently appeared. Elholm and al [6] have studied in a non-symetrique spiral casing with rectangular section and constant thickness. A vortex was noticed inside the volute and its evolution with the flow was followed up. The measurements in the modified geometry (with rectangular sections) to facilitate the access of measurements, are also performed by Deojeda and al [5] in pump configuration. Others performed analysis comprise only the flow in the runner and at the runner outlet.

Part of losses is due to unsteady wake of runner and of stay and guide vanes. The analysis of the interface runner-stator was performed as completely manner, very recently, by Casey and al [3]. By LDV and PTV measurements they succeed to analyse the flow in exit of runner and its interaction with the director system of a centrifugal pump. In a similar approach to the following one here, but in a centrifugal pump configuration, they determines the wake of the runner and the rate of turbulence for several operating points.

For measurements performed in a complete classic configuration of a turbine-pump a collaboration INPG, E.D.F. and NEYRPIC has be performed. The development of a measure system by LDV associated to the pressure measurements (see Mesquita and al [1]) has allowed the performance of a comprehensive study program of the tridimensional flows pattern in the fixed parts of a turbine - pump industrial pattern.

An experimental database on the flow in a turbine - pump tests pattern has been thus constituted. The measurements in the spiral casing, in turbine regime - see Mesquita and al [2], and in pump regime - see Ciocan and al [4], have been performed. Complementary to these references, this paper refers to the flow between stay and guide vanes, in pump regime.

The results of these measurements have served as boundary layers and validation data for a Navier - Stock 3D calculation coupled with a turbulence k - ε model, in permanent regime in the same spiral casing. This calculation has given very good results which were presented in Gomes and al [10].

The data shown in this paper will serve as boundary layers conditions for the same calculation carried out in unsteady regime and as the validation data for a coupled calculation runner - stator.

3. Experimental facility and technical of measurements

For the carrying out of the measurements, an industrial test pattern Francis Nq75 turbine - pump has been specially instrument for to allow the measurements performed by L.D.V. - see Mesquita and al [1], Ciocan and al [4]. The adaptations allow the carrying out of LDV measurements, by windows, at two levels: between the stay and guide vanes and in the 9 sections of the spiral casing. The interfaces for circumferentieles static pressure measurements and for pressure probes have been also provided. The place of these interfaces is shown in the figure 1.

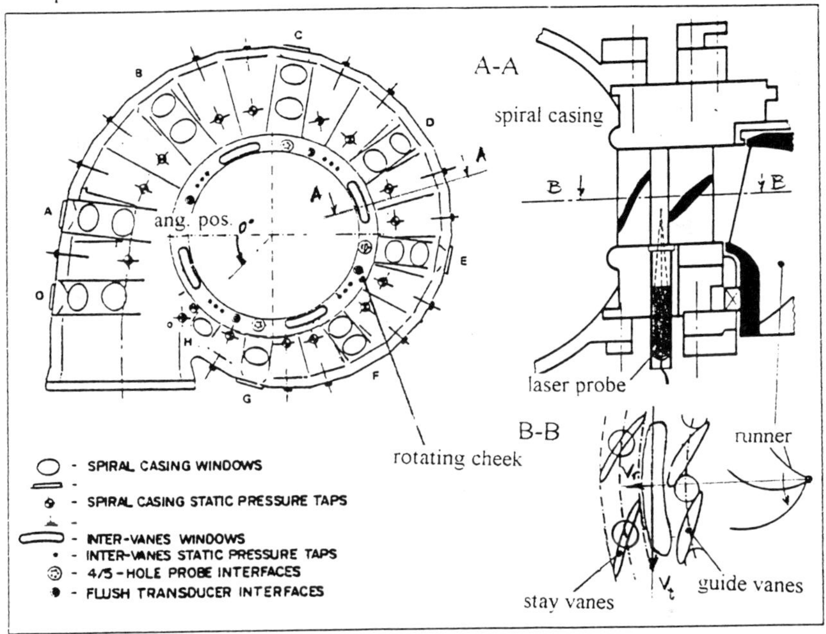

Figure 1 The spiral casing geometry and location of the measurements

For the access of the measurement between 23 stay vanes and 24 guide vanes, 4 windows in optical glass (BK7), arranged to 90°, placed on a mobile cheek have been provided. The rotation possibility of the cheek allows to realise measurements in a cylindrical section between the stay and the guide vanes. An interface for the displacement of the carriage which sustains the laser probe has been provided on the cheek.

The carriage for the displacement of the laser probe has been realised solidary with the pump body to decrease the vibrations effects of the spatial stability from the measurement point. The precision of the displacement is of 0.01 mm. The reference position for the sinking of the measurement volume has been obtained by the crossing of beams on the windows.

The LDV system used is a DANTEC system with two components, in the back scattered light and transmission by optical fibre, with a laser 5 W argon - ion source. The optical characteristic of the system are: laser wave lengths 488 and 514.5 nm, probe diameter 14 mm with a 30 mm beam expander, focal length 100 mm, fringe spacing 3.86 μm, volume of measurements 0.15 mm of x,y, and 2.4 mm of z. The signal processing for the Doppler frequency was accomplished by a spectrum analysis.

A management electronic of the impetus was performed to record the impulse delivered by a coder with photoelectric cells. The coder delivers 3 kind of signals: a reference impetus 0 per one tour and two others, dephased of 90°, 2500 time by tour. The 0 serves for the time basis reference, the second serves for to determine the time for each measurement and the last serves to determine the rotation direction. These signals are synchronised, sent to B.S.A., and used for the synchrony analysis. The clock signal 0 serves also of the runner angular position as compared to the fixed parts of the pattern, by means of marks with the help of a precision stroboscope.

The uncertainties of the laser measurements were estimated according to the method of Mofat and al [7] to 2%.

The total pressure measurements were performed with a miniature 4 holes probe (dimension 3 mm x2.4 mm). After the calibration, this probe allows the deduction of the three components of the velocity vector. The uncertainties for the probe are +/- 0.75 deg for the angles, - /+ 0.85% for the total pressure and +/- 2.52% for the mean velocity. All pressures (which pressure probe and from pressure tapes) have been measured with a Desgranges and Huot electronic balance with a precision of +/- 1mb.

4. Results of measurements

The further shown measurements represent part of the data base constituted on this pattern linked to flow between the stay and guide vanes, in pump regime. The flows analysis in the spiral casing volute in the turbine regime has been performed in Mesquita and al [2] and in pump regime by Ciocan and al [4].

The measurements have been performed for five differed flow rates which correspond to two guide vanes openings. For the nominal opening of 23 mm, 0.8Qn, 0.9Qn, Qn, 1.2Qn and the point of maximal efficiency corresponding to the opening 20 mm has been performed.

4.1. STATIC PRESSURE

The results of these measurements are shown in figure 2.

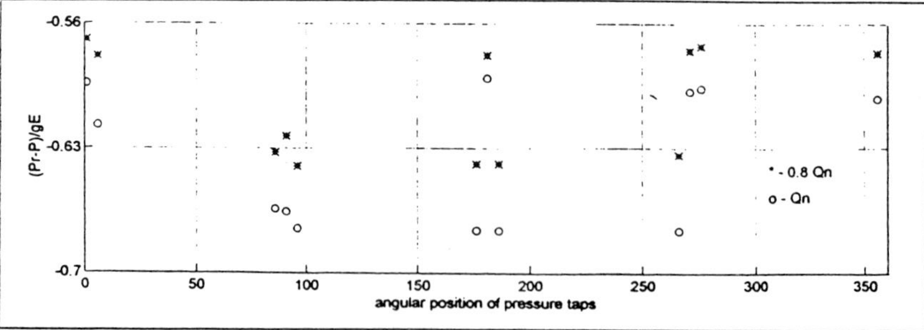

Figure 2. Circumferential static pressure evolution

One can notice that over a period 15°, corresponding to a canal between the guide vanes, the pressure evolution is important and represents the guide vanes wake.

4.2. TOTAL PRESSURE

The pressure probe has shown, according to figure 3, a vertical distribution of the bidimensional and non - uniform velocity. Indeed the component as per the height (z) is negligible as against the two others - from 1% to 3% of the total velocities. The static pressure profiles are uniforms as per the height but a variation is noticed according to the azimuth position and the operating point - see figure 3.

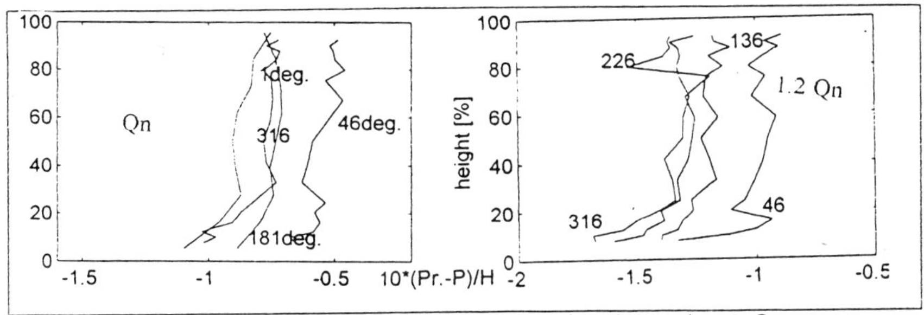

Figure 3. Total pressure evolution for Qn and 1.2 Qn

4.3. MEAN VELOCITY

Two velocity components, for each measurement point, were measured and the corresponding average and root-mean-square were calculated. The two components: radial and circumferential are considered according to the figure 1.

The means have been performed on 7000 acquisition points. One can notice the non - uniformity vertical distribution of the velocity field which was noticed previously by total pressure probe. Following an interval of 15° one can notice a wake. zone (the velocity decrease), as well as maximum velocities in the proximity of the wall zones - see figure 4.

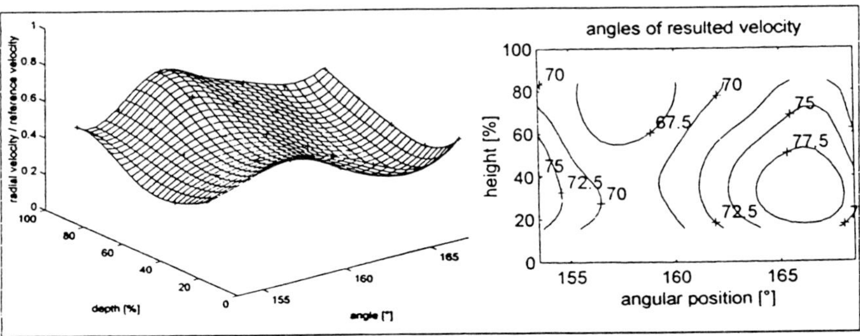

Figure 4. Velocity and angles of resulted velocity distribution in a canal between guide vanes

Moreover we find a periodic distribution as per the azimuth position, but with the different flow rates in guide vanes canal, for a same functioning point. The non - uniformity of the velocity field is emphasised by its removal from the nominal point. The distribution of the resulted velocity angle is the same with that found by total pressure probes.

The same vertical distribution can be noticed at the exit of stay vanes - see figure 5.

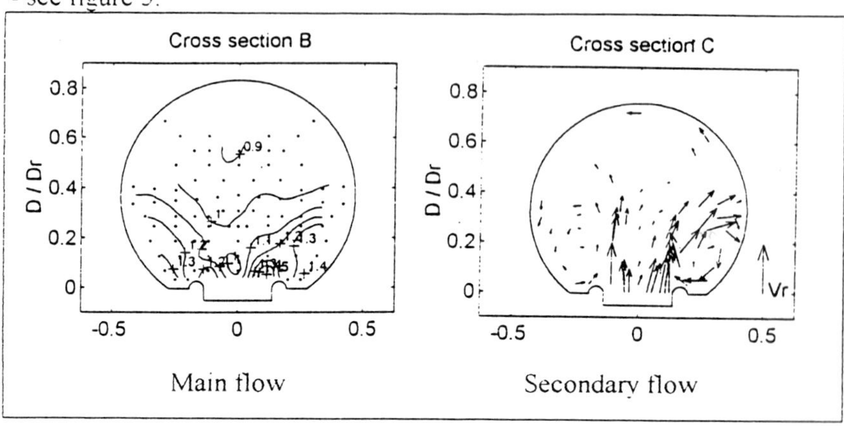

Figure 5. Example of the velocity distribution in the spiral casing

4.4. PERIODIC UNSTEADY VELOCITY SYNCHRONISED WITH RUNNER ROTATION

In the periodic turbulent flows, the statistical averages depend, besides the space variables, on the angle of phase Φ. The idea of this approach is therefore to separate the determining part, linked to the runner wake - the periodic fluctuation of the velocity from the totality of the velocity fluctuations. This technique, called synchronisation means, allows to represent the velocities according to the runner phase. Therefore we shall have:

$$v(i) = \bar{v} + v^{\sim} + v'$$

where: v(i) is the instantaneous velocity of a particle; $\bar{v}$, the mean velocity ($\Sigma v(i)/N$); $v^{\sim}$, synchrony unsteady velocity; v', turbulent unsteady velocity with $< v'^2 > = 0$. To calculate $v^{\sim}$ we take into account a time interval (window), small enough against the periodic flow variations - see figure 6.

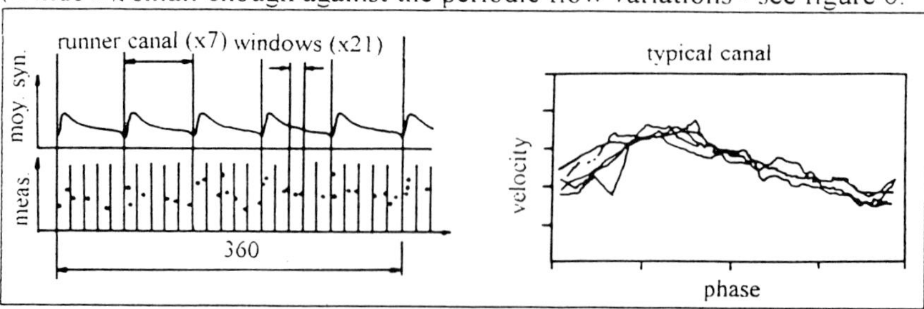

Figure 6. Synchrony unsteady velocity calculation

For a turbomachine, the flow being periodic, this window becomes an interval of angle corresponding to an angular variation of the runner position (therefore a cannel becomes a runner canal). Consequence of several tests, 21 windows were chosen per canal, which correspond to a runner rotation angle of 2.44°. On the other hand the points number of each window has to be sufficient to have a correct average. Thus, the tests have been performed and the best compromise between the points acquired number and the acquisition time has been ~300 points by window. The good flow symmetry between runner canals has allowed to use a representative canal, called typical canal - see figure 6.

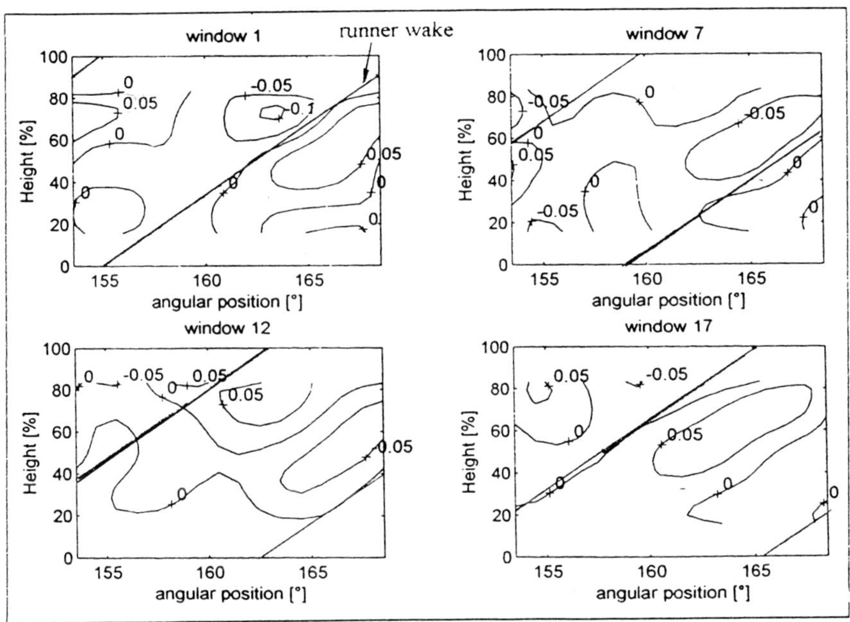

Figure 7 v~ evolution function of runner phase

4.5. TURBULENCE INTENSITY

The calculation of the turbulence intensity was performed by carrying out the isotropy hypothesis according to the 3 directions. Since the size order of the root-mean-square for the two measured is the same, the root-mean-square for the 3rd component (which was not measured by LDV) was considered the arithmetic average of the other two. This calculation can be carried out by and without overlooking v~ due to the runner.

The root-mean-square of the mean velocity can by calculate as:

$$\sigma^2 = [\Sigma(v(i) - \bar{v})^2]/(N-1), \text{ overlooking } v^{\sim}$$

or

$$\sigma^2 = [\Sigma(v(i) - \bar{v} - v^{\sim})^2]/(N-1)$$

Therefore the turbulence intensity was considered:

$$k = 3(\sigma_r^2 + \sigma_t^2)/4$$

Where: σ is root-mean-square; N - number of measurements for each point; σ_r - root-mean-square of radial velocity; σ_t - root-mean-square of circumferential velocity; k - turbulence intensity. There is between the two calculations a difference of up to 10% for the same space distribution.

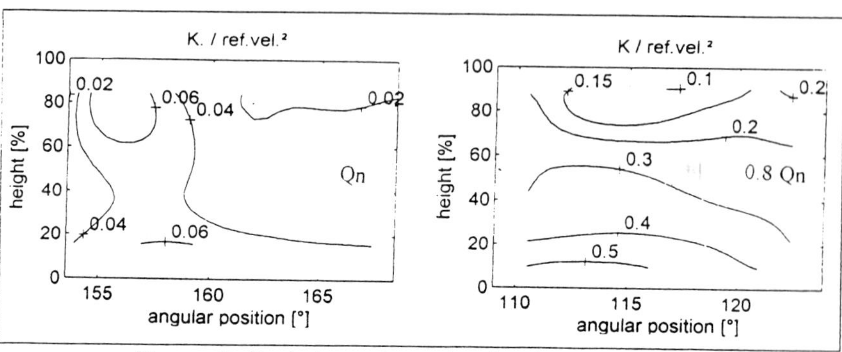

Figure 8. Turbulence intensity for Qn and 0.8 Qn

Mention should be made of the fact that any supply difference between the runner canal adds a false information to the calculation of the turbulence intensity.

5. Discussion

The distribution of the velocity and pressure field allow to analyse the flow at this level. Firstly the velocity field is marked by the guide vanes wake, which gives it a periodical evolution following the angular position. It finds the same characteristic in the static pressure evolution. To this a flow rate evolution between guide vanes canal is added which can be a pressure gradient effect induced at this level by the spiral casing geometry. This can be noticed in the static pressure profiles evolution already found by Sideris and al [9] which also show the influence on the spiral casing form on the velocity at the runner outlet profile, imposed by the pressure distribution in the spiral casing. The effect of the guide vanes wakes is also confirmed by the distribution of the angle of the resulting velocity.

The distribution of the velocity profile following the height, with the on - velocity up and down is in accordance with the distribution the runner outlet velocity profile. This non - uniformity increases with the removal from the nominal point.

This velocity distribution, which is transmitted to the entry of the volute, has a very large importance for the secondary flow formation in the volute. Thus in the tongue zone the velocity excess of the upper part of the section crushes the inferior component and creates a vortex which will be transported by the flow. Moreover the secondary flow in all sections of the spiral casing has as origin this velocity distribution. For details see Ciocan and al [4].

The results of the numerical simulation which uses a uniform velocity profile as against this distribution type can give a very different distribution of flow rate as per the azimuth position. In figure 9 the mean velocity evolution following the section for a calculation performed under the two conditions (with the good boundary layers this calculation gives the very good results - see Soares and al [10]) is shown.

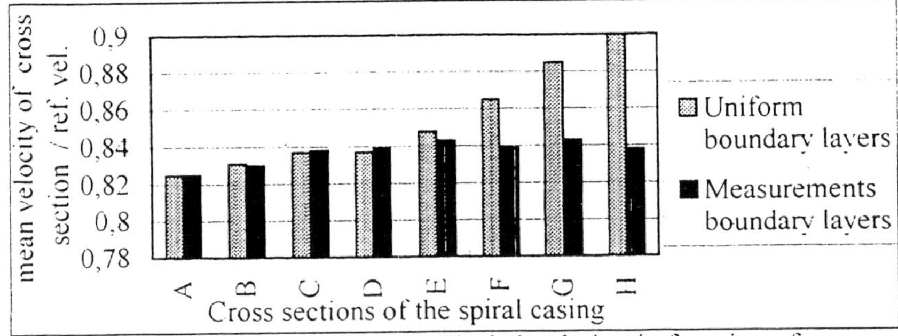

Figure 9 Results of the numerical simulation in function of boundary conditions

From the point of view of the synchrony component of the runner, one notices the propagation of the runner wake (the inclination of the synchrony wake is the same as that the angle of the runner trailing edge) up to this level (1.33 outlet runner diameter) - see figure 7. This is not the case of the configuration whiteout stay and guide vanes where the wake is crushed by the main flow in the spiral casing. This wake is due to the difference of the velocity between the suction side and the pressure side of the blade at the runner outlet. The turbulence level is raised enough and has the same intensity for the two velocity components. The turbulence increases strongly with the removal from the nominal point (which is in agreement with Casey and al [3]) - see figure 8. The obtained error by overlooking the determining part of the velocity synchrony component in the turbulence calculation is of 10%.

6. Conclusion

Despite the difficulty of the access, the complexity of the pump regime and the large complexity of adjustment for the measurement equipment to be implemented (very raised costs and implementation duration of the several months), the detailed data bank was achieved on the flow in the fixed parts of the turbine-pump tests pattern. That has allowed a precision analysis of the steady and unsteady flow, and the turbulence intensity of the flow between stay and guide vanes. That allows also to explain the secondary flow sources in the spiral casing.

Even the comparisons are difficult to carry out and they risk to be inconclusive taking into account the determining influence of geometrical characteristics of each machine (runner and spiral casing geometry), it is considered that there are representative results for global comparisons of qualitative order or of physical explanations. The 3D Navier-Stokes unsteady calculations with this data bank as validation conditions, will allow a detailed loss analysis.

8. Acknowledgements

We acknowledge with gratitude EDF (Electricité de France), NEYRPIC (GEC ALSTHOM) and Région Rhône-Alpes for their collaboration and financial support of this project.

9. References

1. Amarante Mesquita, A. L., (1993) Experimental Techniques for the Flow Characterisation in Hydraulic Turbomachines Models, *Proceedings of the 2nd Meeting of the Latin-American Division of the International Association of the Hydraulic Research - IAHR*, Vol. 1, pp. 193-209, Ilha Solteira, Brazil.
2. Amarante Mesquita, A. L., Philibert, R., Very, A., and Kueny, J. L., (1994) Laser Velocimetry Measurements in a Pump Turbine Spiral Casing, *Proceeding of the 19th Symposium of the International Association of the Hydraulic Research - IAHR*, pp. 395-404, São Paulo, Brazil.
3. Casey, M.V., Eisele, K., Zhang, Z., Gulich, J., Schachenmann, A., (1995) Flow analysis in a pump diffuser: LDA and PTV measurements of the unsteady flow, *Laser Anemometry, FED-Vol. 229, ASME, pp. 89-99*
4. Ciocan, G.D., Amarante Mesquita, A. L., and Kueny, J. L., (1996) Three-Dimensional Flow Patterns in a Pump Volute at Nominal and Off-Design Flow Conditions by LDA Measurements, *Proceedings of the 6th International Symposium on Transport Phenomena and Dynamics in Rotating Machinery - ISROMAC - 6*, Vol. 2, pp. 229-237, Honolulu, USA.
5. Deojeda, W., Flack, R.D., Miner, SM., (1996) Laser Velocimetry Measurements in a Double Volute Centrifugal Pump, *Proceedings of the 6th International Symposium on Transport Phenomena and Dynamics in Rotating Machinery - ISROMAC - 6*, Vol. 2, pp. 131, Honolulu, USA.
6. Elholm, T., Ayder, E., and Van der Braembussche, R., (1992) Experimental Study of the Swirling Flow in the Internal Volute of a Centrifugal Pump, *Journal of Turbomachinery*, Vol. 114, pp. 366-372.
7. Mofat, R. J., (1985) Using Uncertainty Analysis in Planning of an Experiment, *Journal of Fluid Engineering*, Vol. 107, pp. 173-178.
8. Shuliang, C., Qian, H., and Ruchang, I., (1986) Experimental Study on Flow Characteristics and Hydraulic Losses in a Pump Turbine Volute in Pump Operation, *Proceeding of the 13th Symposium of the International Association of the Hydraulic Research - IAHR*, paper N°. 46, Montréal, Canada
9. Sideris, M. T., and Van der Braembussche, R., (1987) Influence of a Circumferential Exit Pressure Distortion on the Flow in a Impeller and Diffuser, *Journal of Turbomachinery*, Vol. 109, pp. 48-54.
10. Soraes Gomes, F., Amarante Mesquita, A. L., Ciocan, G., and Kueny, J. L., (1994) Numerical and Experimental Analysis of the Flow in a Pump-Turbine Spiral Casing in Pump Operation, *Proceedings of the 17th Symposium of the International Association of the Hydraulic Research - IAHR*, pp. 249-258, Beijing, China

SELF-SUSTAINED OSCILLATION OF GAS-LIQUID FLOW IN A CENTRIFUGAL PUMP WITH SEMI-OPEN IMPELLER

J. KUROKAWA, J. MATSUI, H. TAKADA, and T. HIRAYAMA
Yokohama National University
156, Tokiwadai, Hodogaya-ku, Yokohama, 240 Japan

1. Introduction

When a centrifugal pump is operated at very low discharge with low suction head, the air dissolving in water merges to become bubble and stays inside an impeller channel. The staying bubbles accumulate to form a large bubble zone and decrease pumping head. This phenomena is called "air locking", and often arises when air-rich water is used in a pumping system, such as a water supply system of a building.

The air locking phenomena was also raised when air is supplied forcedly into a suction pipe instead of using air-rich water and low suction head. However, a sever self-sustained oscillation was encountered under the same operating conditions, when a semi-open type impeller was used. In this case the pump had stable head-capacity performance and was operated under non-cavitation condition.

There are many types of self-sustained oscillations caused by gas-liquid flow in pumping system, such as low frequency oscillation accompanying cavitation in [1], [2], the oscillation due to time lag between air inlet and impeller inlet as in [3], and so on. These phenomena is accompanied with severe oscillation of a suction pipe. However, the present oscillation is raised only in a semi-open impeller with no suction pipe oscillation.

In order to elucidate the mechanism of the present oscillation, the instantaneous measurements of blade-to-blade pressure were performed and the parameters influencing the oscillation were determined together with the theoretical considerations.

2. Nomenclature

p	: static pressure	ρ	: density of water
r, r_2	: radius and impeller radius	U	: impeller tip speed

Q, q : volume flowrates of water and air, respectively (under suction pressure)
x : distance along vane center line from leading edge
ϕ : discharge coefficient, defined as $Q / (A_2U)$ (A_2: impeller outlet area)
ψ : head coefficient, defined as $H / (U^2/2g)$ (H: pump head)
ψ_s : coefficient of static pressure, defined as $(p - p_{s1}) / (\rho U^2/2)$
τ : power coefficient, defined as $P_w / (\rho A_2 U^3/2)$ (P_w: shaft power)
subscripts s1, s2 : 900mm and 16mm upstream of impeller inlet, respectively
d : discharge (400mm downstream of impeller outlet)

E. Cabrera et al. (eds.), Hydraulic Machinery and Cavitation, 391–400.

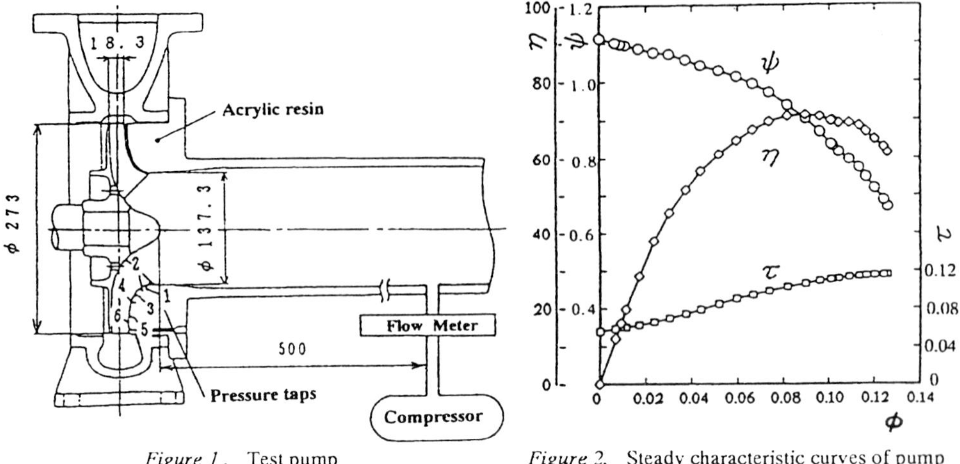

Figure 1. Test pump

Figure 2. Steady characteristic curves of pump

3. Experimental Apparatus and Dimensions

The pump tested is shown in Fig. 1. The impeller is a semi-open type, of which specific speed is 230[m, m^3/min, rpm], blade outlet angle β_2=25°, radius r_2=137mm, blade number 5, and tip clearance 1mm. The suction cover and the discharge pipe are made of acrylic resin so that the bubble behavior can be observed. On the suction cover six pressure holes, No. 1~6 shown in Fig 1, are drilled and semi-conductor type pressure transducers (natural frequency of 10 kHz) are mounted for instantaneous measurements.

The test apparatus consists of a suction pipe of 2m length, a pump, a discharge pipe of 10m length and a reservoir. The compressed air is supplied at 500mm upstream of the impeller inlet, and is separated in the reservoir. A swirl stop is mounted at 600mm upstream of the impeller inlet. The pump rotational speed was varied from 500 rpm to 1800 rpm, corresponding Reynolds number ranging from 9.8×10^5 to 3.5×10^6.

4. Experimental Results and Discussions

4.1 PUMP PERFORMANCE

The pump tested has stable characteristic curves as shown in Fig. 2 with a negative gradient of head-capacity curve over the whole operating range.

When the pump is operated at very low discharge and air is supplied into a suction pipe, the pumping head gradually decreases and begins to oscillate, as shown in Fig. 3. Not only pump head ψ but also discharge coefficient φ, discharge pressure

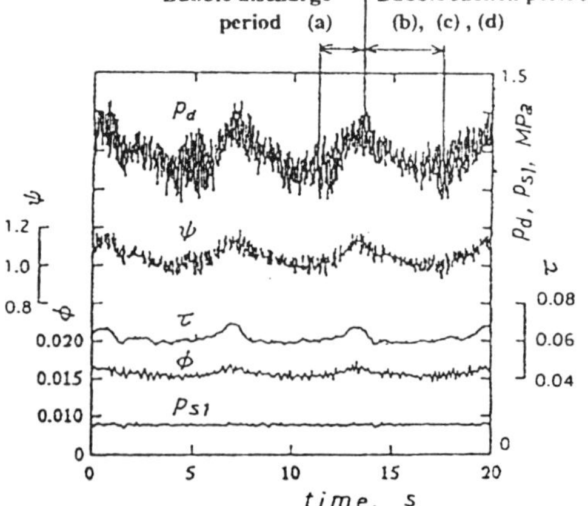

Figure 3. Oscillation of pumping head and others (Q/q = 0.014)

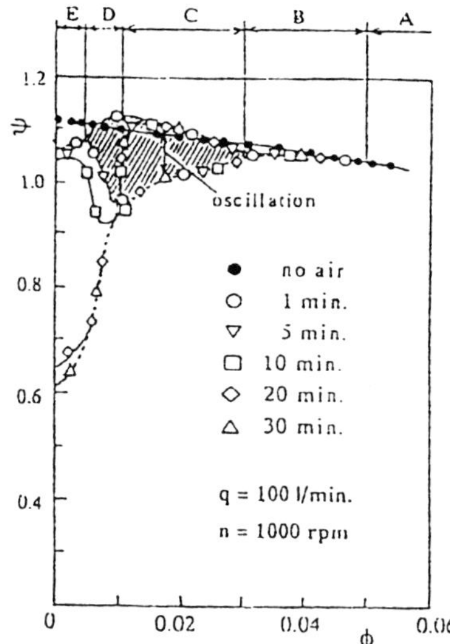

Figure 4. Change in head curve with time

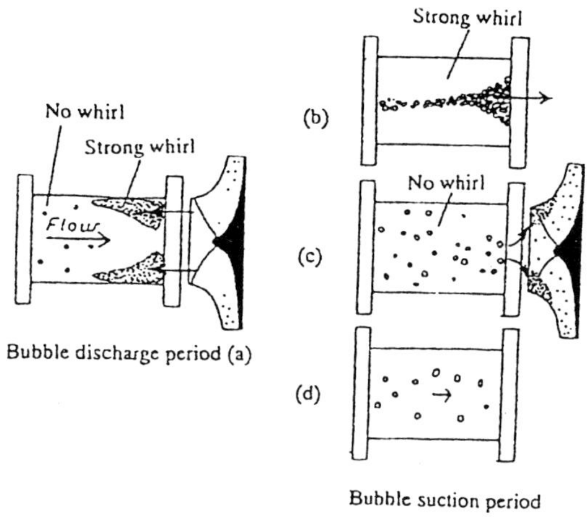

No whirl
Strong whirl
Flow
Bubble discharge period (a)
Strong whirl
(b)
No whirl
(c)
(d)
Bubble suction period

Figure 5. Sketch of bubble behavior in a suction pipe while oscillating (Pattern C)

p_d and shaft power coefficient τ oscillate with the period of 6.7s and in the same phase. However, the pressure p_{s1} at 900mm upstream of the impeller inlet does not oscillate.

Time dependence of head curve is shown in Fig. 4. Both the maximum and the minimum values are plotted when oscillation occurs. After 30 minutes from the start of air supply , the oscillation is still maintained only in the discharge range C shown in Fig. 4, and in the other range the head curve comes to be the dotted line showing a steady operation. From Fig. 4 it is recognized that there exists following five patterns ;

A: Pump operates steadily without decrease in pumping head, though a bubble zone exists in an impeller channel and many small bubbles discharge at the pump outlet.

B; Flow pattern is almost same as A, but the blade channel is full of small bubbles, the pumping head slightly decreases and the head-capacity curve becomes horizontal.

C; Sever oscillation occurs with periodical bubble discharge from the impeller inlet to the upstream suction pipe. The oscillation becomes larger with a decrease in pump discharge. Pressure fluctuation of this pattern is shown in Fig. 3 and the sketch of the flow behavior is illustrated in Fig. 5.

The bubbles accumulating in the impeller are suddenly discharged to the upstream along the casing wall with strong whirl, and liquid single flow enters into the impeller in the pipe center (Fig. 5(a)). Many small bubbles spreading in the pipe then gather to the pipe center due to a strong whirl, grow to large bubbles, form a large group of bubbles (Fig. 5(b)) and then come into the impeller. As the inlet whirl then gradually disappears, the residual bubbles spread to a whole pipe and come into the impeller together with those from the upstream (Fig. 5(c), (d)). The flow pattern again returns to Fig. 5(a).

Throughout one cycle of oscillation, small bubbles flows out steadily at the pump outlet, though the discharge pressure p_d fluctuates largely as shown in Fig. 3.

D; The impeller inlet flow is separated to liquid flow and gas flow. Soon after the start of air supply the oscillation occurs between two curves shown in Fig. 4, but after

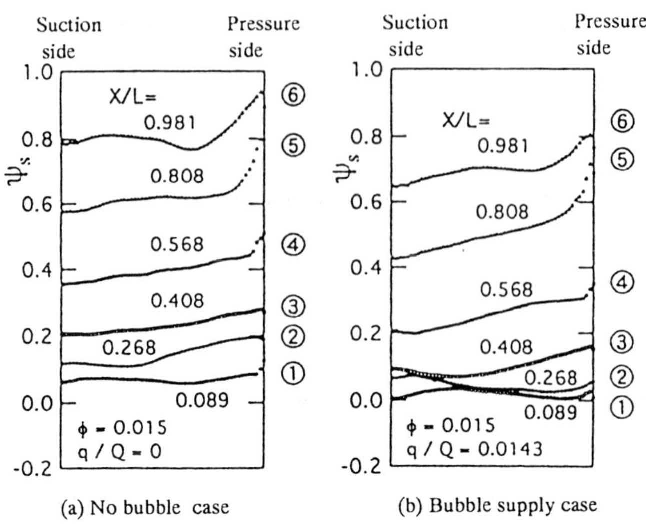

(a) No bubble case (b) Bubble supply case

Figure 6. Blade-to-blade pressure distribution (ϕ =0.015, 1000rpm)

5 minutes it ceases and the head curve comes to the dotted line. The amplitude of oscillation decreases with a decrease in pump discharge.

E; Bubbles accumulate in the impeller inlet region without oscillation, and spread to occupy the upper part of suction pipe, which decreases the pumping head gradually.

4. 2 INSTANTANEOUS BLADE-TO-BLADE PRESSURE DISTRIBUTIONS

In the case of sever oscillation (pattern C), the shaft power also oscillates, which implies that the blade loading fluctuates. The instantaneous pressure of one impeller channel is then measured and was averaged over 50 rotations at each measuring point. The averaged values of 50-rotations didn't show significant difference from those of 200-rotations. The blade-to-blade pressure distributions thus obtained are shown in Fig. 6, in which the liquid single flow case in Fig. 6(a) is compared with the gas-liquid flow case in Fig. 6(b).

Comparison of Figs. 6(a) and 6(b) reveals that air supply decreases the pressure in the whole impeller channel, when the oscillation occurs. A remarkable difference caused by air supply is seen in the measuring points ①, ②and③. This shows that bubbles accumulate around the impeller inlet region and decreases the blade loading. However, the pressure at ⑤ is seen to recover largely, which shows that the blade loading near the outlet becomes larger than that in the no air case.

It is then concluded that the head drop due to air supply is caused by the loss of blade loading around the impeller inlet due to the accumulation of bubbles.

4. 2. 1 *Bubble discharge period and bubble suction period*

The above-described results are obtained by the 50-rotations-averaged values, corresponding to 3 cycles of oscillation, but the instantaneous blade-to-blade pressure distribution must be different from the averaged one especially in the bubble accumulating period. The data averaging period is thus divided to the bubble discharge period and the bubble

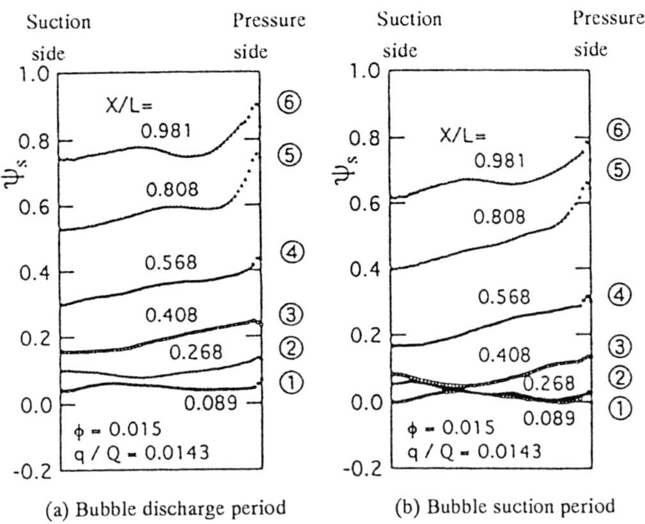

Figure 7. Comparison of instantaneous blade-to-blade pressure distribution (ϕ =0.015, 1000rpm)

suction period by manually switching the data acquisition with observing the phenomena, and the results are compared in Figs. 7(a) and (b).

Comparison of Figs. 7 with Fig. 6 reveals that the instantaneous blade-to-blade pressure profile at bubble discharge period (Fig. 7(a)) is almost same as that of no air case (Fig. 6(a)) and that at bubble suction period (Fig. 7(b)) is almost same as that of the one-cycle averaged (Fig. 6(b)). That is to say, the bubble suction period occupies large part of one cycle of oscillation, and after that bubbles are rapidly discharged to the upstream. In the latter case the impeller is filled only with liquid and attains the maximum pumping head which is equal to that of no air case.

4. 2. 2 *Influence of Air Flowrate on Blade-to-Blade Pressure Distribution*

When the air flowrate ratio is increased, the instantaneous pressure distribution between blades changes as shown in Fig. 8, of which lower figures show the net pressure rise along a blade obtained by subtracting the pressure rise due to centrifugal force.

The upper figures reveal that the pressure decrease is considerable especially around the leading edge, and that the pressures at ①, ② and ③ decreases in order with an increase in air flowrate ratio, while the pressure at ⑥ changes little. This shows that the bubble accumulation zone gradually spreads to the downstream from the leading edge.

In the lower figures in Fig. 8, the pressure difference between pressure side and suction side expresses the blade loading and decreases rapidly in the upstream half while increases in the downstream half with an increase in air flowrate ratio.

Comparison of Figs. 8(a) with 8(b)~(d) also reveals that the air supply decreases the pressure especially in the blade pressure side. Visual observation revealed that bubbles accumulate much more in the pressure side. These two results reveals that the bubbles accumulating in the pressure side near the leading edge decreases the blade loading and makes the main flow deviate to the blade suction side. The deviation of main flow should decrease the pressure recovery in the suction side and the pressure at the downstream

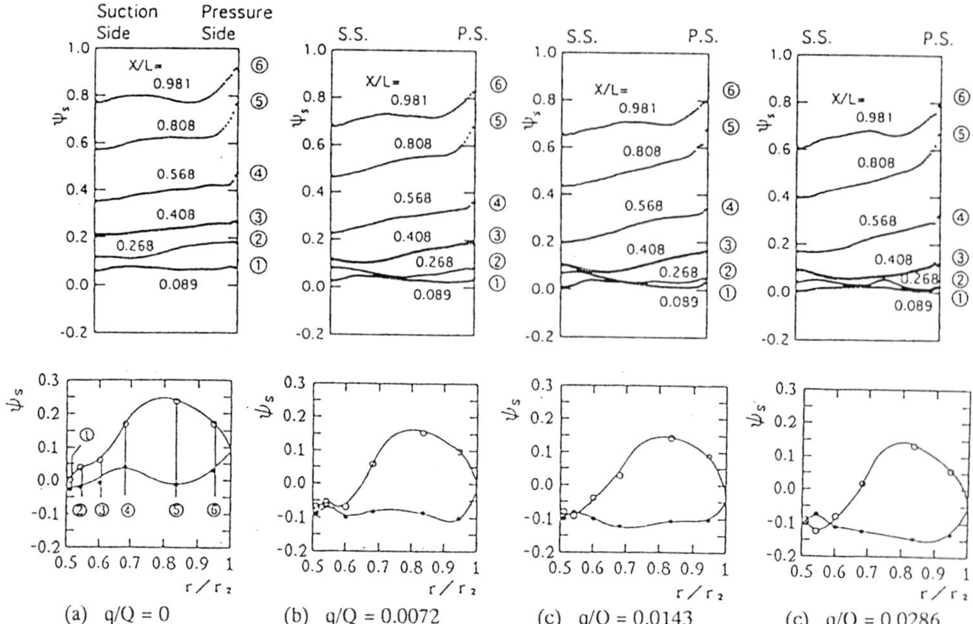

Figure 8. Change in instantaneous blade-to-blade pressure distribution by the change of air flowrate ratio (ϕ =0.015, 1000rpm)

of bubble accumulation zone should recover rapidly due to sudden expansion of main flow, as is shown in Fig. 7. Thus the bubbles accumulate mainly in the blade pressure side at very low discharge, though they accumulate in the suction side at BEP as in [4].

Though sever oscillation is induced at impeller inlet region, the pressure p_{s1} at 900 mm upstream of the impeller is kept constant as shown in Fig. 3. This suggests that the flow oscillation is limited to the pump inlet region and influence little to the upstream suction pipe. To make this point clearer, the variation of pressure p_{s2} at 50mm upstream of the impeller inlet is shown in Fig. 9. The pressure p_{s2} is seen to oscillate in the same phase with the maximum value equal to that of no-air case, the minimum value equal to the upstream pressure p_{s1}. This pressure fluctuation is due to a strong periodical whirl of the recirculation flow which exists near the suction pipe wall of pump inlet.

4. 3 CHARACTERISTICS OF PRESENT OSCILLATION

The period and the amplitude of present oscillation depend largely on the air flowrate and the pump rotational speed. In order to reveal the characteristics of the present oscillation, the variations of amplitude and frequency are shown in Figs. 10 (a) and (b), when the pump rotational speed and the flowrate ratio q/Q of air to liquid are varied.

When the air ratio is increased under the same rotational speed, the amplitude ratio $\Delta\psi/\psi$ of pumping head becomes larger and the frequency becomes smaller. In this case the visual observation revealed that the bubble accumulation zone along a blade becomes larger and both the scale and the strength of oscillation become larger with an increase in air flowrate ratio. When the pump rotational speed is increased, both the amplitude and

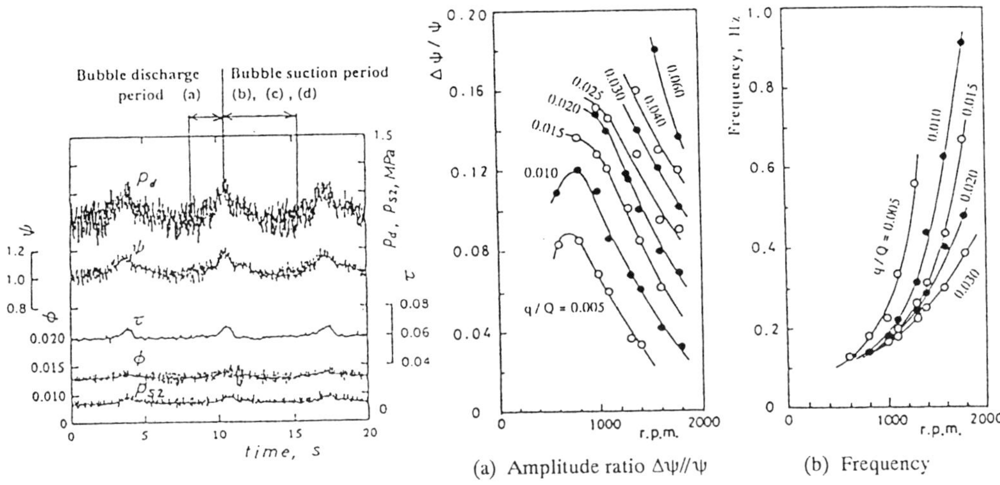

Figure 9. Fluctuation of p_{s2}

(a) Amplitude ratio $\Delta\psi/\psi$ (b) Frequency

Figure 10. Influence of rotational speed and air flowrate ratio

the frequency become smaller resulting in smaller-scaled oscillation.

The present oscillation is one of the unstable phenomenon accompanying gas-liquid flow in a pumping system. The characteristics of the present oscillation is that it occurs in a very low discharge range(10~30% of the designed) in which large recirculation flow should be induced if there were no oscillation, and that it occurs only by supplying more than only 0.5 % of air to liquid. The length of recirculation flow zone in a suction pipe is then considered to influence the phenomena, but the removal of the swirl stop at 600mm upstream in the suction pipe revealed no significant change in the phenomena.

Another remarkable characteristics of the present oscillation is that it occurs only in a semi-open type impeller and doesn't occur in a closed type impeller. The tip clearance was thus considered to influence the phenomena, but the zero clearance test by attaching a rubber plate at the impeller tip revealed no significant change in the phenomena.

5. Theoretical Considerations on the Mechanism of Present Oscillation

In order to elucidate the mechanism and the cause of the present oscillation, the stability analysis based on Gleitzer [5] is applied to the present pumping system.

In the pumping system shown in Fig. 11(a) under non-cavitation condition, it is assumed that the fluctuations of pressure p, liquid flowrate q and gas flowrate q_G (these notations are different from the preceding chapters) be small compared with the mean pressure P and the mean flowrates Q, Q_G of liquid and gas, respectively. Under these assumptions a perturbation method can be introduced to linearize the equation. It is also assumed that the air flowrate at pump outlet is constant and subscripts 1 and 2 are given to the pump suction side and the discharge side, respectively.

Considering that the bubble flow rate is small compared with the liquid flowrate, and that the valve resistance is dominating in the discharge pipe, the linear equations of motion as for the fluctuation terms in the suction pipe, the discharge pipe and the pump are expressed as follows in the non-dimensional form;

$$p_1 = -L\dot{q}_1 \tag{1}$$
$$p_2 = R_V q_2 \tag{2}$$
$$p_2 - p_1 = R_P q_1 - \dot{q}_1 + R_G q_G \tag{3}$$

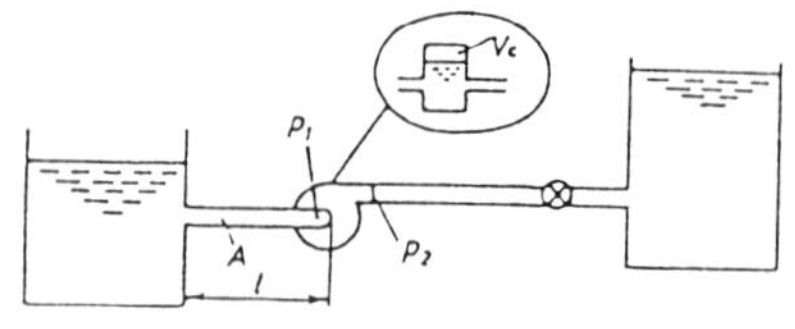

(a) Pump system

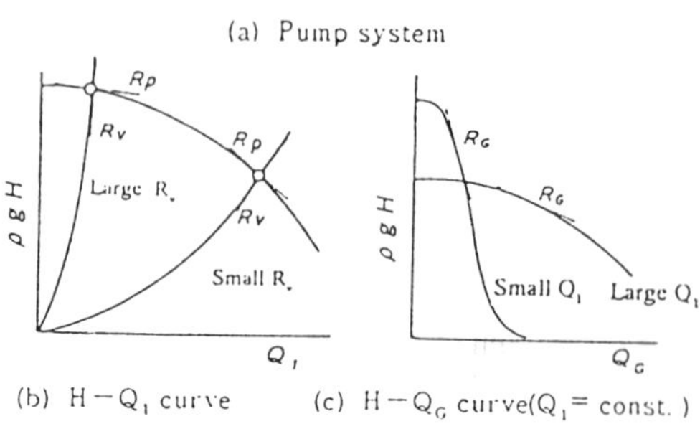

(b) H−Q₁ curve (c) H−Q_G curve(Q₁= const.)

Figure 11. Model and characteristic curves of pumping system and system resistance

Here, the pressure p is normalized by $\rho U^2/2$, the flowrate by AU and the time t by $2L_p/U$, where U, A and L_p are the impeller tip speed, the sectional area of pipe and the equivalent length of pump, respectively. The notation $\dot{q}_1$ denotes the differentiation of q_1 as for time t. L is the length of the inlet recirculation zone (or the length of suction pipe if there is no reverse flow) normalized by L_p. The parameters R_p, R_V and R_G are the gradients of the characteristic curves of the pump system shown in Figs. 11 (b) and (c), and are defined as

$$R_p \equiv d(2gH/U^2)\,/\,d(Q_1/AU),\quad R_V \equiv d(2gH_V/U^2)\,/\,d(Q_1/AU),\quad R_G \equiv d(2gH/U^2)\,/\,d(Q_G/AU)$$

where H and H_V are the pumping head and the system resistance(valve), respectively.

Putting the air volume accumulating in an impeller as V_c and its fluctuation as δV_c normalized by AL_p, the following equations are obtained in the non-dimensional form;

$$\dot{V}_c = q_2 - q_1 + q_G \tag{4}$$
$$\delta V_c = C_P p_1 + M_b q_1 + M_G q_G \tag{5}$$

where $C_p \equiv \partial V_C / \partial p_1$, $M_b \equiv \partial V_C / \partial Q_1$, $M_G \equiv \partial V_C / \partial Q_G$

The motion of a bubble is to be determined from the balance of forces acting on a bubble. Here for the simplicity, the fluctuation of the bubble volume at an impeller inlet is assumed as follows in the non-dimensional form;

$$q_G = \alpha p_1 + \beta q_1 \tag{6}$$

where $\alpha \equiv \partial Q_G / \partial P_1$, $\beta \equiv \partial Q_G / \partial Q_1$

Equations (1) to (6) are all linear with constant coefficients. Hence the solutions are of the form e^{st}. For stability the real part of s must be negative. Substituting expressions of the form e^{st} for the perturbations, the growth rate must satisfy the following equation, if there are to be non-tribial solutions to the resulting system;

$$s^2 + as + b = 0 \tag{7}$$

where $a = -\{\alpha + (M_b + \beta M_G)/L + (1 + 1/L + \alpha R_G)/R_V\} / (C_p + \alpha M_G)$

$b = -\{1 - \beta - (R_p + \beta R_G)/R_V\} / L(C_p + \alpha M_G)$

Generally speaking, the negative valve of b gives static instability, while the negative value of a gives dynamic instability if b is positive. In the present pumping system the parameters in the above-equations are considered to be as follows;

$$C_p < 0,\ M_b < 0,\ M_G > 0,\ R_p < 0,\ R_V > 0,\ R_G < 0,\ \alpha < 0,\ \beta \sim 0$$

Thus the denominators of a and b are negative and b takes a positive value, but the quantity a can be positive or` negative depending on the numerator of a. After all for the dynamic instability must satisfy the following equation.

$$\alpha + M_b/L + (1 + 1/L + \alpha R_G)\ /R_V < 0 \tag{8}$$

The first two terms are negative, and the last term in bracket is positive. The self-sustained oscillation could thus be caused depending on these parameters. If the liquid flowrate is large, the system resistance R_V becomes small, resulting that the system is stable. With an increase in system resistance, the flowrate decreases and the last term becomes very small, which satisfies Eq. (8), resulting in an unstable system. When the flowrate ratio of air to liquid is large, the value αR_G becomes large, resulting in dynamic stability.

From the above-described theoretical considerations, it is concluded that the present self-sustained oscillation could be caused by the unbalance between the system resistance rate R_V and the loss rate of blade loading R_G under thé influenced of the behavior M_b of the bubble accumulating zone.

One of the other remarkable characteristics is that the oscillation occurs only in a semi-open type impeller. The careful observation revealed that the bubbles in an impeller channel become very small-sized, and accumulate mainly near the blade pressure side. This might be caused by the existence of recirculating flow zone induced by flow separation of blade suction side, as shown by the three-dimensional structurein in Fig.13(a). When the bubble zone further grows, the recirculation flow zone might be suppressed and disappear,because the recirculation flow of a semi-open type impeller is very unstable due to the lack of the outer shroud wall.

The visualization of flow behavior using tufts attached on casing wall revealed that the recirculation flow disappears instantaneously and at the same time the bubble accumulating zone rapidly grows to the upstream, and that the casing wall just in front of the impeller inlet is covered with air layer rotating much more slowly than the impeller.

When inlet recirculation flow disappears, both the meridional and the tangential velocities rapidly decreases at an impeller inlet, which results in large angle of attack at blade inlet. A large-sized recirculation flow is thus induced again, and the bubbles are suddenly discharged to the upstream with the recirculation flow. During the period from bubble discharge to bubble gathering to the pipe center, the liquid flowrate increases at impeller inlet, and the pumping head increases due to the recovery of

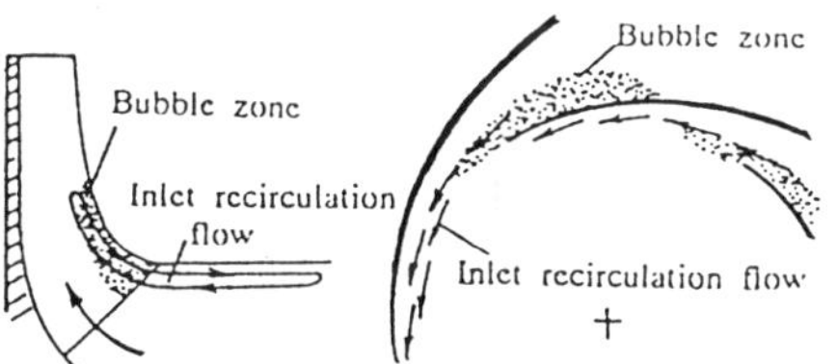

(a) Recirculation flow region and bubble zone

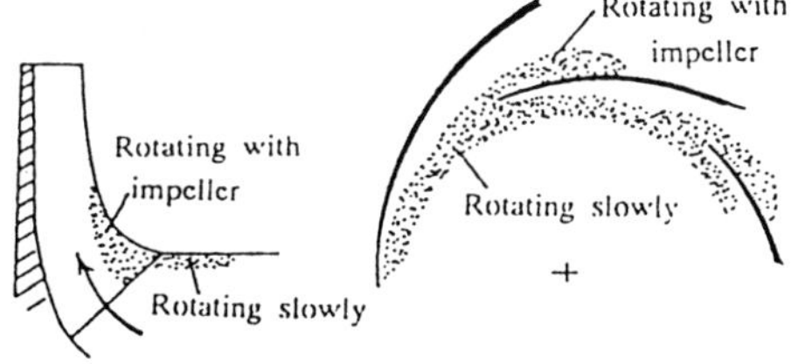

(b) Just before the bubble dicharging

Figure 12. Sketch of bubble behavior and recirculation flow at pump entrance

blade loading, which results in positive slope of head-capacity curve. On the contrary in a closed type impeller the recirculation flow might form a steadily staying bubble zone which coexists stably with a recirculation flow.

From the above-described considerations the disappearance of the recirculation flow due to the growth of bubble zone might be responsible for the present oscillation, and such difference in the bubble behavior between a semi-open type impeller and a closed type could be expressed by the difference in α and M_b in Eq. (8).

When a pump is operated near the shut-off point, the impeller inlet recirculation flow develops to be so strong and large-sized that the growth of bubble zone cannot cause the recirculation flow to disappear. The pumping system thus does not come into a dynamic instability but rapidly come into a static instability, resulting in a large head-drop.

6. Conclusions

A low frequency oscillation caused by a semi-open impeller at low discharge operation in gas-liquid flow is studied. The characteristics of the oscillation are determined experimentally and its mechanism is studied theoretically by use of a linear stability analysis.

The measurements revealed that the head drop is caused by the accumulation of bubbles along a blade pressure side around the leading edge. Theoretical considerations suggest that the bubble accumulation in blade-to-blade channels could be responsible for the present oscillation. The theoretical model provides one possible explanation as for the excitation of the oscillation, that it is caused by dynamic unbalance between the head drop rate due to bubble accumulation and the system resistance rate under the influence of flow interference between the bubble accumulating zone and the recirculation flow.

The reason why the present phenomena occurs only in a semi-open type impeller might be that the recirculation flow at impeller inlet is so unstable that the growth of bubble accumulating zone near the blade leading edge suppresses the recirculation flow and makes it disappear, which rapidly causes a large recirculation flow.

Acknowledgements

The present authors would like to express their sincere gratitude to Dr. K. Yamamoto, Dr. M. Aoki and Mr. I. Ichiki of Ebara Research Co. Ltd. for their valuable suggestions and financial support in the present study.

References

(1) Yamamoto, K., Instability in a Cavitating Centrifugal Pump, *JSME International Jr.*, Ser. 2, Vol. 34, No. 1(1991), pp. 9-17.

(2) Brennen, C. and Acosta, A. J., The Dynamic Transfer Function for a Cavitating Inducer, *Trans. ASME*, Jr. Fluid Dynamics, Vol. 98, No. 2(1976), pp. 182-191.

(3) Kaneko, M. and Ohashi, H., , Self-Excited Oscillations of a Centrifugal Pump System under Air/Water Two Phase Flow Condition , *Proc. IAHR Symp.*(Amsterdam), Vol. 2 (1982), Paper No. 36.

(4) Ichiki I., Airlocking Phenomena at Partial Load Operation in Centrifugal Pump, *Ebara Research Report* (in Japanese), Vol.151(1991-4), p. 24.

(5) Gleitzer, E. M., The Stability of Pumping Systems, *Trans. ASME,* Jr. Fluid Engineering, Vol. 103, No. 2(1981), pp. 193-242.

MEASUREMENTS IN THE DYNAMIC PRESSURE FIELD OF THE VOLUTE OF A CENTRIFUGAL PUMP

J.L. PARRONDO, J. FERNÁNDEZ, C. SANTOLARIA & J. GONZÁLEZ
Área de Mecánica de Fluidos, Universidad de Oviedo
ETSIIG. Campus de Viesques. 33271 Gijón. Spain.

Abstract

This paper presents an experimental investigation into the dynamic pressure field existing in the volute of an industrial centrifugal pump in order to characterize the interaction phenomena between impeller and volute. For that purpose, pressure signals were obtained simultaneously at different points of the volute casing by means of two miniature fast-response pressure transducers. Particular attention was paid to the pressure fluctuations at the passing blade frequency, regarding both amplitude and phase delay relative to a reference point. The analysis of the dependence of the pressure fluctuations on both flow-rate and position along the volute clearly indicates the leading role played by the tongue in the impeller-volute interaction and the increase of the amplitude of the dynamic forces in off-design conditions.

1. Introduction

Centrifugal pumps with volute casing are usually subjected to static thrust in both the axial and radial directions. Radial thrust is particularly great in off-design conditions, because, for both low and high flow-rates, the presence of the volute provokes a non-uniform pressure distribution around the outlet of the impeller (Stepanoff 1957, Csanady 1962). Additionally, centrifugal pumps can also present non-steady radial forces, which are usually associated with the frequency of rotation or with the passing-blade frequency (and with their harmonics). Excitation at the frequency of rotation may be provoked by impeller whirling, when the impeller has an orbital motion coupled to the rotation, due to shaft misalignment for instance. An impeller with small manufacturing imperfections will also present some excitation at that frequency. Excitation at the passing-blade frequency (frequency of rotation multiplied by the number of blades of the impeller) is a consequence of the finite thickness of the blades, which causes flow disturbances in the volute associated with the passage of each blade. In the case of centrifugal pumps with vaned diffusers significant excitation may also exist at the vane passage frequency (frequency of rotation multiplied by the number of vanes) and at the blade-vane passage frequency (passing-blade frequency multiplied by the number of vanes).

E. Cabrera et al. (eds.), Hydraulic Machinery and Cavitation, 401–410.

The effects of those excitation mechanisms may be substantially modified by rotor-stator interaction phenomena. Previous work in this field was mostly focused on axial machinery, either compressors (Gallus et al. 1980) or turbines (Dring et al. 1982). Regarding centrifugal pumps, Adkins & Brennen (1988) investigated both theoretically and experimentally the hydrodynamic forces arising from impeller whirling in a pump with volute casing, and found that those forces could be destabilizing under many operating conditions. Arndt et al. (1990) investigated the interaction of a centrifugal impeller with several vaned diffusers of different geometry by measuring pressure fluctuations in both impeller and diffuser. More recently Chu et al. (1993), also Katz (1996), used particle image velocimetry to measure the velocity distribution in the near-tongue region of the volute of a centrifugal pump, together with pressure and noise measurements. They concluded that dominating phenomena are associated with the interaction of the blades with the tongue, and that even a small increase in the space between impeller and tongue (up to 20% of the impeller radius) causes significant reductions in noise levels.

This paper presents an experimental study of the dynamic interaction between the volute and the impeller of an industrial centrifugal pump, with particular attention to the excitation at the passing-blade frequency. Pressure measurements were taken simultaneously at different positions of the volute casing by means of miniature fast-response pressure transducers. The amplitude and phase delay of the pressure fluctuations obtained for different positions and flow-rates clearly show that blade-tongue interaction is a primary contributor to the fluctuating pressure field in the volute.

2. Experimental Equipment

The pump tested in this investigation was a Worthington EWP-65-200, a centrifugal pump with axial suction and volute casing, equipped with an impeller of 190 mm in outside diameter (Figure 1). Other dimensions of the impeller were: diameter at inlet hub, 67 mm; diameter at inlet tip, 88 mm; discharge width, 16.8 mm; number of blades, 7; and angle of blades at impeller output, 29°. The diameters of the suction and impulse pipes were 107.1 mm and 64.2 mm respectively. Figure 2 shows the performance curves of this pump when driven at 1500 r.p.m. (Fernández & Santolaria 1993), with a best efficiency point at about 48 m^3/h. Manufacturer's data at 2900 r.p.m. indicated a maximum efficiency of 76% for a rate of 93 m^3/h at 43 m head. According to similarity laws, both best efficiency points nearly agree.

The pump was connected in closed-loop to a reservoir with a capacity of 10 m^3. The set-up was designed according to BS-5316/2 specifications. Flow-rate was regulated by means of a butterfly valve located at the reservoir inlet, together with a pinch valve at the outlet of the pump for fine adjustment. The flow-rate was measured with a previously calibrated orifice plate, connected to an inclined piezometric mercury manometer. Measurement uncertainty was estimated to be less than 4% for flow-rate values greater than 20 m^3/h. The pump was driven by a DC-motor governed by a regulation device that maintained a rotational speed of 1500 ± 1.0 r.p.m.

3 mm diameter pressure taps were located every 30° along the volute, at 7 mm

Figure 1. Centrifugal pump drawing and location of the pressure taps.

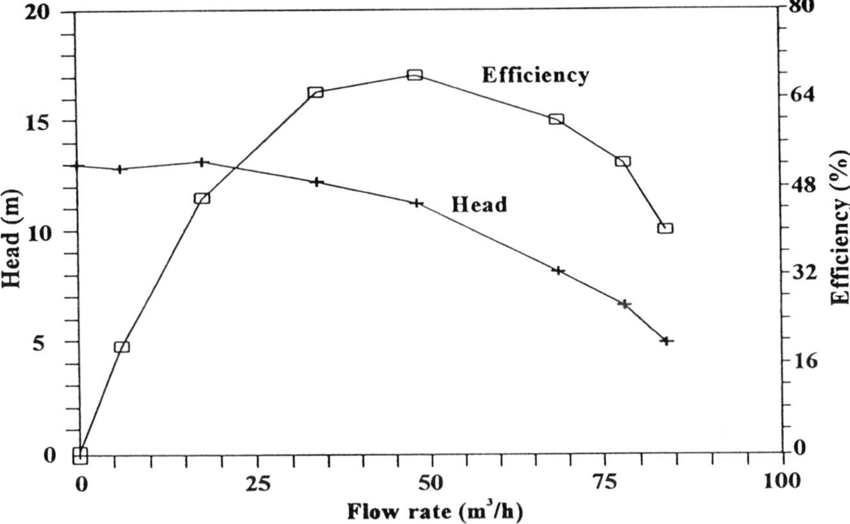

Figure 2. Performance curves of the pump tested at 1500 r.p.m.

from the outlet of the impeller (Figure 1). Static pressure at those positions could be obtained with a Kistler 4043A10 piezo-resistive pressure transducer and a K-4601 current amplifier, which provided absolute pressure values with an uncertainty of less than 0.5% (according to manufacturer's data). Fluctuating pressure could be also measured in the pressure taps along the volute by means of two Kistler 601A miniature fast-response piezo-electric pressure transducers, each of which was connected to a K-5007 charge amplifier (combined uncertainty of less than 1.5% according to manufacturer's data). Pressure signals were processed with an HP-3562A two-channel

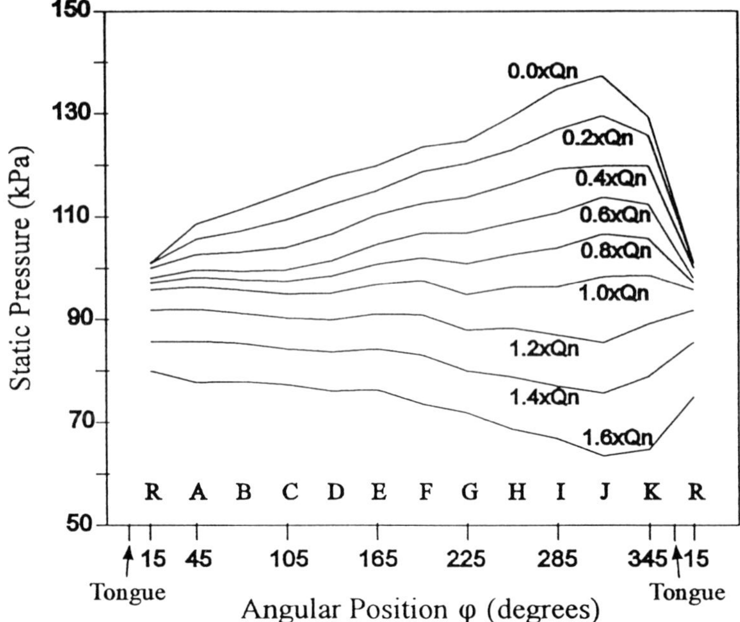

Figure 3. Static pressure distribution around the impeller outlet, for different flow-rates (Qn=48 m³/h).

dynamic signal analyzer, which permitted either time or spectral averaging of the signals and the obtaining of various spectral functions, including power and cross-power spectra, and coherence.

3. Static Pressure Measurements

A preliminary series of tests was conducted to measure the static pressure along the volute, as a function of the flow-rate. For such purpose each of the pressure taps was connected to a collector chamber by means of a flexible hose with a switching valve. Proper operation of these valves permitted the pressure at the taps to be transmitted to the collector, where the piezometric pressure transducer was installed (Fernández & Santolaria 1993). The flow-rate was progressively increased from zero to the maximum value achievable in the set-up, and for each flow-rate the switching valves of the different pressure taps were sequentially operated. Measurements were taken some time after each valve switching to allow for regularization of the pressure in the collector.

Figure 3 shows the circumferential distribution of the static pressure so obtained for different values of the flow-rate; angular position φ is zero at the edge of the tongue and increases when rotating anti-clockwise. It may be observed that the static pressure along the volute is quite uniform for flow-rates around the best efficiency point. This is a

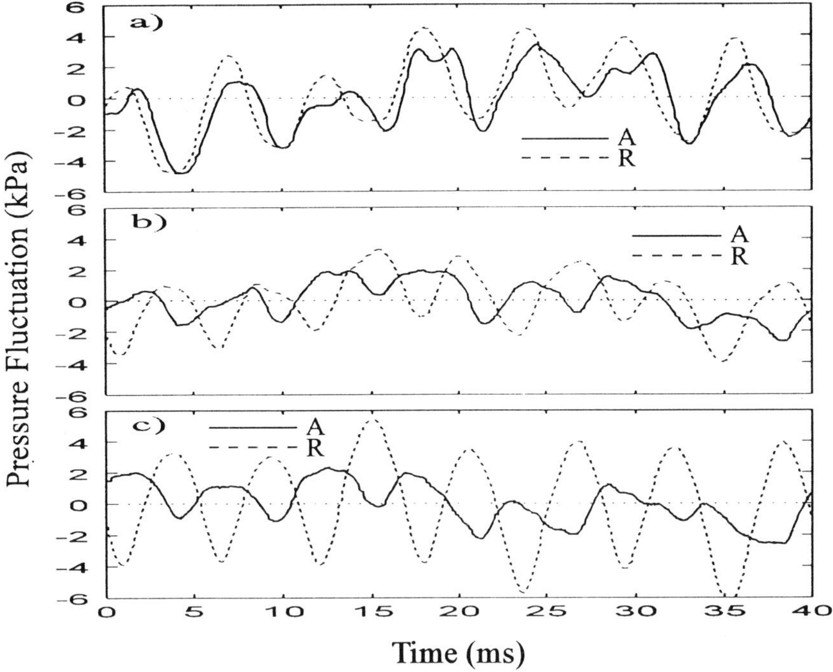

Figure 4. Pressure signals at positions 'A' and 'R' during one complete revolution. Flow-rate: a) 17.1 m^3/h; b) 48.0 m^3/h; c) 67.2 m^3/h.

foreseeable result, since an efficient design of the volute is associated with a minimum radial thrust.

In off-design conditions, however, pressure is non-uniform and achieves a maximum for low flow-rates or a minimum for high flow-rates at about the position $\varphi = 315°$. This result agrees well with the trends indicated in classical texts (Stepanoff, 1957). Such non-uniform pressure distribution is due to the direction of the absolute velocity at the outlet of the impeller, variable with the flow-rate, and the subsequent incidence of the flow on the volute. This interaction between impeller and volute may be said to be stationary, as opposed to the dynamic interaction referred to below.

4. Dynamic Pressure Measurements

For the fluctuating pressure measurements, both piezoelectric pressure transducers were connected to two of the pressure taps. A small adaptor was used so that the sensitive membrane of the transducers was 10 mm from the internal surface of the volute, thus ensuring that any disturbing resonance phenomena inside that cavity would only occur well above the range of frequencies of interest in this study. Simultaneous

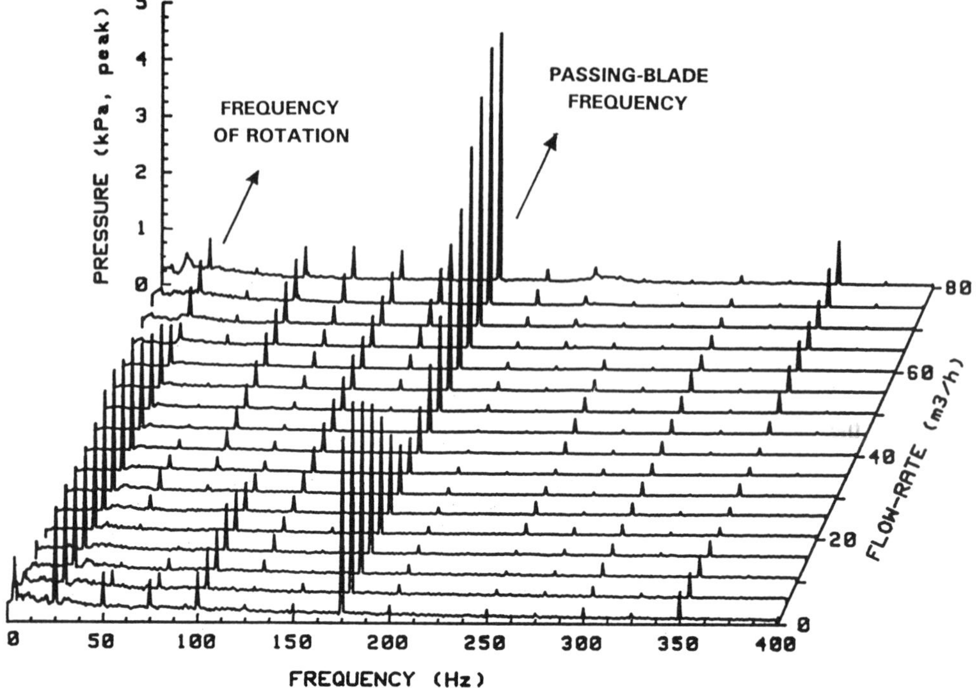

Figure 5. Pressure power spectrum at position 'R' as a function of flow-rate.

capture of the signals from both transducers permitted not only the amplitude but also the phase delay between the corresponding fluctuations to be obtained.

Figure 4 shows the evolution of the pressure at positions 'A' and 'R' (see Figure 1) during one complete revolution of the impeller for three different values of the flow-rate. These curves are the average of 100 trials, with each sample being triggered by the signal from an optical tachometer oriented to the pump shaft. A total of 7 hills and valleys can be appreciated in these curves, corresponding to the 7 blades of the pump as they pass in front of the pressure transducers. Such modulation is due to the finite thickness of the blades and the corresponding non-uniformity of the flow around the outlet of the impeller. However, it is clear from Figure 4 that: 1) the shape and the amplitude of the pressure fluctuations depend on the position of the transducers for a given flow-rate; 2) both shape and amplitude depend on the flow-rate for a given transducer position; and, 3) the time (or phase) delay between the pressure fluctuations at two positions depends on the flow-rate.

All this behaviour is a consequence of the dynamic interaction between impeller and volute, which is the principal objective of this study. A systematic series of experiments were conducted for which one of the transducers was located permanently at the tap with $\varphi=15°$, the one referred to as 'R' in Figure 1, whereas the other transducer was progressively moved from position 'A' to position 'K' (Figure 1). For each pressure tap position and for stepped increments of the flow-rate, after some time for flow settling, the analyzer performed spectral averages of 100 trials of the in-coming signals (for 5 minutes) and obtained their power spectrum, cross-power spectrum and coherence.

Whereas the power spectrum provides a direct measure of the amplitude of the fluctuations at each frequency, the phase of the cross-power spectrum provides a measure of the phase delay between the fluctuations of two signals. The coherence function permits an evaluation of the degree of correlation between them at each frequency. Figure 5 shows the evolution of the power spectrum of the pressure at the reference position ('R'), with respect to the flow-rate. As expected (see Section 1), the predominant spikes correspond to the frequency of rotation of the impeller (25 Hz) and in particular to the passing-blade frequency (175 Hz), together with their respective harmonics. As stated above, the present study was focused on the excitation at the passing-blade frequency, as provoked by the finite thickness of the blades.

Since the power spectrum of the pressure at position 'R' was obtained several times for a given flow-rate (one series of measurements for each of the other pressure taps), examination of the spike amplitude at a given frequency permitted the checking of the maintenance of flow conditions between one test to another. The variations so observed at the passing-blade frequency were always less than 10% and usually less than 3%. At least two series of tests were conducted for each pressure tap position, on different days, and good repeatability was found with respect to both amplitude and phase delay at the passing-blade frequency. Phase delay variations were always less than 4%, referring to 360°. Coherence at the passing-blade frequency was always greater than 0.9 and usually greater than 0.99.

5. Experimental Results at the Passing-Blade Frequency

Figure 6 shows the amplitude of pressure fluctuations at the passing-blade frequency as a function of the flow-rate for different circumferential positions along the volute. The behaviour observed is very different from that of the stationary pressure distribution (Figure 3). In general the pressure fluctuations are greater in off-design conditions, i.e. for both low and high flow-rates. The maximum values, which correspond to the region of the tongue (position 'R') for very high flow-rates, are about 6% of the corresponding stationary pressure. Some of the curves exhibit a clear minimum, at a flow-rate dependent on the tap position ('R', 'B', 'D', 'F'...), and some other curves, however, are much flatter ('A', 'C', 'E'...). The continuous crossing of the curves of Figure 6 suggests the existence of node and anti-node positions along the volute, variable with the flow-rate, which is a typical effect of wave resonance phenomena.

Figure 7 shows the phase delay ψ between the pressure fluctuations at each pressure tap and those at position 'R', after subtraction of the nominal phase associated with the time lag that one blade needs for moving from one position to the other. This nominal phase, which is not dependent on the flow-rate, may be shown to be equal to $210 \cdot i$ degrees, for a position separated $30 \cdot i$ degrees from tap 'R'. Interestingly, most of the curves of Figure 7 exhibit a very similar behaviour: increasing the flow-rate results in

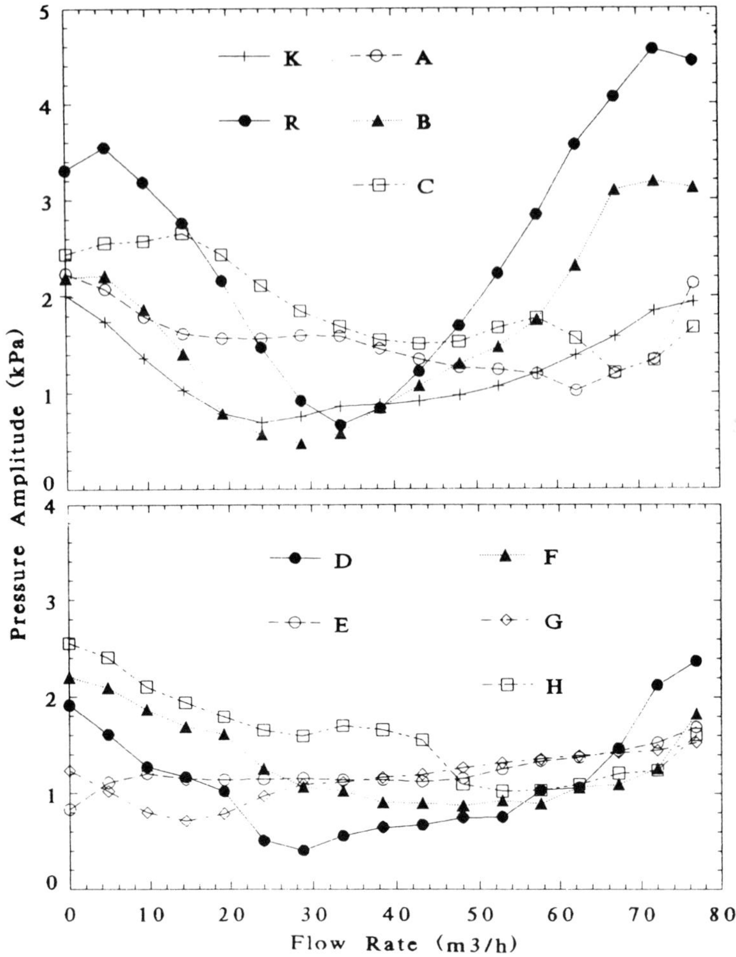

Figure 6. Amplitude of pressure fluctuation (passing-blade frequency) as a function of flow-rate for several locations along the volute.

a reduction of the phase delay ψ, from about 180°-240° for the lower range of flow-rates to values surrounding 0° for flow-rates between 85% of the best efficiency point and the maximum available flow-rate. This means that the pressure fluctuation peaks are usually not in phase with the passage of the blades. Position 'B' is the only one to follow a different trend, which might be attributed to the existence of a node in its vicinity for small flow-rates.

The large relative phase delay at small flow-rates actually corresponds to low absolute phase delays, i.e., the pressure fluctuations around the impeller are synchronized with the passage of the blades in front of the tongue. This result indicates the great influence of the tongue-blade interaction on the dynamic pressure field existing

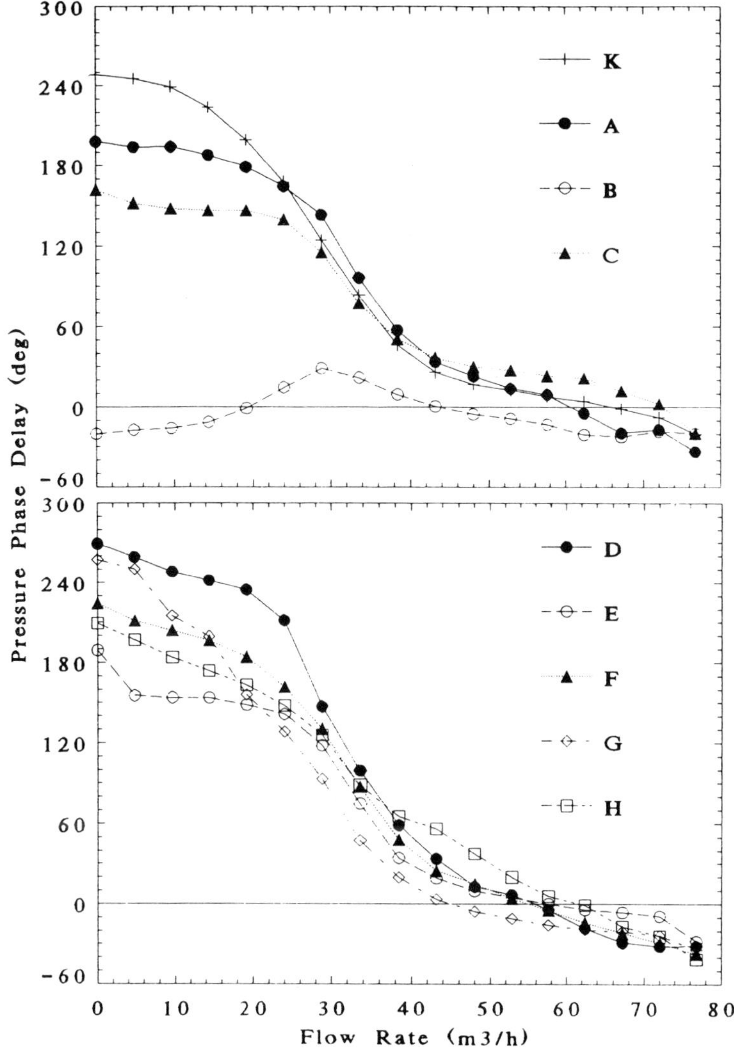

Figure 7. Pressure phase delay (at the passing-blade frequency) relative to position 'R' as a function of flow-rate for several locations along the volute.

within the volute for the lower range of the flow-rate, which agrees with the results of previous studies (Chu et al. 1993, Katz 1996). For increasing values of the flow-rate such influence is progressively reduced and partially counterbalanced by the pressure difference between both sides of the blades. The best efficiency point is also the point for which the dynamic excitation is minimum: minimum pressure fluctuations along the volute and phase delay $\psi \approx 0°$.

6. Conclusions

A systematic series of tests were conducted to measure the dynamic pressure field in a number of positions along the volute of a centrifugal pump. The analysis focused on the pressure amplitude and phase delay at the passing-blade frequency, depending on both angular position and flow-rate. The results obtained show that the distribution of pressure fluctuations greatly depend on the flow-rate. In general the amplitude of the fluctuations is greater in off-design conditions, in particular in the near-tongue region, which suggests the monitoring of the pressure in that zone as an effective means of observing the operation of the pump, for predictive maintenance applications. Minimum dynamic excitation approximately corresponds to the best efficiency point, for which the pressure fluctuations are approximately in phase with the passing of the blades. For low flow-rates, the interaction between impeller and tongue is dominant in the generation of the dynamic pressure field, as indicated by the analysis of the phase delay existing between the pressure fluctuations at different positions.

Acknowledgements

The authors gratefully acknowledge the financial support of the Dirección General de la Investigación Científica y Tecnológica (Spain) under Project TAP-94-0653 entitled "Optimización de técnicas de diagnosis y control de turbomáquinas".

References

Adkins, D.R. & Brennen, C.E. (1988) Analyses of hydrodynamic radial forces on centrifugal pump impellers, *ASME Journal of Fluids Engineering* **110**, 20-28.

Arndt, N., Acosta, A.J., Brennen, C.E. & Caughey, T.K. (1990) Experimental investigation of rotor-stator interaction in a centrifugal pump with several vaned diffusers, *ASME Journal of Turbomachinery* **112**, 98-108.

Chu, S., Dong, R. & Katz, J. (1993) The effect of blade-tongue interactions on the flow structure, pressure fluctuations and noise within a centrifugal pump, in G. Caignaert (ed.), *Pump Noise and Vibrations*, CETIM, Paris, pp.13-34.

Csanady, G.T. (1962) Radial forces in a pump caused by volute casing, *ASME Journal of Engineering for Power* **84**, 337-340.

Dring, R.T., Joslyn, H.D., Hardin, L.W. & Wagner, J.H. (1982) Turbine rotor-stator interaction, *ASME Journal of Engineering for Power* **104**, 729-742.

Fernández, J. & Santolaria, C. (1993) Esfuerzos de origen fluidodinámico en una bomba centrífuga con rodete recortado, *Actas del I Congreso Iberoamericano de Ingeniería Mecánica* **2**, AEIM, Madrid, pp.187-193.

Gallus, H.E., Lambertz, J. & Wallman, T. (1980) Blade row interaction in an axial flow subsonic compressor stage, *ASME Journal of Engineering for Power* **102**, 169-177.

Katz, J. (1996) On the relationship between volute geometry, flow structure, pressure fluctuations and noise in a centrifugal pump, *Flow in Radial Turbomachines*, Lecture Series 1996-01, Von Karman Institute for Fluid Dynamics.

Stepanoff, A.J. (1957) *Centrifugal and Axial Flow Pumps*, Wiley, New York.

FUNCTIONAL MODELLING OF PUMP VOLUTE GEOMETRY

P.R. THACKRAY, R.D. JAMES
Hydraulic Design Engineer, Kvaerner Boving Ltd, Doncaster, UK.,
Senior Lecturer, The University of Hull, UK

Abstract

The task of hydraulic design is usually undertaken using traditional techniques, coupled with many simplifying assumptions. This can lead to long design lead-times and under-utilisation of performance potential. New computational approaches are now required. In this resolve, some considerations of linking intelligent computer technology with fundamental aspects of hydraulic design, in particular pump volute geometry, are discussed. The paper firstly reviews volute flowfield measurements and the effects of geometry, and goes on to discuss some recent experiences in developing an experimental system to provide a functional modelling facility.

1. Background

The desire to understand the real flow characteristics in spiral casings of pumps and turbines stems from an intense need to produce higher hydraulic efficiencies. Early theoretical treatments of pump outflows by Kucharski and others focused on the impeller and assumed perfect fluid flow; the effects of blade curvature were often simplified, and volute effects were ignored completely [1]. Photographic flow studies by Fischer and Thoma [2] in 1932 using dye injection into an open impeller pump with glass sides showed recirculation at low flows. Binder and Knapp [1,3] determined that radial velocities and pressure distributions were non-uniform even at design flow. In 1956 Bowerman and Acosta [4] demonstrated that an impeller operates differently with different volute spirals. Efficiency variations of around 5% were reported for three log-spiral volutes of vane angles varying by 9°. The shape of the volute was also found to be prominent in determining the overall efficiency of the pump. In 1960, Iversen *et al* [5] presented a theoretical pressure distribution analysis which included an estimate for mixing losses. Csanady [6] obtained similar results by conformal mapping of the shed flow cascades.

In the 1960's, Worster [7,8] provided basic empirical guidelines for impeller-volute interaction and geometry. Potential flow theory was used to superimpose a source and free vortex to model the volute flow. The shape of the volute section was found to strongly affect pump performance, and the magnitude and position of the best efficiency flow point. A partly-trapezoidal section provided a smoother pressure distribution around the volute and marginally higher efficiency than a comparable

E. Cabrera et al. (eds.), Hydraulic Machinery and Cavitation, 411–418.

rectangular section. This was also confirmed in tests by Pekrun [9] and others. The shape and angle of the tongue was found by Hira and Vasandani [10] to affect the magnitude and position of the best efficiency flow point. Kurokawa [11] showed that a 2-d ideal frictionless flow analysis could still be used to accurately predict volute flow characteristics over a wide range of off-duty conditions. Imaichi [12] used a potential flow analysis accounting for the effect of vortex shedding to estimate unsteady forces; these were found to agree well with empirical formula. Streamline plots derived by streak photography by Brownell and Flack [13] showed stagnation points and zones of flow separation in the volute; streamlines were also found to deform in proportion to the magnitude of off-design flowrate.

The advent of laser technology meant that non-intrusive measurements of velocity fields became possible. Normalised non-rotating flow profiles taken by Thomas *et al* [14] in 1986 using laser velocimetry showed velocities at the impeller periphery up to 30% higher than the average volute core velocity. The nature of the volute flow was found to depend strongly on impeller outlet angle. Miner *et al* [15] showed that the momentum flux in the volute is always asymmetric, even at design flow. Volute tangential velocity profiles were found to be in close agreement with a free vortex flow, and the jet-wake phenomenon from the impeller was seen to rapidly dissipate within the volute. The turbulent flow structure of the volute was recently measured by Dong *et al* [16] using laser velocimetry. Phase-averaged vorticity distributions showed a flow structure strongly dependent upon geometry and flowrate.

Today many of the hydrogeometric features can be determined and optimised at an early stage of the design process. However, most hydraulic pump design is essentially empirically-based and therefore sub-optimal. Volute geometry is of fundamental importance in minimising losses and achieving an optimal outflow field. In addition to impeller design improvements, problems remain in modelling an optimum volute profile. Because of the large number of variables and design steps involved, a decision making process is obviously required. The following discussion relates to some recent experiences [17] in this field of turbomachinery design, focusing on the use of artificial intelligence (AI) to capture and model required hydraulic design functionality.

2. Modelling Design Intent

Nearly all existing computer-aided design systems suffer from an inability to work intelligently with the designer. On their own, these systems are not knowledgeable, nor gain knowledge, and may accept erroneous inputs without checking against some reference base for conformance and accuracy. A system which provides some inference capability is known as an intelligent CAD system (ICAD). Ideally such a system should be able to understand the intentions of the designer, detect errors, suggest alternatives, etc, by referring to a design knowledge-base. Currently there is very limited experience of geometrical reasoning in engineering design [18,19], and few examples are known of such working applications in the turbomachinery field. In the complex design cycle of a new pump or turbine, the conceptual design stage is currently in greatest need of rationalisation.

Decision making in a design process involves a sequence of conformance or alteration steps, performed within a framework of constraints or requirements. Along with a suit of alteration rules, a specific method for knowledge representation is required, ideally as an intelligent CAD system which is capable of interacting with the designer. "Intelligence" in this context describes the ability to solve problems by applying inference rules from a knowledge-base. This knowledge-base should consist of a matrix of rules and facts represented in a specific language. Amongst the number of different types of inference known, only deductive inference has been studied in sufficient detail for use as a general processing mechanism [20,21].
Design rules can be formulated in a knowledge-base by use of a procedural or declarative computer programming language. However the problem now lies in codifying the knowledge and testing and verifying the output. Additional problems exist with control of the inference process; this will affect the solving process as the characteristics of the designer's thought process will be replicated in the way their decision-making is undertaken. This thought process has been simulated on computers in simple terms using declarative sequences, although the inference mechanism must be regulated. This can be undertaken using a metamodel approach [18], ie, developing functions which govern rules about rules and interdependencies.

3. Functions and Symbols

In order to perform intelligently, ICAD systems must be able to integrate individual design operations into a design process. Such a system can achieve this by processing the input and transferring information internally from one operation to another. Each operation, or function, is defined independently and modelled where possible in terms of other functions. Different forms of information such as procedures, specifications, databases and other knowledge must be used together, coupled with a good user interface. The use of a declarative (or quasi-declarative) language greatly facilitates the modelling of entities and their functionalities in this manner [22]. Simple inferences can be modelled, for example:

if a and b represent parameters such

$$a = f(b) \text{ and } b = f(c), \text{ then } a = f(c) \text{ may be inferred} \tag{1}$$

This concept can be extended to include basic hydraulics rules, for example:

if (Ns<20) then (use symmetric volute rules + Stepanoff progression rules) or,
if ((mixing losses>5%) then ((reduce sidewall angles) and (increase blend radius))) else (modify impeller outlet width) (2)

Such instructions may also be processed iteratively, for example:

if (flow area tolerance < 0.5%) and (low flow separation) then (accept profile) otherwise (repeat by adjusting geometry per sub-rules) (3) or recursively :

$$f\,(\text{rules }(\text{area}_{0\rightarrow 360°})) = f((\text{rule area}_0)+(\text{rules b.c.d....})) \tag{4}$$

The following section relates to some recent experiences of an experimental system which was developed and implemented on a trial computer-aided design system [17]. Because of its interactive qualities, and its ability to process function-argument instructions [23], a LISP-type environment was specifically chosen to construct the knowledge base. Programmes were written in AutoLISP language, operating interactively with AutoCAD™ system.

4. Modelling Functions using AutoLISP

Hydrogeometric information can be represented symbolically using lists. A list consists of groups of elements treated as one expression and stored implicitly in memory; this facilitates the modelling of rules and relationships between parameters. AutoLISP algorithms consist of blocks of lists, arranged as constituent functions. The programmer-designer can also fashion algorithms to manage the inference process, for example cyclically or by conditional branching. Such functions may process any number of arguments. These consisted of variables, commands, constants, user inputs, rules and other functions. During computation, the evaluator reads a sequence of instructions, evaluates it and returns a result which performs a particular task.

5. Experimental System

5.1 INPUT PROCESSING

To deal with relationships between two or three dimensional geometric objects, the knowledge of analytical geometry which defines points, lines, curves, planes, intersections, transformations, blends and fillets was defined as lists. Mathematical knowledge is required implicitly to define blending forms and other trigonometric relationships of the hydraulic profiling functions. This information was supplied by the hydraulic designer. In its basic form, some 10^3 lines of AutoLISP were required to describe a simple impeller-volute design. The process was initiated by firstly evaluating user-inputs to establish fundamental hydraulic geometry and expected performance parameters for the impeller. A predicted throat area A_{th} subsequently formed the basis for required volute geometry.

5.2 HYDRAULIC FUNCTIONS

Individual hydraulic parameters were represented using functions. For example, the hydraulic function called *slipf* was used to define impeller slip factor μ :

```
(defun slipf ()
(setq μ (- 1(/(* pi (sin β2))z))))                (5)
```

AutoLISP's (setq) function was used to define geometric constructions, either locally or globally. Along any volute periphery S, control functions were used to define points $S=f(p_1, p_2, p_3 ...)$ in terms of system x, y, θ coordinates. Points were joined using lines, arcs, curves etc, into closed complex areas A_s, and exported graphically (or as datafiles). A typical peripheral function used was:

(defun periphery (S))= (area p_1, p_2, p_3 ......p_1), where (setq p_1 rule a), (setq p_2 rule b).... (setq p_n rule n) (6)

Self-predictive-corrective algorithms generated complex volute profiles, iterating successively on area A_v until a suitable profile S_v was formed, meeting active hydrogeometric criteria. Two types of rule were entered into the knowledge base: (a) hydraulic assignments, parameters of head, flow, impeller characteristics etc, and (b) geometric parameters such as desired ratios between impeller outlet/volute inlet dimensions, sidewall angles, aspect ratios, fillets, swept centres, tangency allowances etc. Some geometric variables were only partly defined, providing flexibility during iteration. Friction losses, turbulence losses, radial forces and other parameters were estimated concurrently.

Theoretically, any number of sections S_n may be defined along the volute path, but values of $n<10$ proved adequate for small pump designs. To achieve an accuracy of outflow area $A_v < 0.1\%$, (acceptable for most low Ns pumps), typically only four or five iterations per section S_v were required, each with a computing time of some 15 seconds operating on a 486-PC. For higher Ns pumps, smaller convergence tolerances, therefore higher flowarea accuracy, could be set at a cost of longer solving times. The remaining volute path was developed recursively:

$$\text{(defun 3-dvolute } (V_{Ath \rightarrow 0})) \qquad (7)$$

During recursive operation, rules were only accessed once per cycle, thus requisitioning memory and computation only as needed. This allowed the program to tackle problems of varying size. Devised section areas were calculated internally by the foster CAD system and verified separately.

Because the modelling process describes the hydraulic profile in terms of inputs and modifying operations, this readily facilitates the handling of changes. Thus any of the inputs may be altered, and the process repeated parametrically. An example of a typical profile output is shown in Figure 1, and a typical AutoLISP listing is shown in Figure 2.

5.3 PARADOXES

During development, problems occurred with non-convergence and instability caused either by self-intersection of curves, oscillation or overplotting. For complex area shapes, non-convergence was offset by use of double iteration, this also reduced some convergence times. Average-damping was also used to force convergence. A compromise was apparent between definition of adequate number of control points along bounding curves, yet providing limited flexibility for modification purposes.

6. Summary

The method demonstrated a successful application of an artificial intelligence process during the hydraulic design phase of a pump. It was found that the codification procedure determines how effectively design knowledge can be modelled. Functionality has been derived from the knowledge-processing system by development of an additional processing mechanism on top of the conventional computing mechanism. In order to integrate into a mainstream turbomachinery design process, an ICAD system must:

- provide means of easily updating/entering new knowledge,
- be able to work interactively with the design engineer,
- be linked to analytical facility (FEA, CFD systems),
- have means of testing and verifying outputs,
- provide the knowledge-base with ability to adapt generatively.

Such methods can help design engineers during conceptual phases, thereby helping to shorten design lead times. Further work is recommended in the development of a tandem AI-CFD system; this would help integrate fundamental fluid parameters deeper into the root design process, and reduce the reliance of empiricism.

References

1. Binder, R.C., Knapp, R.T.: Experimental Determination of the Flow Characteristics in the Volutes of Centrifugal Pumps, *Trans. ASME*, HYD-58-4, 1936.
2. Fischer, K, Thoma, D.: Investigation of the Flow Conditions in a Centrifugal Pump, *Trans. ASME*, Vol 54., HYD-54-8, 1932.
3. Knapp, R.T.: Centrifugal Pump Performance as Affected by Design Features, *Trans. ASME*, April 1941.
4. Bowerman, R.D., Acosta, A.J.: Effect of the Volute on Performance of a Centrifugal Pump Impeller, *Trans. ASME*, July 1957.
5. Iversen, H.W., Rolling, R.E., Carlson, J.J.: Volute Pressure Distribution, Radial Force on the Impeller and Volute Mixing Losses of a Radial Flow Centrifugal Pump, *Trans. ASME, Jnl Eng for Power*, April 1960.
6. Csanady, G.T.: Radial Forces on a Pump Impeller Caused by a Volute Casing, *Trans. ASME, Jnl.Eng. for Power*, October 1962.
7. Worster, R.C.: The Interaction of Impeller and Volute in Determining the Performance of a Centrifugal Pump, *BHRA paper RR679*, November 1960.
8. Worster, R.C., The Flow in Volutes and its Effect on Centrifugal Pump Performance, *Proc. IMechE*, Vol. 177, 1963.
9. Pekrun, M., Flörkemeier, K.H.: Experimental Investigation on Volutes of Centrifugal Pumps with Radial or Tangential Diffusers, *Pumpentagung Karlsruhe '78*, September 1978, Germany.
10. Hira, D.S., Vasandani.,V.P.: Influence of the Volute Tongue Length and Angle on the Pump Performance, *Jnl. of the Institute of Engineers India*, Part M.E. No.56.
11. Imaichi, K., Tsujimoto, Y, Yoshida, Y.: A Two-dimensional Analysis of the Interaction Effects of a Radial Impeller in Volute Casing, Proc. *IAHR Symposium*, Tokyo 1980.
12. Kurokawa, J.: Theoretical Determination of the Flow Characteristics in Volutes, *Proc. IAHR Symposium*, Tokyo 1980.
13. Brownell, R.B., Flack, R.D.: Flow Characteristics in the Volute and Tongue Region of a Centrifugal Pump, ASME paper 84-GT-82, 1984.
14. Thomas, R.N., Kostrzewsky, G.J., Flack, R.D.: Velocity measurements in a Pump Volute with a Non-Rotating Impeller, *Intl. Jnl. Heat & Fluid Flow*, Vol.7 No.1, 1986.

15. Miner, S.M., Beaudoin, R.J., Flack, R.D.: Laser Velocimeter Measurements in a Centrifugal Flow Pump, *Trans. ASME, Jnl. of Turbomachinery*, Vol.111, 1989.
16. Dong, R., Chu, S., Katz, J.: Quantitative Visualisation of the Flow within the Volute of a Centrifugal Pump, FED-Vol 107, *General Topics in Fluids Engineering*, ASME 1991.
17. Thackray, P.R.: Computer-Aided Analytical Design of Centrifugal Pumps, *Ph.D. Thesis*, Hull University, UK, 1994.
18. Tomiyama, T., Yoshikawa, H.: Requirements and Principles for Intelligent CAD Systems, *Knowledge Engineering in CAD*, Elsevier 1985.
19. Simmons, M.K.: Artificial Intelligence for Engineering Design, *Computer-Aided Engineering Journal*, April 1984.
20. Jakiela, M.J., Papalambros, P.Y.: Design and Implementation of a Prototype Intelligent CAD System, *Trans. ASME Jnl. Mech. Trans. and Automation in Design*, June 1989.
21. Ohsuga, S.: Toward Intelligent CAD Systems, *Computer-Aided Design*, Vol.21, No.5. June 1989.
22. Gevarter, W.B.: The Languages and Computers and Computers of Artificial Intelligence, *Computers in Mechanical Engineering*, Nov. 1983.
23. Winston, P.H., Horn, B. K. P.: LISP, Addison-Wesley Publishing Co., 1981.

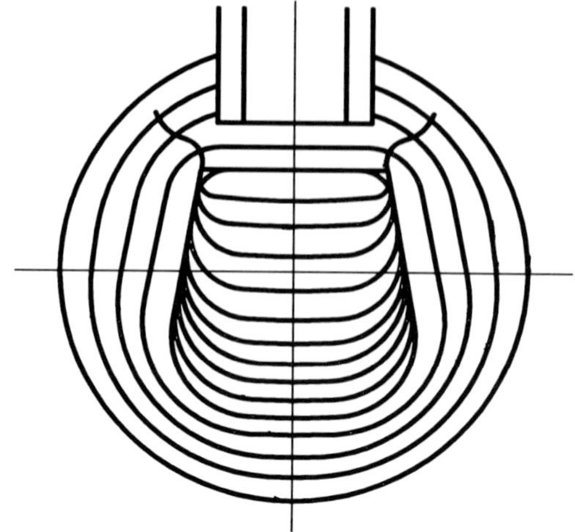

Figure 1. Typical profile output (2-d superimposed).

```
(defun blendfunct1()
(if(< rar 0.57) (setq rfu 0.95)(setq rfu 1.00)) (if(< rar 0.57) (setq rfu bl1)(setq rfu bl2)))
(defun volute1()
(while(or(>=a1(* flowarea 1.001))(<= a1 (* flowarea 0.999)))
(if(< impb2b3 1.5) (setq rule10a rule10b)(setq rule10a rule10c))
(command "erase" t1a "")
(setq thrht (* thrht (sqrt(/ flowarea a1))))(setq thr1 (+ thrht nt))(setq deltax (/ (+ x1 x2) 2))
(setq orig2 (polar orig 1 (dtr 0) deltax))(setq origtop (polar orig1 (dtr 90) thrht))
(setq cltop (polar origtop (dtr 0) deltax))(setq cenc4l (polar orig 1 (dtr 0) x4))(.....)
(setq v1 (*(dist cendis c1)(sin(-(angle cendis c1)(dtr 90)))))
(setq rd1 (*(+(*(-(/ disd 2) rd1) rar) rd1)rfu))(setq x1 (+ x1 (* v1 rar)))
(setq v2 (*(dist cendis c2)(cos(angle cendis c2))))
(setq rd2 (*(+(*(-(/ disd 2) rd2) rar) rd2)rfu))(setq x2 (- x2 (* v2 rar)))(.....)
(command "pline" t4b t1a "")(setq t41 (entlast))
(command "pline" t3a t4a "")(setq t34 (entlast))(.....))
(command "pedit" k21 "y" "j" k22 t32 k33 t34 k44 t43 t41 k45 t46 k47 t48 k11 "" "x")
(command "area" "e" k11)(setq a1 (getvar "area"))
)
(if (< flowcon flowtol)((mixloss)(frictloss)(blend)(tabulate))(accrls))
)
```

Figure 2. Typical AutoLISP computer listing.

ANALYSIS OF FLOW MEASUREMENTS IN THE IMPELLER AND VANED DIFFUSER OF A CENTRIFUGAL PUMP OPERATING AT PART LOAD.

Michel TOUSSAINT
CNAM, Laboratoire de la chaire de Turbomachines, Paris.

François HUREAU
CETIM, Laboratoire d'Hydraulique Industrielle, Nantes.

Abstract

In a recent european contract, four impellers with different specific speeds have been equiped with transparent front shrouds in order to obtain optical access to the flow between the blades. The presented results are the last analysis of internal flow measurements in one of these impellers (Ns32) fitted with a vaned diffuser. They are especially related with the interaction between impeller and diffuser, e.g. the phenomenon of flow separation, wakes, and backflow inside the pump. Unsteady velocity measurements were carried out in the hydraulic laboratory of CETIM with the help of laser doppler velocimetry technique. The data fields have been treated by the CNAM-LEMFI in order to give instantaneous view of the flow between the blades. From these pictures, a three dimensionnal movie has been built, that allows to explain some details in the behaviour of centrifugal impeller between 0.2Qn and Qn.

1. Introduction

If a centrifugal pump operates under its nominal flow rate (Qn), there comes a moment when the "incoming flow" fails to fill simultaneously all the channels of the impeller thereby generating counterflows at the impeller inlet and outlet. The experimental results given in this paper show for example how the flow "swings" from the shroud side to the hub one when the flow rate is reduced, and where are located the backflow areas inside the channels of the impeller. The comparison of results between vaneless and vaned diffuses gives some explanation on the nature of separated flows in this part of a rotodynamic pump. These detailed flow measurements complete other experimental results achieved on centrifugal pumps in Europe [2,3,4,11,12,13,14,18]. They should be extensively used to calibrate the computational fluid dynamics tools actually used in the design methodology of pump (see for exemple [1,5,6,7,10]).

E. Cabrera et al. (eds.), Hydraulic Machinery and Cavitation, 419–427.

2. Experimental setup.

2.1. TEST EQUIPMENT

The test engine is a centrifugal pump which main characteristics at nominal point are: speed 1450 rpm, flowrate 136 l/s, static head 30 m, mechanical power 62 kW. The flow at the outlet of the impeller is drived in a main collector linked to the pipe with 4 branch'lines. This configuration give an axisymetric behaviour of the flow within the pump, in the case of vaneless diffuser. The geometrical data of the centrifugal impeller (6 blades) and the vaned diffuser (8 blades) are given on figures 1 and 2, with the locations of velocity measurements. The test rig, that has been built to fulfill the different tasks for the investigations, is more completely described in reference [17].

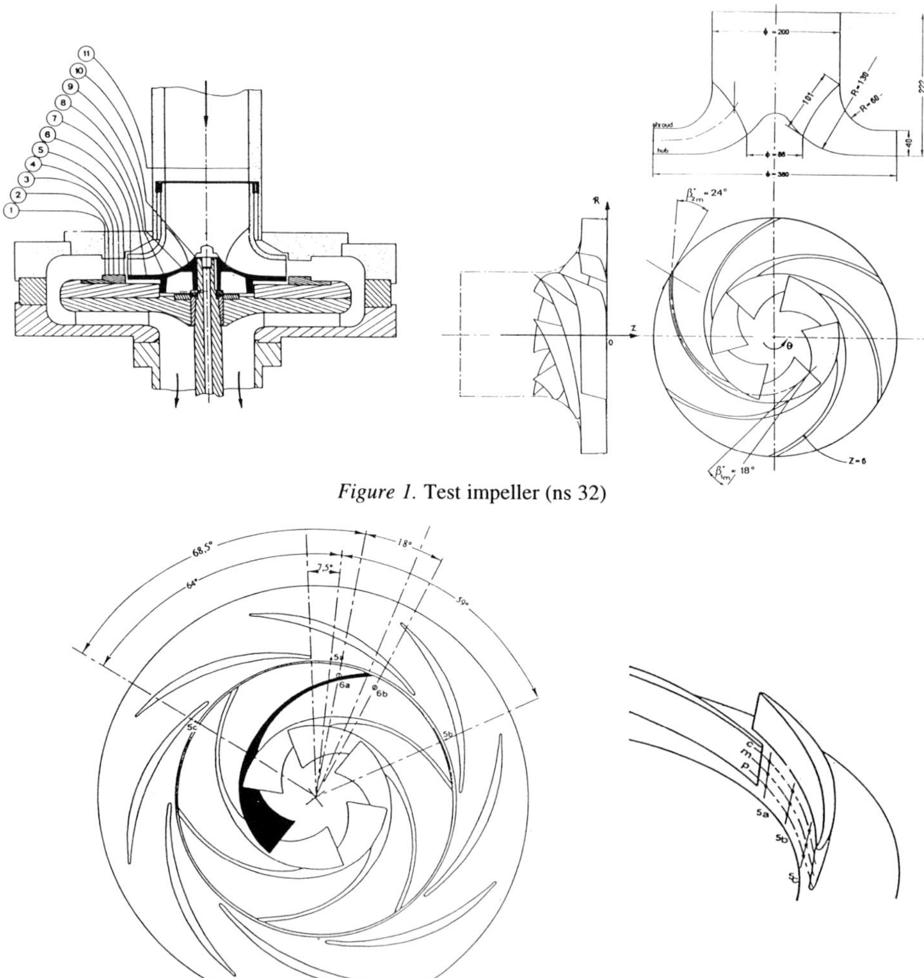

Figure 1. Test impeller (ns 32)

Figure 2. Vaned diffuser

2.2 INSTRUMENTATION.

In addition to conventional apparatus such as a test bed (rotational speed, torque, delivery rate, etc.), the equipment includes a slip ring collector allowing the measurement of pressure provided by built-in sensors. The scale model is also equiped to measure pressure fluctuations upstream and downstream from the impeller. The upper part of the model and the impeller flange are made of perspex to enable observation of any pre-rotation, recirculation or cavitation phenomena. Eleven glass ports, laid out as shown in Figure 1, provide access to impeller velocity measurements using the laser velocity measurement technique. Very fine particles (1 to 10 μm) of titanium dioxide act as tracers within the flow. The system used is a Doppler laser velocimeter with two interference fringe networks, including a COHERENT Argon laser source (6W output for all lines) and a TSI optoelectronic unit. Each reading of the two velocity components of these particles simultaneously leads to the input of the impeller's angle position using an optical encoder.

3. Analysis of experimental results.

3.1. RECIRCULATION IN THE IMPELLER.

The recirculation phenomena affecting the impeller and the diffuser in a plain diffuser type pump are summarised in the two drawings in Figure 3. It is noticeable that at delivery rates of between Qn and 0.61 Qn, slight signs of recirculation occur simultaneously in proximity to the impeller inlet and inside the diffuser. When the rate decreases, recirculation continues to spread, occurring in both the inlet and outlet of the impeller and much of the plain diffuser at very low delivery rates. Recirculation affecting the inlet originates at some distance in the impeller pipe. It takes up approximately 50% of the channel height and remains localised along the inner surface of the blade. This pocket of recirculation seems to "wear itself out" as it travels against the flow to reach the impeller outlet, which is logical since part of the recirculating fluid must be carried along as a result of the viscosity and set back on the right course by the "normal" delivery flow. These various observations confirm that recirculation at the impeller inlet is completely independent of the recirculation observed at the impeller outlet.

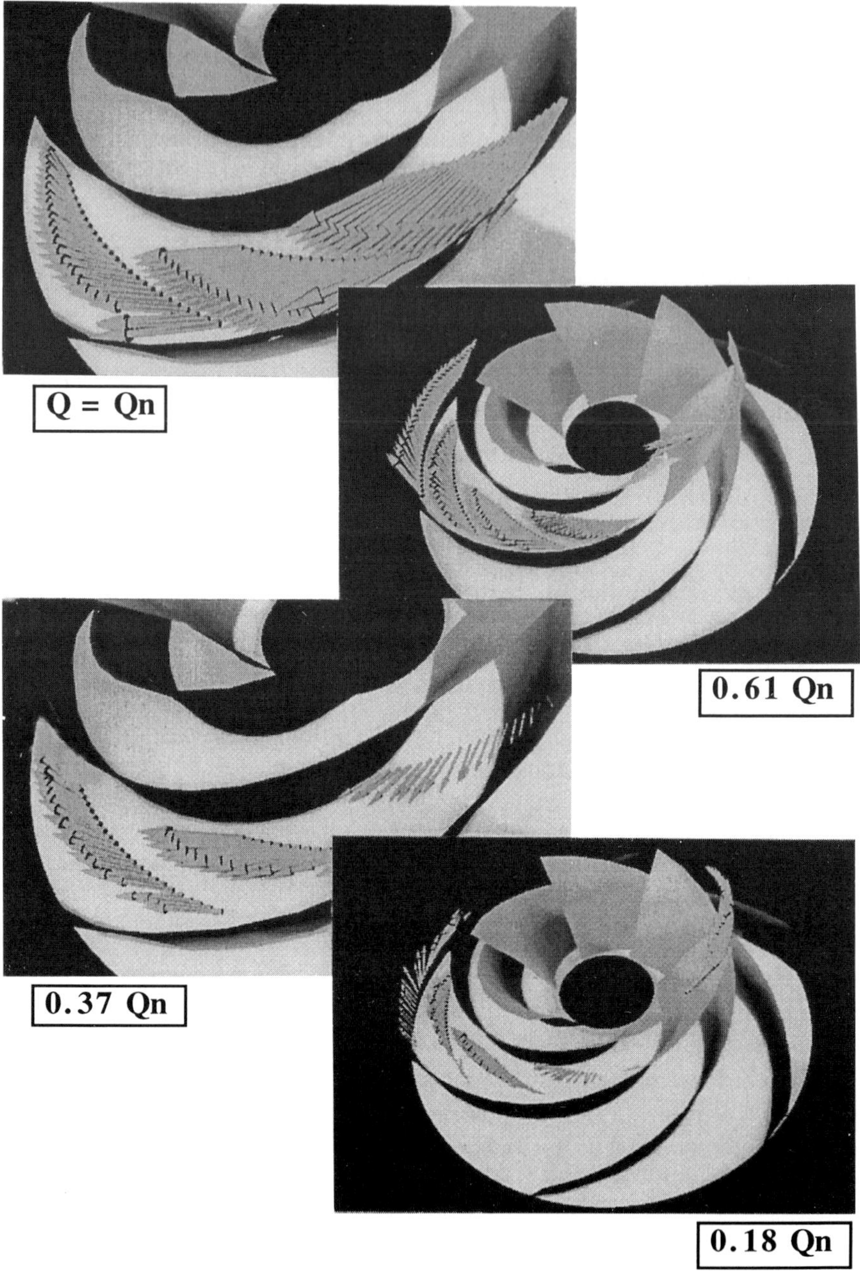

Figure 3: Flow measurements in the impeller.

3.2. INFLUENCE OF THE VANED DIFFUSER.

Although it is true that the relative flow in a centrifugal impeller may be "almost steady" [15] [16], this is not the case in an industrial operation where the impeller is always followed by a diffuser, return pipe or volute. The interaction between the fixed parts (diffuser vanes or volute tongue) and moving parts usually leads to a turbulent flow that is unsteady in both a Galilean referential system and in an impeller-linked relative reference system. Research carried out into this type of flow [8] [9] makes analysis of the results difficult (it is always somewhat specious), since the subsequent processing of pinpoint measurements of velocity and pressure values in the channels between the vanes of a turbomachine impeller always requires the use of extensive computer and graphics facilities. Using 3D images we shall, in this article, present an analysis of the flow in the area of interaction between impeller and diffuser, in a centrifugal pump operating with a vaned diffuser.

A comparison of the various velocity profiles recorded in the impeller in configurations with either plain or vaned diffuser shows that the vaned diffuser has an almost negligible influence on the flow in the inter-vane channels, except near the impeller outlet. Figure 4, which gives a comparison of the profiles recorded on axis 7 in the two operating configurations, confirms this initial analysis. As regards recirculation at the inlet and in the impeller, the presence of a vaned diffuser makes no notable difference compared to the system operating with a plain diffuser.

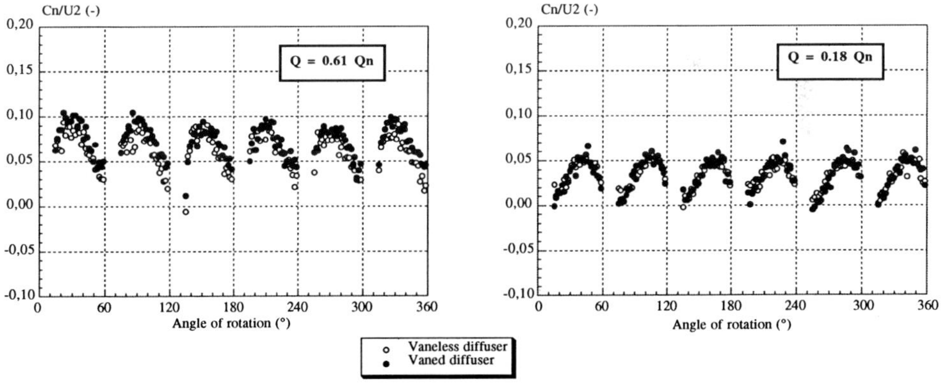

Figure 4: Comparison of velocity measurements with and without vaned diffuser.
Axis 7 - R=0.75 R2 - middle - Cn: normal to axis of measurement

Near the diffuser inlet the influence is more important, especially at part load operation. The figure 6 gives at three flow rates the radial component of velocity on Axis 6 (R=0.92 R2) with and without vaned diffuser. Taking into account the disymetry of flow around the impeller, two locations of measurements on the same axis are given (6a and 6b; see on figure 2)

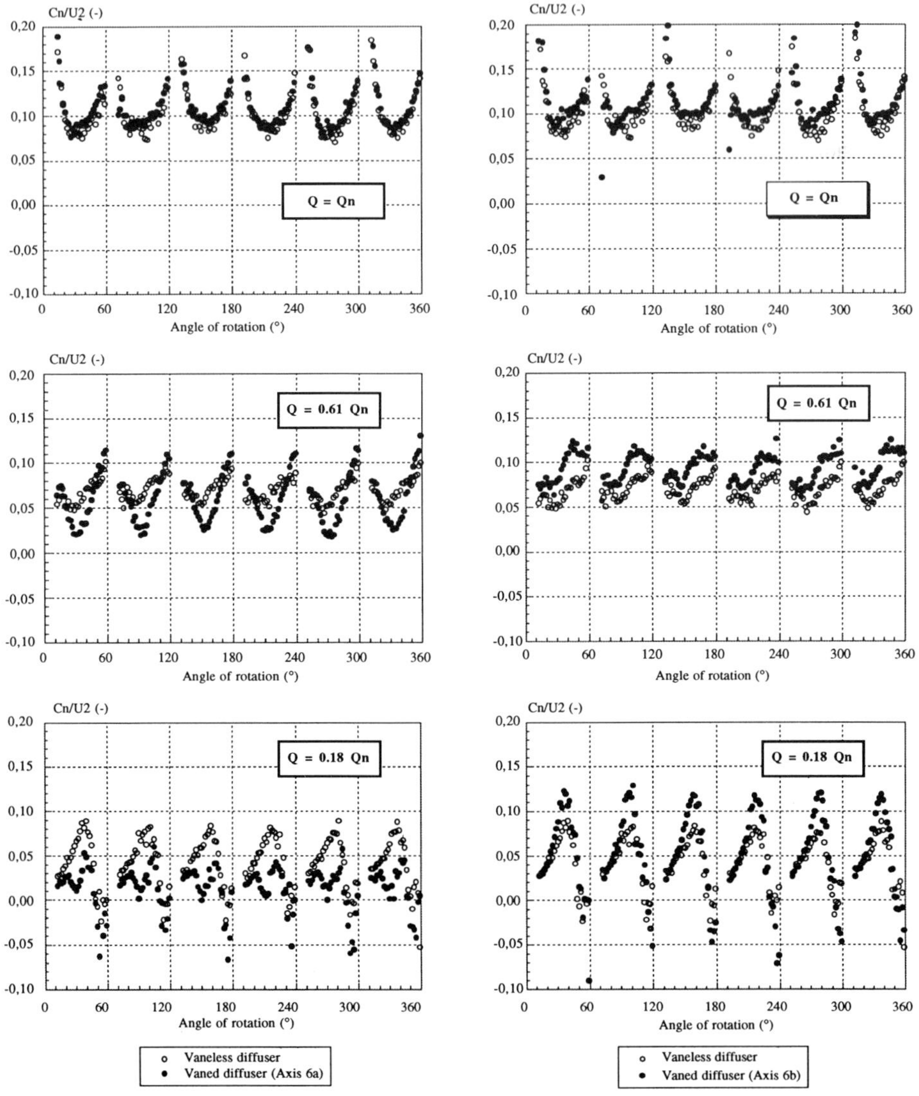

Figure 5: Comparison of velocity measurements with and without vaned diffuser. Axis 6 - R=0.92 R2 - middle (b/b2=0.5) - left: axis 6a; right: Axis 6b

The three-dimensional and unsteady phenomena which occur when the impeller vanes pass in front of the diffuser vanes may be given in vector form. This operation was carried out with support of the LEMFI Laboratory in Orsay using IRIS Explorer software on a Silicon Graphics workstation. An animated film was built using these 3D images.

The film explains the following phenomena in particular:

- at the nominal rate and slightly below (up to 0.9 Qn), the flow is particularly "stable" in the area around the diffuser inlet. The velocity deviations are small over a period of time and distribution is almost uniform between the diffuser vanes.
- at 0.82 Qn, separation occurs in the corner between the bottom of the vane and the concave side of the diffuser vane. This separation is resorbed as the impeller vane approaches and creates a flow running into this area.
- at 0.61 Qn, separation occurs on the shroud side (top of the diffuser vane). This separation is located between the leading edge of the vane and the upper flange and it acts as a result of the significant wake generated at the impeller outlet at this flow rate.
- at 0.37 Qn, the separation again occurs at the bottom of the diffuser vane, owing to recirculation inside the impeller which completely alters the axial profile of the velocity field.
- at 0.18 Qn, the separation phenomenon at the leading edge of the diffuser vane is further accentuated. It occupies the full height of the blading when an impeller vane passes in front of the diffuser vane.

This analysis of the flow between the rotor and stator highlights the two main characteristics of the unsteadiness inside a pump (see ref [1]):

- potential interaction: this corresponds to the encounter between 2 distinct flows. Even in the case of an inviscid flow, a spiral flow passing in front of a flow between fixed channels leads to periodical fluctuations in local velocity values.

- wakes interaction: because of the viscosity of the water, separation arises in the impeller and generates a backwash in the outlet along the concave side of the vanes. The density of the wake varies depending on the pump operating condition. The wake reaches the diffuser vanes and creates further periodic local disturbances.

In this type of machine it may be assumed that the two types of interaction are of similar size. Both phenomena are superposed giving rise to considerable fluctuations in velocity, mainly near the leading edge of the diffuser vanes. These local hydrodynamic disturbances spread further along the flow and cause loud noises, mainly due to acoustic propagation.

4. Conclusions

An analysis of the test results obtained for a centrifugal impeller with or without vaned diffuser has enabled us to emphasise the following points.

* The phenomenon of the recirculation beginning inside the impeller leads to complete instability of the flow between the vanes and causes a significant increase in the overall level of velocity fluctuation. The three-dimensional aspect becomes predominant and everything occurs as if the relative flow in the impeller formed an eddy as a result of the effect of the various curvatures of the sides. Separation appears on the concave side of the vane near the front flange where the radius of curvature is the smallest. At the impeller outlet, the recirculation phenomena in the impeller is superimposed on the separation phenomena affecting the vaneless diffuser, and this complicate analysis at this particular spot. Reverse flows in the impeller are only visible at a very low flow rate (near the trailing edge they appear on the outer surface of the vanes) whereas at or near the nominal rate recirculation can already be observed in the diffuser.

* By comparison, the measurements taken with a vaned diffuser show that the recuperating part at the impeller outlet has very little effect on the flow inside. The area near the leading edge of a diffuser vane is the region where the flow is most disturbed when the rate decreases, in both the impeller and the diffuser. In particular, when the rate reaches the critical recirculation value in the impeller, it is observed that the profiles suddenly fail to match the delivery flow.

This study marks an end to the analysis of periodic flow in the ns32 impeller. All the results obtained will serve as a database for the definition of computation codes. Work on the subject is currently being carried out by the SHF's Turbomachine Working Group - Internal flows and related performances. Furthermore, the turbulent aspect of the velocity field may be dealt with in the course of further research work.

References

1. BERT P.F., COMBES J.F., KUENY J.L. (1996) *Unsteady flow calculation in a centrifugal pump using a finite element method,* XVIII IAHR Symposium on Hydraulic Machinery and cavitation, Valencia, Sep. 16-19.

2. BOIS G., RIEUTORD E. (1990) *Etude de l'écoulement en sortie de roue de pompe centrifuge depuis le débit nominal jusqu'au débit critique de recirculation,* Compte rendu MRT n° 88 H 0636, Octobre 1990.

3. CAIGNAERT G., BARRAND J.P. (1989) *Débits critiques de recirculation: synthèse de l'ensemble des résultats expérimentaux sur les roues SHF essayées en air (ENSAM Lille) et en eau (INSA Lyon, EPF Lausanne, Hydroart Milan)* 20èmes journées de l'Hydraulique, SHF Lyon, 4-6 Avril.

4. CASEY M.V., EISELE K., ZHANG Z., GULICH J., SCHACHENMANN A. (1995) *Flow analysis in a pump diffuser, part 1: LDA and PTV measurements of the unsteady flow,* ASME Paper, FED Summer Meeting, Symposium on Laser Anemometry.

5. CASEY M.V., EISELE K., MUGGLI F.A.., GULICH J., SCHACHENMANN A. (1995) *Flow analysis in a pump diffuser, part 2: Validation of a CFD Code for steady flow*, ASME, Numerical simulations in turbomachinery, Vol.227 pp.135-143
6. COMBES J.F., GRIMBERT I., RIEUTORD E. (1991) *An analysis of the viscous flow in a centrifugal pump with a finite element code*, Fluid Machinery Forum, 1st ASME-JSME Fluids Engineering Conference, Portland, Oregon, June 23-27.
7. CROBA D. (1993) *Modélisation de l'écoulement instationnaire dans les pompes centrifuges. Interaction roue et volute*, Thèse Doctorat, INP Grenoble, Juillet 1993.
8. HUREAU F, KERMAREC J., FOUCHER D. (1993) *Etude de l'écoulement instationnaire dans une pompe centrifuge fonctionnant à débit partiel*, 1st International Symposium on Pump Noise and Vibrations, Clamart, France, July 7-9, pp.
9. HUREAU F., KERMAREC J., STOFFEL B., WEISS K. (1993)*Study of internal recirculation in centrifugal impellers*, ASME, 2nd Pumping Machinery Symposium, Washington, D.C, June 20-24, FED Vol. 154, pp. 151-157.
10. LONGATTE F. (1996) *Analyse phénomènologique du couplage rotor-stator dans les pompes centrifuges. Etude sous Fluent d'une pompe 2D avec diffuseur aubé*, LEGI-CREMHyG, Mai 1996.
11. MOREL Ph. (1993) *Ecoulements décollés dans une roue de pompe centrifuge- Conception et réalisation d'un banc d'essai. Analyse des pressions pariétales*, Thèse Doctorat, Université de Lille, Decembre 1993.
12. PAONE N., RIETHMULLER M.L., VAN DEN BRAEMBUSSCHE R.A. (1989) *Experimental investigation of the flow in the vaneless diffuser of a centrifugal pump by particle image displacement velocimetry*, Experiments in Fluids 7, Springer Verlag 1989, pp 371-378
13. STOFFEL B., WEISS K. (1994) *Experimental investigations on part load phenomena in centrifugal pumps*, World Pumps, October 1994, pp 46-50
14. STOFFEL B., LUDWIG G., WEISS K. (1992) *Experimental investigations on the strucure of part-load recirculations in centrifugal pump impellers and the role of different influences*, XVIth IAHR Symposium, Sao Paulo, Brazil, September 14-18.
15. TOUSSAINT M. (1993) *Contribution à l'étude des recirculations dans les pompes rotodynamiques.* Thèse de doctorat, Université P et M Curie Paris VI, Paris, Décembre 1993.
16. TOUSSAINT M. (1994) *Etude des recirculations dans les pompes centrifuges roue Ns 32 diffuseur lisse*, Rapport interne CNAM/CETIM, Paris, Septembre 1994.
17. TOUSSAINT M., HUREAU F. (1993) *Etude expérimentale de l'écoulement instationnaire dans une pompe centrifuge fonctionnant à débit partiel*, IAHR, 6th International Meeting, The behavior of Hydraulic Machinery under steady oscillatory conditions, Lausanne, Sept. 1993.
18. UBALDI M., ZUNINO P., BARIGOZZI G., CATTANEI A. (1994) *An experimental investigation of stator induced unsteadiness of centrifugal impeller outflow*, ASME, La Haye, Paper 94-GT-105.

INFLUENCE OF THE BLADE ROUGHNESS ON THE HYDRAULIC PERFORMANCE OF A MIXED-FLOW PUMP. A VISCOUS ANALYSIS

S. UNDREINER, E. DUEYMES
EDF-DER, 6 quai Watier
78400 Chatou, FRANCE

1. Abstract

The aim of this work is to quantify the blade roughness influence on the hydraulic performances of a mixed-flow pump. In this way the flow in a hydrodynamic tunnel equipped with a blade profile was analysed by measurements performed by the CREMHyG and numerical simulations with the code N3S (developed by 'Electricité de France'). The roughness parameter z_0 used in the simplified rough wall law was calibrated from the measured roughness for three Reynolds numbers and three roughnesses. The application to the pump flow was obtained for the scale model conditions (0.41). For the prototype pump conditions, a calibration was suggested but has to be refined by some other simulations.

2. Nomenclature

Q_n	nominal flow rate	x_d	separation point abscissa
k	measured roughness	z_0	roughness parameter
u_*	shear velocity	U_t	tangential velocity on the wall
ν	kinetic viscosity	ρ	density
L	lenght scale (chord for the pump)	U_c	neck speed

$U^+ = \frac{U_t}{u_*}$ $\quad y^+ = \frac{y u_*}{\nu}$ $\quad z_0^+ = \frac{z_0 u_*}{\nu}$

3. Introduction

A shift of the hydraulic performances of the pump was detected during some specific tests that EDF performed on a test-ring. After a certain number of running hours, the pump head increased of about 5% (figure 1) of the nominal value predicted by the designer. Several hypothesis were analysed to explain this phenomenon. The last one retained was based on a possible modification of the blade roughness. Some theoretical investigations have been performed by computational modelling. The aim was to get a better understanding of the flow structure in the impeller and to know how it can be modified by the roughness. First computations have been performed with an inviscid flow code [1]. At nominal flow rate the results have pointed out an important load on the trailing edge of the pressure side. Indeed, a low velocity level had appeared which

E. Cabrera et al. (eds.), Hydraulic Machinery and Cavitation, 428–437.

could provoke a separation. The computations were completed by a Navier-Stokes approach with the code N3S [2]. Several simulations have been performed: at nominal flowrate, at below and above the nominal flowrate (respectively 0.9 and 1.2 Qn) for a smooth surface. The viscous simulations have given better predictions than the inviscid flow computations, particularly in case of non-nominal operations (0.9 and 1.2 Qn). The comparisons with the measurements were satisfactory. The viscous analysis was completed by a qualitative modelling of the roughness influence on the boundary layers. The results have confirmed the correlation between an important head evolution and a blade roughness modification. In an extreme situation, an evolution on the suction side from a rough to a smooth surface provokes an important increase of the head (about 4%). However it was not yet possible to quantify the relationship between the roughness and the head of the pump: the roughness parameter in the wall law did not represent the right roughness.

The target of this work was to evaluate the roughness corresponding to the parameter used in N3S. In this way an experiment had been performed by CREMHyG [3]. First the perspectives were to analyse the iinfluence of the roughness on the separation point position, second to calibrate the roughness parameter. The results from several N3S simulations were compared to the measurements.

This work is decomposed into six parts:

⇒ in the first part we will describe the CREMHyG experiments and the most important results,

⇒ in the second part, we will briefly describe the N3S code,

⇒ the third part concerns the laws of the wall commonly used in industrial codes for smooth or rough walls,

⇒ the fourth part is the N3S modelling of the water tunnel flow (CREMHyG) and the comparison with the measurements,

⇒ in the fifth part we will suggest a transposition of the hydrodynamic tunnel calibration to the pump flow before concluding in a sixth part.

4. The CREMHyG experiments

The hydrodynamic tunnel was equipped with a blade profile in order to partly reproduce the velocity profile on the pressure side of the pump blade, near the hub (it was obtained from the Euler simulations [1]). The geometry is presented figure (2): the 0.691m length corresponds to the pump chord where the velocity profile is reproduced, the roughness can be adjusted only in the first 0.324m. Three velocities at the neck and three roughnesses are studied: Uc=3, 8 and 24m/s and k=0, 3, 7μm. The separation point was determined either by the pressure gradient cancellation or by means of velocity measurements. From these experiments it appears that: the separation point goes upstream when decreasing the Reynolds number or increasing the roughness (table 1). The roughness influence on the separation point position is more important for the high velocities (Uc=24m/s)than for the low.

Uc (m/s) \ k (μm)	0	3	7
3 m/s	150 (± 2)	148 (± 2)	148 (± 2)
8 m/s	151 (± 2)	148 (± 2)	148 (± 2)
24 m/s	172 (± 2)	166 (± 2)	162 (± 2)

Table 1 Evolution of the separation point abscissa (in mm) with velocity and roughness (CREMHyG [3])

5. The numerical method

The viscous computations are performed with the industrial code N3S (release 3.1) developed by 'Electricité de France'. A finite-element method solves the unsteady Navier-Stokes equations for an incompressible flow. The time discretization is based on a fractional step method which solves successively an advection step, a diffusion step and a generalized Stokes problem. The boundary conditions are based on a non-homogenous Dirichlet condition in the inlet (for the velocity and the turbulent quantities) and a null constraint in the outlet.

6. The wall law

The turbulence model used for these studies is the standard k-ε model associated to a wall law. For a smooth wall N3S solves the law of Reichardt [4] valid for values of y+ ranging from 5 to 200. The rough wall law was established for a pipe flow [5]:

$$U^+ = \frac{1}{\kappa}\ln(y^+) + B - \Delta B(z_0^+) \qquad (1)$$

where B=5.5

ΔB is a constant which depends on the roughness and the shear-stress; for a sand roughness wall:

$$\Delta B = \frac{1}{\kappa}\ln(1 + 0.3 z_0^+)\,. \qquad (2)$$

This law can be simplified for a fully rough flow ($z_0^+ > 60$):

$$U^+ = \frac{1}{\kappa}\ln(\frac{y}{z_0}) + 8.5 \qquad (3)$$

In the previous study on the pump flow with the N3S code the rough wall law was the simplified one with different constants ($\kappa = 0.42$ instead of 0.4):

$$U^+ = \frac{1}{\kappa}\ln(\frac{y}{z_0}) + 8.23 \qquad (4)$$

In our case, since z_0^+ can be inferior to 60, the z_0 parameter can no more represent the physical roughness. A calibration is then necessary.

7. The hydrodynamic tunnel simulation

The flow in the hydrodynamic tunnel of the CREMHyG [3] was simulated for the three neck velocities (3, 8 and 24 m/s). The roughness parameter of the simplified law of the wall (equation 4) was calibrated to the measured roughness (0, 3 and 7μm) essentially at 24m/s velocity.

A study of the finite-element mesh dependance showed that it was possible to use the same mesh for the three velocities. Indeed we did not modify the refinement for the rough simulations. The mesh includes 25492 P2 nodes.

First we studied the flow on a smooth surface to check the ability of N3S to predict the separation point evolution with the Reynolds number. The results are given in the table (2).

U_c (m/s)		3	8	24
x_d (mm) N3S		173	185	195
x_d (mm) Experiments		150 (±2)	151(±2)	172 (±2)
error		15.3 %	22,5%	13.4%
Evolution of the separation point abscissa (mm) compared to the reference case 3m/s	N3S	-	+12	+22
	Exp	-	+1 (<5 and >-3)	+22 (<26 and >18)

table 2 Comparison of computations/measurements for a smooth wall

The N3S predictions are globally correct for the separation point position and for the relative evolution with the Reynolds number. The comparison computations/measurements concerning the velocity profiles confirms this tendancy (figure 3 for the 24m/s velocity). The deviation can be essentially explained by the turbulence modelling (k-ε model and wall law). This approach can not take into account the presence of the adverse pressure gradient and then overestimates the separation point abscissa. Indeed the wall law is no more valid in a recirculating region.

The simulations for the rough walls cases have been performed with the simplified rough wall law. The modelling with the general law (equations 1 and 2) did not give any satisfactory results. Computations of a rough pipe with this law have shown that it was not adapted to the recirculating flow.

The main difficulty was to define a calibration criterion of the z_0 parameter with the experimental roughnesses. The only possibility was to consider the following hypothesis: the measured relative separation abscissa point evolution from the smooth to the rough surfaces had to be reproduced by N3S. Therefore, the abscissa that N3S has to predict is determined and z_0 is computed by successive simulations. With this method we obtained the following calibration table 3.

measured roughness k (m)	0	$3\ 10^{-6}$	$7\ 10^{-6}$
z_0 (m) Uc=24m/s	$5\ 10^{-6}$	$2\ 10^{-5}$	$5\ 10^{-5}$
z_0 (m) Uc=3m/s	$2\ 10^{-4}$	$3\ 10^{-4}$	$3\ 10^{-4}$

Table 3 Calibration of the z_0 parameter at 3 and 24 m/s

For the smooth surface (k=0μm) the z_0 parameter was calibrated with the N3S smooth computations.

The evolution of the z_0 parameter is more important for the higher Reynolds number: z_0 is multiplied by 10 between 0 and 7 μm at the 24m/s velocity and by 2/3 at the 3m/s velocity. This behaviour corresponds to the experimental measurements, table 1.

The suggested calibration shows that the z_0 parameter is a function of the physical roughness and the neck speed. This means in adimensional terms that z_0/L depends on the Reynolds number and k/L where L is a length scale.

In table 3, it appears that:

- the z_0 parameter evolution with the roughness is stronger at the high Reynolds number values,
- the z_0 parameter evolution with the Reynolds number is stronger at low roughness (for the smooth surface).
- the z_0 parameter tends toward a limit value at each Reynolds number.

8. The transposition to the pump flow

The aim is to quantify the roughness parameter z_0 used for the study in the pump flow [1]. The problem is now how to apply the previous calibrations in the water tunnel for the pump flow?

In table 5, the Reynolds numbers are given for each case: the hydrodynamic tunnel, the prototype pump, and the scale model conditions; for the pump flow the Reynolds number is based on the blade chord and the mean velocity at the leading-edge.

		Reynolds number
Water tunnel	3m/s	$2\ 10^6$
	8m/s	$5.3\ 10^6$
	24m/s	$1.6\ 10^7$
pump: scale 1 (hot water)		$1.5\ 10^8$
pump: scale 0.41 (cold water)		$3\ 10^6$

Table 5 Reynolds number for each configuration

8.1. The scale model

In table 5 it appears that the Reynolds number of the flow in the scale model is comparable to the 3m/s velocity in the water tunnel. It is then possible to apply the water tunnel calibrations to the scale model conditions : the ratio z_0/L (L is the length scale) is equal. The results are given in table 6.

k (µm)	3
z_0 (m)	10^{-4}

Table 6 Transposition to the scale model flow

However, scale model flow simulations have shown that:

- ♦ the roughness effects are opposite to those obtained for the prototype pump conditions (the head increases with the roughness),
- ♦ the previous calibration needs a finer analysis of the water tunnel calibration applicability to the scale model flow.

8.2. The scale 1

With results concerning only two Reynolds numbers it is difficult to extrapolate the z_0 values for the pump flow corresponding to 3 and 7 µm: the pump Reynolds number is 10 times higher than the Reynolds number in the hydrodynamic tunnel. In figure (4) we tried to represent from our calibration the z_0 evolutions with the roughness k for two Reynolds numbers. The values z_{01} and z_{02} in figure (4) were used for the pump simulations [2]. If we extrapolate the difference between the velocities 3 and 24 m/s to the difference between 24m/s and the pump flow conditions, we can suggest that the $z_0=10^{-5}$ (m) corresponds to a maximum of 7µm. However, the Reynolds number effects have to be taken into account. This means that for high values the Reynolds number has less influence on the flow. It would be interesting to check this behaviour by a water tunnel simulation at the pump flow Reynolds number. In such case a refined mesh would be necessary.

9. Conclusion

The aim of our work was to quantify the blade roughness influence on the hydraulic performance of a mixed-flow pump. A previous study has shown that a significative head decrease can be correlated with an increase of the blade surface roughness. Our work consisted in a calibration of the roughness parameter (z_0) used for the previous study to a real roughness.

In this way the CREMHyG has performed an experiment in a hydrodynamic tunnel. The roughness effects on the boundary layer separation was quantified by surface conditions and velocities measurements. The experimental results were compared to the N3S simulations.

The results obtained with N3S in the smooth conditions are satisfactory in regard to the simple wall modelling (law of the wall). This means that N3S can predict a correct evolution of the separation abscissa. The deviation between the measured and calculated abscissa is about 15%.

Several difficulties were analysed before suggesting a calibration first in the water tunnel and second in the pump. The main difficulty was to define a calibration criterion in the hydrodynamic tunnel. The presence of a recirculation put the general law of rough wall in the wrong track. Therefore we had to consider the following hypothesis: the relative evolution between the experimental and the numerical results has to be respected for the different velocities. The second difficulty concerned the extrapolation of our suggested calibration to the pump (scale one) flow. Firstly, the Reynolds numbers between the pump and the hydrodynamic tunnel conditions are completly different (a ratio of 10). Secondly, more experimental results are needed to correctly take into account the Reynolds number effects: for example an intermediate case between 8 and 24 m/s.

Despite these difficulties, we suggested a calibration for the pump flow (scale 1) and the scale model.

In case of the scale model the roughness of 3μm can be linked to $z0=10^{-4}$(m). This calibration would be reviewed considering the last simulations results: the flow is extremely sensitive to the roughness in opposite to the flow in the water tunnel (at similar Reynolds number).

In case of the prototype pump the parameter $z_{02}=10^{-5}$ (m) would correspond to a roughness of 7 μm at the most and $z_{01}=10^{-6}$ (m) to 3μm. Therefore a roughness variation of 4μm could provoke a head decrease of 2%. The measured and calculated head evolution is given figure (5).

This study have shown the complexity of modelling the roughness effects by means of a wall law. Several supplementary analysis can be envisaged:

- a direct calibration for the scale model by comparisons between N3S computations and measurements,
- a finer comparison of the flow structures between the prototype pump and the scale model conditions (Reynolds number effects),
- an improvement of the rough wall laws.

10. References

[1] **I. GRIMBERT.** *Dérive des charactéristiques hydrauliques d'une roue de pompe; premières analyse de l'écoulement interne*. Internal report EDF/DER HP-41/91.17

[2] **V.MOULIN**. *Analyse complémentaire de l'écoulement dans une roue de pompe avec le code N3S.* Internal report EDF/DER.HP-41/94/007.

[3] **C. REBATTET**. *Etude expérimentale de l'influence de la rugosité sur le comportement de la couche limite dans une veine*. Rapport CREMHyG 1994.

[4] **J.P CHABARD**. *Projet N3S de Mécanique des Fluides, Manuel théorique de la version 3.* Internal report EDF/DER HE-41/91.30B.

[5] **F.M. WHITE.** *Viscous Fluid Flow.* McGraw-Hill, Inc, Second Edition.

11.Figures

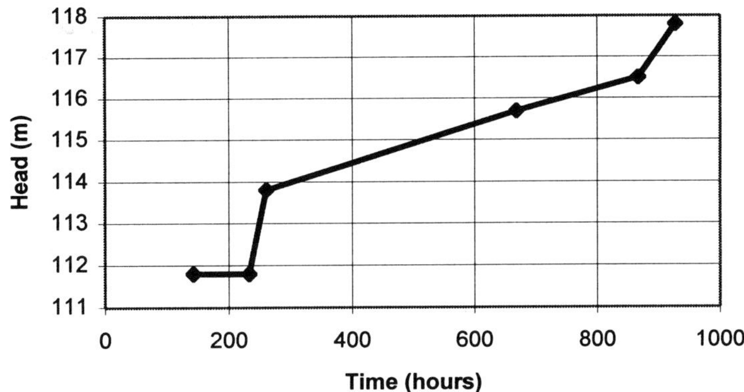

Figure 1 measured head evolution in time during the fifth test serie

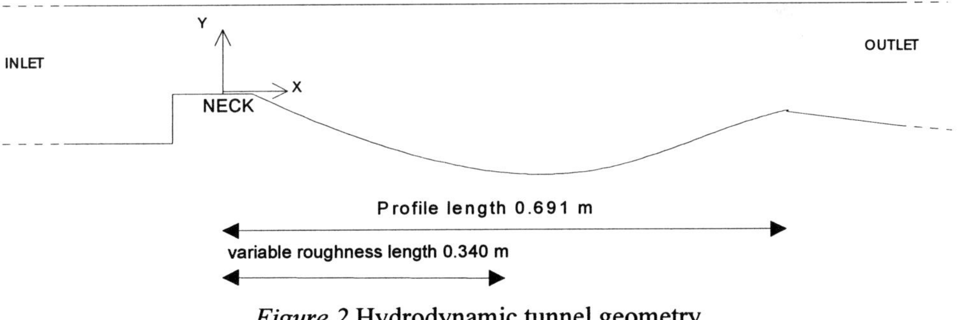

Figure 2 Hydrodynamic tunnel geometry

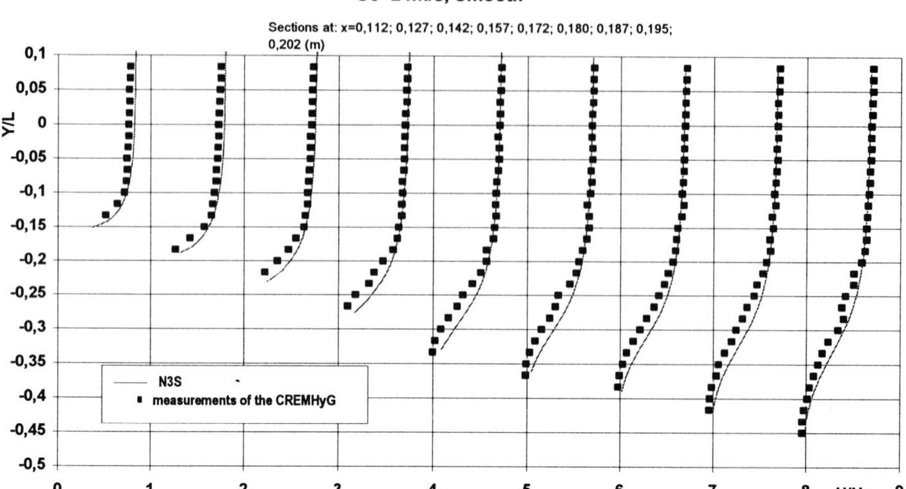

Figure 3 Comparison measured/calculated velocity profiles(Uc=24m/s)

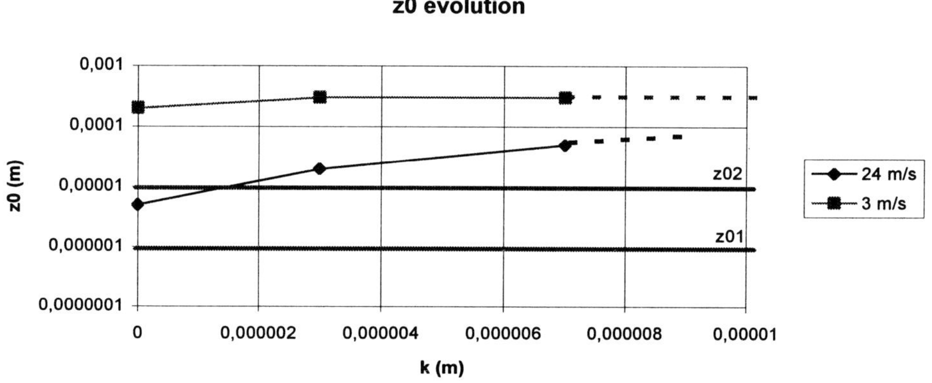

Figure 4 z0 evolution (water tunnel)

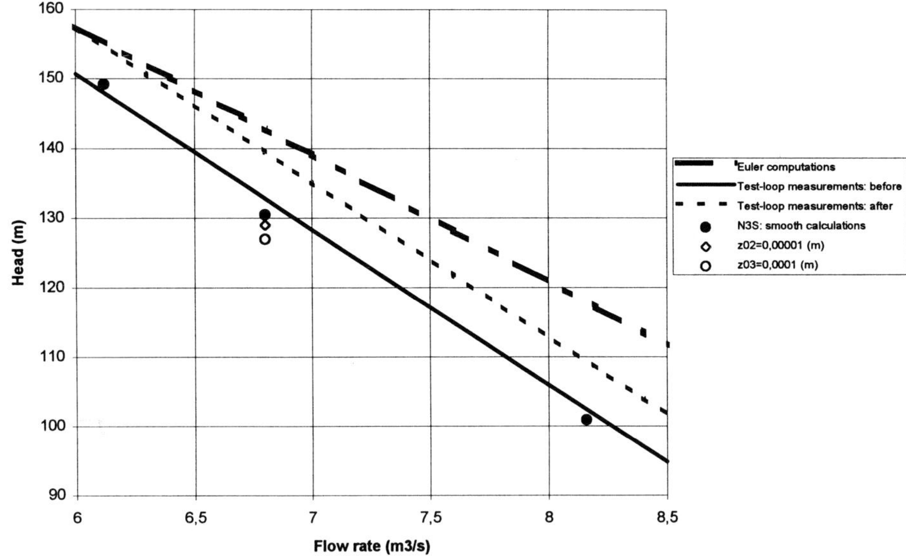

Figure 5 Head of the mixed-flow pump with the flow rate (mesaurements and computations)

IMPROVEMENT OF PERFORMANCE OF CENTRIFUGAL PUMPS BASED ON COMPUTATIONAL AND THEORETICAL METHODS AND EXPERIMENTAL DESIGN

A.F. Vinokurov,
(J-S-C "POMPA", Moscow region, Russia)
A.V. Volkov, G.M. Morgunov, S.N. Pankratov
(Moscow Power Engineering Institute, Moscow, Russia)

Introduction

The problems of quality performance improvement of centrifugal pumps are actual and long-range problems. This type of pumps is widely used in all branches of national economy. Improvement of various technological processes requires modernization of old centrifugal pumps or manufacture of new ones able to provide improved performance.

The problem of raising efficiency of a pump unit , same as improvement of its anticavitation characteristics, belong to the main tasks which attract attention of research workers and designers of pump equipment.Increase of pump efficiency by some per cent, and in other cases - by fractions per cent, considering high driving capacities or a big number of operating hours/year of the pump unit, results in significant energy saving. Improvement of anticavitation characteristics increases reliability of the pump, its ecological safety during operation.

Modernization of the pump, in contrast to making a new hydraulic machine,allows to lower very significantly the level of expenditures on development and use existing foundations, main pipelines, piping adjacent to the pumps, electrical equipment. Besides, necessity of modernizaton also arises in those cases when the pump operation does not agree with the characteristicof the network for which it functions. As a result, the hydraulic machine operates at non-optimum conditions. It means it is necessary to bring operating conditions in line with the optimum point of the pump operational characteristic. Fabrication of a new impeller only, in accordance with the conditions of the network required, makes it possible to get a considerable saving.

Sustantial Part

A successful settlement of the problems under consideration is achieved by using more precise computational methods of research on hydraulic machines allowing to carry out with a high degree of authenticity an analysis of work processes going on in the liquid passage part of the pump for a wide spectrum of integrated and local characteristics, and apply optimization methods to design phases.

E. Cabrera et al. (eds.), Hydraulic Machinery and Cavitation, 438–444.

1D-3D method (one-dimensional inverse and three-dimensional direct hydrodynamic problems) developed in the Moscow Power Engineering Institute (MPEI) for design and study of liquid passage parts of vane-type hydraulic machines makes it possible to reach a high level of outer characteristics of pump equipment.

1D problem of synthesis is a one-dimensional optimization problem using a theory of optimal control. Its solution results in initial approximation of a space image of the median surface of the vane obtained from the condition of minimum loss in the impeller as well as in the feeding and discharging devices. A subsequent transition to a corporeal image of the vane is accompanied with transmission of geometric information for an analysis problem.

3D problem [3] account of a space flow blades of systems hydraulic machines cvazi-establishment unsound by a viscous flow is based on integrated representation 3D of the theory of fields. The given approach permits on order to lower dimension of soluble problems.

The problem is realized in three steps: calculation of potential flow in the lattice, calculation of vortex-type barotropic flow in the vane system, approximate account of viscosity with limitation of the subdomain of the turbulent layer near the wall. As a result of its solution between the vanes, and on the vanes, tangential pressures, integrated hydrodynamic lattice parameters, effective circulation of the impeller , friction coefficients , in particular, inductive resistance, cavitation, etc.

The above mentioned method was applied to modernization of centrifugal pump type " K 50-32-125 " produced serially in Russia by joint-stock company " POMPA ". The pump did not provide operating conditions required by ISO 2858. Research on the original version of the passage part as well as analysis and synthesis of the new (modernized) passage part of the pump was carried out using this computational-theoretical method.

On fig. 1 are submitted meridianal of a projection and plans of impellers the original version and modernized variants.Index "1" corresponds to the original version, index "2" - to the modernized one. Reception such geometry sizes as: R1, R2, b_1,b_2 and the construction of the new plan of a blade was executed with use 1 D - problem.

The main variant of a impeller did not provide required settlement parameters,therefore at perfection of initial variant a problem of maintenance required parameters of quality with simultaneous maintenance of higher power qualities of a new pump. Moreover on conditions of technology produce of a impeller , restriction was in addition exhibited: perpendicular of a leading disk of a axis of rotation. In connection that new a impeller should be produced from sheet have become were in addition imposed restrictions on a structure of a blade - blade of constant thickness

The comparative analysis of initial and modernized variants was is carried out on the basis of 3 D of a method MPEI. On fig. 2,3 dependences relative speed W on a working surface of a blade for pluging (i=1) and peripheral (i=10) areas meridianal projection of a impeller. We shall note, that variant "2" has downturn of significances of speed in the

second half of a blade. Show downturn of size of speed, can in turn up the reason of reduction of losses of friction in a impeller and by that increases a efficiency of a pump.

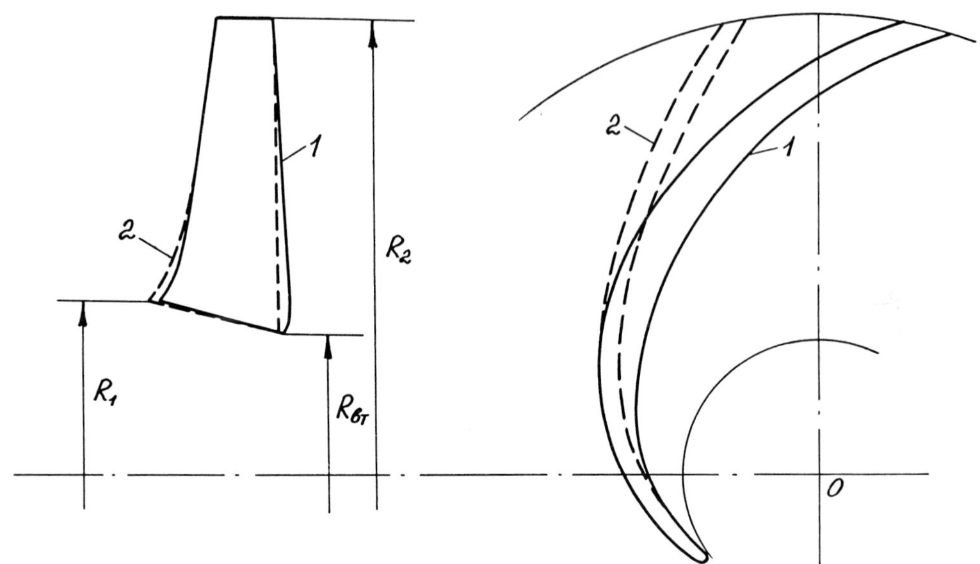

Figure 1. Meridianal of a projection and plan impellers

On fig. 4,5 of dependence change of the circulation Gs along a blade, where s - relative length of a blade. Dependences agreement modernized to variant ("2") have higher significance, that permits to assume about a increase of significance head H develop modernized ("2") a impeller.

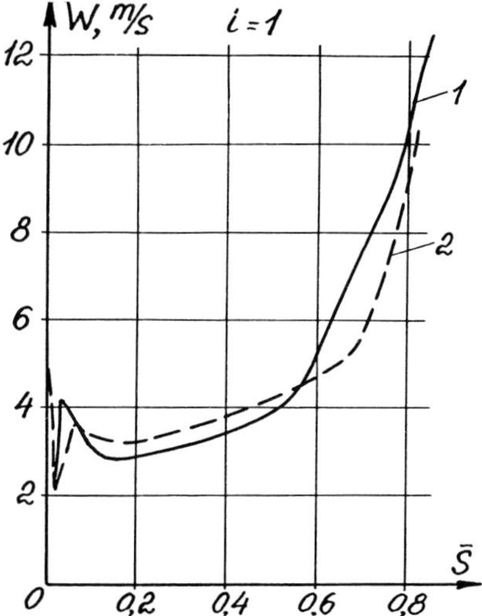

Figure 2. Dependences relative speed i=1.

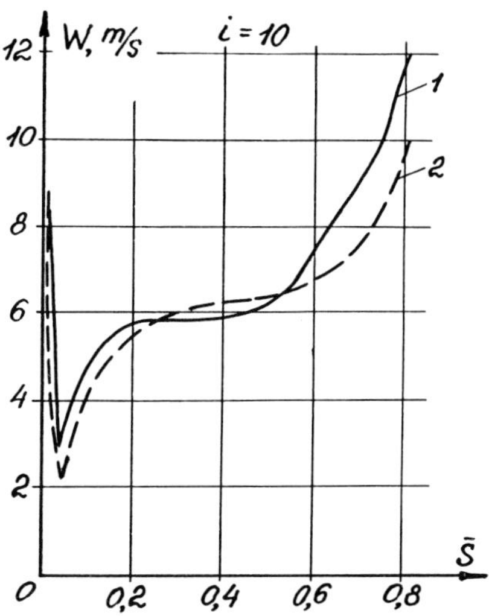

Figure 2.Dependences relative speed i=10.

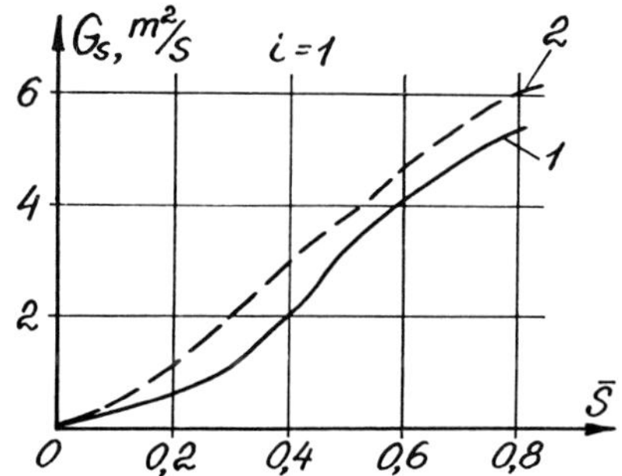

Figure 4. Dependences change of the circulation along a blade i=1.

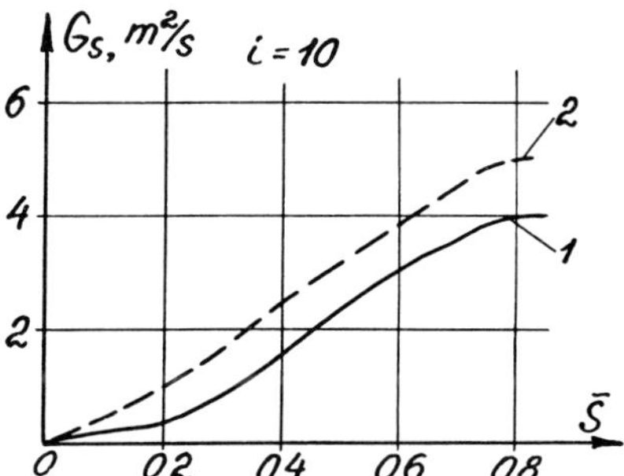

Figure 5. Dependences change of the circulation along a blade i=10.

In table 1 the comparative analysis of main integrated the characteristics received as a result of accounts.

Table 1. Comparison of the integrated characteristics of researched pumps

N	Gs m^2/s	Hfi %	Hi %	He %
1	8.95	1.7	10.0	20.0
2	9.56	1.2	6.6	13.3

, here Hfi - loss in a distant trace, stipulated by non-uniformity change of circulation on scope of a blade, Hi - loss on contours of a wheel, He - sum losses of friction on contours and blades of a impeller. Submitted data permit to make the assumptions that variant "2" has higher iency than initial variant, roughly on 6-7 %.

Fig.6 shows the results of action research on the original and modernized pumps. The modernized version of the pump provides necessary working ability of the hydraulic machine: at flow rate Q= 12.5 m^3/hr it develops head H=20 m as required by the standard. Moreover, it a higher efficiency which results in significant energy saving. The relation presented here confirms effectiveness of this approach to centrifugal pump improvement.

Conclusion

At modernization of a pump were simultaneously decided and technological problems. Use of a more improved process of impeller fabrication (welded and stamped version) made it possible to reduce the quantity of cast iron which is not always welcomed by manufacturers because of arising ecological problems. This decision also reduced considerably the cost of the design due to a lower number of machining, treatment and inspection operations during impeller manufacture. The above mentioned improvements of the pump design made it possible to increase reliability of the pump unit operation and make the hydraulic machines produced by joint- stock company " POMPA " more competitive.

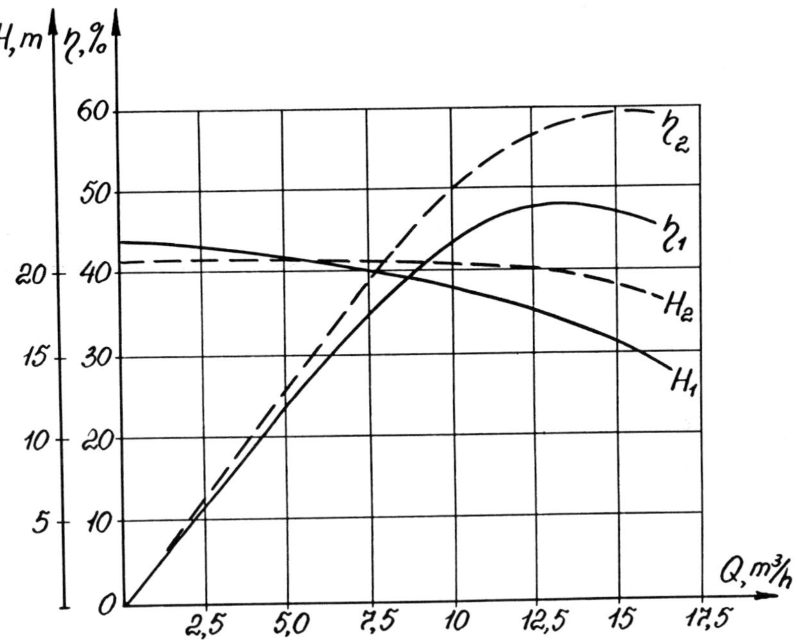

Figure 6. Results of experimental investigation.

References

1. Morgunov G.M.,Volkov A.V.,Gorban V.M. Combined 1D-3D cad method of hydromachine water pessages design..- "Mathematics, computer, control and investments",Int. Conf.,Moscow,15-19 february, 1993, p. 58.

2. Vinokurov A.F.,Volkov A.V. , Kuzmin S.V. One of combine of the methods designing flowing of parts
centrifugal of pumps.- Moscow power engineering institute 1992. N 652. p.52-60. (in Russian)

3. Morgunov G.M. Calculation unseparetion of the flow spacing blades of systems with the account viscous. - Izv. AS USSR, Engineering and transport, 1985, N 1, p.117-126. (in Russian)

LIQUID-PARTICULATE TWO-PHASE FLOW IN CENTRIFUGAL IMPELLER BY TURBULENT SIMULATION

WU Yu-lin, DAI Jing and MEI Zu-yan
Tsinghua University
Beijing, CHINA 100084

OBA Risaburo and IKOHAGI Toshiaki
Tohoku University
Sendai, JAPAN 980-77

Abstract

In the present work, turbulent liquid-particulate two-phase flows at the particle dilute concentration through a centrifugal impeller has been simulated by using the K- ε –Ap turbulence model. In the present numerical treatment, the finite volume method, based on the SIMPLEC algorithm, is applied in a body fitted coordinate system. The liquid-particle two-phase flow turbulence model used in this study can correctly predict essential features of this flow in centrifugal impellers at dilute concentrations.

1. Introduction

Sevaral theoretical and experimental techniques for investigation of multiphase flows are available. Basically, there are two fundamental theories for two-phase flows, namely, the macroscopic continuum mechanics theory and the microscopic kinetic theory. Marble (1963) was one of pioneer researchers to describe basic equations of two-phase flows. From the sixties to seventies, general governing equations of two phase flows for the two fluid model and the mixture model based on the continuum mechanics theory were set up by Drew (1971), Ishii (1975) and others by using different average methods. In the mixture model, it is assumed that there exists only one type of mixture in fluid-dispersed flows, and that the whole flow space is full of the mixture. The mixture flow is described by using a set of unique characters. Roco & Reinhart (1980) applied this model to calculate the liquid-particle mixture flow through centrifugal pump impellers .

In the two-fluid model, the dispersed phase is treated as a pseudo-fluid. In the Eulerian approach, the flow of the dispersed phase is described by conservation equations of mass, momentum and energy in continuum mechanics. In this model, there exists the slip of parameters between the carrier fluid and the dispersed phase. Based on the two-fluid model, increasing interest in the prediction of turbulent multi-

E. Cabrera et al. (eds.), Hydraulic Machinery and Cavitation, 445–454.

phase flows has been noticed during the last twenty years, such as Danon et al (1974), Al Taweel & Landau (1977), Genchiev & Karpuzov (1980), Melville et al (1979), Sharma & Crowe (1978), Michaelides & Farmer (1984) and Shuen et al (1983). Danon et al (1974) tried to develop the one-equation turbulence model by adding a length scale determined experimentally to the turbulent kinetic energy equation for calculating two-phase round jet-turbulent-flows. Al Taweel & Landau (1977) deduced an energy spectrum equation at the transient frequency band, adding a supplement dissipation term to consider appearance of the dispersed phase. Calculated results show that the increase of volumetric density of the dispersed phase causes the turbulence intensity at the high frequency band decrease. Genchiev & Karpuzov (1980) proposed that a sink term has to be added to the single phase turbulent kinetic energy k equation to simulate the influence of the disperse phase. These authors applied the two-equation turbulence model to calculate two-phase turbulent flows. Two-equation turbulence models also have been proposed for dilute fluid-particle flows by Elghobashi & Abou-Arab (1983) and Crowder et al (1984). Algebraic and one equation turbulence models have been suggested for the dense liquid-solid interaction by Roco & Shook (1983). Bertodano et al (1990) applied the two-fluid model combined with the Reynolds stress turbulence model to analyze water-bubble flows. Dai & Wu (1993) have used the k- ε -a-kp liquid-particulate two-phase turbulence model to calculate this flow through centrifugal impellers at dilute concentration. Most of existing models are suitable only for relatively dilute mixtures, in which the particle-particle collision effect and the fluctuation energy interaction between the liquid and the particulate phases are negligible. During the last decade, there has been an interesting development in modeling dense solid-liquid flows, as shown, in References of Ma & Roco (1988), Ahmadi & Abu-Zaid (1990) and Gidaspow et al (1991). As mentioned above, Roco & Shook (1983) applied the macroscopic continuum theory and the one-equation turbulence model with double averaging to calculate dense liquid-particle flows. A probabilistic microscopic model for solving shear motion of spherical particles dominated by friction and lubrication force was proposed by Ma & Roco (1988) to slow granular flow and high dense coal slurry. Gidaspow et al (1991) have developed a computer code that solves a generalization of N-S equations for multiphase flows. Particulate phase viscosity and pressure were derived by using the mathematical techniques of the dense gas kinetic theory.

2. Governing Equations

For incompressible liquid-particle flows through a centrifugal impeller, the continuity-, the momentum- and the turbulent character-equations for both the liquid and the particulate phases in 2D Cartesian coordinate system, fixed on the impeller rotating with an angular velocity ω (rad/s), can be written as follows:

2-1 LIQUID PHASE

(a) Continuity equation

$$(u_j)_{xj} = 0 \tag{1}$$

(b) Momentum equations

$$\{\rho\, u_i u_j - \mu_e[(u_j)_{xi} + (u_i)_{xj}]\}_{xj} = -p'_{xi} + \rho_p (u_{pi} - u_i)/\tau_{rp} - 2\rho\, e_{ilk}\, \omega_l\, u_k \tag{2}$$

where

$$\tau_{rp} = \rho_p d_p^2 (1 + Re_p^{2/3}/6.0)^{-1} / 18\mu \quad Re_p = |u - u_p|\, d_p/\nu$$

$$p' = p - \rho\, \omega^2 r^2/2$$

The subscript xj represents the derivative with respect to xj. The term $2\rho\, e_{ilk}\, \omega_l\, u_k$ represents the Coriolis force.

(c) Equation of turbulent kinetic energy k

In order to calculate turbulent flows in a practical way, a turbulence model is necessary for simulating turbulence process in the flows. In this study, the two-phase turbulence model, particularly, the k- ε –Ap turbulence model is used for simulating the liquid-particulate turbulent flow in centrifugal impellers. The k- ε –Ap model is actually the k- ε turbulence model for liquid flows combined with an algebraic turbulence model for particulate phase flows. The equation of the turbulent kinetic energy k is indicated as follows:

$$\{\rho\, k u_j - (\mu_e/\sigma_k)\, k_{xj}\}_{xj} = G_k - \rho\, \varepsilon \tag{3}$$

where $G_k = \mu_e [(u_i)_{xj} + (u_j)_{xi}] (u_i)_{xj}$

(d) Equation of turbulent energy dissipation rate ε

$$\{\rho\, \varepsilon\, u_j - (\mu_e/\sigma_\varepsilon)\, \varepsilon_{xj}\}_{xj} = C_{\varepsilon 1}\, \varepsilon/k\; G_k (1 + C_{\varepsilon 3} R_f) - C_{\varepsilon 2}\, \rho\, \varepsilon^2/k \tag{4}$$

where $Rf = (k/\varepsilon)^2 (w/r^2) (wr)_r$ represents the effect of rotation on liquid turbulence, w is the circumferential velocity component.

(e) Boussinesq eddy viscosity μ_t

$$\mu_t = C_\mu k^2/\varepsilon \tag{5}$$

Constants in the k- ε turbulence model are taken as:

$C_\mu = 0.09$ $\sigma_k = 1.0$ $\sigma_\varepsilon = 1.3$ $C_{\varepsilon 1} = 1.44$ $C_{\varepsilon 2} = 1.92$ $C_{\varepsilon 3} = 0.8$

2.2 PARTICLE PHASE

(a) Continuity equation

$$[\rho_p u_{pj} - \nu_p/\sigma_p(\rho_p)_{xj}]_{xj} = 0 \tag{6a}$$

The continuity equation (6a) may be expressed in the form of the particle number n_p, that is,

$$[n_p u_{pj} - \nu_p/\sigma_p(n_p)_{xj}]_{xj} = 0 \tag{6b}$$

where $\rho_p = m_p n_p$, and m_p is the mass of a particle.

(b) Momentum equations

$$\{\rho_p u_{pi} u_{pj} - \mu_p[(u_{pi})_{xj} + (u_{pj})_{xi}]\}_{xj} = \nu_p/\sigma_p[u_{pi}(\rho_p)_{xj} + u_{pj}(\rho_p)_{xi}]_{xj} + \rho_p/\tau_{rp}(u_i - u_{pi}) - 2\rho_p e_{ilk}\omega_l u_{pk} + 0.5(\rho_p \omega^2 r^2)_{xi} \tag{7}$$

(c) Eddy viscosity of particulate phase ν_p

The algebraic turbulence model of the particulate phase is used in this computation. So the eddy viscosity ν_p of the particulate phase can be expressed in the following Hinze-Tchen equation:

$$\nu_p/\nu_t = kp/k = (1 + \tau_{rp}'/\tau_t)^{-1} \tag{8}$$

where the particle dynamic response time τ_{rp}' and the fluctuation time of liquid turbulence τ_t are shown in the following forms:

$$\tau_{rp}' = \rho_p d_p^2/(18\mu); \qquad \tau_t = l/u' = (1.5)^{0.5} C_\mu^{3/2} k/\varepsilon$$

The governing equations of motion for the two-dimensional relative steady incompressible liquid-particle flow through such an impeller, expressed in the BFC system (ξ, η), can be represented by the model transport conservation equation (Wu et al 1995 b).

3. Boundary Conditions

The boundary conditions of liquid flows through the impeller for turbulence computation are the same as those used in the reference by Wu et al (1995 a). Only additional boundary conditions of particle turbulent flows are imposed.

The variables in the particulate phase also possess zero normal gradient specification everywhere except the inlet. At the inlet, particle velocity components have the same values as those in the liquid flow. And the bulk density in the particulate phase at the inlet can be easily determined according to the known volumetric concentration.

4. Solution Procedure

Discretization of the governing equations (1), (2) and so on in the form at the BFC system are performed, by using the same finite volume approximation as used in Ref. of WU et al (1995a). The second order central differencing is used for approximating the diffusion and source terms. For convective terms, the hybrid differencing scheme is employed. Solutions of incompressible turbulent flows are nonlinear and strongly coupled. The SIMPLEC algorithm is applied to drive the pressure field and the velocity field to be divergence free and convergence. The present numerical method for solving liquid-particulate flows contains the following steps:

(1) Estimate the initial velocity and the pressure field of a single liquid phase flow in the computational domain.

(2) Solve for the velocity, the pressure and the turbulent characters k and ε in the single phase flows, by using the SIMPLEC method, to get coarse convergence solution of the flow.

(3) Based on the single phase flow solution, we solve for the particulate turbulent flow field, by using the algebraic particulate phase turbulence model.

(4) Calculate the interaction terms between the liquid- and the particulate-phases to get some source terms in the control equations for the liquid flow.

(5) Solve again the momentum equations, the turbulent energy and its dissipation rate equation for single phase flow, but including the interaction terms.

(6) Set the new variables and return to step (3); repeat the process until it converges.

5. Calculated Results

Calculation of the liquid-particulate two-phase turbulent flow has been carried out for the same model slurry pump as that used for experiments in the references by Wu et al (1992, 1993), by using the Cray Super Computer in the Institute of Fluid Science of Tohoku University, JAPAN. The geometry and the inlet flow parameters of the impeller, as well as the grid system of computational domain have been shown in the references by Wu et al (1995 a and b). The reliability of the computational method and computer program in the present work can be obtained by comparison between the calculated results of the pure water flow through the same impeller and the experimental results by using the PIV technique (Wu et al 1995 a) and also by comparison between the calculated results and the experimental data for a rectangular region with a backward-facing step (Wu et al, 1995 b).

Figs. 1, 2 and 3 show the results at the following conditions:

Volumetric concentration of solid particle phase C_V= 0.1

Particle diameter	d_S= 0.1 mm
Particle density	ρ_S= 2.65 g/cm3
Inlet radial velocity	u_{r1}= 6 m/s
Rotation speed	n = 1450 rpm

Figs. 4, 5 and 6 show the results at the conditions:

Volumetric concentration of solid particle phase C_v= 0.1
Particle diameter d_s= 1.0 mm
Particle density ρ_s= 2.65 g/cm3
Inlet radial velocity u_{r1}= 6 m/s
Rotating speed n = 1450 rpm.
From these calculations, the following remarks can be obtained.

(1) According to the comparison between Figs. 2(a), 2(b) and 5(a), 5(b), as well as figures in the reference by Wu et al (1995 b), the existence of solid particles in the impeller has a little influence to the liquid flows at dilute solid concentrations, that is, C_v is smaller than 0.1. In the single liquid phase flow (Figs. 1 (a, b)), the flow recirculation occurs near the pressure side and by the trailing edge, resulting from the large expansion of transverse passage sections, caused by the large blade outlet angle $\beta_{b2} = 34^o$ and the small value of blade number, N_b=4. This phenomenon was verified by the PIV observation results in the reference by Wu et al (1995 a).

But the existence of particle phase makes the maximum value of relative velocity at the leading edge on the pressure side lower slightly, e.g., at the single-phase flow W_{max} = 13.27 m/s (Ref. Wu et al 1995 b), but at the two-phase condition of C_v= 0.1 and d_s= 0.1 mm, W_{max}= 12.24 m/s (Fig. 2(a)), and at the condition C_v= 0.1 and d_s= 1.0 mm, W_{max}= 12.84 m/s (Fig. 5 (b)). Also the existence of particles makes the area of the recirculation and small velocity near the pressure side outlet different. For example, at a condition of C_v= 0.1 and d_s = 0.1 mm, that is, at a small particle condition (Fig. 2 (a)), the area is getting smaller a little than that at the single-phase flow. And at the condition of C_v=0.1 and d_s=1.0 mm, that is, at a large particle condition (Fig. 5 (a)), the area is slightly larger for the large diameter particles possessing greater velocity than that of liquid flows and making the liquid flow circulation more severely.

The existence of particles also affects to the pressure developed by the impeller. The larger the particle diameter is at the same concentration and the same density conditions, the lower the pressure in the liquid flow field is at the impeller outlet. For example, at the single-phase flow P_{max} is 4.99×10^5 Pa, in the two-phase flow with d_s = 0.1 mm, P_{max} is 4.85×10^5 Pa and at the condition with d_s = 1.0 mm, P_{max} is 4.68×10^5 Pa.

(2) The velocity field in the particulate phase is different from that of the liquid phase (Figs. 1(b), 2(b), 4(b) and 5(b)). The velocity difference between these two phases takes place mainly near the outlet of the impeller. That is, the relative particle velocity components at the outlet are higher than those in the liquid flow. The phenomenon agrees with the measurements by using LDV by Carder et al (1991). The results also agree with the observation of individual particle movement by Herbich & Christopher (1963) and by Wu et al (1992, 1993). That is, the particles are not appreciably slowed down even at the large cross-sectional area of the impeller discharge ring, and their absolute tangential velocities do not follow those of the transporting liquid.

(3) An increase in particle diameter results in a marked difference between the particle velocity and the liquid one. From comparison between Figs. 1(b) and 4(b), it is very clear that the particle relative velocity Ws near blade pressure side at the outlet under the condition of d_s = 1.0 mm (Fig. 4(b)) is either larger than that of the liquid flow (Fig. 4(a)), or larger than that of particles with d_s = 0.1 mm in Fig. 1 (b). Because the particle with a larger diameter as a pseudo-fluid in the present model keeps more inertia. This phenomenon also agrees with the observation of individual particles by Wu et al (1993). And an increase in the volumetric particle concentration results in the similar effects.

(4) The existence of particles in the liquid flow makes the turbulent kinetic energy k of the liquid phase lower. This trend will be severe with an increase in particle diameters at the same volumetric concentration. In the single-phase flow, the maximum value of k, i.e., k_{max} is 8.13 m2/s2, but in the two-phase flow with d_s=0.1 mm, k_{max} is 7.03 m2/s2 and at d_s=1.0 mm condition, k_{max} is 6.54 m2/s2 , which is much lower than that in single-phase flow.

6. Conclusions

(1) The liquid-particle two-phase flow turbulence model used in this study, the k- ε -Ap turbulence model, can correctly predict essential features of this flow in centrifugal impellers at dilute concentrations.

(2) The velocity field of particulate phase is different from that of the liquid phase in centrifugal impellers, especially near the outlet ring.

7. References

Ahmadi, G. and Abou-Zaid, S. (1990) A Rate-dependent thermodynamical model for rapid granular flows, *Int. J. Non-Newtonian Fluid Mechanics*, Vol. **35**, 15-3.

Al Taweel, A.M. and Landau, J. (1977) Turbulence modulation in two-phase jets, *Int. J. Multiphase Flow*, Vol. **3**, 341-351.

Bertodano, M.L., Lee S-J., Lahey, R.T. and Drew, D.A. (1990) The prediction of two-phase turbulence and phase distribution phenomena using a Reynolds stress model, *Trans. ASME, J. Fluid Engng.*, Vol. **112**, 107-113.

Carder, T., Masbernat, O. and Roco, M.C. (1991) Liquid-solid slip velocity around a centrifugal pump impeller, *ASME FED*-Vol. **118**, *Liquid-Solid Flows*, 101.

Crowder, R.S., Daily, J.W. and Humphrey, J.A.C. (1984) Numerical calculation of particle dispersion in a turbulent mixing layer flow, *J. Pipelines*, Vol. **4**, **3** , 159-170.

Dai, J. and Wu, Y.L. (1993) Numerical simulation of turbulent liquid-particle flows in centrifugal impeller, *Proc. Int. Symp. on Aerospace and Fluid Science*, Sendai, 404-410.

Danon, H., Wolfshtein, M. and Hetsroni, G. (1974) Numerical calculations of two-phase turbulent round Jets, *Int. J. Multiphase Flow*, Vol. **3**, 223-234.

Drew, D.A. and Segal, L.A. (1971) Averaged equation for two-phase flow, *Studies in Applied Math.*, Vol. **50**, 205-231

Elghobashi, S.E. and Abou-Arab, T.W. (1983) A two-equation turbulence closure for two-phase flows, *Phys. Fluids*, Vol. **26**, pp. 931-938.

Genchiev, Zh.D. and Karpuzov, D.S. (1980) Effects of motion of dust particles on turbulence transport equations, *J. Fluid Mech.*, Vol. **103**, 833-842.

Gidaspow, D., Jayaswal, U.K. and Ding, J. (1991) Navier-Stokes equation model for liquid solid flows using kinetic theory, *ASME* FED-Vol. 118, *Liquid-Solid Flows*, 165-172.

Herbich, J. B. and Christopher, R. J. (1963) Use of high speed photography to analyze particle motion in a model dredge pump, *Proc. IAHR Congress*, London, 89-98.

Ishii, M. (1975) *Thermo-fluid dynamic theory of two-phase flow*, Eyrolles, Paris.

Ma, D.N. and Roco, M.C. (1988) Probabilistic three-dimensional model for slow shearing particulate flow: wet friction, *ASME* FED-Vol. **55**, 53-60.

Marble, F.E. (1963) Dynamics of a gas containing small solid particles, *Proc. of 5th AGARD Combustion and Propulsion Colloquium*, Pergamon Press, London.

Melville, W.K. and Bray, K.N.C. (1979) A model of the two-phase turbulent jet, *Int. J. Heat Mass Transfer*, Vol. **22**, 647-656.

Michaelides, E.E. and Farmer, L.K. (1984) A model for slurry flows based on the equation of turbulence, *ASME* FED-Vol. **13**, *Liquid-Solid Flows and Erosion Wear in Industrial Equipment*, 27-32.

Roco, M.C. and Reinhart, E., (1980) Calculation of solid particles concentration in centrifugal pump impeller using finite element technique, *7th Hydrotransport*, 359-376.

Roco, M.C. and Shook C.A. (1983) Modeling slurry flow: The effect of particle size, *Canadian J. Chem. Engng.*, Vol. **61**, **4**, 494-504.

Sharma, M.P. and Crowe, C.T. (1978) A novel physico-computational model for quasi one-dimensional gas particle flows, *Trans. ASME, J. Fluid Engng.*, Vol. **100**, 343-349.

Shuen, J.S., Chen, L.D. and Faeth, G.M. (1983) Prediction of the structure of turbulent particle laden in rounded jet, *AIAA J.*, Vol. **21**, 1480-1483.

Wu, Y.L., Xu, H.Y. and Gao, Z.Q. (1992) Experimental study on particle motion in slurry pump impeller, *J. Tsinghua University*, Vol. **32**, **5**, 52-58, (in Chinese).

Wu, Y.L., Xu, H.Y., Dai, J. and Wang, L. (1993) Experiment on solid-liquid two-phase flow in centrifugal impeller, *Proc. Int. Symp. on Aerospace and Fluid Science*, Sendai, 379-387.

Wu Y.L., Sun Z.X., Oba, R., Ikohagi, T. (1995 a) Study on flows through centrifugal impeller by turbulent simulation, *ASME* FED-**222**, 63-68.

Wu, Y.L., Dai, J., Oba,R., Ikohagi, T. (1995 b) Turbulent flow simulation through centrifugal pump impeller at design and off-design conditions, *Proc. of 2nd ICPF*, Tsinghua University, Beijing, 155-167.

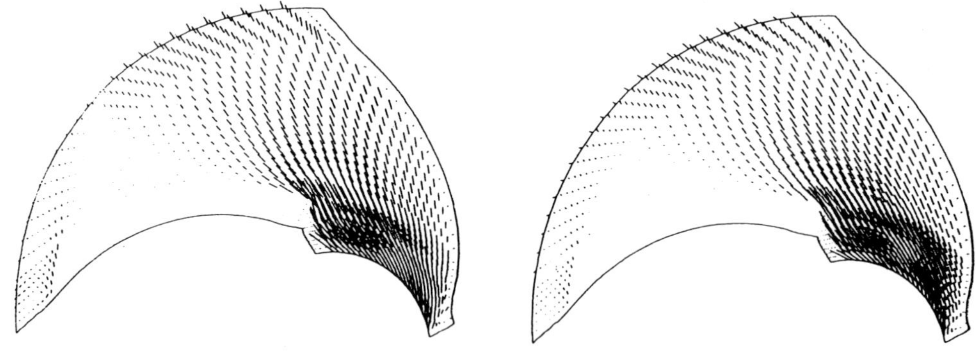

(a) Water Flow in Two-Phase Flow Condition (b) Particle Phase Flow

Fig. 1 Velocity Vectors at Condition (u_{r1} =6m/s, n=1450rpm, C_v=0.1, d_s= 0.1mm and ρ_s=2.6 g/cm3)

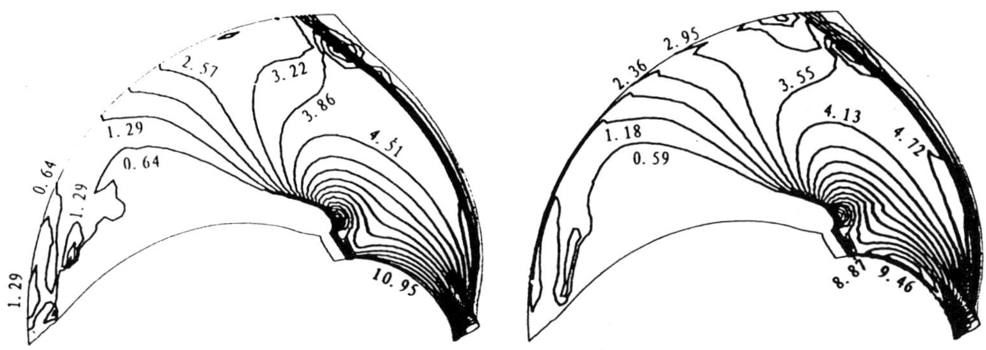

(a) Water Flow (W_{max} = 12.24 m/s) (b) Particle Flow (W_{pmax} = 11.13 m/s)

Fig. 2 Relative Velocity at Condition (u_{r1} =6m/s, n=1450rpm, C_v=0.1, d_s= 0.1mm and ρ_s=2.6 g/cm3)

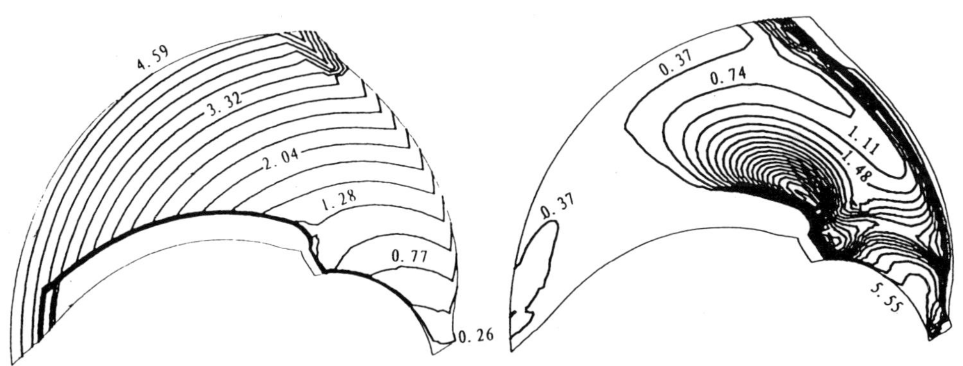

(a) Pressure Contour (P_{max} = 4.85 × 10^5 Pa) (b) Turbulent Kinetic Energy Contour (k_{max} =7.03 m^2/s^2)

Fig. 3 Calculated Results at Condition (u_{r1} =6m/s, n=1450rpm, C_v=0.1, d_s= 0.1mm and ρ_s =2.6 g/cm^3)

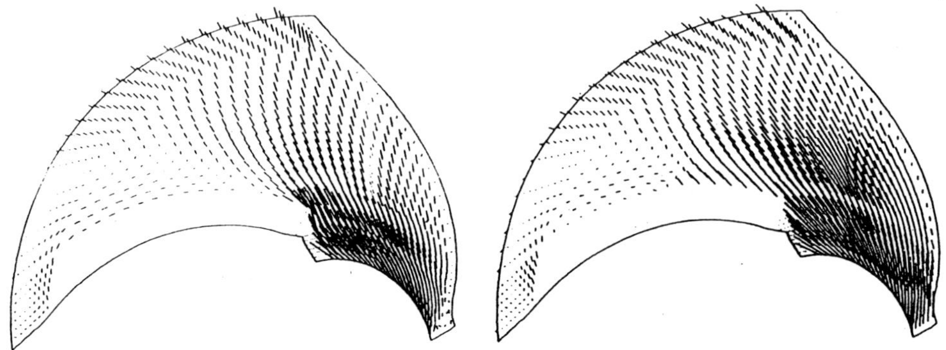

(a) Water Flow in Two-Phase Flow Condition (b) Particle Phase Flow

Fig. 4 Velocity Vectors at Condition (u_{r1} =6m/s, n=1450rpm, C_v=0.1, d_s=1.0mm and ρ_s =2.6 g/cm^3)

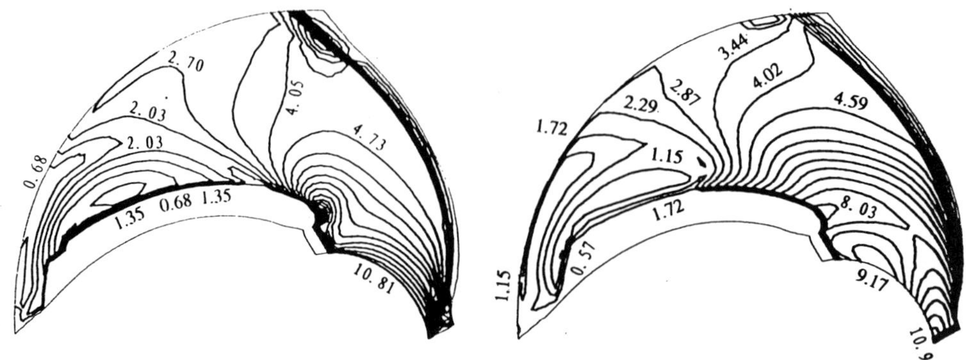

(a) Water Flow (W_{max} =12.84 m/s) (b) Particle Flow (W_{pmax} = 10.90 m/s)

Fig. 5 Relative Velocity at Condition (u_{r1} =6m/s, n=1450rpm, C_v=0.1, d_s= 1.0mm and ρ_s =2.6 g/cm^3)

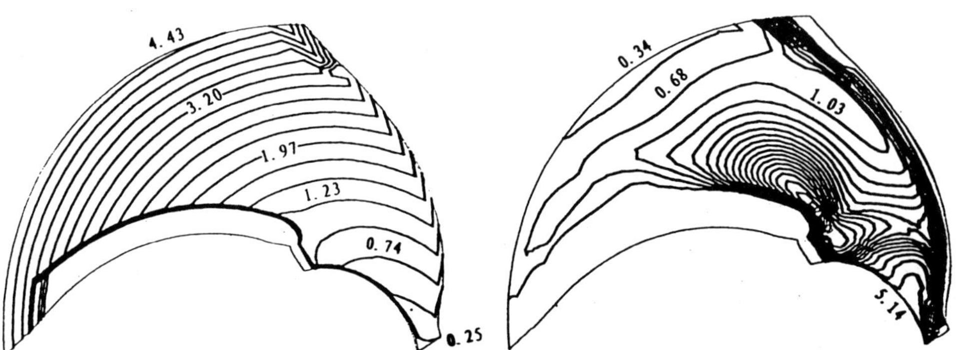

(a) Pressure Contour (P_{max} = 4.68 × 10^{5}Pa) (b) Turbulent Kinetic Energy Contour (k_{max} = 6.51 m^2/s^2)

Fig. 6 Calculated Results at Condition (u_{r1} =6m/s, n=1450rpm, C_v=0.1, d_s= 1.0mm and ρ_s =2.6 g/cm^3)

APPLICATION OF THE METHOD OF KINETIC BALANCE FOR FLOW PASSAGES FORMING

M. BENIŠEK, S. ČANTRAK,
Faculty of Mechanical Engineering, Belgrade, Yugoslavia
B. IGNJATOVIĆ
"HPS Djerdap", Kladovo, Yugoslavia
D. POKRAJAC
Faculty of Civil Engineering, Belgrade, Yugoslavia

Definition of flow field boundaries is a common problem in hydraulic engineering. Their shape should enssure stable fluid flow, without separation and transient phenomena. Such problems are often solved using either the acquired experience or a large number of expensive experiments, with many trials, finally leading to the proper solution.

The inner fluid current is at hydrodynamic equilibrium. It contains the zones of "sound flow" and boundary layer zones. Besides them, there may be some zones of fluid at rest, or moving slowly, separated from the sound flow by the discontinuity surface. These zones are called "dead water". Although they interact with the "sound flow", their flow patterns are completely different. The "dead water" zones are formed if the fluid cannot follow the rigid flow boundaries. Separation happens either due to boundary layer thickening or because of significant fluid inertia. The former case can be avoided by boundary layer suction, but the latter one cannot. It was named "inertial separation" by Strscheletzky (1957). In the rotating fluid, close to the rotation axis, swirling core is formed, which can be treated as "dead water" compared to the "sound flow" between the core and the boundary layer at the pipe walls. In the "sound flow" region, the influence of viscosity can be neglected, flow can be assumed homogeneous and the fluid ideal and incompressible.

The problem of fluid boundaries shaping, which shall ensure the stable fluid flow with minimum undesirable phenomena was studied by Strscheletzky (1957) and (1969). He developed the method of kinetic balance, based

E. Cabrera et al. (eds.), Hydraulic Machinery and Cavitation, 455–463.

on the Euler flow equation for ideal, incompressible fluid

$$\frac{d\vec{C}}{dt} = \vec{F} - \frac{1}{\rho} gradp. \tag{1}$$

Assuming conservative volume forces ($\vec{F} = gradU$), for the elementary mass $dm = \rho dV_i$, of the fluid, from (1) the momentum equation can be written as

$$\rho \frac{d\vec{C}}{dt} dV_i + \rho gradU dV_i + gradp dV_i = 0 \tag{2}$$

where $\vec{C}$ and p are local velocity and the pressure for the elementary volume dV_i.

Zone of fluid flow V_i has its characteristic flow mode ("sound flow" od "dead water"). The whole flow domain contains n elementary volumes V_i $(i = 1, 2, 3 \ldots n)$. The "sound flow" and "dead water" zones are separated by the surfaces of zero, second or higher order of discontinuity, Strscheletzky (1969)

Virtual work of forces acting on the fluid in the volume V_i, at the moment t, for virtual displacement $\delta\vec{r}$ is

$$\int_{V_i} \rho(\frac{d\vec{C}}{dt}\delta\vec{r})dV_i + \int_{V_i} \rho(gradU\delta\vec{r})dV_i + \int_{V_i} (gradp\delta\vec{r})dV_i = 0. \tag{3}$$

For the whole flow domain V

$$\sum_{i=1}^{n} \left\{ \int_{V_i} \rho(\frac{d\vec{C}}{dt}\delta\vec{r})dV_i + \int_{V_i} \rho(gradU\delta\vec{r})dV_i + \int_{V_i} (gradp\delta\vec{r})dV_i \right\} = 0. \tag{4}$$

Since the integrals are additive, and $\sum_{i=1}^{n} V_i = V$ from (4) it follows that

$$\int_{V} \rho(\frac{d\vec{C}}{dt}\delta\vec{r})dV_i + \int_{V} \rho(gradU\delta\vec{r})dV_i + \int_{V} (gradp\delta\vec{r})dV_i = 0. \tag{5}$$

The equation (5) states Lagrange's principle of virtual work. Flow equilibrium in the volume V, at the moment t, is achieved, when the sum of virtual works of the forces acting on the fluid equals zero.

Introducing kinetic energy dE_k and potential energy dE_p, of the elementary mass dm

$$dE_k = \frac{1}{2}c^2 \rho dV, \tag{6}$$

$$dE_p = U dm + p dV, \tag{7}$$

from the equation (5) it follows that

$$\int_{t_1}^{t_2} \delta \int_V (dE_k - dE_p) dt = 0. \tag{8}$$

The equation (8) states the balance condition for the fluid in motion. Non-viscous and incompressible fluid is in equilibrium if the difference between kinetic and potential energy is at minimum.

Introducing the realistic assumption that **total energy in the fluid stream does not change after virtual displacement** i. e. $\delta(dE_k + dE_p) = \delta(dE_t) = 0$, the equilibrium condition (8) is expressed as the variation of the sum of **integrals of action** I_i) formed for the characteristic flow domain zones V_i

$$\delta I = \sum_{i=1}^{n} \delta I_i = \sum_{i=1}^{n} \delta \int_{s_1}^{s_2} \rho \int_{V(1)} \vec{c} d\vec{s} dV_i = 0, \tag{9}$$

where: V_i - i-th fluid flow region, $\vec{c}$ - local flow velocity, dV_i - elementary volume, bounded by the inflow A_{ei} and outflow A_{oi} control surfaces (fig. 1) as well as by the given boundaries, s_1, s_2 - respective positions of the fluid particle at the moments t_1 and t_2 with $d\vec{s} = \vec{c}dt$

The equation (9) also states the equilibrium condition for the fluid in motion. It is more convenient for the solution of practical problems than (8). If one of the flow region boundaries changes, the action integral changes as well. The most convenient flow domain boundary, from the series of trial shapes, is the one which has minimum value of action integral.

In many practical problems the inner flow consists of only one main sound flow region, and one or more closed secondary flow regions, which are separated from the main flow by the free boundaries. For the ideal, non-viscous fluid flow, this boundaries are the discontinuity surfaces i.e. vorticity dissipative layers in the real fluid. In "dead water" zones the fluid is at rest or moves very slowly. Variational conditions can be applied to the sound flow regions only (action integral for the "dead water" equals zero). In the region V_1 having total discharge Q, from the equation (9) it follows that

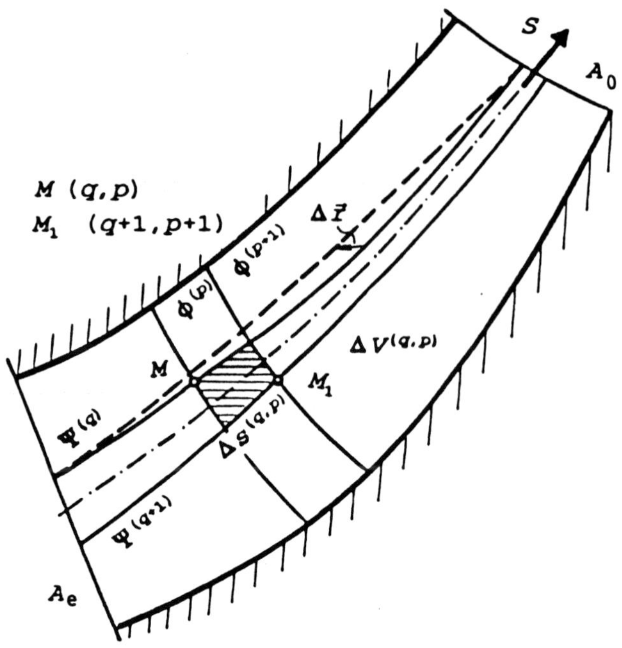

Figure 1: General view of flow passage and virtual streamline displacement

$$\delta I = \delta \int_{s_1}^{s_2} \int_{V_1} \rho \vec{c} dV d\vec{s} = 0. \tag{10}$$

Analytical solution of the equation (10) exists only for the special cases. For that reason, the grapho-analytical or numerical solution is used. The equation (10) is applied to the small, but finite elements $\Delta V^{(q,p)} = \Delta V(x^{(q,p)}, y^{(q,p)}, z^{(q,p)})$ of q-th stream tube ($q = 1, 2, \ldots, m$). Each $\Delta V^{(q,p)}$ is divided into p ($p = 1, 2, \ldots, k$) elementary volumes. The distance between the two respective positions $s^{(q,p)}$ and $s^{(q,p+1)}$, along the mean streamline of the q-th stream tube (fig. 1). The action integral I is approximated as:

$$I = \rho \sum_{q=1}^{m} \sum_{p=1}^{k} \vec{c}(q,p) \Delta V^{(q,p)} \Delta s^{(q,p)} \tag{11}$$

where $\vec{c}(q,p)$ is local flow velocity corresponding to the mean streamline of the q-th stream tube divided into k parts.

Practically, shaping of flow field boundaries reduces to one boundary variation, while the others are fixed, as well as the control surfaces. For each case, the action integral is computed, and the one with minimum integral is adopted as the final solution.

To illustrate kinetic balance method application, some flow field boundaries forming examples are shown.

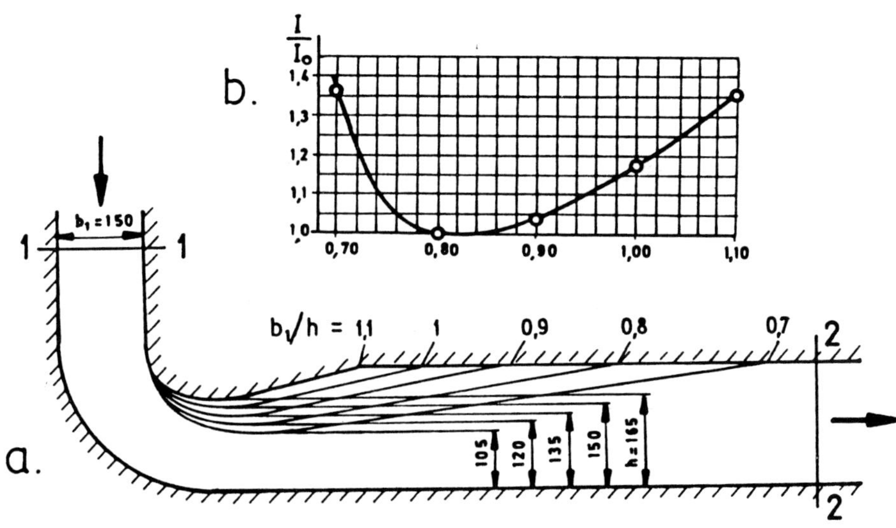

Figure 2: Difusor with parallel lateral walls

Example 1 **Forming of the inner (convex) curved contour of the diffusor with parallel lateral walls** is presented in Strscheletzky (1957). The form of diffusor model is shown in figure 2a. The inner (convex) boundary form is varied as defined by the ratio $h/b_1 = 1.1; 1.0; 0.9; 0.8; 0.7$. The other flow passage contours are fixed. For every flow passage form (h/b_1) the equipotential lines and streamlines are determined, and the action integrals calculated, using the expression (11). The change of dimensionless action integral values I/I_0 is shown in figure 3.b . For the value $h/b_1 = 0.815$, the function I/I_0 has minimum, so this contour was found optimum. The experiments presented in Strscheletzky (1957) have confirmed such a theoretical result.

Example 2 **Forming of the aeration duct of the bottom outlets of the Haditha dam in Iraq**, Benišek (1982). The shape of the aeration duct is shown in figure 3.a. The inner boundary forms and positions are defined by the value $r/b_u = 1.57; 1.37; 1.14$ The other contours are fixed. The calculation results of the dimensionless integral I/I_0 are shown in figure 3.b. For the value $r/b_u = 1.37$ (case II) the function I/I_0 has minimum.

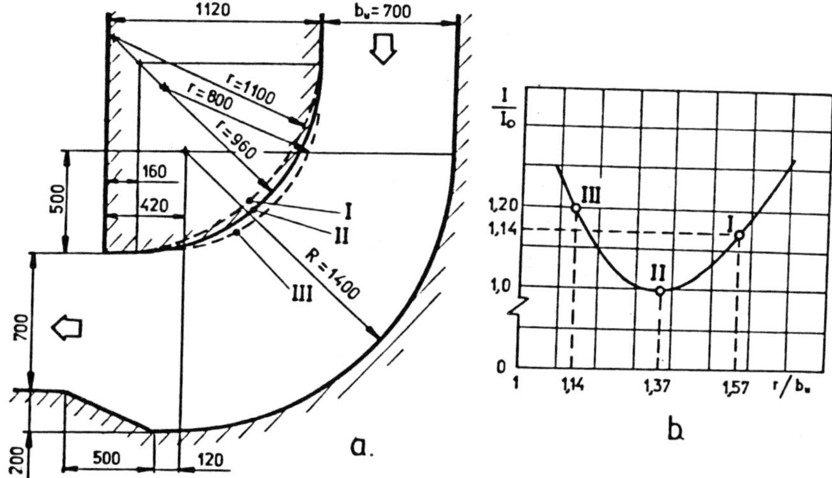

Figure 3: Aeration duct of the bottom outlets

Example 3. **The analytical determination of the intake structure lower contour of the additional hydro-turbine plant on the spillway of the "HEPS Djerdap I"**, Benišek (1993), Ignjatović, (1994) The lower contour of the intake structure is determined by given method. The considered cases are shown in figure 4.a, denoted by numbers 4 to 10, for various head water levels $\nabla 69.5; \nabla 65.0; \nabla 63.0$. The computation results are shown in figure 4.b. The minimums of dimensionless action integral I/I_0, for various head water levels, are denoted by the arrows. The satisfactory contours are between 6 and 8. These results were confirmed by the experiments as well – Teodoresku (1990).

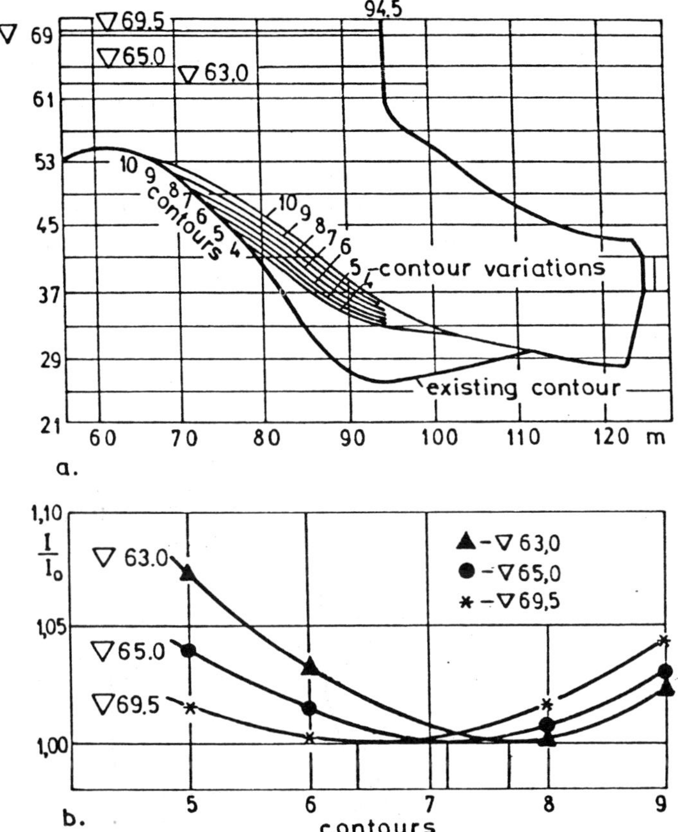

Figure 4: Intake structure contour of the additional hydro turbine plant on the spillway of "HEPS Djerdap I"

Example 4. **Shaping of flow boundaries under the inlet bellmouth of the axial pump in the pumping station "Gradištanski rit"**, Benišek (1992).

Larger capacity axial pumps were installed in the pumping station "Gradištanski rit". For the existing suction level in the sump, their operation caused some undesired phenomena. Shaping of the sump was analysed, in order to provide some necessary conditions: prevent rotational flow, establish uniform velocity field at suction pipe mouth and prevent air-entraining vortices. Besides some other changes, it was necessary to form the fluid region below the larger bellmouth diameter. Figure 5a shows the shape of axy-symetrical fluid flow domain between the suction bell and the guide cone. The shape of the bell is already defined, and it was kept constant, while the inner concave boundary of suction cone was varied. For the three shapes of suction bell contour – curves A, B and C, with $R/h_u = 2.42; 1.91; 1.64$ the action integrals I/I_o were computed for axy-symetrical fluid flow domain between the suction bell and the guide cone. Results of computations are shown in figure 5b. For $R/h_u = 1.91$, corresponding to the curve B the function I/I_o was found to be at minimum. The suction cone was built in the shape of the curve B. Air-entraining vortices in the sump have not appeared and water level in the sump was very quiet.

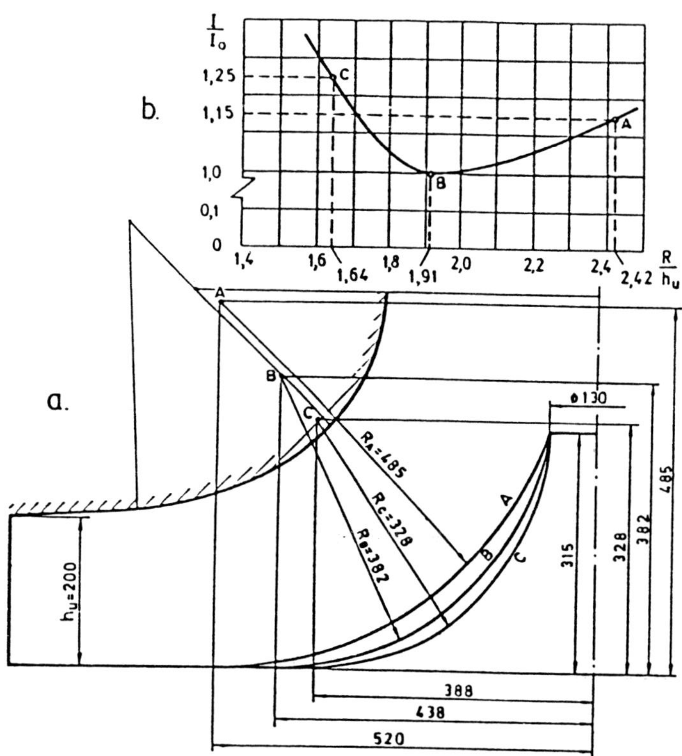

Figure 5: The flow passage between bellmouth and guide cone of the "PS Gradištanski rit"

Conclusion

On the basis of the results obtained in this study, the following conclusions can be derived:

- The presented method of kinetic balance is valuable tool for analytical determination of optimum shape of fluid flow boundaries.
- The method is simple. It requires computation of flow field – streamlines by any method for non-viscous fluid flow solution. The potential flow solution is probably easiest to use.
- The influence of viscosity, which was neglected, should be checked by laboratory measurements for the final solution.
- The number of cases which should be tested in the laboratory decreases significantly by the application of this method.

References

Strscheletzky, M. (1957) Ein Beitrag zur Theome des hydrodynamic Gleichgewichts von Stromungen, *Voith Forschungund Konstruktion*, Heft 2, Aufsafz 1

Strscheletzky, M. (1969) Kinetisches Gleichgewicht der Innenstromungen inkompressibler Flussigkeiten, *VDI-Z*, reihe 7, Nr **21**

Benišek, M. and Čantrak, S. (1982) Forming the aeration ducts of radial gates, Haditha project - Iraq Bottom outlets, Report of Faculty of Mechanical Engineering, Belgrade

Benišek, M., Pokrajac, D. and Čuković, B. (1993) The analytical upper and lower contour forming of the intake structure of the additional hydro-turbine plant on the spillway of the "HEPS Djerdap I, Report of Faculty of Mechanical Engineering, Belgrade

Ignjatović, B., Mandić, D., Benišek, M., Savić, Z. et al (1994) Problems of hydrounits on the spillway of HPP DJERDAP I, Nacional energy conference – CNE'94, Neptun, Romania

Teodoresku, I. et al (1990) Study of the intake structure model of the additional HPP IRON GATE I, Report of ISPH, Bucharest

Benišek, M., Čantrak, S. and Vuković, I. (1992) Study of the "PS Gradištanski rit" sump design, Report of Faculty of Mechanical Engineering, Belgrade

DYNAMICS OF LARGE HYDROGENERATORS

Experimental and Theoretical Aspects to Calculate Guide Bearings Dynamic Coefficients

BRITO, G.C. (1), WEBER, H.I. (2) and FUERST, A.G.A. (2)
(1) *Itaipu Binacional - Itaipu Hydroelectric Power Plant*
Laboratory of the Superintendence of Maintenance
Foz do Iguaçu, PR - CEP 85856-970 - Fax: +55 45 522 12 73
(2) *Unicamp - Campinas State University - Centre of Technology*
Campinas, SP - CEP 13083-970 - Fax: +55 19 239 76 40 - Brazil

Abstract
Based on experimental data, this paper shows that the mathematical models of large hydrogenerators, used to preview the dynamic behavior at the design phase or as a predictive maintenance tool at the operation phase, must include the effects of the stiffness and damping forces of the oil film in the guide bearings.
Based on an analytical study of the Reynolds equation, this paper presents a simplified procedure to determine the oil film stiffness and damping coefficients. This procedure is applied to the Itaipu generating units guide bearings and some practical aspects are examined.

1. Experimental Aspects

The following experimental aspects were established with basis on the data achieved by commissioning, monitoring and special vibration tests performed on the 18 generating units 700 MW Francis turbine type of the Itaipu Power Plant [1]. This knowledge must be taken into account in rotordynamic simulation of large hydrogenerators and their modelling.

1.1 SHAFT VIBRATION SEVERITY

Comparing shaft vibration severities measured on the 18 units at the same operating condition (load, water head and thermal condition), it was observed that there are differences of 600% of such values, despite the fact that the units are split in two groups of identical design and have had identical procedures for manufacturing, assembling, balancing, etc.

The shaft vibrations were measured in two directions, one 90 degrees from the other. It was verified that despite in most of the generating units both

E. Cabrera et al. (eds.), Hydraulic Machinery and Cavitation, 464–473.

vibrations have equivalent amplitudes, some units present vibration in one direction as approximately half of the other direction, indicating a significant static load in these cases.

1.2 ESTIMATION OF OIL FILM STIFFNESS

Oil film stiffness of some guide bearings were estimated using data achieved on the rotor balancing at the commissioning time. For some units the coefficients obtained in such way are very close to the values estimated from the design. However, in other hydrogenerators were detected large discrepancies between experimental and theoretical values.

1.3 EXCITATION OF NATURAL FREQUENCIES

When some units are in a narrow band near to 80% of the nominal load, therefore above the range of low load vortex in the draft tube, there were observed excessive shaft vibrations at the turbine guide bearing (up to 750 μmpp). Such vibrations occurs always in two different frequencies, in the range of three to four times the nominal rotating speed. It is most interesting to observe that both frequencies may change up to 20% from a test to another. Such a phenomenon is still being investigated, but these vibrations are very probably due to the excitation of rotor natural frequencies.

1.4 CHANGES IN BEARING CLEARANCES

The clearances adjusted during assembling vary due to several factors, mainly influenced by thermal expansions. The difference between the journal and pad support ring temperatures can reach some degrees centigrades, which will cause expressive changes in the clearance and in the hydrogenerator bearing dynamic coefficients. Tests have shown that thermal expansions are strongly affected by the bearing cooling water temperature. As this water is taken directly at the spiral case, its temperature changes up to 15°C from summer to winter. Therefore the units present seasonal variations in bearing clearances.

1.5 SEASONAL VARIATIONS ON SHAFT VIBRATION

Shaft vibration and bearing temperatures monitoring has shown that there are significant periodic variations on such magnitudes, with seasonal characteristics. This behavior is originated by the seasonal changing on the bearing cooling water temperature, as described above.

1.6 DYNAMIC BEHAVIOR AFTER A START UP

Shaft vibration monitoring on a generating unit, since its start up at environment temperature until its thermal stabilization seven hours later, demonstrated that vibration severity decreases exponentially to less than the half of the initial value. Such a behavior is due to a significant reduction on the bearing clearances.

1.7 BEARING BRACKET DEFORMATIONS

Significant discrepancies among clearances were detected, caused by deformations in the bearing. Such deformations are originated mainly by localized heating of the bearing bracket near to the generator phases and neutral outputs, by the action of eddy currents. This behavior was confirmed by a Finite Elements Method - FEM analysis on the bearing bracket.

1.8 LUBRICANT VISCOSITY

It was verified that the guide bearings pads temperatures, and consequently the temperatures of their oil films, may differ up to 20°C from a bearing side to another. The oil film dynamic coefficients are directly proportional to the lubricant dynamic viscosity, which mainly depends on its temperature. Therefore, it is very important to take into account the actual oil film temperatures when determining dynamic coefficients.

2. Theoretical Aspects

Mathematical models are essential tools for today's design of rotating machinery. A well elaborated numerical description of intended devices allows a previous knowledge of static and dynamic behavior of the later installed machinery. By this, the design time and costs can be reduced and safety aspects can be improved at an early stage of planning.

Likewise, there is a need for modeling already installed mechanical structures to determine and optimize operational parameters, as well as trade with maintenance aspects. The maintenance optimization extends the operation intervals to increase productivity and prolongs machine life.

The ideas of on-line monitoring and incipient failures diagnostic through correlating vibration signals to a base line or simulated data, is based on the philosophy named predictive maintenance approach, which is already in use in less complex industrial processes. To do that it is desirable a mathematical model which describes the points of interest as good as possible.

The main goal in creating a numerical image of the hydrogenerator dynamic behavior is to help the maintenance team in detect and diagnostic incipient failures, through vibration monitoring. This image contains a huge number of dynamic influences like a structure with natural frequencies and resonances depending on the rotation speed, the gyroscopic effect, electric excitation at the generator and hydrodynamic forces at the turbine, as well as the interactions at turbine seals and hydrodynamic bearings. For such reasons, as well as to find more important influence factors for future modeling, an experimental identification is necessary to validate modelling and understand and estimate the machine dynamics.

This paper describes a procedure to determine stiffness and damping coefficients by means of a simplified calculation, derived from an analytical study of Reynolds equation. Despite being simplified, this study represents adequately the hydrodynamic bearings commonly used in large hydrogenerators: segmented type, with more than six concentric pivoted pads. For this bearing type the pad radius is approximately equal to the sum of journal radius and the bearing clearance, which allows to represent the oil film thickness along the bearing length by a linear function.

3 Dynamic characteristics of oil film

When a journal supported by hydrodynamic bearings vibrates under the influence of external forces, the oil film pressure oscillates. This oscillation gives rise to stiffness and damping forces in the oil film, which influence the critical velocities and vibration amplitudes of the rotor. The coefficients that represent such stiffness and damping effects are defined from forces resulting from the oil film pressure, derived from the Reynolds equation.

Let $F_X=F_X(X,X',Y,Y')$ and $F_Y=F_Y(X,X',Y,Y')$ be the components of the resulting force from oil film hydrodynamic pressure between the journal and pads, in X and Y directions respectively. The velocities of the displacements of the journal centre, X and Y, are $X'=\partial X/\partial t$ and $Y'=\partial Y/\partial t$.

The force $\boldsymbol{F}=\{F_X,F_Y\}^T$ is non-linear for each variable. However, it can be approximated around the point of the journal static equilibrium with adequate precision, using the Taylor Series, if the vibration displacements and velocities are sufficiently small. The vectorial expression for the oil film force, disregarding terms of power equal or greater than two, is: $\boldsymbol{F}=\boldsymbol{F_0}+\boldsymbol{\Delta F}$, where $\boldsymbol{\Delta F}=\boldsymbol{k}\{\Delta X,\Delta Y\}^T+\boldsymbol{c}\{\Delta X',\Delta Y'\}^T$ and $\boldsymbol{k}$ and $\boldsymbol{c}$ are given by Equation 1.

$$\boldsymbol{k}=\begin{bmatrix}\dfrac{\partial F_X}{\partial X} & \dfrac{\partial F_X}{\partial Y}\\ \dfrac{\partial F_Y}{\partial X} & \dfrac{\partial F_Y}{\partial Y}\end{bmatrix}=\begin{bmatrix}k_{XX} & k_{XY}\\ k_{YX} & k_{YY}\end{bmatrix}\quad,\quad \boldsymbol{c}=\begin{bmatrix}\dfrac{\partial F_X}{\partial X'} & \dfrac{\partial F_X}{\partial Y'}\\ \dfrac{\partial F_Y}{\partial X'} & \dfrac{\partial F_Y}{\partial Y'}\end{bmatrix}=\begin{bmatrix}c_{XX} & c_{XY}\\ c_{YX} & c_{YY}\end{bmatrix}\qquad(1)$$

The subscript $\boldsymbol{0}$ on $\boldsymbol{F_0}$ denotes the force at the point of the bearing static equilibrium, i.e., the force originated only by the oil wedge pressure, caused by the static load. The symbol $\boldsymbol{\Delta}$ indicates that the dynamic force and vibration amplitudes are small. The coefficients k_{ij} and c_{ij} are respectively the oil film stiffness and damping coefficients in the directions $i,j = X,Y$.

To calculate the coefficients defined above, it is necessary to differentiate the force acting on the oil film with relation to the displacements and velocities of the relative vibration between journal and pads, in order to determine stiffness and damping coefficients, respectively.

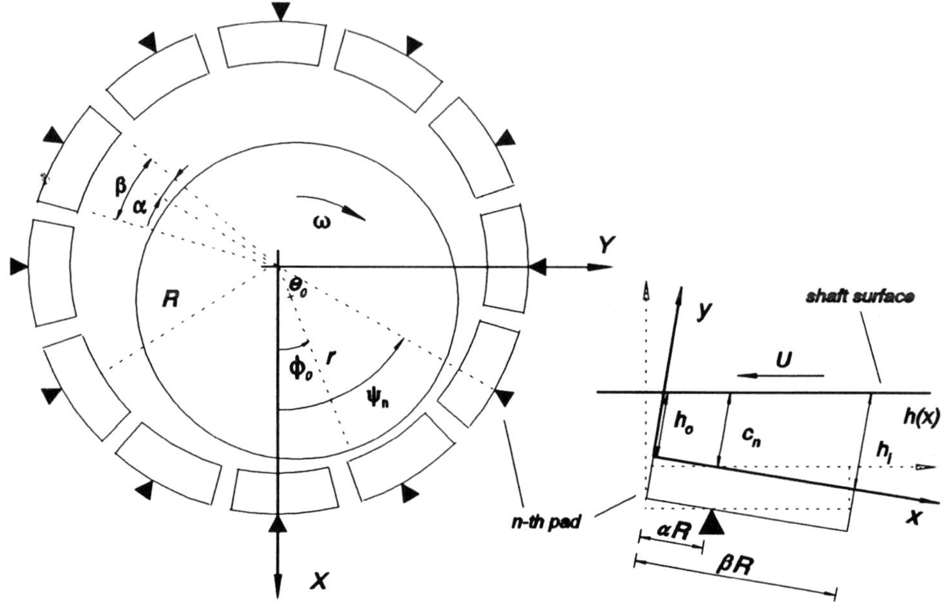

Figure 1. Sketch of a large hydrogenerator guide bearing

4. Oil Film Pressure: Unidimensional Analysis

The Reynolds equation for hydrodynamic bearings can be derived from the movement equation of an oil film element in a laminar flow between parallel plates, in a steady state condition. It is given by:

$$\frac{\partial}{\partial x}\left(\frac{h^3}{\mu}\frac{\partial p}{\partial x}\right) + \frac{\partial}{\partial z}\left(\frac{h^3}{\mu}\frac{\partial p}{\partial z}\right) = -6\,U\frac{\partial h}{\partial x} - 12\,X'\cos\theta - 12\,Y'\sin\theta \qquad (2)$$

Disregarding the axial losses ($\partial p/\partial z=0$), the pressure distribution over the surface of the n-th pad is determined analytically by integrating the Reynolds equation twice with respect to x. The integration constants are obtainded by the boundary conditions for the pressure ($x=0$, $p(x)=0$) and ($x=\beta R, p(x)=0$).

The hydrodynamic pressure bends the pivoted pad, assuming an equilibrium position defined by $\eta=h_i/h_o$, the relationship between the oil film thickness at the leading edge (h_i) and at the tailing edge (h_o) of the pad. To determine such position, it should be remembered that the torque originated by the hydrodynamic pressure at the left and at the right side of the pivot point must be equal in this condition. By this, the following equation for η is obtained [1]:

$$2\ln\eta\ [(1+2\eta)+\frac{\alpha}{\beta}(\eta^2-1)]+(\eta-1)(\eta+5)-4\frac{\alpha}{\beta}(\eta-1)^2=0 \qquad (3)$$

By the previous equation can be noted that η is a constant which depends exclusively on the pad geometry. Once the pad length βR and pivot point αR are determined, η is obtained by solving Equation 3 and it is valid to all bearing pads.

5. Oil Film Pressure: Two Dimensional Analysis

In a finite bearing there are axial losses, which can not be disregarded ($\partial p/\partial z \neq 0$). The effect of this flow in z, the width direction, is considered by using an approximated solution for the complete Reynolds equation. This solution is derived from the specific case where the bearing pad width is much smaller than the bearing length. For such a case the pressure variation in x direction can be disregarded ($\partial p/\partial x = 0$) and the Reynolds equation can be solved analytically again. It can be shown that the pressure has a parabolic distribution in the z direction.

The proposed solution has the following form [1]:

$$p(x,z)=\frac{-6\mu U}{\beta R}\,\frac{(\eta-1)}{(\eta+1)}\,\frac{x\,(x-\beta R)}{h^2(x)}\ [-\left|\frac{2z}{L}\right|^m+1] \qquad (4)$$

where coefficient m depends on the pad geometry. For an infinitely wide pad, m is infinite, which means a constant pressure distribuition along the pad width. For the infinitely short pad, m is equal 2 and the pressure distribution in the z direction is parabolic. In fact, such idea was used before in cylindric bearings [4], where m is determined in relation of bearing dimensions, by an empirical function.

In the present paper another procedure to determine m is used for a finite bearing. The basic idea is to put Equation 4 into the complete Reynolds equation. If $p(x,z)$ actually were the solution to that equation, this procedure would result in a null equality. However, in this case there will be a residue, a function of m and bearing dimensions. The coefficient m is determined [1] by minimizing the total residue, obtained by integrating the residue over the pad surface. This procedure leads m to:

$$m=-\frac{1}{2}+\frac{1}{2}\sqrt{1+12\,(\frac{L}{\beta R})^2} \qquad (5)$$

6 Oil Film Dynamic Coefficients

The stiffness coefficients are determined by differentiating the components of the resulting force (F_X and F_Y) with respect to the displacements X and Y. This forces are obtained by the integration of the pressure over pad surface. Thereby, the following equations are reached [1]:

$$k_{XX} = \frac{\Gamma_0}{c_0} \sum_{n=1}^{n_s} \frac{1 + \cos 2\psi_n}{\chi_n^3} \qquad k_{XY} = \frac{\Gamma_0}{c_0} \sum_{n=1}^{n_s} \frac{\sin 2\psi_n}{\chi_n^3}$$

$$k_{YX} = \frac{\Gamma_0}{c_0} \sum_{n=1}^{n_s} \frac{\sin 2\psi_n}{\chi_n^3} \qquad k_{YY} = \frac{\Gamma_0}{c_0} \sum_{n=1}^{n_s} \frac{1 - \cos 2\psi_n}{\chi_n^3} \tag{6}$$

where:

$$\Gamma_0 = \frac{m}{m+1} \frac{6\mu UL[(\eta-1)\alpha+\beta]^2}{(\eta-1)^2} [- \ln\eta + \frac{2(\eta-1)}{(\eta+1)}] (\frac{R}{c_0})^2 \tag{7}$$

The damping coefficients are determined by differentiating F_X and F_Y with respect to the velocities X' and Y'. Thereby, the following equations are obtained [1]:

$$c_{XX} = \Lambda_0 \sum_{n=1}^{n_s} \frac{[C_{\beta\eta} \cos\theta_n + S_{\beta\eta} \sin\theta_n] \cos\psi_n}{\chi_n^3}$$

$$c_{YX} = \Lambda_0 \sum_{n=1}^{n_s} \frac{[C_{\beta\eta} \cos\theta_n + S_{\beta\eta} \sin\theta_n] \sin\psi_n}{\chi_n^3}$$

$$c_{XY} = \Lambda_0 \sum_{n=1}^{n_s} \frac{[C_{\beta\eta} \sin\theta_n - S_{\beta\eta} \cos\theta_n] \cos\psi_n}{\chi_n^3}$$

$$c_{YY} = \Lambda_0 \sum_{n=1}^{n_s} \frac{[C_{\beta\eta} \sin\theta_n - S_{\beta\eta} \cos\theta_n] \sin\psi_n}{\chi_n^3} \tag{8}$$

where $\theta_n = \psi_n - \alpha$ and the constants Λ_0, $C_{\beta\eta}$ and $S_{\beta\eta}$ are determined as follows:

$$\Lambda_0 = \frac{m}{m+1} \frac{6\mu L [(\eta-1)\alpha + \beta]^3}{(\eta-1)^2} (\frac{R}{c_0})^3 \tag{9}$$

$$C_{\beta\eta} = \frac{-\beta\eta-\beta\eta\cos\beta-2\eta\sin\beta}{\beta\eta(\eta+1)} + \left[\frac{-(\eta^2+1)\delta^2}{(\eta+1)} + \frac{2\ln\eta}{(\eta-1)}\right]\cos\delta +$$

$$+ \left[\frac{-(\eta^4-2\eta^3+2\eta-1)\delta^3}{6(\eta^2-1)} + 3\delta - \frac{2\eta\,\delta\ln\eta}{(\eta^2-1)}\right]\sin\delta \tag{10}$$

$$S_{\beta\eta} = \frac{2\eta+\beta\eta\sin\beta-2\eta\cos\beta}{\beta\eta(\eta+1)} + \left[\frac{-(\eta^2+1)\delta^2}{(\eta+1)} + \frac{2\ln\eta}{(\eta-1)}\right]\sin\delta +$$

$$+ \left[\frac{(\eta^4-2\eta^3+2\eta-1)\delta^3}{6(\eta^2-1)} - 3\delta + \frac{2\eta\,\delta\ln\eta}{(\eta^2-1)}\right]\cos\delta \tag{11}$$

7 Procedure Validation

The dynamic coefficients of the Itaipu generating units upper guide bearing were calculated using two methods, i.e., by using Equations 6 and 8 and the Dynko software. Dynko was elaborated by the 'Arbeitsgruppe Machinendynamik' at Kaiserlautern University - Germany, in order to determine oil film dynamic coefficients using the FDM - Finite Difference Method.

In Figure 2 the results are presented in a graphical form, where the dimensionless dynamic coefficients ($K_{ij} = c_0\, k_{ij} / W$ and $C_{ij} = \omega\, c_0\, c_{ij} / W$) are plotted against the Sommerfeld Number ($So=(\mu\, U\, L\, r^2) / (\pi\, W\, c_0^2)$). It can be observed that the results of both calculation methods have a good agreement. Moreover, it was verified that there is a constant for a given bearing that matchs the stiffness curves perfectly.

8 Conclusions

The experimental observation has shown that the guide bearings oil film forces have a strong influence on large hydrogenerators dynamic behavior. Therefore, such forces must be considered to get a reliable response with the mathematical models of those machines.

The expressions to calculate the oil film dynamic coefficients present a good agreement with the traditional time consuming method. Despite simplified, such expressions can be used to study several important phenomena that occur in large hydrogenerators (bearing deformations, clearances variations, different viscosities), which are usually disregarded. It is important to notice that the influences given by those phenomena are much more significant than those obtained by the simplification.

A future works will improve the description of coefficient m, by defining it as a function of bearing length. It will be also studied the bearing of non concentric type, looking for a procedure to linearize the oil film thickness function.

9 References

1. Brito, G.C. and Weber, H.I. (1996) *Dynamic Behavior of Large Hydrogenerators Guide Bearings* - Manuscript of M.Sc. Thesis (in Portuguese) - Unicamp - Campinas State University - Brazil

2. Fuerst, A.G.A. (1993) *Modellierung einer vertikalen Wasserkraftmaschine in Xavantes - Brasilien* - Diplomarbeit - Universität Kaiserslautern - Germany

3. Ohashi, H. (1991) *Vibration and Oscillation of Hydraulic Machinery* - Hydraulic Machinery Book Series - Avebury Technical - England

4. Varga, Z.E. (1971) *Wellenbewegung, Reibung und Oeldurchsatz beim segmentierten Radialgleitlager von beliebiger Spaltform unter Konstanten und Zeitlich veränderlicher Belastung* - Dissertation Nr. 4734 - ETH - Juris Druck + Verlag - Zurich

11. List of notation

c_n	Clearance at the n-th pad	α	Angle between the pad pivot point ar the origin of (x,y) co-ordinates
c_0	Nominal clearance		
e_0	Journal eccentricity	β	Angle between pad extremities
h	Oil film thickness ($h=h(x)$)	δ	Auxiliary variable $\delta=\beta/(\eta-1)$
L	Pad width	ϕ_0	Journal attitude angle
n	Pad numbering ($n=1..n_s$)	η	Relationship between the oil fil thickness at the input (h_i) and the outp (h_o) of the pad ($\eta=h_i/h_o$)
n_s	Number of bearing pads		
p	Oil film pressure ($p=p(x,z)$)		
r	Journal radius	μ	Dynamic or absolute viscosity
R	Pad radius	ψ_n	Angular position of the n-th pad piv point
t	Time		
X,Y	Journal center displacements	θ	Angular position of a point on the pa surface ($\theta=\theta_n+x/R$, $\theta_n=\psi_n-\alpha$)
X',Y'	Journal center velocities		
U	Journal surface velocity	χ_n	Dimensionless clearance $\chi_n=c_n/c_0$
W	Static load acting on journal	ω	Journal rotating speed

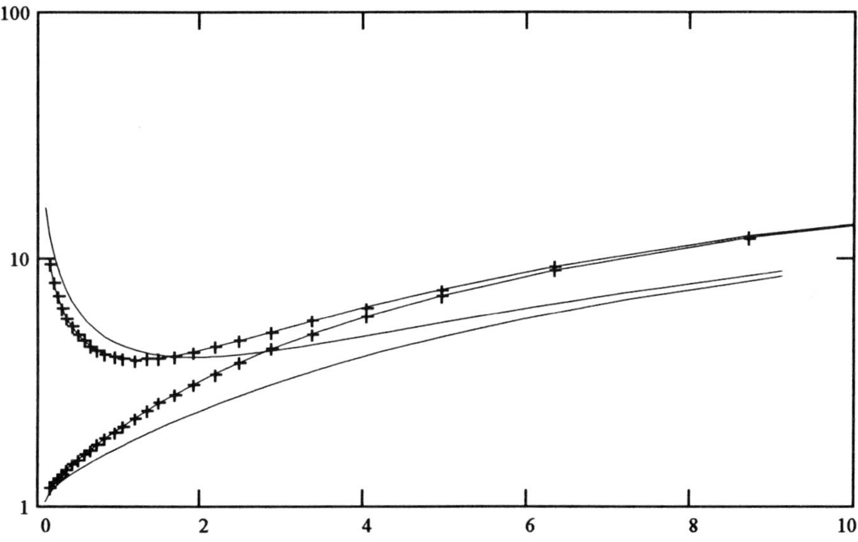

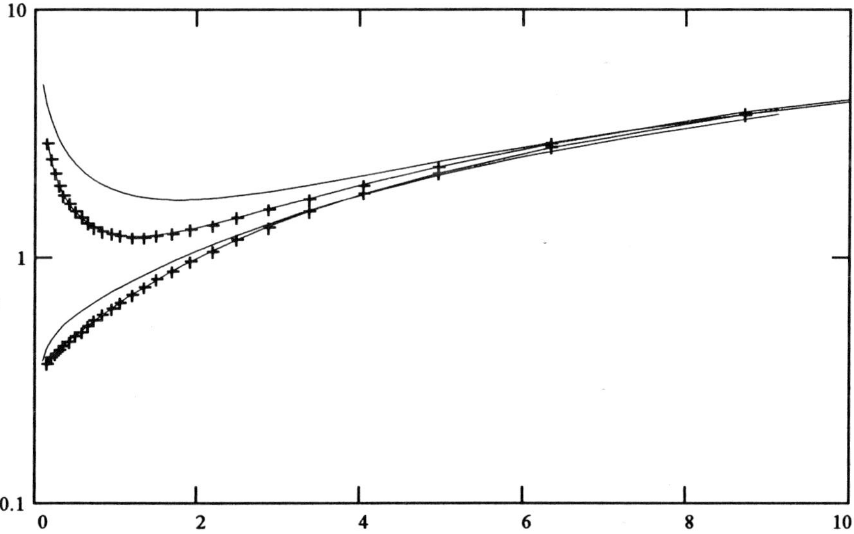

Figure 2. Comparing stiffness (upper) and damping (lower) coefficients: Equations 6 and 8 - crossed lines versus Dynko Program - solid lines.
($K_{XX} > K_{YY}$, $C_{XX} > C_{YY}$, $K_{XY} = K_{YX} = 0$, $C_{XY} = C_{YX} = 0$)

INSTABILITIES IN A FLOW-CONTROL VALVE

CIGADA A.(*), GUADAGNINI A.(**) and ORSI E.(**)
(*) Dip. Meccanica (**) D.I.I.A.R.
Politecnico di Milano, P.zza L. da Vinci, 32, 20133 Milano (Italy)

1. Abstract

This paper is about some instabilities that have been observed during the testing procedure of a flow-control valve. It was observed that the complete closure of the valve was very often reached after a transient state with oscillating behaviour of its mobile element whose displacement was recorded by means of proximity sensors, facing on opposite sides the moving body of the valve. The outputs have been collected by two signal conditioning devices and sent to a dual channel spectrum analyser. Measurements have also been performed through a servo-accelerometer which has been fixed to the pipe. The influence of the pipe length on the behaviour of the system was studied. The oscillations recorded by the proximity sensors are fairly well reproduced by a series of square waves, with a number of low amplitude, strongly damped, higher frequency contents. During a single test the frequency of the oscillations is not a constant but increases in its central part. The phenomenon cannot be related to one of the natural frequencies of the piping system, that are much higher. The observed frequency depends on the length of the adduction pipe.

2. Introduction

Some devices, even though very simple, in a complex piping system may cause noticeable instabilities (e.g. Brunone *et al.*, 1995; Fanelli, 1976; Föllmer and Zeller, 1980; Green and Wood, 1980; Orsi, 1994; Tanda and Zampaglione, 1991). A very careful analysis has to be performed to fully identify the dynamics of these phenomena and to propose an interpretation which has to be checked by experimental data. We will be concerned with the testing procedure of a flow control valve, that is designed to stop the flow in a pipe when the flow rate reaches a given value. These elements are of great interest especially for safety reasons in industrial processes. The structure of the valve, as reported in *Figure* 1, is very simple. It was mounted on a steel pipe line, with 50 mm internal diameter and some preliminary tests allowed to obtain characteristic curves as a function of the orifice diameter δ, of the initial opening h and of the spring preload F_S. In standard operating conditions the device is like an orifice in the main pipe of diameter D with transversal section $s = (\pi\ d\ \delta)$, d being the stem displacement

E. Cabrera et al. (eds.), Hydraulic Machinery and Cavitation, 474–483.

$(0 \leq d \leq h)$. The dynamic force of the fluid is opposed to the preload of the spring. When a fixed calibration limit is trespassed, the flow rate should instantaneously go to zero. We have performed a laboratory study of the behaviour of such a device. When the flow rate is progressively incremented and the conditions at the valve calibration limits are reached, some instabilities arise: the complete closure of the valve is very often reached after a transient state with oscillations of its mobile element. Therefore, the laboratory tests have been focused on the analysis of this transient state.
The displacement of the mobile element was recorded by means of proximity sensors, facing on opposite sides the moving body of the valve. The outputs have been collected by means of two signal conditioning devices and then sent to a dual channel spectrum analyser capable of directly summing the two inputs. Measurements have also been performed through a servo-accelerometer which has been fixed to the pipe. The servo-accelerometer output has sometimes been recorded together with the one coming from one of the proximity sensors to eventually get the relation between the two. During this first phase the influence of the pipe length, L_P, on the observed behaviour was studied: tests were performed for $L_P = 2$ and 400 m, the sampling rate f_S was varied between 50 and 1280 Hz and the phenomenon was recorded until the complete closure of the valve was obtained.

3. Experimental set - up

The way to approach the problem passed through the measuring of the movements of the valve head inside its body. An adequate accuracy level had to be achieved, but the main worries were about the eventual disturbances on the oncoming flow due to the measuring transducers, which could affect the phenomenon to be studied to such an extent to make it disappear. The choice had to take into account the fact that the adopted sensors had to work in water, too. The most attractive solution looked the one which employed some non-contact eddy current sensors. However, as it was not possible to fix one of these sensors in front of the plate of the valve head, due to the already mentioned disturbing fears, some alternative solution had to be looked for. This solution is shown in *Figure* 1: a cone has been screwed and glued to the valve stem, allowing for a radial positioning of the eddy current pick-up, which has been fixed drilling a hole in the tube carrying the valve. The conicity has been chosen in such a way to adjust the head stroke to the probe measuring field. The sensor has been put downstream the valve throat and it is "hidden" by the guide plate beading, to limit the effects due to its presence to a very small consequence. To prevent the plate from rotating and therefore leaving the probe exposed to the free stream, a steady pin has been inserted between the valve body and the plate. During the first laboratory tests to check the efficiency of the whole system, it was observed that one probe alone was not enough to get a satisfactory accuracy. In fact, as a certain backlash is observed between the guide and the stem, a movement of the head, relative to the valve seat, is always allowed. Therefore, the single probe output can't distinguish between the axial and the radial movements of the stem. A suitable way for separating these two effects is to fix another probe facing the cone at the opposite side of the same diameter where the first probe was. The two measurements combined together give the possibility of detecting the needed part of the signal out of the global measurement. The outputs of the two probes, properly conditioned, have been sent to a dual-channel spectrum analyser, for both data storing and analysis. A further step was the measuring system calibration. Calibration curves have been obtained by means of a slip table moved by a micrometric screw allowing for an accuracy of 1/100 mm. The combination of component errors in the overall system accuracy had to be pointed out. Calibration has been performed following two different procedures, therefore giving the needed values to perform an uncertainty analysis (e.g. Doeblin, 1990). The first way is to simply establish a link between the displacement superimposed to the valve head and the average change of the probe outputs, with respect to the initial rest position. This gives the calibration curve of *Figure* 2, expressing the along-stream stem motion as a function of the average voltage readings. The second way passes through the eddy current probe calibration and the cone angle according to the following:

$$d = \frac{(\Delta V_1 + \Delta V_2)}{2k_P} \frac{1}{tg\,\alpha} = \frac{\Delta V}{k_P} \frac{1}{tg\,\alpha} \qquad (1)$$

where ΔV_i are the probe outputs [V], ΔV is the average of the probe outputs [V], k_P is the eddy current calibration factor [V/mm], and $tg\,\alpha$ is the cone angle. Uncertainties are bound to each term of the equation. The eddy current probes have a fixed gain and their calibration is given by the manufacturer for a certain number of materials. A common calibration value is 7.874 V/mm (200 mV/mils) with a resolution of a few μm for AISI 4140: the cone screwed to the stem has therefore been made up of this steel; anyway, as the probe faced a not-flat surface and as water was between the probe and the cone, some doubts arose about the possibility of a significant calibration change. The cone angle has been built so that $tg\,\alpha = (1.5/12 \pm 0.1/12)$ mm/mm. As several calibration curves were available, it has been decided to carry out an estimation of both k_P and $tg\,\alpha$ by means of a least square method minimising the difference between eq. (1) and the superimposed stem displacements, to get reliable values of the uncertainty bound to both k_P and $tg\,\alpha$.

The identified values are $tg\,\alpha = 1.6 / 12$ mm/mm, and $k_P = 8.635$ V/mm, thus allowing for an estimation of the uncertainties related to $tg\,\alpha$ and k_P (which have been fixed close to the difference between the known theoretical values and the identified ones). The $tg\,\alpha$ value is coherent with the a-priori supposed uncertainty, the estimation of k_P suggests a conservative uncertainty limit of $\pm$ 1 V.

The following uncertainties have been therefore assumed: $unc(\Delta V) = \pm 0.01$ V, $unc(k_P) = \pm 1$ V/mm, $unc(tg\,\alpha) = \pm 0.1/12$ mm/mm. A sensitivity analysis has been carried out to point out the weight of the calibration error $unc(k_P)$ as compared to the others. This leads to the calculation of the following derivatives coming from eq. (1):

$$\frac{\partial d}{\partial \Delta V} = \frac{1}{k_P\, tg\,\alpha} \qquad (2)$$

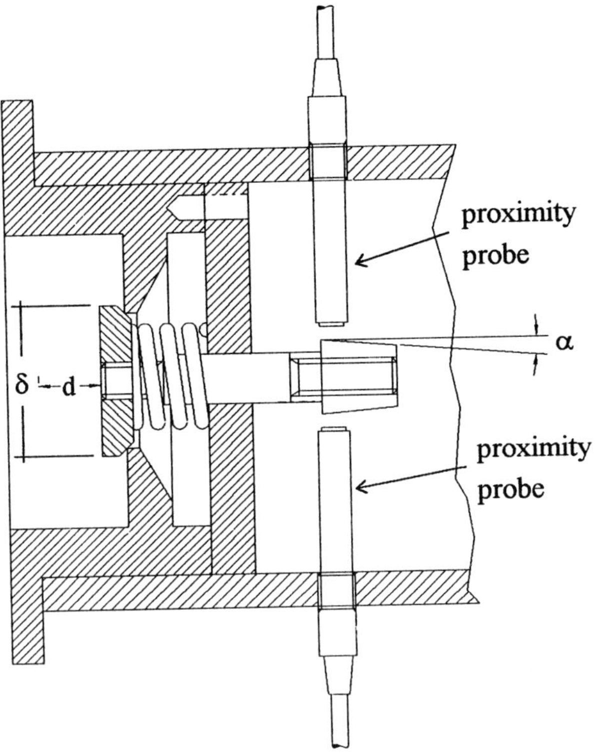

Figure 1. View of the experimental set - up for the valve.

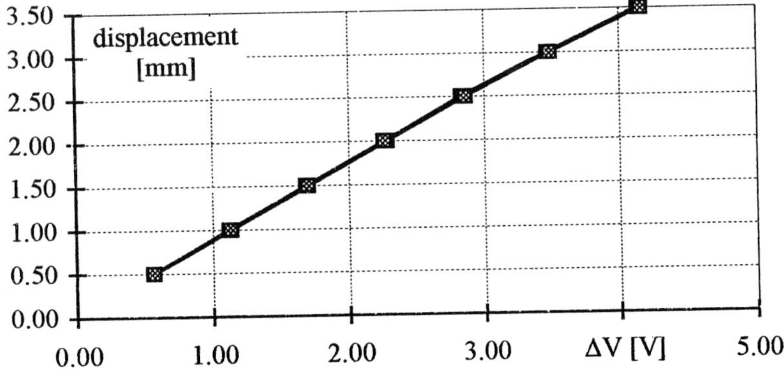

Figure 2. Calibration curve for proximity sensors.

$$\frac{\partial d}{\partial k_p} = \frac{\Delta V}{tg\,\alpha}\left(-\frac{1}{k_p^2}\right) \tag{3}$$

$$\frac{\partial d}{\partial\, tg\,\alpha} = \frac{\Delta V}{k_p}\left(-\frac{1}{tg^2\alpha}\right) \tag{4}$$

An example concerning one of the calibration curves is shown in Table 1.

TABLE 1. Results of the calibration procedure

(1) [mm]	(2) probe 1 [V]	(3) probe 2 [V]	(4) ΔV [V]	(5) d [mm]	(6) ΔV *err* [mm]	(7) k_p *err* [mm]	(8) tg α *err* [mm]	(9) Ea % [-]	(10) Esq % [-]
0.50	0.60	0.53	0.57	0.49	0.01	-0.06	-0.03	0.21	0.14
1.00	1.21	1.03	1.12	0.97	0.01	-0.12	-0.07	0.20	0.14
1.50	1.75	1.63	1.69	1.47	0.01	-0.18	-0.10	0.20	0.14
2.00	2.25	2.28	2.27	1.97	0.01	-0.24	-0.14	0.20	0.14
2.50	2.85	2.83	2.84	2.47	0.01	-0.31	-0.18	0.20	0.14
3.00	3.55	3.38	3.47	3.01	0.01	-0.37	-0.21	0.20	0.14
3.50	4.05	4.23	4.14	3.60	0.01	-0.44	-0.26	0.20	0.15

Column (1) is the superimposed head displacement, (2) and (3) are the probe outputs referred to the initial position, (4) is their average value, (5) is the stem displacement obtained by means of eq. (1), with the identified k_P and tg α values, (6) is the error linked to the transducers outputs, (7) is the one due to k_p and (8) the one due to tg α; (9) is the absolute percentage error and column (10) is the percentage root square error. It is shown that even a broad uncertainty in the value of k_P does not affect significantly the stem motion measurement. The overall system accuracy has been fixed to 0.15% of the actual reading and any further improvement of the measurement set-up was not considered to be necessary.

4. Results and discussion

In the following some preliminary results are presented. They are about some time histories of both the two eddy current probes and the servo-accelerometer fixed to the pipe, when the valve instability phenomenon was exhibited. As it was thought that the frequency of the valve motion depended upon the length of the upstream pipe, we decided to perform the measurements twice, with two different pipe lengths (L_P = 2 and 400 m), to look for a frequency-pipe length link.

As an example of the experimental data, in *Figures* 3a and 4a the output of one of the two eddy current probes, that is the valve stem and head motion, is reported as a function of time, for L_P = 400 m (sampling frequency f = 128 Hz). The spectra of the sensor outputs are then calculated ; some of the routines were modified from Press *et al.* (1992). In *Figures* 3b and 4b the spectra H of the collected data are shown, in order to give the value of the average valve motion frequency : for these two conditions the observed main frequency is about 1 Hz. It has to be noted that the recorded motion is a kind of a square wave: the valve body hits the seat and then it suddenly goes back to its full-opened position. At the end of each these two movements a damped high-frequency transient is exhibited. It is due to the seat, the stem and the spring stiffness, although they are not significant to the description of the observed instability. Some remarks can be drawn from these time histories. One thing to be underlined is that the first harmonic frequency of the stem motion is not constant: although spectra of the acquired signals can be performed, as shown in *Figures* 3b and 4b, the observed peak has to be considered an average value; all the same, these values are significant because the above mentioned changes of frequency are not so remarkable. However, the acquired data systematically show that at the beginning and towards the end of the stem alternative motion the frequency tends to decrease. The second thing to be remarked is that the square cycles are not symmetrical in the sense that the rest times of the valve head at the closed and opened positions are different. These two facts may be considered as a general trend, as they have always been observed over more than 50 tests performed. Concerning the time domain signals, it is noted that some deviations from a regular square wave may be due to the fact that the stem is sometimes subjected to some random flow irregularities.

Figures 5, 6 and 7 show some records for the short pipe, at different sampling frequencies. In this case, the two eddy current probe outputs have been averaged to give the movements of the stem along the mean direction of the stream. The general features of the phenomenon are similar to those observed for the long pipe. The frequency tends to slow down at the beginning and at the end of the unstable motion and the square cycles are not symmetrical. Going to the spectra of the time histories (*Figures* 5b, 6b, 7b), it is clear that the shortening of the pipe is linked to a higher value of motion frequency, which is now between 16 and 18 Hz. The higher order peaks do not deserve particular attention, being multiple of the 16 ÷ 18 Hz harmonic.

A further check on the measured values is withdrawn by means of a servo-accelerometer fixed at 1 m upstream the valve. Its output has sometimes been recorded together with the one coming from one of the proximity sensors to eventually get the relation between the two. *Figures* 3a and 4a shows the recorded time history together with one of the two eddy current probes output. As only two channels at a time could be recorded, and as it was clear from the first tests that the stem movement was the full stroke, it has been decided that one eddy current probe output could be considered enough to have a reference for the comparison with the servo-accelerometer signal. Concerning the acceleration measurement, the signal to noise ratio is worse, due to the fact that the servo-accelerometer senses also the higher order vibration modes of both the pipe and the supporting structure (which, however, are well separated from those of the stem motion); all the same, it is possible to get the peak of the valve instability out of the accelerometer signal.

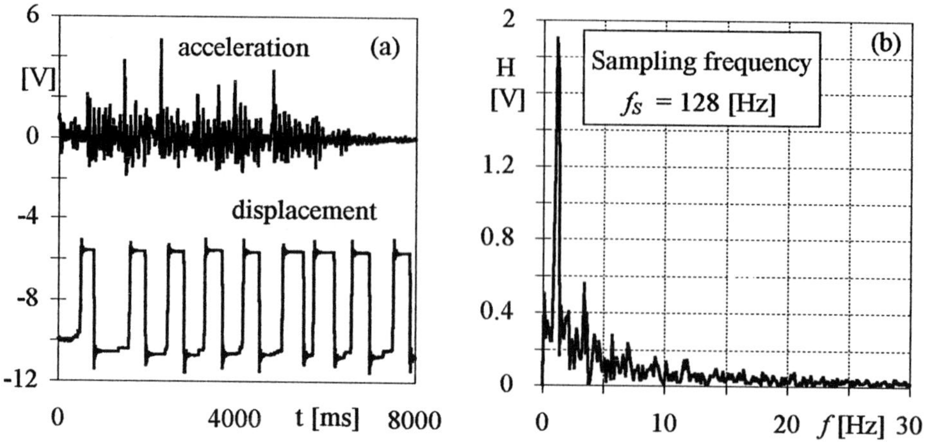

Figure 3. (a) Example of output from proximity sensor and servo-accelerometer (L_P = 400 m; total sampling time T = 8000 ms) and (b) spectrum of the displacement time history.

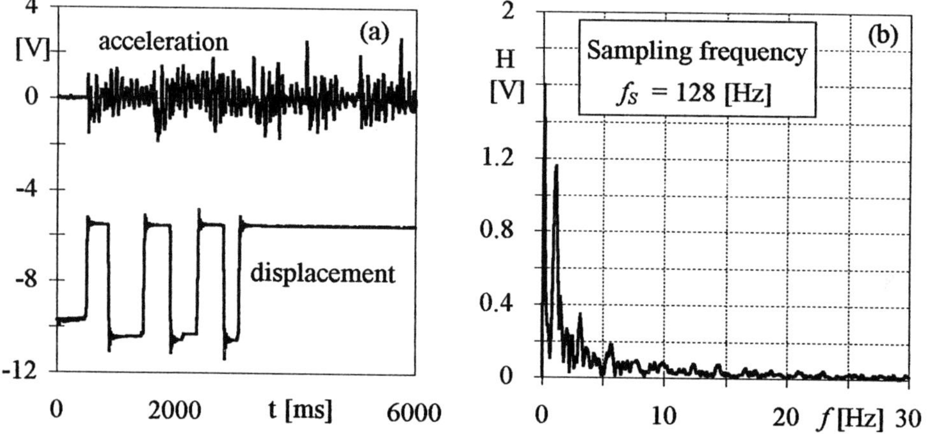

Figure 4. (a) Example of output from proximity sensor and servo-accelerometer (L_P = 400 m; total sampling time T = 8000 ms) and (b) spectrum of the displacement time history.

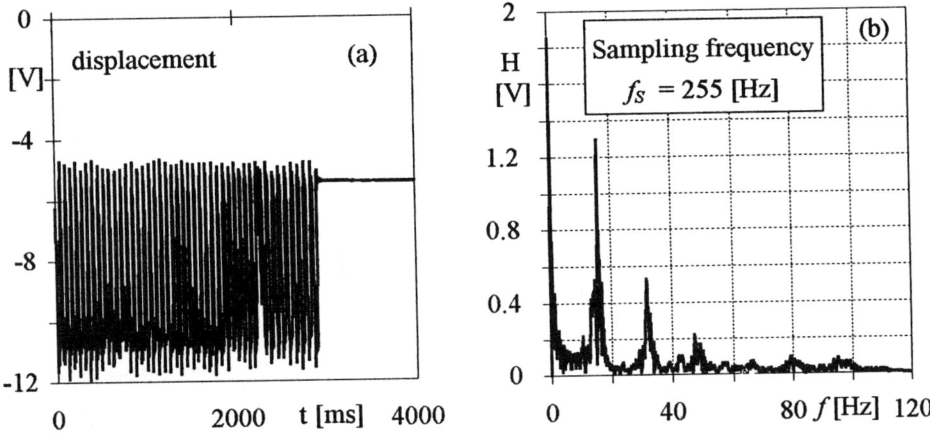

Figure 5. (a) Example of output from proximity sensor (L_P = 2 m; total sampling time T = 4000 ms) and (b) associated spectrum.

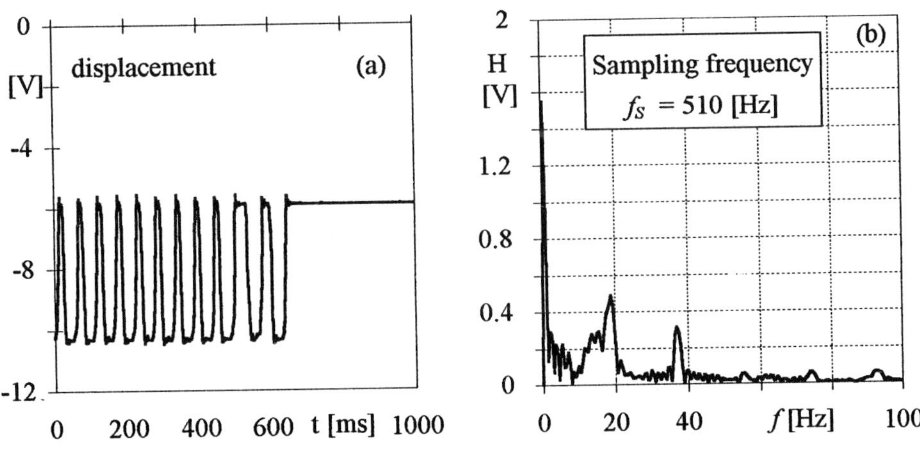

Figure 6. (a) Example of output from proximity sensor (L_P = 2 m; total sampling time T = 2000 ms) and (b) associated spectrum.

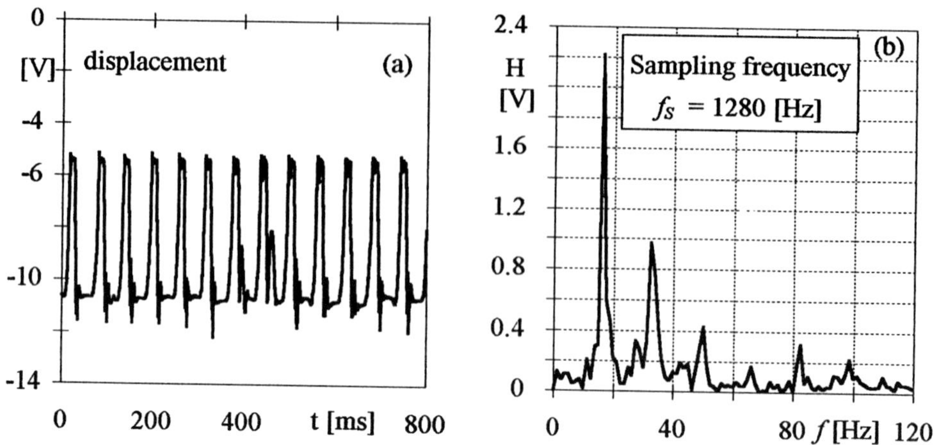

Figure 7. (a) Example of output from proximity sensor (L_P = 2 m; total sampling time T = 800 ms) and (b) associated spectrum.

From the readings of the accelerometer it is clear that the phenomenon cannot be related to one of the natural frequencies of the piping system, that are higher.

From a hydrodynamic point of view two tentative interpretations of the observed instabilities may be proposed. The first one is related to some analogies with a water hammer process : this may cause the motion of the mobile element of the valve, as it is subjected to the opposing hydrodynamic and spring forces. The second one is related to the behaviour of the system as the valve is reaching its closing position: as the flow section narrows, the fluid velocity increases and the local pressure drops. This action adds to the flux momentum. When the valve is closed, the spring force is opposed only to the force given by the fluid column in the upstream pipe, so that it might be possible that the valve opens again and an instability is generated. This observations are somehow consistent with the experimental evidences reported by Föllmer and Zeller (1980). At the present it is not possible to fully accept or cast aside any of these two schemes. Further experiments are needed, with different pipe lengths ; it is also important to properly measure the pressure values immediately upstream and downstream the valve.

5. Conclusions

A series of laboratory tests have been performed to study the oscillating behaviour that has been detected during the testing procedure of a flow-control valve. This transient phenomenon may assume a remarkable importance as it may be caused and maintained by several reasons which have to be thoroughly investigated. The displacement of the mobile element was recorded by means of proximity sensors ; moreover, a servo-accelerometer has been fixed to the pipe. The sampling rate was varied between 50 and

500 Hz and the phenomenon was recorded until the complete closure of the valve was obtained. The influence of the pipe length, L_P, on the observed behaviour was analysed. The time length of the transient conditions varies; in one case a persistence of the phenomenon was observed to last several minutes. During a single test the frequency of the oscillations is not a constant but it tends to increase in its central phase. The observed frequency, f, depends on the length of the adduction pipe: it ranges between 16 and 18 Hz for L_P = 2 m and it is about 1 Hz for L_P = 400 m. This behaviour suggests some analogies with a water hammer process and this problem is currently under investigation, also with the recording of the pressure values upstream and downstream the valve.

6. References

Brunone, B., Golia, U.M., and Greco, M. (1995) Effects of two-dimensionality on pipe transients modeling, *J. of Hydr. Eng.* ASCE, **121**(12), 906- 912.

Doeblin, E. O. (1990) *Measurement Systems: Application and Design*, Mc Graw-Hill, New York.

Fanelli, M. (1976) Bibliographie raisonnee sur : “Les resonances hydrauliques dans les circuits industriels”, ENEL - CRIS, Milano.

Föllmer, B. and Zeller, H. (1980) The influence of pressure surges on the functioning of safety valves, *Proc. of the Third Int. Conf. on pressure surges*, Canterbury, England, paper J2, 429-444.

Green, W.L. and Wood., G.D. (1980) The stability of direct acting spring loaded relief valves taking into account the upstream conditions, *Proc. of the Third Int. Conf. on pressure surges*, Canterbury, England, paper G4, 45-62.

Orsi, E. (1994) Su un fenomeno di instabilità in valvole a limitazione flusso (in Italian), *Proc. of the Seminar* “Moto vario nei sistemi acquedottistici”, Bari, 72-75.

Press, W., Flannery, B., Teukolsky, S. and Vetterling, W. (1992) *Numerical Recipes: The Art of Scientific Computing* (Second Edition). Cambridge University Press, New York.

Tanda, M.G., and Zampaglione, D. (1991) Analisi di fenomeni di risonanza in un sistema distributore al termine di un'adduttrice a gravità: primi risultati (in Italian), Giornata di studio per la celebrazione del centenario della nascita di Girolamo Ippolito. Napoli.

STUDY OF STAYVANE VIBRATION BY HYDROELASTIC MODEL

Jean-Loup DENIAU
GEC-ALSTHOM NEYRPIC
82,Avenue Léon Blum BP 75, 38041 Grenoble Cedex, FRANCE

A method using hydroelastic model of stayvane, completed by mechanical calculations is presented. Comparison with measured stress value on a prototype is made. Some results obtained with different profiles are given.

1. Introduction

Usual method described to avoid stay vane cracking in hydraulic turbines consists in checking that frequency of vortex shedding is inferior to the first natural frequency of the vane.

Shedding frequency is a function of boundary layer characteristics near the separation points. Many workers have developed a Universal Strouhal Number Sh, based on the spacing h between the separation streamlines, and the velocity V along them.

$$\boxed{fk = Sh \times \frac{V}{h}} \qquad (1)$$

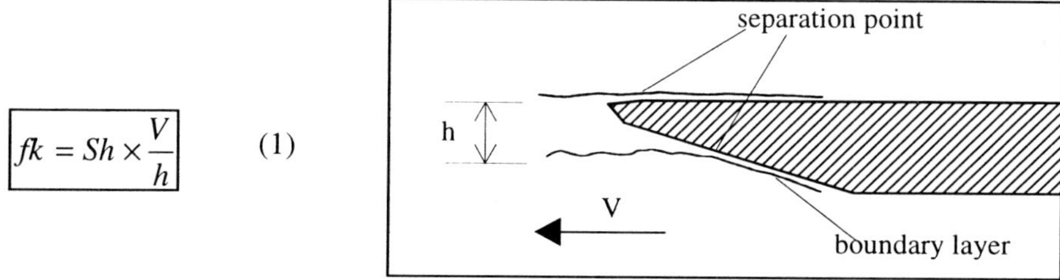

FIG. 1. Illustration of wake thickness

In this formula, Strouhal number Sh is independent of geometry.

Sh is difficult to evaluate, and different values were proposed. But from published works, and from our experimental studies, we can conclude in all cases :

$Sh < 0.28$

Evaluation of velocity V and wake thickness h is made by hydraulic calculations in stayring.

Then, the method of justification is to check :

E. Cabrera et al. (eds.), Hydraulic Machinery and Cavitation, 484–493.

$$f1 > K \times 0.28 \times \frac{V}{h} \tag{2}$$

where :

f1 = frequency of the first eigenmode (in general a bending mode)

K = safety factor to take into account :

- uncertainty in calculations of V , f1 and h.
- the lock-in phenomena (modification of vortex frequency when fk and f1 are near).

This precedent method is usual for turbines under great and mean head.
However, for turbines with low head, this approach can induce a design with thick stay vanes (to have high natural frequencies), and not profiled trailing edge (to have high wake thickness and thus low vortex frequencies).
Consequently, this design can induce worse hydraulic performances, and is generally inadmissible.

An other method is generally used in this case. This method consists in profiling trailing edge in order that vortices are very weak.
This method has generally been used with a empirical approach to solve eventual critical cases where vortex induced vibrations has occured.

Unfortunately, at this time, no reliable calculation method exists to evaluate vibration stresses induced by trailing edge shedding. To palliate this, we can have recourse to experimental method. However, study of vortex shedding requires hydroelastic models.

2. Conditions for hydroelastic simulation

A hydroelastic model requires following conditions (using classic symbols) :

- a good representation of real flow near the trailing edge of the profile, and particularly of the flow convergence which exists in a stay ring.

- a strict reproduction of ratio between elastic forces in the structure, and inertial fluid force.

$$\frac{E_{model}}{(\rho_{structure})_{model} \times V^2_{model}} = \frac{E_{proto}}{(\rho_{structure})_{proto} \times V^2_{proto}} \tag{3}$$

- the added mass effect must be correctly represented :

$$\frac{(\rho_{fluid})_{model}}{(\rho_{structure})_{model}} = \frac{(\rho_{fluid})_{proto}}{(\rho_{structure})_{proto}} \tag{4}$$

and lateral confinement must be similar.

Easiest method is :

- to use similar materials and fluids in model and in prototype. Mechanical boundary conditions at the ends of profile have to be reproduced.

- to do tests with flow velocity equal at velocity in the prototype.

In these conditions, stresses in model are equal to stresses in prototype.

3. Problems to solve

Maximal stresses in stayvanes are usualy limited in a little zone near attachment (a few mm at the model scale).
Example of axial stress distribution in stayvane (without stress concentration):

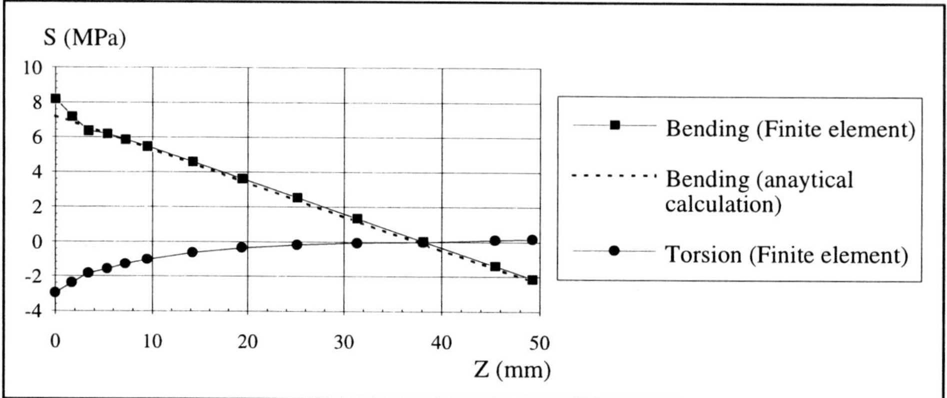

FIG. 2. Evolution of stress with axial distance at the attachment (elementary loads in bending and torsion at the middle of a 150mm height stayvane)

The direct measurement of stresses at ends is practically impossible.

4. Proposed methodology

Based on simultaneous use of finite element calculations, and hydroelastic simulations Procedure is summarised in schematic representation of fig. 3.
A finite element model (using approximately 900 quadratic elements in a half stay vane), gives end stresses during calibration .

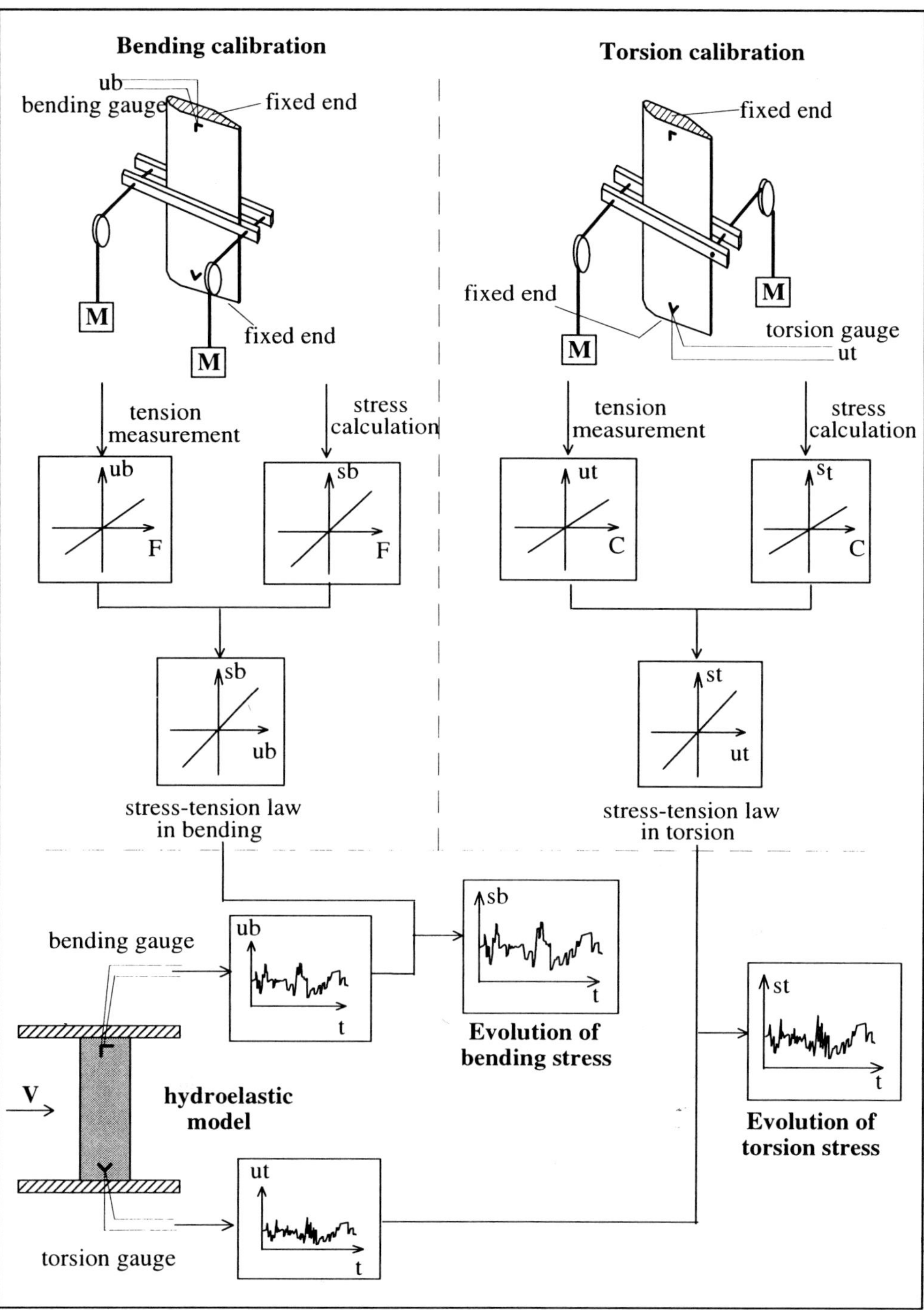

FIG. 3. Time evolution of stresses : schematic procedure

5. Presentation of a model for method's validation

Tests for validation of the method were done with a simple model of a stayvane, in a closed channel (120 mm width, 150 mm height). Variable angles of incidence i are possible. A modifiable lateral wall permits to reconstitute the flow convergence which exists in a stayring.
Incident flow velocity can be regulated between 0 and 22 m/s.

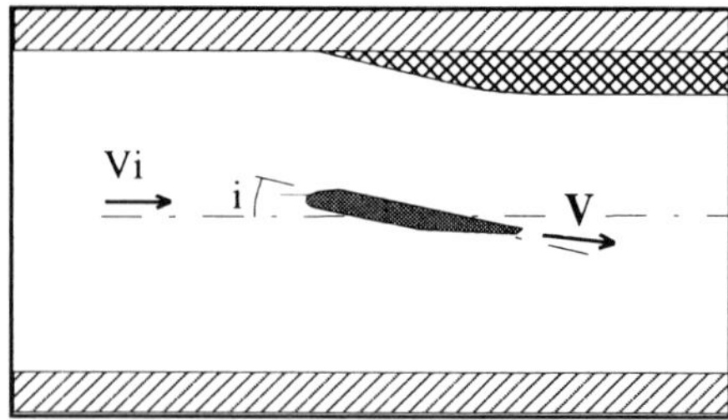

FIG. 4. Geometry of hydroelastic model

6. Main general results obtained

Tests were done with 5 different profiles :

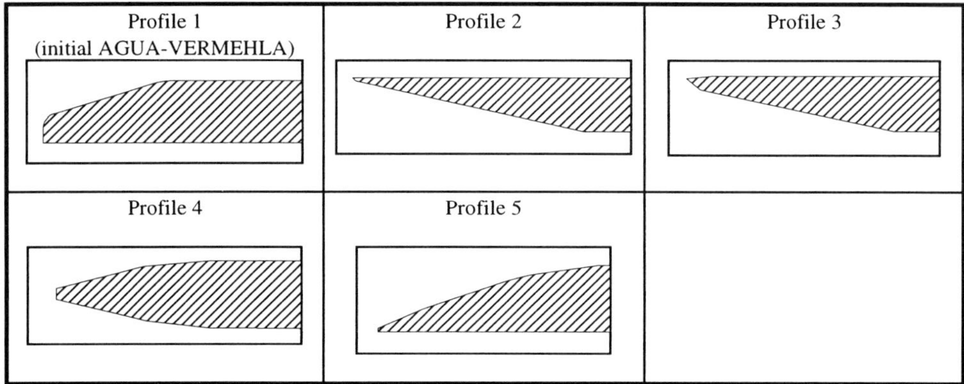

FIG. 6. Trailing edge geometry of different tested profiles.

• All tests show a very good agreement between measured and calculated values of first natural frequencies in bending and torsion mode. Example for profile 2 :

TABLE 1. Natural frequencies for profile 2 (in Hz)

	bending mode 1		torsion mode 1	
	in air	in water	in air	in water
measure	1508	1100	3460	2900
analytical calculation	1527	1116	3378	2831
EF calculation	1572	1147	3429	2873

Appendice gives relations used for analytical calculations. Element finite calculations are achieved using the same model than for stress calibration (part 4).

• For reduction factor of frequency induced by added mass, tests give a excellent concordance between mesured and calculated values (see appendice). Example for precedent profile 2 :

TABLE 2. Reduction factor of frequency from air to water (profile 2)

	bending mode 1	torsion mode 1
measurement	0.729	0.838
calculation	0.731	0.838

• Tests show a good concordance between peak to peak value of measured stresses in attachment of a prototype stayvane, and in the correspondent hydroelastic model (profile 1) :

In the prototype.

Stresses measured at AGUA-VERMEHLA (before modification) in 1979 : Smax = 82 MPa peak to peak for V estimated between 11 and 13 m/s at the trailing edge.

In the hydroelastic model.

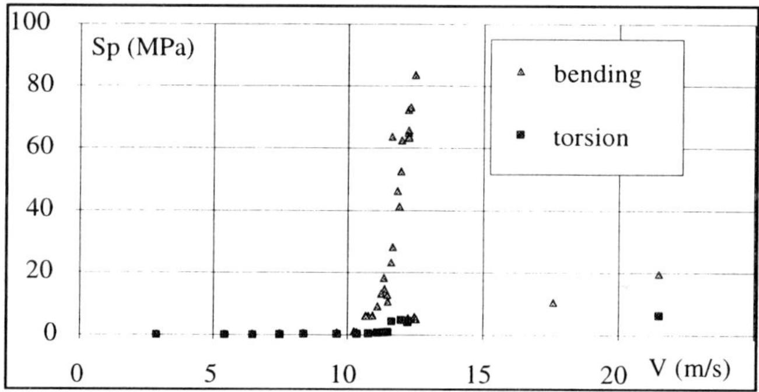

FIG.7. Profile 1 with i = 0°

7. Some results obtained with different other profiles

Following figures give, at the same scale, peak to peak stress values obtained during hydroelastic tests :

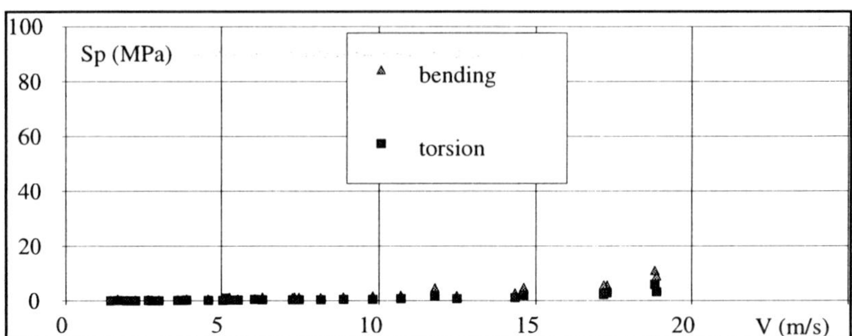

FIG.8. Profile 2 with i = 14°

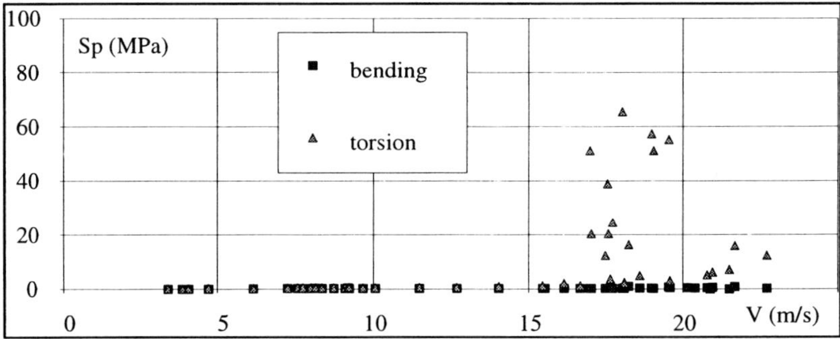

Fig. 9. Profile 3 with i = 14°

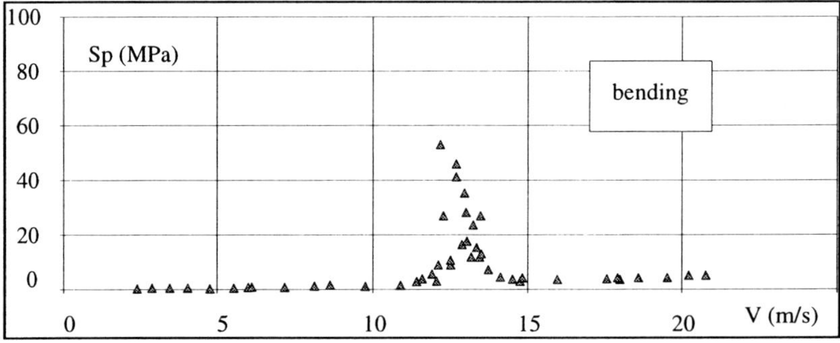

FIG. 10. Profile 4 with i = 0°

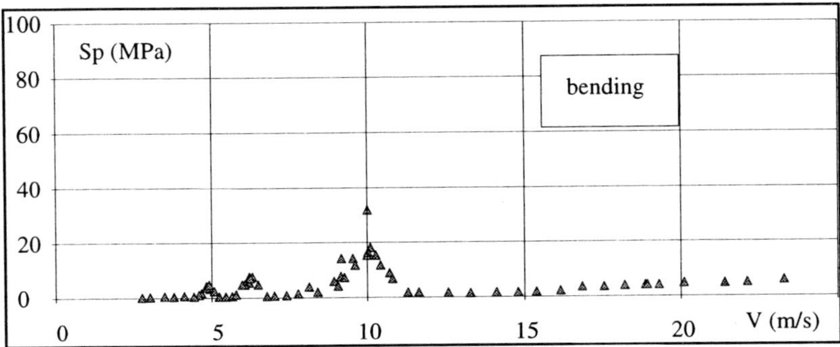

FIG. 10. Profile 4 with i = 14°

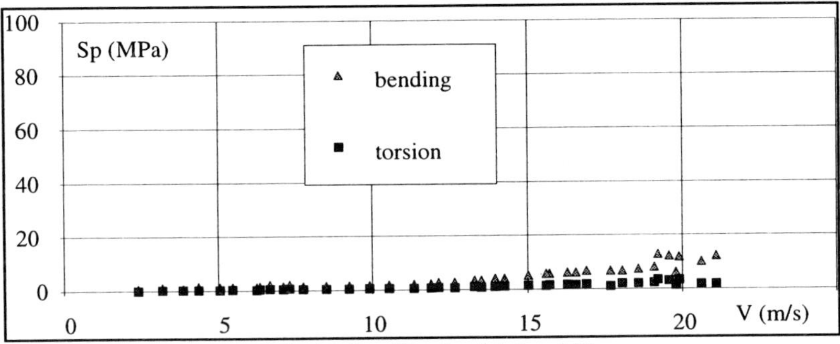

FIG. 11. Profile 5 with i = 14°

During resonance, stress amplification factor can reach values between 50 and 100.

During tests at i=14° with profile n°4, a strange phenomenon occurs : three successive resonance at the first bending mode appear. Only the third resonance at flow speed of 10 m/s is forecasted by classical design method (relation (1) in part 1.).
Following figures illustre this phenomena :

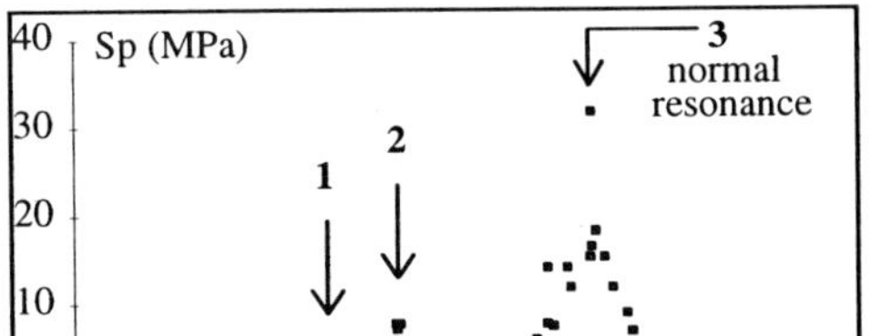

FIG. 12. Profil 4 : Peak to peak stress value in bending mode

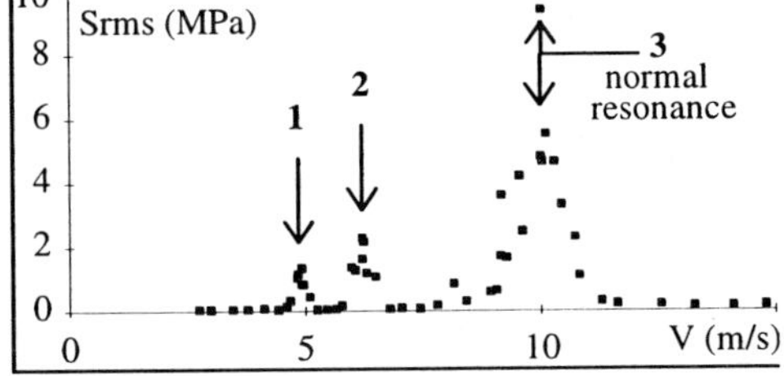

FIG. 13. Profil 4 : Efficace stress value in bending mode at first natural frequency (1100 Hz)

8. Conclusions

Tests described in this paper have permitted to verify possibility of a experimental approach in the field of mechanical structural response at vortex shedding.

For each stayvane profile to study, these tests have to be completed by hydraulic calculations, to check that wake caracteristics behind the trailing edge are very similar in model and in stayring. Calculations have to take into account the effect of double cascade (stayvane and guidevane).

Some strange phenomena occured with a symetrical trailing edge, which exhibit three succcessive resonance at the first natural frequency.

Appendice : analytical calculation of natural frequencies

Natural frequencies of vibration can be calculated assuming stay vane are beams fixed at two ends.

For example, the frequency of the first bending mode is calculated by :

in air : $$f_b = \frac{22{,}35}{2 \times \pi \times H^2} \times \sqrt{\frac{E \times Im}{\rho \times S}}$$

where : *H* height of stay vane
Im minimal moment of inertia
S area of cross section
ρ steel density
E modulus of elasticity

in water : $$f_b\,\text{water} = \frac{1}{\sqrt{1 + \frac{\pi}{4} \times \frac{\rho\text{water}}{\rho} \times \frac{l^2}{S}}} \times f_b\,\text{air}$$

where l = length of cross section
ρ_{water} = water density

For the natural frequency of the first torsion mode :

in air : $$f_t = \frac{1}{2 \times H} \times \sqrt{\frac{G \times J}{\rho \times Ip}} \times \frac{1}{\sqrt{1 - \frac{2}{\beta \times d}\,\text{th}\left(\frac{\beta \times d}{2}\right)}}$$

where : $$J = \frac{S^4}{4 \times \pi^2 \times Ip}$$

Ip = polar moment of inertia

$$\beta \times d = 2{,}148 \times \frac{H}{\left(2 \times \frac{S}{e} - 1\right)}$$

Precedent formula takes into account no warping of beam.

In water : $$f_t\,\text{water} = \frac{1}{\sqrt{1 + \frac{\pi}{128} \times \frac{\rho\text{water}}{\rho} \times \frac{l^4}{Ip}}} \times f_t\,\text{air}$$

STUDY OF DYNAMIC BEHAVIOUR OF NON-RETURN VALVES

P. FRANÇOIS
Industrial Hydraulic
Technical Center for Mechanical Industries
BP 7617
F - 44076
NANTES Cedex 03

1 - Introduction

Although fluid carrying industrial installations have long been insured against non-return valves malfunctions, attention must be drawn to the risk of operating failures involved by the use of valves not well suited to the installation they are fitted to.

As a result of CETIM's research in this field, non-return valves operation can be characterized which facilitates the selection of a non-return valve for a given installation.

2 - Purpose of study

Non-return valves are essential oil pipe systems protecting devices. Their "one-way" function prevents many damages such as the draining of an installation in case of supply pressure decrease, back flows, unexpected liquid mixtures, liquid losses, etc .

E. Cabrera et al. (eds.), Hydraulic Machinery and Cavitation, 494–503.

However, they must be suitably designed for the pipe system they are fitted to because damages may happen in transient conditions when the stable fluid flow rate decreases suddenly and reverses on valve closure.

Subsequent pulsations influenced by the valve characteristics may then cause destructive water hammers (upstream and downstream pressure surges) likely to damage the valves and fittings.

Working in close collaboration with the valve manufacturers and the AFIR (Valves Industrial French Association), the CETIM's industrial valve committee has set out a method aimed at characterizing non-return valves operation during that stage. Current tests are required to provide an approved non-return valve closure characterization procedure giving results transposable from one installation to another.

3 - Test equipment on an experimental schakle

The test rig allows to simulate a pump shutdown with back pressure development in downstream pipework.

The final test arrangement and components are shown in Figure 1.
The test line is composed of the following elements (from upstream to downstream) :

- a pump shutdown which can be synchronized with the operation of a number of valves,
- a quick opening discharge valve which is mounted next to the test valve.

Its opening can be simulated when a shutdown signal is sent to the pump, thus eliminating the influence of the motor pump set and pipework inertia between the pump and the valve during the tests.

Then, there is :

- the test valve,
- a dynamic pressure transducer, pressure range 100 bar, frequency range 1 kHz,
- a fast acting electromagnetic flowmeter working in direct and reverse flow, frequency range 100Hz.

Finally, the test line ends by the operation of the quick-acting valve at the tank outlet which can be synchronized with the pump shutdown and the opening of the upstream valve.

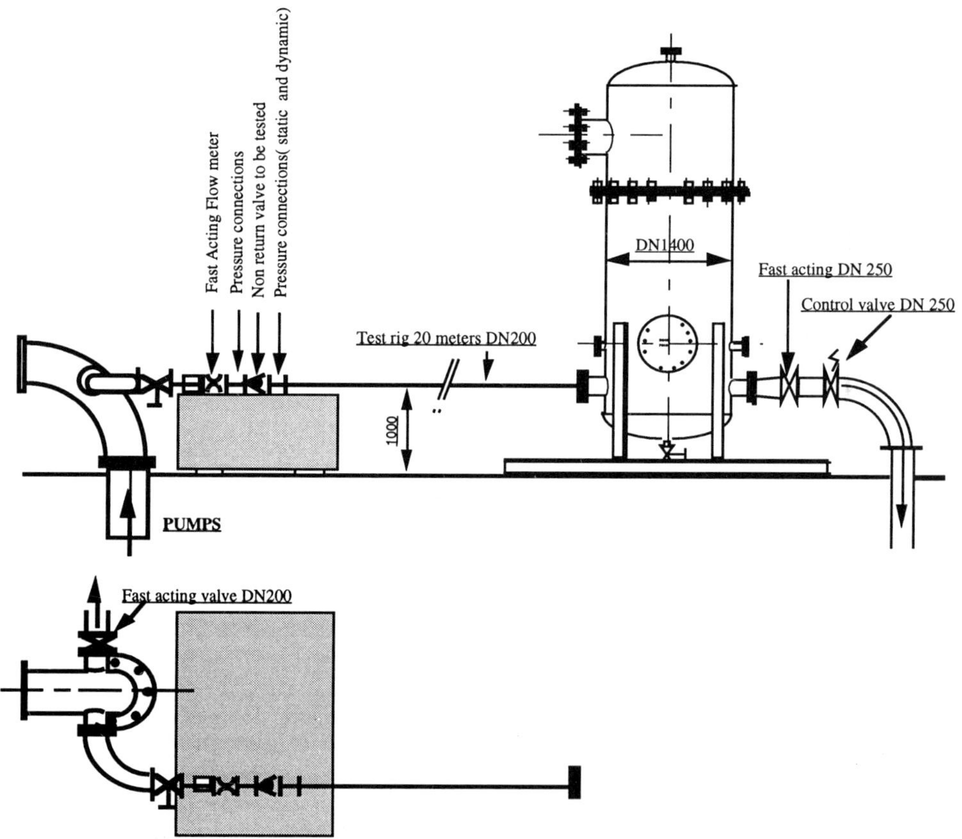

Figure 1 : Non return valves test rig diagram

Test results are recorded by a data-acquisition system and the corresponding curves are plotted automatically. The tests are carried out in a laboratory which has received the COFRAC/Essai agreement for several programs about the hydraulic materials (valves, pumps, turbines) operating under quality control system.

When the pump is shut down, the flow rate, the pressure before and after the non-return valve are recorded simultaneously as a function of time (figure 2).

Flow rate fluctuations (dv/dt) and return rate (Vr) are determined from a graph similar to Figure 2. The non-return valve dynamic characteristic plot is obtained by changing the test loop operating parameters (downstream pressure).

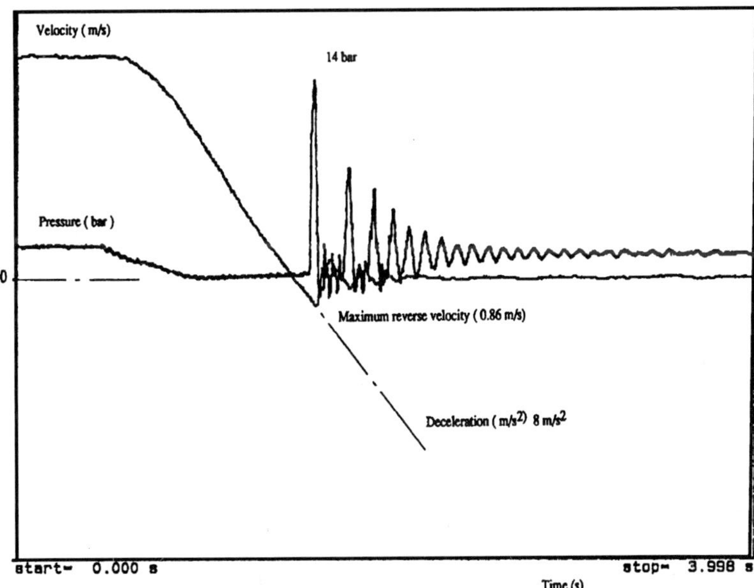

Figure 2 : Valve behaviour when closing

4- Use of results

In addition to the non-return valve closing time, two parameters are of particular importance.

The reverse flow rate maximum velocity Vrmax (if the valve closes after the direction of flow has reversed) governs the magnitude of pressure peaks according to the following Joukowski relationship:

$$\Delta P = \rho \, a \, Vrmax \qquad (1)$$

where ρ is the flow density (kg/m^3) and a the pressure waves propagation velocity (m/s).

The wave speed in the fluid-pipe system is given by :

$$a^2 = \frac{1}{\rho} \frac{K}{1 + DK/Ee} \qquad (2)$$

with K is the bulk modulus of the fluide (Nm^{-2}), E the pipe Young's modulus (Nm^{-2}), D the diameter of the pipe and e its thickness.

The slop of velocity-time curve is the average deceleration which in turn is an indicator of backflow severity.

Test results are usually sketched to represent the reverse flow rate maximum velocity Vrmax versus flow deceleration dv/dt curve (Figure 3) which is the non-return valve characteristic [1].

As a matter of fact, when an initial stable rate of flow runs through a wide open valve which means that initial V velocity is greater than Vc which controles the valve wide opening, dv/dt is the sole variable that governs Vrmax for a given valve [2].

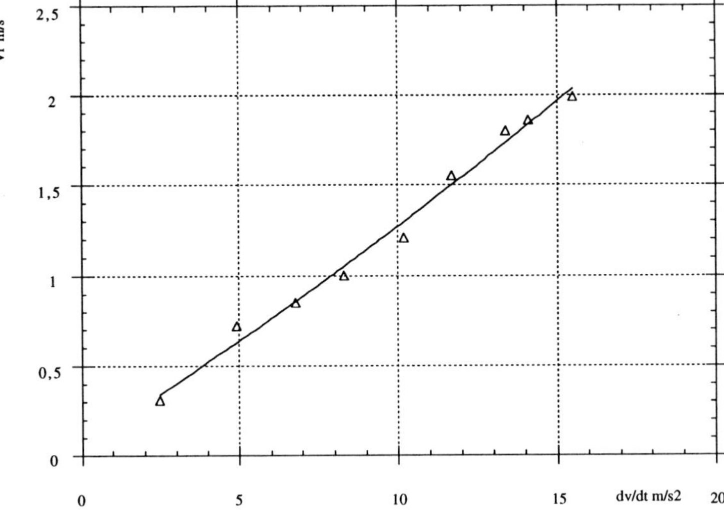

Figure 3 : Dynamic characteristic for a moving ball type non return valve DN 200.

The ideal valve would close before the reversal of the direction of flow regardless of deceleration value. The dynamic characteristic of the valve would coincide with the horizontal axis.

The representation on the same graph of characteristics of 6 widely different valve types is very helpful to compare their respective performances (Figure 4).

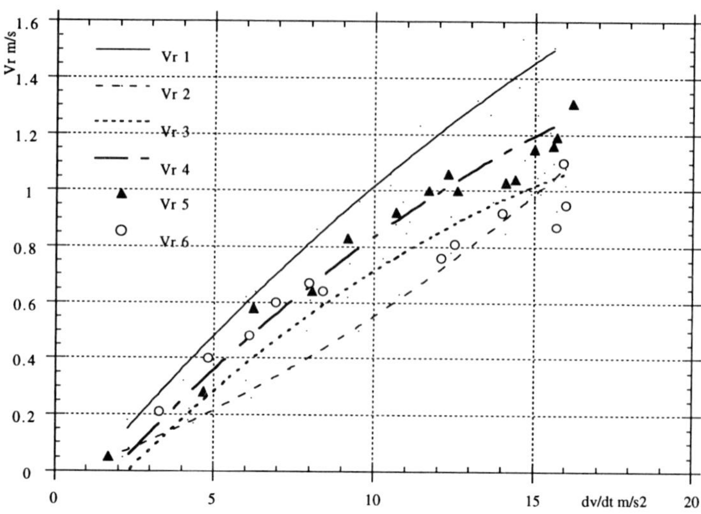

Figure 4 : Dynamic characteristics for several duo check split disk valves DN 200

5 - The selecting of a non-return valve

The first information required to select a non-return valve for a particular installation is the predictable deceleration of the fluid column due to a pump failure, a pipe break, etc.

If the pump inertia is ignored, and in a simple system, a first approximation of dv/dt is given by:

$$dv/dt = gH/L \qquad (3)$$

where g = gravity acceleration, H = pump discharge height and L = pipe length. We are currently building a model integrating the various parameters relative to the pump inertia and the pressure drop.

In addition, the tests are aimed at validating the model built.
Then, it is necessary to evaluate the maximum pressure allowable in the installation in case water hammers are generated as it will be used in equation (1) to determine the reverse flow rate maximum velocity Vrmax.

The selection of the best suited non-return valve shall rely on the study of a variety of valves characteristic curves. The final choice shall not necessarily be the highest performing valve but certainly the one that is best suited to the installation under consideration.

In addition to pulsation-induced water hammers, a certain amount of fluid is lost upstream the installation which may be a more stringent criterion in some specific cases.

6 - Setting of test configuration

During the tests, the validation of Joukowski's equation has always been our preoccupation, because this is the first calcul for the use of dynamic characteristics.

We have noticed a good connection between the Vr measured speed and the pressure in the case that the test line is homogenous, that is to say when diameters of flowmeter, non return valve and the installation are equal.

When the reverse flow is too little, we use this relation (1) to determinate the reverse velocity.
As it was difficult to measure the twist speed with the relation, because of a too weak. We have decided to assert ourselves some conditions : the test line and the non return valve will have the same diameters.

Moreover, we dispose 3 flowmeters with 100, 200 and 250 mm diameters. We limit to a 250 mm non return valve for diameters which don't enter in the flowmeters range, we realise test with two different flowmeters.

From this point, we have noticed the divergent and convergent influence on the dynamic characteristic, which can be met in industrial using (Figure 5).

In this case, it is primordial to define the installation. The non return valve dynamic characteristic is single, but the junction of a divergent or convergent pipe leads to an another non return valve with different characteristics.

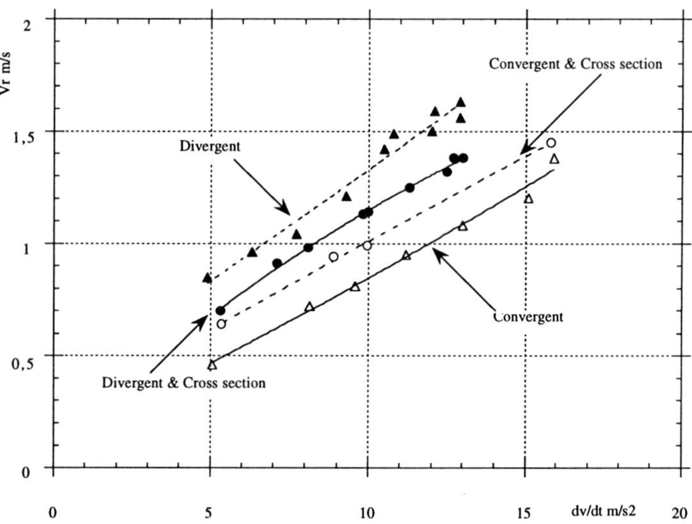

Figure 5 : Valves dynamic characteristics, a duo-check split disk DN 150

7 - Conclusion

For a given installation, i.e. deceleration $\frac{"dv"}{dt}$, pressure, wave propagation velocity "a" and maximum allowable pressure surge are defined, one can select the non-return valve best suited to these parameters, provided its operating characteristics are known.

CETIM's Industrial Hydraulics Departement test rig is designed to test pipes up to 250 mm, to deliver flow rates up to 4000 m^3/h for a 4 bar pressure and to create deceleration from 10 m/s^2 to 20 m/s^2 against tried diameters.

8 - References

1 - PROVOST GA (1982) The dynamic Characteristics of non-return valves, 11e Symposium AIRH AMSTERDAM, 1-7

2 - THORLEY A-R-D (1983) Dynamic response of check valves, Procedings 4e int. pressure Surges, 231-242.

OPTIMUM HYDRAULIC DESIGN OF TWO-WAY INLET CONDUIT OF WANGYUHE PUMPING STATION

LU LINGUANG, ZHOU JIREN
Agricultural College,Yangzhou University, China

ZHANG RENTIAN
Jiangsu Surveying And Design Institute of Water Conservancy, China

ABSTRACT

The numerical analysis of 3-D turbulence flow through the two-way inlet conduit of is executed. In the light of the analysis the conduit is finally optimized by considering all geometrical parameters which influence the conduit shape. The hydraulic characteristics of two pump systems with different inlet conduit are compared with model tests, which show that the better characteristics are achieved for that with the optimized conduit.

1. INTRODUCTION

Wangyuhe Pumping Station will be built between the Yangtze River and the Tai Lake in China, where 9 axial flow pump sets are going to be installed, the diameter of each pump being 2.5 meters. The station is required to pump water in two directions: one for drainage and the another for irrigation. The alteration of the directions can be brought about only by regulating the four gates: two located in the inlet conduit and two in outlet conduit, respectively, as shown in Fig.1. Thus several hydraulic constructions could be saved which are otherwise necessary for the direction controlling.
A two-way inlet conduit is employed by the pumping station (See Fig.1).

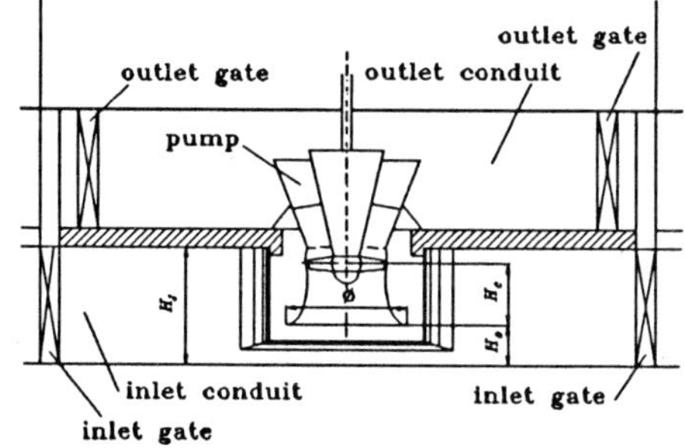

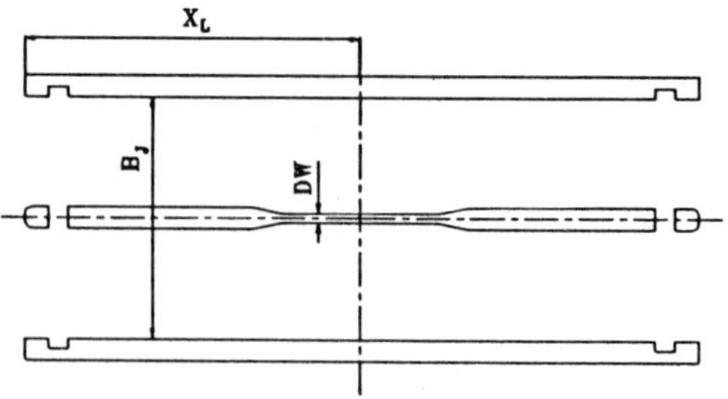

Fig.1 Configuration of two-way conduit

E. Cabrera et al. (eds.), Hydraulic Machinery and Cavitation, 504–513.

Due to the absence of study for the hydraulic design of the two-way inlet conduit before, two problems could be occurred to the inlet conduit. First, the flow field provided for the pump by the conduit may be non-uniform, which can reduce the efficiency of the pump system. Second, there probably be eddies in the non-working side of the inlet conduit, which may induce vortex band leading to severe oscillation of the pump.

In order to ensure the pump to be operated with safety and high efficiency it is necessary to optimize the hydraulic design of the conduit. The numerical analysis of 3-D turbulence flow through the two-way inlet conduit of Wangyuhe Pumping Station is executed, based on which the effects of all geometrical parameters of the conduit on the hydraulic characteristics are investigated one by one. Finally, the comparisons of model tests are conducted between two pump systems with the same model pump and outlet conduit but different inlet conduit. The calculated results of flow pattern in the inlet conduit and the hydraulic performance of the systems are experimentally validated.

2. MATHEMATICAL DESCRIPTION OF THE FLOW FIELD

2.1 GOVERNING EQUATIONS

The steady, Reynolds averaged Navier-Stokes equations for flows through the two-way inlet conduit, along with the closure standard k-ε model, can be written in Cartesian coordinates as:

(1) Continuity equation

$$\frac{\partial u_i}{\partial x_i}=0 \tag{1}$$

(2) Momentum equation

$$u_j\frac{\partial u_i}{\partial x_j}=f_i-\frac{1}{\rho}\frac{\partial p}{\partial x_i}+\frac{\partial}{\partial x_j}[(\nu+\nu_t)(\frac{\partial u_i}{\partial x_j}+\frac{\partial u_j}{\partial x_i})] \tag{2}$$

(3) k-equation

$$\frac{\partial u_i k}{\partial x_i}-\frac{\partial}{\partial x_i}[(\nu+\frac{\nu_t}{\sigma_k})\frac{\partial k}{\partial x_i}]=P_r-\varepsilon \tag{3}$$

(4) ε-equation

$$\frac{\partial u_i \varepsilon}{\partial x_i}-\frac{\partial}{\partial x_i}[(\nu+\frac{\nu_t}{\sigma_\varepsilon})\frac{\partial \varepsilon}{\partial x_i}]=\frac{(C_{\varepsilon 1}\cdot\varepsilon\cdot P_r-C_{\varepsilon 2}\cdot\varepsilon^2)}{k} \tag{4}$$

where x_i,u_i and f_i, with the suffix i=1,2,3, represents the Cartesian coordinates, the Cartesian velocity components and the Cartesian force components, respectively. P_r, the generation ratio of the turbulent energy, may be mathematically expressed by

$$P_r = \nu_t \left(\frac{\partial u_i}{\partial x_j} + \frac{\partial u_j}{\partial x_i}\right)\frac{\partial u_i}{\partial x_j} \tag{5}$$

in which ν_t stands for the dimensionless turbulent viscosity which is associated to k and ε by the relation:

$$\nu_t = C_\mu \frac{k^2}{\varepsilon} \tag{6}$$

The values of the empirical constants appearing in the above equations is given as follows[1]:

$C_\mu=0.09$, $\sigma_k=1.0$, $\sigma_\varepsilon=1.3$, $C_{\varepsilon 1}=1.44$, $C_{\varepsilon 2}=1.92$

2.2 BOUNDARY CONDITIONS

2.2.1 *Upstream B.C.*

The upstream boundary is set at the inlet section of the suction sump, where the approach flow could considered as having a logarithmic velocity profile:

$$u = \frac{1}{\kappa} u_* \ln\left(\frac{z}{d_0}\right) \tag{7}$$

where κ is the Von Kármán constant, u_* is the friction velocity, z is the vertical distance of the centre of the grid element to the bottom of the suction bay and d_0 is the roughness length of the bottom. The friction velocity may be obtained from:

$$u_* = \frac{\kappa q}{d\left(\ln\frac{d}{d_0} - 1\right)} \tag{8}$$

where q is the design discharge per unit width and d is the water depth in the suction sump.

The imposed profile of k and ε are of the forms:

$$k = \frac{u_*^2}{\sqrt{C_\mu}}\left(1 - \frac{z}{d}\right) \tag{9}$$

$$\varepsilon = \frac{u_*^3}{\kappa}\left(\frac{1}{z} - \frac{z}{d}\right) \tag{10}$$

2.2.2 *Downstream B.C.*

The outlet boundary is set at the inlet of the impeller chamber of the pump, where an absence of flow circulation over the cross section and a zero velocity gradient normal to the section are given[2,3,4].

2.2.3 *Solid Wall B.C.*

According to 'the law of the wall'[5], the following relation can be applied:

*b89Wâα Ç

$$u_* = \frac{\kappa u_w}{\ln(\frac{z_w}{d_w})} \tag{11}$$

in which z_w and u_w are the distance from the concerned element to the wall and the total velocity parallel to the wall in that element, respectively. The values of k and ε in the element are calculated via:

$$k = \frac{u_*^2}{\sqrt{C_\mu}} \tag{12}$$

$$\varepsilon = \frac{u_*^3}{\kappa Z_w} \tag{13}$$

2.2.4 *Free Surface B.C.*

The free surface may be considered as a symmetry plane for the velocity components and for the turbulence energy when wind shear is neglected. For the dissipation, ε, the used boundary conditions reads:

$$\varepsilon = \frac{(\sqrt{C_\mu} k_s)^{1.5}}{\kappa (Z_s + C_{BE} d)} \tag{14}$$

where z_s and k_s are the distance from the centre of the concerned element to the free surface and the turbulence energy in that element, respectively[5], and C_{BE}=0.07 is an empirical constant[6].

3. OBJECTIVE FUNCTIONS

According to the design conditions for the impeller of an axial pump, it is necessary for the pump to be operated with maximum efficiency that the velocity distribution over the inlet section of the impeller chamber should be uniform and the velocities should have no any radial component. Correspondingly, the two objective functions[7] are employed for the following optimum hydraulic computations.

4. BASIC FLOW PATTERN IN THE CONDUIT

The basic flow pattern calculated is shown in Fig.2, from which it can be seen that the pump intakes water through the bellmouth. The flows entering the pump are converged to the bellmouth from all the directions: the front, the two sides and the rear of the bellmouth. Obviously, the velocities from different directions are not uniform: that from

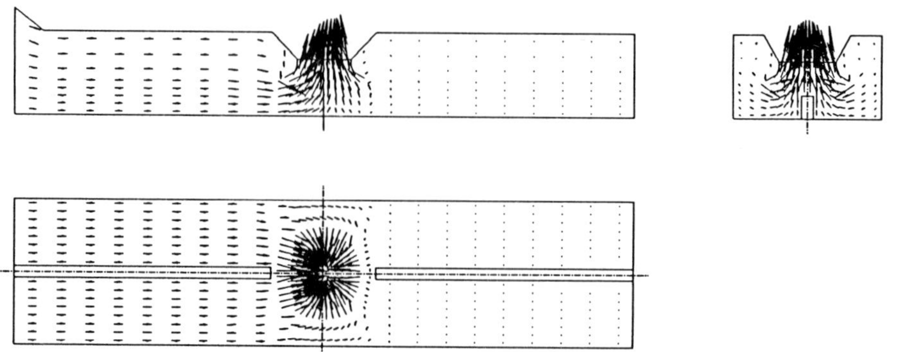

Fig.2 Basic flow pattern in the two-way inlet conduit

the front are the largest, while that from the rear are the smallest. This is the most important feature of the intake form in the two-way inlet conduit. It reveals that to obtain full uniform velocity distribution for the pump is difficult or even impossible. What we can do is to equalize the flow flux in all the directions as far as possible by hydraulically optimizing the design of the conduit.

The velocity distribution is clearly shown in the model test by the colour wires stuck on the inlet section of the bellmouth and the floor under the bellmouth. The flow pattern observed is essentially consistent with that calculated.

5. RESULTS OF THE OPTIMUM COMPUTATIONS

5.1 THE FIRST PART OF THE COMPUTATIONS

In the light of the numerical analysis, the hydraulically optimum computations can be executed. The undetermined parameters for the first part of the computation, in which the bellmouth is made of cast iron, so its outer shape is the the same as its inner one, are: clearance of bottom floor, H_B, diameter and height of the bellmouth, D_L and H_c, thickness of the baffle plate, DW, height, length and width of the inlet conduit, H_j, X_L and B_j (see Fig.1). The single-factor comparison method is applied in the optimum calculations. The schemes with different geometry parameters are principally taken in order according to the influence extent of the parameters on the flow field in the inlet conduit. The number and the relative geometry parameters of computational schemes in the first part are listed in Table 1, where scheme 11 is the original inlet conduit. The effect of the parameters upon the objective functions may found in Table 2 ~ Table 8. Following comments could be made based on the results:

(1) The clearance of bottom floor effect the degree of velocity uniformity obviously (See Table 2). The original value seems to be too small, so the centre of the pump impeller has to be raised to make the clearance increase to 0.84D (D is the diameter of the impeller).

(2) The diameter of the bellmouth has notable effect on the objective functions. From Table 3 it can be seen that the original one 1.44D is suitable.
(3) The height of the bellmouth influences markedly on the objective functions, too. Table 4 shows the larger the height, the better the inlet flow field.

Table 1 Number and relative geometry parameter of computation schemes

No.	X_L	H_j	B_j	H_B	H_c	ϕ	DW
11	5.40D	1.48D	2.60D	0.68D	0.56D	1.44D	0.5m
12	5.40D	1.48D	2.30D	0.68D	0.56D	1.44D	0.5m
13	5.40D	1.28D	2.60D	0.84D	0.56D	1.60D	0.2m
14	5.40D	1.28D	2.75D	0.84D	0.56D	1.44D	0.2m
15	5.40D	1.48D	2.60D	0.84D	0.56D	1.44D	0.5m
16	5.40D	1.28D	2.60D	0.84D	0.56D	1.44D	0.5m
17	5.40D	1.28D	2.60D	0.84D	0.56D	1.20D	0.2m
18	5.40D	1.28D	2.60D	0.84D	0.56D	1.44D	0.2m
19	5.00D	1.28D	2.60D	0.84D	0.56D	1.44D	0.2m
20	5.40D	1.28D	2.60D	0.84D	0.46D	1.44D	0.2m

Table 2 Effect of clearance of bottom floor

No.	H_B	u_{max}(m/s)	u_{min}(m/s)	$\bar{u}$(m/s)	V_u(%)	$\bar{\theta}_{max}$(°)	$\bar{\theta}_{min}$(°)	$\bar{\theta}$(°)
11	0.68D	5.52	2.56	3.98	83.60	90.0	60.0	78.0
15	0.84D	5.47	2.64	3.99	85.23	90.0	64.0	78.4

Table 3 Effect of diameter of bellmouth

No.	ϕ	u_{max}(m/s)	u_{min}(m/s)	$\bar{u}$(m/s)	V_u(%)	$\bar{\theta}_{max}$(°)	$\bar{\theta}_{min}$(°)	$\bar{\theta}$(°)
16	1.44D	5.34	2.04	3.98	85.77	90.0	65.7	78.1
13	1.60D	5.09	1.47	3.93	86.25	90.0	59.0	74.9
17	1.20D	5.11	0.33	3.93	75.60	90.0	64.1	76.4

Table 4 Effect of height of bellmouth

No.	H_c	u_{max}(m/s)	u_{min}(m/s)	$\bar{u}$(m/s)	V_u(%)	$\bar{\theta}_{max}$(°)	$\bar{\theta}_{min}$(°)	$\bar{\theta}$(°)
18	0.56D	5.10	1.77	4.02	87.79	90.0	64.6	77.4
20	0.46D	5.06	0.33	3.80	84.33	90.0	52.7	71.1

Table 5 Effect of thickness of baffle plate

No.	DW	u_{max}(m/s)	u_{min}(m/s)	$\bar{u}$(m/s)	V_u(%)	$\bar{\theta}_{max}$(°)	$\bar{\theta}_{min}$(°)	$\bar{\theta}$(°)
16	0.5m	5.34	2.04	3.98	85.77	90.0	65.7	78.1
18	0.2m	5.10	1.77	4.02	87.79	90.0	64.6	77.4

Table 6 Effect of width of inlet conduit

No.	B_j	u_{max}(m/s)	u_{min}(m/s)	$\bar{u}$(m/s)	V_u(%)	$\bar{\theta}_{max}$(°)	$\bar{\theta}_{min}$(°)	$\bar{\theta}$(°)
18	2.60D	5.10	1.77	4.02	87.79	90.0	64.6	77.4
14	2.75D	5.37	2.64	3.99	88.14	90.0	63.9	77.1

Table 7 Effect of height of inlet conduit

No.	H_j	u_{max}(m/s)	u_{min}(m/s)	$\bar{u}$(m/s)	V_u(%)	$\bar{\theta}_{max}$(°)	$\bar{\theta}_{min}$(°)	$\bar{\theta}$(°)
15	1.48D	5.47	2.64	3.99	85.23	90.0	64.0	78.4
16	1.28D	5.34	2.04	3.98	85.77	90.0	65.7	78.1

Table 8 Effect of length of inlet conduit

No.	X_L	u_{max}(m/s)	u_{min}(m/s)	$\bar{u}$(m/s)	V_u(%)	$\bar{\theta}_{max}$(°)	$\bar{\theta}_{min}$(°)	$\bar{\theta}$(°)
18	5.40D	5.10	1.77	4.02	87.79	90.0	64.6	77.4
19	5.00D	5.24	1.82	4.02	88.04	90.0	64.6	77.1

(4) From the results in Table 5, the thickness of baffle plate also has obvious effect on the functions and should be as thin as possible. In fact, the reason to found a baffle plate under the bellmouth is to prevent vortex bands from forming in the inlet conduit, and there is no requirement for thickness of the plate. Considering the smallest thickness of concrete pouring in the construction site, 0.2m is acceptable.
(5) In a certain range, the width, height and length of the inlet conduit have a little effect on the functions (See Table 6, Table 7 and Table 8). The values of the three basic dimensions of the pumping station greatly influence on the capital investment, so they had better to be determined more considering other factors. For example, the length may be decreased only if the arrangement of the upper structure of the pumping station is permissible.

5.2 THE SECOND PART OF THE COMPUTATIONS

To further reduce the velocities near the bellmouth the conduit width has been widened from the original value 2.6D to 2.8D in the second part of the computations. This change, however, does not mean any increasing of capital expenditure, because the widened part of the width is realized by thinning width of the supporting pier from 1.1 m to 0.6 m and remaining the width of 1.1 m in the vicinity of the gate grooves.

In order to avoid any possible vibration of the bellmouth the it is decided to be made from cast iron to concrete. In this part of the computations the purpose is mainly to compare two types of outer shape of the bellmouth: bell-like and cylinder outer shape. The calculated flow fields in the two inlet conduit are shown in Fig.3 and Fig.4, respectively. The calculated flow field reveals that the flow in the conduit with bellmouth of bell-like outer shape is easy to be induce large eddies with inclined axis in the non-working side of the conduit (See Fig.5(a)). These eddies complicate the flow between the bellmouth and the non-working side. Here flows entering into the pump and the eddies twist with each other, which makes it be possible for a vortex band to be produced. The flow in the non-working side of the conduit with bellmouth of cylinder outer shape, however, is quiet: no eddy can be found there (See Fig.5(b)), which is validated by the model test of the pump system[9].

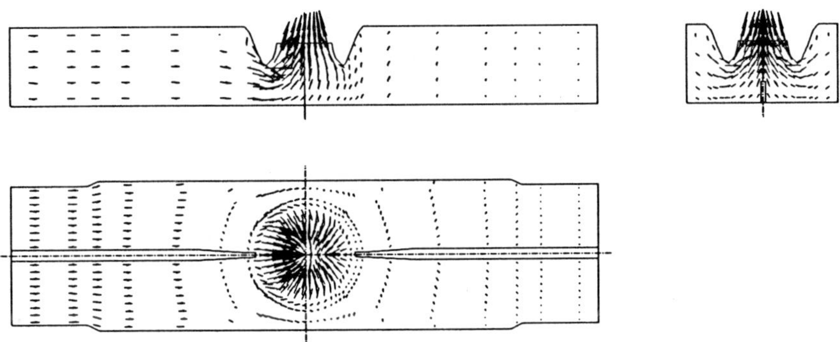

Fig.3 Flow field in the conduit with bell-like outer shape of bellmouth

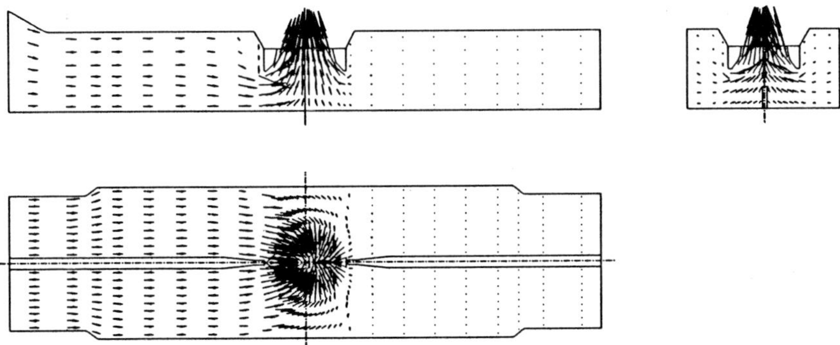

Fig.4 Flow field in the conduit with cylinder outer shape of bellmouth

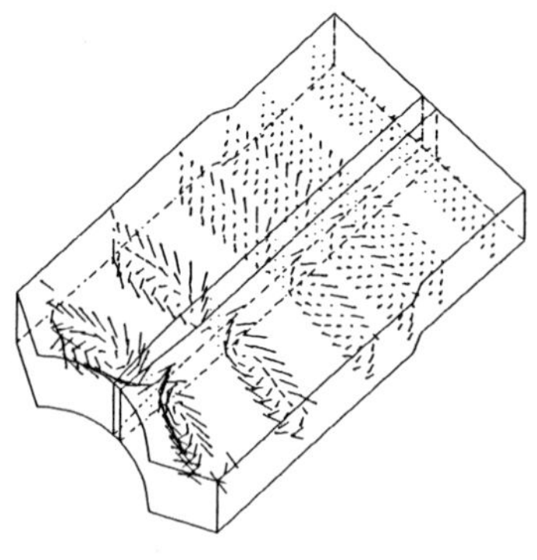

(a) Bell-like outer shape

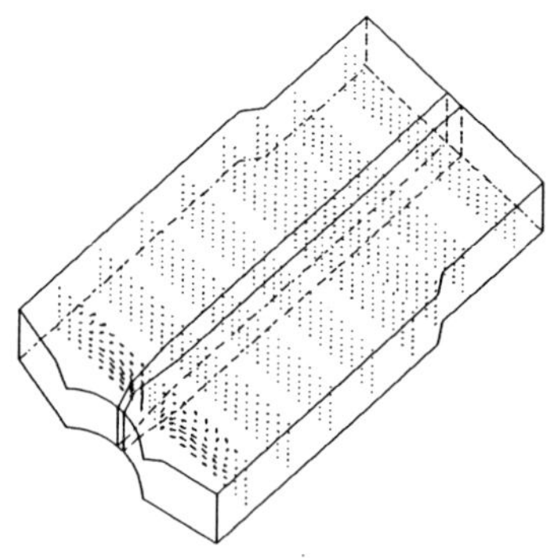

(b) Cylinder outer shape

Fig.5 Flow field in the non-working side of the two-way conduit

Difference of hydraulic characteristics between the two pump systems with different bellmouth of outer shape may be compared according to the computation results listed in Table 9. The outer shape of bellmouth hardly effect the objective functions.

Table 9 Effect of outer shape of bellmouth

outer shape of bellmouth	u_{max}(m/s)	u_{min}(m/s)	$\bar{u}$(m/s)	V_u(%)	$\bar{\theta}_{max}$(°)	$\bar{\theta}_{min}$(°)	$\bar{\theta}$(°)
bell-like	3.96	2.66	3.45	90.48	90.0	63.58	78.11
cylinder	3.98	2.73	3.45	90.78	90.0	66.50	77.76

6. EXPERIMENT COMPARISONS

Model tests with the two pump systems have been finished. The model pump and the outlet conduit are the same in the pump systems, while the inlet conduit is different: one is original and another is optimized. A brief comparison between the hydraulic characteristics of the two pump systems at the optimal operating point is given in Table

Table 10 Comparison between hydraulic characteristics of the two pump systems

	inlet conduit	Q(m^3/s)	H(m)	P(kW)	η(%)	n_s	β(°)	reference
1	original	0.299	2.463	11.78	61.34	1269	+2	[8]
	optimized	0.2986	2.03	9.36	63.51	1466	0	[9]
2	original	0.311	2.848	15.04	57.76	1161	+4	[8]
	optimized	0.3164	2.21	11.08	61.85	1416	+2	[9]

10, where H, Q, P, η, n_s and β is the head, discharge, power, efficiency, specific speed of the pump system and the blade angle of the pump, respectively.

Measured results in Table 10 show that under the condition of the discharge Q being equal the hydraulic characteristics of the pump system for one with the optimized inlet conduit are improved as indicated below:

(1) The head H at the optimal operating point decreases 0.46m for the first group of data and 0.64m for the second group of data, respectively, and the specific speed n_s increases correspondingly from 1269 to 1466 for the first group of data and from 1161 to 1416 for the second group of data, which enable the new pump system to be more suitable to the feature of the special low head of the pumping station.

(2) The efficiency of the pump system at its optimal operating point increases 2.17% for the first group of data and 4.09% for the second group of data.

7. CONCLUSIONS

By the help of the hydraulic optimization, the efficiency of the pump system at the optimal operating point is increased by 2%~4% while the head is decreased by 0.46m to 0.64m.

The geometry parameters clearer to the inlet section of the pump, such as the hight and diameter of the bellmouth, the clearance of the bottom floor, the thickness of the baffle plate, have more notable effect on the hydraulic characteristics of the pump system, so they need more carefully to be optimized.

The outer shape of the bellmouth greatly influence the flow field in the non-working side of the conduit. The outer shape of cylinder is much better than that of bell-like which is easy to lead eddies.

REFERENCES

1. Jin.,Z. (1989) Numerical Solution to the Navier-Stokes Equations and Turbulence Models, *Hohai University Press*, pp54.

2. Song,C.C.S., He,J.M. and Chen,X.Y.(1991) Calculation of Turbulence Flow Through a Francis Turbine Runner and an Elbow Draft Tube, *International Power Generation Conference*, San Diego, CA.

3. Vu,T.C. and Shyy,W.(1990) Navier-Stokes Flow Analysis for Hydraulic Turbine Draft Tubes, *Journal of Fluid Engineering*, Vol.112, pp199-204.

4. Vu,T.C. and Shyy,W.(1990) A Comparative Study of Three Dimensional Viscous Flows In Semi and Full Spiral Casings, *IAHR Symposium*, Belgrade Yugoslavia.

5. W.Rodi,W.(1980) Turbulence models and their application in hydraulics, *IAHR Section on Fundamentals of Division II: Experimental and Mathematical Fluid Dynamics*, Delft, The netherlands, pp44-46.

6. Hossain,M.S.(1980) Mathematical Modelling of turbulent buoyancy flows, Ph.D.Thesis, University of Karlsruhe.

7. Lu,L. and Booij,R.(1994) Numerical Determination of A Hydraulically Optimized Suction Box of A Large Pumping Station, *IAHR Symposium*, Beijing China.

8. Jiangsu University of Science and Technology (1993) Report of Model Test for Pump System of Wangyuhe Pumping Station (1), pp8.

9. Jiangsu University of Science and Technology (1996) Report of Model Test for Pump System of Wangyuhe Pumping Station (2), pp10.

FLOW ANALYSIS FOR THE INTAKE OF LOW-HEAD HYDRO POWER PLANTS

A. RUPRECHT, M. MAIHÖFER, E. GÖDE

Institute for Fluid Mechanics and Hydraulic Machinery
University of Stuttgart
Pfaffenwaldring 10, 70550 Stuttgart, Germany

Abstract

In low-head hydro power plants often severe flow problems arise at the inlet region. Since model tests are mostly too expensive, especially for small power plants, numerical flow analysis is introduced. In this paper the numerical analysis of the flow behavior is shown for the intake region of two small power plants. For the first power plant the problem was that due to a disturbed flow in the trash rack, causing severe head losses, the power output was too low. By suggesting a relative simple modification of the geometry the problem could be cured. In the second case investigated there were oscillation problems initiated by a vortex shedding at the separation pier. Again, by changing the shape of the pier based on the results of the flow analysis, this problem could be avoided.

1. Introduction

Hydro turbines today obtain a very high hydraulic efficiency. Great effort is focused on all parts of the machine (stay vanes, wicket gates, runner, draft tube) in order to get the best power output. This is also the case for low-head turbines, which are frequently installed in rivers. In contrast to the high level turbine design, the shape of the intake region often is often not optimized in terms of flow behavior. This may have different reasons:

- the power plant is only of secondary importance, since the geometry is dominated e. g. by ship movements or flood resistance,
- the power plant is integrated into existing civil work arrangements or
- simply in order to achieve a cheap construction.

A poor design, however, can either result in severe energy losses, which can be caused by flow separations or by an oblique inflow to the trash rack. Or it can cause operational problems e. g. by unsteady vortex structures. Often minor changes in the geome-

E. Cabrera et al. (eds.), Hydraulic Machinery and Cavitation, 514–523.

try can increase either the power output of the plant considerably or it can lead to smooth turbine operation. Therefore it is necessary to pay the same attention to the intake as to the other components.

Hydraulic model tests of the entire intake region are very expensive and time consuming as well and are usually not applicable for smaller power plants. Therefore a numerical method is developed for the analysis of theses types of flow. In this paper the intake geometries of two small low head power plants are investigated using this numerical method. The existing flow problems were analyzed and modifications of the geometry are examined to cure the problems.

2. Basic equations and numerical methods

A viscous incompressible flow is assumed, either two-dimensional or three-dimensional and either steady or unsteady, depending of the problem. The calculations are based on the Reynolds-averaged Navier-Stokes equations combined with the eddy viscosity assumption of Boussinesq. The momentum equations are obtained to

$$\frac{\partial U_i}{\partial t} + U_j \frac{\partial U_i}{\partial x_j} + \frac{1}{\rho}\frac{\partial P}{\partial x_i} - \frac{\partial}{\partial x_j}\left[(\nu + \nu_t)\left(\frac{\partial U_i}{\partial x_j} + \frac{\partial U_j}{\partial x_i}\right)\right] = 0 \qquad (1)$$

all equations are written in tensor notation assuming the summation convention of Einstein (summation on repeated indices). The continuity equation is given by

$$\frac{\partial U_i}{\partial x_i} = 0 \qquad (2)$$

The turbulent viscosity ν_t is calculated from a Prandtl mixing length formulation

$$\nu_t = |\omega| \cdot l_m^2 \qquad (3)$$

where $|\omega|$ is the amount of vorticity and l_m is the mixing length.

The mixing length formulation is preferred instead of the frequently used k-ε model, since it is relatively simple and robust and therefore reduces the computational effort. In addition the k-ε model in unsteady flows often shows too high dissipation and some times suppresses vortex shedding. Especially the standard k-ε model with logarithmic wall functions causes severe problems, since the separation of the boundary layer is not predicted accurate enough by wall functions.

For the flow predictions it is assumed that the free surface is of minor influence to the global flow structure and therefore is assumed to be constant. The trash rack cannot be considered exactly in the calculation, because its resolution by a computational grid would cause an extreme computational effort. Therefore the trash rack is introduced in form of a "porous" medium with an additional local pressure loss. The pressure loss is calculated from the empirical formulas of Kirschmer [1] (straight inflow) and Spangler [2] (oblique inflow).

The simulation is performed by the finite-element code FENFLOSS, which is developed at the university of Stuttgart. FENFLOSS uses a segregated solution algorithm for the momentum equations combined with a modified Uzawa type pressure correction. The linear systems of equations are solved by an ILU-preconditioned conjugate gradient algorithm (BICGSTAB2). For details the reader is referred to [3,4].

3. Applications

3.1. VERIFICATION

Since it is complicated, time consuming and therefore very expensive to do field measurements and because the calculations are often used to predict the flow behavior before the power plant is built, a simple model test rig has been constructed in order to compare the calculation results with some flow visualizations in the experiment. The test rig, shown in fig. 1, consists of 3 inlet gates and 3 turbines, which can be opened or closed in order to obtain different flow situations. In the test rig no trash rack has been considered. A two-dimensional calculation has been applied.

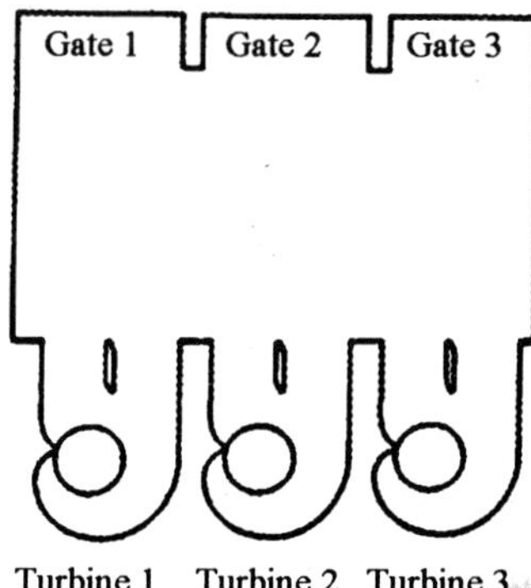

Fig. 1: Geometry of the test rig

In fig. 2 the streamlines of the calculation and the experiment are compared for a straight inflow. In this case the inflow enters through the middle gate and all three turbines are open. One can observe, that there are two recirculation regions behind the closed gates. The flow to the middle turbine is nearly undisturbed, whereas the flow to the left and to the right turbine show a strong turn on the piers.

In Fig. 3 the behavior for strong oblique inflow conditions are presented. The flow enters the test rig by the right gate and the right turbine is closed. Therefore a vortex occurs behind the closed gates and in front of the closed turbine. The flow to middle and the left turbines is highly disturbed and flow separations at the piers can be observed. These flow separations can cause severe losses and unsteady phenomena within the turbine and should be avoided if possible.

The comparison between the experiment and the calculations shows a good agreement. The structure of the flow can be predicted quite accurately. Therefore the computational procedure can be applied for the prediction of field conditions and to search for solution of the arising problems.

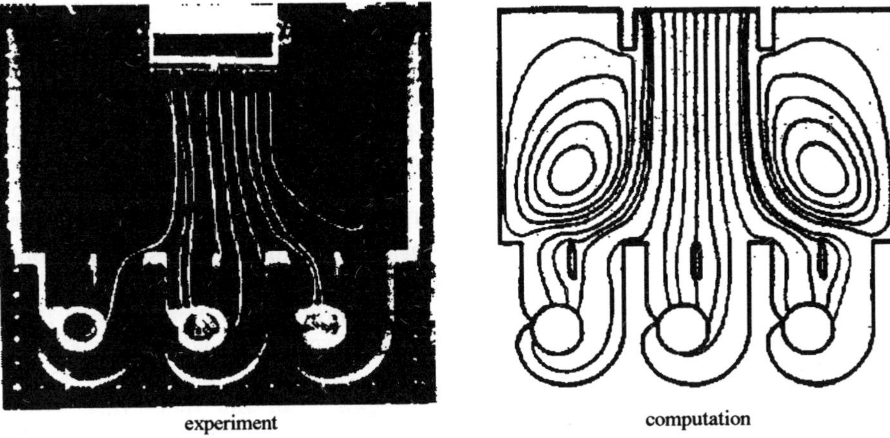

Fig. 2: Comparison of experiment and calculation for straight inflow conditions

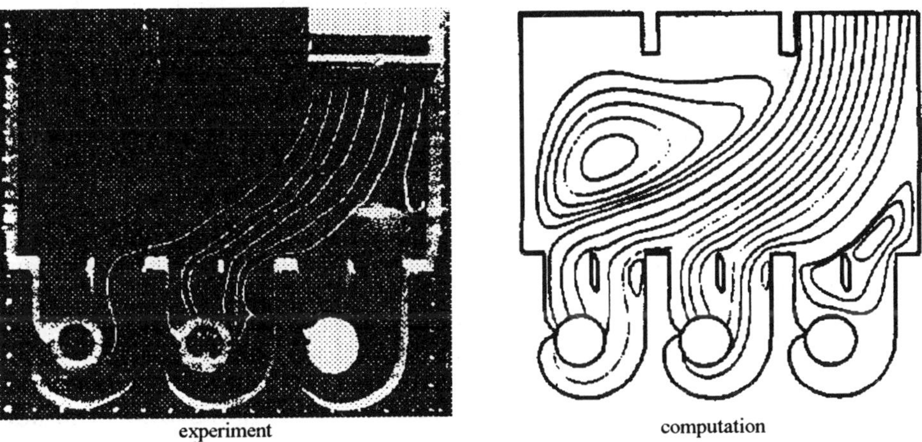

Fig. 3: Comparison of experiment and calculation for oblique inflow conditions

3.2. APPLICATION 1

As a first example a small hydro power plant, schematically shown in fig. 4, is presented. Next to an existing weir a new power plant has been build. After commissioning it was found that the power output did not reach the estimated level. It was assumed that the problem is caused by the shape of the inlet. It could be observed that the flow separates at the sharp corner when going into the channel to the turbine, leading to a disturbed flow field in the trash rack cross-section.

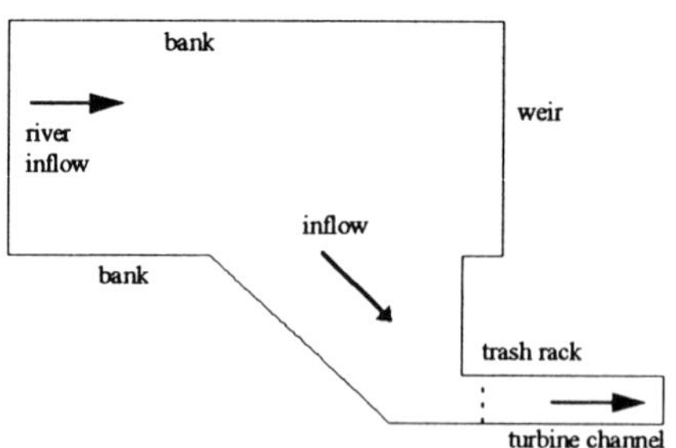

Fig. 4: Geometry of application 1

Since the water depth is nearly constant the investigation is carried out two-dimensional. In order to assure the results and to show the influence of secondary motion also a three-dimensional calculation has been carried out.

The flow behavior described above has already been found by the two-dimensional flow analysis. The streamlines are shown in fig. 5. It clearly can be observed that there exists a flow separation which extends into the trash rack. This leads to a locally higher velocity in the trash rack cross-section, fig. 6, and therefore to higher losses. Looking to the flow angle of attack, fig. 7, it can be seen that the flow enters the trash rack with an up to 10° wrong inflow angle. This also causes an increase of flow losses. It has to be pointed out, that the trash rack itself makes the flow more uniform. A simulation without trash rack leads to a flow angle deviation up to 20° and a more disturbed velocity profile. But this equalization of the flow causes additional losses.

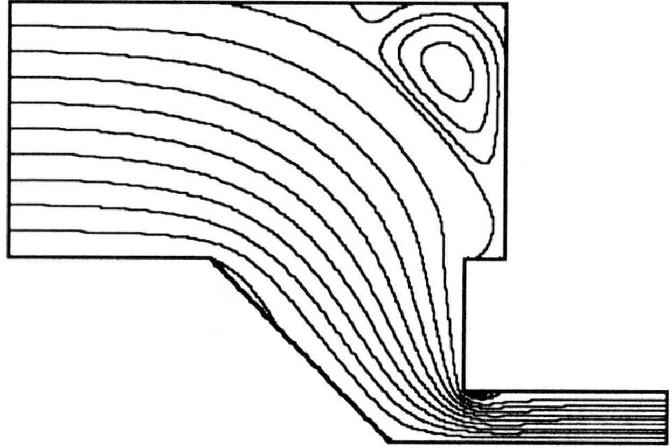

Fig. 5: Streamline plot

Due to the disturbed flow in the trash rack cross-section the losses in the trash rack are approximately 3 times higher than for a uniform flow. Since the velocity in the channel is rather high anyhow, losses in the trash rack are severe and cause the drop in thepower output.

In order to be sure that there are no mayor three-dimensional effects dominating the flow behavior a 3D flow analysis has been carried out. In fig. 8 the velocity profile in the trash rack cross-section is shown. Again it can be observed that there is also the recirculation region initiated by the sharp corner. Except the influence of the boundary layer at the bottom the flow shows a two-dimensional behavior. Therefore a 2D calculations is sufficiently accurate.

The high energy losses in the trash rack mean, that if it is possible by changing the inflow geometry in such a way, that the velocity profile in the trash rack is quite uni-

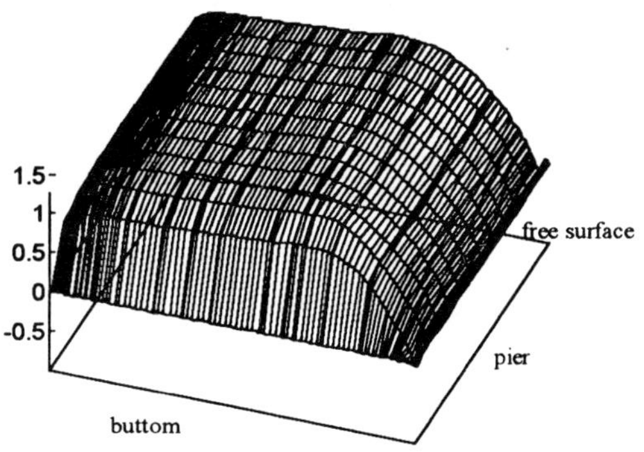

Fig. 8: Velocity profile in the trash rack cross-section

form, a considerable increase of power output can be achieved. Since it would be very expensive to reconstruct the power plant, it was tried to cure the flow situation only by minor changes of the geometry. The shifting of the trash rack cross-section further down-stream had to be avoided, because it is too expensive to move the trash rack inclusive the automatic cleaning devices.

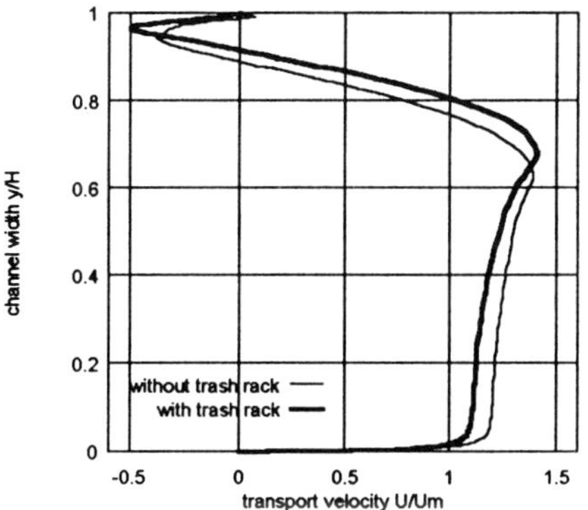

Fig. 6: Velocity distribution in the trash rack cross-section

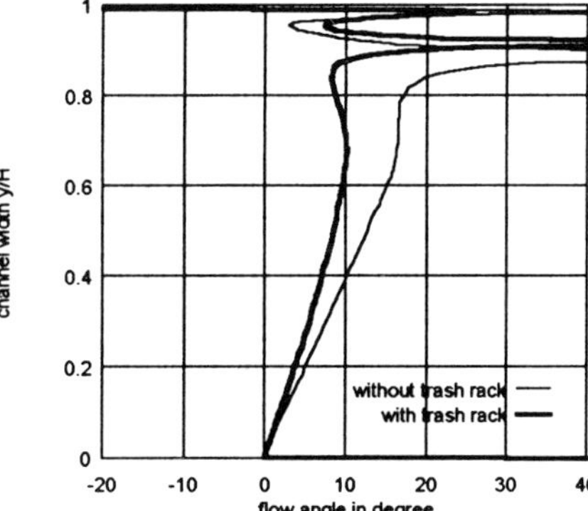

Fig. 7: Flow angle in the trash rack cross-section

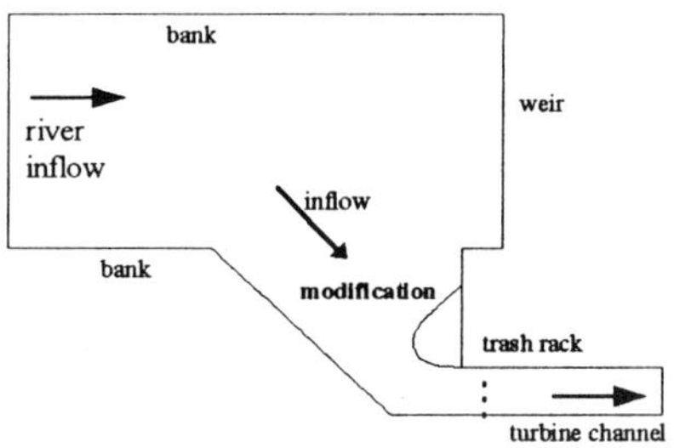

Fig. 9: Modified geometry

By giving the inflow section a round shape, see fig. 9, the flow behavior could be improved. Fig. 10 shows the streamlines for the modified contour. It can be seen, that the flow enters the trash rack quite undisturbed. There is no longer a flow separation at the inlet to the turbine channel. This can be seen more clearly in fig. 11 and fig. 12, where the velocity distribution and the flow angle in the trash rack cross-section are shown. One can observe that the velocity profile is quite uniform and the maximum angle of attack at the trash rack is lower than 2°. Therefore the head losses in the trash rack are reduced remarkably.

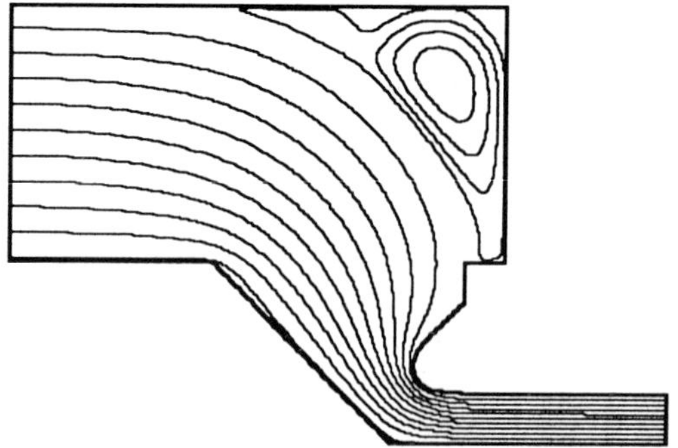

Fig. 10: Streamline plot of the modified geometry

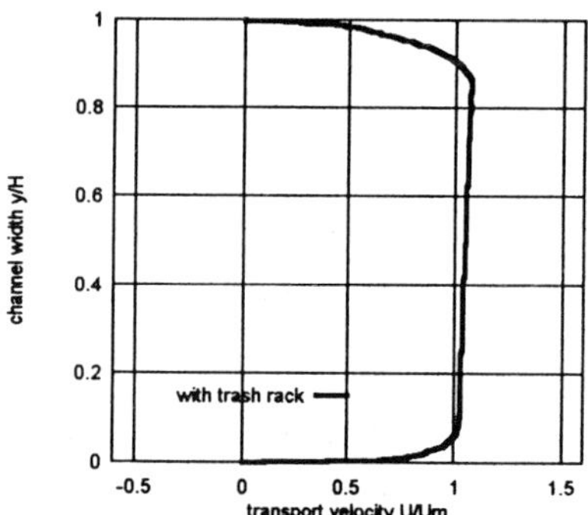

Fig. 11: Velocity distribution in the trash rack cross-section, modified geometry

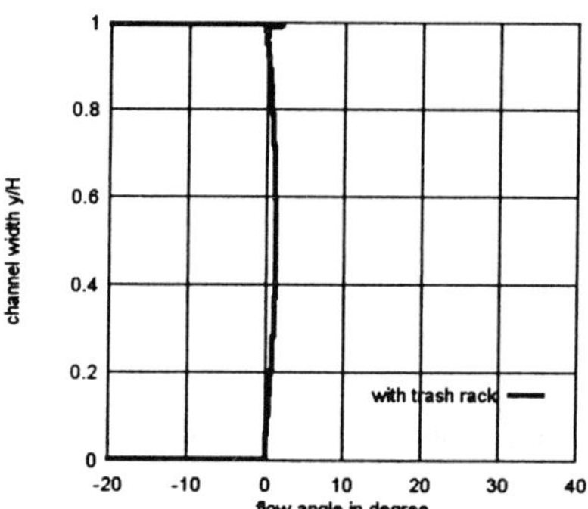

Fig. 12: Flow angle in the trash rack cross-section, modified geometry

3.3. APPLICATION 2

As a second example another small hydro power plant with two turbines, shown in fig.13, is presented. In this plant one turbine shows strong oscillations during "steady" operation, which causes severe problems to the bearing (water lubricated bearings). These oscillations are initiated by vortex shedding from the separation pier. Since the water depth nearly remains constant in the whole intake region and strong secondary flow behavior is not expected, a two-dimensional simulation is applied.

The calculations show, that the flow is unstable. A steady flow simulation did not converge. Therefore, an unsteady flow analysis has been carried out. The flow patterns can be seen in fig. 14, where the streamlines for different time steps are presented.

It can be observed that vortices are shedding from the separation pier and move towards the turbines. When reaching the turbine inlet channel on the inner side the vortices disappear but the strong velocity changes causes the role up of smaller vortices in the channel. These vortices move down-stream to the turbine. The unsteady behavior can also be seen in fig. 15. The variation in time of the velocity is shown for three different locations of the intake is shown. It can be seen, that the flow is periodic with an oscillation time of about 1 minute.

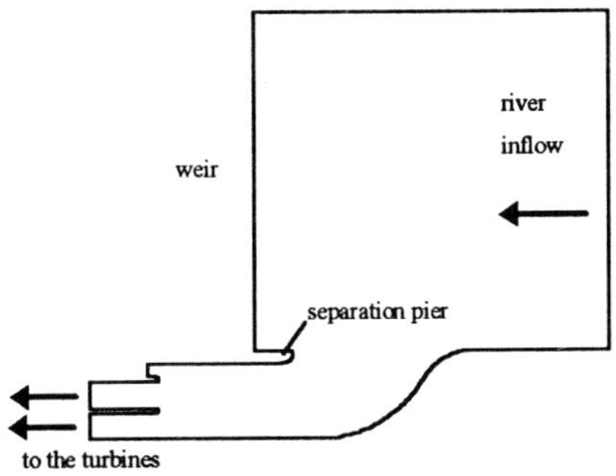

In order the cure the problem, the geometry of the pier was modified. In fig. 16 the streamlines for the new contour is shown. With this modification the flow remains stable and no oscillations occur anymore. After this modification have been built by the end of last year it was observed that the oscillation problems do no longer exist and the turbines are running smoothly.

Fig. 13: Geometry of application 2

4. Conclusions

Numerical flow analysis has been applied in order to evaluate the flow in the inlet region of different low-head power plants. By comparing the calculation with a flow visualization on a small test in the laboratory it was shown that the characteristics of the flow can be predicted with sufficient accuracy.

The application of the numerical procedure to two different existing power plants is shown. In both of the plants severe operating problems exist concerning the flow in the intake region. In the first case the energy losses in the trash rack were extremely high due to a highly disturbed velocity field. In the second plant oscillation problems existed caused by vortex shedding at a separation pier. In both cases the flow problems could be analyzed and understood and appropriate changes in the geometry could be found to cure the problems.

It has been shown that the numerical simulation can be used as a tool for the design of the inlet region of hydraulic turbines. It can predict the characteristics of steady state flows as well as unsteady vortex shedding. By applying this investigation in advance of the construction many operational problems could be avoided. Since the effort of the numerical flow simulation is rather low compared to model tests, it should be justifiable to analyze the flow behavior even of very small power plant in advance.

References

[1] Kirschmer, O., "Untersuchungen über den Gefälleverlust an Rechen", Mitteilungen des Hydraulik Instituts der TH München, 1-5 (1926-1932).

[2] Spangler, J., "Untersuchungen über den Verlust an Rechen bei schräger Zuströmung", Mitteilungen des Hydraulik Instituts der TH München, 2, 1928.

[3] Ruprecht, A. et al. "Einsatz der numerischen Strömungsmechanik in der Entwicklung hydraulischer Strömungsmaschinen", Mitteilungen Nr. 9, Inst. f. Strömungsmechanik und Hydraulische Strömungsmaschinen, Universität Stuttgart, 1994

[4] Ruprecht, A., "Finite Elemente zur Berechnung dreidimensionaler, turbulenter Strömungen in komplexen Geometrien", Mitteilungen Nr. 3, Inst. f. Strömungsmechanik und Hydraulische Strömungsmaschinen, Universität Stuttgart, 1989.

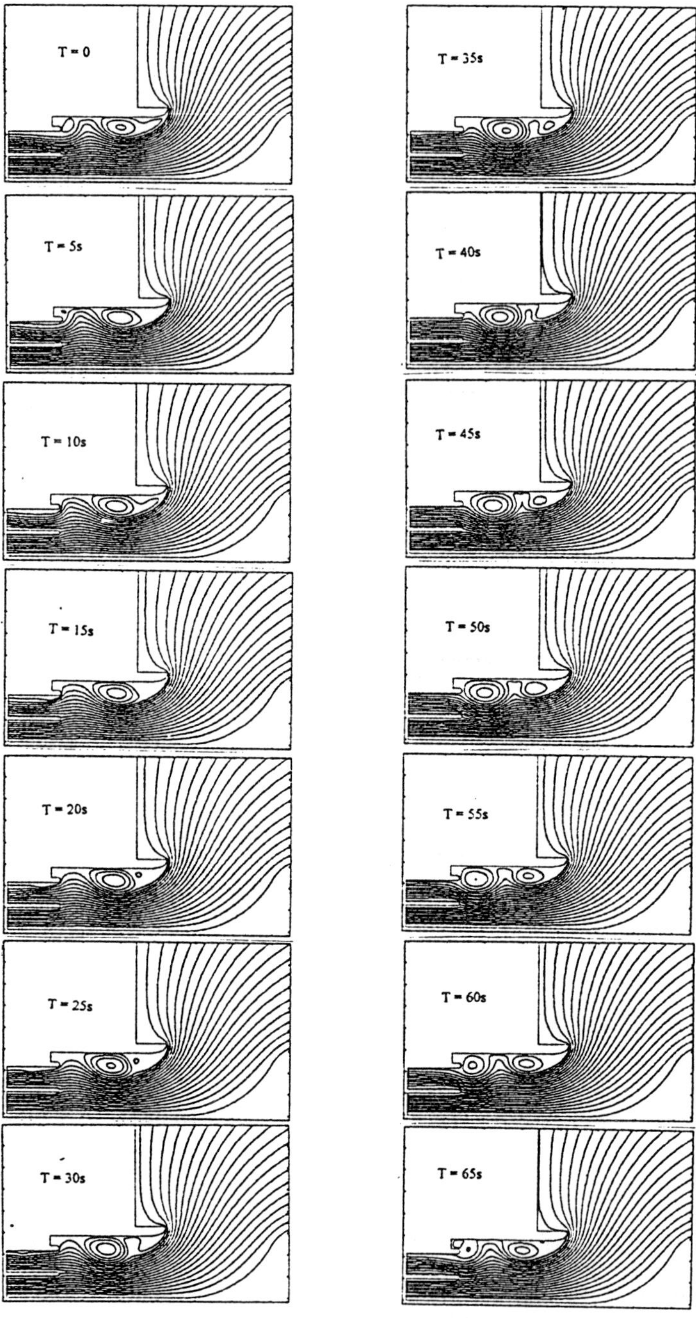

Fig. 14: Streamline distribution for different time steps

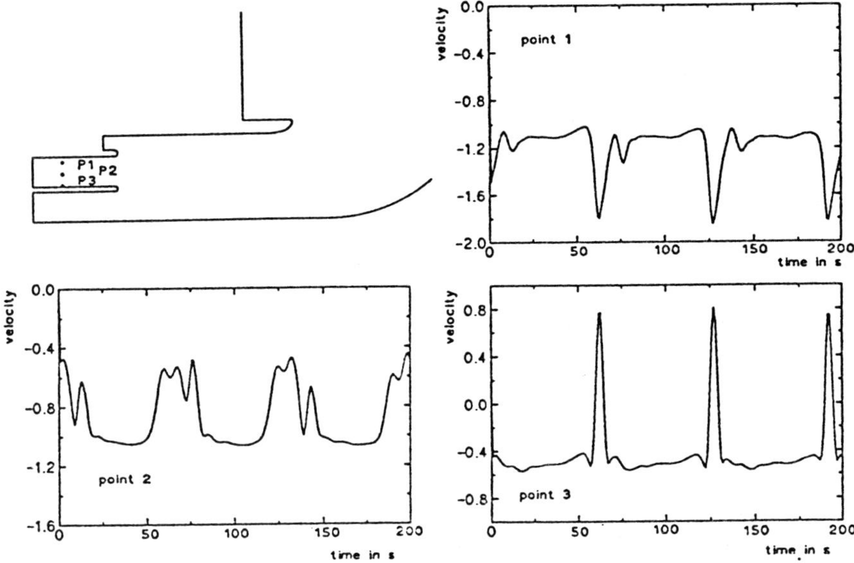

Fig. 15: Velocity variation in time for three different locations

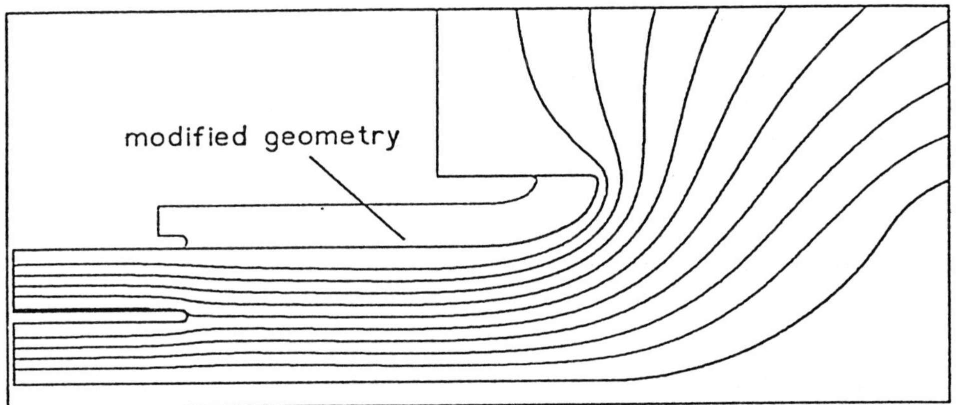

Fig. 16: Streamline distribution for the modified geometry

EFFICIENCY ALTERATION OF FRANCIS TURBINES BY TRAVELLING BUBBLE CAVITATION

Experimental and theoretical study

CH. ARN, PH. DUPONT AND F. AVELLAN
IMHEF-EPFL
33, av. de Cour, CH–1007 Lausanne, Switzerland
Email: christophe.arn@imhef.dgm.epfl.ch

Abstract.
The setting level of a hydraulic machine, specially for low head machines, is decided with respect to the possible alteration of the efficiency due to the cavity development. This alteration can easily be noticed by following the evolution of the efficiency η as a function of the Thoma number σ leading to the so-called $\eta - \sigma$ cavitation curves. Observation of the cavity extent in the flow passage of the runner allows to associate the drop of efficiency with a particular type of cavity development.

However, depending on the type of cavities this drop cannot be very easily explained. Obviously, for a leading edge attached cavity corresponding to high head operating points, the presence of the vapour phase on the blade suction side limits the pressure at the vapour tensile strength value which causes the flow alteration. In the case of travelling bubble cavitation, corresponding to the outlet cavitation at the nominal head, previous experiments with a 2-D NACA profile show that the modification of the mean pressure field is mainly due to the bubble dynamics. The aim of this paper is to present the results and the analysis of two experiments intending to explain the influence of the nuclei content on the mean pressure field correction due to the bubble dynamics.

1. Introduction

Travelling bubble cavitation takes place for the design value of the head, at the throat of the Francis runner flow passage, close to the outlet and corresponds to a low flow angle of attack. The development of this type of cavitation, responsible of an efficiency alteration of the machine is very sensitive to the content of cavitation nuclei and to the value of the setting level. For this reason, this setting level is determined with respect to this type of cavitation. The physical phenomenon underlying such an efficiency

E. Cabrera et al. (eds.), Hydraulic Machinery and Cavitation, 524–533.

alteration of the turbine is not yet very clear. Previous experiments [9] performed with a NACA009 profile mounted on a hydrodynamic balance show us the hydrodynamic loads to be dependent on the cavitation coefficient σ and on the content of the cavitation nuclei. We can so deduct that these two parameters are responsible for that pressure modification on the blade as the pressure distribution is generating the lift.

An approach to study this influence is to consider that the pressure field modification is mainly due to the bubble dynamics [8] [10] [7]. If we consider the travelling bubble like a non moving sphere in expansion in a still fluid, the computation of the pressure field around the bubble is possible. Indeed, the relative velocity between the bubble and the fluid is very low. The potential theory of incompressible flows allows to class this case as the determination of the pressure generated by a potential $\phi = -\dot{R}R^2r^{-1}$ where R is the radius of the sphere. The expansion of this sphere generates a radial pressure field around. The Bernoulli equation leads to the expression of the pressure field p. Thus, the pressure on the blade can be obtained by superposing the two bubble potential fields according to the image superposition technique, the wall being a flow surface. The Rayleigh-Plesset model allows us to determine the evolution of the bubbles radius, which is necessary to compute the radial pressure field. In the case of the NACA 009 bidimensional blade whose pressure field is computed by a Navier-Stokes finite elements code, the evolution of the bubbles along streamlines is determined solving the Rayleigh-Plesset equation using a fifth order Runge-Kutta method. One can see an example on Figure 1 with a value of the cavitation coefficient σ equal to 0.43. The evolution of the pressure computed in a point of the wall is also reported.

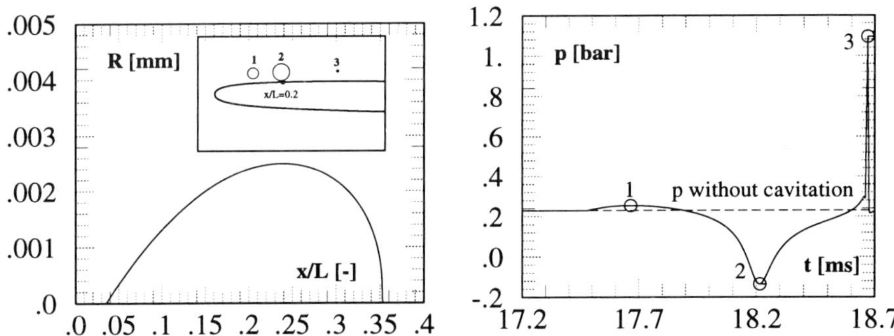

Figure 1. Bubble radius evolution along a streamline in a NACA009 blade and pressure generated at $x/L = 0.2$. $\sigma = 0.43$.

Obviously, this method is valid for the case where the bubbles remain relatively spherical which corresponds to high values of the cavitation coefficient σ. Moreover, the amplitude of the pressure are higher as the measured one [3]. Thus, this model allows qualitatively a good approach of the physical phenomenon. For lower values of σ, the size of the bubbles is greater than the layer containing the activated nuclei. Thus, the shape of the bubbles have a trend to be flattened up to form an hemisphere [1]. This trend depends strongly to the Weber number. In this case, the pressure on the wall under the bubble is naturally the vapour pressure value p_v. If the nuclei content is great enough to obtain the cavitation saturation, the pressure becomes constant at the value of p_v. Then we can define three different zones of travelling bubble cavitation influence. The first one is the part of the blade where the bubbles are spherical. The pressure modification is there mainly due to the bubble dynamics. The second zone is the region where the bubbles becomes hemispherical and the pressure under the bubbles are limited by the vapour tensile strength. The last one is the part of the blade where a saturation of the development of the bubbles is obtained and where the pressure on the blade is constant and equal to the value of p_v. The aim of this paper is then to present the results of an experiment intending to confirm these different points by measuring the pressure distribution on a two dimensional blade and on a Francis runner with travelling bubble cavitation.

2. Experimental set-up

2.1. MEASUREMENTS ON THE NACA 009 BLADE

The test are carried out in the IMHEF high speed cavitation tunnel [5]. The experimental hydro-foil is a 2D NACA 009, 100 mm long and 150 mm wide, truncated at 90 % of its chord length. 18 piezoresistive absolute pressure transducers are distributed on the suction side of the blade, as shown on Figure 2. The measurement range covers 0 to 100 bar. Each transducer is supplied by an independent current source and its output pressure signal is separately amplified and band-pass filtered. Data acquisition is performed with the help of two digital transient recorders with a 12 bits resolution per sample. The first one (Lecroy 8212a) allows simultaneous data acquisition of 32 signals at a maximum sampling frequency of 5 kHz whereas the second one ensures simultaneous data acquisition of 12 channels at a maximum sampling rate of 1 MHz (3 Lecroy 6810 Modules) the static calibration of the pressure transducers is performed with the blade mounted in the test section by varying the static pressure in the tunnel from 0.3 to 10 bar. To achieve a dynamic calibration of the pressure transducers, a special technique is developed to generate a pressure impulse in the test section [6]. The transducers output are compared to the output of a 601 Kistler pressure transducer mounted in the test section too. The main result is a good concordance up to 15-20 kHz.

A control of the nuclei content is performed during all the experiment. Indeed, the travelling bubble cavitation is impossible without the injection of cavitation nuclei. The nuclei are generated by an expansion of air-saturated water in a series of injection modules [2]. By varying the number of these modules, one can obtain the required quantity of cavitation nuclei

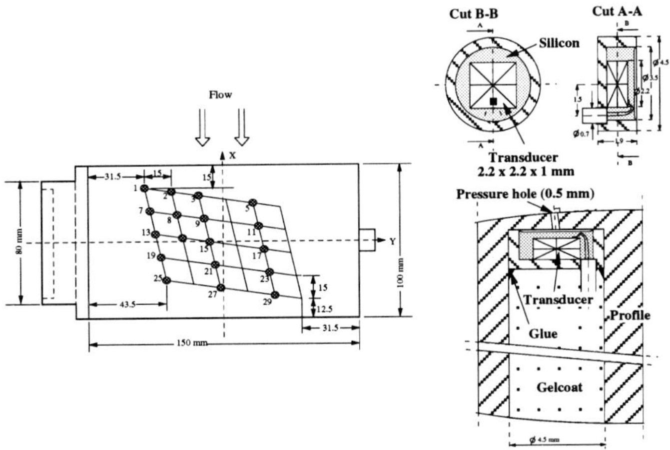

Figure 2. NACA 009 hydro-foil equipped with 15 transient pressure transducers. Pressure transducers mounting

in the test section. The distribution of these nuclei is measured by a cavitation nuclei counter [4] which use the Venturi effect for the detection of the nuclei explosive growth.

2.2. MEASUREMENTS ON THE FRANCIS RUNNER

The experiment is conducted with a Francis runner model which has a specific speed $\nu = 0.33$. It represents a standard case of a Francis turbine design. The tests of the model are performed on the high performance research test rig of IMHEF. This closed loop test rig covers a range of flow rates up to 1.5 cubic meters per second, with a maximum net head of 60 meters. The primary quantities as the torque, the flow rate, the head, the angular velocity and the water temperature are continuously recorded to determine the power and the efficiency of the machine with an overall accuracy better than 0.1 percent.

Two blades of the runner are equipped with each five absolute transient pressure transducers similar to these mounted on the 2D blade. The position of the transducers is described on Figure 3. The acquisition set-up is the same as the other experiment except the conditioning electronics connected to the on-board transducers. This electronics is placed in the head of the runner, it has an amplifier with a remotely variable gain and a multiplexer that allows the scanning of all the transducers. Signals are converted to frequency to avoid electro-magnetic disturbances. They are brought out from the rotating part through a slip ring collector and finally reverted to voltage signals which are digitized on the waveform recorders.

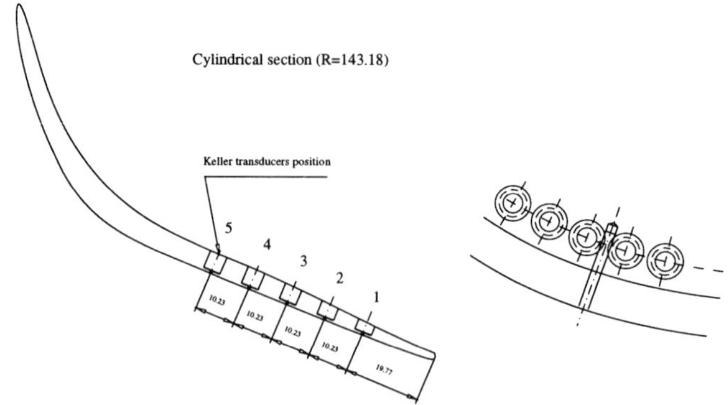

Figure 3. Transducers position on the two blades of the Francis runner

3. Results and discussion

3.1. NACA 009 BLADE

For the case of the 2D blade, the cavitation coefficient $\sigma = \frac{p_{ref}-p_v}{1/2\rho C^2}$ is defined by a relation similar as the definition of the local cavitation factor $\chi_E = \frac{p_{ref}-p_v}{\rho E}$ for the turbomachines where C is the velocity and E the energy. A previous experiment conducted with the same blade mounted on a five components hydrodynamic balance give us the evolution of the lift coefficient c_z as a function of this cavitation coefficient σ [3]. We can observe an increase of the lift from the inception of the travelling bubble cavitation. Then, after a maximum value corresponding to $\sigma = 0.32$, the lift falls. These results are reported on the Figure 4.

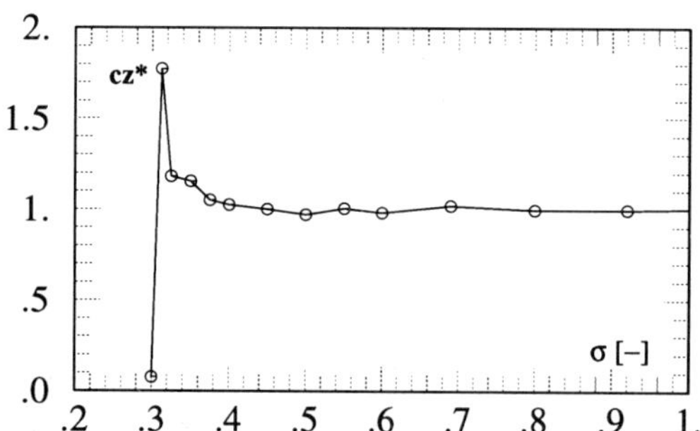

Figure 4. Measured lift coefficient as a function of the cavitation coefficient σ.

An approach to explain this lift drop phenomenon is to measure the pressure distribution at the same operating points, for an upstream velocity of 20 m/s and with cavitation nuclei injection in order to obtain the saturation condition. These distributions are reported on Figure 5 and compared with the result of a Navier-Stokes computation. The frames corresponding to each σ are too illustrated on this Figure. One can observe a good agreement between the pressure distribution measured for a high value of σ and the computed distribution. This provides a certain confidence in the results. From the inception of travelling bubble cavitation, a depression appears in the zone where the size of the bubbles is maximum ($\sigma = 0.4$ and $x/L = 0.3$). In this condition, the bubbles stay relatively spherical. For a value of $\sigma = 0.375$, the bubbles become hemispherical and the depression extends over the suction side but doesn't go below the vapour pressure value. For lower values of σ, the measured pressures correspond in all positions to the vapour pressure. Indeed, the travelling bubble cavitation saturation is reached as shown in the different frames on Figure 5. The main noticing is that the maximum lift corresponds to the situation where the value of $-\sigma$ reaches the plateau of the pressure coefficient distribution. In fact, this is mainly the position of the point where the pressure is minimum which determine the behavior of the hydraulic characteristic in function of the σ. In the case where the $c_{p,min}$ value is near the trailing edge of the blade, which corresponds to low flow angle of attack, only a lift drop can be appear since the vapour pressure goes to limit the suction of the blade. In the other case where the $c_{p,min}$ is reached near the leading edge, an increase of the lift is generated by the development of travelling bubble cavitation in the zone where the pressure is higher as the vapour pressure. The drop appears only when the vapour pressure is reached on the suction side.

3.2. THE FRANCIS RUNNER

Generally, the pressure distribution in a Francis runner at the best operating point corresponds to the first case that we described in the previous section. The flow angle of attack is low and the minimum pressure is localized on the outlet of the blade. The tests are conducted at this operating point for three different heads: 10 m, 15 m and 20m. The results of a flow computation in the case of a head of 15 m are presented on Figure 6. One can observe on this figure that the minimum of the pressure is localized in the outlet of the blade for each computation mesh line except in the zone close to the runner band (k=19). Based on the conclusion of the previous described experiment, the efficiency must directly decrease with the cavitation coefficient. All the more so since the whirl has no effect at the best operating point. The cavitation coefficient usually used in hydraulic turbomachines is the Thoma number $\sigma = NPSE/E$.

The cavitation curves obtained are illustrated on Figure 7. The values of the efficiency are related to the cavitation free values. We can determine the value of σ_0 which is the highest value of the Thoma number where the efficiency is modified. This value is the same for all the tests with nuclei injection and is equal to 0.068. The values of χ_E corresponding to this Thoma number are equal to 0.024, 0.022 and 0.02 for the three test heads. The com-

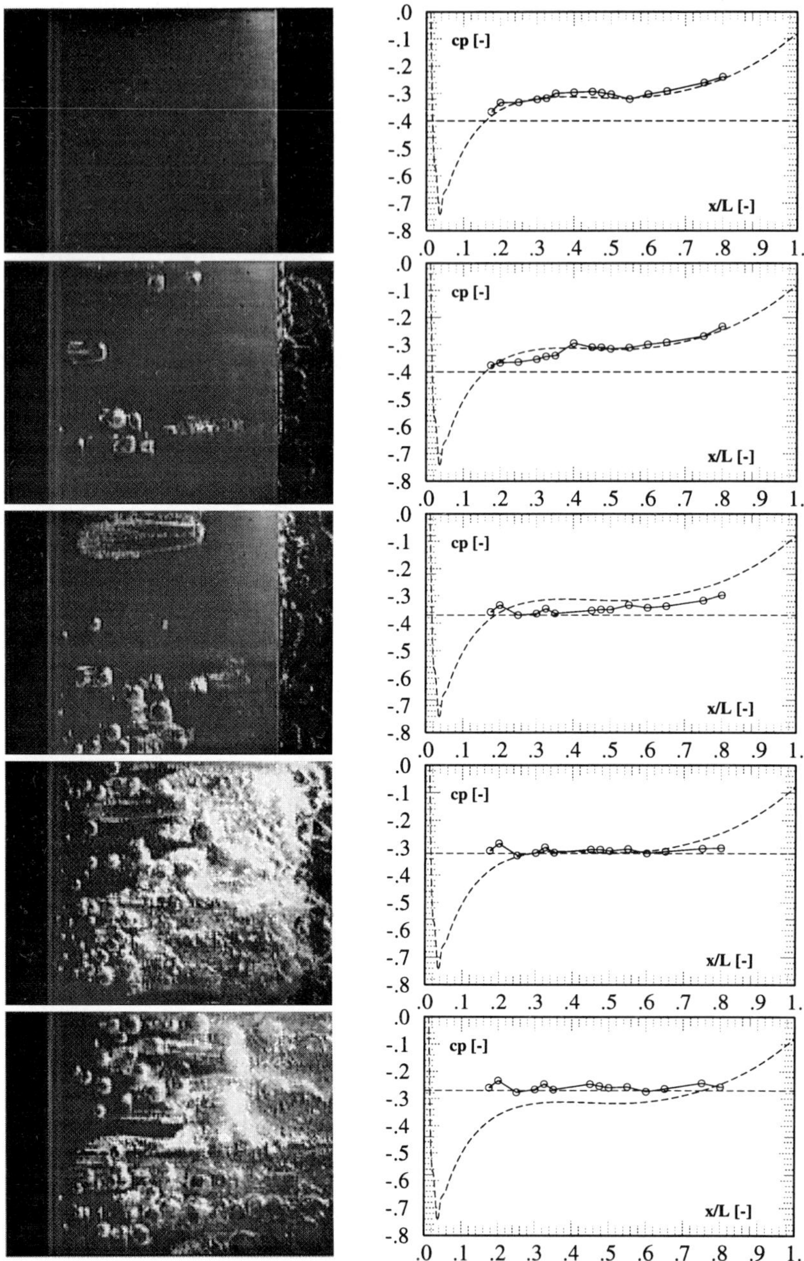

Figure 5. Measured and computed pressure distribution on the NACA 009 blade for the following values of σ: 1.0, 0.4, 0.375, 0.325, 0.275. (From top to bottom). The horizontal dashed line represents the $-\sigma$ value

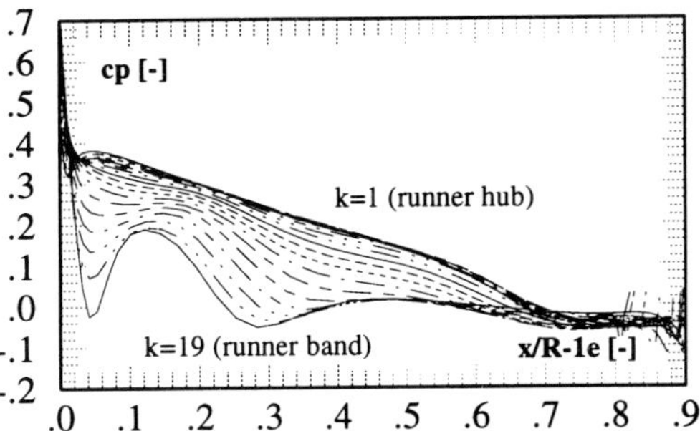

Figure 6. Results of a Navier-Stokes flow computation at the best operating point. The reference pressure is the computation domain outlet pressure. H=15 m

parison with the pressure coefficient computed by the Navier-Stokes code shows us that the beginning of the efficiency drop appears when the all the values of the minimum pressure coefficient are lower as $-\chi_{E,0} = -0.022$. At this state, it corresponds to an important development of the outlet cavitation. For the case corresponding to the computed one, the first bubbles are developed for a value of the Thoma number of 0.085. Moreover, based on the computed results, we can determine the maximum value of the Thoma number where the vapour pressure is reached on the blade. The minimum pressure coefficient on the blade provides this value of σ which is 0.11. Thus, we have a difference of the σ from 0.04 between the point where the vapour pressure is reached on the blades and the beginning of the efficiency drop. The development of travelling bubble cavitation must be then relatively established before the efficiency falls.

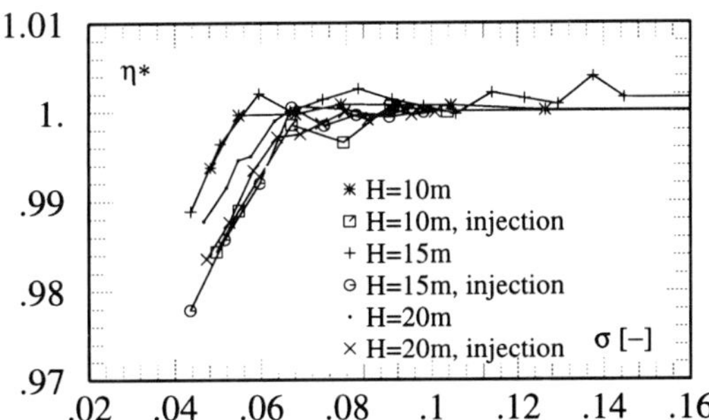

Figure 7. Normalized $\sigma - \eta$ curves for the Francis runner tests.

As in the case of the 2D blade, we can plot the results of the pressure measurements with the computed distribution at the corresponding computation mesh line. The Figure 8 presents these results where we can observe that the correspondence between the measurements and the computation is very good for the case without cavitation. For lower values of the Thoma number, the pressure distribution is modified by the development of travelling bubble cavitation. The different pressure measurements show us a decrease of the pressure until the vapour pressure value. The development of the bubbles is due to the upstream pressure drop localized close to the runner band. However, the main performance of the runner decrease since the development of the travelling bubble cavitation take effect mainly at the outlet of the blade.

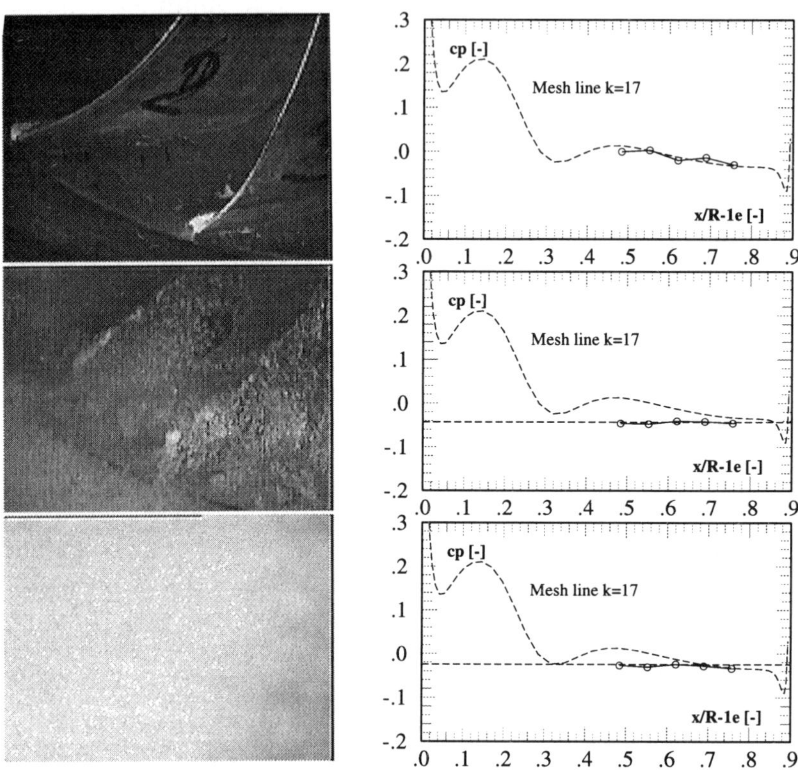

Figure 8. Measured and computed pressure distribution on the Francis runner blade for the following values of σ: 0.43, 0.07, 0.06. H=15 m. (From top to bottom). The horizontal dashed line represents the $-\chi_E$ value.

4. Conclusion

Based on these two experiments, the effects of the travelling bubble cavitation on the turbomachines performances are better described. The efficiency modification appears mainly when the saturation of the cavitation is reached. The saturation causes a modification of the pressure distribution by limiting the pressure field to the vapour pressure value. The part of the machine where the shape of the bubbles stay spherical is not really influenced in term of performances. The acoustic emission caused by the bubbles growth or collapses is an other problem which justify the study of the acoustic pressure field around the active bubbles. However, in order to obtain a modeling of the pressure distribution modified by travelling bubble cavitation, it is necessary to determine the different zones of the blade where the bubbles are spherical, hemispherical and the cavitation saturation is reached. The size and the position of these different regions depends naturally to the cavitation nuclei concentration and to the operating point of the machine. This is the reason why the study of the bubbles evolution close to a wall in a variable pressure field is important for the characterization of the performances alteration in a hydraulic turbomachine. The way to this characterization in the case of the efficiency alteration by travelling bubble cavitation is certainly the study of a reliable model of the bubble evolution in a runner, and then the modification of its shape including the case of the saturation. The effects of the bubbles on a pressure field being now relatively known and described, this is the knowledge of the distribution of these different travelling bubble cavitation development on the runner that will be make possible a prediction of the efficiency drop in the Francis turbines due to this type of cavitation.

5. Acknowledgment

The authors are particularly grateful to the members of the IMHEF Cavitation research group and to its technical staff. This research is financially supported by the PSEL "Fonds Suisses pour Projets et Etudes de l'Economie Electrique".

References

1. Y. Kuhn de Chizelle et al. Observations and scaling of travelling bubble cavitation. *J. Fluid. Mech.*, 293:99–126, 1995.
2. C. Brand et al. The imhef system for cavitation nuclei injection. Sao Paolo, 1992. AIRH.
3. Ch. Arn et al. Experimental and theoretical study of the 2d blade lift alteration by traveling bubble cavitation. ASME, August 1995.
4. F. Avellan et al. Theoretical and experimental study of the inlet and outlet cavitation in a model of francis turbine. pages 38–55, Stirling, August 1984.
5. F. Avellan et al. A new high speed cavitation tunnel for cavitation studies in hydraulic machinery. volume 57, pages 49–60. ASME, 1987.
6. F. Pereira et al. Dynamic calibration of transient sensors by spark generatd cavity. Symposium of Bubble Dynamic and Interface Phenomena, September 1993.
7. J.P. Franc et al. *La Cavitation, Mécanismes physiques et aspect industriels.* Presses Universitaires de Grenoble, 1995.
8. S. Kumar et al. A study of pressure pulses generated by traveling bubble cavitation. *J. Fluid. Mech.*, 225:541–564, 1993.
9. B. Gindroz. *Lois de Similitude dans les essais dans les Essais de Cavitation des Turbines Francis.* PhD thesis, EPFL, 1991.
10. J.T. Daily R.T. Knapp and F.G. Hammit. *Cavitation.* Mac GRaw Hill, New York, 1970.

CAVITATION EROSION PREDICTION ON FRANCIS TURBINES-PART 1 MEASUREMENTS ON THE PROTOTYPE

P. BOURDON, M.FARHAT, R. SIMONEAU
Hydro-Québec
1800 boul. Lionel Boulet, Varennes, Québec, Canada, J3X 1S1
F. PEREIRA, P. DUPONT., F. AVELLAN
IMHEF/EPFL
33 av. de Cour, CH 1007 Lausanne, Switzerland
J.-M. DOREY
Electricité de France
6, Quai Watier, 78401 CHATOU Cédex, France

1. Abstract

In the process of developing tools for cavitation erosion prediction of prototypes from model tests, 4 on board aggressiveness evaluation methods were tested on a severely eroded blade of a 266 MW Francis turbine. These are pressure, pit counting, DECER electrochemical and vibration measurements. All methods provided coherent results on the blade mounted measurements. The test program provided understanding of the heterogeneous erosion distribution of the prototype blades and quantitative data for comparison in subsequent tests on the model of the machine.

2. Introduction

The prediction of cavitation erosion of a prototype turbine from model tests requires that measurement tools be available to characterize the aggressiveness of the cavitating flow in both scales. To develop such tools, IMHEF, Electricité de France and Hydro-Québec pooled their resources in an ambitious research program involving various measurement techniques both on the prototype and on the model of a 266 MW Francis turbine with a well documented cavitation erosion history. Preliminary measurements on this prototype (1) had proven to be incomplete but also very encouraging. A more ambitious test program was conceived with improvements in the array of sensors utilized, sample mounting methods, data acquisition systems and hydraulic test conditions. The test program was divided in two parts, the first with pressure sensors mounted on the suction side of blade #4, a well eroded blade, to identify the type of cavitation present and its aggressiveness in terms of pressure pulses. In the second, polished metallic samples were mounted in place of the sensors for pitting studies. This experimentation program took place in June 1995.

E. Cabrera et al. (eds.), Hydraulic Machinery and Cavitation, 534–543.

3. Test set-up

3.1 MECHANICAL INSTALLATIONS

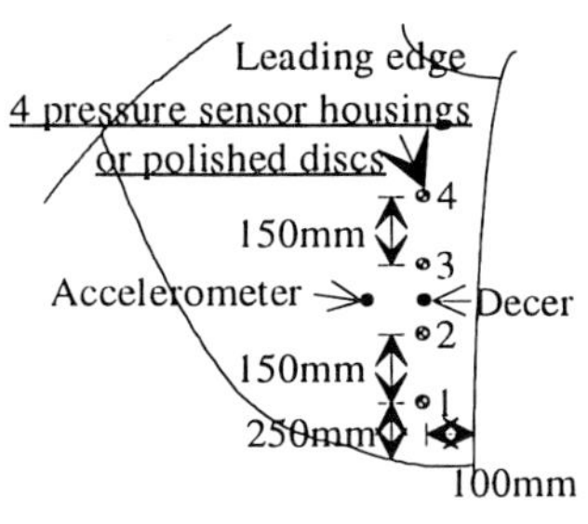

Figure 1. Blade 4 sensor and disc positions

This machine has the particularity that the erosion is much worse on about half of the blades and is localized towards the trailing edge near the blade to band fillet. Blade #4 was retained for testing as it had been used in the preceding tests because of its severe erosion history. Consequently, six 45 mm diameter holes were counterbored in this blade as shown in Figure 1 on the low pressure side. An intricate drilling and shimming scheme was developed by IMHEF to insure perpendicularity of the hole axis with the complex blade surface and optimal flush mounting of the various sensor housings. All housings or sample mounting supports were securely bolted down to the blade from the pressure side. Figure 2 shows the magnetic drill mounting base positioned in place as well as some of the already bored holes.

Holes were also bored above the draft tube access door to mount a dynamic pressure sensor, pass cables for 5 underwater accelerometers and cables to immersed spark plugs for the dynamic calibration of the set-up in watered conditions. A cylinder housing the custom designed onboard data acquisition electronics was welded in place in the runner nose cone. Cables were run from the sensors to this unit through a stainless steel tube welded to the pressure side of bade 4 and covered with an hydraulically smooth buildup of resin to minimize turbulence. Signal and control cables were then brought out to the stationary world through the hollow turbine-alternator shaft using the air injection path and a double slip ring arrangement fixed to the end of the alternator exciter shaft. A threaded hole was also machined at the lower turbine guide bearing to receive a piezoelectric force exciter for sinusoidal sweep calibration of the transmissibility function between the guide bearing and 5 underwater accelerometers mounted on the runner blades.

Figure 2. Drill mounting base

3.2 SENSORS

For the pressure tests, 4 of the blade holes (# 1-4) received stainless steel sensor housings bearing each 4 piezoresistive pressure sensors allowing to measure both static and dynamic pressures. Each housing contained 2 400 bar and 2 1000 bar Keller sensors. These housings were replaced with optically polished 316L stainless steel discs for the pitting tests. The two remaining holes were occupied respectively by a DECER electrochemical cavitation erosion sensor and a 4kHz bandwidth damped piezoresistive accelerometer. These were used in both the pressure and pitting tests.

In addition to the onboard sensors mentioned above, four high frequency accelerometers monitored the cavitation impacts at the lower guide bearing while another was installed above the draft tube access door along with a Kistler 701 wide band dynamic pressure sensor. A tachometric probe allowed to synchronize data acquisition on multichannel wide band data acquisition systems. The test program included calibration of the setup through the measurement of transmissibility functions both in air and in water between the blades and measurement points at the lower guide bearing. This step is required to determine absolute cavitation aggressiveness levels from remotely measured vibration or pressure data. For this purpose, 5 accelerometers were mounted on blades 3, 4 ,7 10 and 13 on the suction side at a position corresponding to the location of the DECER probe on blade 4 shown in Figure 1. Direct and reciprocity impact and direct spark generated bubble calibration techniques were utilized. Two underwater spark plugs were used to generate bubbles at two locations close to the suction side of blade 4. A sine sweep technique was used in a reciprocity mode by exciting the structure with controlled force at the lower guide bearing.

3.3 DATA ACQUISITION SYSTEMS

For the purpose of this program, IMHEF designed an onboard data acquisition system with two variable gain 100kHz bandwidth data channels and two front end remotely controlled channel multiplexers. This system allowed to scan the 16 blade mounted pressure sensors as well as the 4kHz bandwidth accelerometer. Signals were converted from voltage to frequency for transmission through the dual slip ring arrangement to avoid noise pickup in the alternator environment. The signals were then converted back to voltage with a frequency to voltage converter. Extra tracks on the slip rings were used to carry an RS 232 control link, the DECER electrochemical work and auxiliary potentials and the power supply current for the onboard electronics.

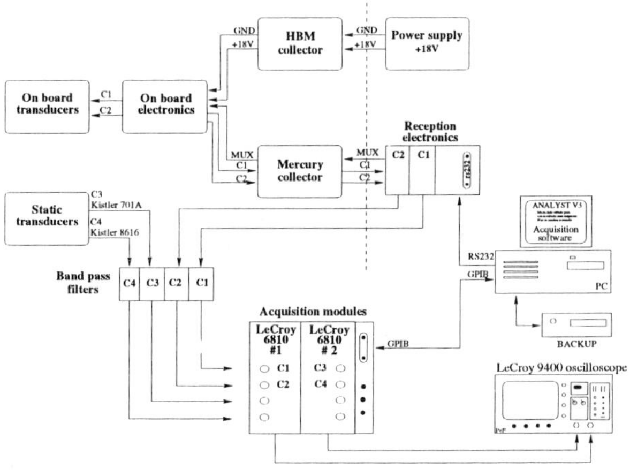

Figure 3. IMHEF data acquisition system.

Data acquisition was assured by two separate systems provided by IMHEF and Hydro-Québec. The first system is based on 2 LeCroy 6810 acquisition modules configured to acquire each 2 input signals with 12 bit resolution at sampling frequencies up to 5MHz. A 1 Mbyte memory in each module is shared between the two input channels. The acquisition is controlled via a GPIB port by a PC running Analyst software developed at IMHEF in the ASYST environment. On line monitoring of signals from this system is performed with a LeCroy DO9400 oscilloscope. Signals are first passed through antialiasing filters before acquisition. This system is illustrated in Figure 3.

The second includes a 9 channel HP3565S 13 bit 100kHz bandwidth data acquisition system controlled by LMS Fourier Monitor software, a Nicolet 500 4 channel 12 bit 10 Mhz digitizer with a 1 megasample memory on each channel and a Sony PC216A 16 bit resolution digital tape recorder used in an 8 channel 10 kHz bandwidth configuration.

Both systems were utilized for the dynamic calibration of the test setup and for acquisition during the tests. The instrumentation was completed with a bank of programmable bandwidth and gain analog filters and envelope detectors. A potentiostat circuit conditioned the DECER signal which was recorded on an analog chart recorder as well as on the SONY tape recorder. An audio amplifier and loudspeaker combination allowed to assess audibly on site the impulsive nature and the intensity of the cavitation signals perceived by the blade mounted accelerometer.

4. Test conditions

Conditions for the pressure and pit counting tests appear in the following table.

TABLE 1. Test conditions

Test Identification	Guide vane opening (%)	Power (MW)	Downstream level (m)	Duration (h.min)
		Pressure tests		
1	78	228	205.7	6.0
2	90	259	205.8	3.0
3	78	228	207.4	4.0
4	90	255	207.6	2.15
		Pitting tests		
M1	78	227	206.2	.41
M2	78	223	207.4	.41
M3	90	253	207.5	.40
M4-1	85	247	205.7	.17
M4-2	80	234	205.6	.16
M4-3	75	219	205.5	.15
M4-4	70	201	205.5	.15
M5	72	206	205.9	.40

In order to evaluate the effect of the variation of the cavitation index on the cavitation development, tailwater levels were raised in the downstream reservoir over a two week period to allow testing of the machine with high downstream levels with all available machines operating. These conditions had not been possible in the preliminary program (1). Low tailwater testing was done by operating only the test unit in the powerhouse.

5. Test results

5.1 CALIBRATION

Before running the operating tests on the prototype, an extensive calibration program was performed to characterize the transmissibility function (ratio of output acceleration power spectrum to input force power spectrum) between the attacked areas on the blades and the monitoring points either on the blades or at the lower guide bearing. Measurements were performed with the runner in air and then in water. The linearity of the structure was verified by reciprocity measurements using instrumented hammer excitation techniques with the machine in air. The average transmissibility function with the runner under water was then evaluated by impacting at the lower guide bearing monitoring point at 90 degrees from upstream on the spiral case upstream side and measuring simultaneously the response on 5 blades. The result of this measurement is shown in Figure 4. This function was then used to infer acting forces on the blades from acceleration measurements during subsequent testing.

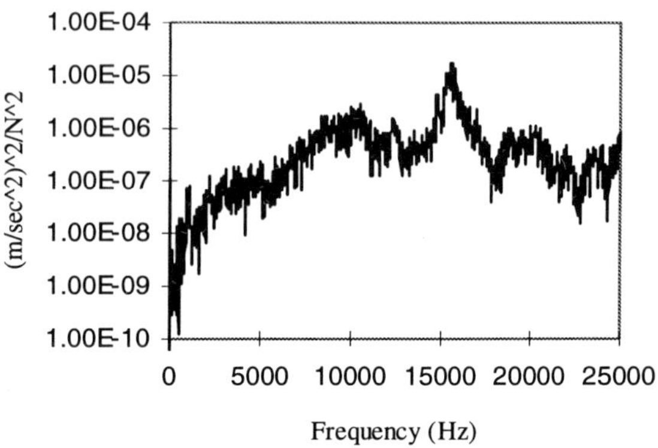

Figure 4. Lower guide bearing to blade transmissibility.

5.2 PRESSURE TESTS

The first pressure test performed with a low tailwater level produced cavitation conditions on blade 4 such that violent implosions were localized in the area of the two downstream pressure sensor housings. This was later confirmed in pitting test M1 under similar conditions and by the severe marking of the stainless steel sensor housings after the pressure tests. As a result, the 8 sensors in the most downstream positions and 2 more on the next upstream housing were progressively put out of service as the time exposure increased. The remaining sensors exhibited a main

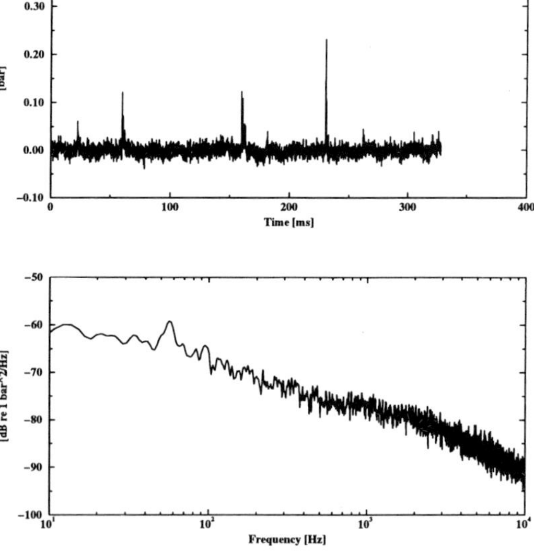

Figure 5. Typical pressure sensor time trace and average Power Density Spectrum

component in their power spectral density spectra around 60 Hz which corresponds to the guide vane passing frequency for a rotating observer. An example is given in Figure 5. This result suggests the existence of a cavitation cloud whose pulsation is modulated by the pressure variations generated by passing in front of the guide vanes. This in turn commands forced vortex shedding in the closure area of this cloud and impacts on blade 4 at this rate. This is confirmed by the envelope analysis of the high frequency acceleration of the blade which shows a strong component at or near this frequency. In pressure test #3, this hypothesis was supported by the remaining upstream sensors which were still operational. As the downstream level was raised, the cavitation cloud moved in the upstream direction in such a way that sensors on. housing 4 measured vapor pressure while those on housing 3 showed intense activity with enormous fluctuations which occasionally saturated the frequency to voltage converter. The pitting tests confirmed these indications as disc 3 was the most severely marked under the same hydraulic conditions.

5.3 PITTING TESTS

5.3.1 Pit counting

The 1993 test program had seen the most exposed glued discs torn away in the flow and the impossibility of testing with high downstream levels to evaluate the effect of this parameter on the cavitation development. As mentioned earlier, these two elements were corrected in this program. After the pressure tests, the pressure sensor housings were replaced with polished disc holders bolted down to the blade and the discs were replaced after each of the roughly 40 minute exposure periods. An exception to this was made in test M4 which consisted of 4 successive 15 minute exposures at 85, 80, 75 and 70% guide vane opening in one sequence with a low downstream level. Three of the pressure test conditions were repeated, the 90% low downstream level test was skipped as it had been performed in 1993, the sweep test was added and was followed by a final point at 72% guide vane opening.

The results of these tests are summarized in Figure 6 which shows all of the disc samples along with a 10X micrograph of typical pitted areas on those discs that were marked by the cavitation impacts. The size of the pits can be appreciated by the 1 mm references which are shown below samples 2M1 and 3M2. The first digit in the sample number corresponds to its position on the blade (see Fig. 1). Downstream is left and upstream right. At the left is indicated the guide vane opening and downstream level for each test. From this Figure it can be clearly seen that the largest pits are observed near the trailing edge of the blade in tests M1 and M4 with the low downstream level. Smaller pits occur on discs 3 or 2 when the downstream level is raised as in test M2 or the guide vane opening is reduced as in test M5. Opening the guide vanes to 90% has for effect to move the pitting area to disc 4 towards the trailing edge or even beyond as no pits are observed on upstream discs 2 to 4. The sweep test shows the worst pitting on disc 2, the result of the combined marking during the 80 and 75% guide vane opening exposure.

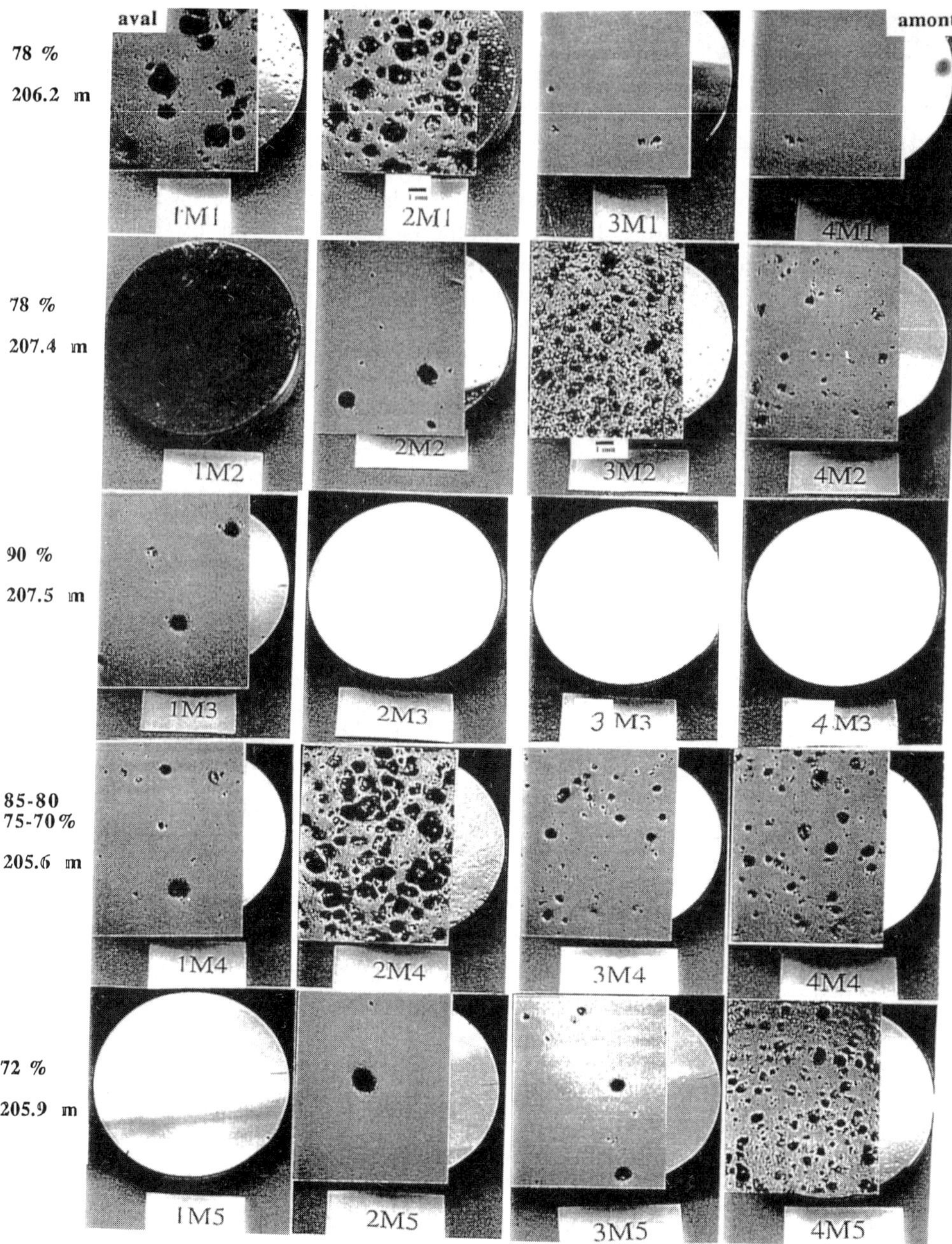

Figure 6. Pitted discs and 10X micrographs of pitted areas.

These results are consistent with those of the previous campaign and show the combined effect of the increasing velocity with larger guide vane opening which promotes a more voluminous cavitation development and the adverse effect of a higher downstream level which contracts the cavitation development and forces implosions on the blade in the closure area of the cavitation cloud in the pressure recovery region.

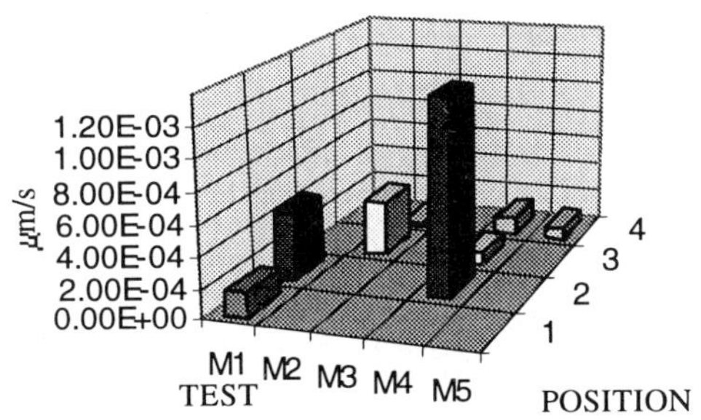

Figure 7. Volume pitting rate vs test and disc location

Since larger developments generate bigger shedded vortices, larger pits are found closer to the trailing edge of the blade. These observations are summarized in Figures 7 and 8 which show respectively the Volume pitting rate V_d (total volume of pits per unit area and time) and the characteristic radius R_v (average of pit radius weighted by V_i, the pit volume). The data for these figures are obtained by scanning the pitted samples with an UBM Laser profilometer and processing the output files with the "Adresse" software developed for EDF by CREMHyG (2).

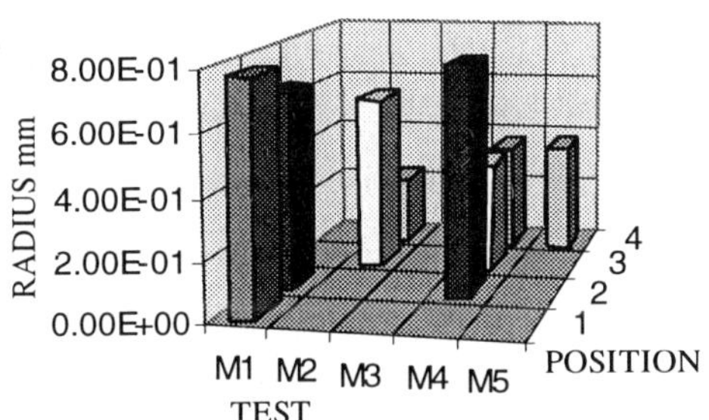

Figure 8. Characteristic radius Rv of the pits vs test and disc position on blade 4.

These Figures clearly show that the largest Volume pitting rates are recorded in tests M1 and M4, the sweep test, in position 2 closer to the trailing edge. This is due to conditions at 80 and 75% guide vane opening which probably both contribute to pitting at this location. Individual pit diameters approaching 2 mm, the largest we have ever observed, were found in test M1. These pits are found downstream where the larger and more energetic vortices implode. This is consistent with the notion of erosive power (3) where the potential energy of each cavity scales with its volume. In test M3 at 90% guide vane opening, no pits were recorded on samples 2 to 4 and very few on sample 1. Under these conditions implosions occur on the blade towards the trailing edge or in the flow beyond.

5.3.2 DECER electrochemical erosion detection

The DECER erosion detector located between samples 2 and 3 produced a localized maximum erosion rate at 75% guide vane opening in sweep test M4 with a low downstream level and a next higher value at 78% with a high downstream level in test M2. This is illustrated in Figure 9 which shows the DECER erosion current in μa in relation to the guide vane opening and downstream level. This current is proportional to the actual erosion rate on the surface of the titanium sensor.

The initial pressure tests, the pit counting tests and the DECER measurements all indicate a variable cavitation development on blade 4 affected both by guide vane opening and downstream level with maximum aggressiveness in the 75 to 80% span of guide vane opening.

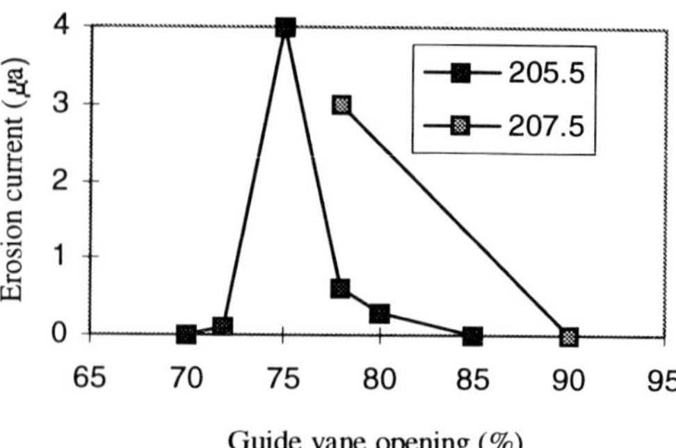

Figure 9. DECER erosion current vs guide vane opening

5.3.3 Vibration measurements

Vibration and pressure measurements were made during all tests. The blade mounted accelerometer was retained for these latter tests. The results are summarized in Figure 10. The three curves represent data normalized to the maximum observed value of each parameter during the low downstream level pressure and pit counting tests. The free symbols represent similar data for the high downstream level tests of both types.

Impacts were found to occur on the blade at or near a 60 Hz rate which corresponds to the guide vane passing frequency (20 guide vanes and 3 Hz rotation frequency) as indicated by the blade mounted accelerometer high frequency amplitude modulation envelope. The envelope of high frequency acceleration band pass filtered in a third octave bandwidth centered on 4 kHz was analyzed. Similar indications were obtained with the 0 degree upstream accelerometer at the lower guide bearing. In this case the 15 to 30 kHz bandwidth was retained for analysis. These results agree with those of the 1993 tests. The three curves present respectively with diamonds, squares and crosses the Mean Square Value of modulation of the high frequency acceleration on the blade and at the lower guide bearing and the MSV of pressure pulsation above the draft tube access door. These parameters were calculated at 20 +/- 1 and 2 times the runner rotation frequency over a quarter of a Hz bandwidth at each of these 5 frequencies.

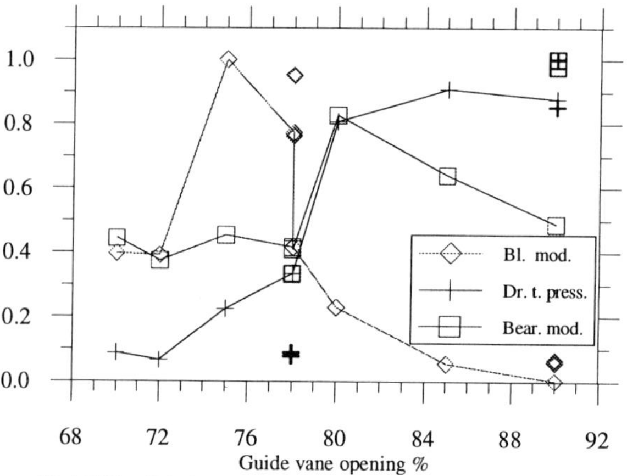

Figure 10. MSV of blade and bearing acceleration modulation and draft tube pressure around the guide vane passing frequency.

These curves clearly show that maximum aggressiveness on blade 4 occurs around 75 to 80% G.V. opening depending on downstream level. A higher downstream level displaces the high intensity points towards the greater G.V. openings. The indications at the lower guide bearing are different with maximum values occurring at 80% G.V.

opening for low downstream levels and 90% for high downstream levels. The guide bearing sensor modulation envelope is in good agreement with the indications of the draft tube pressure sensor. As opposed to the blade mounted accelerometer which is affected essentially by local events, the response of this pressure sensor, like the guide bearing accelerometer, is determined by the overall runner cavitation performance.

6. Discussion and conclusions

The availability of tools to characterize the aggressiveness of cavitation at the prototype level has been demonstrated in this test program. Localized pressure, pitting, DECER electrochemical erosion detection and vibration measurements have all concurred to characterize the cavitation behavior of blade 4 of this Francis prototype. Increasing guide vane openings or lower downstream levels produce a larger cavitation development and greater aggressiveness inasmuch as the implosions do not occur beyond the trailing edge of the blade. Overall vibration measurements at the guide bearing and pressure measurements at the draft tube reveal maximum aggressiveness at the maximum tested guide vane opening and downstream level. This suggests that all blades do not behave identically under the same hydraulic conditions from a cavitation point of view and that maximum overall aggressiveness occurs under high flow velocity and downstream levels which produce the best conditions for high erosive power of the cavitating flow. A set of blades with high erosion, of which blade 4 is typical, go into cavitation early and are followed later by the others when hydraulic conditions are more severe. This explains the heterogeneous blade erosions that have historically been observed on this machine. Because of this behavior, the transposition of these results to the model level may not be simple. However, the tools required to characterize quantitatively cavitation aggressiveness at the prototype level have been demonstrated and data is available for comparisons with those obtained in the model test program (4). In closing, the authors would like to thank Louis Bezençon and Georges Jotterand of IMHEF and Pierre Lavigne and Jacques Larouche of IREQ for their fundamental contributions to the success of this program.

7. References

1. Bourdon, P., Simoneau R., and Dorey, J-M., "Accelerometer and Pit Counting Detection of Cavitation Erosion on a Laboratory Jet and a Large Francis Turbine", Proceedings of the *XVII IAHR Symposium*, Beijing, China, 1994, Vol. 2, pp. 599-615.
2. Fortes -Patella R and Reboud, J.L., "Analysis of cavitation erosion by numerical simulation of solid damage", Proceedings of the *XVI IAHR Symposium*, Sao Paulo, Brazil, 1992, Vol. 2, pp.617-626.
3. Farhat, M., Pereira, F. and Avellan F., "Cavitation Erosion Power as a Scaling Factor for Cavitation Erosion in Hydraulic Machines", Proceedings of the *ASME Symposium on Bubble Noise and Cavitation Erosion in Fluid Systems*, ASME Fed-Vol. 176, pp. 95-104, New-Orleans, USA, Dec. 1993.
4. Caron, J.-F., et al., "Cavitation Erosion Prediction on Francis Turbines, Part II: Model Tests and Flow Analysis", paper to be presented at the XVIII IAHR Symposium, Section on Hydraulic Machinery and Cavitation, Valencia, Espania, September 1996.

DETERMINATION OF CRITICAL CAVITATION LIMIT IN THE PRESSURE CONTROL DEVICES

A. CASTORANI[1], G. DE MARTINO[2], U. FRATINO[1]
[1] *Water Eng. Dep. - Polytechnic of Bari (Italy)*
[2] *Hydraulic and Environmental Eng. Dep. - University of Naples (Italy)*

Abstract

This paper describes a method for predicting the cavitation intensity of hydraulic devices by means of the analysis of recorded pressure signal. Infact, it seems possible to get in detail the behaviour of the observed variable through the determination of some frequencies of the pressure fluctuations produced by flow turbulence and by the cavitation impulses. It has been found that it is possible to define the numerical value of the critical cavitation limit by means of the determination of a global index as the root mean square of the pressure fluctuations and in which way the spectral signal analysis proves this approach.

1. Introduction

The determination of the performances of the pressure control devices must necessarily consider a careful analysis of the cavitation characteristics to assure their correct operations in every working conditions [*4*].
Therefore, in view of the modern guidances, the evalutation of the risks involved in the development of the cavitation phenomena has become a crucial moment of the design analysis. From this point of view, using the results of previous experiments, the laboratory procedures for their determination has been partly standardized, as proved by the standards used in several countries [*1*].
The used methodologies are based on the quantification of several secondary effects produced by the cavitation and, as a rule, these approachs are based on the measurement and evalutation, in a relative way, of the vibrational and acoustic parameters produced by the cavitation. Moreover, a method suitable to correlate the cavitation with the hydraulic variables that identify the fluid motion seems closer to a lagrangian problem approach.
Therefore, during an experimental research designed to point out the characteristics of several methods for predicting the cavitation intensity and carried out both on a Larner Johnson ported valve of 100 mm diameter and some 200 mm diameter orifices of several shapes, it seemed useful to examine a new approach based on pressure recordings.

E. Cabrera et al. (eds.), Hydraulic Machinery and Cavitation, 544–553.

2. The purpose of the research

Using some experimental results obtained from conventional analysis techniques, the temporal behaviours of the recorded pressure in several points both upstream and downstream the device location have been analized to characterize the pressure signal so that informations on the cavitation characteristics have been obtained[1].
The choice has been confirmed, in a preliminary study [5], by dimensional considerations that make possible an approach based on the evalutation of the pressure fluctuations produced on the fluid mass. Infact the description, through a dimensional analysis, of the phenomenon produced by the fluid flowing in a control section, allows to define the variables by the following relation:

$$\phi_1(V, \Delta P, P_m, \bar{r}, \rho, \mu, \tau, \varepsilon, \Delta\gamma, C_p, \Theta, P_w, T_r, D, \psi, shape) = 0 \tag{1}$$

where, the indipendent variables ρ, V, D and Θ being chosen and applying the π theorem, it leads to:

$$\phi_2(Eu, Ne, Ne_r, Re, We, Fr_d, Ma, Ne_w, Ec, St, \psi, shape) = 0 \tag{2}$$

that is the general formulation of the relation describing the fluid motion through a pressure control device.
The mathematical shape of the equation *(2)* can be modified to show the influence of the dimensionless cavitation parameter σ. Infact, σ being a pressure forces ratio as:

$$\sigma = \frac{P_m - P_w}{P_m - P_v} \tag{3}$$

it can be written as:

$$\sigma = \frac{(Ne - Ne_w)}{Eu} \tag{4}$$

from which it seems correct to use σ instead of Ne_w, the first one being just a linear combination of the other one.
Downstream of this mathematical position and the dimensionless parameters arised by hydraulic and bubble dynamics considerations, suitable to a lagrangian approach, being neglected we have:

$$\phi_3(Eu, Ne, Ne_r, Re, Ma, \sigma, St, \psi, shape) = 0 \tag{5}$$

expression in which the dimensionless parameters have a different weight according to the boundary conditions.
In the case that the fluid motion, in absolute turbulence conditions, produces the development of cavitation phenomena and modifying the equation so that the cavitation index σ becomes the dependent variable, it is possible to obtain:

$$\sigma = \chi(Eu, Ne, Ne_r, Re, Ma, St, \psi, shape) \tag{6}$$

that shows that σ is function of loss coefficient through the Euler number (*Eu*), of the mean system pressure and therefore of its relation with the vapor pressure of the water (*Ne*), of the pressure fluctuations around the mean value (Ne_r), of the Reynolds number (*Re*), of the Mach number (*Ma*), of the bubble permanency time in low pressure zones (*St*), of the opening degree (ψ) and of the valve shape.

---

[1] Several researchers previously have been performed many tests to correlate the pressure signal to the cavitation phenomena induced by the hydraulic devices and the results of these researches have been very useful [*2, 6, 9*].

In view of the previous relations an experimental research in a hydraulic device subjected to cavitation phenomena and for which it is required the design of whole similitude model, must necessarily consider the necessity to respect simultaneously all the analogies of the dimensionless parameters in the equation *(5)*, including the one of the cavitation index σ.
In the real situation, this requirement prevents to operate in a full respect of the similitude, determining the impossibility to use a different scale for the model, unless we accept several simplifying assumptions.
Therefore, it is necessary to neglect some dimensionless parameters defined in the equation *(5)* and to respect, in addition to σ, only the conditions imposed by the device geometry (the opening degree ψ and the shape) and by the Euler number.
In view of these restrictions, it is very critical to suggest experimental researches using scale models that bring unavoidably to introduce some distorsions between prototype and model and that determine those uncertainties, known as like scale effects, which many researcher have studied for a long time [*1, 5, 7, 8*].

3. Experimental apparatus

The experimental set-up is composed by two parallel 100 mm and 200 mm diameter pipelines on which the hydraulic devices have been installed, supplied by a electropump assuring a 0,9 MPa pressure value and discharges over 100 l/s, (*fig. 1*).
The valve used in the tests is a control Larner Johnson ported valve which presents, at its end, a moving steel cylinder with sixteen parabolic-shaped openings that produce, in full opening position, a ψ value equal to 0,92. The four, 200 mm diameter and 19 mm thickness, steel sharp-edged orifices have 4, 5, 9 and 13 radial symmetry holes respectively. The 4 and 9 holes orifices have a ψ value equal to 0,40, whereas the 5 and 13 hole orifices have a ψ value equal to 0,16.

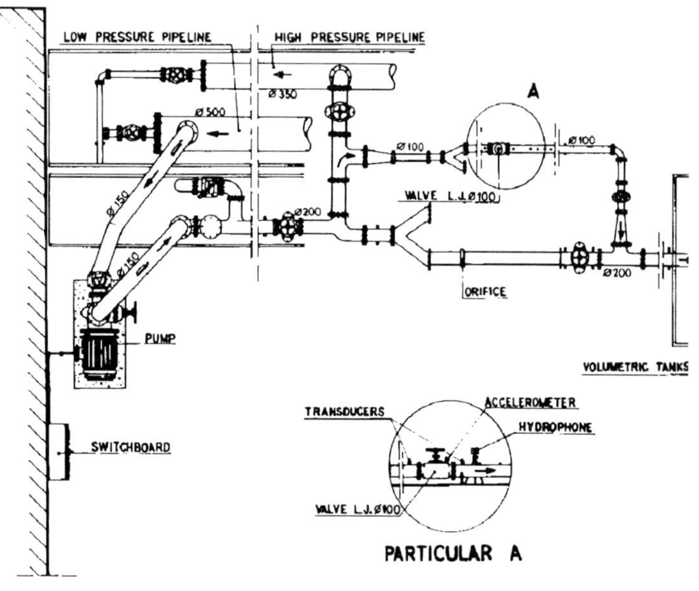

Fig. 1 - Experimental set-up

Preliminarily to the development of the cavitation research, on the 100 mm and on the 200 mm diameter pipelines, several tests have been performed to evaluate the Darcy-Weisbach friction factor in order to subtract the calculated friction loss between the two pressure taps from the measured gross pressure drop.
The pressure taps used for evaluating the gross pressure drop are located one diameter upstream and ten diameters downstream the device location and the pressure drop has been measured by a mercury differential manometer and by Bourdon tube pressure gauges. The discharge value has been valutated by a measuring orifice plate located on the delivery pipe and by some volumetric tanks.
In the laboratory circuit have been located several parallel slide gates, upstream and downstream the device location, for settling the desired pressure and discharge values in every test condition.
Therefore, during the laboratory tests, a pressure value equal to 0,5 MPa has been fixed upstream the valve and a pressure value equal to 25 Kpa downstream the orifices for making the experimental results indipendent from possible uncertainties due to the pressure scale effects that have been evaluated in previous tests [*5*]. In a preliminary study, several tests in steady condition have been performed to define, for the valve, the curve of the loss factor versus the opening degree and, for the orifices, the loss coefficient values *[5]*.
The evalutation of cavitation intensity has been performed using several sensor types: a Larson & Davis 820 sound level meter and a Bruel & Kjaer 4384 accelerometer for the vibration measurements. For measuring the pressure fluctuations, two pressure transducers, having 200 Hz maximum frequency response, have been used. The pressure measurements have been performed in several points along the pipeline, located at 1, 3, 5 and 10 diameters downstream the device location.
These tests, in addition to provide informations about the cavition intensity, allowed to define the downstream section in which the pressure recovery occurs and the normal hydraulic gradeline is re-established.

4. Analysis of results

The experimental research, performed using the traditional cavitation techniques, provided the results in agreement to the conclusions of other researchers, proving the validity of these approaches [*3*, *5*].
The reliability and the suitability of the test results has been proved, evaluating the mechanical and acoustic parameters produced by cavitation phenomena, but the limitation to use the laboratory procedures has been partially overcome when the problem has been studied through a detailed analysis of pressure signal. In fact, this approach seems to identify a typical behaviour of the pressure signal when the critical cavitation limit is reached. This circumstance is very meaningful because σ_c represents the universal design limit for a cavitation free device [*1*].
The recorded signal processing has been carried out according to two different ways: the first one by means of a global evaluation of the pressure signal and a second one in which the components and the frequencies involved in the cavitation inception have been defined for getting a new analysis procedure.

4.1. GLOBAL ANALYSIS

In this first phase, the evalutation of the signal characteristic has been carried out through a global analysis, verifying preliminarily the correctness of the study by means of the invariability of the mean and root mean square values versus the acquisition times. The root mean square value of the pressure signal, evaluated just upstream and downstream the device location, shows however that, for a fixed constant value of the upstream pressure, this quantity has a different behaviour starting from a certain σ value. An inspection, made to recognize the σ values for which the disjunction occurs, brings, for all investigated opening degrees, to identify it in a position close to the one conventionally defined, in the researches performed by accelerometer and sound level meter, as critical cavitation index σ_c[2]. So, this remark suggests a physical interpretation of this index, through a dimensionless variable known as Newton number, connected with the turbulence conditions of the flow and defined by:

$$Ne_r = \frac{\sqrt{\dfrac{\sum_{i=1}^{N}(P_i - \overline{P})^2}{N-1}}}{\rho V^2} \tag{7}$$

All the graphs, plotting Ne_r versus σ, are similar to that in *fig. 2*, that is referred to the valve. From the figure, infact, it is possible to point out, close to the σ_c value, a clear divergence of the two curves. Hence, this behaviour of Ne_r confirms the role of the dynamics of the turbulence on the cavitation characteristics of the flow, as shown in *(5)*.

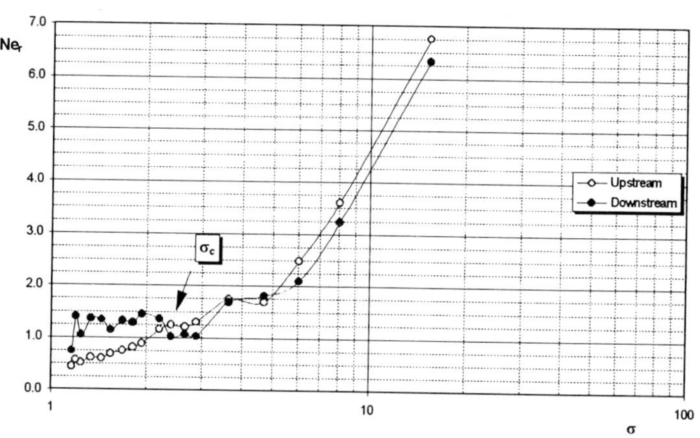

Fig. 2 *- Dimensionless index Ne_r versus σ in the 100 mm valve (C=25%)*

This behaviour has been found again, with the same characteristics, when the analysis has been extended to the 200 mm orifices, and particularly to the 5 and 13 holes ones characterized by a smaller ψ ratio. The orifices, on which many tests have been carried out in different flow conditions, show a strong tendency to cavitate and the Ne_r-σ curves, determined at the same pressure taps location, have no further doubts about the possibility that this occurence is somehow due to the shape of the streamlines in the

---

[2] In the technical literature [*2, 3, 4, 5*], the cavitation intensity is defined by means of four conventional levels called incipient cavitation, critical cavitation, incipient damage cavitation and chocking cavitation that represent the boundary marks between cavitation intensities associated to different risks for the hydraulic system.

outlet section of the device (*fig. 3*)[3]. The confirmation of this behaviour, obviously, defines a new methodological approach to the problem, expecially for its versatility and, moreover, permits to join a physical significance to a fundamental index in the evalutation of cavitation performance of a device, indipendently from a graphical settlement.

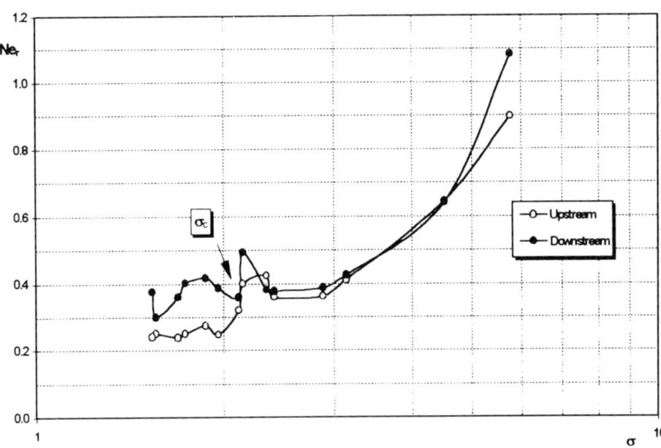

Fig. 3 - Dimensionless index Ne_r versus σ (13 holes 200 mm orifice)

4.2. SPECTRAL ANALYSIS

Before describing the signal processing methods, it is important to underline that the frequency analysis of the recorded signal points out that the acquisition time was sufficient to estimate the phenomenon [*5*]. Settling the correctness of the pressure measurements, the problem is to identify the frequencies involved in the flow turbulence and in the inception of the cavitation phenomena. From this point of view, in the preliminary phase of the investigation, in absence of cavitation phenomena, a comparison between the pressure spectra in different sections was carried out, determining an essential agreement between the spectra relating to the upstream and downstream sections of the device. Hence, the turbulent nature of the flow doesn't appear to bring any changes during the passage through the section in which there is the local pressure drop, expecially if this one is of slight amount, so the spectral peculiarities of the pressure signal, mainly due to the pump and system characteristics are virtually the same. This investigation has been extended to the measurements performed along the downstream pipeline too, showing a similar behaviour.

However this situation changes clearly when, decreasing σ, the cavitation inception occurs with the development and subsequent implosion of vapor cavities. The amplitude spectra, at the same σ, at the section just downstream the valve location have a different trend from the ones defined at the upstream section. This trend is shown in *fig. 4*, where are reported the spectra for a σ value, equal to 2,77, less than σ_c. The two spectra appear different and the downstream one has components higher for frequencies less than 20 Hz, particulary in the ranges [6-12 Hz] and [16-20 Hz]. The component values in these frequency ranges, infact, are much higher than those defined, at the same frequencies, upstream. From this analysis, later improved by other measurements, it appears that, in presence of cavitation, some changes in the spectral

[3] The cavitation tests on the orifices carried out using the traditional techniques by accelerometer and by sound level meter show σ_c values in agreement with those defined by this plots.

structure of the pressure fluctuations components in the downstream section occur. On the other hand, it is likely that the same character of the phenomen, by means of the number and the power of the induced implosions, gives rise to a singular sequence of time of the frequency intervals that, from time to time, are modified.

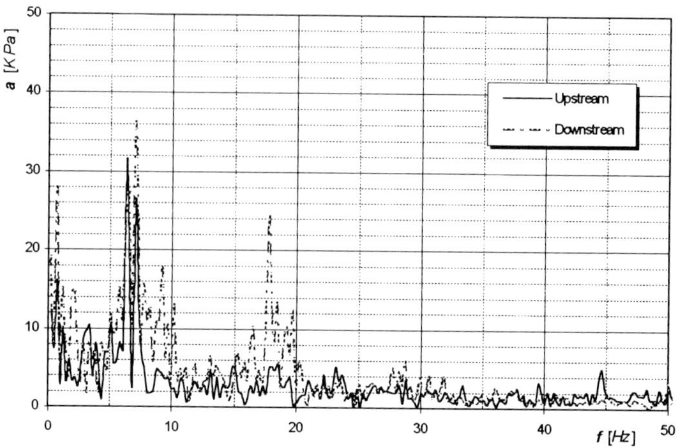

Fig. 4 - Amplitude spectra upstream and downstream the 100 mm valve (C=40%)

In a general way, if the upstream pressure measurements define the turbulence of flow and the vibrations originated by the pump, it is possible to observe that the flow turbulence, when the discharge increases, involves, step by step, higher frequencies producing, after a first meaningful growth of the lower frequencies, a general rise of the spectrum in the whole frequency field. The behaviour of the amplitude spectrum just downstream the device section is instead very different. In fact, if for $\sigma<\sigma_c$, the spectral character of the pressure signal between the upstream and downstream section is similar, except for negligibile differencies due to the hydrodynamic characteristics of the jet flowing from the device, when the discharge increases, the observed differencies increase reaching high values through a meaningful growth of the spectrum, mainly for lower frequency values. This behaviour can be explained as an effect due to the character of the flow, thanks to the large development of turbulent eddies and to the pulses on the fluid mass generated by the increased number of vapor bubbles implosions. The graph in *fig. 5*[4], referred to the valve, explain this behaviour; infact, it is possible to see that, for light cavitation, there is a slight increase of the spectrum for frequencies lower than 10 Hz and higher than 16 Hz, effect on which the cavitation phenomenon plays an important role, because of the reoccurrence of the involved frequencies varying either Re or the valve opening degree. When the discharge increases, the phenomenon raises in the above mentioned ranges, producing, as a result, the raising of the closer frequencies, whereas it fades totally in the other ones. So, because of the increased fluid compressibility due to the formation of the vapor bubbles, the pressure pulses, coming either from the flow dynamics or from the implosions of the bubbles, cause fluctuations of low frequency; in substance the smaller stiffness of the fluid mass provokes a shift downwards of the involved frequencies. The behaviour of the chocking spectrum improves the previous remarks, infact, for this

---

[4] The σ values are equal to 16,59 for no cavitation condition, to 2,56 for light cavitation, to 1,86 for heavy cavitation and to 1,30 for chocking cavitation.

cavitation condition, it is possible to observe a further shift down of the involved frequency.

In this case, the pulses are returned from the fluid mass by means of a limited number of low frequency harmonics, to point out an increased absorbing capacity of the stresses. The hydraulic behaviour of the orifices confirms the above-mentioned conclusions, even if, in different frequency ranges, because of the different geometry.

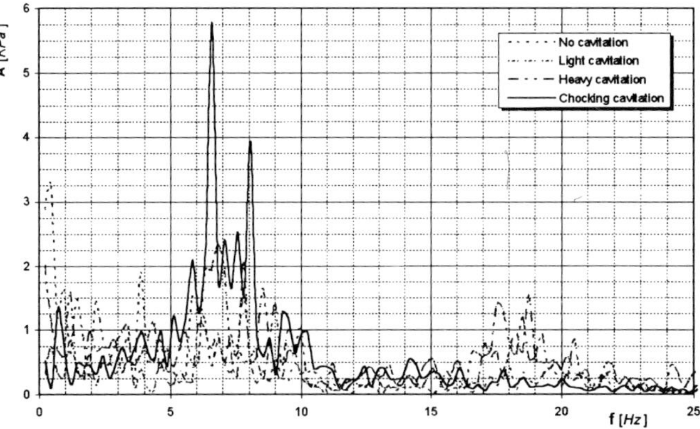

Fig. 5 - Pressure spectra downstream the 100 mm valve for several σ

In view of the previous considerations, it is possible, through a detailed study of the spectral behaviour of the pressure fluctuations downstream, to make a careful analysis of the temporal evolution of the cavitation. Infact, limiting the study to the 20 Hz maximum frequency, a variable representative, by means of a trapezoidal integration, of the spectral behaviour is defined. This variable, called $\Sigma_{[a;b]}$, represents the spectrum area of the observed frequency range $[a;b]$ and has the following expression:

$$\Sigma_{[f_1;f_2]} = \int_{f_1}^{f_2} a(f)\, df \tag{7}$$

in which $a(f)$ is the amplitude value at the f frequency and f_1 and f_2 are the frequencies that define integration extreme terms.

The chosen integration intervals are [0-20 Hz], [15-20 Hz] e [5-10 Hz] and, through graphs like in *fig. 6*[5], it is possible to set out their peculiarities. The figure shows that the curves have a bell-shaped trend, typical of the cavitation measurements. The curve of the integral values in the [0-20 Hz] range spectrum has obsviously higher values than the other two and defines the critical cavitation limit where the curve has the abrupt change of slope, whereas the other two curves have substantially the same trend, even if the one defined in lower interval is, on the average, higher.

This behaviour confirms some of the above mentioned considerations, showing, in the defined frequency range that, besides an initial constant amplitude values, there is an increase involving first the higher frequencies and then the lower ones. In short, it is possible to see that, for σ value less than σ_{ch}, the observed variables decrease quickly to underline a sudden drop of the cavitation intensity, that evolves in a hydrodynamic condition of fully chocked flow where the flowing jet is surrounded by a vapor cavity instead of water.

---

[5] In the same figure the incipient, critical and chocking cavitation limits are represented, determined previously by accelerometer and by sound level meter.

The last mention is for the analysis made for defining the way in which the pressure fluctuations evolve along the pipeline varying the hydrodynamic conditions. The spectra comparison shows that they have similar shape when the flow conditions define light phenomena of cavitation, unless slight differencies due to the way in which the mean stress transmission occurs. Infact, it is possible to find a complete agreement of the involved frequencies, while the amplitudes have a decreasing behaviour when the X/D ratio increases.

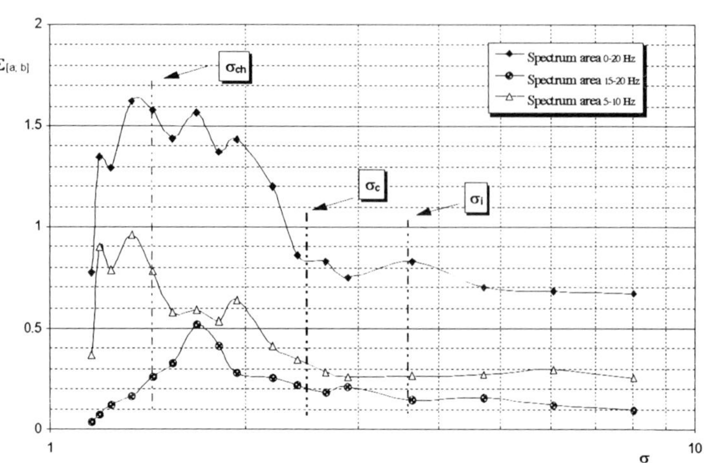

Fig.6 - Plot of $\Sigma_{[a;b]}$ versus σ for the 100 mm valve (C=20%)

For that purpose, , for the valve, the $\Sigma_{[5\text{-}10\ Hz]}$ value is plotted versus the pressure taps position in different cavitation conditions (*fig. 7*). From the figure, the behaviour of this variable varying the pressure tap location is clear. It, for pressure tap fixed, increases in a monotone way, while decreases, step by step, moving downstream. Thus, for σ fixed, the cavitation phenomena show a maximum intensity value just downstream the device section and have a lower one when the X/D ratio increases, because of expected longitudinal variability of the cavitation intensity.

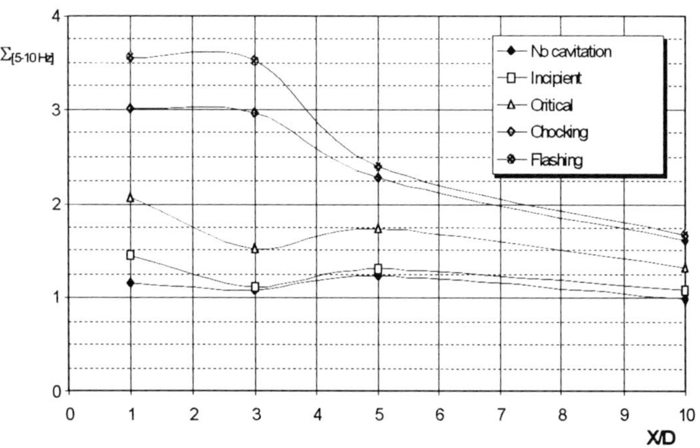

Fig. 7 - Plot of $\Sigma_{[5\text{-}10\ Hz]}$ versus X/D at several σ for the 100 mm valve (C=30%)

This remark gives further proof to the theory that states the existence of a close relation between the cavitation intensity and the pressure fluctuations spectra, confirming that the pressure measurements, by means of the spectral analysis, are an important source for defining the cavitation development.

5. Conclusions

The described method for detecting the cavitation intensity appears of great interest and, in the future, of possible large use. In fact, the evalutation of the pressure signal allows, through a global analysis, the physical definition of the critical cavitation index. At last, the pressure fluctuation spectra show that the pressure fluctuations intensity increases when the cavitation intensity raises and that the pressure fluctuations, although with a random trend in magnitude and frequency, have a clear drop when the opening degree of the valve increases and, at the same opening degree, going from upstream to downstream.

Notation

C	Stroke, in per cent, of the closing element of the valve	
C_p	specific heat at constant pressure	(J kg^{-1} K^{-1})
D	diameter	(m)
f	frequency	(Hz)
$\overline{P}$	expected value of the pressure	(N m^{-2})
P_i	instantaneous pressure values	(N m^{-2})
P_m	absolute upstream pressure	(N m^{-2})
P_w	vapor pressure of the water	(N m^{-2})
$\overline{r}$	root mean square of pressure fluctuations	(N m^{-2})
T_r	bubble permanency time in low pressure region	(s)
V	mean pipe velocity	(m s^{-1})
$\Delta\gamma$	difference in specific weight across an interface of different densities	(N m^{-2})
ΔP	pressure drop	(N m^{-2})
ε	bulk modulus of water	(N m^{-3})
μ	viscosity of water	(N m^{-2} s)
ρ	density of water	(N m^{-2})
σ	cavitation index	
τ	surface tension	(N m^{-1})
Θ	temperature	(K)
ψ	opening degree or area ratio	

$$Eu = \frac{\Delta P}{\rho V^2} \qquad Ne = \frac{P_m}{\rho V^2} \qquad Ne_r = \frac{\overline{r}}{\rho V^2} \qquad Re = \frac{\rho VD}{\mu} \qquad We = V\sqrt{\frac{\rho D}{\tau}}$$

$$Fr_d = \frac{V}{\sqrt{\frac{\Delta\gamma}{\rho}D}} \qquad Ma = V\sqrt{\frac{\rho}{\varepsilon}} \qquad Ne_w = \frac{P_w}{\rho V^2} \qquad Ec = \frac{V^2}{C_p\Theta} \qquad St = \frac{D}{VT_r}$$

References

1. **ANSI-ISA**: "*Control valve Capacity Test Procedure*" S75.02, Research Triangle Park, NC, 1988.
2. **Anton L.**: "*The effects of turbulence on cavitation inception*" La Houille Blanche n°5, 1993.
3. **Castorani A., De Martino G., Fratino U.**: "*Metodologie di valutazione delle caratteristiche cavitanti delle valvole di regolazione*" XXV Convegno di Idraulica e Costruzioni Idrauliche, Torino, 1996.
4. **Di Santo A., Fratino U., Piccinni A.F.**: "*Criteri di scelta delle valvole di regolazione al servizio dei grandi adduttori*" Convegno A.I.I. Macchine ed Apparecchiature Idrauliche, Baveno (NO), 1993.
5. **Fratino U.**: "*Effetti cavitanti nelle valvole di regolazione*" PhD Thesis, Università di Napoli, 1996.
6. **Li S.C., Zhang Y., Hammitt F.G.**: "*Characteristics of cavitation bubble collapse pulses, associated pressure fluctuations and flow noise*" Journal of Hydraulic Research n°2, IAHR, 1986.
7. **Tullis J.P.**: "*Hydraulics of Pipelines*" Wyles and Sons, New York, 1989.
8. **Tullis J.P.**: "*Cavitation Guide for Control Valves*" prepared for U.S. Nuclear Commission, 1993.
9. **Vigander S.**: "*Wall pressure fluctuations in a caviting turbulent shear flow*" Cavitation in Fluid Machinery, ASME, New York, 1965.

STABILITY OF AIR CAVITIES IN TIP VORTICES

A. CRESPO, F. CASTRO*, F. MANUEL AND D. H. FRUMAN **

E.T.S.I. Industriales, Universidad Politécnica de Madrid
José Gutiérrez Abascal, 2, 28006 Madrid, Spain
** Universidad de Valladolid, Spain*
*** ENSTA, Groupe Phénomènes d'Interface, Palaiseau, France.*

Abstract: A model is presented to interpret the behavior of air cavities formed in the tip vortex issued from hydrofoils in a water tunnel, when air is injected. These cavities have the ability to move upstream and reach a stable position near the hydrofoil. From detailed measurements of the velocity components the conditions for the cavities to appear are examined, and compared to the model results. The proposed model, that is based on the equilibrium balance of the cavity, and takes into account that it is closed downstream by a reentrant jet developing in a Lamb vortex, can explain aspects of the behavior not contemplated by another models.

1. Introduction

If air is fed into a water tunnel containing a finite span wing, it may be captured by the low pressure region appearing in the vortex generated at its tip (Figure 1). Cavities in tip vortices have been observed by Meijer (1981) in an open-channel, and Manuel et al. (1987) in a closed-channel of circular cross-section; Meijer (1981) termed them swimming cavities because of their ability to move upstream. Escudier and Keller (1983) associated these cavities to the vortex-breakdown phenomenon, and obtained analogous shapes by pumping air in the region downstream of the vortex-breakdown; however, for the experiments reported here the vortex core remains smooth and no irregularities resembling vortex-breakdown are observed prior to air injection; nevertheless, the analytical predictions made by Keller et al. (1985) based on the two-stage breakdown hypothesis of Escudier and Keller (1983) coincide with those presented here in the appropriate limiting situations. Vapor cavities can also occur in tip vortices depending on the value of the minimum pressure reached on the vortex axis and the air contents in the water (Arndt et al. (1991)). For large air content values (termed weak water), the inception cavitation numbers at a given Reynolds number become weakly dependant of the lift coefficient (Arndt and Keller (1992)). It appears that, for water containing large amounts of free air, capture of nuclei is the dominant effect and that ventilated cavities are formed prior to vapor cavities occurrence. It has been noticed by Arndt and Keller (1992) that in weak water the inception process is highly intermittent, as it also happens in some cases for the phenomenon presented here.

The investigation of the behavior of ventilated cavities shape and intermittency, is

E. Cabrera et al. (eds.), Hydraulic Machinery and Cavitation, 554–563.

interesting from three points of view : first, there is no need to operate under reduced pressure conditions, second, interesting information can be gathered to understand the behavior of large natural (vapor) cavities and, third, ventilated cavities behavior can be associated with some aspects of the vortex breakdown process. Besides, ventilated cavities occur very often in blades and hub vortices of propellers operating near the free surface, surface-piercing propellers, turbines torches, etc.

As it is discussed by Manuel (1981), Manuel et al. (1987) and Escudier and Keller (1983), two characteristic transition regions can be identified in the cavities of Figure 1. The first region contains the stagnation point at the front, and can be considered as a transition from the unperturbed flow to a uniform cylindrical cavity. The second region is a sudden constriction of the first cavity, followed by a wavy pattern. The second transition region has been studied by Manuel et al. (1987) using an analogy with water wave theory in which hydraulic jumps, progressive waves, and solitary waves move along the interface; similar results for progressive waves were obtained by Ackerett (1930), Uberoi et al. (1972), and Keller and Escudier (1980).

The model presented here to study the first transition region is a continuation of a previous one proposed by Manuel et al. (1992). It is assumed that there is conservation of mass, momentum and Bernoulli's constant. In the closure region there is a reentrant jet that contributes to the overall momentum balance. According to observations, the cross section of the viscous core of the vortex and the area of the reentrant jet are assumed to be small compared with the cross section of the cavity, and this area is small compared to the cross section of the tunnel. Keller et al. (1985) assume that in the interface the velocity is zero, and that there is a thin rotational layer near the interface where the velocity changes from zero to its uniform value outside; Escudier and Keller (1983) suggest that this layer may grow by viscous diffusion or become unstable, thus explaining the various spiral forms of vortex breakdown. In our model it is assumed that in the vicinity of the stagnation point the fluid at the interface is at rest and is dragged along by viscous diffusion of the rotational layer, until it reaches a velocity of the order of that of the external irrotational flow. Experimental evidence of the existence of a reentrant jet at the back of the cavity has been accumulated. The appearance of the jet may be a consequence of the instability of the shear layer as a result of the inflexion point required to satisfy the zero normal velocity gradient condition at the interface.

The model predicts a criterion for the cavity to be steady in the flow as a function of the flow conditions upstream of the cavity and the section of the reentrant jet. In contradiction to Escudier and Keller`s (1983) model, in which the cavity will only occur for a given combination of vortex core diameter and maximum azimuthal velocity, our model predicts that there is a range of equilibrium conditions for the cavity depending on the section of the reentrant jet. This is in agreement with the experimental observation showing that stable positions all along the downstream diffusing vortex are possible. The model retains radial variations of the axial velocity, because an overshoot near the axis has been observed experimentally, and uses the Lamb vortex for swirl velocity, because it fits the experimental results much better than the Rankine vortex used by Keller et al. (1985) and Manuel et al . (1992).

Experiments have been conducted in two water tunnels, ENSTA and ETSII, with rectangular (ENSTA) and circular (ETSII) test sections respectively, using hydrofoils of rectangular planform (ENSTA and ETSII), and elliptic planforms (ENSTA), that have been described by Manuel et al. (1992); here, only the results needed for comparison and interpretation of the model will be presented.

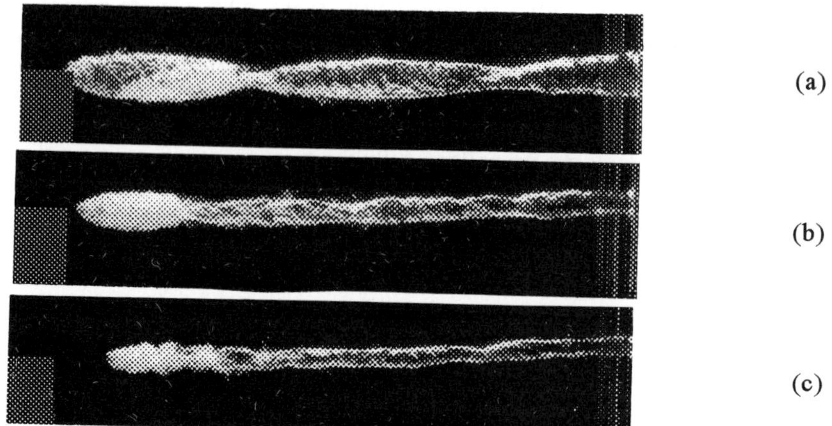

Figure 1. Cavity at three different times (a: 1'06", b: 2' 30" and c: 6') after stopping the air supply. Rectangular planform, 10° incidence angle, W_1=5 m/s. ENSTA tunnel.

2. Experimental results

2.1 CAVITY OBSERVATIONS

Injected air is trapped by the vortex and a cavity, whose leading edge nearly coincides with the transition between the test section (of constant area) and the diffuser (of variable area and adverse pressure gradient) develops. In the case of elliptical planform the bubble remains in the diffuser, whatever the flow conditions. However, if the incidence angle of the foil is above a critical value the cavity is able, occasionally and in a fraction of a second, to move back and forth from the diffuser to the wing. For rectangular planforms and large enough values of water velocity, air flow rate, and incidence angle, the bubble is able to move upstream in the main test section, and, under some circumstances, to get in contact with the foil. By modifying the test conditions, the leading edge of the cavity can be stabilized directly in contact with the foil or at a short distance downstream; however, the range of flow conditions necessary to maintain the cavity at a certain distance from the wing is very narrow: changes in the incidence angle of the order of 1° can send the cavity back to the diffuser or attach it to the wing. Under some circumstances the successive waves pinch off and another reentrant jet occurs.

The cavity in the tip vortex of the rectangular foil can be maintained within the test section even if the air supply is stopped; for a test performed in the ENSTA water tunnel this process lasted about ten minutes. In the absence of injected air the size of the leading bubble and the satellite cavity decrease, and the leading edge moves downstream in a quasi-steady manner, as shown by the three photographs of Fig. 1. With continuous air injection rate an analogous behavior is observed: for larger air injection rates the cavity becomes larger and gets closer to the wing. The reduction of the size of the cavities is related to the continuous draining of the air inside the cavity, while the downstream displacement of the bubble is associated to the force balance over the bubble.

It should also be pointed out that the leading edge of the cavity oscillates around its stable position, with increasing amplitudes and decreasing frequencies as the cavity becomes smaller and moves downstream.

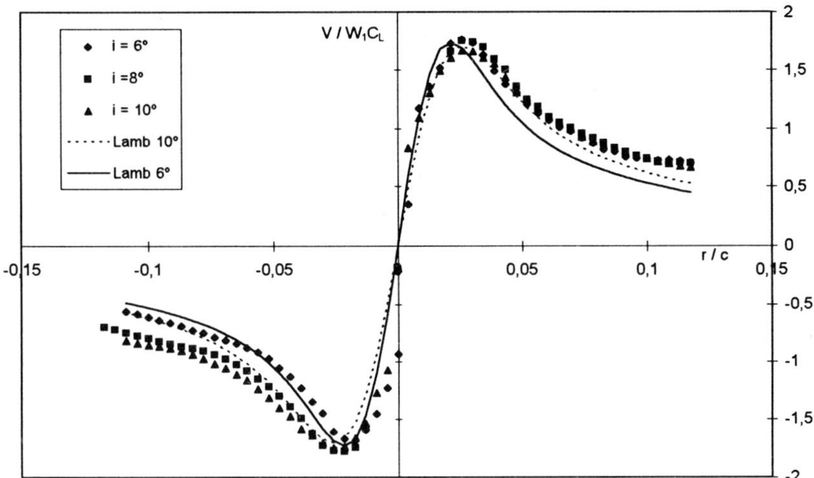

Figure 2. Radial distribution of azimuthal velocities normalized with $C_L W_1$ at three different incidence angles. Rectangular planform, 3 chords downstream. Comparison with Lamb vortex for i= 10°. ENSTA tunnel.

2.2 VELOCITY FIELD

Figure 2 shows, for tests performed in the ENSTA tunnel, a typical radial distribution of the azimuthal velocities, V_1, normalized with the free stream velocity, W_1, and the lift coefficient, C_L, for the vortex generated by the rectangular planform wing operating at three different incidence angles, 6°, 8° and 10°, at a location situated three chords downstream of the trailing edge. It is noteworthy that all the curves fall almost in the same one showing that the normalization is appropriate. The profiles follow the usual behavior: solid body rotation within the vortex core and potential velocity distribution outside it. The swirl velocity profile is quite well adjusted by a Lamb vortex (see equations (2) and (3)), as also shown in Figure 2.

The radial distribution of axial velocity has a typical overshoot around the vortex axis, whose diameter is slightly larger than of the viscous core (Manuel et al. (1992)).

The value of the ratio of the maximum azimuthal velocity, V_{MAX} to the axial velocity at the axis, W_0, including the effect of the overshoot, is an important parameter of our model to determine cavity stability,

$$k = \frac{V_{MAX}}{W_0} . \quad (1)$$

The values of k as a function of the downstream distance, are displayed in Figure 3, for conditions under which the cavity, appears either in a stable position for the rectangular planform or moving back and forth for the elliptic planform. The parameter k in general decreases with distance, except for the rectangular foil of ENSTA.

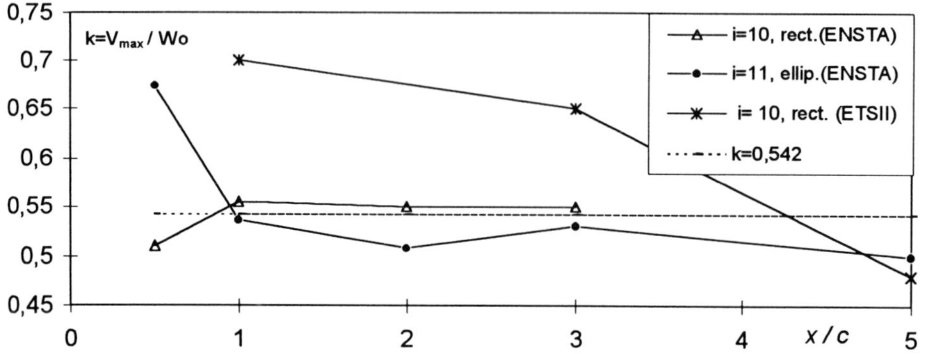

Figure 3. Variation of k with downstream distance. Comparison with analytical prediction. Measurements for both elliptical and rectangular wings at several downstream distances, i=11° and 10° respectively. ENSTA and ETSII tunnels.

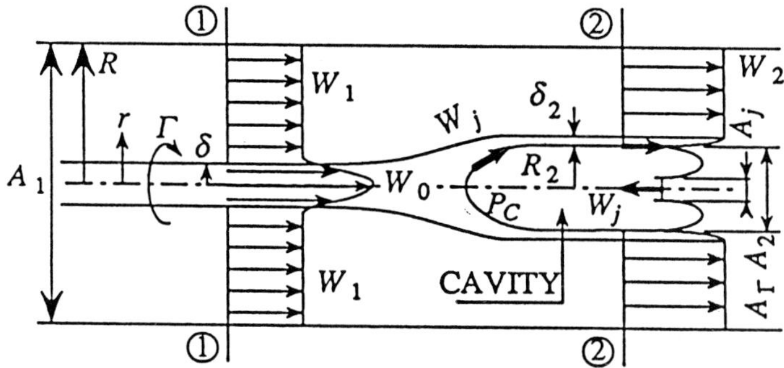

Figure 4. Schematic showing the geometry of the domain under consideration.

3. Proposed model

The geometry of the portion of the flow for which the model will be developed is shown schematically in Figure 4. In section 1 the approaching flow has a uniform axial velocity, W_1, outside a vortex of radius δ and intensity Γ; inside the vortex the approach velocity has an overshoot. Azimuthal velocities are supposed to behave as in a Lamb vortex,

$$V = \frac{\Gamma}{2\pi r}(1 - e^{r^2/\delta^2}) \; . \tag{2}$$

The maximum azimuthal velocity and the corresponding radial position are,

$$V_{MAX} = 0.638\frac{\Gamma}{2\pi\delta} \quad \text{at} \quad r = 1.12\delta \ . \tag{3}$$

In the following δ and Γ will be estimated from experiments using equations (3). Figure 2 shows that the experimental values of the swirl velocity are fitted quite well by equation (2) when equations (3) are satisfied.

The pressure distribution is obtained by integration of the following equation,

$$\frac{dP}{dr} = \frac{\rho V^2}{r} \ , \tag{4}$$

that, using equation (2), gives

$$P_1 = P_0 + \frac{\rho\Gamma^2}{4\pi^2\delta^2}\left(\log 2 - \frac{1}{2X^2} + \frac{e^{-X^2}}{X^2} - \frac{e^{-2X^2}}{2X^2} - E_1(X^2) + E_1(2X^2)\right) , \tag{5}$$

where, $X=r/\delta$, P_0 is the pressure at the axis, and E_1 (X) is the exponential integral. For r (or X) going to infinity a constant value of the pressure is reached,

$$P_{1\infty} = P_o + \frac{\rho\Gamma^2}{4\pi^2\delta^2}\log 2 = P_o + 1.703\rho V_{MAX}^2 \ , \tag{6}$$

where equation (3) has been used. By integration of the pressure field, equation (5), across the upstream section 1 we obtain the force,

$$\frac{F_1}{\pi R^2} = P_0 + 1.703\rho V_{MAX}^2\left[1 - \frac{1}{\log 2}\left(\frac{\delta}{R}\right)^2 \log\left(\frac{R}{\delta}\right)\right] . \tag{7}$$

Where it has been assumed that $\delta << R$ and only a first order correction term has been retained.

In the downstream section the flow is supposed to be irrotational, with constant axial velocity, and a circulation equal to that upstream, outside a thin rotational shear layer of thickness δ_2. As the areas of the two rotational regions should be of the same order, because only the fluid particles inside the rotational core of radius δ in region 1 are the ones that can be in the rotational layer of thickness δ_2 in region 2, the value of δ_2 is such that $\delta_2 = O(\delta^2/R_2)$. It will be shown later that the radius of the cavity is much larger than δ, and much smaller than the radius of the test section; the ordering of the different lengths involved is as follows (see also equation (18)):

$$\delta_2 \ll \delta \ll A_j^{1/2} \ll R_2 \ll R \ , \ \delta_2 = O(\delta^2/R_2) \ , \ A_j = O(\delta^2 \log(R_2/\delta)) \ . \tag{8}$$

For the particular case of Figure 1, typical values are δ= 1 mm, R_2= 10 mm and R= 100 mm; the jet area is smaller than that of the cavity, and slightly larger than that of the upstream vortex. As the shear rotational layer is very thin, it can be assumed that over most of region 2 the flow is irrotational and the azimuthal velocity is given by $V=\Gamma/(2\pi r)$ and on using equation (4) the pressure distribution over region 2 is obtained:

$$P_2 = P_C - \frac{\rho\Gamma^2}{4\pi^2}\frac{1}{2}\left(\frac{1}{r^2} - \frac{1}{R_2^2}\right), \tag{9}$$

where is P_C the cavity pressure, and by integration we obtain the force on region 2,

$$F_2 = P_C\pi R^2 - 2.457\rho V_{max}^2\pi\delta^2\left(\log\frac{R}{R_2} - \frac{A_2}{2\pi R_2^2}\right), \tag{10}$$

where equation (3) has been used.

It is assumed that the pressure in the cavity is equal to the stagnation pressure along the axis of the upstream region,

$$P_C = P_0 + \frac{1}{2}\rho W_0^2 . \tag{11}$$

Manuel et al., (1992), consistently with the fact that there is a reentrant jet in the back of the cavity, assumed that the constant value of the velocity in the interface is the jet velocity, thus obtaining a lower value of the pressure in the cavity; however, it is not clear how this assumption can hold at the tip stagnation point. It is more reasonable to assume that, if the cavity is long enough for the viscous forces to be significant, the external flow will drag along the stagnant flow in the interface until it reaches a finite velocity, of the order of that of the reentrant jet.

For the control volume of Figure 4, we have conservation of mass,

$$W_1 A_1 + W_j A_j = W_2 A_2 , \tag{12}$$

and the conservation of momentum,

$$F_1 + \rho W_1^2 A_1 = F_2 + \rho W_2^2 A_2 + \rho W_j^2 A_j , \tag{13}$$

where the effect of the overshoot has been neglected because it is of order δ^2. It is also assumed that the jet of area A_j and velocity W_j, that enters in the cavity in the back, produces a force directed upstream, and its mass is incorporated to the liquid again through the interface. It is also assumed that there is conservation of stagnation pressure in the outer irrotational region, and in particular, near the wall, $R >> R_2 >> \delta$,

$$P_{1\infty} + \frac{1}{2}\rho W_1^2 = P_{2\infty} + \frac{1}{2}\rho W_2^2, \tag{14}$$

where, from equations (9) and (3),

$$P_{2\infty} = P_C + 1.228\rho V_{max}^2 \left(\frac{\delta}{R_2} \right)^2 . \quad \textbf{(15)}$$

First, is considered the case in which all higher order terms, according to what is indicated in equation (8), can be neglected. Then, $A_1 = A_2$, from (12) $W_1=W_2$, and from (14) and (15) $P_{1\infty} = P_{2\infty} = P_C$. Then, from equations (6) and (11), it is obtained that,

$$k = \frac{V_{MAX}}{W_o} = 0.542 \; . \quad \textbf{(16)}$$

The momentum equation has not been used, and is satisfied identically. The basis of this result is quite simple: the leading point of the cavity is a stagnation point, where the head of the axial velocity just cancels the pressure minimum of the vortex. Manuel et al. (1992) performed a similar analysis using a Rankine vortex and obtained for k a value of $1/\sqrt{2}=0.707$ instead of 0.542. This value is also obtained using the two-stage transition model for breakdown of Keller et al. (1985), with a Rankine vortex and without considering overshoot (see also Escudier, 1988, the value of k that he defines differs by a factor 2 from the one defined in equation (1)). Spall et al. (1987) give a criteria for breakdown based on a Rossby number, Ro, that is related to k by Ro=0.57/k, without considering overshoot; they find that, for Reynolds numbers based on core radius larger than 100, there is breakdown when $k>k_c$, where $k_c=0.877$; this result is based on experimental and numerical research both of them and another authors. The comparison of the result of Spall et al. (1987) with equation (16) shows that larger values of vortex intensity are needed for the appearance of vortex breakdown than for the formation of the cavity. This is consistent with the fact that, if there is no air injection, no vortex breakdown appears in the single phase for the value of k given by equation (16).

If higher order terms are retained, equation (16) is modified to give

$$k = \frac{V_{MAX}}{W_o} = 0.542 \left(1 + 3.406 \; \frac{W_1^2}{W_0^2} \left(\frac{R_2}{R} \right)^2 \right) . \quad \textbf{(17)}$$

This means that, as the radius of the cavity increases the maximum azimuthal velocity necessary for the cavity to be stable also increases; equation (16) is then a lower limit for V_{MAX}. From the momentum equation (13), that so far has not been used, the following relation is obtained for the radius of the cavity as a function of the upstream kinematic variables and the jet characteristics:

$$\left(\frac{R_2}{R}\right)^4 = \frac{1}{\log 2}\left(\frac{W_0}{W_1}\frac{\delta}{R}\right)^2 \log\left(\frac{R_2}{\delta}\right) + 2\frac{W_j}{W_1}\left(1+\frac{W_j}{W_1}\right)\frac{A_j}{A_1} \quad . \tag{18}$$

This equation was first obtained by Manuel et al. (1992) without the first term in the right hand side, and a minus sign in the brackets of the last term. The last term represents the effect of the jet; if this term is zero, this equation gives the value of the cavity radius, R_2, as function of the vortex radius, δ, and equation (17) gives the azimuthal velocity, V_{MAX}, necessary for the cavity to be stable; there is then a relationship between δ and V_{MAX} that has to be satisfied; this means that for a vortex whose properties change with distance, stability will only be possible at one position. On the other hand, if we retain the last term of equation (18), the cavity can be at different positions of the vortex: if the cavity gets smaller, equation (17) shows that the bubble should move to another position where V_{MAX} should also be smaller, in this position δ will have a new value, and equation (18) can still be satisfied if the jet characteristics are appropriate. For the two terms in (18) to be of the same order, the area of the jet should be: $A_j=O(\delta^2\log(R_2/\delta))=O(\delta_2 R_2\log(R_2/\delta))$ (as anticipated in equation (8)), slightly smaller than that of the vorticity region close to the interface. For the elliptic wing the radius of the vortex is smaller than the for the rectangular wing, then A_j is also smaller, and the last term of equation (18) may not be large enough to provide the stabilizing effect previously mentioned. Besides, for very small δ all the right hand side of equation (18) becomes very small, the radius of the cavity, R_2, is also very small (this is observed experimentally for elliptic wing), and equation (17) shows that the cavity will only appear for values of k so close to the limiting value given by equation (16), that it is not possible to observe it experimentally.

Equations (17) and (18) indicate that the minimum value of k for the cavity to be stabilized in the test section is 0.542; in figure 3, it is compared with the measurements corresponding to incipient cavity appearance (stable cavity for rectangular wings and oscillating cavity in elliptic wings) in the test section. Since inequalities (8) hold for all the experiments, it is expected that k should be close to this value, whenever the cavity is stable; this is in agreement with the experimental observation that for the cavity to be stable in the test section, the range of variation of the incidence angle is very small. For the rectangular wing in the ENSTA tunnel, k is close to 0.542 all along the vortex. For the elliptic wing at a short distance from it, k is considerably larger than 0.542, and further downstream is very close to 0.542, but as it was already mentioned the cavity is never stabilized and moves back and forth. For the rectangular wing in the ETSII tunnel k gets close to 0.542 about 4 chords downstream, whereas the experimental observations show that the cavity remains stable about 1 to 2 chords downstream of the wing.

4. Conclusions.

Ventilated cavities formed in the tip vortex issued from finite span hydrofoils have different behaviors depending on their planforms. Experiments show that the cavities are generally formed by a leading bubble with a reentrant jet followed by a satellite cavity with longitudinal waves whose amplitude and length decrease with the distance downstream. Moreover, depending on the wing and the flow conditions, the cavity formed initially in the transition between the test section (of constant area) and the diffuser (of increasing area and adverse pressure gradient) can or can not move upstream and, sometimes, get in contact with the foil.

By considering the equilibrium of the leading bubble with a reentering jet, a model is proposed that gives the range over which the ratio of the local maximum tangential velocity to the free stream velocity has to be maintained to allow for cavity stability. Comparison of the model prediction and experimental results show a good agreement and can explain some aspects of the observed behavior.

Further experimental and theoretical work should be conducted to investigate the behavior of the tip vortex and its interaction with the ventilated cavity for a greater range of flow conditions. Further test should also provide information on the cavity characteristics, in particular the length at which instabilities occur in the interface

5. Acknowledgments

The authors wish to express their deep appreciation for the partial support received from the Spanish (grants n°86-B, A.I. and PB 89-0184 of the DGICYT) and French (grants n° 92/205 A.I. and DRET 88/1038) governments.

6. References

Ackerett, J. (1930) *Uber Stationäre Hohlwirbel*. Ing. Archiv. pp 399-402.

Arndt, R.E.A., Arakeri, V.H. and Higuchi, H., (1991) *Some observations of tip vortex cavitation*, Journal of Fluid Mechanics, Volume 229, pp 269-289.

Arndt, R.E.A. and Keller, A.P., 1992, *Water quality effects on cavitation inception in a trailing vortex*, Journal of Fluids Engineering, Volume 114, pp. 430-438.

Escudier, M.P. (1988) *Vortex Breakdown: Observations and Explanations*. Progress in Aerospace Sciences. Volume 25, N° 2, pp 189-229. 1988

Escudier, M.P. and J. J. Keller. (1983) *Vortex Breakdown: a Two-Stage Transition*. AGARD CP N° 342. Aerodynamics of vortical type flows in three dimensions, paper 25.

Keller, J.J., Egli, W. and Exley, J. (1985) *Force- and loss-free transitions between flow states*, J. Appl. Math. Phys. 36(6), 854-889.

Keller, J.J. and Escudier., M.P. (1980) *Theory and Observations on Hollow Core Vortices*. J. Fluid Mech. Vol. 99, part 3, pp 495-511.

Manuel. F. Ph. D thesis. (1981) Universidad Politécnica de Madrid.

Manuel, F., Crespo, A., Castro, F. (1987) *Wave and Cavity Propagation along a Tip Vortex Interface*. Physico Chemical Hydrodynamics, Vol. 9, N° 3/4, pp. 611-621.

Manuel, F., Crespo,A.,Castro,F., Fruman,D.H., Beuzelin,F., Gaudemer, R. (1992) "*Ventilated cavities in tip vortices*." Proceedings of the Institution of Mechanical Engineers. IMechE 1992-11. International Conference on Cavitation. Cambridge .

Meijer, M.J. (1981) *Swimming Vortex Cavities*. Euromech Colloquium 146. Institut de Mechanique de Grenoble.

Spall, R. E., Gatski, T. B., and Grosh, C. E. (1987) *A criterion for Vortex Breakdown* Phys. Fluids Vol. 30 (11), pp 3434- 3440.

Uberoi, M. (1972) *Stability of a Coaxial Rotating Jet and Vortex of Different Densities*. Phys. Fluids, Vol. 15.

CAVITATION EROSION PREDICTION ON FRANCIS TURBINES-PART 3 METHODOLOGIES OF PREDICTION

J.M. DOREY, E. LAPERROUSAZ
Electricité de France
6, Quai Watier, 78401 CHATOU Cedex, France
F. AVELLAN, P. DUPONT
IMHEF/ EPFL
33 Avenue de Cour, CH 1007 Lausanne, Switzerland
R. SIMONEAU, P. BOURDON
Hydro-Québec
1800 boul. Lionel Boulet, Varennes, Québec, Canada J3X 1S1

Summary: In the frame of a joint research programme between EDF, Hydro-Québec and IMHEF, different methods are investigated to predict cavitation erosion on Francis turbines from model. They are based on measurement of pitting, pressure fluctuations and acceleration. The measurement techniques have been detailed in Part 1 and Part 2. The present article describes essentially the theoretical and practical aspects of the methods and discusses the results obtained until now from the model and prototype tests. The first analysis shows that the methods proposed are suitable to measure cavitation aggressiveness on model and on prototype, and that the level on the model is several orders of magnitude smaller than on the prototype. To adjust transposition laws, a more complete set of data is needed.

1. Introduction

Despite many efforts in the past, cavitation erosion still remains an unsolved problem regarding the acceptance tests on model for hydraulic turbines including Francis turbines. The different methods used today to fix acceptable tailwater level from cavitation model tests (to add a " safety margin" to the efficiency drop tailwater level, to fix a certain acceptable cavity length, to use paint erosion tests or so)do not give full satisfaction since they involve empirical experience and subjective considerations.

As this problem remains essential for sizing and design of hydraulic turbines, EDF, HydroQuébec and EPFL have conducted an important research programme on this subject, including tests on a 260 MWe Francis Turbine and its model. The programme has been previously presented [1]. The measurements of cavitation intensity on the prototype and on the model are described in Part 1 and Part 2.

E. Cabrera et al. (eds.), Hydraulic Machinery and Cavitation, 564–573.

2. Cavitation aggressiveness and model-to-prototype similitude

2.1. CAVITATION AGGRESSIVENESS

It must be reminded that cavitation erosion is the result of accumulation of impacts due to the collapse of vapour "structures" near the wall. So cavitation erosion can be split in two different mechanisms :

- first, the hydraulic mechanism: local depression in the flow generates vapour structures, growing and then collapsing, leading to minute impacts on the wall;
- second, the damage mechanism: under cavitation impacts the wall material is damaged, ending with mass removal.

The interface between these two distinct phenomena is called the "cavitation aggressiveness", which is the impact loading applied on the wall by successive collapses. In a first approach, it can be set that cavitation aggressiveness is the pure consequence of the first mechanism and the input of the second.

So the cavitation aggressiveness can be defined as a set of impacts striking the wall, each impact being characterized by its pressure P_i, its duration t_i, and the characteristic size L_i of the surface stroked. Although the mechanisms of collapse are not yet fully elucidated, it is admitted now that these parameters are in the range of gigaPascal for P_i, microsecond for t_i, and 0.1-1 mm for L_i. Up to now, no on-line instrumentation is able to measure cavitation aggressiveness, and the methods proposed hereafter are based on indirect measurements.

2.2. MODEL TO PROTOTYPE SIMILITUDE

Here we remind only the main features of the reasoning leading to scaling laws. More argued statements of these laws can be found in [2]. Between a prototype (reference size: L_{proto} , speed : N_{proto}) and its model (reference size: L_{model} , speed : N_{model}), the hydraulic mechanism can be theoretically ruled by scaling relationships:

- The extend of cavitation area: the well known Sigma similitude between model and prototype theoretically ensures that similar areas are under vapour pressure;
- The size of vapour transient cavities: since cavitation extent is similar, the size of cavities is also similar, i.e. proportional to the scale;
- The rate of production of vapour cavities: if they are produced by a pocket attached on the wall, or by a vortex shedding, their rate of production follows a constant Strouhal number. If the vapour structures are generated by nuclei, this rate will be proportional to the number of nuclei per volume unit;
- The potential power of vapour structures: it is the product of their potential energy (i.e. the volume multiplied by the pressure that forces the collapse minus vapour pressure) by their rate of production. Since size, pressure and rate can be scaled, this also can be scaled.
- The cavitation impacts on the wall: the impact being considered as a shock, the impact pressure P_i due to the collapse of a vapour cavity can be scaled by $\rho c \mathbf{V}$

(ρ = water density, $\mathbf{c}$ = sound velocity in water, $\mathbf{V}$ = reference flow velocity), while the impact size L_i and the time t_i can be scaled by the size of the vapour cavity i.e. by the homology factor.

With these considerations in mind, aggressiveness can be transposed from model to prototype.

On the contrary, the loss of mass under cavitation aggressiveness cannot be ruled by similitude laws. Up to now, too little is known on the processes involved in material damage to lay down a simple modelisation.Experimental data are still needed. These may be provided by devices such as water jet [4] or vortex generator [8], or others.

3. Prediction methods and results

3.1. PREDICTION USING PITTING MEASUREMENTS

3.1.1 *Requirements for pitting tests*

This method is based on the analysis of cavitation pitting obtained after a short duration test (less than an hour) on polished samples made of soft metal (pureCu, pure Al, Ag or so) mounted on the model. The test is run in hydraulic similitude with investigated prototype operating conditions, including cavitation similitude.

To get the highest cavitation aggressiveness, it is important to reach the highest head possible on the model platform. This will be favourable for transposition, to get a sufficient pitting, and to diminish extraneous effects such as damping due to air content, and viscous effects. It is often impossible to reach the same head on the model as on the prototype, but a head of about one half to one quarter the prototype one seems to be a good range. Air content in the water of the loop has to be reduced as low as possible to avoid any damping effects on the collapse.

Preliminary tests are required to fix test duration that must be long enough to get a representative pitting , but not too long to avoid excessive overlapping of pits.

3.1.2. *Techniques and apparatus*

The easiest way to put soft polished metal in the erosion area is to use samples mounted in the blades. If this technique has been successful on the prototype, unfortunately, on the model, preliminary tests with pure aluminium samples shows undesirable cavitation pitting behind small discontinuities (less than 0.2 mm height) between sample and blade. Electrolytic deposition of metal directly on the blades is found safer but it rules out the use of Al (poor quality of deposits) and Cu (variable hardness of deposits). In this case, silver has been found to be suitable.

As for pitting measurement, we used a UBM laser profilometer that provides a mapping of the surface. After levelling, pitting can be extracted by a software named "Adresse" that determines the values of pit depth h_i, pit radius R_i, and pit volume V_i for each pit. This process is detailed in [3]. This software calculates also the pressures and sizes, and perfoms the transposition from model to prototypeaccording to the laws set out hereafter.

3.1.3. *Transposition procedure*

After a test duration T, pitting obtained is measured on a surface $S_{a\ mod}$ and provides a set of pits $(h_i, R_i)_{mod}$, each one being characterized by its depth h_i and radius R_i. From the pits geometry and the material properties (simple shear stress $\mathbf{S_0}$, elastic Young modulus **E**, Poisson modulus ν, sound velocity $\mathbf{C_s}$), pressure P*i* and size L*i* of the corresponding impact can be deduced using the following relationship obtained by analysis of elastoplastic deformation [5] [6]:

$$\frac{R_i}{L_i} = f_1\left(\frac{C_l}{L_i} \cdot dt\right) \qquad \text{and} \qquad \frac{h_i}{L_i} = \frac{S_0}{E} \cdot \left(\frac{P_i}{S_0}\right)^m \cdot A^{1-m}$$

where $A = f_2\left(\frac{C_l}{L_i} \cdot dt, \nu\right)$, $m = f_3\left(\frac{C_s}{L_i} \cdot dt, \frac{C_l}{L_i} \cdot dt\right)$, $\mathbf{C}_l$ is the sound velocity in the water and **dt** is the time duration of the impact. $\mathbf{f_1}$, $\mathbf{f_2}$ and $\mathbf{f_3}$ are determined by numerical simulation of plastic deformation, $\mathbf{C_s dt / L}_i$ and $\mathbf{C_l dt / L}_i$ can be assumed to be constant [7]. So the relationship can be written in the following form :

$$(P_i, L_i) = f_{mat}(h_i, R_i)$$

where $\mathbf{f_{mat}}$ depends only of the material.
This provides the set of impacts $(P_i, L_i)_{mod}$ obtained on the model.

Given : - the scale ratio $\lambda = L_{proto} / L_{mod}$, and
- the speed ratio $n = N_{proto} / N_{mod}$

of the model, this set $(P_i, L_i)_{mod}$ is then transposed to prototype $(P_i, L_i)_{proto}$ by means of the following scaling laws:

$$P_{i\ proto} = P_{i\ mod} \cdot \lambda \cdot n \quad \text{and} \quad L_{i\ proto} = L_{i\ mod} \cdot \lambda$$

This deduced aggressiveness $(P_i ,L_i)_{proto}$ is supposed to be applied on a surface:

$$S_{a\ proto} = S_{a\ mod} \cdot \lambda^2 \quad \text{during a time} \quad T_{proto} = T_{mod} / n$$

These relations are deduced from the similitude considerations mentionned above, with the hypothesis of constant Strouhal number for production of cavities.
With the inverse relationship $\mathbf{f_{mat}}^{-1}$ between pressure pulse and pit, one can calculate pitting $(h_i, R_i)_{proto}$ that would be produced by this aggressiveness $(P_i, L_i)_{proto}$ on the prototype, on the prototype material. A volume pitting rate can be inferred:

$$V_d = \Sigma V_i / (S_{a\ proto} \cdot T_{proto})$$

where V_i is the volume of pit *i* ($V_i = f(h_i, R_i)$).
Finally, erosion on the prototype is deduced using correlations between pitting and mass loss obtained in laboratory (high pressure cavitating jet or venturi) for a range of impulse pressures and materials:

$$Er = f(V_d, \text{material})$$

3.1.4. *Pitting Results*

The pitting obtained on the prototype is described in Part I and in a previous paper[4]. What must be noticed here is the very large size of the larger pits: up to 2 mm diameter despite the hardness of the material (316L stainless steel). This confirms the assumption on scale of pits proportional to the size of the machine.

2 mm

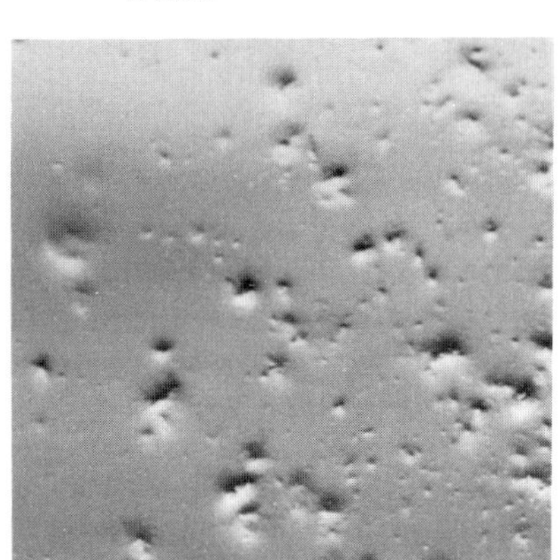

Figure1: typical pitting on the prototype on 316L SS (130 HB)

0.2 mm

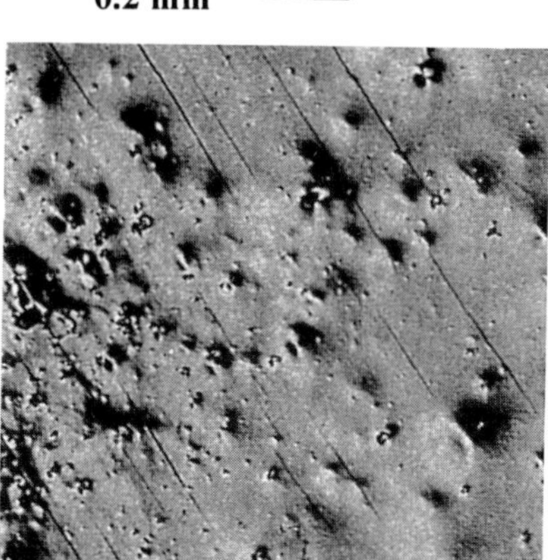

Figure 2 : pitting obtained on the model on pure aluminium (16 HB)

On the model, preliminary tests have been carried out with aluminium samples flush mounted on two blades. After a short duration, a good pitting was obtain, even if on some samples parasite cavitation was observed. This can be seen in figure 2.

Nevertheless, these preliminary tests are not strictly comparable to the prototype tests since, as explained in Part 2, the model is not exactly representative and the cavitation behaviour is different from the prototype, what forbid direct comparison.

3.1.5. *Tentative transposition*

Despite this fact, an interesting exercise can be carried out using the prototype results : from measurements on stainless steel on sample 2 (one of the more severely damaged, $\sigma = 0.120$, $\phi = 0.27$, vane opening 78%) one can use the inverse procedure to predict what is expected on the model. The test lasted $T_{proto} = 30$ mn. The sample 2 was analysed on $S_{proto} = 100\ mm^2$ and gives the following results:

- number of pits : 193
- total volume of pits: $V_{tot} = \Sigma V_i = 4.35\ 10^8\ \mu m^3$

From the pits set (h_i, R_i, V_i) the following values are calculated:

- mean weighted depth: $h_v = \Sigma(h_i.V_i)/\Sigma V_i$
- mean weighted radius: $R_v = \Sigma(R_i.V_i)/\Sigma V_i$
- volume pitting rate: $V_d = V_{tot}/(S_{proto}.T_{proto})$

These values are listed in table 1, as for the successive steps of the transposition.

Then, the mechanical characteristics of 316 L stainless steel beeing $S_0 = 400$ MPa, E = 200 GPa, C_s = 5800 m/s, the inferred pressures P_i and size L_i of impacts are calculated. They range from 1.9 to 4.5 Gpa on the prototype. Transposition is then applied. Since the model scale is $\lambda = 14.66$ and n = 0.1125, pressures are divided by 1.65, while sizes are divided by 14.66. The corresponding surface and duration are:

$$S_{mod} = S_{proto} / \lambda^2 = 0.465 \text{ mm}^2 \quad \text{and} \quad T_{mod} = T_{proto} \cdot n = 3.36 \text{ mn}.$$

Once aggressiveness is transposed, one can calculate the pitting on the model, for a given material with the software "Adresse". On the same material (316L SS), the pitting obtained is described in column 2 of Table 1. What is significant is that, while the radius R_v decreases almost proportionally to the scale, the depth is divided by more than four time the scale. Most of such pits would be hard to measure. The weakness of the aggressiveness on the model justifies the use of a softer material.
The same aggressiveness (set (P_i,L_i)) applied on pure Al ($S_0 = 100$ MPa, E = 50 GPa, $C_s = 5000$ m/s) leads to the pitting described in column 3 of Table 1. Here, the pitting is very important since many pits are deeper than large. This seems to be unrealistic. It can be compared with the pitting obtained during one of the preliminary tests on the model ($\sigma = 0.07$, $\phi = 0.27$, vane opening 78%)). After a test duration of 15 mn, a 9 mm^2 surface of one of the most eroded samples has been analysed and the results are given in column 4 of Table 1. The pitting is much lower, pits are less deep and larger than expected but the similitude conditions are not respected (specially σ similitude), so that no conclusion can be drawn up to now. More experimentation is currently carried out, using silver coating of the blades. At the present time, the results show clearly that pitting measurements on model can be obtained, and that transposition has to be adjusted.

Table 1 : Results of transposition from prototype pitting tests

	Prototype tests on 316L SS	Transposed to the model on 316L SS	Transposed to the model on Al 59	Model test on Al 59
Duration (mn)	30	3.36	3.36	15
Surface (mm^2)	100	0.465	0.465	9
hv (μm)	22.2	0.207	36	1.9
Rv (mm)	0.66	0.045	0.045	0.17
V_d (μm/s)	1.29 10^{-3}	1.07 10^{-4}	2.05 10^{-2}	6.2 10^{-4}
(mm/hour)	4.66 10^{-3}	3.87$10^{-4}$	0.074	0.223 10^{-2}

3.2. PREDICTION USING VIBRATION ANALYSIS

This method aims to measure the amplitude of shock forces applied on the blades by cavitation. Since direct measurement is not possible, a method is proposed to infer these forces from the signal of an accelerometer mounted outside the machine. To avoid extraneous noises from mechanical phenomena or flow, forces are calculated in a high frequency range. The transfer function from the blades where cavitation occurs to the measurement spot can be determined by two different means : spark generator and instrumented impact hammer. From energy considerations, it can be said that inferred forces F_{mod} obtained on the model can then be transposed to the prototype using the following formula:

$$F_{proto} = f(\lambda,n) \, . \, F_{mod}$$

where λ is the scale ratio between prototype and model, and **n** is the speed ratio.

That such a relation or another may exist rests on the following considerations. Developed forces for similar flows on model and prototype should scale with the energy of cavities, i.e. their volume. The rate of collapse of these cavities on the runner must also be considered as this parameter may vary between model and prototype and influences the inferred forces values.

With the results of both campaigns in hand, we identify scaling law on the basis of the calibrated values of the mean square inferred high frequency forces. From the knowledge of the damaged areas on the prototype, the inferred forces values can be converted to erosion rates using erosion data obtained on a laboratory jet.

Transmissibility functions (ratio of response acceleration power spectrum to input force spectrum) were measured between the lower guide bearing and the eroded a reason the blades by a reciprocity instrumented hammer impact method. An average response was established by measuring the response of 5 blades on the prototype and 4 on the model with the runner under water in both cases. The 0-25 and 0 -100 kHz frequency ranges were utilized respectively on the prototype and on the model. The useable frequency ranges where adequate coherence (>.8) between the response and the excitation were realized were 0-16kHz on the prototype and 0-80kHz on the model. The need to measure the transmissibility function with the runners in water was confirmed by the significant differences between values measured with the runners in air or in water. On the prototype, smaller values were seen with the runner in water, as anticipated, due to the coupling of vibration energy from the runner blades to the water while the opposite was true on the model where increased coupling from the lower guide bearing to the runner is achieved with the presence of water in the runner crown to bearing gap.

Forces were inferred on the prototype in the .8 to 11.296 kHz as had been done but this time with the transmissibility function measured with the runner underwater. This produces substantially higher force levels than estimated in the previous measurement campaign. On the model, the 20 to 35 kHz range was utilized because of interfering high frequency vibration at the lower guide bearing generated by a 90° 1 to 1 gearbox linking the model shaft to the generator. This underestimates the real force values by a factor of at least 2 or 4. The presence of these undesirable vibrations is inconsistent with reciprocity measurement approach and will be eliminated in the future tests. The results of the prototype inferred force calculations are summarized in figure 3 with maximum values observed on the prototype at 90% guide vane opening with high downstream levels. Figure 4 shows the correlation with volume pitting rate. The highest level on the model is also obtained at 90%

guide vane opening but the inferred forces per unit area on the model are four to five times much weaker than on the prototype.

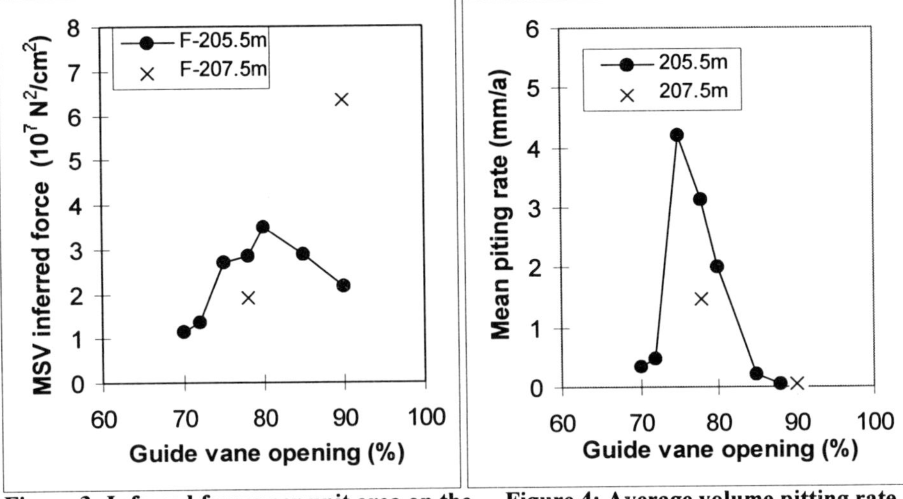

Figure 3: Inferred forces per unit area on the proto (lower guide bearing acceleration)

Figure 4: Average volume pitting rate on 4 disks for same tests on the proto.

Cavitation characteristics on the model however appear to differ slightly from those observed on the prototype. The high frequency acceleration amplitude modulation spectra at the lower guide bearing and in the crown of the runner show that impacts occur at the guide vane passing frequency as on the prototype but modulation components of similar amplitude are also present at the blade passing frequency. This confirms that flow conditions are not fully homologous to those of the prototype as discussed in Part 2.

3.3. PREDICTION USING ELECTROCHEMICAL PROBE DECER

This method based on electrochemical effects during cavitation erosion can give a good representation of cavitation aggressiveness through the activation current delivered by the titanium electrochemical probe which is directly proportional to the actual erosion rate [8]. This localized on-line measurement on both prototype and model is a complement to pitting results to provide the influence of operating parameters on cavitation aggressiveness. On the prototype, this method gives similar results than the others methods, with maximum erosion near optimum opening The maximum current measured is in the range of 3 μA for a 7.5cm^2 of a grade 2 Titanium. (This corresponds to 20 mm/year erosion on Ti grade 2 according to laboratory jet tests [8]). On the model, the maximum current level detected on Ti49 is 0.05μA on a surface of 0.32 cm^2. This is very low and correspond to 0.16 mm/year of Titanium erosion (according to laboratory jet tests [8]). Since the resistance of Ti 49 is ten times lower than Ti grade 2, it means that cavitation erosion found on the model seems to be about 1000 less than on the prototype.

3.4. PREDICTION USING CAVITATION EROSIVE POWER

It has been set previously (Avellan *et al*) that the cavitation erosive power could be evaluated from energy dissipated by collapses of transient cavities and their production rate. This power can be expressed as :

$$P_{er} = \frac{1}{2} K \cdot \rho \cdot \left(c_{p\max} + \sigma\right) \cdot C_{ref}^3 \cdot St \cdot \frac{V_c}{L_c}$$

where $\mathbf{c_{p\ max}}$ is the maximum pressure coefficient in the recovering pressure region downstream the main cavity, σ the cavitation coefficient, $\mathbf{C_{ref}}$ a velocity reference, **St** the Strouhal number corresponding to the production rate of the transient cavities, $\mathbf{L_c}$ the main cavity length, $\mathbf{V_c}$ the transient cavities volume and **K** a scaling constant. This parameter can be calculated on the model using visualisation to determine $\mathbf{V_c}$, $\mathbf{L_c}$ and **St**, and transposed from the model to the prototype.

Hydraulic similitude and sigma similitude will ensure that $\mathbf{c_{p\ max}}$, σ and **St** are the same on the model and on the prototype. Then it can be assumed that $\mathbf{L_c}$ is proportional to the reference length **L**, and that $\mathbf{C_{ref}} \propto N \cdot L$ (**N** : rotational speed)

As for $\mathbf{V_c}$, two hypothesis can be proposed:

- if it is assumed that $\mathbf{V_c}$ is proportional to $\mathbf{L^3}$, then the ratio between erosive power between model and prototype can be written:

$$\frac{P_{proto}}{P_{model}} = \left(\frac{C_{ref\ proto}}{C_{ref\ model}}\right)^3 \cdot \frac{V_{c\ proto}}{V_{c\ model}} \cdot \frac{L_{c\ mode}}{L_{c\ protol}} = \left(\frac{N_{proto}}{N_{model}} \cdot \frac{L_{proto}}{L_{model}}\right)^3 \cdot \left(\frac{L_{proto}}{L_{model}}\right)^3 \cdot \left(\frac{L_{proto}}{L_{model}}\right)^{-1}$$

$$r = \frac{P_{proto}}{P_{model}} = \left(\frac{N_{proto}}{N_{model}}\right)^3 \cdot \left(\frac{L_{proto}}{L_{model}}\right)^5 = n^3 \cdot \lambda^5$$

(notice: this is equal to the classical ratio on machine powers)

Since $\mathbf{n} = 0.1125$and $\lambda = 14.66$ then in this case: $\mathbf{r} = 964$

- if the following law, established in a previous statistical study [10] is applied:

$$V_{cr} = k \cdot L_c^{2.6} \cdot C_{ref}^{-0.4}$$

then the ratio becomes:

$$r = \frac{P_{proto}}{P_{model}} = \left(\frac{N_{proto}}{N_{model}}\right)^{2.6} \cdot \left(\frac{L_{proto}}{L_{model}}\right)^{4.2} = n^{2.6} \cdot \lambda^{4.2}$$

then in this case: $\mathbf{r} = 269$

These two results emphasises clearly that, even if some uncertainty remains on the transposition laws, the cavitation intensity is much higher on the prototype than on the model. Then, to predict erosion on the prototype from transposed power, experimental correlations from reference data have to be used.

4. Conclusions

The proposed procedures to predict cavitation erosion from model tests take advantage from a campaign on a large Francis Turbine and its model.
Particular methods are set from each instrumentation type: pitting, vibration, electrochemical probe and erosive power. Results allow to confront them to actual data and to assess the hypothesis on transposition of cavitation aggressiveness. At the present time, since the programme is not completed, only partial conclusions can be drawn. Obviously the four methods proposed are able to quantify cavitation aggressiveness on the model. They all clearly show that cavitation aggressiveness is much smaller on the model than on the prototype, and this is due to less energetic collapses, due to the reduction of both pressures and sizes of impacts.
As for a definitive assessment of transposition and prediction processes, the completion of the programme is needed for a thorough synthesis.

References:

[1] E. Laperrousaz et al, 1994 "Prediction cavitation erosion in Francis turbines on the basis of scale model testing", *Proc. 17th IARH Symposium*, Beijing, China.
[2]Y. Lecoffre, P. Grison;, J.M. Michel , 1986 "Prevision de l'érosion de cavitation pour les turbomachines " *Proc. 14th IARH Symposium*, Montreal, Canada.
[3] J.M. Dorey, R. Simoneau ,P. Bourdon, M. Farhat, F. Avellan 1994 "Quantification of cavitation aggressiveness in three different dvices using accelerometer, DECER, and pitting measurements" , *Proc.Second International Symposium on Cavitation*, Tokyo, Japan.
[4] P. Bourdon, R. Simoneau, J.M. Dorey, 1994 "Accelerometer and pit counting detection of cavitation erosion on a laboratory jet and a large Francis tu rbine" *Proc. 17th IARH Symposium*, Beijing, China.
[5] R. Fortes-Patella, J.L. Reboud, J.M. Dorey, 1991 "Simulation of cavitation impact damageon an elastoplastic solid", *Proc. ASME Cavitation and Multiphase Flow Forum*, Portland, USA.
[6] R. Fortes-Patella ,J.L. Reboud , 1992 "Analysis of cavitation erosion by numerical investigation of solid damage" *Proc. 16th IARH Symposium* , Sao-Paulo, Brasil.
[7] R. Fortes-Patella ,J.L. Reboud, 1995 "A new approach to evaluate cavitation erosion power" *Proc. International Symposium on Cavitation*, Deauville, France.
[8] R. Simoneau ,P. Bourdon, M. Farhat, F. Avellan, J.M. Dorey, 1993 "Cavitation erosion, impact intensity and pit size distribution of jet and vortex cavitation", *Proc. ASME annual winter meeting, Bubble noise and cavitation erosion in fluid systems* , New Orleans, USA.
[9] R. Simoneau, 1995 "Cavitation pit counting and steady state erosion rate" *Proc. International Symposium on Cavitation*, Deauville, France.
[10] F. Pereira, Ph. Dupont, F. Avellan, 1995 "A statistical approach to the study of transient erosive cavities on a 2D profile" *Proc. ASME annual meeting* , South Carolina, USA.

CAVITATION EROSION PREDICTION ON FRANCIS TURBINES. PART 2 MODEL TESTS AND FLOW ANALYSIS

PH. DUPONT, J.-F. CARON, F. AVELLAN
EPFL, IMHEF/LMH
33 Av. de Cour, CH-1007 Lausanne, Switzerland
P. BOURDON, P. LAVIGNE, M. FARHAT, R. SIMONEAU
Hydro-Québec
1800 Boul. Lionel Boulet, Varennes, Québec, Canada J3X 1S1
J.-M. DOREY, A. ARCHER
Electricité de France, DER
6, quai Watier, F-78401 Chatou, France
E. LAPERROUSAZ
Electricité de France, CNEH
Savoie-Technolac, B.P. 26, F-73370 Le Bourget-le-Lac, France
M. COUSTON
GEC-Alsthom, Neyrpic
75, rue Général-Mangin, B.P. 75, F-38041 Grenoble, France

ABSTRACT Different measurement techniques have been used to detect cavitation on a Francis turbine model. The results are compared to those obtained on the prototype and presented in the first of this series of articles. The runner model used for that study is build on the basis of a geometrical recovery of one of most eroded blade of the prototype. The results of the different measurements are presented and commented by comparison with prototype measurements. This comparison leads to a proposal of the physics which should be involved in transposition laws for the prediction of prototype erosion from cavitation model tests. The consequences of such scaling laws, as well as their application to the prototype and model results, are part of the third facet of this work.

1. INTRODUCTION

The problem of cavitation erosion on Francis turbines leads IMHEF, Hydro-Québec, Electricité de France (EDF) and GEC Alsthom-Neyrpic to be involved together in a large and ambitious research program on this topic. Preliminary measurements using various techniques have been performed on the prototype of a 266 MW Francis turbine and are fully described in Part I of this work.

The present paper is concerned with the model tests performed at IMHEF. A runner model is manufactured using the data provided by the geometry recovery of one of the prototype blades [1]. The model measurement campaign, similar to the prototype one, is

E. Cabrera et al. (eds.), Hydraulic Machinery and Cavitation, 574–583.

described. As for the prototype measurements, the model runner is equipped with different transducers in order to follow the cavitation development. The results of hydraulic characteristics, vibratory monitoring, pit counting, pressure pulses and electro-chemical probes measurements are reported. Additionally, a flow analysis is presented using both Euler and Navier-Stokes calculation codes.

One of the goal of the present study is to define, on the one hand, the best way to perform cavitation tests on model to be as close as possible to the cavitation behavior of the prototype, and one the second hand, the transposition laws to apply to predict cavitation erosion on prototype from the model tests and the way to measure it. The measurements show important differences between model and prototype cavitation development. These differences are commented and analyzed. This leads to propose possible explanations about the physics involved in these scale effects.

In the third part of this series of articles, the similitude and the differences between results of both campaigns are used to give recommendations for the prediction of cavitation erosion from model tests.

2. GEOMETRY RECOVERY AND DESIGN OF THE MODEL RUNNER

In order to reproduce the cavitation pattern of the prototype on the model, one of the most eroded blades, namely blade number four, as well as the blade to blade channel corresponding to its suction side have been measured on site. Three techniques were used and compared with each other [1]: the classical template technique, a 3D laser interferometer and a portable 3D digitizer. Because of slight discrepancies between the blade and the blade to blade channel results, it was decided to manufacture the model upon the geometric data of the selected blade. This choice would have lead to bigger outlet openings on the model runner then the measured ones on the prototype. It has been so decided to slightly extend the model blades in order to keep these openings of the runner identical to the prototype. From this result, a model of the runner with a 300 mm diameter has been machined using a five axes CNC machine by GEC Alsthom-Neyrpic. To control the possible differences of the flow between model and prototype, flow calculations on the two geometries have been performed.

3. FLOW CALCULATION

The analysis of the flow in the model and prototype runner is done using both Euler and Navier-Stokes calculation codes. The geometry recovery of blade four is first used to generate the meshing of the prototype runner. The operating condition followed corresponds to the best efficiency point. The comparison of the results of the different calculation codes used with this prototype geometry is given in left part of Figure 1.

This flow analysis shows that the cavitation development seems to be due to a local bulge of the blade, close to the band and at the third of the blade chord length. This local bulge creates a local underpressure located just upstream of the eroded area of the prototype. Moreover, changing the blade to band fillet size of the tested model runner at suction side has shown to change the cavitation inception. It is so suspected that both corner flow along the blade to band fillet together with this local underpressure is responsible for the unexpected erosion observed on the prototype. The influence of the model blade elongation is investigated comparing the pressure distribution calculated for both prototype and model blade geometry. The right part of Figure 1 shows the elongation to slightly increase the pressure level at runner outlet close to the band.

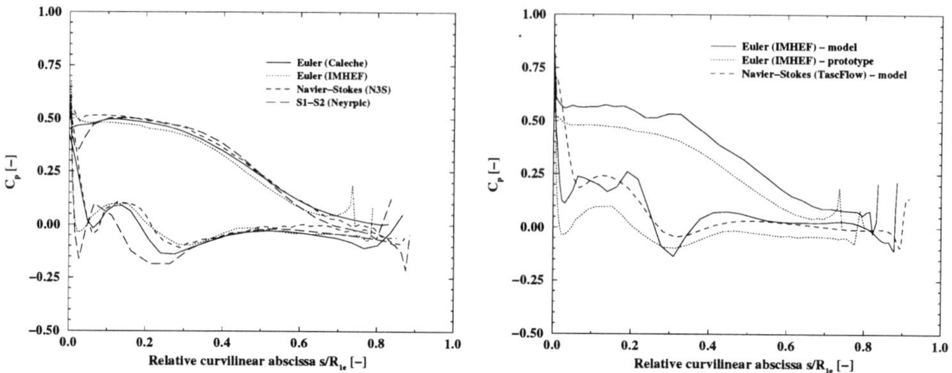

Figure 1: Pressure coefficient distribution along the blade close to the band.
On left: Prototype blade #4 geometry - Comparison of different codes
On right: Comparison between prototype and model geometry results

4. MODEL TESTS SET-UP

The model tests are performed on the high performance research test rig of IMHEF (PF III). This closed loop facility allows to reach flow rates up to 1.5 cubic meters per second, with a maximum net head H_n of 60 meters. A first series of tests are performed using a set of blades without any instrumentation to compare the hydraulic performances of the model with the prototype measurements.

After this first step, 6 of the 13 blades are replaced with instrumented ones. Two blades (2 & 8) are equipped each with five miniature static and dynamic pressure transducers, with a useful 15 to 20 kHz upper frequency determined by a preliminary dynamic calibration. These sensors allow to measure the pressure fluctuations acting on the blade due to cavitation. Two other blades (4 & 10) are instrumented with a total of five DECER electro-chemical cavitation erosion probes, which provide a real time monitoring of the erosion rate. Finally, two blades (6 & 12) are equipped with samples of soft material, namely pure copper, for the pit counting experiment. The locations of the first four transducers (pressure, electro-chemical or samples), are identical to those used on the prototype. Moreover, due to the extra length of the model blades, a fifth location is available downstream of the others. These locations correspond to the erosion area found on the prototype, close to the trailing edge of the blades and near to the band to blade fillet. To prevent any disturbance of the flow due to the mounting of the transducers, it was decided to alternate instrumented and non instrumented blades.

Figure 2: Conditioning electronics

The hub of the runner is especially designed to allow a sealed volume large enough to install the conditioning electronics connected to the onboard transducers (See Figure 2).

This electronics has an amplifier with a remotely variable gain and a multiplexer that allows the scanning of all the transducers, with two simultaneous outputs. Voltage signals are converted to frequency to avoid electromagnetic disturbances. They are brought out from the rotating part through a slip ring collector and finally reverted to voltage signals which are digitized on high speed waveform recorders. In this same sealed volume is placed an accelerometer, fixed to the runner blade ring. This accelerometer replaces for the model tests the one placed on the blade for the prototype tests.

As for the prototype campaign, in addition to the onboard sensors, four accelerometers are used to monitor the cavitation impacts at the lower guide bearing; another one is fixed on the upstream flange of the draft tube cone. The angular position of the last accelerometer is perpendicular to the meridian plane including a dynamic pressure transducer, fixed in the wall of the draft tube cone. The data acquisition systems are essentially the same as those utilized in the prototype program described in part 1 of this series of papers with the addition of a Metrum RSR 512 16 channel digital data recorder.

5. TEST CONDITIONS

Two operating points are followed during the model tests, corresponding to 78% and 90% of the maximum guide vanes opening. To ensure identical runner inlet flow angles on both model and prototype, model guide vanes opening and discharge coefficient $\varphi_{\bar{1}e}$ are chosen to be equal to the prototype campaign values. Two different model heads, 13 and 51 meters, corresponding respectively to rotational speeds of 800 and 1600 RPM, are investigated to check the influence of the head on the cavitation development. For these two heads, sigma coefficient investigations are done to detect cavitation inception. This inception is determined according to the first pressure peaks detection, corresponding to the first cavity collapses.

Table 1: Operating conditions for the model tests.

TEST	MA01	MAM1	MAM2	MAM3
Guide vanes opening [%]	78	78	78	90
$\sigma_{\bar{1}e}$ [-]	0.20	0.08	0.07	0.10

To check the influence of the water nuclei content on the cavitation development, tests were done with and without nuclei injection, using the IMHEF nuclei injection system [2] . If overall cavitation performances are affected by this nuclei content, no influence has been observed on the specific cavitation pattern developing along the blade to band fillet. These specific results are presented in an other paper [3]. To avoid any water quality effects, all measurements are performed after a two hours degassing period.

The operating conditions used for the pit counting and vibrations tests, given on Table 1, are similar to the prototype campaign. The sigma values on model are chosen to obtain cavity development in agreement with erosion observed on prototype.

6. MEASUREMENTS RESULTS

6.1 HYDRAULIC CHARACTERISTICS

6.1.1 Efficiency tests

The model results obtained compares very well with those previously obtained with the original model tested by Neyrpic. These hydraulic characteristics, transposed at prototype scale, corresponds quite well to the on site measurements even if two local maximum efficiency are observed on the latest.

6.1.2 Cavitation tests

The $\sigma-\eta$ cavitation curves have been performed for the model, and compared to the prototype cavitation behavior. Figure 3 shows the model $\sigma-\eta$ cavitation curve for a 90% guide vanes opening, and the two sigma values corresponding to the downstream levels achieved during the prototype campaign. The cavitation development on the model corresponds to the location of the erosion on the prototype and confirms the local underpressure revealed by the calculation. This seems to indicate that, even if the model is built on the basis of only one blade geometry recovery, the cavitation behavior of the prototype is well reproduced. On the other hand, the comparison of the cavitation index for a given cavity extent, observed on the model and estimated on the prototype from the pitting tests, shows an important deviation. For an equivalent cavitation extent, the corresponding cavitation index is lower on the model than on the prototype. This last point needs a deeper analysis, in the light of a local cavitation coefficient notion presented here latter. It has also been noticed a influence of the fillet size between suction side blade and band on the cavitation inception. Decreasing the size of these fillets, cavitation inception index is notably increased.

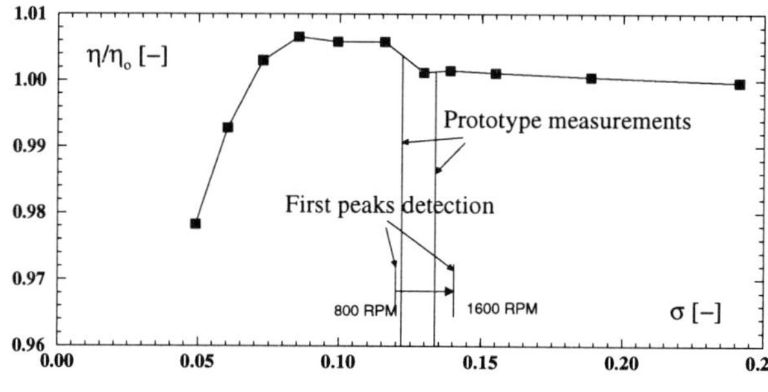

Figure 3: $\sigma-\eta$ cavitation curve at α=90%.

6.2 VIBRATORY MONITORING

6.2.1 Dynamic calibration

To identify an average transmissibility function (ratio of response acceleration to input force autospectra) between the blades and 2 radialy mounted accelerometers at the lower guide bearing, reciprocity instrumented hammer impact techniques [4] are applied between each of these two monitoring points to excite 5 Kistler miniature accelerometers mounted on runner blades. Each measurement sequence consists of 32 radialy applied hammer impacts at the guide bearing. A frequency range of 100 kHz with 64 Hz of resolution is used. The quality of the measurements is excellent as indicated by the coherence function between the input force spectrum and the underwater blade response acceleration spectrum shown in Figure 4. The transmissibility function is then used in a reciprocity mode to infer forces on the blades from the accelerations measured at the lower guide bearing during the tests.

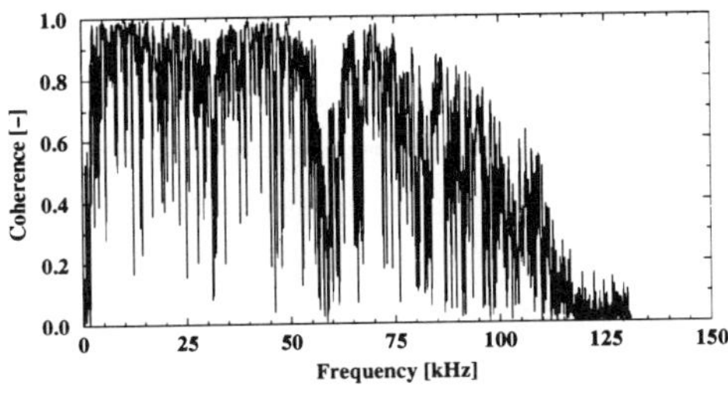

Figure 4: Coherence function between input force and underwater blade response acceleration spectrums.

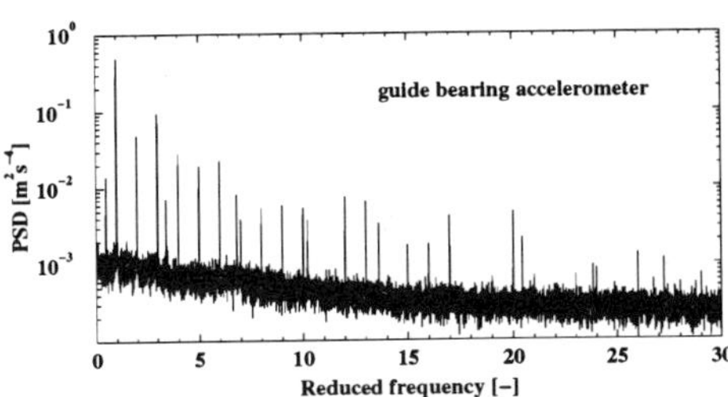

Figure 5: Power spectral density of the 20-35 kHz accelerations signal envelope at the guide bearing.

A 20 to 35 kHz frequency range is utilized to avoid high frequency interference caused by a 90 degree translating gearbox and to minimize artificial inferred force amplitude caused by the lower resonance frequency of the guide bearing accelerometers in comparison to that of those used on the blades.

6.2.2 *Measurements*

In the test cases, 128 blocs of 4096 points are acquired with Hanning weighting and their autospectra averaged to minimize statistical measurement errors on the random data. So, a 90% confidence interval ranging better than +/- 1 dB is achieved at each frequency. Forces acting on the blades are inferred from these autospectra with the transmissibility functions. Forces inferred from the lower guide bearing monitoring accelerometer in the 20 to 35 kHz frequency range are listed in the following table.

Table 2: Inferred forces

TEST	MA01	MAM1	MAM2	MAM2	MAM3
Inferred forces (N^2)	533.6	2130.7	2101.3	1843.8	5222.9

High frequency acceleration amplitude modulation studies are also performed on data recorded digitally on the Metrum data recorder. The modulation of the 20-35 kHz acceleration at the different monitoring points is studied by filtering the data with a band pass elliptical filter prior to the envelope detection.

Power spectral densities with Hanning weighting are evaluated with a frequency resolution of 0.0625 Hz. The frequency axis is normalized to orders of rotation. The high frequency vibration envelopes are seen to be mainly modulated by two sets of frequencies centered around the guide vane passing frequency 20X and the blade passing frequency 13X. The power at these frequencies +/- 1 and 2 times the rotation frequency are extracted. The power is integrated over a 0.4 Hz bandwidth around each of these frequencies. An example of these PSD is given in Figure 5 showing the forcing effects of these frequencies on the cavitation dynamics.

6.3 PIT COUNTING MEASUREMENTS

Three pitting tests (MAM1-2-3) have been performed at the highest head, to ensure the highest pitting rate. These tests are done using Al 59 samples mounted on two different blades. The pitting results are shown on Figure 6 for MAM3 test and blade #6. The exact flush mounting of the samples is rather difficult due to the blades size. A misalignment has induced side effects corresponding to highest local pitting related to the leading edge of samples, as shown on Figure 6.

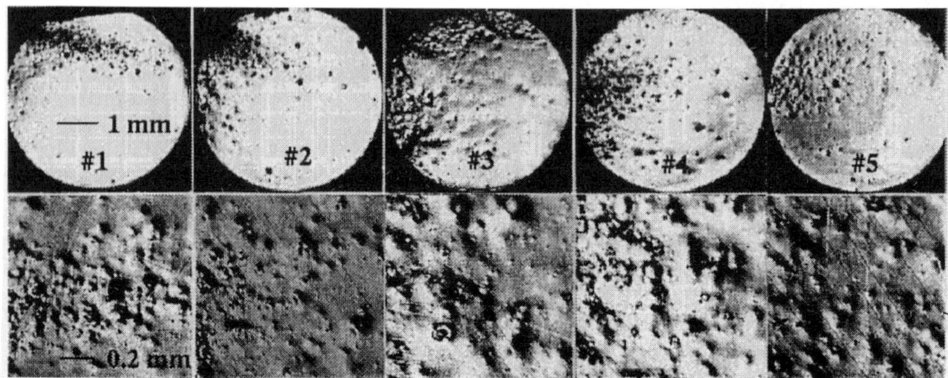

Figure 6: Pitting results on blade #6 for test MAM3 (Flow direction from left to right)

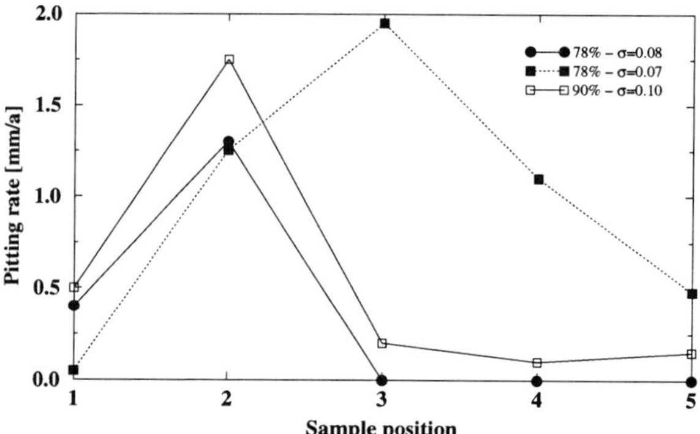

Figure 7: Volume pitting rate for MAM1-2-3 tests using Al 59 samples.

In spite of that, the mean radius of the biggest pits is 0.2 mm, which corresponds to about the tenth of the pits obtained on prototype. This is coherent with the geometric scale ratio equal to 14.67. The volume pitting rates are compared for the three tests on Figure 7. As for the prototype measurements, the maximum pitting is moving closer to the trailing edge of the blade when decreasing the sigma value. As well, a equivalent pitting rate is obtained at the same location and for a highest sigma value for a guide vane opening of 90% compared with 78%.

6.4 PRESSURE PULSES MEASUREMENT

Pressure fluctuations due to cavitation development are measured on two blades on five locations, indicated by dots on Figure 8. These unsteady pressure transducers are able to detect high frequency pressure variations as shock waves due to cavity collapses. This measurement technique is shown to be very efficient to detect the beginning of cavitation. On Figure 8 the incipient cavitation is detect at an opening of 90% for a σ value of 0.132. Using visual observations, this limit is set for a σ value of 0.120.

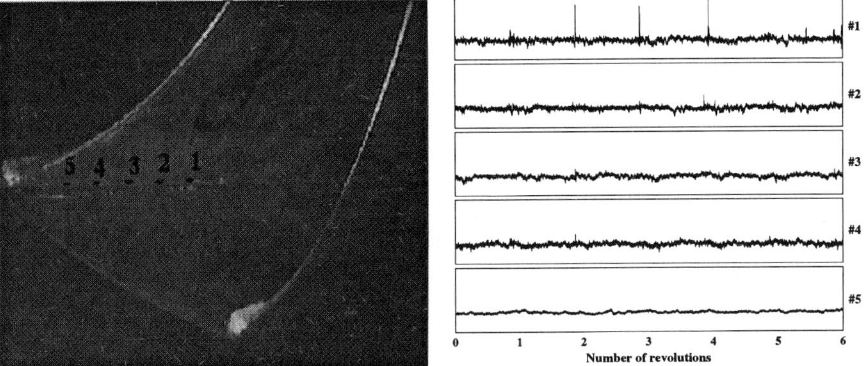

Figure 8: Cavitation pattern visualization and instationnary pressure time signals at five different locations on blade #8 (Operating conditions: α=90%, σ=0.132, H_n=50 m).

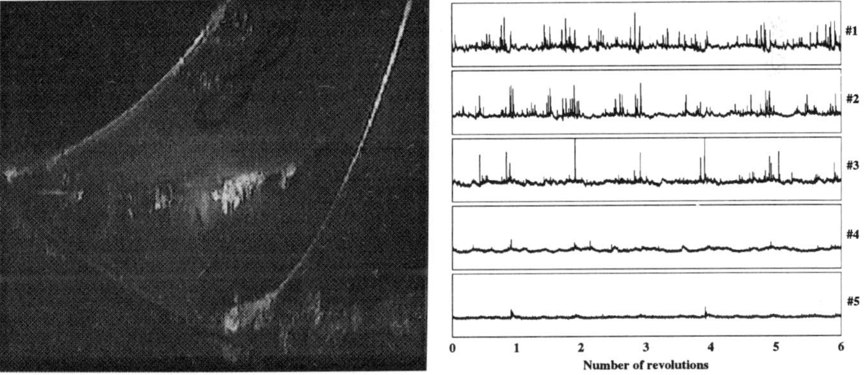

Figure 9: Cavitation pattern visualization and instationnary pressure time signals at five different locations on blade #8 (Operating conditions: α=90%, σ=0.100, H_n=50 m).

For developed cavitation, pressure measurements gives good information about the modulation of the collapses. On Figure 9 for a σ value of 0.100 at 90% of guide vane opening, a displacement of the collapses closer to the trailing edge is observed. A larger number of small collapses occurs close to pressure transducers #1 and #2, while fewer but larger collapses occur close to pressure transducer #3. This shows the biggest vapor cavities to be convected farer downstream to the local minimum pressure area corresponding to there generation point. It can be observed on Figure 9 the strong modulation of the biggest peaks at each revolution. This will lead for a static observer to vibrations frequency modulation at blade passage frequency as shown by the vibratory monitoring. On the contrary, the guide vane passage frequency is more difficult to detect on pressure fluctuations, may be due to the more intermittent characteristics of this modulation.

6.5 ELECTRO-CHEMICAL PROBES MEASUREMENTS

Because of the small size of the electro-chemical probes and the weakness of the cavity collapses on the model, most of the followed operating points are under the detection threshold of this technique. Only the most erosive situations lead to measurable values corresponding to erosion rates more then three orders of magnitude weaker then those observed on the prototype.

7. ANALYSIS

In order to compare the cavitation development on the model and on the prototype, a local cavitation factor χ_E is used instead of the usual Thoma number σ. This local cavitation factor is defined as:

$$\chi_E = \frac{p_{\bar{1}x} - p_v}{\rho E}$$

where ρ is the mass density, E is the energy, p_v corresponds to the vapor pressure and $p_{\bar{1}x}$ is the local static pressure at the runner outlet.

Using the well known Bernoulli equation between the runner outlet and the draft tube outlet, and taking into account the losses in the draft tube, one can written:

$$\chi_E = \sigma + \frac{1}{Fr^2} \frac{Z_{ref} - Z_{\bar{1}x}}{D_{\bar{1}}} - \frac{C_{\bar{1}x}^2}{2E} + \frac{E_{rd}}{E}$$

where σ is the Thoma number, Fr the Froude number and E_{rd}/E the relative energetic losses in the draft tube. Knowing the flow rate coefficient $\varphi_{\bar{1}}$ at the whirl free operating point (φ_o), the energy coefficient at the runner outlet $\psi_{\bar{1}}$, these relative losses can be expressed as a function of the flow coefficients:

$$\frac{E_{rd}}{E} = k_{rd} \frac{\varphi_{\bar{1}}^2}{\psi_{\bar{1}}} + k_{dcin} \left[\frac{\varphi_{\bar{1}}}{\varphi_o} - 1 \right]^2 \frac{1}{\psi_{\bar{1}}}$$

where k_{rd} is the viscous loss coefficient and k_{dcin} the residual kinetic energy loss coefficient. Using a model geometrically homologue to the prototype and investigating corresponding flow conditions, the differences in cavitation behavior between model and prototype are due to both Froude and Reynolds effects. The influence of the head on the incipient cavitation is investigated using the pressure transducers. The incipient cavitation index for a net head of 13 and 51 meters, corresponding to a rotational velocity of respectively 800 and 1600 RPM, is indicated on Figure 3. One can noticed that if, at the smallest model head, the first pressure peaks are detected on the model at a σ value of 0.120, lower then the lowest prototype measurement, the incipient cavitation at the highest model head corresponds to a σ value of 0.140, higher then the highest prototype measurement. As the influence of the Froude term is small, this difference is explained by the modification of the relative losses between the two model heads.

8. DISCUSSION AND CONCLUSIONS

The model, build on the basis of one of the most eroded prototype blade (#4), shows a cavitation development similar to the one suspected on prototype, but occurring for lower σ values. This difference is attributed to two main reasons. The first one is the geometry used to build the model. The cavitation behavior at the suction side of a given blade is not only due to its eigen geometry but also to the blade to blade channel geometry. In that meaning, the model is closer to the cavitation developing on the previous blade (#3), which is one of the less eroded blade on the prototype. The second reason is the scale effects between model and prototype. As shown by the measurements done at 800 and 1600 RPM, Reynolds effects on incipient cavitation are observed. The usage of a local cavitation coefficient leads to much better comparison, but needs to be able to evaluate the different losses on both model and prototype.

The difficulty to compare cavitation erosion on model and prototype is increased by the difficulty to evaluate it on a model. One of the goal of this study is to test different measurement techniques to be used on model to predict erosion on the prototype. The vibratory technique is shown to be well suited to follow developed cavitation, and can be used to quantify erosion rate on model. It is with no doubt the easiest technique to be applied on both model and prototype, because it does not required onboard equipment. It seems on the contrary to be not sensitive enough to detect cavitation inception. The pitting technique on soft material samples gives good information about the localization and the rate of the erosion but it is very sensitive to local geometry defaults. Pressure peaks detection using unsteady pressure transducers is shown to be efficient to detect incipient cavitation as well as the modulation of the collapses. It is on the contrary difficult to be used to quantify the erosion because it is difficult to relate the peaks amplitude to an energy, the location of the collapse being unknown. Finally, electro-chemical probes technique, which is able to give on line erosion rate on a prototype, needs to be more sensitive to be used on model.

In closing, the authors would like to thank IMHEF technical staff for their contributions to the success of this project. They are also indebted to Jean-François Combes from EDF/DER, for the Navier-Stokes (N3S) and Euler (Calèche) flow calculations.

9. REFERENCES

[1] Dumoulin, Ch., Avellan, F., Henry, P. [et al.], « Geometry recovery of a Francis runner prototype at site », Proceedings of the 17th IAHR Symposium, section on hydraulic machinery and cavitation, 15-19 September, 1994, Beijing, China

[2] Brand, C., Avellan, F., « The IMHEF system for cavitation nuclei injection », Proceedings of the 16th IAHR Symposium, section on hydraulic machinery and cavitation, 14-18 September, 1992, Sao Paolo, Brasil.

[3] Arn, Ch., Dupont, Ph., Avellan, F., « Efficiency alteration of Francis turbine by travelling bubble cavitation: Experimental and theoretical study », Proceedings of the 18th IAHR Symposium, section on hydraulic machinery and cavitation, 1996, Valencia, Spain.

[4] Bourdon, P., Simoneau, R., Avellan, F., « Erosion Vibratory Fingerprint of Leading Edge Cavitation for a NACA Profile and a Francis Model and Prototype Hydroturbine », Proceedings of the ASME Symposium on Bubble Noise and Cavitation Erosion in Fluid Systems, pp. 51-67, New-Orleans, Dec. 1993

IMPACT OF VAPOUR PRODUCTION AND CAVITY DYNAMICS ON THE ESTIMATION OF THERMAL EFFECTS IN CAVITATION

Daniel H. FRUMAN
Ecole Nationale Supérieure de Techniques Avancées
Groupe Phénomènes d'Interface (GPI)
91125 Palaiseau Cedex, France

Jean-Luc REBOUD, Benoît STUTZ
Laboratoire des Ecoulements Géophysiques et Industriels
Institut de Mécanique de Grenoble (LEGI - IMG)
B.P. 53, 38041 Grenoble Cedex 9, France

Abstract

Vapour production through cavitation extracts heat from the fluid surrounding the cavity and creates a temperature difference between liquid and vapour. This thermal effect is particularly significant in cryogenic liquids. Estimates of the temperature difference can be made provided : a) the rate of vapour production required to sustain a given cavity, and b) an appropriate model for the heat exchange at the interface of the cavity are known. The vapour production is usually estimated by assuming that it is equal to the non condensable gas flow rates necessary to sustain ventilated cavities of equal geometry. It has been shown that experimental results previously obtained can be quite satisfactorily predicted if the interface is assimilated to a rough flat plate through which the amount of heat necessary to generate the vapour is being fed. In order to obtain data for cavities developed over walls whose geometry and pressure gradient are analogous to those of turbopump inducers, and to achieve a better precision, tests were conducted in a specially designed cavitation loop operating with Freon 114. It is shown that the experimental results are well predicted by the proposed model.

1. Introduction

Vapour production through cavitation extracts heat from the fluid surrounding the cavity and creates a temperature difference between liquid and vapour. This thermal effect is particularly significant in cryogenic liquids such as those utilised in the turbopumps of the European space launcher Ariane and preferentially occurs in the inducer stage. Estimates of this thermal effect can be made provided : a) the rate of vapour production required to sustain a given cavity, and b) an appropriate model for the heat exchange at the interface of the cavity are known. The vapour production is usually estimated by assuming that it is equal to the non condensable gas flow rates necessary to sustain ventilated cavities of equal geometry. However, this globalised vapour production does not tell much about how it is distributed along the interface and if vaporisation is indeed

E. Cabrera et al. (eds.), Hydraulic Machinery and Cavitation, 584–593.

restricted to the interface. To predict the heat exchange at the interface an appropriate model has to be selected and its suitability checked against existing experimental results. Previous estimations (Fruman *et al.*, 1991, Fruman and Beuzelin, 1992) have shown that the experimental results obtained by Hord (1973) can be quite satisfactorily predicted if the interface is assimilated to a rough flat plate through which the amount of heat necessary to generate the vapour is being fed. For that purpose, the vapour is assumed to be uniformly distributed along the interface (rough wall) and the relative roughness has to be properly tuned.

In order to obtain data for cavities developed over walls whose geometry and pressure gradient are analogous to those of turbopump inducers, and to achieve a better precision, tests were conducted in a specially designed cavitation loop operating with Freon 114 ($C_2Cl_2F_4$), developed with the support of the Agence Française de l'Espace (CNES) and the Société Européenne de Propulsion (SEP) and operated by the Centre de Recherches et d'Essais de Machines Hydrauliques de Grenoble (CREMHyG). This investigation was coupled with a research program, conducted at GPI and LEGI and sponsored by the Agence Française de l'Espace (CNES), which consisted of determining:

- the morphology and dynamics of vapour cavities,
- the rate of growth of vapour cavities (Larrarte *et al.*, 1995),
- the void fraction and the two phase flow velocities within the vapour cavities (Stutz and Reboud, 1994),
- the air flow rate required to sustain a ventilated cavity of equal length,
- the morphology and dynamics of ventilated cavities (Larrarte *et al.*, 1995).

Nomenclature

$\boldsymbol{B}$	non dimensional temperature difference
B	cavity width
C_Q	flow coefficient
c_p	specific heat at constant pressure
C_f	flat plate friction coefficient
L	latent heat of vaporisation
ℓ	cavity length
p_{ref}	reference upstream pressure in the Venturi test section
P	Prandtl number
Q	vapour production or air injection flow rate
$Re_{x,\ell}$	Reynolds number computed with the distance $x(\ell)$
T_{ref}	reference temperature in the Venturi test section
T_p	wall (interface) temperature
T_∞	free stream temperature
u_b	velocity at the edge of the viscous boundary layer
U_∞	free stream velocity for a flat plate or velocity on the cavity interface
V_{ref}	reference velocity in the Venturi test section
V_g	mean velocity of the vapour (air) uniformly distributed over the cavity interface
x	distance along the flat plate (interface)
ΔT	temperature difference ($T_\infty - T_p$, $T_{ref} - T_p$)
ΔT_{visc}	increase of temperature due to viscous dissipation
α	void fraction
ε	roughness of the flat plate (interface)
φ_p	heat flux transferred to the flat plate (interface)
ρ	liquid specific mass
ρ_v	vapour specific mass
σ	cavitation number

The results have shown that, for equal length, the vapour and ventilated cavities have a different morphology and dynamic behaviour. In spite of these differences, the estimation of the mean vapour production rate can be made, with a good degree of accuracy, by using the air flow rate necessary to sustain the ventilated cavities. The above mentioned model (Fruman *et al.*, 1991, Fruman and Beuzelin, 1992) was applied to predict the thermal effect for cavities obtained in experiments carried out at CREMHyG in the Freon 114 test loop. The predicted and measured difference between the temperature of the circulating liquid and the temperature in the cavity (at the interface) compare very favourably.

2. Experimental

The Freon 114 experimental facility (called Plateforme d'Essais de Cavitation de Liquides à Effet Thermodynamique (PECLET)) is a closed loop operating with a reference pressure, obtained by pressurising, up to 35 bars, a reservoir with gas Nitrogen. To prevent gas dissolution into the circulating liquid, a flexible membrane separates the liquid and gas phases. The loop is fitted with a test section having the shape of a two dimensional Venturi, designed in such a way that the pressure distribution is analogous to the one existing on the upper surface of an inducer blade (Kueny *et al.*, 1991). The throat section of the Venturi is 44 mm width and 43.7 mm height and the velocity can be varied between 16 and 45 m/s. Temperature can be varied and controlled from 20 to 40°C. Under these experimental conditions, it is possible to achieve cavities having a maximum length of 12 cm before choking. Temperatures on the Venturi wall next to the cavity were measured using five micro-thermocouples situated at 3, 20, 38, 75 and 115 mm from the throat. One reference temperature was measured 20 cm upstream. Mean cavity interfaces were visualised with a video camera and the length determined by image processing (Kueny *et al.*, 1991).

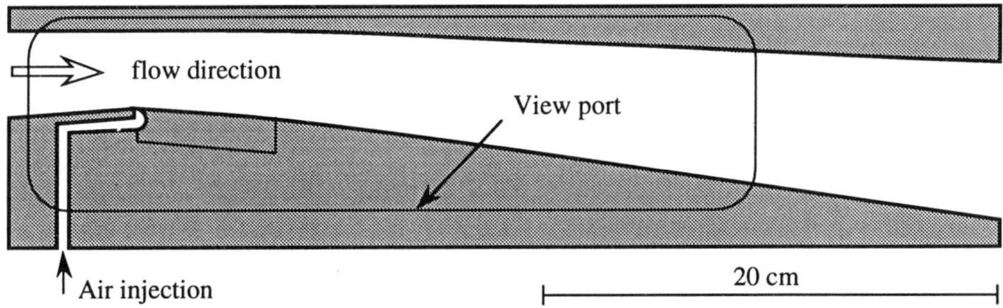

Figure 1 : Schematic of the Venturi in the test facilities.

Tests with air ventilated cavities were conducted in a facility having the same test section and operating with water and pressurised by air, Figure 1. However, maximum throat velocity was only 20 m/s and the pressure was set in order to avoid vapour production at the ventilated cavity interface. Air flow rate was measured using either a turbine flow meter or a rotameter. The same loop was used to perform vapour cavitation tests. Void fraction and phase velocity in the cavity were measured using specially designed optical probes (Stutz and Reboud, 1994, 1996). Vapour and gas flow rates were obtained by integration of the velocity profiles (Stutz, 1996). Independent tests with ventilated and vapour cavities were conducted using an hydrofoil with a sharp leading edge and water as circulating fluid (Larrarte *et al.*, 1995).

3. Proposed model

Since over the interface of the cavity the pressure gradient is nearly zero and the curvature of the interface is small we assume that the heat exchange is analogous to the one occurring at a solid flat plate with zero pressure gradient from which heat is being removed in the amount required for the vapour production given by the flow coefficient C_Q,

$$C_Q = \frac{Q}{U_\infty B \ell} = \frac{V_g}{U_\infty} \tag{1}$$

where Q is the air flow rate, B an ℓ are the width and length of the two dimensional cavity, U_∞ is the free stream velocity and $V_g = Q/B\ell$ is the mean velocity of the vapour assuming that it is brought to the cavity through the whole interface. We will also assume that the flow is turbulent and that the surface behaves as a rough plate. The validity of the above assumptions can only be verified if the results are favourably compared with available data. This is the purpose of section 5.

It can be shown (Brun *et al.*, 1970) that the wall temperature on a flat plate over which a heat flux, φ_p, is being transferred by the fluid is given, for Prandtl numbers $P \approx 1$, by,

$$T_p = T_\infty + \frac{\varphi_p}{\frac{1}{2}\rho c_p U_\infty C_f}\left(1 - \frac{u_b}{U_\infty}(1-P)\right) \tag{2}$$

where T_∞ is the temperature of the liquid, ρ its specific mass, c_p is the specific heat at constant pressure, C_f is the local friction coefficient, u_b is the velocity at the edge of the viscous boundary layer and,

$$\varphi_p = -\rho_v L V_g = -\rho_v L C_Q U_\infty \tag{3}$$

where L is the latent heat of vaporisation and ρ_v the specific mass of the vapour phase. For a rough plate at very high Reynolds numbers, the friction coefficient is given by,

$$\frac{1}{2}C_f = 0.00695\left(\frac{x}{\varepsilon}\right)^{-1/7} \tag{4}$$

where ε is the equivalent roughness. Finally, the ratio u_b/U_∞ is assumed to be equal to that of a smooth plate and given by,

$$\frac{u_b}{U_\infty} = \frac{2.1}{Re_x^{0.1}} \tag{5}$$

Combining (3) to (5) and (2), we obtain,

$$T_p = T_\infty - \frac{\rho_v L C_Q}{0.00695\,\rho c_p}\left(\frac{x}{\varepsilon}\right)^{1/7}\left[1 - 2.1\,Re_x^{-0.1}(1-P)\right] \tag{6}$$

A better data correlation of Hord's results was obtained by using a roughness (in meters), varying with the Reynolds number, of the form (Fruman and Beuzelin 1992),

$$\varepsilon = 2.2\, Re_x^{-0.5} \tag{7}$$

4. Experimental results

4.1. THERMAL EFFECT RESULTS

The cavity length, ℓ, as a function of the cavitation number,

$$\sigma = \frac{p_{ref} - p_v(T_{ref})}{\frac{1}{2}\rho V_{ref}^2} \tag{8}$$

where p_{ref}, V_{ref} and T_{ref} are the reference pressure, velocity and temperature measured upstream of the throat, p_v is the vapour pressure, is plotted in Figure 2 for a variety of test conditions in the PECLET facility. For comparison purposes, results obtained in the water loop have been also indicated. They show that the cavity length increases very sharply for σ values of less than 0.60. The difference between the Freon and the water results can be explained because of the thermal effect. Figure 3 shows the temperature difference, ΔT, between the free stream temperature and the temperature measured on the Venturi wall, as a function of the distance to the throat, non dimensionalized with the cavity length (50 mm), for three reference temperatures and flow velocities. Temperatures decrease linearly with distance whatever the experimental conditions. Maximum differences between the reference temperature and the wall temperature, ΔT, are presented as a non dimensional coefficient $\boldsymbol{B}$,

$$\boldsymbol{B} = \frac{\Delta T\, \rho\, c_p}{\rho_v\, L} \tag{9}$$

for a variety of test conditions as a function of σ, Figure 4. In all these tests, ΔT is comprised between 1 and 5°C.

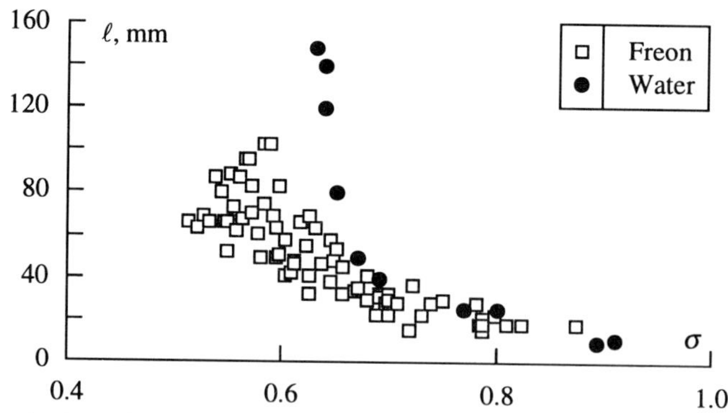

Figure 2 : Cavity length as a function of cavitation number in the PECLET and water facilities.

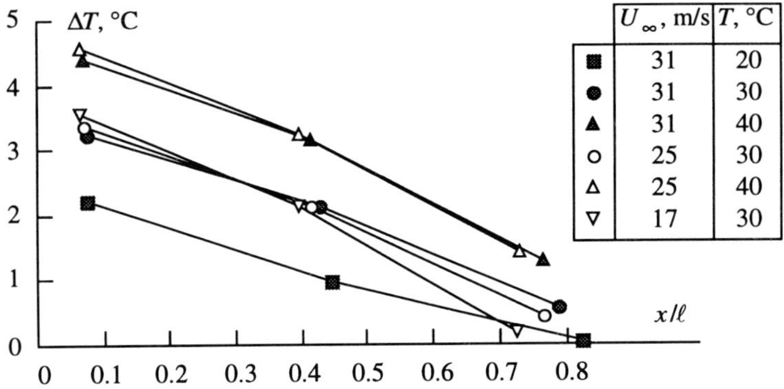

Figure 3 : Temperature difference as a function of distance to the throat for ℓ = 50 mm.

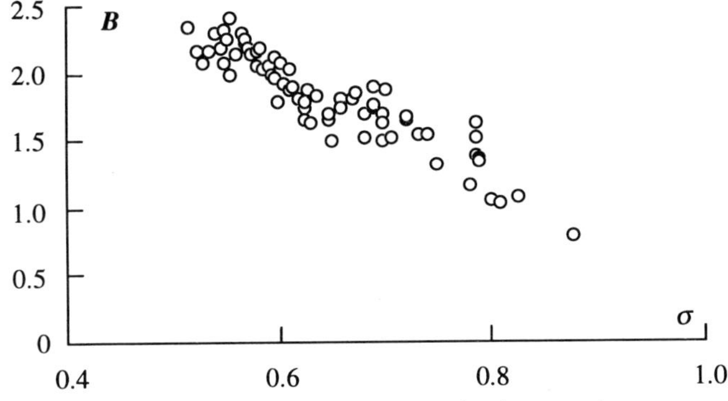

*Figure 4 : **B** factor versus cavitation number.*

4.2. FLOW COEFFICIENT FOR VENTILATED CAVITIES

The flow coefficients and the Reynolds numbers were computed using as the free stream velocity the one on the cavity interface given by the Bernoulli's equation, $V_\infty = V_{ref}(1+\sigma)^{1/2}$. The mean flow coefficient is independent of Reynolds numbers in the range tested, much smaller than those usually encountered in practical applications and in the PECLET loop, and has a value of $C_Q = 3.85 \times 10^{-3}$, with a standard deviation of 0.45×10^{-3}. The latter is consistent with the precision of the measurement of the cavity length, estimated to ±10 percent and due essentially to the fluctuations of the cavity closure. Stutz (1996), has conducted velocity and void fraction measurements within a 8 cm long ventilated cavity. They show, Figure 5a, that near the leading edge of the cavity the void fraction peaks at the interface while becoming nearly constant over the cavity height downstream. The velocities are in the direction of the main flow on the upper half of the cavity thickness and show a backflow in the lower half, Figure 5b. This backflow justifies the homogenisation of the void fraction by the intense mixing so created. The air (vapour) flow rate was computed by integrating, over the cavity thickness at different stations, the product of the velocity and the void fraction. In spite of the difficulties associated with this type of measurements and the precision of the integration, conducted with a limited number of data points, the comparison with the injected flow rate is very much satisfactory, Figure 6. It has to be pointed out, however, that the experimental C_Q is in the lower range as compared to earlier experiments. This may be due to the estimate of the cavity length made by the experimenter.

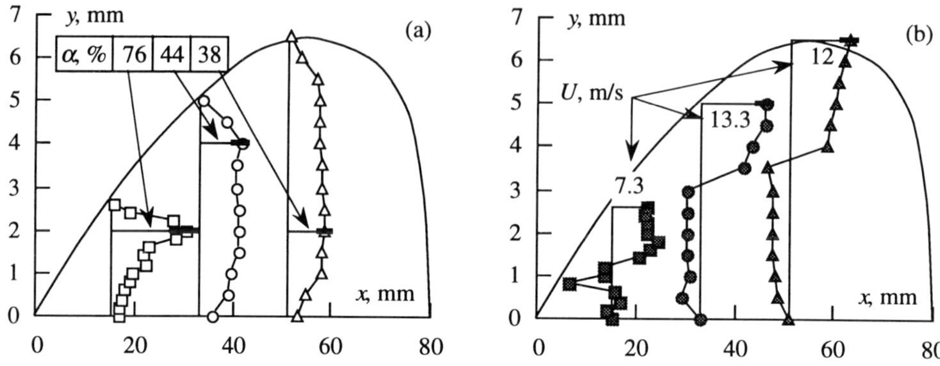

Figure 5 : Void fraction (a) and velocity (b) profiles at three stations for V_∞ = 14 m/s and a ventilated cavity length of 80 mm.

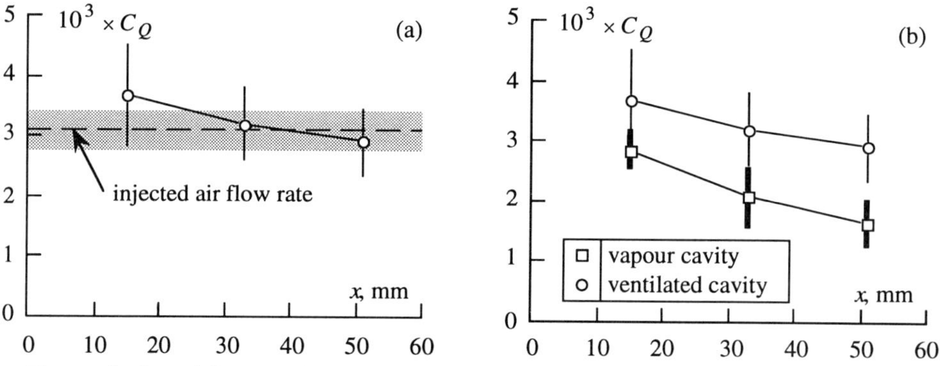

Figure 6 : Local flow coefficients obtained by integration of the product of the void fraction and the velocity from Figure 5 and 7.

4.3. FLOW COEFFICIENT FOR VAPOUR CAVITIES

For the geometry of the Venturi used in the ventilation tests, Stutz (1996) performed velocity and void fraction measurements for a 8 cm long natural stable cavity, Figure 7. As compared to the ventilated cavity situation, the void fraction do not show a significant augmentation near the leading edge in the vicinity of the interface. Further downstream, the void fraction is larger near the wall than near the interface. The velocity profiles are surprisingly close to the ones of the ventilated cavity. In terms of vapour flow coefficients, Figure 6b shows that they are slightly below the non condensable gas flow coefficients but close enough to justify "a posteriori" the hypothesis of the entrainment theory (Holl *et al.*, 1975).

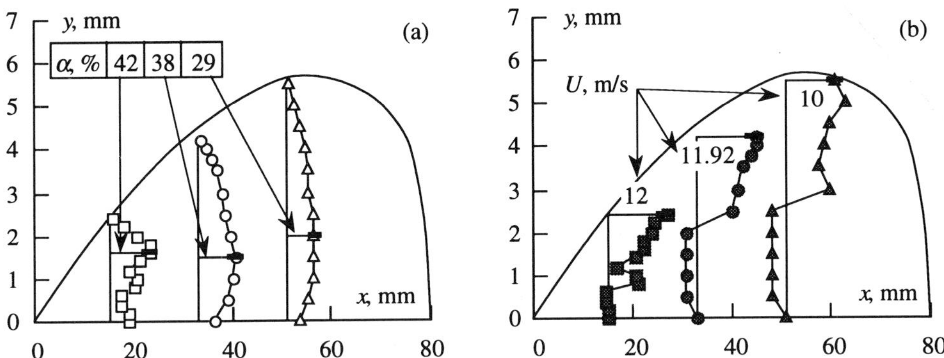

Figure 7 : Void fraction (a) and velocity (b) profiles at three stations for $V_\infty = 14$ m/s and a vapour cavity length of 80 mm.

For unsteady nearly periodic cavities, Larrarte *et al.* (1995) have shown that the vapour production can be assimilated to the cavity growth rate during the initial phase. Assuming that there is no vapour production during the phase of cavity detachment and cloud convection, the mean vapour flow rate during the total duration time of a period gives a flow coefficient 4 to 6 times larger than the one of the stable cavity. However, Stutz and Reboud (1996) have shown that the mean void fraction of these cavities during the growth phase is around 20 percent. If this void fraction is taken into account, the vapour flow coefficient is reduced (by a factor of about five) and the flow coefficient estimates are then close to the ones for stable cavities.

4.4. PRELIMINARY CONCLUSIONS

The results presented above show that the hypothesis of the entrainment theory, assuming that the vapour flow rate can be estimated by the non condensable gas flow rate required to sustain a ventilated cavity, are quite well justified by the experiments if due account is taken of the difficulty of conducting very precise measurements in vapour cavities. The methodology developed by Fruman and co-workers to estimate the thermal effect will be now applied to the flow conditions of the PECLET loop tests.

5. Comparison of experimental and predicted temperature differences

By using expressions (6) and (7) and making $x = \ell$, the temperature differences were computed with $C_Q = 3.85\times10^{-3}$ The results are shown in Figure 8a as a function of the experimental ΔT. It can be seen that the predicted values of ΔT are much larger than the experimental ones. Among the many factors which can justify these differences, a screening investigation has firstly shown that the vapour phase characteristic parameters in expression (6), calculated for the reference temperatures instead of the cavity temperatures, are the most significant. Secondly, it has been observed that because of the high velocities of the liquid flow (20 m/s < V_∞ < 50 m/s), the heat produced by viscous effects might not be neglected. According to Eckert and Drake (1972), in the case of turbulent boundary layers the increase of temperature induced by viscous effects can be estimated (from experiments) by,

$$\Delta T_{visc} = \frac{U_\infty}{2\,c_p}\,P^{1/3} \tag{10}$$

A computation was then performed using the vapour phase parameters at the cavity temperature and taking into account the heat provided by viscous effects by subtracting ΔT_{visc} given by equation (10) to the right term of equation (6). The results are plotted in Figure 8b. Here the agreement is very satisfactory and gives an "a posteriori" justification to the assumption sustaining the model. However, it must be stressed that the assumption that the vapour production is uniformly distributed over the whole cavity interface is not justified by the results presented in Figure 6 showing that the flow coefficient has reached its maximum value at the first measuring station downstream the cavity leading edge. Moreover, the size of the adopted roughness can not be justified either on physical grounds and has to be considered as an adjustable parameter whose physical meaning is yet to be found.

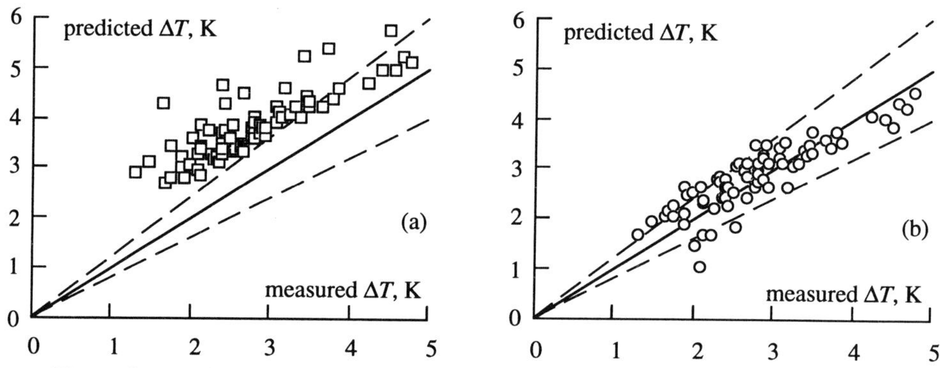

Figure 8 : Predicted versus measured temperature difference for (a) values of the parameters at the free stream temperature and (b) corrected values of the parameters for the vapour phase.

6. Conclusion

Results of tests conducted in two cavitation loops having the same geometric Venturi type test section are presented. In one of the test loops the working fluid is Freon 114 and allows to determine the temperature reduction due to the vapour produced to feed a natural cavity. In the other loop, operating with water, tests were conducted with ventilated and natural stable cavities to determine the non condensable gas and vapour production rates by performing void fraction and velocity measurements within the cavities.

A model that allows the computation of the temperature reduction by assuming that : i) the vapour production is equal to the air injection necessary to sustain a ventilated cavity of equal mean length, ii) the vapour production is uniformly distributed on the whole surface of the cavity interface, and iii) the interface behaves as a rough flat plate through which heat is fed to produce the vapour, is presented.

The experimental results show that the hypothesis of equality between gas flow rate and vapour production rate is rather well justified. Comparison between the experimental temperature differences (liquid/interface) and the values obtained using the proposed model show that reasonable estimates can be made provided the liquid properties and the vapour properties are computed respectively for the free stream and for the interface temperature, and due account is taken of the temperature increase produced by viscous dissipation.

Acknowledgements

The authors acknowledge the financial support of the Agence Française de l'Espace (CNES) and express their gratitude to Dr. G. Albano for his continuous interest on this project. The design, construction and operation of the PECLET facility was made possible thanks to funds and engineering assistance provided by the Société Européenne de Propulsion (SEP).

References

Brun E.A., Martinot-Lagarde A., Mathieu J., 1970, Mécanique des Fluides 3, Dunod, Paris.

Eckert E.R.G., Drake R.M., 1972, *Analysis of Heat and Mass Transfer*, Mc Graw-Hill Book Co.

Fruman D.H., Benmansour I, Sery R., 1991, Estimation of the Thermal Effects on Cavitation of Cryogenic Liquids, Cavitation and Multiphase Flow Forum, *ASME FED*-Vol. 109, pp. 93-96.

Fruman D.H., Beuzelin F, 1992, Effets thermiques dans la cavitation des fluides cryogéniques, *La Houille Blanche* n° 7/8, pp. 557-561.

Holl J.W., Billet M.L., Weir D.S., 1975, Thermodynamic Effects On Developed cavitation. *Trans. ASME, J. Fluids Engng*-Vol 97, pp. 507-514.

Hord J., 1973, *Cavitation in liquid cryogens, Volume II-Hydrofoil.* NASA CR-2156

Kueny J.L., Reboud J.L., Desclaux J., 1991, Analysis of Partial Cavitation by Image Processing and Numerical Prediction, Cavitation '91 Symposium, *ASME-FED*-Vol 116, pp. 55-60.

Larrarte F., Pauchet A., Bousquet P., Fruman D.H., 1995, On the Morphology of Natural and Ventilated Cavities, Cavitation and Multiphase Flow Forum, *ASME FED*-Vol. 210, pp. 31-38.

Stutz B., 1996, Analyse de la structure diphasique et instationnaire de poches de cavitation, Doctoral Thesis, I.N.P.Grenoble, January 12th, 1996.

Stutz B., Reboud J.L., 1994, Experimental Study of the Two Phase Structure of Attached Cavitation, Cavitation and Multiphase Flow Forum, *ASME FED*- Vol. 194, pp. 41-46.

Stutz B., Reboud J.L., 1996, Experiments on unsteady cavitation, *Experiments in fluids,* to be published.

AERATION VERSUS CAVITATION IN DAM SPILLWAYS: SELF-AERATION AND ARTIFICIAL AERATION (AERATORS)

Ramón GUTIERREZ SERRET
Civil Eng. Ph.D. Head Hydraulic Structures Div., Centre for Research and Experimentation, Ministry of Public Works, Transport and Environment. P° Bajo Virgen del Puerto 3, Madrid, SPAIN.

Abstract
This paper describes the most important aspects of the effect that flow aeration has in combating the damage caused by cavitation. An explanation is also given of how to characterise these flows when the air enters the flow in a natural way, as well as giving an account of some of the artificial aeration devices that are placed on spillways when natural aeration proves to be insufficient.

1.- Aeration versus cavitation.

Of the variety of corrective measures that can be taken to rectify the damage caused by cavitation - use of more resistant materials, modification to desgins, etc. - flow aeration occupies an extremely important place. The reasons for this circumstance are summarised below:

- The collapse pressures are lower when the flow contains air.
- The presence of air brings about a considerable reduction in the velocity of the shockwaves generated during the collapse stage; it has been demonstrated that air concentrations [$C = Q_a/Q_a + Q_w$] of 8% cause tenfold reductions or even greater. Furthermore, they attenuate the domino effect on the collapse of adjacent cavities, which further reduces the overall damage.
- The cavities containing air, whose size is greater than those without air, cause a union between the two types, giving rise to larger cavities, the collapse of which either does not cause damage or reduces the extent thereof (when the size is greater than 0.1 mm., virtually no damage is caused).
- Pressures which are much lower than atmospheric pressure are prevented, so cavitation is much less likely to occur.

Where damage resulting from cavitation is concerned, reference must be made to the danger caused by shear flows, because the instantaneous fluctuation of the pressure [P'(t)]
can determine that, momentanously and intermittently, water vapour pressure is reached without the average pressure [P] having dropped below atmospheric pressure [P = P + P'(t)], which therefore causes cavitation.

This situation can occur in the high velocity zones (generally above 25-30 m/s) of spillways, close to the concrete surfaces, especially if these surfaces are irregular in places, in baffles , stilling basins or when gates are partially open. Fig. 1 shows these localities and some others where cavitation may take place, even when shear flows are not involved.

E. Cabrera et al. (eds.), Hydraulic Machinery and Cavitation, 594–603.

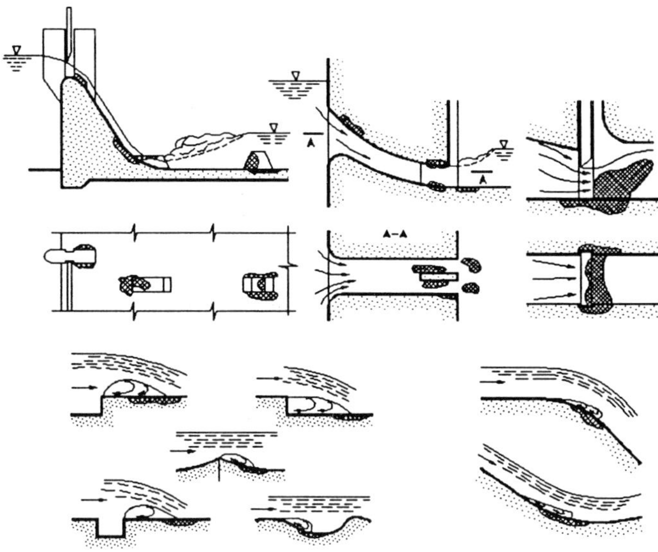

Fig.1.- Location of spillway cavitation.

Among the different indexes that are used to quantify cavitation risk, one of the most widely used, especially where dams are concerned, is the "cavitation number (σ)[1]". Its critical values (σ_{crit}), below which the possibility of damage exists, have been the subject of a great deal of research work and, for normal spillway finishes, the values have been determined as ranging from 0.25-0.20 (Falvey 1990; ICOLD 1992)[2].

However, other researchers propose specific formulae for obtaining σ_{crit} (Arndt 1981) or give values of this parameter in the case of isolated iregularities, offsets, holes, gate zones, sudden expansions, etc. (Kolman 1969, Rouse 1966, Hamilton 1983, Falvey 1990, etc.). There are also authors who use other cavitation indexes in the case of dams (Tullis 1990, Wang and Chow 1979, Ball 1963 and Johnson 1963)[3].

The damage mainly depends on the following factors: depression, exposure

---

[1] Cavitation number : $\sigma = P-P_v/(\rho v^2/2)$.

P,v : pressure and velocity in the proximity of the cavitation zone.

P_v : water vapour pressure.

[2] Values of σ_{crit} have been determined between 0.2 and 17, the latter value being for shear flows (Falvey 1982, 1990; ICOLD 1992; Jansen 1988)

[3] Gutiérrez Serret 1994 presents a summary of the data provided by all these authors.

time, velocity and aeration degree, surface finish tolerance and the resistance of the material. With respect to such factors, some considerations of a practical nature are:

- The damage must be rectified as soon as it has been detected, because its initial progress is extremely rapid, although it does slow down once it is developed.
- The admissible velocity is closely related to the degree of flow aeration and the surface finish, to the extent that greater air concentrations allow for higher velocities for the same finish tolerance.
- Velocities of 25-30 m^3/s can be regarded as dangerous or even less (12-15 m^3/s if there is no flow aeration, Chanson 1993).
- Current practise tends to combine velocity and aerated flow with finishes that are not too smooth, which are not difficult to achieve and which are long-lasting.
- It has been shown that, it is generally the case for the finishes that are currently used, air concentrations of 7-8% in the proximities of the surfaces, prevent damage being caused for the velocities indicated, or even for velocities that are somewhat greater (30 m^3/s).
- It has also been shown that isolated irregularities cause a lot more damage than a high uniformly distributed roughness. Falvey (1990), proposes three tolerances of surface finish, as a function of the cavitation number, depending on whether or not the flow is aerated (Table C1).

FINISHING TOLERANCES			TOLERANCE - CAVITATION N° - AERATION		
TOLERANCE (T)	OFFSET Max. Height (mm)	SLOPE (V:H)	CAVITATION N° (σ)	TOLERANCE WITHOUT AERATION	TOLERANCE WITH AERATION
T_1	25	1: 4	> 0.6	T_1	T_1
T_2	16	1: 8	0.4 - 0.6	T_2	T_1
T_3	12	1: 16	0.2 - 0.4	T_3	T_1
			0.1 - 0.2	Modify design	T_2
			< 0.1	Modify design	Modify design

C.1.- Tolerances depending on cavitation N° and whether or not aeration exists (Falvey 1990).

2.- Natural aeration.

Now that the importance of aeration in the fight against damage caused by cavitation has been explained, a description will be given of the way to characterise natural aerated flows.

The natural aeration mechanism in high velocity flows, is a result of the highly turbulent nature of such flows, this being the cause of considerable agitation on the surface which, in turn, provides some water particles with enough kinetic energy to overcome the effects of gravity and water surface stress; such particles are outwardly projected in the form of drops and surface waves, part of which (mainly the drops), perforate the current when they return, and form air bubbles.

This process transforms the water into a biphasic an bicomponent fluid - water and air - and modifies the flow structure, in such a way that three zones can be created longitudinally, depending on whether air entrainment has not commenced (zone without aeration), is underway but a balance does not yet exist between the air entering and the water leaving (gradually aerated zone), or whether this balance has been reached (uniform aeration zone). Four areas can exist transversely, depending on whether the air or the water is predominant: surface, mixture, intermediate or without air. Both structures can be seen in figure 2.

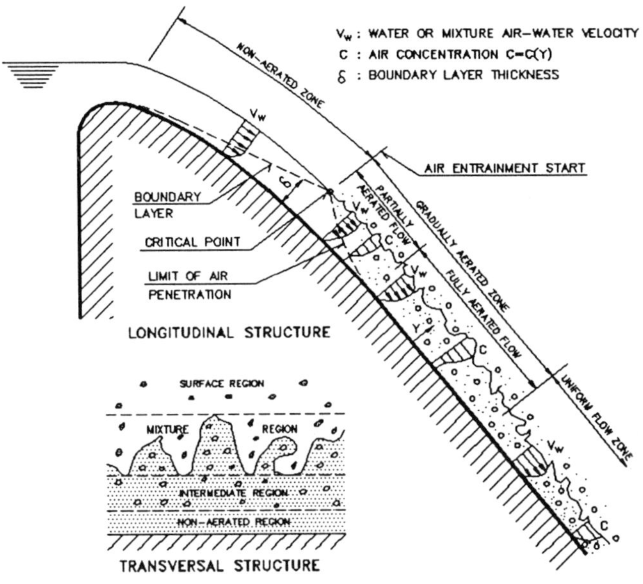

Fig. 2.- Natural aeration. Longitudinal and transversal flow structure.

Aeration theoretically starts at the point where the Boundary Layer reaches the curent surface (Critical Point), but in practice, the irregularities of both surfaces -nappe and Boundary Layer- determine the existence of a strip in which zones without air co-exist with others where entrainment is already underway. It is necessary to know the profile of both surfaces and to use the Boundary Layer theory, in order to locate the Critical Point, and this has been studied by several authors. However, the results obtained reveal considerable differences, so it is advisable weigh up the results.

The "gradual aeration zone" is characterised by a longitudinal variation in the air concentration. According to Wood (1991), the average concentration and the depth throughout the chute [C(x); $h_w(x)$], can be determined by integrating the energy equations for the water-air mixture by numerical methods[4] (finite

[4] When the chute width and its slope are constant, the equations are simplified. The air continuity equation (1) can be directly integrated, thereby obtaining C(x); that for the energy (2), with which $h_w = h_w(x)$ is calculated, the finite differences method is still used for solving (Gutiérrez Serret, 1994).

differences) . Once C(x) and $h_w(x)$[5] are known, the rest of the parameters that characterise the flow can be obtained - concentration at the bottom (C_o), velocity, emulsified flow (h_a) - by applying the expressions that are indicated later on for the "uniform aeration zone". These equations are expressed as follows:

- Air continuity.

$$\frac{d\overline{C}}{dx} = (1-\overline{C}) \left[\frac{(V_T)_u \cos\alpha}{q_w} (k_e\overline{C}_u - K_T\overline{C})(1-\overline{C}) + \frac{\overline{C}}{b}\frac{db}{dx}\right] \qquad (1)$$

- Energy for the water-air mixture.

$$\frac{dh_w}{dx} = \frac{\sin\alpha\cdot(1+h_w\frac{d\alpha}{dx}) + E\frac{q_w^2}{gh_w^2}\frac{1}{b}\frac{db}{dx} - I}{\cos\alpha - \frac{E}{g}\frac{q_w^2}{h_w^3}} \qquad (2)$$

in which:
x : Distance above the chute axis from the Critical Point.
b : Chute width, generally variable; b = b (x)
α : Angle of chute with horizontal, generally variable; $\alpha = \alpha(x)$
q_w;q_a: Specific water/air discharge (m^3/s.m).
h_w: Equivalent depth.
$\overline{C}$: Average concentration in each section; $\overline{C} = q_a/q_a+q_w$.
V_e: Inflow velocity of air into the current.
V_T: Limit velocity of the bubbles in turbulent flows,
[0.17 m/s (Wood 1991) $<V_T<$ 0.40 m/s (Chanson 1992)]. $K_e = V_e/(V_e)_u$;
$K_T = (V_T)/(V_T)_u$[6]. Subindex u: Values $\overline{C}$, V_e and V_T in the uniform aerated zone.
E : Corrective coefficient of the kinetic energy[5]; g : Apparent gravity.
I : Energy line slope: $I \simeq q_w^2 \cdot f_w / 4gh_w^3$ (Darcy-Weisbach)
f_w and f: Roughness Coefficient "equivalent flow" and flow without aeration

[5] Equivalent flow: theoretical situation in the hypothesis in which the water and the air flow separately, but at the same velocity that the water-air mixture would flow at (enclosed figure).

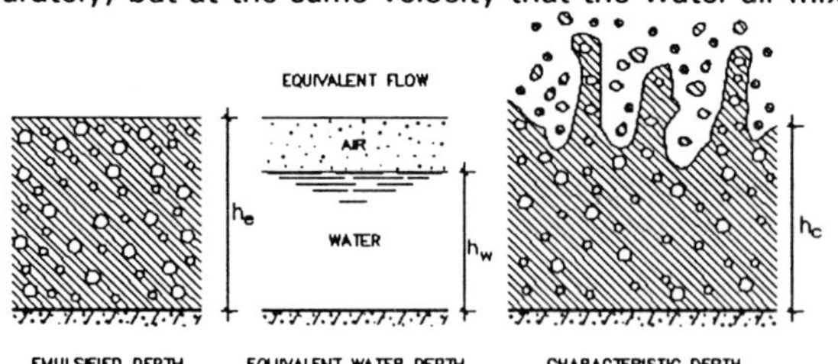

h_c is usually adopted up to heights with an air concentration of 90% (h_{90}).

[6] The following is usually considered when integrating these equations: $K_e = K_t = 1$ y $E = 1$

$$\frac{f_w}{f} = \frac{1}{1+10\overline{C}^4} \tag{3}$$

In the "uniform aerated zone" it is of interest to know the average concentration ($\overline{C}$), the bottom concentration (C_0), the emulsified depth (h_c) and the mixture velocity.

Numerous formulae have been proposed for calculating the average concentration ($\overline{C}$) and they are not all consistent. Those of Gangadharaiah et al. (1970) and Hager (1991) have been selected from these, and their expressions are:

- Gangadharaiah et al. (1970):

$$\overline{C} = 1 - \frac{1}{1 + \Omega\, F_w^{3/4}} \tag{4}$$

in which:

F_w:"Equivalent Fronde number"; $F_w = V/\sqrt{g\, h_w}$

Ω: Constant depending on the roughness and the shape of the channel. $\Omega = 1.3\text{-}5\cdot n$ rectangular and $\Omega = 2.16\cdot n$ trapezoidal (n: n° Manning).

- Hager (1990):

$$\overline{C} = 0{,}75\ (\text{sen}\ \alpha)^{0{,}75} \tag{5}$$

The bottom concentration (C_0) can be calculated, using the formulae of Rao and Gangadharaiah (1971) or Hager (1991), amongst others. Hager's expressions are:

$$C_o = 1.25\left(\frac{\pi}{18D}\alpha\right)^3; \qquad 0^{\circ} \leq \alpha \leq = 40^{\circ} \tag{6}$$

$$C_o = 0.65 \cdot \sin \alpha; \qquad 40^{\circ} \leq \alpha \leq 80^{\circ} \tag{7}$$

The emulsified flow (h_e), or to be more precise, the characteristic depth (h_c), is:

$$h_c = h_w/(1 - \overline{C}) \tag{8}$$

3.- Artificial aeration. Aeration devices (Aerators).

As we have already said, when there is a risk of damage due to cavitation and the natural air entrainment is not sufficient ($\sigma > 0.2\text{-}0.25$ and $C_o < 8\%$), artificial aeration has, since the 60's and 70's, been one of the most widely used solutions for dealing with the cavitation problem. Aeration devices (one or more[7]) are usually constructed down the length of the chute, and these aerators suck the air into the flow.

Its typologies are varied (Fig.3), but they are all combination of four basic elements: ramp, offset, groove and aeration duct, so that the advantages these elements can be enhanced and the operation can be made more satisfactory. They are usually located 30 to 100 m. apart, and the chute's slope changes are suitables locations.

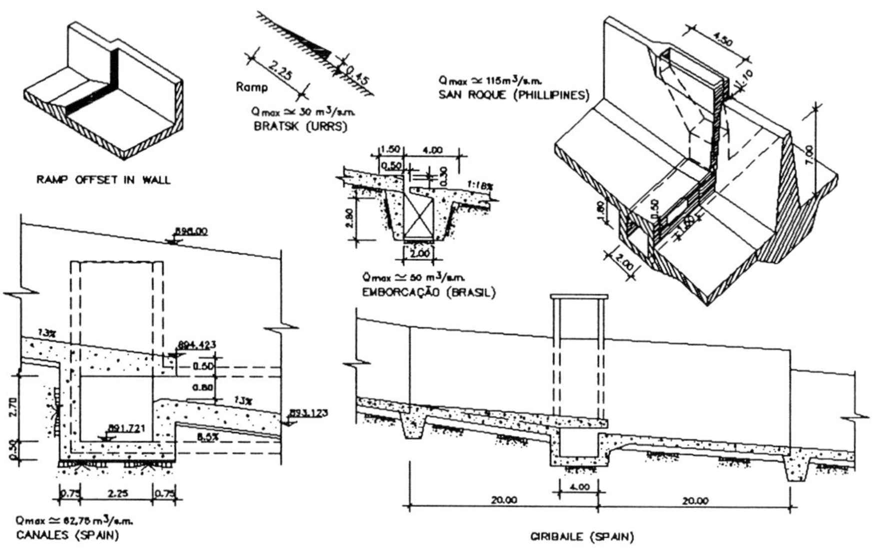

Fig.3.- Aerators. Usual designs.

Ramps provide significant quantities of air for slight water discharges, but they are by no menas ideal when the variability of such discharges is high, and they also cause disturbance to the flow; their height ranges from 0.1 to 1 m. and the angles are from 5° to 15°. **Offsets** cause less disturbance and work well when the discharges are high, heights of 0.5 to 2 m. or even less, are common. **Grooves** distribute the air evenly at the bottom, but a risk of flooding exists when the discharges are slight; heights generally range from 0.2 to 2 m. and widths from 1 to 2 m. **Aeration ducts** are normally rectangular and are designed for air velocities

[7] Spillways with slopes gentler than 20°, usually require more than one aerator to be placed down the length of the chute, but when the slope is steeper than 30°, one single aerator will suffice at the head of the chute. No generalisations can be made about intermediate slopes.

ranging from 30 to 45 m/s and, in exceptional circmstances, from 90 to 100 m/s, the maximum depressions being from 0.5 to 2 w.c.m. (water colunm metres).

The basic operation of aerators (See Fig. 4), is due to the pressure difference between that of the atmosphere and the cavity formed in the aerator, which sucks air in from outisde (air demand). Furthermore, the flow turbulence is greater, bringing about an increase in air entrainment at the upper surface of the nappe. according to the mechanism described, the flow structure in an aerator is made up of five zones (Fig 4.): "approach", "transition", "aeration", "impact" and "bottom aerated flow".

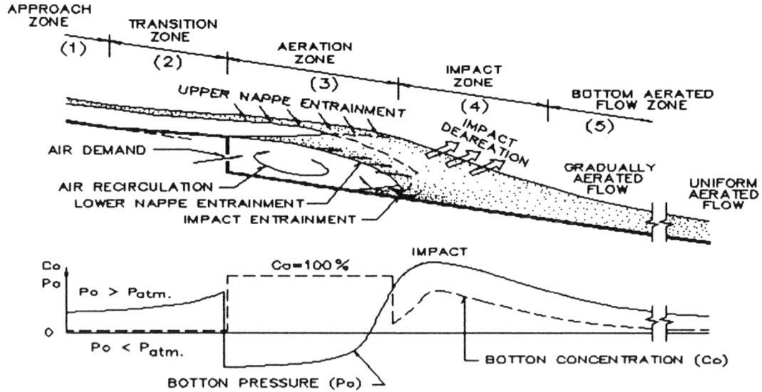

Fig.4.- Artificial aeration. Longitidinal structure of the flow.

The flow conditions in the "approach zone" determine to a large extend the way the aerator operates. A pressure increase, with respect to the hydrostatic, greater friction and an increase in turbulence, all occur in the "transition zone". Air suction takes place in the "aeration zone", and it rejoins the flow via the lower nappe of the cavity, and there is also an increase in entrainment at the upper nappe.

Furthermore, in addition to these aeration processes in the "impact zone", caused by the existing turbulence and pressure gradients, the following occur: air entrained by impact, recirculation of this element in the cavity and later de-aeration. Downstream, in the "bottom aerated zone", the mixture flows in the same way as for natural aeration, a "gradual aerated zone" appearing first, and this is followed by a "uniformly aerated zone", if the chute is long enough.

Quantification of all these entrainments, has been carried out by several authors (see summary by Gutiérrez Serret 1994, 1995), but information is still not complete, especially where air entrainment through the upper surface of the nappe and by impact are concerned, and as regards the de-aeration that takes place in the zone.

4.- Computer program for positioning of aerators. Example.

In accordance with what has been indicated, and on the basis of the gradually aerated flow equations (Eq. 1 and 2.), a computer program has been developed which makes it possible to determine the aerator positioning requirements.

It ought to be pointed out, that the results, should not be considered as more than an approach to the problem, and in the light of this, they must be compared with experience gained to date, as well as with laboratory tests and even prototype.

The program is structured into three stages. At the First stage (A), an analysis is conducted into the way the spillway operates without aerators (natural aeration). Stage (B) determines the position of the first aerator, if necessary, where σ>0.2-0.25 and C_o<8%, and the aerator must be designed. At Stage (C), the depth (h_i) and the average concentration C_i) downstream from the impact zone are taken as initial flow conditions (Chanson, 1992; Falvey 1990) and, equations (1) and (2) are used again to obtain $\sigma=\sigma(x)$ and $C_o=C_o(x)$; a second aerator is then placed if $\sigma <$ 0.2-0.25 and C_o < 8%, and the process is repeated, downstream from this second aerator.

By way of example, figure 5 shows the application of the Giribaile dam (Spain).

BIBLIOGRAPHY.

- ARNDT, R.E.A. (1977). "Recent Advances in Cavitation Research". Advances in Hydroscience, Ven Te Chow ed., Academic Press, Vol. 1.
- CHANSON, H. (1993). "Self-Aerated Flows on Chutes and Spillways". ASCE, Jour. Hydr. Div. vol.119, HY2, pp. 220-243.
- FALVEY, h. (1980). "Air-Water Flow in Hydraulic Structures". Engineering Monography Nº 41, Bureau of Reclamation. U.S.A.
- FALVEY, H. (1990). "Cavitation in Chutes and Spillways". Eng. monogr.42. Bureau of Reclamation. EEUU.
- GANGADHARAIAH, T., LAKSHMANA RAO, N.S. & SEETHARAMIAH, K. (1970). "Inception and Entrainment in Self-Aerated Flows". ASCE, Jour. Hydr. Div. vol.96, HY7, pp. 1549-1565.
- GUTIÉRREZ SERRET, R. (1994). "Aireación en las Estructuras Hidráulicas de las Presas: Aplicación a los Aliviaderos". Tesis doctoral. Esc. Téc. Sup. de Ing. de Caminos, Canales y Puertos. Madrid. España
- GUTIÉRREZ SERRET, R. & PALMA, A. (1995). "Aireación en las Estructuras Hidráulicas de las Presas: Aliviaderos y Desagües Profundos". Premio José Torán, Comité Español de Grandes Presas Madrid.
- HAGER, W. (1991). "Uniform Aerated Chute Flow". ASCE, Jour. Hydr. Div. vol. 117. HY4, pp. 528-533.
- IAHR. (1991). "Air Entrainment in Free-Surface Flows". Design Manual Nº 4, Balkema Ed. Netherlands.
- ICOLD. (1992). "Spillways. Shockwaves and Air Entrainment". Boletín Nº 87.
- LECOFFRE, Y (1994). "La Cavitation". Ed. Hermès. Paris.
- MATEOS, C.(1987). "Aireación y Cavitación en Desagües". Curso sobre Comportamiento Hidráulico de los Desagües de las Presas. CEDEX, Ministerio de Obras Públicas y Transportes. Madrid, España.
- MAY, R. (1987). "Cavitation in Hydraulic Structures: Ocurrence and Prevention". Report Research SR79, Hydraulic Research Wallingford. Oxfordshire UK.
- PINTO, N.L. de S. & NEIDERT, S.N. (1983). "Evaluating Entrained Air Flow through Aerators", Water Power and Dam Construction, vol. 35, pp. 40-42.
- WOOD, I. (1985). "Air Water Flow. Keynote Address". XXI IAHR Congress, Melbourne, Vol.6 pp. 18-21.

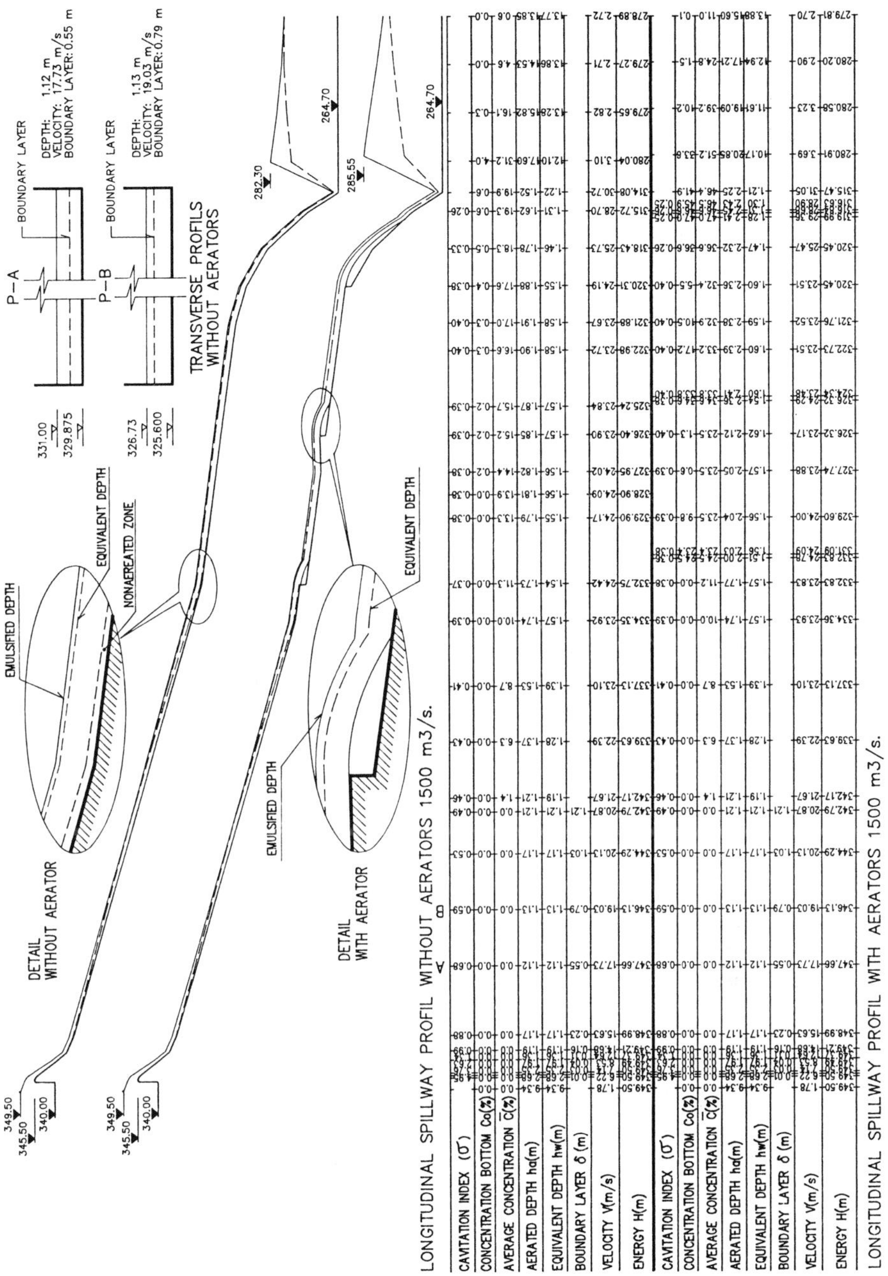

Fig 5.–Aerator Location. Design Process. Application to the Giribaile Dam Spillway (Spain).

LEADING EDGE CAVITATION IN A CENTRIFUGAL PUMP: NUMERICAL PREDICTIONS COMPARED WITH MODEL TESTS

R. HIRSCHI, PH. DUPONT AND F. AVELLAN
IMHEF-EPFL
33, av. de Cour, CH-1007 Lausanne, Switzerland
E-mail: roland.hirschi@imhef.dgm.epfl.ch

AND

J.-N. FAVRE, J.-F. GUELICH AND W. HANDLOSER
Sulzer Pumps P.O. Box 65
8404 Winterthur, Switzerland

Abstract. The aim of this paper is to present the results obtained with a 3-D numerical method allowing the prediction of the cavitation behaviour of a centrifugal pump and to compare this prediction to model tests.
The proposed method consists in assuming the cavity interface as a free surface boundary of the computation domain and in computing the single phase flow. The unknown shape of the interface is predicted by an iterative procedure matching the cavity surface to a constant pressure boundary (p_v). The originality of the presented method is that the adaptation process is done apart from the flow calculation, allowing to use any available code.

1. Introduction

Cavitation behaviour prediction of hydraulic machines, such as inception ψ_{c_i} and standard ψ_{c_s} cavitation coefficients, partial cavity length and its associated performance drop is of high interest for the manufacturers.
When upgrading existing hydraulic installations or designing new geometries, the cavitation guarantees are often the main limiting features. A precise prediction of this phenomenon by numerical simulation is then essential. In recent years, models to predict cavitation development have been refined and applied with success on isolated profiles, for steady [3] and unsteady flows[2][7]. Some 3-D models, based on S1/S2 and Euler flow computation were also developed, but without taking into account the 3-D turbulent and viscous effects of the cavity on the flow [8].
To present the accuracy of the new proposed method, numerical calculations are performed on a centrifugal pump with a commercial Navier-Stokes code (*TASCflow*). A first analysis enables an intrinsic verification of the numerical results given by the solver and make sure they are physically meaningful. Once this analysis is done, particular operating points are chosen to predict the impeller cavitation behaviour.

E. Cabrera et al. (eds.), Hydraulic Machinery and Cavitation, 604–613.

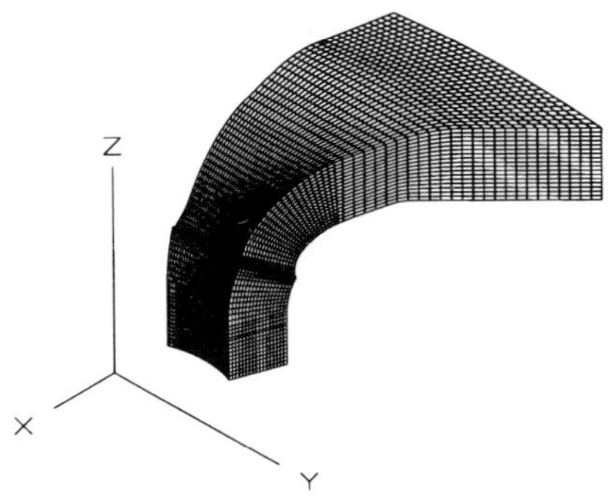

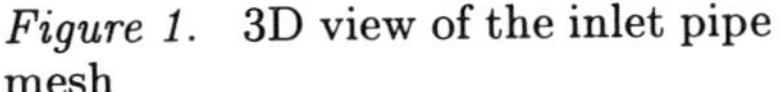

Figure 1. 3D view of the inlet pipe mesh

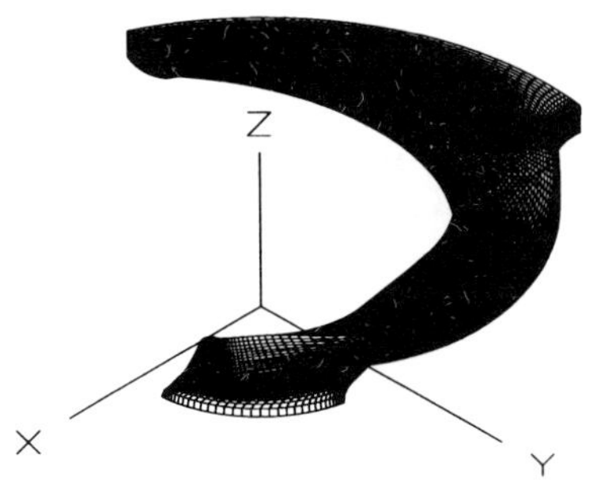

Figure 2. 3D view of the impeller mesh

Cavitation free flow computation is performed, in order to define the hydraulic performances and the cavitation inception coefficient ψ_{c_i}, which is predicted using the minimum pressure coefficient calculated over the blade. Then, the cavity lengths as well as the initial cavity shapes are established as a function of the cavitation coefficient using the *Rayleigh-Plesset* equation. Finally, the cavitation coefficient ψ_{c_s}, corresponding to the beginning of the head drop, is determined by a direct computation of the pump flow, taking into account the flow modification due to the cavitation sheet. The comparison with the measurements is done defining a local cavitation coefficient, based on the ψ_c cavitation coefficient measured during the model tests [1] [4].
The cavitation behaviour prediction of the impeller is only possible if the flow field at its inlet is well known. For that reason, the inlet pipe flow computation is performed in order to define the velocity field and the evolution of the absolute flow angle at its outlet. In Figure 1 is represented the inlet pipe mesh and in Figure 3 the flow field characteristics at its outlet. The Figure 3 brings to the fore the quality of the resulting absolute flow angle repartition. The angular variations in comparison with the ideal angle of 90 degrees are lower than 0.1 degree. Considering these results, one can assume the velocity field to be axial at the outlet of the inlet pipe. This condition will be used as an inlet condition of the impeller flow computation.

2. Single phase flow computation

Eight operating points are calculated using a Reynolds averaged Navier & Stokes finite volume code. The turbulence is taking into account using the well known $k - \varepsilon$ model. As the resolution of the near wall region would require a large number of nodes, a common approach is to model this region using "wall-function". This way allows to relate the near wall tangential velocity to the wall shear stress by means of a logarithmic relation.
The mesh of Figure 2 is a single structured bloc with 95 nodes in the flow direction, 41 from suction to pressure side and 17 from hub to shroud.

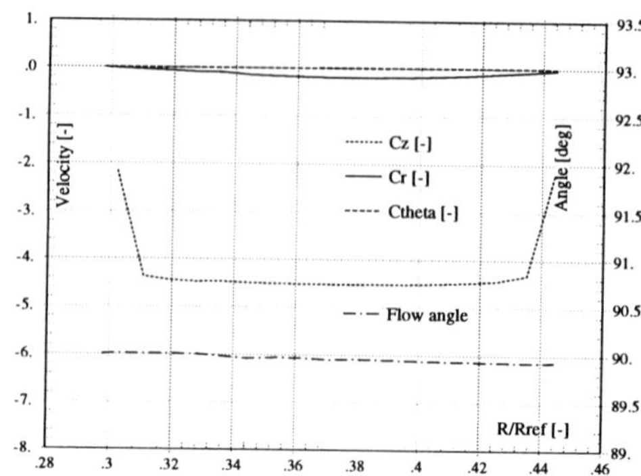

Figure 3. Velocity and absolute flow angle survey at the inlet pipe outlet

3. Cavity length

If one assumed that the dynamic behaviour of the cavity is similar to the bubble one, then the use of the *Rayleigh-Plesset* equation [6], which gives the evolution of a bubble radius depending of the pressure field acting around, allows to take into account the cavity dynamic behaviour. Especially, cavitation sheet beginning as well as its closure is relatively well predicted using the bubble explosion and collapse location. Previous experiments in a test section with a 2-D profile have shown this approach to be in most of the cases surprisingly accurate to predict these two points, even if the cavity thickness needs to be corrected.

4. Initial cavity shape

The proposed method used for the computation with the cavitation sheet has the great advantage to be independent of the flow calculation code. However, the iterative deformation process could be quite time consuming if beginning with the solid surface of the blade. In order to estimate an initial cavity shape, the *Rayleigh-Plesset* equation is used. As illustrated in Figure 4, the initial cavity shape is supposed to be the envelope of the bubble evolution.

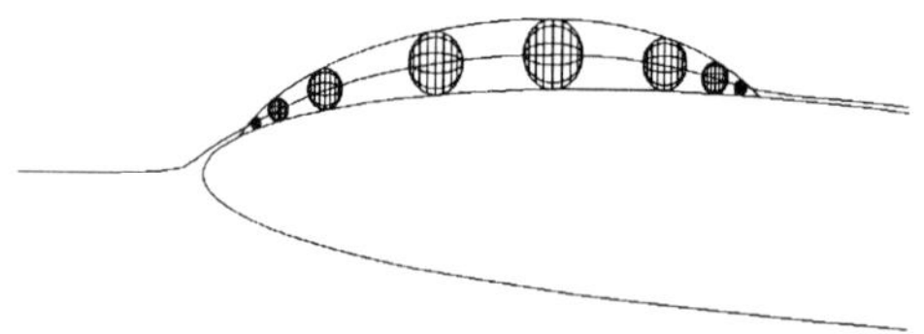

Figure 4. Initial cavity shape = bubble evolution envelope

5. Deformation method

Generally, for 2-D flow computations, one can observe small differences between the initial cavity shape obtained with the method described in the previous section and the deformed one (see Figure 5). But for 3-D problems, the deformation method is necessary in order to take into account the 3-D effects of the cavity on the flow.
So, the cavity shape adaptation is determined according to the pressure distribution obtained with the flow computation at the previous iteration. If one considers t the subscript corresponding to the calculated values at the previous iteration, the expression of the modified cavity thickness e at the abscissa i is given by:

$$e_i^{t+1} = e_i^t + C_2(c_{p_i}^t + \chi_c) \tag{1}$$

where C_2 is a coefficient depending on the *Reynolds* number and a relaxation coefficient C_1.

$$C_2 = f(R_e)\frac{\left[2 - 2^{(1-C_1)}\right]}{1 + \left|\frac{\partial^2 e}{\partial s^2}\right|_i^t} \quad \text{and} \quad C_1 = \begin{cases} 1 & \text{if } \left|c_{p_{i-1}} + \chi_c\right| > S_{cp} \\ \frac{\left|c_{p_{i-1}} + \chi_c\right|}{S_{cp}} & \text{otherwise} \end{cases} \tag{2}$$

The coefficient $\left|\frac{\partial^2 e}{\partial s^2}\right|_i^t$ allows us to avoid oscillations in regions where high thickness gradients occurs, such as the closure of the cavity. S_{cp} corresponds to the root mean square of the difference δ_i^t between the pressure coefficient at the abscissa i and the local cavitation coefficient value χ_c, on all cavity nodes, and is given by:

$$S_{cp} = \sqrt{\frac{1}{fp - dp}\sum_{i=dp}^{i=fp}\left(\delta_i^t - \overline{\delta}^t\right)^2} \quad \text{and} \quad \delta_i^t = c_{p_i}^t + \chi_c \tag{3}$$

with $\overline{\delta}^t$ the average value of δ_i^t on all cavity points. The cavity beginning does not always correspond to the leading edge, so the deformation starts at the abscissa i where the following condition is respected:

$$\left|\frac{\partial^2 e}{\partial s^2}\right|_i^t > 1 \tag{4}$$

The Figure 5 shows an application of this method on a 2D profile.

6. Results

6.1. CAVITATION INCEPTION

The cavitation inception condition, in terms of local cavitation coefficient and static pressure, is given by the condition $\chi_c \leq -Cp_{min}$ where Cp_{min} represents the minimum value of the calculated static pressure coefficient in the impeller [5]. Non dimensionalized, this condition is given by:

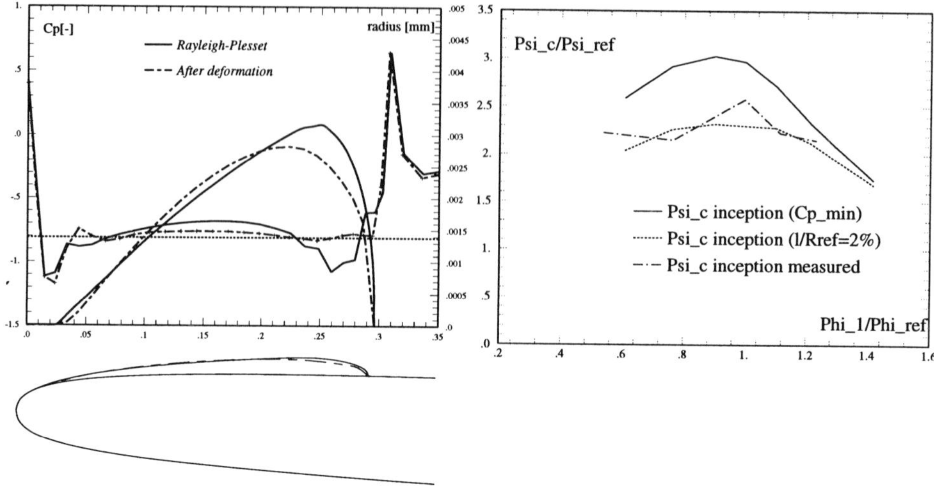

Figure 5. Cavity deformation on a 2D profile

Figure 6. Cavitation inception prediction compared with measurements

$$\psi_c \leq -Cp_{min} + [1 + k_{rd}]\,\frac{R_1^4 A_1^2}{R_{\bar{1}}^4 A_{\bar{1}}^2}\phi_1^2 \tag{5}$$

On the Figure 6 are compared the cavitation inception curve obtained with the minimum pressure coefficient calculated, the same curve obtained with the cavitation inception criterion defined as a cavity length equal to 2% of the outlet radius of the impeller and the measurements curve obtained with this same criterion. These results show that the measurements results are very well predicted by this method. A other point to notice, especially in such a pump of a small specific speed, is the leakage flow between the impeller and the casing on shroud side: an accurate estimation of this leakage flow, in order to calibrate the real flow inside the impeller, is essential for an useful comparison.

6.2. CAVITY LENGTH

Out of the shock less entry, which corresponds to zero incidence, the $|Cp_{min}|$ increases on the suction side for low flow rate or on the pressure side for high flow rate. The direct consequence is a cavity inception and development. In Figure 7 are reported the calculated cavity lengths on five streamlines, from hub to shroud, for four operating points. Remember that the cavity length corresponds to the bubble collapse calculated by the *Rayleigh-Plesset* equation. The comparison with the measurements, corresponding to the square points, shows the assumption made in paragraph 4 (cavitation sheet dynamic behaviour = bubble dynamic behaviour) to be valid.

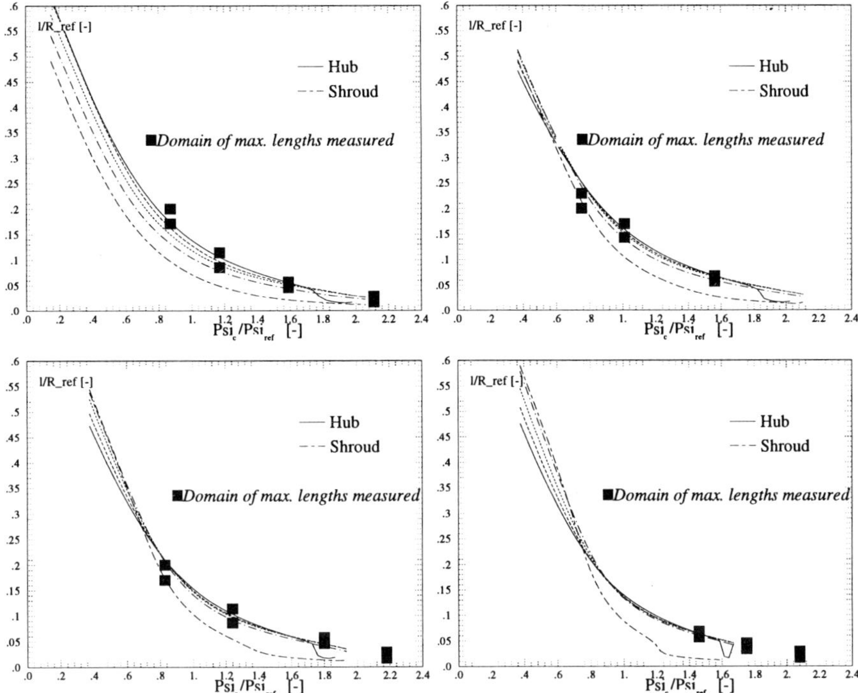

Figure 7. Predicted cavity lengths compared with measurements

6.3. CAVITY DEFORMATION

Once the initial cavity determined, its shape is iteratively modified until the calculated pressure at its interface is constant and equal to the vapor pressure. The pressure coefficient distributions (grid lines 2=hub, 9, 13 & 16=shroud) obtained after the flow computation without cavitation and for a ψ_c/ψ_{ref} value of 0.265 are compared in Figure 8. The cavity effect on the pressure distribution corresponds to the constant pressure region on the suction side. In this example, the pressure loss at the leading edge is partially compensated by the pressure increase near the cavity closure region.

6.4. HEAD DROP COMPUTATION

The comparison of the torque, obtained by the pressure and the viscous forces integration on the blades and impeller side walls, to the moment of momentum in the relative frame of reference allows to quantify the energy loss due to the cavity. In term of torque this is expressed by:

$$\vec{T_t} = \vec{T_f} + \vec{T_r} \tag{6}$$

Where $\vec{T_t}$ represents the transferred torque, $\vec{T_f}$ the provided torque and $\vec{T_r}$ the loosed torque.

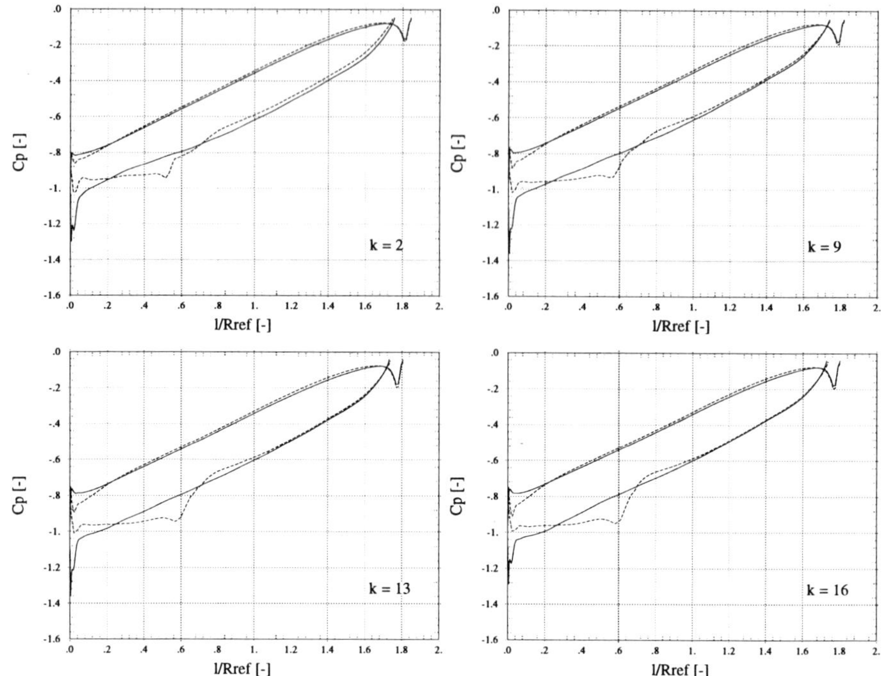

Figure 8. Pressure coefficient distribution without cavitation and for $\psi_c/\psi_{ref} = 0.265 : \phi_1/\phi_{ref} = 1.21$

Transferred power The transferred torque given by the impeller is:

$$\vec{T_t} = \int_{S_{imp}} \vec{r} \wedge (p\vec{n})ds + \int_{S_{imp}} \vec{r} \wedge (\overline{\overline{\tau}}\, \vec{n})ds \tag{7}$$

where S_{imp} represents the blades and impeller side walls, p the calculated pressure, $\vec{r}$ the position vector and $\overline{\overline{\tau}}$ the constraints tensor, which correspond to the molecular and turbulent viscosity. The transferred energy can be expressed under a non dimensional number ψ_t through the transferred power:

$$P_t = \vec{T_t} \cdot \vec{\Omega} = \rho Q E_t \quad \rightarrow \quad \psi_t = \frac{2E_t}{\omega^2 R_1^2} \tag{8}$$

Provided hydraulic power The expression of the provided power is given by the moment of momentum equation in the relative frame of reference. Let $\vec{T_f}$ to be the torque representative of the provided moment of momentum:

$$\vec{T_f} = -\int_{S_{tot}} \vec{r} \wedge (\rho\vec{w})\ \vec{w}\vec{n}ds + \int_V \vec{r} \wedge \left(\rho\vec{S}\right) dv + \int_{S_{tot}} \vec{r} \wedge (\overline{\overline{\tau}}\, \vec{n})\, ds \tag{9}$$

where $\vec{S}$ corresponds to the centrifugal and Coriolis effects, and where S_{tot} represents all the domain surfaces, delimited by the inlet and outlet of the impeller. It is then possible to express the provided energy in the form:

$$P_f = \vec{T_f} \cdot \vec{\Omega} = \rho Q E_f \quad \rightarrow \quad \psi_f = \frac{2E_f}{\omega^2 R_1^2} \tag{10}$$

In the Table 1 are reported, for three relative ψ_c values, the energy coefficient representative of the provided energy (ψ_f) and the transferred energy (ψ_t). For a quite constant transferred energy value, one remarks that the provided energy is decreasing with the pressure level ψ_c at the impeller inlet section. If one plots the calculated relative head evolution, which corresponds to the provided energy, versus the ψ_c value (Figure 9), it can be noticed that the ψ_{c_s}, which corresponds to the head drop beginning, coincides very well with the measured one. The results exposed in Figure 9 show that the calculated impeller head is over-estimated in comparison with the measurements. However, it has to be taken into account the fact that the presented measurements are representative of the entire pump, with the fixed parts (diffuser, inlet pipe,..). It is then reasonable to think that this over-estimation is mainly due to the losses in the fixed parts of the machine, which are not taking into account by the computation.

ψ_c/ψ_{ref}	$\psi_f[-]$	$\psi_t[-]$
cav. free.	1.129	1.28
0.383	1.12	1.285
0.265	1.1	1.28

TABLE 1. Impeller provided and transferred energy as a function of ψ_c ($\phi_1/\phi_{ref} = 1.21$)

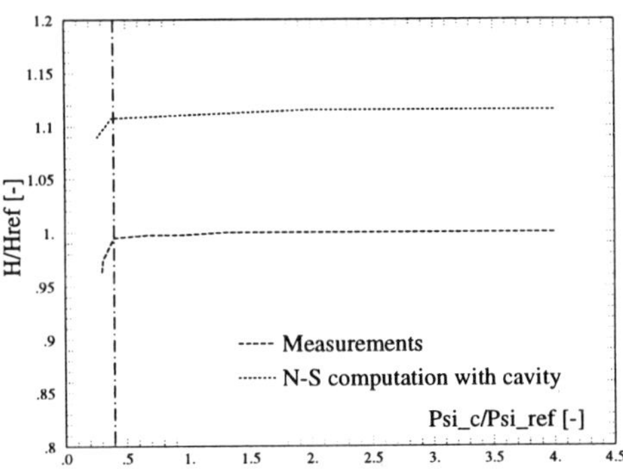

Figure 9. Predicted and measured head drop

7. Conclusion

Considering an impeller geometry, its behaviour and its main cavitation characteristics can be described with the results of a Navier-Stokes 3-D computation, using models of cavitation developments. The predicted behaviour of the cavitation flow in a centrifugal pump, using the proposed method, compares very well with the measurements.

It was brought to the for that the knowledge of the impeller inlet velocity field is essential for the cavitation behaviour prediction. Therefore, a flow computation of the inlet pipe was first performed.

The pressure level corresponding to cavitation inception is determined for all calculated operating points and compared with measurements. The mean standard deviation between the measured and calculated points is lower than 4%.

The cavity lengths computation using the *Rayleigh-Plesset* equation provides very good results. Indeed, the measured cavity lengths behaviour is faithfully reproduced by the calculation, and this for all measured operating points.

The prediction of the head drop beginning is done by a direct flow computation taking into account the cavitation sheet. The very good results obtained with this method proves its efficiency to predict head drop due to cavitation. As a conclusion, the results analysis of this study demonstrates that the presented method provides an interesting tool for the 3-D cavitation development prediction in hydraulic machines.

Figure 10. 3D view of the cavitation sheet

8. Acknowledgments

The authors are particularly grateful to the members of the IMHEF Cavitation research group. We wish to thank *SULZER Pumps* which made possible the publication of this paper and the staff of the test-rig who performed the measurements. This work is financially supported by the Swiss Federal "Commission d'Encouragement à la Recherche Scientifique", *SULZER Pumps, Hydro-Vevey S.A.* and *Rolla S.P. Propeller.*

9. Nomenclature

e	:	thickness	(m)
k_{rd}	:	coefficient of viscous losses	
p	:	static pressure	(N/m^2)
p_v	:	vapor pressure	(N/m^2)
Cp	:	pressure coefficient	$Cp = \frac{p-p_{\bar{1}}}{\frac{1}{2}\rho U_{\bar{1}}^2}$
E	:	energy	(J/kg)
H	:	head	(m)
P	:	power	(W)

Q	:	flow rate	(m^3/s)
R	:	impeller radius	(m)
S, A	:	area	(m^2)
T	:	torque	(Nm)
V	:	volume of the delimited domaine	(m^3)
$\vec{w}$	:	relative velocity	(m/s)
$\vec{r}$	:	position vector	(m)
$\vec{n}$	:	normalized vector	(m/s)
$\vec{\Omega}$	:	vector of the angular rotation speed	(s^{-1})
α	:	absolute flow angle	$(°)$
ω	:	angular rotation speed	(s^{-1})
ρ	:	density	(kg/m^3)
χ_c	:	local cavitation coefficient at impeller inlet	$\chi_c = \frac{p_{\bar{1}} - p_v}{\frac{1}{2}\rho U_{\bar{1}}^2}$
ϕ_1	:	flow rate coefficient	$\phi_1 = \frac{Q}{\pi\omega R_{\bar{1}}^3}$
ϕ_{ref}	:	design flow rate coefficient	
ψ_1	:	energy coefficient	$\psi_1 = \frac{2E}{\omega^2 R_1^2}$
ψ_{ref}	:	design energy coefficient	
ψ_c	:	net positive suction energy coefficient	$\psi_c = \frac{2NPSE}{\omega^2 R_{\bar{1}}^2}$

Subscripts

I	:	high pressure section of the machine (pump : volute outlet)
1	:	high pressure section of the impeller (pump : impeller outlet)
$\bar{1}$	:	low pressure section of the impeller (pump : impeller inlet)
$\bar{I}$	:	low pressure section of the machine (pump :inlet pipe inlet)
x	:	reference to a particular streamline
i	:	reference to a domain node in the flow direction
t	:	reference to the previous time step

References

1. F. Avellan. *Cavitation tests of hydraulic machines: Procedures and instrumentation.* ASME Winter Ann. meeting, 1993.
2. Y. Delannoy and J.L. Kueny. *Two phase flow approach in unsteady cavitation modeling.* ASME, Toronto, 1990.
3. Ph. Dupont and F. Avellan. *Numerical computation of leading egde cavity.* ASME Summer Ann. meeting, 1991.
4. Ph. Dupont, E. Parkinson, and W. Walther. *Cavitation development in a centrifugal pump: numerical and model tests predictions.* ASME Winter Ann. meeting, 1993.
5. J.P. Franc, F. Avellan, B. Belahadji, L. Brianon-Marjollet J.Y. Billard, D. Frchou, A. Akarimi D.H. Fruman, J.L. Kueny, and J.M. Michel. *La cavitation.* Presses universitaires de Grenoble, 1995.
6. R.T. Knapp, J.T. Daily, and F.G. Hammit. *Cavitation.* Mac Graw Hill, New-York, 1970.
7. A. Kubota, H. Kato, and H. Yamaguchi. *A new modelling of cavitating flows: a numerical study of unsteady cavitation on a hydrofoil section.* J.F.E, 1992.
8. T. Maitre, J.L. Kueny, P. Geai, and A. Von Kaenel. *Numerical prediction of three-dimensionnal partial cavitation in a rocket turbopump inducer.* ASME Summer Ann. meeting, Washington, 1993.

THE RELATION BETWEEN EROSION RIPPLES ON THE WETTED SURFACE OF HYDRAULIC TURBINE AND INSTABILITY WAVES IN THE TURBULENT BOUNDARY LAYER

SHEHUA HUANG * LIANGJUN CHENG **

* Department of River Engineering, Wuhan University of Hydraulic & Electric Engineering, Wuhan, 430072 P.R.China

** Department of Electric Power Engineering, Huazhong University of Science & Technology, Wuhan, 430074 P.R.China

Abstract The perturbation theory is used to analyze the hydrodynamic instability of the viscous sublayer in the turbulent flow near the wall region. Based on the relation between the viscous sublayer and the coherent structure in the turbulent boundary layer, a simplified turbulent coherent structure model, or one for the unstable wavelike motion of the Kelvin-Helmholtz vortex layer with approximated continuous distribution under the action of perturbation, is proposed. This model has been used to study the motion of dilute particles in the turbulent boundary layer, and it has been found that the coherent structure plays a dominant role in the motion of dilute particles.

Keyword erosion ripples, turbulent coherent structure, hydraulic turbine, solid-liquid two phase flow.

1. Introduction

The operation of hydraulic machinery in silt-laden flow always results in a serious abrasion damage of its wetted surface. Of various silt-abrasion patterns of wetted surface in hydraulic machinery, ripples are most common. Research of erosion ripples formation is the first step to study any other more complicated erosion pattern. Because ripples often appear in flat wetted surface, it is reasonable to consider that the structures of solid-liquid two phase boundary layer and motions of particles in it have critical effects on formation of erosion ripple on the rigid surface. It is well known that boundary layer theories are very complex and in development now, so not to say solid-liquid two phase boundary layer theories. For this reason, most of research information on erosion ripple formation in hydraulic machinery is based on experiments[1]. In this paper, being combined with theory of turbulent coherent structure, the properties of instability wave in turbulent boundary layer and particle motion in it have been analyzed for the first time, and some meaningful results have been obtained.

E. Cabrera et al. (eds.), Hydraulic Machinery and Cavitation, 614–621.

2. Similarities between Turbulent Boundary Layer and Transition Region of Laminar Boundary Layer

According to boundary layer theories, laminar boundary layer and turbulent boundary layer are different from each other. The remarkable distinction between them is that there is different distribution of average velocity normal to boundary. Based upon the Prandtl's thoughts, bounded flow field can be divided into two areas, that is, boundary layer flow area and potential flow area. In the boundary layer, flow pattern depends on the ratio of viscous force to inertial force. When Reynolds number is small, flow is laminar, while Reynolds number is large and viscous force near boundary is not negligible yet, flow appears turbulent. If Reynolds number is between the two conditions mentioned above, boundary layer flow pattern presents a transitional state in which the laminar flow and turbulence coexist. Generally, when we consider that laminar flow and turbulent flow are completely two different flow pattern, the emphasis of thinking is on the difference between them, but at the same time it will be unfavorable for the similarity of them to be neglected. Based on the mention above, what follows is the investigation of similarities of turbulent boundary layer to laminar boundary layer, which will be conducive to understanding flow structure of turbulent boundary layer and the formation of erosion ripples on the surface of rigid boundary.
In turbulent boundary layer, no matter how large the Reynolds number is, there are always three layers of main flow structure, that is, viscous sublayer, transitional layer and logarithm law's region, although the thickness of sublayer will become thin along with the increase of Reynolds number. Now suppose that viscous sublayer, transitional layer and logarithm law's region respectively have a certain similarity to laminar flow, that is, in turbulent boundary, velocity distribution of viscous sublayer is similar to parabolic velocity distribution of laminar flow, while velocity distribution in core region of turbulence corresponds to logarithm law velocity distribution of turbulent flow. There is no doubt for the similarity of the latter. As for the former, since velocity distribution of viscous sublayer is usually considered to be straight line, turbulent boundary layer flow over horizontal half infinite flat plate shown in Figure 1 being taken as example in which viscous sublayer flow is discussed, the similarity of sublayer flow to laminar flow is verified as follows.
In Figure 1, suppose streamwise velocity U is steady and uniform along flow direction, then

$$\frac{\partial u}{\partial t} = \frac{\partial u}{\partial x} = 0, \qquad \frac{\partial p}{\partial x} = 0.$$

Corresponding equations of motion are:

$$v\frac{\partial u}{\partial y} = \nu\frac{\partial^2 u}{\partial y^2}, \qquad (1)$$

$$\frac{\partial v}{\partial y} = 0. \qquad (2)$$

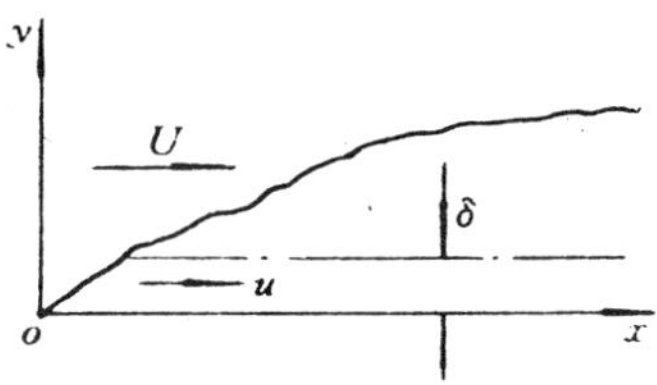

Figure 1 flow in sublayer region

From(2) one can get $v=v_0$, or $v=0$, here v_0 being a constant. Take $v=v_0$, then substitute it into equation (1) and integrate the equation:

$$u = C_1 + C_2 \exp(v_0 y / \nu). \tag{3}$$

Provide boundary conditions as:

$$y=0,\ u=0; \qquad y=\delta,\ u=U.$$

With above boundary conditions one can determine the constant C_1, C_2 in (3) and get:

$$u = U[1-\exp(v_0 y / \nu)]/[1-\exp(v_0 \delta / \nu)]. \tag{4}$$

Let $\bar{u} = u/U$, $R = v_0 y / \nu$, $\bar{y} = y/\delta$, then equation (4) becomes

$$\bar{u} = [1-\exp(R\bar{y})]/[1-\exp(R)]. \tag{5}$$

In addition, Taylor series of e^x is

$$e^x = 1 + x + x^2/2! + x^3/3! + \cdots,$$

Since the value of $(R\bar{y})$ is small here, as approximation, the forth three terms of the series can be taken, that is

$$\exp(R\bar{y}) \approx 1 + R\bar{y} + (R\bar{y})^2/2! \quad , \tag{6}$$

Substitute (6) into (5) and get

$$u = \frac{R\bar{y} + (R\bar{y})^2}{\exp(R) - 1} \quad . \tag{7}$$

It can be seen from (7) that streamwise velocity distribution in viscous sublayer is parabolic. Because v_0 is very small, $R \to 0$, then the limit of (7) for $R \to 0$ is $\bar{u} = \bar{y}$. Therefore, the common character of velocity distribution in viscous sublayer is the parabolic distribution with its limit for $R \to 0$ being straight line. The character of velocity distribution mentioned above is shown in Figure 2 .

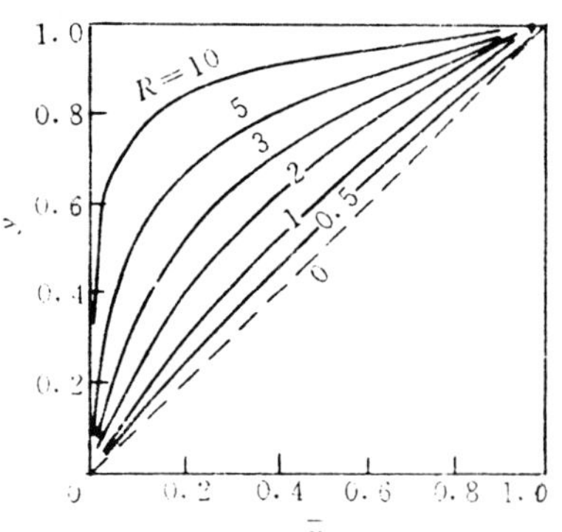

Figure 2 velocity distribution in viscous sublayer

Through above analysis, the similarity of viscous sublayer to laminar flow has been verified. Further more, taking account of the character that in turbulent boundary layer, viscous sublayer, intermediate sublayer and logarithm law zone coexist and interact with each other and result in flow instability, It is reasonable to consider that there are some similarities between laminar transitional boundary layer and turbulent boundary layer at least in the respect of flow instability. At this point, Blackwelder[2] also point out that for turbulent boundary layer, the coherent structures associated with bursting process near the wall region is similar to the instability waves of laminar transitional boundary layer flow over flat plate or concave plate in quite a few important aspects.

3. Analysis of Stability of Viscous Sublayer

Based upon linear instability theory, in this section, flow stability of viscous sublayer in turbulent boundary layer is analyzed. Suppose two dimensional viscous sublayer is disturbed by some external perturbation which can be represented as stream function with a form of Tollimen -Schlichting travelling waves :

$$\phi = \phi(\bar{y})\exp[i\alpha(\bar{x} - \bar{c}t)]$$

The linear perturbations meet the Orr-Sommerfeld equation :

$$(\bar{u} - \bar{c})(\phi''' - \alpha^2\phi) - \bar{u}''\phi = (\frac{i}{\alpha R})(\phi^{(4)} - 2\alpha^2\phi'' + \alpha^4\phi) \quad , \tag{8}$$

where δ is the thickness of viscous sublayer, λ is wave length of perturbation, $\alpha = 2\pi\delta/\lambda$ is wave number, $\bar{u} = u/U$ is non-dimension phase velocity and U is external flow velocity, $R = U\delta/\nu$ is Reynolds number of viscous sublayer, while all derivatives are with respect to $\bar{y} = y/\delta$.

On boundary there are non-slip definition:

$$\phi(0) = \phi'(0) = 0 \tag{9}$$

For convenience of analysis, assume that flow regions outside viscous sublayer are completely turbulent flow with characteristic Reynolds number being very large, the flow can be regarded as non-viscosity flow. So Orr-Somerfield equation reduces to non-viscosity differential equation. At the outside boundary of viscous sublayer($y = \delta$), there is,

$$\phi'' - \alpha^2\phi = 0, \qquad (\bar{y} \geq 1). \tag{10}$$

General solution of (10) can be write as:

$$\phi = C\exp(\pm\alpha\bar{y}). \tag{11}$$

where a solution corresponding to signal "+" is not consistent with physical meanings and will be omitted .Differentiate (11)and we have

$$\phi' + \alpha\phi = 0 \qquad (\bar{y} \geq 1). \tag{12}$$

Hence there have already been four boundary conditions, that is, (9), (10) and(12) for O-S equation. On the numerical method to get the solution of the equation (8), one can refer to literature[3]. Here are the numerical results quoted from literature [4] (see Table 1).

Table 1 Relations between Reynolds number (R) and the parameters of neutral stable perturbation waves

$\bar{c}$	0	0.10	0.20	0.30	0.40	0.50	0.60	0.70
α	0	0.05	0.118	0.213	0.342	0.537	0.809	1.066
R	∞	2.415×10^5	1.292×10^4	2124.9	558.2	191.9	70.0	33.4

Due to the character that in viscous sublayer Reynolds number is approximately constant, let $R=136$, then the parameters of Tollimen-Schlichting travelling wave in neutral stability can be interpolated from Table 1:

$$\bar{c} = c/U \approx 0.52, \qquad \alpha = \frac{2\pi\delta}{\lambda} \approx 0.625,$$

Namely $\lambda \approx 10\delta$.

If amplitude Y of perturbation is the same order of magnitude as the thickness δ of viscous sublayer, approximately there will be $\lambda \approx 10Y$. Although this result is very close to the experimental values of literature [6] in which the ratios of wavenumber of erosion ripples to its amplitudes are changed from 9.2 to 10.5, the mechanism of the ripple formation are not simply attributed to solid particle motion in viscous sublayer, because the thickness of viscous sublayer is not only very thin (10^{-2} cm order of magnitude), but also the flow velocity in it is very slow. In fact, suspended particles' diameters in rivers are greater than or equal to the above magnitude, so in viscous sublayer there are not enough energy is to control particle's motion.

On the other hands, the effects of boundary's roughness on the instability of viscous sublayer should be taken into account. Assume that there is not flow separation in viscous sublayer, then the micro-irregularity of boundary is bound to be reflected into flows of viscous sublayer. So in viscous sublayer, streamline patterns should be similar to the shape of micro-irregularities of rigid boundary. We express boundary roughness function Δ as Fourier series:

$$\Delta = \sum_{k=1}^{\infty} \alpha_k \exp(ikx) \tag{13}$$

To sum up, it is reasonable to consider that in the viscous sublayer turbulent boundary layer, small and instable wavelike disturbance exists and can be written as Fourier series following:

$$\phi = \sum_{k=1}^{\infty} \phi_k(\bar{y}) \exp[ik(\bar{x} - \bar{c}t)] \tag{14}$$

4. Turbulent Coherent Structures near the Wall and Dilute Particle's Motion in it

The analysis in paragraph 2 indicate that viscous sublayer has no direct effect on particle's motion in turbulent boundary. So we can imagine that as a trigger of turbulence near the wall, it is possible for turbulent coherent structure to control the solid particle's motion. Although coherent structure is very complex, in order to investigate particle's motion in the boundary layer, Here is a simplified model of coherent structure.

As shown in Figure 3, turbulent boundary layer flow is roughly divided into two section, i.e. , viscous sublayer and logarithm law's region. In turbulent boundary layer, a section I--I having been chosen arbitrarily, it is obvious that distribution of average velocity within viscous sublayer, as mentioned in paragraph 1, is parabolic while that in the other section, i.e., turbulent core region observes logarithm law. Under limit conditions, the velocity at the interface between viscous sublayer and "logarithm law " region is discontinued and corresponding, the velocity gradient normal to streamwise direction is infinity. Practically although velocity should be continuous due to viscosity and there is a intermediate layer between viscous layer and logarithm law region, the velocity gradient is very great in this intermediate layer, so it is possible for large scale spanwise vortex to generate in the intermediate layer. We now describe it with spanwise K-H vortex layer which are nearly distributed continuously along stream wise direction. It is well known that K-H vortex is extremely instable. Hence the small disturbance which always lies in viscous sublayer will have effects on the K-H vortex layer, Providing that K-H vortex layer waver at limited amplitude under act of small disturbance, and the waves are similar to (14) but its amplitude is amplified. For convenience of analysis, the coordinates of position for vortex center are accepted in followings instead of stream function, that is,

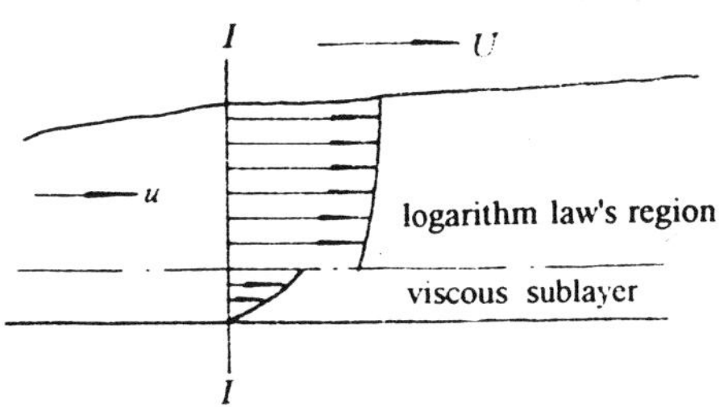

Figure 3 flow sketch in turbulent boundary

$$f(x,t)=\sum_{k=1}^{\infty} A_k \exp[ik(x-ct)] \tag{15}$$

When the vortex layer wavers into high velocity region, it will be accelerated. otherwise it decelerated. For any vortex line, there are high velocity section and slow velocity section alternatively along vortex line in the spanwise direction, and due to action of acceleration and deceleration, the vortex line is twisted and stretched with "horseshoe" shape or "hairpin" pattern structure being formed. At last vortex structure breaks off into random turbulence. After that, the above quasi-periodic process begin again. It is necessary to point out that the above process doesn't last very long and is often covered by the random characteristics of turbulence ,So, for a long time coherent structure is not understood well. We think that K-H vortex layer and its wave like movement is the main feature of coherent structure near wall ,with other characters of coherent structure being derived from it. What follows are application of above simplified model of coherent structure to analysis of particle's motion in the turbulent boundary. Under the condition of dilute particles phase the existence of particles hasn't great effects on the flow structure in liquid phase boundary layer. Further more, if collision between particles is neglected, dilute particles motion will be similar to that of a single particle. For a particle the diameter of which is d, when moving near the simplified coherent structure as shown in Figure 4, it will be drawn into the vortex layer by pressure gradient. Suppose the particle moves at the same streamwise velocity as the vortex layer, but at the direction normal to wall the particle is forced to vibrate under the action of coherent structure. If we take the effects of particle's inertia into account, the harmonic wave which governs the particle's vibration will be only one which wave number is the same as or near to particle's harmonic frequency. So particle vibration at balance point can be expressed as:

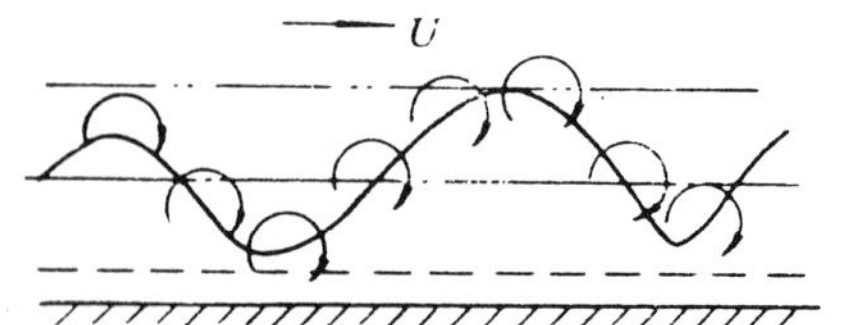

Figure 4 simplified model of coherent structure

$$f_d(x,t) = \sum_{k=1}^{\infty} A_{kd} \exp[iK_d(x-ct) - \delta_d] \tag{16}$$

Where K_d is the wave number corresponding to particle's harmonic frequency, A_{kd} is amplitude corresponding to K_d, δ_d is phase lag due to velocity difference between fluid and particle.

Given that particle is in balance point when $t=0$, that is,

$$x_0 = 0, \qquad y_0 = 0 \quad ,$$

then after the time interval of t the displacement of the particle becomes

$$x=ut \quad , \tag{17}$$

$$y = \text{Real}\{f(x,t)\}$$
$$= A_{kd} \cos[K_d(x-ct) - \delta_d] \tag{18}$$

Substituting (17)into (18)to eliminate parameter variable t, one obtains

$$y = A_{Kd} \cos[K_d(1-c/u)x - \delta_d] \ . \tag{19}$$

From equation (19), it is indicated that the trajectory of particle motion in turbulent boundary layer is a cosine wavelike curve. Since solid wall is always flat plate or regular curved surface, particle's motion result in impingement on the wall, and the effect of impact is attempted to have boundary shape suitable to particle's motion. At last surface ripples have arisen.

4. Conclusion

Through analysis of the similarities between turbulent boundary layer and laminar transitional boundary layer, in this part, it is revealed that the erosion ripple formation on wetted surface of hydraulic turbine is association with the instability wavelike motion of fluid near wall region of turbulent boundary layer. Based on the relation between viscous sublayer and coherent structure, a simplified turbulent coherent structure model that an unstable wavelike motion of the Kelvin-Helmholtz vortex layer with approximately continuous distribution along flow direction under the action of perturbation is proposed. This model is applied to analyze dilute particles' motion in turbulent boundary layer, and the conclusion is obtained that the trajectories of particles are sinusoidal curve, the amplitude and wave length of which mainly depend on external flow velocity, streamwise velocity of Kelvin-Helmholtz vortex layer, particle's diameter and boundary roughness. The quantitative relation between these parameters remains to be investigated further.

Acknowledgment This work is supported by doctoral fund for institutes of higher learning, National Education Committee of China .

References

1. Karimi, A., Schmid, R. K., Ripple formation in solid liquid erosion. *Wear*, 156(1), 1992. 33-47.
2. Blackwelder, R. F., Analogies between transitional and turbulent boundary layers. *Phys. of Fluids*, 26(10), 1983. 2807-2815.
3. Lin C. C., Theory of hydrodynamic stability. Cambr. Univ. Press. 1955.
4. Hou H. C., *Basic problems of river dynamics.* Hydraulic and Electric Power Press, Beijing. 1982.
5. Stephen, K. H., Coherent motions in the turbulent boundary layer. *Annu. Rev. of Fluid Mech.* 23. 1991. 605-639.
6. Yu W. G., A study of the abrasive erosion appearance of fish-scale or ripples in silt-laden flow. Thesis of MS. degree, IWHR in Beijing. 1987.

ACOUSTIC METHOD AND ITS APPLICATIONS ON MEASURING AND JUDGING CAVITATION OF HYDRAULIC TURBINE

LIU KEHUANG, YANG CHUN
Yangtze River Scientific Research Institute
Wuhan. China

1. Introduction

In past 15 years, acoustic methods on judging cavitation has been adopted in vacuum model tests and prototype observations in hydraulic structures in the institute.The researches has been made on the physical machanism and acoustic characters of cavitation, and its measuring and judging methods . This paper will give some of the researches related with hydraulic turbines.

1.1. CAVITATION TYPES AND DAMEGES OF HYDRAULIC TURBINE

Applying Bernoulli's equation to upper and lower pool levels of a hydraulic turbine, the fellowing equation can be obtained as cited in [1] and [2]:

$$\left(\frac{P_k}{r}-\frac{P_d}{r}\right)/H=\left(\frac{P_a}{r}-H_s-\frac{P_d}{r}\right)/H-(W_k^2-V_k^2)/2gH-(V_2^2-W_2^2)/2gH$$
$$-\eta_v V_2^2/2gH-h_{k-2}/H\pm\triangle h_{k-2}/H \qquad (1)$$

in which P_a , P_k and P_d are the atmospheric, minmum and vapour pressures, respectively; H is the total water head and H_s is the suction head of the turbine; W,U and V are relative, tangential and absolute velosities, respectively; H_{k-2} is the height of the minmum pressure point k to the exit point 2;H_{0-2} is the height of the point 2 to the point 0 at the centre line of the turbine; ΔH_{k-2} is the head loss between the point k and the point 2; η_2 is the hydraulic efficiency of the tailrace.

At $P_r/r = P_d/r$, incipient cavitation occures, then

$$\left(\frac{P_a}{r}-H_s-\frac{P_d}{r}\right)/H=(W_k^2-V_k^2)/2gH-(V^2-W_2^2)/2gH+\eta_v V_2^2/2gH$$
$$+(h_{k-2}/H-\triangle h_{k-2}\pm h_{0-2}/H \qquad (2)$$

in which,let σ_y be the left term, which is called as cavitation index of turbine facility or external character cavitation index; let σ_T be the right term of the equation, which is called as inner character cavitation index. Let σ_1 , σ_2 , σ_3 and σ_4 be the first, the second, the third and the rest items of the right of equation, which represent the cavitation indexes of the point k, the point 2, the tailrace and the runner struc-

E. Cabrera et al. (eds.), Hydraulic Machinery and Cavitation, 622–631.

ture. Then the equation 2 becomes

$$\sigma_y = \sigma_T = \sigma_1 + \sigma_2 + \sigma_3 + \sigma_4 \qquad (3)$$

The equations [2] or [3] are called cavitation incipience equation, which describes cavitation incipience state of a turbine. For $\sigma_y > \sigma_T$, no cavitation occures.

It is indicated that turbine cavitation may have different types such as vane cavitation (represented by σ_1), cavity cavitation (by σ_2), vortex cavitation (by σ_3), cavity and vortex related cavitation (by σ_4). Those types have some relationship among each other. Turbine cavitation may cause three dameges such as erosion, vibration and energy efficiency decreasing, which may cause safety problem or high cost.

1.2 TURBINE CAVITATION STUDY

Up to now,equation [2] can not be solved accuratly.Turbine cavitation study is mainly depended on model tests. In model studies, it was often used to change Pa/r in σ_y that means to change vaccum, and to determine the cavitation incipient index σ_i or ctitical index σ_k by measuring energy coefficients. According to the measured indexes and coefficients, the suction height and install elevation of the turbine are designed. This method is called as external character method.

The method is fail to reveal The turbine cavitation type and feature. It only focucuses on one of the three possible cavitation damages. Therefore, many turbines designed by the method have different cavitation erosion and sonic vibration problems around the world. On other hand, it is difficult to determine the suction height Hs reasonablly by this method, that is a important coefficient for cost. Fur thermore, this method can not be used in prototype studies.

1.3 CAVITATION MEASURING TECHNIQUE

As mentioned above, it is very important to develop cavitation measuring technique in studing turbine cavitation.Some turbine experts even indicated as citied in [3] that advanced cavitation measuring technique would be one of the keys for large hydraulic turbine development.

Acoustic method is a modern technique in measuring and judging cavitation. In past 15 years,it has been used in more than 10 hydraulic projects including spillways,navigation locks, and hydropower plants both in models and prototypes in my insititute [4], [5]. Based on our practice and experience, acoustic method has the following advantages:

Theoretical reliability. It can reveal types, locations, characters and damage potential of cavitation.

High sensitivity. It can make the judgement comprehensively from time, frequency, amplitude and energy of underwater noise, which ensures the judgement being accurate and reliable.

Convenient to use. It is convenient to use both in model and prototype with little requirement of working environment and conditions.

2. Acoustic Method And Principle

2.1 PRINCIPLE OF ACOUSITIC METHOD [6] [7] [9~12]

In any non-steady pressure process, there are underwater noises in a flow field. The pressure in the field will be

$$P(x,y,z,t) = P_o(x,y,z,t) + P'(x,y,z,t) \tag{4}$$

in which P' is the sonic component of the pressure, The dynamic sonic sources equation in a fluid field is

$$\nabla^2 P' - \frac{1}{C_o^2} P' = -q' + \nabla \overline{f}' + \frac{\partial \tau_{ij}^2}{\partial x_i \partial x_j} \tag{5}$$

The three items on the right side are the main sonic radiant sources, which represent the sources caused by the volume variation, force fluctuation and turbulence of the fluid in turns.The three sources have the charactiristics of single pole,dipole and quadrupole, which radiant efficiencies are in proportion to the first power, the third power and the fifth power of Mach number. Generlly Mach number in water flow is much less than unit so that the sonic radiant energy efficiency of volume variation is the highest.The cavitation bubble is the most significant singlepole sonic source, so that it can be easily measured by acoustic method.

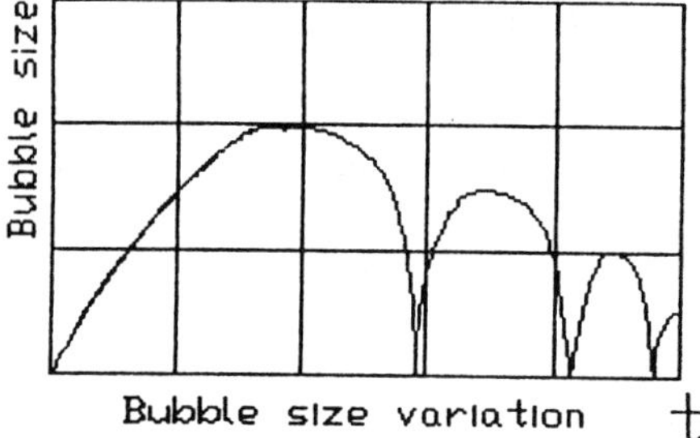

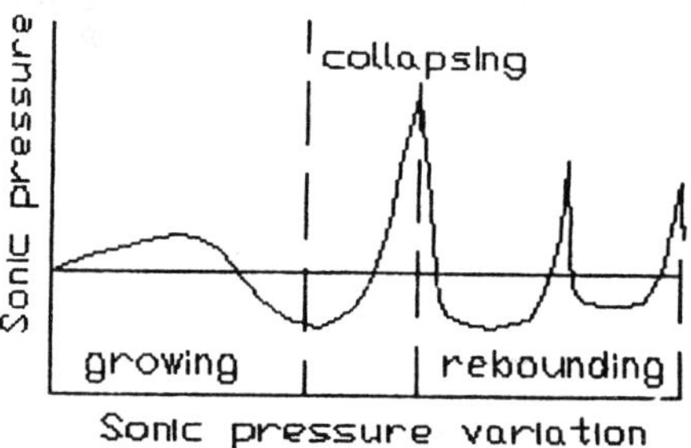

Figure 1. A ssingle bubble development

2.2 CAVITATION NOISE CHARACTER OF A SINGLE BUBBLE

Generally, the process of a single cavitation bubble can be devided into 4 stages as growing,compressing,collapsing and rebounding as shown in Figure.1. The process can be described as follow: a critical gas nucleu expands at certain dynamic pressure or sonic pressure and forms a bubble.If the bubble size was larger than the critical gas nucleu or expanded in a low pressure zone, its vibration frequency would change.If the frequency was lower than that of dynamic pressure, the bubble would only vibrate non-linely, which means cavitation incipience. If the bubble moved to a high pressure zone, the bubble would be compressed with vibrating frequency increasing. If the frequency was higher than that of dynamic pressure, the bubble would collapse and rebound immediately, which is cavatation development.

By analyzing the bubble character, it is indicated that the size of a bubble can not be compressed to zero,but a bubble may collapse and rebound to minmun size with maximum pressure.The smaller the size is compressed, The higher the shock wave pressure is produced, which will radiate intense cavitation noise.

The cavitation noise have the following charactiristics.

In Time Domain. For steady pressure,collapsing time of a bubble is about 75 persent of that of growing. For non-steady pressure, with dynamic pressure increasing, the time of callopsing and growing and the size of the bubble are increased.

In amplitude domain (sonic pressure). During a bubble growing and collapsing, there are a hegetive low frequency vibrating component and a positive sharp peak of sonic pressure. It is indicated by analyzing the sonic pressure equation that the positive peak occurs at the time of the bubble compressed to its minmum size. If the gasous pressure rate within the bubble was larger than 0.2, the positive peak value would be less than that of the negetive.If the rate was less than 0.2, the peak would hundreds times of that of the negetive.

In frequency domain. During a bubble growing, the sonic pressure is negetive with low frequency. During a bubble collapsing, the pressure appears pluses with high frequency components increasing significantly. The feature of pulses determines, the attenuation character of the high frequency componenents. It can be proved theoretically that, for a sharp pulse, a right angle pulse, and a triangle pulse a trapezoid pulse, the spectra attenuate in a rate of 0 dB/oct, 6 dB/oct, 8~10 dB/oct, 10~12 dB/oct, respectively, after vibration modulation of bubbles with dominant distributed radius. This is one of the basis of judging with or without, which type of, and damage potanlial of cavitation.

For small gas content, the sonic pressure appears shock wave. Its spectrum has low attenuate rate,which indicates vapour cavitation with high cavitation erosion potontial. For large gas content (in case of gas was cavitation or vortex cavitation), the sonic pressure spectrum has high attenuate rate, which means little cavitaion erosion but may cause sonic vibration.

In sonic energy. The sonic radiant energy is almost less than 1 persent of the potontial energy of bubble in growing. However, the energy ratio of radiant and potential of the bubble is far larger in collapsing than in growing.For air content rate of a bubble being less than 0.5 persent, The ratio is about 67 persent. For the rate being 0.5 to 10 persent, the ratio is 66 to 15 persent. For the rate larger than 10 persent, the ratio is less than 15 persent.

In acturally flow field, the underwater noise is determined by the character of averaged dominant bubbles,which is basically similar to that of a single bubble but with wider frequency band.Therefore, the acoustic charactiristics of a single bubble is the basis of the acoustic method of judging cavitation.

2.3 MEASURING AND ANALYZING SYSTEM

The measuring and analyzing system consists of hydrophones, signal amplifiers, a tape recorder, a FFT dynamic singnal analyzer, a PC and output equipment. To measure reliablly underwater noises and to judge cavitations, the basical requirements of the system are multichannels, high sensitivity, wide frequency band (200HZ to 250KHZ), good directivity (to receive singnal from a given direction but the whole flow field), convenient usability, and high data process ability.

2.4 JUDGING METHODS BY CAVITATION NOISE

By using underwater noises, some methods have been used in our cavitation studies both in models and prototypes such as methods of sonic wave pattern, correlation function, spectrum and its attenuation, sonic energy, shock wave, air supply, and 3D display, ect. They gave satisfactory results and good accordance with field cavitation inspections, felled vibrations and other simultaneous measured data as pressures, vibrations and swings, air entretment and load variations. The four methods used in this paper are as fellow:

Spectra level difference. Because of the difference of bubble air content rate,the sonic energy caused by bubble collapsing may be times or hundreds times of that bubble growing. The sonic energy level difference may be up to 10dB to 20dB between both cases.Therefore, this method can be used for comparing different sonic processes such as testing condition with "backgrowd" in a vaccum tank, different operation processes, and testing processes with different vaccum. In practice, a Fourier spectrum is used for comparing sonic energy, which called sonic pressure spectrum level method. The spectrum is got from one third octave band analyzing and frequency band modifing. In the spectrum level comparison, the larger the different in high frequency componants of the spectra, the more intense the cavitation.
Spectrum Attenuation. According to a single bubble character and our practice, within 2 to 3 octaves, if the spectrum attenuation rate is larger than 8 dB/oct, there is no cavitation, if the rate is about 6 to 7dB/oct, there is vortex type or gasous cavitation,which would not cause erosion but might produce sonic vibration;if the rate is less than 6 dB/oct, there is developed cavitation. The less the rate, the more intense the cavitation.
High Frequency Shode Ware. This method is used to analyze high frequency characters of sonic preaaure. (frequency higher than 10KHZ). If the spectrum attenuation was very small near to zero in the frequence band, There would occure sharp pulses in sonic pattern, which would mean shock wave type cavitation developed.
Sonic Enercy. Sonic energy can be obtained by three ways. One is to get tatol sonic enercy level directly from the one third octave band analyzing. Another is to sum up the componants of different frequence from the spectrum. The third is to take the value of correlation function at $\tau = 0$. The sonic enercy method can be used to comparing different conditions.

Above methods can be used individually or combinedly.The shock wave method may used to judge vapour cavitation collapsing and rebounding with high frequency. The spectrum attenuation method is suitble to reveal gasous cavitation or cavatation incipience with low frequency.To ensure a reliable judgement,different methods may be used as comparing to analyze whole process of cavitation development.

3. Applications In Hydraulic Turbine Study

3.1 APPLICATION IN A FRANCIS TURBINE MODEL

The diameter of turbine runner is 250mm.14 operating cases were tested with water head being H_{max}=198m, H_p=162m, H_{min}=135m in prototype. The opening ratios of the guide blade were full open, 60 and 30 persents,. and that in accordance with optimum efficiency and 5 persent output restraint curve of the turbine.

In each case, the test was conducted with different σ_y by changing vaccum in the model. Four hydrophones were installed at the top cover Z_{13}, the base ring Z_0, the entrance and exit of the vertical cone of the tailrace Z_{21} and Z_2 . High speed photographing, video recording, and pressure fluctuation measuring were simultaneously taken for comparision.

Four types of cavitation were revealed as vane cavitation of blade, vetical axis vortex cavitation, eccentric vortex cavitation and cavity cavitation. For better understanding, the cavitation of blade is graded into four developing stages as , , and , which in turns describle the cavitation bubbles looking like a thin foil, sheets, a cloud and a intermittent vortex band.

Figure 2 gives the relationships of the measured sonic energy $Z(0)+\Sigma Z(f_i)$ with cavitation index σ_y for four cases. The following table gives the test conditions and results of critical cavitation indexes.

The critical indexes were determined by sonic energy method. For gasous or vortex cavitation, the sonic energy would increase with σ_y to a maximum value and then would decrease sharpdly. the reaso is that with this type cavitation developing, the air content of the bubble is increased. At a certain air concentration the sonic energy would be absorbed, that caused sonic signals becoming " weak ". Therefore,the inflection point before the maxmum value will accordete to the critical index. For vapour type cavitation, the sonic energy would increase with σ_y to a inflection point and then it would increase rapidtly with a steeper gradient, because of sonic energy character of this cavitation development. Hence, the inflection point is accorded to the critical index.

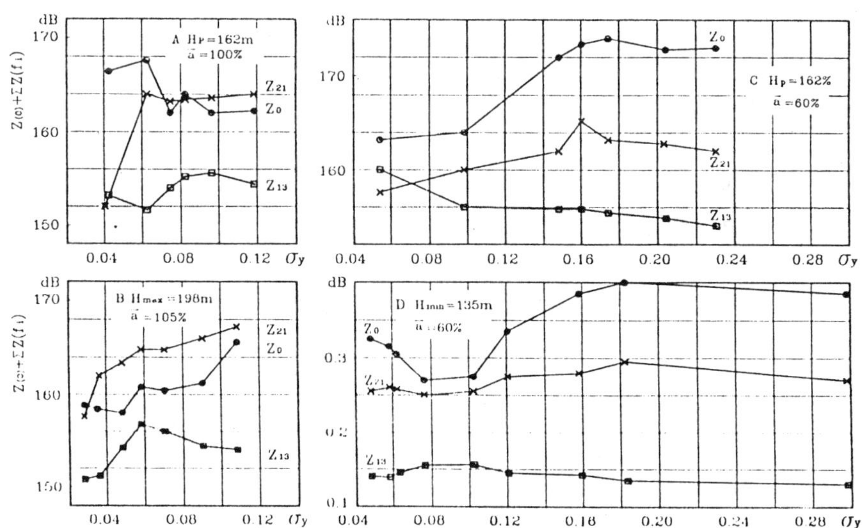

Figure 2. Relation ship of sonic energy with vaccum

For cases in above table, the cavitation type and development can be described as follow.

Tabie. Criticai Caoiiation Index of onic Energy Method anei eotranai characteristics Method

Test No	Test Condition			onic Energy Method						ebtetnai characieristis method	
	Hp(m)	a (%)	6r	6k			6i			6k'	6i'
				Z_{21}	Z_0	Z_{13}	Z_{21}	Z_0	Z_{13}		
1	162	100	0.041~0.119	0.062	0.083	0.090	>0.119	0.094	>0.119	0.064	0.119
2	198	105	0.048~0.127	>0.127	>0.127	0.078	>0.127	>0.127	>0.127	0.068	0.102
3	162	60	0.037~0.212	0.140	0.154	<0.08			>0.212	0.053	
4	135	60	0.049~0.299	0.183	0.183	0.102			0.183	0.062	0.092

Test No.1 For σy = 0.119, vane cavitation in grade occured at the upper middle part and the lower part of the blade. With σy decreasing, the cavitation at the upper part would tend to develop into middle of the blade, and the cavitation at the lower part would develop much intensely. For σy = 0.062, the cavitation at the lower part developed into grade , that there was full of cavitation cloud.

Test No.2 For σ = 0.127, vane cavitation in grade occured at the middle and the lower parts of the blade entrance side; vortex cavitation appeared at the vertical cone, which sized like a "rugby" ball. With σy decreasing, the vane cavitation moved up wards and the vortex cavitation developed looking like a " sylender ". For σy = 0.048,the vane cavitattion at the upper middle part of the blade back moved into the head and the middle of blade; and those at the lower part developed into grade.The vortex cavitation also developed with larger size, looking like a "gyro".

Test No.3 For σy = 0.212, vane cavitation in grade occured at the exit of the blade back, and in grade at the head of the blade.Some small vortex band cavitation appeared at the exit of blade. With σy decreasing, the cavitation at the exit remained in grade , but at entrance of the blade back, the cavitation developed into grade and grade . The vortex band appeared in the cone. For σy = 0.087, the flow field was fully covered will vortex band, the vane cavitation could not be recognized.

Test No.4 For σy = 0.29, No cavitation occured at the blade, some small vortex bands could be seen occasionally in the cone. For σy = 0.183, vane cavitation in grade occured at the lower part of the entrance of blade; and vane cavitation in grade appeared at the upper and the middle parts of the blade, the vortex band developed into a larger size. With σy decreasing, the cavitation at the blade entrance side developed into grade .The eccentric vortex bands became more intense and finally there were full of cavitation clouds the field.The vane cavitation could not be recognized.

With the cavitation description and the critical indexes given in the table, the followings were indicated.

The critical cavitation indexes determined by the sonic energy method were good accordance with the cavitation bubble development and characterstics.

The cavitation at the tailrace is gasous type vortex cavitation, the cavitation at the blade is varpour type cavitation, and the cavitation near the base ring is combined of the above 2 types on the test conditions. The results of cavitation judgement by the method are basically accordance with the principle and practice of hydranulic turbine cavitation.

The sonic energy method could measure and judge the type and location of critical cavitation index, but the external method has only one critical index, which does not accordant with the prinicple and practice of turbine cavitation.

The critical index in external method is less than that in the sonic method, which is not safe for turbine design. This has been proved by many turbines in operation.For example,if vane cavitations in grade appeared at the blade back,which was caused by uniform negitive pressure, the left force would increase, that means higher energy efficiency. However, when the energy efficiency decreased, the vane cavitations would have much intensly developed. Therefore, if the critical index was determined according to one persent of energy efficiency decreasing, it would be too small to be safety.

3.2 APPLICATION IN A FRANCIS TURBINE

There are 4 Francis turbines installed in Geheyian hydropower plant, Hubei, with capacity 300MW for each unit. The water heads are designed as H_{max}=125m, H_p =103m and H_{min}=87m. The turbines 1# and 2# are made in Canada, 3# and 4# in china. A prototype observation was conducted at water head H=115m on the turbine 1#. The purposes of the observation were to study the hydraulic steability of the turbine, the transient character of the intaking system, the efficiency of the turbine,and safety problems as cavitation and vibration, ect.The observations included pressure and pressure fluctuation, underwater noise, air suppliment and air noise, energy output and discharge (by water hammer method),vibration and swing, ect. Before the observation, the turbine system was dewatered for inspecting cavitation erosion. This paper will focus the cavitation problem only.

Three hydrophones were installed outside the steel plant, in which Z_1 was at the scroll case as the sonic " background ", Z_2 was at the top cover for measuring sthe cavitation noise from the runner,Z_3 was at the entrance of the tailrace to measuring the vortex cavitation noise.

3.2.1 Sonic Energy Method

Figure 3 gives the sonic energy variations with the turbine output. It indicated:

The everage sonic energy at the tailrace was 10dB larger than that at the scroll case. The maximum sonic energy occured at 60 persent output (177MW). This is a gasous type vortex cavitation, that might cause sonic vibration. Other measurement conformed the pressure fluctuation was 0.6HZ in frequency, which accorded to 1/3 frequency of the runner being the same of the vortex band frequency.The fluctuation value was upto 4.34 persent of the total head, which was slightly larger than the value guaranted by the maker. Meanwhile, the swing amplitude of the main shaft reached maximum value 0.283mm, which exceeded the swing allowance 0.25mm by the maker. The maximum air noise was also occured in the tailrace and the runner chamber. At the condition of full output, retalive large sonic energy was also measured, that was caused by the runner cavitation as mentioned below.

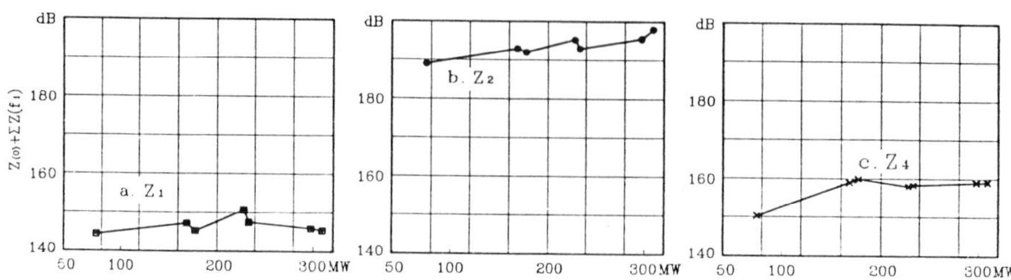

Figure 3. Sonic energy vs output

The sonic energy measured by Z_2 at the top cover was 30dB to 40dB larger than that at the scroll case Z_1. The higher the output load, the larger the energy. It was indicated that there would be vapour type cavitation occured with bubbles collapsing and rebeunding,which might cause cavitation erosion on the runner.Cavitation inspection by dewatering the turbine proved there were cavitation erosions at beginning stage on the blade of the runner.

3.2.2 High Frequency Shock Wave Method

By 1/3 octave band analyzing at 30 frequecies from 125HZ to 100KHZ, the noise spectra for frequency higher than 10KHZ were shown in Figure 4, it revealed:

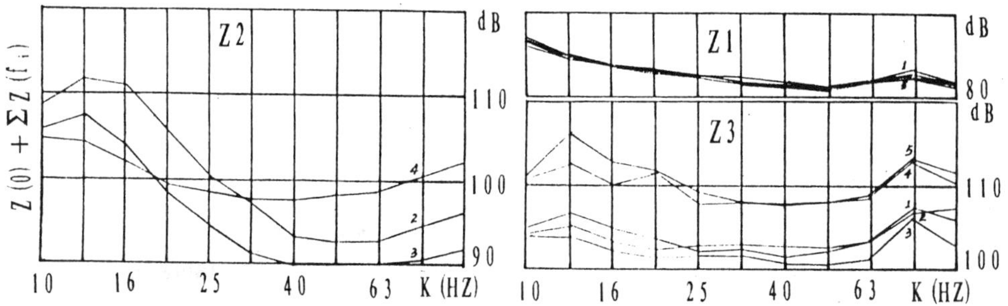

Figure 4. Spectra of underwater noise

At the scroll case (Z_1), the components in high frequency of the spectra were small and did not vary with output, which means no cavitation.

At the tailrace (Z_3), for outputs 177MW and 228MW, the spectrum attenuation was larger than 10 dB/oct at frequency band of 2KHZ to 40KHZ, which belong to flow turbulence noise. The sonic pressure for 177MW was more intense than for 228MW, which indicated that vortex cavitation occured at the tailrace. This judgement is in accordance to that by sonic energy method. For output 297MW, the sonic pressure was the most intense in the conditions and its spectrum attenuation was small, which indicated cavitation with character of high frequency shock wave.

Comparing the results in sonic energy method, the cavitation noise came from the exit side of the blade where high frequency cavitation occured.

At the top cover (Z_2). The spectrum components in high frequency band, were 20dB higher than that at the scroll case (Z_1), except output 75MW, the spectrum attunuation was very small, which revealed cavitation developed much intensively.

4. Conclusion

It is reliable therotically and practively to measure and judge turbine cavitation by accoustic method. Futhermore, the method is more reasonable and practive than the external characteristics method, which could reveal the different types,locations and danage potentials of cavitation. However, the method is still preliminary studied. Further study and practice are underway.

5. References

[1] CHENG LIANGUN (1981) HYDRAULIC TURBINE, China Mechanical Industry Publishing Company

[2] Н.И.Пылаев, Ю.У.Эделъ (1974) Кавитация в Гицротурбинах

[3] LIU KEHUANG (1993) Brief Intruduction Of Hydraulic Prototype Observation Development In YRSR1 In Past 10 Year, National Symposium On Safely Monitoring Technique on Large Dams, Hunan, China

[4] LIU KEHUANG (1988) Measuring and Judging Cavitation by Underwater Noise, National Symposium on cavitation and Its Erosion, Xinjiang, China

[5] LIU KEHUANG (1991) Approches to Discernment of Cavitation by Acoustic Characteristics, International Symposium on Modelling Cavitation Phenomena, Wuhan, China

[6] DONADL ROSS (1976) Methanics Of Under Water Noise

[7] HE ZOUYONG, ZHAO YUFANG (1981) Acoustic Principle, National Fefence Industry Publishing Company

[8] SUN QINGLONG (1982) Spectrum Analyse and Measurement, People's Telecommunication Publishing Company

[9] LIU KEHUANG (1988) Prototype Study On Sonic Vibration and Its Elimination of No.2 Ship Lock in Gezhouba Project, National Symposium on Hydraulics, Cheng Du, China

[10] HUANG BOMIN, LIU KEHUANG (1991) Operation Experiment Sumary on the 170Mw Huge Turbine of Gezhou ba Hydropower Plant, International Symposium On Hydraaulic Research In Nature And Laboratury

NUMERICAL SIMULATION FOR DILUTE SANDY WATER FLOW IN PLANE CASCADE

X. B. LIU
Sichuan Institute of Technology
Chengdu, Sichuan, P.R. China

Q.C. ZENG
University of Queenland
Brisbane QLD 4072, Australia

L. D. ZHANG
Sichuan Institute of Technology
Chengdu, Sichuan, P.R. China

The flow of dilute sandy water (sand carrying capacity less than 1.0kg/m^3) in plane cascade is numerically solved. The collision efficiency is defined. The influence of the Reynolds number based on the characteristic length of the cascade, the angle of attack and the sand diameter on the collision efficiency are discussed. The trajectories and impinging action of sand particles moving in plane cascade flow are calculated by including the effects of the wall, the boundary layer, the wake, the sand particle size and the collision of the sand particle-wall.

1 INTRODUCTION

There are many hydraulic turbines operated in silt laden rivers in the world. The presence of solid particles in the flow often cause abrasive degradation of certain parts of the hydraulic turbine which are especially exposed to the flow. The flows in hydraulic turbine gate vane and runner are cascade flows, therefor, the study of the cascade flow is very important. In this paper, the flow of dilute sandy water in plane cascade is numerically solved. The

E. Cabrera et al. (eds.), Hydraulic Machinery and Cavitation, 632–640.

collision efficiency is determined. The influence of the Reynolds number based on the characteristic length of the cascade, the angle of attack and the sand diameter on the collision efficiency are discussed. In this solid-liquid flows, Eulerian equations are used for the description of the liquid phase dynamics, whist the solid particles are treated within the Lagrangian framework. The trajectories and impinging action of sand particles moving in plane cascade flow are calculated by including the effects of the wall, the boundary layer, the wake, the sand particle diameter and the collision of the sand particle-wall. The following assumptions are implied in the work:

(a) The particles are spherical. The physical properties of each phase are constant.

(b) The particle-particle interactions are neglected.

(c) The flow is steady.

2 BASIC MOTION EQUATIONS

2.1 Particle Motion Equations

Tchen (1947) extended the equation for the slow motion of a spherical particle in a stagnant fluid derived by Basset, Bousinesq and Oseen to the case of a particle suspended in an unsteady velocity field. Here, the modified Tchen's equation of motion for a solid particle in arbitrary flow field by Liu (1994, 1995) in the i direction is given as:

$$\frac{dV_{pi}}{dt} = \frac{1}{K_m + S}\left[\frac{3}{4d}C_D\left|\vec{V}_f - \vec{V}_p\right|(V_{fi} - V_{pi}) + \frac{3}{2d}K_B\sqrt{\frac{\nu_f}{\pi}}\int_{-\infty}^{t}\frac{\frac{dV_{fi}}{d\tau} - \frac{dV_{pi}}{d\tau}}{\sqrt{t-\tau}}d\tau\right.$$

$$+\frac{3}{4}C_M\Omega_i \times (V_{fi} - V_{pi}) + \frac{6}{\pi d}K_S\left|\nu_f\frac{\partial V_{fj}}{\partial x_i}\right|^{1/2}(V_{fj} - V_{pj})\mathrm{sgn}\left(\frac{\partial V_{fj}}{\partial x_i}\right)$$

$$\left. +(1+K_m)\frac{dV_{fi}}{dt} - \nu_f\nabla^2 V_{fi} - (1-S)g_i\right] \tag{1}$$

where, V denotes the velocity; $\Omega_i = \omega_{pi} - 0.5\nabla \times V_i$ (ω_p is the angular speed of particle freely rotating); d is the particle diameter; t, τ denote the time; S is the particle-fluid density ratio (ρ_p / ρ_f); g is the gravitational acceleration; x_i are the components of coordinate; ν is the kinematics viscosity; sgn is the sgnum function; $K_m(\approx 0.5)$, $K_B(\approx 6.0)$, $K_S(\approx 1.615)$ and $C_M(\approx 1.0)$ are the coefficients of the virtual mass, the Basset, the Saffman lift and the Magus lift forces (In main streams, the Saffman lift and the Magus lift are negligible), respectively. For dilute liquid-particle flows the particle motion drag coefficient C_D can be expressed as:

$$C_D = \begin{cases} 24/\mathrm{Re}_p & \mathrm{Re}_p = \left|\vec{V}_f - \vec{V}_p\right| d/\nu_f < 1 \\ (24/\mathrm{Re}_p)(1+0.15\mathrm{Re}_p^{0.687}) & 1 \le \mathrm{Re}_p \le 1000 \\ 0.44 & \mathrm{Re}_p > 1000 \end{cases}$$

the subscripts i and j are the coordinate tensors. f and p refer to the fluid and the particle, respectively.

In fact, the solution of equation (1) is very difficult. The calculation flow area is generally divided into two regions: a. the main flow region (① region in Figure.1), b. the boundary layer flow and the wake flow region (② region in Figure.1). While it is a well-known fact that a spining sphere in a moving fluid experiences a lateral lift force due to its rotation (Magus effect), it is perhaps not so generally appreciated that a similar effect is present when a sphere moves through a viscous fluid in shear flow. The formar mechanism is known as "slip-spin" motion while the latter is known as "slip-shear" motion. At low Reynolds number, Saffman [1965] showed that, for a freely rotating particle, unless the rotating velocity is very much greater than the rate of shear, the lift force caused by the particle rotation is less by an order of magnitude than that caused by the shear. Recently, using the laser photography technique, the result of the particle motion research shows that the particle do not rotate at most areas in the flow field.

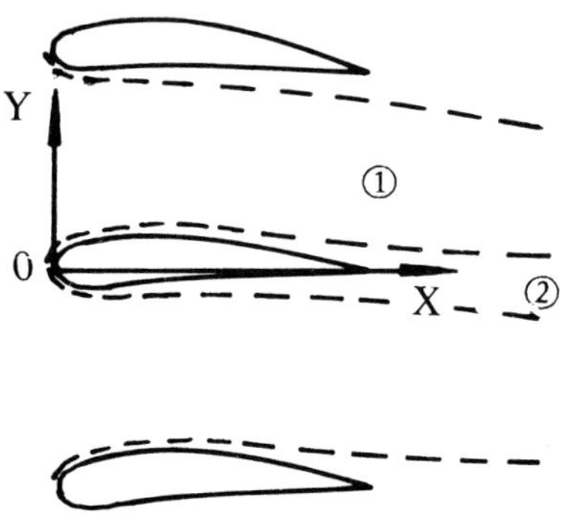

Figure.1 The coordinate of particle motion

Hence, the Saffman lift and the Magus lift as well as $-\nu_f \nabla^2 V_{fi}$ terms in equation (1) can be neglected for ① region, and the particle reduced equation may be expressed as

$$\frac{\mathrm{d}V_{pi}}{\mathrm{d}t} = \frac{1}{\tau_p}(V_{fi} - V_{pi}) + \frac{6\beta}{d}\sqrt{\frac{\nu_f}{\pi}} \int_{-\infty}^{t} \left(\frac{\mathrm{d}V_{fi}}{\mathrm{d}\tau} - \frac{\mathrm{d}V_{pi}}{\mathrm{d}\tau} \right) \frac{\mathrm{d}\tau}{\sqrt{t-\tau}} + \beta \frac{\mathrm{d}V_{fi}}{\mathrm{d}t} \qquad (2)$$

where, τ_p is the particle response time, $\beta = 3/(2S+1)$.

For ② region, we consider the coordinate system which is that the *x0*-coordinate is the surface distance measured from the O point and the *y0*-coordinate is measured normal to the surface. Because of effects of viscous and the particle rotating velocity is very low in the boundary layer. Thus, In this region, the effect of the Magus lift force and the component of Saffman lift force in the *x0*-direction are neglected. Hence, in ② region, the Lagrangian equations of particle motion in the *x0* and *y0* directions can now be expressed as, respectively,

$$\frac{dV_{px0}}{dt} = \frac{C_{x0}}{\tau_p}(V_{fx0} - V_{px0}) + \frac{6\beta}{d}\sqrt{\frac{\nu_f}{\pi}}\int_{-\infty}^{t}\left(\frac{dV_{fx0}}{d\tau} - \frac{dV_{px0}}{d\tau}\right)\frac{d\tau}{\sqrt{t-\tau}} + \beta\frac{dV_{fx0}}{dt} - \frac{2}{3}\beta\nu_f\nabla^2 V_{fx0} \quad (3)$$

$$\frac{dV_{py0}}{dt} = \frac{C_{y0}}{\tau_p}(V_{fy0} - V_{py0}) + \frac{6\beta}{d}\sqrt{\frac{\nu_f}{\pi}}\int_{-\infty}^{t}\left(\frac{dV_{fy0}}{d\tau} - \frac{dV_{py0}}{d\tau}\right)\frac{d\tau}{\sqrt{t-\tau}} + \beta\frac{dV_{fy0}}{dt} - \frac{2}{3}\beta\nu_f\nabla^2 V_{fy0}$$

$$+\frac{6.46}{\pi d}\beta\left|\nu_f\frac{\partial V_{fy0}}{\partial x0}\right|^{1/2}(V_{fy0} - V_{py0}) \quad (4)$$

where, C_i is the component of correction factors which account for the effects of the wall on the drag. We adopt the following expressions for C_{x0} and C_{y0} :

$$C_{x0} = (1 - \frac{9}{16}\delta + \frac{1}{8}\delta^2 - \frac{45}{246}\delta^4 - \frac{1}{16}\delta^5)^{-1}, C_{y0} = 1 + \frac{9}{8}\delta + \frac{81}{256}\delta^2 \quad (5)$$

where $\delta = d/(2y0)$, *y0* denotes the coordinate normal to the wall.

2.2 Fluid Motion Equation

Before the particle equation (1) of motion is solved, the velocity and the pressure distributions of fluid fields must be obtained according to the particle equation (1) of motion. For this cascade flow, the velocity and the pressure distributions of fluid fields are calculated with the aid of a 2-D numerical approximation solving the N-S equations.

3 PARTICLE REBOUND MODE

After impacting a solid boundary, the magnitude and direction of the particle rebounding velocity is dependent on the particle material, the surface material, and on the impact conditions. The experimental studies were carried out in special erosion tunnels, which were designed to include the aerodynamic effects in the rebound characteristics. The following empirical relations (Tabakoff and Hamed 1977) for the rebound-to-impact restitution rations were used in the trajectory calculates:

$$\begin{cases} V_{py02}/V_{py01} = 1.0 - 0.4159\beta_1 - 0.4994\beta_1^{\ 2} + 0.292\beta_1^{\ 3}, \\ V_{px02}/V_{px01} = 1.0 - 2.12\beta_1 + 3.0775\beta_1^{\ 2} - 1.1\beta_1^{\ 3} \end{cases} \tag{6}$$

The rebounding velocity magnitude V_2 and the rebounding angle β_2 are calculated using the following expressions:

$$V_{p2} = \sqrt{V_{py02}^{\ 2} + V_{px02}^{\ 2}}, \ \beta_2 = \mathrm{tg}^{-1}(V_{py02}/V_{px02}) \tag{7}$$

where, V_{py0} and V_{px0} represent the particle velocity components normal and tangent to the solid surface, and the subscripts 1 and 2 refer to the conditions before and after impact, respectively. In the above equations, β is the angle between the velocity and the tangent to the surface.

4 DEFINITION OF COLLISION EFFICIENCY

To a great extent, the collision efficiency η indicate the erosive wear rate of the duct surface. Hence, the collision efficiency η is an important parameter. The definition of the collision efficiency η is that at unit time the total number N_I of the particle impacting the duct surface is divided by the total number N_V of the particle in the flow field. That is

$$\eta = \frac{N_I}{N_V} \tag{8}$$

5 NUMERICAL SIMULATION

5.1 Boundary and Initial Conditions

(a) Boundary of fluid phase:

The coming flow velocity $V_{f\infty}$ of the fluid at infinity is given. The velocity of the fluid at the wall can be written as:

$$V_{fi} = 0 \tag{9}$$

(b) Initial conditions of particulate phase:

The relations of the particle and the fluid initial velocity components can be expressed as:

$$V_{pi0} = e_0 V_{fi0} \tag{10}$$

where e_0 denotes a initial tracking coefficient of particle to fluid. The larger particle size and the weightier particle, the smaller e_0. If the particle impact the solid wall, the reflection of the particle must be considered in terms of particle rebound model.

5.2 Nnmerical Simulation Method of Particle Motion

To simulation the motion of suspended particle, the particle initial position and velocity need to be given. A fourth-order Runge-Kutta scheme is used to integrate numerically the Lagrangian particle equations of motion. The Basset history term in the equation is singular at $\tau = t$ and its discretization needs particular attention, to include its effect on the particle motion accurately, it is discreatized in the following manner,

$$\int_0^t \left(\frac{dV_{fi}}{d\tau} - \frac{dV_{pi}}{d\tau} \right) \frac{d\tau}{\sqrt{t-\tau}} = 2\sum_{k=0}^{K-1} \left\{ \frac{1}{\sqrt{\Delta t}} \left[V_{fi}(k+1) - V_{fi}(k) - V_{pi}(k+1) + V_{pi}(k) \right] \cdot (\sqrt{K-k} - \sqrt{K-k-1}) \right\} \tag{11}$$

where, Δt is the time step, *V(k)* is the velocity at time $k\Delta t$, $K\Delta t$ is the time duration t.

Also using the Runge-Kutta integration technique, the trajectories of aparticle can be found from its velocity components,

$$\frac{dx_i}{dt} = V_{pi} \tag{12}$$

Knowledge of particle velocity and trajectories make possible extrapolation of incidence velocities and angles at the impact locations. Then erosive wear rates may be determined with the erosive wear model.

6 NUMERICAL RESULT AND DISCUSSION

In this study, the cascade spacing T of the plane cascade is 0.5m. The wing chord length is 0.4m. The coming flow impact angles α_0 are given as -10°, -5°, 0°, 5°, 10°. The particle sizes d are 50μm, 100μm, 500μm, 800μm,

1mm. The sandy particle density ρ_p is 2650kg/m^3. The water density ρ_f is 1000kg/m^3. The water kinematics viscosity of the water ν_f is $1.006 \times 10^{-6}\text{m}^2/\text{s}$. The particle initial position X_0 is -0.5m. The initial tracking coefficient e_0 of particle to fluid is 0.9. The coming flow Reynolds Re_L are 7.0×10^5, 1.4×10^6 and Re_L is defined as

$$\text{Re}_L = \frac{V_{f\infty} L}{\nu_f} \tag{13}$$

where L is the characteristic length of the body (in this paper, L is the largest thickness of the profile, that is 0.07m)

Figure 2 shows the motion of particles in the cascade flow.. It can be clearly seen from the figure that the impact region mainly occurs in the profile head and the suction pressure side of the profile tail for the larger positive impact angle, and in the profile head and the pressure side of the profile tail for the larger negative impact angle. The vortex flow near the profile tail leads to increasing collision of particles.

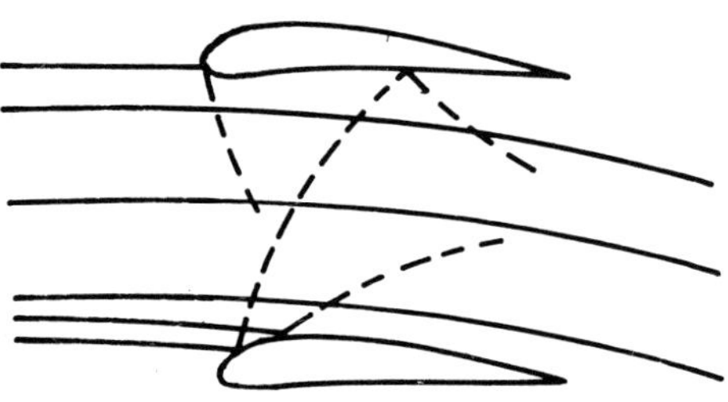

(a) the coming flow impact angle α_0 =0°

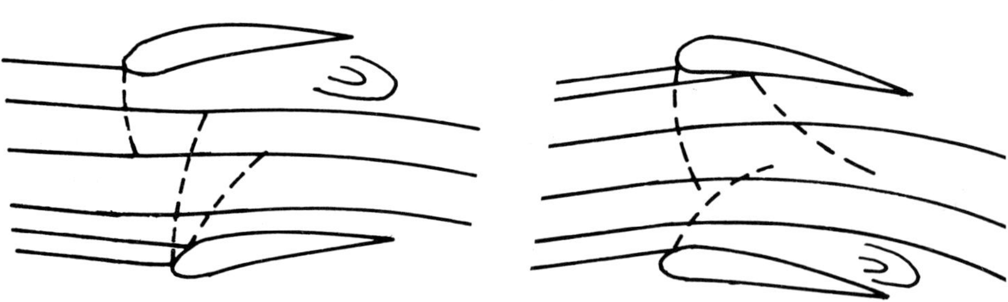

(b) the coming flow impact angle α_0 =-10° (c) the coming flow impact angle α_0 =10°

Figure. 2 The motion of particles in the cascade flow for the coming flow impact angle α_0 = 0° ,-10° , 10°

$\text{Re}_L = 1.4 \times 10^6$, d=500μm, - - - the particle rebounding trajectories

Figure 3 shows the effects of the coming flow Reynolds, the coming flow impact angle and the particle size on the collision efficiency. It can be clearly seen from the figure that the collision efficiency increase with the coming flow Reynolds, the absolute value of the coming flow impact angle and the particle size.

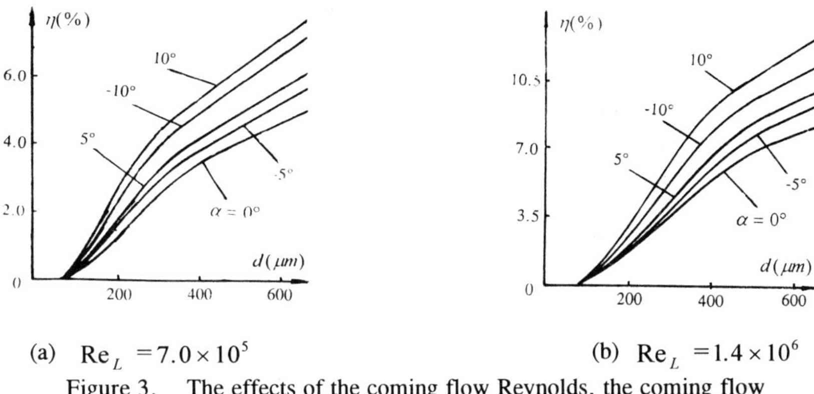

(a) $Re_L = 7.0 \times 10^5$ (b) $Re_L = 1.4 \times 10^6$

Figure 3. The effects of the coming flow Reynolds, the coming flow impact angle and the particle size on the collision efficiency

Figure 4 shows the particle trajectories in the cascade flow for the coming flow Reynolds $Re_L = 1.4 \times 10^6$, the coming flow impact angle $\alpha_0 = 5°$, the same particle initial position and motion characteristic as well as the different particle sizes. As particle size increases, the deviation of the particle trajectory to the water streamline is larger. The particle trajectories are the same as the water streamlines for the particle sizes less than 50μm. It is clear that particles near suction pressure side having larger size tend to devitate towards the suction pressure side and particles near pressure side having larger size tend to remove the pressure side.

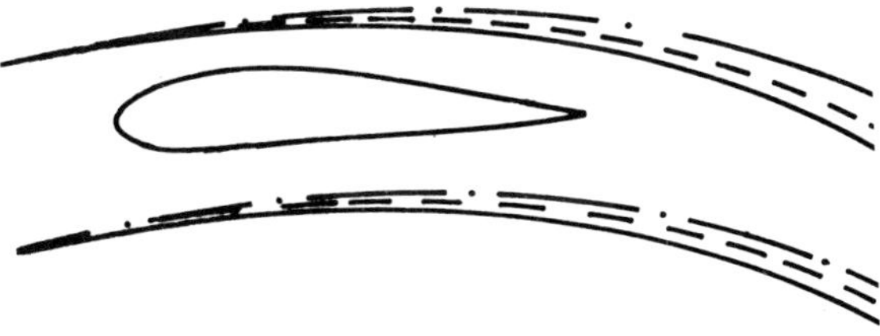

Figure 4. The particle trajectories in the cascade flow $Re_L = 1.4 \times 10^6$, $\alpha_0 = 5°$,

——— streamline of water,

– – – – particle trajectories for d=500μm,

— · — particle trajectories for d=800μm

7 CONCLUSIONS

In this paper, a flow of dilute sandy water in plane cascade is numerically solved. The collosion efficiency is defined. In this solid-liquid flows, Eulerian equations are used for the description of the liquid phase dynamics, whist the solid particles are treated within the Lagrangian framework. The trajectories and impinging action of sand particles moving in plane cascade flow are calculated by including the effects of the wall, the boundary layer, the wake, the sand diameter and the collision of the sand particle-wall. The wear distributions of the through-flow duct surface in solid-liquid flows may be also calcaulated with the aimd of this paper's numerical simulation method and the wear model. The wear region predicted from calculation results show good agreement with that from field observation in the hydraulic turbomachinery.

REFERENCES

Liu, X.B. and Cheng, L.J. (1994) On the particle motion in a flow field in hydraulic turbomachines, J. Huazhong University of Science and Technoloy, 22, 10-16.

Liu, X.B. (1995) Study of solid-liquid two-phase turbulent flow and wear in hydraulic turbomachinery, Ph.D. Thesis, J. Huazhong University of Science and Technology, Wuhan.

Saffman P.G. (1965) The lift on a small sphere in a slow shear slow, J. Fluid Mech, 1965, 22, 385-341.

Tabakoff, W. and Hamed, A. (1977) Aerodynamic ffect on erosion in turbomachines, JSME and ASME, paper No.70, 392-401.

Tchen, C. M. (1947) Mean value and correlation problems connected with the motion of small particles suspended in a turbulent fluid, Ph.D. Thesis, Delft , Hague.

REVIEW OF RESEARCH ON ABRASION AND CAVITATION OF SILT-LADEN FLOWS THROUGH HYDRAULIC TURBINES IN CHINA

MEI Zu-yan and WU Yu-lin
Tsinghua University
Beijing, CHINA 100084

1. Introduction

Silt content is very high in many rivers in China. A large number of hydraulic turbines working in these rivers encounters serious problems of silt abrasion and aggravated cavitaion. In lighter cases machine performances are impaired and in serious cases structural damages may result. In recent years, Chinese hydraulic engineers are faced with the serious challenge of solving the problems of silt-laden flow in the many hydro projects on the Huanghe (Yellow River) and the Changjiang (Yangtze River) and many other silt-laden rivers. Attention has been focused on the cavitation erosion rate in the sediment carrying flow, that is, the effects of silt concentration on the cavitation behavior and intensity of cavitation erosion. This report covers the major results of field observation and analysis of extensive laboratory research in recent years.

2. Effect of Sand Content on the Cavitation Process

2.1 INCIPIENT CAVITATION NUMBER

The existence of silt in water flow has a definite influence on the inception of cavitation and its behavior in bubble growth. Liu, Y.X. (1983) did the first tests to study the incipient cavitation number in silt-laden flow in China. His experiment was carried out in a water tunnel with a test section of 80 × 20 mm^2 and the flow velocity of 12 to 23 m/s to measure the cavitation bubbles following a cavitation generator (a cylinder of diameter D = 15 mm). Tests were made with sand from the Huanghe with grain size of 0.014 - 0.018 mm and sand content of 1 - 87 kg/m^3. The incipient cavitation number is defined as

$$K_i = 2\,(P_o - P_v)/(\rho\, U_o^2) \qquad (1)$$

where P_o and U_o are the reference pressure and velocity at the inlet of the water tunnel. In the experiments, for different length of cavitation bubbles λ there is a definite status cavitation number K_m under a certain condition. Fig. 1 shows the experimental results of variation of incipient cavitation number obtained by observing the bubble inception

E. Cabrera et al. (eds.), Hydraulic Machinery and Cavitation, 641–650.

and results of variation of the status cavitation number for maximum sound intensity of cavitation noise. The inception cavitation was determined by measurement of the cavity length, here $\lambda = 0.5$ mm, resulting in a radio $\lambda /D = 3\%$. Another group of experimental results was obtained by using a NASA 4412 airfoil in another water tunnel with plastic sand of $d_{50} = 0.165$mm and density $\rho = 1070$ kg/m^3, as shown in Fig. 2. From these two experimental results, it is very clear that there exists a critical concentration of $S_{cr} = 10\text{-}13$ kg/m^3 at which the incipient cavitation number K_i reaches a maximum value. For S less than 10-13 kg/m^3 the sand plays a role of inducing cavitation whereas for S greater than 10-13 kg/m^3 it has a retarding effect.

2.2 CAVITY LENGTH

In clear water, the cavity length does not change for a given cavitation number. But for the test with silt-laden flow, the status cavitation number K_m for a certain cavity length changes with variation of sand concentration, as shown in Fig. 3. Fig. 4 shows the same results of variation with S at a fixed value of K_m at 2.2 - 3.0. Fig. 5 is the results of the same experiment as that for Fig. 2. For a definite cavity length, the critical concentration S_{cr} (for which the cavitation number is highest) may be 5-8 kg/m^3 for Fig. 3 and 10-13 kg/m^3 for Fig. 5.

2.3 CAVITATION FLUCTUATION

There is a critical concentration value S_{cr} for the variation of cavity fluctuation length ΔLmax as shown in Fig. 6, where the S_{cr} is 10-13 kg/m^3 for NASA 4412 airfoil.

3. Interaction between Cavitation and Silt Abrasion

3.1 GENERAL UNDERSTANDING OF THE INTERACTION

If the surface of a metal is first pitted to sponge-like by cavitation, silt will easily cut or impound on roughened surface and aggravate damage. On the other hand, when metal surface are scratched by silt, the rise and dents will induce cavitation to occur which in turn intensifies the attack. In a silt-laden flow, cavitation and abrasion is thus seen to mutually induce. Some well known facts about cavitation erosion and silt abrasion are (Mei, 1993):

a. The viscosity of particle-laden flow is higher than clear water. As the critical vaporizing pressure is raised, inception of cavitation is retarded.

b. Intensity of cavitation erosion is related to flow velocity in the 6th power or higher.

c. Intensity of sand abrasion is roughly proportional to the 3rd power of velocity.

d. There is an evident incubation period with cavitation erosion. The rate of damage may be identified into stages of acceleration, deceleration and equilibrium. Sand abrasion has no incubation period, damage is essentially proportional to time.

e. Damage to metal surfaces by interaction of cavitation and abrasion is much more serious than each attack alone. A rough indication may be: silt abrasion alone causes 4 times weight loss than by cavitation in clear water, combined action causes 16 times weight loss than by cavitation alone.

It is also well known that varying grain size of sand causes different degree of abrasion damage while the other properties such as shape of grain, hardeness and graiding being essenstially equal. A very rough general idea is that there may be a critical size of sand of 0.05 mm diameter, above which it will cause substantial damage of operation of machines. Some researchers however contend that this "critical size" should be much lower, to around 0.02 mm (Chen, B.E., 1993)

3.2 OVERALL OBSERVATION OF SILT ABRASION ON TURBINES

An analysis was made on abration data of turbines from domestic and abroad. The product of operating head H and concentration of harmful sediments S_d (taken to be $d > 0.05$ mm) which essentially represents the energy of particles, is selected to be a measure of abrasion intensity. As an analysis (Chen, C.J. and Zhang Z.Q., 1990) shows, $H \times S_d = 7$ is roughly the dividing line between moderate damage and serious damage. It thus may be conducted that in hydro power station designs the condition of $H \times S_d < 7$ must be assured.

3.3 TESTS ON CAVITATION EROSION IN SILT-LADEN FLOWS

The weight loss rate of specimens due to cavitation varies with the sand condent in the flow. Cheng, Z.J. (1990) indicated a number of results of abrasion and erosion tests for particle-laden flows. The tests with specimen experiments in water tunnel, with rotating disc or magnetostriction apparatus show there is a critical concentration for silt damage, i.e., at which value the loss weight of specimens reaches a maximum, as shown in Tab. 1. Fig. 7 shows typical results of cavitation erosion tests using various concentration of fly-ash and bakelite powder (Rong,X.Y., 1984). It is obvious that in a liquid containing some kind of solid particles, there must be a critical concentration and an equivalent concentration. When the latter is reached, the average rate of weight loss will be equivalent to that in clear water cavitation test. But in different kinds of liquid, solid particles and specimen materials will have defferent critical concentration values. Rong, X.Y. (1984) has conducted cavitation erosion tests with rotating discs in an effort to determine the role of particles in the process of cavitation erosion. He used uniform plastic granules in order to avoid abrasion by particle thereby yielding only the results of cavitation erosion. Cheng, G.M. (1991) has also conducted similar cavitation erosion tests to determine the erosion rates using magnetostriction apparatus on liquid contaning various solid particles such as natural sand from the Huanghe, quatz sand, bakelite powder, fly-ash and plastic balls.

3.4 DIFFERENT TYPES OF EROSION BEHAVIOR

Wu, P.H. (1994) pointed out that there may be three types of erosion process in the silt-laden flow. Most of the results listed in Tab. 1 belong to type A of erosion where a critical concentration S_{cr} exists at which the cavitation erosion loss rate is maximum.

This type of results may reflect the pure cavitation damage of specimen at silt-laden flow. Test results of some Soviet scholars in the 1960's and some Chinese researchers in recent years (Zheng, D.Q., 1988) show that the critical concentration S_{cr} is not at the maximum as in type A, but accurs at the minimum loss rate. That is, the loss rate of type B erosion process will be larger for both the denser and lighter concentrations than that for the critical concentration condition. A third kind of erosion process is type C which has a loss rate that continually increases with the increase of the solid concentration, as shown in Ref. of Jin (1994). Figs. 8 and 9 show this type of the loss rate in which the damage on the specimen is caused purely by solid abrasion, if there is very little cavitation erosion under certain conditions.

TABLE 1. Critical sediment concentration of mixture in cavitation erosion tests

No.	Test method	Particle material	ρ_s	S_{cr}	d_{50}	References
1	MSA 15.5kHz	Huanghe sand	2400	5.0	0.016	Liu Y.X., 1983
2	Tunnel foil	Plastic	1070	8.0	0.165	Cheng L.J. 1990
3	MSA 19.6kHz	Fly ash	2150	107.5	0.023	Jin T.L. 1994
4	MSA 19.6kHz	Bakelite	1048	74	0.023	ditto
5	Rotating disc	Huanghe sand	2400	15	0.03	Wang Z.G. 1986
6	Tunnel cylinder	Huanghe sand	2400	2.89	0.03	ditto
7	Rotating disc	Plastic	1070	2.0	0.25	Rong 1984
8	Rotating disc	Plastic	1070	3.6	0.94	ditto
9	Rotating disc	Plastic	1070	5.0	2.1	ditto
10	MSA 19.19kHz	Plastic	1070	42.8	0.135	Han D. 1990
11	MSA 19.19kHz	Huanghe sand	2400	128	0.135	ditto
12	MSA 19.19kHz	Huanghe sand	2400	128	0.45	ditto
13	MSA 19.6kHz	Quartz	2600	3.07	0.037	Cheng G.M. 1991
14	MSA 19.6kHz	Quartz	2600	5.27	0.10	ditto

4. Model Experiments of Turbines in Silt-Laden Flows

Results of two tests of model turbines conducted in closed circuit test stands with silt-laden flow have been reported (Huang S.H., 1994). Fig. 10 shows the variation of relative cavitation coefficients σ_c / σ_{co} versus sand concentration from a test of an HL001-LJ-25 model turbine with 24 guide vanes and 14 stay vanes designed for the Liujiaxia Power Station. This test used a head 15 m with guide vane opening ao = 14 mm and the unit speed n_{11o} = 68.75 rpm and the unit discharge Q_{11o} = 600 l/s at the design condition

TABLE. 2 Test data of variation of σ_c / σ_{co} (Fig. 10)

Test	S_V %	S_W %	Sand type	Density	d_{50} mm	σ_c / σ_{co}
1 ○	0.0	0.0	Clear water	1.00		1.0
2 △	1.37	1.47	Plastic	1.07	0.25	1.20
3 ∨	2.01	2.15	Plastic	1.07	0.25	0.86
4 ×	0.35	0.85	Huanghe sand	2.37	0.06	1.04

As shown in Fig. 10, a critical concentration of $S_{cr} = 1.5$ kg/m3 exists for the maximum cavitation coefficient σ_c which was measured by the energy method according to IEC Standard 7.3.1.4.. But Fig. 11 shows a similar test result of a model turbine with parameter n_{11o}= 70.8 rpm and Q_{11o}= 990 l/s. In this test, with sand density of 2614kg/m3, cavitation coefficient σ_{cmax} were determined by using the maximum acoustic energy measurement of cavitation noise (Lu et al, 1992). Lu who performed this test believed there were shortcomings in using the energy method for determining the critical cavitation coefficients of a turbine in turbid flows since it could not be augmented by observation of the cavitation pattern.

5. Parameters of Turbines Operating in Silt-Laden Flows

In order to reduce silt abrasion and cavitation erosion of the turbine, the properties of the flow must be taken into consideration in the hydraulic design of turbines.

5.1 CHARACTERISTIC VELOCITY

In China, it has been sugggested (Tong, 1993) the peripheral velocity U_2 and the relative velocity W_2 at the outlet of turbine runners should be limited to:

For $S_w < 5$ kg/m3 U_2 or $W_2 < 38$ - 40m/s.
For $S_w < 12$ kg/m3 U_2 or $W_2 < 34$ m/s.

Needless to say, the recommended limiting velocities are dependent on the runner material. From observation in actual operation, the limiting velocities for 1Cr18Ni9Ti plate steel is about 25 m/s. From rotating disc tests with Huanghe sand, the critical velocities for A3, 20SiMn, 0Cr13Ni4Mo and 0Cr13Ni5Mo steels are 34 m/s (damage will increase rapidly above this velocity), and from a different testing source, for 0Cr13Ni4Mo steel is 38 m/s.

5.2 SPECIFIC SPEED

For these characteristic velocity values, the relation between the specific speed of turbines operating in silt-laden rivers versus turbine head are expressed in Fig. 12 for Kaplan turbines and in Fig. 13 for Francis turbines. Statistical data are also plotted on these curves (Tong, 1993).

5.3 UNIT SPEED AND UNIT DISCHARGE

The rated rotating speed is normally selected according to the weighted averaged head of the power plant. Under clear water conditions, the rated speed n_{11r} is usually taken to be 1.0 to 1.2 times that of the optimum value n_{11o}. In silt-laden flows, it is recommended that n_{11r} be the same as n_{11o}. Suggested unit speed and unit discharge are listed in Table 3 for Kaplan turbines operating on silt-laden rivers and in Table 4 for Francis turbines:

TABLE 3. Unit speed and unit discharge for Kaplan turbines

He (m)	5.0	10.0	15.0	20.0	25.0	30.0	35.0	40.0
n_{11r} (rpm)	175	160	146	134	123	116	112	111
Q_{11r} (m^3/s)	3.26	2.4	1.96-2.05	1.7-1.77	1.355-1.52	1.237-1.32	1.065-1.145	1 - 1.07

TABLE 4. Unit speed and unit discharge for Francis turbines

He (m)	35	50	70	100	120	150	180	200	250	300	350	400
n_{11r} (rpm)	79.4	71.7	67.8	65.7	64.3	61.2	59.7	59.2	57.2	55.9	55.2	54.6
Q_{11r} (m^3/s)	1.584	1.36	1.128	0.887	0.772	0.637	0.529	0.47	0.37	0.30	0.247	0.208

5.4 GUIDE VANE PITCH CIRCLE DIAMETER

The height of guide vanes for turbines of low specific speed is normally quite short. In order to satisfy the inlet circulation before the runner inlet, the curvature of the guide vanes must be quite large, here the cross section of the flow passage through the guide vanes is relatively narow. These conditions will cause high flow velocity in the guide vane passages that will cause serious abrasion on the surfaces of the vanes in silt-laden flows. Hence the pitch circle diameter Do of guide vanes will have to be made larger than conventional values, and according to Tong (1993):

General range	Do = (1.15 - 1.26) D_1
For high specific speed turbines	Do = (1.16) D_1
For low specific speed turbines	Do = (1.18 - 1.20) D_1
Turbines in silt-laden flows	Do > (1.2) D_1

For operation in dense concentration, beside using a larger pitch circle diameter and smaller surface curvature, lesser thickness for the guide vanes should also be considered.

6. Conclusions

Experiments on cavitation erosion and silt abrasion have been conducted in test facilities such as the magnetostriction apparatus, rotating disc rigs as well as silt-laden water tunnels or model turbine test stands in China in recent years. The perfomance characteristics of hydraulic turbines operating in silt-laden flows have been fully analyzed. Some suggestions on the selection of hydraulic parameters of these turbines have been derived from these results.

7. References

Chen, B.E. and Li, W.G. (1990) Approximate calculation of effect of solid material on performance of pump, *Pump Technology*, **2**, 12-15, (in Chinese).

Chen, C.J. and Zhang, Z.Q. (1989) On wear of turbines in hydro stations, *Hydro Electricity*, **2**, 34-39, (in Chinese).

Cheng, G.M. (1991) Experimental research on cavitation erosion and sediment abrasion in silt-laden flow, *Master Thesis, IWHR*, (in Chinese).

Cheng, Z.J. (1990) Experimental study on the silt critical concentration of erosion and abrasion in silt-laden flows, *Technology on Hydraulic Engineering and Hydro Power*, **2**, 57-63, (in Chinese).

Cheng, Z.J. (1990 a) Trends of abrasion and cavitation mechanism research on hydraulic machinery in China, *Proc. Abrasion and Cavitation in Hydraulic Machinery*, 8-12, (in Chinese).

Han, D. and Huang, J.T. (1990) Model test research on vibratory cavitation and silt abrasion, *J. of Hydraulic Engr*. 1990, **1**, 34-37, (in Chinese).

Huang, S.H. and Cheng, Z.J., (1994) Experimental study on critical cavitation coefficient of hydraulic turbines in silt-laden flow, *Hydraulic Machinery of Technology*, **5**, 9-12, (in Chinese).

Jin, T.L. (1994) Recent advancement of cavitation erosion and wear in sand-laden water, *Proc. 17th IAHR Symposium*, Beijing, 555-556.

Liang, J.G. (1993) Problems in parameter selection of hydraulic turbines in dense silt-laden hydro station, *Proc. Abrasion and Cavitation in Hydraulic Machinery*, 48-53, (in Chinese).

Liu, Y.X. (1983) Effects of silt-laden flows on cavitation, *J. of Hydrauic Engineering*, **5**, 55- 58, (in Chinese).

Lu, L. et al (1992) Measurement of incipient cavitation coeffient of model hydraulic turbines in silt-laden flow, *IWHR Research Report*, (in Chinese).

Mei, Z.Y. (1993) A survey of experiences on hydraulic turbines working in silt-laden rivers, *Proc. of 4th Asian Int. Conf. on Fluid Machinery*, Suzhou, Invited lecture.

Rong, X.Y. (1984) Cavitation and cavitation damage in silt flow and its interaction with sediment abrasion, *Master Thesis, IWHR*, (in Chinese).

Tong, W.M. (1993) Selection of objective parameter of hydraulic turbines on silt-laden river, *Proc. Abrasion and Cavitation in Hyydraulic Machinery*, 30-40, (in Chinese).

Wang, Z.G. and Yao G. (1986) Cavitation and abrasion of the high speed flow with dilute concentration of small size silt, *J. of Hydraulic Engineering*, **10**, 63-67, (in Chinese).

Wu, P.H. (1991) Discusion on the solid critical concentration and the damage under the combined action of cavitation and silt abrasion, *Proc. of 10th Symposium of Hydro Power engineering*, Beidaihe, 176-183, (in Chinese).

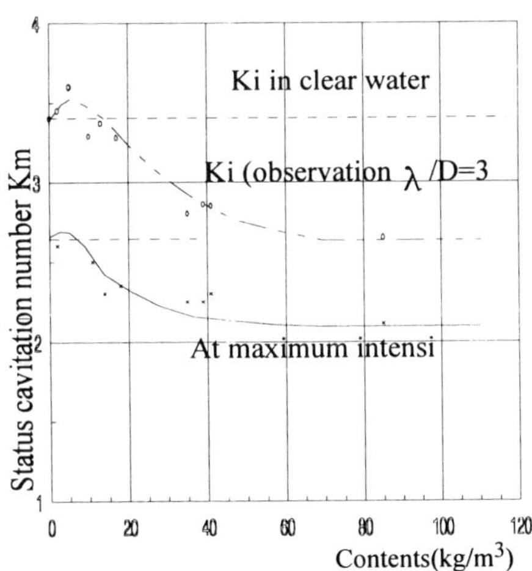

Fig. 1 Effects of solid content on incipient cavitation number for a cylinder

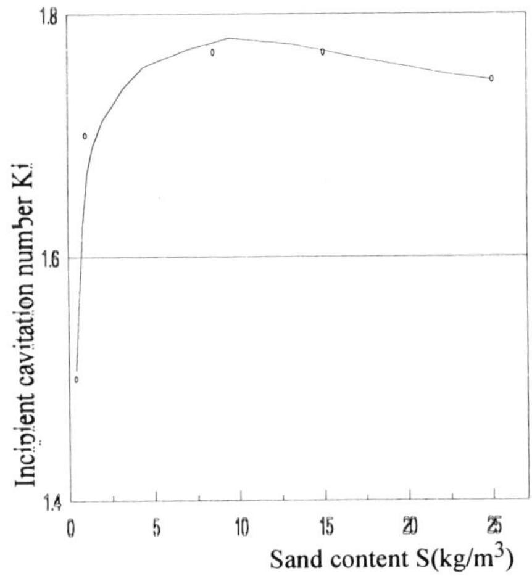

Fig .2 Relation between sand contents and K_i for NASA4412

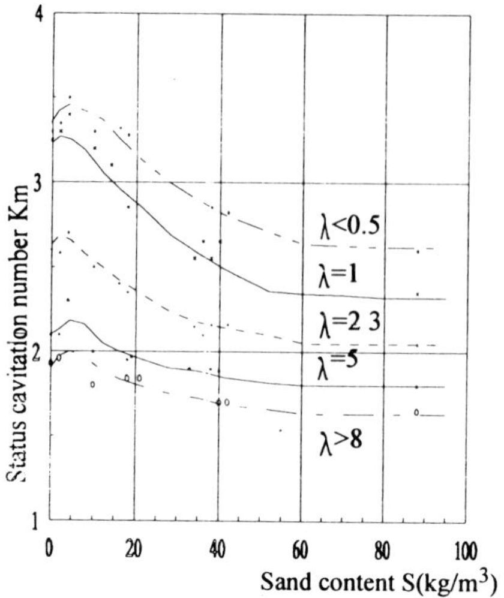

Fig. 3 Effects of sand content S on $K_i m$ for different length λ

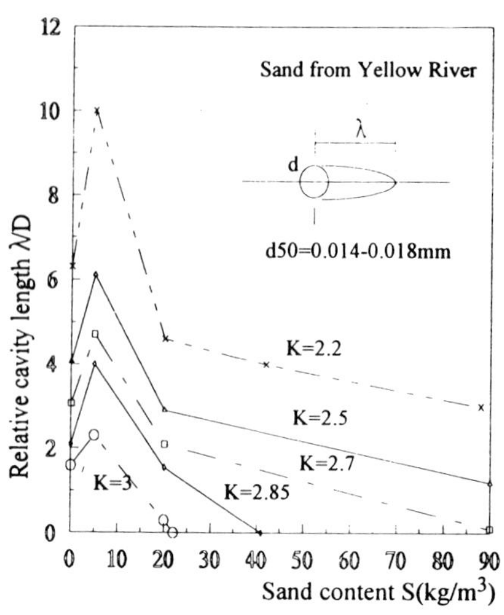

Fig. 4 Relation between sand content S and cavity length

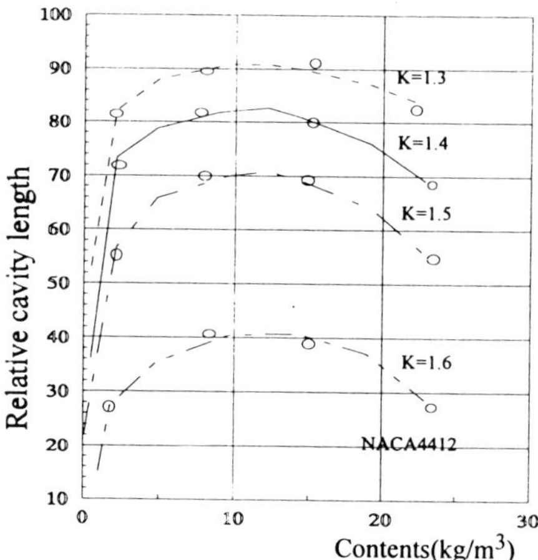

Fig. 5 Relation between sand content and cavity length for NASA 4412 (d_{50}=0.165mm, ρ_s=1070(kg/m3), α=4°)

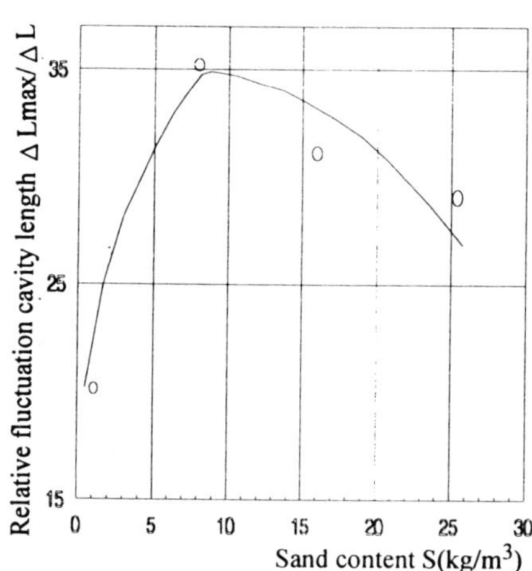

Fig. 6 Relation between sand contents and relative fluctuate cavity length (NACA4412, α=4° , d_{50}=0.165mm, ρ_s=1070(kg/m3))

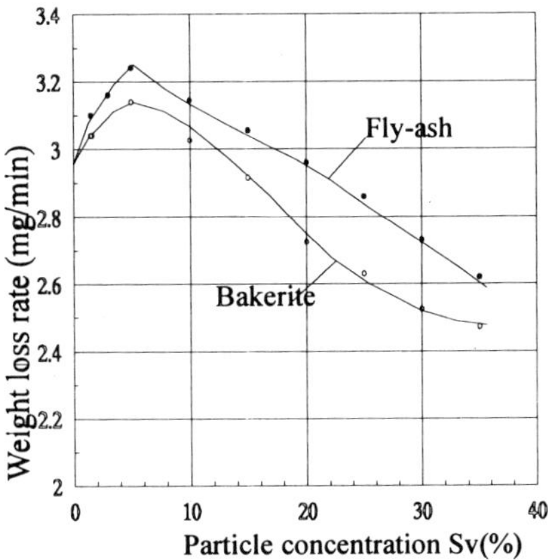

Fig. 7 Comparison of weight loss rate versus particle concentration

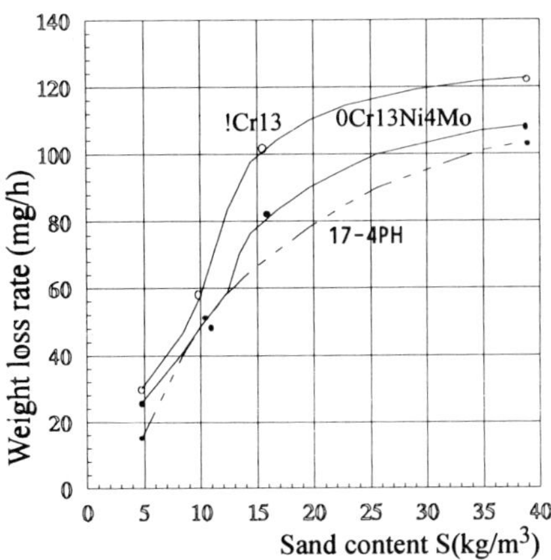

Fig. 8 Type-C loss weight rate of sand erosion

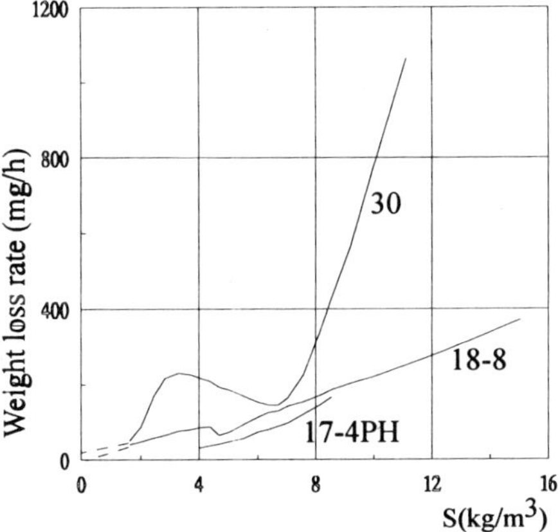

Fig. 9 Similar type-C loss weight rate

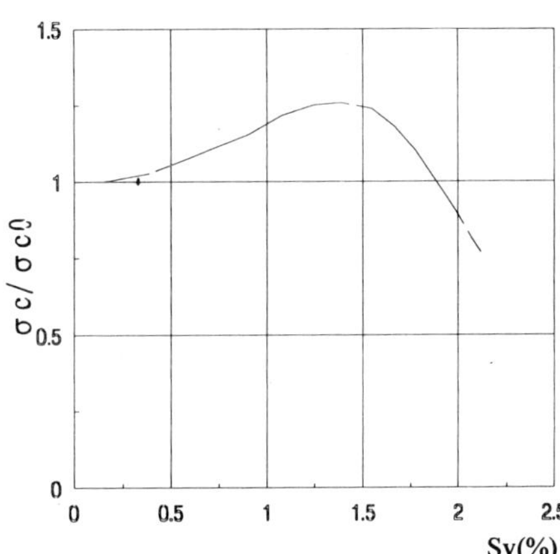

Fig. 10 Relation between critical cavitation coefficient and sand concentration

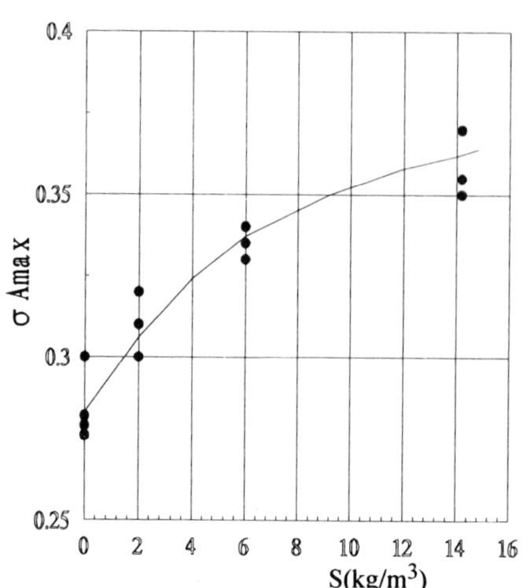

Fig. 11 Effects of concentration on incipient cavitation coefficient

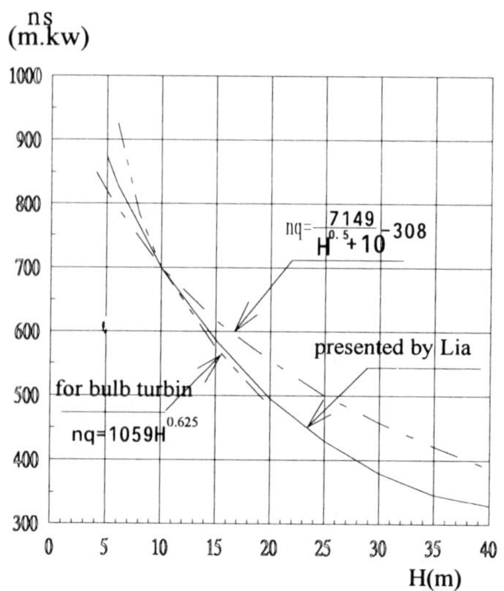

Fig. 12 Relation between ns and effective head for Kaplan turbines

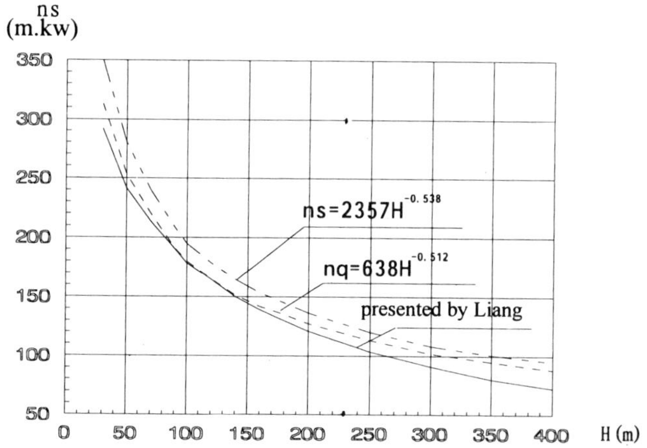

Fig. 13 Relation between ns and effective head for Francis Turbines

EFFECT OF THE LEADING EDGE DESIGN ON SHEET CAVITATION AROUND A BLADE SECTION

Jean-Luc REBOUD
Laboratoire des Ecoulements Géophysiques et Industriels (LEGI)
Institut de Mécanique de Grenoble,
B.P. 53, 38041 Grenoble cedex 9, France

Claude REBATTET
Centre deRecherches et d'Essais de Machines Hydrauliques (CREMHyG)
Institut National Polytechnique de Grenoble,
B.P. 95, 38402 Saint Martin d'Hères cedex, France

Philippe MOREL
Société Européenne de Propulsion (SEP)
Forêt de Vernon, B.P. 802, 27207 Vernon cedex, France

Abstract

An experimental study of the cavitation behaviour of two-dimensional hydrofoils simulating a section of an inducer blade has been performed in the CREMHyG large cavitation tunnel. Two leading edge shapes, chosen at the request of the SEP to approach rocket engine inducer designs, have been tested with respect to the development of sheet cavitation. Pressure distributions along the foil suction side and the tunnel walls have been measured for different cavit lengths. Cavity shapes have been provided by visualisations under laser sheet light and image processing. Laser anemometry (LDA) around the leading edge and total pressure measurements along the foil suction side have been performed to characterise the effects of cavitation on the liquid flow. A numerical model of cavitating flows has been improved using those experimental results. Numerical predictions have been performed, showing a good agreement with experiments on a large range of flow conditions.

I. Introduction

To improve their operating range with respect to low inlet pressure, rocket engine turbopumps are always designed with an inducer stage on which cavitation occurs. The cavitation behaviour of the inducer is mainly responsible of the performance of the whole pump and determines the fuel and oxidiser tank static pressure limitations. The inducer stage of the liquid hydrogen pump of the Vulcain motor, equipping the European launcher Ariane 5, has been designed by the SEP (Société Européenne de Propulsion. Such inducers have been tested on the water high speed pump test rig of the CREMHyG

E. Cabrera et al. (eds.), Hydraulic Machinery and Cavitation, 651–660.

(Centre de Recherches et d'Essais de Machines Hydrauliques de Grenoble). Sheet cavities attached to the suction side of the blades have been observed (Bernard *et al.* 1988). To study more in detail that kind of cavitation, two-dimensional hydrofoils simulating a mid-span section of an inducer blade have been tested in the CREMHyG's large cavitation tunnel (Kueny *et al.* 1991). In parallel, numerical models for the prediction of cavitating flows have been developed, involving two dimensional (Rowe and Blottiaux, 1993, Kueny *et al.* 1991), quasi three dimensional (Kueny *et al.* 1988) and three dimensional formulations (Maitre *et al.* 1993).

In the present study, two leading edge shapes, chosen at the request of the SEP to approach rocket engine inducer designs, have been compared with respect to the development of sheet cavitation. Pressure measurements along the foil suction side and the tunnel walls have been performed for the different cases. In the same time, visualisations under laser sheet light and image processing provided the cavity shapes. To characterise the effects of cavitation on the global behaviour of the liquid flow around the hydrofoil, laser anemometry (LDA) around one leading edge and total pressure measurements along the foil suction side using a 4 holes probe have been performed, under cavitating and non cavitating conditions. A two dimensional numerical model of cavitating flows has then been improved using the experimental results.

2. Experimental set up

The experimental study of the cavitation behaviour of two-dimensional hydrofoils simulating a mid-span section of an inducer blade has been performed in the CREMHyG's large cavitation tunnel (Fig. 1).

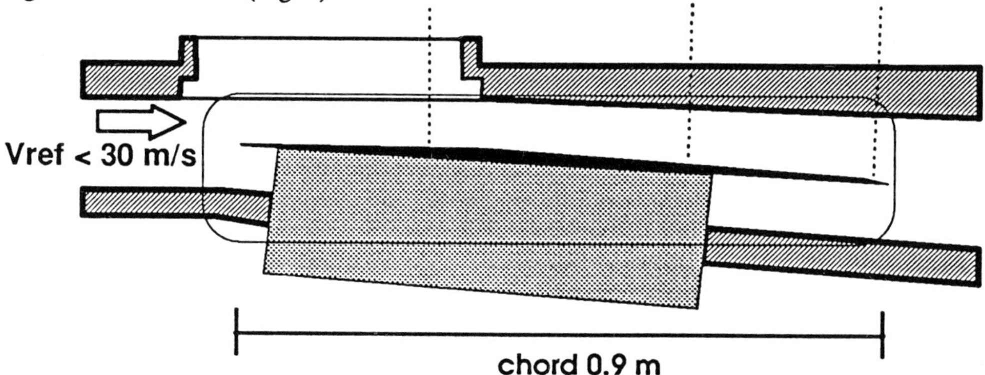

Figure 1 : CREMHyG's large cavitation tunnel (width = 0.12 m)

The upper and lower walls of the cavitation tunnel have been designed to generate along the suction side of the foil a pressure distribution as close as possible to the one existing along the inducer blade in non cavitating condition. The comparison of the pressure distribution has been performed using two dimensional and quasi-three dimensional perfect fluid flow computations (Kueny *et al.* 1991). The hydrofoil chords are 0.9 meter long and their relative thickness is smaller than 1.5%. The foils are held by a 5 mm thin streamlined leg, through which pressure taps have been drilled. The reference flow velocity Vref and static pressure Pref are measured in the upstream rectangular section

Sref (137x120 mm^2), located 200 mm upstream the foil leading edge. Those reference values are used to define the cavitation number σ = (Pref - Pv) / 0.5 ρ Vref ^{2} , where Pv is the water vapour pressure. The maximum flow velocity Vref may be larger than 30 m/s. No marked effect of the amount of air dissolved in water has been observed on developed cavitation. During the present experiments, the air content of the water has been kept near saturation.

The behaviours of two different leading edges have been compared with respect to the development of sheet cavitation. Their shapes are reported on Figure 2. The first leading edge presents a parabolic shape, about 3 mm long, creating with the suction side a sharp ridge. It simulates a blade whose upstream end has been bevelled. The second one is sharper and its rounded upstream end, where the cavity detachment occurs, presents a small curvature radius close to 0.2 mm. The two profiles will respectively be named hereafter "bevelled leading edge" (BLE) foil and "sharp leading edge" (SLE) foil. The second hydrofoil is 3 mm shorter than the first one and its relative thickness is a little smaller, but the two suction sides are identical.

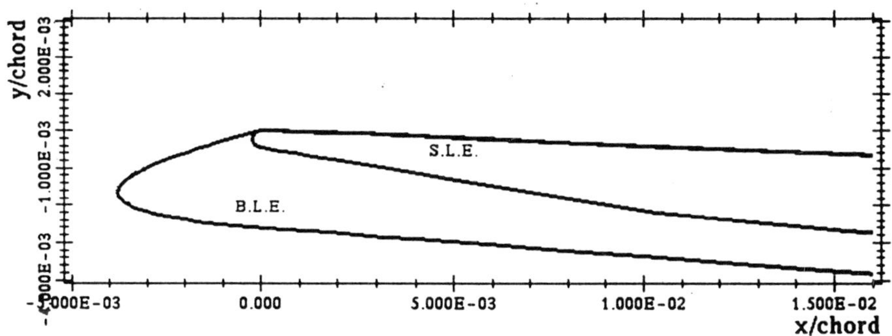

Figure 2 : Sharp and bevelled leading edges (4.5 degrees angle of attack).

Two different angles of attack have been tested : 4.1 and 4.5 degrees. They have been chosen to provide flow conditions close to those predicted in a blade to blade channel of the inducer at its nominal operating point. Because of the small ratio (0.137/0.9) between the tunnel height and the chord length, the 0.4 degrees difference has a great influence on the cavitation behaviour.

Pressure measurements have been performed along the foils suction side (10 taps) and the tunnel walls (3 taps in the reference section, 7 along the lateral and lower walls and 4 along the upper wall). Pressures taps are connected with 1 mm diameter Rilsan pipes to piezo resistive *DRUCK 910* transducers. The transducers calibration is performed by comparison with a *Desgranges and Huot* standard of pressure. The accuracy obtained is then better than +/- 0.1% of the 5 bars range.

The flow rate is measured by a *BEN* electromagnetic flowmeter, with a relative accuracy of 0.2 %. The combination of inaccuracies on pressure and velocity measurements ensures on the pressure coefficient, Cp = (P - Pref) / 0.5 ρ Vref ^{2}, and on σ a global uncertainty of +/- 0.02 at Vref = 10 m/s.

Cavity shapes have been obtained using a laser sheet light and recording images with a CCD video camera. Images have then been processed on a micro computer equipped with a 512x512 *PCoeil* processing board. In situ calibration is performed using a graduated stem held at different position through the pressure taps. A resolution of 0.4 mm / pixel is obtained, which give an accuracy of +/- 5 to 10% on the cavity thickness estimates.
Velocities have been measured using a *DANTEC* two components LDA system. Total pressure measurement have been performed along 3 axes using 4 holes probes, connected to the *DRUCK 910* pressure transducers. Velocities have been deduced from the total and static pressure measurements. Comparisons with LDA data acquired along the same axes have been used to validate the method under non cavitating and cavitating conditions.

3. Numerical model

A two dimensional model has been improved using the experimental results. That model is based on the coupling of : (a) a potential flow code taking into account the cavity as a mean streamline at vapour pressure, (b) turbulent boundary layer computations and (c) a simple model for the losses generated in the cavity wake.
The potential flow code has been developed from the blade to blade flow model of the quasi three dimensional model (Kueny et al. 1988). It involves the finite element method and is based on a stream function formulation. In the two dimensional case, an obstruction factor comprised between 0 and 1 can be taken into account to simulate a partial obstruction of the flow passage.
Turbulent boundary layers are computed along the tunnel and hydrofoil walls using a finite volume discretisation and a mixing length hypothesis as turbulence model (Cebeci and Bradshaw, 1977). Displacement thicknesses along the tunnel walls and profile sides are subtracted to the tunnel sections. Owing to the low curvatures of the tunnel a simplified method can be applied : it takes into account in the transverse sections a mean obstruction factor, growing with the distance from the upstream section Sref.
The vapour cavity surface is considered as a free streamline at constant vapour pressure, which detachment point is imposed. That point is located at the ridge of the BLE foil, and at the rounded nose of the SLE foil using the empirical criterion of smooth separation (free streamline tangent to the body at the separation point). The cavity length is a given parameter of the numerical model whereas the cavitation number σ is unknown. Downstream the cavity, the streamline smoothly joins the hydrofoil surface. In that cavity wake, one part of the displacement thickness can be considered as singular head losses. Their effect is taken into account by modifying the obstruction factor of the meshes located downstream the cavity. It may then affect the flow around the hydrofoil through the Kutta-Joukowski condition, written as the equality of static pressure on the two sides of the trailing edge.

4. Experimental results

4.1 PRESSURE DISTRIBUTIONS AND CAVITY SHAPES

Pressure measurements along the foil suction side and the tunnel walls have been performed for the different cases at 3 flow velocities : Vref = 10, 16 and 20 m/s. In non cavitating conditions at the two angles of attack, the two hydrofoils generate quite similar pressure coefficient distributions (Figure 3). Because of the small curvature of the cavitation tunnel and hydrofoils, the static pressures measured appear to be quite uniform in the cross sections. The effect of the difference of leading edge design is not significant, even for the first pressure tap located 47 mm downstream. At the 4.5 degrees angle of attack, LDA measurements show that a small separation bubble probably occurs just downstream the leading edge of the SLE foil.

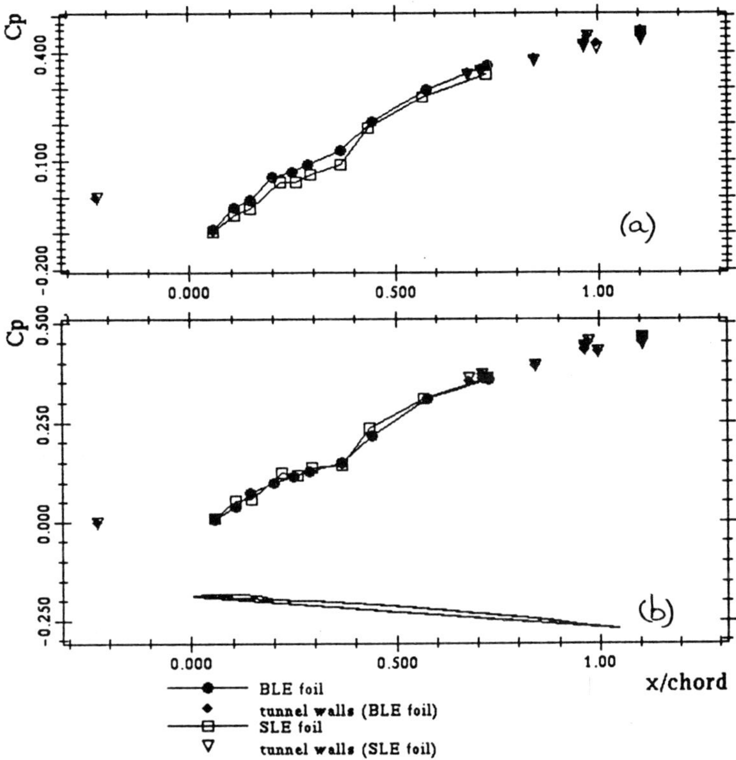

Figure 3 : Pressure coefficient distribution along the foils suction side and tunnel walls, without cavitation. Vref = 16 m/s, angle of attack : (a) 4.5 degrees (b) 4.1 degrees.

Cavities varying from 5 to 20% of the chord length have been studied. The pressure coefficients measured within the cavities are close to $-\sigma$, as can be seen on Figure 4 for cavities about 15 cm long. The main effect of the leading edge design is observed for the 4.5 degrees angle of attack. In that case, the cavitation number σ at given cavity length is more than 25% greater on the SLE hydrofoil (Figure 5). On the other hand, at reduced angle of attack (4.1 degrees) the results are quite similar on the two hydrofoils. No marked effect of the flow velocity can be observed in the studied range, the cavitation number at given cavity length never varying more than 0.02.

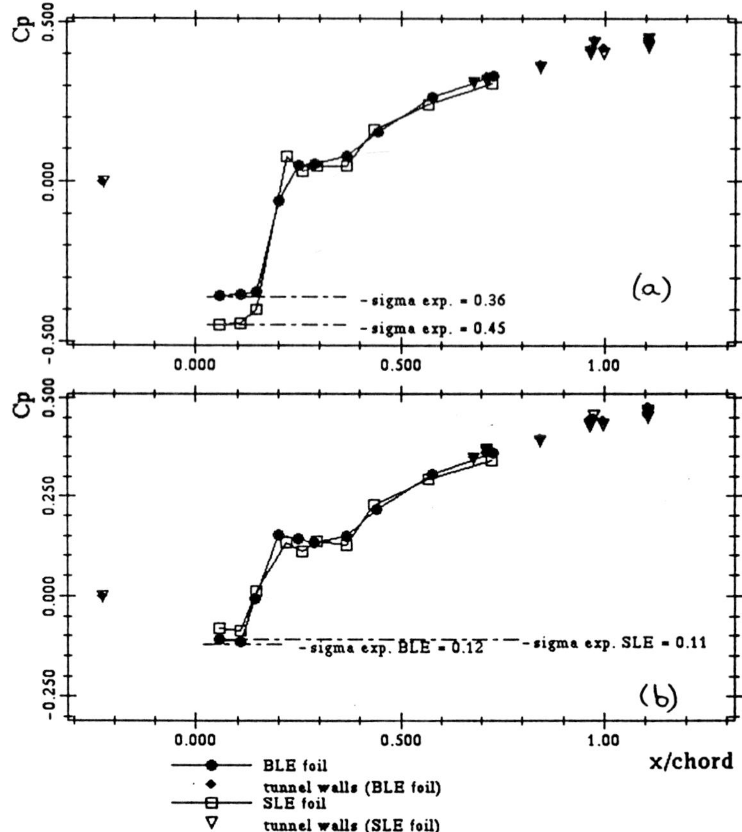

Figure 4 : Pressure coefficient distribution along the foils suction side and tunnel walls. Cavity length = 14 cm, Vref = 16 m/s, angle of attack : (a) 4.5 degrees (b) 4.1 degrees,

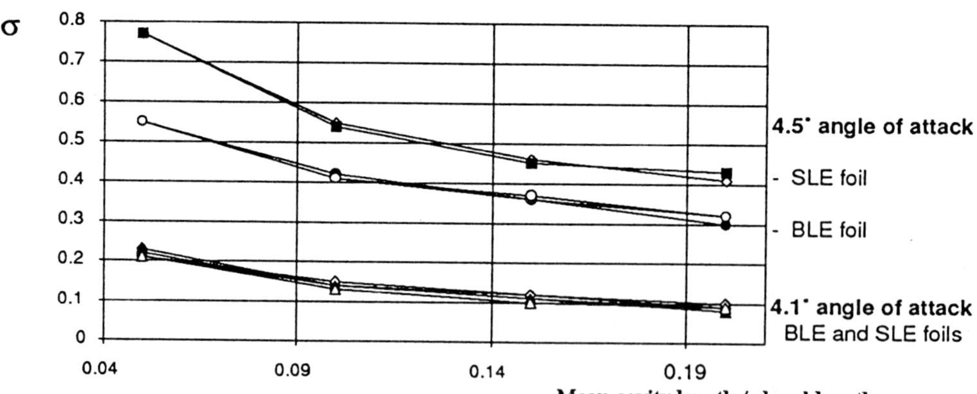

Figure. 5 : Cavitation number vs cavity length (Vref = 10, 16 and 20 m/s)

Visualisations under laser sheet light and image processing provided the cavity shapes. The greater value of σ observed at the 4.5 degrees angle of attack on the SLE foil is associated with cavity thickness increasing faster near the leading edge. The maximum relative cavity thickness is close to 7% of the cavity length. Under stroboscopic light an unsteady behaviour can be observed, large vapour clouds being shed by the main cavity. At the 4.1 degrees angle of attack, the cavities obtained are much thinner, their relative thickness always being smaller than 4%. In this case, the cavities always appear smooth and quasi steady.

4.2. EFFECT OF THE LOSSES GENERATED BY CAVITATION

Total pressure measurements along the BLE foil suction side have been performed using the 4 holes probe. Cavitating and non cavitating cases have been compared to characterise the effects of cavitation on the global behaviour of the liquid flow. Flow velocity profiles have been deduced from the total pressure given by the probe and the static pressure measured along the tunnel walls. The velocity profiles obtained at Vref = 10 m/s have been reported on Figure 6. The cavity is 14 cm long (15% of the chord length). Its shape obtained by image processing has also been drawn. One can notice the propagation along the foil suction side of losses generated by cavitation. The displacement thickness associated to the boundary layer along the foil side is increased in the cavitating case, whereas the boundary layer along the tunnel upper wall remains unchanged. The displacement thickness increasing due to the cavity can be estimated from the velocity profiles between 10 and 20% of the cavity thickness. Those results are consistent with those obtained in venturi type ducts using LDA measurements (Merle and Delannoy, 1994). The interaction of such losses with the trailing edge flow probably leads to a modification of the flow distribution on both sides of the foil, with a small decrease of the flow rate in the suction side channel and of the angle of attack in the leading edge region.

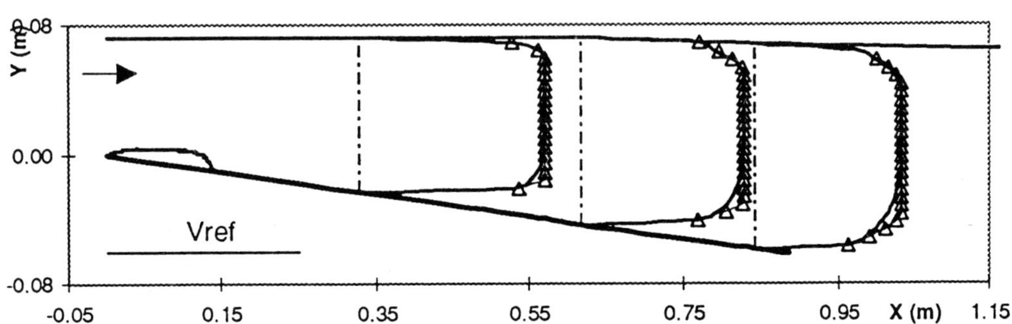

Figure 6 : Velocity profiles from total pressure measurements in non cavitating (white points) and cavitating (bold lines) conditions. BLE foil, 4.5 degrees angle of attack, cavity length 14 cm, Vref = 10 m/s.

LDA measurements have been performed around the SLE foil leading edge. On Figure 7, the difference between the velocity fields measured in cavitating and non cavitating cases has been reported. It can there obviously be seen that the cavitation induces a deviation of one part of the flow rate from the foil suction side toward the pressure side. That result is consistent with our previous remark.

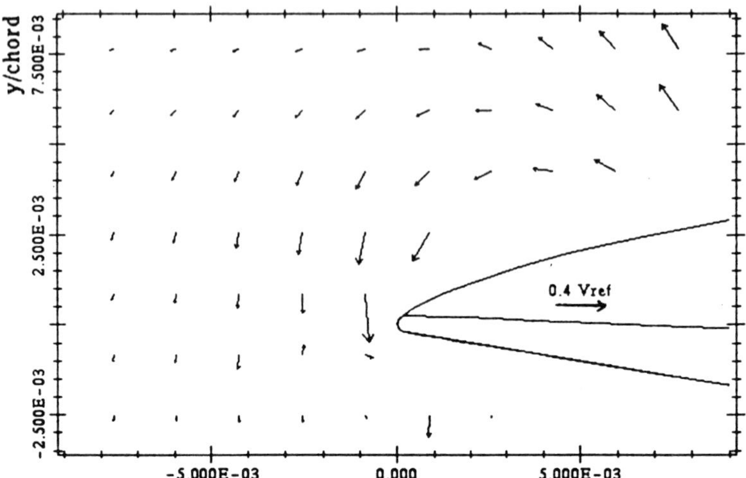

Figure 7 : Differential velocity field : difference between cavitating and non cavitating cases. SLE foil, 4.5 degrees angle of attack, cavity length 14 cm, Vref = 10 m/s

5. Comparison between experimental and numerical results

Comparisons between numerical and experimental results have been performed on a large amount of flow configurations. In the non cavitating cases, a good agreement has been found on the pressure distributions, provided that the obstruction effect of the boundary layers are taken into account. In the cavitating cases, it can be observed that the losses generated by cavitation have a great influence on the predicted value of σ at given cavity length. For example, on the BLE foil for a cavity length of 14 cm and a reference velocity of 16 m/s, the experimental value of σ is 0.36. The predicted σ is equal to 0.42 without taking into account such losses and 0.38 while subtracting in the cavity wake a displacement thickness equal to 20% of the cavity thickness, as estimated experimentally (Figure 8).

Such results are confirmed when applied to other flow configurations as illustrated on Figure 9. For cavities being 10% to 20% of the chord long, the cavitation number is predicted with an error smaller than 0.04 for the 2 foils at the 2 angles of attack. The worse results are obtained for short cavities at the 4.5 degrees angle of attack, σ being underestimated, and for the BLE foil at the 4.1 degrees angle of attack, σ always being overestimated. On the one hand, our simplified model of losses generated by cavitation has certainly to be improved in the case of short or thin cavities. On the other hand, it seems that a better description of the geometry in the detachment region is needed : rounding the sharp ridge on the mesh of the BLE foil will provide a better description of the real geometry and the prediction of the cavitation number might then be improved.

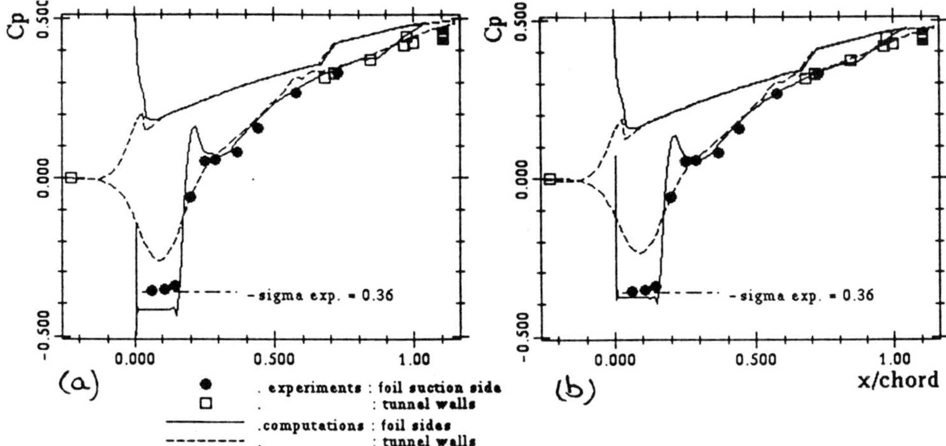

Figure 8 : Pressure distribution : BLE foil, angle of attack = 4.5 degrees, Vref = 16m/s .
- experimental data : black circles (foil suction side) and white symbols (tunnel walls)
- numerical results : full lines (foil sides) and dotted lines (tunnel walls) :
(a) without taking into account the losses generated by the cavity : σ = 0.42
(b) taking into account the losses generated by the cavity : σ = 0.38

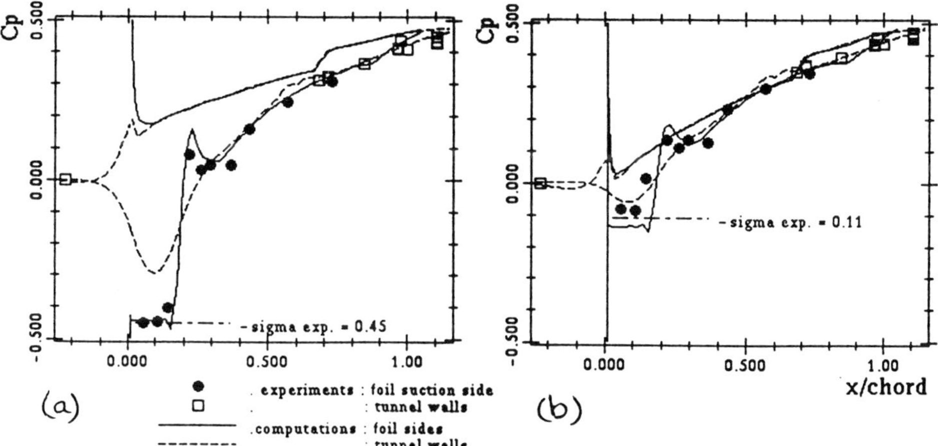

Figure 9 : Pressure distribution taking into account losses generated by the cavities. Vref = 16m/s, cavity length 14 cm, SLE foil : (a) 4.5 degrees (b) 4.1 degrees.

6. Conclusion

That experimental and numerical study brought out the following remarks :
- on our two dimensional thin hydrofoils the effect of the leading edge geometry may be an important factor of the cavitation behaviour. That effect depends on the angle of attack, but the bevelled leading edge always gave equivalent or smaller cavitation

numbers than the sharp one for the two angles tested. That result is probably no valid for an higher angle of attack if the cavity detachment shifts upstream to the ridge on the BLE foil. It can also not be extended without caution to three dimensional blades because of possible 3D effects around the leading edge.
- losses generated by cavitation have been observed along the foil suction side using total pressure measurements in cavitating and non cavitating cases. Their interaction with the trailing edge flow slightly modify the flow distribution around the foil, what has also been observed on velocity measurements around the leading edge.
- a numerical model of steady cavitating flows has been modified to take into account such losses. The predictions of the cavitation number σ for the different hydrofoils, angles of attack and cavity lengths have been improved by that modification.
The experimental data obtained in those different flow conditions provide useful test cases for improvement and validation of different numerical codes. For example, taking into account the vapour clouds shedding observed experimentally with an unsteady two-phase flow model (Reboud and Delannoy, 1994) may be one of the future way to predict the behaviour of cavitating inducers.

Acknowledgements

The authors acknowledge the Agence Française de l'Espace (CNES) and the Société Européenne de Propulsion (SEP) for the financial support of this project. The CREMHyG's large cavitation tunnel has been designed by J. Genevey and the experimental study has been performed with the help of M. Riondet, R. Morel and J. Baudoin.

References

Bernard R., Rebattet C., Martignac F., Desclaux J. (1988) Essais hydrauliques de l'inducteur hydrogène du moteur fusée Vulcain, *La Houille Blanche,* n° 7-8, 1988.

Cebeci, Bradshaw (1997) *Momentum transfer in boundary layers*, Mc Graw Hill.

Maitre T., Kueny J.L., Geai P., Von Kaenel A. (1993) Numerical predictions of three-dimensional partial cavitation in a rocket turbopump inducer, 2nd ASME Pumping Machinery Symposium, Washington.

Merle L., Delannoy Y. (1994) Flow above natural and ventilated cavities, Cavitation and Multiphase Flow Forum, *ASME-FED*, vol. 193.

Kueny J.L., Reboud J.L., Desclaux J. (1991) Analysis of Partial Cavitation by Image Processing and Numerical Prediction, Cavitation '91 Symposium, *ASME-FED*-Vol 116, pp. 55-60.

Kueny J.L., Schultz J.L., Desclaux J. (1988) Numerical prediction of partial cavitation in pumps and inducers, *proc. 14th Symposium of the IAHR* , Trondheim.

Reboud J.L., Delannoy Y. (1994) Two-phase flow modelling of unsteady cavitation, *proc. 2nd Int. Symposium on Cavitation*, Tokyo, April, 5-7, 1994, pp. 39-44.

Rowe A. and Blottiaux O. (1993) "Aspects of Modeling Partially Cavitating Flows", *J. Ship Research*, Vol.37, n° 1, pp. 34-48.

HYDRAULIC MACHINERY AND CAVITATION

VOLUME II

HYDRAULIC MACHINERY AND CAVITATION

Proceedings of the XVIII IAHR Symposium on Hydraulic Machinery and Cavitation

VOLUME II

Edited by

E. CABRERA, V. ESPERT and F. MARTÍNEZ
Polytechnic University of Valencia, Spain

Polytechnic University of Valencia
Hydraulic and Environmental Eng. Dept.
FLUID MECHANICS GROUP

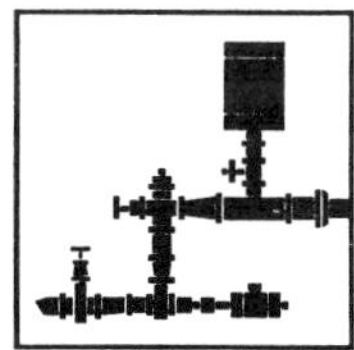

Co-sponsors:

KLUWER ACADEMIC PUBLISHERS
DORDRECHT / BOSTON / LONDON

A C.I.P. Catalogue record for this book is available from the Library of Congress.

ISBN 0-7923-4209-7 (Volume II)
ISBN 0-7923-4208-9 (Volume I)
ISBN 0-7923-4210-0 (Set of 2 Volumes)

Published by Kluwer Academic Publishers,
P.O. Box 17, 3300 AA Dordrecht, The Netherlands.

Kluwer Academic Publishers incorporates
the publishing programmes of
D. Reidel, Martinus Nijhoff, Dr W. Junk and MTP Press.

Sold and distributed in the U.S.A. and Canada
by Kluwer Academic Publishers,
101 Philip Drive, Norwell, MA 02061, U.S.A.

In all other countries, sold and distributed
by Kluwer Academic Publishers Group,
P.O. Box 322, 3300 AH Dordrecht, The Netherlands.

Printed on acid-free paper

Printed in the Netherlands

TABLE OF CONTENTS

Acknowledgements xvii
Preface xix
Foreword xxi

VOLUME I

INVITED LECTURES

The hydroelectricity in the world-Present and future.
Brekke, H. 3

Analysis of transients caused by hydraulic machinery.
Chaudhry, M.H. 17

Some present trends in hydraulic machinery research.
Fanelli, M. 23

Rapid prototyping of hydraulic machinery.
Schilling R., Riedel, N., Bader, R., Ascherbrenner, T., Weber, Ch., Fernandez, A. 40

Fluid transients in flexible piping systems.
A perspective on recent developments
Wiggert, D.C. 58

SYMPOSIUM CONTRIBUTIONS

GROUP 1 HYDRAULIC TURBINES. ANALYSIS AND DESIGN

1. 1 NUMERICAL ANALYSIS OF COMPONENTS

A decision aid system for hydraulic power station refurbishment procedure.
Bellet, L., Parkinson, E. Avellan, F., Cousot, T., Laperrousaz, E. 71

A three dimensional spiral casing Navier-Stokes flow simulation.
Bruttin, C.H., Kueny, J.L., Boyer, B., Héon, K,. Vu, T.C. and Parkinson, E. 81

Tip clearance flow in turbomachines- experimental flow analysis.
Ciocan, G.D. and Kueny, J.L. 91

Numerical prediction of hydraulic losses in the spiral casing of a Francis turbine.
Drtina, P. and Sebestyen, A. 101

Improvements of a graphical method for calculation of flow on a Pelton bucket.
Hana, M. 111

Flow behavior and performance of draft tubes for bulb turbines.
Kanemoto, T.,Uno, M., Nemoto, M., and Kashiwabara, T. 120

Performance analysis of draft tube for GAMM francis turbine.
Kubota, T., Han, F., and Avellan, F. 130

Modelling complex draft-tube flows using near-wall turbulence closures.
Ventikos, Y., Sotiropoulos, F. and Patel, V.C. 140

1.2 NUMERICAL ANALYSIS AND DESIGN OF HYDRAULIC MACHINERY

Fluid flow interactions in hydraulic machinery.
Aschenbrenner, T., Riedel, N. and Schilling, R. 150

Numerical optimization of high head pump-turbines
Buchmaicher, H,. Quaschnowitz, B., Moser, W. and Klemm, D. 160

Numerical hill chart prediction by means of CFD stage simulation for a complete Francis turbine.
Keck, H. Drtina, P. and Sick, M. 170

Development of a new generation of high head pump-turbines, Guangzhou II.
Klemm, D., Jaeger, E.U., and Hauff, C. 180

Development of integrated CAE tools for design assessment and analysis of hydraulic turbines.
Massé, B,. Pastorel, H. and Magnan, R. 190

Design and analysis of a two stage pump turbine
Mazzouji, F. , Francois, M. Hebrard, F., Houdeline, B. and Bazin, D. 200

Study of high speed and high head reversible pump-turbine
Nakamura, T., Nishizawa, H., Yasuda, M., Suzuki, T. and Tanaka, H. 210

Analysis of the performance of a bulb turbine using 3-D viscous numerical techniques
Qian, Y., Suzuki, R. and Arakawa, C. 220

Analysis of the inlet reverse flows in a pump turbine using 3-D viscous numerical techniques
Qian, Y., Suzuki, R. and Arakawa, C. 230

Importance of interaction between turbine components in flow field simulation
Riedelbauch, S,. Klemm, D. and Hauff, C. 238

From components to complete turbine numerical simulation
Sabourin, M., Labrecque, Y. and de Henau, V. 248

Validation of a stage calculation in a Francis turbine
Sick M., Casey, M.V. and Galpin, P.F. 257

Simulation of flow through Francis turbine by LES method
Song, C.C.S., Xiangying, Ch., Ikohagi,T., Sato, J., Shinmei, K. and Tani, K. 267

Simulation of flow through pump-turbine
Song, C.C.S., Changsi, C., Ikohagi,T., Sato, J., Shinmei, K. and Tani, K. 277

1.3 LOSS ANALYSIS AND SCALE EFFECTS

The scale effect in Kaplan turbines. New relationships for the calculation of scalable and non-scalable hydraulic losses and of the coefficient V.
Anton, I.M. 284

Analysis of losses in hydraulic turbines.
Brekke, H. 294

Scaling-up head discharge characteristics from model to prototype.
Couston, M. and Philibert, R. 304

Recent development of studies on scale effect
Ida, T., Kubota,T. , Kurokawa, J. and Tanaka, H. 313

Prediction of scalable loss in Francis runners of different specific speed
Kitahora, T., Jurokawa, J., Matumoto, M. and Suzuki, R. 323

Scale effect of jet interference in multinozzle Pelton turbines
Nakanishi, Y. and Kubota, T. 333

Further development of step-up formula considering surface roughness
Nichtawitz, A. 342

Numerical simulations of jet in a Pelton turbine.
Nonoshita, T. , Matsumoto, Y., Kubota,T. and Ohashi, H. 352

An assessment of the loss distribution in Francis turbines
Suzuki, R., Qian, Y., Kitahora, T. and Kurokawa, J. 361

GROUP 2. HYDRAULIC PUMPS

Unsteady flow calculation in a centrifugal pump using a finite element method.
Bert, P.F., Combes, J.F. and Kueny, J.L. 371

Steady and unsteady flow pattern between stay and guide vanes in a pump-turbine
Ciocan, G., Kueny, J. L. and Mesquita, A.L.A. 381

Self-sustained oscillation of gas-liquid flow in a centrifugal pump with semi-open impeller
Kurokawa, J., Matsui, J., Takada, H. and Hirayama, T. 391

Measurements in the dynamic pressure field of the volute of a centrifugal pump
Parrondo, J.L., Fernández, J., Santolaria, C. and González, J. 401

Functional modelling of pump volute geometry
Thackray, P.R. and James, R.D. 411

Analysis of flow measurements in the impeller and vaned diffuser of a centrifugal pump operating at part load.
Toussaint, M. and Hureau, F. 419

Influence of the blade roughness on the hydraulic performance of a mixed-flow pump. A viscous analysis
Undreiner, S. and Dueymes, E. 428

Improvement of performance of centrifugal pumps based on computational and theoretical methods and experimental design
Vinokurov, A.F., Volkov, A.V., Morgunov, G.M. and Pankratov, S.N. 438

Liquid-particulate two-phase flow in centrifugal impeller by turbulent simulation
Wu, Y., Dai, J., Mei, Z., Oba, S. and Ikoagi, T. 445

GROUP 3. HYDRAULIC ELEMENTS. DYNAMIC CHARACTERIZATION AND HYDRAULIC BEHAVIOUR.

Application of the method of kinetic balance for flow passages forming.
Benisek, M., Cantrak, S., Ignatovic, B. and Pokrajac, D. 455

Dynamics of large hydrogenerators.
Brito, G.C., Weber, H.I. and Fuerst, A.G.A. 464

Instabilities in a flow-control valve.
Cigada, A., Guadagnini, A. and Orsi, E. 474

Study of stayvane vibration by hydroelastic model.
Deniau, J.L. 484

Study of dynamic behaviour of non-return valves.
François, P. 494

Optimum hydraulic design of two-way inlet conduit of Wangyuhe pumping station
Linguang, L., Jiren, Z. and Rentian, Z. 504

Flow analysis for the intake of low-head hydro power plants
Ruprecht, A., Maihofer, M. and Gode, E. 514

GROUP 4. CAVITATION AND SAND EROSION

Efficiency alteration of Francis turbines by travelling bubble cavitation.
Arn, Ch., Dupont, Ph. and Avellan, F. 524

Cavitation erosion prediction on Francis turbines - Part 1. Measurements on the prototype.
Bourdon, P., Pfarhat, M., Simoneau, R., Pereira, F., Dupont, P., Avellan, F. and Dorey, J.M. 534

Determination of critical cavitation limit in the pressure control devices.
Castorani, A., De Martino, G. and Fratino, U. 544

Stability of air cavities in tip vortices.
Crespo, A., Castro, F., Manuel, F. and Fruman, D.H. 554

Cavitation erosion prediction on Francis turbines - Part 3 Methodologies of prediction.
Dorey, J.M., Laperrousaz, E., Avellan, F., Dupont, P., Simoneau, R. and Bourdon, P. 564

Cavitation erosion prediction on Francis turbines - Part 2. Model test and flow analysis.
Dupont, P.H., Caron, J.F., Avellan, F., Bourdon, P., Lavigne, M., Farhat, M., Simoneau, R., Dorey, J.M., Archer, A., Laperrousaz, E. and Couston, M. 574

Impact of vapour production and cavity dynamics on the estimation of thermal effects in cavitation.
Fruman, D.H., Reboud, J.L. and Stutz, B. 584

Aireation versus cavitation in dam spillways: self-aeration and artificial aeration (aerators).
Gutiérrez , R. 594

Leading edge cavitation in a centrifugal pump: Numerical predictions compared with model tests
Hirschi, R., Dupont, P.H., Avellan, F., Favre, J.N., Guelich, F., and Handloser, W. 604

The relation between erosion ripples on the wetted surface of hydraulic turbine and instability waves in the turbulent boundary layer
Huang, S. and Cheng, L. 614

Acoustic method and its applications on measuring and judging cavitation of hydraulic turbine.
Kehuang, L. and Chun, Y. 622

Numerical simulation for dilute sandy water flow in plane cascade
Liu, X.B., Zeng, Q.C. and Zhang, L.D. 632

Review of research on abrasion and cavitation of silt-laden flows through hydraulic turbines in China
Mei, Z. and Wu, Y. 641

Effect of the leading edge design on sheet cavitation around a blade section
Reboud, J.L., Rebattet, C. and Morel, P. 651

VOLUME II

GROUP 5. HYDRAULIC TRANSIENTS AND CONTROL SYSTEMS RELATED WITH HYDRAULIC MACHINERY AND PLANTS

Optimal closure of a valve for minimizing waterhammer.
Abreu, J., Cabrera, E., García-Serra, J. and López, P.A. 661

Qualitative flow visualizations during fast start-up of centrifugal pumps.
Barrand, J.P. and Picavet, A. 671

Analysis of a numerical model for the oscillatory properties of a Francis turbine group.
Cattanei, A., Capozza, A., and Molinaro, P. 681

Transients analysis and dynamic criteria for HPP exploatation
Gajic, A., Pejovic, S., Krsmanovic, L.J. and Stojanovic, Z. 691

Modelling a protection device in a low pressure lifting system.
Giustolisi, O. and Mastrorilli, M. 701

Dynamic compression of entrapped air pockets by elastic water columns.
Guarga, R., Acosta, A. and Lorenzo, E. 710

Generalization of pump station boundary condition in hydraulic transient simulation.
Izquierdo, J., Iglesias, P., Espert, V. and Fuertes, V. 720

Analysis of unsteady characteristics of flows through a centrifugal-pump impeller by an advanced vortex method.
Kamemoto, K., Kurasawa, H., Matsumoto, H. and Yokoi, Y. 729

Expert system for analysis of pumped storage schemes.
Koelle, E., Andrade, J.G.P. and Luvizotto Jr., E. 739

Model-based analysis of active PID-control of transient flow in hydraulic networks.
Lauria, J.C. and Koelle, E. 749

Simulation of transients in pressurized hydraulic systems with visual tools.
Martínez, F., Izquierdo, J., Pérez, R. and Vela, A. 759

Dynamic behaviour of governing turbines sharing the same electrical grid
Nielsen, T.K. 769

Prediction of natural frequencies in a hydro power plant supplying an electric network by itself having a known load type
Raabe, J. 779

Modelling and practical analysis of the transient overspeed effect of small Francis turbines.
Ramos, H. and Betamio, A. 789

Simulation of turbine governing in time domain.
Stuksrud, D.B. 799

Parametrical modelling of power characteristics of the Francis and Kaplan hydraulic turbines.
Tolea, M.F. and Kueny, J.L. 809

Unsteady frictions in pipelines.
Vennatro, R. 819

GROUP 6. OSCILLATORY AND VIBRATION PROBLEMS IN HYDRAULIC MACHINERY AND POWER STATIONS

Swirl flow in conical diffusers.
Dahlhaug, O.G. 827

Experimental investigation of vortex core in reverse swirl flow from Francis runner.
Furuie, Y., Mita, H. and Hosoi, Y. 835

Hydraulic oscillation analysis using the fluid-structure interaction model.
Gajic, A., Pejovic, S. and Stojanovic, Z. 845

Francis turbine surge: discussion and data base.
Jacob, T. and Prenat, J.E. 855

Non-stationary flow in reversible Francis turbine runner due to wakes trailing the guide vanes.
Jernsletten, J. 865

The swirling inlet flow effects on the pressure recovery of a low head water turbine draft tube.
Kikuyama, K., Hasegawa, Y., Augusto, G., Nishibori, K. and Nakamura, S. 875

Two kinds of whirl on fixed-blade propeller type turbine.
Léonard, F. 885

Self-excited hydraulic oscillations dued of unstable valve behaviour. A case study.
Mateos, C., Pérez-Andújar, T., Andreu, M. and Cabrera, E. 895

An experimental study on fins. Their role in control of the draft tube surging
Nishi, M., Wang, X.M., Yoshida, K., Takahashi, T. and Tsukamoto, T. 905

Model for vortex rope dynamics in Francis turbine outlet.
Pedrizzetti, G. and Angelico, G. 915

Vortices rotating in the vaneless space of a Kaplan turbine operating under off-cam high swirl flow conditions.
Pulpitel, L., Skotak, A. and Kontnik, J. 925

Experimental investigation of frequency characteristics of draft tube pressure pulsations for Francis turbines.
Qinghua, S. 935

On the suppression of coupled liquid/pipe vibrations.
Tijsseling, A.S. and Vardy, A.E. 945

Unsteady hydraulic force on an impeller due to rotor-stator interaction in a diffuser pump.
Tsukamoto, H., Uno, M., Qian, W., Teshima, T., Sakamoto, K., and Okamura, T. 955

Swirling flow with helical vortex core in a draft tube predicted by a vortex method.
Wang, X.M. and Nishi, M. 965

GROUP 7. EXPERIMENTAL INVESTIGATIONS RELATED WITH HYDRAULIC MACHINARY AND ITS APPLICATIONS

Friction loss of rough wall passage in a turbomachinery
Akaike, S. 975

Redesign of sharp heel draft tube- Results from tests in model and prototype.
Dahlbäck, N. 985

Model and prototype draft tube pressure pulsations
Kerkan V., Bajid, M. Djelic, V. Lipej, A. and Jost, D. 994

Fish bypass system impact upon turbine runner performance at Rocky Reach dam.
Lang, A. and Christman, B. 1004

LDV measurements in an impeller-generated turbulent jet developing in a new coflow.
Peterson, P., Larson, M and Jönsson, L. 1014

Inline radial force measurement of turbine runners
Riener, J., Egger, A. and Schnur, G. 1024

Different types and locations of part-load recirculations in centrifugal pumps found from LDV measurements.
Stoffel, B. and Weiss, K. 1034

Turbulent 3D flows near the impeller of a mixed-flow pump
Wang, B. and Hellman, D.H. 1044

GROUP 8 PRACTICAL APPLICATIONS OF THE HYDRAULIC MACHINERY

Overload in Kaplan turbines of Salto Grande hydropower complex.
Baccino, M. and Bonecarrere, E. 1053

Full sized tests on a french main coolant pump under two-phase flow.
Huchard, J.C., Bore, C. and Dueymes, E. 1063

Computer simulations of dynamic performance of 400 Mw adjustable speed pumped storage units.
Kita, E., Nakagawa, H., Kuwabara, T. and Harada, M. 1073

Performance comparison of nuclear reactor recirculation pumps tested under large reynolds number difference
Saiki, K., Ikura, T., Matsumoto, K., Komita, H., Kobayashi, M., Saito, T. and Tanaka, H. 1083

Performance of Candu heat transport pumps under two-phase flow conditions.
Samarasekera, H. and Kumar, A.N. 1093

Interdependence of draft tube and tailwater flow in bulb turbine power plants.
Schneider, C.H., Knapp, W. and Schilling, R. 1103

Study of hydraulic transients using the bond graphs method.
Tiago Filho, G.L. 1113

GROUP 9. MONITORING, PREDICTIVE MAINTENANCE AND REFURBISHMENT.

Numerical flow analysis of a Kaplan turbine.
Jost, D., Lipej, A., Oberdank, K., Jamnik, M. and Velensek, B. 1123

18-Paths acoustic flowmeter and transducer protrusion.
Lévesque, J.-M., Néron, J. and Tran, C.M. 1133

Step-up in rehabilitation: not a myth, a science
Mahé, B., de Henau, V. and Sabourin, M. 1142

Super system: a hydroelectric unit condition monitoring system in operation at Hydro-Quebec.
Mossoba, Y. 1152

Application of rotor response analysis to fault detection in hydro powerplants.
Nascimento, L.P. and Egusquiza, E. 1162

The placement exploration of diagnostic measuring points on operating unit equipment.
Liu, X.T. 1172

OPTIMAL CLOSURE OF A VALVE FOR MINIMIZING WATER HAMMER

ABREU, J.,
Univ. de Coimbra. Dep. de Engenharia Civil. Portugal

CABRERA, E., GARCIA-SERRA, J., LOPEZ, P.A.
Univ. Politécnica de Valencia. U.D. Mecánica de Fluidos.Spain

Summary

Due to its relevance, many authors have paid a lot of attention to determine the optimal closure of a valve. As some of the results have in practice difficult implementation, the problem remains open.

This paper deals with the different methods of analysis, and highlight their hypothesis, results and reliability. By the simultaneous use of the rigid and the elastic models, the influence of the inertia and the elasticity of the system, regarding the optimum closure, are considered. Finally, some practical conclusions are pointed out.

1. Introduction.

The use and operation of valves in pressurized hydraulic systems is essential. Their manoeuvres introduce pressure surges, which must be supported by the all system. Regardless of the protective devices used in practice, it is of a paramount interest to know, given a closure time Tc and, in order to minimize the surge, in which way should be carried out the manoeuvre. In practice good regulations requires short Tc, whereas from the instalation's safety point of view, longer times of Tc are needed. Then a compromise solution for Tc is adopted. Once Tc has been defined, the optimal closure manoeuvre that we are looking for, should give rise to the minimum overpressure ΔH_{max}.

Many contributions can be found in the literature. Ruus (1957, 1966), and Cabelka and Franc (1959) deal on the "optimum valve closure arrangement", concept that is later extended to "valve stroking" (Streeter, 1963). These techniques often impose a rather rigorous valve movement that requires accurate and complex driver systems, out of interest due to its scarce reliability. Valve stroking presents, as relevant characteristic, that, after the manoeuvre, no residual surge remain into the pipe.

Other approaches look for optimization techniques introduced by the expanding fields of operations research. Driels (1975) studied to minimize the pressure rise following a two stage valve closure, using a simple search algorithm, in order to find when the speed of the valve actuator should change. Contractor (1983) applied the dynamic programming technique, being the objective function the pressure rise at the valve. Later Contractor (1985) applied the Box-Complex method in a two/three stages valve closure.

Goldberg and Karr (1985) suggested an optimal valve closure movement, called quick stroking. Finally Azoury et al. (1986), and in order to optimize the manoeuvre, suggest a continuous power funtion for the inherent valve characteristics.

This paper faces the problem from both rigid and elastic models point of view. The first one, better known as mass oscilation, gives a good physical understanding of

E. Cabrera et al. (eds.), Hydraulic Machinery and Cavitation, 661–670.

the problem and, in slow transients, provides reasonable results. A later application of the waterhammer theory, as is colloquially called the elastic model, allows to evaluate the contribution of the system's elasticity. Ideal (frictionless) and real systems, are considered in both cases, in order to know about the friction influence into the final solution.

2. Pipe and boundary conditions parameter.

The system shown in Fig. 1 is considered. A simple and uniform pipeline, with a valve at its downstream end, is discharging water coming from the reservoir into free air, i.e. a penstock supling water to a Pelton turbine.

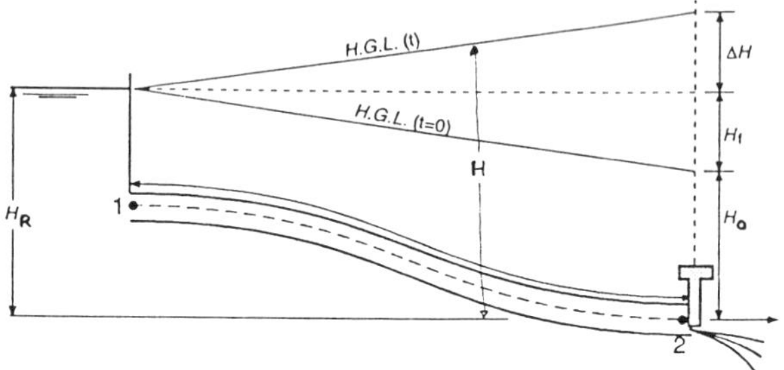

Fig. 1. Basic considered installation.

Waterhammer pressures, due to a flow control manoeuvre, depend both on the valve characteristics and on the pipeline response. This unsteady one dimensional fluid flow in closed pipes is described by two quasi-linear hyperbolic partial differential equations (Chaudhry 1987, Wylie and Streeter 1993). The dynamic behaviour of a simple uniform pipe is completely defined (Abreu et al., 1996) by three parameters: Inertance,I, Pipeline Resistence, R, and Elastic Time.,T_e. Are given by:

$$I=\frac{L}{g\,A} \qquad R=f\frac{L}{2\,g\,D\,A^2} \qquad T_e=\frac{2\,L}{a}$$

being a, the waterhammer wave velocity, and A, L, D and f the cross-sectional area, length, diameter and friction factor of the pipe, respectively.

These parameters can be gathered into two nondimensional parameters:

$$\rho = \frac{T_W}{T_e} = \frac{I\,Q_o}{T_e\,H_R} = \frac{a\,V_o}{2g\,H_R} \qquad (1)$$

$$h_f = \frac{H_f}{H_R} = R\frac{Q_o^2}{H_R} \qquad (2)$$

in which $T_w = I\,Q_o/H_R$ is the water starting time of the conduit and the suscript o refers to the initial steady state conditon. It is assumed that the reservoir level, H_R, remains constant during the transient, whereas the valve boundary is defined by the steady-state orifice equation, i.e.

$$Q = (C_d\,A_v)\,\sqrt{2g\,H} \qquad (3)$$

beingC_dA_v the valve's area times the discharge coefficient, and H the head loss valve.

The behaviour of a valve is characterized throughout its "inherent valve schedule". It can be found ploting the flow throughout the valve at differents openings versus the stem position at a constant pressure drop. Each valve has its own inherent characteristic. Generally is represented by τ versus time according the equation:

$$\frac{Q}{Q_o} = \tau \sqrt{\frac{H}{H_o}} = \frac{C_d A_v}{(C_d A_v)_o} \sqrt{\frac{H}{H_o}} \tag{4}$$

as can be seen, i.e., in Chaudhry (1987) or Wylie and Streeter (1993). Thus, if $H = H_o$, and in absence of system-valve interaction, the boundary condition will be:

$$\frac{Q(t)}{Q_o} = \tau\ (t) \tag{5}$$

Nevertheless, when the valve is part of the whole system, it does not operate at a constant pressure drop, and then the "effective" valve schedule differs from the "inherent" one.

The dimensionless parameter defining the relative time of closure is given by:

$$\Gamma = \frac{T_c}{T_W} = \frac{g\,H_R}{L\,V_o}\,T_C \tag{6}$$

We claim the use of Γ, instead of the classic Allievi's parameter, θ = Tc/Te, because Γ has a significant relevance, as inertia parameter, into the resulting flow rate ("effective valve schedule").It is worth to be noted that these parameters are related by $\theta = \rho\ \Gamma$.

3. Optimal closure valve schedule, -rigid model-, in a frictionless system.

From this approach we obtain simplified, but closed results, valids for the understanding of the phenomenon. A later inclusion of the friction allows to evaluate its influence.

From the Control Theory, and assuming that the body behaviour is under the dynamic fundamental Newton's equation, it is well known that the most efficient way for stopping a moving body in a given time Tc, is to apply a constant effort during this interval. According this, the pressure law to be introduced by the valve in order to get an optimal closure, is the one gived by Fig. 2.

The rigid model gives, as required overpressure:

$$\Delta\ H^{opt}_{max} = \frac{LV_o}{gT_C} \quad or \quad \Delta h^{opt}_{max} = \frac{\Delta H^{opt}_{max}}{H_R} = \frac{1}{\Gamma} \tag{7}$$

From (4), the closure schedule to be followed by the valve can be determined by:

$$\tau(t) = \frac{V}{V_o}\sqrt{\frac{H_o}{H}} \tag{8}$$

being needed an initial and instantaneons closure step equal to $\Delta\tau = 1 - \tau_{o,\,opt}$, given by:

$$\tau_{0,opt} = \tau\ (t = o) = \sqrt{\frac{H_o}{H_o + \Delta H^{opt}_{max}}} = \sqrt{\frac{1}{1+\Gamma}} \tag{9}$$

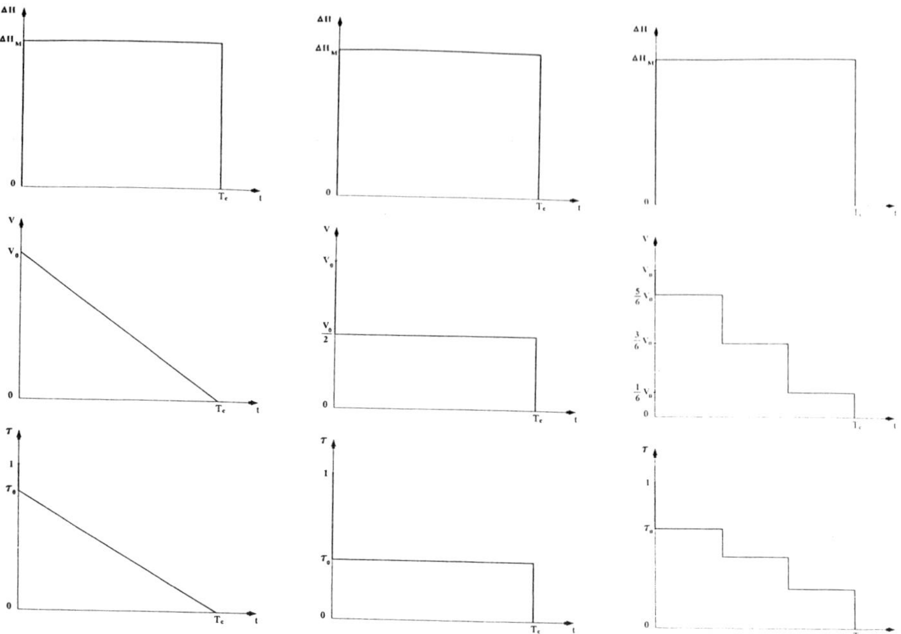

Fig 2. Theoretical optimal closure characteristic
a) Rigid model; b) Elastic model (Tc = 2L/a); c) Elastic model (Tc = 6L/a)

According the rigid model theory, if a constant ΔH_{max}^{opt} is applied, the flow velocity decreases in a linear way. Then, and according (8), $\tau(t)$ is between t = 0 and t= T_c also linear. The global manoeuvre (Fig. 2.a) given by an instantaneous closure plus a linear one, will be called in the following "optimal closure". Although fundamented into the rigid model, this procedure has been the starting point of the work done by Ruus (1957) and Cabelca and Franc (1959).

From (7) and (9), we get:

$$\Delta h_{max}^{opt} = \frac{1}{\tau_{0,\,opt}^2} - 1 \qquad (10)$$

Greater inertias ($\Gamma \rightarrow 0$), requires greater Δh_{max}^{opt} ,and greater discontinuities in the optimum closure as well, as can be seen from (7) and (9).

For non optimal instantaneous partial closures, the following can be stated :
a) $\tau_0 \prec \tau_{0,\,opt}$. In this case the maximum pressure is reached at t = 0, being its value:

$$\Delta h_{max} = \frac{1}{\tau_0^2} - 1 > \Delta h_{max}^{opt} \qquad (11)$$

quite similar to the expression (10).
b) $\tau_0 \succ \tau_{0,\,opt}$. Now the maximum pressure appears at t = T_c, being its value equal to:

$$\left. \Delta h_{max} = \frac{1}{2} k^2 + \sqrt{\frac{1}{4} k^4 + k^2} \quad ; \quad k = \frac{\tau_0}{\Gamma} \right\} \qquad (12)$$

that for τ_0 = 1 (full linear closure) gives rise to Johnson's expression (Johnson, 1920).

In Fig.3a) and 3b), these results can be seen.

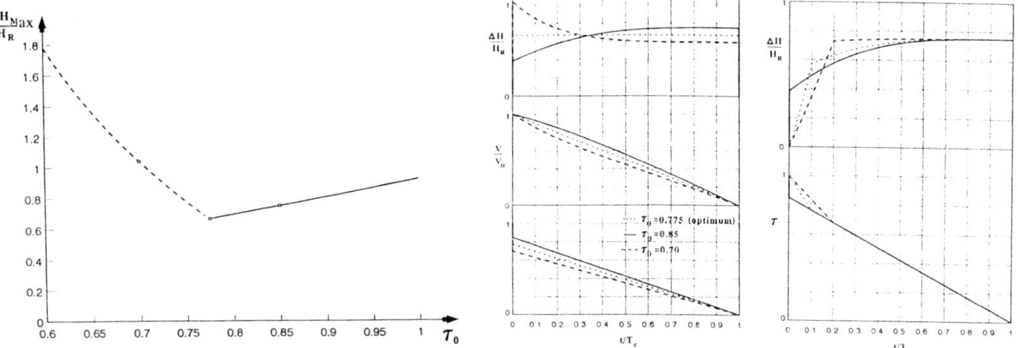

Fig 3.- Two-stage valve closure. Rigid model ($\Gamma = 1.50$)
a) τ_o versus maximum relative pressure head rise
b) Instantaneous plus linear closure valves (with $\tau_{0,opt}$= 0,775, τ_0= 0,7 and τ_0= 0,85)
c) Two stage closure valve (τ_o = 0,85, t_p = 0,1 T_c and t_p = 0,2 T_c)

From Fig 3.a) we can conclude that it does not worth an instanteneous closure with $\tau_o < \tau_{0,opt}$, because it is sharper than the optimal manoeuvre, but with a higher overpressure. Then only instantaneous closures with a final value of $\tau_o > \tau_{0,opt}$ will be considered. In this last case and according Fig. 3c), the same peak pressure can be obtained with a two stages linear closure than with the instantaneous -non optimal-, plus linear closure, one. We have, depending on time where the slope of the closure valve change, t_p, different transient answers but, in any case, the final pressure remains the same.

In summary, Fig 3 highlight clearly the interest of the two stages linear closures, widely used in practice.

4. Optimal closure valve schedule from the elastic model point of view, in a frictionless system.

When elastic effects are considered, the closure valve schedules and the velocity laws as well, are given by Fig.2b) and 2c). The quick stroking technique (Goldberg and Karr, 1987) is based into these laws. As far as the valve stroking frictionless (Streeter, 1963) and the valve stroking with friction (Streeter, 1967) concerns, the initial and instantaneous closure of an optimal procedure is replaced by a manoeuvre that at t = 2L/a gives the maximum overpressure at the valve point and a uniform flow along the pipe. Later on (t > 2L/a), the valve is operated in order to mantain the pressure constant ($\partial H/\partial t = 0$) and that, from the continuity equation, implies $\partial Q/\partial x = 0$, remaining the flow uniform without any residual surge into the pipe.

Taking into account the elastic effects, optimal laws represented by Fig. 2b) and 2c), give as optimal overpressure:

$$\Delta h^{opt.}_{max} = \frac{1}{\Gamma - \frac{1}{2\rho}} \tag{13}$$

that, as can be seen, is similar to the obtained from the rigid model, but now assuming a closure time equal to:

$$T'_c = T_c - \frac{L}{a} \tag{14}$$

that cannot be considered surprising taking into account the transient generated by a valve stroking closure.

It is worth to be noted here that, as any optimal law implies partial instantaneous manoeuvres, in practice they have a rather complex implementation. By this reason it is helpfull to calculate the increment of the overpressure obtained when, with the same closure time T_c , we apply a linear inherent law $\tau = \tau$ (t), instead of an optimal one.

The difference, in relative terms, can be stablished according

$$\delta = \frac{\Delta h_{max} - \Delta h^{opt.}_{max}}{\Delta h^{opt}_{max}} \tag{15}$$

that appears in Fig. 4, with regard to the basic parameters ρ and Γ. Then:

a) In low friction systems, with significative elastic and inertial effects -small values of γ and ρ-, a simple linear closure gives results close to the optimal ones.

b) The difference between both policies increases with the inertia and the rigidity. This gap can be important from, say, $\Gamma < 3$ and $\rho > 3$.

c) In the low inertia domain (i.e. $\Gamma > 3$), the opposite behaviour can be find: the difference increases with the elasticity of the system ($\rho < 3$). In any case, with a deviation under 6%, this change of tendence is more qualitative than quantitative.

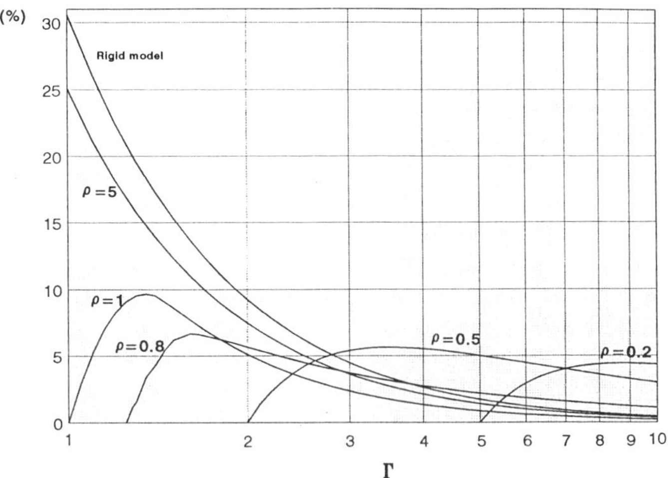

Fig. 4. Relative diference of linear versus optimal valve closure.

5. Optimal closure schedule for real systems, from the rigid model point of view

Like before, a constant overpressure step, ΔH, is imposed as system´s answer. The law τ(t) to be imposed and the answer, in velocity terms, V(t) are quite the same than before,because their linear parts are now a little bit concaves.

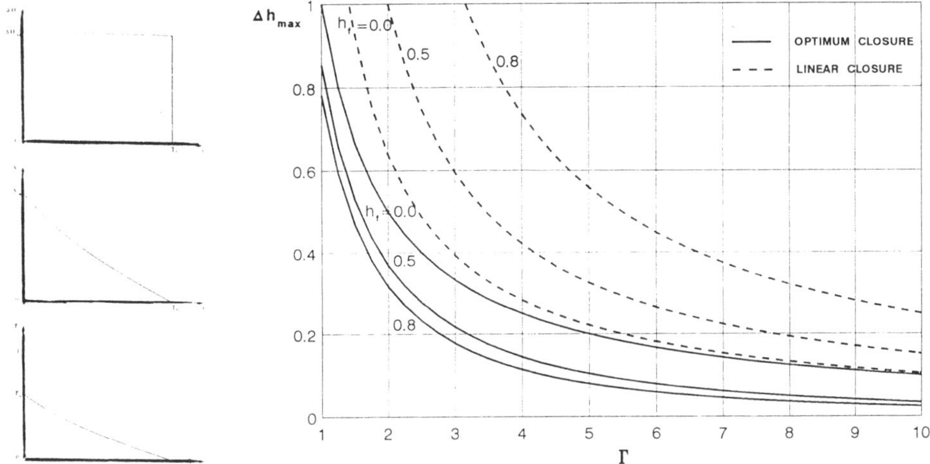

Fig. 5.- Rigid model with friction a) Teoretical optimum closure characteristics.
b) Linear closure versus optimum closure

In Fig. 5b), values of Δh_{max} , for different closure types and friction systems, can be obtained. The friction parameter, $h_f = H_f/H_R$, play now a relevant role not being, then, Δh_{max}^{opt} only Γ dependent as before was. Results from an optimal manoeuvre, -full lines in Fig 5b)-, can be obtained from:

$$\Gamma = \frac{1}{\sqrt{h_f \; \Delta h_{max}^{opt}}} \tan^{-1} \sqrt{\frac{h_f}{\Delta h_{max}^{opt}}} \tag{16}$$

that is the equation (5.6) in (Wylie and Streeter, 1993), with V = 0 for t = T_c, and assuming an optimum initial instantaneous closure ($\Delta\tau = 1 - \tau_{0,opt}$).

Due to the friction system-boundary coupling phenomenon that modify the "inherent characterisitc" for the valve, given rise to the "effective characteristic" (Abreu et al. 1996), the parameter h_f plays a relevant rol in the final value of Δh_{max}. This influence increases with the coupling effects, shaping concave the final characteristic of the valve.

After a first impression, could be thought that the overpressure Δh_{max} would decrease with the increasing friction. But the results, as can be seen in Fig 5b), are just the opposite because of the coupling phenomenon. In fact the only way to minimize the overpressure is a prompt devanishment of the velocity law with a convexe shape profile, just the opposite that the one due to the coupling effect. That requires.

$$V = V_o \left(1 - \frac{t}{T_C}\right)^m \; with \quad m > 1 \tag{17}$$

that, according the rigid model herein considered, gives as the overpressure evolution Δh:

$$\Delta h = m \frac{1}{\Gamma} \left(1 - \frac{t}{T_c} \right)^{m-1} - h_f \left(1 - \frac{t}{T_c} \right)^{2m} \tag{18}$$

The term representing the friction effect, contributes to minimize Δh(t) and takes more relevance with the increasing values of m. In fact more concavity implies a faster initial reduction of the velocity. This analysis, even being ideal, underline the relevance of the velocity profiles in the Δh(t) evolution.

Fig. 6 is equal to Fig 3, but taking now into account friction. Main conclusions, are:

a) The optimum closure, -an inital and instanteneous closure plus a concave closure, gives rise to a constant, and minimum, value of the overpressure Δh_{max}^{opt}.

b) If, in the optimum schedule, the second stage of the closure is linealized, the results are better than any other two stages linear closure.

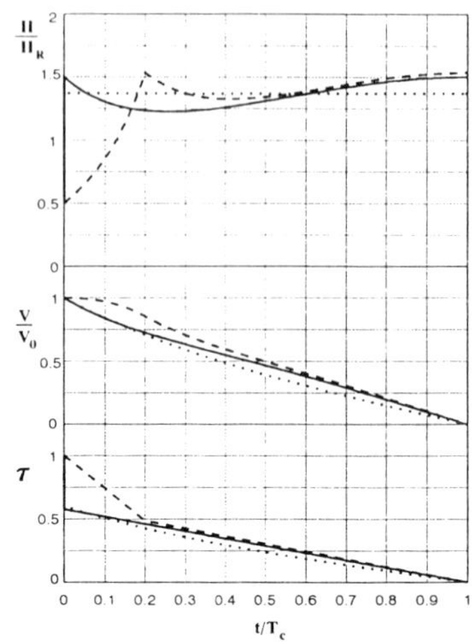

Fig. 6.- Two stages closure valve (Rigid model).

6. Optimal closure schedule of a valve, with friction and elastic effects.

As in practice both friction and elastic effects are present, conclusions to be showed in the following are the most important ones. Because it is also a theoritical one, the instantaneous closure is not taked into account. In its turn the valve stroking procedure is considered. Fig 7 depicts three diferent closures, with their respective answers. Clearly can be seen that the most simple and practical closure, the linear one, gives rise to the maximum ΔH_{max}, and the best results are supplied by the valve stroking (called optimum in Fig.7).

Basic parameters in Fig 7 and Fig 8 are: Γ = 2, ρ = 2.5 and h_f = 0,5. With reference to this last figure (Fig 8a)), it is very important to notice the influence of the coordinate values (t_p,T_p) where the change of slope is produced in the final ΔH_{max}. In Fig. 8 b) the existence of an optimum well defined value can be seen. In order to show how much can be reduced, with an instantaneons closure, the maximum pressure, this case is also considered. Results are not relevant.

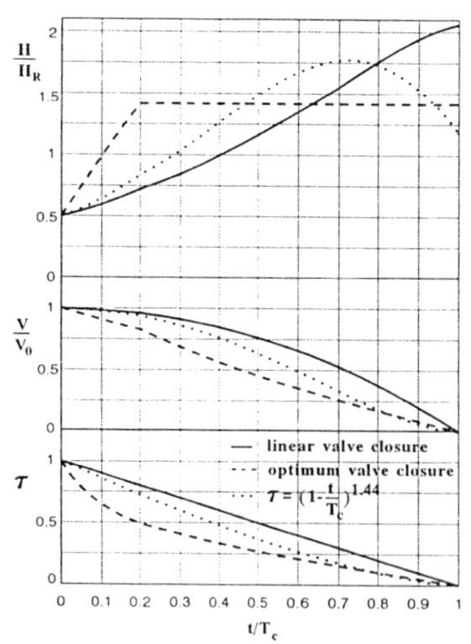

Fig. 7. Comparison of different valve closures policies.

When a two stages valve closure is decided as solution, the Fig 8 b) highlight the relevance of a correct selection of (t_p, τ_p) in the final value of ΔH_{max}. By this reason, it is more usefull to carry out this kind of study than to look for an optimization model that, later, has no practical viability.

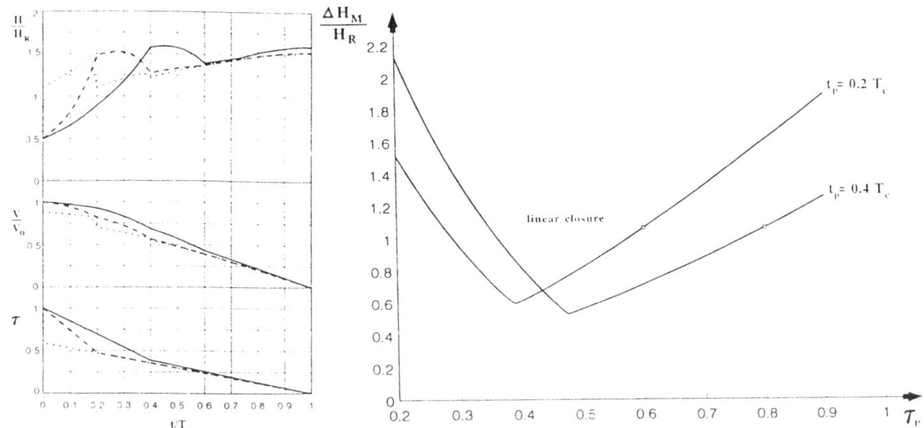

Fig. 8.- a) Comparison between three valve motions.
b) Maximum pressure versus break point coordinates (t_p, τ_p)

7.Concaves valve closures monoeuvres.

Azoury et al. (1986) suggest an inherent concave closure law, according:

$$\tau = \left(1 - \frac{t}{T_c}\right)^m \text{ ,with } m>1 \qquad (19)$$

that, regardingless its reliability, the main question concerning this law is to determine the optimum value of m for a given parameter h_f. According the rigid model, Fig 9 gives the answer that, against the Azoury's opinion, we claim is also valide when the elastic effects are considered, except in very extremal cases ($\rho<1$).

And the same can be said, regarding the influence of Γ, now in coincidence with the Azoury's opinion: its value has not significance in the m_{opt} selection, which, practically, only depends on h_f. In any case, as can be seen in Fig. 7, the answer obtained from this closure, even adopting m_{opt} ($h_f = 0{,}5 \Rightarrow m_{opt} = 1{,}44$) is not so good as the valve stroking one, although improve the results of the linear closure.

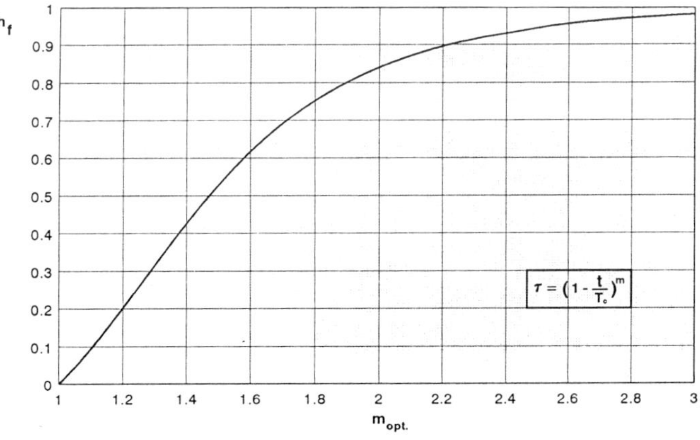

Fig. 9.- Optimum inherent valve closure coefficient.

8. Conclusions.

From this paper, the following conclusions can be stated:

a) In order to minimize the overpressure due to a T_c valve closure, the optimum policie is to mantain constant along this periode ΔH_{max}. That can be achieved with an initial and instantaneous partial closure, from $\tau = 1$ to $\tau = \tau_{opt}$, and a second closure between $t = 0$ and $t = T_c$, from $\tau = \tau_{0,opt}$ to $\tau = 0$. The shape and characteristics of this second phase closure depens on the rigid or elastic model (continuous or by steps), and on the ideal or real system (linear or convex) .

b) The quick and the valve stroking, being the optimum policies in real and elastic systems, have not been implemented in practice, due to its complexity.

c) When the wider used two-stages bilinear closure is adopted, it is very important to select the coordinates (t_p, τ_p) where the change of slope is introduced. That can be done from a search procedure.

d) The inherent closure proposed by Azoury, with a continuons and convexe shape, even for the optimum m_{opt} exponent, does not improve the results of the valve stroking, although have better behaviour than the linear closures.

e) The most practical closure, the linear one, supply better results for low h_f, but mainly when the elastic and the inertia effects are important (small values of ρ and Γ).

f) The dynamic programming (Contractor, 1983), not provide an optimum policie, due to the fact that the monotonocity property is not satisfied by the recursion proposed.

g) The simplest rigid model allows a better understanding of the problem considered.

References

Abreu, J., Cabrera, E., Izquierdo, J. and García-Serra (1996). Flow modeling in pressurized systems. From waterhammer to steady state. Submited to Journal of Hydraulic Engineering, ASCE.

Azoury, P.H., Baasiri, A.M. and Najm, H. (1986) Effect of valve-closure schedule on water hammer. Journal of Hydraulic Engineering, ASCE, vol. 117, 890-903.

Cabelka, J. and Franc, I. (1959) Closure characteristics of a valve with respect to water hammer. Eighth Congress of the International Association for Hydraulic Research, Montreal, Canada, 6A1-6A23.

Chaudhry, M.H. (1987) Applied hydraulic transients, Ed. Van Nostrand Reinhold Company, New York

Contractor, D.N. (1983) Valve stroking to control waterhammer transients using dynamic programming. Proceedings applied mechanics bioengineering and fluid engineering conference. Fluids Eng. Division. ASME, Vol. 4, 77-81.

Contractor D.N. (1.985). Two and Three Stage Valve Closures to Minimize Waterhammer Transients. Proceedings of the 1985 Pressure Vessels and Piping Conference, ASME, vol. 98-7, 357-361.

Driels, M. (1975) Predicting Optimum Two-Stage Valve Closure. ASME Paper No. 75-WA/FE-2.

Goldberg, D. E. and Karr, C.L. (1987) Quick Stroking: design of time-optimal valve motions. Journal of Hydraulic Engineering, ASCE, Vol. 113, 780-795.

Hohnson, R.D. (1920) Discussion of Gibson's paper on Pressures in Penstocks caused by gradual closing of Turbine Gates. Trans A.S.C.E. Vol. 83, 707-740, 754-762.

Ruus,E. (1957) Bestimmuing vor Schiess Funktionen welche den kleinsten Wert des maximalen Druckstosses ergeben. Thesis toward the Doctor of Engineering Degree. Tech. University of Karlsruhe. Germany.

Ruus, E. (1966) "Optimum rate of closure of hydraulic turbine gates". A merican Society of Mechanical Engineers. Fluid Engineering Conference, Denver.

Streeter, V.L. (1963) Valve Stroking to Control Waterhammer. Jorunal of the Hydraulics Division, ASCE, Vol. 93, HY3, 81-98.

Wylie, E.B. and Streeter, V.L. (1993) Fluid Transients is Systems, Ed. Prentice Hall, New Jersey (U.S.A).

QUALITATIVE FLOW VISUALIZATIONS DURING FAST START-UP OF CENTRIFUGAL PUMPS

J.P. BARRAND, A. PICAVET
Laboratoire de Mécanique de Lille
Ecole Nationale Supérieure des Arts et Métiers
8, Boulevard Louis XIV
59046 LILLE Cedex

INTRODUCTION

The main uses of pumps in hydraulic plants require quasi-steady state operation. However, some applications such as rocket engines include severe transient phases. The approach assuming that, at any time, all flow conditions are identical to those occurring during steady-state working is no more sufficient. Thus several laboratories have started to study the behaviour of turbomachines working under unsteady conditions ([SAI82], [TSU82]).

The Mechanics Laboratory of Lille, France, has been working for several years on the topic of fast start-up of centrifugal pumps. This paper intends to present the results of investigations focusing on the flow which takes place at the inlet and inside of a centrifugal pump during a fast start-up.

1 - EXPERIMENTAL PLANT

The experiments have been performed on the DERAP test rig (Demarrage Rapide de Pompes Centrifuges) which is set up in the Laboratory of Mechanics of the Ecole Nationale Supérieure d'Arts et Métiers of Lille [BAR93].

The pump is driven by an asynchronous electric motor by the means of an electromagnetic clutch. The rotation speed of the engine is measured with a photo-electric cell. An eccentric is mounted on the shaft of the impeller and two magnetic sensors are pointed at it. The recording of their signals allows to calculate the instantaneous rotation speed of the machine.

The pump is inserted in an hydraulic circuit, the pipes of which are connected to a vessel and are instrumented with piezo-electric pressure sensors and with an electromagnetic flowmeter. The circuit can be set in two different configurations which are represented on the figures 1.a and 1.b. The pressure sensors are identified by numbers 1 to 6 in each case.

E. Cabrera et al. (eds.), Hydraulic Machinery and Cavitation, 671–680.

Two casings are available for the machine. The first one can be instrumented with piezo-electric pressure sensors in order to provide unsteady pressure measurements at the outlet of the impeller. The pump barrel can be instrumented in the same way (see figures 1.c and 1.d).

The second casing, as well as the upper shroud of the impeller, is made of plexiglas and allows visualization of the flow in a part of the inlet pipe and at the outlet of the impeller (see figure 3).

2 - DESCRIPTION OF THE EXPERIMENTS

The test described in this paragraph is the fast start-up of the pump, the circuit being in its long configuration, towards an operating point of 2900 rpm and 24 m^3/h, which is the nominal operating point of the pump. The results corresponding to this experiment are represented on the figures 2.a, 2.b and 2.d.

At the beginning of the operation, the motor is running at its maximum rotation speed. The clutch is engaged and the speed of the pump rises from 0 to 2650 rpm in 0.3 second then, after the synchronization of the clutch, goes on growing more slowly until 2900 rpm (see figure 2.a). The head provided by the pump grows the same way but the energy is at first absorbed by the inrtia of the fluid, then by the losses in the circuit. Thus, the growth of the flowrate is delayed with respect to the evolution of the other parameters (see figure 2.b).

On the figure 2.d are represented p_{inlet}, pressure measured at the inlet (with the transducer 3), p_{outlet}, measured at the outlet (with the transducer 4) of the pump and $p_{outlet} - p_{inlet} = \rho g H$, H being the head provided by the pump. The pressure measurements in the pipes are also necessary to calculate the real evolution of the flowrate because of the time constant of the electromagnetic flowmeter.

If we assume $p^*=p+\rho gz$ constant and equal to 0 in the tank, we can write the mass oscillations equation between the vessel and the inlet or the outlet. Assuming one dimensional flow, this leads to:

$$p_{inlet} = -\rho \frac{L_{inlet}}{S} \frac{dQ_v}{dt} - K_{inlet} Q_v^2 \qquad (2.1)$$

$$p_{outlet} = \rho \frac{L_{outlet}}{S} \frac{dQ_v}{dt} + K_{outlet} Q_v^2 \qquad (2.2)$$

In these expressions, the inertial terms are much higher than the loss terms as long as the flowrate is small. The high transient value of $\frac{dQ_v}{dt}$ explains why the inlet pressure decreases very fast and reaches a minimum before growing towards its final steady value. We find the lengths of the pipes in the equations (2.1) and (2.2). This shows that the configuration of the circuit is of a great importance in such transient problems and explains why the plant was intended to be set in two different configurations.

3 - COMPARISON BETWEEN EXPERIMENTAL RESULTS AND MONODIMENSIONAL MODEL BASED CALCULATIONS

Ghelici [GHE93] has developped a monodimensional model of unsteady flow inside the impeller of a centrifugal pump. Assuming that the number of blades is infinite and writing the momentum equation for liquid particles, without consideration of hydraulic losses, leads to:

$$\rho g H_i = \rho u_2 C_{u2} + \rho \frac{R_2^2 - R_1^2}{2\tan\beta}\frac{d\omega}{dt} - \rho \frac{\ln(\frac{R_2}{R_1})}{2\pi b \sin^2\beta}\frac{dQ_v}{dt} \qquad (3.1)$$

That means that the head provided by the impeller in unsteady conditions is the sum of a quasi-steady term and of an inertial and an angular acceleration rate. It can be extended, in a first attempt, to the whole machine and the quasi-steady term can be identified to the expression approaching, after Sedille [SED67], the steady characteristic of the pump. That leads to:

$$\rho g H = p_{outlet} - p_{inlet} = a\omega^2 + b\omega Q_v + c{Q_v}^2 + K_1\frac{d\omega}{dt} - K_2\frac{dQ_v}{dt} \qquad (3.2)$$

a, b and c are determined from the steady characterization tests of the machine. The equation (3.2) can be associated with equations (2.1) and (2.2). We obtain a differential equation the unknown quantity of which is Q_v. K_1 and K_2 are estimated from the experimental results, using a least squares procedure. K_{inlet} and K_{outlet} are deduced from the final steady value of the records of p_{inlet} (transducer (3)) and p_{outlet} (transducer (4)). Using the records of the rotation speed N allows us to solve the differential equation and to calculate Q_v then p_{inlet} and p_{outlet}.

The results are represented on the figures 2.c and 2.e. The agreement between theoretical and experimental results appears to be rather good, especially for the final values of the different quantities. The measured and calculated heads are also represented versus Q_v in the H-Q diagram (see figure 2.f). The calculations are performed using terms estimated from experimental results. However, some difference during unsteady working conditions exists. We can therefore say that the model is not sufficient and that the flow is certainly much more complicated than assumed. Further experimentation is necessary to identify the real phenomenons that occur during a fast start-up of the pump.

4 - STUDY OF THE INTERNAL FLOW

4 - 1: Visualization of the flow at the inlet of the pump

As one of the casings of the pump is made of plexiglas, visualizations of the flow during fast start up were provided using the wool threads technique. The colouring trickle couldn't be aplied because of the unsteady features of the problem. Three thin metallic frames are introduced in the inlet pipe (see figure 3). One of them is set very close to the entry of the pump (see figure 4.a). The threads are tied on the transversal bar of the frame so as to allow them to swivel. The working pump is then shot using a camcorder. The sampling frequency is only of twenty four pictures per second, which is enough to get reliable and useful information about unsteady phenomena lasting one half of a second.

The first test concerns steady state working. The pump runs at a constant speed and the regulation valve is progressively closed. When the flowrate becomes lower than the critical recirculation flowrate, we see the wires on the first frame changing their direction to indicate the existence of a tangential velocity component. Then we see the frame twisting because of the high level of this tangential velocity. It reveals the occurrence of a prerotation at the inlet of the pump, which is a classical result for machines operating at partial flowrates [SEN80]. The other frames and wires, which are located farther from the pump, show that the prerotating field tends to occupy the whole plexiglas duct when the regulation valve comes close to shutting.

A fast start up of the pump is then performed. The camcorder is placed so as to see the three frames as well as the impeller nosepiece, in order to detect the beginning of the movement. As the machine accelerates strongly for 0.3 second and as the flowrate grows for nearly 0.5 second, we get ten to twelve pictures during the transient working. These pictures have been digitized and printed (see figures 4.a to 4.j).

On the figure 4.a, the pump is motionless. On the figure 4.b, the rotation has started. When the flow starts to grow, we see the wires on the first frame leaving their axial direction and indicating the existence of a tangential velocity component (see figure 4.e). Then the frame twists (see figures 4.f, 4.g, 4.h and 4.i) indicating that the level of this tangential component is very high. The wires tied on the other frames don't swivel. When the flowrate gets its steady value, the wires and the frame come back in their initial position (see figure 4.j).

So we clearly see the transient existence of a prerotating velocity field during the start-up of the pump and, more precisely, at the beginning of the growth period of the flowrate. As explained in the paragraph 2, the evolution of the flowrate is delayed with respect to the rotation speed because of the inertia of the liquid in the circuit. Therefore, the flow behaves as if there were a succession of partial flowrates operation conditions. Lefebvre and Barker [LEF95] evoke the occurrrence of such phenomena after the results of measurements of a pump operating under unsteady conditions.

4 -2: Visualization of the flow in the diffusor

Thin metallic shafts on which are tied wool threads are introduced in the volute. When the pump is running in steady conditions, the direction of the threads is nearly tangential. The recording of a fast start-up corresponding to the same test as in the former paragraph is then performed using the camcorder. On the figure 5.a, the pump is motionless. We see that, at the very beginning of the rotation (see figures 5.b and 5.c), the angle which exists between the direction of the flow and the impeller outlet section is very important. The flowrate is at once almost equal to 0. When it starts to grow, nearly $1/10^{th}$ of a second after the clutch has been engaged, the direction of the threads becomes tangential again (see figure 5.d). This confirms the presence of vortex in the diffuser and in the volute.

CONCLUSION

Visualizations have provided very useful qualitative results about the flow occurring at the inlet and inside a pump starting-up very fast. As the prerotative velocity field is certainly linked with internal effects, such as stall, a visualization of the internal flow with a fast camcorder would be very interesting to perform.

References

[BAR93] J.P. Barrand, N. Ghelici, G. Caignaert: Unsteady Flow During the Fast Start-up of a Centrifugal pump. ASME FED-Vol.154, *Pumping Machinery*, 1993.

[GHE93] Naceur Ghelici: Etude du régime transitoire de démarrage rapide d'une pompe centrifuge. PhD Thesis. Université des Sciences et Technologies de Lille. 1993.

[SED67] Marcel Sedille: *Turbo-machines hydrauliques et thermiques*. Ed MASSON. 1967.

[LEF95] P.J. Lefebvre - W.P. Barker: Centrifugal Pump Performance During Transient Operation. *Journal of Fluids engineering*, Vol. 117, pp. 123-128, March 1995.

[TSU82] H. Tsukamoto - H. Ohashi: Transient Characteristics of a Centrifugal Pumpduring Starting Period. *Journal of Fluids Engineering*, Vol. 104, pp. 6 - 14, March 1982.

[SAI82] Sumio Saito: The transient Characteristics of a Pump during Start-Up. *Bulletin of the JSME*. Vol. 25, N°201, March 1982.

[SEN80] M. Sen: Study of inlet flow of centrifugal pumps at partial flowrates. Ph. D. Thesis. Université Libre de Bruxelles, June 1980.

Notations

p: static pressure

L: length of pipes

N: rotation speed

R: radius of the impeller

K_{inlet}, K_{outlet}: global head loss coefficients

Q_v: flowrate

S: section of pipes

C: absolute velocity of the fluid

β: angle between the curve describing the shape of the blade and the tangent

H_i: indicated head of the pump

ω: angular speed

u: ωR, peripheral speed

b: width of the blades

Marks

inlet, outlet: refers to the inlet and the oulet of the pump

1, 2: refers to the inlet and the outlet of the impeller

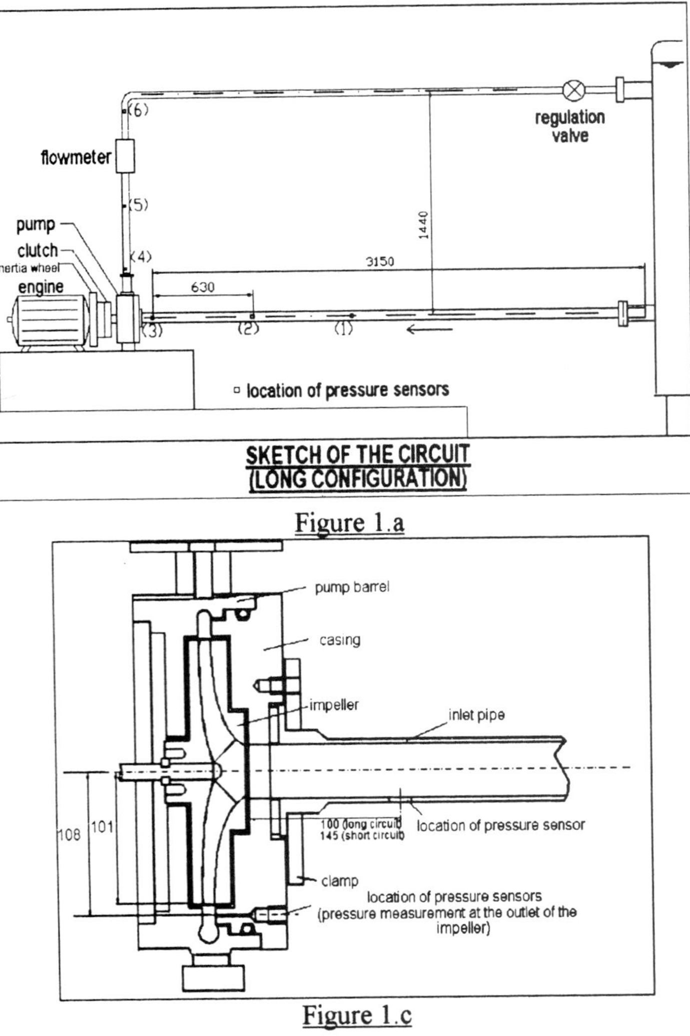

Figure 1.a

Figure 1.c

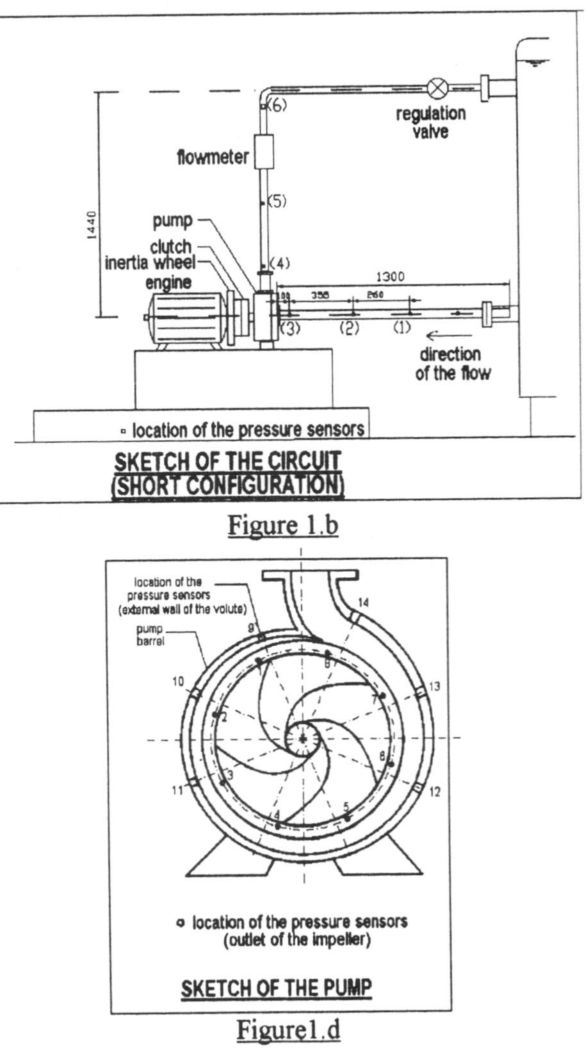

Figure 1.b

Figure1.d

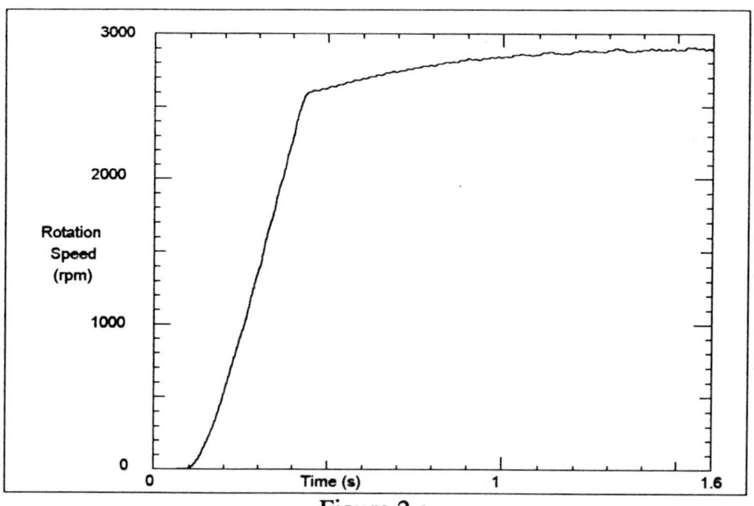

Figure 2.a

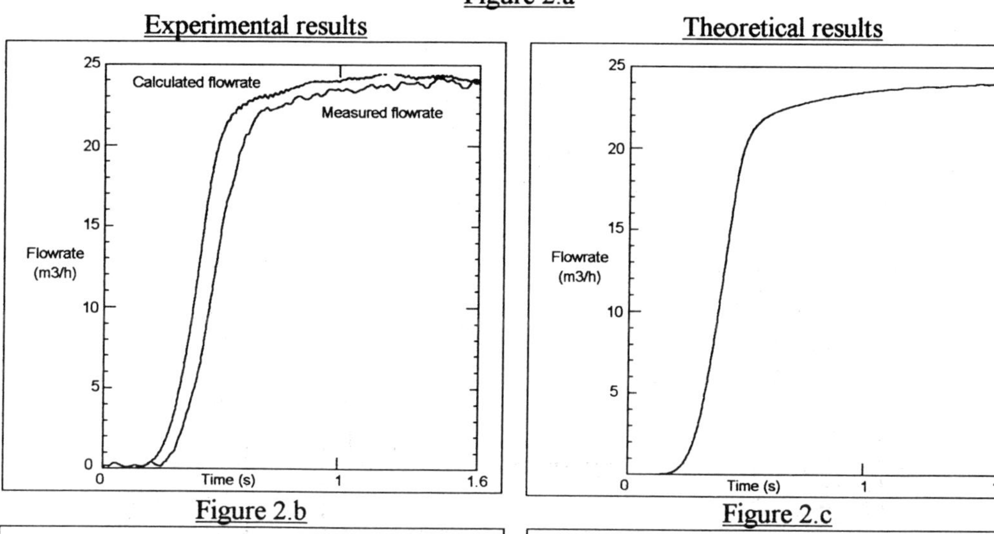

Figure 2.b

Figure 2.c

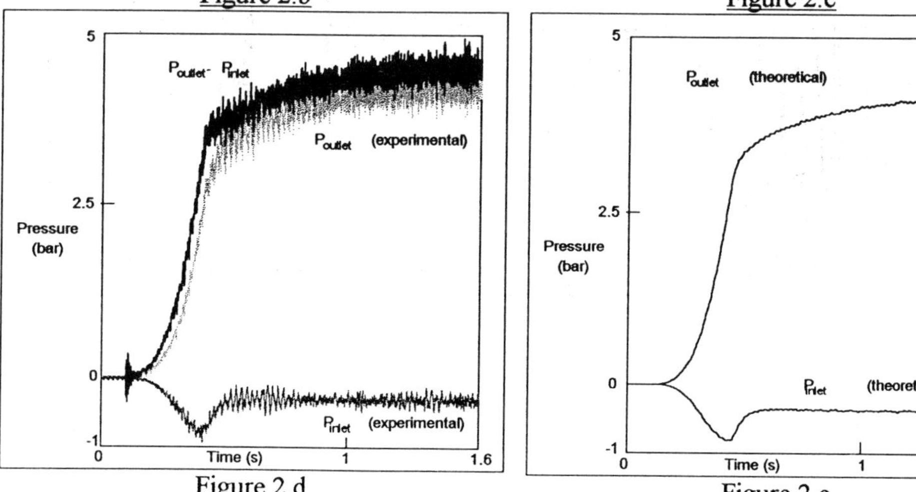

Figure 2.d

Figure 2.e

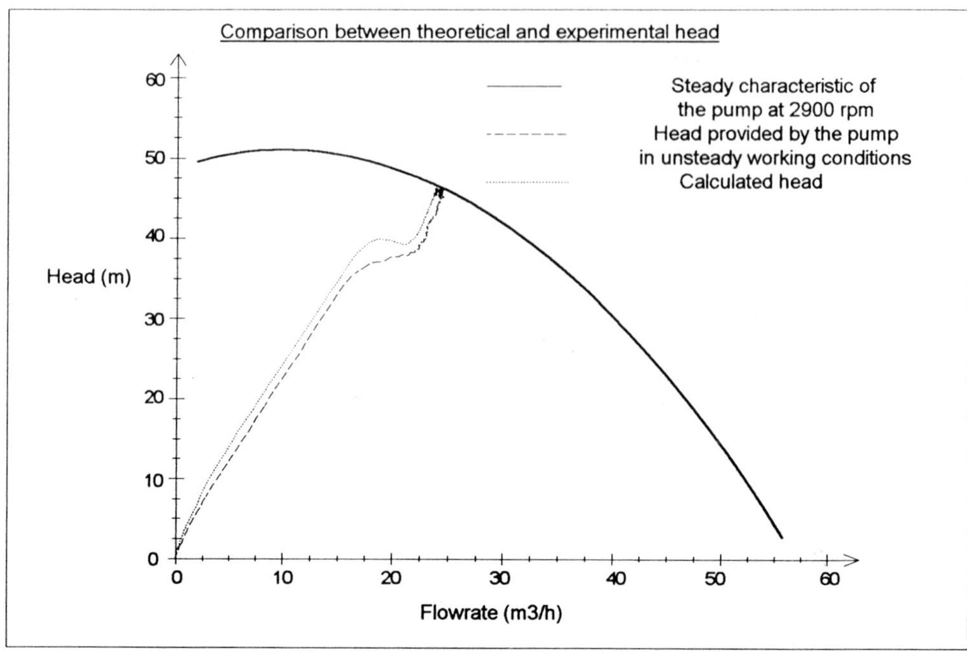

Figure 2.f

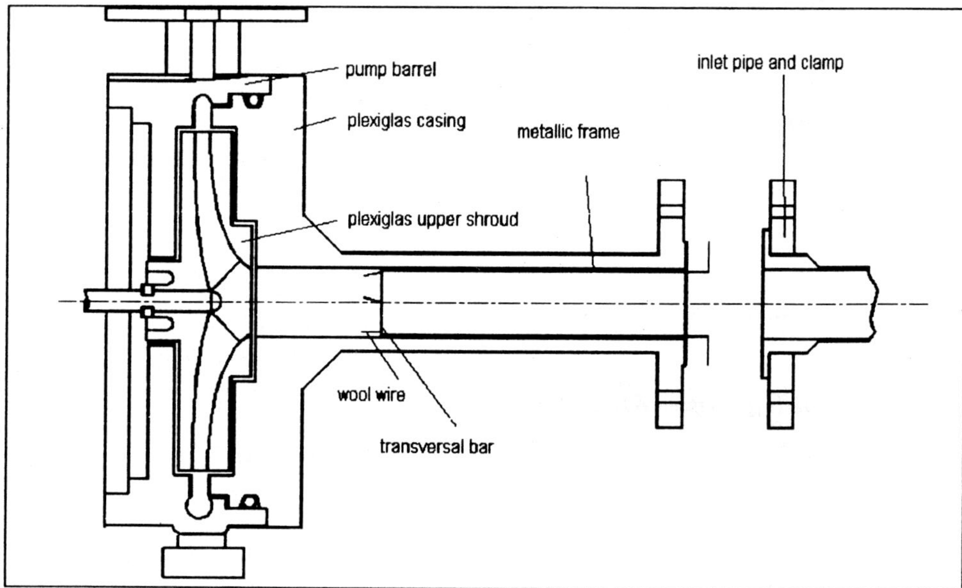

Figure 3

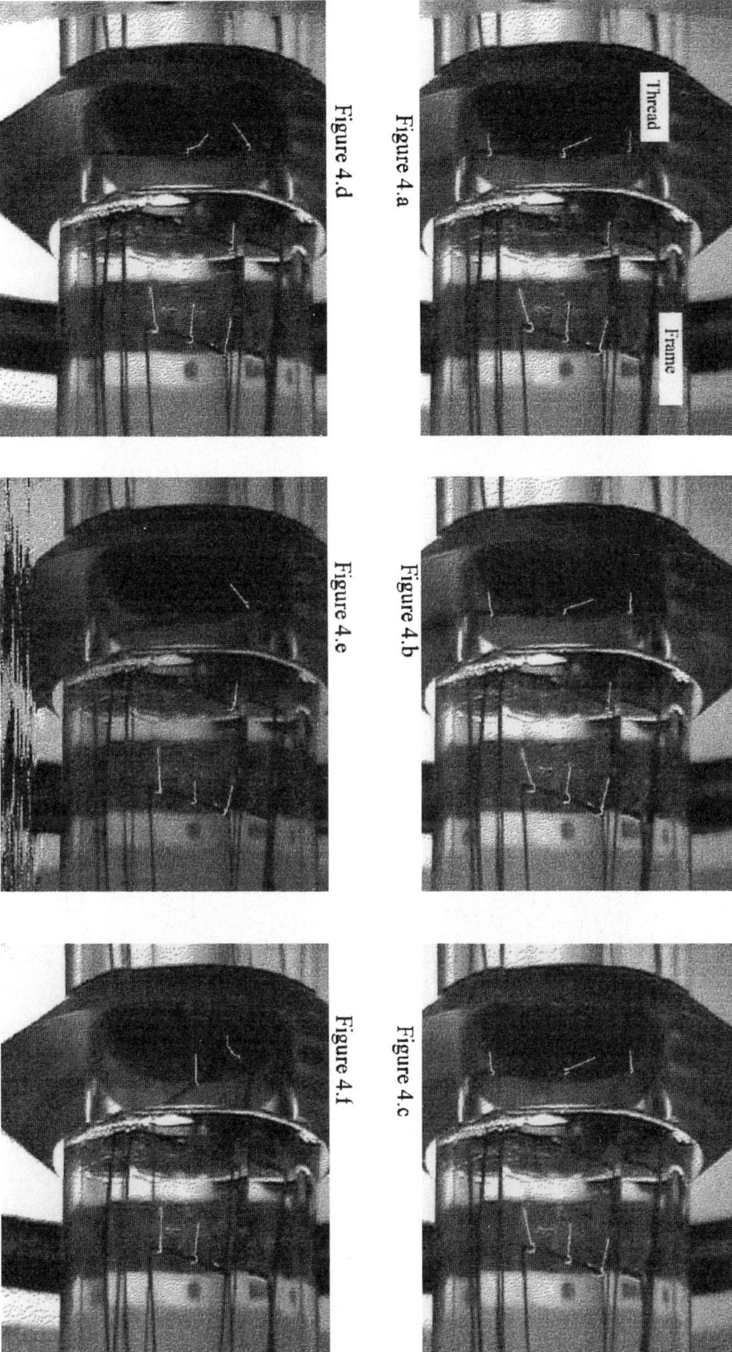

Figure 4.a

Figure 4.b

Figure 4.c

Figure 4.d

Figure 4.e

Figure 4.f

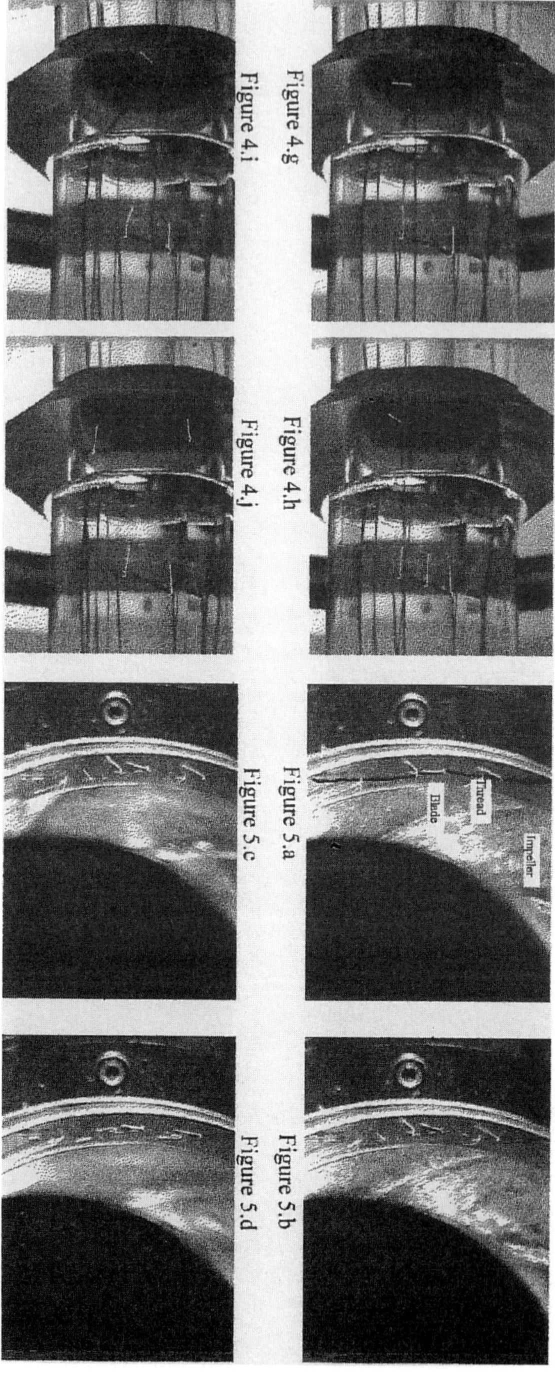

Figure 4.g

Figure 4.i

Figure 4.h

Figure 4.j

Figure 5.a

Figure 5.c

Figure 5.b

Figure 5.d

ANALYSIS OF A NUMERICAL MODEL FOR THE OSCILLATORY PROPERTIES OF A FRANCIS TURBINE GROUP

A. CATTANEI
Università di Genova
Genova, Italy

A. CAPOZZA
CISE
Milano, Italy

P. MOLINARO
ENEL-DSR-CRIS
Milano, Italy

1. Introduction

The present work represents an assessment of the authors' several years of study on the application of the transfer matrix method to hydropower plant components. It is part of the strong research effort promoted by ENEL-DSR-CRIS on the oscillatory properties of hydropower groups equipped with Francis turbines, particularly important in presence of cavitating vortex rope (CVR).

The paper concerns the study and the implementation in a computer code of a model for a group composed by a Francis turbine, an alternator and a speed governor. The model has been developed by the same authors [1] and its final result is the determination of the turbine hydraulic impedance. The fundamental characteristic is maximum generality and rigourousness at the same time so that most of the possible interactions with speed governor and alternator (a priori not neglectable) may be taken into account. This kind of approach has been imposed by the lack of knowledge about the problem and has led to a rather complex model whose simplification will be the next step of the work.

2. Brief Description of the Model

2.1 MAIN CHARACTERISTICS AND BASIC EQUATIONS

Dealing with oscillatory phenomena, the usual decomposition in mean and harmonic fluctuating part is performed; thus each state variable F(t) (incoming and outcoming discharges Q_P and Q_D, net head H_P-H_D, rotational speed Ω, driving and resisting torques M_m and M_e, wicket gate opening angle θ, etc.) has the following structure:

$$F(t) = F^o + F'(t) = F^o + \Delta F \exp(j\omega t) \tag{1}$$

F^0 being the mean value, F'(t) the fluctuating time dependent part and ΔF the complex amplitude of F'(t). Since the present work deals with small oscillations, the mean values will be considered known quantities and will be input data for the program.

E. Cabrera et al. (eds.), Hydraulic Machinery and Cavitation, 681–690.

The whole system (fig.'s 1-2) is composed by the Francis turbine, the speed governor and the electric part (alternator, excitation system and tie line). For a complete representation, three ports (with three independent variables) should be considered:

1) volute inlet-penstock outlet — hydraulic variables H_P, Q_P;
2) runner outlet-draft tube inlet — hydraulic variables H_D, Q_D;
3) transmission line-electric network — electrical variables V_B, I_B;

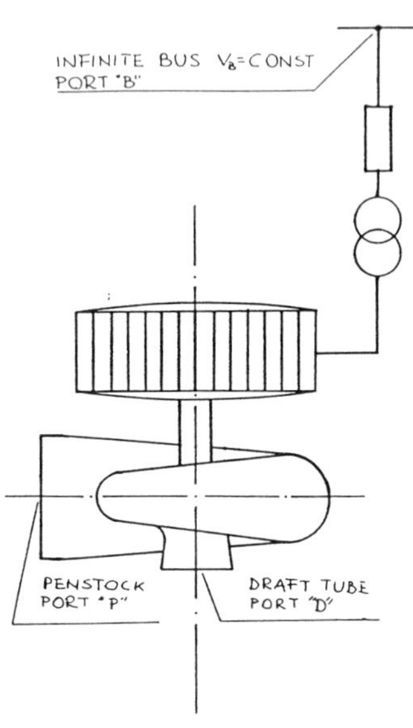

Figure 1. Francis turbine group.

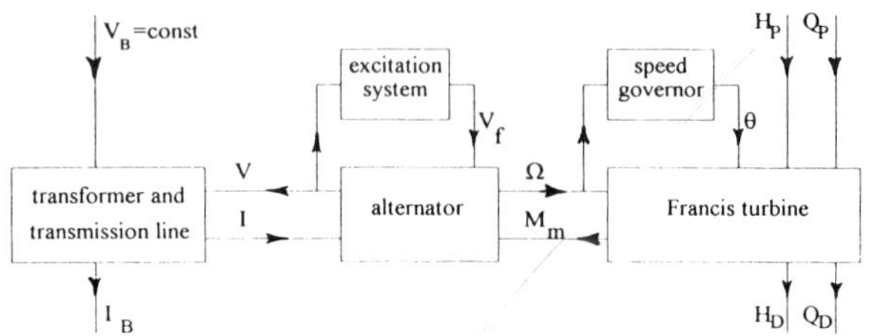

Figure 2. Structure of the system.

Focusing on the group local behaviour, the voltage V_B is assumed constant and only two independent variables remain. Then, fluid compressibility and blade cavitation are neglected so that incoming and outcoming discharges equal reducing to one independent variable only, the discharge oscillation ΔQ. Thus the turbine is represented by its hydraulic impedance only and each oscillating quantity F is expressed in the form:

$$\Delta F = F_{,Q}\,\Delta Q \tag{2}$$

$F_{,Q}$ being the transfer function between F and Q. The determination of the various transfer functions requires a lengthy manipulation of the equations of each module. The procedure is reported in [1]; only the resulting equations are reported here.

<u>Francis turbine:</u>

$$\begin{pmatrix} \Delta Q_D \\ \Delta(H_P - H_D) \\ \Delta M_m \end{pmatrix} = \begin{pmatrix} 1 & 0 & 0 \\ (H_P - H_D)_{,Q} & (H_P - H_D)_{,\Omega} & (H_P - H_D)_{,\vartheta} \\ M_{m,Q} & M_{m,\Omega} & M_{m,\vartheta} \end{pmatrix} \begin{pmatrix} \Delta Q_P \\ \Delta\Omega \\ \Delta\vartheta \end{pmatrix} \tag{3}$$

The transfer functions in eq.n (3) are obtained taking the 1-D, non-permanent equations of transport of kinetic energy and moment of momentum, expressing velocity as a function of discharge, flow angle and cross section, making the small oscillations approximation and numerically integrating the equations on the turbine conduits. Two remarks are necessary:

1) three equations are obtained whose unknowns are the turbine five state variables oscillations ΔQ, $\Delta(H_P-H_D)$, $\Delta\Omega$, ΔM_m, $\Delta\theta$. A further "internal" oscillating variable, the flow angle, appears.

Since no prime principle may be employed to determine it, some relations are required to express it as a function of the state variables. Within the blade passages the flow angle is assumed to equal the blade angle; within the annuli it is obtained by means of moment of momentum conservation along the absolute trajectory.

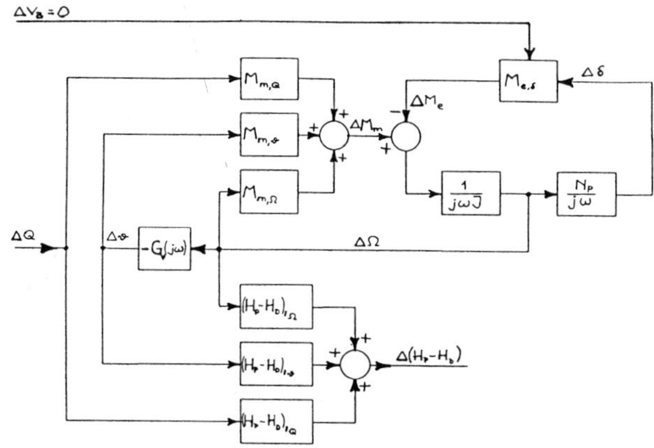

Figure 3. Block scheme of the system.

Thus, while the meridional component V_m of velocity oscillates in phase with respect to the discharge, the tangential component V_u oscillates with a delay time T given by the ratio of the upstream volume of the annulus to the mean discharge:

$$T = \frac{Vol(r_o - r)}{Q^o} = \pi b \frac{r_o^2 - r^2}{Q^o} \tag{4}$$

$$\Delta V_u(r) = \frac{r_o}{r} \Delta V_u(r_o) \exp(-j\omega T) \tag{5}$$

r_o being the radius of the outlet section of the upstream blading imposing the moment of momentum, r the actual radius and b the channel heigth;

2) numerical integration of the transport equations requires the knowledge of the turbine geometry (blade angle, cross section, volume distributions along the machine and wicket gate angle). It is a major difficulty and will be treated in the next section.

Speed governor:

$$\Delta\vartheta = -G_v(j\omega)\Delta\Omega \tag{6}$$

where $G_v(j\omega)$ is the transfer function of the governor:

$$G_v(j\omega) = K_v \frac{1 + j\omega T_2}{1 + j\omega T_1} \tag{7}$$

Alternator:

$$\Delta M_e = M_{e,\delta}\,\Delta\delta \tag{8}$$

Eq.n (8) relates the electric torque M_e to the electric angle δ. Many different forms exist for the transfer function $M_{e,\delta}$. A fourth order model has been employed allowing rotor damper windings, excitation system, and tie line effects to be taken into account.

Mechanical equilibrium:

$$\Delta\Omega = \frac{1}{j\omega J}\left(\Delta M_m - \Delta M_e\right) \tag{7}$$

J being the moment of inertia of the rotating parts.

Relation between angular speed and electric angle:

$$\Delta\delta = \frac{N_p}{j\omega}\Delta\Omega \tag{8}$$

Connections and feedbacks between the various modules are represented in fig. 3.

3. Capabilities and Limitations of the Model

The present model stems from the needing for a rational and complete approach to the problem. Thus the model must be capable to take into account as many effects as possible. The result is a model rather complex, requiring many input data sometimes difficult to know, and its simplification will be a future step of the work.

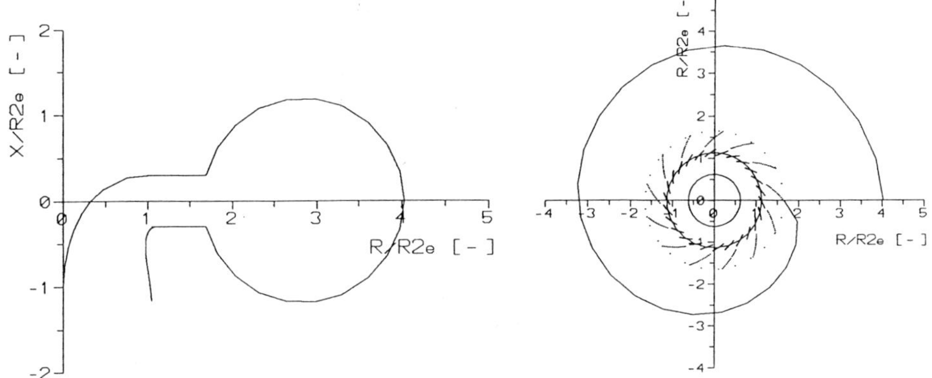

Figure 4. Pescara turbine geometry.

The input data are the mean values of the state variables of the group (wicket gate opening, discharge, turbine efficiency, electrical power, output voltage, power factor, etc.), a dozen of time constants, gains and electrical reactances for electrical part and speed governor, and the turbine geometry.

TABLE I. *Pescara I salto* turbine characteristics

net head H	27.8 m	(ψ=0.66)
discharge Q	25.16 m^3/s	(φ=0.334)
power P	5800 kW	
speed n	300 rpm	
specific speed n_s	358	

The turbine geometry determination is surely troublesome and may lead to practically unbearable difficulties; to circumvent this problem a computer code has been written which determines the required quantities starting from the main turbine dimensions.

The main dimensions, when not known, are determined by means of statistical correlations [2-3]. As an instance a geometry is represented in fig. 4 which has been generated to perform the computations showed in the next chapters (the lengths are non-dimensionalized by means of the runner trailing edge radius). It corresponds to the turbine of the plant of *Pescara 1° salto* whose characteristics are reported in table I.

4. Model Analysis

The input data are numerous and therefore an extensive analysis of the model, that is a parametric study, would be required. The possible combinations of the input data are too many for a complete study to be reported here so only the most interesting results will be shown. The analysis will first regard the effect of the turbine mean working point on the turbine impedance and then the effects of the interaction with the electric part. This requires expressing the head oscillation as an explicit function of ΔQ and is easied making reference to the block scheme of fig. 3. Since the interaction takes place only in case of angular speed oscillation, this has to be expressed as a function of ΔQ. Imposing the equilibrium of rotating parts (third line of eq.n (3) and eq.n's (6-8)) yields:

$$\frac{d\Omega}{dQ} = \frac{M_{m,Q}}{\left(\frac{N_P}{j\omega} M_{e,\delta} + j\omega J\right) - \left[M_{m,\Omega} - G_v(j\omega) M_{m,\vartheta}\right]} \tag{9}$$

Eq.n (9) governs the oscillation of Ω. There are two feedbacks on the driving torque: a direct one acts through the transfer function $M_{m,\Omega}$ and an indirect one (due to the action of the speed governor) acts through $M_{m,\theta}$. Making system with the second line in relation (3) $\Delta\Omega$ and $\Delta\theta$ may be eliminated yielding the following expression for the turbine impedance Z:

$$Z = \left(H_P - H_D\right)_{,Q} + \left[\left(H_P - H_D\right)_{,\Omega} - G_v(j\omega)\left(H_P - H_D\right)_{,\vartheta}\right]\frac{d\Omega}{dQ} \tag{10}$$

In eq.n (10) two distinct contributions to the turbine impedance may be individuated. The first one, due to the term $(H_P-H_D)_{,Q}$, represents the hydraulic impedance in a strict sense. The second one is the term in brackets which represents the effect of the interaction with the electric part and may be considered as a perturbation superposed to the first term: it is null in absence of interaction (i.e. Ω does not oscillate). Two ways result for the interaction: it may take place through the term $(H_P-H_D)_{,\Omega}$ (direct way) and through the term $(H_P-H_D)_{,\theta}$ (indirect way).

4.1 EFFECT OF THE TURBINE MEAN WORKING POINT ON THE TURBINE IMPEDANCE

The behaviour of the group depends significantly on the turbine mean working point due to the non-linearity arising from the presence of the mean values of discharge and wicket gate opening angle in the small oscillation equations. This dependance has been evidenced performing computations for twelve different mean working points; they have been obtained combining several values (φ= 0.15; 0.20; 0.25; 0.307, ψ= 0.75; 1.10; 1.45) of the non dimensional discharges and head. The working points are represented on the hill diagram of fig. 5. The resulting behaviours of the turbine impedance in absence of angular speed oscillation, i.e. the term $(H_P-H_D)_{,Q}$ alone, are represented in fig. 6 (ν is ratio of the pulsation ω to the mean rotational speed Ω^0). Experimental results of previous works and qualitative considerations [5-8] show a typical trend of the turbine impedance; in absence of interaction with the electric part three main contributions may be considered:

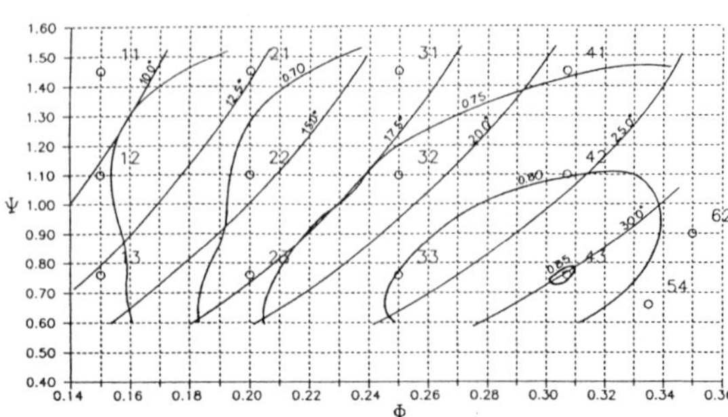

Figure 5. Pescara turbine hill diagrams.

1) a "steady state characteristic" $Re[Z]_{\omega=0}$ (that is the slope of the curve H/Q in steady conditions at constant wicket gate opening), main contribution to Re[Z];
2) the inertia of the fluid contained in the turbine, main contribution to Im[Z];
3) the delay in angular momentum transport inside the annuli, the "distortion effect".

The superposition of these effects is clearly recognizable in all the plottings of fig. 6; the mean working point affects strongly both $Re[Z]_{\omega=0}$ and the distortion (wavy behaviour of Re[Z]) but more weakly Im[Z]. In particular increasing the discharge at constant net head causes $Re[Z]_{\omega=0}$ to diminish while increasing the net head at constant discharge (that is closing the wicket gate) causes $Re[Z]_{\omega=0}$ to increase; this is in agreement with the slope of the steady characteristics obtainable from the elementary theory of turbomachinery.

A qualitative explanation may also be found for the variation of the repetitivity ("frequential period") of $Re[Z]_{\omega}$ with the mean discharge: looking at eq.n's (4-5), which give the structure of the delay term, one sees that increasing Q^0 the time delay T decreases, that is the frequential period increases. Such kind of analysis, if applied to the effect of the wicket gate mean opening angle, would lead to conclusions opposite to the results: infact opening the wicket gate at constant discharge causes the extention of the distributor-runner annulus to decrease and thus the time delay should also decrease with an effect analogous to that of the discharge. The numerical results show an opposite trend due to the presence of integral terms in the equations not taken into account in the elementary analysis.

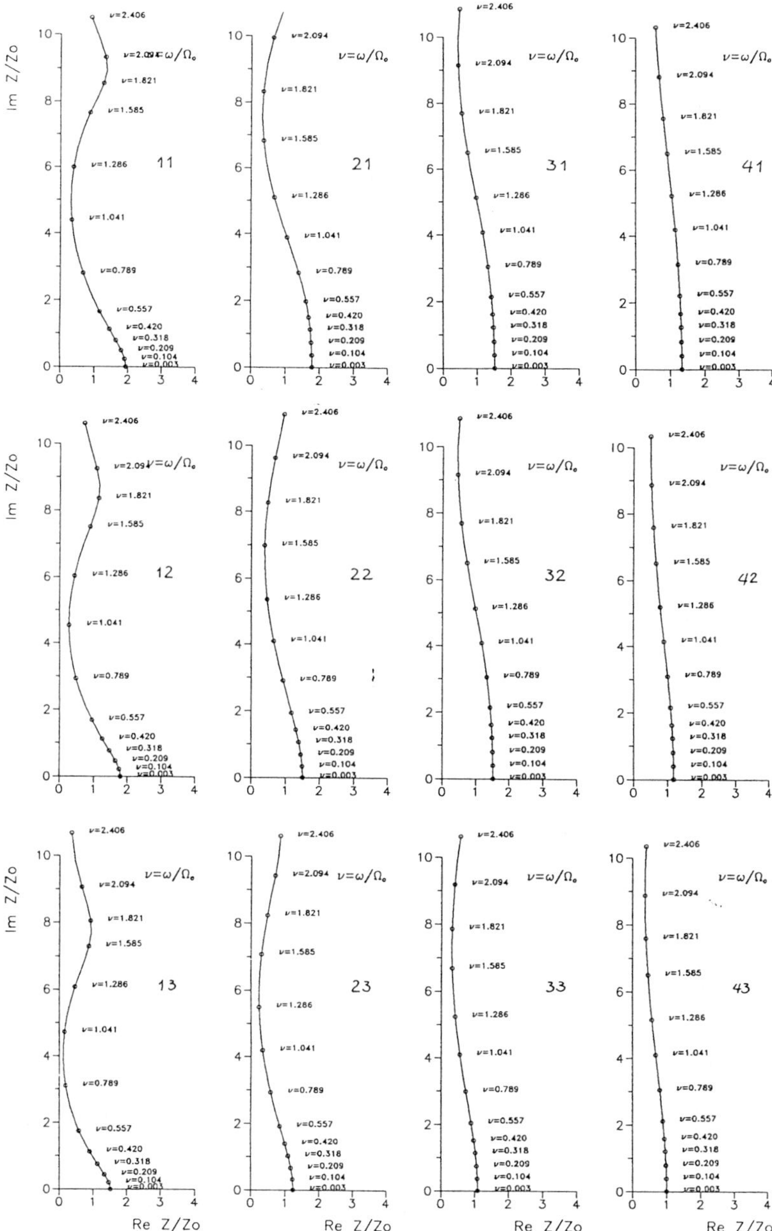

Figure 6. Effect of mean working point on turbine impedance.

4.3 INTERACTION WITH THE ELECTRIC PART.

As already stated, the interaction with the alternator takes place through the rotational speed oscillation through two different ways. With reference to the turbine mean working point 23, in fig.'s 7-9 some examples of turbine impedances will be shown, chosen in such a way to allow estimating the relative importance of the two effects. In present case all the terms appearing in eq.n (10) are retained and the behaviour is represented by the lines with symbols while the behaviour in absence of oscillations (the "reference" case) is represented by the lines without symbols.

In fig. 7 the turbine impedance in presence of both speed governor and alternator rotor damper windings is shown for typical values of the parameters of electric part and control systems. The interaction with the electric part appears in terms of a node in the plot of the turbine impedance and is significant only in a small range of pulsation (few radians per second) around a value of the order of 6 rad/s, that is the typical frequency of the electromechanical oscillations (generator resonance). Out of this range, rotational speed oscillations are negligible and no difference may be observed between the two curves. Depending on the relative values of the various gains and time constants the node may also assume the shape of a cusp or a "knee".

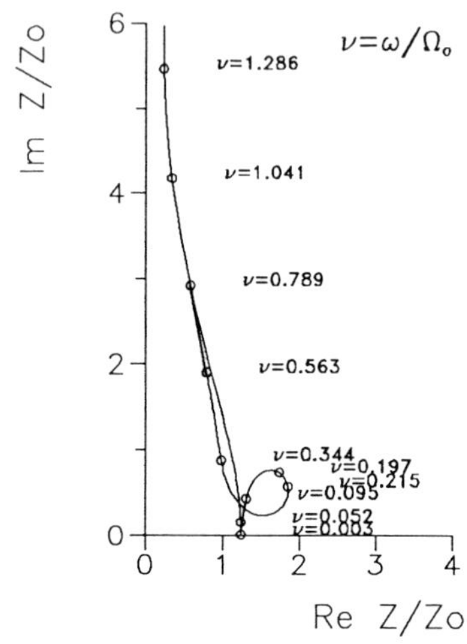

Figure 7. Electric part effect: both s.g. and d.w. present.

Similar trends (presence of irregularities within a limited frequency range) have been already observed during experimental researches [7-8]. Due to the proximity with the frequency range typical of the CVR oscillations (about 1/3-1/4 the rotational speed), an interaction with the electrical part may be expected. In order to estimate the relative influence of the various ways of interaction some calculations have been performed in which either the speed governor or the rotor damper windings have been eliminated.

In fig.'s 8-9 the turbine impedances are represented which are relative to the two cases in which both the speed governor and the damper windings are absent and only the damper windings are absent. The case with presence of damper windings and absence of speed governor has not been represented since no difference from the reference behaviour (lines without symbols) could be detected just looking at the curves. This, together with the fact that in presence of both speed governor and damper windings (fig. 7, line with symbols) the node is clearly present suggests that the interaction with the electric part mainly takes place through the speed governor and that the direct influence of the rotational speed on the turbine impedance is much smaller. The same conclusions may be obtained comparing fig.'s 8 and 9 in which the rotational speed oscillations are enhanced by the absence of damper windings: fig. 8

shows the effect of the rotational speed oscillation in absence of speed governor, and in fig. 9 the superposed effect of the speed governor is shown.

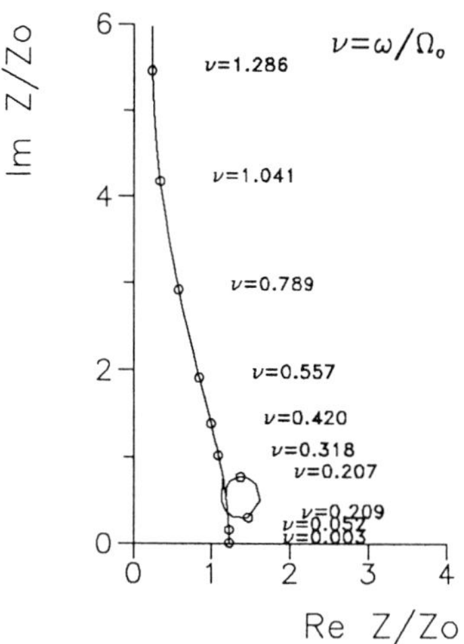

Figure 8. Electric part effect: both s.g. and d.w. absent

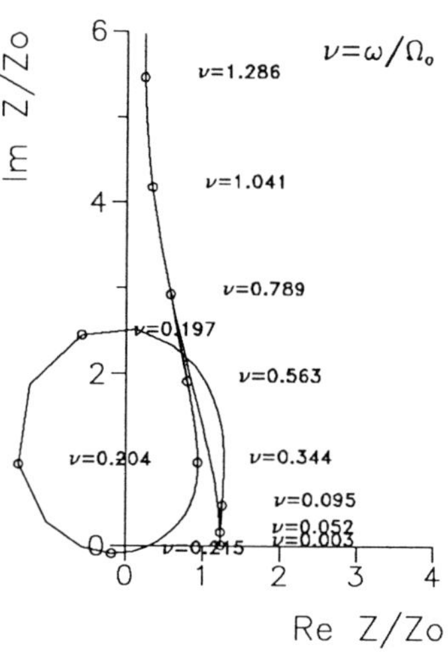

Figure 9. Electric part effect: s.g. present and d.w. absent.

With the only sake of showing how other electric parameters may influence the turbine impedance, in fig. 10 the effect of a higher tie line reactance value (twice as big as in fig. 7) is presented, still following the usual conventions. Since the tie line links the alternator terminals voltage to the network one, the bigger the reactance the weaker the link and the bigger the possibility of local oscillations of the group: the qualitative behaviour is the same as in fig. 7, but the node consequent to the electromechanical oscillation is broader.

As a general consideration one can say that all the parameters (power factor, rotating masses inertia, etc.) which may influence the electromechanical resonance will also affect the turbine hydraulic impedance.

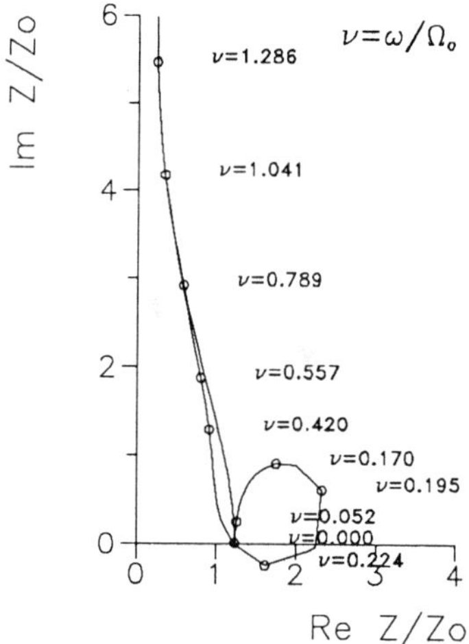

Figure 10. Effect of tie line reactance.

5. Conclusions

A theoretical model for the transmissive properties of a Francis turbine hydropower group has been implemented in a computer code and the first calculations indicate that the model is very promising: the results show a good agreement with previous experimental works and theories. It is not easy to achieve this by means of a completely theoretical model.

Particularly the influence of the turbine working point and some effects of the interaction with the electric part have been identified. This result is due to the fact that the model is rigorous and allows all possible effects to be taken into account.

On the other hand rigorousness means complexity and thus other work has still to be done before the present model be easily employable and reliable; mainly two other steps should be taken, both presenting difficulties.

First, a parametric study of the model is necessary in order to understand which input data are fundamental and which may just be estimated; in this way the physics of the phenomenon could be deepened and the model could also be simplified. Due to the big number of input variables a careful planning of the calculations is required.

Second, the code should be validated with reference to an existing plant but practical reasons make it difficult to provide a complete set of experimental data: harmonic oscillations of the state variables should be produced and measured at various frequencies.

6. List of References

1. Cattanei, A., Capozza, A.; "A Transfer Matrix Model for a Francis Turbine Group Interacting with Hydraulic Circuit and Electrical Network", XVII IAHR Symposium, Beijing, 1994.
2. de Siervo, F., de Leva, F.; "Modern Trends in Designing and Selecting Francis Turbines", Water Power and Dam Construction, August 1976.
3. Lugaresi, A. Massa, A; "Designing Francis Turbines: Trends in the Last Decade", Water Power and Dam Construction, November 1987."
4. Bovet, T.; "Contribution a l'étude du tracé d'aubage d'une turbine a réaction du type Francis", (in French), Information technique Charmilles, n. 9, 1962; (also ASME Paper 61-WA-155).
5. Doerfler, P.; "Operating Problems of Francis Turbines Half Load Surge", XI IAHR Symposium, Amsterdam, 1982.
6. Angelico, G., Fanelli, M.; "La macchina idraulica come sede di trasferimento di fenomeni pulsanti", (in Italian), ENEL-DSR-CRIS internal report, 1986.
7. Fanelli, M., Siccardi, F.; "Réponse d'une turbomachine Hydraulic à des fluctuation des paramètre dynamiques du circuit ", (in French)
8. Ng, S. L., Brennen, C.; "Experiments on the Dynamic Behaviour of Cavitating Pumps", ASME Jrn. of Fluids Eng., June 1978.

TRANSIENTS ANALYSIS AND DYNAMIC CRITERIA FOR HPP EXPLOATATION

GAJIĆ A., PEJOVIĆ S., KRSMANOVIĆ LJ., STOJANOVIĆ Z.
Faculty of Mechanical Engineering Belgrade, Yugoslavia

Abstract

Transient regimes in hydro power plants and hydraulic systems are: normal operating conditions, emergency operating regimes and catastrophic cases. The list of analyses which must be carried out and some backgrounds needed for these computations are discussed. This list is vital for economical design, construction and operation of hydraulic systems.

The paper deals with characteristics of turbomachines in pump, turbine and dissipation operating zones. Some methods of four quadrant representation are briefly described; Suter and modified Suter curves are particularly pointed out. The methods of interpolation and extrapolation of performance curves is shown.

Hydraulic transients analysis (waterhammer, hydraulic oscillations and stability), extreme values of stresses and the pressure fluctuations along the system and other characteristics important for the safety of the plant, should be determined. Comparing machine vibrations to the recommended statistical data, exposure rate can be estimated and probability of vibration appearance considered.

1. Introduction

1.1 GENERAL

Hydraulic transients describe the disturbances in a fluid caused during a change from one steady state to another. The principal components of the disturbances are pressure changes caused by the propagation of pressure waves throughout the hydraulic system.

In the 1960s computer based solutions of hydraulic transients in pumping systems and hydro power plants were developed. The numerical solution is usually based upon the method of characteristics which converts partial differential equations of motion and continuity into four ordinary differential equations. These equations are then expressed in a finite difference form and their solution is carried out on a digital computer. This method, or any other method, also requires the governing equations or performance characteristics of each component in the system in order to solve the boundary conditions Chaudhry (1979), Pejović (1977), (Pejović *et al.*, 1987), (Wylie *et al.*, 1993). Computer based solutions of steady-oscillatory motion were developed in frequency domain by the use of procedures borrowed from electrical transmission-line theory.

E. Cabrera et al. (eds.), Hydraulic Machinery and Cavitation, 691–700.

The forced oscillatory motion and the resonating characteristics of systems may be easily identified Chaudhry (1979), Pejović (1979), (Wylie *et al.*, 1993).

1.2 DANGERS OF TRANSIENT REGIMES

Transient regimes are inevitable; a hydro pumping and a power plant must be started up, switched off, undergo changes of load, etc., not to mention human errors, equipment breakdowns, earthquakes, etc. It is obvious that there are many different transient situations. The engineer is only concerned with those that might endanger the safety of a plant and its personnel, cause some damage or result in operational difficulties of any kind. To set apart the cases which deserve his attention, the engineer must first clearly define what are the consequences of transient regimes he fears the most in a plant. These might be:

(i) maximum pressures; (ii) occurrence of local vacuum; (iii) cavitation; (iv) vibration; (v) too strong oscillations of the masses.

Maximum pressures during transient regimes may destroy pipelines, valves or other components, causing considerable damage and sometimes loss of human lives. In less drastic cases, strong pressure surges may cause cracks in internal lining, damage connections between steel and concrete, destroy or cause deformations to equipment (such as air valves, penstock valves or gate valves). Sometimes the damage is not noticed at the time, but results in intensified corrosion which, combined with repeated transients, may cause unexpected collapse of the penstock.

Vacuum is usually avoided, even at a high cost, because it creates stresses and strains which are very different to those produced during normal operating regimes. Vacuum may cause the collapse of thin- walled penstocks or reinforced concrete sections which were not designed to withstand such strains. This operational difficulties may arise in pumping systems, hydro power plants, oil-hydraulic transportation pipelines as well as in water cooling systems of thermal and nuclear power plants etc. Groundwater, often heavily polluted, or gases, may be sucked into the pipeline or tunnel during a vacuum phase of a transient regime. This is, obviously, a great danger occuring in public water supply systems. Dissolved gases are released from the water whenever the local pressure drops considerably and these gases may promote the corrosion of any steel sections, with consequent damage (Wylie *et al.*, 1993).

Cavitation occurs when the local pressure is lowered to the value of the vapour pressure at the ambient temperature. All the gases within the water are released and the water starts to evaporate when the local vacuum is 96-97 %. When the pressure waves reach this point, water is condensed instantly, leaving behind space filled only with very rarefied gases. The water column enters this space and collides with whatever is on the side of the cavity. In this case both vacuum and strong pressure surges are present which may result in a substantial damage. The main problem with regard to cavitation is that its effects cannot be estimated by the analysis with any reasonable degree of accuracy, because the various parameters are difficult to determine during the design stage.

Hydraulic vibrations, if strong enough, cause damage of pipelines, tunnels, measuring and control equipment crumble concrete and may damage any internal lining in the tunnel. If resonance occurs, the whole system may easily be destroyed.

The engineer has to consider carefully all the potential dangers for the plant, estimating the weak spots. Only then should he embark upon the analysis since this usually means a considerable expense of both time and money.

The questions the engineer needs to ask are:

(i) is the safety of the plant endangered during any transient regime?

(ii) is some kind of protection required and if so, which one?

(iii) what is the most dangerous operation for the plant and is the adopted protection adequate?

2. Typical transient regimes in HP plants

The scope of analysis for a feasibility study or a detailed design is as follows:

2.1 NORMAL OPERATING CONDITIONS

The hydro power plant should be designed to operate without difficulties during any of the following conditions:

(i) normal steady state conditions either turbining, pumping or speed no-load; (ii) standstill to turbine operation and loading; (iii) turbine operation to synchronous compensation and return to turbine operation; (iv) stopping from turbine operation; (v) synchronous compensation to closure; (vi) standstill to pump operation and loading; (vii) stopping from pump operation; (viii) pump operation to synchronous compensation and back to pump operation; (ix) synchronous compensation to closure; (x) pump to turbine operation; (xi) load rejection during turbine operation followed by rapid closure of the wicket gates; (xii) power failure during pump operation followed by rapid closure of the wicket gates.

2.2 EMERGENCY OPERATING REGIMES

In emergency cases the plant is generally submitted to higher strains than in normal transient situations, but nevertheless should be able to withstand these and remain undamaged and operational. A few characteristic cases are given below:

(i) machines with variable closing speed of the wicket gates. One servomotor mechanism may maintain the maximum closing rate to the end of the wicket gate closure, thus creating a shock in the system; (ii) plants with synchronous relief valves. One of the valves may fail to open; (iii) plants with several units. The wicket gates of one unit may remain open during the closing-down of the whole plant. This machine has to be closed by its penstock valve. (iv) gate in the inlet structure. In low head plants this gate may malfunction, either by closing down permanently or not completely; (v) gate in draft tube. In some low head plants this gate is a protection of a runaway condition and may fail to operate; (vi) runaway regime. This is caused by a machine remaining open after a load rejection while the other machines in the station close down normally. The water forces the unit to enter the runaway zone regardless of the initial conditions. This dangerous state is terminated by closing down the inlet gates, or the draft tube gate, or the emergency spool valve closing the wicket gates (vii) unsuccessful start. Rapid start up of the whole plant followed by failure of one machine which has to be closed down immediately. (viii) unfavourable set of events. Usually start up or closing down of the whole plant, before the transients, caused by an earlier operation, had not died out, results in amplification of the pressure surges. Initial and boundary conditions should be the most unfavourable. For example, the water elevation in the surge tanks and/or air vessels should be either maximum or minimum depending upon which is worst for the given case.

2.3 CATASTROPHIC CASES

There are disastrous cases when the plant is endangered by simultaneous failure at two or more safety devices, or by an extremely unfavourable series of events. Some of those cases are:

(i) Machines with variable closing speed of the wicket gates. Two or more servomotor mechanisms maintain the maximum wicket gate closing speed over the closure period; (ii) plants with relief valves. Two or more relief valves fail to open during a rapid closure; (iii) plants with several units. The wicket gates of two or more units remain stuck in the fully open position; the corresponding penstock valves close normally; (iv) runaway. One or more units go to runaway because all devices in their branches fail; (v) instantaneous closure of wicket gates. Wicket gates close in a time less than $2 \cdot L/a$; (vi) instantaneous closure of a valve or gate; (vii) column separation and reverse waterhammer; (viii) resonance in the wicket gates. Oscillatory movements of its blades with a frequency equal to one of the natural frequencies of the system; (ix) auto-oscillations provoked by leakage on an imperfectly closed valve. Usually caused by a ball valve; (x) breakdown of the penstock. For instance a crack in the penstock near to the powerhouse; (xi) earthquake. Water within the system will act as a piston during an earthquake.

2.4 SAFETY

Safety coefficients: K1 (related to the ultimate strength of material) and K2 (related to the yield limit) should have the following values:

Table 1: Safety coefficients

		K1	K2
1.	Normal operation	4 to 5	1.5 to 2.2
2.	Emergency conditions	2 to 3	1.4 to 2.0
3.	Catastrophic cases	1	1

The plant is not usually designed to withstand the catastrophic events because the costs would be prohibitive. If some damage happen to the plant in catastrophic circumstances it is regarded as a "vis major" case.

2.5 IMPORTANT WARNING

The above list is rather general. This list should be compiled in each particular case, taking into account specific features of the plant. Special care should be devoted to the small pumping and power plants because there is a strong tendency to reduce the costs of analysis and design to the minimum. The incidents in future plant practice are dangerous, and it is much better to entrust this simplified and reduced analysis to a well-trained and experienced specialist who can help to prevent the incidents in future plants even when working within narrow constrains of time and money.

3. Machine data for transient computations

The performance characteristics of a machine, defined in four quadrants in the coordinate system with the abscissa the unit speed and the ordinate unit discharge and unit torque, describe its behaviour in all possible operating regimes. The data for the construction of these curves are obtained from the result of laboratory investigations during stationary regimes.

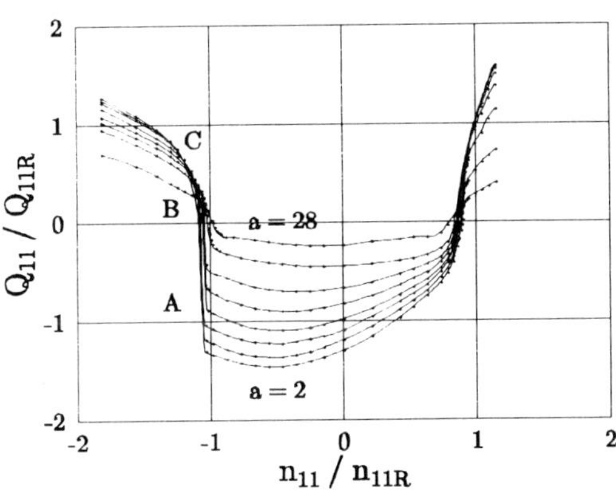

Figure 1: Four-quadrant turbine characteristics

Turbine characteristics representation for all regimes of operation has been the subject of several papers (Marshall *et al.*, 1965). Those characteristics have to be known for several guide vanes openings. When turbine characteristics have to be determined for a non-measured guide vane opening, they are determined by interpolation.

The starting point for the characteristics representation analysis have been Suter curves. As the authors consider Suter curves to be well known, they are not discussed here in detail. The characteristics interpolation should be performed in such a manner as to obtain smooth curves that resemble original characteristics in form. Several investigators Gajić (1983), Martin (1982), ...have tried to interpolate turbine characteristics by different kinds of polynomials, like Lagrange and Hermitte polynomials, but the results obtained were not satisfactory.

This indicated that Suter curves had to go through an additional transformation before a successful interpolation could be performed. This transformation had also to overcome the following inconvinience imposed by Suter transformation: namely, when the relative guide opening is close to 0, the value of W_H for $\Theta = 3 \cdot \pi/2$ becomes very great, tending towards infinity.

One of possible transformations was suggested earlier in Martin (1982). The abscissa in these graphs is:

$$\theta = \pi + \arctan \frac{v}{\alpha / a_R} \tag{1.1}$$

while the ordinates are:

$$WH = \frac{h}{\alpha^2 + v^2 / a_R^2}, \tag{1.2}$$

and

$$WM = \frac{\beta}{\alpha^2 + v^2 / a_R^2}. \tag{1.3}$$

where a_R is the relative opening given as:

$$a_R = \frac{a}{a_F} \tag{1.4}$$

where a is the guide vanes opening and a_F is the maximum guide vanes opening.

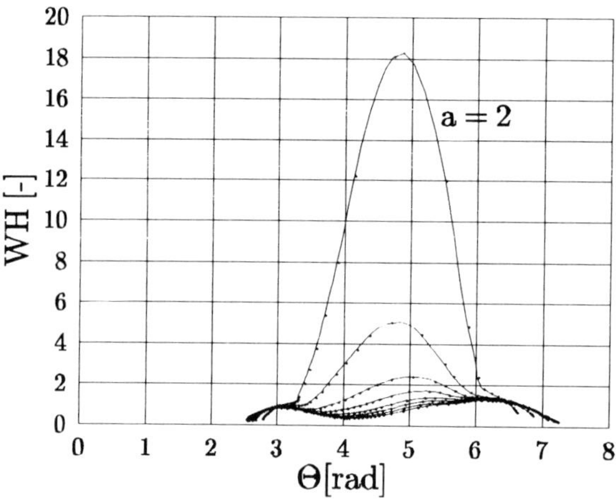

Figure 2: Suter representation of turbine characteristics

By this method the aforementioned shortcoming of Suter representation is overcome, and the curves become closer to each other and more uniform, although this improvement is not equal for all regimes of operation.

The opinion of the authors of this paper is that the referencing of the guide opening to maximum guide vanes opening is not neccessarily the best solution.

Namely, this maximum opening is not a characteristic of the turbine itself, but is determined by the conditions of turbine operation in the plant.

It seems reasonable, thus, to use a^*, the guide vanes opening for the optimal turbine regime, instead of a_F, as a^* is a characteristic of a turbine itself. Furthermore, this approach is more compatible with the method of calculation of v, h, α i β, as they are obtained by reduction of Q, H, n i β, respectively, to the corresponding values in the turbine optimum operating regime. Furthermore, from Fig.2 in Martin (1982), it is obvious that dependence of v upon $a_R = a/a_F$ is not quite linear.

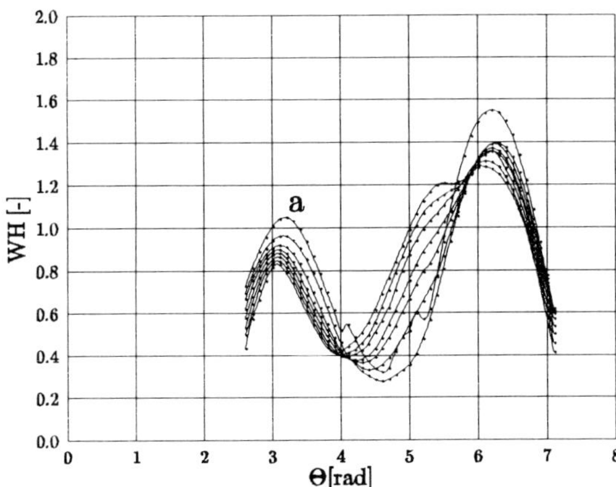

Figure 3: Unit curves transformed with $\gamma = 0.8$

Another field for a possible improvement of Martin's formulae is the value of the exponents associated with the relative guide vanes opening in Eqs (1), (2) and (3). As the inclusion of this relative opening itself alleviates the abovementioned shortcoming of Suter method of characteristics representation, it is not mandatory that this exponent in Eq.(1) should equal 1, or 2 in Eqs (2) and (3). Another value could be used instead, which would further facilitate characteristics interpolation, i.e,

$$\theta = \pi + \arctan \left[\frac{v}{\alpha / a_r^{\gamma}} \right], \tag{1.5}$$

and

$$\frac{h}{\alpha^2 + v^2 / a_r^{2 \cdot \gamma}}. \tag{1.6}$$

where γ is a value specific for a turbine, and $a_r = a/a^*$.

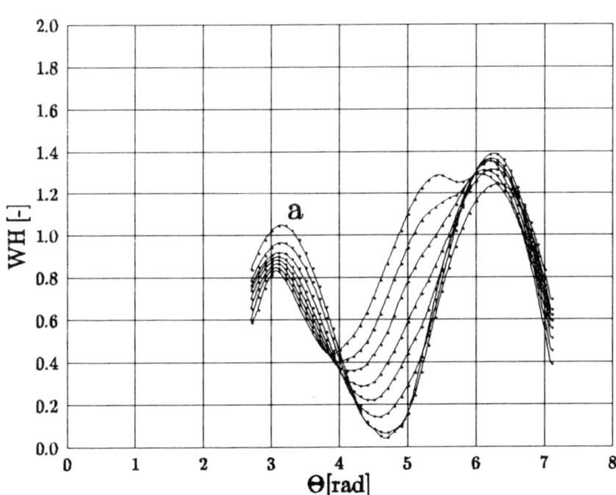

Figure 4: Unit curves transformed with $\gamma = 1.2$

Determination of the optimum γ for a turbine is based upon the following considerations. It is supposed that the characteristics interpolation is easier when the curves are more closely and more uniformly spaced. Therefore, the appropriate γ should be determined so that curves are as much uniformly spaced as possible, not only in certain operating regimes, but for the whole zone of the unit operation.

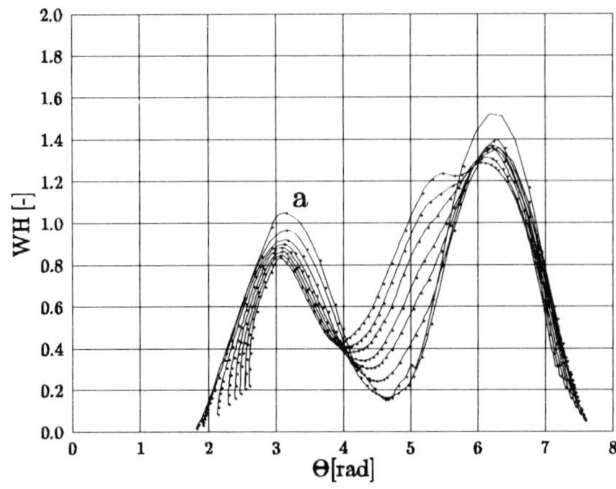

Figure 5: Unit curves transformed with $\gamma = 0.92$

It is, therefore, necessary to introduce a certain criterion for uniformity. In an ideal case, these differences should be identical; however, in reality they should be as close as possible to some mean value. The measure of non-uniformity should then be the mean standard deviation of the differences.

The most uniform distribution for a selected value of v would then be the one with the least mean deviation.

This approach should be extended to a whole range of unit operation, while the total mean deviation can be used as a measure of distribution uniformity.

The procedure decribed above can be divided into the following steps:

(i) a range of γ values is determined, as $0 < \gamma < 2$;

(ii) a value of γ is selected;

(iii) curves for a whole range of turbine operation are transformed with the selected value of γ;

(iv) a range of θ values is divided into segments, as to obtain an array of significant θ values;

(v) for each θ selected, first the differences between the curves are determined, and subsequently their aritmetic mean value and the mean deviation;

(vi) step (v) is repeated for all values of θ in the array, to cover the whole range of operation; in each repetition sum of the mean deviations is updated, until the total mean deviation is determined;

(vii) steps (iii)-(vi) are performed for all selected values of γ.

The value of γ that results in the least total mean deviation is considered to be the optimum for the given unit.

Figs 3, 4 and 5 ilustrate the results obtained with the γ values of 0.8, 1.2 and 0.92, respectively; the most favourable results have been obtained for the last γ value.

4. Vibration severity

The vibrations of the transient state has recently become very important. In addition to the vibration in the steady state condition, the transient operation of hydraulic machines (pumps, turbines, reversible pump-turbines) such as: starting, stopping, load rejection, over speed, runaway, etc. must be considered. Fig. 6 shows the allowable vibration amplitudes and the exposure rate R for the transient operation.

The exposure rate is expressed in dimensionless terms and indicates the ratio of the vibration time to the total operation time (probability). An exposure rate of $1 \cdot 10^{-3}$ would indicate operating in the vibrating condition for about 9 hours per year. An exposure rate of $1 \cdot 10^{-6}$ represent the operation for 30 seconds per year. The operation at vibration amplitudes above the latter curve should not be permitted at any time. Single excursions above these amplitudes should also be prevented if possible Rund (1980), Ohashi (1991).

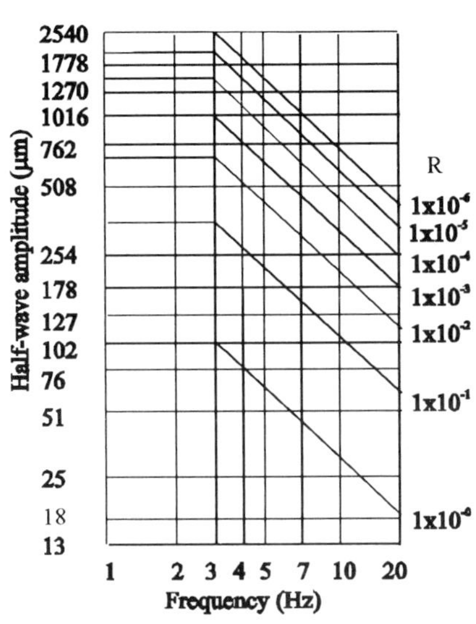

Figure 6: Vibration amplitude of normal and transient operation

The amplitudes of about one-half of those given by the exposure rate of $1 \cdot 10^{0}$ would be considered "excellent", while those on the curve would be "normal". "Strong" vibrations would be indicated by the curve for an exposure rate of $1 \cdot 10^{-3}$, and "severe" vibrations would be near the curve for the exposure rate of $1 \cdot 10^{-6}$.
The usual case of transient operation will produce vibration amplitudes in the "strong" range. The components unit will generally be undamaged through the life of the unit. Operation at higher amplitudes, or for a longer total exposure may lead to severe bearing damage or serious damage to the appurtenant members through structural failure.

The costs of such possible damage may be expanded along with the exposure rates and a probable risk when evaluating possible remedial measures for reducing excessive vibration amplitudes.

5. Conclusions

The usual case of transient operation will produce vibration amplitudes in the "strong" range. Operation at higher amplitudes, or for longer total exposure may lead to serious damage to appurtenant members through structural failure. The costs of such possible damage may be expanded along with the exposure rates and a probable risk when evaluating possible remedial measures for reducing excessive vibration amplitudes. By calculating, measuring and analysing the hydraulic transients (waterhammer, hydraulic vibrations and stability) extreme values of machine vibrations and the pressure along system are evaluated.

Comparing machine vibration to the recommended statistical data exposure rate could save a lot of money and time in improving the operation reliability and reduce maintainance costs.

6. Acknowledgments

Grateful acknowledment is expressed to the Serbian Foundation for Sciences for supporting investigations Grant S.2.06.21.142 Faculty of Mechanical Engineering Belgrade.

7. References

Chaudhry, H. M. (1979) *Applied Hydraulic Transients*, VAN, New York.

Gajić, A. (1983) *A Contrubution to the Investigations of Unsteady Phenomena in Hydro-Electric Power Plants*, (in Serbian), Ph.D. Dissertation, University of Belgrade, pp.79-85

Krivchenko, G. I., Arshenevsky, N. N., Kvyatkovskaya, E. V., Klabukov, V. M. (1975) *Hydraulic Transients in Hydroelectric Power Plants*, (in Russian), Moscow

Marshal, M., Flesch, G., Suter, P. (1965) Calculating the Waterhammer in Storage Pump Installations with Electronic Digital Computers, *Proceedings of the Int. Symp. on Waterhammer in Pumped Storage Projects*, ASME, Chicago, pp.67-75

Martin, S. (1982) Transformation of Pump-Turbine Characteristics for Hydraulic Transient Analysis, *Proceedings of the 11th IAHR Symposium, Amsterdam*, Vol.2, pp.30.01-30.15

Ohashi, H. (Editor) (1991) *Vibration and Oscillation of Hydraulic Machinery*, Hydraulic Machinery Book Series, Avebury Technical, England

Pejović, S. (1977) *Waterhammer and Hydraulic Transient Analysis*, (in Serbian), Beograd

Pejović, S. (1979) *Hydraulic Vibrations of Hydraulic Systems*, (in Serbian), Beograd

Pejović, S., Gajić., A., Obradović, D. (1983) Forms of Pump-turbine Characteristics in Water Hammer Calculation, *Proceedings of ASME Winter Annual Meeting*, Boston

Pejović, S., Boldy, A.P., Obradović, D. (1987) *Guidelines to Hydraulic Transient Analysis*, Technical Press, England

Rund, F. O. (1980) Vibration Criteria for Transient Operation of Hydroelectric Units, *Proceedings of IAHR Symposium on Hydraulic Machinery and Cavitation*, Tokyo, pp. 517-526

Suter, P. (1966) Darstellung der Pumpencharacteristik für Druckstosserechungen, *Sulzer-Forschungsheft*, pp.45-48

Wylie, B. E., Sweeter, L. V. (1993) *Fluid Transients in Systems*, Prentice Hall, Englewood Cliffs

MODELLING A PROTECTION DEVICE IN A LOW PRESSURE LIFTING SYSTEM

ORAZIO GIUSTOLISI
MICHELE MASTRORILLI
Water Engineering Department, Bari Polytechnic
via E. Orabona, 4 - 70125 Bari

1. Introduction

A *water hammer protection device* in lifting systems has been experimentally studied, especially low pressure ones [1][2][3][7] (see figures n°1 and n°2, pages n°5 and n°6 of the paper). Water hammer, in the experimental tests of the analyzed device, has been caused by the fast water flow interception with a spherical valve located just downstream from the pump of the hydraulic system. Such a device is composed of a generic check valve and a concentrated head loss located at one end of a bypass; the latter must be installed between the upstream reservoir and the delivery pipe of the hydraulic system, just downstream from the section in which water flow cutoff has been generated, with the valve close to the reservoir. The valve allows water to flow from the reservoir to the delivery pipe through the bypass. The positive experimental results regarding the efficiency of this device **bypass-check valve-concentrated head loss** in protecting low pressure lifting systems from water hammer, have been the starting point for studies which have been directed towards the creation of a numerical model of the hydraulic system. From the trials presented in [1][3], the authors have noted that a numerical model, simulating water hammer in a system equipped with the studying protection device, must use theoretical studies from the cavitation and models reproducing it. Previously [2], a simulating model, without considering the possibility that cavitation occurs, was realized. As the new experimental trials show [1][3], having been carried out starting from steady velocities much higher than older ones [2], the previous simulating model can be applied only when cavitation does not occur.

2. Physical Functioning

The protection device, which the authors are studying, has exibited a complex physical running. It has been observed that, when the protection device was operating, cavitation often occured [1][3].

E. Cabrera et al. (eds.), Hydraulic Machinery and Cavitation, 701–709.

Indeed, the protection device, composed of **bypass-check valve-concentrated head loss**, allows water to flow through the bypass in the delivery pipe of the lifting system just after the water cutoff caused by spherical valve V1 (figures n°1 and n°2).
The protection device response, neverthless, has a little time delay due to the time interval that perturbating waves, at the wavespeed of about 1300 m/sec [8], spend to reach the check valve from V1.
Also if the response defer of the protection device was very short, authors observed that it was enough long, being the flow interception by a spherical valve very fast, to allow the reaching of cavitating pressures.
The latter consideration has been supported by experimental diagrams of pressure detected at the pressure outlet just downstream from the bypass insertion [1][3][7].
The observation that inserting the protection device further from the valve V1 (in the point n°3 has been inserted the protection device to detect the experimental trials to validate the numerical model) had caused higher pressure spikes just after the check valve opening (the check valve running was monitored [1][3][7]), due to the sudden compression of air bubbles and the longer response defer of the protection device, suggested the authors to equip the protection device with a concentrated head loss.
The concentrated head loss has the function to reduce high velocities which occur in the bypass due to the sudden check valve opening caused by underpressure waves.

3. Calculating Model

The simulating model has been written starting from results which authors have obtained in [4] and using for the term J the expression [5][6]:

$$J_u = J + \frac{k}{g}\bullet\left(\frac{\partial V}{\partial t} - c\bullet\frac{\partial V}{\partial x}\right) \tag{1}$$

so it has been obtained the system,

$$\begin{cases} H_{i,j} + \dfrac{c\bullet(1+k)}{g}\bullet U_{i,j} = H_{i-1,j-1} + \dfrac{c\bullet(1-k)}{g}\bullet W_{i-1,j-1} + \dfrac{2\bullet k\bullet c\bullet U_{i,j-1}}{g} - \displaystyle\int_{i-1,j-1}^{i,j} J\bullet dx \\ H_{i,j} - \dfrac{c\bullet(1+k)}{g}\bullet W_{i,j} = H_{i+1,j-1} - \dfrac{c\bullet(1-k)}{g}\bullet U_{i+1,j-1} - \dfrac{2\bullet k\bullet c\bullet W_{i,j-1}}{g} - \displaystyle\int_{i+1,j-1}^{i,j} J\bullet dx \\ \dfrac{Y^{\gamma}_{i,j-1}}{Y^{\gamma}_{i,j}} = R^{\gamma}_{i,j} = \dfrac{H_{i,j}}{H_{i,j-1}} \\ \dfrac{1}{R_{i,j}} - 1 = \dfrac{A\bullet\Delta t}{Y_{i,j-1}}\bullet\left(W_{i,j} - U_{i,j}\right) \end{cases} \tag{2}$$

The system (2), with little modifications [4], can specify unsteady flow in all internal sections of the bypass and the delivery pipe of the hydraulic system. It has been necessary to develop a similar system for the computational section x, which correspond to the insertion of the protection device, and for the section of the check valve (and the concetrated head loss) close to the upstream reservoir.

3.1. SYSTEM FOR THE JUNCTION

Looking at the system (2), fixing the positive versus of the velocity into the bypass that goes from it to the junction with the delivery pipe, it has been possible to write:

$$\begin{cases} H_{x,j} + \dfrac{c\bullet(1+k)}{g}\bullet U_{x,j} = H_{x-1,j-1} + \dfrac{c\bullet(1-k)}{g}\bullet W_{x-1,j-1} + \dfrac{2\bullet k\bullet c}{g}\bullet U_{x,j-1} - \int_{x-1,j-1}^{x,j} J\bullet dx = N_1 \\ H_{x,j} - \dfrac{c\bullet(1+k)}{g}\bullet W_{x,j} = H_{x+1,j-1} - \dfrac{c\bullet(1-k)}{g}\bullet U_{x+1,j-1} - \dfrac{2\bullet k\bullet c}{g}\bullet W_{x,j-1} - \int_{x+1,j-1}^{x,j} J\bullet dx = N_2 \\ H_{n,j} + \dfrac{c\bullet(1+k)}{g}\bullet U_{n,j} = H_{n-1,j-1} + \dfrac{c\bullet(1-k)}{g}\bullet W_{n-1,j-1} + \dfrac{2\bullet k\bullet c}{g}\bullet U_{n,j-1} - \int_{n-1,j-1}^{n,j} J\bullet dx = N_3 \\ \dfrac{Y^{\gamma}_{x,j-1}}{Y^{\gamma}_{x,j}} = R^{\gamma}_{x,j} = \dfrac{H_{x,j}}{H_{x,j-1}} \ :\ H_{x,j} = H_{n,j} \ :\ Y_{x,j} = Y_{n,j} \\ \dfrac{1}{R_{x,j}} - 1 = \dfrac{A\bullet\Delta t}{Y_{x,j-1}}\bullet\left(W_{x,j} - U_{x,j} - U_{n,j}\right) \end{cases} \tag{3}$$

In writing (3) the assumption has been made that in the junction there is a single concentrated air pocket which expands and compresses according to a thermodynamic transformation [4]. In this case, therefore, the time variation of its volume is related to three velocities as in the fifth of (3). We can note that the upstream velocity $U_{n,j}$ of the last section of the bypass, if positive, causes a compression of the air pocket. Indeed, the last equation in (3) should be written,

$$\frac{1}{R_{x,j}} - 1 = \frac{A\bullet\Delta t}{Y_{x,j-1}}\bullet\left(W_{x,j} - U_{x,j}\right) - \frac{A_B\bullet\Delta t}{Y_{x,j-1}}\bullet U_{n,j} \tag{4}$$

where it has been supposed the bypass diameter, in A_B, to be different from that of the delivery pipe. The solution of the system (3) has been given by passing through a resolving equation in $R_{i,j}$ [4],

$$\begin{cases} A\bullet R_{x,j}^{(1+\gamma)} + B\bullet R_{x,j} + C = 0 \\ A = 3\bullet H_{x,j-1} \ :\ B = \dfrac{c^2\bullet(1+k)\bullet Y_{x,j-1}}{g\bullet\Delta x\bullet A} - N_1 - N_2 - N_3 \ :\ C = -\dfrac{c^2\bullet(1+k)\bullet Y_{x,j-1}}{g\bullet\Delta x\bullet A} \end{cases} \tag{5}$$

which, substituted in (3), gave the unknown terms.

3.2. SYSTEM TO THE CHECK VALVE

In order to write the mathematical system at the section of the check valve, a previous hypothesis has been made that the valve has an ON/OFF running. This last hypothesis seems possible because the valve, in the experimental trials, opened quickly when the bypass was going to produce its effect of protecting device while, when it closed, the motion was slow [1][2][3][7]; However, the numerical model has resulted satisfactory also with the previous hypothesis which makes it not dependent on the specification of the check valve. Regarding the concentrated head loss close to the valve, it has been assumed, in calculations, to be located in the same section of the valve that can allow us to conglobate in it also the concentrated head loss related to the valve.
For the concentrated head loss the usual equation: $\Delta p = K \bullet V^2$ has been assumed.
Then the calculation in the section of the check valve has been performed:

- writing system (2), with the condition $W_{o,j} = U_{o,j} \geq 0$ [4], **valve opened**;
- writing system (2), with $U_{0,j}=0$, **valve closed**.

In the first case it has been obtained:

$$\begin{cases} H_{0,j} = \text{upstream reservoir level - } K \bullet W_{0,j}^2 = H_m - K \bullet W_{0,j}^2 \\ H_{0,j} - \dfrac{c \bullet (1+k)}{g} \bullet W_{0,j} = H_{1,j-1} - \dfrac{c \bullet (1-k)}{g} \bullet W_{1,j-1} - \dfrac{2 \bullet k \bullet c}{g} \bullet U_{1,j-1} - \int_{1,j-1}^{0,j} J \bullet dx = N \\ W_{0,j} = U_{0,j} \geq 0 \quad \Rightarrow R_{0,j} = 1 \end{cases} \tag{6}$$

and then the solving equation in $W_{o,j}$,

$$A \bullet W_{0,j}^2 + B \bullet W_{0,j} + C = 0 \text{ with } A = K \ : \ B = \frac{c \bullet (1+k)}{g} \ : \ C = N - H_m \tag{7}$$

Studying the coefficeints of (7), being A and B always positive, the system (6) gives a solution if $C \leq 0$ then if $N\text{-}H_m \leq 0$. The last condition represents, in this model ON/OFF, the possibility, related to the hydrodynamic state N coming from the $i+1$ $(i=0)$ section and the upstream reservoir level H_m, that the check valve is open. In the second case it has been obtained:

$$\begin{cases} H_{0,j} - \dfrac{c \bullet (1+k)}{g} \bullet W_{0,j} = H_{1,j-1} - \dfrac{c \bullet (1-k)}{g} \bullet U_{1,j-1} - \dfrac{2 \bullet k \bullet c}{g} \bullet W_{0,j-1} - \int_{1,j-1}^{0,j} J \bullet dx = N \\ \dfrac{Y_{0,j-1}^{\gamma}}{Y_{0,j}^{\gamma}} = R_{0,j}^{\gamma} = \dfrac{H_{0,j}}{H_{0,j-1}} \\ \dfrac{1}{R_{0,j}} - 1 = \dfrac{A \bullet \Delta t}{Y_{0,j-1}} \bullet W_{0,j} \end{cases} \tag{8}$$

Resolved system (8), in this case the physical condition, which makes it valid, is that close to the check valve, beeing $U_{0j}= 0$, is true $H_{0,j} \geq H_m$.
Based on previous considerations, (6) has been adopted for the valve opened with its condition for changing its state; on the contrary (8) has been used for the valve closed with its condition for changing its state.

4. Numerical Results and Conclusions

The computer program to simulate the unsteady flow in the experimental lifting system equipped with the protection device has been based on previous mathematical systems from *DFGM* model. The comparison, reported here, between numerical tests and experimental data has been performed assuming that the insertion point of the bypass was in n.3 (see figure n° 1) because in this case the physical phenomenon is more interesting [1][3][7]; for other values we have assumed:

- void fraction F_V equal to 10^{-5};
- steady velocity equal to *1,66 m/sec*;
- γ variable according to the function $\gamma = 0,75+40 \bullet dY/dt$;
- k [5][6] equal to 0,0369 as derived from experimental data [7];
- K of $\Delta p = K \bullet V^2$ equal to 0-6.

From the numerical results compared with experimental data (see following diagrams) we can note that the previous method, based on *DFGM* extended, is satisfactory to simulate the unsteady flow in this particular physical situation more complex than a usual one. Authors, moreover, give a great importance to the agreement between computer results and experimental trials, because the proposed device has shown itself to be very efficient, based on a wide experimental phase [1][3][7], to protect the lifting system.
From the first and the third diagram in figure n°3, we can note that the model is able to predict, with a good aproximation, pressure spikes which occur when the concentrated head loss (K=0) is not operating.
The second diagram of figure n°3 shows that the numerical model is able to simulate the protection device running also when there was the concentrated head loss effect (K=6). Moreover, authors have found a good general agreement between numerical results diagrams and the same ones of the experimental tests in all the comparations [1][3][7].
This is a satisfactory result considering the hypotesis in the *Discrete Free Gas Model*, the ON/OFF running of the check valve and the $K \bullet V^2$ formula adopted for the concentrated head loss.
Finally, the numerical simulation of the phenomenon has allowed authors to verify some physical considerations about the protection device running and to explain better experimental trials [1][3][7]; for example the high velocities in the bypass just after the check valve opening and pressure spikes due to the sudden air bubbles have been reproduced in the model.

Figure 1. Experimental Lifting System

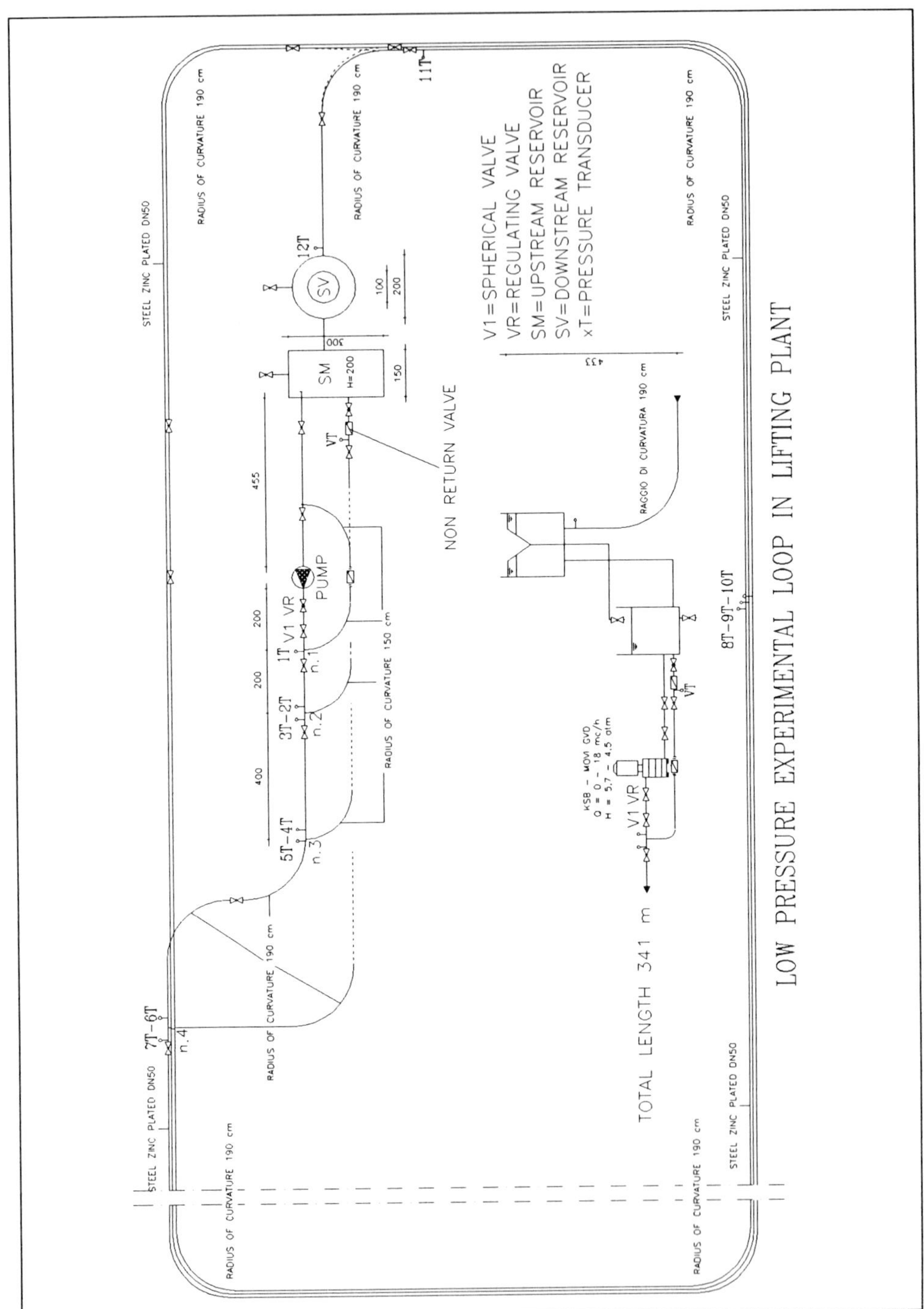

Figure 2. Functional Scheme

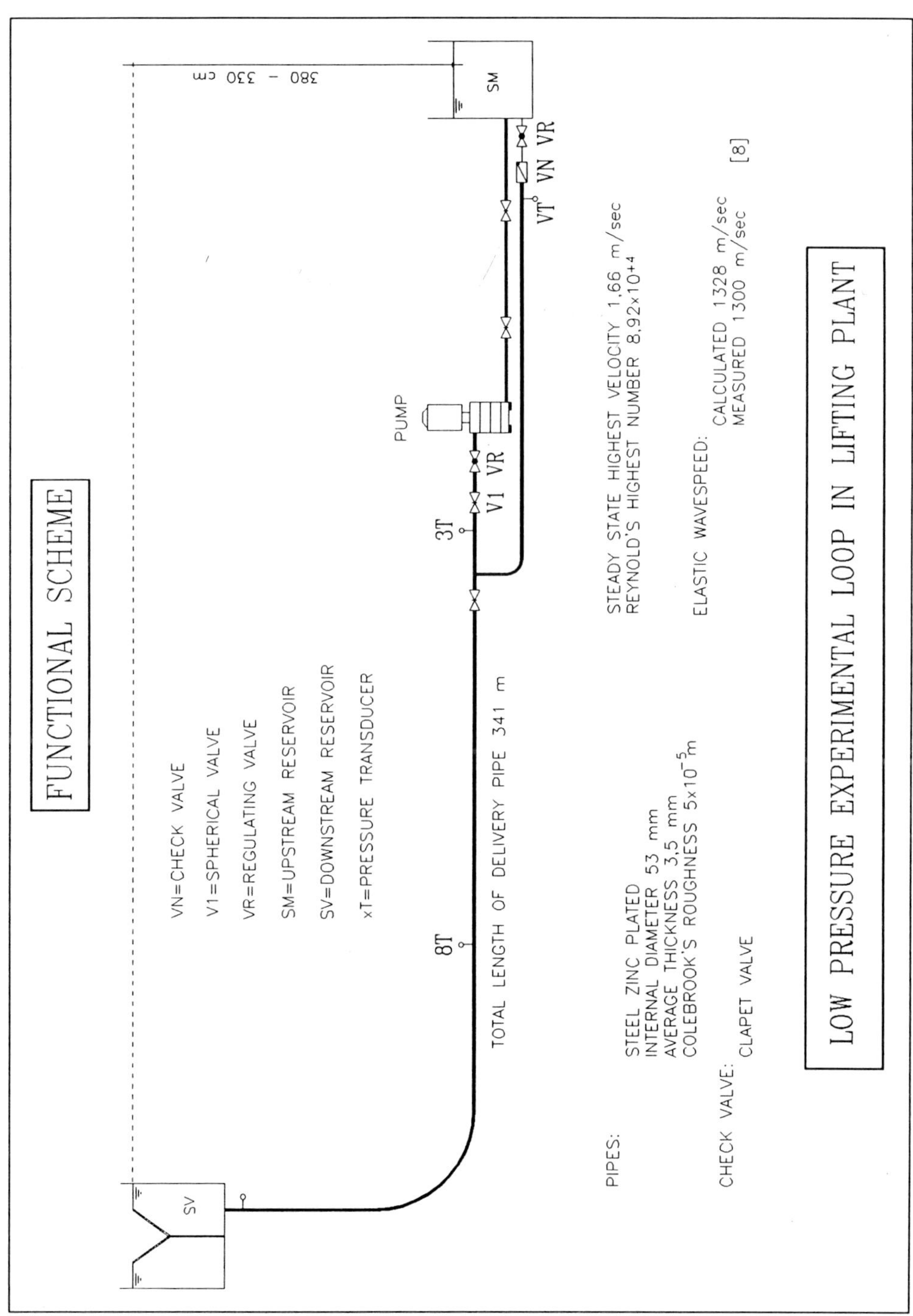

Figure 3. Experimental Trials (hidden line) and Numerical Results (continuous line)

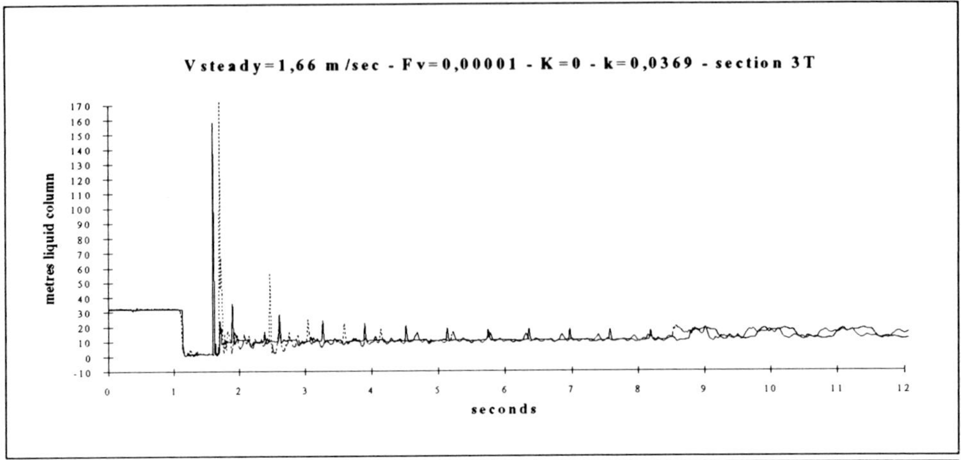

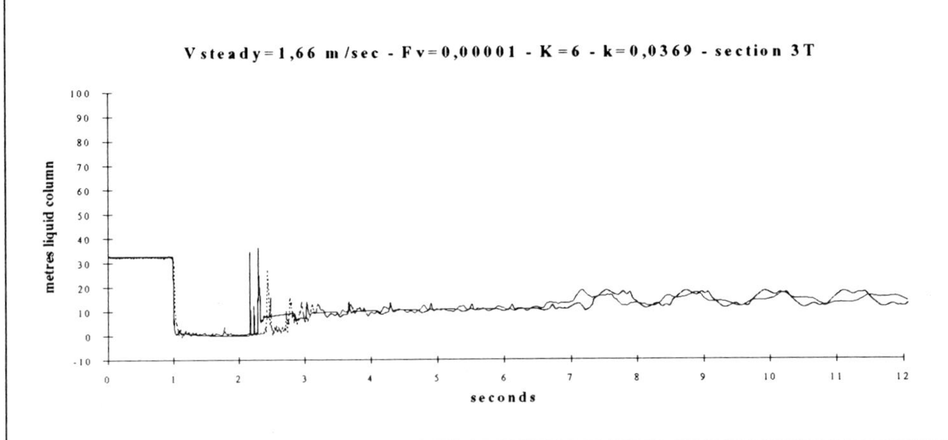

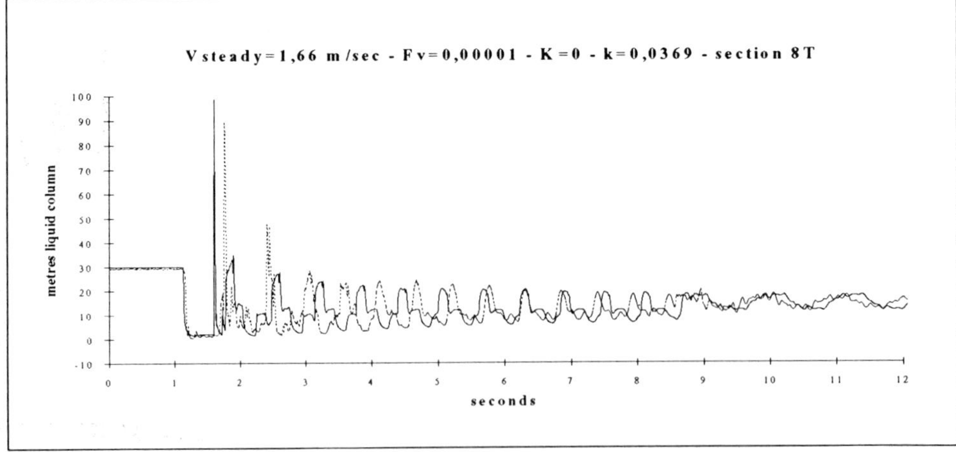

Nomenclature

A, A_B	pipeline area	m^2
c	wave speed	m/sec
g	acceleration due to gravity	m/sec^2
H	piezometric head	m
U,W	upstream and downstream velocities	m/sec
Y	volume of the air pocket	m^3
Fv	void fraction	
$\Delta x, \Delta t$	spatial and temporal discretization	m, sec
H_m	upstream reservoir level	
i,j	spatial and time indices	
n,x	indices of the junction section	
J	friction term from uniform flow	
J_u	friction term in unsteady flow	
k	coefficent of Brunone-Golia-Greco	

References

1. Giustolisi O. (1994): *Dati sperimentali su un metodo di protezione per un impianto di sollevamento a bassa pressione* - Convegno Moto Vario nei Sistemi Acquedottistici, Bari, Italy 30-31 May.
2. Giustolisi O., Mastrorilli M. (1994): *Proposal of Water Hammer Control Device in a low pressure system: study and interpretation of experimental data*- 2nd International Conference on PIPELINE SYSTEMS, Edinburgh, Scotland 24-26 May.
3. Giustolisi O., Mastrorilli M. (1994): *Protezione di un impianto di sollevamento: inserzione di un bypass con una perdita di carico localizzata* - Convegno Moto Vario nei Sistemi Acquedottistici, Bari, Italy 30-31 May.
4. Giustolisi O. (1995): *DFGM model with generic politropic transformation and a varying exponent* - HYDRA 2000, IAHR XXVI Congress, London, United Kingdom, 11-15 September.
5. Brunone B., Golia U. M., Greco M. (1992): *Some Remarks on the Momentum Equation For Fast Transients* - in AA.VV., Hydraulic Transient with Water Column Separation (9th and last Round Table of the IAHR Group), Valencia, Fluid Mechanics Group, Valencia, Spain.
6. Brunone B., Golia U. M., Greco M. (1992): *Modelling for Fast Transients by Numerical Methods* - in, AA.VV., Hydraulic Transient with Water Column Separation (9th and last Round Table of the IAHR Group), Valencia, Fluid Mechanics Group, Valencia, Spain.
7. Giustolisi O. (1995): *Un dispositivo di controllo del moto vario in un impianto di sollevamento a bassa pressione di esercizio: analisi teorica e rilievi sperimenatali* - Tesi di Dottorato in Ingegneria Idraulica VI ciclo Napoli.
8. Giustolisi O. (1994): *Experimental Evaluation of the Elastic Wavespeed in the Unsteady Flow*. 5th International Conference, Hydrosoft '94, Porto Carras, Greece 21-23 September.

DYNAMIC COMPRESSION OF ENTRAPPED AIR POCKETS BY ELASTIC WATER COLUMNS

R. GUARGA, A. ACOSTA, E. LORENZO.
Fluid Mechanics and Environmental Engineering Institute.
School of Engineering, University of Uruguay.
Montevideo, Uruguay.

1. Introduction

The problem of dynamic compression of entrapped air pockets in a pipeline has a great technical importance when designing water pressured pipelines.

C. Martin (1976) raised and solved the problem for the case in which the liquid column can be considered as incompressible and of a constant length. He elaborated diagrams that allow calculating the maximum absolute piezometric head registered in an entrapped air pocket during its dynamic compression by a rigid water column.

E. Cabrera et al. (1992) analyzed the problem for the case in which the liquid column can be admitted as incompressible but its length must be considered variable.

R. Guarga et al. (1994) examined the limit between these models mentioned above, establishing that this limit can be set using the following parameter (the nomenclature of the variables is specified at the end of this paper):

$$\Phi = \frac{\forall_0}{D^2} \frac{|\sin\alpha|}{H_t^*} \qquad (1)$$

In that paper, Guarga et al. conclude that if $\Phi < 0.1$, Martin's diagrams may be used with errors that will not exceed a 15%.

These papers which have been mentioned, allow performing the project of pressured water pipelines with a conceptual tool for the analysis which enables calculating the effects of the possible pressure peaks caused by the dynamic compression of entrapped air pockets.

Nevertheless, these results are excessively conservative when the compressibility of the water column cannot be neglected and that fact usually leads to more expensive designs than actually necessary. Therefore, it is of great technical interest to study the limit until which the compressibility of the water column can be neglected.

E. Cabrera et al. (eds.), Hydraulic Machinery and Cavitation, 710–719.

2. The Basic Problem to be Studied

In Figure 1 the basic problem to be studied is illustrated. The tank (whose free surface remains at a constant level) feeds a pipe of length L (through a valve), with an air pocket at its end.

When the valve is closed, the system does not work. The absolute pressure in the air pocket will be p^*_{go} and its volume $\forall_o$. The phenomenon to be studied begins when the valve is completely opened in a negligible time compared to period $T = 2L/a$ (where "a" is the wave propagation velocity in the pipe).

When the valve is open, a phenomenon takes place which is trivial in the case in which $p^*_{go}/\gamma = H^*_t$. This dynamic phenomenon will be studied hereafter.

3. Equations which Consider Finite the Wave Propagation Velocity.

The equations which rule the mentioned phenomenon are the following:

In the pipe $(0 < s < L)$

$$g\frac{\delta H^*}{\delta s} + \frac{\delta V}{\delta t} + \frac{f\, V|V|}{2\, D} = 0 \tag{2}$$

$$g\,\frac{\delta H^*}{\delta t} + a^2\frac{\delta V}{\delta s} = 0 \tag{3}$$

At the end, where the air pocket is $(s = L)$

$$H^*(L,t) = H^*_g(t) \quad ; \quad t \geq 0 \tag{4}$$

$$\frac{dH^*_g}{dt} = -n\frac{H^*_g}{\forall}\frac{d\forall}{dt} \tag{5}$$

$$\frac{d\forall}{dt} = -AV(L,t) \tag{6}$$

At the begining $(s = 0)$, where the valve is located

$$H_{go}^{*} = H(0,t) \quad if\ t\leq 0 \quad ; \quad H_{t}^{*} = H^{*}(0,t) \quad if\ t>0 \qquad (7)$$

The initial conditions (t = 0) are the following:

$$H_{g}^{*}(0) = H_{go}^{*} \ ; \ \forall(0) = \forall_{o} \ ; \ V(s,0) = 0 \ ; \ H^{*}(s,0) = H_{go}^{*}$$

The unknown functions are H* (s,t); V (s,t) and ∀ (t).

4. Inspectional Analysis of the Equations

So as to carry out the inspectional analysis of the system of differential equations which has been raised, of the initial conditions and of the corresponding border, the following change of variables has to be made:

$$s=\hat{s}\ L;\ \ t=\hat{t}\ t_{R};\ \ V=\hat{V}\ V_{R};\ \ \forall=\hat{\forall}\ \forall_{go};\ \ H^{*}=\hat{H}^{*}\Delta H^{*}$$

where t_R and V_R are a time and a velocity of reference, whose most convenient determination will come about from the inspectional analysis itself. The symbols with ^ correspond to dimensionless variables. If the change of variables is made, and assigning to t_R and V_R the following values:

$$t_{R}=\left(\frac{L\forall_{o}}{g\Delta H^{*}D^{2})}\right)^{1/2} \quad ; \quad V_{R}=\left(\frac{\forall_{o}\Delta H^{*}g}{D^{2}L}\right)^{1/2}$$

a system in the dimensionless variables, controlled by the following dimensionless numbers is obtained:

$$\pi_{1}=\frac{H_{T}^{*}}{\Delta H^{*}};\ \ \pi_{2}=f\frac{\forall_{o}}{D^{3}};\ \ \pi_{3}=n;\ \ \pi_{4}=\frac{a^{2}\forall_{o}}{Lg\Delta H^{*}D^{2}}$$

Consequently, the solutions will take up the following form:

$$\hat{H}^{*}=F_{1}(\hat{s},\hat{t},\pi_{1}..\pi_{4});\ \ \hat{V}=F_{2}(\hat{s},\hat{t},\pi_{1}..\pi_{4});\ \ \hat{\forall}=F_{3}(\hat{t},\pi_{1}..\pi_{4})$$

5. Calculation of the Maximum $\hat{H}^*_g$ Value (i.e. $\hat{H}^*_{gM}$)

When designing water pipes, the main interest is concentrated in the appraisal of the maximum pressures which can be found in the air pocket. Therefore, the maximum values of $\hat{H}^*_g$ must be calculated, in which, according to equation (4), the value of $\hat{H}^*$ is in $\hat{s} = 1$. From these maximum values, the one which matters is the highest, which will be called $\hat{H}^*_{gM}$.

In the maximum values of $\hat{H}^*_g$, we find $d\hat{H}^*_g/dt=0$, wherefore, we derive $d\forall/dt=0$ from equation (5). Consequently, we find with equation (6) that $\hat{V}(1,\hat{t}_M) = 0$. From here we conclude that the instants in which a maximum $\hat{H}^*_g$ takes place, will be a subassembly of the roots of the equation:

$$\hat{V}\,(1,\hat{t},\pi_1..\pi_4) = 0 \tag{8}$$

Consequently, such instants depend only on the four dimensionless numbers from π_1 to π_4. From here we derive that $\hat{H}^*_{gM}$ is also and exclusively a function of these four dimensionless numbers, i.e.:

$$\hat{H}^*_{gM} = F_1(\pi_1..\pi_4) \tag{9}$$

6. The Results

With the purpose of comparing our results later on with those obtained by Martin (1976), the results will be stated using the same variables as this author.

Remembering the four dimensionless numbers shown in (9), it is obvious that π_1 can be built from H^*_t / H^*_{go}. Therefore, it is possible to state H^*_{gM} / H^*_t in the following manner:

$$\frac{H^*_{gM}}{H^*_t} = G(\frac{H_t}{H^*_{go}}, f\frac{\forall_o}{D^3}, n, \delta) \tag{10}$$

where $\delta = (a^2\ V_o) / (L\ g\ \Delta H^*\ D^2)$.

Thus, an expression is obtained which incorporates the same variables that Martin uses and where a new variable "δ" appears. For the cases studied by Martin (an uncompressible column and a rigid pipe) the wave propagation velocity is infinite and "δ" will equally be so.

Consequently we can foresee that since "δ" tends towards infinite, the quotient H^*_{gM} / H^*_t will tend towards the value calculated by Martin when H^*_t/H^*_{go}, fV_o/D^3

and "n" are equal.

Expression (10) is calculated numerically integrating the system of differential equations, of initial conditions and of the border described in section 3. We have set n = 1.2, which coincides with the value used by Martin (1976). The results of the calculation are shown in a graphic manner in figures 2, 3, 4 and 5. In Figure 2 we calculated the cases where $fV_0 / D^3 = 0$ (without any friction) while in figures 3, 4 and 5 we calculated the cases where $fV_0 / D^3 = 0.25, 0.5$, and 1 respectively.

In each of the figures, the values are the ones indicated in Table 1, and also, each figure shows the curve calculated by Martin for the liquid uncompressible column and the rigid pipe (infinite "a").

Table 1.

Figure No	Values of δ
2	1, 10, 10^2, 10^6 (*)
3	1, 10, 10^2, 10^4 (**)
4	1, 10, 10^2
5	1, 10, 10^2

* calculated only for $H^*_t / H^*_{go} = 6$

** calculated only for $H^*_t / H^*_{go} = 5$

From the observation of figures 2, 3, 4 and 5 we may conclude that:

a. Such as was pointed out in the Introduction, Martin's results lead to forecast overpressures which are larger than the ones which come about from incorporating compressibility to the liquid column and elasticity to the pipe (finite "a").

b. As was previously indicated, when "δ" grows, the quotient H^*_{gM} / H^*_t tends towards the values given by Martin's calculations. In Figure 2 the superposition of the dots of Martin's curve and the present approach (finite "a") is obtained with $\delta = 10^6$ and in Figure 3 with $\delta = 10^4$.

7. Experimental Verification

In figure 6 the numeric result and the measurement of a specific case are shown. The experimental device was builded in the Laboratory of Fluid Mechanics and Enviromental Engeneering Institute (Montevideo, Uruguay). The corresponding values for the variables involved are the following: H^*_t=48.06 m, H^*_{go}=10.33 m, a=1157 m/s, L=38.01m, D=0.0143m, $\forall_0$=102 cm^3, f=0.0326. The polytropic coefficient was supposed 1.2.

As can be noticed, the numeric model reproduces the experience adequately.

8. Conclusions

The main conclusion which is derived from the above is the noteworthy effect which comes about from incorporating the acoustic effects (finite "a") in the calculation of overpressures produced by the dynamic compression of an air pocket inside of a hydraulic pressure system. This may lead to significative savings in the designing of aqueducts and liquid pipes in general.

9. Nomenclature Used in this Paper

a = wave propagation velocity (m/s)
D = diameter of pipe (m)
f = Darcy-Weisbach's coefficient
g = gravity's acceleration (m/s^2)
H^* = $p^* /\gamma + z$ (m)
H^*_g = p^*/γ (m)
H^*_{go} = initial absolute head in the gas (m)
H^*_t = $H_t + 10.33$ (m)
L = length of the liquid column (m)
n = polytropic coefficient of the air development
p^* = absolute pressure in the liquid (N/m^2)
p^*_g = absolute pressure in the gas (N/m^2)
p^*_{go} = absolute pressure in the gas at t=0 (N/m^2)
s = coordinate measured according to the pipe's axis, with origin in the valve (m)
t = time (s)
t_M = instant in which the maximum pressure takes place in the air pocket (s)
V = the liquid's mean velocity (m/s)
z = altitude from the middle point in the pipe measured from the level of reference (m)
α = the angle between the axis of the pipe (where the air pocket is located) and the horizontal line (*)

γ = the liquid's specific gravity
ΔH^* = H^*_t - H^*_{go}
$\forall$ = the air pocket's volume

(*) This variable is used when the length variation of the liquid column is considered. In the cases treated in the present paper, that variation is not considered.

10. References

Martin, C., "Entrapped air in pipelines". *Proceedings of the Second International Conference of Pressure Surges*, BHRA, London, England, 1976

Cabrera, E., Abreu, J. y Vela, A., "Influence of the liquid length variation in hydraulic transient". *ASCE, Journal of Hydraulic Engineering*, Vol. 118, N°12, Dic. 1992.

Guarga, R. Lorenzo, E. y Acosta, A., "Aire atrapado en una tubería. Análisis de dos modelos". *Proceedings of the XVI Latin American Congress of IAHR*, Vol 1, pg. 355-367. Santiago, Chile, Nov. 1994.

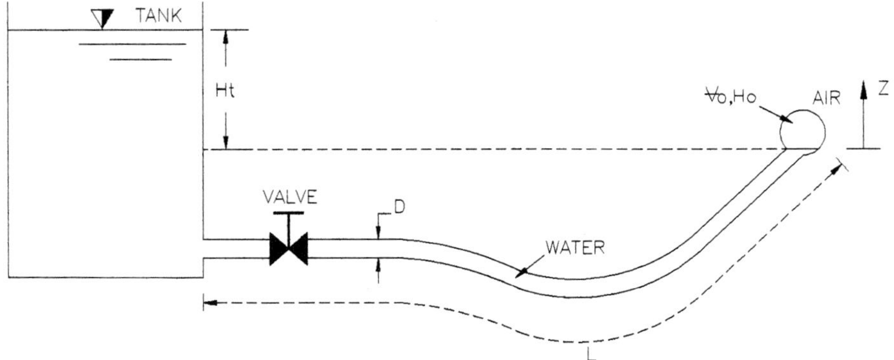

Figure 1 - Scketch of the system analized.

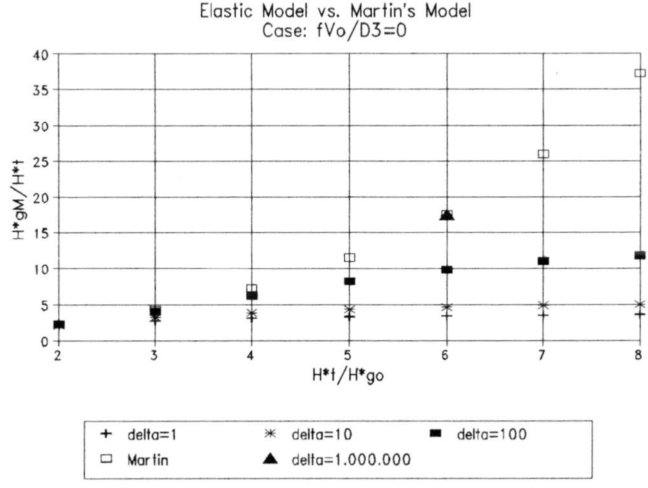

Figure 2 - Compared results from Elastic Model and Martin's Model. Case $fV_o/D^3=0$ and n=1.2

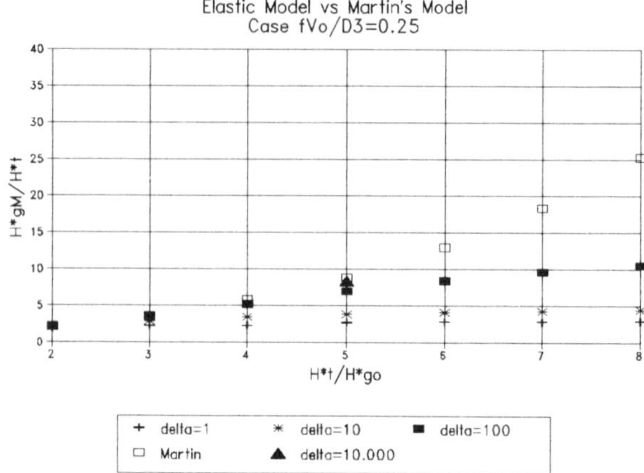

Figure 3 - Compared results from Elastic Model and Martin's Model. Case $fV_o/D^3=0.25$ and n=1.2

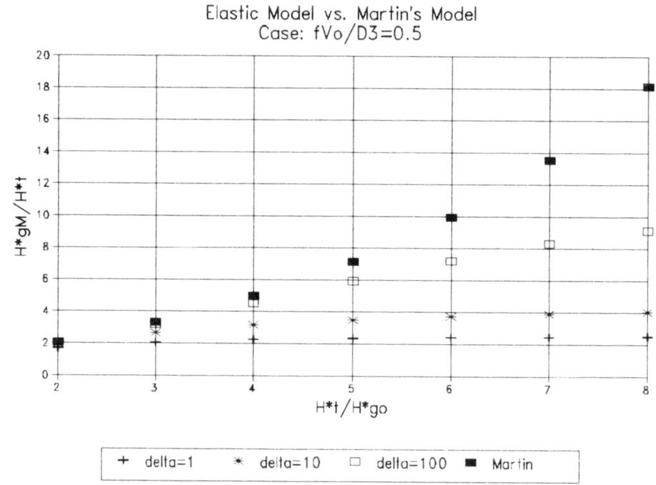

Figure 4 - Compared results from Elastic Model and Martin's Model. Case $fV_o/D^3=0.50$ and n=1.2

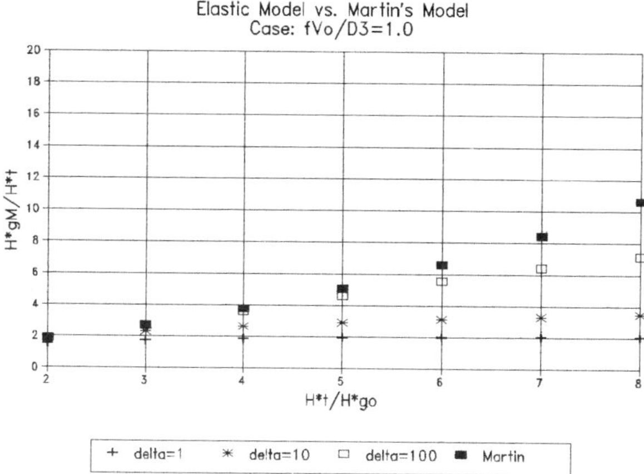

Figura 5 - Compared results from Elastic Model and Martin's Model. Case $fV_o/D^3=1$ and n=1.2

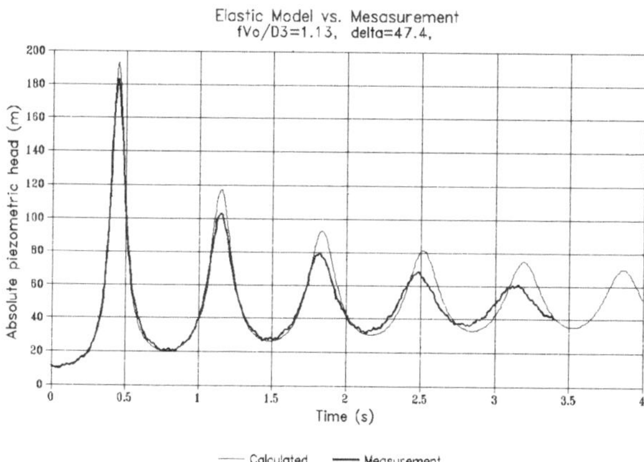

Figura 6 - Record data and numerical result, at s = L.

GENERALIZATION OF PUMP STATION BOUNDARY CONDITION IN HYDRAULIC TRANSIENT SIMULATION

IZQUIERDO, J., IGLESIAS, P., ESPERT, V. AND FUERTES, V.
Unidad Docente Mecánica de Fluidos.
Departamento de Ingeniería Hidráulica y Medio Ambiente.
Universidad Politécnica de Valencia. Apartado de Correos 22012.
46071 VALENCIA (Spain).

Abstract

In this paper, a generalized treatment of the equations describing the behavior of a pumping station is presented. It enables the simulation of many reasonable combinations of elements within a pumping station, by using a single routine and one set of state data suitably maintained. Thus, it represents a unified algorithm massively solved that easily accommodates to virtually every condition in a pumping station. Besides, the interpolating method it uses allows to incorporate manufacturers information into dimensionless curves helping plausibly to produce more real results. The interest of the paper is mainly computational. Nevertheless, the scheme we propose enables the joint consideration of a wide range of elements in a pumping station without the need of including short reaches between them. In this way, it is claimed to represent a good improvement for existing packages to model transient analysis in networks.

1. Introduction

Owners and manufacturers of water distribution systems have to face frequently problems with very different time scales. These problems are accommodated by means of specialized computer programs dedicated to solve the problem from the most simple perspective. Pumps, pipes and valves are sized and located using steady-state analysis. Pump operating values and reservoir capacities are determined by using extended period simulation. And all thickness of pipes, the sizing and location of protection devices are decided by using the rigid o the elastic models. In Abreu et al. (1996a and 1996b) boundaries separating the ranges of applicability of these models are studied for several practical systems, showing that a great deal of unsteady conditions can easily be analyzed by means of the rigid model. Nonetheless, there are situations in which the elastic model is unavoidable. Computer

E. Cabrera et al. (eds.), Hydraulic Machinery and Cavitation, 720–728.

programs dedicated to the elastic model are complex and time-consuming and force to introduce system simplifications than can spoil the results they produce.

It is a well-known fact that two partial differential equations, namely, the continuity equation and the momentum equation, govern the behavior of flow within pressurized pipelines (Chaudhry, 1987, Wylie, 1993). At the same time, simpler models are special cases of the unsteady behavior. The Method of Characteristics (MOC) provides a systematic way of evaluating the dynamic and continuity equations. In this way, the full range of flow conditions (elastic, rigid, quasi-steady and steady) within a pipeline can be modeled by using the MOC (Karney, 1990). However, when trying to adequately represent external boundary conditions such as pumps, reservoirs, in-line valves, surge tanks, pressure relief valves,..., and, especially, practical combinations of these, some difficulties arise. One has either to write specific routines to deal with every particular combination of elements, -what generates a huge amount of code-, or to use short reaches to link the different components of a complex boundary condition in order to work with component-dedicated routines.

In this paper, after reviewing some aspects of the formulation of boundary conditions, we emphasize on the case of a pumping station, as representing the more complex setting, and describe a procedure to deal with practical associations of devices, especially around a pump, in a combined way. It enables the simulation of many reasonable combinations of elements within a pumping station by using a single routine and one set of state data suitably maintained. Thus, it represents a unified algorithm massively solved that easily accommodates to virtually every condition in a pumping station.

2. Boundary conditions

From a computational point of view, boundary conditions may be classified, depending on how many nodal values are needed to describe its hydraulic behavior, as follows:

- Boundary elements depending upon the hydraulic condition in only one point, that is to say, elements in which the flow through them is related to the piezometric head in only one internal point of the system. We will call them *parallel elements* (PE). A valve or orifice discharging to the atmosphere, a reservoir or a surge tank connected to one point of the pipe, are examples of PEs. In a simplified way its constitutive equation can be described by a function of Q_e

$$H(\mathrm{P}) = \mathbf{F}(Q_e) \qquad (1)$$

  Q_e, being the flow through the element and H the piezometric head at the insertion point, P.

- Elements whose hydraulic behavior is defined by conditions in two points P_1 and P_2. In these elements the flow depends upon the pressure gradient, ΔH_e, between the two points. We will note these elements as *series elements* (SE). Examples of SEs are, booster pumps and, in general, in-line elements, such as valves of several types. Their constitutive equation can be given in a simplified way by

$$\Delta H_e = H(\mathrm{P}_1) - H(\mathrm{P}_2) = \mathbf{F}(Q_e) \qquad (2)$$

The equations describing the hydraulic behavior of any of these elements depend on its characteristics and range from simple relations to systems of algebraic and/or differential equations. Compressible flow models generally treat boundary conditions as isolated or localized cases of incompressible flow and use lumped parameters to describe its behavior. These equations, when considered together with the characteristics equations, allow a step by step solution of the problem.

Analyzing, from practical grounds, the combinations of elements within a boundary condition, one can see that a configuration made out of one or several SEs associated with possibly one upstream PEs and another downstream PEs seems general enough.

Let us, then, consider a generic configuration of this type, like in Figure 1. Let A denote the association (perhaps fictitious) of the SEs in the boundary condition and PE1 and PE2 the possible parallel elements linked respectively upstream and downstream to A.

We will use the following notation: Q = flowrate through A; if a rated value Q_r is available, $q = Q/Q_r$ = relative flowrate; Q_{e1} = flowrate in or out PE1; Q_{e2} = flowrate in or out PE2; Q_1 = flowrate upstream the boundary; Q_2 = flowrate downstream the boundary; H_u = piezometric head upstream the boundary; H_d = piezometric head downstream the boundary.

The following equations are relevant to the boundary:

- The characteristics equations

$$\begin{aligned} Q_1 &= C_p + H_u / B_1 \\ Q_2 &= C_n + H_d / B_2 \end{aligned} \tag{3}$$

where

$$B_1 := \frac{a_1}{gA_1}, \quad B_2 := \frac{a_2}{gA_2} \tag{4}$$

being a_i = wave celerity; A_i = pipe cross-sectional area; g = acceleration due to gravity and C_p and C_n are the terms conveying the information from nearby points from the previous time step, according to the MOC. For the purpose of this paper they are constant and are calculated every time step. From (3) one can obtain

$$H_d - H_u = B_2 Q_2 + B_1 Q_1 - (B_2 C_n + B_1 C_p) \tag{5}$$

- The energy equation requires that the difference between the discharge head and the suction head equals the difference between the dynamic head of the pump ΔH_p and the total head loss, ΔH_e, at the resistive components (check valve, delivery valves,...) of A:

$$H_d - H_u = \Delta H_p - \Delta H_e \tag{6}$$

The dynamic head of the pump can be represented using standard methods (see Chaudhry, 1987 and Wylie et al., 1993) by

$$\Delta H_p = H_r h = H_r(\alpha^2 + q^2) f(\theta) \tag{7}$$

where H_r = rated pump head, α = relative pump speed, defined by $\alpha = \omega/\omega_r$, being ω = pump rotational speed and ω_r = rated pump rotational speed, $\theta = \mathrm{atan2}(\alpha/q)$ and $f(\theta)$ the dimensionless head of Suter curves (Marchal et al. 1965).

The head loss across the SE A can be characterized by

$$\Delta H_e = K_e Q|Q| \tag{8}$$

where K_e = joint head loss coefficient of the resistive components of A.

After equating (5) and (6) and substituting (7) and (8) we get

$$H_r(\alpha^2 + q^2)f(\theta) - B_2 Q_2 - B_1 Q_1 - K_e Q|Q| + U = 0 \tag{9}$$

where

$$U := B_2 C_n + B_1 C_p \tag{10}$$

- The torque equation, after integration using a second-order trapezoidal approximation, gives

$$\beta + \beta_0 = -\lambda(\alpha - \alpha_0) \tag{11}$$

being

$$\lambda := \frac{2I\omega_r}{M_r \Delta t} \tag{12}$$

where β = relative torque, ratio between torque M and rated torque M_r; β_0 = relative torque at the beginning of the time step; α_0 = relative rotational speed at the beginning of the time step; I = inertia of the impeller, entrained fluid and rotating parts of the pump and Δt = time step used in the MOC. The dimensionless Suter curve for the torque, $\varphi(\theta)$, allows to write β as

$$\beta = (\alpha^2 + q^2)\varphi(\theta) \tag{13}$$

Combining (11), (12) and (13) one can obtain

$$\lambda\alpha + (\alpha^2 + q^2)\varphi(\theta) + V = 0 \tag{14}$$

where

$$V := \beta_0 - \lambda\alpha_0 \tag{15}$$

The presence of the PEs forces to consider additional equations. For instance, for the PE1:

- The continuity equation

$$Q_1 + Q_{e1} = Q \tag{16}$$

- The constitutive equation of the PEs, of type (1), expressed as

$$H_1 = W_1^{(0)} + g_1(Q_{e1}) \tag{17}$$

where $W_1^{(0)}$ is a constant depending on the values that certain magnitudes inherent to the PE1 take at the beginning of the time step and $g_1(Q_{e1})$ represents the head loss at the junction of the PE1 and the pipeline. Combining the first equation in (3), representing the positive characteristic, with (17) one can obtain

$$B_1 Q_1 + g_1(Q_{e1}) + W_1 = 0 \tag{18}$$

where

$$W_1 = W_1^{(0)} - B_1 C_p \tag{19}$$

and, taking into account the expression for the losses at the intake

$$g_1(Q_{e1}) = K_1 Q_{e1} \left| Q_{e1} \right| \tag{20}$$

and equation (16), one has

$$B_1 Q_1 + K_1 (Q - Q_1) \left| Q - Q_1 \right| + W_1 = 0 \tag{21}$$

An analogous treatment for the PE2 gives

$$B_2 Q_2 + K_2 (Q - Q_2) \left| Q - Q_2 \right| + W_2 = 0 \tag{22}$$

where B_2 and W_2 are defined similarly.

After reordering, equations (9), (14), (21) and (22) can be written

$$\begin{array}{llllllll}
\lambda\alpha & + (\alpha^2 + q^2)\varphi(\theta) & & & & + V & = 0 \\
H_r(\alpha^2 + q^2) f(\theta) & & - K_c Q|Q| & - B_1 Q_1 & - B_2 Q_2 & + U & = 0 \\
 & K_1 (Q - Q_1)|Q - Q_1| & & + B_1 Q_1 & & + W_1 & = 0 \\
 & K_2 (Q_2 - Q)|Q_2 - Q| & & & - B_2 Q_2 & + W_2 & = 0
\end{array} \tag{23}$$

Equations (23), whose coefficients have been defined above, constitute a system of non-linear equations with unknowns α, q or Q, Q_1 and Q_2, that can easily be solved by the Newton-Raphson method. The following facts should be observed.

1) The different terms in (23) are not equally relevant in any circumstance. For instance, the first terms in the first two equations are exclusively for pumps. If the SE has no pump the first equation as well as the second term of the second equation must be ignored. Also, if there is no upstream PE1 or PE2, the corresponding term of losses must not be considered. Moreover, according to the current state of the present devices, some terms can shift between being relevant or irrelevant. As an example, the first equation in not compatible with a pump working on its operating point. The theoretical generality of the present modeling seems to vanish after these considerations. Nevertheless, by defining and suitably managing *state variables* M_{ij}, mimicking the real behavior of the different devices, (23) can be written

$$\begin{array}{llllllll}
M_{10}\lambda\alpha & + M_{11}(\alpha^2 + q^2)\varphi(\theta) & & & & + M_{15} V & = 0 \\
M_{20} q & + M_{21} H_r(\alpha^2 + q^2) f(\theta) & - M_{22} K_c Q|Q| & - M_{23} B_1 Q_1 & - M_{24} B_2 Q_2 & + M_{25} U & = 0 \\
M_{30}(Q - Q_1) & & + M_{32} K_1 (Q - Q_1)|Q - Q_1| & + M_{33} B_1 Q_1 & & + M_{35} W_1 & = 0 \\
M_{40}(Q_2 - Q) & & + M_{42} K_2 (Q_2 - Q)|Q_2 - Q| & & - M_{44} B_2 Q_2 & + M_{45} W_2 & = 0
\end{array} \tag{24}$$

Along the transient, depending on the real state of each device, dictated by the current situation, the different state variables are suitably defined and redefined as 0 or 1, thus accommodating equations (24) to the joint current state of the boundary condition. Also, the other parameters in (24) are suitably accommodated. As an example, let us suppose a pump with check valve, fed by a reservoir and protected by a pressure relief valve. Let us also suppose that the transient evolves from the regime state and is triggered by a power failure. The successive values of the state variables can be expressed as binary numbers of 16 digits corresponding to the 16 coefficients M_{ij} in (24), ordered from left to right (observe that M_{10} is always 1). Some instances are:

- Pump with α given: 01-010011-0010-1000
- Pump running down: 11-010011-0010-1000
- Open check valve: 00-001011-0010-1000
- Closed check valve: 00-100000-0010-0011
- Passive PE: 00-000011-0010-1000
- Active PE: 00-000011-0010-0111

More simple representations can be devised. But, these state variables, that exhibit several redundancies, are shown to be computational efficient. Moreover, these redundancies can be intelligently used to check the correctness of certain states. The *state numbers* of the different devices within the boundary condition are logically operated in order to produce the joint state number that fully describes the ensemble and tailors system (24) to the current scenario.

2) Functions $f(\theta)$ and $\varphi(\theta)$, that are relevant only in the presence of a pump, may be handled under any of the existing ways of working with differentiable expressions for functions given by points. Specifically, we claim that the use of cubic splines is a good choice. Given a digitization of these curves, a simple routine allows to obtain a straightforward differentiable representation that can be readily used in the Newton-Raphson algorithm. Given any value of the specific speed both the values of the function and its derivative are readily available. In addition, information provided by manufacturers, if available, can be suitably threaded with the digitization of the Suter curves, thus plausibly producing more real results. In case of a boundary condition without pump, coefficients M_{11} and M_{21} are set to zero, making irrelevant its corresponding terms in the first two equations.

3) Three non-existing terms in (23) have been added in (24) and endowed with specific state variables, namely M_{20}, M_{30} and M_{40}. The purpose, -observe they are linear-, is clear. They allow for direct assignment to variables q, Q_1 and/or Q_2, depending on the current state of the ensemble, turning into identities their corresponding equations. For example, if a check valve is closed, the second equation will have coefficients shifted to 100000, thus it will force $q = 0$, what translates correctly the real behavior of the boundary. Similarly, if a PE is inactive or, equivalently, there is no flow in or out of it, Q_1 must equal Q. This is achieved by shifting the corresponding equation coefficients to 1000. Observe that the responsibility of making this assignments could be left to the quadratic terms (the loss terms). Nevertheless, it would prevent from using the Newton-Raphson method. In effect, the Jacobian would have determinant zero and, as a consequence, it would not be locally inversible, what is a necessary condition for the Newton-Raphson method.

4) All the above state variables are not a burden for the procedure. Observe that when an equations is trivialized, -with the purpose of assigning directly a value to certain variable-, it is linearized, as well. It is a well-known fact that the Newton-Raphson method gives the solution after the first iteration when applied to a linear system. It can be checked that, in any case, no more than three iterations are needed and, frequently, the solution is obtained after the first iteration.

5) The presence of certain PEs, such as air vessels, forces some extra elaboration. Let us consider a vertical air vessel as an example of PE1. In this case, equation (19) can be obtained as follows. First, the equation relating the flow in or out of the vessel to the elevation of the water in its interior, being S_1 the vessel cross-sectional area,

$$Q_1 = -S_1 \frac{dz_1}{dt} \tag{25}$$

is integrated by means of the rectangle rule to obtain an approximated value for z as a function of z_{10} and Q_{10}

$$z_1 = z_{10} - \frac{\Delta t}{S_1} Q_{10} \tag{26}$$

Then, the geometrical characteristics of the vessel allow for the calculation of the volume of air within the vessel

$$\forall_1 = S_1(hc_1 - z_1) \tag{27}$$

where hc_1 is the vessel height.
The evolution of the air within the vessel, assuming polytropic exponent n, gives the absolute pressure of the air

$$p_1 = p_{1R}\left(\frac{hc_1 - z_{1R}}{hc_1 - z_1}\right)^n \tag{28}$$

Suffix '1R' indicates regime values.
The energy equation can be written

$$H_1 = z + hb_1 + z_1 + \frac{p_1}{\gamma} - h_v^* + g_1(Q_1) \tag{29}$$

where z = elevation of the insertion point, hb_1 = distance from the vessel base to the axis of the pipe, h_v^* = absolute atmospheric pressure. The first terms in this expression are $W_1^{(0)}$ in (19). Solving system (23) allows one to obtain a value for Q_{e1} that is copied into Q_{e10} and used to iterate in (26). When the difference between Q_{e1} and Q_{e10} is smaller than certain tolerance, the final variables assignment is performed.

3. Model validation

Since the intent of this approach is to improve the management of transient simulation models instead of improving their capability to predict measured results from field tests, we will validate it by comparing its performance with a universally validated code. We will choose the SURGE5 package (Wood et al., 1967). It is worth noting here that the interpolation technique used in both codes is different. As mentioned above we use two-dimensional cubic splines to interpolate on the Suter curves, which allow to merge the manufacturer data with the non-dimensional data provided by the Suter curves.

The simple installation in Figure 2 has been used to compare the results. Superimposed graphs of the pressure downstream the pump after a power failure, given by both approaches are shown in Figure 3. It can be observed the close agreement between both models. Under the authors viewpoint it constitutes a fair validation of the model presented in this paper. Other scenarios have also been tested and the deviations between both models have been always negligible from an engineering point of view.

4. Conclusion.

A generalized treatment of the equations describing a boundary condition in transient analysis has been presented. It enables the simulation of many reasonable combinations of elements within a pumping station and other settings as well, by using a single routine and one set of state data suitably maintained. In addition, the two-dimensional-splines-based interpolation technique it uses allows to incorporate manufacturers data to dimensionless data provided by Suter curves. This helps produce more realistic results.

The generalized boundary condition may be implemented in a simple and inexpensive way in transient simulation packages, thus avoiding the need of short reaches, which naturally make a model highly resources-consuming.

Finally, the generalized boundary condition has been tested with the SURGE5 approach and has shown to be in good agreement with this code results.

5. References.

Abreu J., Cabrera E., Izquierdo J. and García-Serra J. (1996a). Flow Modeling in Pressurized Systems, From Water Hammer to Steady State I. Fundamentals. Submitted to the Journal of Hydraulic Engineering of ASCE for possible publication.

Abreu J., Cabrera E., Izquierdo J. and García-Serra J. (1996b). Flow Modeling in Pressurized Systems. From Water Hammer to Steady State II: Examples. Submitted to the Journal of Hydraulic Engineering for possible publication.

Chaudhry, H.M. (1987). *Applied Hydraulic Transients.* Van Nostrand Reinhold, New York, N.Y.

Karney, B. (1990). Energy relations in transient closed-conduit flow. *J. of Hydr. Engrg.,* ASCE, 116(10), 1180-1196.

Marchal, M., Flesh, G. and Suter, P. (1965). The calculation of Waterhammer problems by means of digital computer. *Proc. International Symposium on Water hammer in pumped storage projects.* ASME. pp- 168-188.

Wood , D.J., Dorsch, R.G. and Lightener, C. (1965). Wave plan Analysis of Unsteady Flow in Closed Conduits. *Proc. ASCE J. Hyd. Div.,* Vol 92, No HT2, pp 83-110.

Wylie, E.B. and Streeter, V.L. (1993). *Fluid transients in Systems.* Prentice-Hall, Englewood Cliffs, New Jersey.

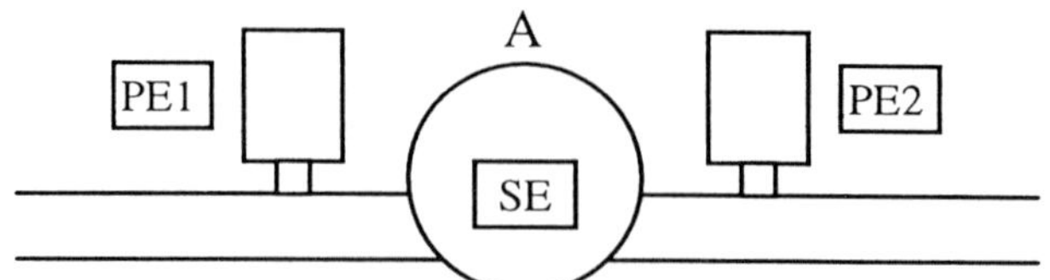

Figure 1. General boundary condition.

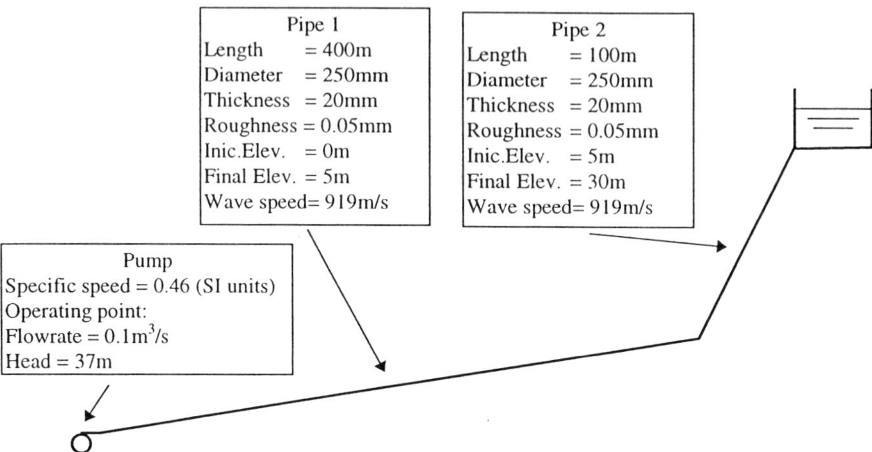

Figure 2. Test installation.

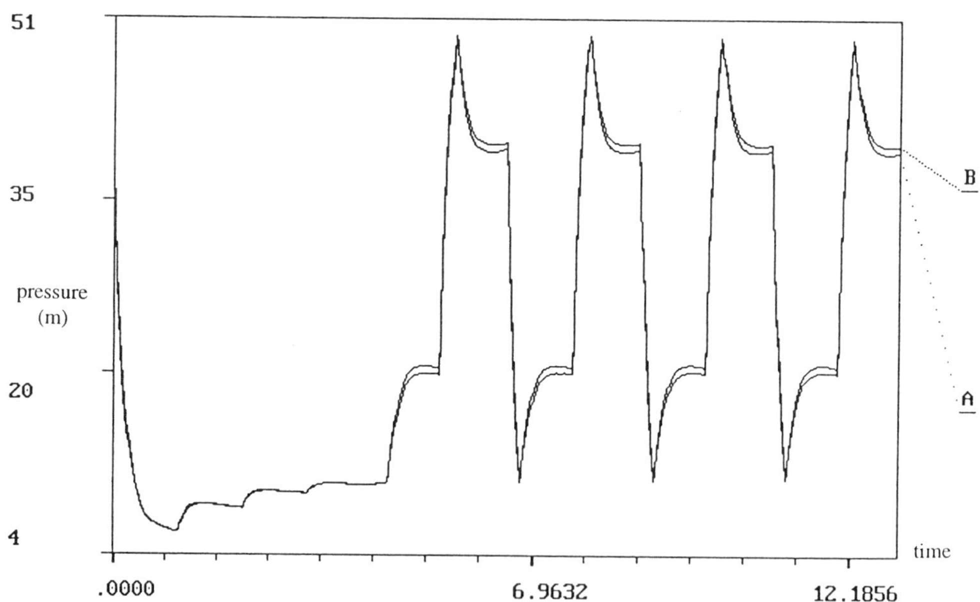

Figure 3. Comparison of results: A-Current approach, B-SURGE5.

ANALYSIS OF UNSTEADY CHARACTERISTICS OF FLOWS THROUGH A CENTRIFUGAL-PUMP IMPELLER BY AN ADVANCED VORTEX METHOD

K.KAMEMOTO, H.KURASAWA, H.MATSUMOTO, Y. YOKOI
Department of Mechanical Engineering and Materials Science
Yokohama National University
156 Tokiwadai Hodogaya-ku Yokohama, 240 Japan

Abstract

In the present study, the applicability of the vortex methods is extended for the analysis of unsteady characteristics and occurrence of unsymmetrical flow through a centrifugal impeller of a pump. In this numerical investigation, two-dimensional unsteady features of the whole flow-field are solved without introducing the periodicity of the blade-to-blade stream, by using an advanced vortex method. As an example of applications of the present method, unsteady flows through a centrifugal impeller consisting of simple configuration of blades were calculated, and the results show how the separation of boundary layer develops into a strong vortex structure and how the vortex is periodically convected downstream and produces an oscillatory flow in a blade-to-blade passage.

1. Introduction

In relation with the problems concerning vibration and noise of turbomachinery, the development of an easy-to-handle numerical technique to predict unsteady flow characteristics of an impeller has been desired since before, for design and improvement of operation-performance. The present study has been carried out in order to develop a useful method for analysis of complex, unsteady flows through an impeller, which does not necessarily need use of a computer with large capacity for the numerical procedures.

In the past thirty years, the vortex methods have been developed and applied for analysis of complex vortical flows and simulation of unsteady flows related

E. Cabrera et al. (eds.), Hydraulic Machinery and Cavitation, 729–738.

with problems in various engineering fields, because they are standing on simple algorithm based on physics of flows, and their numerical stability is usually quite well, and it should be noted here that grid generation in the flow field is not necessary. On the recent vortex methods, Sarpkaya(1989) presented a comprehensive review of various types of vortex methods, and Kamemoto(1995) explained attractive features of the vortex methods in the application into various unsteady separation flows. Gao and Kamemoto(1993) numerically investigated a separated unsteady flow in a centrifugal impeller at a low rate of flow by using a vortex method. In their calculation, periodic boundary conditions for blades were introduced by which blade-to-blade streams were assumed to be the same each other, in order to simplify the problem, and they were succeeded in showing periodic flow separation from leading edges and periodic formation of vortex structures in the blade-to-blade flow.

In the present study, the applicability of the vortex method is extended for analysis of the unsteady characteristics of flow separation and occurrence of rotating stall of various types in flows through a centrifugal impeller of a pump. In this method, two-dimensional unsteady characteristics of the whole flow-field are solved by an advanced vortex method, without introducing the periodicity of blade-to-blade streams. One of the most interesting points in the present method is that unsteady development of separation regions and formation of separation vortices along each blade surface are individually and simultaneously investigated step by step of time. Therefore, the results obtained from the present calculations will provide designers and researchers with helpful information to understand appearance of un-axi-symmetrical characteristics of the whole flow through an impeller which may be thought as a sign of occurrence of rotating stall in a centrifugal impeller at a rather small flow rate. In this paper, the mathematical basis of the present vortex method is explained, and calculated results of typical unsteady flow characteristics of a centrifugal impeller are discussed, by showing flow patterns and variation of head coefficients with time elapsed after the start of impeller-revolution at a constant speed for the three cases of designed-, exceeding- and partial-discharges.

2. Method of numerical analysis

2.1. VORTEX METHOD

The vortex methods have been developed for numerical analysis of incompressible viscous and unsteady flows, and their applicability has been extended to various unsteady flows related with difficult problems in engineering fields. Governing equations of the advanced vortex method are based on the Navier-Stokes equation and the continuity equation which may be written in vector form

$$\frac{\partial \boldsymbol{u}}{\partial t} + (\boldsymbol{u} \cdot grad)\, \boldsymbol{u} = -\frac{1}{\rho}\, grad\, p + \nu \nabla^2 \boldsymbol{u} \tag{1}$$

$$div\, \boldsymbol{u} = 0 \tag{2}$$

If we now take the rotation of Eq.(1) and take account of Eq.(2), the vorticity transport equation for incompressible viscous flow is obtained

$$\frac{\partial \boldsymbol{\omega}}{\partial t} + (\boldsymbol{u} \cdot grad)\, \boldsymbol{\omega} = (\boldsymbol{\omega} \cdot grad)\, \boldsymbol{u} + \nu \nabla^2 \boldsymbol{\omega} \tag{3}$$

where the vorticity $\boldsymbol{\omega}$ is defined as

$$\boldsymbol{\omega} = rot\, \boldsymbol{u} \tag{4}$$

Here, Equations (3) and (4) are the governing equations of the vortex methods. As we can eliminate the vorticity $\boldsymbol{\omega}$ in Eq.(3) by substituting Eq.(4), it is easily understood that the velocity field represented by $\boldsymbol{u}$ can be calculated from Equations (3) and (4), without necessity of solving pressure p. This fact is one of the most attractive features of the vortex methods.

As previously explained by Wu and Thompson (1973) in their study of the integro-differential formulation of the Navier-Stokes equation, mathematical expression for the so-called generalized Biot-Savart law derived from integration of the vorticity definition equation(4) is given as

$$\begin{aligned} \boldsymbol{u}(\boldsymbol{r}) \;\; = \;\; & \frac{1}{\alpha} \int_V \frac{\boldsymbol{\omega}_o \times \boldsymbol{R}}{R^\beta} dV_o \\ & - \frac{1}{\alpha} \int_{S_1 + S_2} \left[\frac{(\boldsymbol{n}_o \cdot \boldsymbol{u}_o) \cdot \boldsymbol{R}}{R^\beta} \right. \\ & \left. + \frac{(\boldsymbol{n}_o \times \boldsymbol{u}_o) \times \boldsymbol{R}}{R^\beta} \right] dS_o \end{aligned} \tag{5}$$

here, subscript $[_o]$ denotes variable and integration at a location $\boldsymbol{r}_o$, and $\boldsymbol{n}_o$ denotes the normal unit vector at a point on a boundary surface. $\boldsymbol{R} = \boldsymbol{r} - \boldsymbol{r}_o$, $R = |\boldsymbol{R}| = |\boldsymbol{r} - \boldsymbol{r}_o|$, and $\alpha = 2\pi, \beta = 2$ for two dimensional flow, and $\alpha = 4\pi, \beta = 3$ for three dimensional flow.

In Eq.(5), the inner product $\boldsymbol{n}_o \cdot \boldsymbol{u}_o$ and the outer product $\boldsymbol{n}_o \times \boldsymbol{u}_o$ stand for respectively normal and tangential velocity components on the boundary surface, and they respectively correspond to source and vortex distributions on the boundary surface. Therefore, once the vorticity distributions are approximated by discrete vortices in the flow field, the configuration of the flow boundary is easily represented by distributions of source and /or vortex elements around it (Boundary Element Methods), which are directly related with the boundary conditions on the surface of the configuration. Eventually, it is concluded that the present vortex method seems essentially suitable for the investigation of moving boundary problems like a rotating impeller or oscillating bodies.

Once we introduce the velocity $\boldsymbol{u}$ estimated from the Biot-Savart law into Eq.(3), the unsteady transportation of vorticity can be analyzed by solving the vorticity

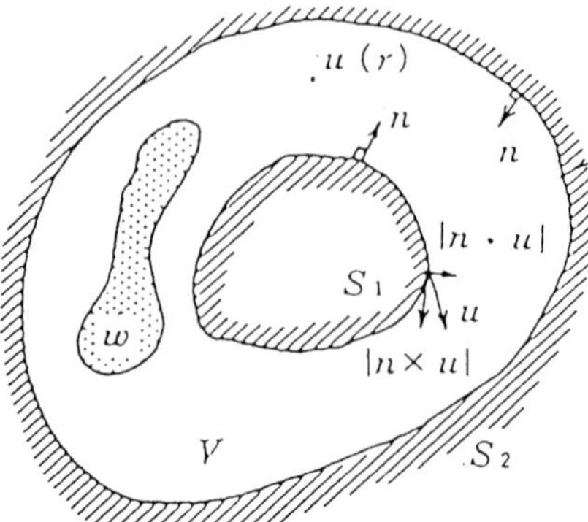

Figure 1. Flow field containing vorticity region.

transport equation. If we consider the vorticity of a fluid particle at a location $\boldsymbol{r}_o$ in the Eulerian coordinates at time t, the rate of variation of vorticity with time is expressed in a Lagrangian form

$$\left(\frac{d\boldsymbol{\omega}}{dt}\right)_o = (\boldsymbol{\omega} \cdot grad)_o \, \boldsymbol{u}_o + \nu \, (\nabla^2 \, \boldsymbol{\omega})_o \tag{6}$$

Therefore, writing the vorticity of the fluid particle at time t as $\boldsymbol{\omega}_o(t)$, we obtain an approximate expression of the transport of vorticity while the fluid particle moves during a small time interval dt as follows.

$$\begin{aligned} \boldsymbol{\omega}_o(t + dt) &= \boldsymbol{\omega}_o(t) + (\boldsymbol{\omega} \cdot grad)_o \, \boldsymbol{u}_o \cdot dt \\ &\quad + \nu \, (\nabla^2 \, \boldsymbol{\omega})_o \, dt \end{aligned} \tag{7}$$

In Eq.(7), the second term in the right hand side is the three-dimensional convection of vorticity, which always becomes zero for two-dimensional flow, and the third term is the rate of viscous diffusion of vorticity. If the Reynolds number of the flow is sufficiently high, the convection term is considered much larger than the diffusion term, and thus, the third term in Eq.(7) may be neglected in the computation. Furthermore, if the high Reynolds number flow is two-dimensional, Eq.(7) is approximated by the simple equation $\boldsymbol{\omega}_o(t + dt) = \boldsymbol{\omega}(t)_o =$ constant. Therefore, if we take a small sectional area ds for the fluid particle and the vorticity is assumed constant in this area, the two-dimensional fluid particle is thought a free vortex element which transports a constant circulation $\Gamma_o = \boldsymbol{\omega}_o ds$.

On the other hand, the motion of the fluid particle and the moving boundary is represented by a Lagrangian form of a simple differential equation.

$$\frac{d\boldsymbol{r}_o}{dt} = \boldsymbol{u}_o \tag{8}$$

Then, the trajectory of the fluid particle over a time step dt is approximately computed from the Euler method or the Adams-Bashforth method as follows.

$$\boldsymbol{r}_o(t+dt) = \boldsymbol{r}_o(t) + \boldsymbol{u}_o \cdot dt \tag{9}$$

$$\boldsymbol{r}_o(t+dt) = \boldsymbol{r}_o(t) + \{1.5\boldsymbol{u}_o(t) - 0.5\boldsymbol{u}_o(t-dt)\}dt \tag{10}$$

If the vorticity layers existing in the flow field are represented with discrete vortex distributions, not only a flow velocity at an arbitrary position but also the convective velocity of each discrete vortex can be calculated from Eq.(5) of the Biot-Savart law. Substituting the velocities into Eq.(7) and Eq.(8) or (9), the vorticity transport and trajectory of each discrete vortex over the time step are estimated, which provide new distributions of discrete vortices corresponding to the vorticity layers transported during the time step. Consequently, the iteration of the above procedure provides the basic scheme of the grid-free Lagrangian simulation of unsteady, incompressible and viscous flow.

In this study, a two-dimensional incompressible flow through a centrifugal impeller was treated. Assuming that thickness of boundary layers around surface of any blades is negligibly thin, we represent the configuration of blades by distributing vortex panels on the surface of each blade and we introduce a radial flow provided by a point source at the center of revolution, where the strength of vortex of each panel can be determined from normal velocity conditions on the surfaces of rotating blades. In order to represent boundary layer separation from both leading edge and trailing edge of a blade, we introduce discrete vortices which are usually called "vortex blobs" near the edges at every time step, where each blob has its viscous core and Gaussian-distribution of vorticity, and strength of circulation of each blob is given by relative velocity at the separation point and time step size used for the calculation.

2.2 THEORETICAL HEAD COEFFICIENT

In this study, pressure distributions along the blade surfaces were calculated from the unsteady Bernoulli's equation for the relative flow, and the torque T acting on the impeller was calculated by integrating moment of pressure force acting on the blade surfaces. Then, a theoretical head coefficient ϕ based on the torque T is estimated as

$$\phi = \frac{T\omega}{\rho Q U_2^2} \tag{11}$$

here, ω, Q and U_2 respectively denote the angular velocity of the impeller, the rate of flow and the peripheral speed of the impeller.

3. Results and discussion

3.1. CENTRIFUGAL IMPELLER

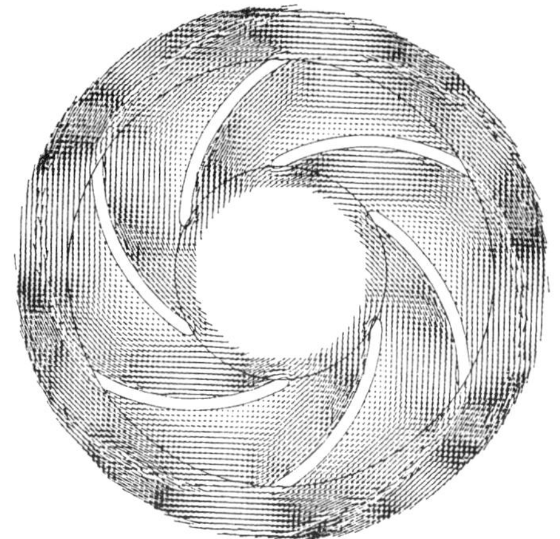

Figure 2. Flow Pattern at t=2.0, Ce=1.0.

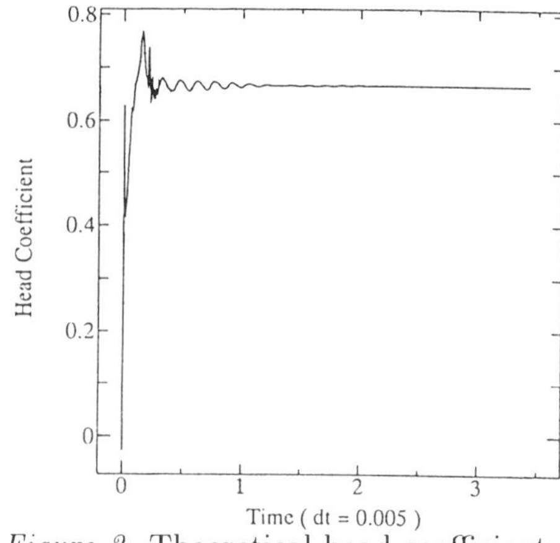

Figure 3. Theoretical head coefficient versus time elapsed (Ce=1.0).

In this study, in order to investigate the essential characteristics of unsteady flow, a two-dimensional impeller consisting of simple configuration was treated, and any effects of guide-vanes and volute casing were ignored. Defining the velocity V_o of the radial direction at the entrance of the impeller for the design point as the reference velocity and the outlet diameter D_2 as the reference length, the inlet diameter D_1 is given as $D_1/D_2 = 0.5$, inlet blade angle $\beta_1 = 25.2^o$, and outlet blade angle $\beta_2 = 30.0^o$. The impeller has six blades whose shape consists of semi-elliptical leading edge and one-circular-arc camber, and the thickness of blade is constant as six percent of the radius of the camber-arc. The angular velocity of the impeller ω and volumetric flow rate per unit breadth of the impeller at designed operation Q_o are normalized as $\widetilde{\omega} = \omega D_2/V_o = 8.5$ and $\widetilde{Q_o} = Q_o/(D_2V_o) = 0.5\pi$ respectively. The Reynolds number was given as $R_e = D_2V_o/\nu = 10^6$, and discharge ratio C_e is defined as $C_e = Q/Q_o$. Normalized non-dimensional time t is defined as $t = TV_o/D_2$, where T is the time elapsed after revolution of impeller starts at constant speed, and so one revolution of impeller takes the non-dimensional time of $t = 0.74$. The size of non-dimensional time step for time-marching calculation was $dt = 0.005$ which corresponds to 2.4 degrees in revolution. In this paper, calculated results corresponding to three cases of discharge ratios of $C_e = 0.6, 1.0$ and 1.5 are shown, and the unsteady flow characteristics at the lower discharge $C_e = 0.6$ are described in detail.

3.2. FLOW CHARACTERISTICS AT DESIGNED OPERATION AND EXCEEDING-DISCHARGE OPERATION

Figure 2 shows instantaneous velocity-vector distribution of the flow through the impeller at $t = 2.0$ in the case of design operation $C_e = 1.0$. It is clarified that flow separation is not detected around any leading-edges of the blades and there are

Figure 4. Flow Pattern at t=2.0, Ce=1.5

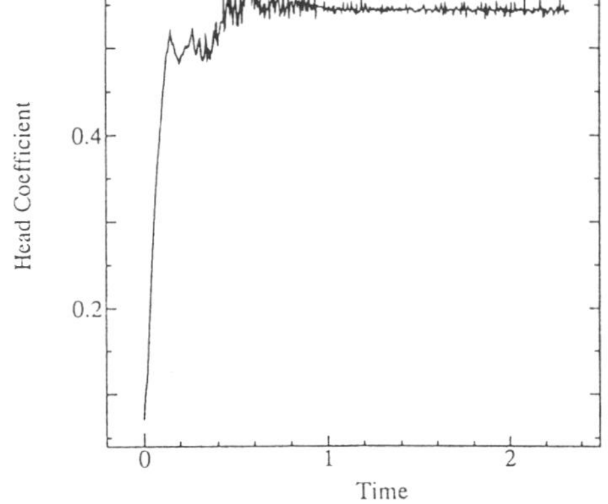

Figure 5. Theoretical head coefficient versus time elapsed (Ce=1.5).

no distinguishable differences among the flows around individual blade, whereas the whole flow field were calculated without introducing the approximation of periodicity of the blade-to-blade stream.

Figure 3 shows variation of theoretical head coefficient versus time elapsed after start of revolution of impeller at the designed discharge. It is interesting that initial oscillation of the head coefficient appears and rapidly damps with increase of time, where the non-dimensional period of the head oscillation is 0.144. It has been confirmed from the present numerical investigation that the oscillation is corresponding to periodic velocity fluctuations of the flow discharged from the impeller which is caused by interaction between initial strong vortices shed from each blade and trailing edges of rotating blades.

Figure 4 shows instantaneous velocity-vector distribution of the flow through the impeller at $t = 2.0$ in the case of larger discharge $C_e = 1.5$. It is seen that flow separation from leading edge of each blade appears and it forms narrow separation bubble over the pressure side surface of each blade. However, there are no distinguishable differences among the flows around individual blades, which means that axi-symmetrical flow is still kept even in the case of discharge larger than the designed.

Figure 5 shows variation of theoretical head coefficient versus time in the case of larger discharge. In this case, random fluctuation of the head coefficient due to flow separation from leading edges becomes dominant, the initial periodic oscillation of head coefficient is not detected clearly.

3.3 UNSTEADY FLOW CHARACTERISTICS AT PARTIAL DISCHARGE OPERATION

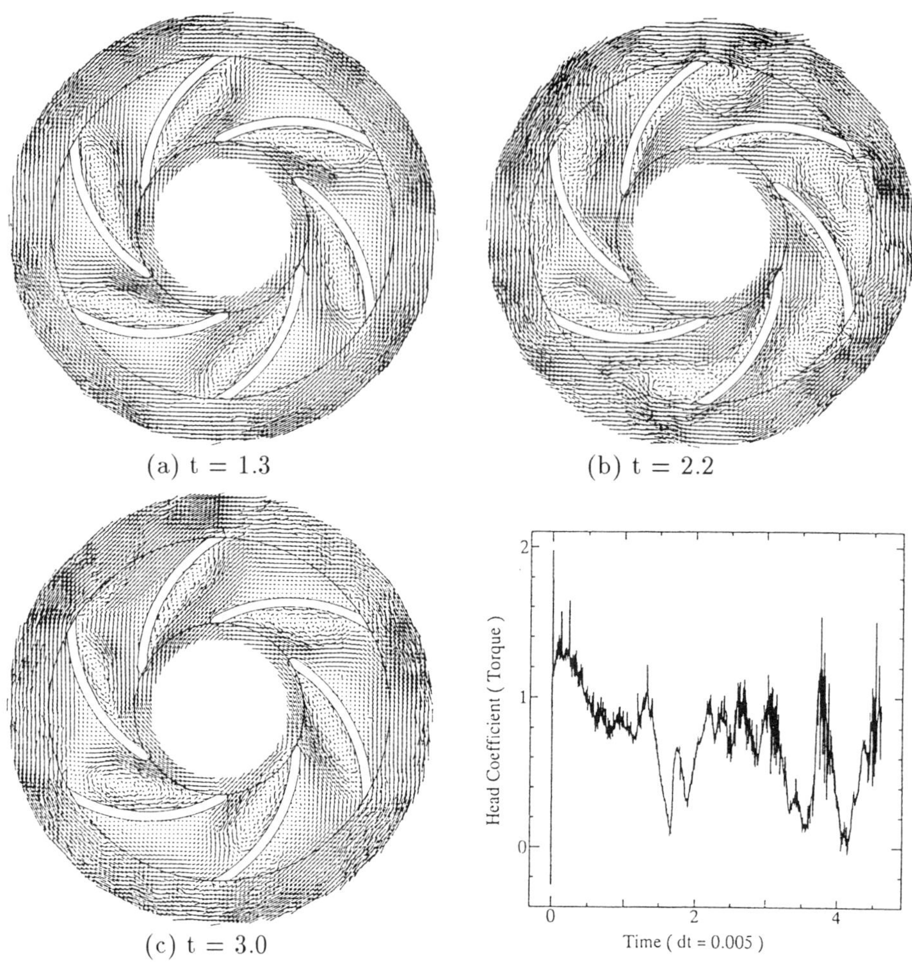

Figure 6. Flow Patterns at Ce = 0.6

Figure 7. Theoretical head coefficient versus time elapsed (Ce = 0.6)

Figure 6 shows a set of instantaneous flow-patterns at different times elapsed in the case of smaller discharge $C_e = 0.6$. It is clearly demonstrated that a separation babble is formed due to the flow separated from a leading edge of each blade and develops into a large vortex structure along its suction-side surface as shown in Fig.6(a), and then the large vortex is shed into the discharge flow and at the same time, a new separation bubble appears downstream of the leading edge as shown in Fig.6(b), and it develops into another large vortex as shown in Fig.6(c). One can find that there exist differences of bubble-scale and vortex-structure among individual blade-to-blade streams, which tend to disturb the axi-symmetric feature of the flow through the impeller.

Figure 7 shows variation of theoretical head coefficient acting on the impeller

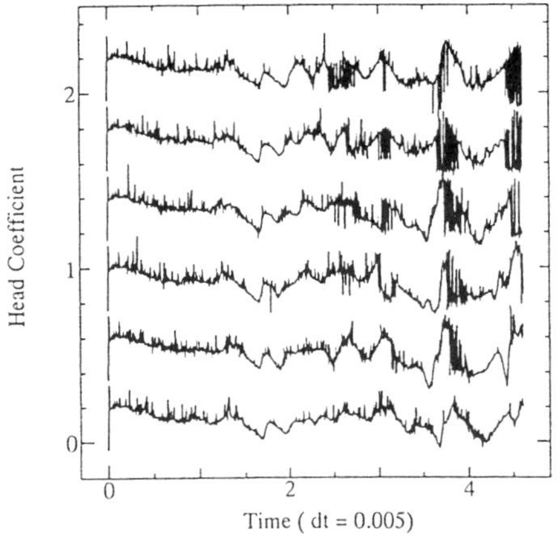

Figure 8. Theaoretical head coefficients of individual blades versus time elapsed (Ce = 0.6).

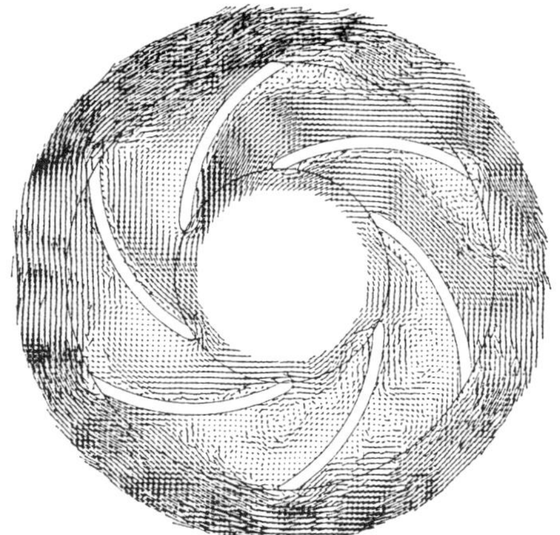

Figure 9. Unsymmetrical flow pattern at t = 4.55 Ce = 0.60.

versus time elapsed after start of revolution of impeller at a smaller discharge $C_e = 0.6$. It is known from this figure that considerable fluctuations of head coefficient appears in this case, which is corresponding to fluctuation of torque of impeller. The cause of the head-coefficient-fluctuation seems to be based on the interaction between the vortex structure in discharged flow and trailing edges of rotating blades, and, however, periodicity of the fluctuation is not clear as far as the present calculation is concerned.

Figure 8 shows variation of individual theoretical head coefficients calculated from pressure distribution on each blade. One can find differences in the aspects of head-coefficient fluctuations among the individual curves.

Figure 9 shows an example of un-axi-symmetrical flow patterns obtained in the present calculation, where non-dimensional time is 4.55. It is very interesting to find that there exist extremely strong vortex structures in two neighboring blade-to-blade streams which seem to block the flows between the blades, whereas there are no severe vortices in the other four streams of blade-to-blade and so the flows around the four blades look like smooth comparatively.

5. Conclusion

In this study, two-dimensional unsteady features of the whole flow-field of a centrifugal impeller are numerically investigated without introducing the periodicity of the blade-to-blade stream, by using an advanced vortex method. The results are pointed as follows.

(1) In the case of design operation, initial oscillation of theoretical head coefficient caused with interaction between initial vortices and trailing edges of rotating

blades appears just after the start of revolution of the impeller, and it rapidly damps with increase of time.
(2) In the case of an exceeding-discharge operation, flow separation from leading edges of blades occurs and narrow separation bubbles are formed along individual pressure-side surfaces.
(3)In the case of a partial-discharge operation, separation bubbles tend to develop into strong vortices along suction-side surfaces of the blades, and according to the unsteady characteristics of flow, severe fluctuation of head coefficient acting on the impeller takes place. At the same time, there exist differences of bubble-scale and vortex-structure among individual blade-to-blade streams, which tend to disturb the axi-symmetric feature of the flow through the impeller.

From the present study, it is confirmed that the vortex method is very convenient for investigation of unsteady characteristics of the flow through a centrifugal impeller. Then, it is expected that this method will be applied extensively to the complex problems of unsteady interaction between impeller-blades and guide vanes and/or a volute casing.

References

Gao, H., Kamemoto, K. 1993. Investigation of a separated unsteady flow in a centrifugal impeller at a low rate of flow by the vortex method. *Proc. 4th Asian International Conference on Fluid Machinery.*:582-587. Suzhou. China.

Kamemoto, K. 1995. On attractive features of the vortex methods. *Computational Fluid Dynamics Review 1995.* edited by M.Hafez and K.Oshima: 334-353. JOHN WILEY & SONS.

Sarpkaya, T. 1989. Computational methods with vortices - the 1988 Freeman scholar lecture. *J. Fluids Engng.* 111:5-52.

Wu, J.C., Thompson, J.F. 1973. Numerical solutions of time- dependent incompressible Navier-Stokes equations using an integro-differential formulation. *Computers & Fluids.* 1:197-215.

EXPERT SYSTEM FOR ANALYSIS OF PUMPED STORAGE SCHEMES

KOELLE, E.
Professor and Consulting Engineer
UNICAMP & Koelle Engineering - Brasil

ANDRADE, J.G.P. & LUVIZOTTO Jr., E
Doctor Assistant Professor
UNICAMP- University of Campinas, São Paulo - Brasil

Abstract

In Pumped Storage Schemes (PSS) it is necessary to know the hydrodynamic personality of the whole installation, in extensive and transient period of operation, in order to prevent inadequate pressure surges and to provide the operational optimization for energy production.

This paper presents a general computer model that provides the engineer with the means to simulate steady, transient and extensive period operation of the installation, considering also oscillatory excitations that induce flow instabilities. To perform the analysis, the complete hydraulic characteristics of the turbine-pump are based on Suter variables, adjusted by Fourier series, and PID feedback controller provides the necessary variation of parameters for stability analysis. Part load operation of machines that induce flow pulsation can be analyzed.

Examples are presented to illustrate the use of the expert system and to show the influence of parameter variations on the installation components.

E. Cabrera et al. (eds.), Hydraulic Machinery and Cavitation, 739–748.

1. Introduction

Several recently presented papers dealing with Pumped Storage Schemes (PSS) have shown the need to obtain the optimization of operational parameters of the plant in order to define the optimum condition of operation [1, 2]. The representation of the characteristics of the hydraulic machine by means of Fourier series using Suter variables has been shown to be most adequate for introduction into computer programs, because of the facility in obtaining any operational condition of the machine[3, 4].

The variation of the governor parameters in extensive time steady state operation simulations has allowed the determination of elements for the Adaptative Control System [5]. The method for the investigation of hydraulic networks personality through a computational program using MOC - Method of Characteristics was consolidated in a recent paper [6]. The personality of a hydraulic network is defined through its dynamic response to a set of stimuli and is represented by the set of nodal heads and flows in the elements of the network in a given period.

The determination of the hydraulic personality is essential in order to predict the dynamic behaviour of hydroelectric plants and becomes critical for pumped storage schemes, because of the multiplicity of operational possibilities when there is more than one machine connected to the same pressure conduit. The influence of the maneuvers done in a machine propagating to other ones must be investigated through the variation of the governor parameters, in order to avoid undesirable pressure surges and operational instability [7]. With the previous investigation, the adequate governor parameters are obtained and this paper shows that the Ziegler-Nichols (**ZN**) method is recommendable for the determination of such parameters, using the computational model in the time domain.

The determination of the transfer functions, in the frequency domain, for the analysis of complex systems, presents many difficulties [8] and requires the linearization of functions. Besides, the matricial methods for writing the network equations, either in the time domain or in the frequency domain, present numerical problems that lead to results that are inconsistent with the physical reality if the matrixes are not well conditioned. For these and other reasons, the investigations in the time domain, which are now eased by the advances in personal computers, allowing high processing velocities, are being optimized in order to allow the analysis and the operational monitoring of hydroelectric plants in real time. The advantages in the use of MOC for analysis in the time domain are illustrated at the end of this paper.

2. Characterization of the PSS Installation

In the simulation of a PSS, the computational model incorporates the dynamic response of the governor [6, 7] and the installation is represented by its elements (ENOS), linked by NODES. The full description of the installation requires the knowledge of the full characteristics of the hydraulic machine and of each of its elements, as illustrated in Figures 1 and 2, in which a pumped storage hydroelectric plant admits two possibilities of physical arrangement: a) System 1, Figure 1: The draft tubes of the three machines are connected to a single pressure conduit; b) System 2, Figure 2: The draft tubes are independent. The shown representation makes possible to establish easily the topology of the network by means of vectors that represent each ENO, containing four elements: 1) the number of the element; 2) the numerical code that indicates its type: pipe, tank, machine, . . . ; 3) the number of the upstream NODE; 4) the number of the downstream NODE.

Thus, in Figure 1, the vector (14, 3, 15, 16) is the pump-turbine (type code: 3), element number 14, between nodes 15 and 16. The set of these vectors defines the complete topological matrix of the installation.

3. The computational program

The writing of the equations of the network operation conditions is based on MOC to simulate the behaviour of the flow in the pipes and through which the equation for Node 9 is obtained:

$$Q_{P_E} = E_N - B_N H_P \tag{1}$$

where Q_{P_E} is the flow in the non-pipe ENO linked to the NODE, E_N, B_N (*node constants*) are calculated for each calculation step, and H_P is the nodal head.

The equation of the node involves the two unknowns Q_{P_E} and H_P that must be found by means of a second equation representing the nodal boundary conditions, that is to say, the equation of the NON-PIPE ENO linked to the nodes of the network.

Figure 3 shows the block diagram of the computational program, with parallel processing, that makes possible to analyze the PSS including the determination of the governor settings. The general writing of the equations for the Turbine-Pump, including the governor, is detailed elsewhere [6, 7, 9].

As one can observe in Figure 3, Module B - Personality Investigation presents the option to determine the governor settings and, in this case, the Ziegler-Nichols method is used. Once the governor settings are determined, it is possible to investigate the other operational conditions (transient, oscillatory and extensive) in order to obtain the personality of the installation, that is to say, the response to a given stimulus.

With the values of E_N and B_N calculated for all the nodes, Module C makes possible the parallel processing for the determination of the instantaneous conditions of operation of the NON-PIPE ENOS, through the equation:

$$H_{P_E} = H_{P_1} - H_{P_2} = E_E - B_E Q_{P_E} \quad \text{in which,} \quad H_{P_E} = H_{P_1} - H_{P_2} \tag{2}$$

in which H_{P_E} represents the head difference between the extreme nodes of the NON-PIPE ENO, and

$$E_E = \frac{E_{N_1}}{B_{N_1}} - \frac{E_{N_2}}{B_{N_2}} \qquad B_E = \frac{1}{B_{N_1}} + \frac{1}{B_{N_2}} \tag{3}$$

are the *ENO constants*, obtained at each calculation step as a function of the conditions in the extreme nodes 1 and 2.

Equation 2, together with the equation of the NON-PIPE ENO makes possible the determination of the values of H_{P_E} and Q_{P_E}, as shown in [9] and, therefore, the personality of the PSS.

4. Determination of the governor settings

There are several methods for determining the governor settings. Using the usual methods (Hovey, Paynter and Chaudhry), Changming and Brekke [10] have shown the advantages of the method proposed by Chaudhry for the determination of governor parameters, because it provides a more rapid response to the analyzed head variations and, also, because it reduces the oscillations during the transient conditions caused by different stimuli.

With the calculation program presented in Figure 3, it is possible to determine the governor settings simulating a step in the power ($\Delta\gamma$) of the machine. If the governor has only the proportional characteristic (**P**), then, by varying the proportional constant (K_P - incorporated in the routine **ZN** of Figure 3), it is possible to obtain the value of $K_P = K_{P_u}$ that causes the permanent oscillatory phenomenon whose period is T_u. These values of K_{P_u} and T_u make possible to determine all the constants of the **PI** or **PID** governor, according to the method proposed by Ziegler-Nichols, namely:
For a PI governor

$$K_P = 0.45 K_{P_u} \tag{4}$$

$$T_i = \frac{P_u}{2} \tag{5}$$

For a PID

$$K_P = 0.47 K_{P_u} \tag{6}$$

$$T_i = \frac{P_u}{2} \tag{7}$$

$$T_d = \frac{P_u}{8} \tag{8}$$

in which K_P, T_i and T_d are proportional constant, integral time and derivative time respectively.

It must be noted that routine **ZN** (Figure 3) makes possible, in this way, to determine consistently the governor settings for all the load conditions, assuming, for each one, $\Delta\gamma/\gamma = 0.1$ for instance.

Figures 4 and 5 illustrate the determination of K_{P_u} and T_u for $\gamma = 0.7$ in the installation of Figure 1. In this installation each turbine is equipped with a **PID** governor. Thus, the calculation program makes possible to obtain the governor settings for all load conditions and, since these ones are associated to the opening (Y) of the guide vanes, it is easy to obtain the functions: $K_P = f_1(Y)$, $T_i = f_2(Y)$ and $T_d = f_3(Y)$, through which the *Adaptative PID Governor* is constructed. It should be noted that in the proposed method, the values of the constants T_w (water time) and T_m (machine time) are already incorporated.

The values of the governor constants determined for $\gamma = 0.7$, for the schemes of Figures 1 and 2, are the same and assume the following values: $K_{P_u} = 5.0$ and $T_u = 5.65s$, resulting in $K_P = 3.35$, $T_i = 2.83s$ and $T_d = 0.71s$.

Since it is suggested that, when there is more than one machine, a feedback term (b_t) should be included, a value of $b_t = 20$ was adopted, thus making a feedback controller.

Figure 6 presents the result of a simulation of load variation in the PSS of Figure 1 with a feedback PID governor, with governor settings $K_p = 3.35$, $T_i = 2.83s$, $T_d = 0.71s$ and $b_t = 20$ and load schedule as follows: Turbine 4: $\gamma_4 = 0.5$ to $\gamma_4 = 0.7$, Turbine 14: $\gamma_{14} = 0.8$ to $\gamma_{14} = 0.7$ and Turbine 19: $\gamma_{19} = 0.8$ to $\gamma_{19} = 0.7$. Figure 7 shows head variations at the nodes 5, 16 and 21.

After some simulations it can be concluded that System 2 is more sensitive to rotation speeds and the guide vanes have more amplitude of movement with the same governor constants. The fact shows that the regulation constants must be adequate for each specific installation.

5. Illustrative example

In order to illustrate the use of the proposed program, several simulations were done on the installations presented in Figures 1 and 2, with the objective of finding the most adequate arrangement from the operational view point, since the arrangement of Figure 1, more economical, could present operational problems due to the interference between the three machines with interconnected draft tubes.

In the definition of the governor constants the **ZN** process was used, as presented. It was also simulated a vortex in the draft tube generated at the frequency of 2.5 Hz, representing, at node 5, an excitation of the type: $Q_{exc} = \Delta Q sin(2\pi f t)$.

The results of this simulation with $\gamma_4 = 0.6$, $\gamma_{12} = \gamma_{17} = 1.0$ are presented in Figure 8 for the installation of Figure 2, where in Equation 3 the value of $\Delta Q = 5\% Q_0 = 7m^3/s$ were assumed.

In Figure 8, for the installations of Figure 2, a total load rejection is promoted in machine 4 and a simultaneous load acceptance in machines 12 and 17.

The presented results and other simulations show that the installation of Figure 1 is adequate and, since it is more economical than the installation of Figure 2, it may be adopted for the installation design.

6. Conclusions and recommendations

The calculation model proposed in this paper (for the general writing of equations, see [6, 7 and 9]) is an Expert System for the analysis and monitoring in real time of hydraulic installations of pressure conduits of any nature.

The determination of the personality of the installation is fundamental in order to guarantee a operation without risk. In analogy with a human being, whose personality may be modified by education, the installation may have its personality modified through the selection of parameters and general arrangement in the design phase.

The same model used for the development of the design is adequate for the control of operation in real time since it allows the use of parallel processing that increases considerably the speed of calculation.

The optimum values of the governor K_p, T_i and T_d may be determined through the simulator that includes the **ZN** method, in which the data of the installation (T_w, T_m) are incorporated in the simulation.

The results obtained in the studies carried out and the advances in personal computers show clearly that simulations in the time domain will predominate over simulations in the frequency domain since they present several advantages in the numeric processing, mantaining all the non-linearities of the installation. The MOC is progressive in time and does not involve matricial calculations that induce unconsistent results.

The expert system presented in this paper is the adequate tool for the determination of the functions that compose the adaptative controller that incorporates, besides the adequate governor settings for each opening of the guide vanes (that is to say, of the generated or demanded load), the optimum rotation, in the case of PSS installations in which there is a cyclo-converter for the variation of the rotation speed of the gyrating parts, as presented in [4 and 5].

The present paper is the result of the authors effort for consolidating the calculation algorithm used in the solutions of consultancy problems in hydraulic installations and made public during the last decade, through several papers and thesis listed in the following bibliography.

Acknowledgments

The authors would like to thank to *Fundação de Amparo à Pesquisa do Estado de São Paulo - FAPESP* and *Conselho Nacional de Pesquisa - CNPq* for funding part of these studies.

7. Bibliography

[1] Chen Nai - Xiang et al, (1994) A New Representation of the Characteristic of Pump-Turbine and Calculation of Transients of Complex Operating Conditions, *IAHR, XVII Symposium* - 15-19 September,Beijing, China - pg 1487-1496.

[2] Huvet, Patrick, et al, (1994) Optimization of the Pump-Turbine parameters for the Tianhuangping Pumped Storage Project, *IAHR, XVII Symposium* - 15-19 September,Beijing, China - pg 1005-1016.

[3] Andrade,J.G.P. and Martin, C.S. (1992) Representation of Pump-Turbine Characteristics using Fourier series, *Proceedings of the International Conference on Unsteady Flow and Fluid Transients* - September, Durham, UK - pg 171-180.

[4] Luvizotto, E. and Koelle, E. (1992) The analytic representation of the characteristics of hydraulic machines for computer simulations, *Proceedings of the International Conference on Unsteady Flow and Fluid Transients* - September, Durham, UK - pg 181-190.

[5] Andrade, J.G.P., Koelle, E. and Gonçalves, M.N.F. (1994) Operational Optimization of Hydroelectric Power Plant using Fourier Series to Represent the Pump-Turbine Characteristics, *IAHR XVII Symposium* - September, Beijing, China - pg 1309-1320.

[6] Koelle, E., Luvizotto, E. and Andrade, J.G.P. (1996) Personality Investigation of Hydraulic Networks using the MOC - Method of Characteristics, *7th International Conference on Pressure Surges and Fluid Transients in Pipelines and Open Channels*, 16 - 18 April, Harrogate, UK.

[7] Andrade, J.G.P., Koelle, E., Luvizotto, E. (1995) Pressure Pulsation in Power Storage Scheme due to improper governor settings and flow instabilities, *7th International Meeting of the IAHR Work Group on the behavior of Hydraulic Machinery on Unsteady and Oscillatory Conditions*, 5 - 7 September.

[8] Doerfler, P. (1986) A cross impedance method for frequency-domain representation of oscillations in power plants with meshed waterways, *Proceedings of the 5th International Conference on Pressure Surges*, 22 - 24 September, Hannover - F.R. Germany, pg 169-176.

[9] Almeida, A.B. and Koelle, E. (1992) *Fluid Transient in Pipe Networks*, Computational Mechanics Publication and Elsevier Applied Science.

[10] Changming, Ma and Brekke, H. (1992) Numerical simulation of hydraulic turbine and its governor transients, *Proceedings of the International Conference on Unsteady Flow and Fluid Transients* - September, Durham, UK - pg 191-197.

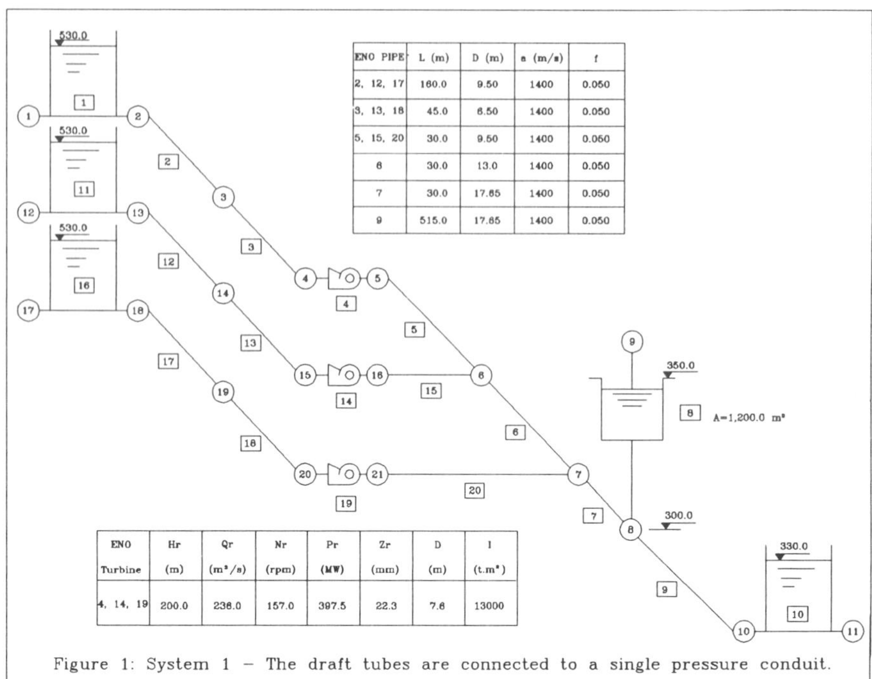

ENO PIPE	L (m)	D (m)	a (m/s)	f
2, 12, 17	160.0	9.50	1400	0.050
3, 13, 18	45.0	8.50	1400	0.050
5, 15, 20	30.0	9.50	1400	0.050
6	30.0	13.0	1400	0.050
7	30.0	17.65	1400	0.050
9	515.0	17.65	1400	0.050

ENO Turbine	Hr (m)	Qr (m³/s)	Nr (rpm)	Pr (MW)	Zr (mm)	D (m)	I (t.m²)
4, 14, 19	200.0	236.0	157.0	397.5	22.3	7.6	13000

Figure 1: System 1 – The draft tubes are connected to a single pressure conduit.

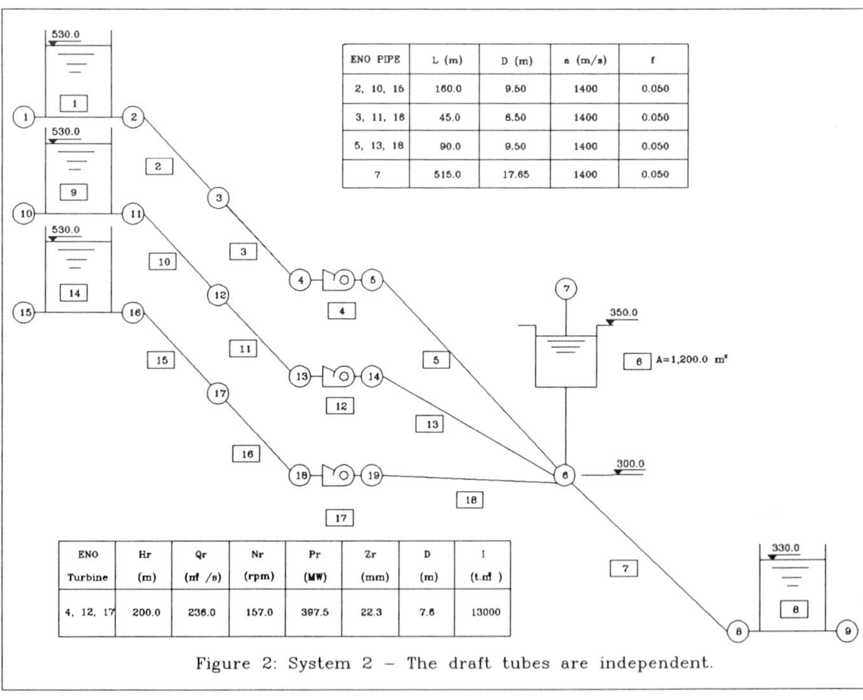

ENO PIPE	L (m)	D (m)	a (m/s)	f
2, 10, 15	160.0	9.50	1400	0.050
3, 11, 16	45.0	8.50	1400	0.050
5, 13, 18	90.0	9.50	1400	0.050
7	515.0	17.65	1400	0.050

ENO Turbine	Hr (m)	Qr (m³/s)	Nr (rpm)	Pr (MW)	Zr (mm)	D (m)	I (t.m²)
4, 12, 17	200.0	236.0	157.0	397.5	22.3	7.6	13000

Figure 2: System 2 – The draft tubes are independent.

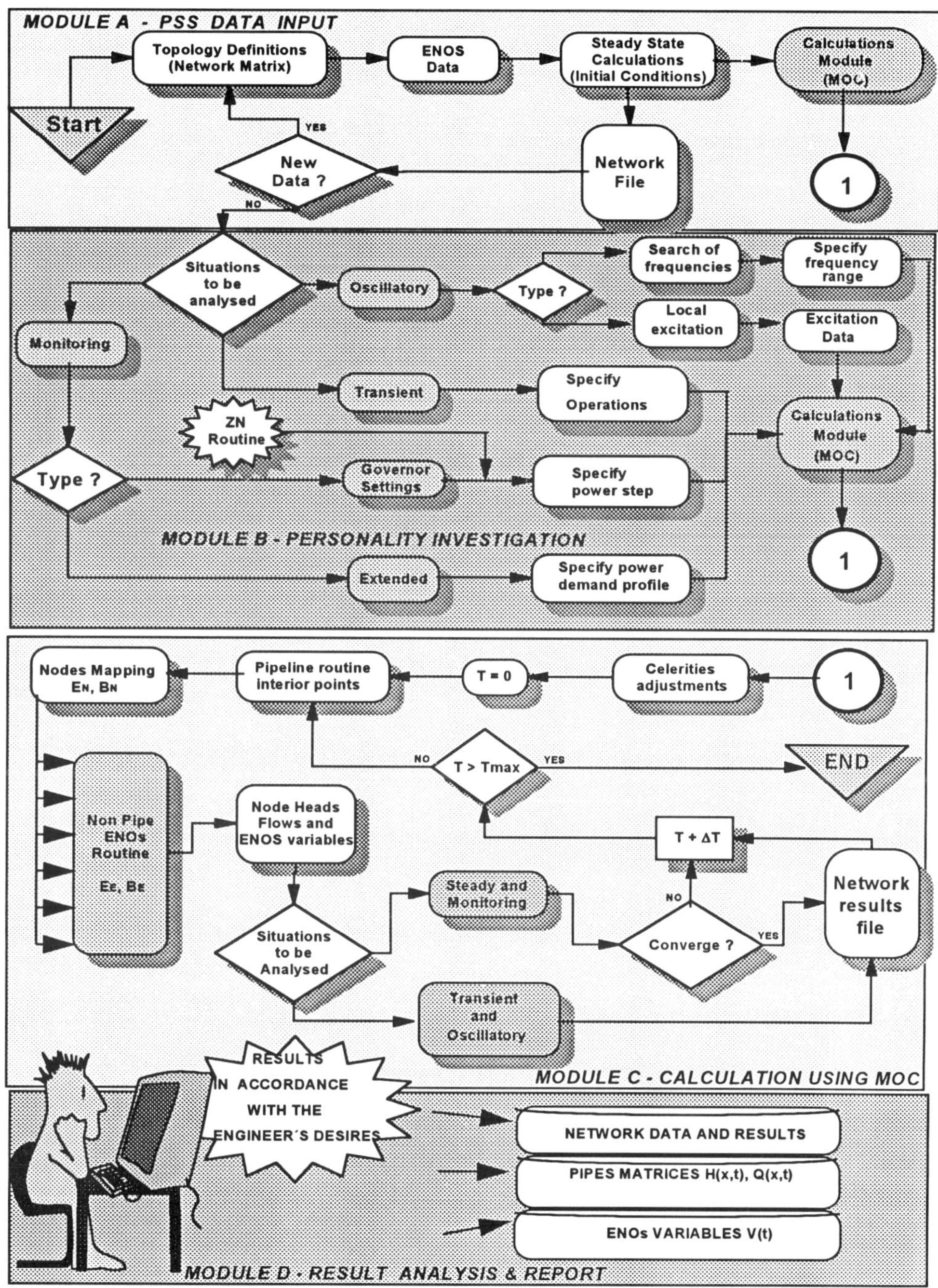

Figure 3. General Program Flow Chart

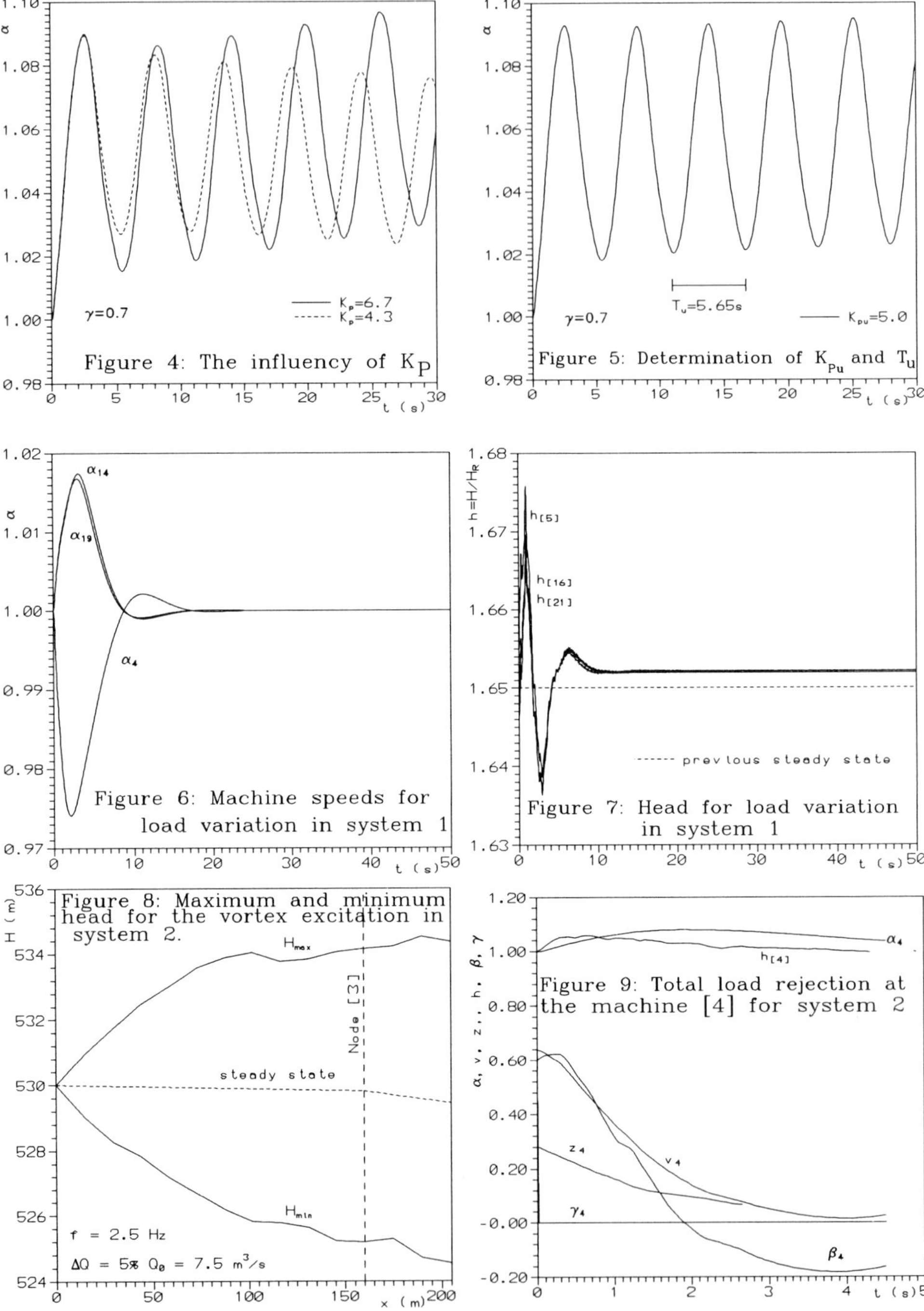

Figure 4: The influency of K_P

Figure 5: Determination of K_{Pu} and T_u

Figure 6: Machine speeds for load variation in system 1

Figure 7: Head for load variation in system 1

Figure 8: Maximum and minimum head for the vortex excitation in system 2.

Figure 9: Total load rejection at the machine [4] for system 2

MODEL-BASED ANALYSIS OF ACTIVE PID-CONTROL OF TRANSIENT FLOW IN HYDRAULIC NETWORKS

JOSÉ CARLOS LAURIA
Escola de Engenharia Mauá
Estrada das Lágrimas, 2035
09580-900 São Caetano do Sul – BRAZIL
Fax: +55 (0) 11 743-0707 – E-mail: maua@eu.ansp.br

EDMUNDO KOELLE
Consulting Engineer
Fax: +55 (0) 11 842-8270 São Paulo – BRAZIL

ABSTRACT – The transient-flow response to the action of automatic valves driven by the working fluid and under PID-control is modelled and simulated so as to investigate its potential and capabilities as an alternative arrangement for the control and suppression of hydraulic transients in pipeline systems.

1. Introduction

Active PID-controllers are the newest dimension in fluid transients and pressure surge. These devices have been employed for more than five decades to solve process control problems and they are used on 95 percent of existing control loops (Coghi, 1994). Until recently, almost all PID controllers were modelled in linear terms, but now nonlinearities are being incorporated to their study (Paruchuri & Rhinehart, 1995). Besides, there is a growing interest in applying the PID control to maintain a desired operating condition in hydraulic networks (Almeida & Koelle, 1992).

Open-loop transient-flow simulations are nowadays carried out with large level of accuracy as result of the association of the method of characteristics and digital computers. For closed-loop, valve-controlled flows, however, the procedures for complete analysis have not been clearly defined yet. There is an emphasis on the stability investigation of the control loop, assuming the valve as a "black-box" and considering indirectly the elasticity effects by a time delay.

Poll (1993) investigated the action of automatic control valves for specifying the pressure pattern in transient flow and showed that, in some cases, the conventional PID-arrangement does not present adequate performance. Conversely, little is done on its complementary part, the flow control.

Given the high level of interest in active control of hydraulic transients, a study programme has been carried out by the authors to couple hydraulic-transient analysis and control theory to study the response of the flow to the action of fluid-driven valves under PID-control to set the system flowrate. Emphasis is placed on the structural knowledge of the control as a prerequisite to experimental investigation.

E. Cabrera et al. (eds.), Hydraulic Machinery and Cavitation, 749–758.

The present work summarises the modelling of the problem and reports on the main results of an extensive numerical study on the operating behaviour, in the time domain, of valves for active control, so as to outline their controllability limits.

2. General Considerations

To start out a systematic analysis of the valve role in the automatic control, so that a comprehensive understanding can be achieved, it is necessary to reduce the installation arrangement to its essentials. The simplest case of active control in hydraulic systems, with practical significance, is a gravity-driven flow along single-diameter pipe, having level variations in the upstream reservoir as input and using PID-control to maintain the flowrate at the set point by means of a valve powered by the working fluid.

Besides the liquid properties (ρ = mass density; μ = coefficient of viscosity), numerous parameters are required to fully describe the problem. Firstly, the available head H. With respect to the pipe they are: L = length; D = diameter (or A = area); f = friction factor, and a = wavespeed. The PID-controller action is expressed by: k_p = proportional gain; t_i = integral time, and t_d = derivative time.

Koelle (1991) develops a general model for control valves that has formed the solution of several particular cases. Of these, the present work is an application of Koelle's proposal, substantiated by Kubie (1982), to model the active flowrate control.

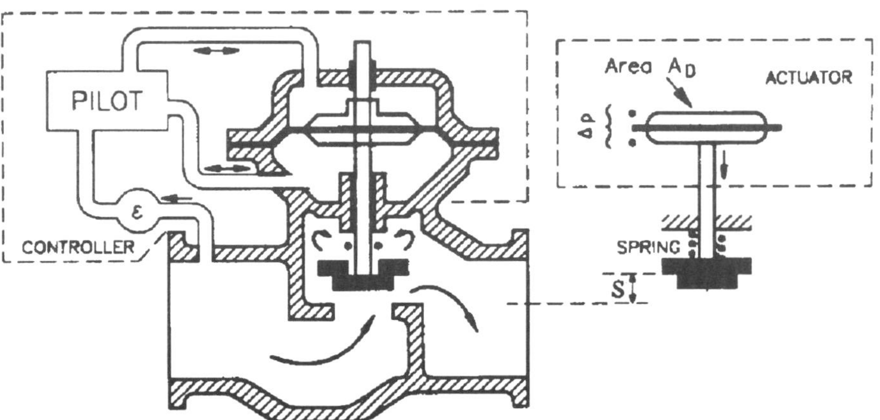

Figure 1. Schematic representation of the control valve

The automatic control valve (fig. 1) is represented by: d = nominal diameter; m_T = total disc-mass (including added mass); c = spring constant; λ = hydraulic-friction factor associated to the gap between the body and valve disc; D_v = valve-disc diameter and σ = spring initial tension.

A deviation-sensing device compares the flowrate to the set point and sends a signal to the controller. Via proper pipework, the pilot control system transmits, according to a prescribed manner, hydraulic power from the fluid stream to the main-valve actuator. A pressure difference (Δp) on the actuator diaphragm (area A_D) causes the movement of the valve disc required to control the flowrate.

3. Transient-Flow Analysis

The method of characteristics has proven to be the most accurate technique for the numerical analysis of hydraulic transients. The success of applying it to non-linear and time-dependent problems stimulated its adoption to simulate the flow behaviour subject to the constraints of the complex boundary condition for the case in point.

There are several references (e.g., Chaudhry, 1987, Thorley, 1991 or Tullis, 1989) on the application of the method of characteristics to the analysis of transient flows. It was adopted, though, the structured procedure to solve steady, transient and oscillatory flows based on the "Node Equation" as set forth by Almeida and Koelle (1992).

According to the definition sketch in figure 2, the continuity condition for a NON-PIPE ELEMENT (ENO) linked to the node yields the NODE EQUATION

$$Q_{P_E} = E_N - B_N H_P \tag{1}$$

where the unknowns are the node head, H_P, and the flowrate, Q_{P_E}. The known values E_N and B_N are got from the installation topology, the hydraulic characteristics for the flow in pipes and the values of the state variables corresponding to a previous time step. If only pipe elements are connected to the node, the corresponding head is

$$H_P = \frac{E_N}{B_N} \tag{2}$$

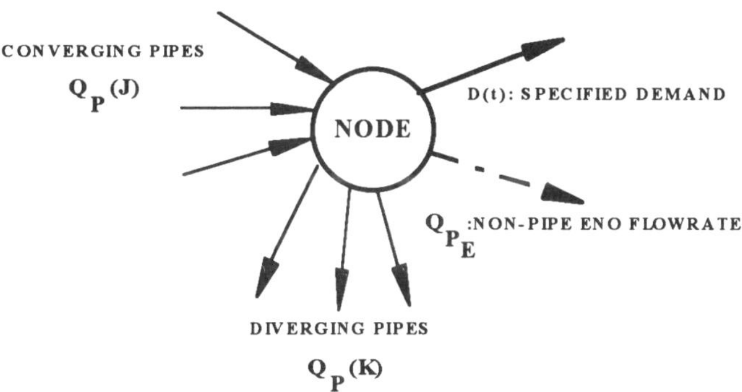

Figure 2. Node in a hydraulic installation

For an automatic control valve (fig. 3), applying twice the node equation for the upstream and downstream node, it results

$$H_{P_E} = H_{P_1} - H_{P_2} = E_E - B_E Q_{P_E} \tag{3}$$

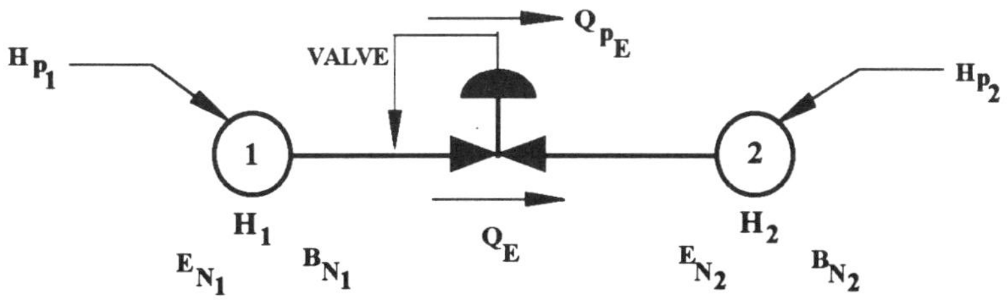

H_{P_1} VALVE Q_{P_E} H_{P_2}
1 2
H_1 Q_E H_2
E_{N_1} B_{N_1} E_{N_2} B_{N_2}

Figure 3. Automatic control valve in a hydraulic installation

In equation (3), E_E and B_E represent the coupling between the valve and the network, according to the valve operating schedule. They are nodal values given by

$$E_E = \frac{E_{N_1}}{B_{N_1}} - \frac{E_{N_2}}{B_{N_2}} \tag{4}$$

$$B_E = \frac{1}{B_{N_1}} + \frac{1}{B_{N_2}} \tag{5}$$

To calculate the state variables Q_{P_E} and H_{P_E} one more expression is necessary. It comes from the valve dynamics related to the PID-controller action (Almeida & Koelle, 1992). As a consequence, however, an additional equation is required because the aperture is an unknown in the present case of fluid-driven valves.

As a boundary condition in transient flow, the valve is assumed as a variable-area orifice that causes a head drop expressed in terms of the loss coefficient. Due to difficulties in obtaining data from tests in transient flow conditions, experimental information related to steady-state regime from Wing Jr. (1960) is adopted as the valve-dissipative coefficient (k_v) against stem displacement (s), or reduced aperture $A_r = s/s_{max}$, as can be seen in figure 4.

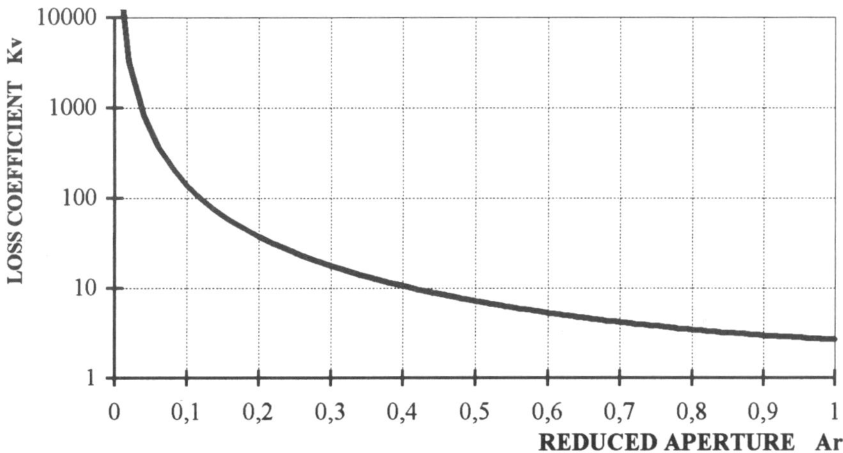

Figure 4. Loss-coefficient curve for the valve considered

4. Valve Dynamics

As result of the parameters identification, the equation of motion for the valve disc in the vertical direction can be written as

$$F_H - F_g - F_\lambda - F_S - F_\sigma + F_C = m_T \frac{d^2 s}{dt^2} \tag{6}$$

where: F_H = hydraulic force; F_g = gravity force; F_λ = hydraulic friction in the clearance between the body and valve disc; F_S = spring elastic force; F_σ = spring initial force; and F_C = controller force.

For a globe-like type of valve, Kubie (1982) assumes, as first approximation, that the hydraulic force corresponds to the pressure difference between the upstream and downstream position of the valve. The pilot system, commanded by an ideal PID-controller, defines the controller force, which results from the action of this same pressure difference on the actuator diaphragm plus a term required for the pilot system to operate only in transient conditions.

As the interest here is in the detailed analysis of the valve response, for adequate numerical evaluation, reduced values of the computational time step must be defined. It has to be of the order of 0.2 to 0.1 of the valve-disc natural period, which, in general, is a fraction of the pipeline period 4L/a (Koelle, 1992).

As a consequence of the discretization, a squared time-interval takes place in the denominator of the acceleration term in equation (6). The participation of the inertia term, therefore, is overemphasized due to the necessity of small time steps and severe problems of inconsistency of the solution can occur.

Although somewhat limited in its scope, a parametric study of the present problem shows that the inertia term has a minor participation in the course of the events (Lauria, 1994). Thus, a weighting factor must be associated to it.

Ideally, weighting factors should be determined from theoretical arguments, but it has been only possible to evaluate its value in a qualitative way via parametric analysis. After a careful numerical study, it became clear that this term is unimportant in the equation of motion and the valve can be modelled as a single force balance.

It can be shown that in the majority of cases the friction term F_λ is very small (Kubie, 1982), which holds for the present situation, and it does not deserve to be considered any further.

As to the assumptions for the disc inertia and the frictional effects on valve disc, Thorley (1983) adopt them for modelling the action of check valves, whose door motion can be faster than the disc motion of the control valve considered here.

5. Controller

An ideal PID-controller is supposed to maintain the instantaneous flowrate (Q(t)) in a desired condition (Q_d). The action of the pilot system defines a force given by

$$\Delta p\, A_D = \left(p_u - p_d\right) A_D + \Delta p(0) A_D \tag{7}$$

where $\Delta p(0)$ refers to the condition for the pilot system to be effective only during the transient flow. Besides, p_u and p_d are the up- and downstream pressure in pipe. The controller is described by the algorithm

$$r(t) = k_p \varepsilon + \frac{k_p}{t_i} \int \varepsilon \, dt + k_p t_d \frac{d\varepsilon}{dt} \tag{8}$$

in which $\mathbf{r(t) = \Delta p(t) - \Delta p(0)}$ is the output signal, and $\boldsymbol{\varepsilon = (Q(t)/Q_d) - 1}$ represents the relative deviation.

To discretize the PID-controller equation the choice is for central-difference approximation because the error inherent to it is smaller than the error for forward and backward-difference formulae (Smith, 1978).

Associating the equation for the valve-disc motion, the head-loss expression for the valve and the control algorithm, it results a set of two equations whose unknowns are the instantaneous flowrate and the valve reduced-aperture and which has to be solved in a trial-and-error procedure. Koelle and Poll (1992) set forth a solution for the case of pressure control with PID regulators, on which the following procedure is based.

The flowrate for a previous time step is adopted as initial guess in the solving process because, in consequence of the relatively small time steps required for the problem analysis, it is not expected significative variation in flowrate from one time interval to another. Afterwards, the expression for the provisional aperture is iterated until the desired accuracy is reached; then a new provisional flowrate is evaluated. The process is finished when two successive flowrates agree to a specified tolerance.

The Newton-Raphson method proved to be inadequate for solution of the above set of equations, presenting severe oscillating behaviour and no convergence. The roots were calculated using the bisection method. For simulations performed in a 486-PC-microcomputer, no convergence was reached for an accuracy better than six significant figures and 300 to 400 iterations were required for convergence.

6. Single Pipe Simulation

Taking into consideration the simplest case of active control as mentioned previously, extensive and detailed numerical simulations were conducted to configure the limits of applicability of the control valve modelled.

Figure 5 exemplifies a typical result for the simulation of the flow. It shows the flow response to a 10% step-change in the upstream-reservoir level and a wavespeed of 1000 m/s. On the representation of the simulation it can be seen that the oscillating pattern gradually reduces after $t = 90$ s, and eventually the desired flowrate is set in a constant fashion at $t = 140$ s. The example corresponds to a situation where the PID-controller tuning provides a stable definition of the set point. Depending on the adjustment of parameters, however, no controllability can be achieved.

As it is shown in figure 5 the PID-controller tuning associated with the area ratio A/A_D cause a high-frequency movement of the valve disc during the initial period of the control action. The diaphragm in the actuator, in its turn, is subjected to repeated fluctuating loads of high frequency and, as consequence, its life is significantly reduced or it even can fail by fatigue, which frequently occurs in the field during the operation of such devices. In these conditions the diaphragm is the weaker point in control-valve arrangement and the main responsible for control failures (Koelle, 1991).

In addition, despite the fact that the valve inertia is negligible in terms of force balance in normal conditions, oscillation patterns as the one described in figure 5 can have a frequency close to the natural frequency of the mechanical system of the valve, leading to the occurrence of resonance. Severe transient conditions can be induced by sudden valve closure. It is required, therefore, the introduction of a damping effect in the system to stabilize the valve-disc movement and to prevent resonance (Koelle, 1992). It is a known fact that PID-arrangements are prone to instability. Many times, there is only a narrow margin for specifying the control parameters so as to assure controllability (Luyben, 1990). This feature could be clearly verified during the simulations performed, pointing out that the model agrees with the actual behaviour of actual PID-controllers.

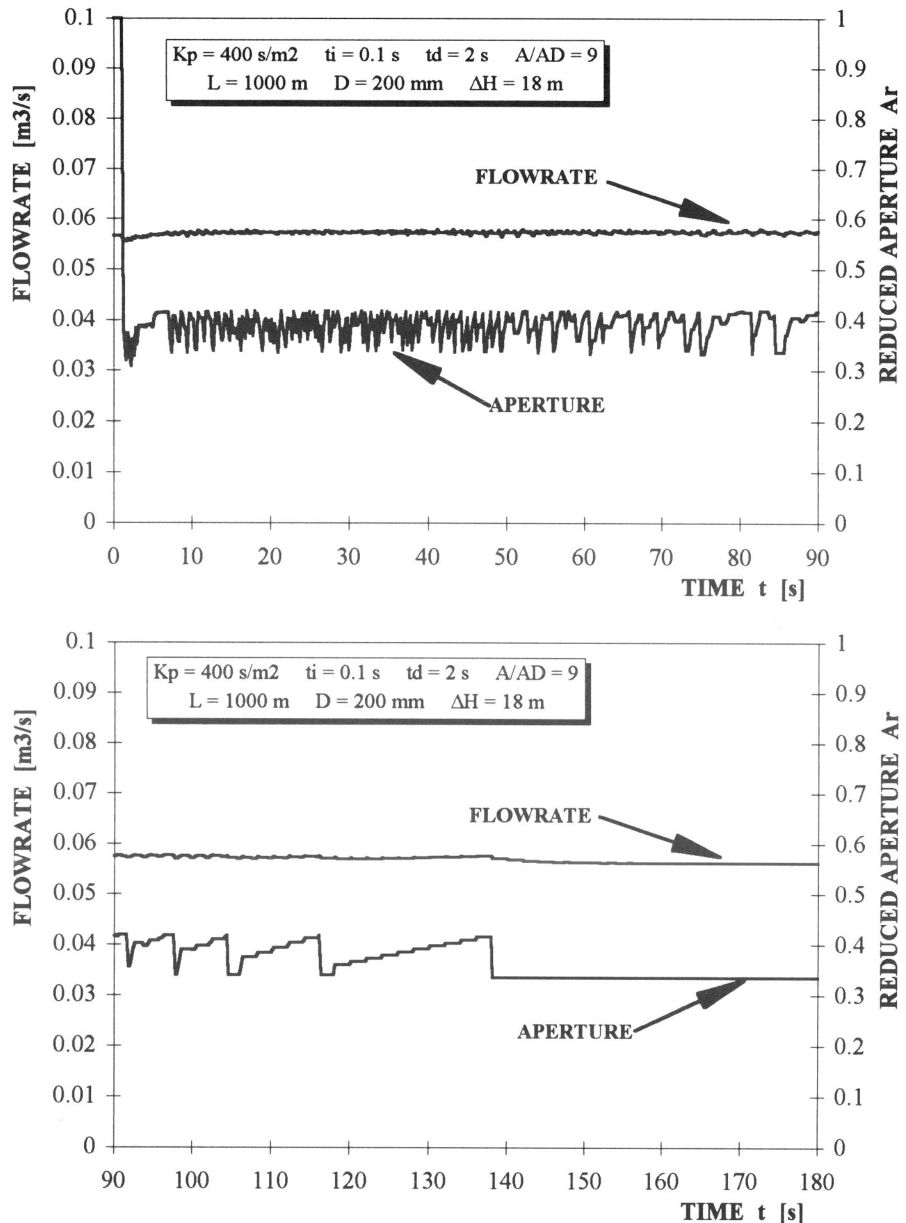

Figure 5. Flow response to valve action for setting the flowrate

There is a shortage of experimental data on transient flows under PID-control. In a search through the open literature it was found in Catheron (1951) information related to industrial flow processes (fig. 6) that are somewhat similar to the problem treated here. A certain degree of correlation exists between both, making it possible to conclude that, at least in qualitative terms, the model presented here is representative and justifiable.

7. Application to Networks

The dynamic behaviour of fluid-driven control valves is largely influenced by the form the piping system defines the disturbance spectrum in transient flow.

Within the scope of hydraulic transients, the study of the flow control in networks is of higher importance, presenting great difficulties for generalization, though. To check out the applicability of the model in broader terms and to provide comparison with single-pipe installations, a specific configuration of network is picked up.

Figure 7 schematizes the network analysed and defines its topology. What is more, it is composed according to the application of the node equation, as mentioned on item 3.

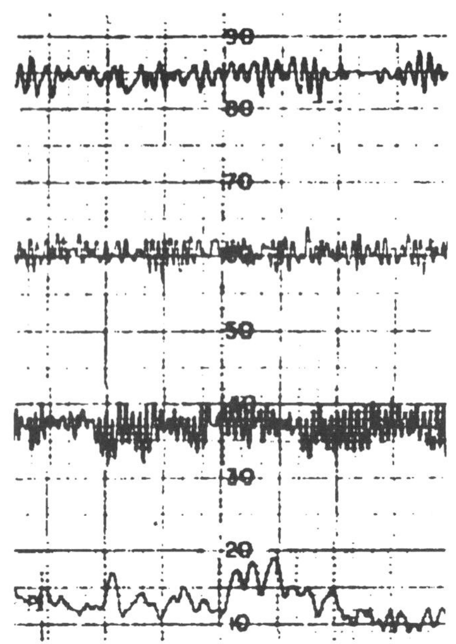

Figure 6. Similar recorded flow patterns (from Catheron, 1951)

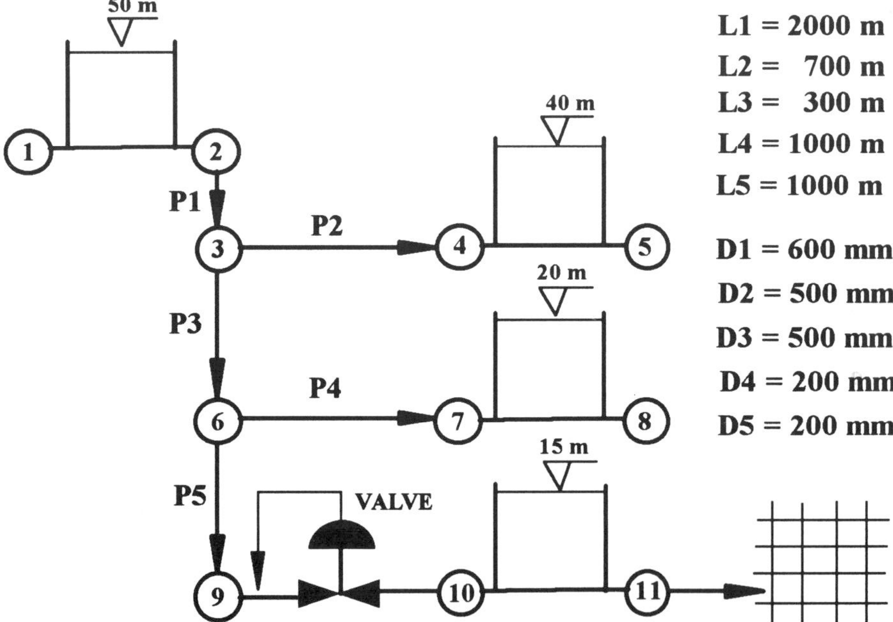

Figure 7. Network schematic

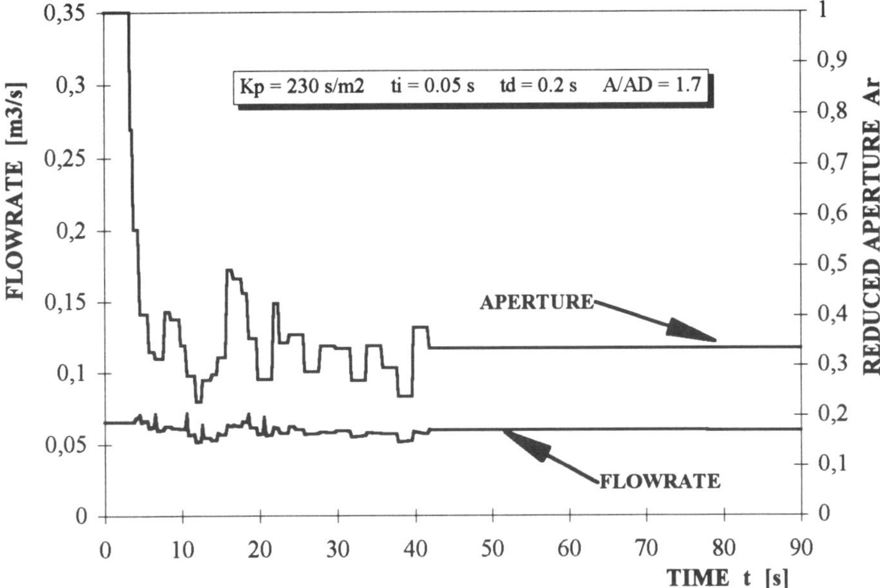

Figure 8. Flow response to valve action in a network

For the same disturbance as in the single-pipe example, the simulation shown on figure 8 for the mentioned network corresponds to the structure of a flow response to a typical PID-controller tuning where the steady-state condition is reached in a relatively short time span, at the expense of a certain offset. On the other hand, the reduction of the deviation would cause an oscillatory pattern for a wider period, as it is the intrinsic characteristic of the PID control.

In relation to the single-pipe situation, the flow pattern in networks causes the controllability to be more critical, requiring smaller integral times (which improves stability) and, mainly, smaller ratios of pipe area to diaphragm area because they represent a significant factor of amplification or attenuation of disturbances.

It was verified in both situations that the larger the value of the proportional gain, the smaller the offset but the more oscillatory the response of the flow becomes. What is more, it was identified the existence of a maximum value of the proportional gain beyond which the flow control will go unstable.

The simulations also showed that growing values of the derivative time have the feature of speeding up the corrective action. However, this procedure can stress the characteristic tendency to oscillation of phenomena with short time constants, as transient flows. The flow response can be very sensitive to small variations in its value, an effect that is also identified by Guevara and Carmona (1990) in studying the application of variable speed drivers under PID control to the regulation, starting and stopping of pumps.

8. Conclusions

Extensive work can be found in the current literature on PID-control theory, but only in few instances the valve is treated as a non-linear component, and even less on its application for active control of flowrate.

The model described here reproduces the intrinsic characteristics of the PID-control. At the present level of knowledge, the association of the model with computer simulations using the method of characteristics, allows the investigation in the physical domain, at least qualitatively, of the PID-control of valves driven by the working fluid, so that computer simulations provide a valuable tool to gain knowledge and understanding to support the conduction of experimental work on the subject.

Effective control to achieve desired goals depends strongly on understanding the dynamic system that is to be controlled. The shape of the flow-response curve varies to a considerable degree according to the ratio of pipe area to diaphragm area and the settings of the PID-controller. The offset and the settling time following a disturbance in the system depend critically on their relative values.

9. References

Almeida, A. B. de & Koelle, E. (1992) *Fluid Transients in Pipe Networks*, Computational Mechanics Publications, Southampton.

Catheron, A. R. (1951) Factors in precise control of liquid flow, *Instruments* **24 (6)**, 705-710.

Chaudhry, M. H. (1987) *Applied hydraulic transients*, Van Nostrand Reinhold, New York.

Coghi, M. A. (1994) Analysis and tuning of control loops (in Portuguese), *Automação & Controle* **4 (3)**, 36-40.

Guevara, Y. & Carmona, R. (1990) Unsteady and steady flow control on pumping systems, *IEEE Transactions on Industry Applications* **26 (5)**, 954-960.

Koelle, E. (1991) Transient behaviour of control valves in hydraulic networks, *9th Round Table of the IAHR Group on Hydraulic Transients with Water Column Separation* – Valencia, Spain, 425-439.

____ (1992) Control valves inducing oscillatory flow in hydraulic networks, *International Conference on Unsteady Flow and Fluid Transients* – Durham, United Kingdom, 343-352.

____ & Poll, H. G. (1992) Dynamic behaviour of automatic control valves (ACV) in hydraulic networks, *16th Symposium of the IAHR – Section on Hydraulic Machinery and Cavitation* – São Paulo, Brazil, 127-139.

Kubie, J. (1982) Performance and design of plug-type check valves, *Proc. Institution of Mechanical Engineers* **196 (3),** 47-56.

Lauria, J. C. (1994) *Investigation of the transient response of automatic control valves in hydraulic installations* (in Portuguese). Doctor's Thesis, Univ. of São Paulo. São Paulo.

Luyben, W. L. (1990) *Process modelling, simulation and control for chemical engineers* – 2nd. edition, McGraw Hill, New York.

Paruchuri, V. P. & Rhinehart, R. P. (1995) Model-based flow control boosts accuracy, eases tuning, *InTech* **42 (4),** 52-56.

Poll, H. G. (1993) *Dynamic behaviour of control valves in hydraulic networks* (in Portuguese). Master Thesis, Univ. of São Paulo. São Paulo.

Smith, G. D. (1978) *Numerical Solution of Partial Differential Equations: Finite Difference Methods* – 2nd edition, Oxford University Press, Oxford.

Thorley, A. R. D. (1983) Dynamic response of check valves, *4th International Conference in Pressure Surges* – Bath, England, 231-242.

____ (1991) *Fluid Transients in Pipeline Systems*, D. & L. George, Herts.

Tullis, J. P. (1989) *Hydraulics of Pipelines*, John Wiley & Sons, New York.

Wing Jr., P. (1960) Practical determination of control valve C_v. *ISA Journal,* **7 (9),** 90-94.

SIMULATION OF TRANSIENTS IN PRESSURIZED HYDRAULIC SYSTEMS WITH VISUAL TOOLS

F. MARTINEZ[1], J. IZQUIERDO[1], R. PÉREZ[1], A. VELA[2]
[1]*U.D. Mecánica de Fluidos. Universidad Politécnica de Valencia Camino de Vera, s/n. E-46071 VALENCIA (Spain)*
[2]*Departament de Tecnologia. Universitat Jaume I Campus de Penyeta Roja, E-12071 CASTELLÓ (Spain)*

Abstract

In contrast with classical methods used to analyze different transient phenomena due to the start-up, shut-off or normal operation on the pressurized hydraulic systems, in which hydraulic machinery is usually present, a different approach based on the use of a visual tool is presented. This tool uses the object oriented language to build the system model by means of block diagrams. Many additional features make a powerful tool to simulates dynamic systems in a very clear and easy way this new software product.

1. Introduction

The study of hydraulic transients in pressurized systems such as hydroelectric power plants, either during start-up or shut-off maneuvers or during normal operation is of great interest from the perspective of both design and exploitation. As a consequence, numerous papers have been devoted to simulate these events by using the rigid and elastic models and both in the time and frequency domains.

Currently, every transient event has its own mathematical formulation, but no completely general tools have been developed yet. In order to solve a particular problem, recourse is made of mathematical libraries or specific software packages.

In this paper we intend to enhance the potentiality of a commercial package for the study of transients in hydroelectric power plants and other hydraulic systems. VisSim (Visual Solutions, 1994) is a visual tool written in an object-oriented language (OOL), and devised for the simulation of the behavior of general dynamic systems.

Even though the program has been devised to solve systems of non-linear ordinary differential equations (ODEs) in the time domain, allowing a posteriori response analysis in the frequency domain, the use of the z-transform enables to solve water hammer

E. Cabrera et al. (eds.), Hydraulic Machinery and Cavitation, 759–768.

equations by the MOC, too. Moreover, it is possible to use vectorial links in order to build models directly in the frequency domain for the study of resonance phenomena. Thus, every type of transient in a hydraulic system can be analyzed by using this tool.

Among other applications we have to quote here the following: determination of design parameters for a desired response; sensitivity analysis of design variables and control parameters; validation of the model used in the simulation of a specific phenomenon, and, summing up, its use as a support tool in decision making.

In order to show the potentiality of the new analysis tool, in this paper several cases are analyzed and the corresponding models presented. As an introduction example, we first present the study of oscillations in surge tanks using an ODE formulation. Next the analysis of the water hammer in the penstock of a hydroelectric power plant after a full load rejection is introduced. The use of the z-transform is now needed in order to consider the fluid compressibility.

2. Characteristics of the software package used

VisSim is a visual programming environment for system design and simulation, oriented to simulate the time evolution of dynamic systems governed by linear and non-linear ODEs. Using a visual programming language, the whole system is modeled in the form of block diagrams. It is possible to select from over 75 graphical blocks, grouped in annotation, arithmetic, Boolean, integration, linear, nonlinear, random generator, signal consumer, signal producer, time delay, trascendental and DDE. The different blocks are then wired them together to build the global model. Parts of the model can be integrated in compound blocks for a better understanding of the structure of the model, being multi-layer block nesting supported. Finally, the possibilities for building models can be enlarged with add-on modules such as Analyze for carry out the stability analysis of linearized parts of the system, RT for real-time and data acquisition, Neural-Net for developing neural network models, C-Code generator, etc.

With regard to their simulation capabilities, until seven integration algorithms are available, including adaptive schemes. Besides, the program supports implicit static solvers based on the Newton-Raphson method or others to solve for one or more unknowns every time step, during the simulation period. VisSim includes also the possibility of optimizing the response for one or more parameters, choosing among Powell, Polak-Rivière or Fletcher-Reeves methods as solvers. It is also valuable, from the point of view of simulation capabilities, the interactive or background operation modes, real time simulation, automatic restart option, snapping states, etc.

But the most interesting aspect for designers and education purposes is the interactive simulation on the screen. Several display blocks can show the results in plots (including time history, XY plots, discrete and frequency domain graphics, log or semi-log scales and other features), meters, strip-charts, display boxes, etc. So, the response of the system can be checked instantly for different parameter values or model configurations.

Finally, the most recent version of VisSim 2.0 supports animation, filters, vector arithmetic operations, C expressions, etc. In the same way that OOL are gaining access to program developers, new tools like VisSim or other similar products based on OOL will be commonly used in the future to solve a broad range of mathematical problems.

3. Transients in surge tanks

The study of hydraulic transients when surge tanks are used to protect the tunnel against waterhammer waves produced in the penstock of a hydroelectric power station by load changes on a turbine, constitute one of the most classical problems of transients studied in literature (Jaeger, 1977).

But except for the hypothesis of a frictionless system, no analytical solution of the governing equations is known in the time domain. For a simple surge tank and taking friction in the tunnel into account, the dynamic and continuity equations define a closed system of ODE to be solved for the tunnel flow Q_t, and the surge tank water level z unknowns, given the change in the turbine flow Q_{tur}. The transient equations can be written as (Chaudhry, 1987):

$$\frac{dQ_t}{dt} = \frac{g A_t}{L} (- z - c Q_t |Q_t|) \tag{1}$$

$$\frac{dz}{dt} = \frac{1}{A_s} (Q_t - Q_{tur}) \tag{2}$$

being g = gravity, A_t= cross-sectional area of the tunnel, L= tunnel length, c= tunnel friction coefficient and A_s = cross-sectional area of the surge tank.

Figure 1 shows the block diagram for an illustrative example corresponding to a full load rejection in a power plant. The sketch of the installation and the parameters of the model are given on the illustration. The initial flow is assumed to be Q_{to} = 48 m^3/s, being the corresponding initial water level in the surge $z_o = - c \cdot Q_{to}^2 = - 15.4$ m. The block diagram is self-explaining. We only will point out the integration block 1/S, which includes the initial value of the corresponding signal. Note that each variable can have only one signal entry, but can be repeated anywhere for convenience and clarity of the diagram. The best way to build block diagrams is by expliciting one variable for equation, an attaching to that variable the output signal of the block diagram which represents the corresponding equation. Figure 1 also shows the results of the transient in the time domain and in an XY plot. The damping is due to the tunnel friction, and the maximum level, 24.8 m over the reservoir level, is reached in the first oscillation at 280 sec from start.

Next a more complex case is modeled using some of the nonlinear blocks provided by VisSim. The new example refers to a differential surge tank, which sketch and notation are shown in figure 2. The governing equations now depend on the state of the system.

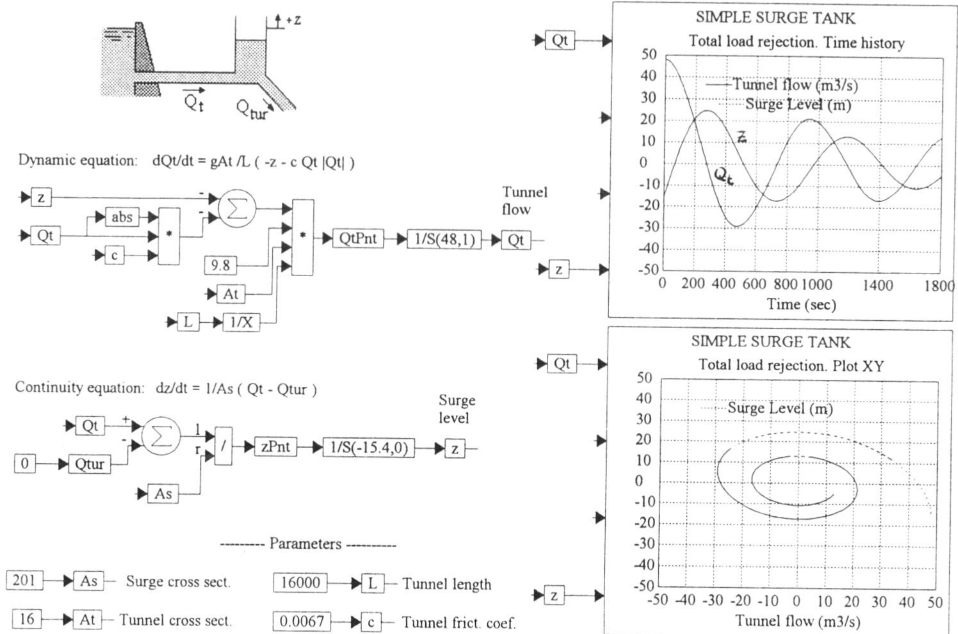

Figure 1. Block diagram for a simple surge tank and output results for a full load rejection

The dynamic equation (1) is preserved, but variable z must be substituted for z_r, which represents the water level into the riser, while $z < z_{r,max}$. If the riser spills into the tank, z_r must be substituted for $z_{r,max}$ = const. Finally if the water in the outer tank floods the riser, i.e. $z>z_{r,max}$, then z_r will be substituted for the unique variable z. During the phase of descent z and z_r will be again distinguished if $z < z_{r,max}$.

The continuity equation at the junction of the tunnel and the tank is now expressed as:

$$Q_t = Q_s + Q_r + Q_{tur} \tag{3}$$

in which Q_r = flow in or out of the riser, and Q_s = flow in or out of the outer tank. The value of Q_s depends upon the difference of the water levels in the riser and in the tank, and upon the size and characteristics of the orifice at the bottom of the tank. It may be computed from the following equation:

$$Q_s = \pm C_d A_{orf} \sqrt{2g|z_r - z|} \tag{4}$$

in which C_d = coefficient of discharge of the orifice, and A_{orf} = cross-sectional area of the orifice. If $z_r > z$ then Q_s will be positive, and will be negative if $z_r < z$.

To compute the variation of liquid level in the riser and in the tank, the following equations must be applied, while $z < z_{r,max}$:

$$\frac{dz}{dt} = \frac{1}{A_s} Q_s \quad , \quad \frac{dz_r}{dt} = \frac{1}{A_s} (Q_t - Q_s - Q_{tur}) \qquad (5, 6)$$

During the ascending motion equations (1), (4), (5) and (6) form a closed system of four ODE to be solved for Q_t, Q_s, z and z_r. When the riser spills into the outer tank the system is uncoupled because $z_r = z_{r,max}$ = const. The variation of level in the outer tank will then be given by:

$$\frac{dz}{dt} = \frac{1}{A_s} (Q_t - Q_{tur}) \qquad (7)$$

In this case, equation (4) only discriminate which part of the flow entering the outer tank will flow through the orifice and which part will spill from the riser. If the riser is flooded, equation (4) has no effect because the whole flow enters through the riser, and equation (1) in terms of the unique variable z, together with equation (7), will form a closed system of two ODE equations to be solved for Q_t and z, like for a simple tank.

During the motion of descent we will have again four equations from the moment $z < z_{r,max}$, and the flow through the orifice will be reversed. The size of the orifice must be enough to avoid the surge drainage.

This complex dynamics can be modeled by VisSim using the 'merge block' which selects between two input signals depending on a boolean control signal, and the 'resetIntegrator block' which integrates the input signal with an optional reset capability. Figure 2 shows the general block diagram for a differential tank, with a compound block for each of the main equations. The parameters of the example are the same than those for the previous ones, except for the total surge cross-section which has been divided into 151 m^2 for the outer tank and 50 m^2 for the riser. The height of the latter has been limited to 15 m. Also new values have been added for the orifice parameters.

The block diagrams of the continuity equations deserve some explanation and are shown in figures 3 and 4, respectively. In the one corresponding to the outer tank the 'merge block' selects between equation (5) for $z_r < z_{r,max}$ and equation (7) when spilling, while the reset integrator block fuses together variables z and z_r when the riser is flooded. In the block diagram of the riser continuity equation, the reset integrator keeps up z_r equal to $z_{r,max}$ while spilling, i. e. $z < z_{r,max}$, $z_r \geq z_{r,max}$ and $Q_r > 0$. But if the riser is flooded or the flow is reversed (motion of descent) z_r will keep on evolving. Finally, in case of flooding of the riser, the total cross-section must be used for the continuity equation.

The results for a total load rejection are shown as well in the figure 2. At 50 sec the riser starts to spill, and the flow through the orifices Q_s reaches its maximum value of 20 m^3/s. At 210 sec the riser is flooded, Q_s vanishes, and the maximum level of 17 m is reached in the tank at 280 sec. At 340 sec the riser surfaces again, its water level goes down faster than the one in the outer tank, and Q_s is reversed. The minimum level is reached at 550 sec and is higher than the initial level. This is approximately maintained while the level of the outer tank becomes equal and during this short period the flow into the tunnel is supplied exclusively through the orifices ($Qt = Q_s$).

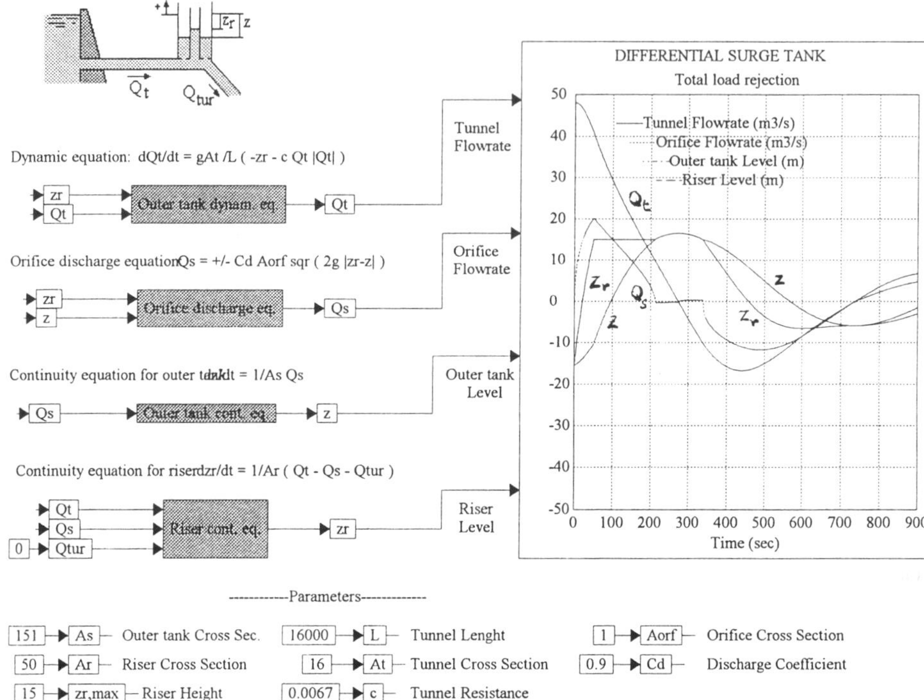

Figure 2. **Block diagram for a differential surge tank and output results for a full load rejection**

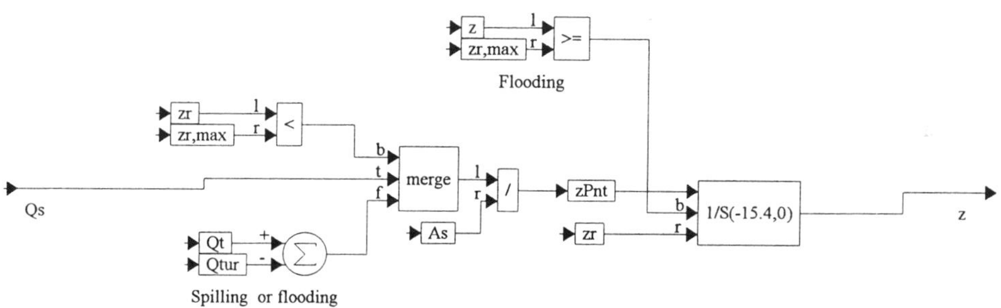

Figure 3. **Block diagram of the continuity equation for the outer tank**

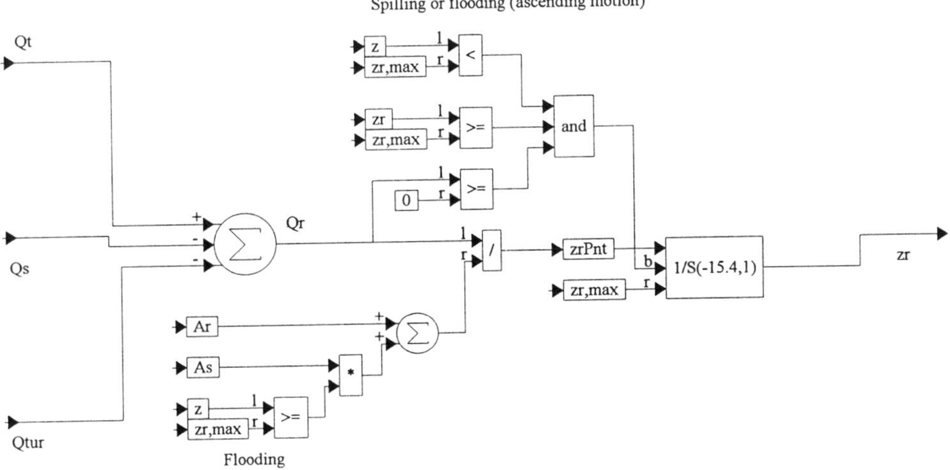

Figure 4. Block diagram of the continuity equation for the riser

From this example, new cases can be developed using the facilities provided by the OOL and the block diagram languages. For instance, setting the initial values of z, z_r and Q_t to zero, and forcing Q_{tur} to be constant (usually 50 % of the full load), start up conditions can be analyzed. By introducing a new compound block to model the governor behavior (e.g. constant power) the stability of the tank can be checked (there exists an add-on of the program to build the locus root for a linearized model automatically obtained over some selected blocks at the stopped simulation time). Finally the optimization capabilities can be used to size the surge tank parameters (diameters, orifice constants, etc) in order to limit the maximum level, to avoid tank drainage or to obtain some desired stability.

4. Water hammer in hydroelectric power plants

Although VisSim is not at first oriented to simulate the behavior of dynamic systems governed by partial differential equations, the problem of waterhammer in one-dimensional systems can be solved by using the Method of Characteristics (MOC) and the z-transform. As known, the z-transform is used to relate output signals with input signals in dynamic processes that evolve in discrete time or for continuous signals sampled in discrete time. VisSim includes a 'Sample&Hold' block to sample continuous signals and other one to simulate the essentials of control actions in discrete systems, the 'Unit Delay' block, which mathematically is represented by z^{-1}. This block is intended for modeling a digital delay in a continuous simulation, and works as follows: the value contained in the buffer is copied to the output tab at each pulse of the clock;

simultaneously the current value of the input signal is stored in the unit buffer. The clock, always necessary to carry out a digital control, is modeled by wiring a pulseTrain block to the Boolean input connector tab of the unitDelay block. If the period of the clock is set equal to the computational time interval Δt of the MOC, the unitDelay block can be used to transfer the values of H and Q at the different calculation points from time t to time $t+\Delta t$.

Lets us summarize the principles of the MOC and show through an illustrative example, how VisSim can work to solve waterhammer problems. Compressible flow within a pipe can be modeled by a well-know set of PDE of hyperbolic type (Chaudhry, 1987). Along its characteristics these equations simplify to a set of ODE, that can be solved numerically by dividing the pipe into reaches and selecting a time step satisfying the CFL condition. The numerical integration of these equations produces a set of algebraic equations for flowrate and head at each point P separating two contiguous reaches:

$$Q_P = C_p - C_a H_P \quad , \quad Q_P = C_n + C_a H_P \tag{8, 9}$$

where:

$$C_p = Q_A + C_a H_A - R\Delta t Q_A |Q_A| \tag{10}$$
$$C_n = Q_B - C_a H_B - R\Delta t Q_B |Q_B| \tag{11}$$
$$C_a = gA/a \quad , \quad R = f/(2DA) \tag{12}$$

By solving simultaneously equations (8) and (9) for Q_P and H_P one can obtain the current state of flow and head at point P. However, at boundaries only one of the equations (8) and (9) is valid and special boundary conditions are needed to determine the conditions at the boundaries at time $t+\Delta t$. Usually one of the boundary conditions contains the perturbation that is causing the transient, and the other one determines the way in which the waves are reflected.

The translation of the previous procedure into a block diagram for the well known problem of a linear valve closure downstream of a single pipe, having a reservoir at the upstream end, is shown in figure 5. This case can be understood as a simplification of the study of the waterhammer in a penstock due to a total load rejection in the turbine. Three calculation points P have been considered. The parameters of the system are shown in the same figure. The time step for this example is 0.5 sec, which have been assigned as time interval for the pulse train block and as simulation time step. The leftmost blocks allow to obtain the values of constants C_p and C_n associated to each point at time t. The rightmost blocks are intended to obtain the values of H_P and Q_P at time $t+\Delta t$, as the solutions of equations (8) and (9). In fact, only C_{p1} and C_{n3} are combined to obtain H_{P2} and Q_{P2}. For the calculating H_{P1} and Q_{P1} the constant C_{n2} is used, together with the boundary condition at the reservoir, $H_{P1} = 100$ m in this case. To calculate H_{P3} and Q_{P3} the constant C_{p2} is used together with the boundary condition at the valve. The closure law has been defined as $Q = 1 - 0.2\ t$ m^3/s for the sake of simplicity, being the closure time 5 sec. Note that the initial value of velocity will be also 1 m/s. because $A = 1\ m^2$.

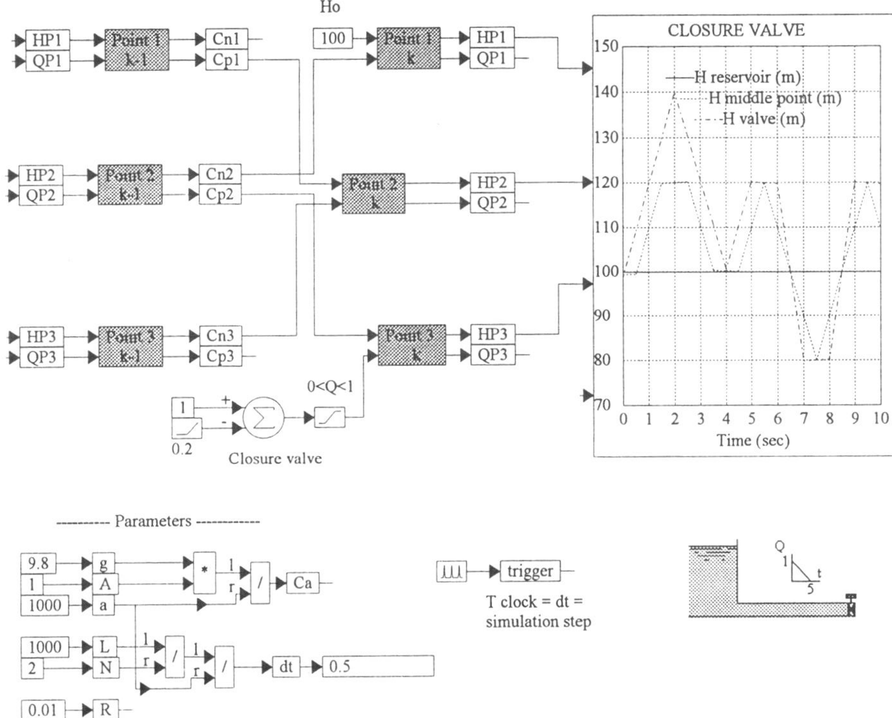

Figure 5. Block diagram for solving a water hammer due to a linear closure valve at downstream. $T_c = 5$ s

The results shown in the plot correspond to the valve and to the middle point of the pipe. The maximum upsurge in the valve is 40 m, and in the middle point 20 m, as expected according to the Michaud formula, due to the low value of the losses (R = 0.01). Figure 6 shows the detailed diagram for the compound block 'Point 1 (k-1)'. It can be checked how the Unit Delay block 1/Z is used to transfer the values of H_{P1} and Q_{P1} obtained at t + Δt to compute the C_{n1} and C_{p1} that will be used in next step.

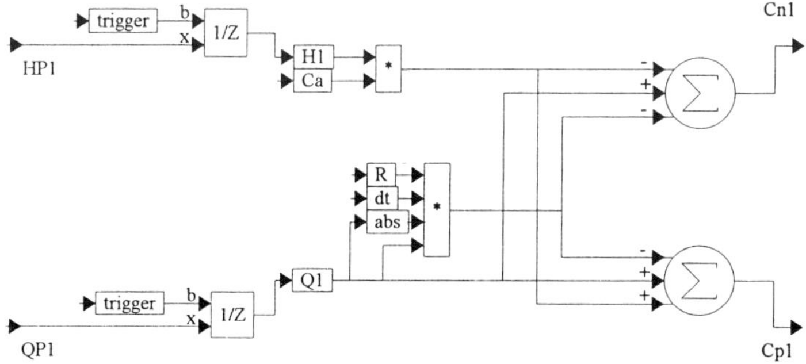

Figure 6. Block diagram for computing C_{n1} and C_{p1} to the next step

More complex examples, in particular as far as boundary conditions concern, can be solved using the same technique. The authors have solved the water hammer for pump stopping due to a power failure using the Suter curves with excellent results. To introduce the Suter curves the 'map block' has been used, which perform piecewise linear interpolated 1- or 2-dimensional table look-ups. Some times the boundary condition includes differential equations. In this case the simulation step must be an integer fraction of the time interval used for the calculus of water hammer. One of the most complex cases of transients in hydroelectric power plants, the calculus of waterhammer due to a partial or total load rejection, taking into account the governor behavior and the turbine characteristics, falls into this category.

5. Conclusions

In the presentation of the paper clarity has been chosen against quantity. The authors have developed many others examples besides the three illustrative ones here presented. Amongst them we would like to emphasize in those regarding the simulation of resonance phenomena. The response in frequency of a system can be solved by the Transfer Matrix Method (Chaudhry, 1987) using compound blocks to characterize the behavior of the components, and resonance frequencies as well as node pressure locations can be found by means of the optimization blocks. On the other hand, a real time model for a pumping station endowed with fixed an variable speed pumps has been developed with success, too.

VisSim offers a new way to study the behavior of hydraulic dynamic systems, and our students are practicing nowadays with this program for educational purposes. On the other hand the new features of the most recent version, such as animation, adds new appealing to this product. But, in our opinion, more efforts must be done in the future by developers to improve the robustness of the product and particularly to improve the convergence of the built-in solvers for both, implicit variable calculation and optimization.

6. References

Abreu, J.M., Guarga, R., Izquierdo, J. (1995) *Transitorios y Oscilaciones en Sistemas Hidráulicos a Presión.*.Valencia

Chaudhry, M.H. (1987), *Applied Hydraulic Transients.* Van Nostrand Reinhold Co. New York

Jaeger, C. (1977), *Fluid Transients in Hydro-Electric Engineeering Practice.* Ed. Blackie. Glasgow.

Visual Solutions (1994), *VisSim User's Guide.* Visual Solutions, Inc, 1994. Westford. MA O1886. http://www.vissim.com/

DYNAMIC BEHAVIOUR OF GOVERNING TURBINES SHARING THE SAME ELECTRICAL GRID

Torbjørn K. Nielsen
Senior Research Engineer
Kværner Energy as, Trondheim, Norway

1. Introduction

A mathematical model of a power generating system containing several turbines, with separate conduit systems, sharing the same electric grid are presented. The purpose of the model is to analyse the dynamic behaviour of the governing turbines during load admittance and load rejectance on the grid with focus on the mechanical and hydraulic side.

The paper describes the differential equations involved in the transference of hydraulic power to mechanical rotating power in the turbines, and from rotating mechanical power to electric power in the synchronous generators. The different aggregates are electrically connected. For each aggregate, both the frequency governor and the voltage governor are modelled.

The turbine model is based on the one dimensional Euler turbine equation as previously reported at the IAHR Symposium in Sao Paulo. The generator model is based on the so called swing equation of the generator describing the dynamics of the angle between a point on the rotor and the synchronous reference frame.

2. Model description

2.1 GENERAL

In order to simulate frequency or load governing in a system with several hydro power plants connected to the same grid, the model must have the dynamic elements that describes reservoirs and conduit system, turbines, generators, frequency governors, voltage governors and the electrical grid.

The reservoirs represent constant pressures as up stream or down stream boundary conditions. The dynamic behaviour of the water in tunnels and tubes are modelled by the continuity equation and the equation of motion for each part and connected into a system of differential equations, transformed and solved by the Method of Characteristics. This method is well documented and will not be looked further into in this paper.

E. Cabrera et al. (eds.), Hydraulic Machinery and Cavitation, 769–778.

To model the electrical grid is a question of finding the grid's properties regarding capacitance and inductance. For the purpose of this paper, the grid is simplified as shown in figure 3.

In the following, the modelling of the turbine, the generator, the frequency governor and the voltage governor will be described.

2.2 THE TURBINE MODEL

The turbines in the system transfers the hydraulic power to rotational mechanical power. In order to regulate the power output, the wicket gate opening (or the opening of the nozzle valves in Pelton turbines) are governed by the frequency governor. The turbine behaviour, dependent upon both pressure and speed of rotation, decides the flow in the system.

The turbine model is based on the one dimensional Euler turbine equation and is previously reported at the IAHR Symposium in Sao Paulo in 1992. By this method, the turbine performance is modelled by use of two differential equations, one for flow and one for the angular speed of rotation:

$$I\frac{dQ}{dt} = \rho g H_e - \rho g H_R \left(\frac{Q}{\kappa Q_R}\right)^2 - \rho s\left(\omega^2 - \frac{H_e}{H_R}\omega_R{}^2\right) \quad (1)$$

$$J\frac{d\omega_t}{dt} = T_t - T_g \quad (2)$$

The turbine torque, T_t, is expressed by:

$$T_t = \rho Q\left(m_s - \frac{\psi m_R}{\omega_R}\omega_t\right)\left(1 - \frac{\Delta H}{H_e}\right) - R\omega_t{}^2 \quad (3)$$

Where:

Q	-	Flow
H	-	Head
ω_t	-	The turbine's angular speed of rotation
I	-	Hydraulic inertia in the turbine element
J	-	Polar moment of inertia
ψ	-	Pressure number
s	-	Self governing parameter
ΔH	-	Hydraulic losses in the turbine
T_g	-	Generator torque
m_s	-	Turbine torque at zero speed of rotation

Subscript:

R	-	Rated
e	-	Effective

The self governing parameter and the hydraulic losses must be adjusted in order to match the performance diagram of the turbine.

2.3 THE GENERATOR

In ordinary programs for stability analysis, the generator is represented only by the polar moment of inertia, J, in the equation above. If several generators are connected to the same grid, the energy transference from rotational mechanical power to electric power must be modelled. However, we are not interested in phenomena like local current transients in the generator windings etc., but only the generator torque's dependency of the electrical load. Therefore a very simplified model is suggested, and it seems to be sufficient for our purpose.

When the electrical load increases, the angle, δ, between the generator stator and rotor, referred to a rotating frame, will increase as illustrated in Figure 1.

The angle must be seen in relation to the grid frequency. With a nominal grid frequency of f = 50 Hz, the angular grid frequency is $\omega_{grid} = 2\pi f$. The synchronous angular speed of rotation of the generator is a function of the number of generator poles, P, according to:

$$\omega_{syncr} = \frac{2}{P}\omega_{grid} \tag{4}$$

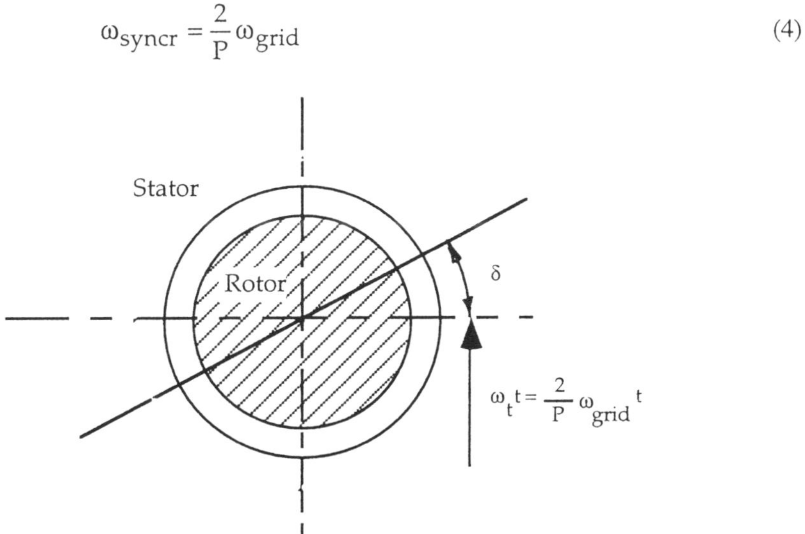

Figure 1 Angle between stator and rotor in a rotating reference frame.

In steady state operation, the angular speed of the turbine must be equal to the grid frequency. During the transient period of a load change, the angle will change according to the differential equation:

$$\frac{d\delta}{dt} = \frac{P}{2}\omega_t - \omega_{grid} \tag{5}$$

where ω_t is the angular rotational speed of the turbine. The angle is in balance when the turbine rotation is synchronous i.e. when:

$$\omega_t = \omega_{syncr} = \frac{2}{P}\omega_{grid} \tag{6}$$

For the transference from mechanical rotating power to electrical power, the voltage, E, is primarily a function of the angular speed of rotation, ω:

$$E = k\phi\omega \qquad \text{or} \qquad \omega = \frac{E}{k\phi} \tag{7}$$

The torque, T_g, is primarily a function of the electric current, I:

$$T_g = k\phi I\cos\varphi \qquad \text{or} \qquad I = \frac{T_g}{k\phi\cos\varphi} \tag{8}$$

where: $k\phi$ - Defines the magnetic flux
φ - Phase angle dependent on the property of the grid.

And according to Ohm's law: $E = R_{grid}I$.

Then the differential equation of the angle can be written:

$$\frac{d\delta}{dt} = \frac{P}{2}\omega_t - \frac{1}{k\phi}E \tag{9}$$

and:

$$E = R_{grid}\frac{T_g}{k\phi\cos\varphi} \tag{10}$$

In the torque equation we must introduce a damping of the angular movement:

$$J\frac{d\omega_t}{dt} = T_t - T_g - m_d\frac{d\delta}{dt} \tag{11}$$

T_g can with good approximation be modelled as a sinus function of this angle according to:

$$\frac{T_g}{T_{gR}} = \frac{\sin\delta}{\sin\delta_R} \tag{12}$$

The function is illustrated on Figure 2.

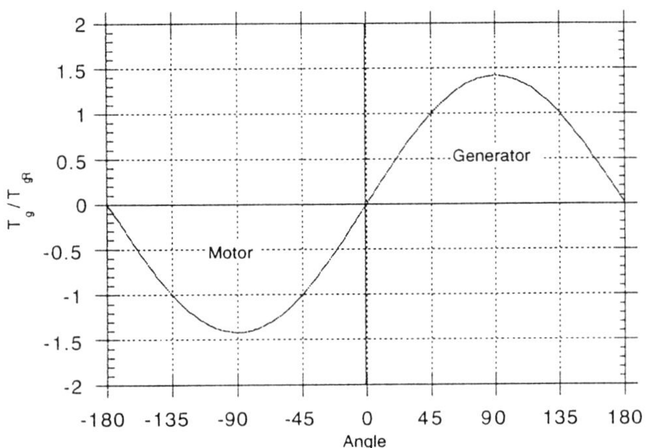

Figure 2 The electrical torque as a function of the angle between stator and rotor.

2.4 SEVERAL GENERATORS CONNECTED TO THE SAME GRID

When several generators are connected to the same grid, the delivered current will be summed up. If the electrical grid between the generators are neglectable the equation for the voltage on the grid, with a total of n generators connected, will be:

$$E = R_{grid} \sum_{i=1}^{n} \frac{1}{k_i \phi_i \cos \varphi} T_{gi}(\delta_i) \tag{13}$$

The differential equation for the angle between rotor and stator for generator no. i will be:

$$\frac{d\delta_i}{dt} = \frac{P_i}{2} \omega_{ti} - \frac{1}{(k\phi)_i} E \tag{14}$$

The sum of the influences from each aggregate decides upon the grid frequency after a change of load. The permanent speed droop is adjustable on the frequency governor. The permanent speed droop means that a certain frequency deviation, dependent on the load, will be allowed. The permanent speed droop also decides how the total load change will be distributed to each of the connected generators, as all synchronous machines connected to the same grid must end up at the same stationary frequency.

2.5 FREQUENCY GOVERNOR

The transfer function for a PI governor with permanent speed droop and servo motor time constant is:

$$H_{reg} = \frac{1+T_D s}{(\delta_b + \delta_t T_D s)} \frac{1}{(1+T_K s)} \tag{15}$$

The equation can be transformed to time regime. The differential equations will be:

$$\frac{dy_K}{dt} = c \tag{16}$$

$$\frac{dc}{dt} = \frac{y_{ref}}{T_K}\left[-\frac{1}{\delta_t}\frac{1}{n_{ref}}\frac{dn}{dt} + \frac{1}{\delta_t T_d}\left(\frac{n_{ref} - n}{n_{ref}}\right) - \frac{(\delta_b T_K + \delta_t T_d)}{\delta_t T_d} c - \frac{\delta_b}{\delta_t T_d}(y_{ref} - y)\right]$$

where:

y	-	Servo motor position
c	-	Servo motor velocity
n	-	Speed of rotation
δ_t	-	Transient speed droop
δ_b	-	Permanent speed droop
T_d	-	Integration time
T_K	-	Servo motor time constant

For simplicity reasons the equations shown are for a PI governor with only one time constant representing the servo motor. The consequence of taking into account additional time constants is that each time constant will give one more differential equation.

2.6 VOLTAGE GOVERNOR

By governing the generator's magnetising power, the flux ϕ will be changed and thereby also the output voltage. The voltage governor is modelled as a PI governor with the equation:

$$\frac{d(k\phi)_i}{dt} = -\frac{1}{\delta_{tg} E_{ref}}\frac{dE}{dt} + \frac{1}{\delta_{tg} T_{dg} E_{ref}}\left(E_{ref} + \frac{1}{\delta_{bg} I_{ref}}(I - I_{ref}) - E\right) \tag{17}$$

The equation contains the derivative of the voltage on the grid, which can be found by taking the derivative of the equation (17) above. If the dynamic properties of the grid is included, the capacitance will decide the derivative of the grid voltage according to the differential equation:

$$C_{grid} \frac{dE}{dt} = \sum_{i=1}^{n} \frac{1}{(k\phi)_i \cos\varphi} T_{gi}(\delta_i) - \frac{E}{R_{grid}} \tag{18}$$

3. Simulations

3.1 POWER PLANT DATA

The model is tested by simulating the system shown in Figure 3. The system contains two water power plants electrically connected to a grid.

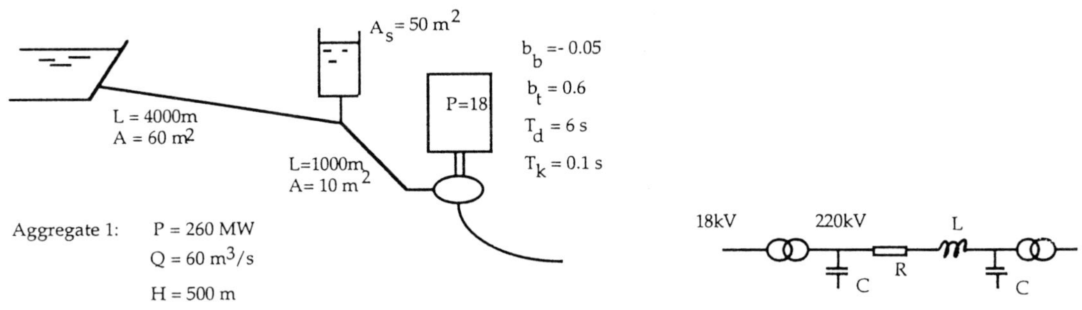

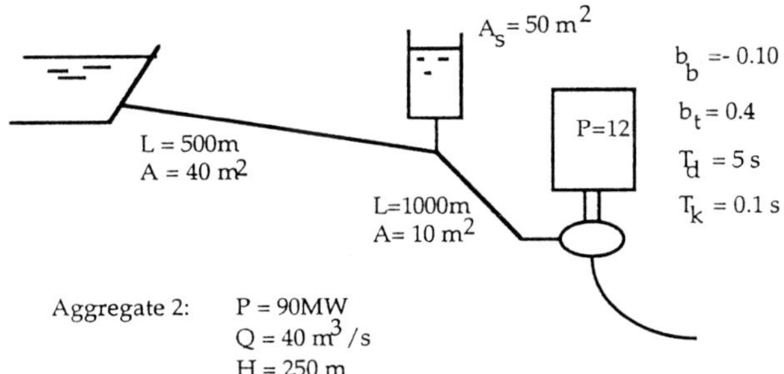

Figure 3 Power generating system used in the simulations.

3.2 SIMULATIONS

The simulation results are shown in the following figures.

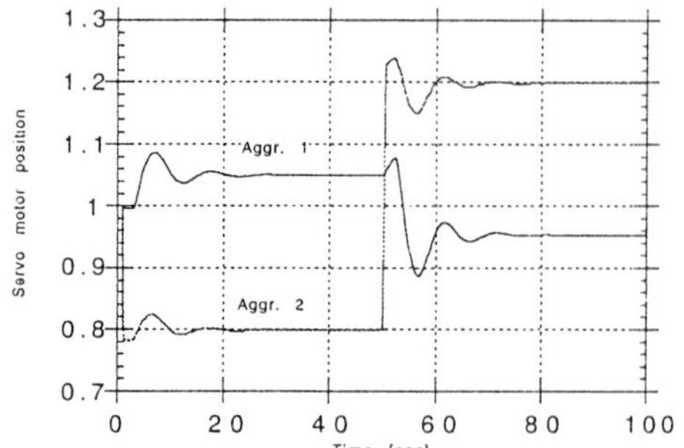

Figure 4

The power output of the aggregate 2 is decreased 20%, and then after 50 seconds, increased up to 20 % above rated. The bigger aggregate 1 will, in order to maintain the frequency, adjust the power output accordingly.

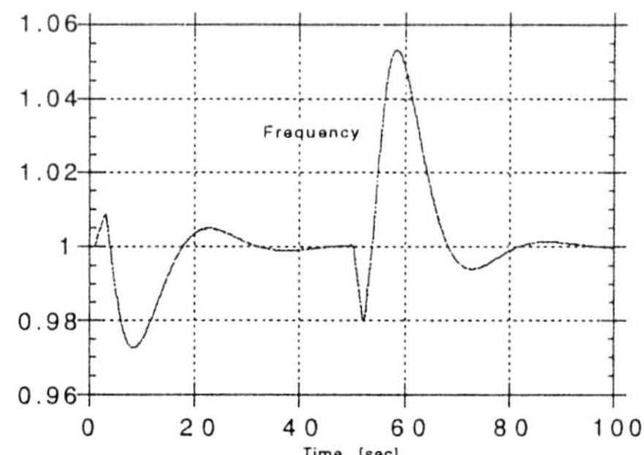

Figure 5

The frequency variation during the load regulation.

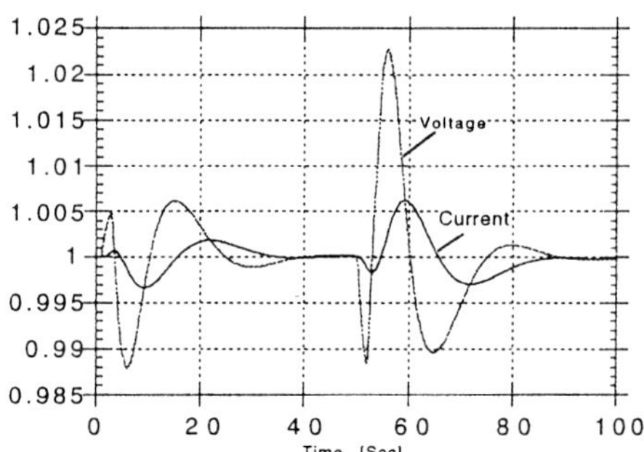

Figure 6

Voltage and current due to the load regulation shown in figure 4.

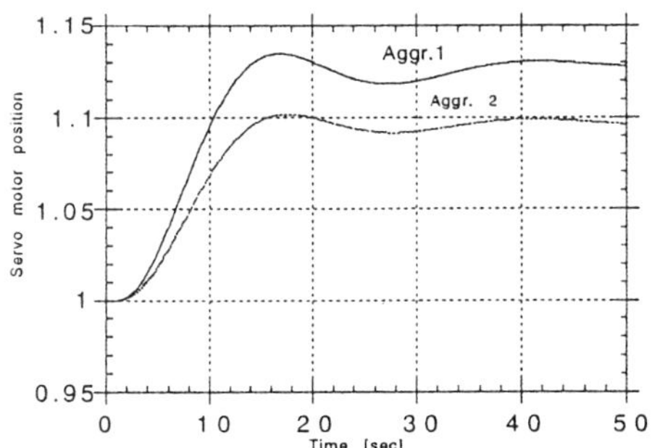

Figure 7

Servo motor position during 10% load admittance. The two aggregates have different permanent speed droops, 0.01 and 0.02 respectively.

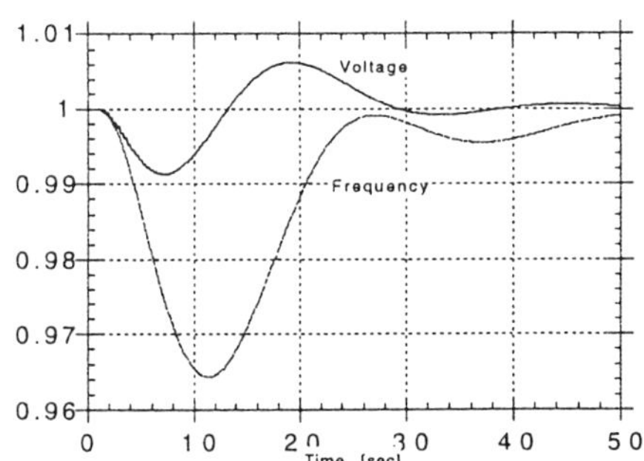

Figure 8

Voltage and frequency variation during 10% load admittance.

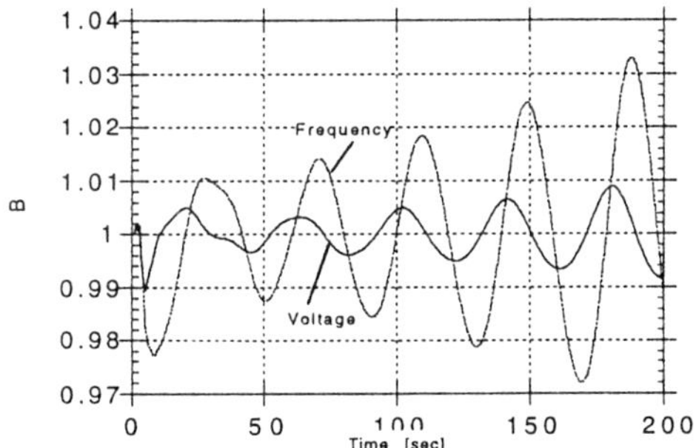

Figure 9

If the shaft crossection of one of the power plants is less than Thoma crossection, the result will be unstable U-tube oscillations resulting in both unstable frequency and voltage governing.

REFERENCES

1. T.K.Nielsen, F.O.Rasmussen: Analytic Model for Simulation of Francis Turbines Implemented in MOC. IAHR-Symposium - Sao Paulo 1992.

2. T.K.Nielsen: Transient Characteristics of High Head Francis turbines. Dr.Ing dissertation, NTH.

3. Fitzgerald, Kingsley, Umans: Electric Machinery, McGraw-Hill Book Company 1990.

PREDICTION OF NATURAL FREQUENCIES IN A HYDRO POWER PLANT SUPPLYING AN ELECTRIC NETWORK BY ITSELF HAVING A KNOWN LOAD TYPE

J.RAABE
Dr.-Ing, Professor Emeritus, Lehrstuhl für Hydraulische Maschinen und Anlagen, Technische Universität, München, Germany
Sonnenstr. 1 D 82049 Pullach/Isar , FR Germany

1. Introduction

Consider a hydro-electric power plant, according to Fig.1, whose pipe system consists of a penstock upstream and a tailwater tunnel downstream having different lengths L_o and L_u , and hence water hammer-conditioned wave passage times $T_{lo}= L_o/a_o$,$T_{lu}=L_u/a_u$,a_o and a_u being the celerities (sonic speeds). The penstock is starting and the tailwatertunnel is ending in a large basin having constant pressure head. The other ends of both the pipe sections are connected with the pressure-and suction-flange of a speed-and gate-controlled Francis turbine having its known hill-diagram. The turbine is driving an alternator (generator) supplying an electric network by itself having a known load type ,e.g. resistors, machine tools, pumps and blowers. This set has a definite start up time T_a . The turbine gate is adjusted by a servomotor, see Fig.1 via a valve activated by a speed deviation meter(here symbolized by a fly-ball pendulum), and fed back by the servomotor piston rod via a dash pot, ensuring a stiff connection of valve-activating lever end and servo motor piston under quick motion here considered. This integral proportional governor by its elastic feedback has a reset time T_i, see [1].

Under fast operation considered it has an adjustable temporary speed droop δ_T ,see [1]. Piping and set are the machine with its servomotor deviation Δs as input,and its speed deviation $\Delta n=n - n_N$ as output,whereas the governor has Δs as input and Δn as output. Shortcircuiting machine and governor, one obtains the control loop of plant considered in this paper with respect to its dynamic behaviour under external disturbance especially under a change of load type within the electric network indicated by change of exponent z relating by $M_g \sim n^z$ generator torque and speed n of the set to one another.

The plant with its head H_N has a linearized pipe -conditioned head loss Δh_l,see Fig.1. The rated flow Q_N is linked to flow velocities c_{No} ,c_{Nu} within both the pipe sections, thus having water hammer linked parameters $a_u c_{Nu}/(gH_N)=K_u$;$a_o c_{No}/(gH_N)=K_o$ .By external disturbances,the plant is assumed to perform oscillations with that small amplitudes, so that relations between plant elements may be linearized. Obviously damping δ and cyclic frequency ω of the natural oscillations, depend

E. Cabrera et al. (eds.), Hydraulic Machinery and Cavitation, 779–788.

on the plant parameters T_i, T_a, T_{lo}, T_{lu}, K_o, K_u, z, δ_T ,already mentioned, together with flow resistance coefficient φ and load factor Λ , being the ratio of torque into rated torque.

The main aim of this paper is to predict from the governing differential equation for the speed deviation ratio $x = \Delta n/n_N$, being of the free oscillator type, whose damping factor and "spring rate" depend on the frequency $-\delta+i\omega$ by transcendental functions , both the components of this frequency in function of the known plant parameters above mentioned, and how to eliminate instabilities occurring by adequate change of these parameters by procedures also shown in [2,3]

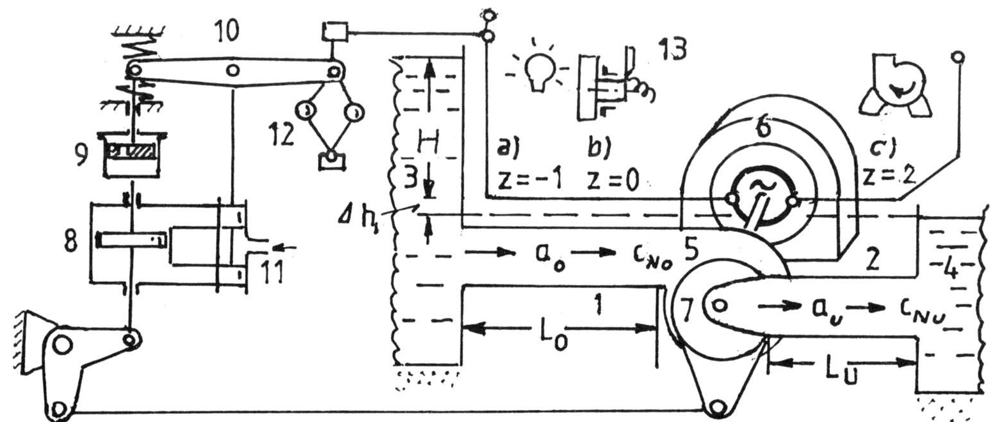

Fig.1 Plant Scheme 1 penstock 2 tailwater duct 3 head water 4 tailwater 5 turbine 6 alternator 7 gate ring 8 servomotor 9 dashpot 10 feedback lever 11 valve 12 tachometer 13 Electric network supplying a) resistive load z=-1 b) machine tools z=0 c) pumps & blowers z=2

2. In- and Out-put of plant elements, governing equation

Equation of motion of the set for a known electric network load type : Input: Head surge ratio $h = \Delta H/H_n$, servomotor position deviation ratio $m = \Delta s/s_M$ .Output: speed deviation ratio x and its rate $\dot{x}$:

$$\Lambda\,[(K_{\eta m} + K_{Qm})m - (1 + z - K_{\eta n} - K_{Qn})x + \frac{3}{2}h\,] = T_a\,\dot{x}\,, \tag{1}$$

in which the K_{ij} are reflecting non-dimensional features of the turbine hill diagram, connected with the working point due to the load factor Λ:

$$K_{\eta m} = \frac{s_M}{\eta_i}\frac{\partial\eta_i}{\partial s}\,,\quad K_{Qm} = \frac{s_M}{Q_{11}}\frac{\partial Q_{11}}{\partial s}\,,\quad K_{\eta n} = \frac{n_N}{\eta_i}\frac{\partial\eta_i}{\partial n}\,,\quad K_{Qn} = \frac{n_N}{Q_{11}}\frac{\partial Q_{11}}{\partial n}\,, \tag{2}$$

where η_i- internal efficiency, Q_{11} - unit flow , T_a - set start up time.

Intervention of piping by water hammer: Our pipe system, with penstock o and tailwater tunnel u , its corresponding water hammer-conditioned parameters K_o and K_u, its linearized pipe friction coefficient φ , being defined by $\Lambda\bar{\lambda}(L/d)c^2/gH_N$($d$-pipe diameter, λ-Weißbach coefficient), its flow surge ratio $q=\Delta Q/Q_N$, and its complex frequency $p = -\delta + i\omega$, has an impedance μ according to

$$\mu \equiv - h/q = K_o \tanh p\, T_{lo} + K_u \tanh p\, T_{lu} + \varphi \quad . \tag{3}$$

Governor relation: Input, speed deviation ratio x; Output , servomotor position deviation ratio m . With isodrome time T_i of dash pot , and temporary speed droop δ_T, the equation of the opened governor circuit reads

$$T_i\, \dot{x} + x = - \delta_T\, T_i\, \dot{m} \quad . \tag{4}$$

The governing differential equation of plant oscillation in terms of x :

$$\ddot{x} + [A - B\, \mu/(1 + K_\eta\, \Lambda\, \mu/2)]\, \dot{x} + [C - D\, \mu/(1 + K_\eta \Lambda\, \mu/2)]\, x = 0, \tag{5}$$

$$A = (\Lambda/T_a)\,[(K_{\eta m} + K_{Qn})/\delta_T + 1 + z - K_{\eta n} - K_{Qn}], \text{(6)} \quad K_\eta = (\eta_{iN}\, H_N)/(\eta_i H), \text{(7)}$$

$$B = (3/2)(\Lambda^2\, K_\eta/T_a)(K_{Qm}/\delta_T - K_{Qn}), \text{(8)} \quad C = \Lambda\, (K_{\eta m} + K_{Qm})/(\delta_T T_a T_i), \text{(9)}$$

$$D = (3/2)(K_{Qm}\, \Lambda^2\, K_\eta)/(\delta_T T_a T_i) \quad . \tag{10}$$

3. Approach to solve (5) by subjecting all elements of plant to same p

To solve the governing differential equation (4) , assume that all the elements of the plant inclusive the piping are oscillating with the same complex frequency , according to , see [2]

$$x = x_o e^{(-\delta + i\omega)t} \equiv x_o e^{p\,t} \quad . \tag{11}$$

Putting this in (5) , one gets the characteristic equation for δ and ω

$$\delta^2 - 2i\omega\delta - \omega^2 + [A - B\mu/(1 + K_\eta \Lambda\mu/2)](-\delta + i\omega) + C - D\,/(1 + K_\eta \Lambda\mu/2) = 0. \tag{12}$$

Splitting of impedance μ according to (3) into real and imaginary part due to penstock o and tailwater tunnel u,according to

$$\mu = \mu_{Ro} + \mu_{Ru} + \varphi + i(\mu_{Jo} \quad \mu_{Ju}) \;, \tag{13}$$

$$\mu_{Ro} = - \frac{K_o \tanh \delta T_{lo}}{\cos^2 \omega T_{lo} (1 + \tanh^2 \delta T_{lo} \tan^2 \omega T_{lo})} \;, \tag{15}$$

$$\mu_{Ru} = - \frac{K_u \tanh \delta T_{lu}}{\cos^2 \omega T_{lu} (1 + \tanh^2 \delta T_{lu} \tan^2 \omega T_{lu})} \;, \tag{16}$$

$$\mu_{Jo} = \frac{K_o \tan \omega T_{lo}}{\cosh^2 \delta T_{lo} (1 + \tanh^2 \delta T_{lo} \tan^2 \omega T_{lo})} \;, \tag{17}$$

$$\mu_{Ju} = \frac{K_u \tan \omega T_{lu}}{\cosh^2 \delta T_{lu} (1 + \tanh^2 \delta T_{lu} \tan^2 \omega T_{lu})} \;, \tag{18}$$

and decomposition of the μ -linked term $\mu/(1 + K_\eta \mu\Lambda/2)$ into real part Re and imaginary part Im , according to

$$\text{Re} = \frac{1}{N} \{(\mu_{Ro} + \mu_{Ru} + \varphi)[1 + K_\eta \frac{\Lambda}{2}(\mu_{Ro} + \mu_{Ru} + \varphi)] + K_\eta \frac{\Lambda}{2}(\mu_{Jo} + \mu_{Ju})^2\}, \tag{19}$$

$$\text{Im} = \frac{1}{N}\{(\mu_{Jo} + \mu_{Ju})[1 + K_\eta \frac{\Lambda}{2}(\mu_{Ro} + \mu_{Ru} + \varphi)] - K_\eta \frac{\Lambda}{2}(\mu_{Jo} + \mu_{Ju})(\mu_{Jo} + \mu_{Ju} + \varphi)\} \;, \tag{20}$$

$$N = [1 + K_\eta \frac{\Lambda}{2}(\mu_{Ro} + \mu_{Ru} + \varphi)]^2 + K_\eta^2 \frac{\Lambda^2}{4}(\mu_{Jo} + \mu_{Ju})^2 \;, \tag{21}$$

results in a splitting of (11) into real and imaginary part yielding the following characteristic equations for ω and δ

$$\delta^2 - \omega^2 - A\,\delta + C - (D - \delta\,B)\mathrm{Re} + \omega B\,\mathrm{Im} = 0 \quad , \tag{22}$$

$$-\,2\,\delta\,\omega + A\,\omega - B\,\omega\,\mathrm{Re} - (D - \delta\,B)\,\mathrm{Im} = 0 \quad . \tag{23}$$

4. The case of undamped plant oscillation $\delta = 0$

Assuming undamped plant oscillations with $\delta = 0$ converts (22) and (23)in

$-\omega^2 + C - D\,\mathrm{Re} + \omega B\,\mathrm{Im} = 0$,(24) $\quad A\,\omega - B\,\omega\,\mathrm{Re} - D\,\mathrm{Im} = 0$. (25)

With δ=0,μ_{Ro} and μ_{Ru} from (15) and (16) are vanishing .Furtheron μ_{Jo} from (17) and μ_{Ju} from (18), with φ neglected,are now reading

$\mu_{Joo} = K_o \tan\omega\,T_{lo}$ (26) $\quad \mu_{Juo} = Ku \tan\omega\,T_{lu}$, (27)

and Re from (19) and Im from (20) together with N from (21) are reading

$\mathrm{Re}_o = \dfrac{1}{N_o} K_\eta \dfrac{\Lambda}{2}(\mu_{Joo} + \mu_{Juo})^2$, (28) $\quad \mathrm{Im}_o = \dfrac{1}{N_o}(\mu_{Joo} + \mu_{Juo})$, (29)

$$N_o = 1 + K_\eta^2 \frac{\Lambda^2}{4} (\mu_{Joo} + \mu_{Juo})^2 \quad . \tag{30}$$

Even if both the relations (24) and (25) are equivalent as characteristic equations , for convenience (25) is used furtheron. Using (22) -(30) (25) yields, after having brought it to the common denominator ,and ommitting the latter as a term never vanishing (31)

$$A\omega + K_\eta \frac{\Lambda}{2} (A\,K_\eta \frac{\Lambda}{2} + B)(K_o \tan\omega T_{lo} + K_u \tan\omega T_{lu})^2 - D(K_o \tan\omega T_{lo} + K_u \tan\omega T_{lu}) = 0 .$$

This is a transcendental equation for the cyclic frequency. For a huge plant the smallest value of ω is of interest . Linearizing the tan-functions $\lim\omega = 0$ brings the following lowest cyclic frequency (32)

$$\omega_{\min} = [1/(K_o T_{lo} + K_u T_{lu})] \sqrt{A - D(K_o T_{lo} + K_u T_{lu}) / [K_\eta \Lambda (K_\eta \Lambda A/2 + B)/2]} \; .$$

Example: Francis turbine of medium specific speed, working at bep,supplying an electric network , having machine tools, impeller pumps & blowers as load, yielding z=0,5;L_o=10³m; L_u=300 m; a_o= 10³ m/s; a_u=900 m/s K_o =1,7; K_u=1,53; δ_T=0,3; T_a=10s; T_i=12s; K_η=1; Λ=1; $K_{\eta m}$=0; K_{Qn} = 0; $K_{\eta m}$=0; K_{Qm}=1, yielding A=0,361; B=0,4167; C=0,027; D=0,0405 The lowest natural frequency of the plant $\omega_{\min}$ = 0,468 1/s. Hence a period $T = 2\pi / \omega$ = 13,4 s.

A vanishing of the radicand in (32) , gives a criterion to suppress any plant oscillation , according to

$$A = D(K_o T_{lo} + K_u T_{lu}) \; . \tag{33}$$

Since the K_{ij} -conditioned terms are variable with the load factor ,the keeping of (33)is questionable.The keeping of orderly oscillations would require A to be larger than the right hand side of (33). Hence at given plant parameters T_{lo},T_{lu} , K_o K_u,K_η , Λ, z , δ_T,$K_{\eta m}$,$K_{Q\eta}$, $K_{\eta n}$,K_{Qm} the dash pot reset time of the governor,has to follow from (34)

$$T_i > (3/2)(\Lambda K_\eta/\delta_T)(K_o T_{lo} + K_u T_{lu})K_{Qm} / [(K_{\eta m} + K_{Qm})/\delta_T + 1 + z - K_{\eta n} - K_{Qn}] .$$

For a set working with given reset time T_i and given parameters T_{lo}, T_{lu} K_i,K_{IJ} a change of load factor $\Delta\Lambda$ and load type by Δz ,requires by

(34) ,after having substituted the unequality sign by an equality sign in the special case of $K_{Qn}=0$, $K_{\eta m}=0$, $K_{\eta n}=0$, $K_{Qm}=ct$, the following change of temporary speed droop in function of load factor change $\Delta\Lambda$ and load type change Δz

$$\Delta\delta_T = \delta_T[\Delta\Lambda/\Lambda - \Delta z/(1+z)] \quad . \tag{35}$$

The general solution of the transcendental equation (31) to determine an arbitrary natural frequency ω requires some consideration .If both the wave passage times T_{lo} and T_{lu} for penstock and tailwater tunnel, respectively , are incommensurable, a practice-oriented solution of (31) does not exist. Any relevant solution of (31) has to assume the ratio T_{lo}/T_{lu} to be rational,e.g. m/n ,m and n being integers. Hence with the new coordinate $y = \omega T_{lu}$,according to fig. 2 ,showing the special case m=3, n=1, in the environce of infinitely many stations , having the abscisses=$ln\pi$,l being integer,(34) is complied by argument $y - l n \pi$ nearly zero. This has something to do with the relative smallness of the parameter D . This first group of solutions may be determined by linearizing $\tan(m/n)(y-ln\pi)=(m/n)(y-ln\pi)$.Hence (32) is converted into a 3rd degree equation of y, namely

$$yA = K_u\left[\frac{K_o}{K_u}\frac{T_{lo}}{T_{lu}}(y-ln\pi)+y-ln\pi\right]\left\{DT_{lu}-K_\eta K_u\frac{\Lambda}{2}\left(\frac{A}{2}K_\eta\Lambda+B\right)y\left[\frac{K_o}{K_u}\frac{T_{lo}}{T_{lu}}(y-ln\pi)+y-ln\pi\right]\right\} . \tag{36}$$

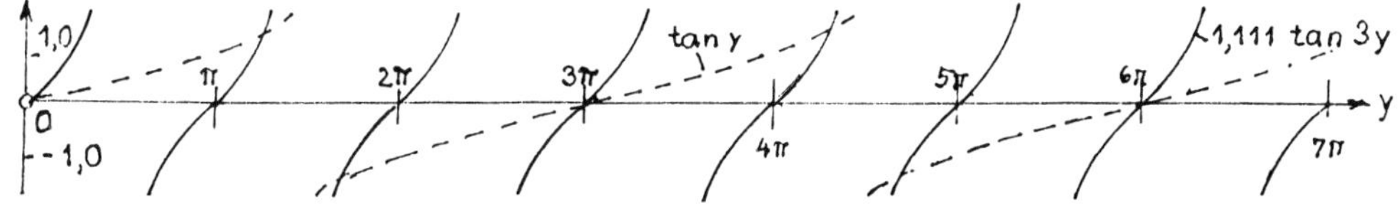

Fig.2. 1st group solution of (36) having transcendental term 1,111 tan 3y+tan y,y=$T_{Lu}\cdot\omega$ · —— 1,111 tan 3y, --- tan y.Smallness of other terms shifts solution close to 3π, 6π....

Fig. 2 shows an example of the case n=1, m=3 . A second group of $l(m-1)$ solutions is in the neighbourhood of the abscisses $l\pi$,.. $l(m-1)\pi$ being more distant from these abscisses than the 1st group solution was from the abscisses $y=l n\pi$, these solutions of the 2nd group solutions have to be calculated by expanding the tan-functions into *Taylor* Series,yielding instead of the 3rd degree equation (36) an equation of considerably higher degree, when practice-oriented accuracy is desired. For the relevant case of a huge plant one may state that from the 2-fold infinity of cyclic frequencies thus obtained only the lowest may be of interest when excluding that,according to (32). Hence the example considered above , for l = 1 results by (36) in ω = 9,42 rad/s, connected with a period T= 0,662 s, already rather unbelievably low for a big plant.

5. The case of damped plant oscillations

For damped oscillations the 2 unknowns δ and ω are following from both the characteristic equations (22) and (23). Starting from (23) ,this reads

$$F(\delta,\omega) = -2\delta\omega + A\omega - B\omega\,\mathrm{Re} - (D-\delta B)\mathrm{Im} = 0 \quad . \tag{37}$$

As it is well known, the total differentials $d\delta$, $d\omega$ now approximated by finite steps $\Delta\delta$, $\Delta\omega$, are, with $F = F(\delta,\omega)$, related to each other by

$$\Delta\delta = -(F_\omega/F_\delta)\,\Delta\omega \quad , \tag{39}$$

where F_ω, F_δ are the partial derivatives of F with respect to ω and δ, respectively. They are obtained by (15) through (21) via

$$F_\omega = -2\delta - B\,\mathrm{Re} - B\,\omega\frac{\partial \mathrm{Re}}{\partial\omega} - (D - \delta B)\frac{\partial \mathrm{Im}}{\partial\omega} + A \, , \tag{40}$$

$$F_\delta = -2\omega - B\,\omega\frac{\partial \mathrm{Re}}{\partial\delta} + B\,\mathrm{Im} - (D - \delta B)\frac{\partial \mathrm{Im}}{\partial\delta} \quad , \tag{41}$$

where $\partial\mathrm{Re}/\partial\omega$, $\partial\mathrm{Im}/\partial\omega$, $\partial\mathrm{Re}/\partial\delta$, $\partial\mathrm{Im}/\partial\delta$ are calculated by the "chain rule" of calculus, e.g.

$$\frac{\partial \mathrm{Re}}{\partial\omega} = \frac{\partial \mathrm{Re}}{\partial\mu_{Ro}}\frac{\partial\mu_{Ro}}{\partial\omega} + \frac{\partial \mathrm{Re}}{\partial\mu_{Ru}}\frac{\partial\mu_{Ru}}{\partial\omega} + \frac{\partial \mathrm{Re}}{\partial\mu_{Jo}}\frac{\partial\mu_{Jo}}{\partial\omega} + \frac{\partial \mathrm{Re}}{\partial\mu_{Ju}}\frac{\partial\mu_{Ju}}{\partial\omega} \quad , \tag{42}$$

in which the derivatives of Re with respect to μ_{Rou} and μ_{Jou} have to be taken from (19) and (21), and the derivatives of Im with respect to μ_{Rou} and μ_{Jou} have to be calculated by (20) and (21), thus bringing

$$\frac{\partial \mathrm{Re}}{\partial\mu_{Ro}} = \frac{\partial \mathrm{Re}}{\partial\mu_{Ru}} = \frac{1}{N}\left\{\left|1 + K_\eta\frac{\Lambda}{2}(\mu_{Ro} + \mu_{Ru} + \varphi)\right| + (\mu_{Ro} + \mu_{Ru} + \varphi)K_\eta\frac{\Lambda}{2}\right\} - \tag{43}$$
$$\frac{1}{N^2}\left[(\mu_{Ro}+\mu_{Ru}+\varphi)\left[1+K_\eta\frac{\Lambda}{2}(\mu_{Ro}+\mu_{Ru}+\varphi)\right]+K_\eta\frac{\Lambda}{2}(\mu_{Jo}+\mu_{Ju})^2 K_\eta\Lambda\left[1+K_\eta\frac{\Lambda}{2}(\mu_{Ro}+\mu_{Ru}+\varphi)\right]\right],$$

$$\frac{\partial \mathrm{Re}}{\partial\mu_{Jo}} = \frac{\partial \mathrm{Re}}{\partial\mu_{Ju}} = K_\eta\frac{\Lambda}{N}(\mu_{Jo} + \mu_{Ju}) - \tag{44}$$
$$\frac{1}{N^2}\left[\left\{(\mu_{Ro}+\mu_{Ru}+\varphi)\left[1+K_\eta\frac{\Lambda}{2}(\mu_{Ro}+\mu_{Ru}+\varphi)\right]+K_\eta\frac{\Lambda}{2}(\mu_{Jo}+\mu_{Ju})^2\right\}K_\eta^2\frac{\Lambda^2}{2}(\mu_{Jo}+\mu_{Ju})\right] \; .$$

The derivatives $\partial\mu_{Ro}/\partial\omega$, $\partial\mu_{Ru}/\partial\omega$, $\partial\mu_{Jo}/\partial\omega$ and $\partial\mu_{Ju}/\partial\omega$ within (42) are obtained from (15) through (18) by setting for the subscripts o or u in the following relations the subscript e

$$\frac{\partial\mu_{Re}}{\partial\omega} = K_e T_{le}\frac{\tanh^3\delta T_{le}\sin 2\omega T_{le}}{\cos^4\omega T_{le}(1 + \tanh^2\delta T_{le}\tan^2\omega T_{le})^2} \quad , \tag{45}$$

$$\frac{\partial\mu_{Je}}{\partial\omega} = K_e T_{le}\frac{1 - \tanh^2\delta T_{le}\tan^2\omega T_{le}}{\cos^2\omega T_{le}\cosh^2\delta T_{le}(1 + \tanh^2\delta T_{le}\tan^2\omega T_{le})^2} .$$

To get $\Delta\delta$ for a known natural frequency ω (of which one has according to the foregoing chapter an infinity of values), we start from ω_o corresponding to $\delta = 0$. For $\omega_1 = \omega_o + \Delta\omega_1$ the corresponding step $\Delta\delta_1$ by (39) becomes

$$\Delta\delta_1 = -(F_\omega/F_\delta)\,\Delta\omega \quad . \tag{47}$$

According to the mean value rule of calculus the ratio F_ω/F_δ shall be adopted for an intermediate station between $\omega_o, \delta = 0$ and ω_1, $\Delta\delta_1$. However in a first approximation, we are taking δ from station ω_1, $\delta = 0$. Hence for the 2nd step with $\omega_2 = \omega_1 + \Delta\omega_2$ and $\delta_1 = \Delta\delta_1$ the corresponding step

$$\Delta\delta_2 = -(F_\omega/F_\delta)(\omega_2,\delta_1)\Delta\omega_2 \quad , \tag{48}$$

and so on . By this procedure the damping coefficient δ of a real plant may be computed in function of cyclic frequency and plant parameters.

6. Damping and frequency of plant when water hammer treated separately

In this chapter an approach is made to get damping and frequency of the plant in function of known plant parameters, save those due to water hammer. However, the latters are accounted in so far as the impedance-conditioned factor $\mu/(1 + K_\eta\Lambda\mu/2)$ in the governing equation of plant oscillation (5) is derived from an impedance according to (3) under the following 2 assumptions, 1) The oscillations of the pipe system are not damped, 2) The pipe system is subject to oscillations having a frequency due to a period being twice the water hammer wave reflection time $2L/a$, being a the celerity and L the length of pipe branch.

Hence the pipe oscillation period of a pipe branch $T = 4L/a$. Moreover without damping, $p = i\omega = i\ 2\pi/T = i\pi a/2L$. Accordingly with wave passage time $T_l = L/a$, the argument in the impedance according to (3), $pT_l = i\,\frac{\pi}{2}$. Hence $\tanh p\,T_l = \tanh i\frac{\pi}{2} = i\tan\frac{\pi}{2} = i\,\infty$. In this case, with $\varphi = 0$, the impedance according to (3) now is reading

$$\mu = (K_o + K_u)\,i\,\infty \quad . \tag{49}$$

Hence the impedance-conditioned parameter in the governing equation (5)

$$\mu/(1 + K_\eta\Lambda\mu/2) = 2/(K_\eta\Lambda) \quad . \tag{50}$$

Putting this in the governing equation this adopts the rather simple form

$$\ddot{x} + [A - 2B/(K_\eta\Lambda)]\dot{x} + [C - 2D/(K_\eta\Lambda)]x = 0 \quad . \tag{51}$$

Now the approach (11) for solution yields two characteristic equations

$$\delta^2 - \omega^2 - [A - 2B/(K_\eta\Lambda)]\,\delta + C + 2D/(K_\eta\Lambda) = 0 , \tag{52}$$

$$-2\delta\omega + [A - 2B/(K_\eta\Lambda)]\,\omega = 0 \quad . \tag{53}$$

From the last equation one obtains a relation for the damping factor

$$\delta = [A - 2B/(K_\eta\Lambda)]/2 \quad . \tag{54}$$

Inserting this in (52) one obtains for the natural plant frequency

$$\omega = \sqrt{C + AB/(K_\eta\Lambda) - [2D/(K_\eta\Lambda) + A^2/4 + B^2/(K_\eta{}^2\Lambda^2)]} \quad . \tag{55}$$

For a real cyclic frequency the radicand has to be positive according to

$$C + AB/(K_\eta\Lambda) > 2D/(K_\eta\Lambda) + A^2/4 + B^2/(K_\eta{}^2\Lambda^2) \quad . \tag{56}$$

Putting the values A, B, C, D, from (6) through (10) in the last relation and accounting for $2K_{Qm} >> K_{\eta m}$ then one obtains the following unequation for the ratio of set start up time T_a into governor reset time T_i

$$T_a/T_i << [\delta_T\Lambda^2/(2K_{Qm} - K_{\eta m})]\ M \ , \tag{57}$$

$$M = \frac{3}{2}K_\eta\left[\frac{K_{Qm}}{\delta_T} - K_{Qn}\right]\left[\frac{K_{\eta m}+K_{Qn}}{\delta_T} + 1 + z - K_{\eta n} - K_{Qn}\right] + \frac{1}{4}\left[\frac{K_{\eta m}+K_{Qn}}{\delta_T} + 1 + z - K_{\eta n} - K_{Qn}\right]^2 + \frac{9}{4}\left[\frac{K_{Qm}}{\delta_T} - K_{Qn}\right]^2 .$$

In a Francis turbine of medium specific speed $K_{Qn} = 0$. If the turbine is working around the bep, then $K_{\eta n} = 0$, $K_{\eta m} = 0$, $\Lambda = 1$, $K_\eta = 1$, (57) reduces to a rather simple rule. (49) may be converted into an equation, after its right hand side has been multiplied by a factor α smaller than 1, say $\alpha = 0{,}8..0{,}9$. Then the simplified version of (57) just mentioned, reads

$$\Lambda^2(1 + z)\left[(3/2)K_\eta + (1/4)\delta_T(1 + z)/K_{Qm}\right] = 2\,(T_a/T_i)\,\alpha \quad (58)$$

Assuming for a proper dynamic behaviour of the plant $(T_a/T_i)\,\alpha$ to be unaltered, and considering also K_{Qm} and K_η to be constant, differentiation of the last equation answers the problem how to change the temporary speed droop δ_T owing to a measured increment of load factor due to the turbine $\Delta\Lambda$ and of load type in the electric network by Δz

$$\Delta\delta_T = -\left[\frac{6\,K_\eta K_{Qm}}{1+z} + \delta_T\right]\left[\frac{2\Delta\Lambda}{\Lambda} + \left[\frac{1}{1+z} + \frac{\delta_T}{6\,K_\eta K_{Qm} + \delta_T(1+z)}\right]\Delta z\right] . \quad (59)$$

If one wants to suppress any oscillation (57) may be used with an equality sign. In this case $\alpha = 1$. To be sure in this respect, one may also reverse the sense of the unequality sign in (56), then $\omega = i\,G$, G being positive and $\exp(i\omega t) = \exp(-G\,t)$ obviously guaranteeing damping. However, the factor $\exp(-\delta t)$ remains in the approach (11). To make sure that no oscillation occurs the damping coefficient has to be set to have a distinct value δ_0. Hence if B, K_η, Λ are given, then A has to be

$$A = 2\,\delta_o + 2B/(K_\eta \Lambda)\,. \quad (60)$$

Inserting in this relation the values of A and B according to (6) and (8) the last equation in terms of turbine hill diagram, load factor, electric network load type and temporary speed droop represents the compliance of a secondary condition for all these parameters in the form

$$\Lambda\left[\frac{K_{\eta m} + K_{Qn}}{\delta_T} + 1 + z - K_{\eta n} - K_{Qn}\right] = 2\,\delta_o\,T_a + 3\,\Lambda^2 K_\eta\left(\frac{K_{Qm}}{\delta_T} - K_{Qn}\right) \quad . \quad (61)$$

Since the parameters K_{ij} of the shell diagram are not changing so much this last relation also gives a hint at how to change the temporary speed droop of the governor in dependence of change in turbine load factor and electric network load type.

Some critical remarks on the result of this chapter, being contradictory to the statements in chapter 3 through 5, will be made. Under item 1) of this chapter undamped pipe oscillations have been assumed, being reflected by the approach $p = i\omega$ within an impedance μ according to (3).

However, as one may find in [1] this formula presumes the complex frequency p of piping to be the same as that of the plant. Since the latter in this chapter was found to be damped, that of the piping, according to (3) should also be damped. In [1] it is shown that the argument of the tanh function in (3) is different in both the pipe sections, whereas in this chapter it is the same. Surely, when moving the gate by the governor, an element of the plant, also the piping is subject to plant oscillations. But in this chapter the decisive feature of pipe oscillation is water hammer reflection time.

To explain these contradictions one may recall chapter 4 and 5 where the plant was found susceptible to an infinite number of natural frequencies, one of which may coincide with (55). The passage of quasi pitch-periodic flow fields in gate and runner vane channel, due to a usually incommensurable number of runner- and gate -vanes, also excites vibrations. Assuming spiral casing and draft tube inlet(at least under off-design load)to be constant pressure vessels, and the vane-less interspace between gate and runner to have constant pressure, water hammer waves are running between gate inlet and runner exit, having a frequency linked to the geometry of both. Interference of these vibrations with those of the piping may lead to resonance or beating.In consequence of annular effect|4|the thickness of wall-near boundary layer decreases with rising frequency,thus enlarging internal loss there. In this boundary layer pressure and velocity fluctuations are nearly in phase,hence linked to a power flux.Outside this layer both the fluctuations , may have up to 90° time lag decreasing power flux to zero. Growing fluctuation amplitudes may occur in volumes having a surplus of power inflow.

Recently Dr. *W. Hudovernik* ,assistant professor at University of *Innsbruck* ,has published a report on transients in pipe systems , mainly comparing measurements on 2 Austrian high head plants, *Silz* and *Kühtai* with calculations. He distinguishes a) Simple systems , in which the path of pressure waves is limited by 2 reflecting faces, e.g. penstock with outlet b) Coupled systems ,having more than two reflecting faces, e.g.penstock, surge tank and gallery.The author also distinguishes with respect to excitation the simple one,having a temporal gradient of flow rate without change of sign, and multiple excitation,having this change. As a whole ,an astonishing agreement is attained in the author's measurements with calculation mainly gained by method of characteristics [5].

7. Conclusions

By linearizing the equation of motion of an hydroelectric set,having its turbine hill diagram with its features K_{ij} ,and its start up time, whose generator is supplying alone an electric network charged by a definite load , indicated by a certain exponent of speed in the generator torque vs speed relation, the turbine being supplemented by a pipe system with its water hammer conditioned characteristics, controlled by a proportional integral speed governor with dashpot reset time and temporary speed droop, the governing differential equation for speed deviation may be established. This governing equation is of the oscillator type, having damping coefficient and"spring rate" , both being on the one hand rational functions of known plant parameters , but on the other hand transcendental functions of the complex frequency wanted of the plant considered. Assuming all components of the plant inclusive the pipe system to be subject to this frequency , and small amplitudes, the cyclic frequency of undamped oscillations may be obtained from a transcendental equation. In the case of commensurable wave passage times of both pipe sections the system is susceptible to an infinite number of natural frequencies. The relevant lowest one results via linearization by a quadratic equation, the next highest one by a 3 rd degree equation. Higher frequencies for a relevant huge plant are of no interest,even if calculable by graph intersection, or numerical by *Taylor* expansion of goniometric functions. Relevant lowest frequency follows in function of plant parameters. Taking into account damping, a stepwise solution is required. Deriving the plant frequency by limiting water-hammer conditioned frequency only to the pipe system , cyclic frequency and damping of the plant result via straightforward method in function of plant parameters. This allows to set criteria for control parameters in function of known plant parameters to attain a definite dynamic behaviour of the plant.

8. References

[1] Raabe,J.(1989) *Hydraulische Maschinen und Anlagen* VDI Düsseldorf
[2] Raabe,J.(1985) *Hydro Power*. VDI Düsseldorf
[3] Raabe,J.(1961) Stabilitätsbetrachtungen an Wasserturbinenreglern m. Berücksichtigung des Druckstosses *Maschinenmarkt* 67 Nr 12 15-23
[4] Schlichting,H.(1958) *Grenzschichttheorie* ,Braun, Karlsruhe
[5] Hudovernik,W.(1995) Kompendium über die dynamische Druckbeanspruchung im Triebwassersystem von Mittel-und Hochdruck-Wasserkraftanlagen. Report Universität Innsbruck, Austria , Inst.für Wasserbau

9. Nomenclature

a	celerity, sonic speed	m/s
A	coefficient	1/s
B	coefficient	1/s
C	coefficient	$1/s^2$
c_N	flow velocity at Q_N	m/s
D	coefficient	$1/s^2$
K_η	dimensionless parameter (2)	
$K_{\eta m}$	defined by (2) $\sim \partial\eta_i/\partial s$	
K_{Qm}	defined by (2) $\sim \partial Q_{11}/\partial s$	
K_{Qn}	defined by (2) $\sim \partial Q_{11}/\partial n$	
$K_{\eta n}$	defined by (2) $\sim \partial\eta_i/\partial n$	
H	head	m
h	$\Delta H/H$ head surge ratio	
ΔH	head surge	m
K_i	$a_i c_{Ni}/(gH)$ water hammer parameter	
Im	imaginary part defined (20)	
L_i	length of pipe section i	
l	integer	
m	$\Delta s/s_M$ servomotor deviation	
Δs	servomotor deviation	m
s_M	servomotor stroke	
N	denominator due to (21)	
n	speed	rev/s
Q	discharge,flow rate	m^3/s
ΔQ	flow rate deviation	m^3/s
q	$\Delta Q/Q$ flow deviation ratio	
T_{li}	wave passage time in pipe i	
T	period $2\pi/\omega$	
T_a	set start up time $(2\pi)^2\Theta n^2/P$	s
P	power	m^2kg/s^3
p	pressure	$kg/(ms^2)$
Re	real part due to (19)	
T_i	dash pot reset time	s
t	time elapsed	s
x	$\Delta n/n_N$ speed deviation ratio	
Δn	speed deviation	rev/s
z	due to M_G=konst n^z	
M_G	generator torque	
Θ	axial mass moment of inertia of the set	kgm^2
Δ	deviation	
δ_T	temporary speed droop	
δ	damping coefficient	1/s
ω	cyclic frequency	1/s
α	factor	
Λ	load factor-torque/rated torque	
η	efficiency (internal)	
μ	impedance = - h/q	

Subscripts

N	rated point
o	due to penstock
u	due to tailwater tunnel
i	internal
J	due to imaginary part
R	due to real part
min	lowest value
o	due to δ = 0

MODELLING AND PRACTICAL ANALYSIS OF THE TRANSIENT OVERSPEED EFFECT OF SMALL FRANCIS TURBINES

HELENA RAMOS, and A. BETÂMIO DE ALMEIDA

Ph.D. in Civil Engineering and Full Professor of Dept. of Civil Engineering

Technical University of Lisbon, I.S.T., (Fax: 351-1-8418150) Av. Rovisco Pais, 1096 Lisboa Codex, Portugal

Abstract

The transient overspeed effect of slow Francis turbine can be a major factor in the design of a small hydroelectric scheme with long penstocks or tunnels.

After a full-load rejection, in a small hydroelectric power plant equipped with low inertia Francis turbine, the runner speed will increase and the runaway conditions can be quickly attained. The hydrotransients induced by overspeed of reaction turbines depends on the turbine type. The specific speed indicates the main characterization of the turbine behaviour in what concerns the flow variation during runaway conditions.

In order to improve the pressure transient analysis since the early phases of the design of the hydraulic circuit, for lack of information about the equipment to be installed, the authors present in this paper a simplified method for reaction turbines modelling that was validated by computational and experimental comparisons.

E. Cabrera et al. (eds.), Hydraulic Machinery and Cavitation, 789–798.

1. Introduction

Transient overspeed of small Francis turbines in hydropower systems with long circuits can be a major factor in overpressures and safety assessment. For a small powerplant connected to a large (infinite) electric grid it is assumed that speed-frequency control stability is guaranteed and that transmission line effect can be neglected, in what concerns the hydraulic transients. Actually, one of the most severe safety problem in what concerns the waterhammer phenomenon occurs after a full-load rejection of small low reaction turbines (with low inertia). In that case the hydraulic system is disconnected from the electric grid and the flow disturbance is due to runner speed variation as well as to wicket closure.

Lack of complete technical data from turbine manufacturers during the early phases of the project enables the use of powerful turbine models (e.g. Suter technique). A simplified computational modelling can be an important tool to overcome this strong limitation to the hydraulic system design and to the specifications of more economic solutions. The authors developped a new type of model based on the dynamic orifice (D. O.) concept which includes the turbo-generator integrated modelling. This model has been validated by experimental tests in laboratory, with turbine models, and in real systems. Comparison with Suter characterization is also very satisfactory.

2. Turbine and pipeline modelling

In hydropower schemes equipped with low inertia turbo-generator units and slow reaction turbines, the runaway conditions can be easily attained during a load rejection and consequently dangerous overpressures can be induced. According to the interaction between the components of the system, dynamic response must be carefully analysed, in order to avoid excessive pressures or turbine overspeed.

The mathematical modelling is based on several basic equations. First equation, the rotating mass equation, applied to each turbo-generator unit:

$$I\frac{d\omega}{dt}+T_G=T_H \tag{1}$$

where T_G is the resistant torque, T_H the hydraulic motor torque, I the total polar moment of rotating mass inertia $\left(I=\frac{WR^2}{g}\right)$ and ω the angular velocity of the rotor.

The polar moment of rotating mass inertia has a great influence in speed variations of turbo-generator. Based on recent manufacturers' data it is possible to have generator inertia values of small power plants equipped with low inertia units and comparing with more powerful normal units (Figure 1) that, in genera,l have greater inertia values.

For low inertia units after a full-load rejection, the wheel speed will increase very fast and the runaway conditions can be easily attained.

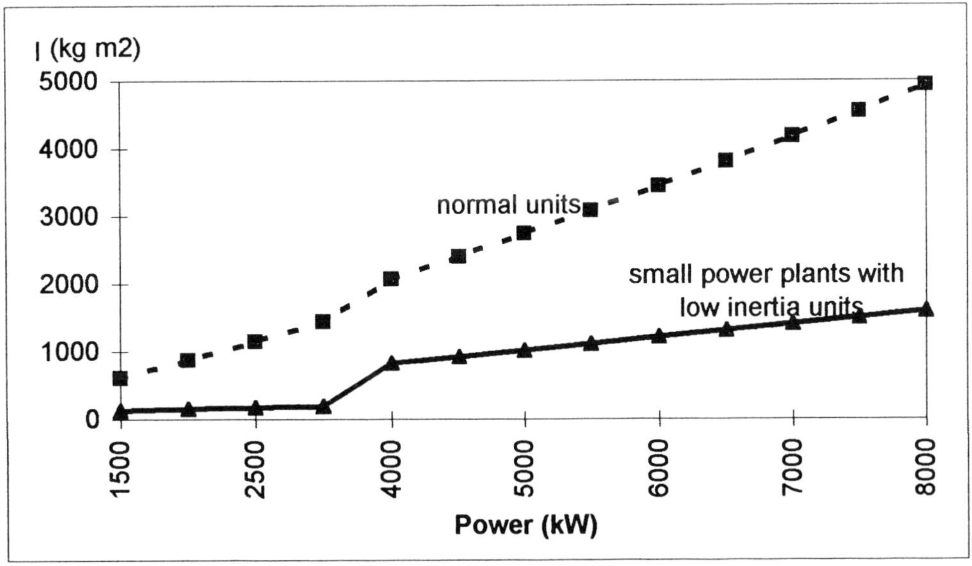

Figure 1 - Generator inertia for rotational speed of 750 r.p.m.

The hydraulic motor torque T_H is obtained by the simplified turbine modelling (Figure 2) based on the dynamic orifice technique or D. O. technique applied to Francis turbines:

$$Q_T = C_g(y)C_s(\alpha_R)\sqrt{H_n} \tag{2}$$

$$T_H = C_t(Q_T, \beta_R, H_n, N, \eta)H_n^{3/2} \tag{3}$$

where

C_g = gate parameter or dimensionless gate opening (=1 open; = 0 closed);

C_s = speed parameter;

C_t = torque parameter;

y = wicket gate opening;

α_R, β_R = D. O. parameters;

η = efficiency;

H_n = net head;

N = rotating speed.

The parameters C_s and C_t are functions based on heuristic solutions deduced in RAMOS, 1995.

The two model parameters α_R and β_R are defined as follows:

$$\alpha_R = \frac{Q_{Rw}}{Q_R} \qquad \beta_R = \frac{N_{Rw}}{N_R} \tag{4}$$

with Q_{Rw} = discharge at runaway speed, Q_R = rated discharge, N_{Rw} = runaway speed and N_R = rated turbine speed.

These two dimensionless parameters are very important in the developed D. O. model. For Francis turbines β_R is about 2 ± 10%; α_R value depends on the turbine type (slow machines will have α_R <1). In Figure 3 values α_R are presented as a function of the turbine specific speed N_s (kW, m) where $N_s = N_R \frac{\sqrt{P_{H,R}}}{H_R^{5/4}}$, with N_R = rated wheel speed (r.p.m.), $P_{H,R}$ = rated turbine power (kW) and H_R = rated net head (m).

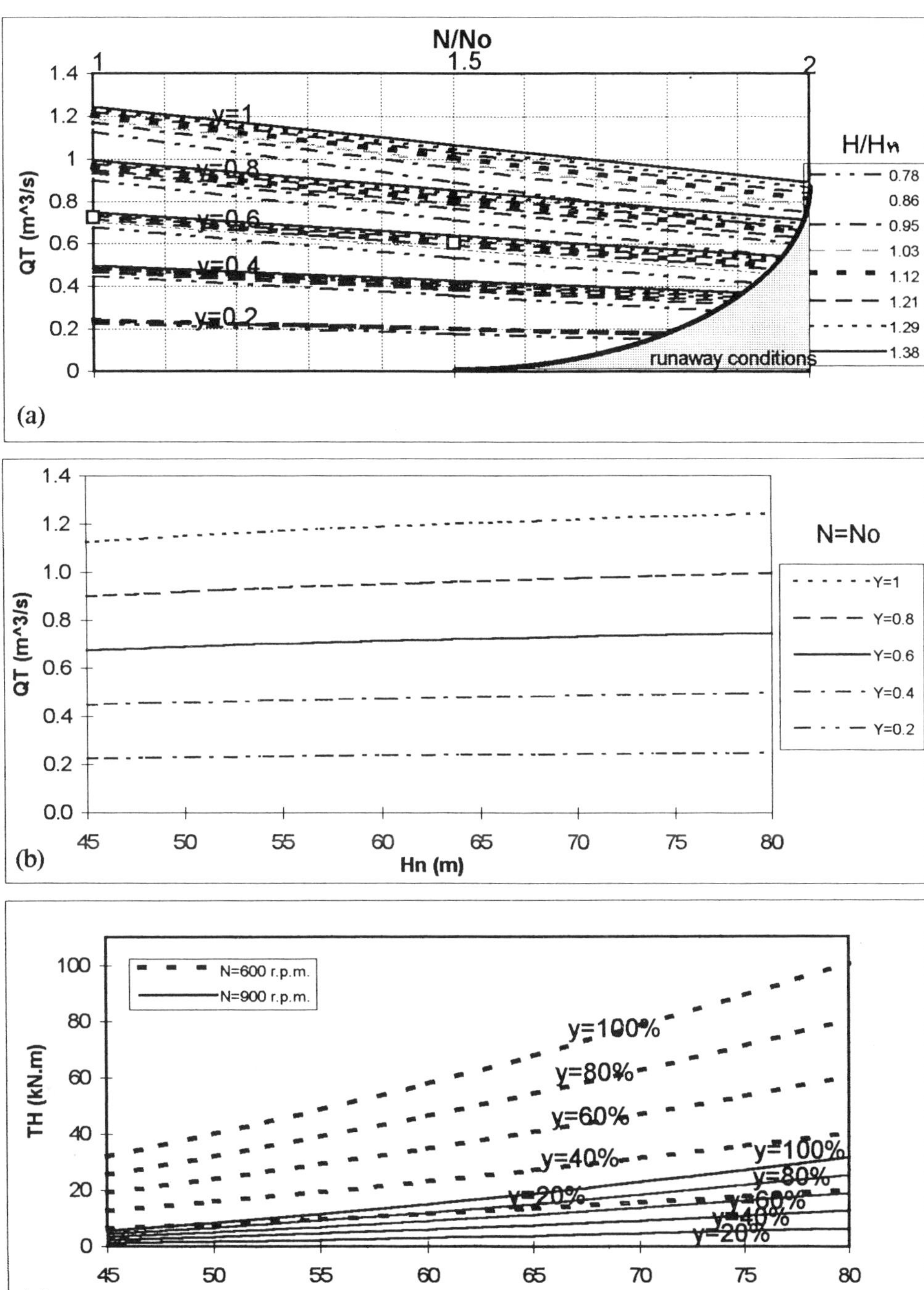

Figure 2 - Characteristic curves for slow Francis turbines based on Dynamic Orifice (D. O.) technique. Example for $\alpha_R = 0{,}65$: a) variation of Q_T with rotational speed (N); b) variation of Q_T with head (H_n) and N = N_R; c) variation of hydraulic motor torque (T_H) for two values of rotational speed ($N<N_{Rw}$)

This information is based on average data from manufactures.

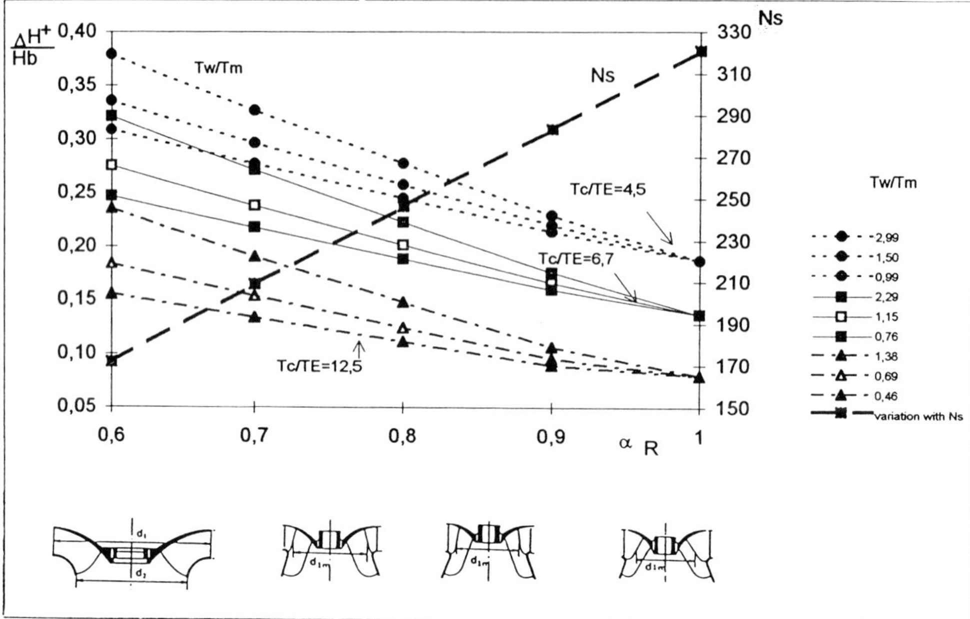

Figure 3 - Francis turbine upsurge after a full load rejection according to the D. O. technique as a function of α_R (N_s), T_C/T_E and T_w/T_m, with $\beta_R = 2.0$

For slow Francis turbines the discharge at runaway speed will be less than the rated discharge. This turbine characteristic will potentially induce a fast discharge reduction in the powerplant.

The equation of rotating masses and the turbine equations are solved together with the well know Method of Characteristics (MOC) equations, for pipe elements, in order to obtain the variations of head, discharge and wheel speed for different values of wicket gate opening/closure:

$$H_p = C_{L,R} \pm B_{L,R}\, Q_p \tag{5}$$

with

L (and signal -) for C^+ line;

R (and signal +) for C^- line;

$$B = \frac{c}{gA};$$

and c = wave celerity.

Different techniques can be chosen for the numerical integration of the friction term which will influence the $C_{L,R}$ and $B_{L,R}$ values.

The turbine and pipe components, as well as other system elements, can be linked together according to the system topology (ALMEIDA and KOELLE, 1992) and the transient response can be simulated.

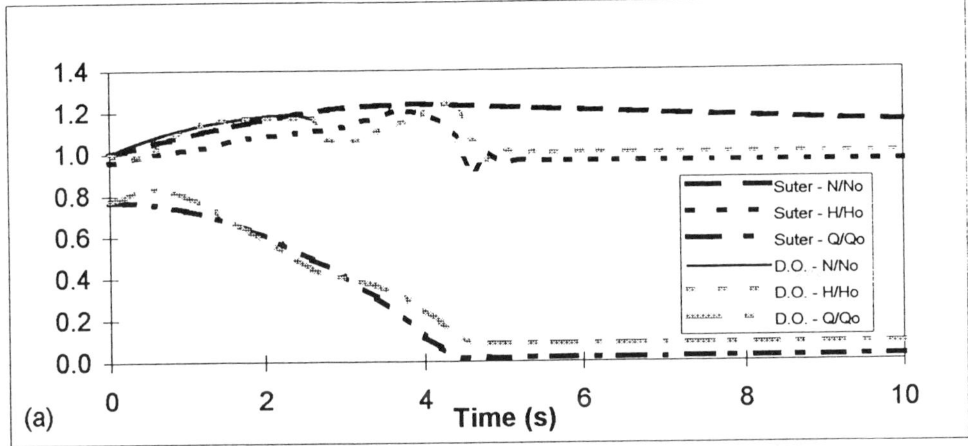

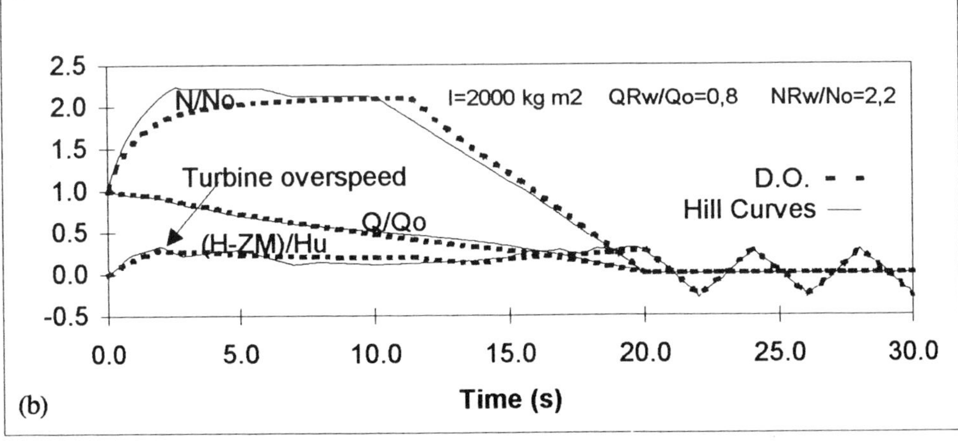

Figure 4 - Comparison of different Francis turbine modelling techniques: a) D. O. technique and Suter parameters; b) D. O. technique and hill curves characterization

3. Comparison analysis and model validation

Comparisons of the simplified Francis turbine D. O. modelling technique with results obtained by Suter technique and by hill curves modelling technique for different case studies are presented on Figure 4: a) shows the good agreement between Suter modelling and the simplified technique; b) shows the comparison with the simulation based on hill curves characterization furnished by a turbine manufacturer.

In Figure 5 a comparison is made between an experimental test at L.N.E.C. made with Francis physical model at the end of pipeline (RAMOS, 1995) and a computer simulation with D. O. model technique.

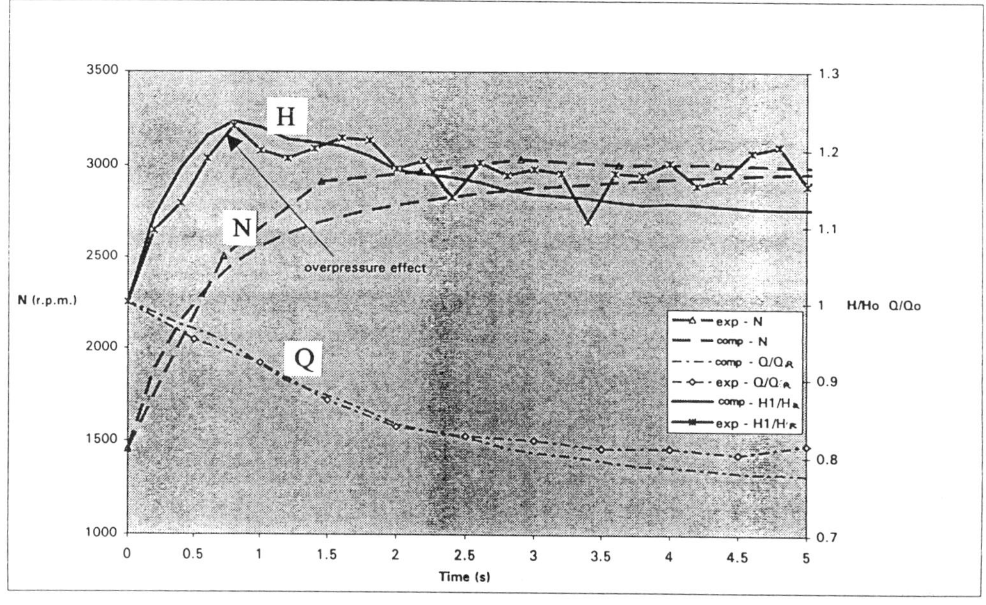

Figure 5 - Comparison between L.N.E.C. experimental test results of turbine overpressure, discharge and wheel speed with computer simulation results based on D. O. technique

4. Conclusions

This simplified analysis allows the improvement of the engineering design support in what concerns, essentially, the different design phases and also the verification of hydraulic special operational conditions, in small power plants. For these types of undertakings the budget and schedule are limited and the economic viability projects depend on the costs and the respective exploitation. So, it is very important to consider new solutions to allow costs and environmental impact minimization e.g. suppressing expensive or environmentally inconvenient protection devices (e.g. surge chambers) using a complete computational system model instead of conservative empirical methods or simplified criteria.

The optimization of the final solutions corresponding to the hydraulic circuit can be achieved after several simulations and their correspondent cost-benefit studies.

This type of turbine model enables an easy implementation of the different types of reaction turbines so that, it is possible to get the best solution with a short computational time spending, much before the real turbine characteristic (in most of the cases very incomplete) are know from the manufacturer and the civil works are already constructed.

5. Bibliography

Almeida, A.B. and Koelle, E. (1992) - *Fluid Transients in Pipe Networks*, C. M. P., Elsevier.

Brekke, H. (1976) - *A Study of the Influence Characteristics on Turbine Governing*. Second International Conference on Pressure Surges, 22nd - 24th, September.

Chaudhry, M.H. (1987) - *Applied Hydraulic Transients*. Second Edition. Van Nostrand Reinhold Company.

Mataix, Claudio (1975) - *Turbomaquinas hidráulicas*. Editorial ICAI, Madrid.

Pejovic, S., Boldy, A.P., Obradovic, D. (1987) - *Guidelines of Hydraulic Transients Analysis*, Technical Press.

Raabe, J. (1985) - *Hydro Power: the design, use, and function of hydromechanical, hydraulic, and electrical equipment*. Dusserldorf. VDI - Verlag.

Ramos, H. (1995) - *Simulation and Control of Hydrotransients at Small Hydroelectric Power Plants* (English abstract and Portuguese version). Ph.D. Thesis. Technical University of Lisbon, Portugal.

Wylie, E., Streeter, V. (1993) - *Fluid Transients in Systems*. Prentice Hall.

SIMULATION OF TURBINE GOVERNING IN TIME DOMAIN

Approximation in frequency domain using the Limited Zero and Pole Method

STUKSRUD, DAG BIRGER
Ph.D. student
The Norwegian University of Science and Technology, Norway

Abstract

In this work there have been developed rational approximations for the elastic pressure/flow transfer function in a pipe. The approximation method is called Limited Zero and Pole Method, and it approximates the function with a selected number of its own zeros and poles. The method was carried out by Li [3] for a system without friction. The approximation has been successfully extended by including a stationary friction term. Originally, the pressure/flow transfer function is a transcendental function, but becomes algebraic and linear through the approximation. This makes it possible to transfer the function from the frequency domain to the time domain, and to use modern automatic control theory for linear systems. A state space formulation are established and the pressure and the flow are simulated in time domain. The simulations are compared with results computed by the well known Characteristics Method, described by Streeter & Wylie [1]. The results shows very good agreement.

1. Introduction

In turbine governing the pressure and flow in front of the turbine are of special interest. In time domain they are described by the Allievi equations, which forms a basis for water hammer calculations in penstocks and tunnels. The equations are partial differential equations, and they are usually solved by the Characteristics Method described by Streeter & Wylie [1], or by other numerical techniques based on lumped parameters. In stability analysis of hydro power plants, the Allievi equations are Laplace transformed into the complex frequency domain. Taken the elastic behaviour of the water column and the friction into account, the transfer function between the pressure and flow in front of the turbine, becomes a transcendental function. This is derived by Brekke [2]. Modern automatic control is based on the concept of state space formulations. This demand that a model of the physical system can be described by a set

E. Cabrera et al. (eds.), Hydraulic Machinery and Cavitation, 799–808.

of first order differential equations in time domain and by a rational transfer function in the frequency domain. Therefor, it has to be found rational approximations for the pressure/flow transfer function to use modern techniques like block diagram and matrix in canonical forms. Today, most commercial programmes for automatic control demands a state space formulation as input. In 1991 Li [3] presented a new technique to approximate the transcendental hyperbolic tangent term in the pressure/flow transfer function. The method was called the Limited Zero and Pole Method. The approximations showed very good agreement with the original function. Li developed his approximations for the case without friction forces. This work extend the model by including a stationary friction term.

2. Fundamental Equations

The Allievi equations describes continuity and momentum for a fluid in a pipe. They are hyperbolic partial differential equations. They can be written on dimensionless form in the following way, shown by Brekke [2]:

$$\frac{\partial q}{\partial l} = \frac{H_0 Ag}{Q_0 a^2} \frac{\partial h}{\partial t} \tag{1}$$

$$\frac{\partial h}{\partial l} = \frac{Q_0}{AgH_0} \frac{\partial q}{\partial t} + \frac{1}{AgH_0} \frac{\pi d\tau}{\rho} \tag{2}$$

The constants in the equations are defined in the nomenclature at the end of the paper. The following assumptions that are made: The flow is turbulent and one dimensional, The pipe is circular, horizontal and rigid. Convective terms are neglected.

Equations (1) and (2) can be Laplace transformed and then solved in frequency domain. By introducing a stationary friction term, K, the transfer function between pressure and flow in front of a turbine becomes [2]:

$$\frac{h}{q} = -2h_w \frac{Z}{S} \tanh\left(\frac{L}{a} Z\right) \tag{3}$$

Where Z is a complex variable:

$$Z = \sqrt{S^2 + KS} \tag{4}$$

h_w is called the Allievi constant:

$$h_w = \frac{Q_0 a}{2 A g H_0} \tag{5}$$

The stationary friction parameter is a linearized frictional constant based on the steady state flow. It can be found by derivating the specific head loss per unit length [2]:

$$K = \frac{4 \lambda Q_0}{\pi d^3} \tag{6}$$

3. Approximation In Frequency Domain

As equation (3) indicates, the transfer function between the pressure and flow, becomes a transcendental function containing a hyperbolic tangent term. The purpose is to find a linear and rational approximation to this function using a method called the Limited Zero and Pole Method, described in [3].

3.1. THE LIMITED ZERO AND POLE METHOD

Equation (3) can be regarded as tree part functions.

$$\frac{h}{q} = h_1 \cdot h_2 \cdot h_3 \tag{6}$$

h_1 is suppose to approximate the hyperbolic tangent term, h_2 shall describe the non linear Z/S function as good as possible and h_3 is a constant term:

$$h_1 \approx \tanh\left(\frac{L}{a} Z\right) \tag{7}$$

$$h_2 \approx \frac{Z}{S} \tag{8}$$

$$h_3 \approx -2h_w + error\left(h_1 + h_2\right) \tag{9}$$

As we can see the constant term, *h3*, also is used to correct error from the other two approximations.

3.1.1. Approximations to h_1

The hyperbolic tangent function can, by introducing $S=j\omega$, be written in the following way:

$$h_1 = j \cdot \tan\left(\frac{L}{a} \omega \sqrt{1 - \frac{K}{\omega} j} \right) = j \cdot \tan(u) \tag{10}$$

The poles can be found by letting the transfer function: $h_1 \to \infty$
This is the case if:

$$u = \frac{\pi}{2}(1 + 2n), \quad n = 0, 1, 2, \ldots \tag{11}$$

The zeros can be found by letting the transfer function: $h_1 \to 0$
This is the case if:

$$u = \frac{\pi}{2}(2n), \quad n = 0, 1, 2, \ldots \tag{12}$$

The transfer function has infinite poles and zeros in the complex plane. Figure 1 shows a qualitative placement of them. Without friction forces the zeros and poles would be laying on the imaginary axis (marginal stability). By combining equation (10), (11) and (12) it is possible to find expressions for the poles and zeros as function of a the counter variable n. The technique is to approximates the function with a selected number of its own zeros and poles, by introducing a finite counter variable N. The following approximation can then be established:

$$h_1 = \frac{\prod_{n=0}^{n=N} \left(S^2 + KS + (2n)^2 \cdot C \right)}{\prod_{n=0}^{n=N} \left(S^2 + KS + (1+2n)^2 \cdot C \right)} \tag{13}$$

Each number of the counter variable n would give two complex conjugated zeros and poles, except the two first zeros ($n = 0$) which both are real. C is a constant called the zero/pole constant:

$$C = \left(\frac{\pi a}{2L} \right)^2 \tag{14}$$

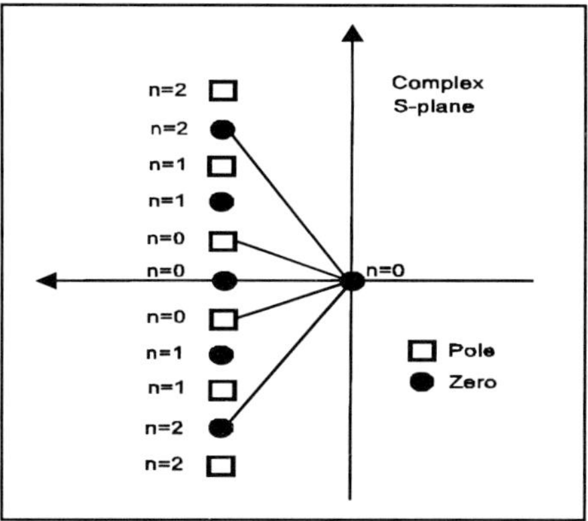

Figure 1. The location of the zeros and poles.

3.1.2. Approximation to h_2

The transfer function h_2 has finite zeros and poles. The best way to approximate this non linear function, as the author sees it, is to use one zero and two poles like:

$$h_2 = \frac{(S+K)}{S^2} \tag{15}$$

Originally the function has two zeros and only one pole. Using they as a approximation would not be satisfactory because of the zero/pole cancelling that will happen.

3.1.3. Approximation to h_3

To correct error introduced from the other two approximations, the author suggest the following function.

$$h_3 = -T_W(N+1) \tag{16}$$

The expression is based on trail and error. T_w is called the pipe constant, given as:

$$T_W = \frac{Q_0 L}{AgH_0} = h_W \frac{2L}{a} \tag{17}$$

3.2. THE FINAL APPROXIMATION

Equations (13), (15) and (16) gives the final approximation for the pressure/flow transfer function:

$$\frac{h}{q}=\frac{-T_W\left(N+1\right)\left(S+K\right)\prod_{n=0}^{n=N}\left(S^2+KS+\left(2n\right)^2\cdot C\right)}{S^2\prod_{n=0}^{n=N}\left(S^2+KS+\left(1+2n\right)^2\cdot C\right)},\quad n=0,1,2,..,N \tag{18}$$

If N=0 the approximation will be of 4th order (three zeros and four poles). The frequency range where the approximation is accurate is extended up to ω = 1.5. If N=1 the function will be of 6th order (five zeros and six poles) and the frequency range is accurate up to ω=3.5. It is possible to extend the range of accurate approximation up to any frequency we want at the price of increasing the order of the system. An approximation of 8th order (seven zeros and eight poles) is normally good enough for turbine regulator design according to Li [3]. The frequency range is accurate up to ω=5.5. Figure 2 shows the amplitude for a 8th order approximation and for the original function. Figure 3 shows the phase. As we can see, the accuracy of the approximation is confirmed. The approximation is given by:

$$\frac{h}{q}=\frac{-3T_W\left(S+K\right)\left(S^2+KS\right)\left(S^2+KS+4C\right)\left(S^2+KS+16C\right)}{S^2\left(S^2+KS+C\right)\left(S^2+KS+9C\right)\left(S^2+KS+25C\right)} \tag{19}$$

Written on canonical form:

$$\frac{h}{q}=-3T_w\frac{\beta_7S^7+\beta_6S^6+\beta_5S^5+\beta_4S^4+\beta_3S^3+\beta_2S^2+\beta_1S}{S^8+\alpha_7S^7+\alpha_6S^6+\alpha_5S^5+\alpha_4S^4+\alpha_3S^3+\alpha_2S^2} \tag{20}$$

Where the canonical constants are given by:

$\alpha_7=3K,\ \alpha_6=\left(3K^2+35C\right),\ \alpha_5=\left(70CK+K^3\right),\ \alpha_4=\left(35CK^2+259C^2\right),$

$\alpha_3=259C^2K,\ \alpha_2=225C^3$

$\beta_7=1,\ \beta_6=4K,\ \beta_5=\left(6K^2+20C\right),\ \beta_4=\left(4K^3+60CK\right),$

$\beta_3=\left(K^4+60CK^2+64C^2\right),\ \beta_2=\left(20CK^3+128C^2K\right),\ \beta_1=64C^2K^2$

Figure 4 shows a block diagram for the linear and canonical transfer function h/q with seven zeros and eight poles (8th order).

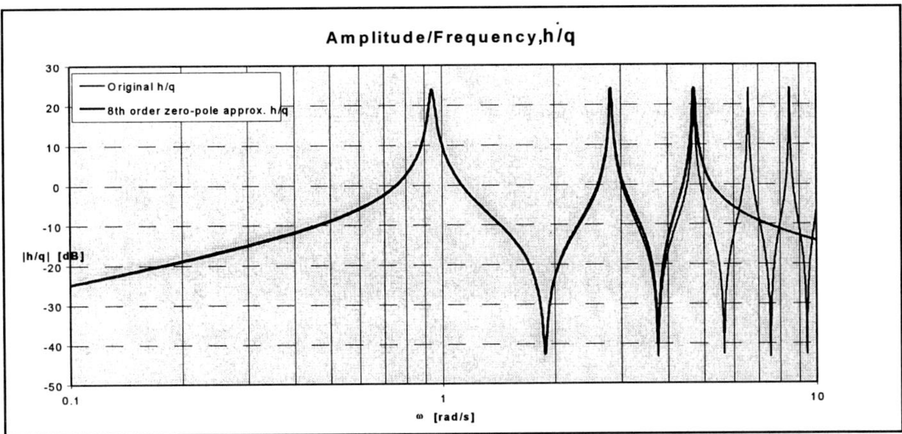

Figure 2. Amplitude for a 8th order approximation (thick line) and for the original function (thin line). ($L=2000$, $D=1.45$, $Q_0=4.51$, $H_0=1000$, $a=1200$, $\lambda=0.013$, $H_w=0.17$, $T_w=0.56$)

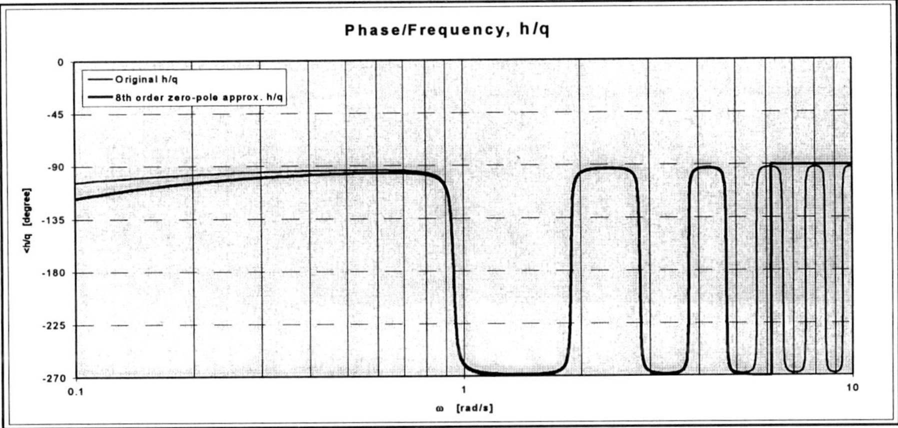

Figure 3. Phase for a 8th order approximation (thick line) and for the original function (thin line).

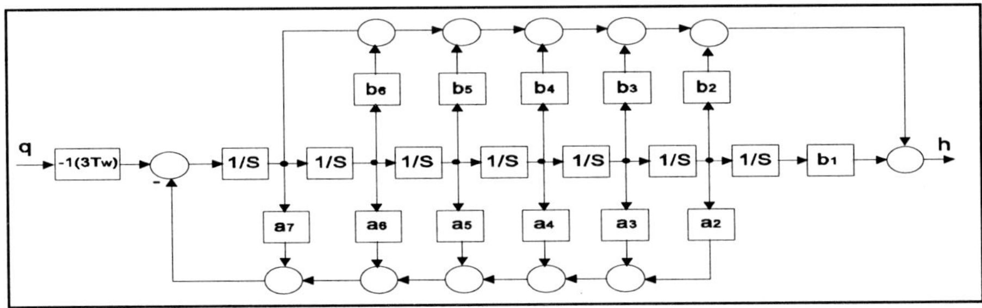

Figure 4. Block diagram for a 8th order approximation to the transfer function h/q.

4. Simulation In Time Domain

The block diagram in figure 4 and the canonical function in equation (20) forms a basis for a state space formulation. The pressure - guide vane transfer function is given by [2]:

$$\frac{h}{y} = \frac{\frac{h}{q}}{1 - \frac{1}{2}\frac{h}{q}} \tag{21}$$

Equation (21) combined with equation (20) results in a transfer function of 8th order, which can be written on canonical controllable matrix form:

$$\begin{bmatrix} \dot{x}_1 \\ \dot{x}_2 \\ \dot{x}_3 \\ \dot{x}_4 \\ \dot{x}_5 \\ \dot{x}_6 \\ \dot{x}_7 \\ \dot{x}_8 \end{bmatrix} = \begin{bmatrix} 0 & 0 & 0 & 0 & 0 & 0 & 0 & 0 \\ 1 & 0 & 0 & 0 & 0 & 0 & 0 & \frac{3}{2}T_w\beta_1 \\ 0 & 1 & 0 & 0 & 0 & 0 & 0 & -\alpha_2 - \frac{3}{2}T_w\beta_2 \\ 0 & 0 & 1 & 0 & 0 & 0 & 0 & -\alpha_3 - \frac{3}{2}T_w\beta_3 \\ 0 & 0 & 0 & 1 & 0 & 0 & 0 & -\alpha_4 - \frac{3}{2}T_w\beta_4 \\ 0 & 0 & 0 & 0 & 1 & 0 & 0 & -\alpha_5 - \frac{3}{2}T_w\beta_5 \\ 0 & 0 & 0 & 0 & 0 & 1 & 0 & -\alpha_6 - \frac{3}{2}T_w\beta_6 \\ 0 & 0 & 0 & 0 & 0 & 0 & 1 & -\alpha \quad - \frac{3}{2}T_w\beta_7 \end{bmatrix} \begin{bmatrix} x_1 \\ x_2 \\ x_3 \\ x_4 \\ x_5 \\ x_6 \\ x_7 \\ x_8 \end{bmatrix} + \begin{bmatrix} 0 \\ -3T_w\beta_1 \\ -3T_w\beta_2 \\ -3T_w\beta_3 \\ -3T_w\beta_4 \\ -3T_w\beta_5 \\ -3T_w\beta_6 \\ -3T_w\beta_7 \end{bmatrix} y(t) \tag{22}$$

Eight first order differential equations has to be solved by a numerical method, as Runge Kutta for instance. This process is a inverse Laplace transformation. x_i is called state vectors. They have no physical meaning unless the last one, which gives the output pressure:

$$h(t) = \begin{bmatrix} 0 & 0 & 0 & 0 & 0 & 0 & 0 & 1 \end{bmatrix} \cdot \begin{bmatrix} x_1 & x_2 & . & . & x_8 \end{bmatrix} \tag{23}$$

$y(t)$ describes the guide vane opening. If the turbine is a Pelton turbine, $y(t)$ can be modelled like a valve. This have been done in this case. The turbine characteristic is neglected (constant performance), though it easily can be included. A time step equal $\Delta t = 0.02$ second is used for comparison with the Characteristics Method. If we choose a pipe length like 2016 meters, the pipe must be divided into 84 parts, according to [1]:

$$nx = \frac{L}{\Delta t \cdot a} = \frac{2016}{0.02 \cdot 1200} = 84 \tag{24}$$

Figure 5 shows the dimensionless pressure $h=H/H_0$ as a function of time. The valve or guide vane opening is closing linearly in 10 seconds. As we can see, the pressure calculated by the Limited Zero and Pole Method shows very good agreement with the Characteristics method, in the closure period. After the valve is closed the pressure calculated by the Limited Zero and Pole Method is much more damped than the curve computed by the Characteristics method. The damping is not physical, it's caused by the approximation terms themselves. An approximation of higher order would not be so damped. The time period between every peak seems to fit exactly. Compared with experimental data, the Characteristics Method gives to little damping [1]. This is a weak point with the method. So, the curves computed by both methods are not reliable after the closure period.

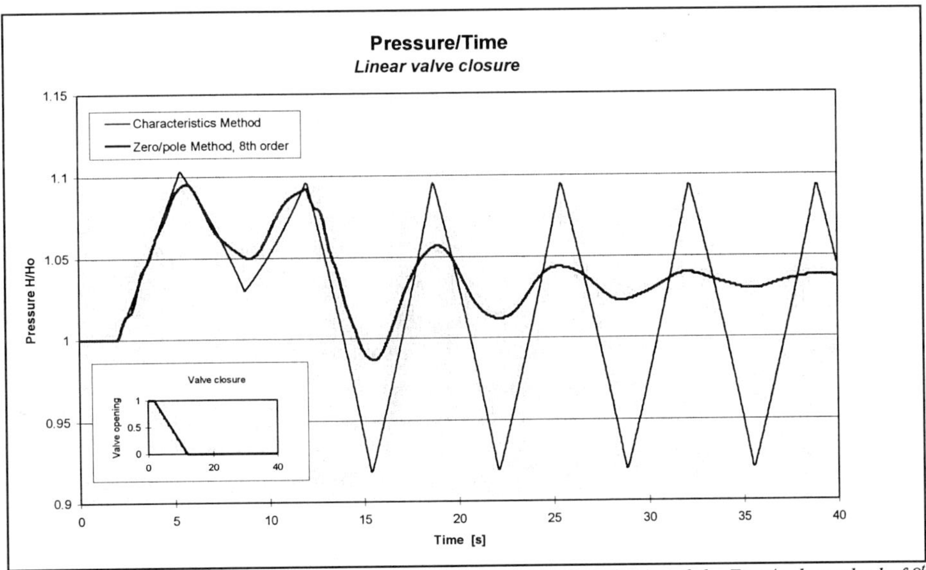

Figure 5. Pressure calculated by the method of characteristics (thin line) and the Zero/pole method of 8th order (thick line). Linear valve (guide vane) closure. ($L=2016$, $D=1.45$, $Q_0=4.51$, $H_0=1000$, $a=1200$, $\lambda=0.013$, $H_w=0.17$, $T_w=0.56$, $\Delta t=0.02$)

5. Conclusion

The purpose of this work was to extend an approximation technique, called Limited Zero and Pole Method, introduced by Li [3]. A stationary friction term has successfully been included in the approximation model. Simulations in frequency domain shows very good agreement with the original transcendental function. The canonical matrix, formed by rational approximations due to block diagrams, are solved in time domain by inverse Laplace transformation using Runge Kutta of 4th order. Comparison with the Characteristics method verifies the accuracy of the approximations. The author is convinced that the model can be extended furthermore by including a dynamic frequency dependent damping term as well. The turbine characteristics may also easily be included in the model.

References

1. Streeter, V. & Wylie, E.: *Fluid Transients in Systems*, Prentice-Hall, Inc. USA 1993

2. Brekke, Hermod, *A Stability Study on Hydro Power Plant Governing including influence from a Quasi non-linear Damping of Oscillatory Flow and from the Turbine Characteristic*. Dr. Techn. Thesis, Oslo 1984

3. Li, Xin Xin,: *State Space Model Formulation of Turbine Regulation Systems*, SINTEF report, Trondheim 1991.

Nomenclature

A	Cross sectional area of pipe	m^2	λ	Darcy-Weisbach const.	-
a	Wave speed	m^2/s	**N**	Integer	-
α_i	Canonical constant	-	**nx**	No. of pipe elements	-
β_i	Canonical constant	-	**n**	Count variable	0,1,..N
C	Zero-pole constant	s^{-2}	$\mathbf{Q_0}$	Stationary fluid flow	m^3/s
d	Diameter of pipe	m	**q**	Dim.less Fluid flow	Q/Q_0
Δt	Time step	s	ρ	Density	kg/m^3
g	Gravity	m^2/s	**S**	Laplace operator	$j\omega$
h	Dim.less Pressure	H/H_0	**t**	Time operator	s
$\mathbf{H_0}$	Stationary pressure	m	$\mathbf{T_w}$	Pipe constant	-
$\mathbf{h_w}$	Allievi constant	-	τ	Shear stresses	N/m^2
$\mathbf{h_i}$	Part function of h/q	-	ω	Angular frequency	s^{-1}
j	Complex variable	-	**x**	State vector	-
K	Stationary friction term	s^{-1}	**y**	Dim.less guide vane	Y/Y_0
L	Pipe length	m	**Z**	Complex function	-
l	Pipe length	m	•	Time derivative	-

PARAMETRICAL MODELLING OF POWER CHARACTERISTICS OF THE FRANCIS AND KAPLAN HYDRAULIC TURBINES

Mugur Florin TOLEA
Jean Louis KUENY
Laboratoire des Ecoulements Geophysiques et Industriels
LEGI - ENSHMG - INPG
BP 53 38041 Grenoble Cedex 9 France

1. Introduction

This paper proposes a new approaches on the mathematical modelling and numerical simulation of the steady-state running of Francis, Kaplan and propeller hydraulic turbines by using the new techniques of parametric identification from experimental data.

The fundamental idea of the parametrical modelling is to establish with the help of the general equation to the energy transfer in turbine, a table of coefficients, with physical meanings, proper to each studied turbine that can characterise the behaviour of the turbomachinery in the area of normal exploitation.

Held account of numerical algorithms and the experimental means used to define this parametrical modelling, the performances of the proposed model (strength and precision of estimation) are perfectly controllable; this fact allow to use this model in a large variety of application as economic simulations of hydraulic power plants, the optimisation of the design and the manufacturing of turbines, the analysis of transitory phenomena and the control theory of hydraulic turbines .

2. General equations of the energy transfer in hydraulic turbines

The energy transfer in a hydraulic turbine can be evaluated if one supposes the former as a complex system that transforms thanks to runner, the available maximal water power in mechanical energy to the level of the turbine shaft. This energy transfer is schematised in the figure 1; the three levels that correspond to the different parts of the structure of the turbine are following:

- theoretical water power provided to the runner, obtained as a difference between available maximal water power and energy losses in the fixed parts of the hydraulic circuit (piping, sluices, tilt, guide-vanes, etc.);

E. Cabrera et al. (eds.), Hydraulic Machinery and Cavitation, 809–818.

- water power available for the energy transfer in the runner, obtained as a difference between theoretical energy provided to the runner and volumetric and hydraulic losses in the runner;
- mechanical energy provided to it the turbine shaft, obtained as a difference between the water power available for the energy transfer and mechanical losses on parts in movement (friction of discs and bearings).

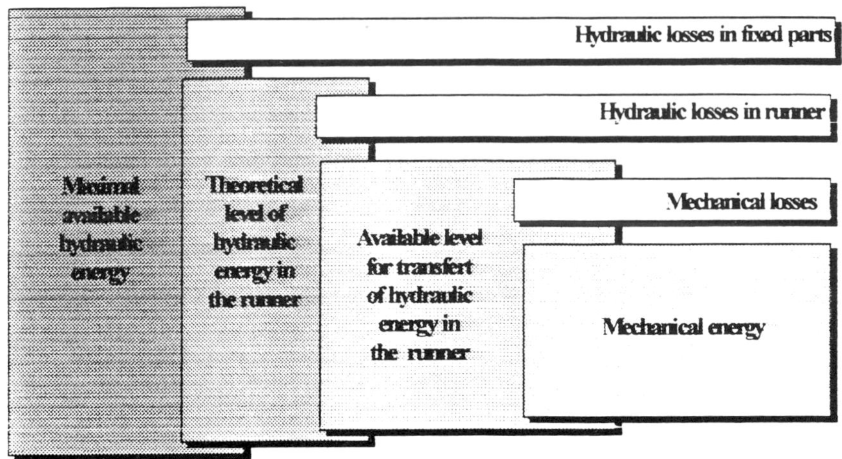

Figure 1. Transfer energy levels in a hydraulic turbine

General equations of the energy transfer in a turbine can write if one considers the specific water power statement to the long of the circuit. The available specific water power to the hydraulic power plant, E_H, insures the level of theoretical energy provided to the runner, E_T, and energy losses over fixed parts of the hydraulic circuit, ΔE_{FP}:

$$E_H = E_T + \Delta E_{FP} \tag{1}$$

The transfer of energy that produced in the runner of a Francis turbine can be evaluated with the Euler relationship for the turbomachinerys. For a given runner geometry , theoretical available energy is:

$$E_T = U_{2e}^2 \left\{ -1 + \frac{|Q_T|}{A_{2e} U_{2e}} \left(\frac{R_{1e}}{R_{2e}} \frac{A_{2e}}{A_{1e}} \frac{1}{k_1 \, tg\, \alpha_1} + \frac{1}{k_2 \, tg\, \beta_2} \right) \right\} \tag{2}$$

This energy level has to insure mechanical energy to it the turbine shaft, E_M, and has compansed the hydraulic losses in the runner ΔE_R and mechanical losses generated by friction of discs and bearings ΔE_M:

$$E_T = E_M + \Delta E_R + \Delta E_M . \tag{3}$$

The three anterior equations can be grouped in one global equation of the energy transfer that is going to allow the parametrical modelling of the steady-state running of hydraulic turbines:

$$E_H = E_M + \Delta E_{FP} + \Delta E_R + \Delta E_M . \tag{4}$$

To characterise the efficiency of the transfer, one recommends utilisation of this equation in efficiency terms, it is to tell:

$$\eta = 1 - \Delta \eta_{FP} - \Delta \eta_R - \Delta \eta_M . \tag{5}$$

3. The parametric estimation of loss efficiency in a hydraulic turbine

Energy losses also in fixed parts of the hydraulic circuit as in the runner can be general considered as the result of the two phenomena: viscous friction's of the fluid during the flow in the turbine and shocks produce by direction changes of the flow in blades.

For fixed parts of the turbine one estimates that the specific energy losses due to viscous friction's of the fluid are proportional with the square of the flow rate of the turbine:

$$\Delta E_{FFP} = K_{FFP} \, Q_T^{\,2} . \tag{6}$$

For the totality of fixed parts of the turbine one considers that the shock losses can be estimated according to the difference between the running flow rate and the nominal flow rate of the turbine:

$$\Delta E_{SFP} = K_{SFP} \left(Q_T - Q_0 \right)^2 . \tag{7}$$

In the runner, the energy losses due to viscous friction's of the fluid can be considered proportional with the square of the relative speed of the fluid in the runner. By constructing speeds triangles for the output section of the runner and takes account the volumetriques losses in runner, one can estimate this type of losses of next manner:

$$\Delta E_{FR} = K_{FR} \cdot W_{moy}^2 = K_{FR1} \cdot \eta_{vol} \cdot U_{2e} \left[\left(\frac{Q_T}{\pi R_{2e}^2 U_{2e}} - 1 \right)^2 + \frac{4 \cdot K_{FR2} \cdot Q_T}{\pi R_{2e}^2 U_{2e}} \right] . \tag{8}$$

For shock losses in the runner, main concentrated in the entry section of the runner, to the level of blade edges, one considers that they depends on the flow rate and the difference between the real relative angle of the flow in the entry section of runner and the relative angle of the blade in the same section. By using speeds triangles , shock losses in the runner can be estimated thus:

$$\Delta E_{SR} = K_{SR}\, Q_T^{\,2}\, (\beta - \beta_0)^2. \tag{9}$$

To use the energy transfer equation under the form (5) in efficiency terms, one introduces the flow rate coefficient

$$\varphi = \frac{Q_T}{\pi\, R_{2e}^{\,2}\, U_{2e}}. \tag{10}$$

and the pressure coefficient

$$\psi = \frac{g\, H}{0.5 \cdot U_{2e}^2}. \tag{11}$$

With these notations one estimates the efficiency losses in the hydraulic circuit of next manner:

- losses in fixed parts due to viscous friction's of the fluid

$$\Delta\eta_{FFP} = h_{FFP}\, \frac{\varphi^2}{\psi}; \tag{12}$$

- losses in runner due to viscous friction's of the fluid

$$\Delta\eta_{FR} = h_{FR1}\, \frac{(\varphi - 1)^2 + h_{FR2}\, \varphi}{\psi}; \tag{13}$$

- shock losses in fixed parts

$$\Delta\eta_{SFP} = h_{SFP1} \left(\frac{\varphi}{\sqrt{\psi}} - h_{SFP2}\, \frac{\varphi_0}{\sqrt{\psi_0}} \right)^2; \tag{14}$$

- shock losses in the runner

$$\Delta\eta_{SR} = h_{SR1}\, \frac{\left(\varphi - h_{SR2}\varphi_0\right)^2}{\psi}. \tag{15}$$

For mechanical losses one considers that the power lost by friction of discs and bearings can be estimated for the area of running with a polynomial of 3-th order in rotation speed of the turbine shaft:

$$\Delta P_M = K_{M1}\, \omega + K_{M2}\, \omega^2 + K_{M3}\, \omega^3. \tag{16}$$

By using the same lessdimensional coefficients, mechanical losses can be estimated thus:

$$\Delta\eta_M = h_{M1}\, \frac{1}{\varphi} + h_{M2}\, \frac{1}{\varphi\, \sqrt{\psi}} + h_{M3}\, \frac{1}{\varphi\, \psi}. \tag{17}$$

Held account of the small variations of the flow rate for constant opening values of the opening guide vanes for a constant head of the turbine, the hydraulic loss coefficients vary only with the value of opening. On the other hand, the mechanical loss coefficients are constant for all openings for a constant head.

This parametric loss estimation is in any case an approximation that allows to make a global manner a robust analysis of the running of a hydraulic Francis or Kaplan turbine. Well on , one can imagine estimations more complicated for all losses but the complexity of the model and the difficulty of identification of its parameters are going to increase strongly and therefore the applicability of the model can be decreased appreciably.

Results of validation of the parametric model on several turbines have demonstrated that these estimations in five terms for losses of efficiency in a turbine are sufficient for all running conditions.

With these considerations, from general equation for the energy transfer (5), equation of the parametric model of the steady state running of a hydraulic turbine becomes:

$$\eta = 1 - h_{FFP}\frac{\varphi^2}{\psi} - h_{FR1}\frac{(\varphi-1)^2 + h_{FR2}\,\varphi}{\psi} - h_{SFP1}\left(\frac{\varphi}{\sqrt{\psi}} - h_{SFP2}\frac{\varphi_0}{\sqrt{\psi_0}}\right)^2 - \\ - h_{SR1}\frac{\left(\varphi - h_{SR2}\varphi_0\right)^2}{\psi} - h_{M1}\frac{1}{\varphi} - h_{M2}\frac{1}{\varphi\sqrt{\psi}} - h_{M3}\frac{1}{\varphi\,\psi} \tag{18}$$

4. Numerical identification of model parameters

The steady-state running of a hydraulic turbines is completely described by the parametric equation (18) and the specific coefficient table [h]:

$$[h] = \begin{bmatrix} h_{M1} & h_{M2} & h_{M3} & h^1_{FFP} & h^1_{FR1} & h^1_{FR2} & h^1_{SFP1} & h^1_{SFP2} & h^1_{SR1} & h^1_{SR2} \\ h_{M1} & h_{M2} & h_{M3} & h^2_{FFP} & h^2_{FR1} & h^2_{FR2} & h^2_{SFP1} & h^2_{SFP2} & h^2_{SR1} & h^2_{SR2} \\ \dots & \dots & \dots & \dots & \dots & \dots & \dots & \dots & \dots & \dots \\ h_{M1} & h_{M2} & h_{M3} & h^N_{FFP} & h^N_{FR1} & h^N_{FR2} & h^N_{SFP1} & h^N_{SFP2} & h^N_{SR1} & h^N_{SR2} \end{bmatrix}. \tag{19}$$

Due to the fact that coefficients of the model are very difficult for calculate theoretically, same impossible, one proposes a method of numerical identification from experimental data [6]. For this reason it is necessary to have a series of complete measures on a turbine to a constant head for a sufficing number of values to the openings of guide-vanes, N.

For each running point determined by measures one writes the equation of the model (18) with the corresponding parameters what suitable have to solve the following system of non - linear equations:

$$\begin{bmatrix} Fobj\left(\varphi_1^1, \psi_1^1 [h]^1\right) \\ \cdots \\ Fobj\left(\varphi_P^1, \psi_P^1 [h]^1\right) \\ \cdots \\ Fobj\left(\varphi_1^N, \psi_1^N [h]^N\right) \\ \cdots \\ Fobj\left(\varphi_P^N, \psi_P^N [h]^N\right) \end{bmatrix} - \begin{bmatrix} 1-\eta_1^1 \\ \cdots \\ 1-\eta_P^1 \\ \cdots \\ 1-\eta_1^N \\ \cdots \\ 1-\eta_P^N \end{bmatrix} = \begin{bmatrix} 0 \\ \cdots \\ \cdots \\ \cdots \\ \cdots \\ \cdots \\ 0 \end{bmatrix}. \tag{20}$$

where F_{obj} elements are non - linear functions similarly that the model equation (18)

The numerical identification of parameters of the model has been realised thanks to special software for a numerical resolution of the system (20) by 3 methods:

- Non-linear Least Squares Implementation [2];
- Sequential Quadratic Programming [3];
- Goal attainment method [1];

All these methods use squared criteria of optimisation to the different functions of error between measured values and predicted values by the model; to increase the strength of the model, the optimisation are oriented to the cancellation of sensitivity functions of parameters. One proposes as initial solution for the table of coefficients [h] the case of an ideal turbine, whose mechanical and hydraulic losses are null.

According to the used numerical identification method, the relative estimation error of the model varies between 0.4 for the goal attainment method and 1.4% for Levenberg -Marquardt method with a medium variation of 5% between values of parameters obtained by different methods [5]. To have the best estimations one has conceived a structure hybridises of calculation for compound the convergence speed of algorithm, the strength of Gauss - Newton algorithm and the controlled precision of goal attainment method.

To have in a minimum time the information indeed on the running of a hydraulic turbine using this method of parametrical modelling one has developed an integrated numerical module that generates automatic measure sequences on a turbine at the same time with the sequences of numerical identification of characteristic coefficients [4].The performances of it this integrated module have been tested with the good results on several hydraulic Francis turbine; considering the strength of the parametric model one estimates the good results on Kaplan and propellers turbines whose tests are in the process to be ended.

5. Validation of the parametric model by numerical simulation

The numerical experiments for the validation of the proposed model have been made to the Researches and Test Center of Hydraulic Machinery of

Grenoble (CREMHYG - France). In this paper one proposes the results of numerical identification of a hydraulic Francis turbine with a specific speed n_S = 250 tr. /min. Tests have been suitable for five openings of the guide-vanes for several constant turbine head. The table of characteristic coefficients of this turbine is the following:

TABLE 1 The coefficients set of tested Francis hydraulic turbine for different opening a of the guide-vanes

Coefficient	$a/a_0=0.6$	$a/a_0=0.8$	$a/a_0=1$	$a/a_0=1.2$	$a/a_0=1.4$
h_{M1}	0.0068	0.0068	0.0068	0.0068	0.0068
h_{M2}	0	0	0	0	0
h_{M3}	0.0002	0.0002	0.0002	0.0002	0.0002
h_{FFP}	0.0017	2.1148	0.9988	0.7538	0.4952
h_{FR1}	0.0058	0.0343	0.0584	0.0335	0.1204
h_{FR2}	0.0046	2.3232	1.4373	3.5951	0.7541
h_{SFP1}	46.1622	5.0480	14.4277	0	0.998
h_{SFP2}	0.79946	0.9622	0.9895	0.9999	0.9972
h_{SR1}	42.9129	25.899	24.1303	18.3514	11.6447
h_{SR2}	0.5952	0.8468	0.9724	1.0646	1.0779

One can consider that the functioning of the studied turbine is described for all running conditions by these coefficients and the parametric equation of the energy transfer (18).

Utilisation of the parametric model proposed in this paper can be oriented to several directions of analysis of a Francis, Kaplan or propeller turbine:

- *simulation of the steady-state running* for different openings of guide-vanes for different turbine head with the plug in account of it the Reynolds effect on the efficiency machinery. Points of the turbine running for nominal head and the nominal opening have been estimated by report to measured values with a global relative error of 0.8% (figure 2). The experiments of parametric identification for this turbine for head that vary between 0.25 and 1.25 around the nominal value, have allowed to study the effect Reynolds on the turbine. We have studied the relative medium variations of coefficients by report to the different head of the turbine; if one uses the apparently Reynolds number to the output section of the runner according to the peripheral speed one can observe that some coefficients remain constant for all conditions of exploitation's (in general mechanical loss coefficients and most of shock losses) and others have "power" laws of variation with this Reynolds number.

- *estimation of the turbine efficiency curves* for the running domain; for the nominal value of the head, one has presented in figure 3 it the efficiency curves of the studied Francis turbine. Global estimation error has been inferior 0.8%.

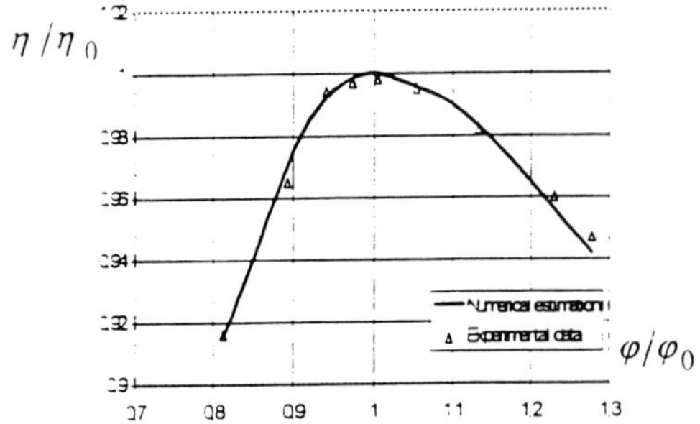

Figure 2. Running characteristic of tested turbine for nominal opening of guide vanes

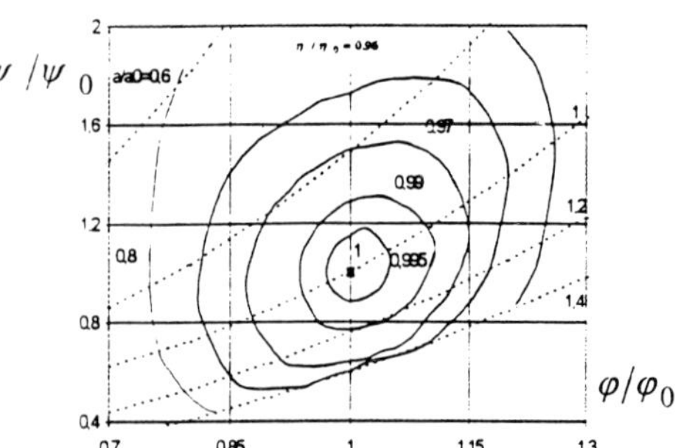

Figure 3. The prediction of efficiency curves of the tested Francis turbine.

- *economic simulation* of a hydraulic turbine of the viewpoint of the energy production: for given water supply conditions, the parametrical modelling allows to make predictions on the energy production to study different strategies of exploitation or refurbishing of hydroelectric power plants.

- *Hydraulic and mechanical loss estimation* in the turbine: thanks to table of characteristic coefficients of a turbine one can estimate for different running conditions, the efficiency loss in the mechanical and hydraulic circuit. For example one presents in the figure 4, the five types of component losses of the model for the nominal opening and the nominal head; in the figure 5 one gives an example of shock loss evolution in the runner for some openings of guide-vanes.

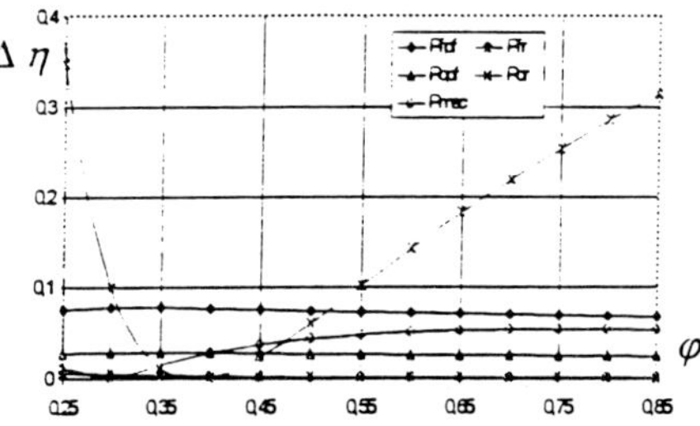

Figure 4. The energy losses estimation of nominal opening of guide vanes.

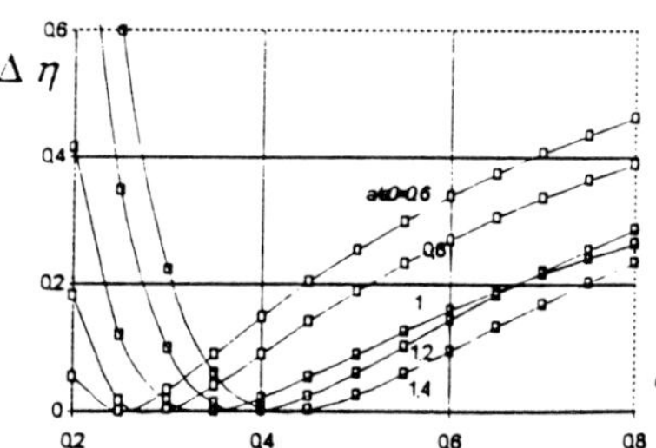

Figure 5. The predicted evolution of shock losses in runner for different openings of the turbine guide-vanes

The hydraulic and mechanical loss analysis with this model gives useful information for the optimisation of the construction and the running of a turbine without being obliged to use expensive and complicated experimental means.

6. Conclusions

The main concept of the parametric model proposed in this paper is to characterise a hydraulic turbine by a table of well mastered coefficients thanks to the general equation of the energy transfer.

The strength of the model comes due to the fact that the former has been conceived by numerical identification from experimental data: the different used calculation diagrams have allowed to obtain the errors of estimation of a turbine behaviour inferior to errors of measure: the precision of estimation of the parametric model is perfectly controllable, with a relative medium value situated around 0.8%.

Using the table of characteristic coefficients and the parametric equation of the model one can simulate all the area of running of a turbine and to estimate the efficiency curves in some conditions of exploitation's. Considering the physical meaning of each characteristic coefficient of a turbine, the parametrical model gives global information on the distribution and the evolution of efficiency losses in the different parts of the turbine; this fact allow the development of the strategies of optimisation of the design and the construction of the turbomachinery. The concept of parametric model allows to be easy to use in transitory analyses in hydraulic installations or in control models of hydraulic turbines regulation for power plants.

7. Mathematical Symbol List

a = guide vanes opening;
E = level energy ;
g = gravitation;
h = turbine parameter;
H = turbine head;
K = proportionallity coefficient;
P = power level;
Q = turbine flow rate;
R = characteristic radius;
U = peripheral runner speed;
W = relative fluid speed in runner;
β = relative angle of the flow;
Δ = loss ;
φ = flow rate coefficient;
ψ = pressure coefficient;
η = turbine efficiency;
ω = runner rotation speed;

Indices:

0 = nominal value;
2e = output runner section index;
FFP = fluid friction in fixed parts;
FR = fluid friction in runner;
M = mechanical nature;
N = number of guide vanes openings;
P = number of experimental data;
SFP = shock in fixed parts;
SR = shock in runner;
T = theoretical estimation

8. Acknowledgements

This paper is due to a collaboration France - Romania and financial support of CEE in the framework of Joule II Project, who's we acknowledge with gratitude.

9. References

1. Fleming, P.J.: Computer-Aided Control System Design of Regulators using a Multiobjective Optimisation Approach, *Proc. IFAC Control Application of Non-linear Programming and Optimisation*, Capry, Italy, pp. 47-52, 1985.

2. Marquardt, D.W.: An algorithm for least-squares estimation of non-linear parameters, *J. SIAM*, **11** (1963), 431-441.

3. Powell, M.J.D.: *Non-linear Optimisation*, Academic Press, New York, 1981.

4. Tolea, M., Kueny, J.L.: Module numerique integre pour identification parametrique des caracteristiques energetiques des turbomachines, *Proceedings PEH 95 Conference*, Nice, France, pp. 221-223, 1995.

5. Tolea, M., Kueny, J.L.: Numerical identification of power losses in turbomachines using a multiobjective optimisation method, *PEH 96 Conference*, Bordeaux, France, 1996.

6. Walter, E., Pronzato, L.: Identification de modeles parametriques a partir de données experimentales, Masson, Paris, 1994.

UNSTEADY FRICTIONS IN PIPELINES

Simulation in time domain using the modified Allievis equations.

VENNATRØ, ROAR
Ph.D. student
The Norwegian University of Science and Technology, Norway

Abstract

This paper presents a friction model to be used in a numerical program to get a solution of oscillatory flow. The model is in time domain. The model is build on Allievis equations, shown in Streeter & Wylie [2]. The intention of this work is to make a model that simulate a transient flow with good accuracy. This is done using a diffusion term as the model for the unsteady friction. This has been done by B.Svingen [3] using a finite element method. The diffusion term added as an extra term in to the Allievis equation. The equations are solved by the method of characteristics. The model is compared with a laboratory test done by Eichinger [1], and shows good agreement with that experiment.

Résumé

Nous présentons un modèle incluant des termes de frottement dans un écoulement oscillatoire,cd modèle est basé sur l'équation d'Allievi avant pour point de départ la conservation de la quantité de mouvement et la continuité. Le but recherché est de simuler l'écoulement en régime transitoire et les variations de pression avec une bonne précision. Comme nous le montrons dans cet article ceci est obtenu en ajoutant un terme de diffusion à celui des pertes stationnaires des équations d'Allievi. Les équations sont résolues par la méthode des caractéristiques, et les résultats montrent une bonne corrélation avec les mesures faites en laboratoire.

1. Introduction

By making tunnels and pipelines in a hydropower plant as smooth as possible, the friction is set to a minimum and the power output is maximised. Since the friction term is stabilising the hydropower plant, minimising the friction could make the power plant unstable in an isolated grid. From this point of view it is necessary to use a model with a dynamic friction term included when calculation of the stability at a sudden and large load changes is executed. The model has to detect the transient behaviour with god accuracy. The equations are build on Allievis equations which is the basis for water

E. Cabrera et al. (eds.), Hydraulic Machinery and Cavitation, 819–826.

hammer calculations. The models is described by two partial differential equations and solved by the method of characteristic.

2. **Fundamental Equations**

The Allievis equations describe continuity and motion of a fluid element in a pipe, equation (1) and (2). The friction term f in the equation of motion (1) is based on the stationary flow $Q_{0,}$ which makes f a constant. If only the stationary friction term is used in a simulation, this will underestimate the dampening of the pressure transient.

$$\frac{\partial h}{\partial x}+\frac{1}{Ag}\frac{\partial q}{\partial t}+\frac{f}{2gDA^2}q|q|=0 \tag{1}$$

$$\frac{\partial q}{\partial x}+\frac{Ag}{a^2}\frac{\partial h}{\partial t}=0 \tag{2}$$

To make a better approach to the transient flow these equations are modified. This is done by adding an extra term to the equation of motion (1). This term describes the diffusion in x, y and z direction as shown in equation (3).

$$\frac{\partial h}{\partial x}+\frac{1}{Ag}\frac{\partial q}{\partial t}+\frac{f}{2gDA^2}q|q|-\frac{\nu}{Ag}\left(\frac{\partial^2 q}{\partial x^2}+\frac{\partial^2 q}{\partial y^2}+\frac{\partial^2 q}{\partial z^2}\right)=0 \tag{3}$$

Since (3) is an one dimensional equation the diffusion in y and z direction will mathematically not be present in the model. Thus the one directional diffusion term has to include all spatial diffusion effects. The proposed model is shown in equation (4) and (5) and it describes the flow in a circular pipe. The pipe is horizontally oriented and completely rigid.

$$\frac{\partial q}{\partial x}+\frac{Ag}{a^2}\frac{\partial h}{\partial t}=0 \tag{4}$$

$$\frac{\partial h}{\partial x}+\frac{1}{Ag}\frac{\partial q}{\partial t}+\frac{f}{2gDA^2}q|q|-k\frac{\nu}{Ag}\frac{\partial^2 q}{\partial x^2}=0 \tag{5}$$

As shown in the equation of motion (5) the losses are modelled with two terms. The first term is the ordinary friction term and the other is a diffusion term in the x direction.

The constant k in equation (5) compensate for the diffusion in y and z directions as shown in equation (3).

3. Solution by characteristics method .

Since a general solution of these partial differential equations is not available, the equations (4) and (5) are solved by the well-known method of characteristics. The method is transforming the partial differential equation into an ordinary differential equations, and these equations are integrated to finite differential equations that are suitable for a numerical algorithm.

3.1. Characteristic equations.

The simplified equation of continuity and motion are identified as L1 and L2. These equations are combined linearly using an unknown term ξ.

$$L = \xi L1 + L2 \tag{6}$$

$$\frac{1}{Ag}\left(\frac{\partial q}{\partial t} + \xi Ag\frac{\partial q}{\partial x}\right) + \xi\frac{Ag}{a^2}\left(\frac{\partial h}{\partial t} + \frac{a^2}{\xi Ag}\frac{\partial h}{\partial x}\right) + \frac{f}{2gDA^2}q|q| - k\frac{\nu}{Ag}\frac{\partial^2 q}{\partial x^2} = 0 \tag{7}$$

The solution of equations (6) and (7) leads to the characteristic equations. This is two pairs of equations C^+ and C^-, one for the positive and one for the negative characteristic. The C^+ equation is shown in (8).

$$\frac{1}{Ag}\frac{dq}{dt} + \frac{1}{a}\frac{dh}{dt} + \frac{f}{2gDA^2}q|q| - k\frac{\nu}{Ag}\frac{\partial^2 q}{\partial x^2} = 0 \tag{8}$$

$$\frac{dx}{dt} = +a \tag{9}$$

The equation for C^- is shown in (10).

$$\frac{1}{Ag}\frac{dq}{dt} - \frac{1}{a}\frac{dh}{dt} + \frac{f}{2gDA^2}q|q| - k\frac{\nu}{Ag}\frac{\partial^2 q}{\partial x^2} = 0 \tag{10}$$

$$\frac{dx}{dt} = -a \tag{11}$$

It is possible to transform these equations into the finite difference form.

3.2. The finite difference form.

The equations are transformed into the finite difference form by integrating along the characteristics [1]. The integral of equation (8) for the positive characteristic is shown in equation (12). The partial derivative $\int_{X_D}^{X_P} \frac{\partial^2 q}{\partial x^2} dx$ is being integrated over a distance Δx in the next time step. This is shown as the distance between P and D in Figure 0.1. The result from the integration of the C^+ equation is shown in equation (13) and the integration of the C^- equation is shown as equation (14).

$$\int_{H_A}^{H_P} dh + \frac{a}{Ag}\int_{Q_A}^{Q_P} dq + \frac{f}{2gDA^2}\int_{X_A}^{X_P} q|q|dx - k\frac{\nu}{Ag}\int_{X_D}^{X_P}\frac{\partial^2 q}{\partial x^2}dx = 0 \tag{12}$$

$$C^+: \quad H_p - H_A + B(Q_P - Q_A) + RQ_P|Q_A| - D_h\left(\frac{dQ_P}{dx} - \frac{dQ_D}{dx}\right) = 0 \tag{13}$$

$$C^-: \quad H_p - H_B - B(Q_P - Q_B) - RQ_P|Q_B| + D_h\left(\frac{dQ_P}{dx} - \frac{dQ_C}{dx}\right) = 0 \tag{14}$$

In the equations the constants are defined as $B = \frac{a}{gA}$, $R = \frac{f}{2gDA^2}$ and $D_h = k\frac{\nu}{Ag}$. The system of equations is implicit and solved with an iteration method.

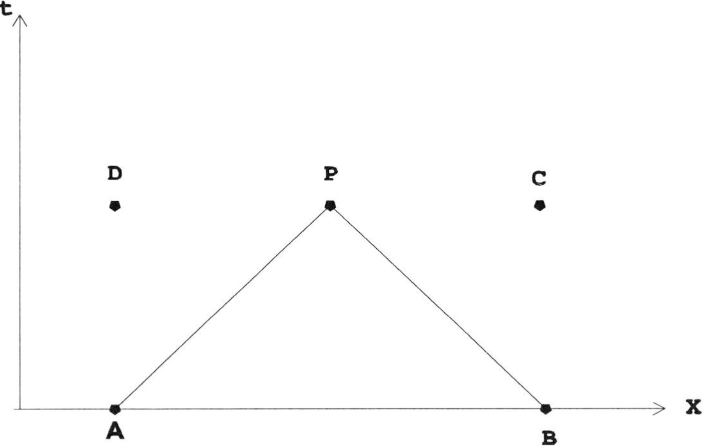

Figure 0.1 Characteristic lines in xt plane.

By solving for $Q_P=Q_j$ and $H_P=H_j$ these equations can be written as equation (15) and (16). In this system of equations j is the index for time and i is the index for length.

$$Q_j = \frac{C_P - C_M + D_h\left(2\frac{\partial Q}{\partial X}\bigg|_{X_i} - \frac{\partial Q}{\partial X}\bigg|_{X_{i-1}} - \frac{\partial Q}{\partial X}\bigg|_{X_{i+1}}\right)}{B_P + B_M} \tag{15}$$

$$H_j = \frac{C_P B_M + C_M B_P + D_h B_M\left(\frac{\partial Q}{\partial X}\bigg|_i - \frac{\partial Q}{\partial X}\bigg|_{i-1}\right) - D_h B_P\left(\frac{\partial Q}{\partial X}\bigg|_i - \frac{\partial Q}{\partial X}\bigg|_{i+1}\right)}{B_P + B_M} \tag{16}$$

The negative and positive characteristics give a solution for Hj and Qj at the upstream and downstream end of the pipe. This leads to equations (17), and (18) at the upstream end of the pipe

$$H_j = H_R \tag{17}$$

$$Q_j = \frac{H_i - C_M + D_h\left(\frac{\partial Q}{\partial x}\Big|_i - \frac{\partial Q}{\partial x}\Big|_{i+1}\right)}{B_M} \tag{18}$$

and equation (19) and (20) at the downstream end of the pipe.

$$Q_j = f(valve) \tag{19}$$

$$H_j = C_P - Q_i B_P + D_h\left(\frac{\partial Q}{\partial x}\Big|_i - \frac{\partial Q}{\partial x}\Big|_{i-1}\right) \tag{20}$$

The term $\frac{\partial Q}{\partial x}$ is discretized using a forward difference scheme for point C and a backward difference scheme for point D shown in Figure 0.1. Forward and backward difference is used to avoid central space difference over the wave front. Central space difference over the wave front would give an numerical error and must be avoided.

4. Simulation

The model is compared with experiments don by Eichinger [1]. The experimental set-up is shown in Figure 0.2.

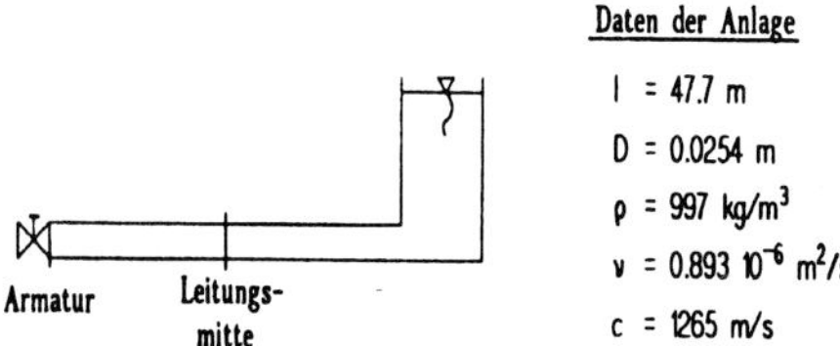

Daten der Anlage

l = 47.7 m

D = 0.0254 m

ρ = 997 kg/m^3

ν = 0.893 10^{-6} m^2/s

c = 1265 m/s

Figure 0.2 Eichingers experimental set -up.

Simulation of the pressure in the pipeline is shown in Figure 0.3. It shows a comparison between simulation of the pressure oscillations with the stationary friction term only and the pressure oscillations with the dynamic friction term included. In Figure 0.4 the results from [1] is shown.

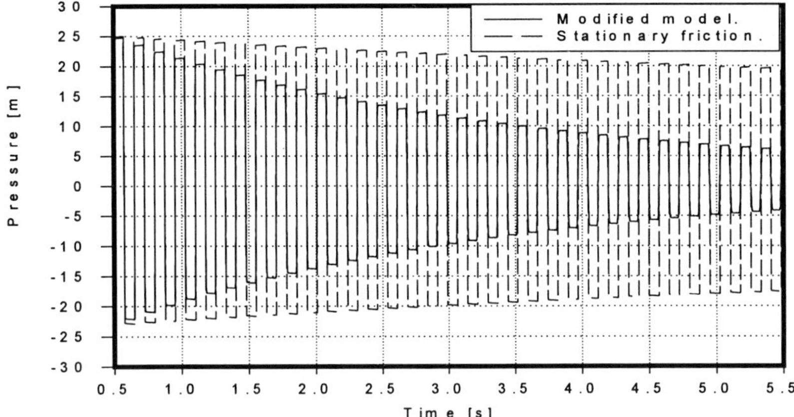

Figure 0.3 Comparison between mathematical models with stationary and dynamic friction.

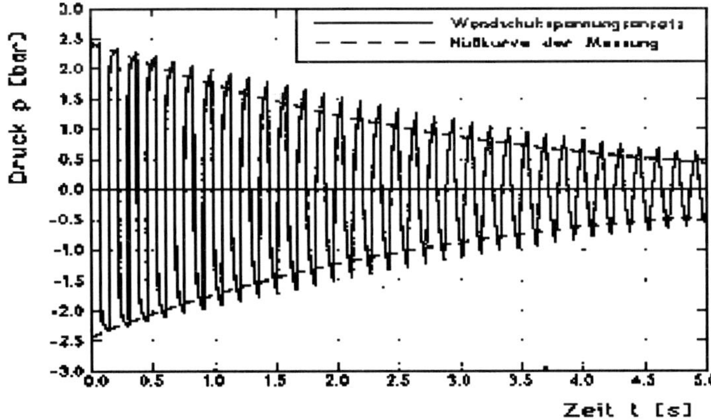

Figure 0.4 Laboratory tests and simulation done in Stuttgart by Eichinger [1]

5. Conclusion

The purpose of this work was to establish an one dimensional model for simulation of transient flow in the time domain. Results presented in this paper shows that the model gives a good correlation with laboratory experiments. Even though the term $\frac{\partial^2 Q}{\partial X^2}$ is the smallest of the diffusion term in Navier Stokes equation, it is used with success as a model of the dynamic friction. The constant k in the motion equation is found by trial and error, and further work will be done to establish a function for this constant.

References

1. Eichinger, Peter, Untersuchung des Reibungsverhaltens bei instationaren Stromungsvorgangen in Rohrleitungen. Von der Fakultat Energietechnik der Universitat Stuttgart zur Erlangung der Wurde eines Doktor- Ingenieurs (Dr.-Ing.) genehmighte Abhandlung.

2. Streeter, V. & Wylie, E.: *Fluid Transients in Systems*, Prentice-Hall, Inc. USA 1993

3. Svingen. Bjørnar, Frequency dependent Friction and Modal Solution of pipe flow by the Finite Element Method. IAHR Work Group 5-7 September Ljubljana 1995

Nomenclature

A	Cross sectional area of pipe	m^2	**f**	Darcy-Weisbach const.	-
a	Wave speed	m^2/s	ν	Kinematic viscosity	m^2/s
D	Diameter of pipe	m	**x**	Length	m
g	Gravity	m^2/s	**k**	constant	-
q	Flow	m^3/s	D_h	Diffusion constant	s/m
h	Pressure	m			

SWIRL FLOW IN CONICAL DIFFUSERS

DAHLHAUG, OLE GUNNAR
Ph.D. student
The Norwegian University of Science and Technology, Norway

Abstract

In order to study swirling, confined, turbulent and non cavitating flow in conical diffusers, a series of 2D LDV and pitot measurements are described. A tangential entry swirl flow inducer was used to generate the swirl flow, LDV was used for the measurements of velocity distribution, and a three hole pitot tube was used for the pressure distribution measurements. The Reynolds number was constant at Re = 5,7 * 10^5 and the Swirlnumber varied from S = 0,1 to S = 0,35. Numerical analysis with FLUENT's RSM, RNG and K-ε turbulence models were carried out. The experimental and numerical results are compared and discussed.

1. Introduction

It is well known that the flow performance of the draft tube has a large influence on the efficiency for low head turbines. Traditionally, the design and development of draft tubes has been based mostly on the experience from existing units and the performance has relied on laboratory model testing, which is very costly, time-consuming and difficult. The design method could be replaced if more efficient computational methods are developed for this type of flow. Examples of flow calculations with some of the best available methods are presented by Charles C.S. Song [3], J.F.Combes [4] and T.C Vu [5]. These calculations show that there exist numerical models that can be applied to this type of flow, but the analytical simulation in the complete range of swirl flow developed in draft tubes are still a great challenge for the developers of computational fluid models. An anisotropic turbulence, the high degree of coupling between the swirl and the pressure field in addition to the influence of geometry makes the modelling of swirling flow very complex. The rotating vortex rope in the draft tube makes the flow two phased and even more complex. Experimental measurements are still scarce, and it will be of great importance to have good measurements for the validation of computational fluid models. In this work the draft tube has been simplified to circular diffusers and bend in order to study the kinetics of this type of flow. The experimental results and numerical analysis presented in this paper are a part of the author's Ph.D.-study which will mainly be concentrated on the experimental results.

E. Cabrera et al. (eds.), Hydraulic Machinery and Cavitation, 827–834.

2. Experimental Details

The experimental test section is shown in Figure 1. The velocity measurements were carried out with Laser Doppler Velocimeter (LDV). The outlet pressure was controlled to ensure that no cavitation occurred. The independence of the Reynolds number for this type of flow is earlier shown by R. Guarga 's [6] work, who studied swirl flow in straight circular section tubes. R.Guarga and his co-workers employed a the Hamilton principle to the swirl flow and demonstrated a functional relationship with the swirl number. In this work, the swirl number S is the focused variable.

$$S = \frac{\int_0^R r^2 \cdot \rho \cdot U \cdot W \cdot dr}{R \cdot \int_0^R r \cdot \rho \cdot W^2 \cdot dr}$$

The variables are defined in the nomenclature. Both axial and tangential mean velocity and fluctuating velocity components were measured. The experimental apparatus consisted of a 2-component LDV system (TSI) with a 300mW Ar-ion laser and probes with focal length of 350 mm . Glass windows were mounted in section A, B, and C (Figure 1) to ensure a homogenous path for the laser beams. The statistical analysis is based on 3000 accepted points in a coincidence mode. For seeding nylon spheres with a diameter of 4 microns were used. A tangential entry swirl flow inducer (Figure 2) generates the swirl flow, and two control valves at the inlet to control and change the swirl level. The cylindrical test section has a diameter of 100 mm at section A and B, 150 mm at section C, and the diffuser between section B an C has an opening angle of 6°. A closed tank with a control valve was used to control the pressure at the outlet. For visual control of the flow, parts of the test section was made in plexiglass. Here, a vortex rope could be observed and measured.

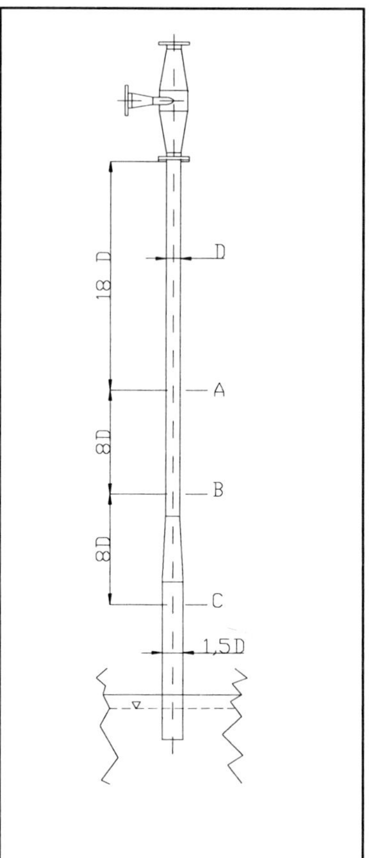

Figure 1. Test section

Figure 2. Tangential entry swirl flow inducer.

3. Numerical Analysis

For the presented numerical analysis FLUENT was used. FLUENT is a commercial general purpose computer program for modelling of fluid flow, heat transfer, and chemical reactions. For this analysis FLUENT's 2D axisymetric scheme which has a finite volume discretisation with second order upwind scheme in space and SIMPLE in time were used on the experimental case described earlier. The governing equations for swirl flow are the full Navier-Stokes equations written in cylindrical co-ordinates with the K-ε, RNG and RSM models for the turbulence analysis. A gridsize of 50 * 500 cells covered the test rig from section A to the outlet. LDV measurements from section A were used as inlet conditions for the analysis. In addition, turbulence intensity was used for the K-ε model. The turbulent intensity is defined as the ratio of the root-mean square of the turbulent velocity fluctuations to the mean flow velocity which for this case were set to 10%. The constants used are adopted from FLUENT's manual and are listed as follows:

$$C_{1\varepsilon} = 1{,}44 \quad C_{2\varepsilon} = 1{,}92 \quad C_{\mu} = 0{,}09 \quad \sigma_k = 1{,}0 \quad \sigma_{\varepsilon} = 1{,}3$$

which is related to the two equations for k and ε:

$$\frac{\partial}{\partial t}(\rho k) + \frac{\partial}{\partial x_i}(\rho U_i k) = \frac{\partial}{\partial x_i}\frac{\mu_t}{\sigma_k}\frac{\partial k}{\partial x_i} + G_k + G_b - \mu\varepsilon$$

$$\frac{\partial}{\partial t}(\rho \varepsilon) + \frac{\partial}{\partial x_i}(\rho U_i \varepsilon) = \frac{\partial}{\partial x_i}\frac{\mu_t}{\sigma_\varepsilon}\frac{\partial \varepsilon}{\partial x_i} + C_{1\varepsilon}\frac{\varepsilon}{k}G_k - C_{2\varepsilon}\frac{\varepsilon^2}{k}\rho$$

where G_k is the generation of k and G_b is the generation due to buoyancy.

The Renormalization Group (RNG) K-ε model differs from the above K-ε model with theoretical functions for constants, new terms in the dissipation rate transport equation, and the turbulent transport equations are extended to permit predictions of swirling flows. A few more differences are added to the RNG model, but are not significant for the analysis presented in this paper. The inlet conditions are the same as for the K-ε model.

The Reynolds Stress Model (RSM) used in FLUENT are based on B.E Launder's [8, 10] assumption to provide closure:

$$\frac{\partial \overline{u'_i u'_j}}{\partial t} + U_k \frac{\partial \overline{u'_i u'_j}}{\partial x_k} = \frac{\partial}{\partial x_k}\left(\frac{\nu_t}{\sigma_k}\frac{\partial \overline{u'_i u'_j}}{\partial x_k}\right) + P_{ij} + \Phi_{ij} - \varepsilon_{ij} + R_{ij} + S_{ij} + D_{ij}$$

where P_{ij} is the stress production rate, Φ_{ij} is a source/sink due to pressure/strain correlation, ε_{ij} is the viscous dissipation, R_{ij} is the rotational term, and S_{ij} and D_{ij} are curvature related terms that arise when the equations are written in cylindrical co-ordinates. The inlet conditions are the same as for the K-ε model. The K-ε, RNG, and RSM models are also described by D.G.Sloan [1], B.E. Launder [8] and in FLUENT's users guide [7].

4. Experimental results compared with numerical results

The presented numerical analysis are at S=0,25. The numerical analysis together with experimental results for mean velocity profiles in axial and tangential directions are presented in Figure 3, 4, and 5.

The four presented velocity-profiles in Figure 3,4, and 5 are results from the LDV-measurements, and the numerical analysis from the K-ε, RNG, and RSM turbulence models respectively. The presented velocities are relative values to the mean flow velocity defined as:

$$U_{rel} = {}^{U}\!/_{U_{mean}}$$

where

$$U_{mean} = {}^{Q}\!/_{A_{inlet}}$$

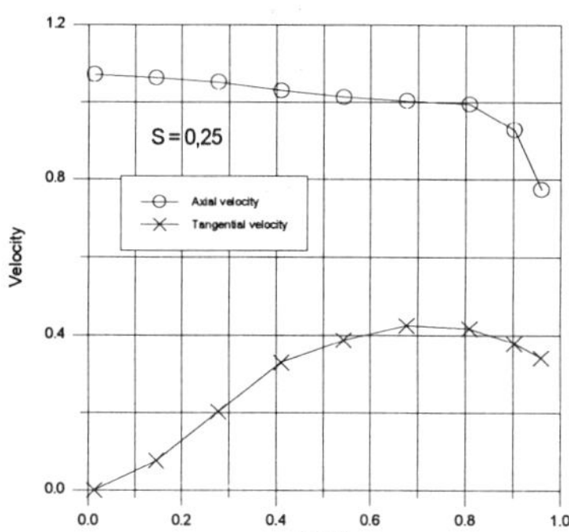

Figure 3. Inlet velocities at section A.

Figure 3 shows the measured inlet velocity profiles at section A. These were used as inlet boundary conditions for the numerical analysis. Figure 4 and 5 show the measured and calculated tangential velocity profiles at section B and C.

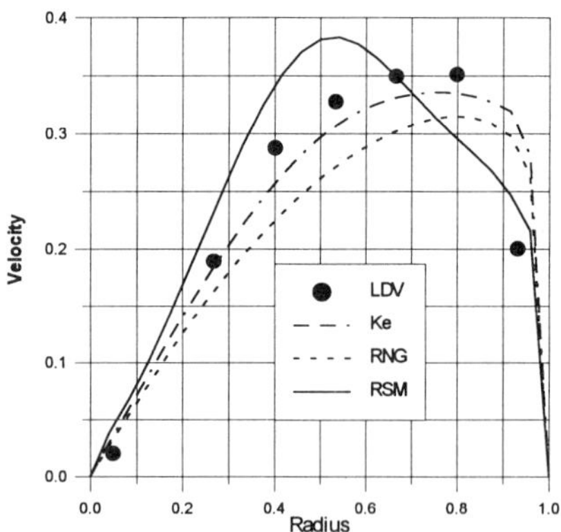

Figure 4. Tangential velocities at section B.

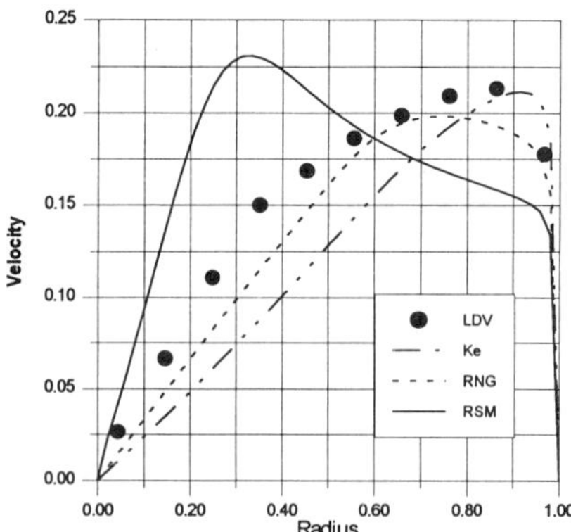

Figure 5. Tangential velocities at section C.

At section B, FLUENT's K-ε and RNG model calculate the velocity profiles fairly good compared to the experimental results. The RSM-model does not give satisfactory results. At section C the K-ε and RNG model tend to keep the forced vortex profile, and the RSM model show more a pattern of a Rankine vortex. Conservation of angular momentum ($r \cdot W = constant$) tends to create a free vortex, in which the tangential velocity W increases as the radius decreases, and with W decaying to zero near r=0 as viscous forces begin to dominate the flow is known as a Rankine vortex. The K-ε model assumes an isotropic exchange of turbulent moment, and in a region where there is a forced vortex the turbulence according to B.E. Launder [10] will tend to restore isotropy as shown by the presented examples. However, in the region with free vortex, the generation of turbulence increase and does not restore isotropy. The computed individual Reynolds stresses by means of the RSM model, provides a better alternative in such cases, but it is influenced by the geometry and sensitive to the grid and boundary conditions.

5. Experimental results

The presented experimental results are at swirl numbers from S=0,10 to S=0,35 These are the mean velocity profiles in axial and tangential directions at four different swirl numbers. In addition the turbulent kinetic energy in the corresponding directions are presented. All presented velocities are relative to mean velocity U_{mean} at the inlet section A.

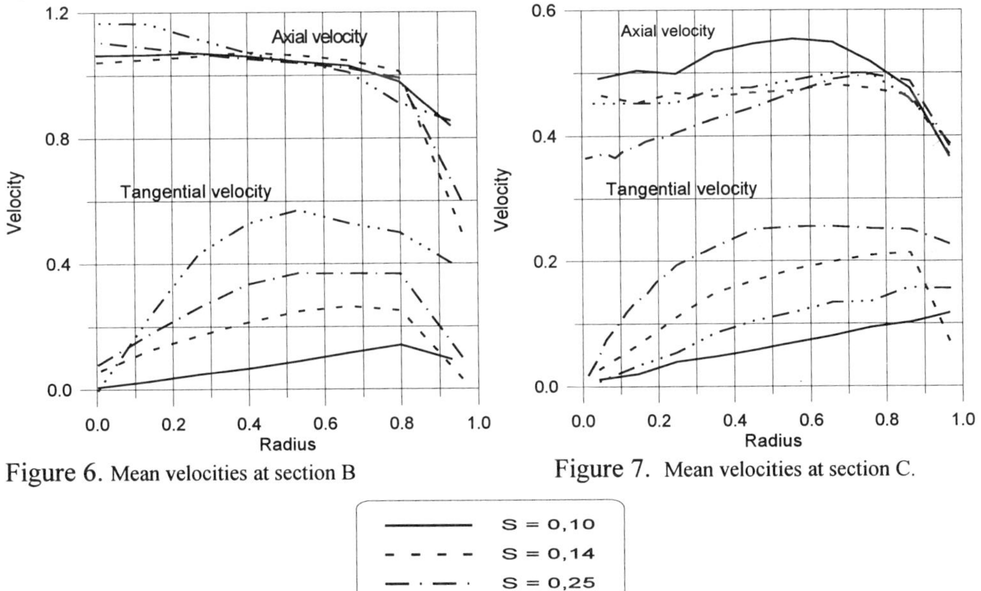

Figure 6. Mean velocities at section B

Figure 7. Mean velocities at section C.

The change of the downstream velocity profiles due to the diffuser are shown as the difference between the profiles in Figure 6 and 7. The axial velocities will change near the centre of the tube due to a reduction of radial pressure gradient downstream which creates an upstream negative axial pressure gradient. The tangential velocity profile tends to keep its forced vortex profile, but with reduced velocity and pressure gradients.

The forced vortex tends to stabilise the flow, and the turbulent kinetic energy in this region will decrease. The turbulent kinetic energy tends to move towards isotropy downstream the diffuser. This is shown in Figure 8 and 9 for different swirlnumber with the ratio B between the turbulent kinetic energy in tangential and axial directions as shown below:

$$B = \frac{\overline{w'^2}}{\overline{u'^2}}$$

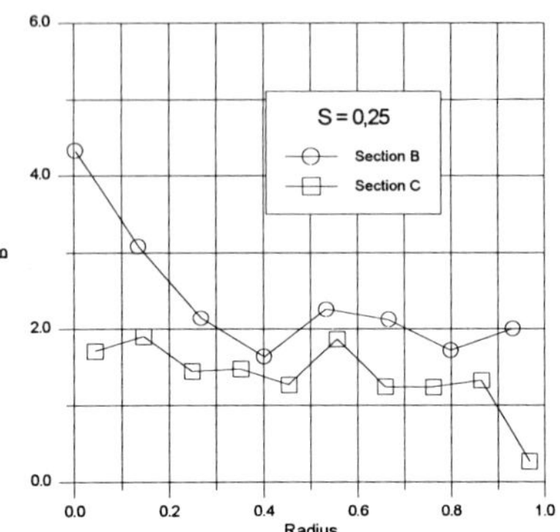

Figure 8. The ratio B at S=0,25.

The third fluctuating component is missing, and it will be investigated by the author in his thesis. This effect is investigated for swirl flow in combustion chambers with air as the working medium with hot-wire anemometer measurements. The investigations shown by A.K.Gupta [2] are both carried out with and without flames, and show that the third fluctuating component v' does not vary much from the tangential fluctuating component w' for swirlnumber 0,3.

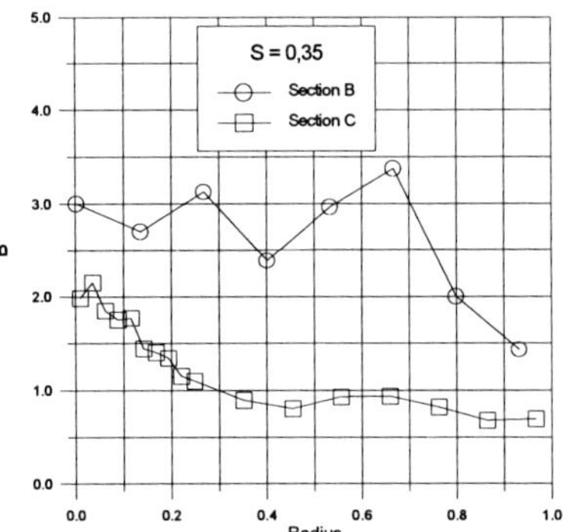

Figure 9. The ratio B at S=0,35.

6. Conclusion

The purpose of this work was to gain experience with swirl flow and to see if numerical analysis can give satisfactory results. The K-ε and RNG turbulence model used at swirlnumber lower than 0,4 gives fairly good results. The tangential velocity profile set up by the tangential swirl flow inducer is very much dominated by a solid body rotation profile, and it does not change much downstream. The forced vortex stabilises the flow and shows a close to isotropic turbulence downstream of a diffuser. Since isotropic turbulence is assumed in a K-ε model, the examples shown in this paper favourites this turbulence model. The velocity profile in axial direction changes due to the large pressure gradient along the axis downstream the diffuser, and the tangential velocity profiles does not change significantly for the inlet profiles and swirlnumbers in these experiments. The fluctuating components in x and θ direction move closer to isotropy downstream of the diffuser, but the component in r direction is not measured in the presented work. The further work of the author's thesis will be to measure the third velocity and fluctuating component in swirl flow in diffusers and bend in order to obtain a better verification of the flow field.

References

1. Sloan David G, Smith P.J, Smoot L.D. *Modelling of swirl flow in turbulent systems.*
2. Gupta A.K., Lilley D.G, Syred N. *Swirl flows*. Abacus Press, (1984)
3. Song Charles C.S, Jianming He, Xiang Ying Chen, *Calculation of turbulent flow through a Francis turbine runner and an elbow draft tube.* (1991)
4. Combes J.F, Verry A, Delorme M, Philibert R, Vanel J.M. *Numerical and experimental analysis of the flow inside an elbow draft tube.* (1990)
5. Vu T.C, Shyy W. *Viscous flow analysis as a design tool for hydraulic turbine components*. (1990)
6. Rafael Guarga F, Jesus Gracia S, Alejandro Sanchez H, Eduardo Rodal C. *LDV and pressure measurements in swirling confined, turbulent and non-cavitating flows.* (1985)
7. FLUENT *Users Guide* Version 4.3 (1995)
8. Launder B.E. *An introduction to single-point closure methology*. (1994)
9. Schuishiro Hirai, Toshimi Tagaki, Massharu Matsumoto. *Predictions of the Laminarization Phenomena in an Axially Rotating Pipe Flow.* (1988)
10. Launder B.E. *Second-Moment closure: Methology and Practice.*

Nomenclature

A	Cross sectional area of pipe	$\mathbf{P_{ij}}$	Stress production
B	Ratio between v' and u'	$\mathbf{\Phi_{ij}}$	Pressure/ strain correlation
$\mathbf{C_{1\varepsilon}}$	Constant	**Q**	Stationary fluid flow
$\mathbf{C_{2\varepsilon}}$	Constant	**S**	Swirl number
$\mathbf{C_{\mu}}$	Constant	σ_ε	Constant
D	Diameter of pipe	σ_k	Constant
$\mathbf{D_{ij}}$	Source term due to curvature	**r**	Radius
ε	Viscous dissipation	**R**	Radius
g	Gravity	**Re**	Reynoldsnumber
$\mathbf{G_b}$	Generation of k	$\mathbf{R_{ij}}$	Source term due to rotation
$\mathbf{G_k}$	Generation of k due to bouyancy	ρ	Density
k	Turbulent kinetic energy	$\mathbf{S_{ij}}$	Source term due to curvature
L	Pipe length	**t**	Time
ν	Kinematic viscosity		
μ_{it}	Turbulent viscocity	$\mathbf{U_{rel}}$	Relative Velocity
$\mathbf{u_i, u_j}$	Fluctuating velocity components	$\mathbf{U_i}$	Velocity
u',v',w'	Fluctuating velocity components in x, r, θ directions	**U,V,W**	Time mean velocity components in x, r, θ directions

EXPERIMENTAL INVESTIGATION OF VORTEX CORE IN REVERSE SWIRL FLOW FROM FRANCIS RUNNER

Yoshinobu FURUIE[1], Hikaru MITA[1], Yutaka HOSOI[2]
[1] *Graduate Student, Tokyo Denki University, Tokyo, Japan*
[2] *Professor, ditto.*

Abstract

It is well known that the precession movement of the helical vortex core generated in the swirl flow from the runner exit of Francis turbine causes the pressure oscillation in the draft tube. This phenomenon which occurs in the part-load operation have been investigated in detail for a long time. At the part-load operation, flow from the runner exit has the swirl in the same direction as that of the runner rotation. On the other hand, when the turbine operates at over-load, flow from the runner exit has the swirl against the runner rotation and the characteristics of the pressure oscillation in such a situation as this is very different in quality from that at the part-load operation. Sometimes this phenomenon raises a hard problem in the actual plant. In the present paper, the behavior of the vortex core and the basic characteristics of pressure oscillation under the conditions of reverse swirl flow are investigated using a small Francis turbine test rig with a cone-type draft tube.

Nomenclature

a	guide vane opening (GVO) [m]	n	runner speed [rpm]
D_e	exit diameter of runner [m]	H	effective head [m]
Q	discharge [m^3/s]	f	surge frequency [Hz]
Δh	surge amplitude [m]	n_{11}	unit speed, $= nD_e/\sqrt{H}$
Q_{11}	unit discharge, $= Q/(D_e^2\sqrt{H})$	f_{11}	unit frequency, $= fD_e/\sqrt{H}$
q	injecting air quantity [m^3/s]		

Subscripts

opt optimum

E. Cabrera et al. (eds.), Hydraulic Machinery and Cavitation, 835–844.

1. Introduction

As for precise prediction of the behavior of draft surge of the prototype through the model test, a large number of similarity factors are required to be considered [1]. It is well known that the draft-surge characteristics is strongly influenced by cavitation and depends on the geometrical shape of the draft tube[2]. In this study, a simplified experiment with a cone-type draft tube is carried out without the effect of cavitation in order to basically investigate the behavior of the vortex core in reverse swirl flow from the runner exit.

2. Experimental apparatus and method

Figure 1 shows the test rig, the tested Francis runner of 48[m, m^3/s, rpm] in specific speed and 0.11 [m] in exit diameter, and a set of the guide vanes. A draft tube made of acrylic resin for the visualization of the vortex core is conical with 8 degrees in cone angle. The four pressure taps on the wall are located every 90 degrees at the same level of $0.19De$ downstream of the runner exit. For arrangement of guide vane opening, the circular disc gauges touched internally on the edges of all of the guide vanes are used for each opening. The induction motor and the torque meter are installed and speed of the induction motor is adjustable with the inverter, and the output of hydroturbine is absorbed by the resistance by regenerative braking.

The test is performed at the test head of 3 m to 10 m. Pressure oscillations are measured with the semi-conductor pressure transducers. Signals from every sensors are simultaneously recorded in a multi-channels digital recorder and all data are analyzed using Fourier transform. Cavitation coefficient is kept high not to affect the draft-surge. The behavior of vortex core in the draft tube is observed with a video camera.

3. Variation of shape of vortex core

In the region of Francis turbine operation represented by coordinates with n_{11}-axis as abscissa and Q_{11}-axis as ordinate, swirl flow from the runner exit has the opposite direction to each other with non-swirl zone which runs from high-n_{11} and high-Q_{11} to low-n_{11} and low-Q_{11} through the vicinity of the optimum operating point. In the region above non-swirl zone, flow from the runner exit has the swirl against the runner rotation.

Figure 2 shows the efficiency hill chart in the n_{11}-Q_{11} coordinates for the tested turbine obtained in advance. In the regions painted out in halftone in this coordinates, the vortex core visualized by air separated from water was observed under the present test condition of non-cavitation. The vortex core which is not visible for lack of the negative pressure in the core enough to separate air from water but actually exists in the swirl flow, was visualized by injecting air. In that case, it was made sure that only few quantity of air ($q/Q \fallingdotseq 2\times10^{-3}$ %) was supplied within the limits which the operating condition and the characteristics of amplitude and frequency do not change. Injected air under the condition

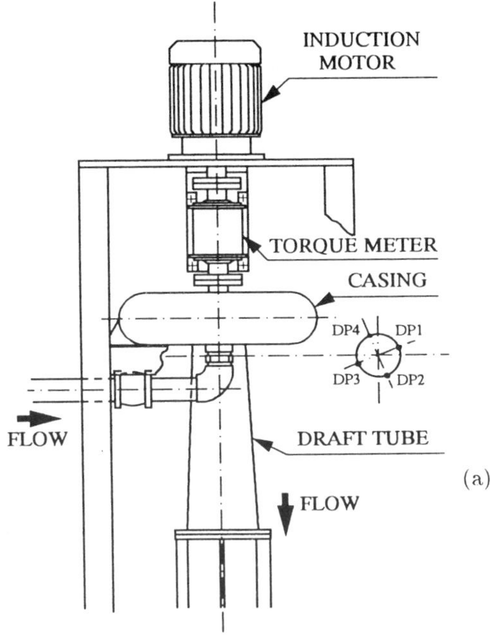

(a)

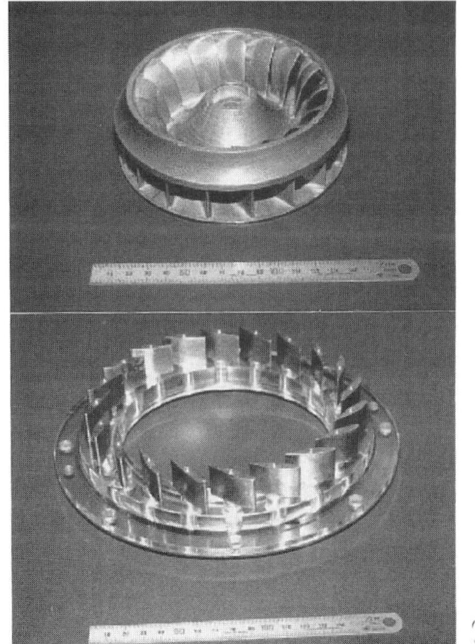

(b)

(c)

Figure 1. (a) Test rig
(b) Tested Francis runner
(c) A set of guide vanes

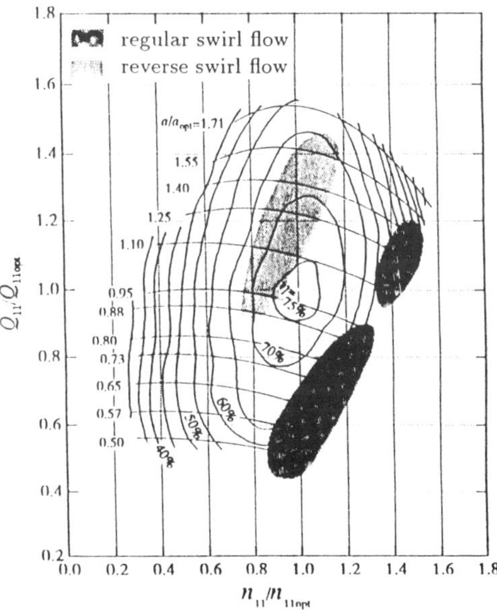

Figure 2. The efficiency hill chart for the tested turbine

of natural suction for the present test rig was actually throttled.

Figure 3 shows the varying shapes of the vortex core with n_{11} varied systematically and with GVO (a/a_{opt} =125 %) constant. The pictures on the upper side show the flow of water from the runner exit without air injection into the draft tube and those on the lower side show the vortex cores with air injection into the central area of the draft tube at the same level as that of the pressure taps. It is found in these pictures that the invisible vortex cores become visualized and the thin vortex cores, even some of invisible cores, become thick with not only single rope but also double ropes in form entangled. The vortex core visualized by air injection does not fix into a certain shape and it is unsteadily variable. In the case of regular swirl flow, however, the behavior of the vortex core when invisible one is visualized is different from the above and it takes the clear shape of the helical rope.

In the vicinity of $n_{11\mathrm{opt}}$ ($\fallingdotseq$ 52 [m, rpm]) as far as this extent of over-gate of 125% in a/a_{opt}, the peripheral velocity component is small, and then the visualized vortex core in a little reverse swirl flow is formed into the shape which is not clearly helical but gnarled. This shape is rather similar to that of the cavitated vortex core at over-load in the vicinity of $n_{11\mathrm{opt}}$ in the actual turbine which often can be seen in the model turbine[2].

Figure 4 (a) shows the pictures of the vortex core observed at both cases of the regular swirl flow and the reverse swirl flow in the present study. It is found that these vortex cores visualized by air injection are similar to the case of an actual model turbine as shown in Figure 4 (b)[3]. In this model test, cavitation coefficient is changed under the conditions that amplitude of the pressure oscillation does not change, so that cavitation coefficient is kept high.

4. Frequency characteristics

Figure 5 shows the two examples of pressure oscillations measured at four taps on the wall of the draft tube and their power spectra, for each of cases of reverse swirl flow and regular swirl flow, respectively. It was also found that the pressure oscillations in both cases are of rotational type. In this connection, it has been reported by Nishi,*et al.* that the synchronous pressure surge did not occur in a conical draft tube using a set of guide vanes instead of a runner, even conducted under the conditions of cavitation[2].

Figure 6 shows the two kinds of the 3-D spectral maps with axes of n_{11}, f_{11}, and$\Delta h/H$ for the pressure oscillations[3] at 73 % and 125 % in a/a_{opt}. In the map at 73 % GVO, it is apparent that the predominant frequencies in the regular swirl is lining in good order. In the map at 125 % GVO, the distribution of frequencies is generally random and a group of the specially predominant frequencies is one in the regular swirl.

Figure 7 shows a plot of two groups of frequencies, i.e., the predominant frequencies for the case of regular swirl and those with the visible vortex core without air injection for the case of reverse swirl, for each of the guide vane openings, against n_{11}. In this figure,

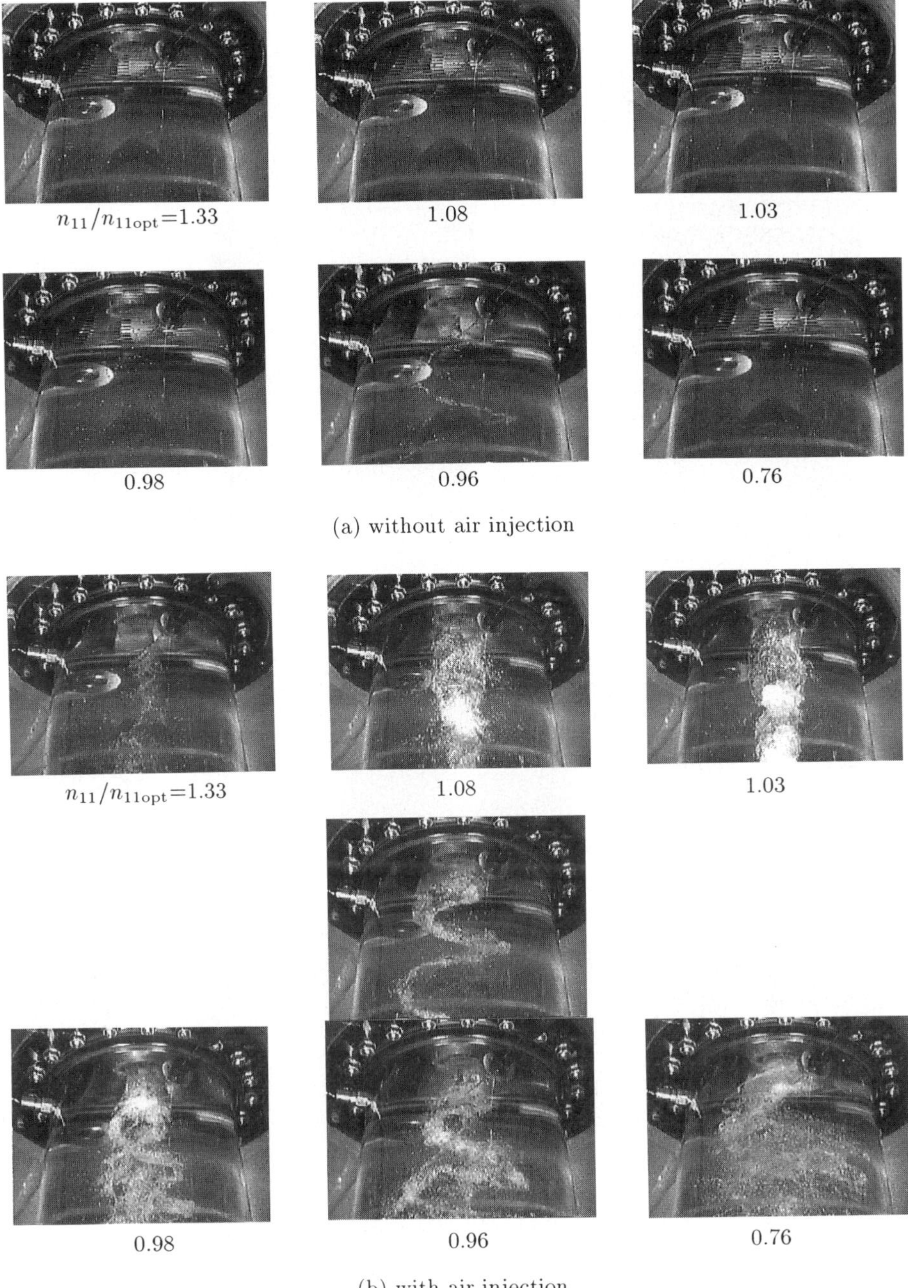

Figure 3. The shapes of the vortex core with varying $n_{11}/n_{11\text{opt}}$ at a/a_{opt}=125%

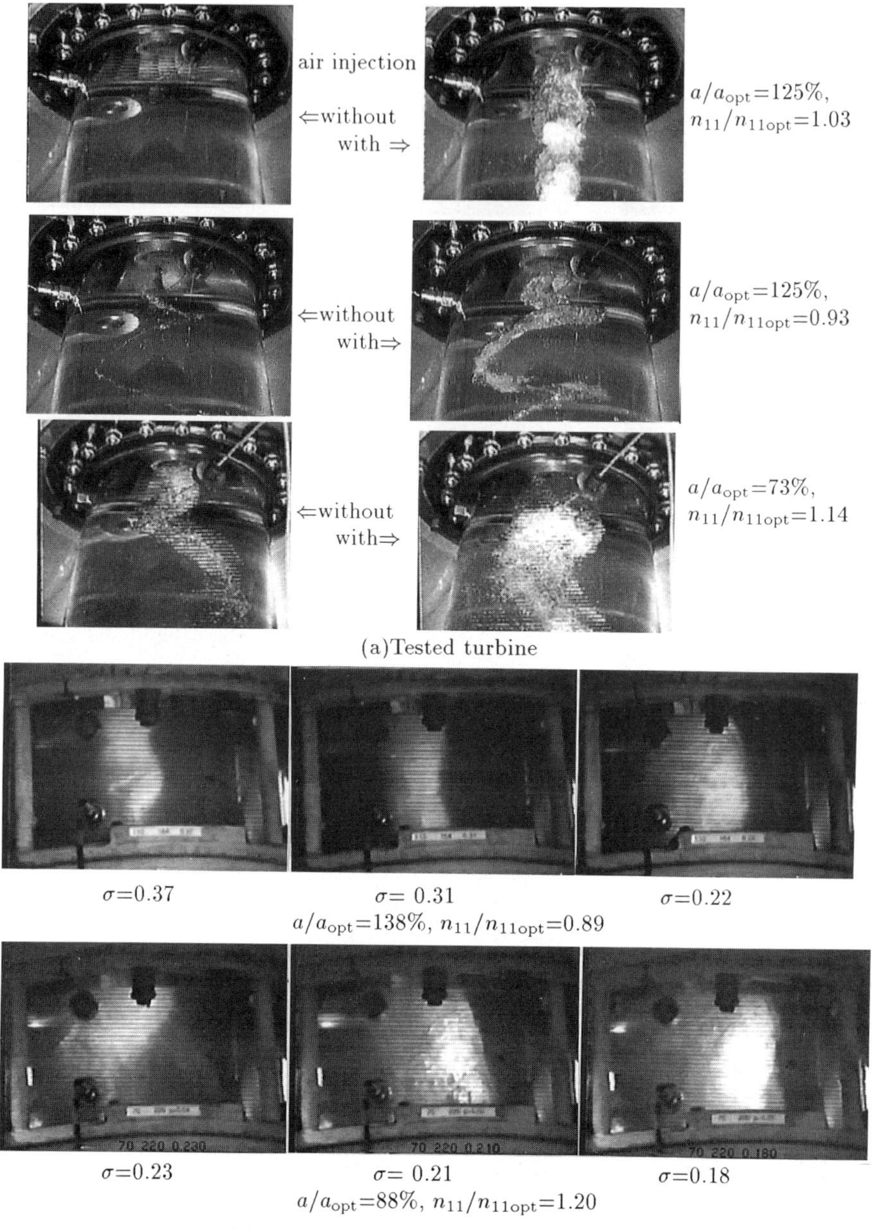

Figure 4. Comparison of the shape of the vortex core between the tested turbine and an actual model turbine (n_{sq} = 59[m, m^3/s, rpm]

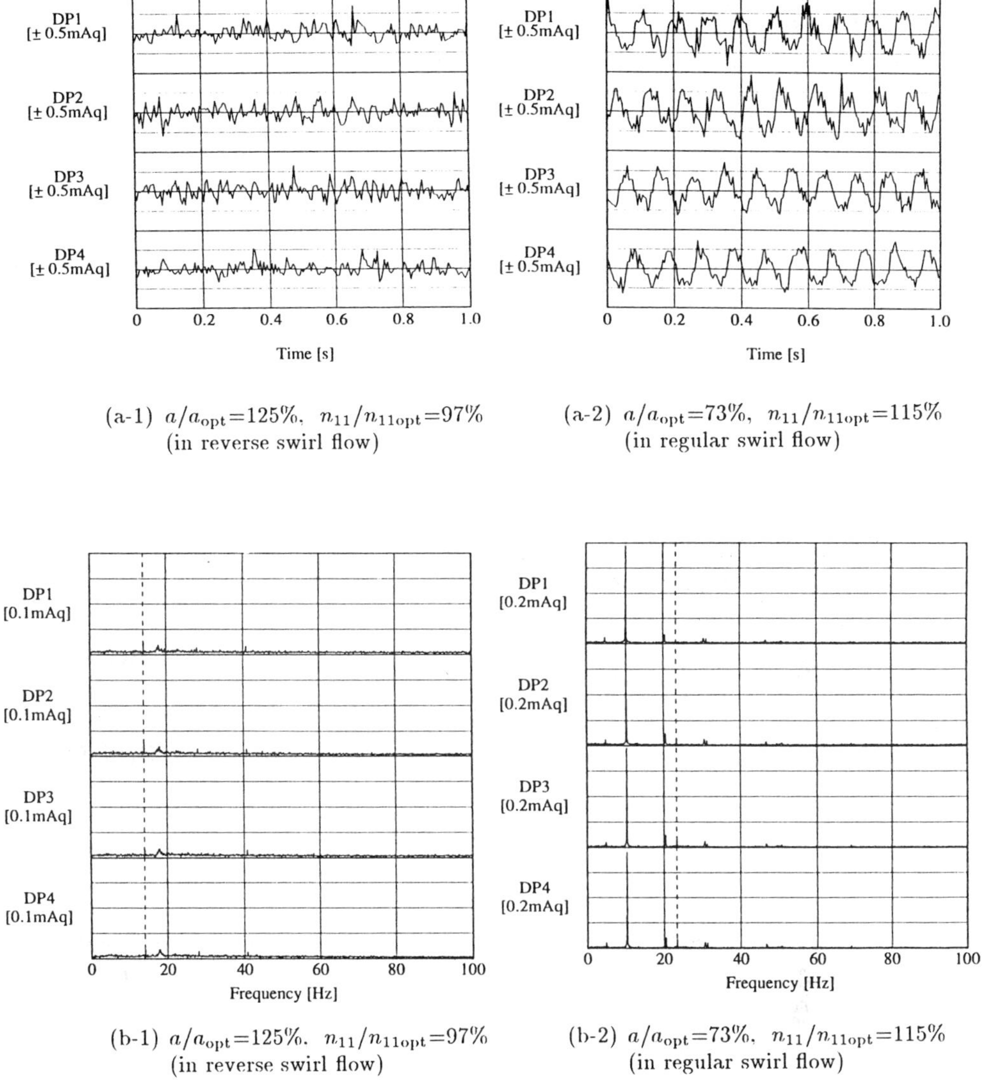

(a-1) a/a_{opt}=125%, n_{11}/n_{11opt}=97% (in reverse swirl flow)

(a-2) a/a_{opt}=73%, n_{11}/n_{11opt}=115% (in regular swirl flow)

(b-1) a/a_{opt}=125%, n_{11}/n_{11opt}=97% (in reverse swirl flow)

(b-2) a/a_{opt}=73%, n_{11}/n_{11opt}=115% (in regular swirl flow)

Figure 5. Pressure oscillations and their power spectra

the case of reverse swirl, for each of the guide vane openings, against n_{11}. In this figure, a group of the lower frequencies than the turbine-rotational frequency belongs to the case of the regular swirl flow and it is represented as $(n_{11}/60)/2.3$, as it is the well-known correlation[4]. Another group of higher frequencies than the above belongs to the case of the reverse swirl flow when the vortex core is visible as it is.

The increase in f_{11} with increase in n_{11} in the regular swirl flow and the decrease in f_{11} with increse in n_{11} in the reverse swirl flow mean that the frequency increases with increase in the peripheral velocity component of the flow from the runner exit. It is, however, found that the tendency of frequency in the reverse swirl flow for each GVO has also an increasing tendency with increase in n_{11} against the above after it reaches a minimum. Amplitude was relatively large at the minimum frequency for each GVO, even though it was small compared with the amplitude in the regular swirl flow.

The fundamental frequency of the pressure oscillation in the regular swirl flow agrees with that of the vortex core revolution as it is also well known[5]. In the reverse swirl flow, however, the frequency of the pressure oscillation does not agree with that of the vortex core revolution. That is, the fact that revolutions of the vortex core is lower than the turbine rotation was clearly observed. In this connection, we have observed the same phenomenon as the above in the model test carried out under the cavitation condition in a manufacturer[3]. It is considered that the pressure oscillation in the reverse swirl flow is not influenced by the precession movement of the vortex core but the swirl flow itself. This phenomenon is one of the most remarkable feature of the reverse swirl flow.

Therefore, it is assumed that unusual tendency of frequency in the reverse swirl flow as shown in Figure 7 depends on the peripheral velocity distribution of the flow from the runner exit. It was observed by the tuft method at the same time that the peripheral velocity component of the flow near the wall was always in the same direction as that of the runner rotation in spite of the reverse swirl flow in the vicinity of the central area around the vortex core. Such a one-directional flow pattern near the wall as this is added to the peripheral velocity based on free vortex motion whether the swirl flow from the runner is in the same direction as that of the runner rotation or not[6]. That is also considered peculiar to the reverse swirl flow which is not found in the swirl flow in the experiment under the condition of the swirl flow generated by only the guide vanes without runner.

We observed further the flow from the runner during the operation with the two tendencies of f_{11} decreasing and increasing through the minimum f_{11} with increase in n_{11} in the reverse swirl flow. In the result, it was found that the peripheral velocity component of the flow near the wall in the direction of the runner rotation decreases according to the decrease in n_{11}. This fact means the one-directional flow pattern near the wall as mentioned above. It is considered that the frequency f_{11} increases with increase in n_{11} when the resultant peripheral velocity of the flow near the wall increases in the direction of the runner rotation beyond a certain degree.

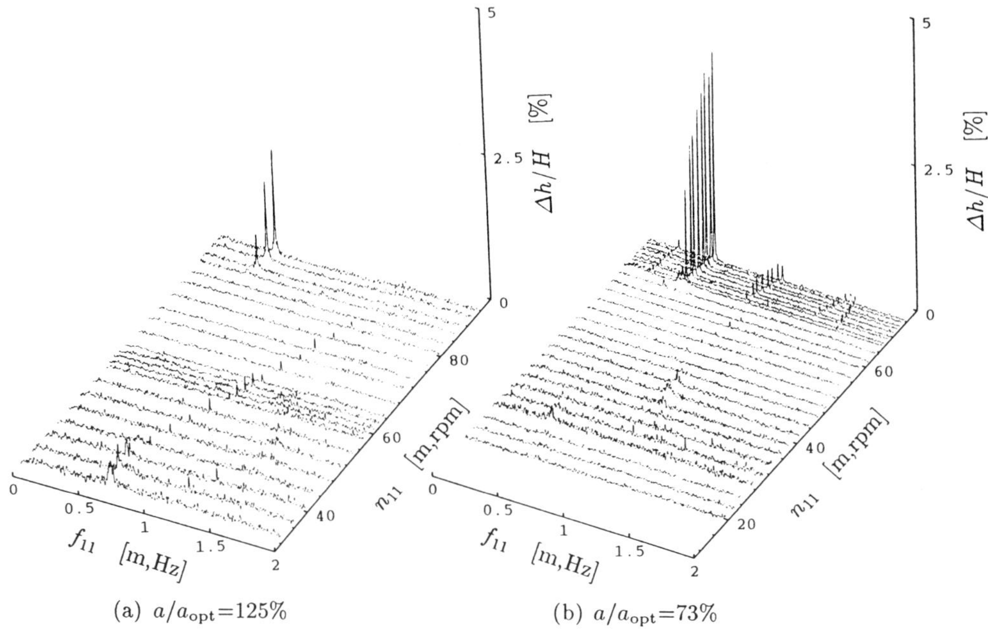

(a) a/a_{opt}=125%

(b) a/a_{opt}=73%

Figure 6. 3-D spectral maps with n_{11}axis

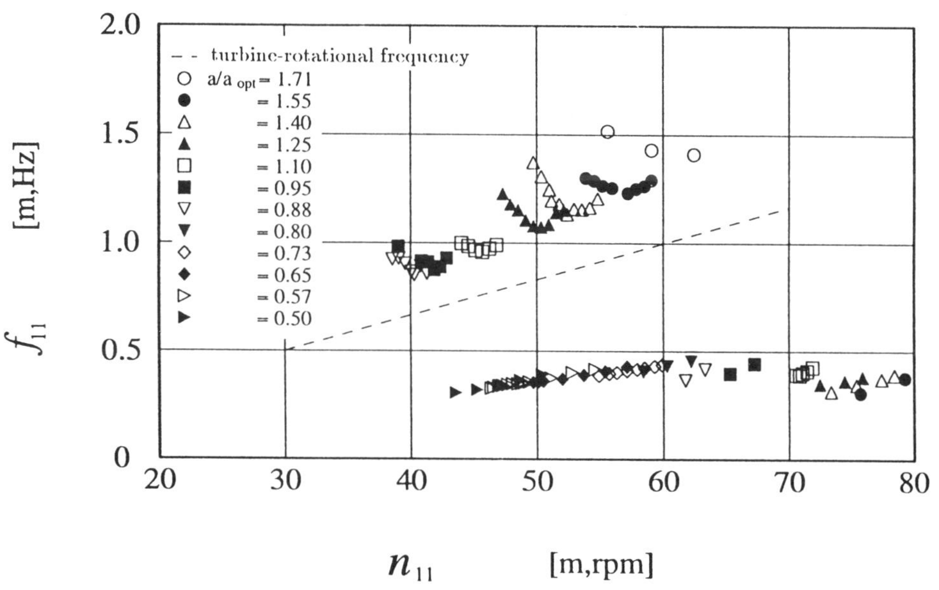

Figure 7. Frequency characteristics

5. Conclusions

As a result of an experiment with a cone-type draft tube to basically investigate the behavior of the vortex core in reverse swirl flow from the runner, the following conclusions are drawn:

1. The vortex core in the reverse swirl flow which is to have the fundamental shape under the condition of non-cavitation was observed by air injection. In comparison between two cases of the regular swirl flow and the reverse one, it is found that the shape of the vortex core of the latter is rarely fixed to that with single rope and is unstable.
2. The frequency of the pressure oscillation does not agree with that of the vortex core revolution in the reverse swirl flow. It is also found that revolutions of the vortex core is lower than the turbine rotation.
3. The frequency of the pressure oscillation in the reverse swirl flow depends on the swirl flow itself but the behavior of the vortex core.
4. It is assumed that the existence of the peripheral velocity component of the flow in the direction of the runner rotation near the draft tube wall makes one of the sharp distinction of the characteristics of the pressure oscillation in the draft tube between the two kinds of cases of the regular swirl flow and the reverse swirl flow.

References

[1] Fisher, R.K., Jaeger, E.U., A contribution for understanding the flow characteristics effecting pressure surge, *Proc. IAHR Symposium, Beijing* (1994), Vol.1, A-1, pp.33-44.

[2] Nishi, M., Kubota, T., Senoo, Y., Surging characteristics of conical and elbow-type draft tubes, *Proc. IAHR Symposium, Stirling*(1984), pp. 272-283.

[3] Hosoi, Y., Experimental evaluation of characteristics of draft tube surging, *IAHR Work Group on the behavior of Hydraulic Machinery Under Steady Oscillatory Conditions, Proc. 7th Meeting,* Ljublyana, Slovenia (1995), A-4.

[4] Hosoi, Y., Characteristics of draft tube surging –A review of recent knowledge–, *Research Reports of The Faculty of Engineering, Tokyo Denki University,* No.37, Tokyo Denki University (1989), pp.27-38.

[5] Guarga, R., Fanelli, M.A., Nishi, M., System Instability Caused by Hydraulic Machinery, *Chapt.6, Vibration and Oscillation of Hydraulic Machinery*, (Ohashi, H., ed.), International Editorial Committee for Book Series on Hydraulic Machinery, Gower Publishing Co., Vermont (1991), pp.210-225.

[6] Hosoi, Y., Characteristics of pressure surge due to whirling water from exit of water turbine runner, *Bulletin of the JSME*, Vol.16, No.99 (1973), pp.1324-1335.

HYDRAULIC OSCILLATION ANALYSIS USING THE FLUID-STRUCTURE INTERACTION MODEL

GAJIĆ A., PEJOVIĆ S., STOJANOVIĆ Z.
Faculty of Mechanical Engineering Belgrade, Yugoslavia

Abstract

The paper presents a fluid transients analysis in hydraulic systems including fluid-structure interaction (FSI). The method is based upon an improved model, that takes into account gravitational and hydraulic resistive forces. The results of calculation applying the FSI model are compared to results obtained by simplified models and to those recently published.

1. Introduction

The mathematical analysis of transient regimes in hydraulic systems is based upon the one-dimensional theory. Fluid-structure interaction (FSI) means that variations of pressure in the system cause deformations of the pipe (axial motion, bending and torsion), and those deformations alter the state of fluid. In order to describe these phenomena analytically, during the last decade investigators have developed several models Wiggert (1987), Obradović (1990), Tijsseling (1993).

The common approach in all these methods is modelling the FSI by means of Poisson coupling, while the interaction of system elements is incorporated by analysis of bends, tees, etc.

Model equations can be solved in time and frequency domains. Solution in time domain offers the information concerning changing of parameters in time; one of most frequently applied is the method of characteristics (MOC) (Wylie & Streeter, 1993).

Solution in frequency domain is used for determination of system eigenvalues, and evaluation of system stability.

E. Cabrera et al. (eds.), Hydraulic Machinery and Cavitation, 845–854.

2. Mathematical model

Some basic assumptions of the mathematical model Wiggert (1987), Obradović (1990) are:

- fluid flow in the pipe is one-dimensional, e.g., all parameters in a cross-section perpendicular to the pipe axis have same values, that change along the pipe axis;
- stresses in the pipe remain in the limits of elasticity;
- pipe wall material is homogeneous and isotropic;
- pipe is thin-walled;
- fluid in the pipe is a liquid.

These assumptions were introduced in order to simplify to a certain extent the analysis of loads, although they did not reduce the applicability of the model. For example, the presence of gases dissolved in the fluid is manifested by reduction of celerity of wave propagation in the fluid. This can be handled very effectively by dividing the pipe into several sections with different wave celerities.

The assumption of the one-dimensional flow in the pipeline does not significantly reduce the precision of computation.

Stresses and strains should not exceed the yield stress limit of given materials. Such an assumption is justified by the fact that elements are very rarely designed to operate in the plastic domain. Most pipes are thin-walled (the ratio of inner tube diameter to the pipe wall thickness exceeds 10, $d/\delta \gg 10$).

It is also assumed that during the deformation tube cross-sections remain plain, while elastic waves have wavelengths much greater than pipe cross-section dimensions; when axial motion is analysed, inertial forces that correspond to motion in the perpendicular direction are neglected.

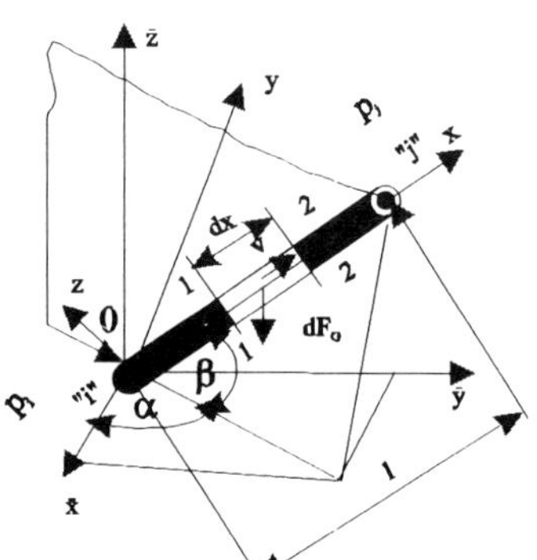

Figure 1: Pipe "i-j" in a global coordinate system

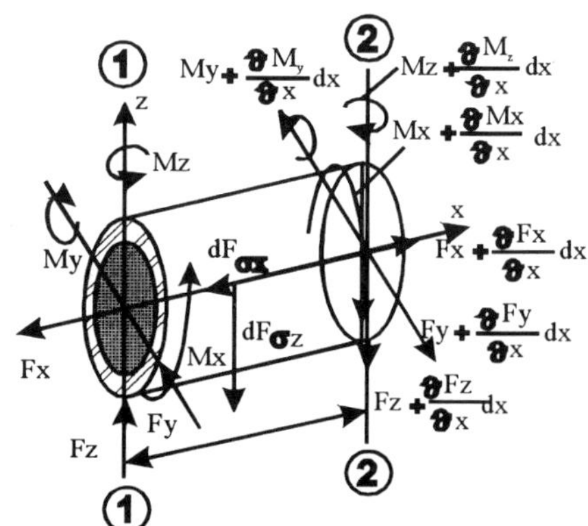

Figure 2: Elementary pipe dx

Figs 1 and 2 present a pipe element of a length dx. Gravitational force acting upon the liquid and the pipe has components in direction of x and z axes. Pipe element is subjected to the action of forces and torques.

Tube and fluid behaviour in a straight tube is described by the model consisting of 14 partial differential equations, forming subsystems (1.1-1.4), (1.5-1.6), (1.7-1.10) and (1.11-1.14), Obradović (1990), (Wiggert *et al.*, 1987).

$$\frac{\partial F_x}{\partial t} - E \cdot A_c \cdot \frac{\partial \dot{u}_x}{\partial x} - 2 \cdot \nu \cdot A_f \cdot \frac{\partial p}{\partial t} = 0 \tag{1.1}$$

$$\frac{\partial F_x}{\partial x} - A_c \cdot \rho_c \cdot \frac{\dot{u}_x}{\partial t} - 2 \cdot \nu \cdot A_f \cdot \frac{\partial p}{\partial x} - g \cdot (A_c \cdot \rho_c + A_f \cdot \rho_f) \cdot \sin\beta + \tau \cdot d \cdot \pi = 0 \tag{1.2}$$

$$\frac{\partial p}{\partial t} + K^* \cdot \frac{\partial v}{\partial x} - 2 \cdot \nu \cdot K^* \cdot \frac{\partial \dot{u}_x}{\partial x} = 0 \tag{1.3}$$

$$\frac{\partial p}{\partial x} + \rho_f \cdot \frac{\partial v}{\partial t} + \frac{4 \cdot \tau}{d} + g \cdot \rho_f \cdot \sin\beta = 0 \tag{1.4}$$

$$\frac{\partial M_x}{\partial t} - G \cdot I_0 \cdot \frac{\partial \dot{\varphi}_x}{\partial x} = 0 \tag{1.5}$$

$$\frac{\partial M_x}{\partial x} - \rho_c \cdot I_0 \cdot \frac{\partial \dot{\varphi}_x}{\partial t} = 0 \tag{1.6}$$

$$\frac{\partial F_z}{\partial t} - G \cdot A_c \cdot (\frac{\partial \dot{u}_z}{\partial x} + \frac{\partial \varphi_y}{\partial t}) = 0 \tag{1.7}$$

$$\frac{\partial F_z}{\partial x} - (A_c \cdot \rho_c + A_f \cdot \rho_f) \cdot \frac{\partial \dot{u}_z}{\partial t} + (A_c \cdot \rho_c + A_f \cdot \rho_f) \cdot g \cdot \cos\beta = 0 \tag{1.8}$$

$$\frac{\partial M_y}{\partial x} - (I_c \cdot \rho_c + I_f \cdot \rho_f) \cdot \frac{\partial \dot{\varphi}_y}{\partial t} - F_z + \frac{1}{2} \cdot (A_c \cdot \rho_c + A_f \cdot \rho_f) \cdot g \cdot \cos\beta = 0 \tag{1.9}$$

$$\frac{\partial M_y}{\partial t} - E \cdot I_c \cdot \frac{\partial \dot{\varphi}_y}{\partial x} = 0 \tag{1.10}$$

$$\frac{\partial F_y}{\partial t} - G \cdot A_c \cdot (\frac{\partial \dot{u}_y}{\partial t} - \frac{\partial \varphi_z}{\partial t}) = 0 \tag{1.11}$$

$$\frac{\partial F_y}{\partial x} - (A_c \cdot \rho_c + A_f \cdot \rho_f) \cdot \frac{\partial \dot{u}_y}{\partial t} = 0 \tag{1.12}$$

$$\frac{\partial M_z}{\partial x} - (I_c \cdot \rho_c + I_f \cdot \rho_f) \cdot \frac{\partial \dot{\varphi}_z}{\partial t} + F_y = 0 \tag{1.13}$$

$$\frac{\partial M_z}{\partial t} - E \cdot I_c \cdot \frac{\partial \dot{\varphi}_z}{\partial x} = 0 \tag{1.14}$$

3. Solution by the transfer-matrix method

In order to prepare equations solving by the transfer matrix method, the expression for shear stress (friction term) has to be linearized, and variations of dependent parameters are analysed as fluctuations around their mean values.

The essential idea of the transfer matrix method is that parameters of the system in a certain cross-section can be determined upon the parameters preceding that cross-section and the transfer matrix of an element between the cross-sections:

$$\{S\}_D = [T] \cdot \{S\}_U \tag{1.15}$$

where: $\{S\}_D$ and $\{S\}_U$ are parameter vectors in downstream and upstream cross-sections respectively, and $[T]$ is the element transfer matrix. This method is also valid for the system as a whole.

The system of equations (1.1)-(1.14) that describe the behaviour of a straight pipe consists of four equation subsystems: (1.1)-(1.4), (1.5)-(1.6), (1.7)-(1.10) and (1.11)-(1.14). These subsystems correspond to the following parameter vectors:

$$\begin{aligned} \{S_1\} &= \{p \; v \; u_x \; F_x\}^T & \{S_2\} &= \{M_x \; \varphi_x\}^T \\ \{S_3\} &= \{u_z \; F_z \; M_y \; \varphi_y\}^T & \{S_4\} &= \{u_y \; F_y \; M_z \; \varphi_z\}^T \end{aligned} \tag{1.16}$$

Transfer matrices of the equation subsystems are T_1, T_2, T_3 and T_4, and for the straight pipe the system transfer matrix T is defined as:

$$T = \begin{bmatrix} [T_1] & 0 & 0 & \\ 0 & [T_2] & 0 & 0 \\ 0 & 0 & [T_3] & 0 \\ 0 & 0 & 0 & [T_4] \end{bmatrix} \tag{1.17}$$

4. Computations

As an illustration of the method, the calculation was performed for the hydraulic system presented in Fig.3 analysed in Svingen (1994) too.

The hydraulic system consists of two straight tubes connected by a bend. At the pipe end 3 there is an open reservoir, and at the end 1 the pipe is open (the water pours out freely). Both pipe ends are fixed. The boundary conditions are the same for both ends: $p = 0$, $\dot{u}_x = 0$, $\dot{u}_z = 0$.

Transfer matrix of the system is:

$$[T_s] = [T_{st}] \cdot [B] \cdot [T_{st}] \tag{1.18}$$

where: $[T_s]$ system transfer matrix, $[T_{st}]$ straight tube transfer matrix, and $[B]$ bend transfer matrix.

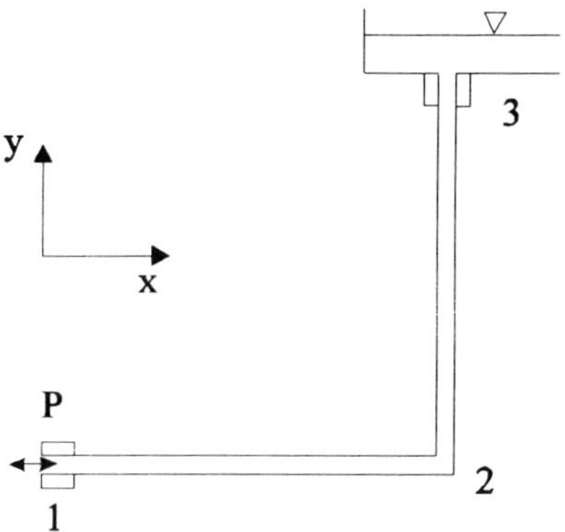

Figure 3: Hydraulic system

Table 1: Characteristics of the system

Pipe density ρ_c:	7900 kg/m^3
Modulus of elasticity E:	$2.1 \cdot 10^{11}$ N/m^2
Poisson ratio ν:	0.3
Pipe outer diameter d_{out}:	82 mm
Pipe int. diameter d_{int}:	80 mm
Length of reaches 1-2 and 2-3 L:	10 m
Liquid density ρ_f:	1000 kg/m^3
Liquid bulk modulus K:	$2.1 \cdot 10^9$ N/m^2

The analysis of a system with a 90° bend in the x-z plane is performed by the simultaneous analysis of the following parameters: p, v, F_x, $\dot{u}_x$, F_z, $\dot{u}_z$, M_y, $\dot{\varphi}_y$.

Transfer matrices of the straight tube are formulated in section 2, and the matrix of a bend in x-z plane is as follows:

$$\begin{Bmatrix} p \\ v \\ \dot{u}_x \\ F_x \\ \dot{u}_z \\ F_z \\ M_y \\ \dot{\varphi}_y \end{Bmatrix}_D = \begin{bmatrix} 1 & 0 & 0 & 0 & 0 & 0 & 0 & 0 \\ 0 & 1 & -1 & 0 & 1 & 0 & 0 & 0 \\ 0 & 0 & 0 & 0 & 1 & 0 & 0 & 0 \\ A_f & 0 & 0 & 0 & 0 & 1 & 0 & 0 \\ 0 & 0 & -1 & 0 & 0 & 0 & 0 & 0 \\ A_f & 0 & 0 & -1 & 0 & 0 & 0 & 0 \\ 0 & 0 & 0 & 0 & 0 & 0 & 1 & 0 \\ 0 & 0 & 0 & 0 & 0 & 0 & 0 & 1 \end{bmatrix} \cdot \begin{Bmatrix} p \\ v \\ \dot{u}_x \\ F_x \\ \dot{u}_z \\ F_z \\ M_y \\ \dot{\varphi}_y \end{Bmatrix}_U \tag{1.19}$$

where A_f denotes the flow cross-section of the pipe.

In the steady state the following condition is met:

$$\begin{Bmatrix} 0 \\ v \\ 0 \\ F_x \\ 0 \\ F_z \\ M_y \\ 0 \end{Bmatrix}_1 = \begin{bmatrix} T_{11} & T_{12} & T_{13} & T_{14} & T_{17} & T_{18} & T_{19} & T_{1,10} \\ T_{21} & T_{22} & T_{23} & T_{24} & T_{27} & T_{28} & T_{29} & T_{2,10} \\ T_{31} & T_{32} & T_{33} & T_{34} & T_{37} & T_{38} & T_{39} & T_{3,10} \\ T_{41} & T_{42} & T_{43} & T_{44} & T_{47} & T_{48} & T_{49} & T_{4,10} \\ T_{71} & T_{72} & T_{73} & T_{74} & T_{77} & T_{78} & T_{79} & T_{7,10} \\ T_{81} & T_{82} & T_{83} & T_{84} & T_{87} & T_{88} & T_{89} & T_{8,10} \\ T_{91} & T_{92} & T_{93} & T_{94} & T_{97} & T_{98} & T_{99} & T_{9,10} \\ T_{10,1} & T_{10,2} & T_{10,3} & T_{10,4} & T_{10,7} & T_{10,8} & T_{10,9} & T_{10,10} \end{bmatrix} \cdot \begin{Bmatrix} 0 \\ v \\ 0 \\ F_x \\ 0 \\ F_z \\ M_y \\ 0 \end{Bmatrix}_3 \tag{1.20}$$

This matrix equation can be used for the determination of system eigenvalues. The state in node 2 can be determined upon the state in nodes 1 and 3, i.e.:

$$\left\{ \begin{array}{c} p \\ v \\ \dot{u}_x \\ F_x \\ \dot{u}_z \\ F_z \\ M_y \\ \dot{\varphi}_y \end{array} \right\}_2 = [T_{st}] \cdot \left\{ \begin{array}{c} 0 \\ v \\ 0 \\ F_x \\ 0 \\ F_z \\ M_y \\ 0 \end{array} \right\}_3 \tag{1.21}$$

The same system was analysed, Svingen (1994), but for the case of frictionless flow. The present authors analysed the influence of friction upon the amplitude of pressure oscillations in node 2 of the hydraulic system and upon the greatest amplitudes of pressure oscillations in the system. Reaches 1-2 and 2-3 have been divided into 8 sections (17 nodes in total).

The procedure was performed for the case of Poisson coupling, and no Poisson coupling. Friction was modelled as non-frequency dependent, e.g.:

$$\tau = \frac{\lambda}{4} \cdot \frac{\rho_f}{2} \cdot (v - \dot{u}_x)^2 \tag{1.22}$$

where: τ - shear stress, λ - coefficient of friction, ρ_f - liquid density, v - flow velocity and $\dot{u}_x$ - pipe motion velocity.

Some of the mentioned quantities required prior computation in accordance with the following expressions:

- reduced modulus of elasticity E^*: $E^* = E/(1-\nu^2) = 2.3076923 \cdot 10^{11}$ N/m^2
- pipe wall thickness δ: $\delta = (d_{out} - d_{int})/2 = 0.001$ mm
- reduced liquid bulk modulus K^*: $K^* = K/(1+\frac{d_{int} \cdot K}{E^* \cdot \delta}) = 1.2152777 \cdot 10^9$ N/m^2
- discharge cross-section area A_f: $A_f = (\pi \cdot d_{int}^2)/4 = 5.024 \cdot 10^{-3}$ m^2
- cross-section area A_c: $A_c = \pi \cdot (d_{out}^2 - d_{int}^2)/4 = 2.512 \cdot 10^{-4}$ m^2
- pipe polar moment of inertia I_0: $I_0 = \pi \cdot (d_{out}^4 - d_{int}^4)/32 = 4.1724477 \cdot 10^{-7}$ m^4
- shear modulus of rigidity G: $G = E/[2 \cdot (1+\nu)] = 8.076923 \cdot 10^{10}$ N/m^2
- pipe axial moment of inertia I_c: $I_c = I_0/2 = 2.0862238 \cdot 10^{-7}$ m^4
- liquid axial moment of inertia I_f: $I_f = \pi \cdot d_{int}^4/64 = 2.0096 \cdot 10^{-6}$ m^4
- linearized friction R: $R = (\lambda \cdot \overline{v})/(g \cdot d_{int} \cdot A_f)$

As $\lambda \approx 0.02$ in many cases in engineering practice, four cases were analysed: $\lambda = 0.00$, $\lambda = 0.01$, $\lambda = 0.02$ and $\lambda = 0.03$. The flow velocity in the stationary state $\overline{v}$ was assumed to be 5 m/s, because flow velocities in hydraulic systems generally lie in the scope of 3-5 m/s.

The results are presented in Figs. 4-9. Parameter p_1 denotes the pressure excitation amplitude at node 1, while p_2 and p_m denote the pressure amplitude at node 2, and the maximum pressure amplitude in the system, respectively. For practical reasons, the pressure amplitudes ratio is given on the logarithmic scale.

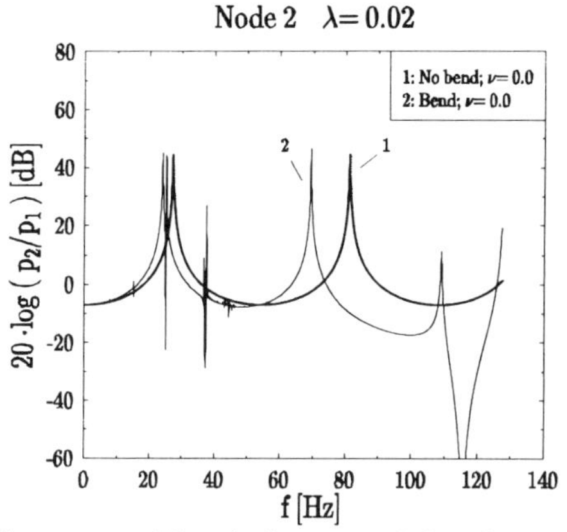

Figure 4: The influence of bend coupling upon pressure amplitude in node 2 (no Poisson coupling, $\lambda = 0.02$)

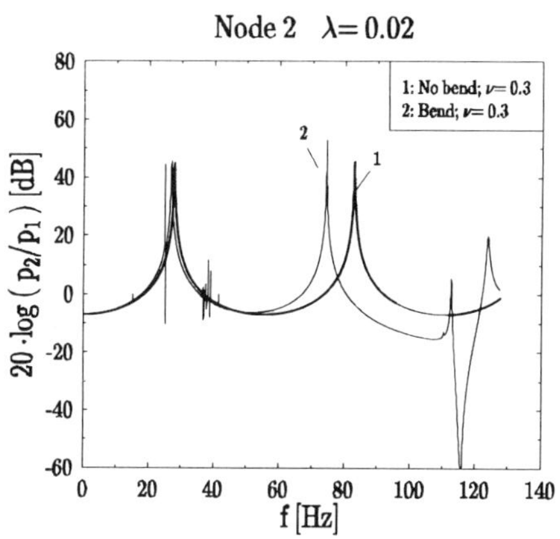

Figure 5: The influence of bend coupling upon pressure amplitude in node 2 (Poisson coupling, $\lambda = 0.02$)

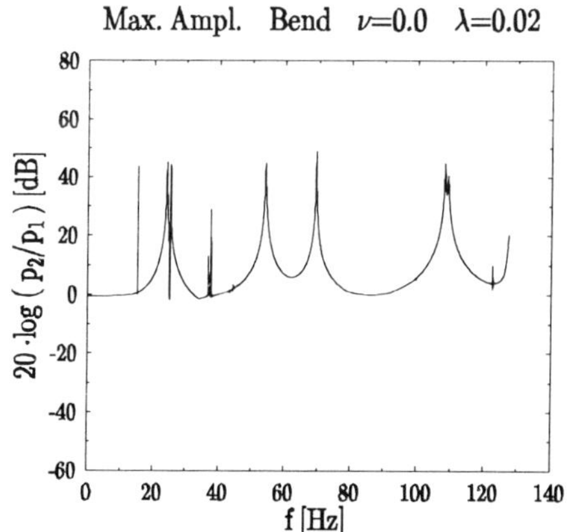

Figure 6: The influence of Poisson coupling upon maximum pressure amplitude in the system (bend coupling, $\lambda = 0.02$)

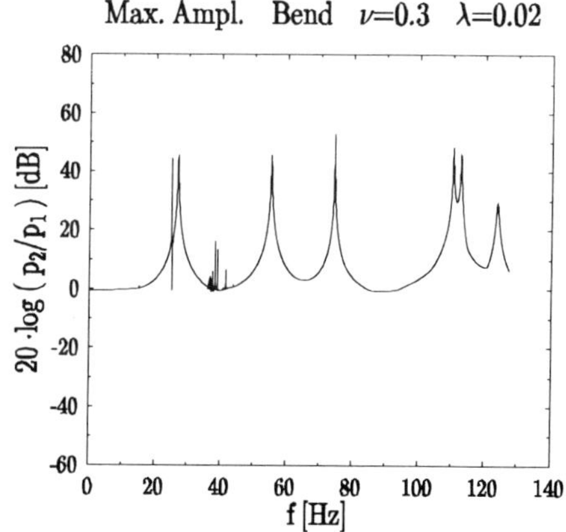

Figure 7: The influence of Poisson coupling upon maximum pressure amplitude in the system (bend coupling, $\lambda = 0.02$)

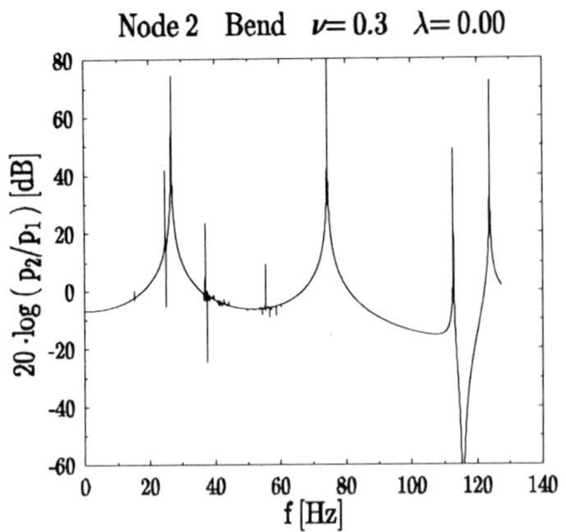

Figure 8: The influence of friction upon pressure variation in node 2 (bend and Poisson coupling, $\lambda = 0.00$)

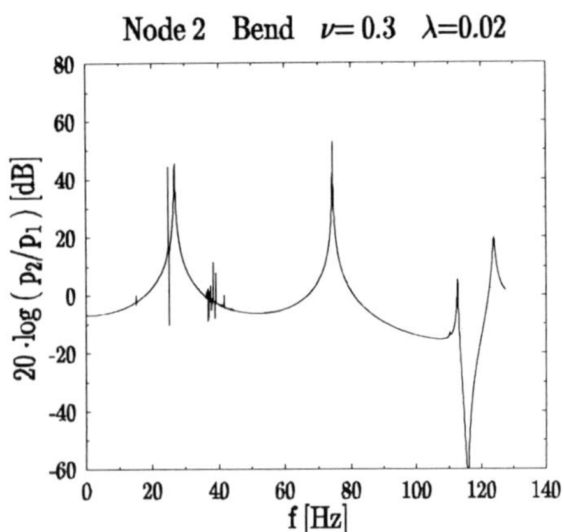

Figre 9: The influence of friction upon pressure variation in node 2 (bend and Poisson coupling, $\lambda = 0.02$)

In Figs 4 and 5 the influence of bend coupling upon pressure variation in node 2 (Fig.3) is presented. In the computation presented in Fig.4 Poisson ratio was assumed to be equal to 0.0. When the bend coupling was neglected, two peaks were observed, at 27.02 Hz and 81.05 Hz. In the case of bend coupling, there are three peaks at 24.01 Hz, 25.21 Hz and 69.30 Hz, the last having the amplitude slightly less than 2 dB higher than the others. Therefore, it can be concluded that the inclusion of bend coupling results in shifting the critical frequencies to the left.

Similar results are obtained in the case of simultaneous inclusion of bend coupling and Poisson coupling (Fig.5). When bend coupling is not taken into account, the critical frequencies are 27.57 Hz and 82.72 Hz; when this is taken into account, the critical frequencies are 24.95 Hz, 26.78 Hz and 74.42 Hz (74.57 Hz). The last one is augmented by about 14 Hz. The critical frequencies are shifted to the left, as in the case of exclusion of Poisson coupling, but to a smaller degree.

In the paper (Gajić *et al.*, 1995) influences of bend coupling and Poisson coupling upon the maximum amplitudes in the system are also presented. The inclusion of bend coupling, without Poisson coupling, into the computation causes shifting of the critical frequencies to the right by about 1 Hz, and the augmentation of corresponding amplitudes by about 1 dB.

Similar observations are true for the case of bend coupling and Poisson coupling (Figs 6 and 7). The critical frequencies are shifted to the right by about 1-3 Hz, and the amplitude augmentations are 1-2 dB at lower frequencies, whereas they amount to 3-6 dB at higher frequencies.

Figs 8-9 present the influence of friction upon the pressure amplitude in node 2 (Fig.3). The influences of bend coupling and Poisson coupling are also taken into account. In Fig.8, when friction is neglected, there are four critical frequencies under 100 Hz. For the sake of comparison, these four frequencies are analysed for the remaining three cases as well. The results are presented in Table 2.

Table 2

Frequency	Pressure amplitude			
	$\lambda = 0.0$	$\lambda = 0.1$	$\lambda = 0.2$	$\lambda = 0.3$
Hz	dB	dB	dB	dB
24.95	41.89	43.17	44.43	36.99
26.78	74.46	52.48	45.59	41.54
74.43	81.85	59.72	52.85	48.81
74.57	40.65	38.37	38.26	37.85

It is evident that the inclusion of friction into computations leads to a decrease in the pressure amplitudes; this decrease is greater for greater frequencies (see frequencies above 100 Hz in Figs 8-9).

5. Conclusions

The results obtained once again indicate that the FSI is important in evaluation of transient regimes in hydraulic systems even in the case of relatively simple systems. This assertion is valid for both Poisson coupling and bend coupling. The influence of friction is important too; this influence is investigated by means of linearized friction R. The parameter R can be used as a similarity factor when this influence is investigated. However, while the influence of friction upon the pressure oscillation amplitudes in a selected point in the system is important, its influence upon the maximum pressure oscillation amplitudes in the system is smaller (due to a variation of the mode shapes). Similar observations can be derived for the influence of coupling (both bend and Poisson). The authors' intention is to try the FSI model presented in the paper with alternate friction models. From the nature of R it can be concluded that if friction coefficient λ is kept constant, the pressure amplitudes in the system can be affected by variying the pipe diameter and the fluid velocity in the stationary state.

6. Acknowledgments

Grateful acknowledment is expressed to the Serbian Foundation for Sciences for supporting investigations Grant S.2.06.21.142 Faculty of Mechanical Engineering Belgrade.

7. Nomenclature

A	m^2	- cross-section area;	φ	rad	- angle of rotation;
ν	-	- Poisson ratio;	d_{int}	m	- inner tube diameter;
δ	m	- pipe wall thickness;	E	N/m^2	- modulus of elasticity;
F	N	- force;	G	N/m^2	- shear modulus of rigidity;
τ	N/m^2	- shear stress;	I_0	m^4	- polar moment of inertia;
B	-	- bend matrix;	K^*	N/m^2	- red. liquid bulk modulus;
L	m	- length;	λ	-	- friction coefficient;
p	Pa	- pressure,	R	s/m^3	- linearized friction;
		pressure amp.;	u	m	- axial displacement;
M	Nm	- torque;	β	rad	- pipe inc. in vert. plane;
ρ	kg/m^3	- density;	S	-	- parameter state vector;
T	-	- transfer matrix.	g	m/s^2	- gravity acceleration;
$\dot{\varphi}$	rad/s	- ang. velocity;	$\dot{u}$	m/s	- velocity of axial disp.;
v	m/s	- flow velocity;	I	m^4	- cross-sect. moment of inertia;
			x, y, z	-	- local coord. system axes;
			$\overline{x}$, $\overline{y}$, $\overline{z}$	-	- global coor. system axes.

8. Superscripts and subscripts

1, 2	- nodes 1 and 2;	$\overline{x}$	- stat. state;	c	- pipe;
D	- downstream;	f	- fluid;	int	- interior;
m	- maximum;	out	- outer;	s	- system;
st	- straight tube;	x, y, z	- coordinates;	U	- upstream.

9. References

Brekke, H. (1984) *A Stability Study on Hydro - Power Plant Governing Including the Influence from Quasi Nonlinear Damping of Oscillatory Flow and from the Turbine Characteristics*, Dissertation, University of Trondheim, Norwegian Institute of Technology, Oslo

Gajić, A., Pejović S., Stojanović Z., Kratica J. (1995) Fluid-Structure Interaction Analysis in Frequency Domain, *Proceedings of WG on the Behaviour of Hydraulic Machinery under Steady Oscillatory Conditions 7th Int. Meeting*, Ljubljana

Obradović, P. (1990) *An Investigation of Pipeline Dynamic Loads Following Transient Regimes in Hydraulic Systems* (in Serbian), Dissertation, Beograd

Pejović, S. (1979) *Hydraulic Vibrations of Hydraulic Plants* (in Serbian), Beograd

Stojanović, Z. (1993) *An Investigation of Dynamic Loads Following Oscillations of Fluid and Structure* (in Serbian), M.Sci Thesis, Beograd

Wylie, E.B., Streeter, V.L. (1993) *Fluid Transients in Systems*, Prentice Hall Inc., New Jersey

Svingen, B. (1994) A Frequency Domain Solution of the Coupled Hydromechanical Vibrations in Piping Systems by the Finite Element Method, *Proceedings of XVII IAHR Symposium*, Beijing, pp.1259-1269.

Tijjseling A.S. (1993) *Fluid - Structure Interaction in Case of Waterhammer with Cavitation*, Dissertation, Delft

Wiggert, D.C., Lesmez, M.L., Hatfield, F.J. (1987) Modal Analysis of Vibrations in Liquid Filled Systems, *Proceedings of ASME Winter Annual Meeting*, Boston

FRANCIS TURBINE SURGE : DISCUSSION AND DATA BASE

Thierry Jacob and Jean-Eustache Prénat
IMHEF-EPFL, av de Cour 33, 1007 Lausanne, Switzerland

0. Background

A system is considered as stable in sustained operating conditions if the response to unavoidable disturbances does not endanger its safety. Normal disturbances of hydraulic turbines are pressure fluctuations [3, 5, 10], influenced by machine design, operating conditions and by the dynamic response of the water conduits and rotating components. They may be associated with mechanical fluctuations of shaft torque, rotational speed, hydraulic load on guide vanes etc. as well as with vibrations.
Low frequency pressure fluctuations are a special threat to stability of operation because they may propagate to the whole piping system and cause hydraulic resonance. They may also excite sub-synchronous free oscillations of the prototype rotating machinery. They occur typically between 0.2 and 3 times the runner rotational frequency.
In Francis turbines, a strong runner outlet swirl develops in off-design conditions and induces flow instabilities in the draft tube. This draft tube surge is the most commonly identified phenomenon among low frequency pressure fluctuations. In addition, draft tube cavitation may change the natural frequencies of the hydraulic system.
In the following pages, we will discuss various aspects of draft tube surges. Our purpose is to establish guidelines for the interpretation of pressure fluctuations. The discussions are based on test results from over twenty turbine models and three prototypes from different manufacturers, with specific speeds n_{QE} ranging from 0.074 to 0.320.

1. Classification of draft tube pressure surges

1.1. OPERATING RANGE

Turbine operation parameters are the energy and discharge coefficients E_{nD} and Q_{nD}. The reference coefficients are at best efficiency. The reference elevation for Thoma number σ is the runner outlet at the band.

$$E_{nD} = \frac{E}{n^2 D^2} \quad ; \quad Q_{nD} = \frac{Q}{nD^3} \quad ; \quad n_{QE} = \frac{Q_{nD}^{1/2}}{E_{nD}^{3/4}} \quad ; \quad \sigma = \frac{NPSE}{E} \qquad (1)$$

Figure 1 is the schematic operating range of a Francis turbine. Constant guide vane angle lines and efficiency contours are set in the hydraulic energy-discharge diagram for a constant rotational speed. The turbine is defined by its best efficiency specific hydraulic energy (a) and discharge (b), and a rated specific hydraulic energy of the project (c), which may be different from best efficiency.
The range of operation is limited by the minimum guide vane angle for sustained operation (d), the maximum guide vane angle (e), the maximum (f) and minimum (g) specific hydraulic energies of the project and by the maximum generator power (h).
Pressure fluctuations in the draft tube are observed at the rated energy coefficient (c), from minimum to maximum guide vane angle. A waterfall diagram displays the Fourier

E. Cabrera et al. (eds.), Hydraulic Machinery and Cavitation, 855–864.

transform amplitude of pressure fluctuations as a function of frequency and discharge coefficient. Pressure fluctuations are made non-dimensional by the specific hydraulic energy. Fluctuation frequencies are made non-dimensional by the rotational frequency.

The following domains are defined within the operating range, see also [3] :

1 very low discharge operation;
2 part load operation;
3 high partial load operation;
4 operation around best efficiency;
5 full-load operation.

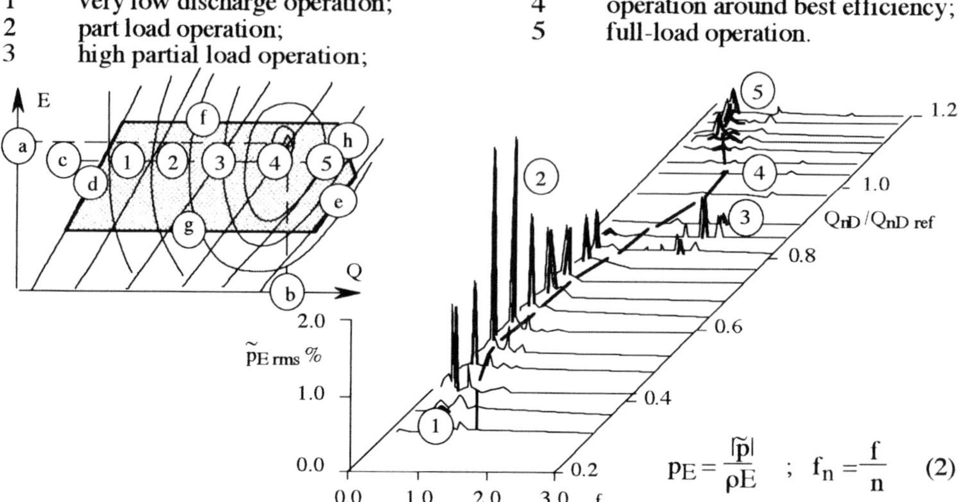

$$p_E = \frac{|\tilde{p}|}{\rho E} \quad ; \quad f_n = \frac{f}{n} \tag{2}$$

Figure 1. Francis turbine hill-chart and waterfall diagram of pressure fluctuations in the draft tube cone

1.2. DRAFT TUBE PRESSURE SURGES

1.2.1. *Part load precession*

Label (2) in the waterfall diagram of figure 1 marks the pressure fluctuations associated with the part load precession. The precession is the motion of a center of rotation around a fixed axis. At part load, the diverging geometry of the draft tube and the draft tube elbow induce a forced precession of the swirling flow at the runner outlet. When Thoma number is small enough, a vapor cavity forms along the vortex center. The helical shape and circular motion of this cavity are evidences of the precession.

The precession generates a rotating field of velocity and pressure fluctuations in the draft tube cone. Defaults in axial symmetry, mostly due to the draft tube elbow, cause a hydraulic excitation at the precession frequency.

Similitude considerations show that the precession frequency can be made non-dimensional by the rotational frequency, and that the excitation magnitude can be made non-dimensional by the turbine specific hydraulic energy. Although the frequency could theoretically vary in a great range, most Francis turbines feature a part load precession frequency of about :

$$\frac{f_p}{n} \approx 0.26 \pm 20\ \% \tag{3}$$

An unusual turbine design can lead to a different precession frequency. Relative frequencies as high as $f_p/n = 0.4$ were found on upgraded turbines.

Pressure fluctuations associated with the part load precession are influenced by turbine design and by the dynamic response of the test circuit or prototype water conduits. If distortions due to dynamic responses of the test circuit and rotating components are small, the order of magnitude of pressure fluctuations at position (1), figure 2, is about :

$$\tilde{p}_{E\,rms} \approx \frac{n_{QE}}{5} \pm 50\ \% \tag{4}$$

Harmonic frequencies of the precession often appear in the spectral data. They usually cause little or no excitation. Multiple vortex precession [12] sometimes occurs at very low discharge, see label (1) in figure 1. Associated amplitudes are generally small.

1.2.2. *Free oscillations of draft tube flow*

A large vapor cavity [6] forms at the runner outlet, due to the residual swirl and the low pressure level. The water plug in the draft tube oscillates against this elastic volume. System constants are the inertia I of the water plug and the cavitation compliance C.

$$f_0 = \frac{1}{2\pi\sqrt{IC}} \quad ; \quad I = \int \frac{dL}{A} \quad ; \quad C = -\frac{\partial V_{vap}}{\partial NPSE} \tag{5}$$

The dotted line in the waterfall diagram of figure 1 marks the frequency of free oscillations. It is smallest at part load and at full load, when the volume of draft tube cavitation is maximum. In this case, the frequency of free oscillations gets very close to the part load precession frequency and draft tube resonance is responsible for the high pressure fluctuation amplitudes.

For a usual draft tube design and a reference Thoma number such as $\sigma \cdot E_{nD} = 1$ for operation at the best efficiency E_{nD}, we can expect the minimum value of the draft tube free oscillations frequency at part load to be close to the precession frequency.

$$\sigma = \frac{1}{E_{nD}} \quad \rightarrow \quad f_{0\ min} \approx f_p \pm 50\ \% \tag{6}$$

For a first approximation, the influences of Thoma number and energy coefficient are :

$$\frac{\partial(f_0/n)}{\partial\sigma} \approx 20 - 35\sqrt{n_{QE}} \pm 50\% \quad ; \quad \frac{\partial(f_0/n)}{\partial(E_{nD}/E_{nDref})} \approx 0.75 \pm 50\% \tag{7}$$

The frequency of free oscillations at full load varies in a wide range according to the turbine design. Its partial derivative to Thoma number is approximately the same as at part load. We could not find a coherent statistic on the magnitude and sign of the partial derivative to energy coefficient.

The water compressibility contributes to the compliance, but is not scalable from model to prototype. This introduces errors in the prediction of free oscillations frequency around best efficiency when the vapor volume is small, or in cases where the draft tube is specially long. In normal cases of part load and full load operation with short draft tubes, the frequency of free oscillations in model tests was not changed significantly by the injection of cavitation nuclei, or even by testing in open circuit mode.

1.2.3. *High partial load instability*

Label (3) in the waterfall diagram of figure 1 marks the pressure fluctuations associated with the high partial load instability, which sometimes occurs between 70 and 90 % of best efficiency discharge. Its physical mechanism is not precisely known. It seems to be connected with pressure waves traveling along the surface of the draft tube vortex cavity, and excited by the passage of the precessing vortex in the draft tube elbow. High partial load instabilities induce large pressure fluctuations in the spiral case and draft tube elbow. They can be responsible for strong vibrations. They are eliminated by air admission flows smaller than 0.1 % of best efficiency discharge.

1.2.4. *Self-amplified pulsations*

Label (5) in the waterfall diagram of figure 1 marks the pressure fluctuations associated with the full load pulsations. When the partial derivatives of cavitation compliance versus discharge and energy have the same sign, the damping of draft tube free oscillations may become negative. This can occur at very low discharges or at full load. Like draft tube free oscillations, the frequency of full load pulsations can be made non-dimensional by the rotational frequency. The amplitude has no real meaning, as this fluctuation is likely to grow catastrophically, depending on the penstock response.

2. Observation of pressure fluctuations

2.1. POSITION OF PRESSURE TRANSDUCERS

The analysis is based on signals from pressure transducers located in the draft tube cone and elbow, where most disturbances are generated. The reference pressure signal is on the inner side of the elbow (1), where amplitudes are often greatest. To allow a separation between rotating and synchronous surges, more transducers are placed at the same elevation, around the draft tube cone (2, 3). The angular position is not important, but immediate interpretation is easier if one is placed on the outer side of the elbow (2).
Spiral case inlet pressure fluctuations (4) are distorted by standing wave effects, characteristic of the test circuit. To evaluate and correct these distortions, additional pressure signals (5, 6) are collected on the feed pipe.
The measurement of torque fluctuations gives a general idea of power fluctuations to be expected on the prototype, even if this observation is not directly transposable.

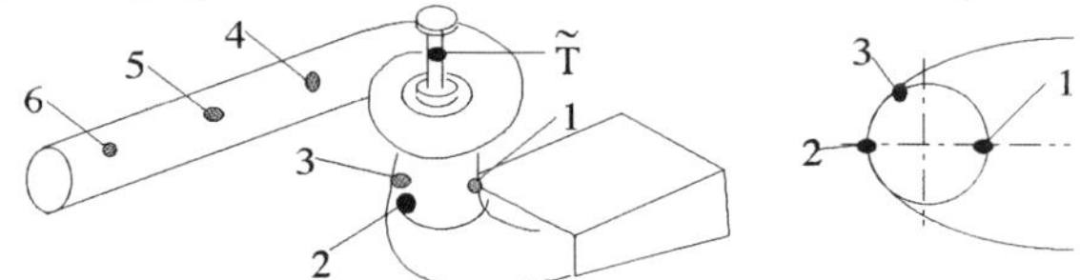

Figure 2. Transducer positions for measurements of stability of operation

2.2. DISCUSSION OF DRAFT TUBE CONE PRESSURE FLUCTUATIONS

2.2.1. *Synchronous and rotating surges*

If we compare the pressure signals from transducers 1, 2 and 3 around a transverse section of the draft tube cone, the magnitudes are the same and the fluctuations are in phase for the draft tube free oscillations, the base frequency of the high partial load instability and for the full load pulsation.
The part load precession and the secondary peak of the high partial load instability have different amplitudes and phase angles within a transverse section of the draft tube cone. These rotating surges can be simplified with an appropriate analysis.

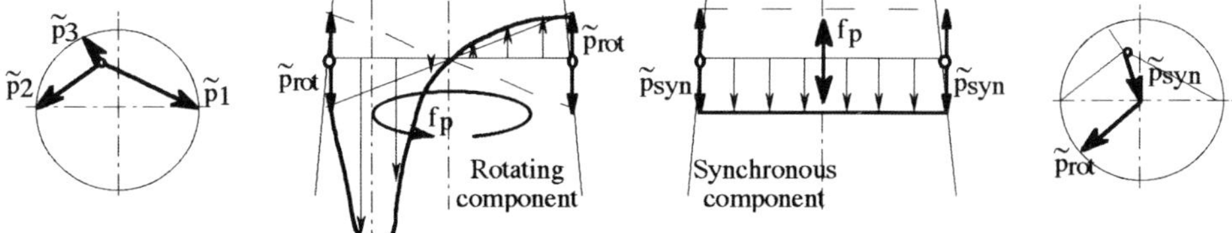

Figure 3. Rotating surge analysis [1]

The observation of pressure fluctuations associated with rotating surges can be quite different, depending on the transducer position. In the example of figure 3, a diagnosis based on the signal from transducer 3 alone would be misleading. Rotating surge analysis can be done on two or more pressure signals from the same circular transverse section of the draft tube cone. The pressure fluctuation is split into two components uniform along the draft tube circumference : $\tilde{p}_{rot}$ is purely rotating and $\tilde{p}_{syn}$ is purely synchronous. With this method, we can compute the pressure fluctuation amplitude at any position around the draft tube cone as the vectorial sum of $\tilde{p}_{syn}$ and $\tilde{p}_{rot}$.
Marginal peaks such as harmonics of the precession cannot be processed this way, but they are less important in the evaluation of stability of operation.

2.2.2. *Discussion of the rotating component of the part load precession*

For a given runner and draft tube cone, $\tilde{p}_{rot}$ is not influenced by Thoma number, by the design of the draft tube elbow or by the length of the feed pipe.

Figure 4 shows the evolution of $\tilde{p}_{rot}$ versus discharge coefficient for different specific speeds. The magnitude of $\tilde{p}_{rot}$ increases with n_{QE}.

The rotating surge analysis is done 0.5 and 1.5 runner diameters below the guide vane center-line. In the top of the draft tube cone, the magnitude of $\tilde{p}_{rot}$ is maximum around 50 % of best efficiency discharge, then drops steadily to zero in the vortex-free region.

In the bottom of the draft tube cone, $\tilde{p}_{rot}$ first rises, then drops around 60 % of best efficiency flow, then goes to a maximum and dies down in the vortex-free region. It does not seem easy to interpolate $\tilde{p}_{rot}$ along the height of the cone, or to tell which elevation is most representative of the precession.

Model measurements of pressure fluctuations must be done at the position where the prototype draft tube is accessible. If an arbitrary elevation must be chosen, it can be one runner diameter below the guide vane center-line, where the model is always accessible.

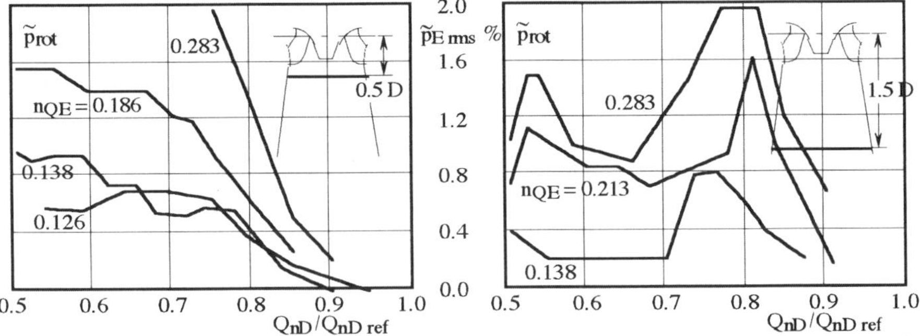

Figure 4. Influence of elevation and specific speed on rotating component of the part load precession

2.2.3. *Discussion of the synchronous component of the part load precession*

$\tilde{p}_{syn}$ was for some time considered as representative of the excitation. Conceptually, it is very close to the passage of the vortex in the draft tube elbow. However, $\tilde{p}_{syn}$, as the pressure fluctuation averaged over the transverse section of the cone, is the sum of the excitation and of the system response, which may vary according to the boundary conditions. Figure 5 shows the pressure fluctuations at the spiral case inlet and $\tilde{p}_{syn}$ for an $n_{QE} = 0.138$ turbine. The correlation between the two amplitudes is excellent.

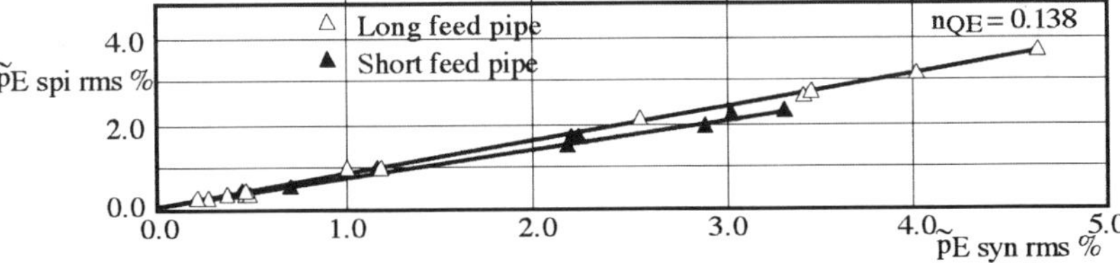

Figure 5. Pressure fluctuation at spiral case inlet and synchronous component of the part load precession

The magnitude of $\tilde{p}_{syn}$ is constant from the top to the bottom of the draft tube cone, over the whole vertical extent of the vapor-filled cavity.

The test was done once with a 6.7 m feed pipe and once with only 3.1 m (19.7 and 9.1 D). The spiral case pressure fluctuation was reduced by 39 % and $\tilde{p}_{syn}$ by 29 % for the 54 % reduction in pipe length. It seems difficult to correct such distortions due to the dynamic response of the test circuit, if $\tilde{p}_{syn}$ is to be used to estimate the excitation.

The gain of $\tilde{p}_{spi}$ to $\tilde{p}_{syn}$ was also found to vary with Thoma number, test head and draft tube depth. No satisfying correlation could be established with the specific speed.

3. Magnitude of hydraulic excitations

3.1. DISTORTIONS DUE TO STANDING WAVES

Local impedances [13] determine the amplitude of pressure fluctuations that develop in a piping system in response to a hydraulic excitation. Pressure signals measured on the feed pipe of a Francis turbine will depend on the position of measurement along the pipe. The amplitude at the spiral case inlet will be influenced by the feed pipe length and wall stiffness. These distortions due to standing waves in the piping system must be corrected for an estimation of the excitation magnitude. The observations of pressure fluctuation amplitudes alone is not sufficient.

The active acoustic power is the energy flow associated with a hydraulic oscillation. It combines pressure and kinetic energies. As it is not influenced by standing waves in the piping, it represents the excitation magnitude more adequately than spiral case inlet pressure fluctuation amplitudes.

The active acoustic power is equal to the real part of the pressure fluctuation times the conjugate of the discharge fluctuation. Its value does not depend on the position along a uniform length of friction-less pipe. It can be positive or negative, indicating the direction of the energy flow. This allows an identification of disturbances generated by the circuit such as feed pump noise, pipe movements etc.

The active acoustic power can also be expressed as the scalar product of pressure and discharge fluctuation vectors, considering their phase shift φ :

$$\mathrm{Real}(\tilde{p}\bullet\tilde{q}^*) = |\tilde{p}|\bullet|\tilde{q}|\bullet\cos\varphi \tag{8}$$

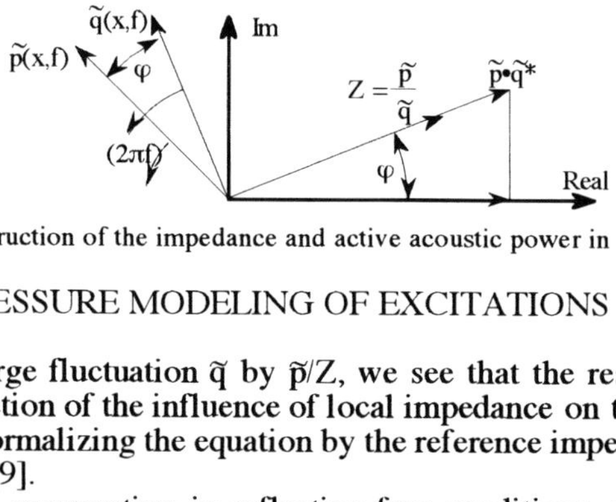

Figure 6. Construction of the impedance and active acoustic power in the complex plane

3.2. SOURCE PRESSURE MODELING OF EXCITATIONS

Replacing the discharge fluctuation $\tilde{q}$ by $\tilde{p}/Z$, we see that the result of acoustic power processing is a correction of the influence of local impedance on the local pressure fluctuation amplitude. Normalizing the equation by the reference impedance Z_0, we obtain a source pressure $\tilde{p}_{src}$ [9].

The source pressure, propagating in reflection-free conditions, would carry the same active power as the actual hydraulic fluctuation :

$$\mathrm{Real}(\tilde{p}\bullet\tilde{q}^*) = \mathrm{Real}\left(\frac{\tilde{p}\bullet\tilde{p}^*}{Z^*}\right) = |\tilde{p}|^2 \bullet \frac{|\mathrm{Real}(Z)|}{|Z|^2} \tag{9}$$

$$\tilde{p}_{src} = \sqrt{Z_0\bullet\mathrm{Real}(\tilde{p}\bullet\tilde{q}^*)} \quad ; \quad \tilde{p}_{src} = |\tilde{p}| \bullet \frac{\sqrt{Z_0\bullet|\mathrm{Real}(Z)|}}{|Z|} \tag{10}$$

The source pressure associated with the part load precession is influenced by the turbine design and operating parameters, but the compensation of reflections due to the circuit reduces the scatter. Between 0.5 and 0.8 Q_{nDref} for operation at E_{nDref} and $\sigma\bullet E_{nD} = 1$, expectable excitation magnitudes are :

$$\tilde{p}_{E\,src\,rms} \approx 0.01 \pm 30\ \% \quad \text{at} \quad f_p \approx 0.26\ n \pm 20\% \tag{11}$$

The amplitude of draft tube pressure fluctuations is correlated with n_{QE}, see figure 4, but this influence is not found in the source pressure estimated at the spiral case inlet.

3.3. DISCUSSION

The use of active oscillatory power to isolate the magnitude of disturbances from the environment response is so widespread in other fields of engineering that it is hardly questionable as a concept. The source pressure formulation is equivalent to the source vectors used in association with a transfer matrix to represent the disturbances generated by a circuit element, when the pressure and discharge components of the vector are related by the reference acoustic impedance Z_0.

The problem with the application of acoustic power methods to turbines is that most disturbances are generated in the draft tube cone and elbow, but the suitable position for the evaluation of the source pressure is the spiral case inlet. There are possibilities of oscillatory power leakage between the emission and the measurement position. The outlet through the draft tube is not a problem : its impedance is imaginary if the draft tube opens in a tank with a sufficiently large free surface, and it is very hard to force acoustic power in this direction. Leakage to the turbine attachments could occur on models with excessive deformations, but this should not be the case.

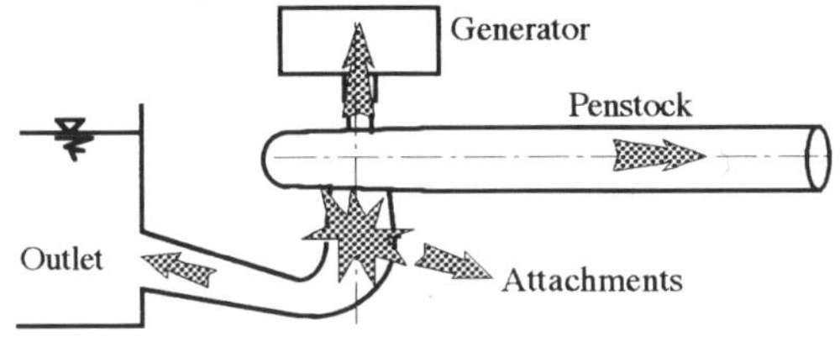

Figure 7. Outflow of oscillatory energy from the turbine

A more likely source of trouble is the exchange of oscillatory power with the rotating machinery, if there are fluctuations of rotational speed. In one case, we could trace discrepancies in source pressure estimations at different test heads to changes in torque fluctuations [9]. A future progress in testing technique will be the use of quick-response tachometers for the measurement of mechanical oscillatory power on the turbine shaft.

4. Prediction of prototype pressure fluctuations

4.1. IMPEDANCE ANALYSIS OF PROTOTYPE INSTALLATION

The eventuality of resonance of the prototype water conduits is investigated in the frequency domain, using linear transmission theories such as impedance methods [13]. The modeling of the plant is usually confined to the piping system, computing impedances from all ends. In the scope of the discussion in (2.3), the electric network and generator should also be included in the impedance analysis [2]. The dynamic model is excited with an arbitrary pulsed discharge at the outlet of one of the runners, and we compute the response in pressure at all units. The absolute amplitude has no meaning at this stage, as the excitation is not realistic.

An important parameter of the simulation is the cavitation compliance C [3, 6], estimated from model tests as in equation (5) and transposed to prototype dimensions and operating conditions. Figure 8 shows the response of a prototype to an arbitrary 1 % pulsed discharge excitation and computed for two values of the cavitation compliance. With C_1, the first resonance frequency is close to the part load precession frequency f_p. The installation of fins on the draft tube wall changed the compliance to C_2. The danger of part load resonance was suppressed. This solution was put to practice on a field case of instability and gave full satisfaction [8].

The cavitation compliance can change very much from one turbine design to another, and it seems difficult to estimate its value from global parameters. The compliance factor at part load and at full load is often in the range of :

$$C_{ED} = \frac{C \bullet E}{D^3} \approx 0.1 \; ... \; 2 \tag{12}$$

In addition to the forced response presented here, impedance methods cover the identification of free oscillation modes, including damping factors. The computed attenuation is generally much smaller than the actual and can be difficult to interpret.

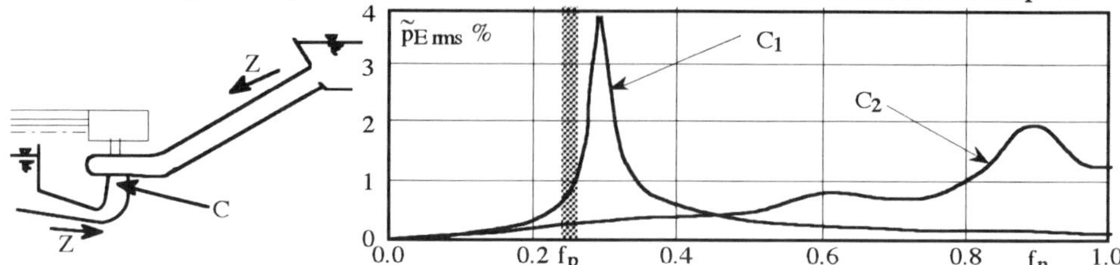

Figure 8. Impedance analysis including cavitation compliance

4.2. TRANSPOSITION OF EXCITATIONS DUE TO THE TURBINE

Excitations due to the turbine are formulated as source pressures, which are transposed to the prototype specific hydraulic energy. Excitation frequencies are scaled to the prototype rotational frequency. The pressure fluctuation $\tilde{p}_{spi}$ at the prototype spiral case inlet must meet two requirements : the local impedance Z_{spi} resulting from the penstock dynamic response and the source pressure $\tilde{p}_{src}$. Reversing equation (10), we obtain :

$$\tilde{p}_{spi\ M \to P} = \tilde{p}_{src\ M} \bullet \frac{(\rho E)_P}{(\rho E)_M} \bullet \frac{|Z_{spi}|}{\sqrt{Z_0 \bullet |Real(Z_{spi})|}} \tag{13}$$

The result of this transposition is compared with field measurements on figure 9. The agreement is good at part load. This method is not suitable for the prediction of full load pulsation amplitudes, as these are self-amplified fluctuations.

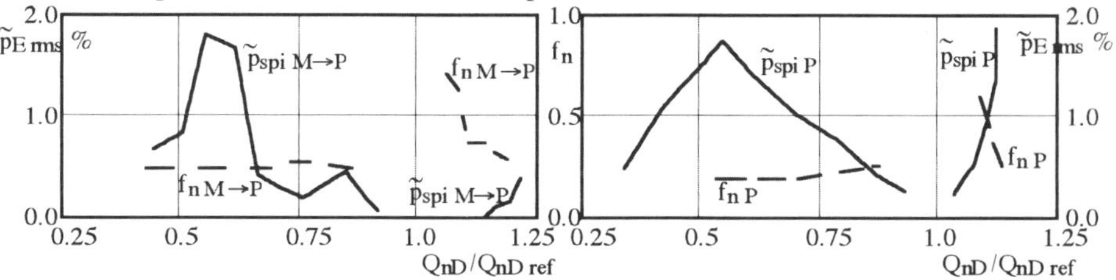

Figure 9. Simulated prototype response to excitations transposed from model tests and field measurement

4.3. DISCUSSION

Hydro power installations are built to minimize energy losses. The real part of a penstock impedance is often very small. From equation 13, we see that errors in the estimation of Real(Z_{spi}) can significantly change the predicted pressure fluctuation amplitude. As the modeling of pipe friction is not accurate, this is a limitation of the method.

Possible exchanges of oscillatory power with the rotating machinery were mentioned in the discussion of 3.3 on the estimation of the excitation magnitude. Accordingly, if the oscillatory power on the turbine shaft can be measured, the resulting excitation power is the sum of the feed pipe and shaft active powers. As active powers are scalar quantities, there are no problems of phase for the addition of the two. A source pressure can then be formulated in the region of the physical excitation, i.e. the draft tube cone and elbow. The penstock and rotating masses impedances combine in a runner outlet impedance from which the local response is computed. At part load, this local pressure fluctuation is the synchronous component of the draft tube surge. The rotating component must still be introduced with due regard to phase for a realistic simulation.

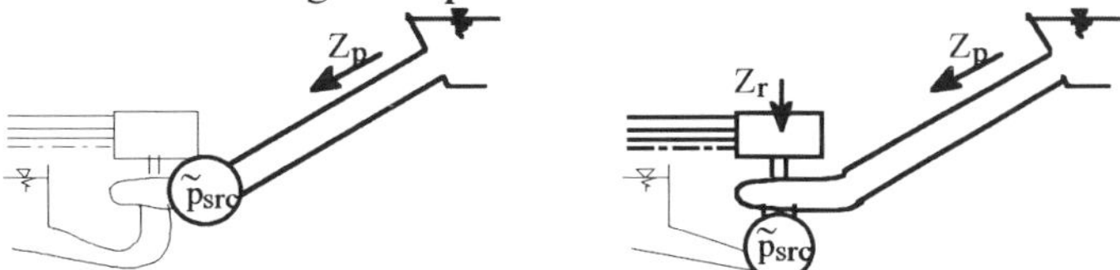

Figure 10. Modeling of responses of the penstock alone or of both penstock and rotating machinery

A prediction of excessive pressure fluctuations makes it possible to optimize stabilizing devices before the prototype is erected.
Structures such as cross-trees, tripods, central cylinders or piles can be installed in the draft tube cone to disorganize swirling flow [7]. These devices reduce the efficiency. In many cases, they were broken after a few hours of operation. Structures in the center of the draft tube cone are effective against the full load pulsation.
Fins on the draft tube wall can reduce the amplitude of pressure fluctuations at part load with small efficiency losses if their position and setting angle are well chosen. In one case, the stabilizing effect of fins was identified as a reduction of the cavitation compliance (figure 8) and the suppression of the synchronous component of the part load precession. The rotating component was not modified [8].
Longer runner cones [11] reduce the reverse flow in the central region of the draft tube cone. In some cases, this reduced the full load pulsation.
Air admission through the runner cone [7] increases the volume of the draft tube cavity. This increases the cavitation compliance and lowers the draft tube free oscillations frequency. Draft tube resonance can be eliminated this way. Air admission can also induce draft tube resonance when the free oscillations frequency without air is higher than excitation frequencies. In some cases, air admission is not effective at full load. It is advisable to have some control over the air inflow valve.
Penstock resonance can be reduced using accumulators or Helmholtz resonators. This can be a good solution for small scale power plants. Active control is considered [4] to govern the spiral case inlet impedance or reduce the excitation power.

5. Conclusions

Most stability problems of Francis turbines can be identified as one of the following :
- Rotating surges related to the part load precession, with possible draft tube resonance;
- High partial load instability, bursts of pressure fluctuations and vibrations;
- Full load pulsation, self-amplified surge.

These phenomena have been described in terms of pressure pulsations. Advanced analysis tools provide the basis of a quick and easy diagnosis of stability of operation. Frequencies and magnitudes can be compared to the typical values listed in this paper.
The excitation magnitude is estimated through the active power of the hydro-acoustic fluctuations in the feed pipe of the model turbine in laboratory test. This method should be improved with the inclusion of oscillatory power exchanges with the generator.
The prediction of prototype pressure fluctuation is done by numerical simulation of the excitation on an impedance model of the plant piping system.

Symbols

A	Cross-sectional area	m^2
C	Cavitation compliance, see equation (5)	$m \cdot s^2$
C_{ED}	Cavitation compliance factor, see equation (12)	
D	Runner diameter	m
E	Specific hydraulic energy of turbine	J/kg
E_{nD}	Energy coefficient, see equation (1)	
f	Frequency	Hz
f_n	Frequency made non dimensional by the rotational frequency	
f_p	Precession frequency	Hz
f_0	Draft tube free oscillations frequency	Hz
I	Inertia of water column	m^{-1}
NPSE	Net positive suction energy of turbine reference elevation at runner outlet periphery	J/kg
n	Rotational frequency of turbine	rev/s
n_{QE}	Specific speed, see equation (1)	
$\tilde{p}$	Pressure fluctuation; for the considered frequency, coefficient of the Fourier transform of the pressure signal	N/m^2
$\tilde{p}_E$	Pressure fluctuation amplitude factor, see equation (2) (rms = effective value)	
$\tilde{p}_{rot}$	Rotating component of draft tube rotating surge	N/m^2
$\tilde{p}_{syn}$	Synchronous component of draft tube rotating surge	N/m^2
$\tilde{p}_{spi}$	Pressure fluctuation at spiral case inlet	N/m^2
$\tilde{p}_{src}$	Source pressure associated with active acoustic power, see equation (10)	N/m^2
Q	Discharge	m^3/s
Q_{nD}	Discharge coefficient, see equation (1)	
$\tilde{q}$	Discharge fluctuation	m^3/s
V_{vap}	Vapor volume	m^3
Z	Hydro-acoustic impedance	$N/m^4/s$
Z_{spi}	Spiral case inlet impedance	$N/m^4/s$
Z_0	Reference acoustic impedance (propagation constant of pipe)	$N/m^4/s$
ρ	Mass per unit volume	kg/m^3
σ	Thoma number, reference elevation at runner outlet periphery	
$..._{ref}$	Reference value (for E_{nD} and Q_{nD} : best efficiency operation)	
$...^*$	Complex conjugate	
$..._M$	Model quantity	
$..._P$	Prototype quantity	
$..._{M \rightarrow P}$	Transposed from model to prototype	
...	Accessory symbols are defined as they appear in the text.	

References

[1] ANGELICO G., MUCIACCIA F., ROSSI G. : Part load behaviour of a turbine : a study on a complete model of hydraulic power plant. IAHR Symposium Montréal (1986)

[2] CATTANEI A., CAPOZZA A. : Analysis of a numerical model for the oscillatory properties of a Francis turbine group. IAHR WG. Ljubljana (1995)

[3] DÖRFLER P. : Francis turbine surge prediction and prevention. Waterpower '85, Las Vegas (1985)

[4] FANELLI M. : Research on off-design behaviour of Francis turbines : an overview of present state, difficulties, open problems, needs and strategies. IAHR W.G. Milan (1991)

[5] FISHER R.K., ULITH P. : Comparison of draft tube surging of homologous scale models and prototype Francis turbines. Voith Research and Construction vol 28e (1982)

[6] FRANC J.P. et al. : La Cavitation, mécanismes physiques et aspects industriels. Presses Universitaires de Grenoble (1995)

[7] GREIN H. : Vibration phenomena in Francis turbines : their causes and prevention. IAHR Symposium Tokyo (1980)

[8] JACOB T., PRENAT J.E., BUFFET G., WINKLER S. : Improving the stability of operation of a 90 MW Francis turbine. Hydropower, Barcelona (1995)

[9] JACOB T., PRENAT J.E., WINKLER S. : Acoustic power analysis for the stability of operation of Francis turbines. IAHR WG, Ljubljana (1995)

[10] KERKAN V., BAJD M., DJELIC V., LIPEJ A., JOST D. : Model and prototype draft tube pressure pulsations. IAHR W.G. Ljubljana (1995)

[11] THICKE R.H. : Methods of controlling turbine draft tube vibrations and stability. Canadian electrical association, hydraulic power section (1980)

[12] WAHL T. : Draft tube surging times two : the twin vortex phenomenon. Hydro Review, feb (1994)

[13] ZIELKE W. : Unsteady one-dimensional flows in complex networks and pressurized vessels: Vibrations and resonance in hydraulic systems. Von Karman Institute for Fluid Dynamics (1980)

NON-STATIONARY FLOW IN REVERSIBLE FRANCIS TURBINE RUNNER DUE TO WAKES TRAILING THE GUIDE VANES

JO JERNSLETTEN
The Norwegian University of Science and Technology
Waterpower laboratory, Trondheim, Norway

Abstract

Experimental and numerical analysis are presented of the unsteady flow in a reversible Francis pump turbine runner due to the effect of wakes trailing the guide vanes. The dynamic fluctuating pressure has been measured with pressure transducers on the inlet part of the runner blades in generating mode in two operational points. A numerical solver of three-dimensional turbulent flow based on weakly compressible Navier-Stokes equations is applied to predict the unsteady flow in the turbine runner. The effect of the guide vanes upstream the runner is modelled with a sinusoidal inlet function which fluctuate the absolute inflow velocity.

Résumé

On présente les analyses expérimentales et numériques d'un écoulement nonpermanent dû aux tourbillons du bord de fuite des aubes directrices. Les fluctuations dynamiques de la pression ont été mesurées à l'aide de capteurs localisés sur le bord d'attaque des aubes de la roue pendant le fonctionnement en turbine et pour deux régimes. On résout numériquement les équations de Navier-Stoke en conservant une faible compressibilité afin de prédire l'instabilité de l'écoulement dans la roue. L'effet des aubages directeurs est simulé en faisant varier sinusoïdalement la vitesse absolue à l'entrée.

1. Introduction

The Weak Compressible Flow (WCF) code, which is used in the numerical analysis is developed by Song and Yuan [9] and Yuan [11]. Song and Chen

E. Cabrera et al. (eds.), Hydraulic Machinery and Cavitation, 865–874.

[10] have calculated the flow in a low head Francis turbine runner and draft tube with the WCF code. Billdal et.al. [1] showed that time depended flow clearly excist in a Francis runner by periodically varying the inflow boundary conditions. They conclude that the interference effect between the runner and stationary components will have a great influence on the overall performance of the turbine. Brekke [2] has discussed the influence of the non-stationary flow caused by the guide vanes on a turbine runner.

2. Basic equations and numerical method

The governing equations of the flow relative to the rotating cartesian coordinate system with z-axis coinciding with the axis of rotation may be written on normalized form

$$\frac{\partial p^*}{\partial t^*} + \overrightarrow{\nabla^*} \cdot (E\,\overrightarrow{w^*}) = 0 \tag{1}$$

$$\begin{aligned} \frac{\partial \overrightarrow{w^*}}{\partial t^*} + \overrightarrow{\nabla^*}(\overrightarrow{w^*}\overrightarrow{w^*}) + \overrightarrow{\nabla^*}(\tfrac{p^*}{2}) - \tfrac{1}{Re}\overrightarrow{\nabla^{*2}}(\overrightarrow{w^*}) + \tfrac{\partial}{\partial x_j^*}(\overline{w_i^{*\prime} w_j^{*\prime}}) = \\ -2\overrightarrow{\Omega^*} \times \overrightarrow{w^*} + \tfrac{1}{2}\overrightarrow{\nabla^*} \times (\overrightarrow{\Omega^*} \times \overrightarrow{r^*})^2 + \overrightarrow{g_z^*} \end{aligned} \tag{2}$$

where p, $\overrightarrow{w}$, $\overrightarrow{\Omega}$, $\overrightarrow{r}$ and $\overrightarrow{g_z}$ are the pressure, the relative velocity, the angular velocity, the radius and the acceleration of gravity respectively. An asterisk denotes normalized variables with respect to the reference scales as following $p^* = \frac{p}{\rho_0 \frac{w_0^2}{2}}$, $\overrightarrow{w^*} = \frac{\overrightarrow{w}}{w_0}$, $t^* = \frac{w_0}{l_0}t$, $\overrightarrow{\nabla^*} = l_0\overrightarrow{\nabla}$, $\overrightarrow{\Omega}^* = \frac{w_0}{l_0}\overrightarrow{\Omega}$ where w_0 and l_0 are the velocity and length reference scales respectively. The bulk modulus of elasticity E is defined as $E = \frac{k}{\rho \frac{w_0^2}{2}} = \frac{2}{Ma^2} = 2\left(\frac{a_0}{w_0}\right)^2$ where $k = \rho a_0^2$ and ρ, a_0 and Ma are the density, the speed of sound and the Mach number respectively. The Reynolds number is defined as $Re = \frac{w_0\, l_0}{\upsilon}$ where υ is the kinematic viscosity. The term $\overline{w_i^{*\prime} w_j^{*\prime}}$ in eq. (2) is the subgrid scale (SGS) turbulence stress. Smagorinsky's SGS turbulence model [8] uses a laminar flow analogy

$$-\overline{w_i^{*\prime} w_j^{*\prime}} = \upsilon_t\left(\frac{\partial w_i^*}{\partial x_j^*} + \frac{\partial w_j^*}{\partial x_i^*}\right) \tag{3}$$

where υ_t is the SGS diffusivity expressed as $\upsilon_t = (C\Delta)^2(2\,S_{ij}S_{ij})^{\frac{1}{2}}$ where C is the SGS coefficient, Δ is the grid size and the strain rate tensor is defined as $S_{ij} = \frac{1}{2}(\frac{\partial w_i^*}{\partial x_j^*} + \frac{\partial w_j^*}{\partial x_i^*})$. In the weakly compressible flow the density ρ is nearly constant. Therefore the contribution from the equation of energy conservation is negligible. The governing equations are solved numerically by using a cell centered finite volume method. The system of equations is solved numerically using MacCormack's [7] explicit predictor-corrector method with second order accuracy in time and space.

The pressure gradient and the velocity gradients normal to the outlet cross section (which is normal to Z-axis) downstream the turbine runner are assumed to be zero

$$\frac{\partial p^*}{\partial z^*} = 0 \qquad \text{and} \qquad \frac{\partial \overrightarrow{w^*}}{\partial z^*} = 0 \tag{4}$$

and the average pressure p^*_{ave} over the outflow boundary is kept constant. The runner blades are solid walls with full slip conditions. The flow conditions upstream the runner are modelled with a inlet function (Jernsletten [6]) defined as :

$$f(t^*, z^*, \theta) = g(z^*) \cdot h(\zeta) \tag{5}$$

The unsteady flow upstream the runner is modelled by forced fluctuations in the absolute inflow velocity $\overrightarrow{v^*} = \overrightarrow{i} \cdot u^*_{abs} + \overrightarrow{j} \cdot v^*_{abs} + \overrightarrow{k} \cdot w^*_{abs}$. The function g in eq. (5) corresponds to the spanwise variation in the absolute inlet velocity. The function h refers to the time and space variation of the velocity components in circumferential direction. Thus

$$\begin{aligned} u^*_{abs}(z^*, \theta, t^*) &= u^*_{0,abs} \cdot g(z^*) \cdot h(\zeta) \\ v^*_{abs}(z^*, \theta, t^*) &= v^*_{0,abs} \cdot g(z^*) \cdot h(\zeta) \\ w^*_{abs}(z^*, \theta, t^*) &= w^*_{0,abs} \cdot g(z^*) \cdot h(\zeta) \end{aligned} \tag{6}$$

where ζ is the phase angle of the fluctuating inflow velocities. The functions h may be expressed as

$$h(\zeta) = 1 - 2A(\cos^m(\zeta) - c_1) \quad , \qquad \zeta = \frac{N_{gv} \cdot \theta}{2} + \frac{2\pi \cdot t^*}{T^*_r} \; , \quad m \succeq 2 \tag{7}$$

where m is an exponent, A is the relative amplitude and the constant c_1 is expressed as $c_1 = \int_0^{2\pi} \cos^m(\zeta)\, d\zeta$ to ensure that the mean value of the function h is equal to unity, $\overline{h} = 1$, T^*_r is the period of one runner rotation, N_{gv} is the number of guide vanes and $\theta = \tan^{-1}(\frac{y^*}{x^*})$ is the polar angle. The relative velocity vector $\overrightarrow{w}$ at the inflow boundary is then calculated from the eq. (6) by $\overrightarrow{w} = \overrightarrow{v} - \overrightarrow{u}$ where the tangential velocity $\overrightarrow{u} = \overrightarrow{i} \cdot u^*_{\tan} + \overrightarrow{j} \cdot v^*_{\tan} + \overrightarrow{k} \cdot w^*_{\tan}$ Hence the relative velocities at the inflow boundary are

$$u^* = u^*_{abs} - u^*_{\tan} \; , \; v^* = v^*_{abs} - v^*_{\tan} \text{ and } w^* = w^*_{abs} - w^*_{\tan} \tag{8}$$

TABLE 1. Main dimensions and technical data for test rig and runner.

Number of stay vanes N_{sv}	12	Best eff. net head $^*H_n$	$16.0\,m$
Number of guide vanes N_{gv}	24	Best rotational speed $^*n$	$508\,rpm$
Number of runner blades N_{rb}	11	Best volumetric flow $*Q$	$0.195\,m^3/s$
Runner inlet diameter D_1	$582\,mm$	Inlet channel width B_1	$52\,mm$
Runner outlet diameter D_2	$338\,mm$	Inlet blade angle β_1	20^0

TABLE 2. Transducer locations on the runner blades in mm.

Measuring point (transducer number)	1	2	3	4
Location side of runner blade	pressure	pressure	pressure	suction
Distance measured from hub	7	-	7	7
Distance measured from shroud	-	7	-	-
Distance from inlet edge	40	40	151	16

Chen [3] measured the velocity distribution in the distributor of a high head Francis turbine with Laser Doppler Velocimetry. The velocity distribution was not uniform in circumferential direction nor in spanwise direction. Therefore will the inflow conditions, given by eq. (8), give a qualitative description of the velocity field because no measurements of the inlet velocities was performed upstream the reversible Francis turbine runner. The parameters $m = 2$, $A = 0.25$ and $N_{gv} = 24$ in eq. (7). The normalized time period $T_r^* = \frac{w_0}{l_0} T_r = \frac{w_0}{l_0} \cdot \frac{60}{n}$. The pressure p^* is extrapolated linearly from interior points in the grid.

The number of grid points in the calculation of unsteady flows in a runner with the number of upstream guide vanes $N_{gv} = 24$ and number of runner blades $N_{rb} = 11$ is $N_{gp} = N_{rb} \cdot N_I \cdot N_J \cdot N_K = 11 \cdot 66 \cdot 19 \cdot 7 = 96558$. All the eleven blade channels are included in the computational domain because of no common divisor between the number of runner blades and the number of guide vanes. No periodical boundary conditions are therefore used.

3. Experimental study

The experimental analysis is made on a reinforced epoxy model runner in scale 1:5.5 . The technical data of the prototype are: $P = 150\,MW$, $H_n = 420\,m$ and $n = 500\,rpm$. More details about the experiments are found in Jernsletten [6]. The main data for the test rig and model runner are given in table 1. The runner inlet and outlet refer to generating mode.

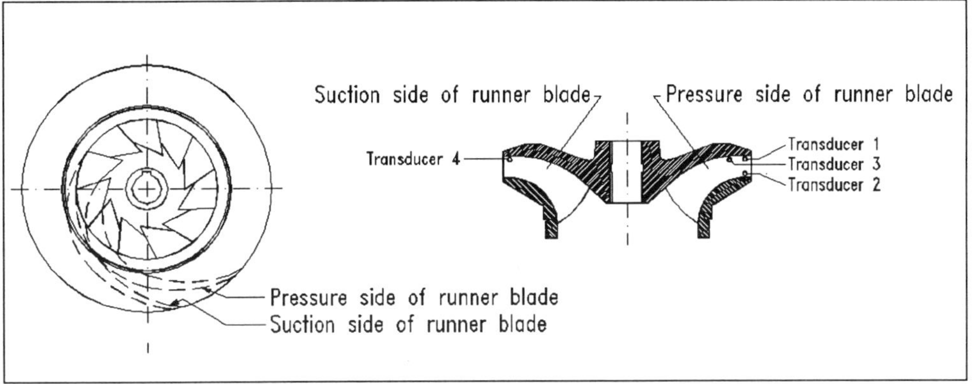

Figure 1. Transducers locations on runner blades

The speed number is defined as $^*\underline{\Omega} = \frac{^*n\pi}{30} \cdot \frac{\sqrt{^*Q}}{(2g^*H_n)^{3/4}} = 0.315$. The specific speed $n_q = n\frac{\sqrt{Q}}{H_n^{3/4}} = 28$. The measurements were performed with pressure transducers mounted flush on the runner blades. The miniature absolute pressure transducers embedded in the runner blades are of resistive full bridge type Kyowa 2 PS-2KA model with $6\,mm$ diameter. Three transducers were mounted on the pressure side and one pressure transducer were mounted on suction side of two opposite blades. In table 2 the locations of the four pressure transducers are specified (see fig. 1).

Two series of measurements are presented for two operational points in generating mode. In the best efficiency operational point the rotational speed $n_{bp} = 516\,rpm$, the flow $Q_{bp} = 0.199\,m^3/s$, the guide vane angle $\alpha_{bp,gv} \doteq 11.8^0$ and the radial distance between guide vanes and runner blades $\Delta r_{bp} = 55.7\,mm$. In the full-load operational point the rotational speed $n_{fl} = 568\,rpm$, the flow $Q_{fl} = 0.210\,m^3/s$, the guide vane angle $\alpha_{fl,gv} = 14.0^0$ and the radial distance between guide vanes and runner blades $\Delta r_{fl} = 53.7\,mm$. The experimental results are compared with the numerical computed results.

4. Results

Numerical predicted averaged flow solutions are presented in fig. 2. The reduced pressure $\underline{p} = \frac{h}{H_n}$ is plotted against position S along theoretical streamlines on both pressure side and suction side of the runner blades, where the pressure $h = \frac{p}{\rho g}$ and H_n is the net effective head. The value of S is equal to $0\,m$ and $0.5\,m$ on the leading edge and the trailing edge of the runner blade respectively. The upper and lower pressure curves correspond

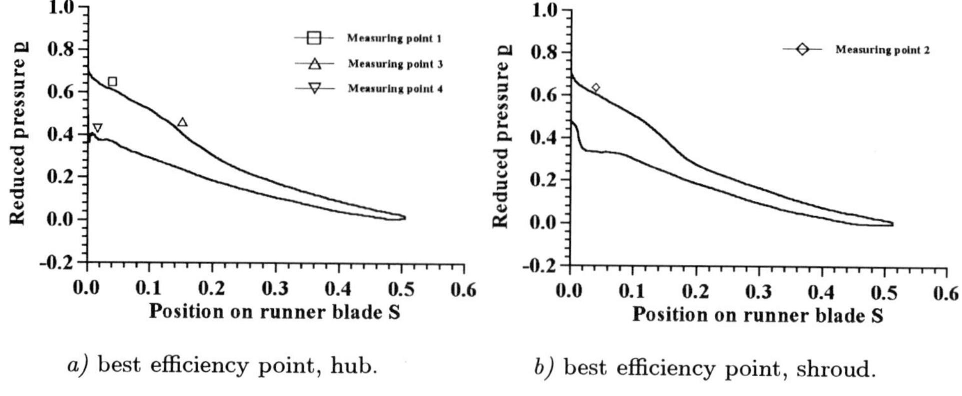

a) best efficiency point, hub. *b)* best efficiency point, shroud.

Figure 2. Averaged reduced pressure at blade sections.

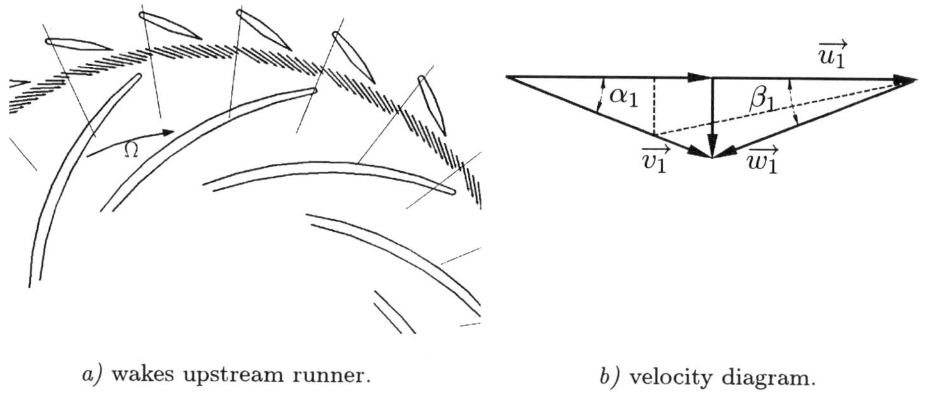

a) wakes upstream runner. *b)* velocity diagram.

Figure 3. Wakes upstream the runner and velocity diagram.

to the pressure side and the suction side of the runner blade respectively. In fig. 2a the averaged numerical predicted reduced pressure $\underline{p}$ along the theoretical streamlines on pressure and suction sides of the runner blade close to hub is presented and compared with the experimental measured pressure values in measuring points 1 (□), 3 (△) and 4 (▽) in the best operational point in generating mode. The locations of the measuring points (transducers) on the runner blades are shown in fig. 1. The agreement between numerical predicted pressure and experimentally measured pressure is good. However the measured values are somewhat larger than the calculated values.

In fig. 2b the numerical predicted averaged reduced pressure curve for $\underline{p}$ along position S close to the shroud on the pressure side is compared

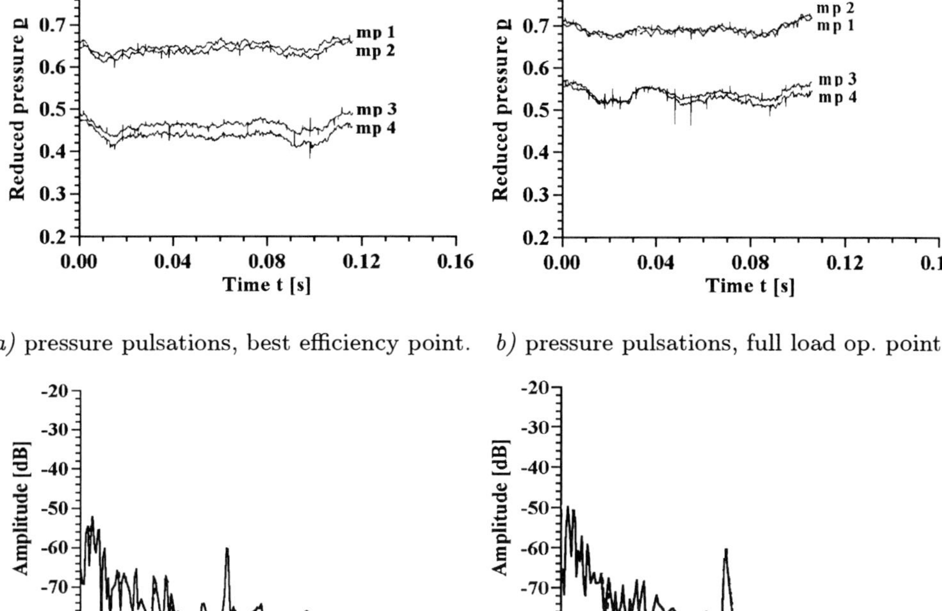

a) pressure pulsations, best efficiency point. *b)* pressure pulsations, full load op. point.

c) FFT analysis m.p. 1 best efficiency point. *d)* FFT analysis m.p. 1 full-load op. point.

Figure 4. Experimental measurements of pressure.

with measuring point 2 ($\diamond$) for the best operational point. Additionally the numerical predicted averaged reduced pressure curve for $\underline{p}$ along the suction side is presented. Also here the agreement is good between the simulation and the experiment. The numerical predicted reduction of pressure $\frac{\Delta p}{\Delta S} \approx 0$ on the suction side of the runner blade between position $S = 0.02$ and $S = 0.10$. This low pressure drop may cause separation and backflow.

However, the experimentally measured timeseries for pressure $\underline{p}$ in the four measuring points are pulsating (see figs. 4a and 4b). Wakes trail each guide vane as shown in fig. 3a and cause fluctuations in the velocity field upstream the runner. The inlet velocity triangle in fig. 3b illustrates the change in relative velocity $\vec{w}_1$ as the absolute velocity $\vec{v}_1$ decrease in the wake. The relative inlet angle β_1 decrease and the relative inflow is directed towards the suction side of the runner blade in the wake. In fig. 3b the dashed vectors denote velocity vectors in the wake and the continous vectors denote velocity vectors between the wakes.

In table 4 the amplitudes of the numerical predicted pulsating timeseries for pressure $\underline{p}$ in the four measuring points at guide vane passing frequency are presented for best operational point and full-load operational point. The numerical predicted pressure amplitudes in measuring point 3 are significant lower than in measuring points 1 and 2 for both operational points. Measuring point 4 shows no significant pressure amplitude.

In the figs. 4a and 4b the experimentally measured timeseries of the pressure $\underline{p}$ in the four measuring points for the best operational point and the full-load operational point are presented. The time period in the figs. 4a and 4b correspond to one runner rotation, i.e. 24 guide vane passing periods. In the best operational point the rotational speed is $n_{bp} = 516$ rpm. The guide vane passing period is therefore $T_{bp,gv} = \frac{60}{24 \cdot n_{bp}} = 0.0048$ sec. and the guide vane passing frequency is $f_{bp,gv} = 206.4$ Hz for the best operational point. In the full-load operational point $n_{fl} = 568$ rpm, $T_{fl,gv} = 0.00440$ sec. and $f_{fl,gv} = 227.2\ Hz$. The fluctuations over longer time periods than the guide vane passing period are due to imbalaces in the spiral casing and the guide vanes of the test rig. Measurements of the guide vane clearance revealed variations in clearance from 0.10 mm to 0.60 mm between different guide vanes. Additionally, if there is an unbalanced inflow from spiral casing this will also cause low frequency fluctuations in time series of pressure.

The FFT-diagram is presented for measuring point 1 in the best operational point in fig. 4c. Significant peaks are observed at frequencies $f = N \cdot f_{rot}$ $(N = 1, 2, .., 24)$ that are multiples of the rotational frequency. Theese peaks are due to the low frequency variations in pressure $\underline{p}$, which are seen in fig. 4a for measuring point 1. The amplitudes of the significant peaks are reduced with increasing N, except for $N = 24$. This is the guide vane passing frequency $f_{bp,gv} = 24 \cdot f_{rot} = 206.4\ Hz$, where a significant peak with amplitude $A = -60.1$ dB is present. For frequencies higher than the guide vane passing frequency the amlitudes are damped because of high internal damping in the reinforced epoxy model runner.In fig. 4d the results of FFT analysis for measuring point 1 in the full-load operational point is presented. Because of higher rotational speed all amplitude peaks are shifted right in this diagram compared with fig. 4c.

In table 3 the experimentally measured amplitudes of the pressure pulsations at the guide vane passing frequency are presented for both operational points and all four measuring points. The pressure amplitude for the guide vane passing frequency is larger on the pressure side than the suction side of the runner blade for both operational points ($A_{mp1}$and $A_{mp2} > A_{mp4}$). The amplitude peak at the guide vane passing frequency is strongly damped downstream on the pressure side ($A_{mp1} >> A_{mp3}$). The ampli-

TABLE 3. Experimentally measured amplitudes in dB at frequency $f_g v$.

Amplitude	$\underline{A}_{mp1}$	$\underline{A}_{mp2}$	$\underline{A}_{mp3}$	$\underline{A}_{mp4}$
Best operational point	−60.1	−62.3	−76.7	−66.9
Full-load operational point	−60.3	−63.2	−74.9	−64.3

TABLE 4. Numerical predicted amplitudes in dB at frequency $f_g v$.

Amplitude	$\underline{A}_{mp1}$	$\underline{A}_{mp2}$	$\underline{A}_{mp3}$	$\underline{A}_{mp4}$
Best operational point	−41.6	−38.9	−55.4	−60.1
Full-load operational point	−45.4	−43.5	−46.4	−60.9

tudes predicted in the measuring points 1 and 2 are approximately 20 dB larger than the experimentally measured amplitudes. This may be because the numerically chosen inlet velocity amplitude is 25% of the total absolute inlet velocity. The value of the inflow amplitude in eq. (7) is obviosly to high when we compare the tables 3 and 4. Because no measurements of the inflow velocity field are performed, the numercal predicted amplitudes are qualitative results. In numerical predicted amplitude on the suction side is strongly damped compared with the pressure side ($A_{mp1}$and $A_{mp2} >> A_{mp4}$). This is not in agreement with the experimental results. Also the relative difference between the amplitudes in measuring point 3 and 4 is not in agreement between eksperimental results ($A_{mp3} < A_{mp4}$) and numerical results ($A_{mp3} > A_{mp4}$). The reason for this may be that a reversible Francis turbine runner has a very strong backward inlet angle. The runner geometry makes it very difficult to avoid skew grid cells in the runner with H-mesh topology. Additionally the grid cells are very coarce.

5. Conclusion

The unsteady Weak Compressible Flow code shows good capabilities in numerical simulations of unsteady flows in turbomachines. The averaged flow calculations were in excellent agreement with experimental results. Unsteady flow calculations showed good qualitative agreement with experimental results. Despite of relative coarse grid with skew cells in the numerical simulations, good qualitative agreement with experimental results are obtained. However, the inflow function should be avoided because this approach leads to qualitative results only. In future works a coupling between a guide vane grid block and a runner grid block should be used instead. Also a draft tube grid block should be included to calculate the performance of the turbine with draft tube.

The experimental results show that there is a great influence from the guide vanes on the flow in the runner. The value of the measured amplitude peak at the guide vane passing frequency is up to $-60\,dB$ close to the leading edge of the runner blades. Significant amplitude peaks are also observed at lower frequencies. However these peaks are not as dangerous as the peak at the guide vane passing frequency due to lower number of cycles during the runners lifetime.

More experimental and numerical investigations should be performed to get more exact knowledge of this phenomenon. The numerical method seems promising and will in the future be extended to deal with direct simulation of the guide-vane/runner/draft-tube interaction.

Acknowledgments

The author thanks Energiforsyningens Fellesorganisasjon and Kvœrner Energy A/S for supporting this work.

References

1. Billdal, J.T., Anderson, H.I., Brekke, H., and Holt, B., "A study on Nonstationary Flow Effects In A Francis Turbine Runner Caused by the Guide Vane Effect", 17th IAHR Symposium, September 1994, Beijing.
2. Brekke, H., "A Review on Oscillatory Problems in Francis Turbine and Simulation of Unsteady Flow in Conduit Systems", 17th IAHR Symposium, September 1994, Beijing.
3. Chen, X., "Theoretical and Experimental Study of Flow Trough the Double Cascade of a Francis Turbine", Ph.D. Thesis, 1992, Norwegian Institute of Technology, Trondheim.
4. Chorin, A.J., 1967, "A Numerical Method for Solving Incompressible Viscous Flow Problems", Journal of Computational Physics 2, pp. 12-26.
5. Eriksson, L.E., "Simulation of Transonic Flow in Radial Compressors", Computer Methods in Applied Mechanics and Engineering 64, 1987, pp. 95-111.
6. Jernsletten, J., "Analysis of Non-Stationary Flow in a Francis Reversible Pump Turbine Runner", Ph.D. Thesis, 1995, Norwegian Institute of Technology, Trondheim.
7. MacCormack, R.W., and Paullay, A.J., "Computational efficiency achieved by time splitting of finite difference operators", 1972, AIAA Paper 72-154.
8. Smagorinsky, J., 1963, "General Circulation Experiments with Primitive Equations", Month Weather Review, Vol. 91, No. 3, pp. 99-164.
9. Song, C.C.S, and Yuan, M, "A Weakly Compressible Flow Model and Rapid Convergence Methods", 1988, ASME Journal of Fluids Engineering, Vol 110, pp. 441-445.
10. Song, C.C.S, and Chen, X.Y., "Calculation of Three-Dimensional Flow in Francis Turbine Runner", 1991, St. Anthony Falls Hydraulic Laboratory, University of Minnesota, Technical Paper 306-Series A.
11. Yuan, M., "Weakly Compressible Flow Model and Simulation of Vortex Shedding Flow About a Circular Cylinder", Ph.D. Thesis, 1988, University of Minnesota.

THE SWIRLING INLET FLOW EFFECTS ON THE PRESSURE RECOVERY OF A LOW HEAD WATER TURBINE DRAFT TUBE

K. KIKUYAMA
Professor
Nagoya Univ.
Nagoya, Japan

Y. HASEGAWA
Assoc. Professor
Nagoya Univ.
Nagoya, Japan

G. AUGUSTO
Graduate Student
Nagoya Univ.
Nagoya, Japan

K. NISHIBORI
Professor
Daido Inst. of Tech.
Nagoya, Japan

S. NAKAMURA
Chief Engineer
Fuji Electric Company
Kawasaki, Japan

Abstract

In this paper, the experimental results concerning the swirling inlet flow effects on the pressure recovery of a divergent passage whose cross-section changes from annular to rectangular were discussed. A developed turbulent flow having a swirling flow component was introduced into the diffuser and the alterations of the flow patterns both in the velocity and pressure distributions in each test cross-sections were carefully examined. The influence of swirling flow component on the axial velocity profile as well as on the streamwise wall pressure distributions was identified for different inlet swirl conditions. To give a detail explanation to the flow behavior and pressure recovery, a flow visualization technique using a tuft grid and surface tuft methods was adopted. Furthermore, based on the measured velocity and pressure, the approximated energy profiles at the best efficiency point of the unit were presented.

1. Introduction

Extensive studies on a small-scale hydraulic plant with a low head site have been made to generate electric power in small rivers with the most economical design. For low head turbines, fluids leave the runner having an appreciable amount of kinetic energy and enter the draft tube, wherein the velocity head at runner outlet becomes 40% to 100% of the effective head [1]. As it is important to optimize the energy conversion by a draft tube, which is an essential part dominating the total efficiency of the generating system, the flows and pressure recoveries within the divergent passage have received considerable attention from the researchers and designers.

In searching for the optimum diffuser performance with a specific configuration, the effects of inlet swirl for a two dimensional diffuser and the non-uniformity of the inlet velocity profiles have been studied widely. Recently, Kanemoto et al.[2] investigated the influence of swirling flow, i.e., free vortex type, in a three-dimensional diffuser using

E. Cabrera et al. (eds.), Hydraulic Machinery and Cavitation, 875–884.

stationary vanes. Research works which involve improvements of conical diffuser with a tailpipe discharge were also performed by several investigators [3],[4]. Inspite of these various works, insufficient information exists for a three-dimensional diffuser with a forced vortex type.

Hence, the present study is concerned with the effects of swirling inlet flow in a three-dimensional diffuser with a cross-section changing from annular to rectangular and its influence on the axial velocity profile as well as on the streamwise pressure distribution obtained for different inlet swirl conditions.

2. Experimental Procedure

The layout of the closed system used for the experimental work is shown in Figure 1. In order to acquire a more vivid explanation of turbulent intensities and the simplicity to measure the velocity by a hot-wire probe, air was used as the working medium instead of water. Air flow generated by a blower was introduced to the settling chamber and discharged into a straight pipe before entering the diffuser. To superpose a tangential flow component, a swirling generator that consists of 12 straight-type vanes was installed at the section just upstream of the divergent channel. The schematic configuration of the divergent channel and the measuring sections are shown in Figure 2.

Two types of measurements were carried out, both of which were guided by a probe traversing mechanism. A cylindrical three-hole pitot tube was used to obtain the pressure distribution and the velocity profile was taken by an I-type hot wire probe in the same test cross-sections.

The experiments were made at the Reynolds number of 48000 based on the inlet properties and four different swirling rates, N=0.0, 0.168, 0.336, and 0.504. The definition of the above parameters are given as follows,

$$R_e = \frac{U_m D}{\nu} \qquad \text{and} \qquad N = \frac{V_0}{U_m} \tag{1}$$

where, D is the pipe diameter, U_m is the average axial velocity at diffuser inlet, and V_0 is the tangential speed of the rotating pipe wall.

In order to express the general strength of swirl for different inlet types of tangential velocity, a general equation of momentum flux ratio, denoted as m, is derived to define the ratio of angular momentum to the axial momentum given as

$$m = \frac{\rho \int r\,|V|\,U\,dA}{\rho\, b \int U^2\,dA} \tag{2}$$

where, U and V are the time-mean axial and tangential velocities respectively and dA is the elemental surface normal to the flow. The dimensionless position is expressed in terms of radial distance, r and the radial difference between the wall and the hub, b.

3. Results and Discussion

3.1. VELOCITY DISTRIBUTIONS

Figure 3 shows the axial and tangential velocity profiles at the inlet region of diffuser, i.e., section M0. It is verified that the longitudinal velocity distribution exhibits almost the same profile despite the different values of rotational rate. The increasing strength of the swirling generator has a weak effect on the axial velocity. But still a remarkable observation is identified particularly when $N=0.336$ and $N=0.504$ wherein, the axial velocity tends to approach the hub due to an increasing pressure gradient caused by the centrifugal force. On the other hand, a forced vortex-type of tangential velocity is seen to dominate in the inlet region with an increasing value of N.

Aside from the influence of axial cone to the components of velocity profiles in the x-z planes of Section M2, the magnitude of longitudinal velocity decreased gradually with an increasing inlet swirl particularly in the central portion.

Far downstream of section M4, the profiles of axial and tangential velocities in the plane of x-y are shown in Figure 4, wherein a strong forced vortex distribution is identified. On the other hand, axisymmetric axial profiles are still generated at the rotational rate of $N=0.336$ and 0.504. A remarkable unsymmetry of axial velocity is observed for $N=0$ to 0.168 and this would be the result of accumulation of stagnant fluids caused by a stall which started to develop on the lower left side of Section M3.

3.2. TURBULENT INTENSITY PROFILES

The r.m.s. values of axial turbulent components at section M0 are shown in Figure 5(a). At the inlet region, the non-dimensional turbulent intensity for $N=0.336$ was significantly increased compared with the obtained values for other different swirling rates. As the presence of moderate swirl in the inlet section is accountable to improve proper mixing where the eddies generated efficiently transmit energy to the wall casing of the diffuser, the increase in r.m.s. value at inlet leads to delay or to prevent the flow separation inside the divergent passage resulting to the increase of pressure recovery.

At far downstream of section M3 and with low swirling rate, as illustrated in Figure 5(b), the values of axial and circumferential turbulent intensities increase particularly in the region where flow separation is present. However, as the rotational rate increases, the region of separation was suppressed and concedes to reduce the magnitude of turbulent disturbance as compared with low swirl.

3.3. PRESSURE DISTRIBUTIONS

In order to evaluate the pressure performance of draft tube with different inlet swirling conditions, the average wall pressure difference is taken from the selected test cross-section of the diffuser. The dimensionless pressure value is defined by the following equation,

$$C_{P_{Ave}} = \frac{p_{Ave} - p_m}{\rho U_m^2/2} \tag{3}$$

where, p_m indicates the wall pressure at Section M0.

The average wall pressure change along the diffuser with different rotational rates (i.e., from N=0 to N=0.84) is shown in Figure 6. In the same figure the pressure change for an ideal fluid is also plotted for comparison. With low rotation rate, N=0.168 the pressure coefficient curve differs largely from other swirling rate conditions, which is due to the existing flow separation on the left side region of diffuser that tends the main flow to become unstable and separate from the channel wall. However, the pressure recovery is achieved continuously in the region upstream of Section M3 for larger values of N. When the rotational rate increased as large as N=0.84, the pressure coefficient at far downstream section is deteriorated compared with N=0.672. This is a result of a low pressure region created at the center with an increasing swirl angle leading to the onset of vortex breakdown which was experimentally verified and theoretically derived in a straight pipe by Harvey [5] and Squire[6] respectively.

Pressure recovery will decrease at swirl angles higher than those necessary to produce optimum performance because of the occurrence of backflow and increase in the wall friction.

3.4. FLOW VISUALIZATION

To verify the occurrence of reverse flow in the divergent passage, a flow visualization technique is adopted using a surface tuft and tuft grid methods. The flow pattern was recorded based on the mobility of the tuft with different rotational rates. When the rotational rate N=0.0, the unstable region was found in the vicinity of Sections M2 and M3. But downstream of this section the flow separation was detected. On the other hand, when N=0.168 the separation was identified on the lower and upper left regions of divergent passage. Although the reverse flow is repressed gradually with an increase of inlet swirl that is when N=0.336, an unstable region was found in one location particularly on the lower left side of diffuser. Further increase of rotational rate, i.e. N=0.672 and 0.84, the reverse flow as well as the unstable region near the wall is suppressed but the movements of the tufts exhibit a flow reversal at the center of diffuser tending the tuft to be entrapped as illustrated in Figure 7.

3.5. ENERGY FLUX PROFILES

In order to obtain a reliable data in approximating the energy fluxes and the flow behavior at section D, four test-ports were added in the vertical and horizontal directions of the same cross-section which are both 45 mm and 40 mm from the centerline respectively.

Assuming the fluid to be incompressible and the kinetic energy of eddies are neglected, the sum of the useful mechanical energy can be simplified as follows:

$$E_{T_D} = E_{K_D} + E_{P_D} \tag{4}$$

where E_{K_D} and E_{P_D} denote the kinetic and pressure energy positioned at section D, which can also be expressed as

$$E_{K_D} = \int_D e_k \, dA \qquad \text{and} \qquad E_{P_D} = \int_D e_p \, dA \tag{5}$$

where,

$$e_k = \frac{1}{2}\rho\, U^2 U \qquad \text{and} \qquad e_p = p\, U\,. \tag{6}$$

Figures 8(a), 8(b), and 8(c) show the approximated graphical representation of kinetic, pressure and total energy flux profiles at section D for low rotational rate, N=0.168. The surfaces were generated by employing the cubic spline interpolation method. For N=0.168, the kinetic energy fluxes are concentrated more on the right side of the outlet region of diffuser. This phenomenon exists due to the separation build-up on the left side portion of the divergent passage from cross-section M2 to M5 until the separated flow reattaches to the wall and tends the flow to deviate on the opposite side of the inlet rectangular section. In the case of uniform axial flow, that is N=0, the kinetic energy profile formed into a ridge at its center, and the core region gradually deteriorates. However, when the swirling rate increases, that is N=0.336 to 0.504, the core region was suppressed and a concave part in the central portion of the flow field is observed,as shown in Figure 9.

The total energy change at section D with different rotational rates is computed by the following equations,

$$\Delta E_T = \Delta E_K + \Delta E_P \tag{7}$$

where;

$$\Delta E_K = \frac{\rho}{2}\int_{M0} |U|^2 U dA - \frac{\rho}{2}\int_D |U|^2 U dA \qquad \text{and} \qquad \Delta E_P = \int_{M0} p\, U dA - \int_D p U dA\,. \tag{8}$$

Figure 10 shows the total energy difference between section D and the inlet region. This verifies that reduction of energy losses in the downstream section of D exists when the strength of peripheral component at the inlet region increases to N=0.336, as the moderate swirl leads to inhibit the separation. Also, by considering the loss effect imposed by the swirling generator, the projected best efficiency point (B.E.P.) for the system performance can be identified as shown in Figures 11(a) and 11(b).

4. Conclusion

An experimental investigation was conducted to determine the influence of inlet swirl on the performance of a three-dimensional diffuser. Velocity and pressure measurements were carried out to identify the flow profiles as well as to determine the diffuser's effectiveness when the inlet swirl is present.

The main results obtained in this study are summarized as follows:

1. The flow inside the diffuser is unstable for low rotation rate of inlet swirl due to the generation of transitory stall. It reduces the pressure recovery of the diffuser due to an increasing loss of kinetic energy.
2. The pressure recovery of the divergent passage can be improved when a moderate tangential component is superposed at the inlet reference region. The optimum energy performance in this experiment is achieved when the rotation rate is N=0.336.

3. The projected best efficiency point of the system based from the present swirling generator configuration exist within the limits of rotation rate, i.e., $0.33>N>0.30$.
4. Further increase of inlet swirl (i.e., $N>0.504$) was found to deteriorate the pressure recovery.

5. References

(1) Kubota, T. and Yamada, S. (1982) Effect of Cone Angle at Draft Tube Inlet on Hydraulic Characteristics of Francis Turbine, *Operating Problems of Pump Stations and Power Plants, IAHR*, Amsterdam.
(2) Kanemoto, T. et al. (1992) Swirling Flow in a Divergent Channel with Annular-Rectangular Cross-Section, *Trans. of JSME*, 58-556, 3552.
(3) Nakamura, I., Ishikawa, K. and Furuya, Y. (1981) Experiments on the Conical Diffuser Performance with Asymmetric Uniform Shear Inlet Flow (The Case of Diffuser with Free Discharge), *Bulletin of the JSME*, Vol.24, No.190, pp.662-671.
(4) Nakamura, I., Ishikawa, K. and Furuya, Y. (1981) Experiments on the Conical Diffuser Performance with Asymmetric Uniform Shear Inlet Flow (Effect of Tailpipe Discharge), *Bulletin of the JSME*, Vol.24, No.196, pp.1729-1738.
(5) Harvey, J.K. (1962) Some Observations of the Vortex Breakdown Phenomenon, *J. Fluid Mechanics*, Vol.14, Part 4, pp.585-592.
(6) Squire, H.B. (1960) Analysis of the Vortex Breakdown Phenomenon, Part 1, Imperial College, London, Dept. of Aero., Report No.102.

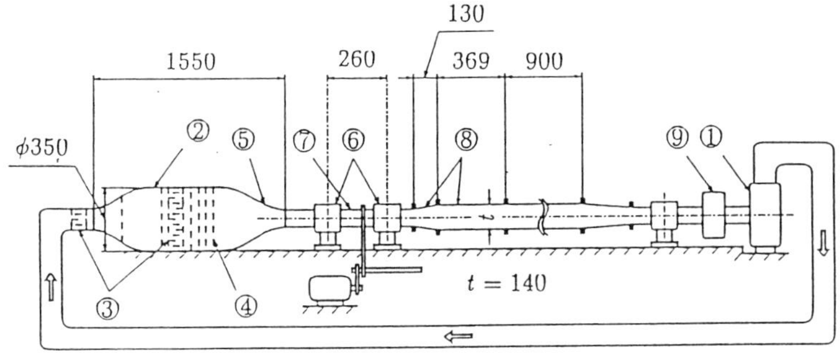

Figure 1. Experimental Apparatus

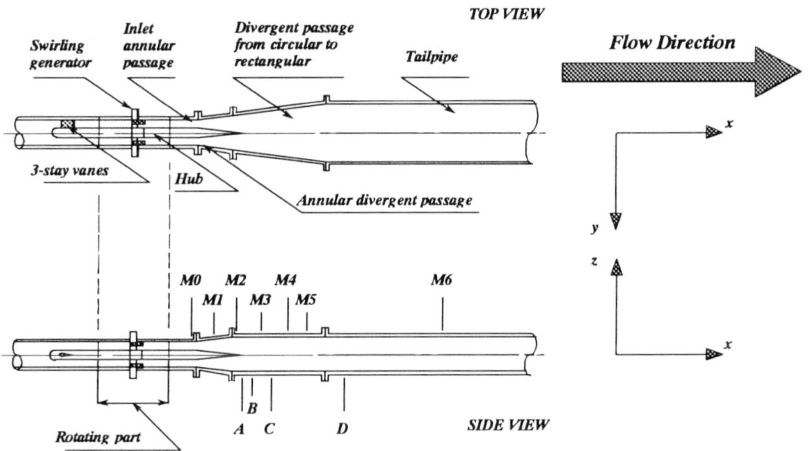

Figure 2. Schematic Diagram of Divergent Channel

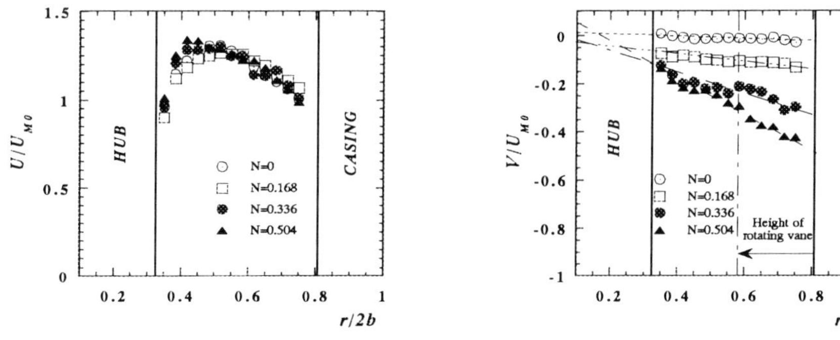

(a) Axial Velocity Profiles (b) Tangential Velocity Profiles

Figure 3. Changes in velocity profiles at section M0 with different rotational rates.

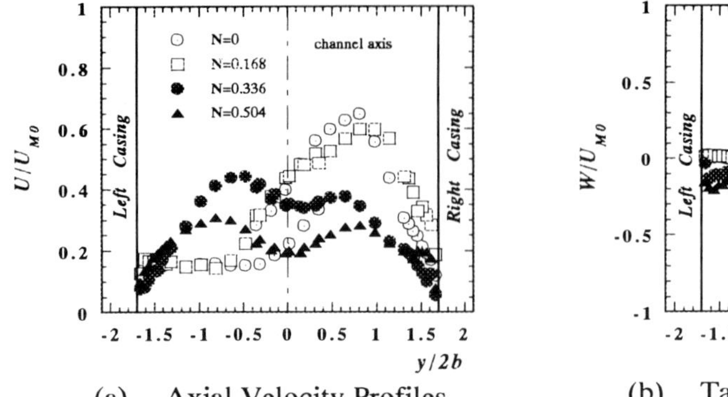

(a) Axial Velocity Profiles

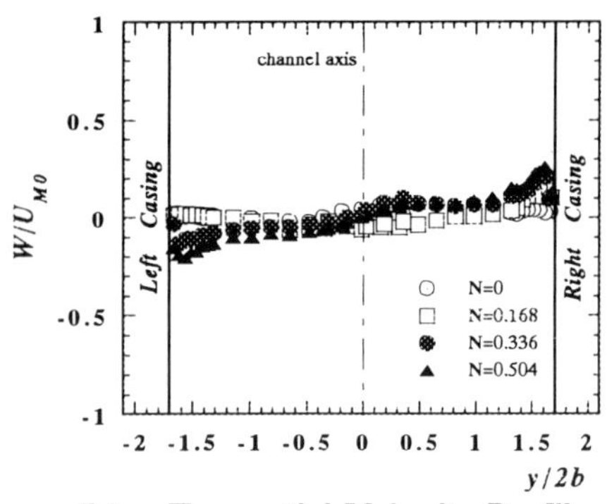

(b) Tangential Velocity Profiles

Figure 4. Changes in velocity profiles with different rotational rates at section M4 parallel to y-axis.

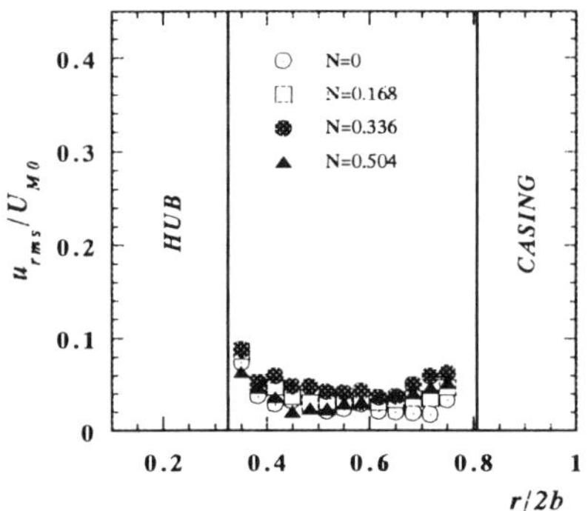

Figure 5(a). Turbulent intensities in axial direction at section M0.

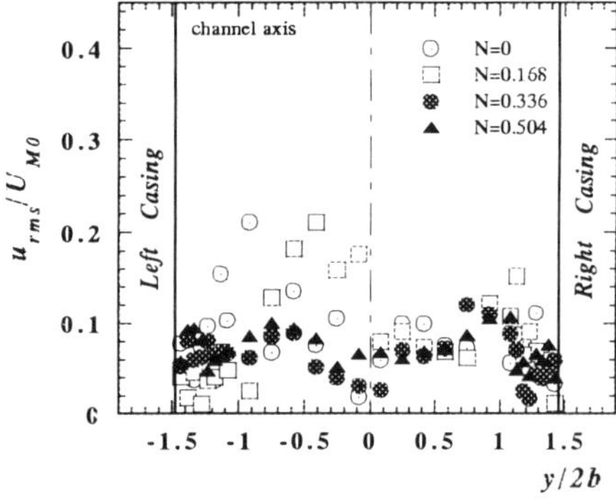

Figure 5(b). Turbulent intensities in axial direction at section M3 along y-axis.

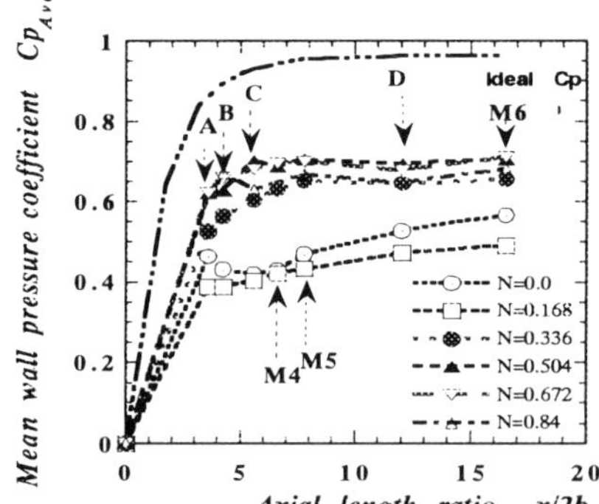

Figure 6. Streamwise pressure distribution with different rates of rotation.

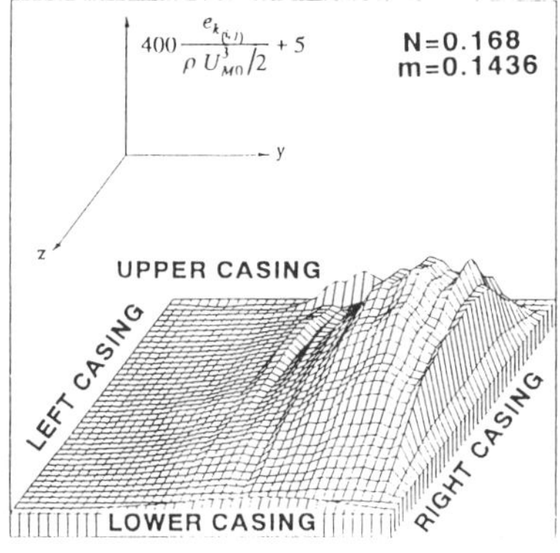

(a) Kinetic Energy Flux

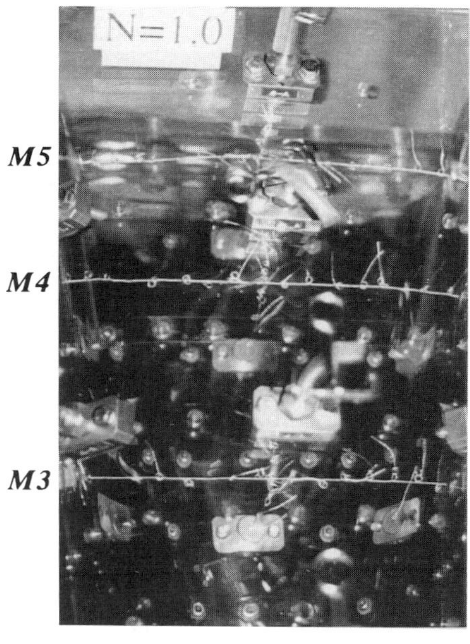

Figure 7. Flow behavior for N=0.84 from section M3 to section M5.

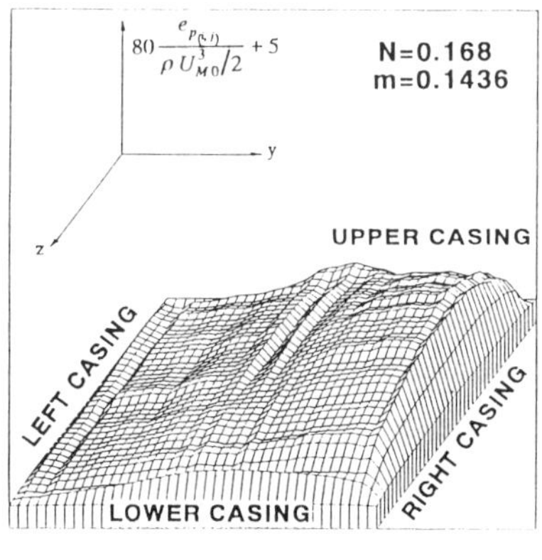

(b) Pressure Energy Flux

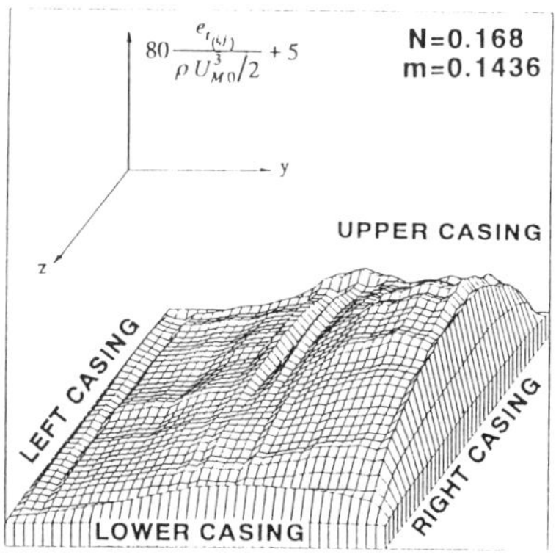

(c) Flux of Total Energy

Figure 8. Profiles of energy flux at section D with a swirling ratio, N=0.168.

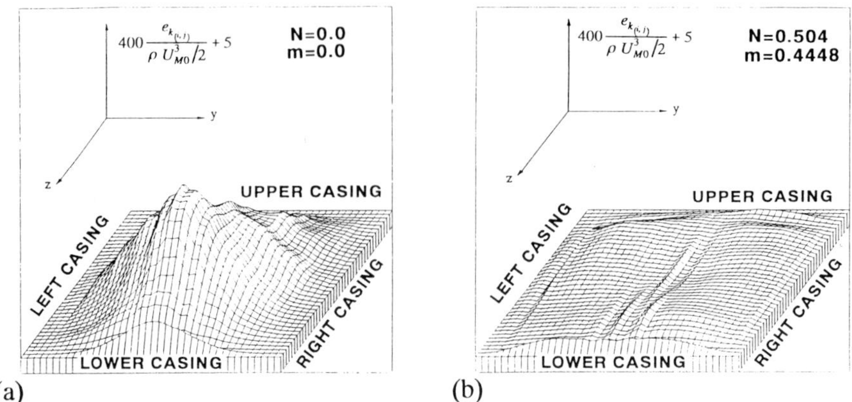

(a) (b)

Figure 9. Kinetic energy flux at section D with a swirling ratio, (a) N=0.0 and (b) N=0.504.

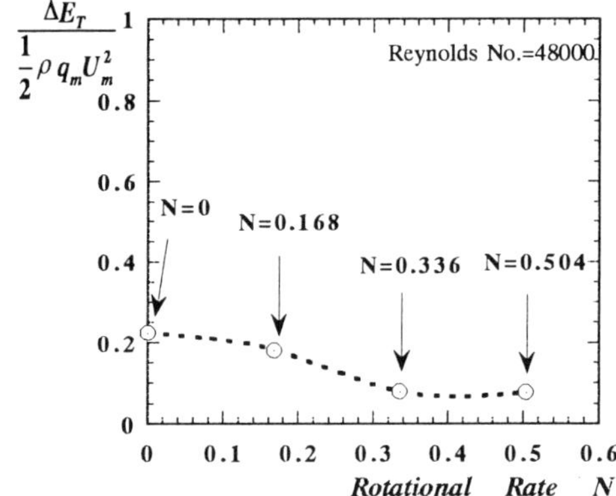

Figure 10. Effects of swirling flow component on the total energy loss between section D and the inlet region for different rates of rotation.

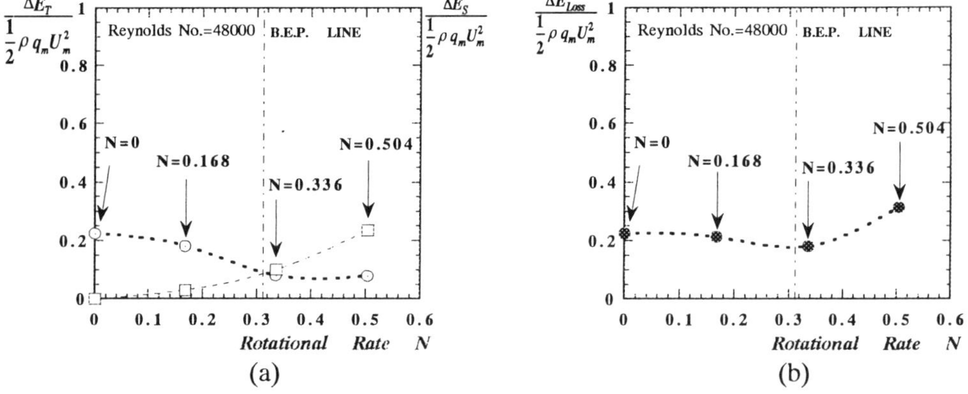

(a) (b)

Figure 11. Projection of B.E.P. as well as energy loss caused by the (a) swirling generator and (b) with the combined effects on the system performance.

TWO KINDS OF WHIRL ON FIXED-BLADE PROPELLER TYPE TURBINE

F. LÉONARD
Hydro-Québec
IREQ, 1800 Blv. Lionel Boulet, Varennes, Qc, Canada J3X 1S1

Two kinds of whirl have been identified on propeller-type generating units. In both kinds, the shaft whirl follows the same direction as the rotation of the machine and the source of the instability seems to be the interaction between the fluid and the structure. The first kind of whirl was observed at 0.33 Hz at plant #3 of Hydro-Québec's Beauharnois powerhouse and, also, at 0.25 Hz at LG-1 and Carillon. It occurred below the peak efficiency operating range. The vibration at 0.33 Hz (0.21 times the rotation speed) affects not only the power output but also the vibration at the guide bearing and the pressure in the penstocks. The second kind of whirl was observed at around 5.14 Hz (3.6 times the rotation speed) at LG-1 and occurred at no load or low load. Measurements at the bearings combined with those of the runner displacements reveal that the shaft undergoes like the second bending mode. Proximity and pressure measurements taken where the blade passes on the discharge ring of the runner at LG-1 shows that the whirl at 5.14 Hz is due to a runner instability.

1. Whirl at 0,33 Hz at Beauharnois powerhouse

Plant #3 at Hydro-Québec's Beauharnois powerhouse has propeller type units built in 1961. Unit #35, which is representative of these turbines, has a rated power of 56 MW and produces 53 MW at peak efficiency for a head of 25.4 m. It rotates at 94.7 RPM (1.579 Hz), has 76 poles, six blades and 24 guide vanes.

When unit 35 is operated at 35 to 45 MW, between the hydraulic rope at low gate openings and the peak efficiency, power swings exceeding 2 MW peak at 0.33 ± 0.02 Hz have been noted. Visual observations of the control cabinets reveal that the generating power at Beauharnois tends to oscillate at a frequency close to 0.33 Hz. This power swing at 0.33 Hz seems to predominate at partial load, above the hydraulic rope, whereas the oscillation at rotation frequency dominates at full load. However, the power swing is only one sign of the phenomenon.

E. Cabrera et al. (eds.), Hydraulic Machinery and Cavitation, 885–894.

The commercial instrumentation used before the SUPER monitoring system [1,2] was developed did not allow the user to distinguish frequencies as low as 0.33 Hz, although it should be mentioned that only unit #35 was instrumented with a SUPER system in 1990. It was only when the SUPER system was applied that a major vibratory side to the phenomenon was noticed. Since then, the utility has been concerned about the possible effects this could have on the increase in fatigue of the machine components. The power swing also affects the power output, the vibration at the guide bearing and the pressure in the penstock. These sinusoidal oscillations are very regular with only a very faint harmonic distortion. As illustrated in Figure 1, this oscillation is present at different amplitudes over a wide operating range from the hydraulic rope at low gate openings to maximum power output. It was even noticed that this oscillation is present in the water level at the intake gate head cover where a pulsation of around 2 cm occurring every 3 s was visible to the naked eye.

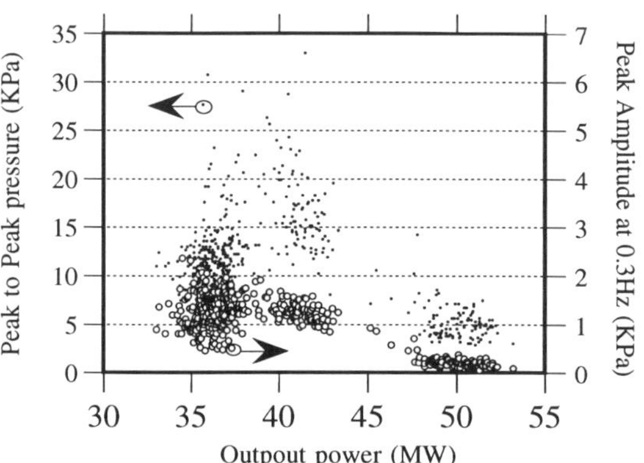

Figure 1. Distribution of the peak amplitude at 0.33 Hz and peak-to-peak oscillation of the pressure at the headcover (vault above the runner) as a function of the power for three years of operation. These measurements were recorded by the SUPER monitoring system [1,2]. Note that the hydraulic rope appeared below 37 MW. The machine was operated below optimum efficiency for many hours because of a defective guide bearing.

Prior to this study, the power swing at 0.33 Hz was attributed to the natural synchronous frequency of the generator [3] or draft-tube surges [4,5]. However, the natural synchronous frequency is higher than 0.33 Hz and, furthermore, a variation in the excitation current at the rotor has no influence on the frequency of the phenomenon at 0.33 Hz, which implies that the latter does not represent excitation of a synchronous resonance of the generator. As for the hydraulic rope, its frequency characteristic is 0.25 to 0.5 times that of the frequency of rotation for this machine. The oscillation associated with the hydraulic rope at low gate opening is irregular both in amplitude and frequency, whereas the phenomenon observed at 0.33 Hz is surprisingly stable. Furthermore, the phenomenon is also present, albeit with a lower amplitude, in the remainder of the operating range (Fig. 1), which means that it cannot be attributed to the hydraulic rope.

The power swing generated by the hydraulic rope phenomenon in the draft tube is well documented but the whirl of a propeller-type runner at a very stable frequency and as low as one fifth of the rotation frequency has never been dealt with in the literature to our knowledge. However, this oscillation phenomenon is not found only at this generating station: a similar oscillation was also observed at 0.25 Hz at La Grande 1 and Carillon which are also equipped with propeller-type turbines. It should be pointed out that it is quite normal that this phenomenon was not observed on reduced-scale models because the nonlinear dynamics of mechanical vibrations does not respect the homologous relationship.

The synchronous resonance of the generator and the hydraulic rope were therefore swiftly eliminated from the diagnostic hypotheses. On the basis of results published by Heskestad and Olberts [6], we also rejected the possibility of Von Karmàn vortex in the penstock or in the gate operating ring; the frequency of the phenomenon increases very little with the flow speed and the hydraulic profiles produce frequencies that are different from 0.33 Hz. Thus several hypotheses were eliminated one by one [7].

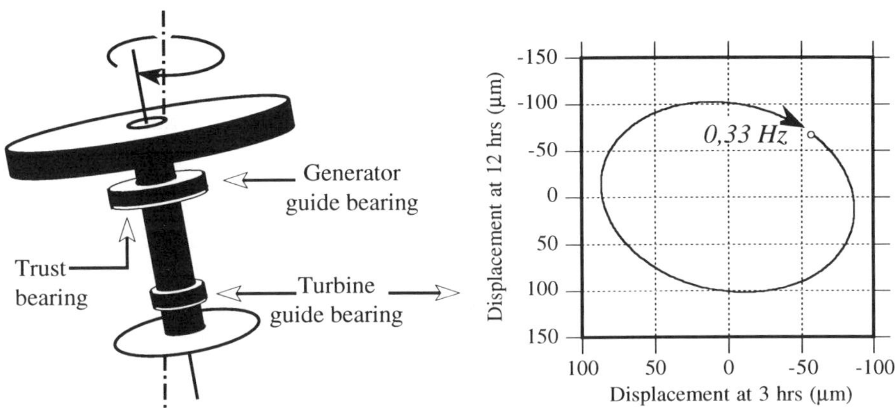

Figure 2. Rigid-body movement of the rotating part of the whirl and orbit mechanism at the turbine guide bearing. At times, the movement of the shaft sweeps the entire bearing gap, namely more than 250 μm.

Vibration measurements at the guide bearings reveal that the shaft is quite stiff in its trajectory in the generator guide bearing and the turbine guide bearing (Fig. 2). The movement seems to be almost circular, at the frequency of 0.33 Hz, in the sense of the rotation and in phase opposition between these two bearings. Similarly, load cells installed on the shoes of the thrust bearing show a rotating movement corresponding to the shaft trajectory [7]. Superposed on this movement, we observed the oscillation of the vertical thrust of the load which generates the power swing. This vertical oscillation distributed over the blade surface and corrected in terms of the penstock sections also gives a pressure oscillation corresponding to that observed in the draft tube and at the headcover. The presence of a strong modulation at 0.33 Hz in the envelope of the acoustic signal generated by the passing of the blades (9.47 Hz), perceived at the draft tube, means that the noise of the flow is strongly modulated by this phenomenon. The statistical scatter in the phase and frequency is minimum for readings of the shaft displacement at the turbine guide bearing, compared to other measurements (power, displacements, stresses, pressure); the frequency driver of this phenomenon is located at the runner. The product of the amplitude of the oscillation pressure by the section of penstock decreases from the head cover toward the water intake; the pressure oscillation signal comes from the runner.

The fact that large pieces of stainless steel over 1 cm thick were torn off from the discharge ring put us on the trail of a mechanism which seems to be related to a fluid-structure interaction at the blade tip. When unit #35 was removed from service on September 11, 1993, it was noted that disbonding of pieces of steel 60 cm by 240 cm wide from the discharge ring of the blade had occurred [8]. These sheets had been welded in 1990 to limit cavitation abrasion. If we assume that the disbonding was due to the passage of a pressure fluctuation accompanying the blades, the pressure measurement in the draft tube is the most likely to contain the vibration signal following the flapping of the metal sections as the blades passed by (6 x 1.579 Hz). This measurement was taken about 3 m from the end of a piece of torn-off metal. Figure 3 gives the results of a study based on three years of accumulated data. The acoustic vibration at 9.45 Hz increased gradually when the unit was operated at 35 MW until April 9, 1992 when the production level was raised to 41 MW. It was then that the vibration level suddenly soared. Metal was therefore torn off especially in the months when the unit was operating at 41 MW, when the whirl was dominant and most stable.

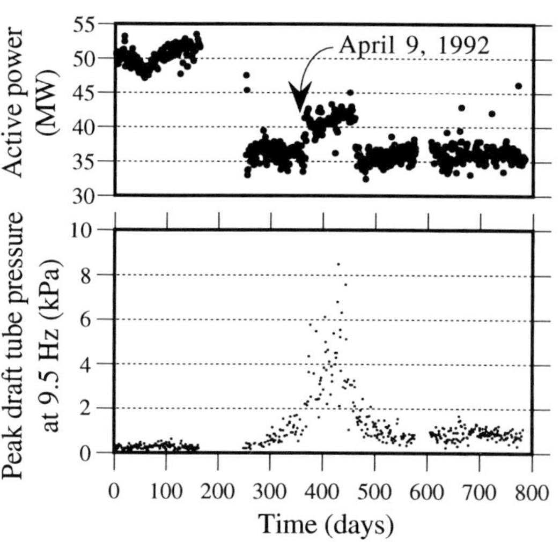

Figure 3. Power and spectral component in the draft tube associated with the flapping of metallic sections as the blades pass. The pull-off is accelerated when the unit was operating at 41 MW, after April 9, 1992. Under these oprating conditions, the vibrations were dominated mainly by whirl which is very stable both in amplitude and in frequency.

Degradation of the discharge ring as well as the different measurements taken indicate a localized excitation source on the periphery of the blades. Since the hydraulic rope is active more in the centre of the runner, there is a strong probability that the latter does not contribute to the excitation mechanism. The regularity of the oscillation in amplitude and frequency can be explained by the participation of a mechanical resonance mode which, added to this excitation source, creates a self-sustained oscillation. Our latest speculations center around a gyroscopic spinning mode where the gyroscopic forces, the pivot and the thrust of the water added to the weight of the spinning part are the leading contributors with the blade tip forces. The guide bearings seem to act as simple abutment that determine the movement of a rigid shaft. The purity and stability of the vibration waveshape seem to be due to a movement of precession following a trajectory determined by the roundness of the bearings and a frequency determined by a substantial inertia under the impulse of a constant force. If such is the case, it is not surprising that this phenomenon has not been seen on the simulation models of manufacturers because they consider guide bearings to be linear springs not abutment. In fact, Ehrich [9] has revealed the existence of subharmonic whirl when a gap is introduced at the guide bearing at the modeling stage.

2. Whirl at 5.14 Hz at LG-1

The generating station at LG-1 is equipped with units from two manufacturers. The turbine-generator units are the 5-blade propeller type with a nominal capacity of 114 MW. They spin at 85.7 RPM (1.4286 Hz), have a head of 27.5 m and share several similarities from the point of view of dimensions. However, the units from manufacturer A have 20 wicket gates and stay vanes, whereas those from manufacturer B have 24 wicket gates and 15 stay vanes. In addition, the former have a pivot located below the rotor, whereas in the latter the pivot is above the turbine guide bearing. These units exhibit whirl around 5 Hz but it is minor for the units from manufacturer B and dominant for the units from manufacturer A. In the latter, the vibration is particularly high at very reduced loads or when the unit is not synchronized. The fairly erratic whirl of the shaft at 5.14 Hz (3.6 times the rotation speed) at the bearing is characterized by a frequency of 8.5 Hz at the pivot mechanism, a vertical vibration of the structure at 18 to 28 Hz, shocks to the discharge ring and oscillation of the draft tube pressure of almost 2 Hz, which excites the synchronous mode of the generator at around 1.4 Hz. The power does not seem to oscillate 5.14 Hz.

A first series of tests with SUPER monitoring system in September 1994 offered a glimpse of the whirl at low load. In the tests with low gate opening, noises of strong metallic banging came from somewhere close by the runner and the discharge ring. From our experience at Beauharnois, we assumed that the blade-discharge ring interaction was at the origin of the whirl observed at LG-1, involving the possible participation of discharge ring displacement. The instability of the envelope of the oscillation of the shaft at 5.14 Hz made it easy to believe that we were in the presence of a more or less random phenomenon. The solutions were to reduce the gap at the turbine bearing or make the discharge ring more rigid opposite the blades. However, before any action was taken, it was recommended to measure the displacement relative of the discharge ring at one selected point with the passing of the blade, its absolute dislacement and the pressure at this same point. Not long afer these recommendatons were made, SUPER recorded the loss of the cone of unit #2 on October 24. The visual inspection that followed this event revealed no sign of friction.

The second series of measurements on the discharge ring [10] provided an opportunity to complete the description of the whirl and it was only after this that we were able to identify the active participation of what seems to be a shaft resonance mode in the whirl. Actually, the deflection in the shaft under the effect of whirl is similar to the second vibration mode of the shaft, as illustrated in Figure 4. The shaft displacement at the turbine bearing appears to be in phase opposition with the displacements at the runner and the generator guide bearing. The presence of a counter-phase trajectory between the runner and the turbine guide bearing is astonishing and implies the presence of marked bending stresses in the shaft and the participation of a mode which is normally found at a higher frequency. It is not easy to relate the whirl frequency with the frequency of the second critical mode of the shaft. After consulting different sources and publications in this field, we put the second mode at around 5.2 times the speed of rotation. However, this value does not correspond in any way to the whirl frequency, although it will be noted that several factors affect the frequency of resonance of the shaft. For example, a flexible rotor with a turbine bearing located near a modal node will have its second mode frequency near the first. The presence of a negative stiffness at the runner due to the blade tip depression can also lower the resonance frequency. Moreover, tests conducted at low openings with and without excitation at the rotor have a frequency identical to that of the whirl: if a mode contributes to the phenomenon, the negative stiffness of the magnetic field barely affects the latter, implying a minimal displacement or a node at the rotor for this mode, thereby confirming the participation of the second mode. Here we need to have more confidence in the measurement than in the calculted estimate of this frequency. The frequency of the whirl is more than 2.5 times greater than the rotation speed and has the form of the second resonance mode of the shaft; this mode therefore very likely plays a role in the whirl.

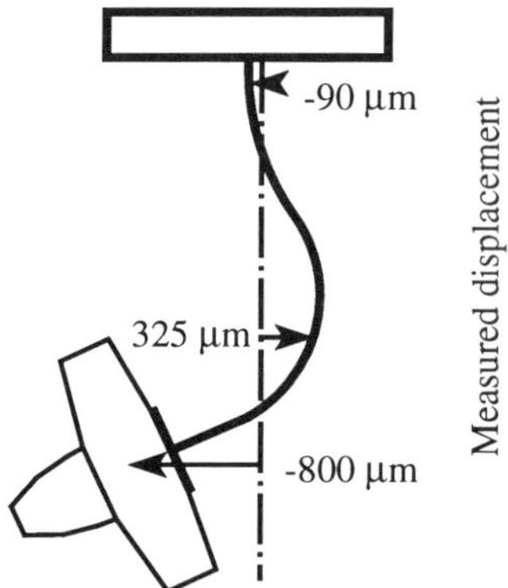

Figure 4. Graphic illustration of the operating deflection shape associated with whirl, with the peak displacement value observed at the alternator guide bearing (top), the turbine guide bearing (middle) and the runner (bottom). At 5,14 Hz, the gyroscopic forces are enormous at the rotor and it is therefore assumed that the shaft is rigidly connected to the rotor with minimum tilt at the rotor. A marked shaft deflection associated with this deflection shape may be seen. The latter resembles the deflection shape of the shaft's second critical speed.

The displacement vibration at the discharge ring is dominated by the passage of the blades, not by whirl, so the discharge ring does not contribute directly to the whirl and there is therefore no point in stiffening it. However, the constant presence of a vibration around 100 Hz can be observed (Figure 5) in all operating regimes, although when whirl is present the vibration is more marked. The predominance of this vibration seems to be due to the excitation of the iron plates of the discharge ring due to the whirl encountering the wake of the wicket gates (5.14 Hz x 20 gates = 103 Hz) and by banging of the discharge ring exciting the resonance frequency of the iron plates (120 Hz). It appears to be the same banging that could be heard during the tests. By banging, we mean three possible phenomena: metallic contact with the runner and the discharge ring, contact between a film of water or hydrodynamic lubrication between the runner and the discharge ring, or a strong hydraulic impulse coming from the runner or the fluid, such as the implosion of a cavitation pocket.

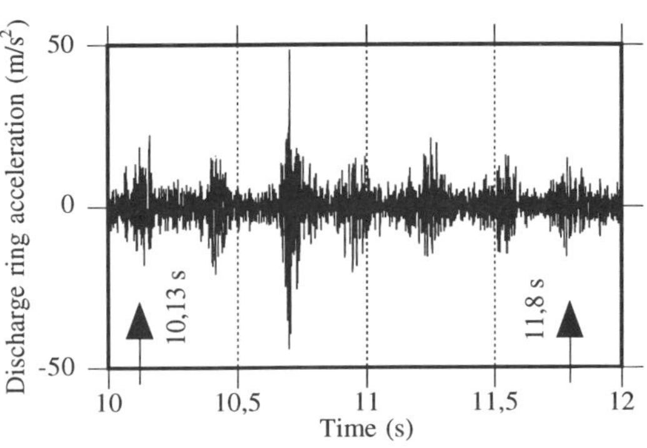

Figure 5. Zoom of a trace showing the presence of beating of the amplitude envelope of the acceleration at the discharge ring with the unit at 30% gate opening and 10 MW over time. There were six beats in 1.67 s, i.e. 3.6 Hz or 2.5 times per turn. The frequency of the modulated acceleration signal varies between 100 and 120 Hz.

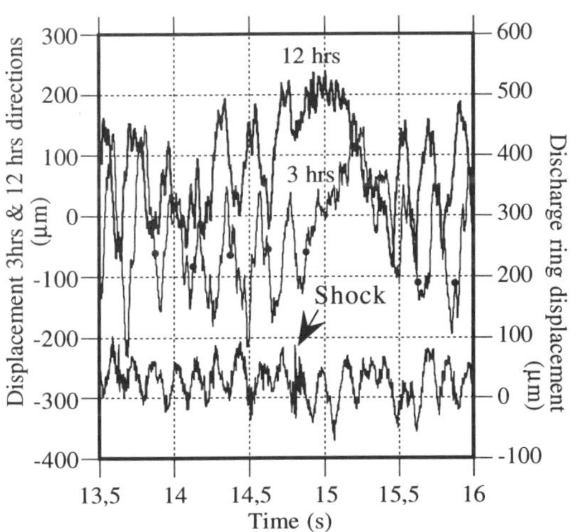

Figure 6. Displacement of the shaft at the guide bearing in the 3 h and 12 h directions and displacement of the discharge ring for unit #7 at LG-1 synchronized to 0 MW. At about 14.8 s, banging appears simultaneously with the stopping of the whirl, then a shaft rebound, followed by start up of the whirl again. The damping of the whirl occurs in a very short instant when the banging occurs. The banging has a lead of 0.05 s on the displacement at the discharge ring because the measurement systems were not started at the same time.

Figure 5 presents an acceleration trace where it appears quite clearly that the amplitude of the discharge ring vibration experiences a repetitive shocks. This measurement was taken at 30% gate opening (g.o.) (10 MW). The frequency of the pulsation was exactly 2.5 times/turn (3.6 Hz)! Furthermore, on some of the peaks the presence of banging can be observed. It should be recalled that if the runner turns at 3.5 times/turn in the direction in which the shaft rotates for a fixed observer, then a turning observer will see the shaft deflection imposed by the whirl at 2.5 times/turn. The reason for this change in frequency originates from the fact that the whirl can be associated with a vector turning at a rate of 3.6 times/turn. Compared to another vector turning in the same direction at a rate of once/turn, it is obvious that the bending rotates at 2.5 times/turn. Since there is no change in frequency for banging between the discharge ring and the shaft, the banging is perceived at 2.5 times/turn. Figure 6 shows that the whirl at 5.14 Hz is damped mainly by the banging at the runner, not by the viscous damping. Banging is therefore the source of damping that limits the amplitude of the whirl. If it were not for the banging in the presence of viscous damping, we would see a stable whirl. Moreover a stable oscillation at 5 Hz is present at a lower amplitude (30 μm peak) on the units from manufacturer B. On these units we suspect that the viscous damping coming from the position of the thrust bearing suffices to avoid having a self-sustained oscillation.

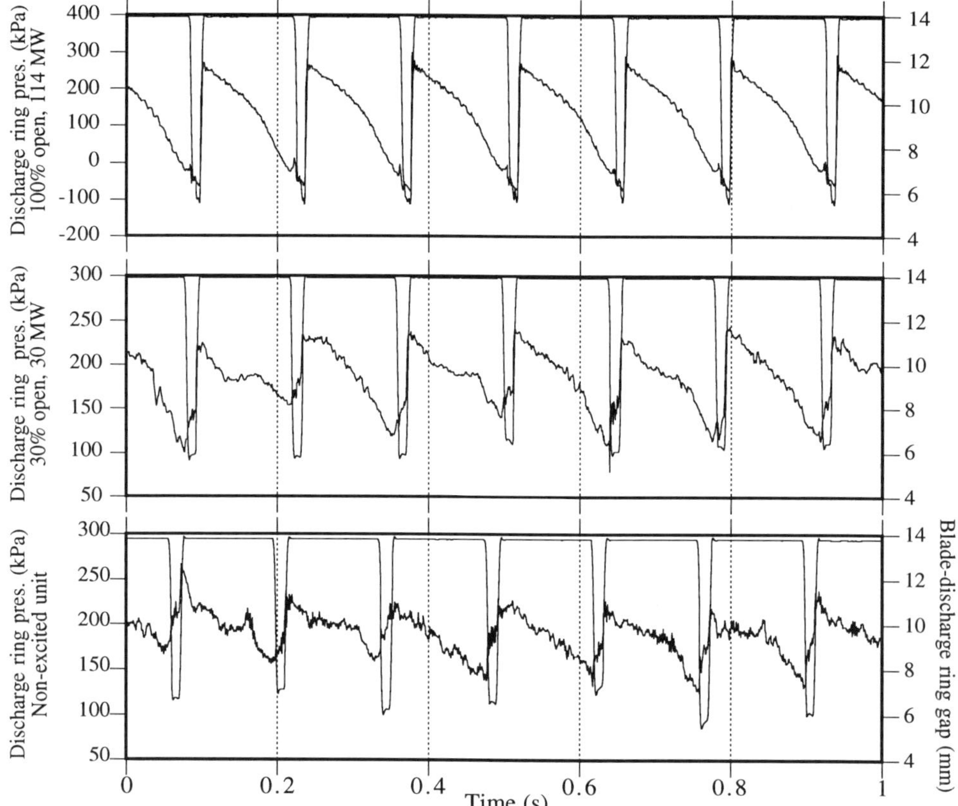

Figure 7. Profile of the pressure at the discharge ring of unit #7 at LG-1 under different operating conditions. At 100% operation, it may be observed that the blades gap is constant and, also, that the pressure has a stable profile from one blade to another, and is close to vacuum when the blade passes.

Figure 7 shows that the pressure at the discharge ring and the gap at the passing of the blades are constant at full load but vary considerably at partial load. On the longer recordings, the blades gaps are fairly well correlated with the blade-tip pressure. The runner acceleration values calculated from the blade-tip pressure fluctuation are of the order of magnitude of the measured accelerations and seem to correspond to an excitation of a damped mode at about 30%. The source of excitation is probably the blade-discharge ring interface.

3. Discussion regarding the whirl drive mechanism

The excitation source seems to be an interaction between the fluid and the structure located at the blade tip in both cases of whirl presented. The pressure measurements, like the tearing off of the steel from the discharge ring of the runner at Beauharnois confirm the presence of a significant depression at the blade tip that acts as a negative stiffness, a source of instability. A negative stiffness tends to force the runner out of its axis against the wall. In the presence of such forces, the runner leans on a point on the discharge ring where the forces are balanced. At least two mechanisms combine to create the forces that displace the runner tangentially to this point of equilibrium. The first, as explained by Hehrich [11] is due to the greater efficiency at minimum gap compared to maximum gap. This lack of equilibrium, associated with the decentering of the runner, forces the runner to be displaced according to a trajectory of the equilibrium of the radial forces acting in the direction of the rotation. It is this first mechanism that is operative in the whirl at 0.33 Hz; acting in an unequal gap this mechanism generates a strong power swing. The second mechanism that we propose is the result of a lack of equilibrium in the forces on either side of the point where the blade trajectory is the closest to the discharge ring (Fig. 8). The absence of a power swing in the whirl at 5.14 Hz shows the dominance of the second mechanism.

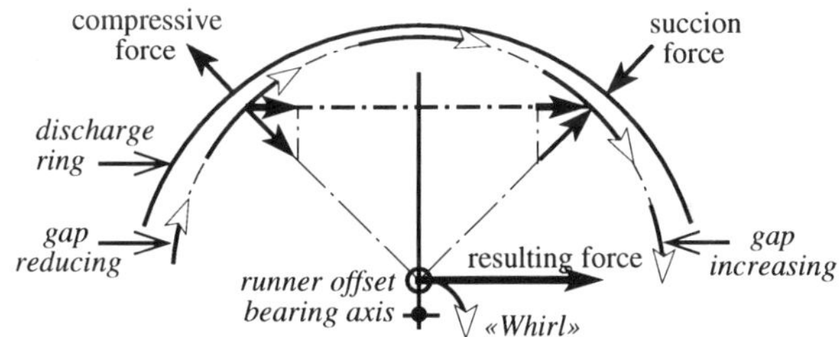

Figure 8. Illustration of a proposed whirl mechanism. The lack of equilibrium in the blade-tip radial forces on either side of the point where the blade trajectory is the closest generates the whirl driving force. The force of attraction between the blade and the wall generated by the high flow speed at the blade tip is not illustrated here. This force generates a negative stiffness which pushes the runner radially toward the wall, decentering it in its discharge ring.

4. Conclusion

These two case studies of whirl prove that the flow at the runner, especially at the blade tip, is a major source of instability at partial load. The presence of a mechanical resonance in the environment of this instability creates a strong self-sustained oscillation which takes the form of a whirl because of the radial symmetry of the instability. The whirl present at very low frequencies does not seem to stress the machine at Beauharnois but at LG-1 the whirl at 5.14 Hz presents a problem because it relaxes or is damped by banging observed at the discharge ring and the deflection associated with this whirl is similar to that of the second vibration mode of the shaft. Since this second vibration mode seems to have its two nodes close to the two guide bearings, protection by vibratory monitoring at the guide bearing is useless. The vibrations on a velocity scale come from the whirl at 5.14 Hz which are more than five times higher than those generated by the hydraulic rope. If we exclude damping by banging of the runner, the self damping of the shaft in its guide bearings is very poor: the addition of dynamic damping would significantly increase this damping and would limit the vibration amplitude to a level similar to that observed at 5 Hz on machines built by the other manufacturer. Another solution would be to modify the gap profile at the blade tip to force a vacuum pocket.

References

1. Pastorel H., Léonard F., Marois D., Watts A., Day H., Roy M. (1990) SUPER: A Computerized Monitoring System of a Turbo-Alternator Hydraulic Unit at Hydro-Québec, *CIGRÉ*, Paris, France, paper 11-202.
2. Léonard F. et al. (1995) Système de surveillance pour la maintenance de groupes hydroélectriques à Hydro-Québec, *Revue européenne. Diagnostic et sureté de fonctionnement*, 5-**3**.
3. Rheingans W. J. (1940) Power Swings in Hydroelectric Power Plants, *Trans. of the American Society of Mechanical Engineering*, **62**, 171-184.
4. Jacob T., Prénat J. E.,Vullioud G., Lopez Araguäs B. (1992) Surging of 140 MW Francis turbines at high load, analysis and solution, *16th Symposium of the IAHR*, 151-160.
5. Falvey H. T. (1971) Draft tube surges: A review of present knowledge and an annotated bibliography, Engineering and Research Center, Bureau of Reclamation, Denver, Colorado, Report REC-ERC-71-42.
6. Heskestad G., Olberts D.R. (1960) Influence of Trailing-Edge Geometry on Hydraulic-Turbine-Blade Vibration Resulting from Vortex Excitation, *Journal of Engineering for Power*, 103-109.
7. Léonard F. (1993) Oscillation de la puissance de groupes hydro-électriques : deux études de cas, Hydro-Québec report IREQ-93-198.
8 Léonard F. (1993) Arrachement au manteau de la roue du groupe 35 de Beauharnois, Hydro-Québec report IREQ-93-313.
9. Ehrich F. F. (1988) High Order Subharmonic Response of High-Speed Rotors in Bearing Clearance, *Journal of Vibration, Acoustics, Stress and Reliability in Design*, 110-**9**.
10. Léonard F., Proulx G., Poudrier R. (1995) Tournoiement à 5 Hz à LG-1, Hydro-Québec report IREQ-95-082.
11. Ehrich F. F. (1988) Self-excited vibration, in Cyril M. Harris (ed.), *Shock and Vibration Handbook*, 3rd edition, McGraw-Hill, New York.

SELF-EXCITED HYDRAULIC OSCILLATIONS DUE TO UNSTABLE VALVE BEHAVIOUR. A CASE STUDY

C. MATEOS[1], T. PÉREZ-ANDÚJAR[1], M. ANDREU[2], E. CABRERA[3]
[1]*Laboratorio de Hidráulica. CEDEX*
Paseo Bajo Virgen del Puerto, 3, E-28005 MADRID (SPAIN)
[2]*Departament de Tecnologia. Universitat Jaume I*
Campus de Penyeta Roja, E-12071 CASTELLÓ (SPAIN)
[3]*U.D. Mecánica de Fluidos. Universidad Politécnica de Valencia*
Camino de Vera, s/n, E-46071 VALENCIA (SPAIN)

Abstract

During the bottom outlet operation of a dam, a single-seal bureau-type valve experienced strong vibration, which in turn induced severe pressure oscillations in the conduit system, with peak to peak amplitudes of up to three times the static pressure.

This paper presents the facts and the frequency domain analysis of the system. Special attention is paid to the gate characterization as internal exciter of the hydraulic oscillation. Analytical results are compared with pressure recordings in order to validate the hypotheses and procedure.

1. Introduction

Any flow disturbance occurring in a pressurized piping system results in pressure waves travelling along it and reflecting at the system boundaries. Generally these waves decay with time and finally vanish due to energy dissipation. However, under certain conditions of the system and/or the flow disturbance, the response may be amplified with time, leading to hydraulic resonance. Two cases may be distinguished. Forced oscillations occur when one of the system boundaries acts as a periodic excitation, causing the hydraulic variables to oscillate with the excitation frequency. Shall this frequency coincide with one of the natural frequencies of the system, resonance will be reached, with considerably large amplitudes of the system response. The second case concerns self-excited oscillations, that can arise if the response of a component in the system is such that flow decreases with increasing pressure, thus causing the system energy to increase in each cycle. In such a situation, the hydraulic variables oscillate at one or more of the system natural frequencies.

Oscillatory flow and resonance can be analysed either in the time domain or in the frequency domain. The analysis in the time domain is carried out by means of the

E. Cabrera et al. (eds.), Hydraulic Machinery and Cavitation, 895–904.

Method of Characteristics. The forcing function, with its specified frequency, must be imposed as a boundary of the system. After an initial transient, the steady oscillatory regime is established, and the amplitudes of the fluctuating hydraulic variables can be determined. This method may not be effective for general studies, as the convergence to the steady oscillatory flow is usually slow. Analysis in the frequency domain assumes that any initial transient has died out, and focuses only in the final steady oscillatory regime. Instantaneous flow and pressure are represented by a purely harmonic oscillation superimposed on the mean flow conditions. By introducing this representation of the flow variables in the equations governing the unsteady flow in any element of the hydraulic system, and linearizing non-linear relationships such as the friction term or head-flow characteristics for valves or turbomachinery, one obtains a set of linear algebraic equations, where the unknowns are the complex-valued amplitudes of the pulsating flow variables. Principle of Superposition applies because of linearity of the resulting equations, so this approach is valid for any periodic forcing function. These equations are then organized by different methods as the well known Impedance (Wylie and Streeter, 1992) and Transfer Matrix (Chaudhry, 1987) methods. Both methods have been widely used and shown effective in the analysis of oscillatory flow and resonance. The Structure Matrix Method, developed for stability analysis of hydropower plants (Brekke, 1984), has latter been applied to analysis of natural frequencies and oscillation modes in hydraulic systems (Li and Brekke, 1990). In the authors' opinion, this latter method presents some advantages, specially when complex systems are dealt with.

Two different general approaches can be used to determine the natural frequencies of hydraulic systems. Frequency Response Analysis is based in the study of forced oscillations. It can be applied either in time or frequency domain analyses. The procedure consists in calculating the system response for a number of frequencies over the range of interest. The natural frequencies would be those resulting in maximum response amplitudes. The other one, known as Free Vibration Analysis (Zielke and Hack, 1972), assumes no external excitation on the system or that it has already withdrawn, and studies the frequency and damping of the system oscillations, as done in modal analysis of mechanical systems. It is only applicable to systems modelled by frequency domain methods. The major feature of this approach is that it can be used to identify system instabilities, provided the system elements are adequately modelled.

Cases of self-excitation induced by leaking valves are widely reported in the literature (see e.g. Erhart, 1979 and Gummer, 1995). If the elastic behaviour and geometry of the valve seal (or the valve itself) are such that leaking area decreases when pressure increases, this may result in a local negative slope of the pressure-flow characteristic for the valve. Such a situation can lead to auto-oscillation, as easily shown by graphical methods (Fanelli, 1974) or theoretical analysis (Guarga, 1995). This paper presents the analysis carried out to confirm the diagnosis of a case of auto-oscillation in the conduit system of a dam. A modal analysis of the hydraulic system is conducted to determine natural frequencies and mode shapes by means of Structure Matrix Method and Free Vibration Analysis. As a leaking gate valve appeared from the beginning as the source of self-excitation, effort has been put in the modelling of its elastic behaviour. This element is then introduced in the global hydraulic system, in order to obtain a solution confirming the system instability due to the valve.

2. Structure Matrix Method

The Structure Matrix Method is a rather simple and elegant way of organizing the linearized equations used for frequency domain analysis of hydraulic systems. Any component (pipes, valves, surge shafts, turbomachinery, etc.) in the hydraulic system is described by means of a matrix equation relating the complex amplitudes of the relevant variables at the nodes that define the component. These element matrix equations are then assembled, as done in matrix methods for structural frame analysis, to obtain a global matrix equation describing the system oscillatory behaviour:

$$\mathbf{A}(s) \cdot \mathbf{h} = \mathbf{q} \tag{1}$$

where the elements of matrix **A** are function of the complex frequency s. In this equation **h** is a generalized pressure vector that contains the amplitudes of the pressure oscillations at all nodes in the system, as well as those of any other mechanical or electrical variable involved. The vector **q** contains the sum of the flow amplitudes at the nodes and reference inputs for non-hydraulic variables. A complete description of component equations and global matrix equation assembly procedure can be found in Li (1988).

2.1. ELEMENT MATRIX EQUATIONS USED IN THIS WORK

2.1.1. *Pipes*

The matrix equation for a pipe is obtained from the Continuity and Momentum differential equations governing unsteady flow in closed conduits. The variables involved and their corresponding positive senses are shown in Figure 1. Note that the flow amplitudes are defined as positive when directed out of the element. This sense convention is valid for any component in Structure Matrix Method.

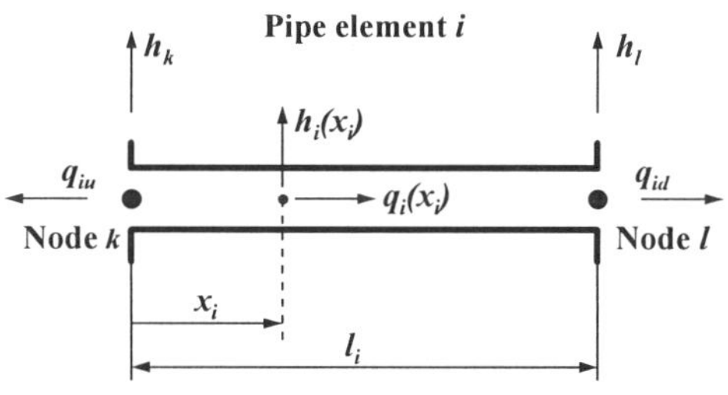

Figure 1. Definition sketch for pipe elements.

According to this, a pipe in a system is represented by the matrix equation:

$$\begin{bmatrix} T_i & S_i \\ S_i & T_i \end{bmatrix} \cdot \begin{bmatrix} h_k \\ h_l \end{bmatrix} = \begin{bmatrix} q_{iu} \\ q_{id} \end{bmatrix} \tag{2}$$

The elements of this matrix depend on the complex frequency and are given by

$$T_i = \frac{-1}{Z_{ci}\tanh(\gamma_i l_i)} \qquad S_i = \frac{1}{Z_{ci}\sinh(\gamma_i l_i)} \tag{3}$$

Z_C and γ are often referred to as Characteristic Impedance and Propagation Constant, respectively. Their values are:

$$Z_c = \frac{\gamma}{Cs} \qquad \gamma = \sqrt{Cs(Ls+R)} \tag{4}$$

where the following parameters are defined for convenience:

$$C = \frac{gA}{a} \quad ; \quad L = \frac{1}{gA} \quad ; \quad R = \frac{fQ_0}{2gDA^2} \tag{5}$$

being A and D the pipe cross section and diameter, a the wavespeed, f the steady state friction factor and Q_0 the mean flow.

2.1.2. *Fixed Valves and Orifices*

The matrix for a fixed valve or an orifice can be derived by linearizing the head loss across it. The following equation is obtained:

$$\begin{bmatrix} -V_i & V_i \\ V_i & -V_i \end{bmatrix} \cdot \begin{bmatrix} h_k \\ h_l \end{bmatrix} = \begin{bmatrix} q_{iu} \\ q_{id} \end{bmatrix} \tag{6}$$

where, denoting by H_0 the head loss for the mean flow Q_0,

$$V = \frac{Q_0}{2H_0} \tag{7}$$

2.1.3. *Surge Devices*

Any component of the system with an oscillating free surface, such as surge or vent shafts, is represented by Continuity Equation when friction and flow inertia are neglected. This can be arranged in matrix form as

$$\begin{bmatrix} 1 & 0 \\ 0 & F_i \end{bmatrix} \cdot \begin{bmatrix} h_k \\ h_l \end{bmatrix} = \begin{bmatrix} q_{iu} \\ q_{id} \end{bmatrix} \tag{8}$$

being

$$F = -s \cdot A \tag{9}$$

In this type of elements, the initial node k is the air above the water surface,then in any case $h_k = q_{iu} = 0$, and therefore the element could be represented by a single equation.

2.1.4. *Flexible Valves*

If the valve opening depends on pressure changes, the valve behaviour is represented not only by the flow-head relationship, but an additional equation must be included relating the valve opening with the pressure variation. After linearization of the first relation, the following matrix equation can be derived for the valve:

$$\begin{bmatrix} -V_i & -Q_y & V_i \\ -Y_h & 1 & Y_h \\ V_i & Q_y & -V_i \end{bmatrix} \cdot \begin{bmatrix} h_k \\ y \\ h_l \end{bmatrix} = \begin{bmatrix} q_{iu} \\ ky_{ex} \\ q_{id} \end{bmatrix} \tag{10}$$

In this equation V is defined as in eqn. (7), and ky_{ex} would represent any external excitation acting on the valve opening y. The other elements in the matrix are:

$$Q_y = \frac{\partial Q}{\partial Y} \tag{11}$$

$$Y_h = \frac{\partial Y}{\partial H} \tag{12}$$

with uppercase letters representing instantaneous values of the corresponding variables.

2.2. FREE VIBRATION ANALYSIS

In free vibration analysis it is assumed that no external excitation acts on the system variables, so all reference inputs in vector **q** in eqn. (1) are zero. Furthermore, because of the sign definition for flow amplitudes, the algebraic sum of flows is zero at the internal nodes of the system. If we further substitute the boundary conditions (q=0 for closed ends and h=0 for open ends and free surfaces), the free oscillation of the system is described by the matrix equation (Li and Brekke, 1990):

$$\mathbf{A}'(s) \cdot \mathbf{h} = \mathbf{0} \tag{13}$$

where $\mathbf{A}'(s)$ is obtained by eliminating in $\mathbf{A}(s)$ the rows and columns corresponding to zero pressure amplitude nodes. This matrix equation has nontrivial solution if

$$\det \mathbf{A}'(s) = 0 \tag{14}$$

Solution of equation (14) yields the infinite complex frequencies of the system

$$s = \sigma + j\omega \tag{15}$$

where ω is the oscillation frequency and σ is referred to as the damping of the system. Positive values of σ will identify unstable conditions with growing oscillation amplitudes.

Mode shapes are a description of the way the system is oscillating, and are represented by the moduli of the oscillation amplitudes along the system. For each of the complex frequencies obtained, a solution of eqn. (13) can be found when one of the variables in vector **h** is set to a fixed value. This solution represents the pressure oscillation amplitudes at the system nodes. The pressure and flow oscillation throughout the system is then obtained from the matrix equations of the pipes, that are the only distributed elements in the system, by means of the following equations:

$$h_i(x_i) = \frac{h_k \,\mathrm{senh}\,\gamma_i (l_i - x_i)}{\mathrm{senh}\,\gamma_i l_i} + \frac{h_l \,\mathrm{senh}\,\gamma_i x_i}{\mathrm{senh}\,\gamma_i l_i} \tag{16}$$

$$q_i(x_i) = \frac{h_k C_i s \cosh\gamma_i(l_i - x_i)}{\gamma_i \operatorname{senh}\gamma_i l_i} - \frac{h_l C_i s \cosh\gamma_i x_i}{\gamma_i \operatorname{senh}\gamma_i l_i} \tag{17}$$

3. Case Study

La Viñuela dam, in the south of Spain, is conceived as a key part of the irrigation and drinking water supply system of the sorrounding region. Due to its hyperannual character, it is necessary to make use of the dead volume of the reservoir. In order to reduce the head losses, as well as for economical reasons, use was made of part of the two bottom outlet pipelines as service conduits. There are three intake pipes in the intake tower. Two of them were conected to one of the bottom outlet conduits, and the lower one to the other. In order to shift the configuration from service intakes to outlet conduits, gate valves were located at different points in the system.

Figure 2 shows an schematic representation of the dam and the part of the conduit system that experienced self-excitation. The valve just upstream the connection between the intake and the outlet conduit is a single seal valve, thus functionning properly only in the normal direction of flow. However, emptying of the bottom outlet with this valve closed can subject the gate leaf to higher pressures in the downstream side. This provoked damage in the gate lintel guides, causing improper sealing of the valve. The leakage flow through the valve induced valve vibration that in turn excited pressure oscillations in the hydraulic system.

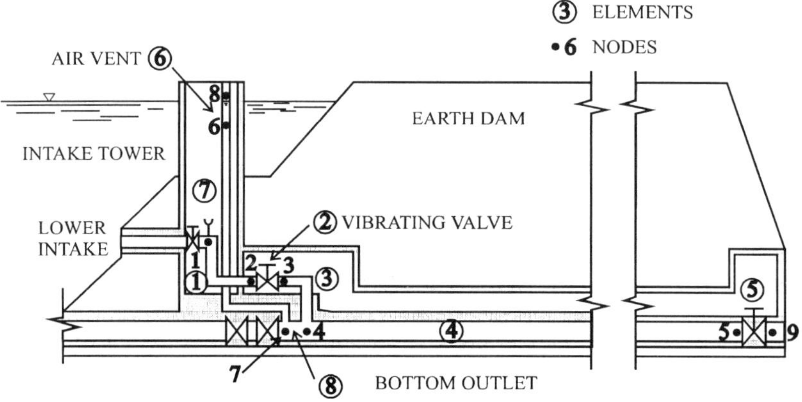

Figure 2. Schematic view of the dam and studied conduit system.

3.1. PROTOTYPE MEASUREMENTS

After detection of valve vibration, an experimental study was carried out at site in order to gain comprehension about the phenomenon (CEDEX, 1994). Pressure transducers (DRUCK PTS·510) were located upstream and downstream the vibrating valve (nodes 2 and 3 in Figure 2), in the intake conduit -at node 1 downstream the air valve and approximately half way between nodes 1 and 2- and in the bottom of the vent shaft. The measurements were conveniently recorded in a magnetic tape recorder (RAKAL).

The operating sequence during the experiments tried to reproduce the manoeuvre conducted during emptying of the bottom outlet after it has been used as service conduit. Gates in the outlet conduit (upstream of node 7) remained closed during the experiment. Valve 5 of Figure 2 was initially closed, and the outlet conduit was isolated from the distribution system by means of other valves not shown in the figure. After closure of both valves in the intake pipeline, valve 5 was lift approximately 5 cm in order to empty the outlet conduit.

Two experiments were carried out with different results. During one of them, no oscillations nor valve vibration were observed. Pressure recordings show the normal evolution of the system due to emptying of the outlet pipeline and vent shaft. It can also be detected an small leak through valve 2, as pressure in node 2 decreases and the air valve opens when pressure at node 1 falls bellow atmospheric.

In the second test, pressure oscillations were experienced in the system, and vibration of valve 2 was observed. According to the pressure recordings, auto oscillation started about 12 s after beginning of the test, when the conduit downstream of valve 2 was still full, with a water level in the vent shaft of approximately 29 m (the initial level was of 51.4 m, due to the static pressure).

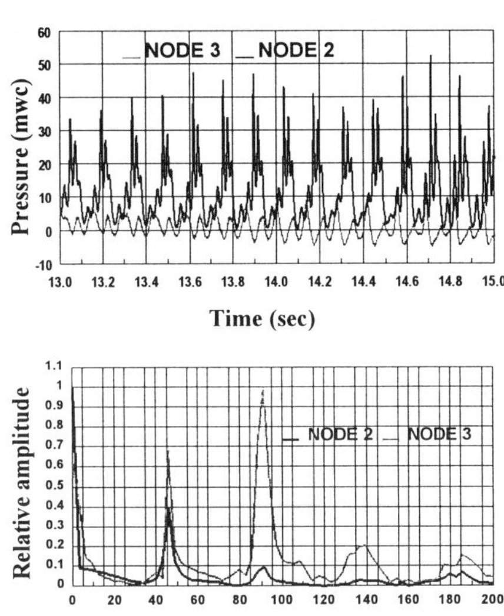

Figure 3. Pressure traces and frequency spectra.

Figure 3 shows pressure traces in the nodes upstream and downstream of the vibrating valve at the first instants after onset of resonance, along with the corresponding frequency espectra. The dominant frequency of node 2 pressure oscillations is clearly of 45.55 Hz, while in node 3 the oscillation shows again this frequency, along with a higher frequency about 91.11 Hz. The initial peak-to-peak maximum pressure amplitude in node 2 was approximately 52 mwc, with a mean value of 11.7 mwc due to the water column between the air valve in node 1 and the valve. After this initial stages, oscillation amplitudes grew up to 115 mwc peak-to-peak, with growing frequencies ranging from the initial 45.55 Hz to 51.82 Hz, due to the decreasing column length in pipe 1. After opening of the by-pass valve the oscillation attenuated and finally vanished temporarily, but, approximately 1 minute later oscillation set on again, reaching maximum pressures higher than transducers full scale output (150 mwc). After that, valve 2 was open to suppress the oscillation.

3.2. FREQUENCY DOMAIN ANALYSIS

After the experimental study, a theoretical analysis was carried out in the frequency domain in order to detect the natural frequencies of the system. The analysis focused in the initial stages of the resonance incident, and the main aim was to confirm the system's instability due to the leaking valve. In order to get reliable results, studies concerning hydraulic and mechanical aspects of the system were made to determine the system parameters needed in the matrix equations:

Hydraulic Analysis. A mass oscillation study of the emptying of the outlet conduit and vent shaft was done to determine the mean flow and water level in the surge shaft at the moment the oscillation started. This analysis included friction and inertia in the vent shaft, as well as local losses at junctions and the valve. The results were checked against pressure recordings, showing good agreement. The leakage flow and discharge coefficient of valve 2 were estimated from the pressure recordings, and wavespeeds for every reach of pipe were calculated according to Thorley (1991), taking into account pipe material, wall thickness and pipe constraints.

Flexible Valve Behaviour. A thorough study of the vibration of the gate valve was made after the site tests (CEDEX, 1994). This study, along with inspection of the valve body (that was substituted by a double-seal one as a remedial measure to the problem) concluded that the lintel guides were damaged due to reverse pressures. This might provoke improper closure and allowed valve movement.

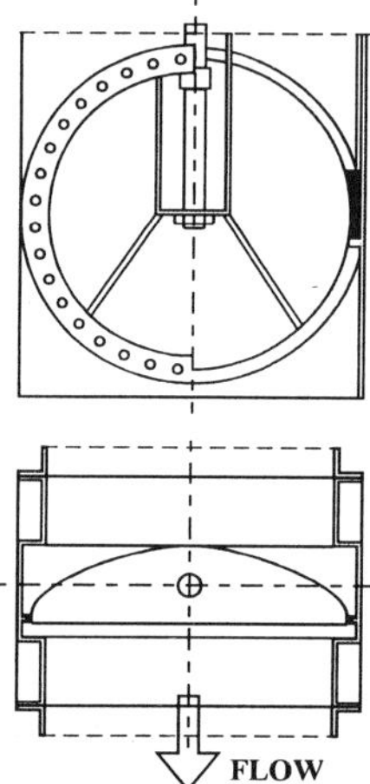

Figure 4. Schematic of the valve

Because of the valve configuration, shown in Figure 4, the Fluctuating-gap Theory (Kolkman, 1984) was applicated as instability mechanism. Under this assumption, the matrix parameters in eqn. (10) are given by:

$$Q_y = \frac{\partial Q}{\partial Y} = \mu L_c \sqrt{2gH_0} \qquad (18)$$

being μ the discharge coefficient (assumed to be constant) and Lc the circumferential lenght of valve, and:

$$Y_h = \frac{\partial Y}{\partial H} = -\frac{\rho g A}{k} \qquad (19)$$

where k is an elastic constant. No mechanical damping is considered in this paper.

In this case, the above mentioned studies concluded that the valve leaf movement could be approximated as a rotation around an axis paralel to the gate rod and located in one of the lintel guides. Then, the elastic constant k was calculated combining flexural and torsional elastic properties of the gate rod, and referring them to the horizontal movement of the valve leaf mass center.

Application of Structure Matrix Method. The system elements and nodes considered are shown in Figure 2. The data used for analysis were: L_1=31.2 m, L_3=8 m, L_4=428 m, L_7=26 m, L_8=10 m, D_1=D_2=D_3=1.2 m, D_4=D_5=D_8=1.6 m, D_6=D_7=0.6 m, a_1=a_3=905 m/s, a_4=a_8=827 m/s, a_7=1012 m/s. A common friction factor f=0.015 was used for all pipes. The mean flow Q_0 was 0.52 m^3/s through elements 4 and 5 and the leakage flow was estimated as 0.01 m^3/s (elements 1, 2 and 3). The head loss is 27.45 m in valve 5 and 1.7 m in valve 2. For the flexible valve, μ=4.6 10^{-3} and k= 296445 N/m were used. After substitution of the boundary conditions in nodes 1, 8 and 9, open to the atmosphere, the global matrix equation for the complete system is:

$$\begin{bmatrix} T_1 - V_2 & -Q_y & V_2 & 0 & 0 & 0 & 0 \\ -Y_h & 1 & Y_h & 0 & 0 & 0 & 0 \\ V_2 & Q_y & T_3 - V_2 & S_3 & 0 & 0 & 0 \\ 0 & 0 & S_3 & T_3 + T_4 + T_8 & S_4 & 0 & S_8 \\ 0 & 0 & 0 & S_4 & T_4 - V_5 & 0 & 0 \\ 0 & 0 & 0 & 0 & 0 & F_6 + T_7 & S_7 \\ 0 & 0 & 0 & S_8 & 0 & S_7 & T_7 + T_8 \end{bmatrix} \cdot \begin{bmatrix} h_2 \\ y \\ h_3 \\ h_4 \\ h_5 \\ h_6 \\ h_7 \end{bmatrix} = \mathbf{0} \quad (20)$$

This equation was solved for the natural frequencies and mode shapes (not shown here for space limitations) using the program Mathematica, which provides good tools for performing the whole analysis (Andreu *et al.*, 1996, Izquierdo *et al.*, 1996).

As the bottom outlet conduit empties after some time, it was also studied the part of the system that remained undergoing oscillation. This system consists only of elements 1 and 2, with node 3 being an open end. The matrix equation for this system is formed by the two first rows and columns of eqn. (20). Table 1 shows the frequencies detected for both systems in the neighbourhood of the measured one.

TABLE 1. Natural frequencies and damping

Complete System		Intake Conduit	
σ	ω	σ	ω
-1.13521	41.2547	1.91254	45.5632
1.58181	45.4228	1.91254	136.689
-1.08748	53.2485		

The results show good coincidence with the measured frequencies. In both systems an unstable condition (σ>0) was found for a frequency close to the measured 45.55 Hz that appeared in the pressure measurements. The analysis of the complete system showed also higher harmonics near the other detected frequency (91 Hz) but with negative values of σ. In order to confirm that these positive values were consequence of the valve flexibility, both systems were analysed considering valve 2 as a rigid one. In that case, very similar frequencies were obtained but all with negative real parts.

4. Conclusion

An incident of self-excited resonance occurred in the conduit system of a dam is presented. After a description of the facts and experimental study, Structure Matrix method is applied to confirm the improper sealing of a valve as the cause of auto oscillation. This paper confirms frequency domain analysis as a simple and effective diagnosis tool in cases of hydraulic resonance, and highlights once again that performing this type of studies during the early design stages could easily avoid operation problems and potential danger in certain types of installations.

5. Acknowledgements

The authors wish to acknowledge the funding and facilities received from the CEDEX for carrying out this research, with particular thanks to its Director Mr. Felipe Martínez. The support given by Iberdrola S.A. to this study, through a research grant to Mr. Andreu, is also acknowledged. And, last but not least, thanks are due to Prof. Brekke for the comments and advices received during the analytical part of this work.

References

Andreu, M., Izquierdo, J. and Iglesias, P.L. (1996) Análisis modal de oscilaciones hidráulicas en sistemas a presión, *Congreso Mathematica 96*, Valencia, 11-12th July.

Brekke, H. (1984) A Stability Study on Hydro Power Plant Governing Including the Influence from a Quasi Nonlinear Damping of Oscillatory Flow and from the Turbine Characteristics, *Dr. Tech. Dissertation*, The University of Trondheim, N-7034 NTH.

CEDEX (1994) Análisis del Funcionamiento de los Desagües Profundos de la Presa de La Viñuela / Málaga, *Internal Report* 41-493-1-038, Madrid.

Chaudhry, M.H. (1987) *Applied Hydraulic Transients - Second Edition*, Van Nostrand Reinhold, New York.

Erhart, R.W. (1979) Auto oscillation in the Hyatt plant's penstock system, *Water Power and Dam Construction*, April, 38-41.

Fanelli, M. (1975) Les phénomènes de résonance hydraulique, *La Houille Blanche* N° 4, 233-253.

Guarga, R. (1995) Resonancia hidráulica. Fundamentos, in Abreu, Guarga and Izquierdo (eds.), *Transitorios y Oscilaciones en Sistemas Hidráulicos a Presión*, U.D. Mecánica de Fluídos, Valencia, pp. 295-345.

Gummer, J.H. (1995) Penstock resonance at Maraetai 1 hydro station, *Hydropower & Dams* Nov., 50-56

Izquierdo, J., Andreu, M. and Fuertes, V.S. (1996) Aplicación del análisis modal a un caso real de oscilaciones hidráulicas, *Congreso Mathematica 96*, Valencia, 11-12th July.

Kolkman, P.A. (1984) Gate vibrations, in P. Novak (ed.) *Developments in Hydraulic Engineering - 2*, Elsevier Applied Science Publishers, London, pp. 55-112.

Li, X. (1988) Hydropower System Modelling by Structure Matrix Method, *HOG Report* 1988:120(A), Norwegian Institute of Technology, Trondheim.

Li, X. and Brekke, H. (1990) Structure matrix method in hydraulic free vibration analysis, *Proc. XV IAHR Symp. - Sec. on Hydraulic Machinery and Cavitation*, Belgrade, Paper T2.

Thorley, A.R.D. (1991) *Fluid Transients in Pipeline Systems*, D. & L. George Ltd, Herts.

Wylie, E.B. and Streeter, V.L. (1993) *Fluid Transients in Systems*, Prentice Hall, Englewood Cliffs.

Zielke, W. and Hack, H.P. (1972) Resonance frequencies and associated mode shapes of pressurized piping systems, *Proc. 1st. Int. Conf. Pressure Surges*, BHRA Fluid Engineering, Cranfield, Paper G1.

AN EXPERIMENTAL STUDY ON FINS, THEIR ROLE IN CONTROL OF THE DRAFT TUBE SURGING

M. NISHI, X. M. WANG, K. YOSHIDA, T. TAKAHASHI
Dept. of Mechanical Engineering, Kyushu Institute of Technology
Sensui-cho 1, Tobata, Kitakyushu 804, Japan

T. TSUKAMOTO
Hydro Power Division, Fuji Electric Co., Ltd.,
Tanabeshinden 1-1, Kawasaki 210, Japan

Abstract

Although the method based on installation of fins has been widely used to alleviate the draft tube surge at part-load operation of a Francis turbine, the corresponding mechanism is still unknown. Thus, the present study was made to clarify the effect of fins on the pressure surge experimentally by analyzing wall pressure fluctuations.

1. Introduction

In the part-load operation of a Francis turbine, a spiral vortex core(or vortex rope) appears in a draft tube and often causes violent pressure fluctuation which is called as the draft tube surge. Since the surge deteriorates the stable operation of a Francis turbine in off-design conditions, many studies have been done to solve such a serious proplem related to unfavorable vibration and noise and to broaden its operating range. Various attempts have been made to develop effective methods for the alleviation of the pressure surge[1]. It is known that the air admission or injection and installation of fins in the inlet cone of a draft tube are popular means. On the other hand, it is also said that they are not always very effective to suppress the pressure fluctuations and a model test should be made beforehand to examine its availability. According to the literature survey, basic studies on the latter are suspected less than those of the former, though there are recent contributions which treated the effect of the fins on the pressure fluctuations[2, 3]. However, it is still expected to make clear how and why the violent pressure fluctuation can be alleviated by using the fins.

In this paper, we demonstrate our experimental study which was made based on the measurement of pressure fluctuations and observation of the cavitating vortex rope simultaneously in a draft tube with and without fins to clarify their effect on the half load surging by analyzing wall pressure fluctuations.

E. Cabrera et al. (eds.), Hydraulic Machinery and Cavitation, 905–914.

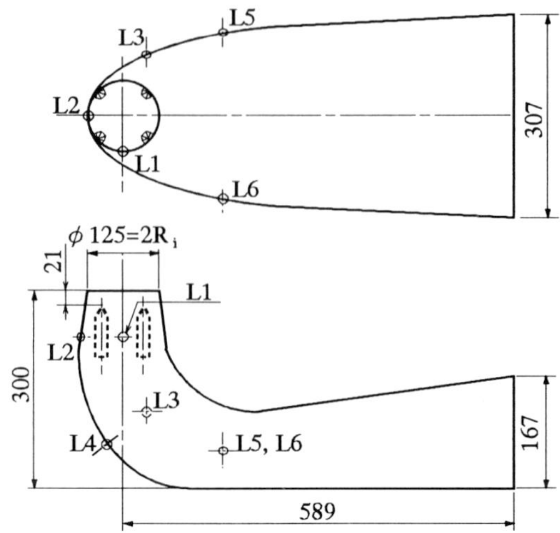

Fig.1 An elbow draft tube

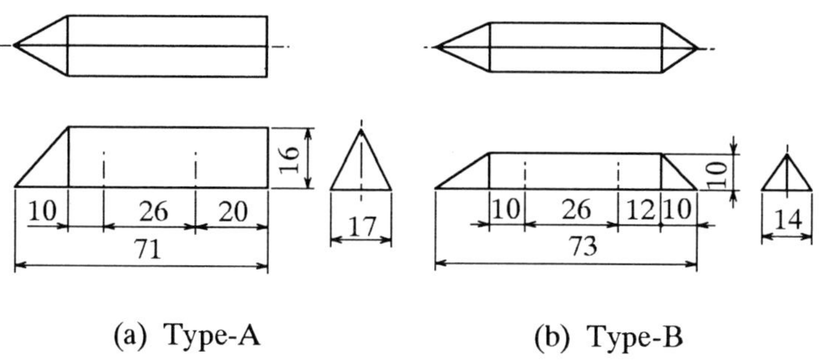

Fig.2 Configurations of fins tested

2. Experimental Apparatus and Method

2.1. AN ELBOW DRAFT TUBE AND FINS

An elbow draft tube made of transparent plastic shown schematically in Fig.1 was used in the present study. It has the following geometries: The inlet cone angle is 9 deg and exit-to-inlet area ratio is 4. Each of L1~L6 in the figure indicates the measurement point where a pressure transducer was installed.

Fig.1 also shows the location of fins mounted on the wall of inlet cone when we investigated their effect on the pressure fluctuation. It is noted that the number was four in all our tests. Regarding the fin's configuration, two conventional types shown in Fig.2 were adopted. The height of Type-A is greater than that of Type-B, and both of them have the triangular cross-section.

2.2. PARAMETERS

As the experiments were conducted using a water test apparatus where the swirl flow was generated by stationary guide vanes[4], the theoretical swirl rate (swirl momentum parameter) and cavitation parameter given by Eqs.(1) and (2) respectively were adopted to specify the test condition for the draft tube. Both of them are determined at the inlet.

$$m_{th} = \Omega / (RM) \tag{1}$$

$$Ki = (H_a - H_s - H_v) / (u_i^2 / 2g) \tag{2}$$

where Ω : angular momentum of swirl flow at the draft tube inlet
R : radius of draft tube inlet
M : axial momentum of swirl flow at the draft tube inlet
H_a : absolute pressure head acting on the tail water surface
H_s : suction head
H_v : vapor pressure head
u_i :axial velocity at the draft tube inlet calculated from the discharge.

To evaluate pressure fluctuation, the following dimensionless quantities $\Delta\psi_{RMS}$ and St are calculated from measured pressures.

$$\Delta\psi_{RMS} = \Delta H_{RMS} / (u_i^2 / 2g) \tag{3}$$

$$St = 2Rf / u_i \tag{4}$$

where ΔH_{RMS} : RMS value of fluctuating pressure head
f : a dominant fundamental frequency of pressure fluctuation

As shown previously, two dominant frequencies are generally found in pressure signals which are detected in the operating range corresponding to flow regime III [5] where a cavitating vortex rope is observed in a draft tube: One is f_{ro} or the rotating frequency due to the vortex-rope precession, and the other is f_s or the natural frequency of the draft-tube vibration system. We obtained these fundamental components from wall pressure measurement at two points together with the method based on the spectrum analysis[6].

3. Results

The effect of fins on the pressure fluctuations was investigated under the conditions where the spiral vortex rope was observed in the inlet cone of the elbow draft tube.

3.1. AMPLITUDE

The cavitation characteristics of amplitude parameter are shown in Fig.3 and Fig.4. The former shows the results obtained in the case of swirl rate m_{th} =1.0, and the latter for m_{th} =1.2 which corresponds to stronger swirl flow than the former flow. Comparison among three cases, i.e., results for the draft tube without fins, those for Type-A fins and the Type-B is made at every measurement point from L1 to L6.

From these results, the following features can be found:

(1) In the higher range of cavitation parameter Ki, $\Delta\psi_{RMS}$ is reduced by using both sets of fins. And it appears that there is little difference in the dimensionless amplitude between Type-A and Type-B in m_{th} =1.0 case, however, Type-A is more effective to suppress the fluctuation than Type-B for m_{th} =1.2.

(2) In the case of m_{th} =1.0, $\Delta\psi_{RMS}$ near the critical Ki, which corresponds to the resonant condition, is suppressed by the installation of both types of fins .

(3) In the case of m_{th} =1.2, however, unfavorable results are recognized in the lower Ki range near the critical. That is, the maximum amplitude couldn't be suppressed by those fins. Moreover, the critical cavitation parameter Ki shifted to the higher value than that of the case without fins.

(4) If $\Delta\psi_{RMS}$ near the critical Ki is concerned, Type-A took the larger value than Type-B.

3.2. FREQUENCY

Concerning the frequency parameters, those results are shown in Fig. 5 and Fig. 6, the test conditions of which correspond to those of Fig. 3 and Fig.4 respectively. In both figures, (a) shows the cavitation characteristics of rotating frequency St_{ro} inside the inlet cone and (b) is for the natural frequency St_s [7]. It is noted that these two components were obtained from pressures measured at L1 and L2.

The following features are observed in these figures:

(1) In every case, St_{ro} is regarded as nearly constant in the test range of Ki.

(2) St_{ro} increases slightly if those fins are installed. The larger the height of fin is, the higher the frequency is.

(3) Considerable change in St_s is observed in the case of m_{th} =1.0 .

(4) St_s for the m_{th} =1.2 case shows almost the same characteristics regardless of fins.

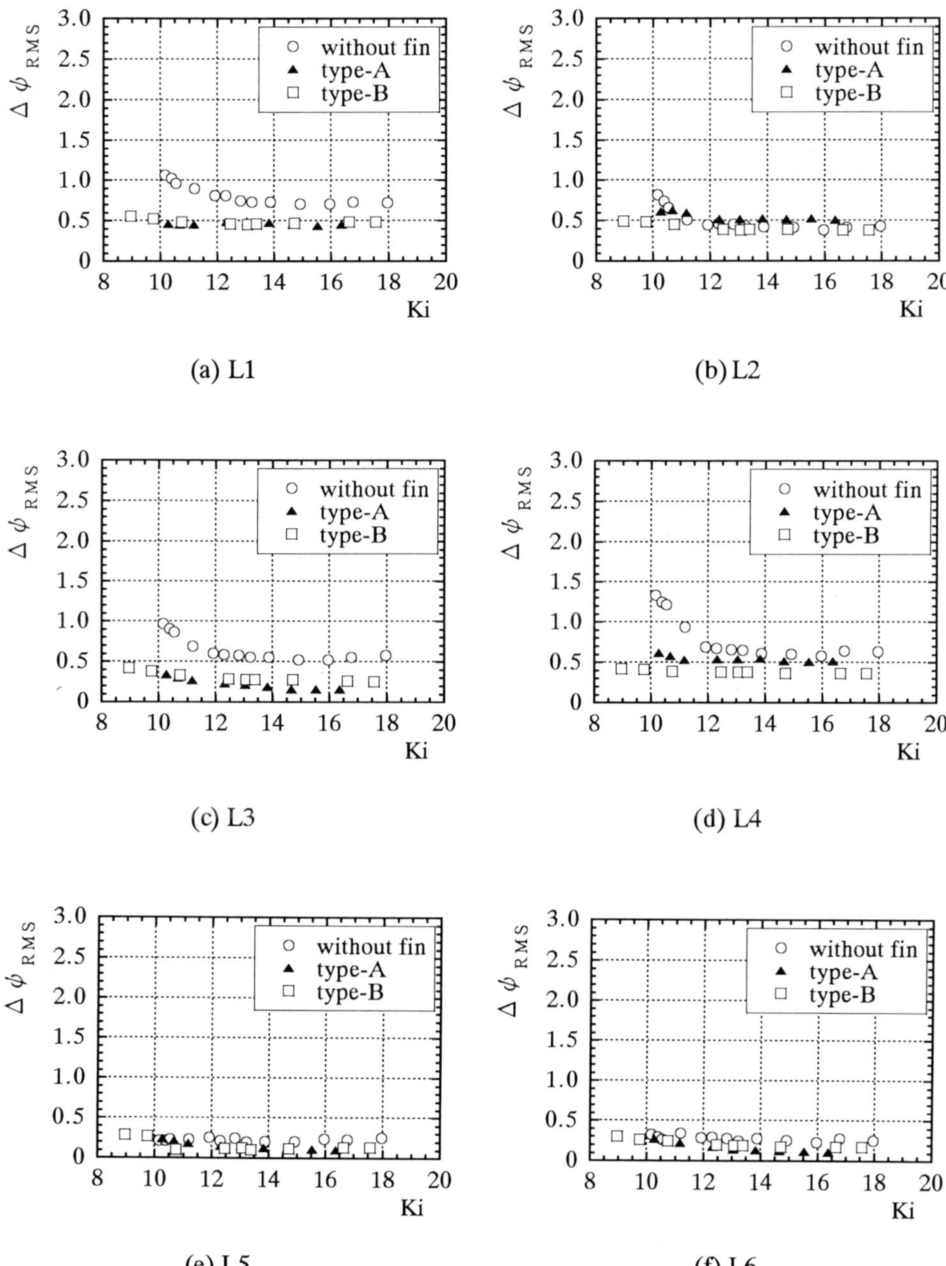

Fig.3 Cavitation characteristics of amplitude parameter $\Delta\psi_{RMS}$ in the case of m_{th} =1.0

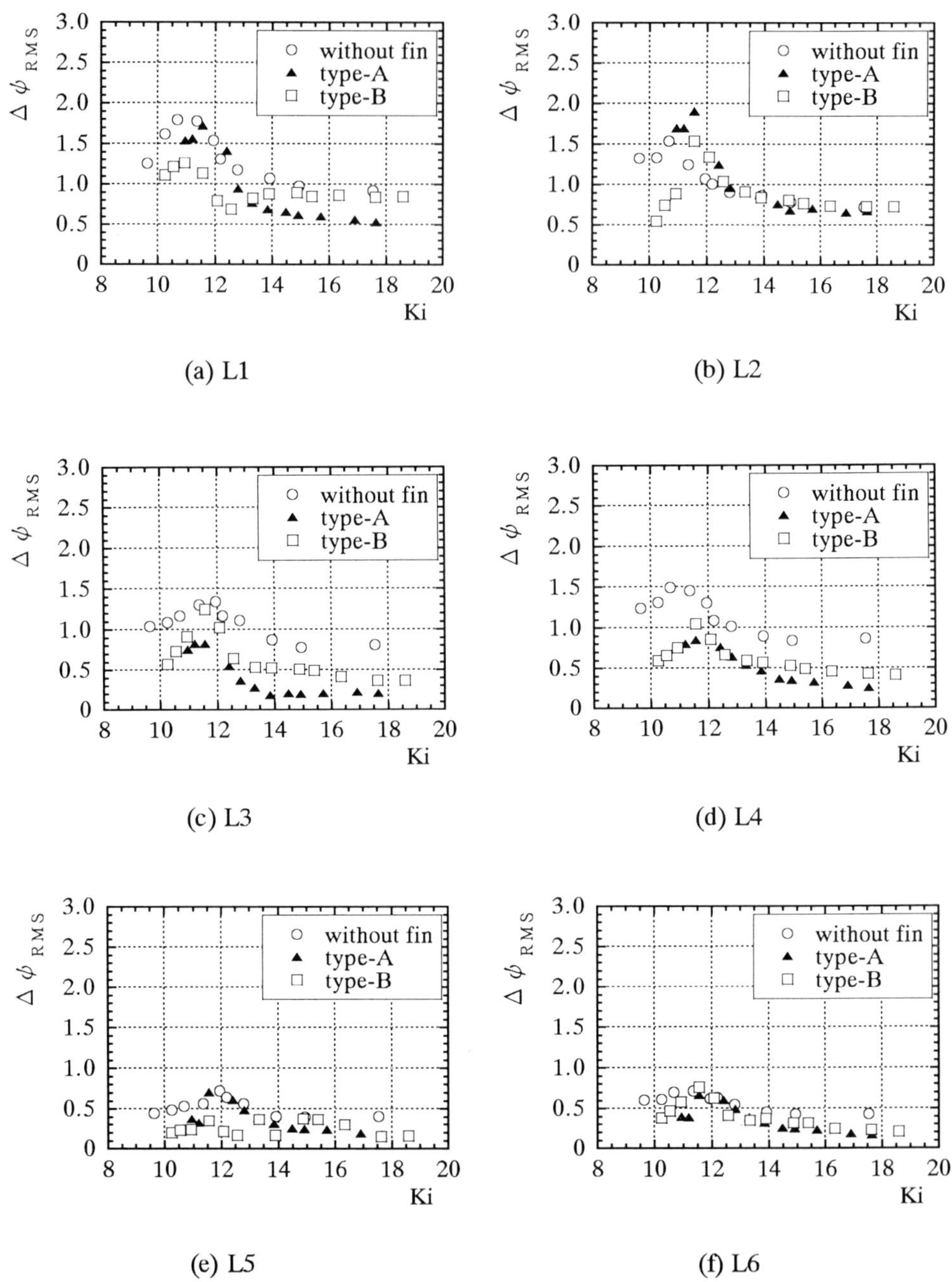

(a) L1

(b) L2

(c) L3

(d) L4

(e) L5

(f) L6

Fig.4 Cavitation characteristics of amplitude parameter $\Delta\psi_{RMS}$ in the case of m_{th} =1.2

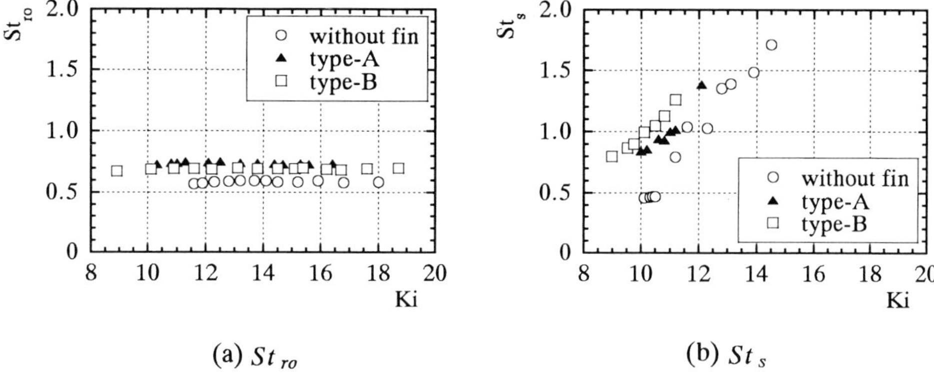

(a) St_{ro} (b) St_s

Fig.5 Cavitation characteristics of frequency parameter St in the case of m_{th} =1.0

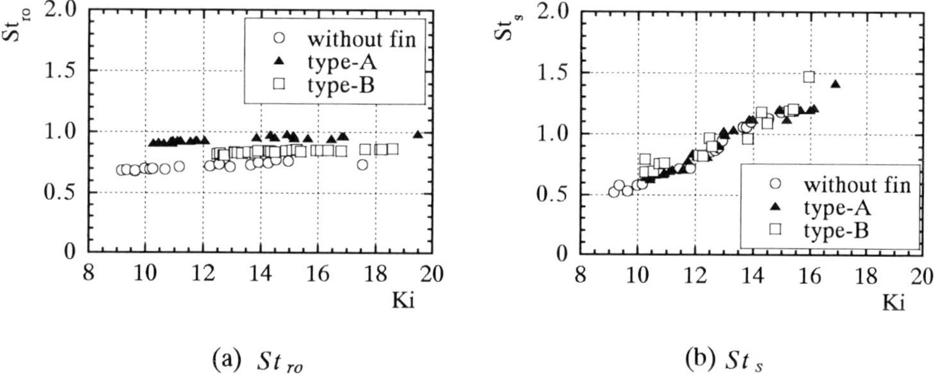

(a) St_{ro} (b) St_s

Fig.6 Cavitation characteristics of frequency parameter St in the case of m_{th} =1.2

4. Discussion

4.1. MODIFIED FREQUENCY PARAMETER

It is easily suspected that the fins will reduce the swirl component in its downstream flow. This is clearly seen in $\Delta\psi_{RMS}$ for the elbow section measured at L3 and L4 in Fig.3 and Fig.4. However, St_{ro} has the nature to not decrease but increase due to the installation of fins. These two results are not inconsistent if the flow pattern in a draft tube was changed by the fins. Considering the effect of fin's height h, we introduce the following modified frequency parameter.

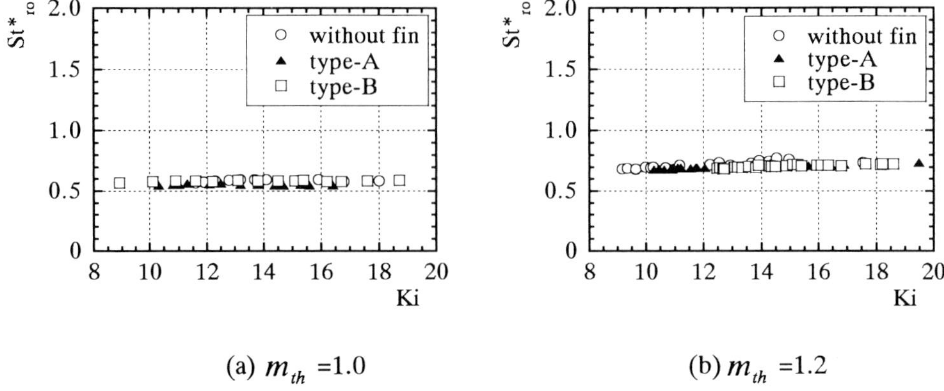

(a) m_{th} =1.0 (b) m_{th} =1.2

Fig.7 Cavitation characteristics of modified frequency parameter $St_{ro}{}^*$

$$St_{ro}{}^* = 2(R-h)f_{ro} / u_i \tag{5}$$

If we make use of this parameter $St_{ro}{}^*$ instead of St_{ro} , those data shown in Fig.5(a) and Fig.6(a) are correlated satisfactorily as shown in Fig.7(a) and (b) respectively.

4.2. ROLE OF FINS

The following explanations were shown regarding the nature of the draft tube surge [7]:
(1) The violent pressure fluctuation called as the draft tube surge having the quasi-synchronous nature (QS type) is caused by the resonance.
(2) This resonance occurs at the condition or the critical Ki where the rotating frequency of the cavitating vortex rope at the elbow section coincides with the natural frequency of draft tube vibration system.
(3) The trigger is attributed to the oscillation of pressure recovery in the foot(downstream diffuser) having the same frequency as the rotating frequency of the rope.

In this section, we mainly discuss on the unfavorable results shown in Fig.4.

If the cavity volume in a draft tube is reduced by the fins, St_s should be greater. This trend is observed in Fig.5(b) for m_{th} =1.0, but not observed in Fig.6(b) for m_{th} =1.2, where the swirl is much stronger than the former. From the observation of cavitation made by a high speed video system, we thought that one of the reasons for the latter case might be another cavities created by the fins. This indicates that we cannot always change the natural frequency St_s by the fins. In such a case, the critical Ki moves to the higher Ki region, since the rotating frequency is increased by the installation of fins, as shown in Fig.4(a) and(b).

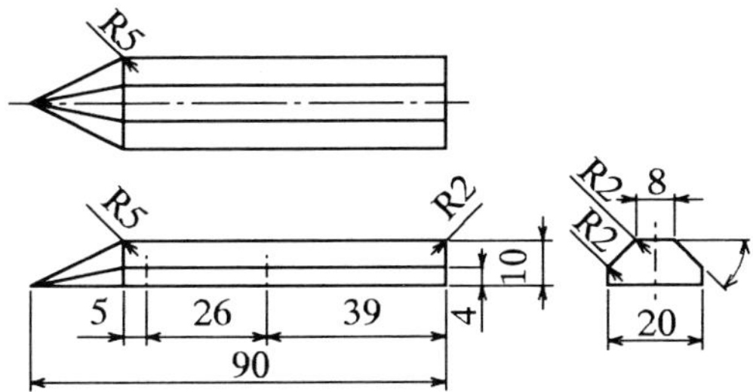

Fig.8 A fin having the trapezoidal cross-section (Type C)

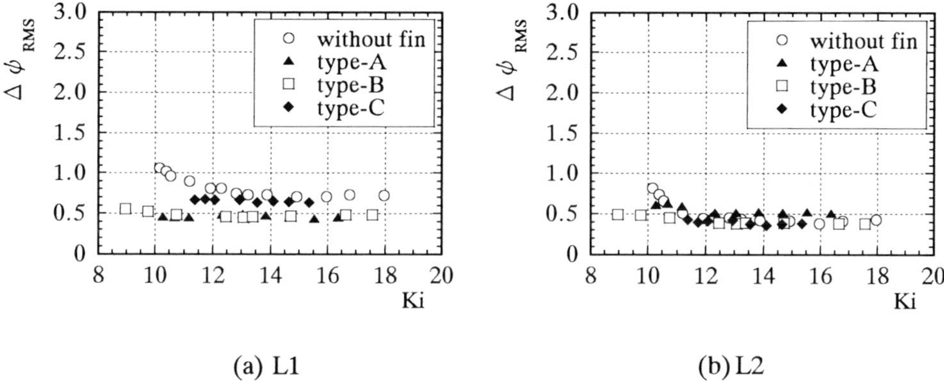

Fig.9 Cavitation characteristics of amplitude parameter $\Delta\psi_{RMS}$ in the case of m_{th} =1.0

It is known that the variation of $\Delta\psi_{RMS}$ in peripheral direction is essentially caused by the oscillation (having synchronous nature) of pressure recovery having the same frequency as St_{ro} (having convective nature). Thus, from the difference between the Type-A and Type-B results in Fig.3 and Fig.4, it is suspected that those fins reduced both swirl and the magnitude of pressure recovery oscillation in the foot in the case of m_{th} =1.0 , but it is emphasized that the type-A fins enhanced the oscillation of pressure recovery by reducing the swirl for the m_{th} =1.2 case. This suggests that severer resonance will be caused by the installation of fins as seen in Fig.4(a) and (b).

5. Further Study

If the fins are used under a cavitating condition, their lives may be affected by the cavitation erosion. Considering this situation, a further study was made to investigate the availability of the different cross-sectional shape from the conventional. As shown in Fig.8, the trapezoidal shape(Type-C) was adopted. The typical results are plotted in Fig.9(a) and (b) which are the reproduction of Fig.3(a) and (b). It is seen that the Type-C is also effective to reduce the amplitude of pressure fluctuation and to suppress the resonance in this operating condition.

6. Concluding Remarks

The following concluding remarks are obtained from the present experimental study on the draft tube fins.
(1) The installation of fins on the inlet-cone wall of a draft tube is basically useful to broaden the part-load operating range of a Francis turbine.
(2) Attention should be paid to the limit of its applicability. Usage of fins in the part load region beyond the limit will be unfavorable due to their enhancement of instabilities.
(3) Rotating frequency increases with the increase in the height of fin.
(4) Reduction of swirl by the fins is related to the natural frequency of the draft tube vibration system with cavitation and the oscillation of pressure recovery in the foot (downstream diffuser). The latter is regarded as the trigger of draft tube surge.
(5) Results shown in this paper will help how to select suitable fins.

7. Acknowledgments

The authors wish to thank Prof. H. Tsukamoto in Kyushu Institute of Technology for his cooperation and Mr. K. Shiramizu in Mitsubishi Heavy Industries, Ltd. for his help in the experimental work.

8. References

1. Grein, H. : Vibration Phenomena in Francis Turbines: their causes and prevention, Proceedings of 10th IAHR Symposium, 1, (1980), .
2. Rocha, G. and Sillos, A. : Power Swing Produced by Hydropower Units, Proceedings of 11th IAHR Symposium, 2, (1982), 44/1-15.
3. Jacob, T., Prenat, J.E., and Buffet, G. : Improving the Stability of Operation of a 90 MW Francis Turbine, Hydropower into the Next Century, Barcelona, 1995.
4. Nishi, M., Kubota, T., Matsunaga, S., and Senoo, Y. : Study on Swirl Flow and Surge in an Elbow Type Draft Tube, Proceedings of 10th IAHR Symposium, 1, (1980), p.557.
5. Nishi, M., Matsunaga, S., Kubota, T., and Senoo, Y. : Flow Regimes in an Elbow-Type Draft Tube, Proceedings of 11th IAHR Symposium, 2, (1982), No.38.
6. Nishi, M., Wang, X., Okamoto, M. and Matsunaga, S. : Further Investigations on the Pressure Fluctuations Caused by Cavitated Vortex Rope in an Elbow Draft Tube, ASME FED-Vol.190, (1994) , pp.63 - 70.
7. Wang, X., Nishi, M., and Tsukamoto, H. : A Simple Model for Predicting the Draft Tube Surge, Proceeding of 17th IAHR Symposium, 1, (1994), pp.95 - 106.

MODEL FOR VORTEX ROPE DYNAMICS IN FRANCIS TURBINE OUTLET

G. PEDRIZZETTI
Dipartimento Ingegneria Civile, University of Firenze
v. S. Marta 3, 50139 Firenze, Italy

G. ANGELICO
ENEL, Centro Ricerche Idrauliche e Strutturali
v. Ornato 90/14, 20162 Milano, Italy

Pressure fluctuations at the outlet of a Francis turbine operating at partial load regimes are evaluated with a computational method based on three-dimensional vortex dynamics. The fundamental frequencies and their dependence on the mean swirl-rate show a good agreement with measured data. Results also indicate the birth of aperiodic regimes.

1. Introduction

The present communication deals with the simulation of the vortex rope present in the outlet of a Francis turbine when operating at partial load regimes. This approach is not devoted to the analysis of a complete turbomachinery system but rather to develop a rigorous mathematical analysis for a local phenomenon of practical relevance. We synthesise the problem calculating the three-dimensional unsteady motion of a vortex filament confined inside a cylindrical wall. In this first analysis we ignore cavitation effects in order to detect those phenomena which can be explained in terms of incompressible vortex dynamics. The system rope-diffuser is assumed as isolated and we neglect any direct interaction with the turbine blades or the diffuser outlet.

The computational model is based on the inviscid dynamics of a thin vortex filament (Moore 1972; Leonard 1985). Such a filament is continuously created at the inlet of the cylinder on the basis of simple arguments which relate the

E. Cabrera et al. (eds.), Hydraulic Machinery and Cavitation, 915–924.

parameters of the turbine with some characteristics of the helicoidal vortex (Fanelli 1989a; Nishi 1991). The vortex is then free to evolve in a three-dimensional way during its roto-translational motion toward the outlet where phenomena of rapid energy dissipation are modelled (Chorin 1990). Pressure fluctuations at the cylindrical wall, for different values of the swirl-rate, are analysed, compared with experimental data, and related with the evolution of the vortex rope.

2. Dynamical system

2.1 VORTEX ROPE GENERATION AT THE DRAFT TUBE INLET

We assume that subsequent filament portions, of helicoidal shape (Rihtarsic et al. 1993; Sirok et al. 1993; Jacob et al. 1993), are continuously generated at the draft tube inlet to mimic the wake at the outlet of the Francis turbine operating at partial load (Fanelli 1989a, 1989b). The filament is "locally" helicoidal and is left free to evolve downstream with no restriction on its shape.

The dimensional parameters of such a model are given by the draft tube radius R, and length L, mean velocity $U(t)$, and vortex circulation Γ, helical radius (or dead zone radius) R_h and pitch P_h. Consider the mean velocity as given by a steady and a fluctuating part

$$U(t) = U_o(1 + f(t)) \tag{1}$$

We make the problem dimensionless by taking R as the unit length and the ratio R/U_o as the unit of time. The draft tube geometry is given by the dimensionless tube length $\ell = L/R$. The present model is not able to predict the vortex characteristics, because the turbine is not modelled at presently, but only its evolution once it has been generated. In order to relate the analysis to actual machinery flow, we consider physical arguments and experimental results for partial load Francis turbine (Fanelli 1989a, 1989b; Nishi et al. 1993).

We consider as input parameters, in addition to ℓ, the mean swirl-rate $\overline{m} = \Gamma/(2\pi R U_0)$ where the vortex circulation Γ is constant in time and can be related to actual machinery parameters and operating point (Fanelli 1989a, Nishi et al. 1993), and the flow oscillations $f(t)$ appearing in (1). The instantaneous swirl-rate is then given by $m(t) = \overline{m}/(1 + f(t))$. We compute the helix radius at the inlet by the relationship (Nishi 1991)

$$m(t) = r_h(t) + 3r_h^3(t) \, . \tag{2}$$

Equation (2) is derived by approximating the result from the principle of minimum kinematic energy and is in good agreement with experimental data. Relation (2) is here adopted because of its simplicity and stability and because it releases us from a definition *a priori* of other parameters (Fanelli 1989a). The helix pitch at the inlet is obtained by the condition that the flow rate must be proportional to the filament strength (Fanelli 1989a)

$$p_h(t) = 2\pi m(t)\left(1 - r_h^2(t)\right) \tag{3}$$

In this way once the swirl-rate m is given, the vortex radius and the pitch can be determined, at least in first approximation, univocally. The relations given by equations (2) and (3) are reported in figure 1.

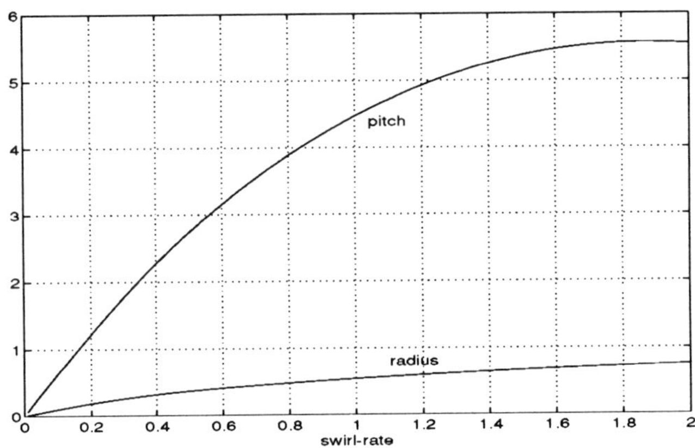

Figure 1. Relations of helical radius and pitch with the swirl-rate as given by eqs. (2, 3).

2.2 VORTEX FILAMENT EVOLUTION

A vortex filament is a distribution of vorticity along a fluid tube whose cross section is much smaller with respect to any characteristic longitudinal length. We assume to deal with an incompressible ideal fluid, then for the Helmholtz and Kelvin laws a thin vortex filament moves with the fluid and maintains its coherence without topological changes, moreover circulation Γ along any fluid contour surrounding the filament is constant both in time and along the filament (Batchelor 1967). Assuming that the filament axis lies over a curve C, then we can introduce a parametric coordinate ξ along the curve such that the point at $\xi = const$ always refers to the same fluid particle. The position of the filament axis is specified by its coordinates along the curve $\boldsymbol{X}_v(\xi, t)$ where t is the time.

The evolution of the filament is computed as that of a material line which is advected by the local velocity at the filament position. The objective is to determine such velocity field as given by the vortex itself with the constrain of the boundary condition of zero normal velocity at the cylinder wall, and the condition of flowing a given discharge along the tube. The velocity at a point x inside the cylinder is given by the superposition of three velocity fields

$$u(x) = u_v(x) + u_b(x) + u_Q(x); \tag{4}$$

where $u_v(x)$is the velocity field given by the vortex filament, $u_b(x)$is the additional irrotational velocity field introduced in order to satisfy the boundary conditions, and $u_Q(x)$ is a uniform field added to the previous ones in order to get the desired discharge.

The velocity induced by a vortex filament in a point x is (Batchelor 1967)

$$u_v(x) = \frac{\Gamma}{4\pi}\int_C \frac{(X_v(\xi) - x)}{\left(|X_v(\xi) - x|^2 + c^2\right)^{\frac{3}{2}}} \wedge \frac{\partial X_v}{\partial \xi} d\xi\,; \tag{5}$$

where the parameter c is called *core* radius and represents a characteristic length scale of the filament cross-section. The explicit presence of the core size in the denominator makes equation (5), known as *cut-off* approximation (Moore 1972), appropriate to calculate the velocity field induced by the filament on itself simply by putting $x = X_v$.

The velocity field (5) does not satisfy the zero normal velocity boundary condition at the cylinder wall. The irrotational contribution $u_b(x)$ due to the presence of the cylinder boundary can be written in terms of potential flow as

$$u_b(x) = \nabla\varphi\,, \tag{6}$$

where the scalar potential $\varphi(x)$ must satisfy the Laplace equation

$$\nabla^2\varphi = 0\,. \tag{7}$$

The boundary conditions for (7) derive by imposing that the normal velocity field given by $u_b(x)$ at the cylinder wall must balance that given by $u_v(x)$. Introducing a cylindrical system of coordinates $\{r, \theta, z\}$ as radial, tangential, and axial coordinates, and calling $u_{v,r}(r, \theta, z)$ as the radial component of $u_v(x)$, the boundary condition for (7) can be written as

$$\left.\frac{\partial\varphi}{\partial r}\right|_{r=R} = -\,u_{v,r}(R, \theta, z), \tag{8}$$

Equation (7) can be solved by separating variables. In this case the solution can be given analytically only for few possible boundary condition at the inlet and outlet of the cylinder. Imposing that the added irrotational flow gives no influence on the discharge in the tube, the solution can be written as

$$\varphi(r,\theta,z) = \sum_{n=0}^{\infty}\Big\{ (a_{n0} cos\, n\theta + b_{n0} sin\, n\theta) + \\ + \sum_{m=1}^{\infty} (a_{nm} cos\, n\theta + b_{nm} sin\, n\theta)\, cos\Big(m\frac{\pi}{L}z\Big)\, I_n\Big(m\frac{\pi}{L}\Big)\Big\}, \tag{9}$$

where $I_n(\cdot)$ is the modified Bessel function of the first kind of order n. Coefficients a_{nm} b_{nm} must be determined by using condition (8). This means, by a computational point of view, that at every time step the radial velocity $u_{v,r}(R,\theta,z)$ is computed on a two dimensional grid at the wall then its double Fourier transform is taken, and the coefficients are directly related to constants in (9). In this way, even though the boundary condition is satisfied exactly only in a discrete number of points, the analyticy of the expansion (9) ensures a continuity of the flow field which could not be achieved by standard discretization methods.

The third contribution to the velocity field entering in (4) is represented by a uniform velocity directed along the tube which ensures the desired value of the flowing discharge. The discharge is constant along the tube even if it can change in time. Assume that, at a certain time, we want a discharge $Q(t){=}\pi R^2 U(t)$, then we compute, numerically, the mean velocity V as given by the *vortex* (generally negative) and *bound* (generally zero) contributions

$$V = \int_0^R \int_0^{2\pi} [u_{v,z}(r,\theta,z) + u_{b,z}(r,\theta,z)] r\, d\theta\, dr \tag{10}$$

and impose

$$u_Q = [\,0, \quad 0, \quad U - V\,]. \tag{11}$$

By a numerical point of view the filament is discretized in a finite number of points along its axis, whose distance must be smaller than the *cut-off* length c, and each point is instantaneously advected by the local velocity (4).

2.3 DISSIPATIVE PHENOMENA

The dissipation of vorticity downstream the draft tube is modelled, in a crude approximation, by continuously removing the final portions of filament as they pass after the outlet of the draft tube.

A dissipative phenomenon may occur during the motion as the filament comes very close to the tube wall. In this case the separated secondary vorticity couples with the primary vortex in a very rapid stretching eventually leading to the destruction of a portion of filament in very short time scale when compared with the evolution of filament axis (Becchi et al. 1992; Pedrizzetti 1992*a*). This is reproduced by eliminating portions of filament as they come closer than the core size to the cylinder wall.

As a further extension of the model, an algorithm to remove tightly folded hairpins has been introduced. This was considered in (Chorin 1990) to model both viscous phenomena and the effect of small scale turbulence, in this way when a sort of knot develops in the geometry of the vortex we assume that a sort of reconnection occurs (Pedrizzetti 1992*b*) and small scale vorticity is rapidly dissipated. The influence of dissipation models is discussed in the results.

3. Results

Numerical simulations have been performed with constant velocity, $f(t) \equiv 0$, at swirl-rate m ranging from 0.2 to 1.2, on a draft tube of length $\ell = 4$. Core radius has been kept fixed at $c = 0.2$. The model appears to be quite stable and in all cases a well developed vortex rope is observed. Its evolution has either rotational and translational components, the nonlinear dynamics of vortex motion leads the vortex to deform from its original helicoidal shape, such a deformation is more relevant for larger values of the swirl-rate. These phenomena are reflected in pressure measurements.

The model is one of inviscid flow and pressure at the wall is computed by the Bernoulli theorem (Batchelor 1967)

$$p(\theta, z, t) - p_i(t) = \frac{\rho}{2}\left(u_i(t)^2 - u(\theta, z, t)^2\right); \tag{12}$$

where the subscript i stands for inlet and ρ is the fluid density. Here we neglected the gravitational hydrostatic and the unsteady contributions. The latter can be neglected in first approximation because the time scales associated with the motion are much larger than one (in dimensionless units) as will be shown below. At the inlet, before entering the tube, the velocity is given by an axial component, equal to the uniform mean velocity, and a swirling component whose ratio with the axial velocity is given by $\overline{m}$, i.e. $u_i^2 = (1 + f(t))^2 + \overline{m}^2$.

Wall pressure oscillates in time in a periodic way, at least for $m \leq 1$, the period of pressure oscillations reflects the roto-translational rope evolution. The fundamental frequencies vary with the swirl rate ranging from about 0.7 to 0.24 dimensionless units, R/U_o, in the studied range of swirl rate. The dependence of frequency with m is reported in figure 2.

These frequency values are in good agreement with experimental observations (Nishi 1991; Jacob et al 1993; Nishi et al 1993); note that Strouhal number is defined in (Nishi 1991; Nishi et al. 1993) as twice the present dimensionless frequency. The observed oscillations do not present any "synchronous" component. These, which act on the whole cross section of the tube at the same time, are observed in the presence of an elbow and are related to impulsive dissipation of portions of the vortex structure as it reaches the elbow or to pulsation of an internally cavitated rope.

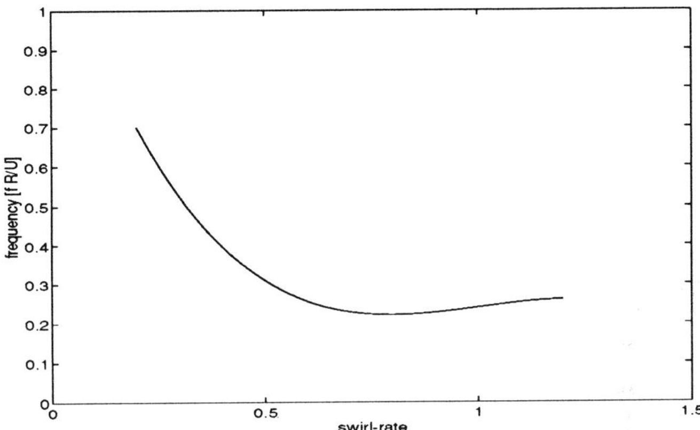

Figure 2. Variation of the fundamental dimensionless frequency of wall pressure fluctuations with the swirl-rate.

Frequency decreases with growing swirl rate reaching a minimum at about $m = 0.8$, and presents a very small increase afterward. These two different behaviours are reflected in the dynamics of the vortex rope. For small swirl rate the initially helicoidal filament decreases its eccentricity downstream in a sort of stable decay reaching an almost rectilinear shape at the tube outlet, on the other hand, when $m > 0.8$ the vortex is more distorted as it reaches the exit reflecting a possibly unstable character. The typical configurations are shown in figure 3 where instantaneous vortex rope is reported for $m = 0.2$ and $m = 1.2$.

Pressure evolution is almost sinusoidal at small swirl-rates showing the dynamics nonlinearities have little influence. For larger values of the swirl-rate the wavy pattern is substantially distorted from sinusoidal showing a more impulsive dynamics and the power spectrum presents several growing peaks indicating a possible tendency toward unpredictable dynamics (Pedrizzetti 1995). Indeed, for swirl-rate $m = 1.2$ the flow evolution looses its purely periodic character and an aperiodic, possibly chaotic, regime establishes.

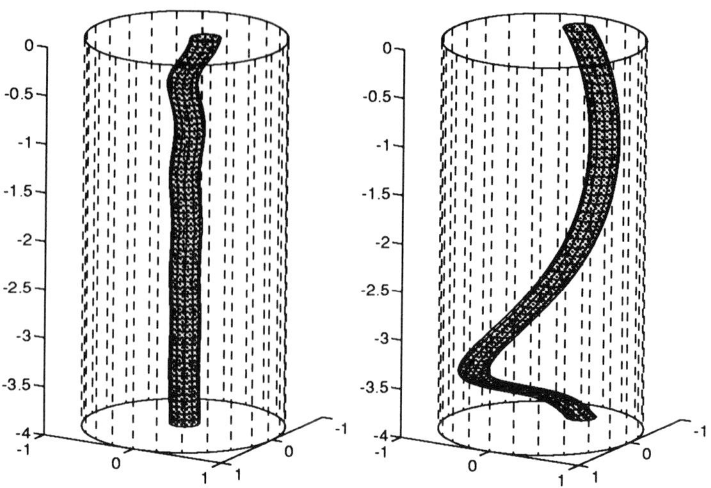

Figure 3. Instantaneous vortex rope, for swirl-rate $m = 0.2$ (left), $m = 1.2$ (right).

This is shown in figure 4 where the time evolution during 20 time units (simulations extend for 100 units) of the pressure coefficient $C_p = (p - p_i)/\frac{\varrho}{2}u_i^2$ is reported for the swirl-rates $m = 1.2$ and $m = 1$. It is worth to notice that the dissipation phenomena, apart from those occurring at the tube outlet, are not present for $m \leq 1$ and that the birth of an aperiodic regime is coincident with the appearance of dissipation in the draft tube.

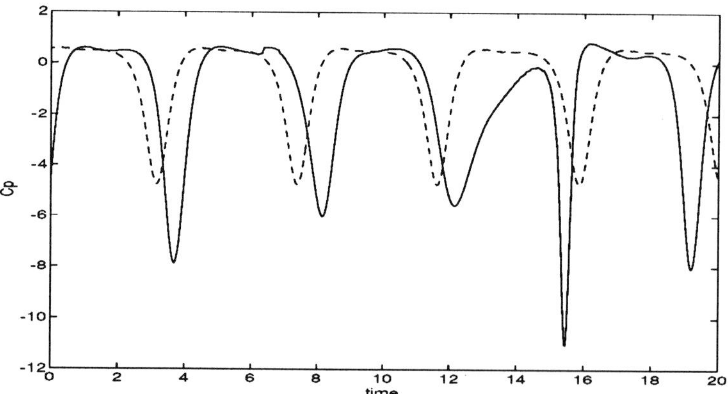

Figure 4. Pressure coefficient time evolution. Pressure is measured at $z = 3$, swirl-rate $m = 1.2$ (continuous line), $m = 1$ (dotted).

4. Conclusions

A model, based on vortex dynamics, has been introduced to reproduce the motion of the three-dimensional vortex rope created at the outlet of Francis turbines when operating at partial load regimes. The diffuser is assumed as rectilinear and the generation-dissipation of vorticity is modelled by simple arguments which mimic, with a minimal representation, the coupling between the draft tube and the turbomachinery.

Results show that the present model captures some features observed in real flow measurements. Pressure fluctuations have been computed at the tube wall, these present fundamental frequencies in quantitative agreement with experimental data. The nonlinear relationship between frequency and swirl-rate has been connected with differences in the actual shape of the vortex rope, further analyses and detailed comparisons are necessary to understand if the different regimes correspond to real differences in the physical behaviour. The growth of the swirl-rate primes relevant internal dissipative phenomena that seem to be responsible for the development of an aperiodic regime.

The model represents a first approximation to the dynamics of real machinery. In particular it should be improved to introduce a more realistic turbine-vortex rope interaction. Cavitation phenomena can be introduced in such a modelling in a straightforward manner by allowing additional degrees of freedom to the vortex filament dynamics. Finally, mathematics can be extended to consider the elbow of the draft tube.

5. References

Batchelor G.K. (1967) *An introduction to fluid dynamics*, Cambridge Univ. Press.

Becchi I., Pedrizzetti G., Frati A., Loru L. (1992) Analisi sperimentale dell'interazione tra strutture vorticose e pareti, *Proc. 23rd Convegno di Idraulica e Costruzioni Idrauliche*, Firenze 31st Aug.-4th Sept., **4**, E.15.

Chorin A.J. (1990) Hairpin removal in vortex interaction, *J. Comp. Phys.* **91**, 1.

Fanelli M. (1989*a*) The vortex rope in the draft tube of Francis turbines operating at partial load: a proposal for a mathematical model, *J. Hydr. Res.* **27** (6), 769.

Fanelli M. (1989*b*) Indirizzi di ricerca per una migliore rappresentazione delle reazioni mutue tra torcia e flusso nella girante Francis, *ENEL-CRIS report* 3818.

Jacob T., Prénat J.E., Randriamamonjy M. (1993) General survey of draft tube cavitation patterns in a medium specific speed Francis turbine, *Proc. 6th Int. Meet.*IAHR work group on "The behavior of hydraulic machinery under steady oscillatory conditions", Lausanne 8-10 Sept.

Leonard A. (1985) Computing three-dimensional incompressible flows with vortex elements, *Ann. Rev. Fluid Mech.* **17**, 523.

Moore D.W. (1972) Finite amplitude waves on aircraft trailing vortices, *Aeron. Quart.* **23**, 307.

Nishi M. (1991) in *Vibration and oscillation of hydraulic machinery*, Oashi H. ed., Avebury Technical, 206-225.

Nishi M., Okamoto M., Wang X. (1993) Evaluation of pressure fluctuations caused by cavitated spiral vortex core in the swirling flow of elbow draft tube, *Proc. 6th Int. Meet.* IAHR work group on "The behavior of hydraulic machinery under steady oscillatory conditions", Lausanne 8-10 Sept.

Pedrizzetti G. (1992*a*) Close interaction between a vortex filament and a rigid sphere, *J. Fluid Mech.* **245**, 701.

Pedrizzetti G. (1992*a*) Insight into singular vortex flows, *Fluid Dyn. Res.* **10** (2), 101.

Pedrizzetti G. (1995) A three-dimensional computational approach to the vortex rope dynamics in Francis turbine outlet, *Proc. 7th Int. Meet.* IAHR work group on "The behavior of hydraulic machinery under steady oscillatory conditions", Ljubljana 5-7 Sept.

Rihtarsic B., Oberdank K., Bajd M., Velensek B. (1993) Experimental flow field analysis in Francis turbine in presence of a vortex core, *Proc. 6th Int. Meet.* IAHR work group on "The behavior of hydraulic machinery under steady oscillatory conditions", Lausanne 8-10 Sept.

Sirok B., Ondracka H., Govekar E., Trdic F. (1993) Experimental description of the complex behaviour of a cavitated vortex core, *Proc. Hydrofractals '93,* Ischia 12-15 October.

VORTICES ROTATING IN THE VANELESS SPACE OF A KAPLAN TURBINE OPERATING UNDER OFF-CAM HIGH SWIRL FLOW CONDITIONS

L.PŮLPITEL, A. SKOTÁK, J. KOUTNÍK
Research engineers
ČKD Blansko, Czech Republic

Abstract

Vortices rotating in the vaneless space of a Kaplan turbine model have been observed under variety of off-cam sluicing conditions. The number of vortices was influenced by the discharge (guide vane opening) and the air admission as well. The phenomena observed could be described in terms known from wave analysis, like modes, stability limits, etc.

1. Introduction

It is well known that a Kaplan turbine operating under off-cam conditions as well as a propeller turbine operating out of the best efficiency point can be subject of severe rotor vibrations. Several publications dealing with this phenomenon have been presented in the past ([1] ÷ [5]). However, the relations between flow patterns and rotor vibrations still seem not to be clear enough. Based on the frequency analysis of both vibrations and pressure pulsations, flow disturbances rotating in the vaneless space of a Kaplan turbine operating under off-cam conditions were predicted [6]. In order to verify the predicted flow pattern, a special model research focused on observation of the flow arising in the vaneless space was carried out [7]. Results of both the observation and the analysis are summarized in this paper. Some hypotheses concerning the nature of flow patterns observed are included in the paper as well.

2. Model tests

The main parameters of the model are:

$H = (8 \div 22)$ m	$D = 400$ mm	$z_R = 4$
$Q = (180 \div 840)$ l/s	$n = (1000 \div 1700)$ rpm	$z_S = 28$

In order to allow observations of flow phenomena upstream of the runner, the model was provided with a transparent discharge ring - Fig. 2. However, the relatively large inlet part of the spiral case appeared to be an obstacle in observations of the vaneless space. Therefore, about a half of the vaneless space could only be observed. The draft

E. Cabrera et al. (eds.), Hydraulic Machinery and Cavitation, 925–934.

tube was provided with two windows. Further, aeration holes leading air into the gap between the runner hub and the head cover cone were provided. Using these holes it was possible to admit or to inject air into the vaneless space. In other words, the flow could take air anywhere on the periphery of the cone downstream end.

2.1. MEASUREMENTS

The model in a horizontal arrangement was operated under the actual head and actual cavitation conditions at a constant large runner blade opening. Several guide vane openings were adjusted during the model tests. The measurement started with the on-cam operation. Then the off-cam operation called sluicing followed. This means a no-load operation allowing a relatively high discharge. The quantities measured including their frequency symbols are given below. Location of some transducers can be seen in Figure 2.

	frequency symbol	
Axial thrust		f_R
Torque on the shaft		f_R
Bending stress in the shaft		f_R
Pressure pulsation in the vaneless space		f_{P1}
Pressure pulsation in the draft tube		f_{P2}

2.2. VORTEX NUMBER DETERMINATION

A stroboscope was used to "stop" the flow pattern appearing in the vaneless space. Then air was admitted or injected into the model to make the flow pattern more visible. Because the angular frequency of the flow pattern in the vaneless space was not known, it was not possible to adjust the stroboscope in advance. In order to avoid a false vortex number determination, simultaneously with the observations, the following rotor vibration analysis was applied to determine the number of circumferential disturbances in another way.

Let us assume that the runner is rotating in the flow field sketched in Figure 1, however, with a different angular frequency than the flow pattern does.

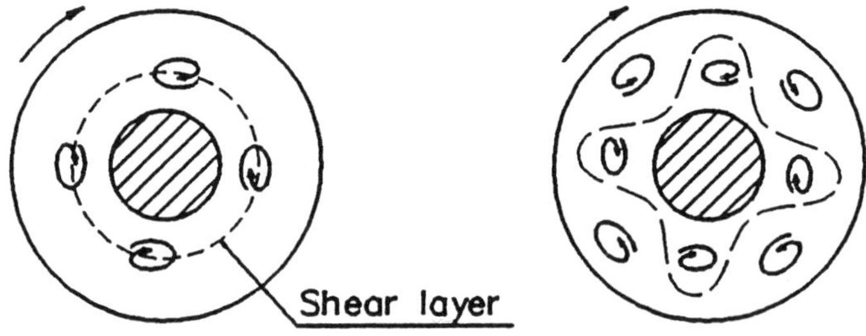

Figure 1. Scheme of a) 4 rotating vortices b) 4 waves

The rotating disturbances induce pressure variations on the stator as well as dynamic loading of the runner blades. The relations among the corresponding frequencies can be summarized in the following way:

$$f_P = z_D \cdot f_0 \tag{1}$$
$$f_B = z_D \cdot (f_n - f_0) \tag{2}$$
$$f_B + f_P = z_D \cdot f_n \tag{3}$$

The resulting mode of rotor vibration forced by the flow disturbances is given by the following criterion well known from rotor dynamics:

$$N = k \cdot z_R - j \cdot z_D \qquad (N = 0, \pm 1) \tag{4}$$

If N = 0, both axial and torsional rotor vibrations are forced. If N = 1, bending rotor vibration or whirling is induced. The relation between the frequency of rotor vibrations and the frequency of the blade excitation is given by Eq.(5), where j should satisfy Eq.(4).

$$f_R = j \cdot f_B \tag{5}$$

Thus, using Eqs (3)÷(5) the number of disturbances can be determined.

3. Results

3.1. FLOW PATTERNS

The flow phenomena observed in the vaneless space can be classified in the following way.

Flow Pattern I. This flow pattern corresponds to the on-cam conditions and is free of any visible disturbances.

Flow Pattern II (sluicing). The back flow core visualized by air can be observed upstream of the runner - Fig. 3. This back flow zone seems to be "squeezed" to the cone and it is of an axisymmetric nature. There are signs that a ring vortex exists on the face of the back flow region. In the axial direction some wave-like modulations of the core can be sometimes seen.

No rotating vortices are present in the vaneless space. There are only vortices shedding from runner blades. The first vortex arises at the vicinity of the corner between the blade leading edge and the runner hub. This vortex is travelling along the suction blade side. The second vortex starts on the blade tip and travels above the blade pressure side. Both vortices remember the flow pattern known from swept aircraft wings [8].

Flow Pattern III (sluicing). Three vortices are rotating in the vaneless space - Figure 4. The vortices seem to be "attached" to the cone and they take practically all air

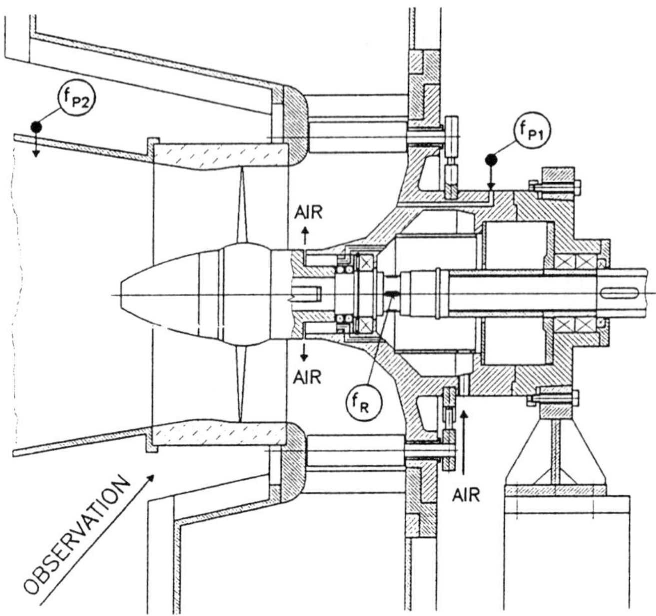

Figure 2. Scheme of the model

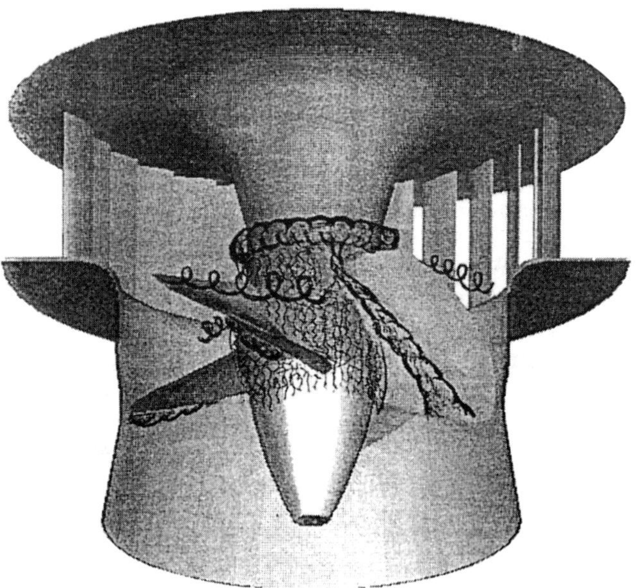

Figure 3. Sketch of flow pattern II (The flow pattern downstream the runner is not sketched)

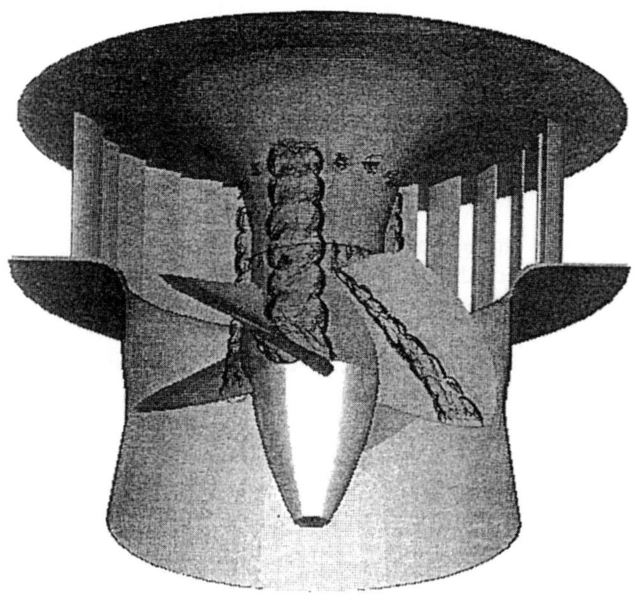

Figure 4. Sketch of flow pattern III (The flow pattern downstream the runner is not sketched)

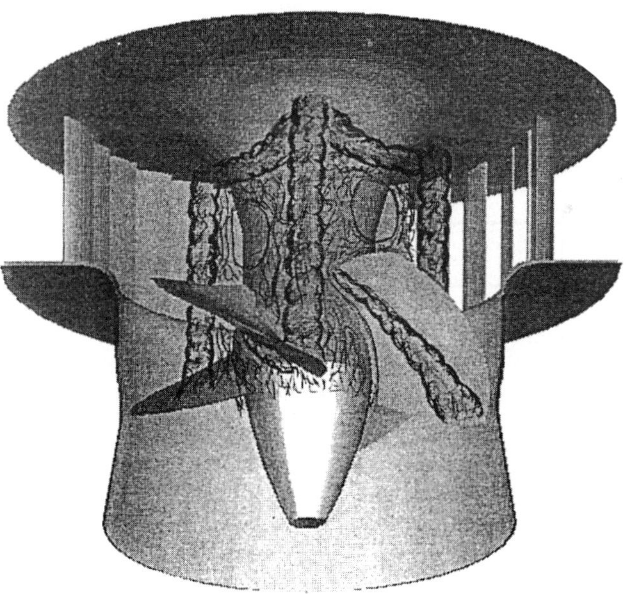

Figure 5. Sketch of flow pattern IV (The flow pattern downstream the runner is not sketched)

admitted into the model. The space between these vortices is free of air bubbles and therefore the cone surface is clearly visible. Air bubbles, if any, are escaping into those "vortex free" regions downstream from the aeration gap. There are some signs that a ring vortex exists in the upstream part of the vaneless space close to the distributor outlet. Because the angular speed of the flow pattern differs from that of the runner rotation, the runner blades are passing the vortices. The origin of the vortices could not be clearly determined.

Flow Pattern IV (sluicing). There are four vortices rotating in the vaneless space - Figure 5. They are "connected" in their upstream part with a "tetragon-shaped" circumferential vortex. The four vortices are not attached to the cone as in the case of flow pattern III. Among the vortices there is a sheet visualized by air bubbles. The sheet is of a tetragonal shape and it has four "windows" through which the cone surface can be seen.

Flow Pattern V (sluicing). Several vortices are arising in the vaneless space in an apparently random manner and the flow pattern picture cannot be stopped using a stroboscope.

3.2 FLOW CONDITIONS

Under no-load flow conditions the speed of the free rotating runner seems to be a parameter which could well characterize the swirling flow in a Kaplan turbine. Therefore, the reciprocal Rossby number Ro was taken into consideration as a swirl parameter which appears to be related to the well known swirl rate M - see [4].

$$1/Ro = \pi \cdot D \cdot f_n / v \tag{6}$$

Flow conditions involving the swirl rates under which the flow patterns mentioned above could be observed are summarized in Table 1.

TABLE 1. Flow conditions

TYPE	a_0 mm	H m	Q l/s	n rpm	Q_{air} l/s	M -	1/Ro -
I	36	8	840	1100	-	-	-
II	16÷18	16	620÷680	1450÷1710	0÷5	1.16÷1.31	5.9÷6.3
III	6÷14	16÷22	200÷600	1000÷1650	0÷10	1.49÷2.93	6.8÷10.8
IV	9, 2	20	375÷380	138÷1460	10÷20	2.14	9.2
V	14	20	580	1650	-	1.49	6.8

3.3. FREQUENCY OF PRESSURE PULSATIONS

The frequency spectra of pressure pulsations measured in the vaneless space can be categorized into two groups. The spectra possess:

- only one clearly predominating frequency peak (without harmonics) corresponding to the flow pattern III or IV - Figure 6a,b,d.

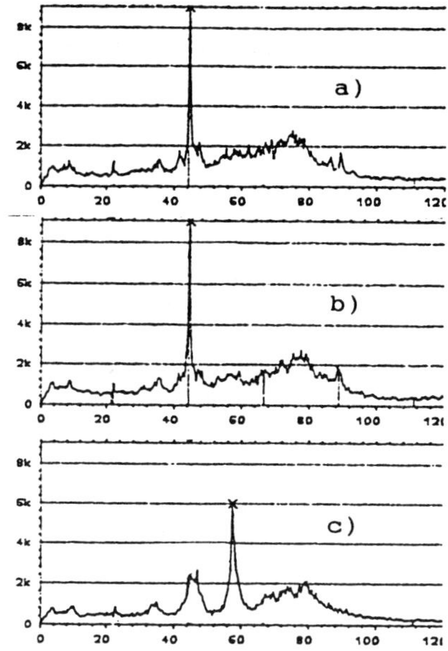

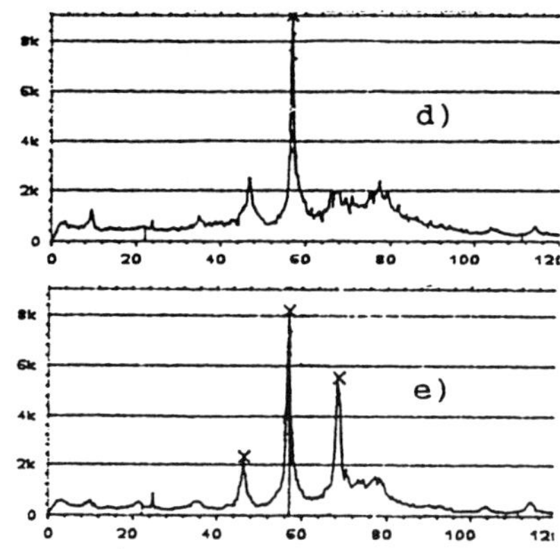

Figure 6. Spectra of pressure pulsations measured in the vaneless space - air effect

	a	b	c	d	e
Q_{air}[l/s]	0	5	10	15	20

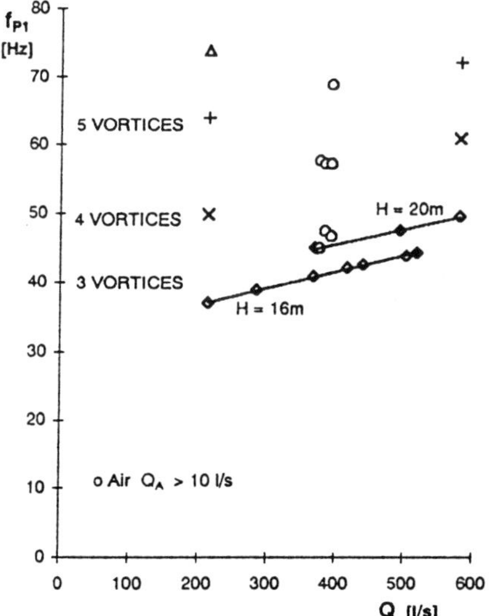

Figure 7. Frequency f_{P1} vs. discharge Q

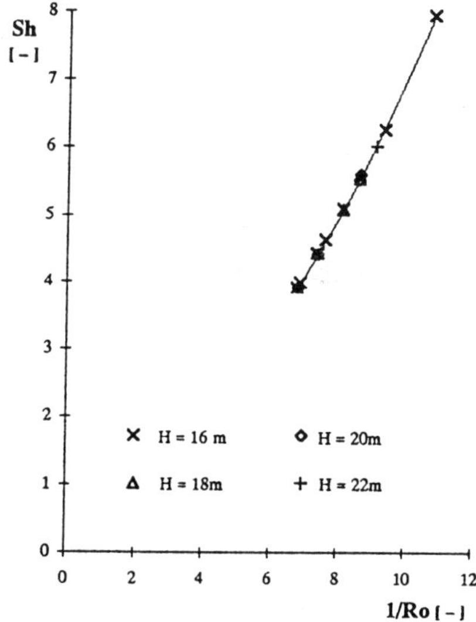

Figure 8. Sh vs. 1/Ro for flow patern III

- two or several frequency peaks (not harmonics) corresponding to the flow pattern V - Figure 6c,e.

The frequency of pressure pulsations measured in the vaneless space depends on the head, discharge and number of vortices - Figure 7. Using the relation Sh vs. 1/Ro the head influence can be eliminated - Figure 8.

3.4. AIR EFFECT

A low amount of air makes flow patterns more visible. Relatively high amount of air, however, can cause a jump from one flow pattern to another one. Figure 6 shows this air effect clearly in frequency spectra of pressure pulsations measured in the vaneless space. The flow pattern III characterized by one frequency peak - Figure 6a - was converted by air into pattern V having 3 frequency peaks - Figure 6e.

3.5. FLOW PATTERNS IN THE DRAFT TUBE

The observation of flow patterns in the draft tube is very difficult. The vortices shedding from the blades mask the whole flow pattern. Based on the frequency analysis of pressure pulsations measured in the draft tube some rotating disturbances of another nature as mentioned above can be expected [6].

3.6. ROTOR VIBRATIONS

As mentioned before, if N = 1, the bending mode of rotor vibrations is forced. Thus, in the case of flow pattern III (3 vortices, j = k = 1, N = 1) the rotor whirls in its rotation direction and the frequency of rotor vibration measured on guide bearings is relatively high (see the following formula in which k, j should satisfy Eq.(4)

$$f_S = N \cdot f_R + f_n = N \cdot (k \cdot z_R \cdot f_n - j \cdot f_P) \qquad (7)$$

and therefore the rotor resonance phenomena can play an important role here. In the case of flow pattern IV (4 vortices, j = k = 1, N = 0) both the axial and the torsional rotor vibrations are forced, however, the rotor resonance phenomena are not to be expected, because the natural frequencies of axial or torsional vibrations are usually higher than the first bending natural frequency. This feature was utilized on the prototype where the flow pattern III was converted into IV using air admission.

4. Discussion

The phenomena observed could be described in terms known from hydrodynamic oscillator theory, such as modes, stability limits, etc. If w, respectively m are taken into consideration as wave numbers in axial, respectively azimuthal directions, the flow patterns can be described as (w, m) wave modes where m is known and w is assumed to be zero

II (0,0) III (0,3) IV (0,4) V (0,3 + 0,4 + 0,5)

Using terms mentioned above, two hypotheses explaining the flow phenomena observed in the vaneless space have been developed.

Hypothesis A. When increasing swirl rate the back flow penetrates along the head cover cone deep into the vaneless space upstream of the runner. The back flow core reaches the head cover and then it expands. At certain critical core radius the mode (0,3) having 3 rotating vortices arises and this one predominates. Under certain discharge conditions several modes are unstable and they are arising in an apparently random manner. A large amount of air injected into the vaneless space changes the stability conditions and it leads to a jump from one mode into another one, e.g. from (0,3) to (0,4).

Thus, the features described above seem to be typical signs of the well known shear layer instability ([8], [9]). It means that the shear layer between the back flow core and the downstream flow becomes unstable - Figure 1a. However, using this hypothesis it is difficult to explain some other features observed, such as

- the ring vortex connecting the rotating vortices,
- pure harmonic character of pressure pulsations (frequency peaks without harmonics).

Hypothesis B. Similarly to the hypothesis A the back flow core becomes unstable under certain swirl rate. However, in this case the whole flow pattern changes its nature. There are vortices in which the water (mixed with air) is running upstream (back flow) and there are vortices (unvisualized) in which the water is running downstream. It means, that to each with air visualised vortex an adjacent unvisualized one takes place. This downstream vortex can not be visualised by air due to high pressure. Using terms known from meteorology we could speak about cyclones (lows) and anticyclones (highs).

Thus, the features mentioned above resemble Rossby waves ([8], [10]) which can be described as (w, m) modes as well - Figure 1b. The tetragon-shaped circumferential vortex could play the role of the jet stream. In addition, waves can produce pressure pulsations of the harmonic nature having spectra with frequency peaks without harmonics.

5.Conclusions

Rotating flow patterns having several vortices originate in the vaneless space of a Kaplan turbine operating under off-cam high swirling flow conditions. The features of these flow phenomena can be described in terms known from wave analysis like modes, stability limits, etc. Thus, the flow patterns observed can be explained as the shear layer instability or waves. These rotating flow disturbances can induce severe rotor vibrations.

6. References

1. Elder, R.A., Dale, J.M., Vigander, S.: Vibrations in a fixed-blade hydraulic turbine unit, *IAHR Symposium, Lausanne* 1968, Paper G2.
2. Eichler, O.: Vibration phenomena on hydraulic axial turbines, *UITAM/IAHR Symposium, Karlsruhe* 1979, Paper B3.
3. Sotnikov, A.A.et al.: Vibrations and pressure pulsations in Kaplan and propeller hydroturbines and their model study, *IAHR Symposium, Amsterdam* 1982, Paper 42.
4. Bettochi, R., et al.: Experimental analysis of the flow in the axial region of propeller - turbine admission ducts, *La Houille Blanche No 7/8* - 1982, pp. 599 ÷ 606.
5. Guarga, R. et al.: Diagnosis of the cause of mechanical vibrations in 135 MW Kaplan turbines at partial load operation, *XXIV IAHR Congres, Madrid* 1991, Paper D-181÷190.
6. Půlpitel, L.: The dynamic behavior of a Kaplan turbine operating under non-standard conditions, *6th International Meeting of the IAHR WG1, Lausanne* 1993, Paper 6-4.
7. Půlpitel, L., Skoták, A., Koutník, J.: Observations of vortices rotating in the vaneless space of a Kaplan turbine model operating under off-cam conditions, *7th International Meeting of the IAHR WG1, Ljubljana* 1995, Paper G - 3.
8. Lugt, H.J.: *Vortex Flow in Nature and Technology,* John Wiley and Sons, Publ., New York 1983.
9. Doerfler, P.: Stability analysis by linear perturbation method for a 2-region vortex model, *2nd International Meeting of the IAHR WG1, Mexico City* 1985, Paper 17.
10. Chen, Y.N. et al.: The Three-Layer-Wave Structure of the Rotating Stall Cell, *VDI Berichte* Nr. 947, (1992), pp.107-32.

List of symbols:

a_0 .... guide vane opening
D .... runner diameter
f_B ....frequency of runner blade excitation
f_P ....frequency of pressure pulsations
f_R .... frequency of rotor vibrations measured on the shaft
f_S .... vibration frequency measured on guide bearings
f_0 .... frequency of flow pattern rotation
f_n .... rotor rotational speed in rps
j, k .... integers, usually $j = 1, 2, 3$; $k = 1, 2$
M .... swirl rate - see [4], [7]
m .... azimuthal wave number
N .... see Eq.(4)
Q, Q_{air} ..... water discharge, air rate
Ro. .... Rossby number, see Eq.(6)
Sh .... Strouhal number $f_P.D/v$
v .... axial component of flow velocity (mean value)
w .... axial wave number
z_R, z_D ...... numbers of runner blades, disturbances

EXPERIMENTAL INVESTIGATION OF FREQUENCY CHARACTERISTICS OF DRAFT TUBE PRESSURE PULSATIONS FOR FRANCIS TURBINES

SHI QINGHUA
Dong Fang Electrical Machinery Co., Ltd.
China

Abstract

This paper presents an experimental investigation of frequency characteristics of draft tube pressure fluctuations for Francis turbines. The different types of vortex ropes in draft tubes were experimentally observed and categorized. The relationship between the surging frequency and the vortex pattern is discussed. The frequency characteristics of pressure pulsations at different measuring locations of the draft tube are also presented.

1. Introduction

Draft tube pressure pulsations for Francis turbines continue to be an important research subject in the hydroelectric industry. For large hydroelectric plants, a remarkable trend is that the stability and safety of operation must be guaranteed in a wide range of operation, even at certain highly abnormal operating conditions such as very low heads and/or extremely small wicket gate openings. The guarantees of pressure pulsations are usually required to be achieved by scale-down model acceptance tests. In some cases, the frequency characteristics of pressure fluctuations are more critical than surging amplitudes because, once the surging frequencies coincide with the natural frequencies of some hydraulic structure components, the resonance will not be avoided.

This paper presents an experimental investigation of frequency characteristics of draft tube pressure pulsations for Francis turbines. The different types of vortex ropes in draft tubes are experimentally observed and categorized. The relationship between the surging frequency and the

E. Cabrera et al. (eds.), Hydraulic Machinery and Cavitation, 935–944.

vortex pattern is discussed. The surging frequencies at different measuring sections of the draft tube are also shown.

2. Test Rig and Measuring Arrangement

The experimental investigation was carried out on a closed-cycle testing rig with a maximum test head of 10m in author's hydraulic research laboratory. The test rig was equipped with a computer-based data acquisition and processing system. The flow rate through the vertical model Francis turbine was measured with Venturi nozzle and the rotational speed was recorded with a digital pulse counter. The power output of the model turbine was measured with a DC electric torque reaction dynamometer. During the test, the test head, the water temperature and air content in the testing circuit remained constant.

The pressure pulsations in the draft tube were measured with a capacitance type pressure transducer mounted on the draft tube wall. The signals from the pressure transducer were processed by computer and the spectrum analysis of pressure fluctuations was carried out with Fast Fourier Transformation (FFT) program. The visual observation of vortex ropes in the draft tube was simultaneously conducted during the test. To observe and photographically record the vortex ropes, the draft tube cone was made of transparent organic glass.

Two model Francis turbines were tested: one with a specific speed of n_s=180m.kW and with a runner diameter (at the intersection of runner band with vane leading edge) of D=250mm; and the other with a specific speed of n_s=170m.kW and also with a runner diameter of D=250mm. The measuring locations for pressure fluctuations are illustrated in Figure 1. The draft tubes of the two model Francis turbines are of different geometrical design parameters.

3. Vortex Pattern

Depending on operating conditions, the vortex rope in the draft tube of a Francis turbine appears with different shapes. Although the vortex shape and the magnitude of pressure pulsations are also related to the suction pressure of a turbine installation and the air content in water, the vortex shape more strongly depends on operating conditions from the viewpoint of the internal flow mechanism of fixed blade type turbines.

Some researchers have simply divided the vortex ropes in draft tubes into two categories, stable ones and unstable ones [1,2]. However, our experimental investigation into the vortex cores in Francis turbine draft tubes shows that the vortex ropes are of various types, especially for the unstable vortices. From our visual observation and photographical recording of vortex ropes, the vortex ropes can be generally divided into the four characteristic types. These vortex ropes are designated as Types I, II, III and IV, as shown in Figure 2.

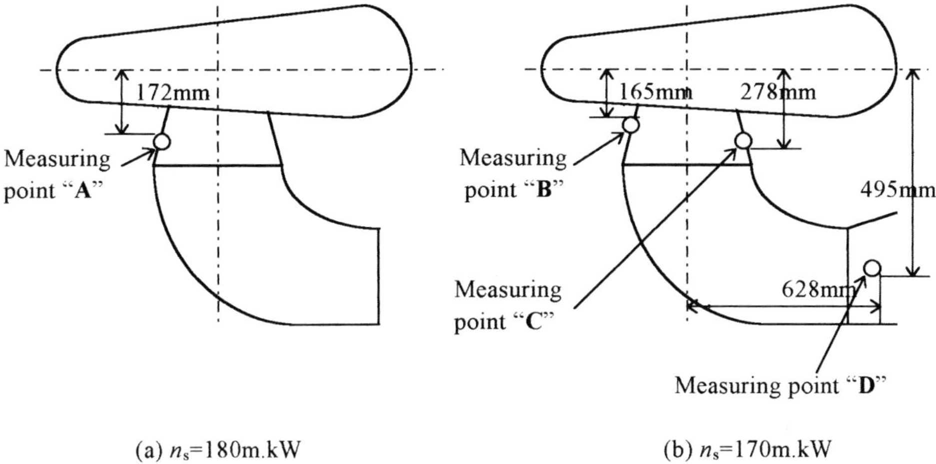

Figure 1. Measuring locations for pressure fluctuations

The Type I vortex rope (Figure 2a) is typical cork screw vortex rope and its shape is somewhat like a twisty snake. This helical vortex rope itself, whose upper end is attached to the runner cone near the turbine axis, rotates around the axis of the vortex core, and towards the downstream side it rotates, with an eccentricity away from the turbine axis, around the turbine axis in the direction of runner rotation. The Type I vortex rope usually results in low frequency pressure pulsations in the draft tube and it is the main excitation source of turbine oscillations and power output swings of the unit in the partial load range.

The Type II vortex rope (Figure 2b) appears somewhat as a round handbag and its upper end is attached to the runner cone near the turbine axis. This vortex rope has a relatively small eccentricity and rotates around the turbine axis in the direction against the runner rotation. This type of vortex core can distinctly be divided into two portions. The upper portion of the vortex rope has a larger diameter while the diameter of the lower portion is relatively small. Because of the large vortex core diameter and vertical swift movement, this vortex rope usually results in large low frequency pressure pulsations. The Type II vortex rope exists in the overload operating range and has a large impact on the overload operation of Francis turbines.

The Type III vortex rope (Figure 2c) is attached to the runner cone in the vicinity of the turbine axis at its origin. This vortex rope is composed of many randomly dispersed vortex filaments and appears somewhat as a sheaf of goatee. The Type III vortex rope is intrinsically unstable. However, this vortex rope usually induces relatively small pressure fluctuations since its appearance is random and dispersed.

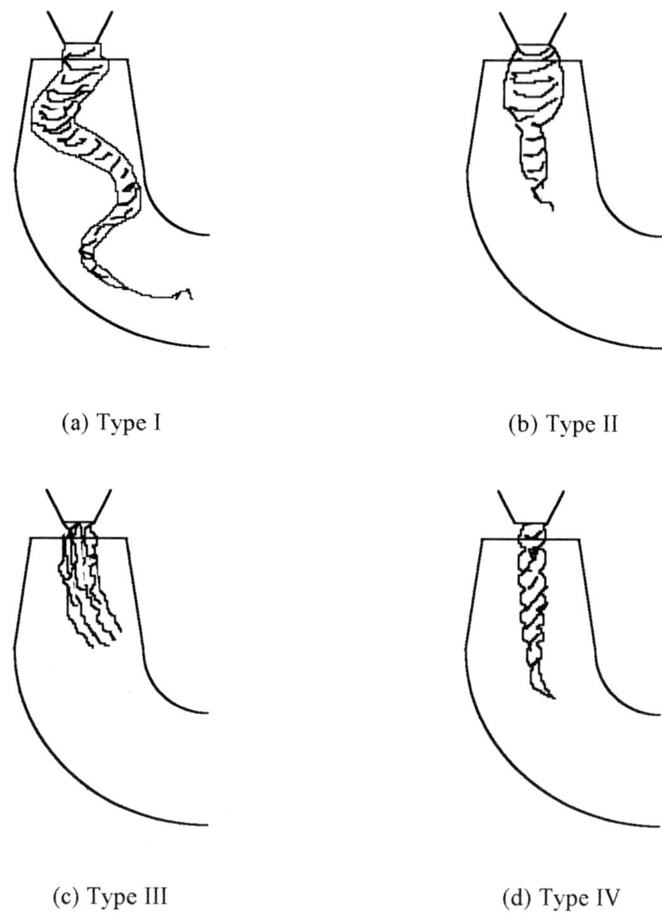

(a) Type I (b) Type II

(c) Type III (d) Type IV

Figure 2. Various types of draft tube vortex ropes

The fourth type of vortex rope (Figure 2d) is typical straight vortex. The upper end of this vortex rope is also attached to the runner cone near the turbine axis and its shape is somewhat like a fluted twist drill. The Type IV vortex rope is stable and scarcely results in low frequency pressure pulsations. Therefore, this vortex rope has a small impact on the operation of Francis turbines.

Figure 3 shows the efficiency hill diagram of the model Francis turbine with n_s=180m.kW, in which the operating zones with different types of vortex ropes are shown. In addition, the operating range free of visual vortex rope in the draft tube is also illustrated.

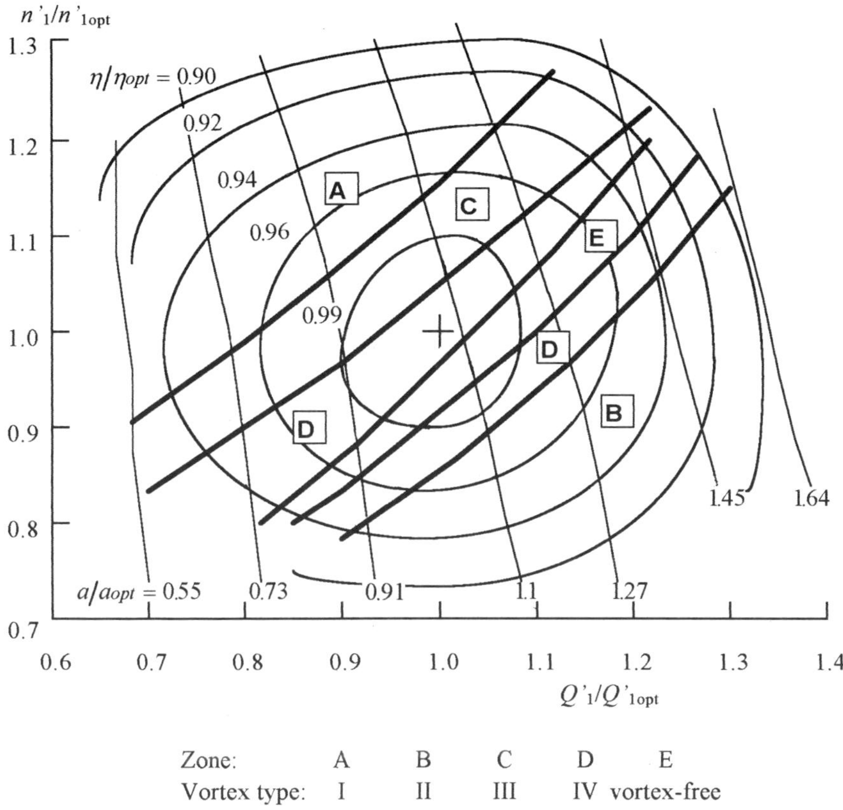

Figure 3. Efficiency hill diagram of the Francis turbine with n_s=180m.kW

4. Relationship between Surging Frequency and Vortex Pattern

The frequency of the draft tube pressure fluctuations measured in the draft tube changes corresponding to each type of the vortex ropes. In other words, different types of vortex ropes possess different dominant surging frequencies.

The model test results show that pressure fluctuations in Francis turbine draft tubes have a wide range of dominating surging frequencies. If the pressure surging frequency and the rotational frequency of the runner are represented by f_v and f_r respectively, the ratio f_v/f_r ranges from 1/20 to 6 for Francis turbine with n_s=180m.kW at the measuring point "**A**" as shown in Figure 1. Depending on the vortex pattern, the surging frequency ranges of the above-mentioned four types of draft tube vortex ropes for the turbine with n_s=180m.kW are illustrated in Table 1.

TABLE 1. Surging frequency ranges of different types of draft tube vortex ropes for the Francis turbine with n_s=180m.kW at the measuring point "**A**"

Vortex type	f_v/f_r
I	1/6 ~ 1/4
II	1/20 ~ 1/4
III	1/3 ~ 1
IV	1/3 ~ 1

The pressure fluctuations induced by the Types I and II vortex ropes are typical low frequency ones. Hazardous pressure fluctuations which may cause severe vibration of turbine, powerhouse structure and/or power swing is mostly induced by these two types of unstable vortex ropes. It can be seen from Table 1 that the surging frequencies of these two vortex ropes are less than the statistical value (f_v/f_r=1/3.6) by Rheingans [3].

The pressure fluctuations induced by the Types III and IV vortex ropes mostly belong to medium frequency ones. Usually, since the pressure surges induced by these two types of vortex ropes are of relatively small oscillating amplitudes, those will not cause serious operational problems.

It should be noted that the pressure pulsation frequency is strongly related to the suction pressure (plant sigma) of a turbine installation. The test results indicate that when the plant sigma is much less than the critical sigma of the turbine, the vortex core grows larger and it takes the form of cavitating dead water core by being filled up inside of the draft tube with turbulence water. A large number of tiny turbulent unsteady vortices are contained in the cavitating dead water core and the dominant frequencies of pressure fluctuations in the draft tube are considerably high. In this case, f_v/f_r is up to 6 for the Francis turbine with n_s=180m.kW.

Figure 4 shows the pressure pulsation frequencies versus the unit discharges at three unit speeds of n'_1/n'_{1opt}=0.8, 1.0, 1.2, measured at the measuring point "**A**" for the Francis turbine with n_s=180m.kW.

The dominating frequencies of draft tube pressure fluctuations for Francis turbines can be classified as four classes, very low frequency, low frequency, medium frequency and high frequency. These frequency classes are respectively designated as Classes A, B, C and D, and their corresponding frequency ranges are illustrated in Table 2.

The Class A frequency rarely appears and occasionally, it is induced by the Type II vortex rope occurring at certain particular conditions. The Class

B frequency is mainly induced by the Types I and II vortex ropes and the draft tube pressure pulsations with this frequency class may cause serious operational hazard.

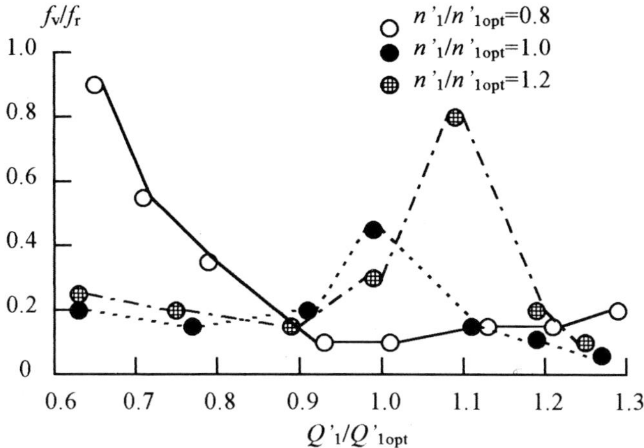

Figure 4. Frequencies of draft tube pressure fluctuations measured at the measuring point "**A**" for the Francis turbine with n_s=180m.kW

The Class C frequency is mainly induced by the Types III and IV vortex ropes, accompanied usually by very small pressure fluctuations. The Class D frequency is caused by the cavitating dead water core filled up inside of the draft tube with turbulent water. Although it is small and appears as high frequency oscillation, the draft tube pressure fluctuations with the Class D frequency may cause high frequency harsh noise.

TABLE 2. Classification of dominating frequencies of draft tube pressure fluctuations

Frequency class	f_v/f_r	Remarks
A	< 1/10	very low
B	1/10 ~ 1/3	low
C	1/3 ~ 1	medium
D	> 1	high

Figure 5 shows the probability analysis for different frequency classes of draft tube pressure fluctuations, measured at the measuring point "**A**" for all testing operating conditions of the Francis turbine with n_s=180m.kW. It can be seen that the probability with Class B is much larger than that with Class

A, C or D. The low frequency pressure fluctuations in the Francis turbine draft tube appear more frequently.

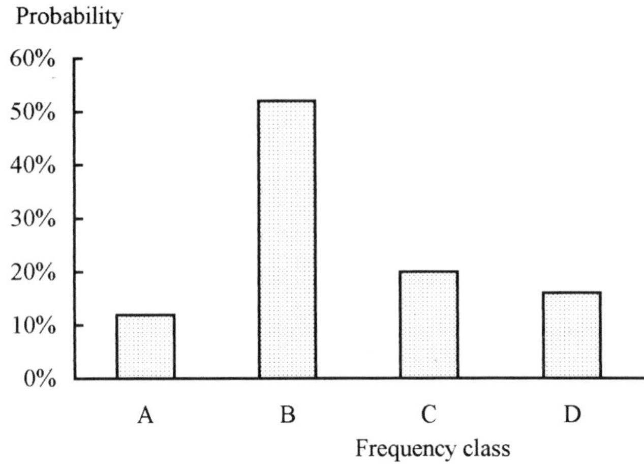

Figure 5. Probability analysis for frequency classes of draft tube pressure fluctuations for the Francis turbine with n_s=180m.kW

5. Surging Frequencies at Different Measuring Locations

The measurement for the pressure fluctuations at different locations of the draft tube was carried out in order to investigate the behavior of draft tube vortex ropes and the influence of the pressure recovery in the draft tube on the frequency characteristics of pressure pulsations. The measurement was conducted on the Francis turbine with n_s=170m.kW and three measuring points were arranged, as shown in Figure 1.

Figure 6 shows the frequency characteristics of the pressure fluctuations at the above-mentioned three measuring points under two unit speeds of n'_1/n'_{1opt}=0.86, 1.14. It can be observed that for the same operating condition, the frequencies of the draft tube pressure pulsations measured at different measuring points differ from one another. Moreover, it seems that the differences are related to the vortex pattern in the draft tube. When the vortex rope in the draft tube is Type II, the frequencies of the pressure pulsations induced by it increase toward the downstream side of the draft tube. The increase in surging frequency may be caused by the recovery of pressure and collapse of cavities in the vortex core downstream of the draft tube. On the other hand, when the vortex rope in the draft tube takes a pattern of Type I, the surging frequencies decrease toward the downstream side. Usually, the Type I vortex rope stretches toward the downstream side of the draft tube, up to the outlet of the elbow tube. In this case, the reduction in surging frequency may be caused by the resistance of water viscosity to the vortex rope procession. Nevertheless, the test results show that the amplitudes of pressure fluctuations at the measuring point "**A**" are larger than those at the measuring points "**B**" and "**C**".

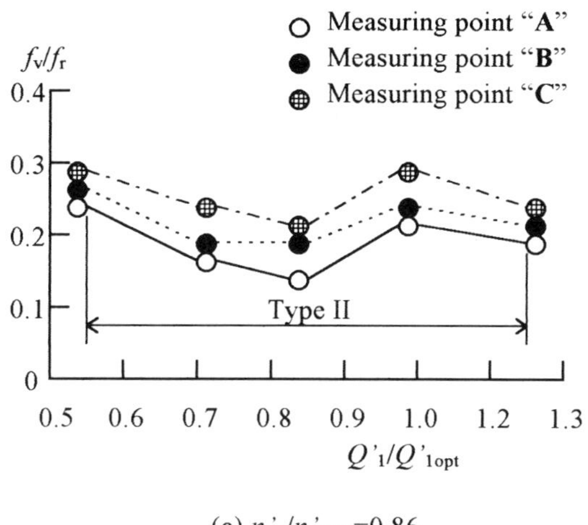

(a) n'_1/n'_{1opt}=0.86

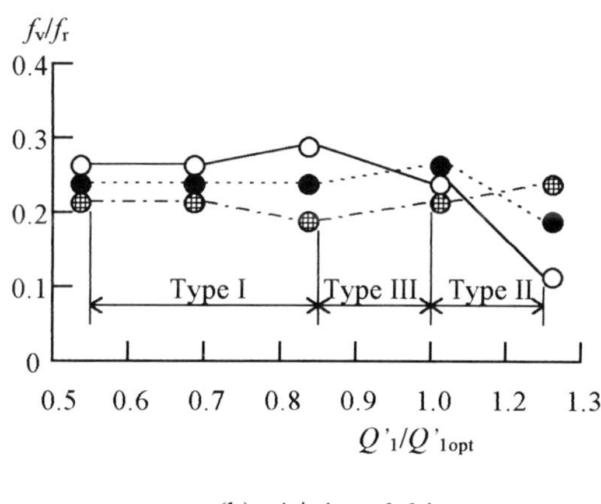

(b) n'_1/n'_{1opt}=1.14

Figure 6. Frequency characteristics of pressure fluctuations measured at different measuring points for the Francis turbine with n_s=170m.kW

6. Conclusions

An experimental investigation into the frequency characteristics of Francis turbine draft tube pressure fluctuations was carried out. The followings are emphasized as conclusions.

1) The vortex ropes appearing in Francis turbine draft tubes can generally be divided into four different characteristic types. Different types of vortex ropes are of different surging frequency characteristics.

2) The typical low frequency pressure fluctuations in Francis turbine draft tubes have a dominant frequency range of f_v/f_r=1/20 ~ 1/4. These pressure fluctuations are related to the unstable vortex ropes occurring in the partial load and overload operating ranges of Francis turbines.

3) At the same operating condition, the frequencies of pressure fluctuations at different draft tube sections differ from one another. The differences may be dependent on the vortex rope types in Francis turbine draft tubes.

The reduction or elimination of the pressure fluctuations in fixed blade type turbine draft tubes is an important industrial task. Up to now, general criteria to judge tolerable degree for each type of vortex rope are not established yet. Therefore, general countermeasures to reduce the vortex intensity or to stabilize the vortex whirl need to be taken.

The draft tube pressure fluctuations may cause severe low frequency vibrations, subjecting the machine components to undue cyclic, alternating stresses. It is more critical to avoid resonance of draft tube pressure pulsations with the natural frequency of the machine components. The frequencies of draft tube pressure pulsations and natural frequencies of the machine components should be adequately separated to be sure of avoiding resonance. Usually, 25 percent separation can provide an adequate margin. Additionally, once the amplitudes of the draft tube pressure pulsations have been determined, design of the machine components should consider the worst cyclic loading.

Acknowledgments

The author wishes to express his gratitude to the Hydraulic Turbine Department of Dong Fang Electrical Machinery Co., Ltd. for supporting this research, with special thanks to Mr. Zhang Fuqin, vice chief engineer and Mr. Peng Junjian, senior hydraulic engineer for their contribution.

References

1. Tanaka, H.: Vibration at off-design operating range of Francis turbines, *Technical Information of Turbine for Three Gorges Hydroelectric Project* (1995), 1-15.
2. Fazalare, R.W.: Trends in selecting and procuring hydro turbines, *Water Power & Dam Construction* **4** (1986), 40-46.
3. Rheingans, W.J.: Power swings in hydroelectric power plants, *Trans. ASME* **62** (1940), 171-184.

ON THE SUPPRESSION OF COUPLED LIQUID/PIPE VIBRATIONS

A.S. TIJSSELING and A.E. VARDY
Department of Civil Engineering
University of Dundee
Dundee DD1 4HN
United Kingdom

ABSTRACT

The vibration of liquid-filled pipe systems can be caused by internal (pressure surges) or external (rotating machinery) sources, or a combination of both. Liquid pulsations can be diminished by devices like an air chamber, whereas pipe vibrations can be reduced by suitable supports. The performance of vibration suppression devices in the system under consideration can be investigated by numerical simulation. This is usually done with conventional waterhammer or pipe-stress computer codes. Present-day software also permits fully-coupled liquid-pipe analyses. The analysis of suppression devices and their interaction with the entire system is an important tool in design and trouble shooting.

The dynamic behaviour of a simple system consisting of one pipe, one liquid column and one suppression device is studied herein by numerical simulation and physical experiment. Liquid-pipe interaction mechanisms are taken into account. A closed water-filled steel pipe is excited by external impact. The suppression device is a short piece of plastic pipe. The closed system prevents rigid-column behaviour of the liquid relative to the pipe, so that the local device does not decrease *amplitudes* of vibration. It does, however, shift the natural frequencies of the system, which is an important means of vibration control. This simple and cheap device may find its application in reducing fatigue problems and in noise reduction.

Keywords: Fluid-structure interaction (FSI), waterhammer, pipe, vibration, suppression.

E. Cabrera et al. (eds.), Hydraulic Machinery and Cavitation, 945–954.

1. Introduction

Vibrations of liquid-transporting pipe systems can be caused by internal (waterhammer, turbulence) or external (earthquake, shaking machinery) sources. In both cases, pressure pulsations in the liquid interact with mechanical vibrations of the pipe walls. Fluid-structure interaction (FSI) phenomena in liquid-filled pipes have been extensively studied since the 1960s (Tijsseling 1996). In all studies, the emphasis has been on measuring and analysing the phenomenon rather than on assessing methods of reducing its consequences.

This paper presents initial results of an investigation into alternative methods of suppressing waterhammer-induced pipe vibrations. The vibrational behaviour of steel pipe systems can be changed greatly by a plastic pipe insert at the right location. The location and dimensions of the plastic insert should be determined from a coupled hydraulic and structural analysis. The aim is to move natural system frequencies away from exciting source frequencies and hence to diminish amplitudes of vibration.

A simple system, namely a single water-filled steel pipe, is investigated herein by physical experiment and numerical simulation. A plastic (ABS) pipe extension is used as a simple and cheap device to change the system's dynamic behaviour. In the experiment, the closed pipe is excited by external impact. In the simulation, a state-of-the-art FSI analysis is utilised.

2. Theory

Full details of the theory underlying the numerical simulations can be found elsewhere (Wiggert *et al.* 1985, 1987; Lavooij & Tijsseling 1991; Kruisbrink & Heinsbroek 1992; Tijsseling *et al.* 1996). A short description is given for guidance.

When a liquid-filled pipe is excited at some location, pressure waves in the liquid and stress waves in the pipe wall are generated and propagate away from the source of disturbance. The acoustic speeds of propagation are approximately

$$c_f = \left(\frac{K}{\rho_f}\right)^{\frac{1}{2}} / \left(1 + \frac{2RK}{eE}\right)^{\frac{1}{2}} \quad \text{and} \quad c_t = \left(\frac{E}{\rho_t}\right)^{\frac{1}{2}} \qquad (1, 2)$$

for pressure and *axial* stress waves, respectively. Section 8 (Nomenclature) gives the meaning of the symbols. Due to the FSI *Poisson coupling* mechanism, disturbances in the liquid pressure/velocity are accompanied by disturbances in the pipe axial stress/velocity and vice versa. In the method-of-characteristics (MOC) approach, this is expressed by the compatibility equations

Liquid

$$\frac{dP}{dt} \pm \rho_f c_f \frac{dV}{dt} - 2\nu \frac{\rho_f}{\rho_t} \{ (\frac{c_t}{c_f})^2 - 1 \}^{-1} \{ \frac{d\sigma}{dt} \mp \rho_t c_f \frac{d\dot{u}}{dt} \} = 0 \tag{3}$$

Pipe

$$\frac{d\sigma}{dt} \mp \rho_t c_t \frac{d\dot{u}}{dt} + \nu \frac{R}{e} \{ (\frac{c_t}{c_f})^2 - 1 \}^{-1} \{ \frac{dP}{dt} \pm \rho_f c_t \frac{dV}{dt} \} = 0 \tag{4}$$

which are valid along paths with directions $dz/dt = \pm c_f$, for Eq. (3), and $dz/dt = \pm c_t$, for Eq. (4), in the distance-time (*z-t*) plane. The equations (3) and (4) are simplified equations without gravity and friction terms. Water-hammer, Eq. (3), and steel-hammer, Eq. (4), equations remain for the hypothetical case where the Poisson ratio ν is zero.

When an acoustic wave reaches a pipe end or junction, it is (partly) reflected. A strong FSI *junction coupling* may then occur. For example, at an unrestrained, massless, closed pipe end, liquid and pipe variables are proportional to each other:

$$A_f P = A_t \sigma \quad \text{and} \quad V = \dot{u} \tag{5, 6}$$

and at an unrestrained, massless, axial junction of two pipes:

$$\{ A_f (V - \dot{u}) \}_1 = \{ A_f (V - \dot{u}) \}_2 \quad \text{and} \quad \{ P \}_1 = \{ P \}_2 \tag{7, 8}$$

$$\{ \dot{u} \}_1 = \{ \dot{u} \}_2 \quad \text{and} \quad \{ A_f P - A_t \sigma \}_1 = \{ A_f P - A_t \sigma \}_2 \tag{9, 10}$$

3. Experiment

The FSI test rig at Dundee University has been extensively used in experiments with: a single pipe (Vardy & Fan 1986, 1989; Tijsseling & Vardy 1996), multi-pipe systems (Fan & Vardy 1994; Vardy *et al.* 1996) and cavitation (Fan & Tijsseling 1992; Tijsseling *et al.* 1996).

The single pipe system has been extended with a short plastic (ABS) pipe (Fig. 1). The maximum allowable pressure for the ABS pipe (class T) is 12 bar. A static pressure of about 6 bar, combined with dynamic pressures of ±3 bar, is sufficient to prevent cavitation and over-pressures. The 0.25 m long ABS extension is screwed onto one end of the 4.50 m long steel pipe, which is struck axially at the other end by a solid steel rod moving at velocity 0.118 m/s. The impact is considered instantaneous and the rod exerts a constant force during the contact time of about 2 ms. Thereafter, the rod and pipe are separated (because of the arrival of stress waves from their remote ends).

Pipe and liquid vibrations are sensed with strain-gauges, piezo-electric pressure-transducers and a laser-Doppler vibrometer, at a sampling rate of 125 kHz. Table 1 gives the essential data on the apparatus.

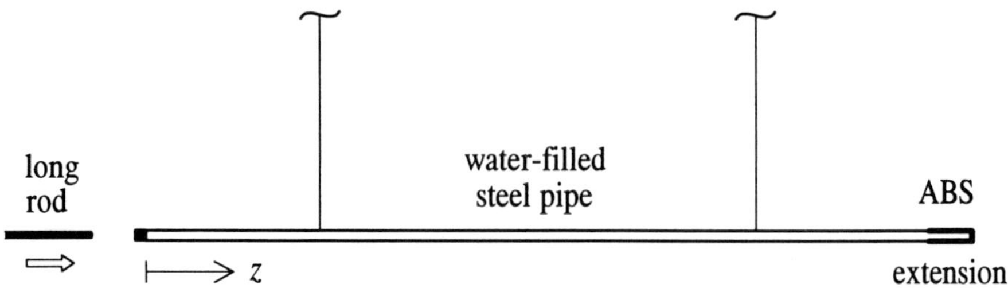

Figure 1. Schematic diagram of experimental apparatus.

Steel pipe	ABS pipe	Water
L = 4502 mm	L = 250 mm	K = 2.14 GPa
R = 26.01 mm	R = 22.8 mm	ρ_f = 999 kg/m^3
e = 3.945 mm	e = 7.4 mm	**Solid steel rod**
E = 168 GPa	E = 2.4 GPa	
ρ_t = 7985 kg/m^3	ρ_t = 1055 kg/m^3	L_r = 5006 mm
ν = 0.29	ν = 0.42	R_r = 25.37 mm
m_1 = 1.312 kg	m_2 = 0.3234 kg	E_r = 200 GPa
		ρ_r = 7848 kg/m^3

Table 1. Input data for simulations.

3.1. STATIC TESTS

The modulus of elasticity (2.4 $\pm$ 0.2 GPa) and the Poisson ratio (0.42 $\pm$ 0.02) of the ABS material were assessed at 22 °C from static compression/decompression measurements on an Instron 1196 test bench using a 200 mm long ABS pipe instrumented with three-way strain gauges at four circumferential positions midlength. During loading and unloading up to 0.4% strain, the test specimen showed almost linearly elastic behaviour (Fig. 2). It is noted that static tests cannot reveal any frequency-dependent behaviour.

The mass density (1055 ± 2 kg/m^3) and the average cross-sectional wall-area (1232 ± 5 mm^2) of the ABS pipe were obtained from immersion tests on an electronic weighing scale. The measured (with a vernier slide gauge) diameter (up to 0.1 mm differences) and wall-thickness (up to 0.2 mm differences axially and up to 0.4 mm differences circumferentially) of the ABS pipe were non-uniform. The relatively large circumferential variation in wall-thickness is assumed to partly explain the differences from average in Fig. 2. The mass of the ABS extension, including threaded thickening (at junction) but excluding end cap, was measured 0.5335 kg. In the numerical simulation, based on Table 1, a value of 0.325 kg was used (concentrated mass at junction neglected).

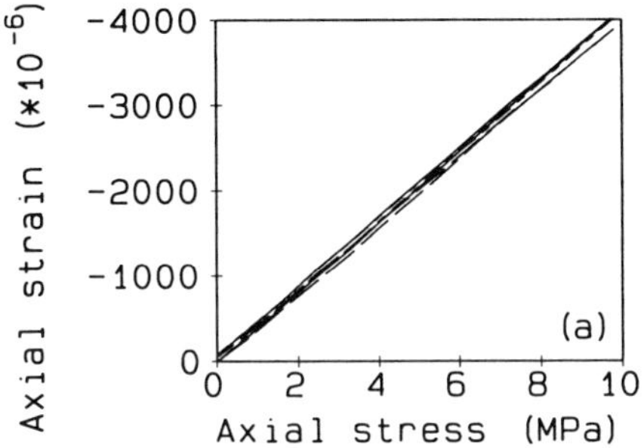

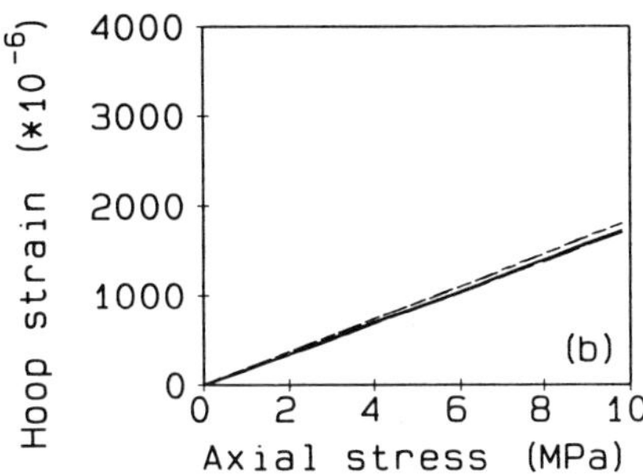

Fig. 2. ABS strain-stress relations obtained from static compression/decompression tests. Circumferential average and individual strains: (a) axial strains, (b) hoop strains.

4. Results

4.1. VALIDATION OF NUMERICAL MODEL

The numerical model outlined in Section 2 has been validated for water-filled steel pipes with closed ends and 90° junctions (see Section 3 for references). Typical results obtained in a 4.5 m long pipe are shown in Fig. 3. Two new elements in the present study are the *axial junction*, represented by the Eqs. (7-10), and the *plastic material* (ABS) of the pipe extension. The masses of the ABS pipe and its steel end cap (m_2) are of the same order of magnitude. The end cap, modelled as a lumped mass, slows down the axial ABS vibration significantly.

The complete system (Fig. 1) consists of two connected pipes, each with its own natural frequency and impedance. It is the liquid pulsations that make the problem interesting. They interact strongly with the vibrating pipes (FSI).

The theoretical wavespeeds in the system, according to the classical formulae (1-2), are c_f = 1354 m/s and c_t = 4587 m/s for the steel pipe and c_f = 574 m/s and c_t = 1508 m/s for the ABS extension. Due to Poisson-coupling-induced axial inertia forces in the pipe wall, the pressure wavespeeds, c_f, are slightly lower: 0.06% in the steel pipe and 0.7% in the ABS pipe (Stuckenbruck *et al.* 1985). Due to Poisson-coupling-

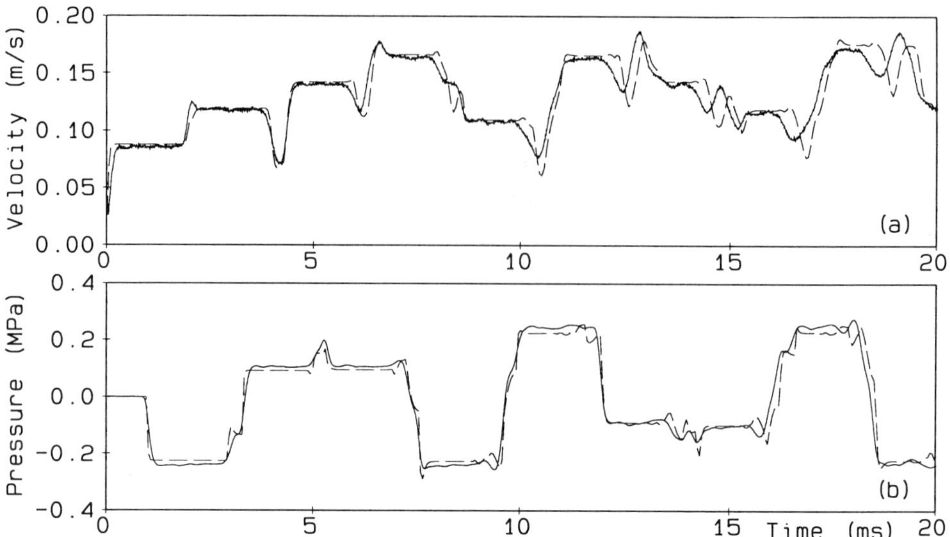

Fig. 3. Comparison calculation (- - -) / measurement (——) in 4.5 m long water-filled steel pipe: (a) axial pipe velocity at z = 0.047 m, (b) liquid pressure at z = 4.5 m.

induced pressure changes, the stress wavespeeds, c_t, are higher: 0.7% in the steel pipe and 8.9% in the ABS extension.

Calculated dynamic pressures, axial strains and axial velocities agree well with the measured transients (Fig. 4). The amplitudes and main frequencies of vibration of both the steel and ABS pipe are predicted accurately. The main frequency of the liquid pulsation is predicted slightly too high. This could result from a small mismatch in wavespeeds or from errors in the assumed effective mass in the junction and boundary conditions.

4.2. PERFORMANCE AS VIBRATION SUPPRESSION DEVICE

A practical and convenient way to assess the influence of plastic inserts on the dynamic behaviour of steel pipe systems is by numerical simulations with a validated computer code. In Fig. 5, numerical results for the pipe with ABS extension (Figs. 1 & 4) are compared with numerical results for a 4.75 m long entirely steel pipe (comparable to Fig. 3). It is seen that the ABS piece shifts natural frequencies and makes wave fronts less steep, both factors being most evident from the pressures in Figs. 5d & 5e, where the main frequency is significantly lower. The amplitudes of fluctuations are roughly the same in both systems, except in Fig. 5c, where the strains in steel are compared with the much larger strains in ABS.

The similarity of amplitudes with and without an ABS piece is a consequence of the short length of the ABS piece relative to the steel pipe. Simulations with a 2.25 m long

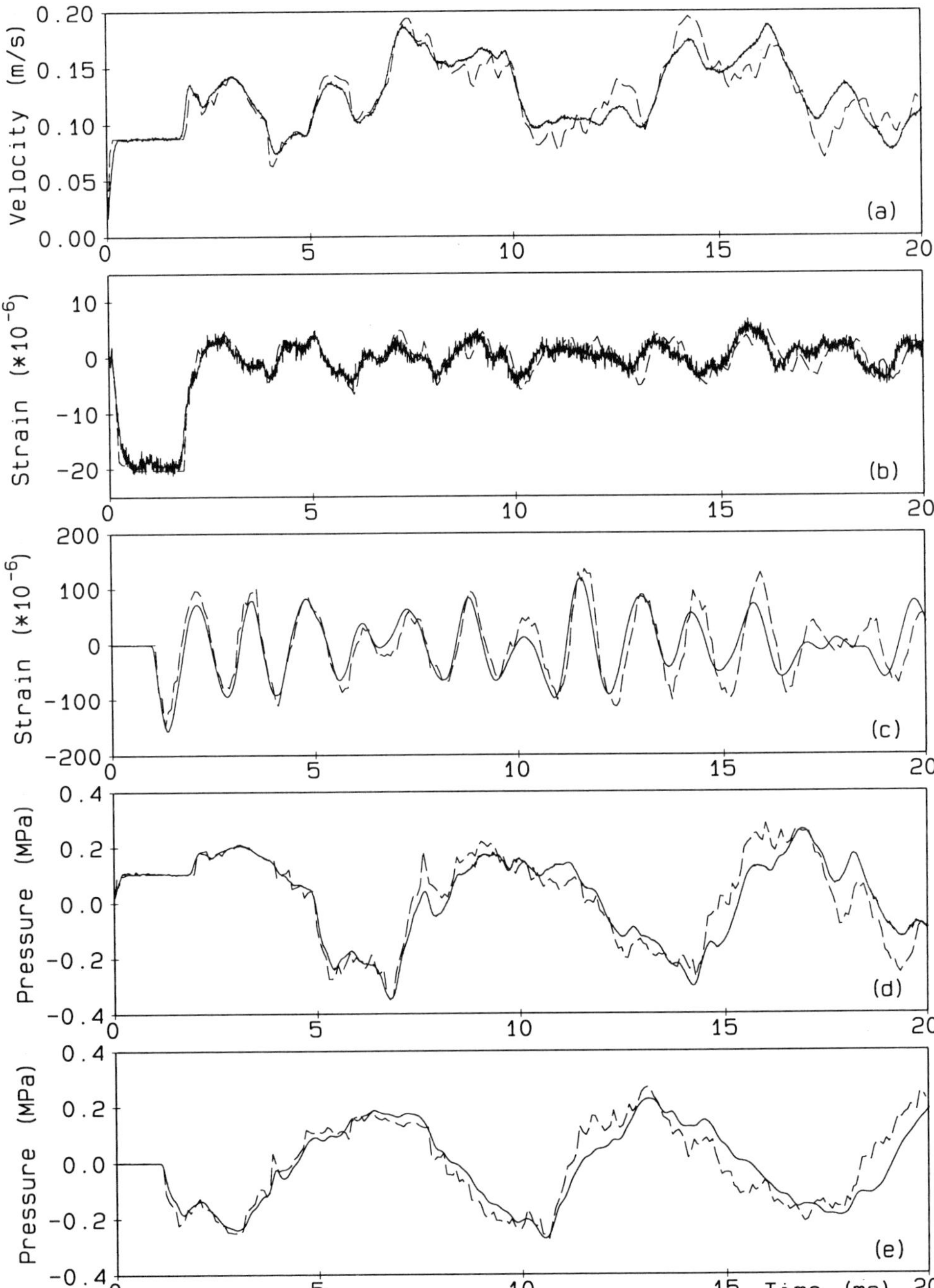

Fig. 4. Comparison calculation (- - -) / measurement (——) in 4.5 m long water-filled steel pipe with 0.25 m ABS extension: (a) axial (steel) pipe velocity at z = 0.047 m, (b) axial (steel) strain at z = 0.574 m, (c) axial (ABS) strain at z = 4.625 m, (d) pressure at z = 0.020 m, (e) pressure at z = 4.750 m.

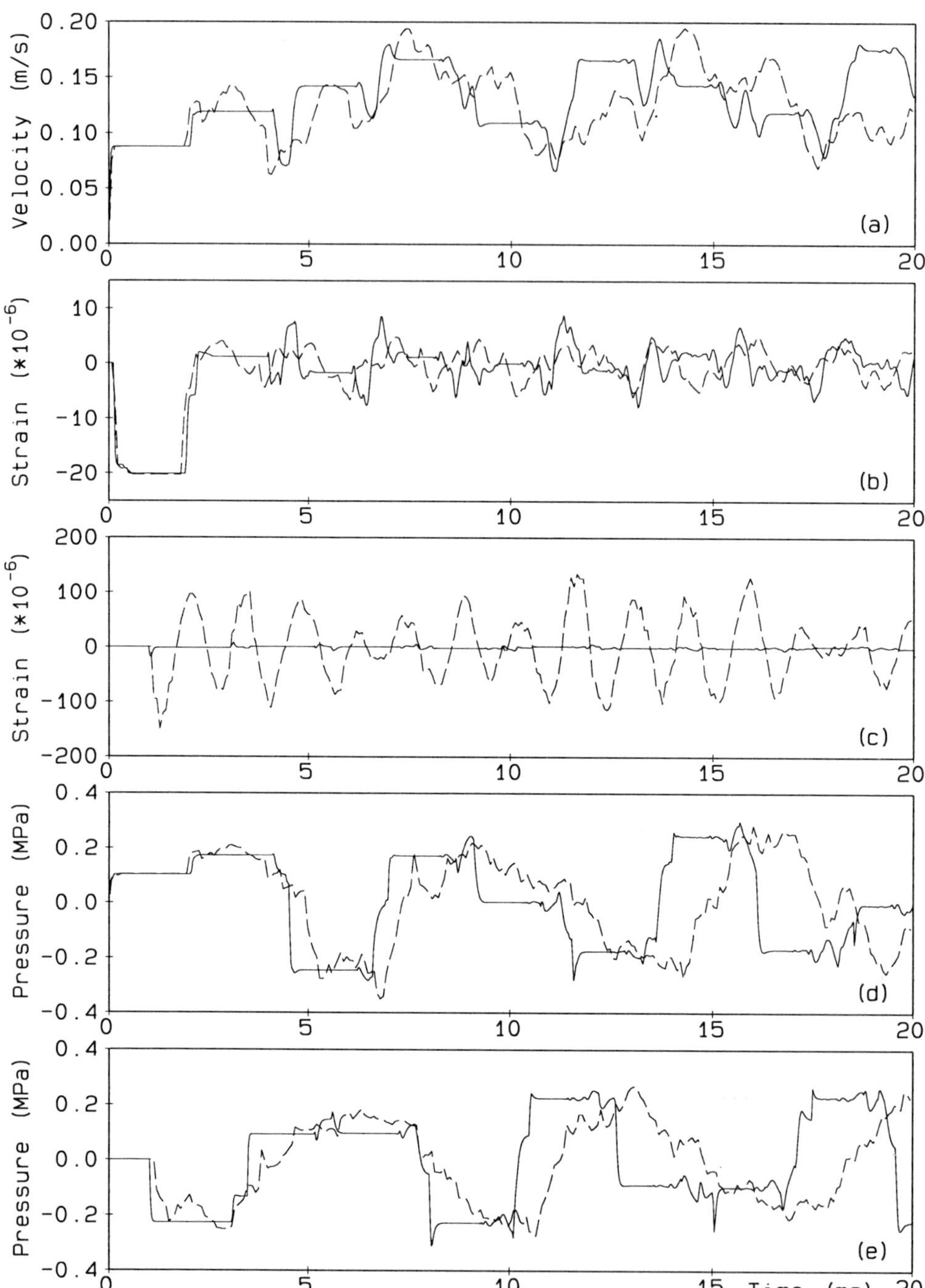

Fig. 5. Comparison 4.50 m steel pipe + 0.25 m ABS pipe (- - -) / 4.75 m steel pipe (——), both systems water-filled, calculations: (a) axial (steel) pipe velocity at z = 0.047 m, (b) axial (steel) strain at z = 0.574 m, (c) axial (ABS/steel) strain at z = 4.625 m, (d) pressure at z = 0.020 m, (e) pressure at z = 4.750 m.

ABS extension (not shown here) predict a 40% reduction in pressure amplitudes, relative to those predicted for a 6.75 m long entirely steel pipe. It is also expected that the ABS pipe would have greater influence if it were located close to the source of excitation.

5. Concluding Remarks

It has been shown that a short plastic extension significantly influences the axial vibration of a water-filled steel pipe. The natural frequencies of the system change and wave fronts become less steep. The amplitudes of vibration are not greatly reduced, when the plastic section is short relative to the steel pipe. Significant amplitude reduction can be obtained with long plastic sections however.

The system herein (Fig. 1) has been successfully simulated with a fully-coupled FSI mathematical model, so that, for the first time, the *junction coupling* (Eqs. 7-10) of two pipes of different materials and with different cross-sectional areas has been validated against experimental data.

Due to FSI *Poisson coupling*, the theoretical axial stress wavespeed in the water-filled ABS pipe is about 9% larger than the value given by classical theory. Other FSI wavespeeds were within 1% of the classical values.

The elastic behaviour of the short plastic (ABS) pipe can be satisfactorily represented by constant values for the Young modulus and the Poisson ratio. Reasonable agreement has been obtained between measurements and theoretical predictions even though visco-elastic, temperature and frequency-dispersion effects have been disregarded.

6. Acknowledgements

This work is part of a research project on the *Suppression of waterhammer-induced vibrations* sponsored by the EPSRC (Grant GR/J 54857) and guided by an FSI Advisory Group consisting of Keith Austin (Flowmaster), Warren Burrower (Greenland Tunnelling), David Clucas (Flowguard), David Fan (JP Kenny, IBM), Anton Heinsbroek (Delft Hydraulics), Simon Pugh (ESDU) and Douglas Warne (EPSRC). Ernie Kuperus and Colin Stark have assisted extensively in the design, construction and execution of the laboratory experiments.

7. References

Fan D. & Tijsseling A. 1992 *Fluid-structure interaction with cavitation in transient pipe flows.* ASME Journal of Fluids Engineering, Vol. 114, No. 2, pp. 268-274.

Fan D. & Vardy A.E. 1994 *Waterhammer including fluid-structure interactions.* Proc. of the First Int. Conf. on Flow Interaction, Hong Kong, September 1994, pp. 439-442.

Kruisbrink A.C.H. & Heinsbroek A.G.T.J. 1992 *Fluid-structure interaction in non-rigid pipeline systems - large scale validation tests.* Proc. of the Int. Conf. on Pipeline Systems, BHR Group, Manchester, UK, March 1992, pp. 151-164, ISBN 0-7923-1668-1.

Lavooij C.S.W. & Tijsseling A.S. 1991 *Fluid-structure interaction in liquid-filled piping systems.* Journal of Fluids and Structures, Vol. 5, No. 5, pp. 573-595.

Stuckenbruck S., Wiggert D.C. & Otwell R.S. 1985 *The influence of pipe motion on acoustic wave propagation.* ASME Journal of Fluids Engineering, Vol. 107, No. 4, pp. 518-522.

Tijsseling A.S. 1996 *Fluid-structure interaction in liquid-filled pipe systems: a review.* Journal of Fluids and Structures, Vol. 10, No. 2, pp. 109-146.

Tijsseling A.S. & Vardy A.E. 1996 *Axial modelling and testing of a pipe rack.* Proc. of the 7th Int. Conf. on Pressure Surges and Fluid Transients in Pipelines and Open Channels, BHR Group, Harrogate, UK, April 1996, pp. 363-383.

Tijsseling A.S., Vardy A.E. & Fan D. 1996 *Fluid-structure interaction and cavitation in a single-elbow pipe system.* Journal of Fluids and Structures, Vol. 10, No. 4 (in press).

Vardy A.E. & Fan D. 1986 *Water hammer in a closed tube.* Proc. of the 5th Int. Conf. on Pressure Surges, BHRA, Hanover, Germany, September 1986, pp. 123-137.

Vardy A.E. & Fan D. 1989 *Flexural waves in a closed tube.* Proc. of the 6th Int. Conf. on Pressure Surges, BHRA, Cambridge, UK, October 1989, pp. 43-57.

Vardy A.E., Fan D. & Tijsseling A.S. 1996 *Fluid/structure interaction in a T-piece pipe.* Submitted for publication in Journal of Fluids and Structures.

Wiggert D.C., Otwell R.S. & Hatfield F.J. 1985 *The effect of elbow restraint on pressure transients.* ASME Journal of Fluids Engineering, Vol. 107, No. 3, pp. 402-406.

Wiggert D.C., Hatfield F.J. & Stuckenbruck S. 1987 *Analysis of liquid and structural transients by the method of characteristics.* ASME Journal of Fluids Engineering, Vol. 109, No. 2, pp. 161-165.

8. Nomenclature

A	cross-sectional area
ABS	acrylonitrile butadiene styrene
c	wave propagation speed
E	Young modulus of elasticity
e	pipe wall thickness
FSI	fluid-structure interaction
K	liquid bulk modulus
L	length
m	mass of end cap
MOC	method of characteristics
P	pressure (cross-sectional average)
R	(inner pipe) radius
t	time
$\dot{u}$	axial pipe velocity
V	fluid velocity (cross-sectional average)
z	axial coordinate (distance along pipe)
ε	axial strain ($\{ \sigma - \nu (R/e) P \} / E$)
ν	Poisson ratio of pipe wall material
ρ	mass density
σ	axial stress

Subscripts

f	fluid
r	rod
t	tube, pipe
1	1st pipe, junction side
2	2nd pipe, junction side

UNSTEADY HYDRAULIC FORCE ON AN IMPELLER DUE TO ROTOR-STATOR INTERACTION IN A DIFFUSER PUMP

H. TSUKAMOTO
M. UNO
Kyushu Institute of Technology
1-1, Sensui-cho, Tobata-ku, Kitakyushu, 804, Japan
W. QIN
Awamura Manufacturing Co., Ltd.
2700, Yomi-cho, Yonago, Tottori, 683, Japan
T. TESHIMA
Mitsubishi Heavy Industries, Ltd.
1-1, Akunoura-machi, Nagasaki, 850-91, Japan
K. SAKAMOTO
Nittetsu Mining Co., Ltd
Shimowake, Suzaki, Kouchi, 785, Japan
T. OKAMURA
Hitachi, Ltd.
603, Kandatsu-machi, Tsuchiura, 300, Japan

1. Introduction

Fluctuations in hydraulic forces on impeller may be caused by interaction between impeller and diffuser vanes/volute casing. Knowledge of the unsteady forces related to the rotor vibration is important to understand rotor dynamics of turbomachinery. There, however, is a limited quantity of data available on the unsteady forces on diffuser pump impeller (Brennen, 1994).

In the present study, unsteady forces on diffuser pump impellers were measured by strain gauges on pump shaft itself as well as the strain beams attached on bearing housing. Unsteady pressures were also measured on the casing wall of diffuser vane passage and volute casing for the test pump with vaned and vaneless diffuser in order to investigate the relations of unsteady hydraulic forces to pressure fluctuations due to interaction between impeller blades and diffuser vanes.

E. Cabrera et al. (eds.), Hydraulic Machinery and Cavitation, 955–964.

2. Nomenclature

b_2	= impeller outer passage width
d_2	= impeller outer diameter
F	= radial force
K	= nondimensional radial force, $F/\{(\rho/2)u_2^2\,\pi d_2 b_2\}$
N	= rotational speed
T	= time required to traverse one rotor pitch, T_i/Z_i
T_i	= period of one revolution, $2\pi/\omega = 1/N$
t	= time
t^*	= nondimensional time, t/T
u_2	= peripheral speed of impeller $\pi d_2 N$
Z	= number of vanes
ρ	= density
σ	= standard deviation
ϕ	= flow coefficient
ϕ_r	= rated flow coefficient
ω	= angular velocity

Subscripts

d	= diffuser
i	= impeller
X, Y	= components in relative system
x, y	= components in stationary frame

TABLE 1. Specifications of test pump

		Pump A	Pump B
Suction diameter	D_s	200 mm	25 mm
Discharge diameter	D_d	200 mm	16 mm × 2
Impeller :			
Outlet radius	R_2	125 mm	49.7 mm
Number of vanes	Z_i	5	5
Diffuser vane :			
Inlet radius	R_3	129 mm	51 mm
Outlet radius	R_4	162.5 mm	70 mm
Number of vanes	Z_d	8	0, 2, 3, 4, 6, 8
Collector:		Volute casing	Fully concentric
Volute width	b_5	76.9 mm	———
Base circle radius	R_5	239.4 mm	78 mm
Rating :			
Rotational speed	N	2066 rpm.	———
Flow rate	Q	6.21 m^3/min	———
Total head rise	H	29.2 m	———

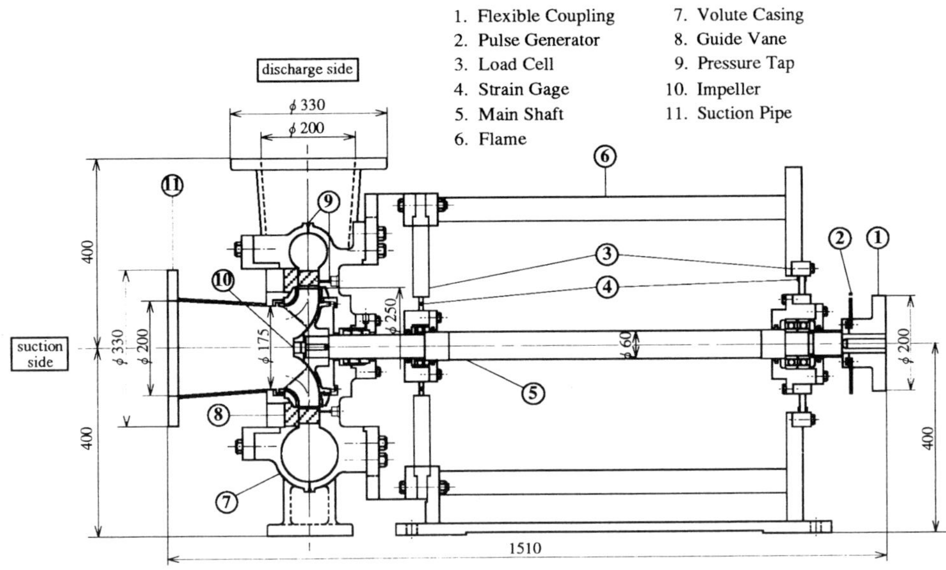

(a) Test pump A

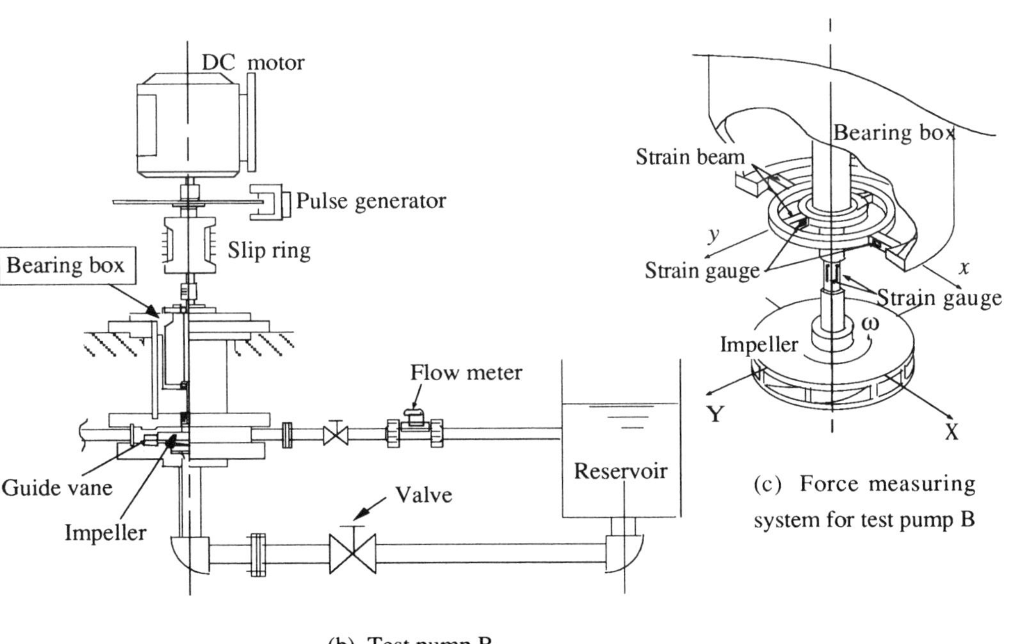

(b) Test pump B

(c) Force measuring system for test pump B

Figure 1. Schematics of test diffuser pumps

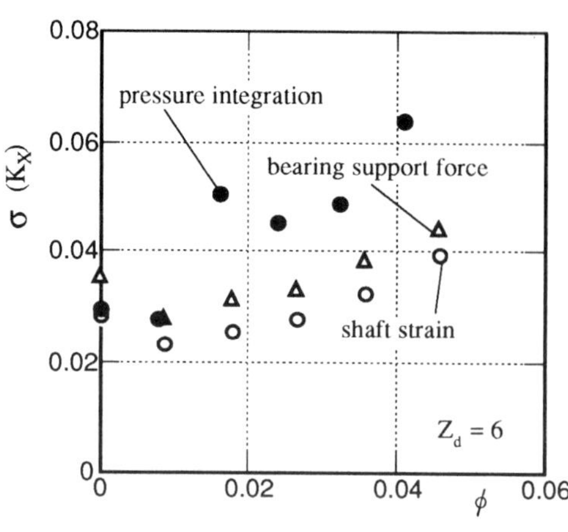

Figure 2. Comparison of radial forces measured by three measurement methods for test pump B

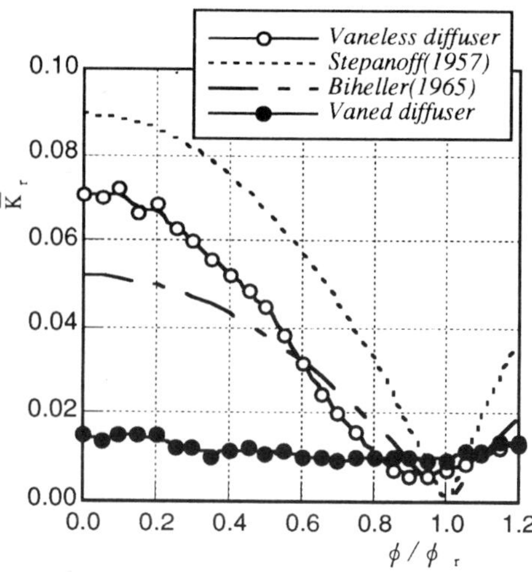

Figure 3. Static radial force as a function of capacity for test pump A. Comparison is made with Stepanoff (1957) and Biheller (1965).

3. Test Equipment and Method

Radial thrust measurements were made for the two test pumps, specifications of which are shown in Table 1. Test pump A is a pump with removable vaned diffuser as well as volute casing (Tsukamoto, et al., 1995). Basic characteristics of unsteady radial thrust were measured by using the specially designed test pump B with fully concentric collector which eliminates the effects of asymmetric volute on radial forces. Figure 1 shows the schematic view of the two test setup with the two test diffuser pumps. The radial forces on the shaft were measured directly by using radial reaction force at bearing assemblies for test pump A and B in the manner similar to Iversen, et al. (1960), as shown in Fig. 2(c) which illustrates schematically the strain-beams for measuring bearing support forces. Test pump B was equipped with strain gauges attached on the shaft neck in two perpendicular directions as can be seen in Fig.2(c). The signals were led to the slip ring and then recorded on data file of computer after A-D conversion. And instantaneous pressures were measured around the impeller exit and were integrated to evaluate the contribution of the discharge pressure to the radial force.

Radial force on the stationary frame is related to that on the rotating relative frame by the following equation.

$$F_X + j\, F_Y = (F_x + F_y)\, e^{j\omega t} \tag{1}$$

Static radial force is defined as time-averaged force on stationary frame. And thus it corresponds to the synchronous component of the shaft strain on the rotating coordinates. The synchronous component of the signal on the relative frame presents the static radial force, as shown in the following equation.

$$F_{XN} + j\,F_{YN} = (F_{x0} + j\,F_{y0})\,e^{j\omega t} \qquad (2)$$

where subscripts N and 0 represent synchronous and time-averaged components, respectively.

4. Test Results and Discussion

Figure 2 shows the comparison of radial thrust data by the present three measurement methods; measurement of bearing support forces, measurement of strain of shaft, and integration of the experimental pressure distributions. The data by pressure measurement show relatively large difference from those by the other two methods. That is because of no consideration of the momentum flux contribution to the radial force. Measured data, however, show qualitative agreement, and thus the present measurements should be acceptable for dynamic hydraulic force.

Figure 3 shows nondimensional static radial force K on the test pump A impeller with vaned and vaneless diffusers at various flow rate. This figure shows also the radial forces calculated by the existing formulae for volute pump by Stepanoff (1957) and Biheller (1965). The experimental results for the pump with vaneless diffuser showed that the resultant forces were maximum at zero flow rate, decreased with increasing flow rate and reached a minimum near the design point ($\phi/\phi_0 = 1.0$). The measured radial forces on the

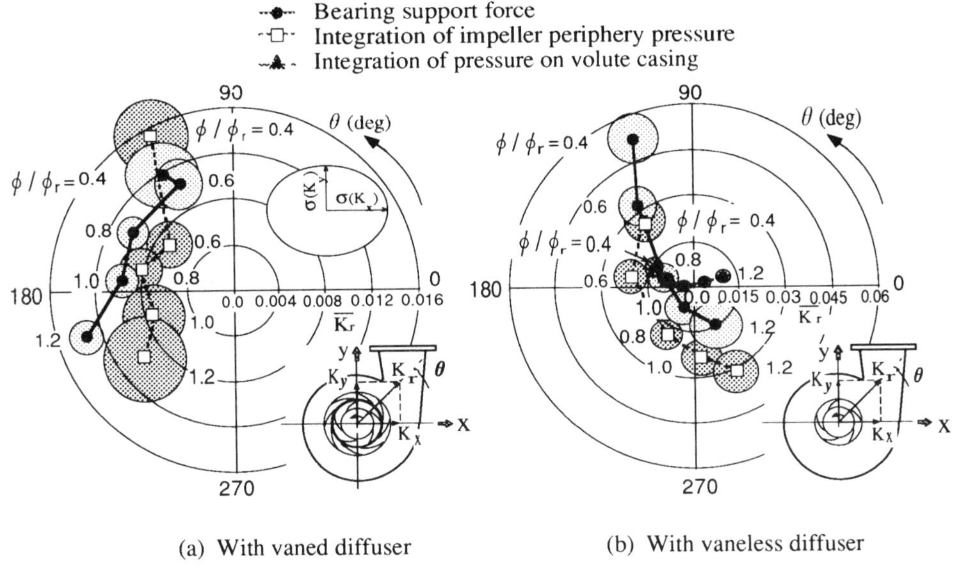

(a) With vaned diffuser (b) With vaneless diffuser

Figure 4. Vector diagrams of time-averaged radial force on test pump A

pump impeller with vaneless diffuser show good agreement with those calculated by existing formulae. On the other hand, the static radial forces on the impeller with vaned diffuser do not show marked variations with flow rate, and the magnitudes of static forces are within the range of data by other researchers (Guelich, et al., 1987).

Figures 4 (a) and (b) present vector diagrams of static radial force on test pump A with vaned diffuser and vaneless diffuser, respectively. The small circles in these figures indicate the standard deviations of radial forces. There is a correlation between the force obtained from the impeller periphery pressure distribution and the measured bearing force. The experimental results show that the direction of the radial force varies with flow rate, as it has been known generally. The radial force for vaned diffuser pump is smaller than that on vaneless diffuser case, as it has been shown by other researchers (Kanki, et al., 1981).

Figure 5 shows the effect of number of diffuser vanes on standard deviation of X-component of dynamic radial force, $\sigma(K_X)$, in test pump B. Larger number of diffuser vanes leads to smaller dynamic radial forces, and the $\sigma(K_X)$ approaches one for $Z_d = 0$ with increase in Z_d.

FFT analysis was made for dynamic radial forces to get dominant frequency for

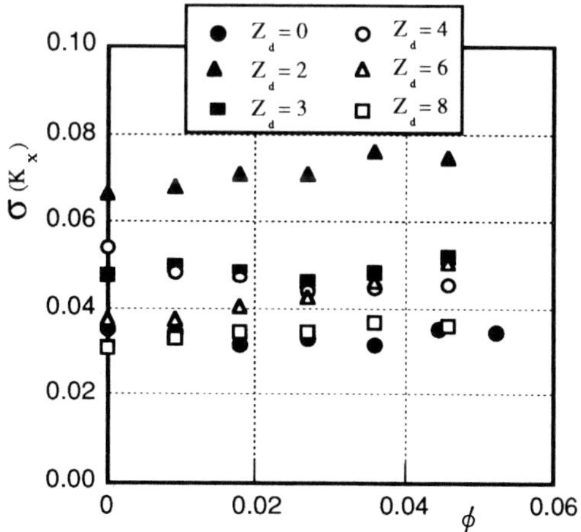

Figure 5. Standard deviation of X-component of dynamic radial force on test pump B impeller as a function of number of diffuser vanes.

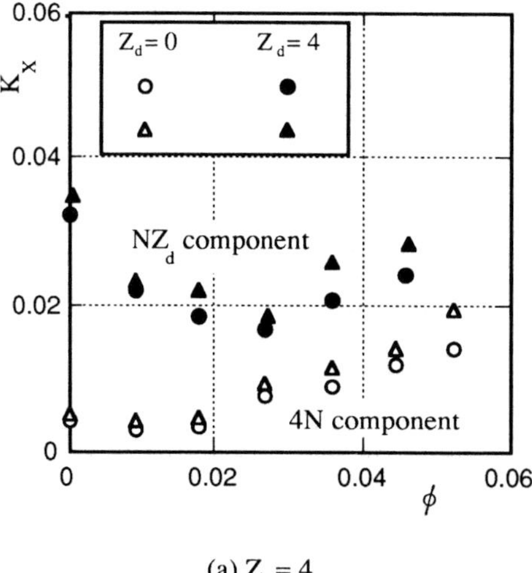

(a) $Z_d = 4$

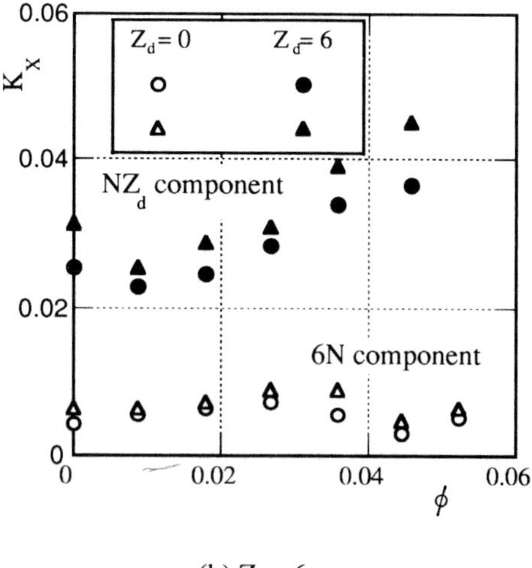

(b) $Z_d = 6$

Figure 6. Predominant frequency component of dynamic force on test pump B impeller

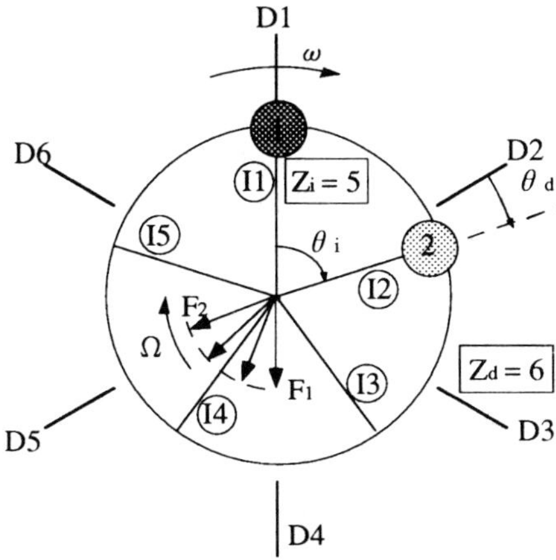

Figure 7. Encounter of rotor vane with stator vane in relative coordinate system; $Z_i = 5$, $Z_d = 6$

TABLE 2. Frequency of dynamic force on pump shaft due to rotor-stator interaction

Z_d	θ_d(deg)	θ_i(deg)	Ω
2	36	- 144	- 4N (= - 2NZ_d)
3	24	144	6N (= 2NZ_d)
4	18	- 72	- 4N (= - NZ_d)
6	12	72	6N (= NZ_d)
8	9	144	16N (= 2NZ_d)

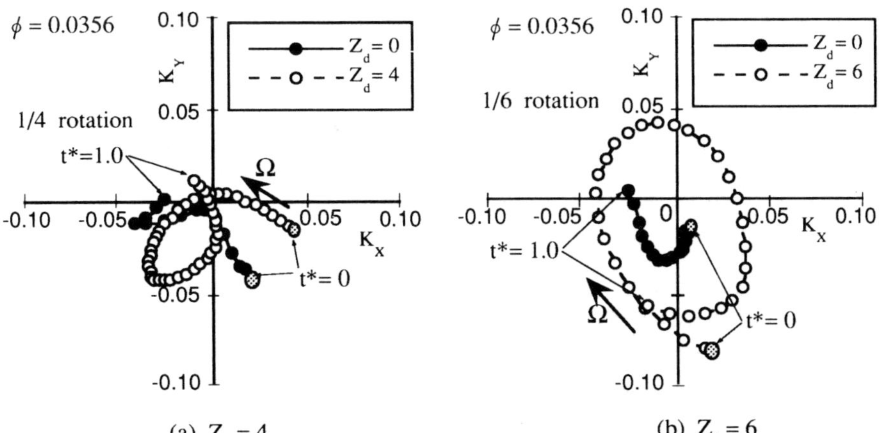

Figure 8. Vector diagram of dynamic force on pump B impeller

dynamic radial forces. The power spectral density function of radial force on impeller showed dominant frequencies relating to the blade passing frequency NZ_d and its harmonics. Figures 6(a) and (b) show the magnitude of dominant frequency components of dynamic forces for $Z_d = 4$ and 6 in test pump B. In these figures the magnitudes of the corresponding component for $Z_d = 0$ for a reference. The variations in radial forces increase with increasing flow rate except for low capacity region. The increase in variations at low flow rate has been known by Hergt and Krieger (1969-1970), and Kawata, et al. (1984).

Figure 7 illustrates schematically the hydraulic interference of impeller blades ($Z_i = 5$) and diffuser vanes ($Z_d = 6$) in the relative coordinates. Impeller blade I1 is first excited by interference with diffuser vane D1, and then impeller blade I2 when the diffuser vane D2 rotates by θ_d. According to the change in interference the radial force vector rotates from F_1 to F_2 with rotating frequency $\Omega = NZ_d$ as shown in this figure.

The frequency of the *n*-th harmonic of the blade passage excitation on impeller due to the interference is

$$f_r = n \cdot Z_d \cdot N \tag{3}$$

On the other hand, the frequency of the *m*-th harmonic of the blade passage excitation in diffuser vane due to the interference is

$$f_s = m \cdot Z_r \cdot N \tag{4}$$

The hydraulic force is composed of a number of discrete partial fluctuations, each of which consists of a wave which propagates in the circumferential direction. If we denote the force with the *k*-th circumferential mode by F_k, the force with the frequency of f_r is expressed as the sum of the modes with various number of modes.

$$F_k = A \cdot \sin(2\pi f_r t \pm k\phi) \tag{5}$$

where, ϕ is the angular coordinate fixed to impeller. The following relation should be derived for the partial force from the analysis of the excitation forces caused by the interference between impeller blades and diffuser vanes (Kubota, et al., 1983):

$$n \cdot Z_d \pm k = m \cdot Z_r \tag{6}$$

where, *n* and *m* correspond to the order of harmonics of the excitation to rotating impeller and stationary diffuser, respectively. Namely, the impeller is excited by the hydraulic force with the frequency f_r propagates in circumferential direction with the frequency f_r / k which satisfies the above relation. In the present test pump B, typical results will be obtained for the fluctuation shown in Table 2. Figures 8 (a) and (b) present loci of radial force vector with the frequency f_r for $Z_d = 4$ and 6 in test pump B.

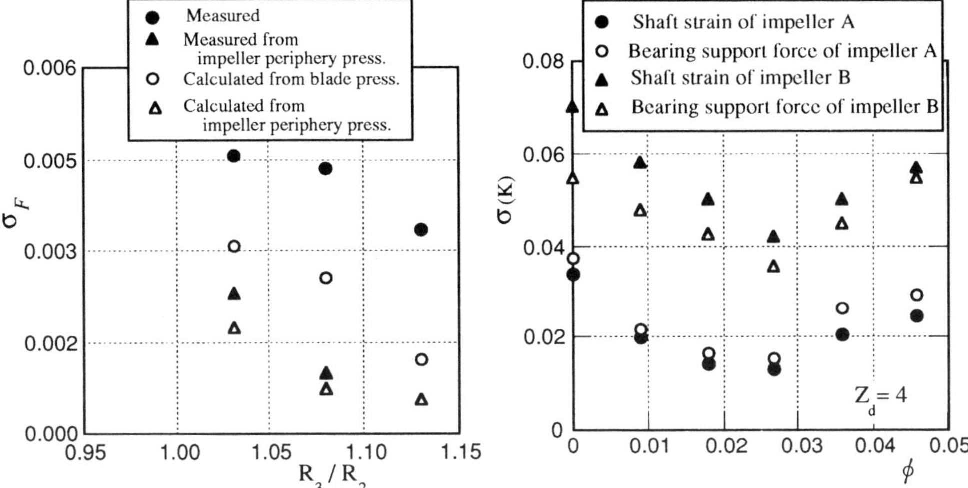

Figure 9. Effect of radial clearance between impeller exit and diffuser inlet on dynamic radial force on test pump A impeller

Figure 10. Effect of flow rate on dynamic radial force on test pump A impeller

The NZ_d component of the force for $Z_d = 4$ propagates in the direction of impeller rotation with rotational speed $\Omega = NZ_d$. These fluctuations correspond to the case in which $1\times 4+1=1\times 5$ in Eq. (6) and thus $m = 1$, $n = 1$, $k = 1$.

On the other hand, the fr component for $Z_d = 6$ propagates in the direction opposite to impeller rotation with rotational speed $\Omega = -NZ_d$. These fluctuations correspond to the case in which $1\times 4+1=1\times 5$ in Eq.(6) and thus $m = 1$, $n = 1$, $k = -1$.

Unsteady hydraulic forces on diffuser pump impeller were calculated by using singularity method (Qin, 1995). The measured unsteady forces are compared with the calculated ones by the integration of calculated impeller blade pressures and impeller periphery pressures. Figure 9 shows the effect of radial clearance between impeller outlet and diffuser inlet on dynamic radial force. The calculated unsteady forces agree qualitatively with the measured ones. The smaller radial clearance leads to larger dynamic force as can be seen in this figure.

Figure 10 presents comparison between calculated and measured dynamic forces as a function of flow rates. While calculated dynamic forces increase with increasing flow rate, measured ones have minimum at rated flow rate.

5. Conclusion

Unsteady forces on a diffuser pump impeller were measured by strain gauges on impeller shaft in the relative frame as well as by strain beams attached on bearing coupling in the stationary coordinate system. Unsteady pressures were also measured on the casing wall of guide vane passage and volute casing for the test pump with vaned and vaneless diffuser in order to investigate the relations of unsteady hydraulic forces to pressure fluctuations due to interaction between impeller blades and guide vanes. As the result of the present study, the following conclusions are derived.

(1) Time averaged static radial forces on the impeller with vaneless diffuser showed good agreement with those calculated by the existing formulae for volute pump.

(2) The time averaged radial forces by direct strain measurements agree well with those by the integration of pressure distribution around the impeller.

(3) The dominant frequencies of unsteady hydraulic forces on test diffuser pump impeller are the blade passing frequency and its higher harmonics, which are the same as the unsteady pressure at impeller exit.

(4) The dynamic radial forces have a frequency component which rotates due to the rotor-stator interaction in circumferential direction.

(5) Larger number of diffuser vanes leads to smaller dynamic radial forces, and the dynamic forces approach one for $Z_d = 0$ with increase in Z_d.

(6) Smaller radial clearance leads to larger dynamic force.

References

Biheller, H. J. (1965) Radial Force on the Impeller of Centrifugal Pumps With Volute, Semivolute, and Fully Concentric Casing, *ASME Journal of Engineering for Power*, **87**, 319-323.

Brennen, C. E. (1994) *Hydrodynamics of Pumps*, Concepts ETI, Inc. and Oxford University Press, 199-204.

Guelich, J., Jud, W., and Hughes, S. F. (1987) Review of Parameters influencing hydraulic forces on Centrifugal Impellers, *Proc. Instn. Mech. Engrs.*, **201-No. A3**, 163-174.

Hergt, P., and Krieger, P. (1969-1970) Radial Forces in Centrifugal Pumps With Guide Vanes, *Proceedings, Institution of Mechanical Engineers*, **184-Pt 3N**, 101-107.

Iversen, H. W., Rolling, R. E., and Carlson, J. J. (1960) Volute Pressure Distribution, Radial Forces on the Impeller and Volute Mixing Losses of a Radial Flow Centrifugal Pump, *ASME Journal of Engineering for Power*, **82**, 136-144.

Kanki, H., Kawata, Y., and Kawakami, T. (1981) Experimental Research on the Hydraulic Excitation Force on the Pump Shaft, *ASME Paper 81-DET-71*.

Kawata, Y., Kanki, H., and Kawakami, T. (1984) The Dynamic Radial Force on the Cavitating Centrifugal Impeller, *Proceedings, The 12th IAHR Symposium, Stirling*, **Paper 3.8**, 305-315.

Kubota, Y., Suzuki, T., Tomita, H., Nagafuji, T., and Okamura, C. (1983) Vibration of Rotating Disc Excited by Stationary Distributed Forces, *Trans. JSME, Ser. C*, **49-439**, 307-312 (in Japanese).

Qin, W. (1995) Study on fluctuating Pressure downstream of a Diffuser Pump Impeller due to Rotor-Stator Interaction, *Ph.D Thesis, Kyushu Institute of Technology*, (in Japanese).

Stepanoff, A. J. (1957) *Centrifugal and Axial Flow Pumps*, John Wiley & Sons, Inc., New York, Second Edition, 116-123.

Tsukamoto, H., Uno, M., Hamafuku, N., and Okamura, T. (1995) Pressure Fluctuation Downstream of a Diffuser Pump Impeller, *ASME Unsteady Flows Forum*, **FED-216**, 133-138.

SWIRLING FLOW WITH HELICAL VORTEX CORE IN A DRAFT TUBE PREDICTED BY A VORTEX METHOD

X. M. WANG, *Lecturer, Kyushu Institute of Technology, Japan*
M. NISHI, *Professor, Kyushu Institute of Technology, Japan*

ABSTRACT

A mathematical model is proposed in this paper to describe the swirling flow with spiral vortex core in a draft tube. In the model, it is assumed that the flow in the circular cross-section is given by the sum of an axisymmetric base flow and that induced by an infinite helical vortex filament in a pipe. The corresponding computer program is developed where a series of the image vortex filaments is considered so as to satisfy the boundary condition at the pipe wall, providing that the flow rate, swirl rate and circulation of incoming flow are prescribed as input values. The velocity distribution in the cross-section and the rotating frequency of vortex core are calculated and compared with experimental results in draft tubes. The reasonable agreement between them is obtained.

1. Introduction

The spiral vortex core is often observed in the draft tube of Francis turbines during their partial load operation. Since the dynamic behavior of the vortex motion is closely related to draft tube surge, many studies have been made to understand its mechanism. As for the numerical investigations, various computer codes based on Reynolds-averaged Navier-Stokes equations have been developed to simulate the flow without or with small swirling component in the draft tube. In the case of strong swirling flow, however, those codes have got some difficulties because the flow field becomes much more complex in the tube where one (or more) precessional spiral vortex core appears. To clarify such distorted velocity distribution in the inlet conical section of draft tube and the rotating speed of the vortex core, a 2D (or two-dimensional) mathematical model, which is called Partially Rolled-up Vortex Model, was proposed by present authors [1].

As an extension of the 2D model, a modified model which takes into account the effect of the helical vortex core on the flow is developed in this paper. And it is assumed that the distorted swirling flow in the cross section is represented by the superposition of the vortex flow induced by a helical vortex filament in a pipe upon an axisymmetric swirling flow called as base flow. The circumferential velocity of the base flow is assumed to have Rankine vortex type distribution, and the axial velocity of base flow is so decided that the superposed velocity satisfies both vorticity transport equation and

E. Cabrera et al. (eds.), Hydraulic Machinery and Cavitation, 965–974.

continuity equation. The applicability of present Q3D (or quasi-three-dimensional) model is examined by comparing the calculated results with experimental data.

2. General Consideration

It is of special significance to understand the flow pattern in the inlet conical section of a draft tube , especially when helical vortex core appears. In most of the cases, the frequency of draft tube surge can be predicted reasonably once the flow field in the inlet section is specified. For numerical analysis of the 3D flow in an elbow draft tube, the flow pattern in the inlet section has to be prescribed as inlet boundary condition.

In order to clarify the behavior associated with the helix effect of the vortex core, we develop a Q3D model based on the similar consideration to the 2D case, where the flow in the inlet conical portion of draft tube is treated as the motion of a helical vortex core rotating in an axisymmetric base flow. The velocity is therefore assumed to be the sum of two components named vortex flow and base flow respectively.

$$\vec{V} = \vec{V}_a + \vec{V}_b \tag{1}$$

$$\Gamma = \Gamma_a + \Gamma_b \tag{2}$$

where $\vec{V}_a$ and Γ_a are velocity and circulation related to the spiral vortex core, and $\vec{V}_b$ and Γ_b are those of base flow.

3. Vortex Flow

Even though it is the usual case that the inlet portion of draft tube has some divergence angle, and the spiral cavitated vortex core does not have an exact helical shape, as a mathematical model, we treat the vortex flow field as that induced by a helical vortex filament in a cylindrical pipe, and both the pipe and the filament have infinite length.

Consider an infinite, right-handed helical vortex core having the radius r_x in a pipe such as shown in Fig.1. The flow induced by the vortex core is irrotational except the core area. To satisfy the boundary condition on the wall, an extra irrotational flow is usually introduced. Several methods have been shown for this purpose. Considering the helicoidal symmetry of the flow, Fanelli[2] reduced the problem into 2D case and calculated the velocity potential by the finite element method. In the case of a filament with finite length, Pedrizzetti[3] treated the similar problem with an analytical method combining with numerical calculation. To develop more convenient method to the irrotational flow, we distribute a series of discrete helical vortex filaments having unknown circulations on a cylindrical surface outside the pipe, which is named image surface $\left(r_a = r_d + r_x;\quad r_a' = R^2/r_a\right)$. Those filaments have the same helical pitch as the original one inside the pipe so that the helicoidal symmetry of the flow is preserved. The velocity

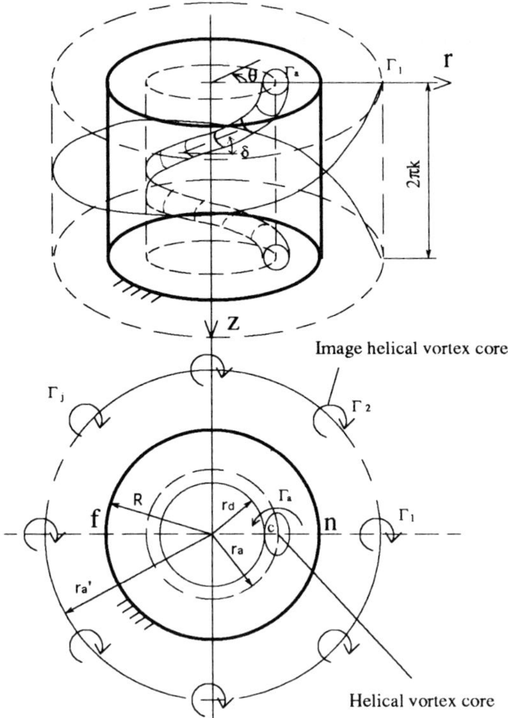

Fig.1 A helical vortex core in a pipe

induced by each filament is given by Hardin's analytical derivation[4]. To calculate the unknown circulation Γ_j (j=1~N) around each vortex filament on the image surface, we examine the boundary condition in an arbitrary cross section considering the helicoidal symmetry. The 3D problem is therefore reduced to a 2D one.

Using the polar coordinate system and choosing a section where the vortex core locates at $(r_a, 0)$, we obtain the radial velocity component at the point (R, φ_i) on the wall

$$v_r{}^i = \frac{\Gamma_a r_a}{\pi k^2} S_4\left(r_a, R, \varphi_i\right) + \sum_{j=1}^{N}\left(\frac{\Gamma_j r_a{}'}{\pi k^2}\right) S_2\left(r_a{}', R, \varphi_{i,j}\right) \tag{3}$$

where $\varphi_j = 2\pi(j-1)/N$ (4) $\varphi_{i,j} = \varphi_i - \varphi_j$ (5) $k = r_a \tan\delta$ (6)

$$S_2(c, r, \phi) = \sum_{m=1}^{\infty} mK'_m(cm/k) I'_m(rm/k)\sin m\phi \tag{7}$$

$$S_4(c, r, \phi) = \sum_{m=1}^{\infty} mK'_m(rm/k) I'_m(cm/k)\sin m\phi \tag{8}$$

I_m and K_m are modified Bessel functions of the m*th* order, and the primes indicate differentiation with respect to the argument. δ is the inclination of the vortex filament.

In order to obtain non-trivial solutions for Γ_j, we choose N-1 points on the wall for the boundary condition as expressed in Eq. (9) and add the condition related to circulation given by Eq. (10)

$$v_r{}^i = 0 \qquad (i = 1 \sim N-1) \tag{9}$$

$$\sum_{j=1}^{N} \Gamma_j + \Gamma_a = 0 \tag{10}$$

Substituting Eq. (3) into Eq. (9), and distributing the boundary points uniformly along the circumferential direction

$$\varphi_i = 2\pi(i-1)/(N-1) \qquad (i = 1 \sim N-1) \tag{11}$$

we can establish a set of equations from Eq. (9) and (10), and obtain the Γ_j by solving it. The circumferential and axial velocity components at a point (r,θ) are expressed as

$$Vu_a(r,\theta) = \begin{cases} \frac{\Gamma_a r_a}{\pi r k} S_1(r_a, r, \theta) + \frac{r_a'}{\pi r k} \sum_{j=1}^{N} \Gamma_j S_1(r_a', r, \theta - \varphi_j) & r \le r_a \\ \frac{\Gamma_a}{\pi r}\left(0.5 + \frac{r_a}{k} S_3(r_a, r, \theta)\right) + \frac{r_a'}{\pi r k} \sum_{j=1}^{N} \Gamma_j S_1(r_a', r, \theta - \varphi_j) & r > r_a \end{cases} \tag{12}$$

$$Vz_a(r,\theta) = \begin{cases} -\frac{\Gamma_a}{\pi k}\left(0.5 - \frac{r_a}{k} S_1(r_a, r, \theta)\right) - \frac{1}{\pi k} \sum_{j=1}^{N} \Gamma_j \left(0.5 - \frac{r_a'}{k} S_1(r_a', r, \theta - \varphi_j)\right) & r \le r_a \\ \frac{\Gamma_a r_a}{\pi k^2} S_3(r_a, r, \theta) - \frac{1}{\pi k} \sum_{j=1}^{N} \Gamma_j \left(0.5 - \frac{r_a'}{k} S_1(r_a', r, \theta - \varphi_j)\right) & r > r_a \end{cases} \tag{13}$$

where

$$S_1(c, r, \phi) = \sum_{m=1}^{\infty} m K'_m(cm/k) I_m(rm/k) \cos m\phi \tag{14}$$

$$S_3(c, r, \phi) = \sum_{m=1}^{\infty} m K_m(rm/k) I'_m(cm/k) \cos m\phi \tag{15}$$

4. Base Flow

If we consider that the flow distortion in the cross section is caused only by the vortex filament in a pipe, an axisymmetric base flow can be obtained by subtracting the flow induced by the vortex filament from the distorted flow field. Assuming that the base flow in the main flow region $(r > r_d)$ to be irrotational and the axial component of vorticity distributes uniformly in the dead water region $(r \le r_d)$, we have the circumferential and radial velocities as Rankine vortex distribution

$$Vu_b(r) = \begin{cases} r\Gamma_b/2\pi r_d^2 & r \le r_d \\ \Gamma_b/2\pi r & r > r_d \end{cases} \tag{16}$$

$$Vr_b = 0 \tag{17}$$

Once the velocity $\vec{V}_a$ induced by the vortex, the circumferential and radial velocity components of base flow, Vu_b and Vr_b, are decided, the axial velocity component of base flow should be decided uniquely. Because the superposition of $\vec{V}_a$ and $\vec{V}_b$ has physical meaning only when the composed velocity $\vec{V}_a + \vec{V}_b$ satisfy the following vorticity equation concerning with vorticities of $\vec{\omega}_a$ and $\vec{\omega}_b$, where viscous effect is neglected

$$D(\vec{\omega}_a + \vec{\omega}_b)/Dt = [(\vec{\omega}_a + \vec{\omega}_b)\cdot\nabla](\vec{V}_a + \vec{V}_b) \tag{18}$$

In the main flow region, both $\vec{V}_a$ and $\vec{V}_b$ are irrotational, so Eq. (18) is unconditionally satisfied. In the dead water region, however, the following condition is required

$$(\vec{\omega}_b \cdot \nabla)\vec{V}_a = 0 \tag{19}$$

Utilizing the feature of the helicoidal symmetrical flow induced by vortex filaments, we have the following equation from Eq. (19)

$$\omega_{bz}/\omega_{b\theta} = r_a \tan\delta / r \qquad r \le r_d \tag{20}$$

where ω_{bz} and $\omega_{b\theta}$ represent the axial and circumferential vorticity components of base flow in the dead water region. From Eqs. (16), (17) and (20), the axial velocity distribution can be expressed as

$$Vz_b = \begin{cases} r^2\Gamma_b/(2\pi r_a r_d^2 \tan\delta) + Cv & r \le r_d \\ \Gamma_b/(2\pi r_a \tan\delta) + Cv & r > r_d \end{cases} \tag{21}$$

The constant C_v can be determined considering the flow discharge condition:

$$\int_0^{r_d} 2\pi r Vz dr + \pi(R^2 - r_d^2)Vz|_{r=r_d} = Q - Q_v \tag{22}$$

$$Q_v = \Gamma_a(R^2 - r_a^2)/2k \tag{23}$$

Q_v represents the flow discharge contributed by all helical vortex filaments and can be calculated by integrating Eq. (13). From Eq. (21) through (23), C_v can be written as

$$Cv = \frac{Q}{\pi R^2} - \frac{\Gamma_a}{2\pi k}\left[1 - \left(\frac{r_a}{R}\right)^2\right] - \frac{\Gamma_b}{2\pi r_a \tan\delta}\left[1 - 0.5\left(\frac{r_d}{R}\right)^2\right] \tag{24}$$

5. Precessional Speed of vortex core

The superposed velocity of $\vec{V}_a$ and $\vec{V}_b$ represents instantaneous flow filed on a cross section, and the flow pattern rotates with the precessional speed of the vortex core. To predict its precessional speed, we deal with the motion of helical vortex filament with the core in a pipe, and express its rotating speed as

$$V_{ro} = Vu_b{}^C + Vu_a{}^C + Vu_{se} \tag{25}$$

$Vu_b{}^c$ indicates the circumferential velocity of base flow at the center point c of the vortex core, $Vu_a{}^c$ the circumferential velocity at point c induced by the vortex filaments on the image surface, and Vu_{se} the self induced circumferential velocity by the original vortex filament inside the pipe, which is calculated based on Ricca's analytical method[5]

$$Vu_{se} = -\frac{\Gamma_a \sin\delta}{4\pi L}\left[\ln\frac{8L}{r_x} + 2\left[I_1(\tau) + I_2(\tau)\right] - 0.25\right] \tag{26}$$

$$L = r_a\left(1 + \tan^2\delta\right) \tag{27}$$

$$\tau = \tan\delta \tag{28}$$

The integrations of $I_1(\tau)$ and $I_2(\tau)$ will be found in reference [5].

6. Application of the Model

When the model is used to pridict the flow field in the inlet section of a draft tube, several parameters such as flow rate, swirl rate and circulation of incoming flow should be given beforehand. Parameters related to the geometry of helical vortex core such as radius of dead water region r_d, radius of the core r_x, and inclination angle δ are also required in the model.

Because the flow pattern in the present study is specified feasibly by one parameter called theoretical swirl rate m_{th} in the experiment whose data is used to examine the present model, we also adopt m_{th} as a principle factor during computation with the model. Then the circulation and other parameters related to vortex core geometry are treated as functions of swirl rate as shown previously[1].

As far as the circulation fractions of Γ_a and Γ_b to Γ in Eq. (2) are concerned, we use the following correlation derived from experimental results [1]

$$Vu_n/Vu_f = k_v m_{th} \tag{29}$$

where Vu_n and Vu_f represent the circumferential velocities at point n and f in Fig.1. It is noted that unlike the 2D's case, no analytical equations for Vu_n and Vu_f can be written. An iterative calculating procedure with given Γ_a and Γ_b to reach equation (29) is needed.

By changing the numbers of the helical vortex filaments and the numbers of accumulating term in Eqs. (7), (8), (14) and (15), we found the choice of N=24, and M=40 was enough to reach stable solutions.

7. Calculated Results and Discussion

The velocity distributions are calculated in the case of m_{th}=0.97. Fig. 2 (a) shows the contour of circumferential velocity, and the numbers indicate the value nondimensionalized by averaged axial velocity V_i. Considerable distortion of the flow field is observed, and the faster velocity region appears in the main flow region near the vortex core. In the Fig 2(b) where contour of axial velocity is plotted, the faster velocity region is found again in the main flow region near the vortex core. A reverse flow region is calculated in the central area of the pipe. Similar features are observed in our measurement with a five-hole pitot probe, and the corresponding results were shown previously [6][7].

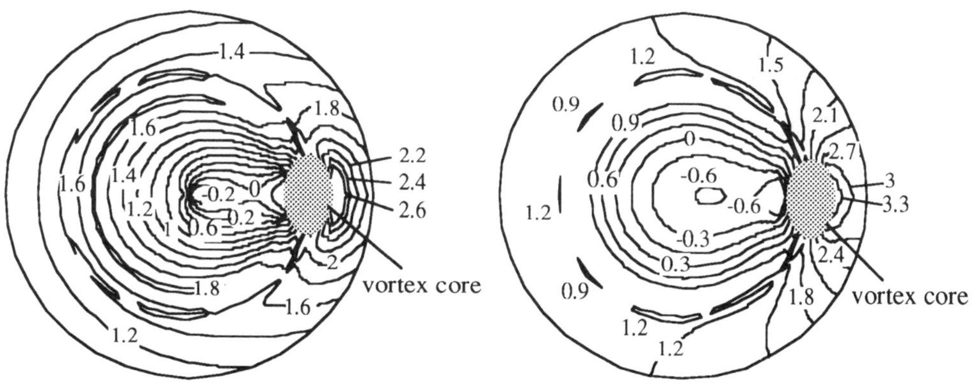

(a) Circumferential component (b) Axial component

Fig. 2 Calculated contour of dimensionless velocity (m_{th}=0.97)

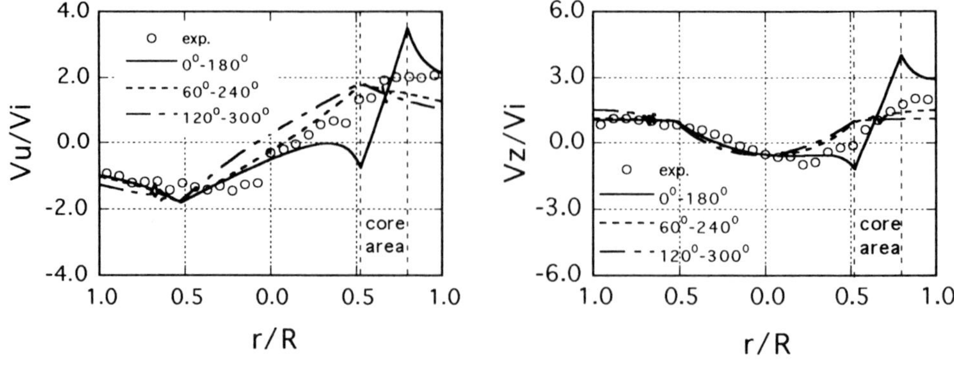

(a) Circumferential component (b) Axial component

Fig. 3 Velocity distribution in meridional sections (m_{th}=0.97)

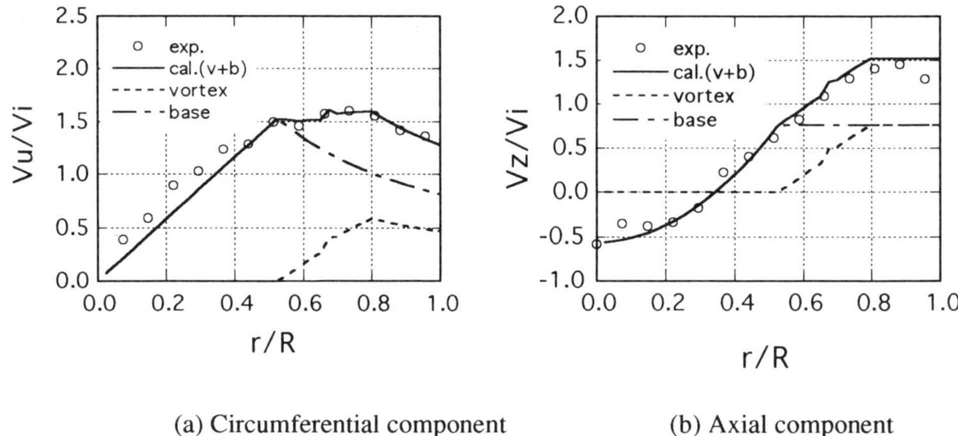

(a) Circumferential component (b) Axial component

Fig. 4 Averaged velocity distribution (m_{th}=0.97)

For further comparison, three diameter lines, one of which includes the vortex core(0^0-180^0), are chosen, and the circumferential and axial velocity distributions on them are shown in Fig 3 (a) and (b) respectively. The experimental results on the diameter line including vortex core are plotted with circular marks. Comparing those marks with the solid curve in Fig. 3(a), we find good agreement between them except the region near the core area. It is noted that the flow inside the vortex core should be re-examined if one intends to know its velocity field. Discrepancy of the axial velocity distribution near the wall is observed in Fig 3(b). It may be due to the irrotational flow assumption in the mail flow region.

Spatially averaged velocity distribution can be obtained by averaging the velocity along circumferential direction, which is equivalent to the time averaged velocity from measurement. The variations of averaged circumferential and axial velocity with respect to radius are shown in Fig 4 (a) and (b) respectively. The calculated results of vortex flow, base flow and superposed flow are represented by dashed, dash dot and solid curves respectively, and experimental results are plotted with circular marks. Satisfactory agreement is obtained between them.

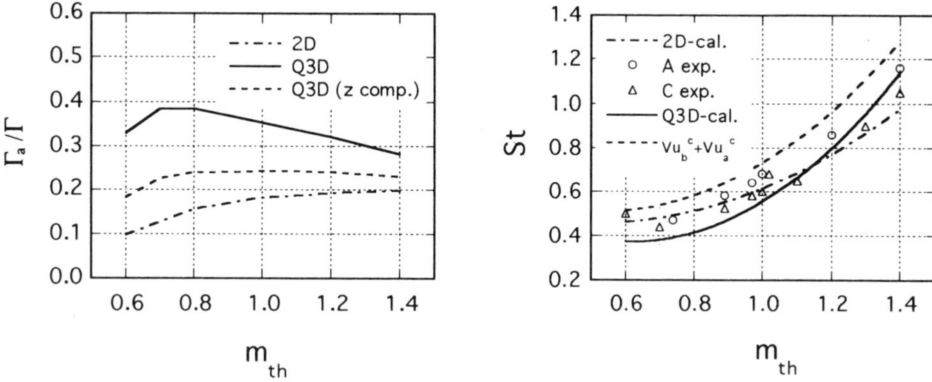

Fig. 5 Circulation fraction around vortex core

Fig. 6 Dimensionless rotating frequency of vortex core

To study the effect of swirling strength on the flow, calculations are carried out for different swirl rates. The circulation fraction around vortex filament to total one entering draft tube are shown in Fig.5 by the solid curve. The dashed curve indicates the axial component of the circulation fraction, and the dash dot curve corresponds to the result from 2D model. Differing from 2D's result, the decreasing tendency of the circulation fraction with increase of the swirl rate is found in Q3D's result. It is also seen that results from present model is higher than 2D's one.

The variation of precessional speed of vortex core with respect to swirl rate is shown in Fig. 6. Dimensionless rotating frequency St $\left(= RV_{ro}/\pi r_a V_i\right)$ is taken as the ordinate, which is calculated from the precessional speed of vortex core. The experimental data from two draft tubes are plotted with circular and triangular marks. The calculated result by Eq. (25) and that obtained disregarding the term of Vu_{se} are shown by solid and dashed curves. By comparison with the experimental results and the calculated result of 2D model which is shown with dash dot line, the calculated results from the present model are found to predict slower rotating speed in the region of smaller swirl rate. It may be due to the inaccuracy of vortex core size, and the higher fraction of circulation as shown in Fig. 5. The tendency of its variation with respect to swirl rate, however, is considered to be reasonable.

8. Conclusions

A Q3D mathematical model based on the partially rolled-up vortex assumption is developed considering the effect of helix of vortex core. The conclusions are summarized as follows:

(1) The swirling flow having the vortex core in the inlet section of a draft tube can be represented by the superposition of the flow induced by a helical vortex filament in a pipe upon Rankine-vortex-type base flow.

(2) The computer program to simulate the distorted swirling flow in a cross section is developed, providing that the flow rate, swirl rate and circulation are prescribed.

(3) The precessional speed of the vortex core is estimated with the consideration of self-induced velocity of the helical vortex core.

(4) The usefulness of the Q3D model is examined by comparing calculated results with experiment data.

(5) The fraction of the circulation around vortex core to total one is estimated less than forty percent by the present model.

9. Acknowledgment

The authors are much indebted to Prof. H. Tsukamoto in Kyushu Institute of Technology and Prof. R. L. Ash in Old Dominion University for their valuable discussions. This study has been partially supported by the Grant-in-Aid for Scientific Research from The Ministry of Education, Science and Culture (Univ. to Univ. Cooperative Research, No. 07045030).

10. References

(1) Wang X.M., Nishi, M. and Tsukamoto, H., (1994) A simple method for predicting the draft tube surge, *17th IAHR Symposium, Section on Hydraulic Machinery and Cavitation, Beijing,* Vol. 1, 95-106.

(2) Fanelli, M.A. (1989) The vortex rope in the draft tube of Francis turbines operating at partial load, *Journal of Hydraulic Research,* Vol.27, No.6, 769-807.

(3) Pedrizzetti, G. (1995) A three-dimensional computational approach to the vortex rope dynamics in Francis turbine outlet, *7th IAHR WG on the Behavior of Hydraulic Machinery under Steady Oscillatory Condition, Ljubljana,* B-1.

(4) Hardin J. C. (1982) The velocity field induced by a helical vortex filament, *Phys. Fluids,* Vol. 25, No. 11, 1949-1952.

(5) Ricca, R. L. (1994) The effect of torsion on the motion of a helical vortex filament, *J. Fluid Mech.,* Vol. 273, 241-259.

(6) Nishi, M., Matsunaga, S., Kubota, T. and Senoo, Y. (1982) Flow regimes in an elbow-type draft tube, *Proc. IAHR 11th Symposium, Amsterdam,* Vol.2, 38.

(7) Wang X. M., Nishi M. and Tsukamoto, H. (1995) Precessional behavior of vortex core in a diffuser with strong swirling flow, *ASME FED*, Vol. 216, 49-54.

FRICTION LOSS OF ROUGH WALL PASSAGE IN A TURBOMACHINERY

S. AKAIKE
Kanagawa Institute of Technology
1030 Shimo-Ogino Atsugi, JAPAN 243-02

1. Introduction

An impeller or a flow passage in a turbomachinery is normally made from casting and the wall surface of the passage is usually not hydraulically smooth on the friction loss. When the friction loss of the passage is estimated by using the friction loss coefficient, it is necessary to consider the effect of the roughness of the passage wall. The friction loss coefficient of the passage is commonly obtained from the well known formulae based on the equivalent sand roughness Ks. However, the roughness of the surface of a prototype turbomachinery is expressed by the arithmetic average roughness Ra. Therefore, it is necessary to determine the correlation between Ks and Ra, or to obtain the formula of the loss coefficient based on Ra.

There exists a considerable literature on the turbulent flow on the rough wall [1] and the mean flow characteristic is almost made clear. The studies on the effect of the wall roughness on the flow loss of a turbomachinery were also performed and the correlation between Ks and Ra was proposed. Nevertheless, it is necessary to reconsider the correlation between Ks and Ra, because there are many differences between the correlation obtained from the previous studies. The detailed discussion concerned with this is presented in the later Chapter.

This study proposes the treatment of friction loss coefficient of such rough wall passage as in a turbomachinery. In the first place, the sharp of a rough surface practically used is examined and the problems of the existing treatment of a rough wall passage are discussed. Based on our studies on the rough wall passages roughened by a sand paper [2] [3], the practical correlation between Ks and Ra is recommended. Moreover, the empirical equation on the loss coefficient based on Ra for the calculation of a rough wall passage is shown.

2. Surface Roughness Practically Used

Figure 1 shows an example of the measuring roughness curves of the surfaces of a runner of Francis hydraulic turbine and a sand paper. The surface roughness profile is commonly not uniform and is changed by a kind of the surface.

E. Cabrera et al. (eds.), Hydraulic Machinery and Cavitation, 975–984.

ISO 468-1982 : Surface roughness defines the surface roughness by the arithmetic average roughness Ra, the ten point height Rz , the maximum height Rmax and so on. The surface roughness of the passage in a turbomachinery is expressed in general by Ra, Rz or Rmax.

TABLE 1 shows a general relationship between Ra and Rz, which is obtained by assuming the two-dimensional roughness, where the roughness elements are arrayed normal in the flow direction. The relationship of Rz/Ra or Rmax/Rz may be varied with the profile of the roughness element. Even if Ra of two different surfaces take the same value, their Rz show usually different values.

TABLE 2 shows the measuring example of surfaces practically used. The casting and grinding surfaces show Rz/Ra$\fallingdotseq$5, while Rmax/Ra of those are different. On the other hand, the roughness profile such machined surface as the milling is varied with the machined direction. Taylor measured the surface roughness on in-service gas turbine two-blades [4], which were taken in the region of the leading edge, the mid-chord and the trailing edge on both the pressure and suction sides for a total of six profiles, respectively. In TABLE 2, the limits of his results are also shown. Rz/Ra of that is nearly the same to the casting surface.

If we consider the roughness profile, it is very difficult to select the reference value of the rough surface which has the effect on the friction loss. It is also pointed out by Taylor [4]. Since the surface roughness in a turbomachinery is prescribed only by the value of Ra under the standard, it is necessary to determine the friction loss coefficient on the basis of Ra for the estimation of the friction loss in a practical rough wall passage.

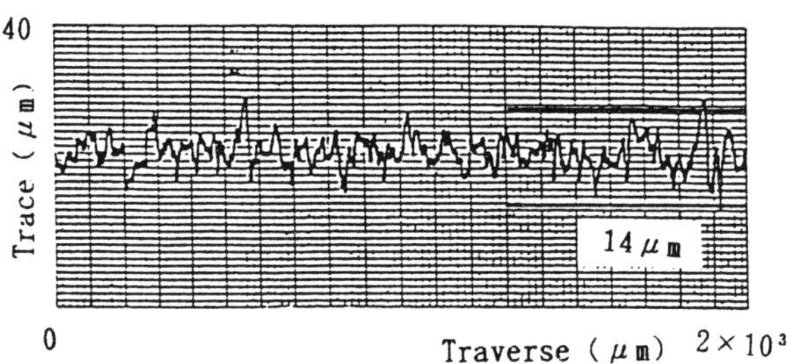

(a) Surface on a runner of Francis hydraulic turbine.

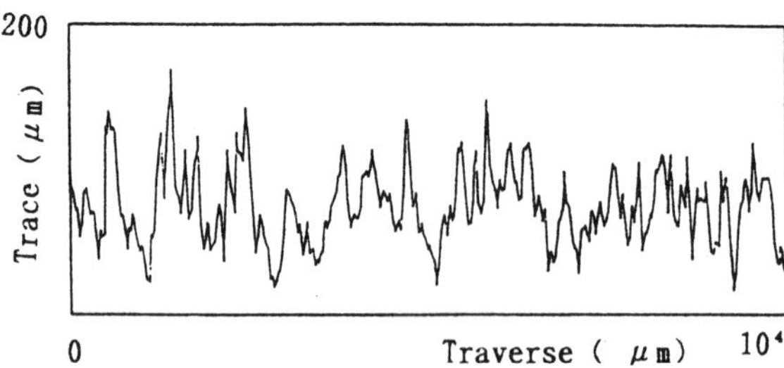

(b) Surface on a sand paper of #180.

Figure 1. Measuring example of rough surface.

TABLE 1. A general relation between Ra and Rz.

Rough surface	Rz/Ra
Square wave	2
Sine wave	3
Triangle wave	4
Machined surface	5
Rasped surface	10
Pulse shape	25

TABLE 2. Measuring example of roughness on surface practically used.

Rough surface	Ra(μm)	Rz/Ra	Rmax /Ra
Casting	9.1	5.80	9.58
	13.6	5.78	9.25
Grinding	3.26	4.44	4.57
Milling(Milling direction)	0.28	32.8	50.0
Milling(Normal direction)	0.32	10.1	14.9
Rasped surface	1.96	9.67	12.9
Sand paper surface	22.7	4.77	
	53.5	5.66	
	108.5	6.72	
In-service gas turbine blades[4]	1.46~6.86	4.48 ~5.39	6.38 ~9.75

3. Existing Method and Problems on Ks

There exists a considerable literature on the friction loss coefficient of the rough wall passage. When the friction loss of the rough wall passage is calculated, in general, the pipe friction loss coefficient λ is determined from the well known Moody chart and so on by using the equivalent sand roughness Ks. This chart is based on the following Colebrook equation.

$$1/\sqrt{\lambda} = -2\log\left[(Ks/4M)/3.71 + 2.51/(Re\sqrt{\lambda})\right] \quad (1)$$

Here, M is hydraulic mean depth and Re is the Reynolds number based on M. Since the friction loss coefficient λ is varied with the value of Ks, it is important to determine the value of Ks. The many studies on Ks of a commercial pipe are performed, but there is a little study on the basis of Ra.

For the two-dimensional rough surface, there exist many studies on the effect of the roughness density on the flow over the rough surface. Bettermann first introduced the roughness density parameter which quantified the effect of the spacing of roughness elements on the flow and was the pitch to height ratio of the traverse square bars [5]. The roughness density parameter was improved by many researchers. Recently, Sigal and Danberg proposed the following roughness density parameter [6].

$$\Lambda s = (S/Sf)(Af/As)^{-1.6} \tag{2}$$

where S/Sf is the ratio of the total surface area to the total roughness frontal area normal to the flow. As is the windward wetted surface area, and Af is the frontal area of an element of roughness. Using equation (2), they correlated many available data and obtained the following correlation of the equivalent sand roughness Ks to the height of an element K for two-dimensional roughness.

$$\frac{Ks}{K} = \begin{cases} 0.003215\,\Lambda s^{4.925} & 1.41 \leqq \Lambda s \leqq 4.89 \\ 8.0 & 4.89 < \Lambda s < 13.25 \\ 151.71\,\Lambda s^{-1.1379} & 13.25 \leqq \Lambda s \leqq 100.0 \end{cases} \tag{3}$$

Ks/K for the three-dimensional roughness is less than that for the two-dimensional roughness.

We assume that equation (3) can be apply to the rough wall where the roughness elements consist of hemispheres or cones and they distribute closely such surface as the casting. Λs of those are generally within the limits of $4.89 < \Lambda s < 13.25$, where the ratio of the height to the radius of the base of the corn is below 0.5. Then, Ks/K=8 is obtained from equation (3). This value is much larger than that of the results as shown in Chapter 4. Therefore, it is difficult to apply this correlation to the surface practically used.

The three-dimensional roughness plate which was modified by a screen was investigated by Furuya *et al.* [7] and Korgsrad *et al.* [8]. The following correlation is obtained by the latter.

$$Ks/K = 3.2 \tag{4}$$

where K is the thickness of the screen and larger than twice the wire diameter of the screen used. This value is also larger than that of the rough wall of the casting.

The above discussion is concerned with the friction loss of completely rough regime. There are some problems in the transition regime, where the flow friction loss becomes to part from the hydraulic smooth condition and to gain the constant value to the Reynolds number. Roberton *et al.* show the relationship of the friction loss coefficient of a smooth to rough transition [9]. They classified the variation of the coefficient to the Reynolds number roughly into three types as shown in Figure 2.

A: Natural roughness after many years or isolated roughness elements ;
Moody chart based on Colebrook formula (1).

B: Sand grains and other angular, closely spaced roughness elements ;
Nikuradse's experiment.

C: Corrugated surface or pulsed surface.

Although the experimental results of Niluradse concerning sand grains show Type B, Moody chart is adapted Type A based on Colebrook formula. For Type B, Koh proposed the following equation for pipe-friction coefficient which is covered from smooth to completely rough regime [10].

$$\frac{1}{\sqrt{\lambda}} = 1.74 - 2.0\log\left[\frac{2Ks}{D} + \frac{48.6\exp\{-0.0212\,Re(2Ks/D)\,\sqrt{\lambda}\text{-}30}{Re\sqrt{\lambda}}\right] \quad (5)$$

Where D is pipe diameter. The results on the roughness for a sand paper show nearly Type C [2] of which the roughness consists of a corrugated surface or pulsed surface. In any event, further study is needed for the transition regime.

The discrete element approach where the effect of a collection of individual roughness elements on the momentum transport processes in the flow is considered, has been investigated by Horge *et al.* [11] against the equivalent sand roughness concept. The differential equations including roughness effects are solved numerically using an iterative marching implicit finite difference procedure. They recommend that this approach is to offer the best possibility for predicting the friction given the roughness geometry and flow conditions. However, there are some problems for the transition regime and it is difficult to relate this approach to the roughness Ra. Then, we seek after the conventional method of which the friction loss can be estimated using a loss coefficient.

4. Correlation Between Ks and Ra

If the correlation between Ks and Ra or Rz is determined, the loss coefficient may be obtained from the existing formulae and the effect of the roughness of the flow passage on the friction loss is considered to a certain extent.

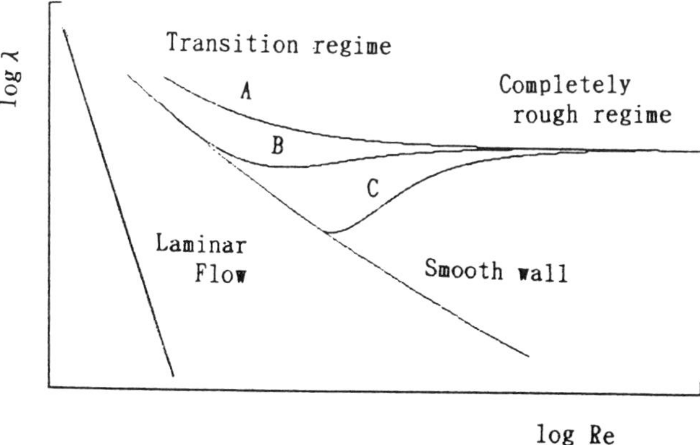

Figure 2. Variation of friction loss coefficient in transition regime [9].

The correlation between Ks and Ra has been investigated by many researchers. Young first introduced the correlation from the experiment on the drag of a blade roughened by a paint at high sub-critical speed [12]. Although, there also are many studies on the boundary layer on the rough wall, a little study investigated the surface of a target of the arithmetic average roughness Ra. Acharya *et al.* obtained the correlation between Ka and Ra of the plate made from casting and the modified plate [13].

In relation to the turbomachinery, Speidel carried out the experiment on a cascade roughened by a sand [14]. Bammert *et al.* investigated the influence of the surface roughness on the performance of a gas turbine and an axial compressor [15]-[17]. The similar studies on an axial compressor were performed by Koch *et al.* [18] and Schaffler [19].

For a hydraulic machines, Fay [20] and Henry [21] introduced the correlation between Ks and Ra in relation to the performance conversion.

We also carried out the experiments on the two-dimensional channel and the square duct roughened by the sand papers [2][3]. For the two-dimensional channel [2], the experiments were carried out for the channel roughened by three kinds of sand paper of the roughness Ra=22.7, 53.5, 108.5 μm by using an air as working fluid. The channel width was 500mm and the height b was varied with from 15mm to 50mm. The equivalent sand roughness Ks was obtained by two ways. One is from the friction coefficient λ by substituting the experimental λ into equation (2) and the other is from the logarithmic velocity distribution. Both results had good agreement.

Figure 3 shows the results of the correlation Ks/Ra of the Ra=108.5 μm to the Reynolds number Re in completely rough regime. Although data scatter from Ks/Ra≒7 to 10, it is noted that the following correlation is reasonable for the completely rough regime. From the similar experiments for the square duct, we confirmed this correlation [3].

$$Ks=9\ Ra \tag{6}$$

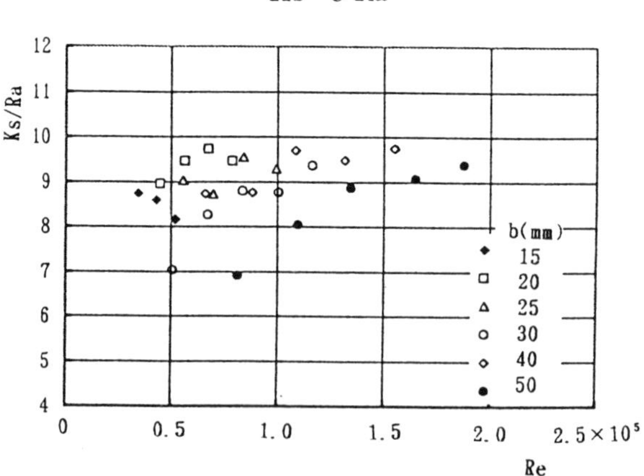

Figure 3. Correlation between Ks and Ra of Ra=108.5 μ m [2].

5. Loss Coefficient Based on Ra

In order to obtain the friction loss coefficient based on Ra, the experiments were carried out for the rough wall roughened by the sand papers by the author *et al.* [2][3]. It seems that the surface of the sand paper is almost the same roughness profile with the casting or the passage in a turbomachinery, because Rz/Ra of the sand paper is about 5~7 as shown in TABLE 2.

The empirical equation for λ based on Ra is obtained from the above experiment[2]. For the completely rough regime,

$$\lambda = 0.038(\mathrm{Ra} \times 10^3/4\mathrm{M})^{1/3} \tag{8}$$

For the transition regime,

$$\lambda = 0.0035(\mathrm{Ra} \times 10^3/4\mathrm{M})^{-2/3}\,\mathrm{Re}^{A} \tag{9}$$

Here, $A = 0.2(\mathrm{Ra} \times 10^3/4\mathrm{M})^{0.6}$

Equations(8) and (9) have the parameter $\mathrm{Ra} \times 10^3/4\mathrm{M}$. Practically, it is better to use this, because Ra is in general expressed by unit μm and the hydraulic mean depth M by unit mm and then we can substitute directly Ra [μm]/4M[mm] into them instead of Ra[mm]$\times 10^3$/4M[mm].

Figure 5 shows an example of the comparison between the experiment[2] and the above equation. In the figure, Prandtl and Kármán formula for the smooth wall, and Colebrook and Koh equations(1) and (5) where Ks is estimated from equation (6) are also plotted.

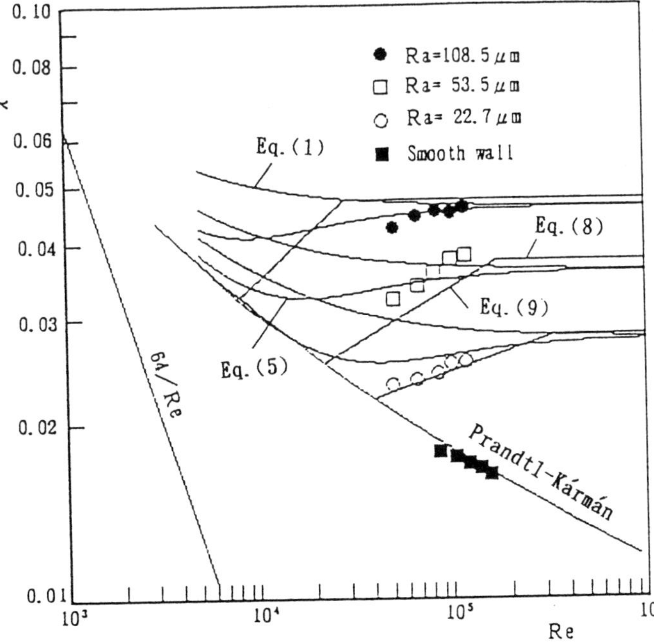

Figure 5. Friction loss coefficient for rough wall passage.

TABLE 3 shows the existing main correlation. However, there are many differences between the correlation obtained from the previous studies. Except for the results of the machined surface of Bammert *et al.*[16], the painted surface of Young[12] and Acharya *et al.*[13] Ks/Ra is about from 5 to 10.

For the completely rough regime, the pipe friction loss coefficient can be calculated from the following well known equation (7).

$$1/\sqrt{\lambda} = 1.14 - 2.0\log(Ks/D) \quad (7)$$

The difference of λ due to the correlation between Ks and Ra is examined as shown in Figure 4. The figure shows λ / λ_0 distribution for Ra=20 μm, where λ is calculated from equation (7) by changing Ck from 1 to 20 for D/Ks=$5\times 10^3 \sim 5\times 10^2$ (D=100mm) and λ_0 is the value Ks/Ra=9. For reference, Ks/Ra obtained from main researcher is marked. The error due to the conversion of Ks by using Equation (6) is below $\pm 10\%$ within Ks/Ra from 5 to 10. Therefore, the correlation of equation (6) is useful for the practical use, although further investigation is necessary for the machined and painted surfaces.

TABLE 3. Existing correlation between Ra and Ks.

Reference	Researchers	Correlation	Rough surface
[12] 1960	Young	Ks = 1.68 Ra	Painted aerofoil
[13] 1986	Acharya	Ks = 6 Ra	Casting
		Ks = 4.2 Ra	Painted surface
[16] 1976	Bammert	Ks = 2.19 $Ra^{0.877}$	Turbine blade
[18] 1976	Koch	Ks = 6.2 Ra	Axial compressor
[19] 1980	Schaffler	Ks = 10 Ra	Axial compressor
[20] 1976	Fay	Ks = (5.26~7.69)Ra	Hydraulic machine
[21] 1980	Henry	Ks = 25 Ra	Hydraulic turbine
[2] 1995	Akaike	Ks = 9 Ra	Sand Paper

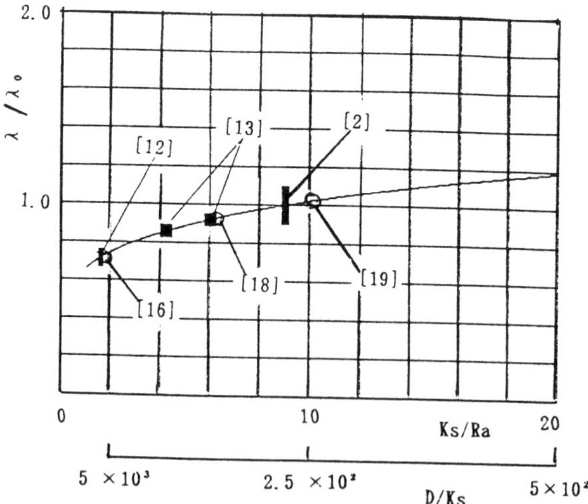

Figure 4. Difference of λ from conversion between Ks and Ra.

For the completely rough wall regime, where the loss coefficient does not depend on Re, the difference of the coefficient between those calculated is small. The comparison between λ_K obtained from equation (7) which is the same with equations (1) and (5) for this regime and λ_R obtained from equation (8) is shown in Figure 6. The coefficients λ_K and λ_R are given for Ra/4M from 10^{-5} to 10^{-2}. The conventional value λ_K obtained from equation (7) by using the correlation equation (6) is large compared with the value obtained from equation (8) for $Ks/4M \lesssim 5 \times 10^{-4}$. However, the difference between those for $Ks/4M \gtrsim 5 \times 10^{-4}$ where λ is large, is small. It is noted that the correlation of equation (8) for the friction loss calculation is practically useful for the completely rough regime.

On the other hand, the friction coefficient for the transition regime does not agree with the conventional equations as shown in Figure 4. The experimental results are less than those calculated from equations (1) and (6). The object of this study is to make clear the friction loss coefficient of a surface which is used in a turbomachinery and is concerned with closely spaced rough element. Therefore, we proposed equation (9) for the transition regime from the experiment on the sand paper surface. Under the present conditions, when we need to obtain the friction loss coefficient directly from Ra, it seems that equation (9) is the only equation for the transition regime. However, further studies on the various surfaces and in the range of the high Reynolds number are needed.

6. Conclusions

In this paper, the treatment of the friction loss coefficient such rough wall passage as in a turbomachinery is discussed and we recommend the following treatment.

(1) For the completely rough regime, the correlation of Ks = 9R is useful for the calculation of the coefficient from the conventional formulae.

(2) For the roughness expressed by the roughness Ra, the empirical equations (8) and (9) are useful for the completely rough and transition regimes.

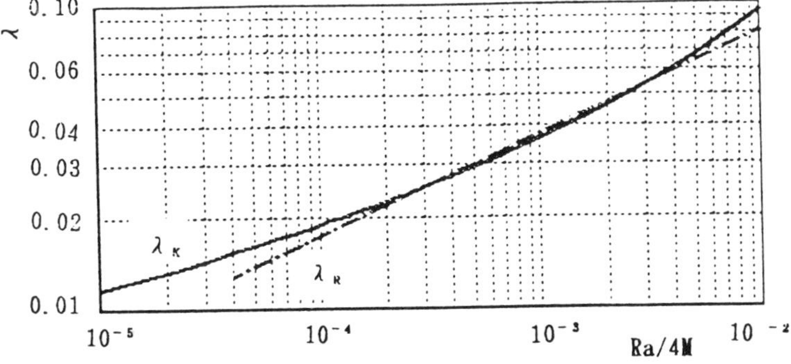

Figure 6. Comparison of friction coefficient for completely rough regime.

References

1. Raupach, M.R., Antonia, R.A. and Rajagopalan, S. : Rough-wall turbulent boundary layers, Appl. Mech. Rev. **44-1**(1991), 1-25.
2. Akaike, S., Nemoto, M. and Nakane, I. : Fully developed turbulent flow in two-dimensional channel with rough wall, Proceedings of 10th Conf. on Fluid Machinery, Budapest(1995), 21-29.
3. Akaike, S., Ohtsuka, N., Nemoto, M. and Nakane, I. : Flow and friction loss in square duct with rough wall, Proceeding of Fifth International Conference of Fluid Mechanics, Cairo **2**(1995), 200-210.
4. Taylor, R.P. : Surface roughness measurements on gas turbine blades, Trans. ASME, J. of Turbomachinery, **117** (1990), 175-180.
5. Bettermann, D. : Contribution a l'etude de la convection force turbulente le long de plaques roughness, Int. J. Heat Mass Transfer, **9** (1966), 153-164.
6. Sigal, A. and Danbergt, J.E. : New correlation rough density effect on the turbulent boundary layer, AIAA J., **28-3**(1990), 28-30.
7. Furuya, Y., Miyata, M. and Fujita, H. : Turbulent boundary layer and flow resistance on plates roughened by wires, Trans. ASME, J. of Fluids Engineering, **88**(1976), 635-644.
8. Korgstad, P.-A., Antonia, R.A. and Browne, L.W.B. : Comparison between rough- and smooth-wall turbulent boundary layers, J. Fluid Mech., **245** (1992), 599-617.
9. Robertson, J.M., Martin, J.D. and Burkhart,T.H. :Turbulent flow in rough pipes, I&EC Fundamentals, **7-2**(1968), 253-265.
10. Koh, Y.-M. : Turbulent flow near a rough wall, Trans. ASME, J. of Fluids Engineering, **114** (1992), 537-543.
11. Horge, B.K., Taylor, R.P. and Coleman, H.W. : Predicting turbulent rough-wall skin friction and heat transfer, Encyclopedia of Fluids Mechanics, Gulf Publishing Co. (1989), 445-468.
12. Young, A.D. : The drag effects of roughness at high sub-critical speeds, J. Royal Aero. Society, **54** (1950), 534-540.
13. Acharya, M., Bornstein, J. and Fscudier, M.P. : Turbulent boundary layers on rough surface, Experiments in Fluids, **4**(1986), 33-47.
14. Speidel, L. : Einfluss der Oberflachenraughkeit auf die Stromungverluste in ebenen Schaufelgittern, Forshung Gebiete Ungen., **20-5**(1954), 129-140.
15. Bammert, K. and Sandstede, H. : Measurements concerning the influence of surface roughness and profile changes on the performance of gas turbine, Trans. ASME, J. of Engineering for Power, **94** (1972) 207-213.
16. Bammert, K. and Sandstede, H. : Influence of manufacturing tolerances and surface roughness of blades on performance of turbine, Trans. ASME, J. of Engineering for Power, **98** (1976), 29-36.
17. Bammert, K. and Woelk, G.U. : The influence of the blading surface roughness on the aerodynamic behavior and characteristic of an axial compressor, Trans. ASME, J. of Engineering for Power, **102** (1980), 283-287.
18. Koch, C.C. and Smith, L.H.Jr. : Loss sources and magnitudes in axial-flow compressor, Trans. ASME J. of Engineering for Power, **98**(1976), 411-424.
19. Schaffler, A. : Experimental and analytical investigation of the effects of Reynolds number and blade surface roughness on multistage axial flow compressor, Trans. ASME, J. of Engineering for Power, **102**(1980), 5-13.
20. Fay, A. : On the high-accuracy scaling-up for pumps and turbines, Conf. 'Pump and Turbine Design and Development', NEL, Paper **2-5**(1976).
21. Henry, P. Influence de la rugosting sur le rendement d'un modele reduit de turbine Francis, Symp. of IAHR, Tokyo **1** (1980), 677-687.

REDESIGN OF SHARP HEEL DRAFT TUBE - RESULTS FROM TESTS IN MODEL AND PROTOTYPE

N. DAHLBÄCK
Senior Research Engineer
Vattenfall Utveckling AB, S-814 26 Älvkarleby , Sweden

Summary

Draft tubes with a sharp heel are quite common in hydro power stations constructed 50 years ago. This design reduced both construction time and investment. We have shown that there is potential for increasing unit performance by a moderate modification of such draft tubes. The redesign has been done with the aid of model tests in a turbine rig. The new design has been installed in a 50 MW hydro unit and an efficiency improvement in the order of 0.5% has been verified through accurate measurements.

Resumé

Il ya nombreux des aspirateurs avec talons acérés dans stations hydroélectrique plusieurs de cinquante ans. Cet design reduit le temps de construction et l´investissement. Nous avons montré un augmentation potentiel de rendement, aprés un changement modéré. Les experiments ont realisé utilisant notre laboratoire de turbines moderne. Le design nouvel est implementé dans un tube d´aspiratuer de Kaplan turbine (puissance 50 MW). Un augmentation environ de 0.5% est verifié aprés essais sur place.

E. Cabrera et al. (eds.), Hydraulic Machinery and Cavitation, 985–993.

1. Introduction

Draft tubes with a sharp heel are quite common in hydro power stations constructed 50 years ago. This design reduced both construction time and investment. The operational properties have also proved to be quite good. However, there is still potential for increasing unit performance by modifying these draft tubes with a moderate input of resources.

Gubin [1] has investigated sharp heel draft tubes and found an efficiency loss of 0.3-2.3% due to the sharp heel.

There are more than 50 sharp heel draft tubes in Sweden in hydro power stations representing 6700 GWh/year electrical generation. A small increase in performance in these power stations represents a considerable economic value. Vattenfall as a power producer was interested in a pilot project demonstration. This pilot project has included both model tests, the construction of steel plates for installation in the prototype and accurate efficiency tests in the prototype before and after the modification of the draft tube.

Full scale tests verifying small improvements noticed in model tests are rare, since they are both costly and hard to perform accurately enough. This paper presents an example of a full scale verification.

The unit used in this project has the following main data; installation year 1949, head 27 m, runner diameter 5.5 m, flow capacity 230 m^3/s and maximum power output 50 MW.

2. Model tests

2.1 TEST RIG AND MODEL DESCRIPTION

Model tests were performed in Vattenfall's own test turbine facility at the Alvkarleby Laboratory. This modern test rig was thoroughly described at an earlier IAHR symposium [2]. Accurate measurements of flow, torque and head results in low total inaccuracy determining efficiency, theoretically less than 0.2%. The earlier presentation [2] also showed that it was possible to detect small differences in efficiency in this test rig. The tests presented here have also used the possibility of easily varying the test head in the test rig.

The model consisted of a 1:11 scale copy of a power station from intake to a tunnel section 20 m down stream of the draft tube outlet. This means that the model Kaplan runner had a diameter of 500 mm. Figure 1 presents a layout of the model set up. This set up has also been used for other tests (not presented here) varying the geometry in intake or tunnel outlet.

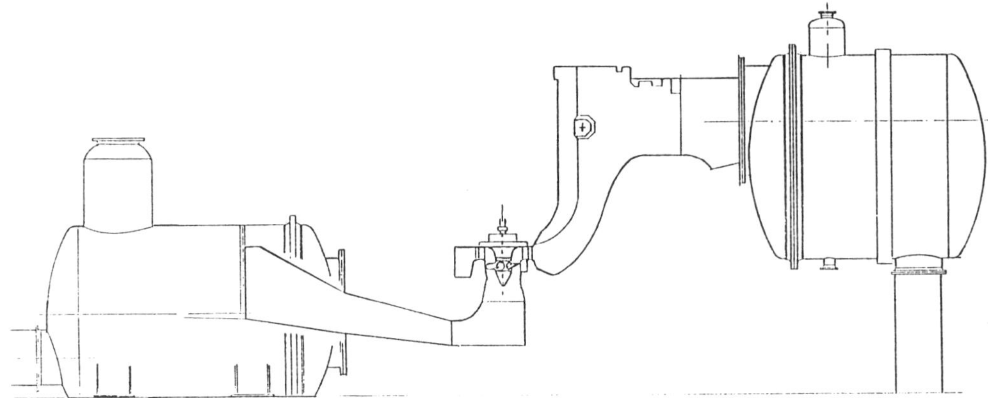

Figure 1 Layout of the model . Size of Kaplan runner = 500 mm.

In addition to the overall measurements of flow, head, torque and speed resulting in efficiency curves, numerous pressure taps were distributed on the draft tube walls. These pressure taps were used to study draft tube pressure recovery and pressure distribution patterns. Windows and other taps in the model made it possible to visualise stream patterns.

2.2 PERFORMED TESTS

Different geometry of the corner in the sharp heel was tested. Pressure field measurements were concentrated on the original design and two redesign options. The tests were also concentrated at three different runner blade angle openings (5°,12.5°, 17,5°). The smallest opening corresponds to best efficiency and the largest opening to nearly maximum flow. At each runner blade angle the guide vane opening was varied, resulting in an efficiency curve.

The model head was varied between 2.2 and 12 m. Still all tests fulfilled the similarity laws in test codes [3] This could be done by simultaneously varying speed and flow. Note that the lowest head, 2.2 m, corresponds to the Froude similarity, i.e. the scaling of the head is the same as the geometric scale of the model.

2.3 MODEL TEST RESULTS

Figure 2 presents efficiency curves for 5 degrees runner blade angle opening , with and without the redesign of the draft tube. Similar curves were obtained for other blade angles. As seen, the shape of the individual blade angle efficiency curves were unchanged at the "left leg" (low flow) but radically increased at the "right leg". We noted an improved overall efficiency, at best combination of blade angle and guide vane opening, in the range of 0.4-0.6% at model head 2.2 m. At higher model heads the tests showed slightly less efficiency improvement. Note that there should be no Reynolds number induced scale effects, due to viscous forces, of efficiency improvements since the flows are the same for the two tests compared. These tests thus imply that there are scale effects not taken care of in the model test codes.

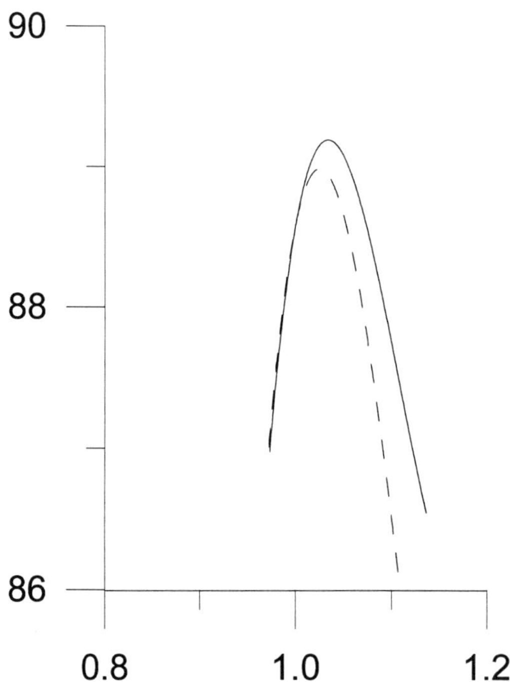

Figure 2 Efficiency curves, turbine efficiency vs nondimensional flow.
Runner blade angle 5 degrees.
Dashed line with original design, solid line with new design.

The pressure measurements at the draft tube wall, were used to calculate a pressure recovery factor ("draft tube efficiency"). These calculations showed that with a redesign "draft tube efficiency" increased from around 50% to around 85%. Please note that the lack of knowledge of velocity fields, especially the rotational component, makes these numbers just a qualitative measure. This is the reason for quote marks around "draft tube efficiency".

Using wall pressure measurements and an interpolation program we have constructed pressure field plots as shown in figures 3 and 4. Please note that the reference level is set to zero at draft tube outlet. Improvement is clearly seen as a lower pressure after the runner. This means a higher net head at the same flow situation, thus a higher output and efficiency.

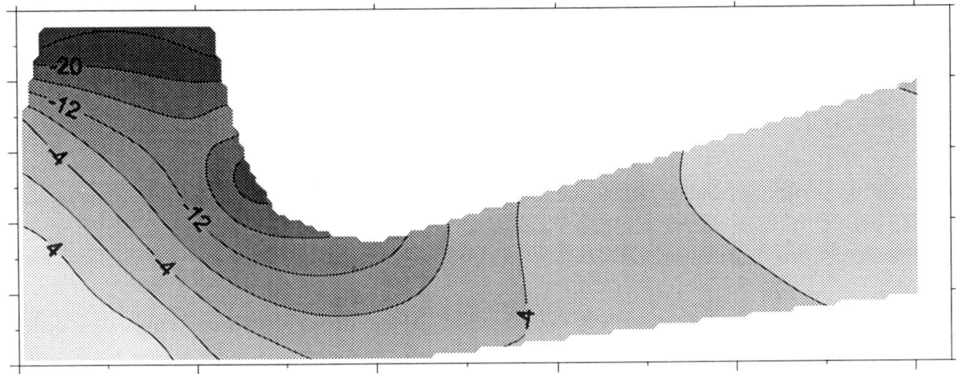

Figure 3 Pressure field plot for original draft tube design.

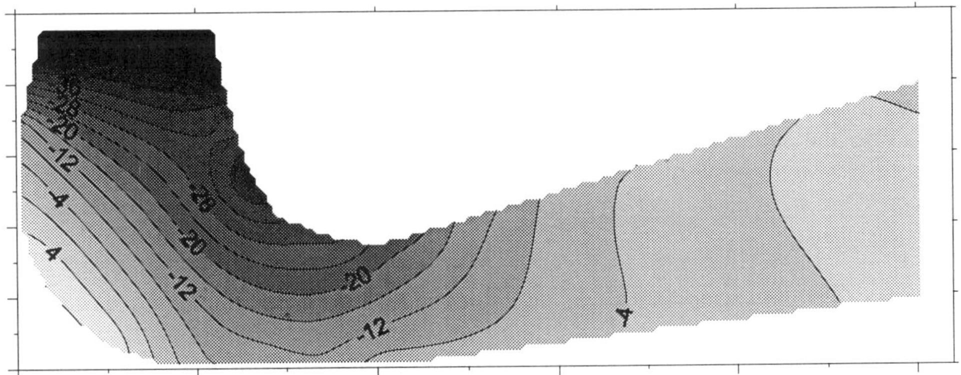

Figure 4 Pressure field plot for new draft tube design.

3. Redesign of Prototype

A steel construction with a shape chosen from model tests was designed to withstand expected pressures and vibrations in the draft tube. The construction consisted of a framework of support bars anchored to the rock penetrating the original draft tube wall. The final shape was made by steel plates welded onto the framework. There will be water on both sides of the steel plates, which is why slots and holes were made to reduce forces due to pressure variations. The construction was 5 m high, 8 m wide and 5 m long . The access opening behind the draft tube gate was 1 m x 3 m, a reason why the steel plates that had been prepared were limited in size. This meant extensive welding work to be done at site.

We chose the steel-plate construction since in a pilot demonstration project we wanted to be able to go back to the original design if there was a major failure. A construction based on concrete would probably have been cheaper and as quickly installed. However, in this case concern is needed for the discharge pit in the bottom of the draft tube.

From this project we also learnt that going into a 50-year-old building demanded some extra effort in researching the exact documentation of anchors and concrete filling areas behind existing draft tube walls.

The cost of design, construction and installation could be lowered when this type of project is repeated. However, that cost in this pilot project will have a pay back time in a few years due to higher production.

4. Verification tests

4.1 METHODS OF VERIFICATION

To verify an expected efficiency improvement in the order of 0.5% there was a need for very accurate efficiency measurements. However, geometry in low head power plants rarely gives opportunities for flow measurements with better than 1% accuracy. This was also the situation for the actual project. Fortunately comparative measurements may give better accuracy. For our verification we used comparisons in three different ways. Test were performed on two different occasions, before and after redesign, two months apart.

The first method was an index test based on Winter-Kennedy differential pressure taps. The repeatability of such a test may be in the order of a few tenths of %. However, it is not proved that the relation between differential pressure and flow will be completely undisturbed by the redesign in the draft tube. Small variations in head may also affect the performance of the unit. It was not possible to plan the tests to give perfect repeatability since the mean river flow (depending on long term weather conditions) also affects the downstream level and head.

In the discharge channel, 800 m down stream from the power plant, a flow measurement system was installed. This acoustic flow measurement system was based on eight paths, along which the velocity was accurately measured. The eight velocities may be calculated to a flow but may also be used directly for comparison between two almost identical flows. Small variation in water level, causing slightly different mean velocity, may be corrected for by the information of velocity profile obtained in the measurements.

To increase accuracy, tests were performed with only one unit running and the others stopped and closed by intake gates. Each discovered leakage was measured or estimated. The test period for a single efficiency value, was as long as 6-8 hours to reduce influence from large scale turbulence and the influence of storage between the unit and the measurement site. The repeatability of velocity measurements is very accurate and the evaluated statistics from the long test run give a number on the large scale turbulence influence less than 0.3%.

The flow measurements in the channel were used for the other two methods of comparative efficiency measurements. One set of test points compared the redesigned unit at best efficiency with a parallel unit. This was done both before and after the redesign. The last method compared the efficiency of the redesigned unit at best efficiency and at 80% load before and after the change.

4.2 VERIFICATION TEST RESULTS

For the unit not redesigned the measurements were separated by two months, and repeated within 0.1% efficiency.

Figure 5 presents index test curves. The overall efficiency improvement for the earlier studied blade angles, is in the range of 0.6-0.8% between best efficiency and 90% load. This is clearly above the inaccuracy of the verification tests. It is also a slightly higher value than the model test showed. No improvement can be seen at maximum output. Within the inaccuracy, a few tenths of %, the different verfication methods showed the same result.

The index tests showed good agreement with model tests regarding the change in individual blade angle efficiency curves. The curves got broader by "lifting the right leg".

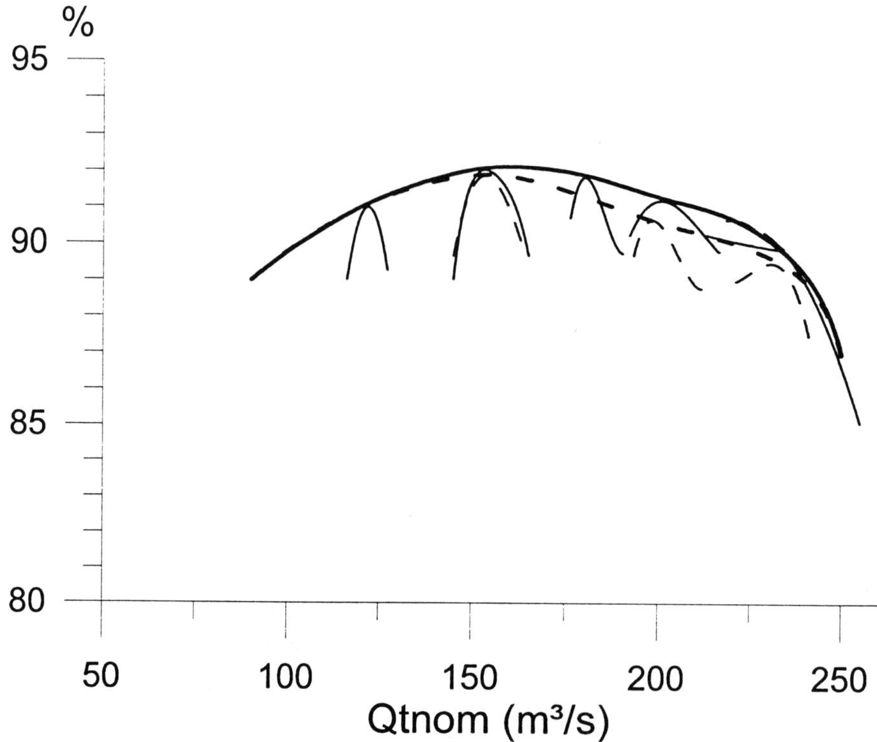

Figure 5. Resulting efficiency curves from index tests (with fixed blade angle curves). Dashed line original design. Solid line after reconstruction.

4.3 EARLIER EXPERIENCE

A few years earlier we did a similar test in a 9 MW hydro power plant. This site had a horizontal stretch in the pen stock well suited for accurate flow measurements. A redesign of the sharp heel draft tube resulted in an efficiency improvement of 1-1.5% from 40 to 90% load. Also for this power plant, model tests showed a slightly lower efficiency improvement.

5. Conclusion

Redesign of a 50-year-old sharp heel draft tube showed improvements of efficiency better than 0.5%. This means that a redesign normally has short pay off time. Small efficiency improvements may be hard to verify in low head power plants. However, the verification methods used here, together with model tests, have clearly shown an efficiency improvement.

The sharp heel draft tube was a competitive design when it was installed. Therefore we may conclude that other draft tubes also can be redesigned, resulting in efficiency improvement.

The relative size between different causes of energy losses in the draft tube is still very little known. The final design of the draft tube is therefore carried out in an experimental set-up.

References:

1. Gubin, M. F. (1973) *Draft tubes of hydro-electric stations*, Amerind publishing Co, New Dehli.

2. Marcinkiewicz, J and Svensson, L (1994) *Modification of the spiral casing geometry in the neighbourhood of the guide vanes and its influence on the efficiency of a Kaplan turbine*, XVII IAHR symposium China , 1994.

3. International standard IEC 995 (1991), *Determination of the prototype performance from model acceptance tests of hydraulic machines with consideration of scale effects.*

MODEL AND PROTOTYPE DRAFT TUBE PRESSURE PULSATIONS

Vladimir KERCAN, M.Sc.
Marin BAJD, PhD.
Vesko DJELIĆ, M.Sc.
Andrej LIPEJ, M.Sc.
Dragica JOŠT, M.Sc

TURBOINŠTITUT
Rovšnikova 7
1210 Ljubljana
SLO

During last several years intensive investigations of draft tube pressure pulsation on various specific speed turbines have been performed. Tests were made in the wide turbine operating range from partial to full load and from runaway to head above optimum. Cavitation and various kinds of air admission influence on the pressure pulsation as well as on the efficiency characteristics were investigated.

All pressure pulsation were investigated on the model and prototype of wide range turbine specific speeds (ns) 159, 213, 340 and 400. Achieved results of the transfer from the model to the prototype has been presented and analysed.

1. Introduction

All producers of Francis turbines daily meet the problem of draft tube vortex appearance and its consequences [1,2,3,4]. Limitations in turbine operation because of the vortex appearance and undesirable consequences on hydraulic system are sufficient reason for its intensive researching. According to such unpleasant and sometimes only hard soluble consequences on the prototypes, the research is directed to model tests and numerical flow analysis with the aim to predict consequences on prototype.

E. Cabrera et al. (eds.), Hydraulic Machinery and Cavitation, 994–1003.

Our vortex researches were orientated to find answers on following questions:

- influence of specific rotational speed on vortex appearance,
- influence of operating regime and cavitation on vortex consequences,
- influence of turbine design of same specific speed on vortex phenomena,
- influence of pressure pulsation measurement position on the results,
- possibilities of the measuring results transfer from the model to the prototype,

To find answers to these questions, the models of 4 specific speeds have been measured in a complete range of turbine operation under different cavitation conditions. Prototype measurements have been made in a wide range of 10 years. This is the main reason for the chosen approach to the processing of the pressure pulsation measuring results at the beginning of 10 years period.

2. Experimental approach to vortex analysis

The complete applicable range of Francis turbines in our vortex phenomena research was covered with 4 different speed models: Ns 159, 213, 340 and 400 respectively [5,6,7,8,9]. Unfortunately all models have not been designed under the same procedure and hence the obtained results have been also influenced by the various turbine design access. However, on the turbine Ns 159 three runners designed upon quite different procedure were investigated in the same spiral case and slightly different draft tube. Three different way designed runners of the model Ns 340 in the same spiral case and draft tube were analysed.

The main task of the research was to establish the general condition of pressure pulsation related to the turbine operating point. All measurements were performed under real prototype cavitation conditions and at wide operating range, even wider than it is realised during the prototype exploitation.

The research program for a basic model Ns 159 is showed on the Fig. 1. Almost the same research program was carried out also on models Ns 213, Ns 340 and Ns 400. Usual dimensionless flow and specific energy coefficients φ and ψ were used for the hill diagram presentation. The dimensionless guide vane opening which defines the turbine geometry is also shown.

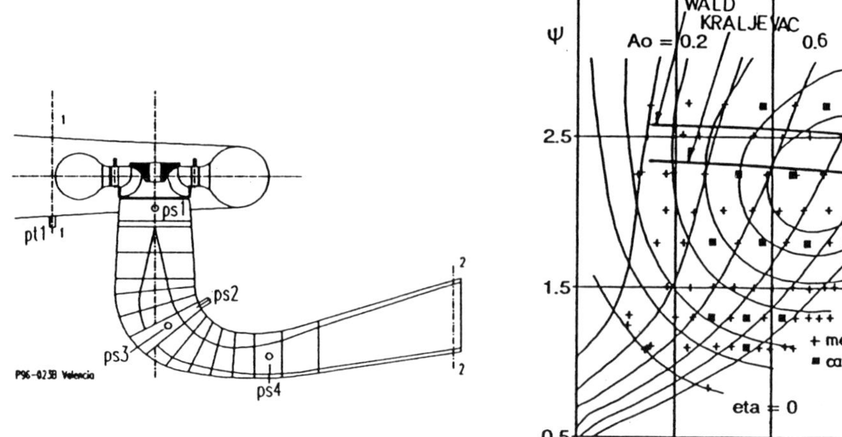

Figure 1. Measuring programme for Ns159 model and prototype tests on HPP Kraljevac and Wald

All models were tested on the Francis turbine test rig under approximately same operating conditions. The test head was between 10 and 16 m, cavitation conditions were achieved by decreasing the suction pressure in the test rig. The special attention was focused on the basic test rig characteristics. The main and by-pass valves were reconstructed to reduce the bearing clearance to minimum in order to reduce their vibrations which caused pressure pulsation in the test rig. Pressure pulsation measurements on 7 test rig positions (measuring points comprehend all characteristic elements of the rig) and in different operation regimes were performed before model tests. It was found out that the pulsation level is relatively low and no characteristic frequency in the frequency range important for vortex investigation was found.

Cavitation influence on the pressure pulsation in different operation regimes was investigated for each model separately. On the Fig. 1 the hill diagram for the model Ns 159 is presented with signed operating points in which model characteristics were measured under different cavitation conditions. It is evident from the diagram, the cavitation influence was measured in a wide operating range which enabled conclusions about the cavitation influence on the model characteristics. The same test programme was carried out on other models.

Pressure pulsation measurements on different positions of flow passage were carried out to investigate the draft tube vortex behaviour in various operating regimes. On all models the positions of pulsation measurements were almost the same: there were four or more pressure taps on the suction side and one or more on the pressure side of the model. The taps positions are presented on the Fig. 1 for the model Ns 159 and for the other models the taps were identical arranged. Positions of all measuring taps are marked dimensionless, relative to the runner outlet diameter 350 mm for all models.

2.1. ANALYSIS OF THE MODEL PULSATION

All pressure pulsation results are presented by means of signal equivalent amplitude calculated on the basis of 98% probability [9]. This amplitude is signed as A98 and normalised with turbine head. The most significant amplitude in a frequency spectrum is signed with A1 and in all diagrams normalised double amplitude 2A1/H is presented. The frequency of most significant amplitude signed with f1 is normalised by the speed (f1/fn), because the speed frequency is characteristic in all operational conditions.

In the Fig.2 results of pressure pulsation measurements on six draft tube positions at optimum specific energy coefficient for the model Ns400 are presented. The highest A98 amplitude appears in the region at app. 80% of optimum flow while the highest amplitude of significant frequency appears in the region at 75% of optimum flow. In the region of strong vortex the frequencies of all taps are practically the same, 0.25-0.30 of rotational frequency. Equivalent amplitude A98 is significant in wider flow range than the amplitude A1. In a wider flow range the pulsation in the draft tube is present only in a narrow range with sharp determined frequency. Due to the fact that amplitude A98 consists of all frequency pressure amplitudes mentioned phenomenon should be taken into account when explaining all analysed results. The pulsation at position PT1 in the turbine inlet has the highest amplitude A98 while the highest amplitude A1 is achieved at the taps PS1 and PS2 - at the inlet of the draft tube. The ratio between the highest amplitude and amplitudes at turbine optimum is between 6 and 8. In the flow region below 40% of optimum flow no significant amplitude A1 exists, in comparison with the highest amplitudes the values decrease for 4-6 times. In the same region amplitude A98 is smaller then the highest one but only for app. 2 times. In this region the pulsations in the draft tube are present but without significant amplitude of vortex frequency, which could caused dangerous oscillations.

Although the influence of cavitation conditions has been investigated at all models and in many operating points, on the Fig.3 the results at only two operating points of the Ns159 model are presented. The point signed No 1 has an intensive vortex, another, No2 is completely vortex free. In the point 1 great influence of cavitation appears at the cavitation coefficient value $\sigma = 0.1$ when both amplitudes, A98 and A1, rapidly increase. In the point 2 till very low σ value, no increase of pulsation amplitudes is observed. The cavitation has no influence on the vortex frequency in the regimes where the strong vortex appears. The frequencies in point 2 are very scattered because the vortex with significant frequency is not developed. Therefore, from point to point, the frequency of the highest amplitude varies and this is the reason for scattering of results and not cavitation conditions. Comparison of various positioned tap amplitudes shows relatively small differences. In any case the differences are much smaller than on the Fig.2 at the model Ns400. In the point of well developed vortex the differences are smaller - well defined vortex is transmitted through all turbine water passages.

The influence of runner design procedure on the vortex appearance has been analysed on the models Ns159 and Ns340, but only results of Ns159 are presented. Due to a prototype requirements the models with 13 and 15 runner blades with the same geometry and only slightly different draft tube extension have been tested.

The third compared runner has 13 blades and the same specific speed but a little lower guide vanes. The pressure pulsation comparison of all three runners at the pressure tap on the draft tube inlet (PS1) and for optimum specific energy coefficient is shown in the Fig. 4. Results of basic 15 blades runner are signed by filled square, of 13 blades runner of the same geometry and different draft tube extension by asterix while of 13

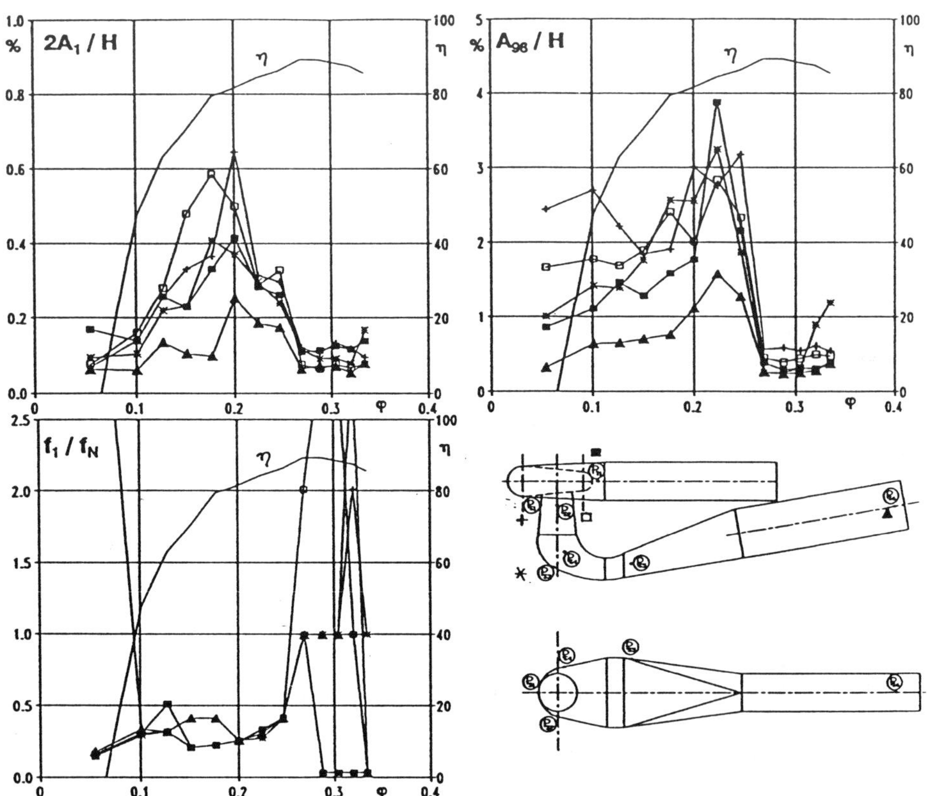

Figure 2. Influence of tap positions on pressure pulsation for various flows at optimum specific energy coefficient measured on the model Ns400

blades runner and lower wicket gate is signed by cross. The difference is unexpected for the same specific speed model and shows great importance of model geometry on vortex consequences.

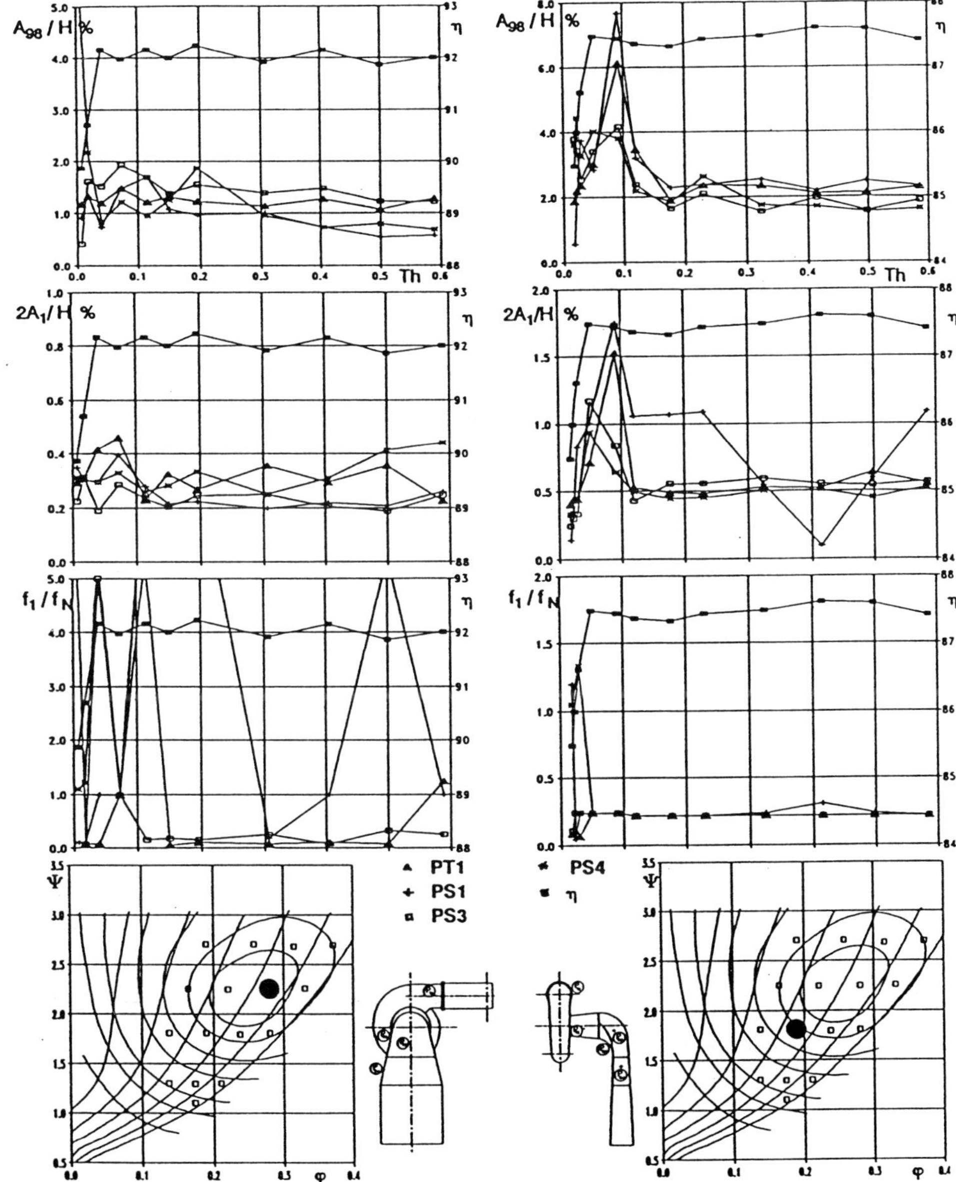

Figure 3. Influence of cavitation conditions on pressure pulsation for model Ns159 in two operating points (with significant vortex and without it)

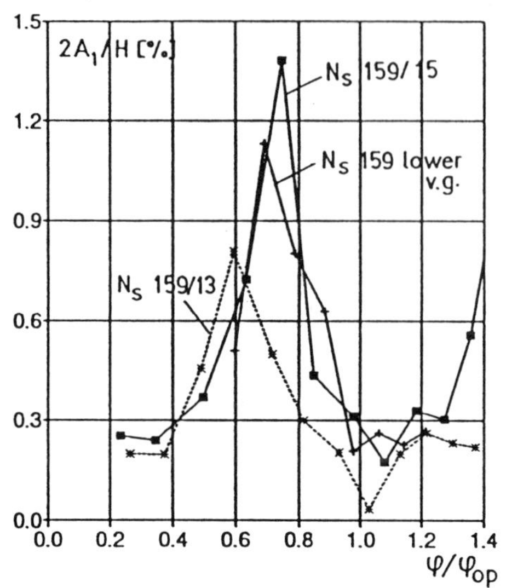

Fig. 4 Influence of runner design on pressure pulsation at the PS1 tap

To determine the influence of the model specific speed on the pressure pulsation amplitudes on the Fig. 5, amplitudes A98 and A1 for tap PS1, at the draft tube inlet, are drawn for optimal specific energy coefficient.

The tests have been carried out at cavitation coefficient of 0.09, 0.10, 0.22 and 0.26 respectively for the models of Ns159, 213, 340 and 400. From the diagrams a great influence of specific speed could be concluded. Unfortunately the relation between pressure pulsation amplitude and specific speed is not well determined. Unexpected high amplitudes of Ns340 model, specially concerning A98 amplitude, could not be easily explained. From our model tests experience we could conclude that the mentioned model is the worse

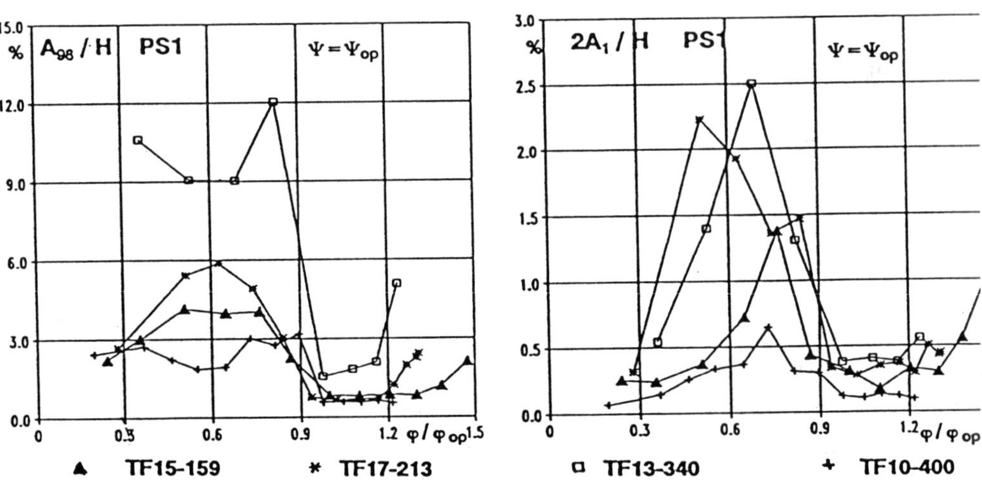

Figure 5. Influence of model specific speed on pressure pulsation at optimum specific energy coefficient and at the draft tube inlet

between all tested models, but more due to its geometry than to it's specific speed. This conclusion is supported by the results of other runner of the same specific speed with significantly smaller amplitudes of pressure pulsation.

2.2. ANALYSIS OF PROTOTYPE PULSATION

The test carried out over a period of many years enable a collection of prototype measurements. Therefore for each tested model one or more prototypes has been tested. Prototype operating conditions do not allow vortex investigations in such wide operating range as the models do. Usually the turbine characteristic measurements have been carried out at one head and under site cavitation conditions.

Hydraulically similar turbines to the model Ns 159 have been investigated in two sites with runner diameter D = 575 mm and D = 1200 mm respectively. The turbine with the runner diameter of 575 mm at the outlet is installed in the power plant Kraljevac, on Adriatic coast. The other one with 1200 mm runner diameter in the power plant Wald in Austria, near Salzburg. The prototype measuring points are presented on the hill diagram on the Fig. 1.

Hydraulically similar to the model Ns 213 is a turbine of HPP Jajce (Bosnia), to the model Ns 340 a turbine of HPP Hemfurth (Germany), while a turbine of HPP Stratos (Greece) is made out of a model Ns 400.

Besides the dynamic characteristics of the mentioned turbines, their energetic characteristics has been measured also, using the thermodynamic method in Wald, current meters flow measurement in Hemfurth and Stratos and index test method in Kraljevac and Jajce.

All prototype pressure pulsation measurements were carried out with piezoresistant transducers in the same way as at the model measurements. The measurement range was of course adjusted to the site conditions.

Results of prototype measurements on PS Wald, hydraulically similar to the model Ns 159, and comparison with model test results for draft tube inlet tap are given on the Fig.6. Prototype pressure pulsation amplitudes are smaller then those of the model. Amplitudes A98 increase for several times in very narrow flow range while model amplitudes are high in a wider range. Prototype turbine operates very smoothly and it is installed very rigidly in the power plant.

The difference between model and prototype amplitudes A1 is much bigger than the difference of A98. Tests on the model and prototype are made at the same cavitation coefficient and therefore the measured difference could not be explained as a cavitation influence.

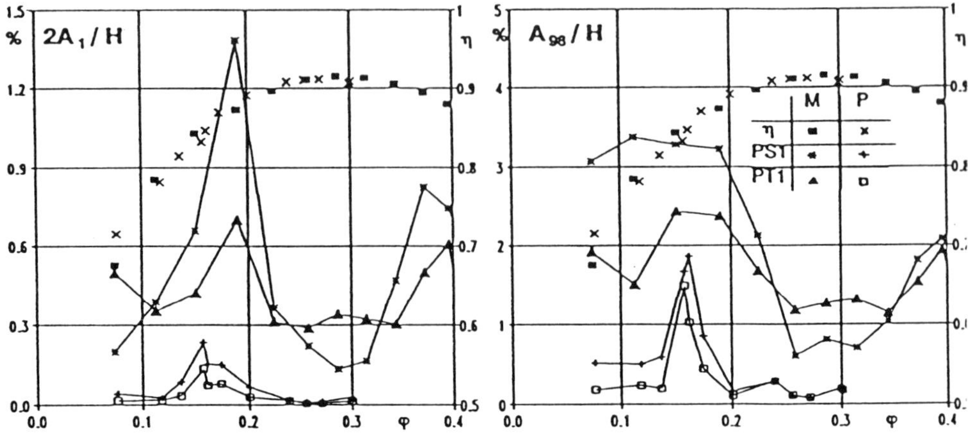

Figure 6. Comparison of pressure pulsations between model Ns159 and prototype HPP Wald

Both tests are performed without air admission. Therefore considering the experience from presented results and all other prototype tests results, no exact conclusion could be accepted for results transfer from the model to the prototype.

4. Conclusion

On the basis of all results obtained from various specific speed models, position of pressure taps, prototypes and operating regimes (in the paper only a smaller part of the results is presented) several conclusions can be derived. Presented results allow discussion about various influences on vortex appearance and consequences. General conclusion of all research is that for each important power plant with Francis turbine detail model tests of pressure pulsation should be performed. Judgement on quality of tested models and prediction of prototype characteristics could be given only from well experienced staff, who has enough experience with model and obviously also with prototype tests and results.

4. References

1. Purdy, C.C. (1979) Reducing power swings of Tarbela's turbines, *Water Power & Dam Construction*, April 1979, 23-27.
2. Bribiesca, S., Paredes, R.B., Paredes, C. (1990) The behaviour of francis unit under partial loads, *Water Power & Dam Construction*, Aug. 1990, 46-52.
3. Guarga, R., Torres, J.J., Solorio, A., Rodal, E. (1986) Comparative study of La Angostura draft tube vortex oscillatory behaviour, IAHR Symposium, Montreal 1986, 6, 11.
4. Fisher, R.K., Ulith, P. (1982) Comparison of draft tube surging of homologus scale model and prototype francis turbines, *Voith Research and Construction*, Vol. 28e, 21.
5. Kercan, V. (1986) Report about francis model turbine test ns159, Technical report 1973, Turboinštitut, Ljubljana.
6. Ranisavljević, M. (1990) Meritve modelne turbine ns213, Technical report 1965, Turboinštitut, Ljubljana.
7. Gregori, J., Kercan, V., Pišljar, M., Hočevar, D. (1987) PS Catalan - dynamic characteristics of model francis turbine, Technical report 1859, Turboinštitut, Ljubljana.
8. Kercan, V. (1986) PS Stratos - report about model acceptance test, Technical report 1738, Turboinštitut, Ljubljana.
9. Kercan, V., Bajd, M. (1988) Computation and experimental investigation of draft tube vortex in francis turbines of various specific speed, IAHR Symposium, Trondheim, 355-364.

FISH BYPASS SYSTEM IMPACT UPON TURBINE RUNNER PERFORMANCE AT ROCKY REACH DAM

ALFRED LANG,
Managing Director
ASTRÖ
Graz- Austria

BILL CHRISTMAN, PE
Principle Civil Engineer
PUD No. 1 of Chelan County
Washington, USA

Abstract

Public Utility District No. 1 of Chelan County, Washington State, USA (Chelan) owns and operates the Rocky Reach and Rock Island Hydroelectric Projects located on the Columbia River. Several species of salmon and steelhead trout migrate through these projects. Downstream migrating juvenile fish pass through either the turbines, where they experience some mortality, or through migration-augmenting spill flows, which are marginally effective and very expensive. Chelan has developed a "Surface Collection" fish bypass system and "Fish-Friendly" turbines at its Rocky Reach Project to mitigate harmful effects to the downstream migration of the juvenile fish. The surface collection system will be installed within the Rocky Reach forebay, supported by the upstream face of the dam, and will encroach upon the generating unit intakes. This paper describes the modeling investigation used to define the effects the surface collector structure will have upon turbine operating characteristics and how that information will be used to balance bypass effectiveness with turbine performance.

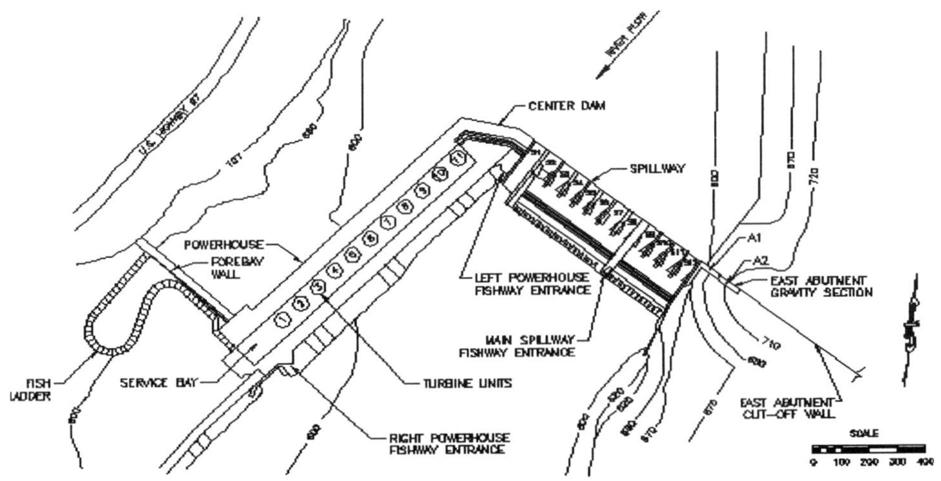

Figure 1: Rocky Reach Dam

E. Cabrera et al. (eds.), Hydraulic Machinery and Cavitation, 1004–1013.

1. Introduction

The Columbia River drains 671.000 km² (259,000 square miles) of Canadian and Northwest United States lands. It is 2044 km (1270 miles) long (745 miles lie within the USA) and discharges into the Pacific Ocean. The Columbia River has the fourth largest hydraulic discharge in North America, the greatest power potential in the USA, and is developed with eleven dams. It also supports several species of salmon and steelhead trout. These anadromous fish migrate upstream as adults to spawn and downstream as juveniles to mature in the ocean. Early in the hydroelectric development of the Columbia River, smolts were forced to migrate downstream through hydroelectric turbines. More recently, spill flows have been used to provide downstream passage around the turbines. At Rocky Reach Dam, where the spillway is located upstream of the powerhouse (Figure 1), spill flows are not significantly effective in attracting smolts away from the turbine intakes. Spill flows are also very expensive in terms of foregone power.

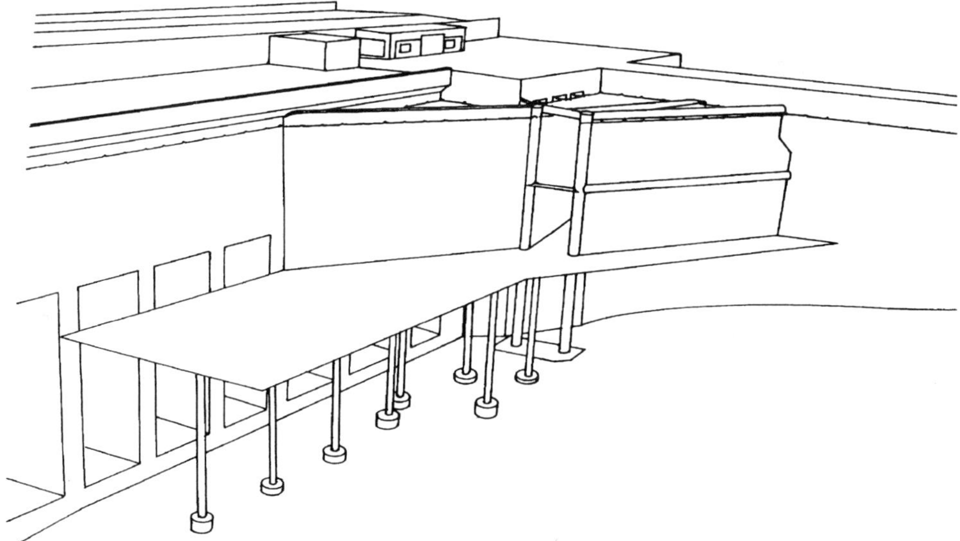

Figure 2: Surface Collecting Fish Bypass System

Chelan's current efforts to mitigate fishery resource degradation at Rocky Reach include development of a Surface Collection Fish Bypass System (Figure 2) and development and installation of Fish-Friendly turbines.

Chelan's goals are to bypass at least 60% of smolts around the turbines with the Surface Collector, improve turbine survival for the smolts which do not utilize the bypass system to the range of at least 96%, and to pursue on/off-site fishery enhancement measures. As a part of Chelan's fish passage decision-making, they will need to determine how many Surface Collector entrances to construct, how deep to make the entrances, and how far the Surface Collector should extend into the project forebay.

These key decisions must take into consideration the balance between expected biological effectiveness and turbine/generator efficiency. The deeper and wider the Surface Collector, the greater the reduction in turbine efficiency.

Fish Passage Features at Rocky Reach Dam

Chelan's fish passage indexing investigations indicated that most smolts enter the upper portions of the turbine intake and therefore pass the turbine in the hub and mid to inner-blade regions. Rocky Reach forebay depth is about 100 feet and the majority of fish migrate within the upper portions of the water column.
Chelan installed and tested a prototype Surface Collection Turbine Bypass System in 1995 and the prototype tests were significantly encouraging.
Although the direct turbine survival rate for smolts is expected to be 96% or higher with newly developed turbine runners, the collector is being developed for two primary reasons. One, indirect turbine mortality, which can be caused by stress, pressure changes, and near-powerhouse predation by avian and other fishery species likely decreases overall turbine passage survival. Secondly, turbine bypass system mortality as measured in the collector prototype was nearly zero. Therefore, Chelan's objective of maximizing overall smolt passage survival at Rocky Reach will be augmented with installation of an effective turbine bypass system.

2. Model Test Objectives

The primary objectives of the model tests are to determine impacts which expansion of the Surface Collector will have upon turbine performance characteristics (e.g. efficiency, cavitation, vibration) and to define a relationship between various Surface Collector lateral/vertical limits and turbine/generator performance.
Efficiency losses have a cost in terms of unproduced available energy. Likewise, blade cavitation and/or excessive machine vibration increase maintenance costs and generating unit outages. On the other hand, increasing the vertical and/or lateral extent of the Surface Collector will increase its biological benefits. Chelan will use the model test results as a part of its efforts to find a balance between power generation cost effectiveness, and its goal of improving overall Rocky Reach fishery impacts to a no-net-loss level of performance. Potential biological benefits not pursued in support of generation efficiency will be made up with other on-site and off-site improvements.

Secondary objectives of the model-test investigation include:

a) Providing guidance as to the Surface Collection system major structural member(s) hydraulically-optimized layout

b) Providing insights to the quasi-Venturi gate final design configuration and hydraulic capabilities

c) Investigating the effects of unbalancing generating unit hydraulic loading. Each of Rocky Reach's generating units consists of three intake bays. Unbalanced hydraulic loading, if it does not unduly influence turbine performance, may be desirable by allowing one-only Venturi gate system per generating unit (thereby reducing Surface Collector construction costs) and/or increasing operational flexibility of multiple Venturi gates.

3. Test Arrangement

The tests were carried out with a 1:20 scale fully homologous model of the existing Kaplan turbines, units C1 to C7. The model runner has five blades with a runner diameter of 356 mm (14 inch).

The main in-flow conditions at Rocky Reach are characterized by the fact that the main turbine flow direction is perpendicular to the river flow direction. In order to simulate the flow condition of the plant, an inlet tank of about 2.4 m x 2.4 m (7.9 ft x 7.9 ft) cross section and 5 m (16.4 ft) length was built, which allowed to install the various anticipated structures in front of the turbine intake (see Figure 4). The bathymetry in the surrounding of the turbine intake and the trash rack were modeled to scale and remained unchanged for all tested configurations. The top wall of the inlet tank simulated the plant head water surface. The inflow to the head tank was equipped with several screens and flow straighteners to provide uniform velocity distribution. The intake part of the homologous model was connected to the head tank and the draft tube was connected to the tailwatertank. From the tailwatertank to the model inlet tank the circuit was closed by a piping line including the flow measuring section and two booster pumps.

The system pressure control unit consisting of an expansion tank, vacuum pump, pressurized air supply and exhaust valve, allows to adjust any absolute pressure from 10 kPa to 500 kPa in the tailwater tank to provide various cavitation conditions.
The two booster pumps of the test stand are connected in series. The main booster pump in the vertical loop is driven by a 300 kW variable speed motor, whereas the second exchangeable one is driven by a 200 kW variable speed motor.

From the main booster pump the water flows into the Venturi-tube inlet vessel and, passing the heat exchanger device and the straightener inside the tank, through the Venturi tube for flow measurement. From there, the water is conducted through the flow distributor to the second booster pump, to the inlet tank and to the model.
The draft tube discharges into the horizontally arranged tailwater tank from where the water flows back to the main booster pump passing the energy dissipator and the flow distributor.

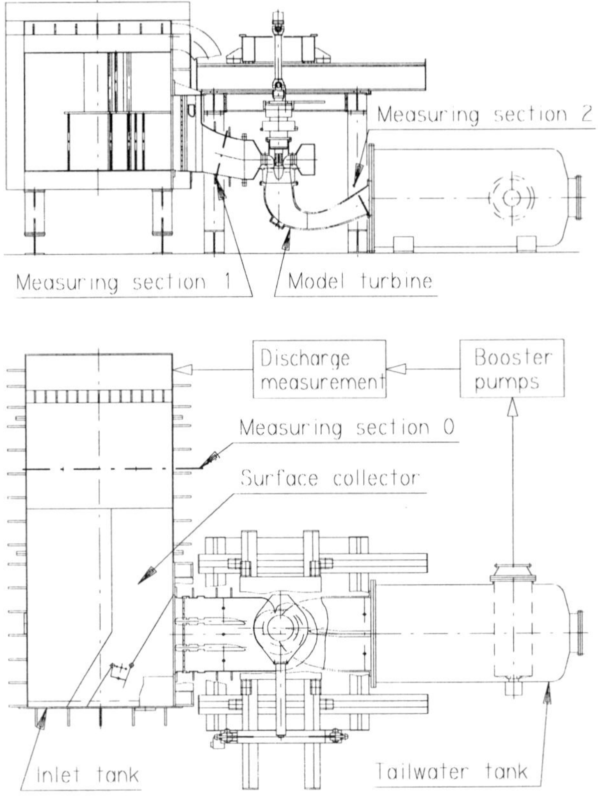

Figure 3: Test Arrangement

The model was mounted on a frame and connected to the hydrostatically supported motor/generator by means of a cardan shaft. The model shaft was set in a double bearing system such that the friction torque caused by bearing and seal is mechanically transmitted to the motor/generator and hydraulic torque was measured directly on the lever arm of the model bearing assembly.

4. Measurement

The tests were carried out with the same standard as applied for competitive tests in a world class hydraulic laboratory. The tests are fully compatible to the latest state of the IEC code and all results can be transformed to the prototype with the premises stated in the IEC codes.

The measurements on the turbine model included head, discharge, torque, rotational speed, suction head and pressure pulsation. Discharge was measured by Venturi tubes, all differential pressures were measured by rotary piston manometers, hydraulic torque was measured by means of calibrated weights and load cell, using hydrostatic support of the dynamometer.

The overall uncertainty of the measurement is approximately 0.2 % of model efficiency. The particular purpose of the test was to compare various configurations and, therefore, the reproducibility being less than 0.1 % in efficiency was the relevant factor.

4.1 TEST PARAMETERS

The main parameters of the tests were performance, cavitation, pressure pulsation and by-pass flow rate. Performance is characterized by efficiency versus power output curves where turbine net head is kept constant. The tests were performed for various runner blade angles and gate openings, creating propeller curves.

Cavitational behavior is described by efficiency versus sigma curves (break curves) for a given operating point, i.e. head, discharge and power output are kept constant. Additionally to the break curves, visual observations were made.
Pressure pulsations were measured in the spiral case, the discharge ring and the draft tube. The evaluation was done by means of FFT analyses. According to the nature of the investigation, emphasis was put on the spiral case pressure fluctuations, particularly on low frequencies.

The bypass flow rate was measured by means of a propeller type current meter. The velocity distribution at the entrance of the bypass system was measured in pretests and a reference point was determined for further measurement, using the correlation between total bypass flow and reference velocity.

In order to determine the model net head, two upstream measuring sections were installed: The first was upstream of the surface collectors in the completely undisturbed flow. The head measurement with this measuring section included all additional losses caused by the surface collector to be investigated. The second measuring section was installed in the turbine intake, shortly downstream of the trashrack. This measuring section was also used for the previously performed competitive tests and, aside from other purposes, served for reference measurements. Efficiency was measured using the differential pressure between the tank measuring section "0" and the draft tube measuring section "2", and comparisons were made between the various configurations. The pressure difference between head tank measuring section "0" and intake measuring section "1" was determined in parallel. The difference between efficiency drop on the one hand and the measured head loss on the other provided an indication about secondary effects on the turbine behavior.

All configurations were tested at one or more runner blade angle, at least at two prototype heads, at non-cavitating conditions and at plant sigma, for full by-pass and small by-pass flow, if applicable.

4.2 TESTED CONFIGURATIONS

The investigated surface collecting system is driven by a hydrodynamic effect. The turbine intake is split by a large horizontal shelf and the main water flow is entering the turbine underneath the shelf creating a lower static pressure due to increased velocity. The decreased static pressure draws water through the primary screen obtaining the attraction flow which guides fish to the by-pass channel. The main task of the investigation was to find a configuration which provides a high attraction water flow and has a low impact on turbine performance. Since, at this point of time, it was not clear how many units will be included in the surface collector system, variations were made, particularly investigating the impact on the turbine adjacent to the one which creates by-pass flow.
The surface collector will only be in operation for several months of the year and will be out of operation for the rest of the year. Therefore, the investigations included the impact on turbine performance for both fully developed by-pass flow and small by-pass flow.

The main geometrical parameters to be investigated were the lateral extention of the shelf, the elevation of the shelf and the shape of the Venturi cowling. Altogether eleven configurations were tested which are described as in Table 4 and Figure 4:

Test No.	Description of Tested Configuration
1	Reference test set: no installations, trashrack
2	Shelf elevation 650 ft, extention 64 ft, no cowling
3	Shelf elevation 650 ft, extention 64 ft, cowling to 630 ft
4	No shelf, roof seal at elevation 650 ft
5	Shelf elevation 630 ft, extention 64 ft, no cowling
6	Shelf elevation 630 ft, extention 100 ft, no cowling
7	Same as no. 6, simulation of adjacent unit
8	Shelf elevation 630 ft, extention 125 ft, no cowling
9	Shelf elevation 630 ft, extention 125 ft, blockage of bay B and C
10	Shelf elevation 640 ft, extention 100 ft, cowling on bay A
11	Shelf elevation 640 ft, extention 100 ft, cowling on bay A, B and C

Table 4: Tested Configuration

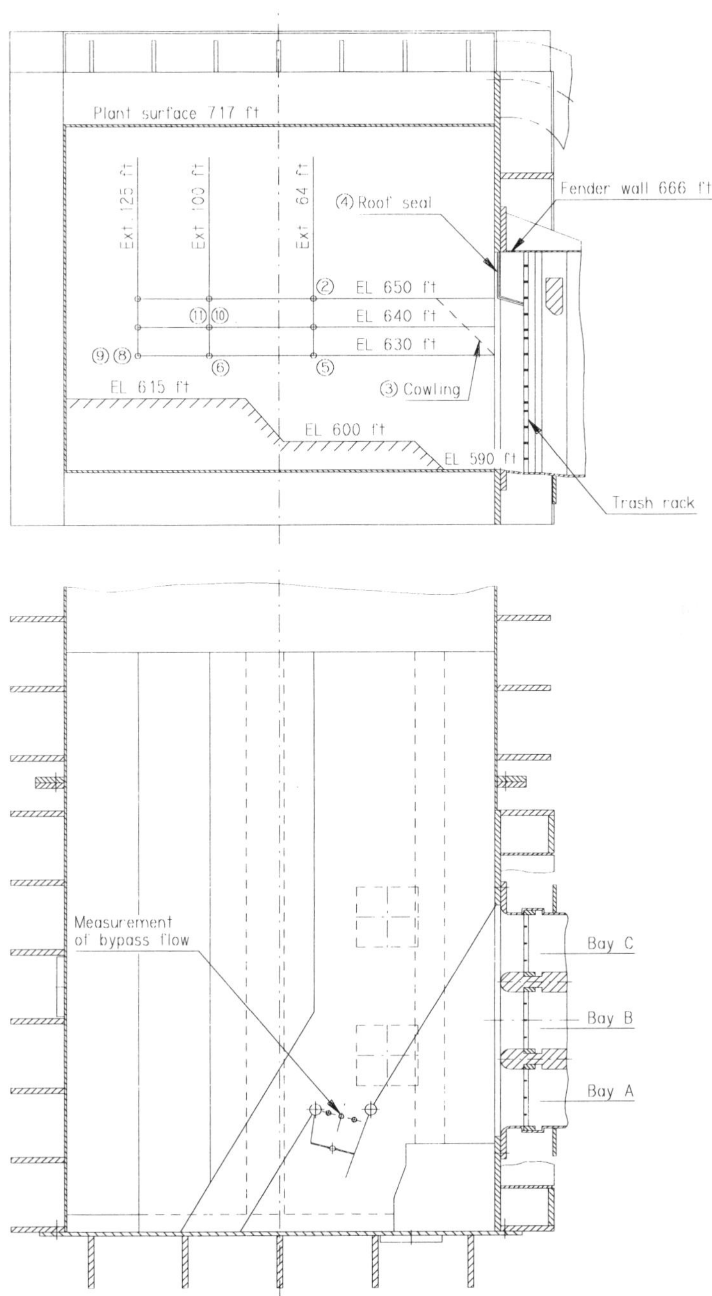

Figure 4: Tested Configurations

5. Test Results

Reference measurements and several other pretests were carried out to make sure that the teststand is functioning properly. As a result of these tests the trashrack losses were determined without other installations. The loss coefficient is approximately 0.9 which is close to the value previously determined in similar tests at Aströ.

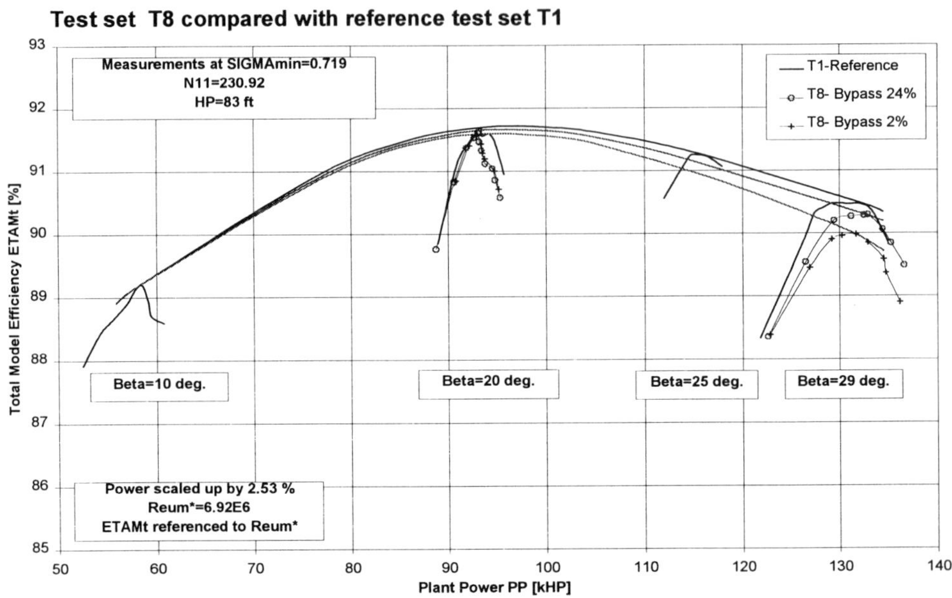

Figure 5: Influence on Efficiency

Comparison between test 2 and test 1 shows no measurable efficiency loss and 6 % by-pass flow. There is a slight Venturi effect which is not causing any losses.
Comparison between test 2 and test 5, where only the shelf elevation was changed shows a decrease in efficiency of 0.2 % and an increase of by-pass flow from 6 to 20 %. Configuration of test 3 has the same area of blockage as test 5. The decrease of efficiency is 0.3 % and the by-pass flow is 18 %. The additional losses are probably due to additional contraction of the turbine inflow caused by th cowling.

Test 6 and test 7 have the same elevation as test 5 but the lateral extention is increased from 64 ft to 100 ft. The difference between test 6 and test 7 is that test 6 has inflow conditions as if the adjacent turbine was in full operation. The by-pass flow is 22 % (test 6) and 24 % (test 7), the maximum efficiency drop is 0.2 % and 0.3 % respectively.

In test 7 where the inflow area is already very restricted, the turbine is sensitive to the amount of by-pass flow. The efficiency drop increases from 0.3 % at full by-pass to 0.8 % at low by-pass.
Test 8 is the same as test 7 with the shelf extended from 100 ft to 125 ft and no change of the results was encountered (see Figure 5).
Test 9 has the same basic configuration as test 8 with bay B and C blocked above the shelf. The by-pass flow is reduced by half and, corresponding to that., the losses in efficiency are increased. As expected, the losses for small by-pass flow remain the same.
Test 10 with medium elevation (640 ft) and medium extention (100 ft) and only one cowling on bay A is one of the likely compromises to be built in the plant. The results are 0.1 % drop of efficiency and 12 % by-pass flow.
Test 11 is similar to test 10 except for cowlings on all three bays. Efficiency drop is 0.2 % and by-pass flow is 20 %.
Test 4, the test of the roof seal showed no influence on the turbine behavior when the intake area is fully blocked from the top down to an elevation of 650 ft.

The loss of efficiency at high power outputs is higher than the equivalent head loss in the intake. This means that there is a secondary influence on turbine performance at large runner blade angles which is not the case for the smaller runner blade angles.

All of the configurations showed an impact only on turbine efficiency. There was no influence on cavitational behavior and pressure pulsation.

6. Future Testing and Conclusions

The model tests are completed and documented. Chelan will test its prototype fish-friendly turbine design with the use of live fish in the spring of 1996, and will test a prototype extended Surface Collector Fish Bypass System during the 1996 spring and summer salmon and steelhead trout outmigration. Chelan expects to combine these results with the model study results in planning for the final Surface Collector configuration. The biologically-based turbine and Surface Collector test results and a tentative reporting of how Chelan has combined the model tests with the biologically-based field tests will be available in the second half of 1996.

LDV MEASUREMENTS IN AN IMPELLER-GENERATED TURBULENT JET DEVELOPING IN A WEAK COFLOW

PER PETERSSON, MAGNUS LARSON, AND LENNART JÖNSSON
Department of Water Resources Engineering
University of Lund, Box 118, S-221 00 Lund, SWEDEN

Abstract: Results are presented from detailed measurements in an impeller-generated turbulent jet developing in a weak coflow. A laser doppler velocimeter was used to measure the mean velocities and normal stresses in three coordinate directions, together with two of the Reynolds shear stresses. An impeller operating at 1800 rpm was placed in a flume with a coflow having a velocity of 0.05 m/s upstream the impeller. The coflow velocity decreased uniformly along the impeller jet as water was entrained into the jet. Complete velocity measurements were made at selected downstream cross sections, including both the zone of flow establishment and the zone of established flow. The mean velocities and the normal stresses approximately displayed self-similar behavior 6*D* downstream the impeller, where *D* is the impeller diameter.

1. Introduction

Fluid mixing is important in many industrial and environmental applications. A mixer is a device consisting of an impeller and a motor that is often used in such mixing operations. The rotating impeller generates a swirling jet that entrains and mixes ambient fluid as it develops downstream, simultaneously as a large-scale circulation is set up in the fluid. Close to the impeller the swirling jet properties are determined primarily by the impeller characteristics (zone of flow establishment, ZFE), whereas further away the jet may be regarded as being generated by a point source of linear and angular momentum (zone of established flow, ZEF). The division between the ZFE and ZEF is typically made based on whether the mean profiles display self-similarity, especially the mean axial velocity profile.

Studies on impeller-generated jets are limited [3],[6],[7]; however, some investigations on swirling jets generated by other means exist [1],[8],[9]. Also, flows similar to impeller jets occur in other fields, such as aerodynamics [5] and naval architecture [4], from which valuable insights might be obtained. The objective of the present study was to determine the three-dimensional velocity field in an impeller-generated jet developing in a weak coflow. A small coflow velocity was employed to avoid the problems with recirculation resulting in a closed tank and to investigate the jet development in a

E. Cabrera et al. (eds.), Hydraulic Machinery and Cavitation, 1014–1023.

flowing ambient, which is of interest in environmental applications or inline mixing [2]. A two-component laser doppler velocimeter (LDV) was employed to measure the velocity field.

2. Laboratory Experiments

A swirling jet was generated with a 1:10 model of the impeller from a Flygt 4501 mixer [3]. The three-blade impeller had an overall diameter of D=0.078 m and a hub diameter of 0.023 m. A rectangular, glass-walled flume with a cross-section of 0.9 x 0.9 m^2 and a total length of 21 m was used in the measurements (see Fig. 1). The water depth in the flume was 0.85 m and a small coflow was employed in the direction of the jet, preventing recirculation in the flume over the distance of observation. A coflow velocity of U_{ao}=0.05 m/s (upstream the impeller) was used in the experiment. The LDV system employed in the velocity measurements was a two-component colorburst-based four-beam fiberoptic system (TSI 90-3) with a 83 mm probe. The impeller was placed as centered as possible in the flume with respect to the focal length of probe, which is 465 mm in water. A traversing system was developed to allow efficient and accurate measurements with the LDV at arbitrary cross sections downstream of the impeller. At each measuring section the probe could automatically be moved in the y- and z-directions (see Fig. 1). The movement of the traverse was controlled by a computer connected to two motors on the traverse via a control unit. Between measuring sections (in the x-direction) the entire traverse system was moved along a rigid fixed rail aligned with the wall of the flume.

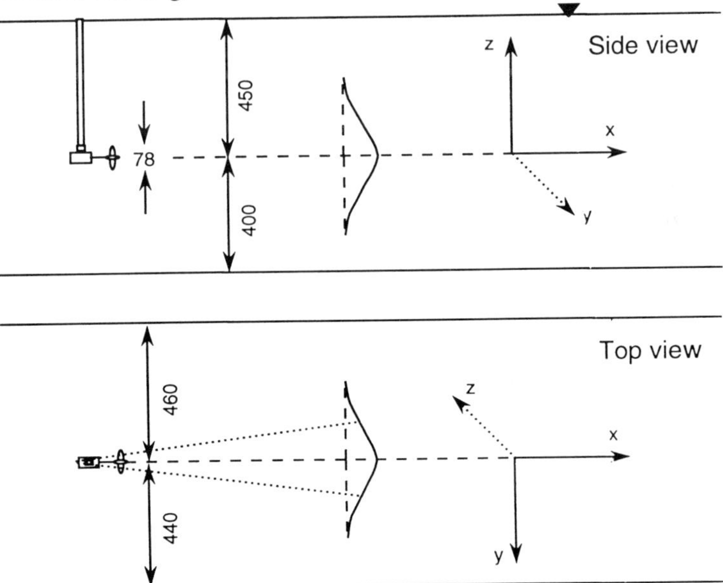

Figure 1. Schematic description of the experimental setup.

The impeller was run at a fixed speed of $N=1800$ rpm during all measurements, and the water was artificially seeded with chalk particles about 5 m upstream the impeller to achieve a satisfactory data rate. In order to obtain the velocity components in all three coordinate directions two sets of measurements were needed at each cross section. Lateral traversing (y-direction) through the jet produced the axial ($U=\bar{U}+u$) and tangential ($W=\bar{W}+w$) velocity components, whereas vertical traversing (z-direction) gave the radial velocity ($V=\bar{V}+v$) together with U once more.

3. Results

Measurements of the mean velocity in the three coordinate directions were performed at selected downstream cross sections together with the corresponding normal stresses and two of the Reynolds shear stresses. The measurement cross sections were located at $0.128D$, $2D$, $4D$, $6D$, $8D$, $10D$, and $12D$ downstream the impeller.

3.1 GENERAL FLOW STRUCTURE

Fig. 2 demonstrates the schematic flow structure in the flume during the experiment. At locations close to the impeller the mean axial velocity profile has a marked trough in the jet center that is mainly an effect of the blocking hub. The trough is rapidly "filled" already a short distance downstream the impeller since the swirl distribution can not maintain such a flow structure. After the maximum mean axial velocity occurs on the jet centerline the velocity profile becomes close to Gaussian and the flow may be regarded as fully established. As the swirling jet develops downstream it entrains surrounding, ambient water causing a marked volume growth, simultaneously as the coflow velocity (U_a) decreases. The entrained water will have a velocity component in the jet direction and thus contribute with momentum to the jet; however, this contribution is negligible in the present experiment.

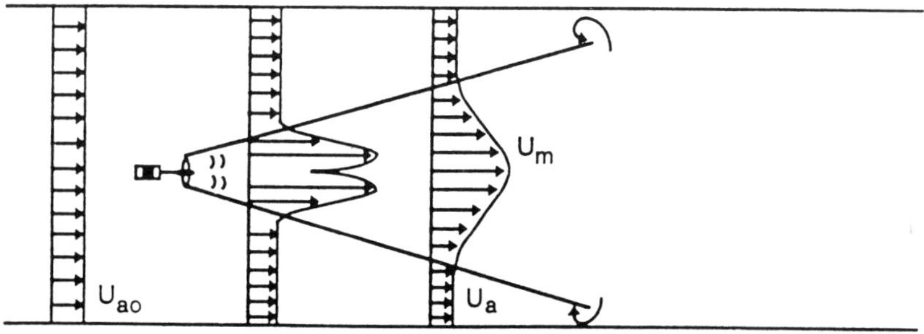

Figure 2. Jet developing in a flume with a coflow.

The continuous entrainment of coflow water reduced the coflow velocity gradually and this reduction was approximately uniform over the cross section. From the original velocity of 0.05 m/s the coflow velocity decreased slowly to close to zero at $12D$. The coflow stabilized the jet path and the deflection or oscillation of the jet was much reduced in comparison with jet development in a closed tank [7]. Close to the impeller a periodicity is introduced in the flow field by the impeller blades that may affect the measurements of mean quantities, especially the turbulence. Spectral analysis was used to investigate the periodicity in the flow and periodic fluctuations were identified between 0-2*D* in the *ZFE*. Limited phase-averaged measurements confirmed the results obtained with spectral analysis. Up to 1*D* the periodicity was found to be marked after which the diffusion of the turbulence from the blade wakes seemed to proceed quickly, and the periodicity had disappeared at 2*D*.

3.2 MEAN VELOCITIES

Close to the impeller (0.128*D*) the results obtained for a coflow are in good agreement with the corresponding experiment for a closed tank [7] where recirculation (counter flow) occurs outside the jet. Thus, in the vicinity of the impeller the ambient flow conditions do not affect the jet development, but the flow is determined purely by the impeller properties. Differences between the jet development in a co- and counter flow were found approximately from 2*D*. The swirl number [1] for the impeller jet was calculated to S=0.26 by evaluating the expressions for the angular and linear momentum flux at 2*D*, where the flow has become axisymmetric, and employing the impeller radius (R_o) as the length scale. This *S*-value corresponds to a weakly swirling jet and the number should be constant going downstream since the momentum flux is conserved; this was confirmed by integrating measured velocity profiles.

Radial distributions of the mean axial ($\bar{U}$), tangential ($\bar{W}$), and radial velocity ($\bar{V}$) normalized with the peripheral blade velocity U_p ($=2\pi NR_o$) are shown in Figs. 3a, b, and c, respectively, from 2*D* to 12*D* (only half the jet is shown due to symmetry and r denotes the radial distance). In the $\bar{U}$-profiles presented the coflow velocity for respective section was subtracted from the measured jet velocity. A distinct trough is found in $\bar{U}$ at 2*D* together with an off-axis peak. At 4*D* the maximum velocity is still found off-axis, although it occurs quite close to the jet axis. Thus, if the ZEF is defined based on the occurrence of a Gaussian profile for $\bar{U}$ this zone will start shortly after 4*D*. At 6*D* the maximum in $\bar{U}$ (U_m) is clearly at the jet centerline.

For $\bar{W}$ two peaks in the radial distribution could be identified close to the impeller (0.128*D*), namely an inner peak produced by the rotating hub and an outer peak produced by the blade-induced swirl [4]. Fig. 3b presents the measured $\bar{W}$-profiles from 2*D* to 12*D*. At 2*D* the two peaks have already merged into a single peak located close to the axis. The swirl distribution changes rapidly between 2*D* and 4*D* with corresponding less influence on the main flow development. The solid-body rotation in the center becomes less concentrated and the peak velocity decays. From 6*D* almost half of the profile shows a weak solid-body rotation and towards the edge of the jet the swirl decays in the form of a free vortex.

Fig. 3c displays profiles of $\bar{V}$, where negative velocity represents an outflow from the jet center. Thus, $\bar{V}$ is directed outwards in the jet center at all cross sections

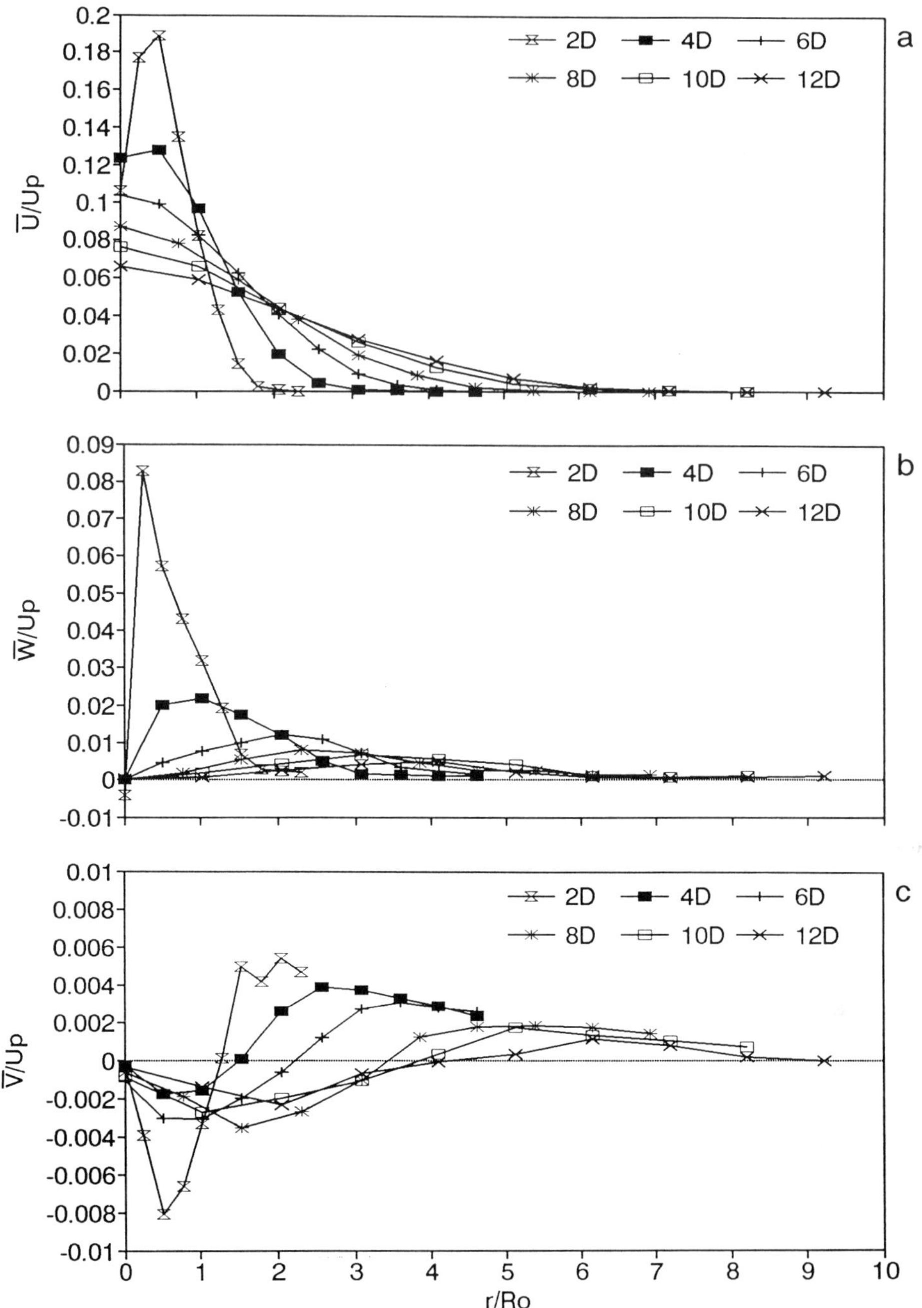

Figure 3. Mean velocity profiles downstream an impeller: a) axial, b) tangential, and c) radial component.

between $2D$ and $12D$. Towards the jet edge the flow direction is reversed and a radial inflow (entrainment) into to the jet may be distinguished. In the jet center values close to zero are obtained (as for $\overline{W}$), which is expected due to symmetry.

3.3 TURBULENCE

Figs. 4a, b, and c display the measured (*rms*) values (normal stresses) for the axial (u_{rms}), tangential (w_{rms}), and radial component (v_{rms}), respectively, normalized with U_p at the locations from $2D$ to $12D$. High *rms* values are observed in the jet core at $2D$ and $4D$ as a result of the initially high turbulence close to the impeller. The off-axis peaks found for all components are a result of the high turbulence produced in the blade wakes and the turbulence generated due to the marked circumferential gradients appearing at these radial locations in the vicinity of the impeller. A rapid general decrease in the *rms* values are noticed from $4D$ to $6D$, whereafter a slower and almost constant decay is obtained for all components. The radial and the tangential components show a peak at the jet center that may be a measurement artefact [6].

Two of the three Reynolds shear stresses ($\overline{uv}$ and $\overline{uw}$) were measured and are presented normalized with $(U_p)^2$ in Figs. 5a and b. For $\overline{uv}$ the present measurements show a smooth profile across the jet and maximum values are obtained near the location of the maximum gradient in $\overline{U}$ (compare Fig. 3a). The radial distribution of $\overline{uw}$ exhibits a complicated development and changes sign between $6D$ and $8D$. At locations close to the impeller $\overline{uw}$ is mainly positive, however, $\overline{uw}$ becomes negative further downstream. Beyond $8D$ the values obtained for $\overline{uw}$ were very small and of the same order as the measurement accuracy and are therefore not presented. Comparing the maximum values for the two measured shear stresses the values obtained for $\overline{uv}$ are about 4 times larger than $\overline{uw}$ at $4D$. Further downstream $\overline{uw}$ is markedly reduced and may be neglected in comparison to $\overline{uv}$.

3.4 SELF-SIMILARITY AND INTEGRAL FLOW DEVELOPMENT

To investigate the approach towards self-similarity $\overline{U}/U_m$ was plotted against $\xi = r/(x+a)$, where a is the distance from a virtual origin to the impeller blades (how a was determined is discussed below). The present measurements indicated that $\overline{U}$ was approximately self-similar from a location where U_m occurred on the jet centerline. A Gaussian curve fitted the $\overline{U}$-profiles well from $6D$ downstream of the impeller, as shown in Fig. 6a. The data from $4D$ were also found to fall on the same curve when normalized with U_m, but since the maximum value was found slightly off-axis this section was not included in the present analysis. The ZEF are therefore assumed to start just downstream of $4D$. If $\overline{W}$ is normalized with W_m (maximum value of $\overline{W}$), a self-similar function for the $\overline{W}$ component can also be fitted reasonably well to the data from $6D$ and beyond (see Fig. 6b; the empirical equation is given in the figure). Finally, the normalized $\overline{V}$-profiles are

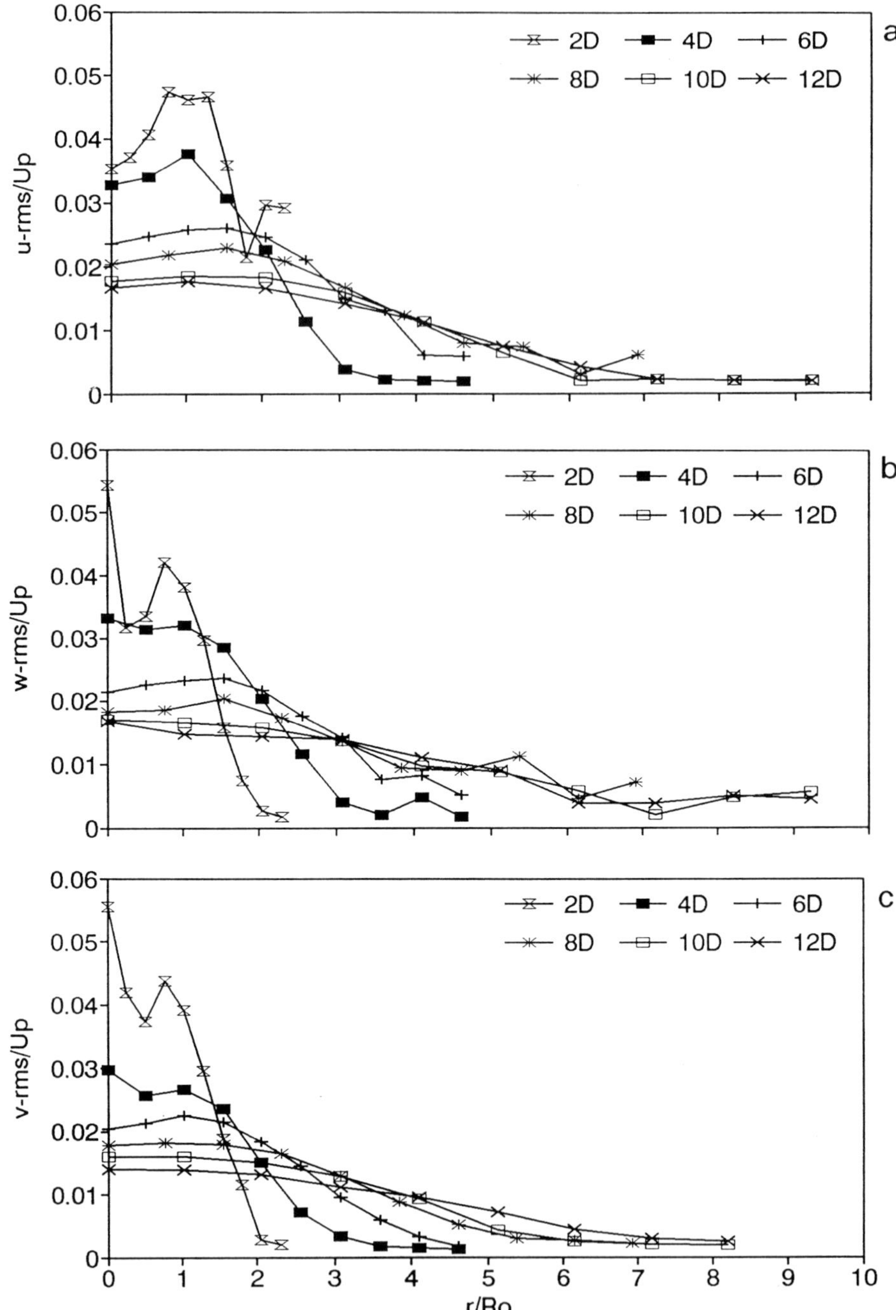

Figure 4. Normal stress profiles (*rms* values) downstream an impeller: a) axial, b) tangential, and c) radial component.

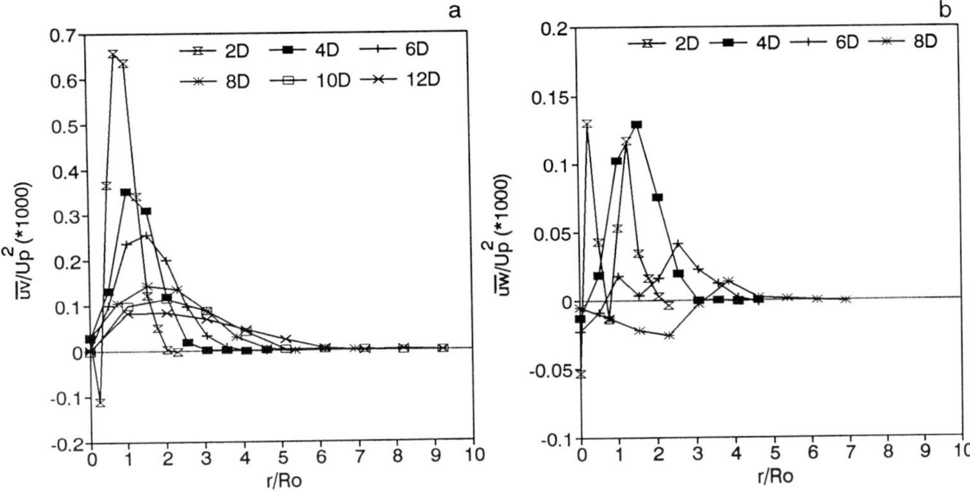

Figure 5. Reynolds shear stress profiles downstream an impeller: a) $\overline{uv}$, and b) $\overline{uw}$.

presented in Fig. 6c. A pattern indicating self-similarity can be seen but the scatter is relatively larger than for $\overline{U}$ and $\overline{W}$, and no curve was fitted to the data.

The radial spread of the jet with distance downstream was determined by plotting the jet width, defined based on the radial distance where $\overline{U}=U_m/2$, as a function of x/D [6]. The jet was observed to spread linearly at a half angle of 5.0 deg, which is a smaller angle than what has been measured for free swirling jets (7.5 deg in [1]). By extrapolating the line describing the radial spread back to the horizontal axis the location for the virtual origin upstream the impeller blades was found to be approximately $4D$. The volume flux Q was determined at downstream cross sections by integrating the measured $\overline{U}$-profiles across the jet, and Q displayed a linear growth over the distance of observation with an entrainment coefficient of $K_e=0.4$ ($K_e=D/Q_o\, dQ/dx$, where Q is the volume flux and index o denotes conditions at the impeller).

4. Concluding Remarks

In the present measurements self-similarity was found for $\overline{U}$ just downstream of $4D$. For free swirling jets generated by other means similarity in $\overline{U}$ was also noted from $4D$ downstream an orifice for weak swirl, whereas for strong swirl similarity was not observed until $10D$ [1]. In [8] and [9] similarity was observed from about $6D$ for jets with moderate swirl. Since the impeller generates a weakly swirling jet the present results are in good agreement with previous studies on swirling jets. The components $\overline{W}$ and $\overline{V}$ displayed tendencies to become self-similar after $6D$, although the scatter was marked. The *rms* values could be successfully scaled with U_m and $\overline{uv}$ scaled well with

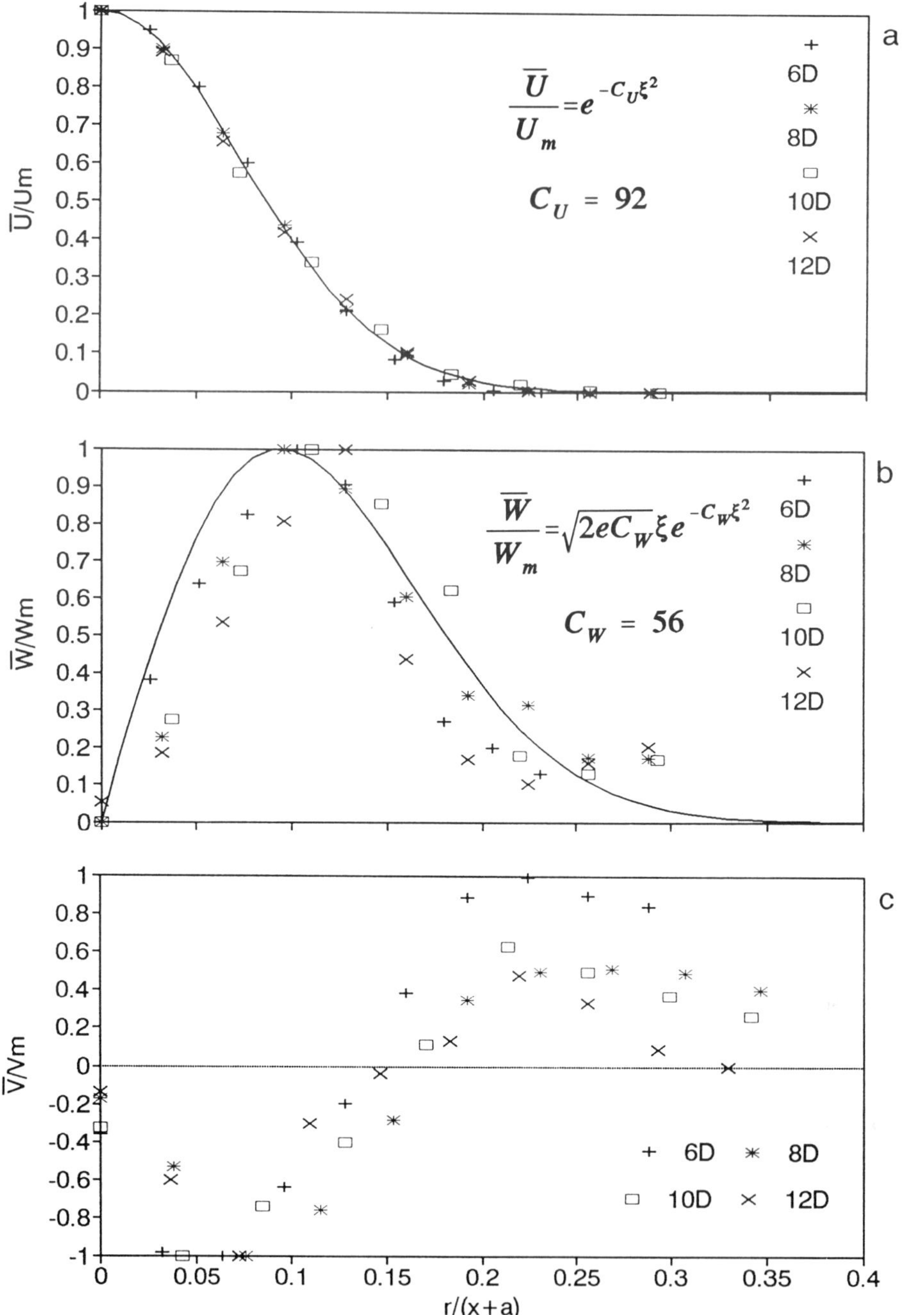

Figure 6. Scaled mean velocity profiles in ZEF downstream an impeller: a) axial, b) tangential, and c) radial component.

$(U_m)^2$, all quantities reaching approximate self-similarity from $6D$. These characteristics of the flow indicate that the jet tends to develop towards a self-preserving state with a balance between the turbulence and the mean flow. Self-preservation can only be verified if the impeller jet could continue to develop undisturbed for a much longer distance than in the present investigation. The jet in the coflow is more narrow than a jet with a corresponding S developing in a stagnant [9] or counterflowing ambient [8]; actually, the jet in the coflow is quit similar to a non-swirling jet (S=0). Thus, the coflow produces a more stable jet development than the case of a free jet or a jet in a return flow, but the radial spread is considerably reduced.

Acknowledgments

This research was supported by the Swedish Research Council for Engineering Sciences (TFR Dnr 93-80). The authors want to thank Magnus Fahlgren at ITT Flygt AB for supplying the experimental impeller setup used in the investigation.

References

1. Chigier, N.A. and Chervinsky, A., 1967, "Experimental Investigations of Swirling Vortex Motion in Jets," *Journal of Applied Mechanics*, 34, 443-451.
2. Duquenne, A.M., Guiraud, P., and Bertrand, J., 1993, "Swirl-Induced Improvement of Turbulent Mixing: Laser Study in a Jet-Stirred Tubular Reactor," *Chemical Engineering Science*, 48(22), 3805-3812.
3. Fahlgren, M. and Tammelin, J., 1992, "The 3-D Velocity Field Near a Submerged Hydrofoil Impeller," *Scientific Impeller*, ITT Flygt AB, 1(1), 49-57.
4. Hyun, B-S. and Patel, V.C., 1991, "Measurements in the Flow Around a Marine Propeller at the Stern of an Axisymmetric Body. Part 1: Circumferentially-Averaged Flow," *Experiments in Fluids*, 11, 33-44.
5. Lepicovsky, J., 1988, "Laser Velocimeter Measurements in a Model Propeller Flowfield," *Journal of Fluids Engineering*, 110, 350-354.
6. Petersson, P., 1996, "Laser Doppler Velocity Measurements in an Impeller-Generated Turbulent Jet," Dept. of Water Resources Engrg., Lund University, Report No. 3195, Lund, Sweden.
7. Petersson, P., Larson, M., and Jönsson, L., 1996, "Measurements of the Velocity Field Downstream of an Impeller," *Journal of Fluids Engineering*, 118, (in press).
8. Pratte, B.D and Keffer, J.F., 1972, "The Swirling Turbulent Jet," *Journal of Basic Engineering*, 93, 739-748.
9. Rose, W.G., 1962, "A Swirling Round Turbulent Jet," *Journal of Applied Mechanics*, 29, 615-625.

INLINE RADIAL FORCE MEASUREMENT OF TURBINE RUNNERS

JOSEF RIENER *Hydraulic Engineer*
ARNOLD EGGER *Measuring Expert*
GERALD SCHNUR *Hydraulic Engineer*

ASTRÖ - Anstalt für Strömungsmaschinen, Graz, Austria

Abstract

Radial forces are important parameters for the design of pump turbines and Francis turbines. Until now the forces were measured with special measuring arrangements. Establishing these arrangements and measuring itself was very time consuming. As the time between start of the project and going into service of the machines should most of the time be as short as possible, these tests were often canceled. Consequently important machine parameters were not available for the design of the rotating parts of the turbine. To overcome this discrepancy, ASTRÖ developed a new measuring system, which allows the change from normal testing to radial force measurement without any delay. The following paper shows details of the new solution as well as first measuring results.

1. Introduction

Hydraulic forces acting on the runner of a hydraulic machine significantly influence the running smoothness. Especially in the case of pump turbines, with their necessary quick change between different operating modes, the runner forces are the limiting values for the changing time. But also for big Francis turbines the measurement of all forces and moments becomes more and more a standard of measurement during development and acceptance model testing. Up to now various measuring systems with different suitability and accuracy are used. For most of the systems a special model arrangement is necessary, and high costs and time consumption often make these tests unfeasible.

Therefore, the idea was to develope a new system, which allows the measurement of steady state and fluctuating forces and moments during model performance testing without any hardware change. This means a combination between a high accurate hydrostatically supported bearing for torque measurement with a suitable radial force measuring system.

E. Cabrera et al. (eds.), Hydraulic Machinery and Cavitation, 1024–1033.

2. Hydrostatically supported model bearing systems

These bearings are the state of the art for characteristic measurement and used in most of the laboratories. Typical hydrostatically supported model bearing systems have a runner shaft and generator shaft, supported with roller bearings in intermediate casings which are supported hydrostatically in an outer casing fixed to the test stand frame. Between intermediate casing and outer casing only contact-free gap seals are allowable. The runner torque is measured as a reaction force on the leverarm with an accuracy better than 0.1 % in steady state conditions. If torque fluctuations are to be measured a torque-meter can be arranged between bearing system and power brake. With these systems also the steady state axial thrust can be measured with sufficient accuracy.

The main features of hydrostatically supported bearings are as follows:

+ high accuracy for steady state torque measurement
+ sufficient accuracy for axial thrust measurement
+ compact design
- insufficient for oscillating force measurement
- complex sealing system

3. Selection of radial force measuring systems

The first step of the combination was to select a radial force measuring system wich allows the implementation into the hydrostatically supported bearing system. The following systems are possible:

a) direct shaft deflection measurement
b) reaction forces between bearing system casing and model casing
c) reaction forces of the shaft bearings
d) force measurement between runner and shaft with piezo force transducers
e) strain gauge equipped sleeve between runner and shaft

The requirements for the force measurement are as follows:

- high natural frequency of the shaft-runner combination
- no influence of shaft seal and bearing stiffness
- no influence of unbalanced rotating masses
- no influence of forces acting on the shaft clutch
- rigid design to make an implementation possible
- maintenance-free signal transmission out of the rotating system

The measurement of the shaft deflection (a) was not possible, because the deflection of the double supported system was too complex. Reaction force measurement between bearing system casing and model casing (b) allways includes the forces acting on the coupling, as well as the connections necessary for the hydrostatic system. Reaction force measurement of the shaft bearings (c) is suitable if only the more important components shall be measured, but not compact enough for implementation. The force measurement between runner and shaft with piezo force transducers (d), a high quality measuring principle, is not compact enough either. The strain gauge sleeve between runner and shaft (e) was found as the best solution for the defined premises.

4. Design of the strain gauge sleeve measuring shaft

ASTRÖ has long expirience with strain gauge sleeve measuring shafts. The oldest design is a horizontal arrangement with two strain gauge sections on two separate sleeves (Fig. 1). The only disadvantage was the application of the strain gauges on the outer surface of the sleeves. Highest quality protection hardly provided sufficiently low influence of the water pressure on the signals. Aging of the protection even enhanced the problem.

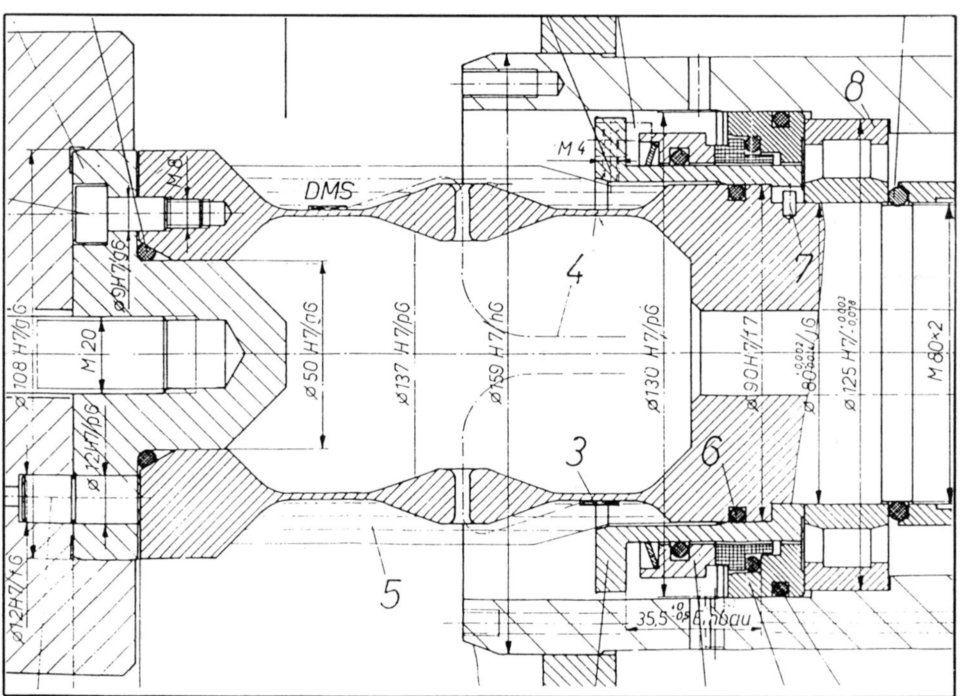

Fig. 1 Two strain gauge section measuring shaft

To eliminate the influence of the cover, the next step was the strain gauge application inside of the measuring sleeve. Therefore, a special application device (Fig. 2) was developed, to allow the application of 24 strain gauges in the exact position inside the hardly accessible bore of the shaft. For gluing with temperable glue only a relative short time between start of application and begin of tempering is available.

Fig. 2 Strain gauge application device

The results with this improvement were excellent, and therfore this design was selected. Due to growing model sizes with higher runner masses, the stiffness of the shaft had to be increased to the highest possible value. This required a large diameter of the shaft between the roller bearings and the shortest possible length between runnerside roller bearing and runner. Also, the principle with two measuring sections was replaced by one measuring section with only one sleeve, after detailed finite element calculations.

5. Instrumentation

An important prerequisit for the realisation was the miniaturisation and the decreasing costs for telemetry components. With this new telemetry system the previously used maintenance-intensive silver slip ring transmitters were no longer necessary. The signals of the 24 strain gauges are led to the amplifier/sender via the hollow shaft and to the receiver/demultiplexer via the antenna. After passing an 8-channel analogous filter the signals are stored on the hard disc of the personal computer via a 16 bit transient recorder. Additionally the angle information of the trigger is stored. For a first information the results can be shown on the test stand computer in the time domain as well as in the frequency domain.

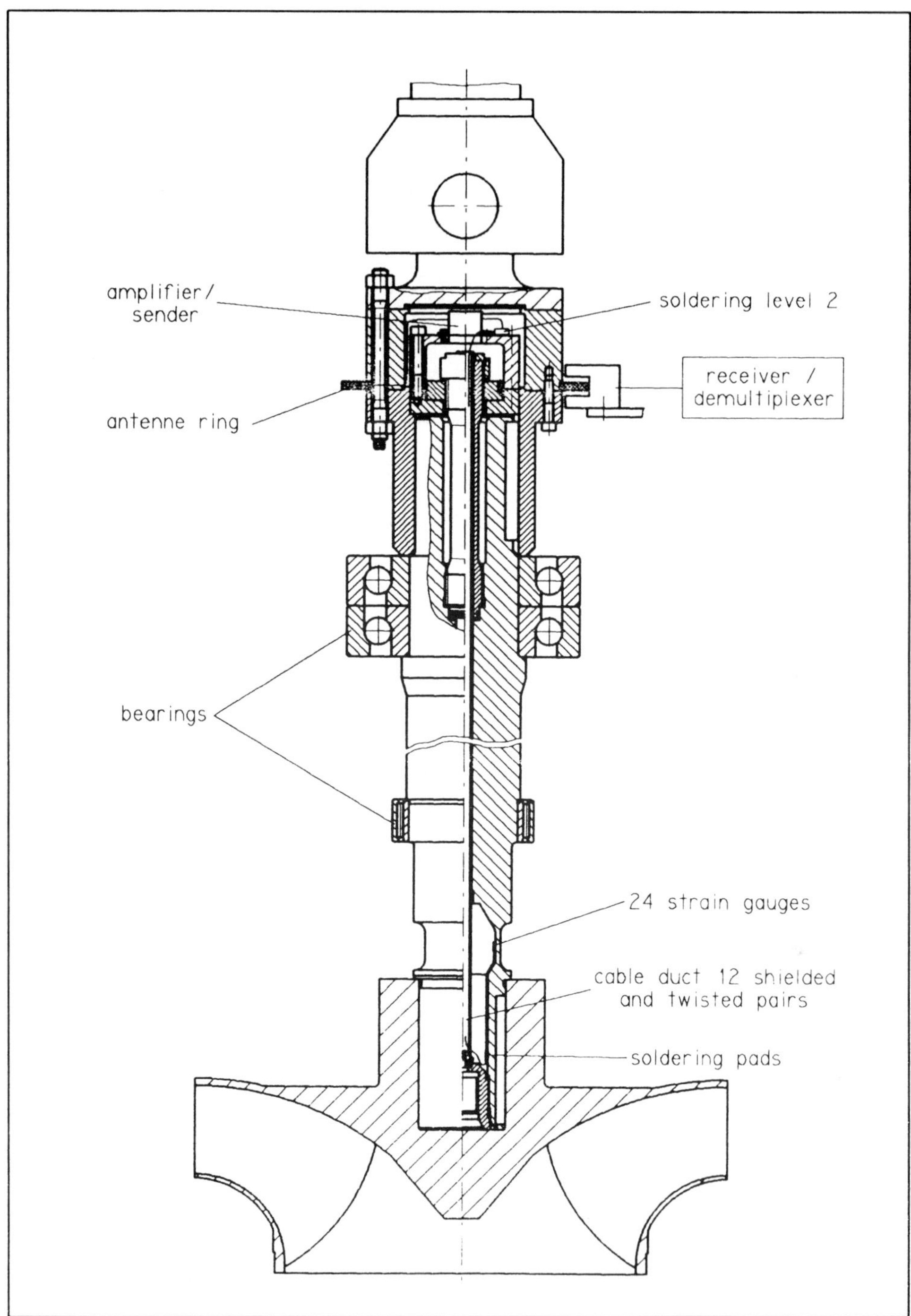

Fig. 3 Measuring shaft main section

6. Evaluation

More extensive measurements shall be evaluated on an external computer. First step is subsampling from the 4000 Hz sample rate to the half of this value. With help of the calibration matrix, force and moment values are calculated from the signals, and the coordinate transformation from the rotating to the steady state system is done. For presentation, the results can be shown in the time domain, in the frequency domain up to 250 Hz and orbit plots. (Fig. 4)

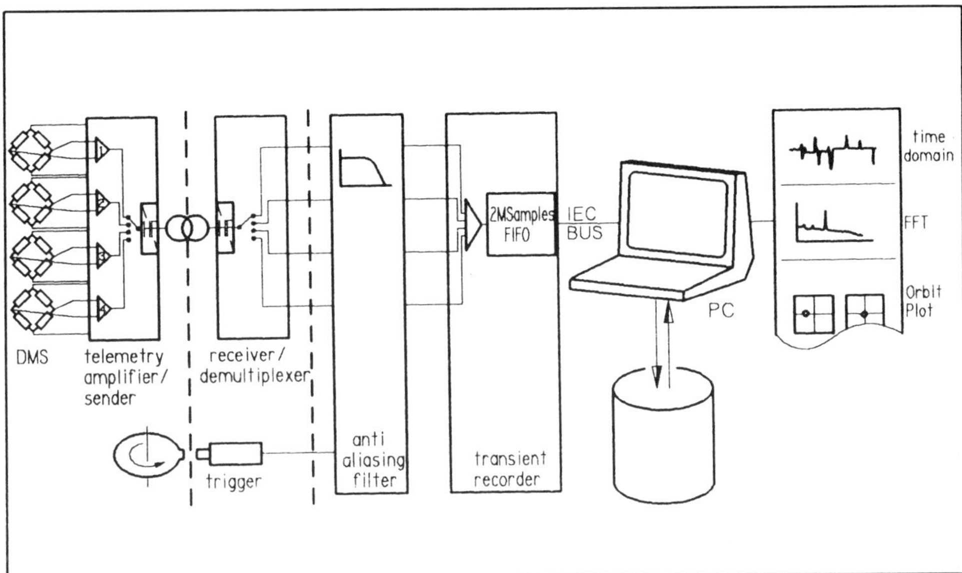

Fig. 4 Data evaluation

7. Calibration

Two types of calibration shall be taken into account, a static and a dynamic calibration. One requirement was the in-situ calibration, where the model is fully mounted on the test stand. The static calibration was done with dismounted draft tube ellbow and runner in air, at low shaft speed.With two load beam devices, one for the radial force (Fig. 5) and one for the radial torque (Fig 6), and load masses, the reference values were applied. With these calibration values the dynamic behavior has to be checked by means of a step-up test. Step-up test means to measure one operating point of the characteristic at various speeds from zero to maximum speed, and to compare the normalized results for the detection of resonance phenomena. Dynamic calibration by means of a shaker on the runner of the water-filled model was not taken into account, because easy realisation is only possible without shaft rotation and, therefore, the important interaction between oscillation and flow is not included in the calbration.

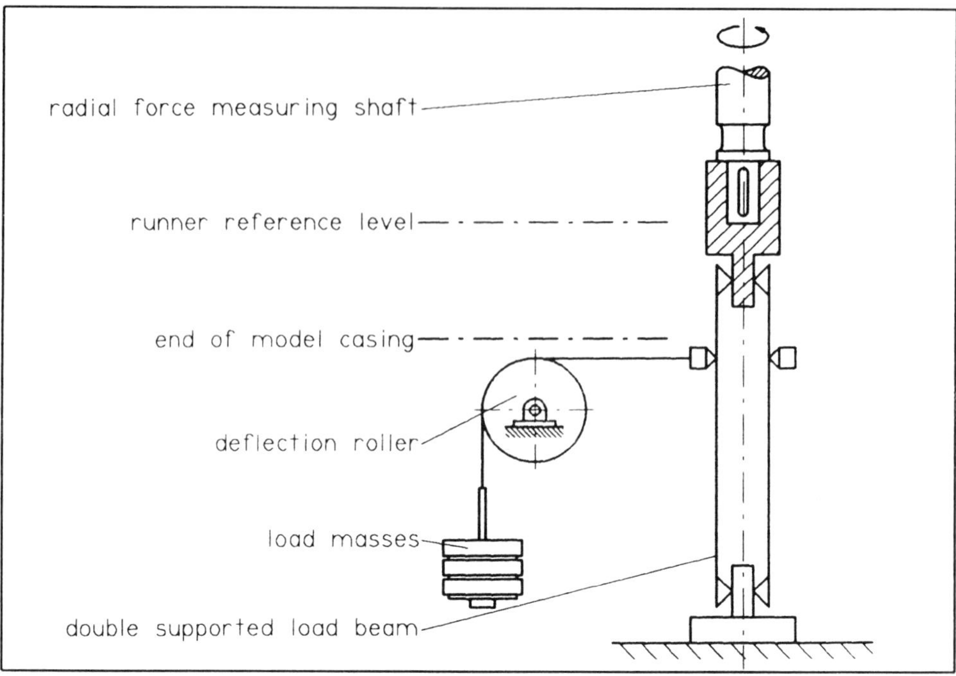

Fig. 5 Radial force calbration device

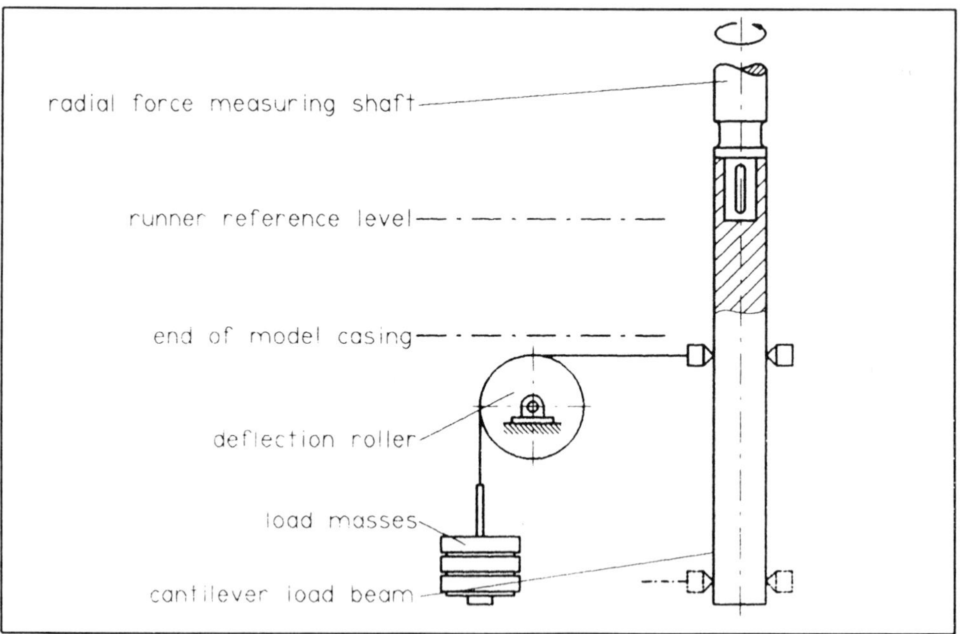

Fig. 6 Radial moment calibration device

8. First experience

During the model acceptance test of a big pump-turbine, the new system was used and, in general, the results were excellent. It was possible to change from efficiency measurement to radial force measurement without any delay. Calbration in presence of the customers' representitives was done within one day. During one day measurements of approximately 25 individual points were possible, enough for determining the runner force characteristic of the machine in the most important regions. Good conformity with comparable data was found. Results in the frequency domain (Fig. 7) and orbit plots (Fig. 8 and 9) are shown as an example.

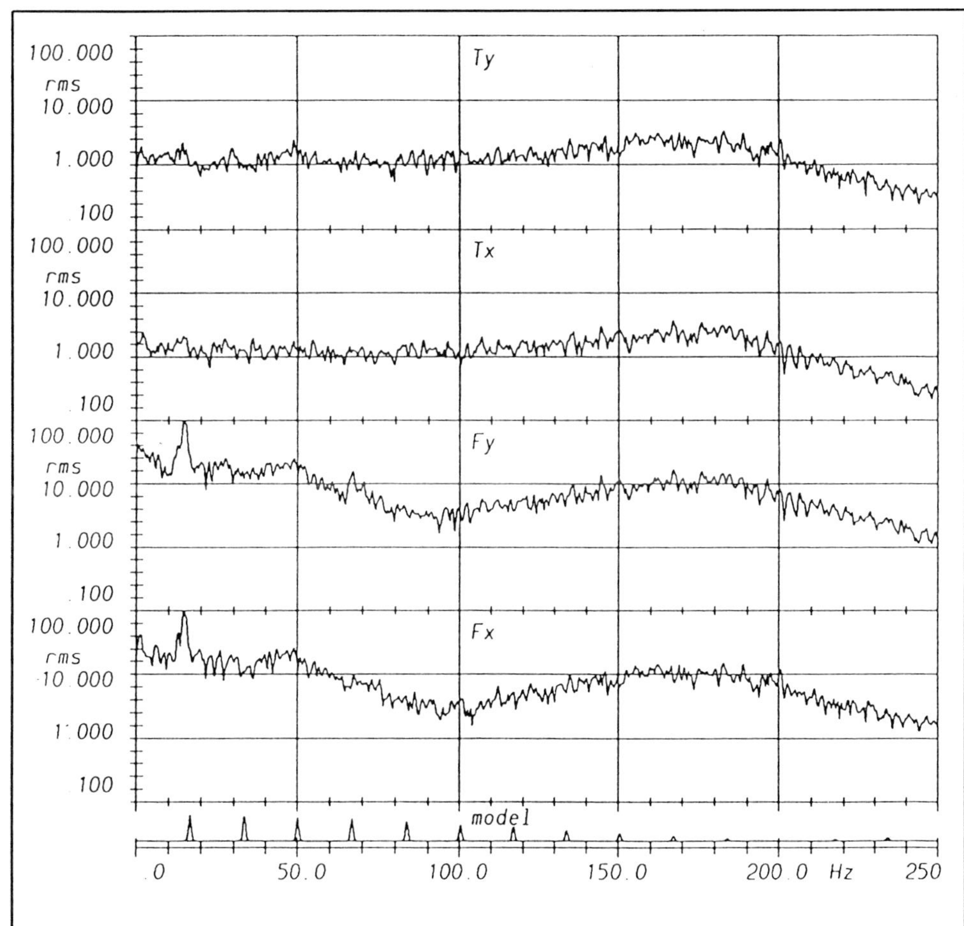

Fig. 7 Force and moment results in the frequency domain

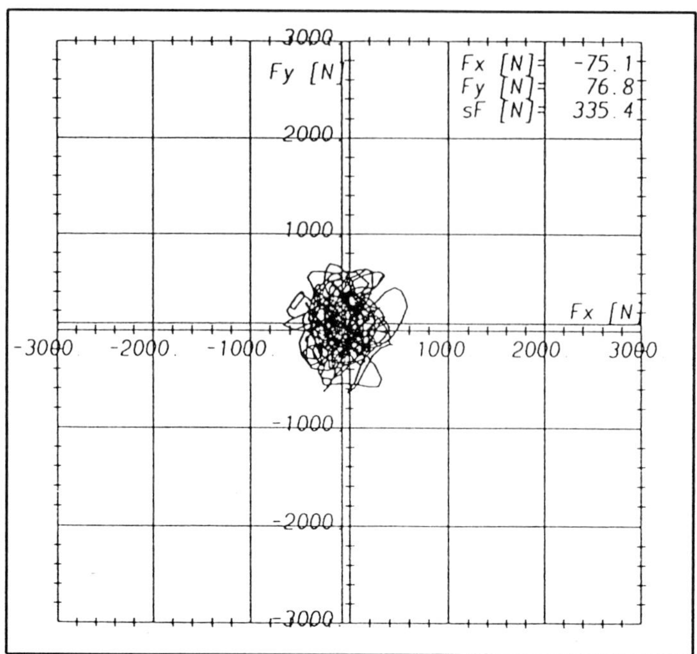

Fig. 8 Orbit plot of the radial force

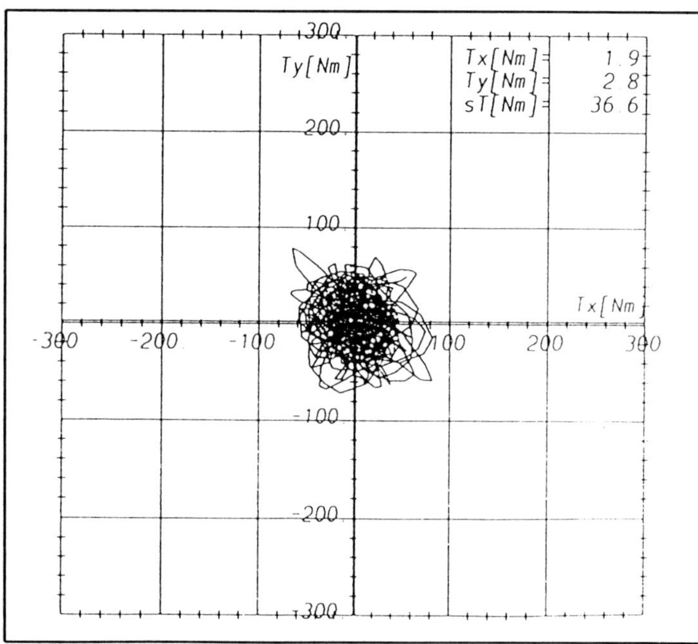

Fig. 9 Orbit plot of radial moment

9. Conclusion

With the introduced system forces and moments acting on the runner can be measured in both steady state and dynamic mode. The measurements can be carried out in course of the usual performance tests as no change of the model arrangement is necessary. This means that important additional information can be obtained in acceptance or competitive tests without causing any delay.

DIFFERENT TYPES AND LOCATIONS OF PART-LOAD RECIRCULATIONS IN CENTRIFUGAL PUMPS FOUND FROM LDV MEASUREMENTS

B. STOFFEL and K. WEISS[1]
Institute for Turbomachinery and Fluid Power
Darmstadt University of Technology
Magdalenenstrasse 4
D-64289 Darmstadt
Germany

Summary

LDV measurements were carried out to analyse the flow field at part load operation of three test impellers having different values of specific speed. On the basis of the measured flow velocities, different types of flow recirculations were identified which can occur in the same impeller from different causes, at different locations and at different values of the flow rate.

1. Introduction

In centrifugal pumps operating at part load conditions, flow recirculations occur if the flow rate is reduced below corresponding critical values. Depending on their extent and intensity, these recirculations impair the pump performance, cause increased noise, vibrations and even damage of pump parts. In combination with cavitation, the inlet recirculation can lead to severe surge phenomena. By these effects, part load recirculations are detrimental for the reliability of pumps and pump plants.

Although the part load recirculations are of considerable importance for many pump applications and have been studied by several researchers (see, e.g., [1] to [3]) the basic physical effects related to the recirculations and their internal structure have not been sufficiently well understood. Therefore, extensive experimental investigations have been carried out at our laboratory in the recent years (see [4] to [8]) with the aim to further improve the understanding of the causes, origins and structures of part load recirculations in centrifugal pumps. Special objectives of these investigations were

- to analyse the part-load flow field (outside and inside the impeller), preferably by using a non-intrusive measurement technique (LDV),

[1] now: WILO, Nortkirchenstrasse 100, D-44263 Dortmund, Germany

E. Cabrera et al. (eds.), Hydraulic Machinery and Cavitation, 1034–1043.

- to identify and separate individual recirculation phenomena and the loca-tions of their origin,
- to interprete the measurement results in respect to the physical causes for the occurrence of recirculations and to develop an appropriate modelization,
- to clarify possible interactions of different types and locations of recirculations.

The measurements were performed at our laboratory as part of a German-French BRITE project supported by the EU. Financial support for the experimental investigations came from the German Pump Manufacturers Association within the VDMA (Verband Deutscher Maschinen- und Anlagenbau) and from the EU in Brussels. The authors wish to express their thank to these institutions. The whole test programme, measured data, their evaluation and interpretation are described in more detail in the Doctoral Thesis of K. Weiss [8].

2. Test Installation and Instrumentation

For the investigations, special test impellers - covering a range of specific speeds - were designed and manufactured, the design being representative for commercial pumps. All of these test impellers had the same nominal flow rate $Q_{AP} = 0.136$ m^3/s at the nominal speed $n_N = 1450$ min^{-1}. Three of the impellers having specific speed values of 20, 32 and 50 min^{-1}, respectively, were investigated at our laboratory. As an important feature, the geometry at impeller inlet (hub and shroud contours, blade geometry) was nearly identical for all test impellers. The inlet diameter was $D_{1a} = 200$ mm, the blade number was $z = 6$ and the blade inlet angle at the shroud was $\beta_{S1a} = 18^o$. Also, the blade outlet angle had the same value of $\beta_{S2} = 24^o$ for the three impellers. The values of the outer diameter D_2 and of the outlet width b_2 of the impellers are given in table 1. The impellers were furnished with a front shroud made of acrylic glass. This enabled the application of the LDV (Laser Doppler Velocimetry) for velocity measurements inside the blading and immediately upstream of the blade leading edges.

The impellers were installed in a special test casing with an exchangeable diffuser downstream of the impeller outlet. The design of the casing aimed at a rotationally symmetrical outer boundary condition at the diffuser outlet. The results presented here were found from measurements for the case of vaneless diffusers. The diffuser dimensions (diffuser width b_3, radial diffuser length L_{Diff}) are also given in table 1.

TABLE 1. Impeller and diffuser dimensions of test impellers

n_q [min^{-1}]	D_2 [mm]	b_2 [mm]	b_3 [mm]	L_{Diff} [mm]
20	500	30	32	30
32	380	40	42	55
50	305	48	50	98

The suction nozzle of the pump casing and the front shrouds of the vaneless diffusers, as well, were also made of acrylic glass to make possible LDV traverses upstream of the impeller inlet and downstream of the impeller outlet. A traversing mechanism assembled with the pump casing served for the positioning of the measuring volume. For some traverses, an underwater mirror inside the water filled casing was used for deflecting the laser beams. A sketch of the test arrangement is given in fig. 1.

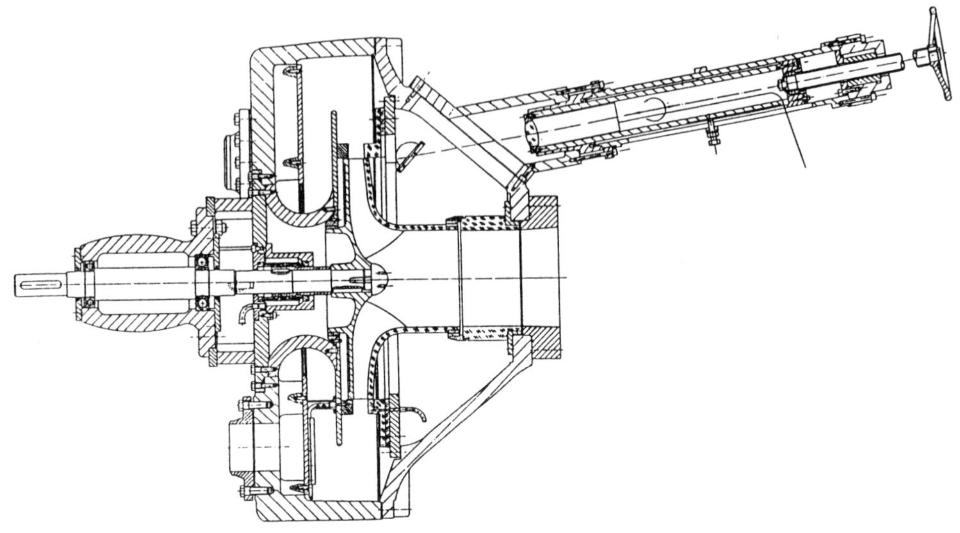

Figure 1. Test installation

For the LDV measurements, a DANTEC system including a 4 W Argon-Ion Laser, a one-component glass-fibre optic probe and a BSA (Burst Spectrum Analyser) was used. The glass-fibre probe was positioned with the aid of the positioning mechanism shown in fig. 1. When evaluating the velocity data, the individual deflection of the laser beams at the (plane or curved) interfaces crossed by the beams was taken into account. For the determination of phase-locked velocity distributions, an angle decoder was connected to the pump shaft.

3. Measurements and Data Evaluation

For all test impellers, velocity measurements were carried out in the range of flow rates from slightly above the nominal one down to extreme part load conditions ($Q/Q_{AP} \approx 0.1$). In respect to the flow phenomena in the impeller, the *impeller flow rate* is relevant. It was determined from the external flow rate (measured by a magnetic flowmeter located in the discharge pipe) by adding the flow rate for axial thrust balancing (measured separately by a second flowmeter) and the sealing gap flow rate (calculated from the sealing gap geometry and the differential pressure across the sealing gap at the impeller suction eye).

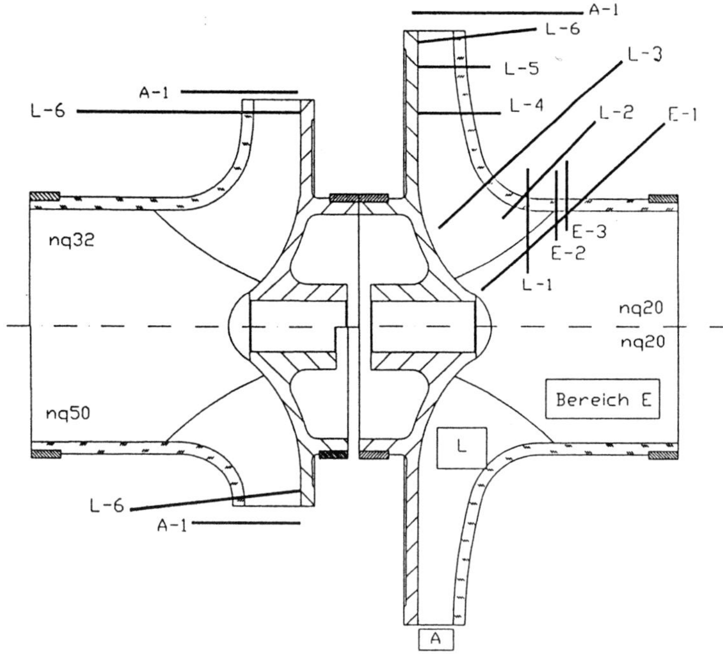

Figure 2. Meridional contours of impellers and location of traversing axes

The meridional cross sections of the different test impellers together with the various traversing axes (marked as black lines) are presented in fig. 2. The axes nominated by "E" are located immediately upstream of the blade leading edges. Traversing axes inside the impeller blading are nominated by "L", and the traversing axis nominated by "A" is located downstream of the blading at a short distance from the impeller outer diameter. Along the axes E-1, E-3, L-3, L-6 and A-1, velocity measurements were made for all three impellers while velocities were measured only on the axes E-2, L-1 and L-2 for the impellers of n_q = 20 and 50 min^{-1} and on the axes L-3 and L-5 for the impeller of n_q = 20 min^{-1}. From the location of the measuring axes, it becomes evident that the most detailed flow analyses were made in the areas near the inlet and outlet of the blading.

By turning the fiber optic probe around its longitudinal axis, two velocity components (orientated perpendicularly to the traversing axis) were determined sequentially for each measuring location. In all cases, one of these components (called $C_{tangential}$) was the circumferential one while the other component (called C_{normal}) was always flush to the meriodional plane and was orientated more or less in the meridional flow direction, depending on the individual traversing axis. No attempt was made to determine the 3rd velocity component and, thereby, the resulting 3-dimensional velocity vector because the flow analysis aimed primarily at the detection of recirculations and at the identification of general flow characteristics.

The two velocity components were evaluated as circumferentially averaged or as phase-locked values or distributions. By combining the measured components together and with the local circumferential velocity of the impeller, information was gained on the relative flow velocity. To support the LDV results and the interpretation of the hub-to-shroud and blade-to-blade velocity distributions, some visualizations of the near wall flow direction of the relative flow inside the impeller channels were carried out. For this purpose, arrays of thin rubber threads were glued to the blade suction and pressure surfaces and to the inner hub and shroud surfaces, as well. Through acrylic windows in the pump casing, the impeller was illuminated by a stroboscopic light source, and the mean orientation and the fluctuations of the threads and the variations with varying flow rate were observed and recorded by a video camera.

For the determination of the onset of recirculation (i.e. backflow) at a measuring location by LDV, the change of sign of the normal component of the velocity served as an indicator. Different "critical" values of the flow rate (with respect to the local onset of backflow) can be defined (see [4],[8]) based on

- the occurence of single negative instantaneous values of the normal velocity component somewhere along the impeller circumference,
- the change of sign of the phase-locked values of the normal component at (at least) one location of the blade-to-blade distribution
- the change of sign of the circumferentially averaged values of the normal component.

4. Results

4.1. ONSET OF RECIRCULATION AT IMPELLER INLET

For all test impellers, the critical value of the flow rate at onset of inlet recirculation (based on the circumferentially averaged normal component) was determined

- at a constant distance of 2 mm from the inner shroud surface on several axes upstream and downstream of the blade leading edges,
- on axis E-3 at several distances from the inner shroud surface.

As a result, it was found that

- the onset of recirculation occurs at an impeller flow rate of $(0.69 \pm 0.02) \cdot Q_{AP}$,
- the origin of the onset of inlet recirculation is situated on the inner shroud surface very near to the position of the blade leading edges in the meridional plane

for all three test impellers. Because of the obvious independance from the differing features of the impellers (specific speed, magnitude of the sealing gap flow rate), the flow rate at onset of the inlet recirculation and the location of its origin seem to be determined by those features/influencing parameters which the test impellers have in common. It has to be mentioned that a somewhat special feature of the test impellers is the relatively long cylindrical part of the shroud. This leads to a moderate pre-swirl (induced by wall friction) in the boundary layer on the inner shroud surface which is superposed with the pre-swirl induced by the swirling flow injected from the sealing gap. But even the non-neglectable differences (among the impellers) in respect to the amount of pre-swirl near the shroud (found from LDV-mesurements on axis E-3) don't cause significant differences concerning the onset of recirculation.

The evaluation of the blade-to-blade distribution of the normal component near the shroud indicated clearly that the origin of this recirculation is located between the middle of the blade spacing and the blade pressure surface. This finding cannot be explained by a blade stall (on the blade suction surface) resulting from an increased incidence angle at part-load, as is commonly supposed. From our experimental findings, it is rather to conclude that the separation of the boundary layer on the shroud surface - as caused by the positive pressure gradient upstream of the blade leading edges which increases with decreasing flow rate - is responsible for the onset of the recirculation. This hypothesis was confirmed not only by a simplified boundary layer calculation in [8], but also by experimental results in [7] which showed that a counter-rotating as well as a co-rotating pre-swirl (generated by injection of a swirling bypass flow upstream of the impeller) were able to suppress the onset of recirculation.

4.2. DEVELOPMENT OF INLET RECIRCULATION

Figs. 3 und 4 show the variation of the (circumferentially averaged) values of the velocity components C_{axial} and $C_{tangential}$ (normalized by the circumferential velocity U_{1a} at D_{1a}) with the relative flow rate Q_{La}/Q_{AP} for a location on axis E-3 near to the shroud surface. For all three impellers, five characteristic ranges of flow rates can be distinguished in these curves. These ranges correspond to different flow situations within the inlet part of the blading area. This was clearly proved by the analysis of LDV results taken on the various traversing axes in this area (E-1 to L-3) and by the flow visualization. The flow situations in the different ranges are as follows:

- range A: No backflow upstream of the leading edges. Weak pre-swirl near the shroud induced by wall friction.
- range B: First instantaneous backflow induced by shroud boundary layer separation. Average axial component still positive, but sharply decreasing with decreasing flow rate.
- range C: Averaged axial component has become negative. From the location of the onset (leading edges), the backflow zone spreads rapidly upstream into the suction line and moderately downstream along the shroud with decreasing flow rate.
- range D: Remarkably increasing tangential component upstream of the blading, its magnitude becoming even larger than U_{1a}.
- range E: Backflow zone fills the whole blade spacing near the shroud. Extension of the recirculation far upstream into the suction line and downstream up to the strongly curved area of the shroud contour.

Especially from the LDV measurements on the axes L-1 to L-3, a second recirculation was identifiable which has its origin in the corner between the blade suction surface and the shroud in the area of strongest curvature of the meridional shroud contour.

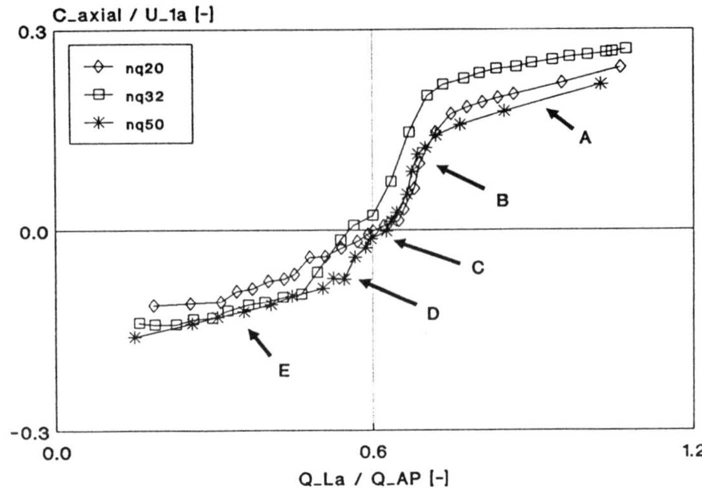

Figure 3. Axial velocity component on axis E-3 at a distance of 4 mm from the shroud

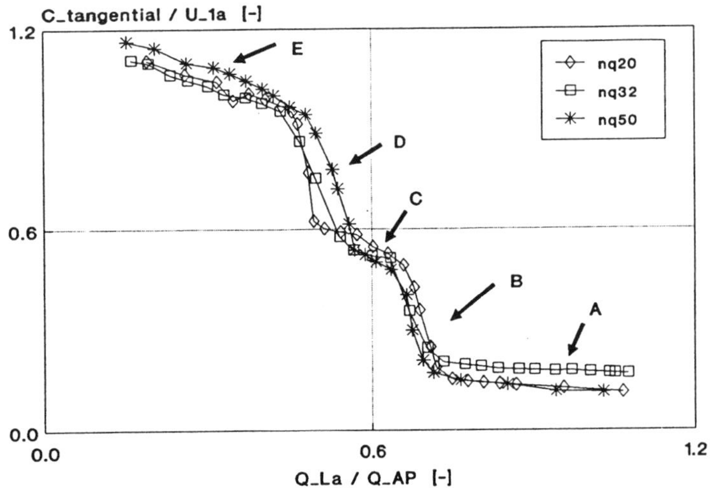

Figure 4. Tangential velocity component on axis E-3 at a distance of 4 mm from the shroud

The onset of this second recirculation occurs at flow rates somewhat lower than $0.6 \cdot Q_{AP}$ depending on the impeller specific speed, i.e. on the meridional shroud contour. This recirculation seems to be induced by the combined effects of wall friction, shroud/blade curvature and Coriolis forces acting on the low energy fluid which moves along the blade suction surface/shroud corner. For $Q_{LA} < 0.6 \cdot Q_{AP}$, the second recirculation zone merges together with the first one. In this situation, the moment of momentum which is transmitted by the blading up to the "point of return" inside the impeller is transported upstream (by the fluid leaving the bladed area in the reverse direction) and induces tangential components $C_{tangential} > U_{1a}$ upstream of the blading.

4.3. RECIRCULATIONS AT AND NEAR IMPELLER OUTLET

Based on the LDV measurements on axis A-1, it can be stated for all three impeller/diffuser combinations that backflow on both diffuser walls exists for flow rates not much smaller than Q_{AP}. On the other hand, for $Q_{La} > 0.3 \cdot Q_{AP}$ no backflow was detectable on axis L-6 (inside the impeller). Therefrom, the recirculation existing at $Q_{La} > 0.3 \cdot Q_{AP}$ is obviously limited to the diffuser area and caused by boundary layer separation on the diffuser walls.

Concerning the backflow phenomena which were observed inside the impellers at flow rates $< 0.3 \cdot Q_{AP}$, a detailed analysis by LDV and flow visualization was worked out. For the impeller of $n_q = 20$ min^{-1}, on axis L-4 (further upstream from the impeller outlet) a backflow sets on near the blade pressure surfaces when the flow rate is slightly decreased below the nominal one. The magnitude of the relative flow velocity W (in the core flow outside the boundary layers) shows a nearly linear increase from blade pressure side to blade suction side and only small variations from hub to shroud. With decreasing flow rate, this distribution of W keeps up its general shape and is only shifted down to smaller values. This leads to a growth of the backflow zone near the blade pressure side (where the W-values become negative) in radial and circumferential direction. This type of recirculation was not found for the other two test impellers and can be explained as a predominantly kinematic (i.e not viscous) effect resulting from the well known model of the "relative vortex" in the rotating frame of the impeller.

Depending on the specific speed, two other types of recirculation were identified on axis L-6 at extreme part load. For all three impellers, backflow occurs near the blade pressure surface. The corresponding velocity distributions are characterized by negative radial components in the middle between hub and shroud and jet-like outflow in the corners of the blade pressure side and the front and back shrouds, respectively. As a simplified model, the combined effects of secondary flow in the blade channels and centrifuging of the boundary layers at the shrouds can serve to explain the measured distributions of recirculating flow. Only for the impeller of $n_q = 50$ min^{-1}, an additional backflow zone occurs near the blade suction surface at nearly zero flow rate which probably results from the steep pressure gradient along the blade suction surface near impeller outlet.

4.3. INTERDEPENDENCIES OF THE DIFFERENT RECIRCULATIONS

For the impellers of $n_q = 20$ and 32 min^{-1}, the recirculations having their origin in the upstream part of the blading (leading edges and curvature zone) merge together with decreasing flow rate. The same holds true for the different types of recirculations which have their origin in the downstream part of the impeller and in the diffuser. But for these impellers no indication was found - even at extreme part load - from our measurements that the recirculations at inlet and outlet join together and form a "global" backflow from the diffuser through the impeller to the suction line.

In contrary, for the impeller of $n_q = 50\ min^{-1}$ (having the shortest radial extent) the measurements of the normal velocity component on the various axes (shown in fig. 5 for $Q_{La}/Q_{AP} = 0.15$) lead to the presumption that the individual recirculations developing at inlet and outlet join together at very low flow rates.

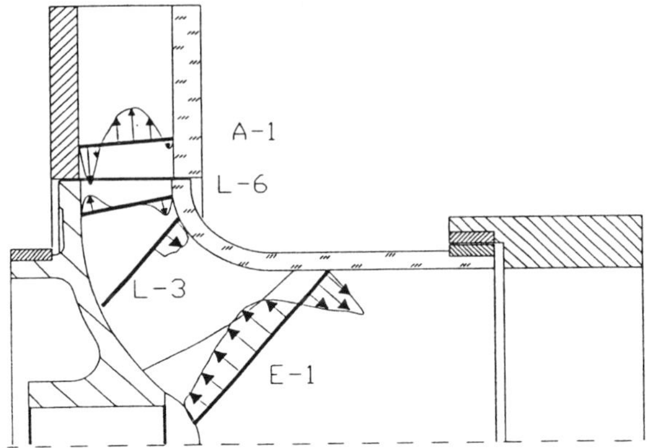

Figure 5. Velocity distributions for the impeller of $n_q = 50\ min^{-1}$ at extreme part load

5. Conclusions

From LDV measurements and flow visualizations on three impellers of different specific speed, different types and locations of part-load recirculations were identified. Part of these were found for all three impellers while other types were observed only for the impellers of the lowest ($20\ min^{-1}$) or highest ($50\ min^{-1}$) value of the specific speed. In fig. 6, five zones (denominated by numbers I to V) are marked as the probable locations where the different types of recirculations have their origin. For each of these zones, different physical causes seem to be responsible for the onset of backflow:

- Zone I: Boundary layer separation from the shroud caused by positive pressure gradient existing at blade inlet.
- Zone II: Combined effects of wall friction, shroud and blade curvature and Coriolis forces acting in the blade/shroud corner.
- Zone II: Negative relative velocities caused by the "relative vortex" in the rotating frame,
- Zone IV: Combined effects of blade channel secondary flow and shroud boundary layer centrifuging near the blade pressure surface; flow separation from the blade suction surface caused by steep pressure increase near blade trailing edge.
- Zone V: Boundary layer separation from the diffuser walls caused by flow deceleration.

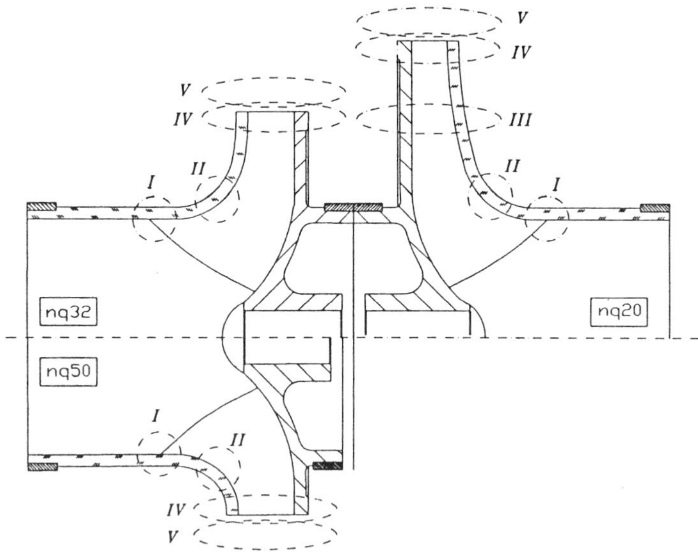

Figure 6. Zones of origin of different types of part-load recirculations

6. References

1 Stoffel, B. and Krieger, P. (1982) Experimental investigations on the energy balance of centrifugal impellers at part load conditions, 11th IAHR Symposium, Amsterdam, pap. 19

2 Schiavello, B. (1983) A diffusion factor correlation for the prediction of the reverse flow onset at the centrifugal pump inlet, 7th Conf. Fluid Mach., Budapest, p. 751-760

3 Sen, H.M. (1983) Prediction of the reverse flow and prerotation in centrifugal pumps, 7th Conf. Fluid Mach., Budapest, p. 805-814

4 Stachnik, P. (1991) Experimentelle Untersuchungen zur Rezirkulation am Ein- und Austritt eines radialen Kreiselpumpenlaufrades im Teillastbetrieb, Doct.Thesis, Darmstadt University of Technology

5 Stoffel, B., Ludwig, G. and Weiss, K. (1992) Experimental investigations on the structure of part-load recirculations in centrifugal pump impellers and the role of different impellers, 16th IAHR Symposium, Sao Paulo, vol. 2, p. 445-454

6 Hureau, J.,Kermarec, J., Stoffel, B. and Weiss, K. (1993) Study of internal recirculation in centrifugal impellers, ASME Fluids Engng. Conf., Washinton D.C., FED-vol 154, p. 151-157

7 Liu, Ch. (1994) Experimentelle Untersuchungen zur Einleitung eines Drallstroms im Teillastbetrieb eines radialen Kreiselpumpenlaufrads, Doct. Thesis, Darmstadt Univ. of Techn.

8 Weiss, K. (1995) Experimentelle Untersuchungen zur Teillastströmung bei Kreiselpumpen, Doct. Thesis, Darmstadt Univ. of Techn.

Turbulent 3D Flows near the Impeller of a Mixed-Flow Pump

Experimental Investigations for a Specific Speed of 120

Baogang Wang, Dieter-Heinz Hellmann
University of Kaiserslautern, Germany

ABSTRACT

Compared to other fluid machines, the flow patterns in a mixed-flow pump are highly complex, due to adverse pressure radial gradients, secondary flows and the flow separation effects in the impeller inlet and outlet.

This paper deals with experimental investigations and their analysis. Special emphasis is laid on the three-dimensional flow fields at the inlet and the exit of a closed mixed-flow impeller to improve design and calculation methods.

By means of a hot wire probe and an angular trigger blade-to-blade measurements of the flow field at the impeller inlet and outlet have been performed from the hub to the shroud within the entire operating range of the pump.

Results of particular interest are the performance curve instability of a mixed-flow pump, the beginning of recirculation at the inlet and the exit of the impeller and the influence of inlet guide vanes in the pump suction side upon the pump performance.

1 INTRODUCTION

Axial and radial impellers have already been subject to exhaustiv studies, but only little is known about mixed flow impellers for their flow field is fully three-dimensional with complex fluid-structure-interactions inside and behind the impeller. Only a few authors [1-8] have studied the flow phenomena of mixed-flow pumps yet, dealing with the improvement of the impeller performance at part load. The knowledge of flow patterns is essential for pump design as well as for flow field calculations. Latest computational fluid dynamic (CFD) techniques allow hydraulic performance predictions to be made based on experimental results.

Performance curve instabilities of a mixed-flow pump are usually caused by reverse flow, if the pump works beyond the best efficiency point (BEP), i.e. irregular flow patterns occur inside and outside of the impeller.

The three-dimensional flow at the exit of a mixed-flow impeller is strongly linked with the impeller geometry (e.g. blade geometry, clearances between casing and impeller). Secondary flows, leakage flows, boundary layers and blade wakes influence directly the pump's efficiency, cavitation, noise and vibration level.

For simplicity the flow entering the mixed-flow pump is usually considered uniform and swirl-free. But in practice the entry flow is influenced by the structure in front of the impeller (e.g. stationary or rotating guide vanes, stiffening ribs and diffuser) or/and the flow rate, especially beyond BEP. A disturbed entry flow may cause cavitation, loss of efficiency and operating instability.

E. Cabrera et al. (eds.), Hydraulic Machinery and Cavitation, 1044–1052.

In order to find the influence of the flow characteristics at the inlet and the exit of a mixed-flow impeller on the performance of the pump, two measuring systems were developed: a global system to measure the performance curve and a local system to measure the flow field in the pump. The hot wire technology is effectively applied for the measurement of the three-dimensional turbulent flow. With the help of an angular trigger, the measuring point can be displaced circumferentially in order to determine the influence of the distance to an impeller blade on the velocity vector.
Further topics are the investigation of the three-dimensional turbulent flow fields and their dependence on the pump operating conditions and the geometry of the suction pipe, especially the inlet guide vanes (IGV). Finally the flow characteristics at the inlet and the outlet of the impeller are discussed.

2. Equipment and Methods

2.1 Test Rig

Fig. 1 shows the test loop used for monitoring the pump performance curve and measuring the flow field.

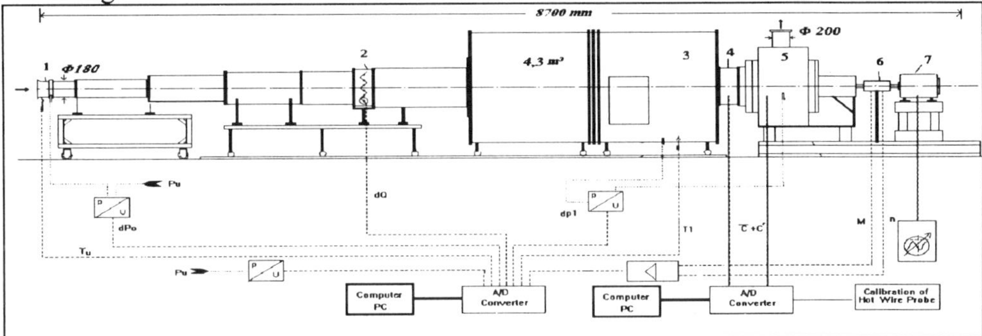

Figure 1

The test loop consists of an inlet nozzle (1) for the measurement of the flow capacity, a throttling device (2) to regulate the flow capacity, a suction tank (3) with the suction pipe (4), a mixed-flow pump (5). The electric motor (7) is linked to the pump with a measuring unit (6) for speed and torque. Atmospheric air is used as test fluid. The design parameters of the pump are:

Speed	1450 rev/min	Impeller inlet diameter	315 mm
Capacity (shockfree)	2070 m³/h	Impeller outlet width	73,5 mm
Nb of blades	4	Impeller outlet diameter	400 mm

2.2 Measuring System

The main devices used for data aqcuisition and processing are:

- PC with I/O-Card
- Error Control
- Calibration-System
- Probe Displacement

For information on the mathematical method and the techniques used for calibration and measurement refer to [9, 10]. Probe positions and coordinate system at the inlet and the exit of the impeller are indicated in Fig. 2.

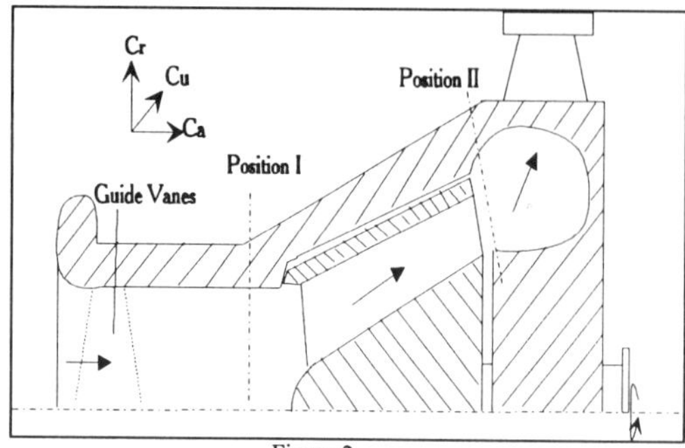

Figure 2

The circumferential position of the probe is exactly determined by the angular trigger; the maximum triggering speed is equivalent to a shaft rotation of 0,5°. The radial position of the probe can be altered in steps of 0,5mm. The axial distance from the probe to the impeller are -60mm (position I) and +10mm (position II) (positive sens corresponding to the direction of flow).

3. Results of Measurements

3.1 Overall Performance

The overall performance of the test pump is shown in Fig. 3, where ϕ, ψ and μ represent the coefficients of pressure, flow rate and shaft power and η is the pump efficiency (air test).

Inlet guide vanes in the suction pipe of the impeller have a positive impact on the head-capacity and efficiency, particularly if the flow coefficient is less than 0,15.

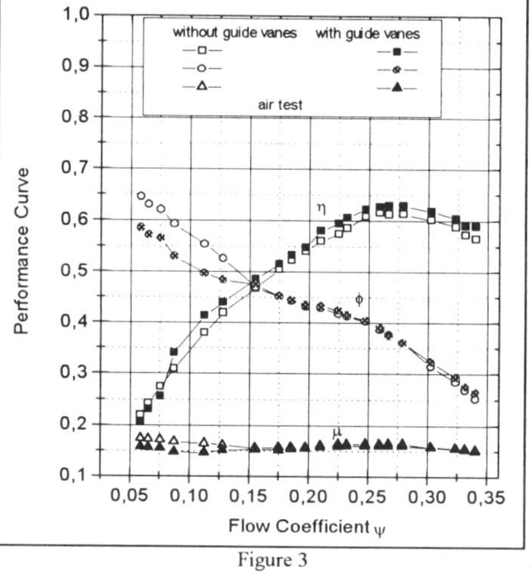

Figure 3

3.2 Impeller Inlet Flow Characteristics

The absolute velocity profils of the three dimensional inlet flow at BEP measured at position I are depicted in figures 4 and 5. The velocity component at each radius from the center to the wall of the suction pipe is obtained by averaging the measurements taken at different circumferential points. The influence of the inlet guide vanes on the entry flow can be obtained by comparing the two figures. For reasons of mechanical stability the IGV are joined in the center, thus the axial velocity (measured between the IGV and the impeller) will be smaller there (radius < 30mm).

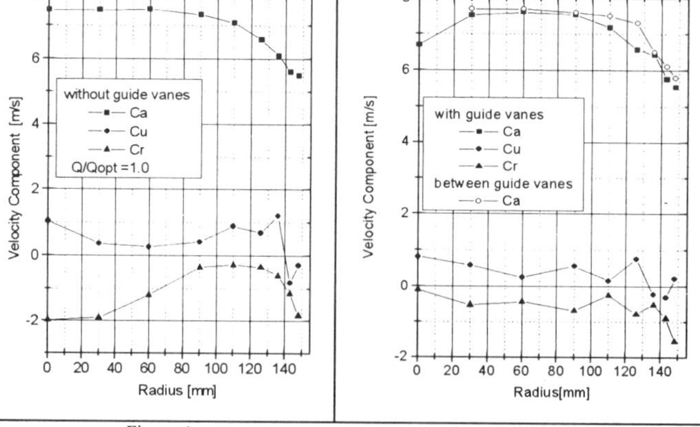

Figure 4 Figure 5

The axial velocity measured between the IGV (Fig. 5) is slightly higher than the axial velocity at position I as the sectional area of the pipe is reduced by the blades of the IGV. The IGV equalize the radial and the circumferential velocity component which become almost constant (but not zero) except for the boundary region.

Figures 6a and 6b monitor the velocities at position I and part load. Two concentric flow regions are visible. The main flow passes through the inner part of the pipe. The outer part contains a vortex referred to as part load swirl indicated by the strong increase of the circumferential velocity. In absence of IGV the vortex region increases to the detriment of the main flow and the circumferential component rises from 5 m/s to 22 m/s (Fig. 6). Obviously there is a clear transition from the main flow to the part load swirl which strongly influences every velocity component and is located at a distinct radius. The IGV increase the main flow area, thus leading to higher efficiency of the pump.

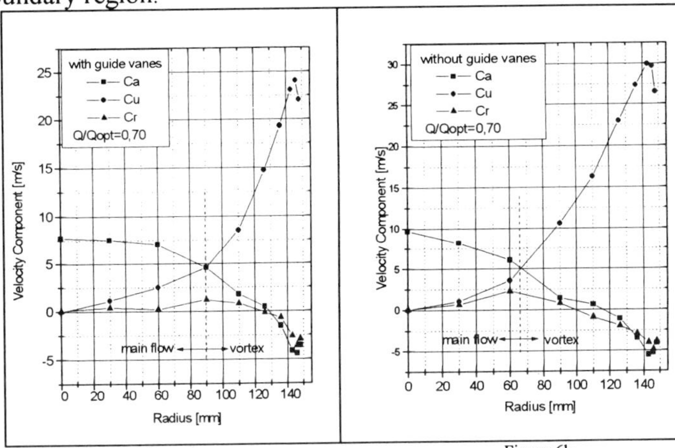

Figure 6a Figure 6b

Figures 7a and 7b show the dependence of the axial velocity on the flow rate at position I. The hub of the IGV causes a velocity decrease of 1-2 m/s in the center (radius < 40 mm) but a velocity increase of 2 m/s (radius> 40 mm). This leads to a more homogeneous inlet flow. The geometry of the IGV proves to be unsuitable at part load (Q/Q_{opt} = 0,2) as the axial component still becomes negative (part load swirl).

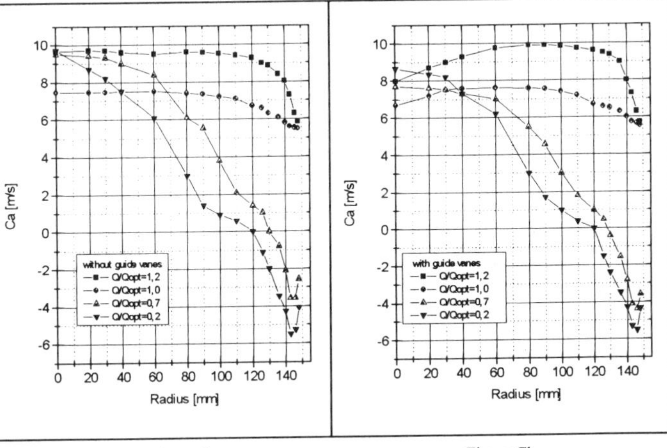

Figure 7a Figure 7b

Fig. 8 is a 3D plot of the axial velocity taken from blade to blade (pos I, with guide vanes, part load). The impact of the blade on the flow pattern is limited to the tip: although the two faces of a blade are substantially different (suction and pressure side) the axial velocity remains constant up to a radius of 130 mm.

3.3 Impeller Outlet Flow Characteristics

The impeller outlet flow was measured directly behind the impeller (position II) at different operating points with IGV. A blade to blade 3D plot at BEP is depicted in

Fig.9. The flow field is stable and smooth; its mean axial velocity is influenced by impeller blades.

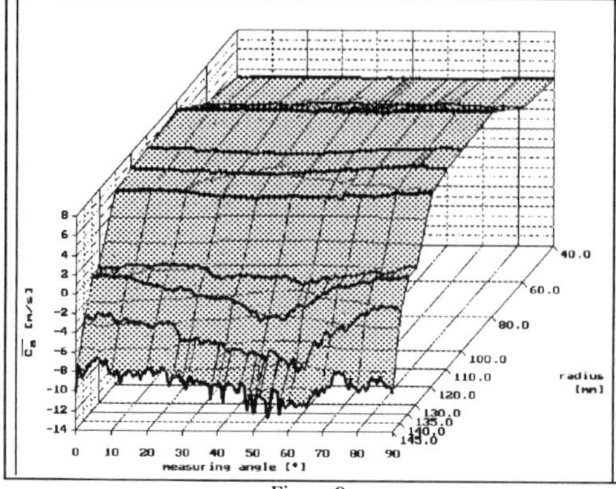

Figure 8

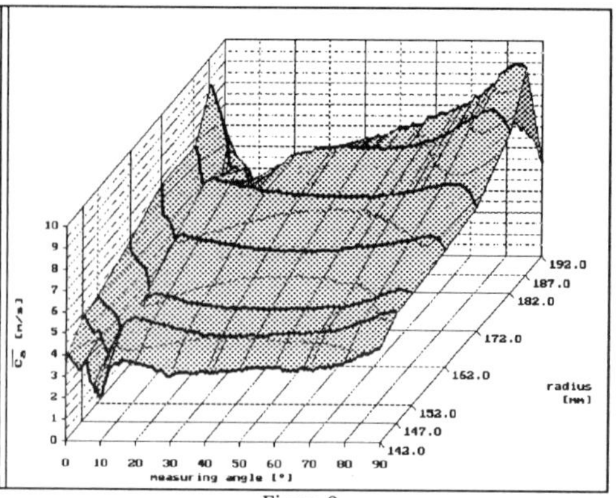

Figure 9

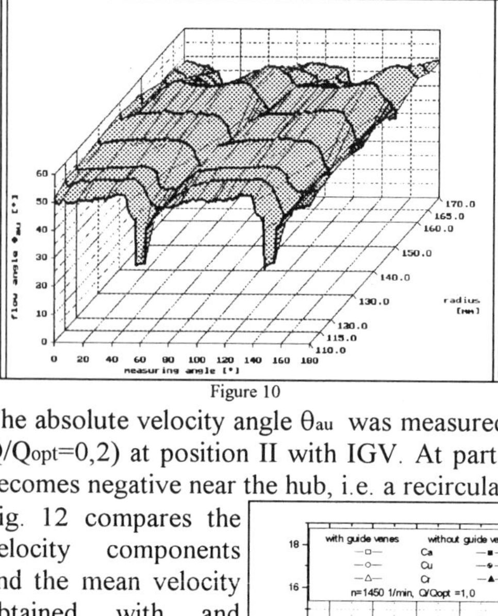

Figure 10

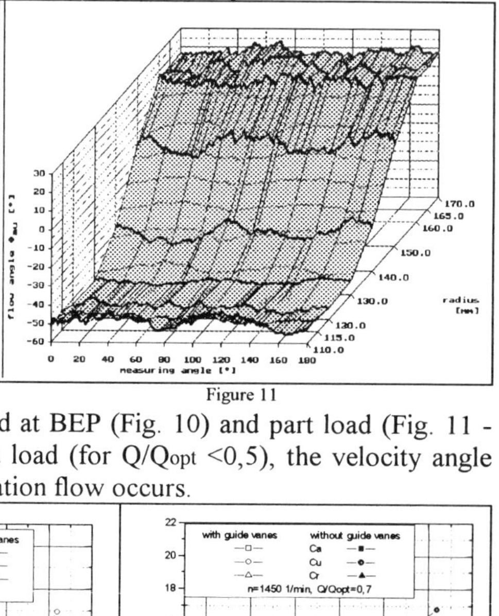

Figure 11

The absolute velocity angle θ_{au} was measured at BEP (Fig. 10) and part load (Fig. 11 - Q/Q_{opt}=0,2) at position II with IGV. At part load (for Q/Q_{opt} <0,5), the velocity angle becomes negative near the hub, i.e. a recirculation flow occurs.

Fig. 12 compares the velocity components and the mean velocity obtained with and without IGV at position II at BEP. The mean velocity (quadratic norm of axial and radial components) remains almost constant. The IGV influence the different velocity components but does

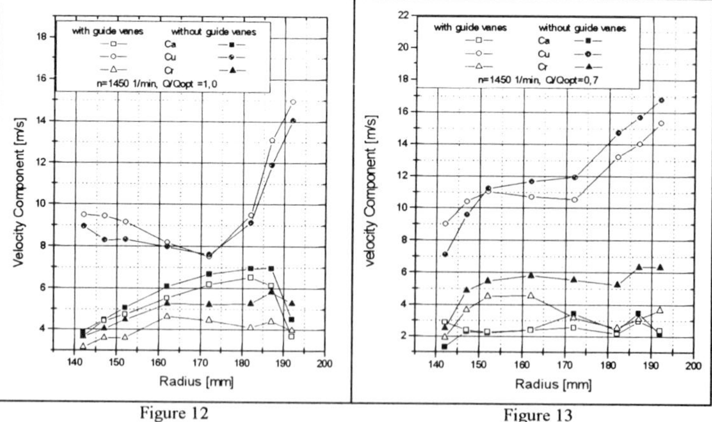

Figure 12

Figure 13

not alter the principal shape of the curves. Fig. 13, taken under the same conditions but at part load (Q/Q_{opt} = 0,7), leads to the same conclusions. The product $u*c_u$ represents the mechanical power output of the pump. By transforming radial velocity into circumferential velocity, IGV have a positive impact on the pump efficiency particularly at part load and even at the design point.

Fig. 14 (one channel of the impeller) and Fig. 15 (four channels) show the distributions of the circumferential velocity component c_u at part and over load and BEP near the hub. At part load Q/Q_{opt} = 0,2 the IGV cause a small decrease (0,4 m/s) of the circumferential velocity. At BEP a positive impact can be observed. At overload the circumferential velocity decreases strongly and contains large pulsations. This instability is caused by an instable entry flow. The flow separation and blade wake flow behind the impeller can be seen in Fig. 15. The axial velocity at various operating points is depicted in Fig. 16 (without IGV) and Fig. 17 (with IGV). The impact of the IGV on the axial velocity is higher compared to the circumferential velocity. The mean value increases and the pulsations (blade wake flow) are remarkably reduced.

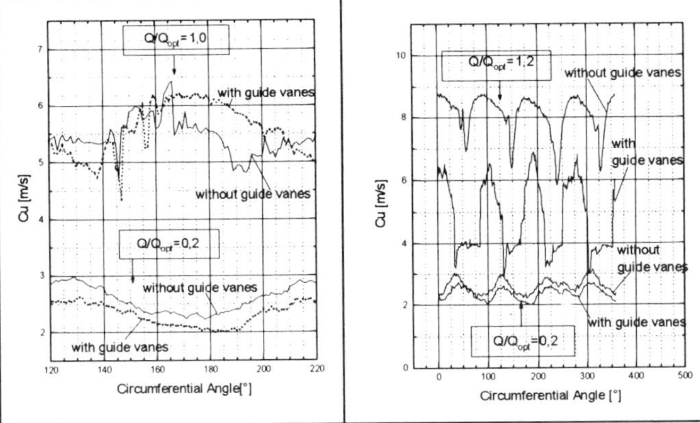

Figure 14 Figure 15

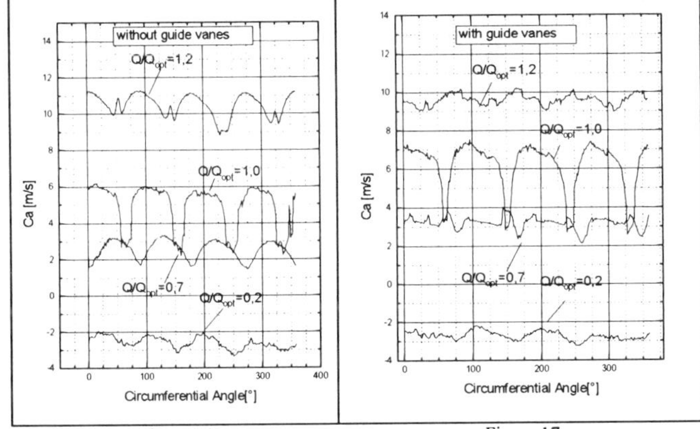

Figure 16 Figure 17

4 TURBULENT FLOW CHARACTERISTICS

Fig. 18 (Q/Q_{opt} = 1,0) and Fig. 19 (Q/Q_{opt} = 0,2) show the distribution of the three

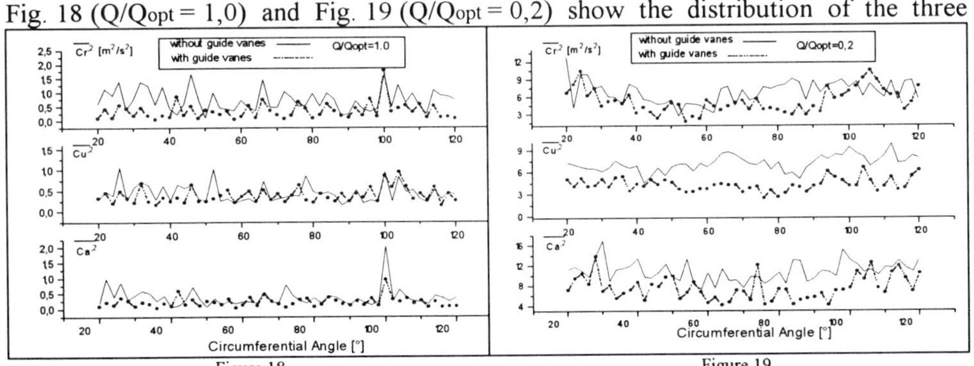

Figure 18 Figure 19

components of turbulent kinetic energy measured near the wall of the suction pipe of the impeller. The IGV provides damping for each velocity component at part load and at the design point. The overall level of turbulences is ten times higher at part load.

Fig. 20 shows the turbulent kinetic energy measured near the hub at position II. At the design point the turbulences near the impeller blades are much higher than in the middle of the channel. At part load the energy distribution is nearly constant. The IGV succeed to lower the turbulences of each velocity component at both operating points.

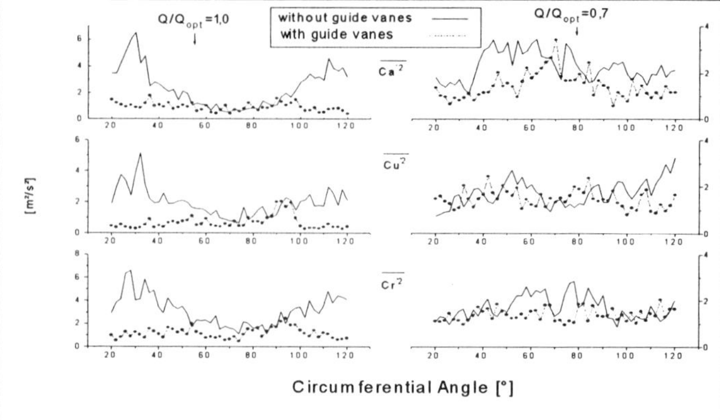

Figure 20

5 INSTABILITY OF PERFOMANCE CURVE

The instability in the performance curve is shown in Fig. 21. Slow variations of the flow rate lead to a hysteresis is the characteristic. Within the range of Q/Q_{opt}=0,61 and Q/Q_{opt}=0,66 the head varies about 4%. In addition there are discontinuities at the end of the range which cause sudden pressure variations. The beginning of the recirculation near the impeller can be derived from Fig. 21.

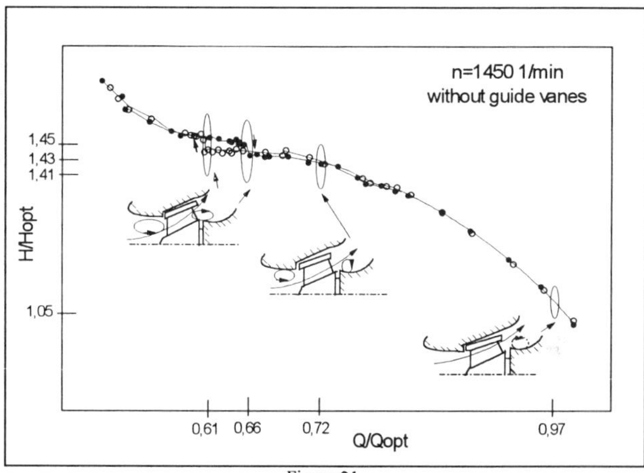

Figure 21

5.1 Unstable Inlet Flow

Figures 22 to 24 give the correlation between the velocity components and the flow

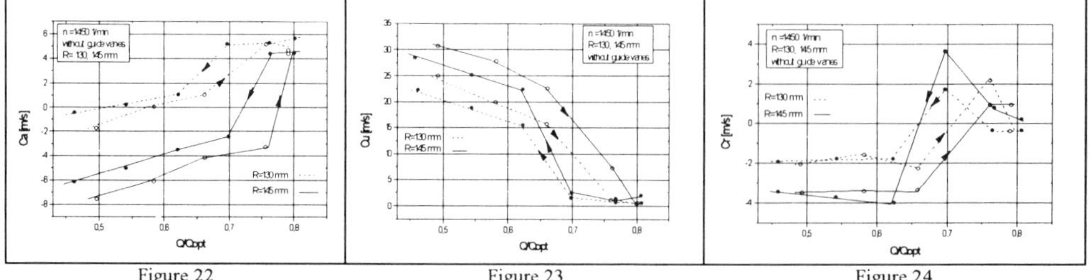

Figure 22 Figure 23 Figure 24

capacity at part load (Q/Q_{opt} = 0,45 - 0,85) measured in proximity to the suction pipe wall at position I. The velocity components depend not only on the flow rate but also on its derivative. The part load swirl occurs at the impeller tip first due to the centrifugal force. There is a hysteresis between the occuring (Q/Q_{opt}=0,72) and the vanishing (Q/Q_{opt} = 0,78) of the part load swirl.

5.2 Unstable Outlet Flow

Fig. 25 shows the three velocity components obtained at position II. The flow rate is varied from Q/Q_{opt} = 0,40 - 1,1. At Q/Q_{opt} = 0,96, the axial component changes the sign which indicates the beginning of recirculation. Reducing the flow rate further, the recirculation flow develops continually. The radial and circumferential components decrease steadily whereas the axial velocity has a negative maximum at Q/Q_{opt} = 0,70 - 0,80. Lowering the flow rate further the three components tend to their minimum with a sudden change of head of the pump (cf Fig. 21). The pressure jump is caused by the outset of the recirculation flow at the impeller outlet.

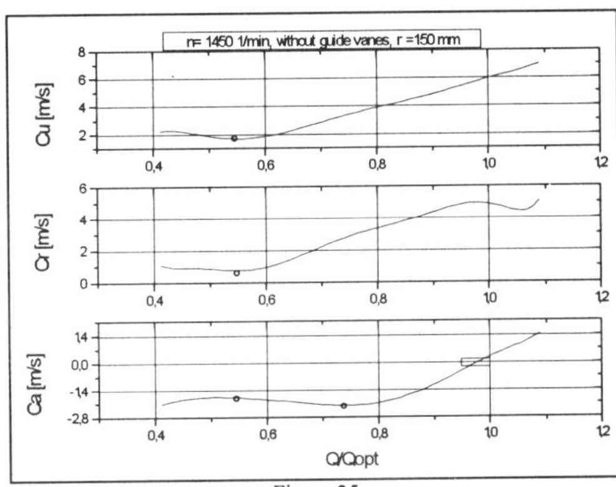

Figure 25

Fig. 26 (Q/Q_{opt} = 0,96) and Fig. 27 (Q/Q_{opt} = 0,865) show the blade to blade developement of the axial velocity component near the hub. The highest increase in speed can be observed in the middle of the channel.

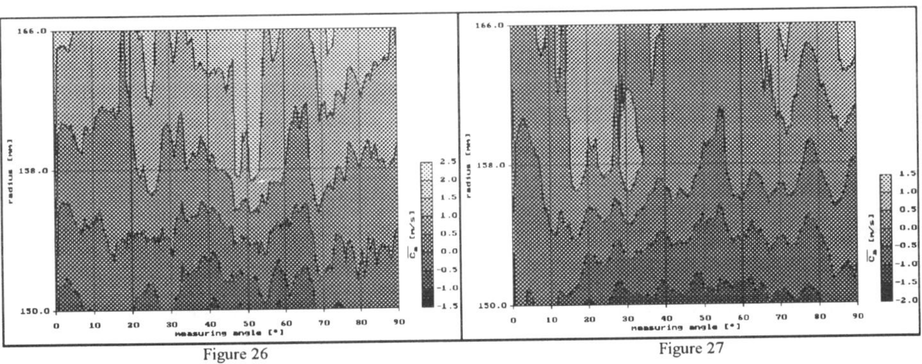

Figure 26 Figure 27

6 CONCLUSIONS

The three-dimensional flow fields in front of and behind the impeller were measured at different pump operating points with and without inlet guide vanes. These investigations brought new aspects about the mixed-flow pump, particularly concerning the influence of the guide vanes and the suction pipe of the pump. The important conclusions can be summarized as following:

- The entry flow of the pump is important and sensitive. It is strongly correlated with operating behaviour of the impeller.

- The measurements of the impeller inlet and outlet flow at part load show the mechanisms of flow recirculation in the pump and their relations to instabilities in the performance curve.
- Inlet guide vanes in the pump suction pipe improve the pump performance effectively in the whole operating range but particularly at part load.
- Inlet guide vanes have a positive influence on each of the three velocity components of the inlet and the outlet flow.
- Inlet guide vanes reduce turbulences and provide a stable spatial energy distribution.

The geometric optimisation of inlet guide vanes will be subject to future studies.

7 NOTATIONS

c_a	[m/s]	axial velocity
c_u	[m/s]	circumferential velocity
c_r	[m/s]	radial velocity
θ_{au}	[°]	angle between axial velocity and circumferential velocity
ψ	$Q/(Au)$	flow rate coefficient
μ	$P/(\rho Au^3/2)$	shaft power coefficient
η	$\rho QHg/P$	efficiency
ϕ	$2Hg/u^2$	pressure coefficient

8 REFERENCES

1. Goto, A. (1992b) The Effect of Tip Leakage Flow on Part-Load Performance of a Mixed-Flow Pump Impeller, ASME Journal of Turbomachinary, Vol. 114, pp. 383-391.
2. Goto, A. (1992a) Study of Internal Flows in Mixed-Flow Pump Impellers With Various Tip Clearances Using Three-Dimensional Viscous Flow Computations, ASME Journal of urbomachinary, Vol. 114, pp. 373-382.
3. Goto, A. (1994) Suppression of Mixed-Flow Pump Instability and Surge by the Active Alteration of Impeller Secondary Flows, ASME Journal of Turbomachinary, Vol. 116, pp. 621- 628.
4. Matthias, H.-B., Käfer, K., Pfingstmann, S. (1995) Investigations of Suction Pipe Flows in Reversible Pumps, Proceedings of The Tenth Conference on Fluid Machinary, pp. 315-324, Budapest.
5. Zhang, K., Sun, J., Wang, Y., Jia, Z. (1995) Optimization of The Mixed Flow Pump, Proceedings of The Second International Conference on Pumps and Fans, pp. 299-306, Beijing.
6. Kanemoto, T., Higuma, T., Huang, J. (1995) Probability of Tandem Cascades for a Mixed Flow Pump Diffuser, Proceedings of The Second International Conference on Pumps and Fans, pp. 307-318, Beijing.
7. Gopalakrishnan, S., Cugal, M., Ferman, R. (1995) Experimental and Theoretical Flow Field Analysis of Mixed Flow Pumps, Proceedings of The Second International Conference on Pumps and Fans, pp. 571-583, Beijing.
8. Kurokawa, J. (1992) Performance Curve Instability of A Diffuser Pump Due to A Rotating Stall and Some Control Methods, Proceedings of 16th Symposium of The IAHR, pp. 723-732.
9. Wang, B., Hellmann, D.-H., Hahn, P. (1995) Examination of The Turbulent Flow at The Impeller Inlet and Exit of An Axial Fan by Using a Triple Hot Wire Probe and An Angular Trigger, Proceedings of The Second International Conference on Pumps and Fans, pp. 531-540, Beijing.
10. Wang, B., Hellmann, D.-H., Hahn, P. (1995) Experimental Determination of The Highly Turbulent Flow at A Revolving Nozzle Exit by Using A Triple Hot Wire Probe and An Angular Trigger, Proceedings of The Tenth Conference on Fluid Machinary, pp. 510-518, Budapest.

OVERLOAD IN KAPLAN TURBINES OF SALTO GRANDE HYDROPOWER COMPLEX

Mabel BACCINO and Enrique BONNECARRERE
Senior Engineers
Comisión Técnica Mixta de Salto Grande
P.O. Box 68036, 50000 Salto, Uruguay
P.O. Box 106, 3200 Concordia E.R., Argentina

1. Introduction

The Salto Grande hydropower complex, in industrial operation since 1982, is located on the reach of the Uruguay river which forms the international border between the Uruguay and Argentina. The complex includes two simetrical power houses located on the river banks and a surface spillway located in the middle of the river. Each powerhouse includes seven 135 MW generating units resulting in a total installed capacity of 1,890 MW. The turbines are of the vertical shaft Kaplan type, of 187,500 CV each, under a rated net head of 25.3 m, and a specified rotation at speed of 75 r.p.m. The runner diameter is 8.5 m.
The Salto Grande Kaplan turbines had a power limit for each head, which was guaranteed by the supplier. This allowed power limit decreased remarkably as the head decreased, which provoked a direct energy loss when the river had abundance of water. In order to increase the generation, studies and tests were carried out with the turbines operating under overload conditions.
This kind of operation resulted in generation with higher powers than allowed by the guarantees, using the water flow that otherwise would have been spilled. The maximum generator capacity of (135 MW) was not surpassed.

2. History

The first complete tests of the turbine operating under overload conditions were carried out in 1989 with the participation of the Institute of Fluid Mechanics and Environmental Engineering, School of Engineering, University of Uruguay and the University of La Plata, Argentina. These tests were carried out for two low heads i.e., 20.4 and 24.4 m.
Vibrations and pressure pulsations were mesured and recorded at different places of the turbine when operating under overload conditions, as the maximum power for each tested head was reached.
The turbine governor is of the electro-hydraulic type. The original design of the governor actuador mechanical cam also covered the zone corresponding to overload. As this zone beyond the guarantees, it was verified that the on-cam relationship under overload conditions corresponded to best efficiency. Therefore, the tests were carried out at the on-

E. Cabrera et al. (eds.), Hydraulic Machinery and Cavitation, 1053–1062.

cam relationship between runner blade angle and guide vane opening.

These tests showed that:

- Vibrations on the head cover, the oil head, and the bearings did not surpass maximum allowed values by standards for large turbines.
- There were no power swings.
- Pressure pulsations on the spiral case, head cover and draft tube elbow did not surpass the maximum allowed values by the manufacturer for the rated conditions.

However, these tests showed that pressure pulsations below the runner under overload conditions, increased considerably for lower heads. It was for this reason that the following studies were undertaken.

Besides these tests, cavitation behaviour at overload conditions was also studied. The experience of more than 10 years of industrial operation showed that the blade metal loss between two maintenance outages (30.000 hs) had been lower than the value allowed for this kind of turbines, always increasing during periods of high mean power generation with high heads (1). The model test data also suggested that cavitation pitting on the power limit curve increased with increasing heads. As overload was only considered for heads lower than the rated head, cavitation pitting did not seem to be an impediment for overload.

3. Study of overload operation using model test data

This paper describes especifically the pressure pulsations below the runner which are associated with the turbine operating under overload conditions. Also the relationship with other operating parameters such as power, efficiency, and flow, comparing the data obtained from the model and the prototype tests, is described.

	Hn (m)	Pt (KW)
Supplier limit	Hn ≥ 25.3	Pt = 137906.0
	25.3 ≥ Hn ≥ 18.8	Pt = 6845.8 x Hn - 35293.9
	18.8 ≥ Hn ≥ 10.6	Pt = 6906.5 x Hn - 36433.5
Maximum overload	Hn ≥ 23.3	Pt = 137906.0
	23.3 ≥ Hn ≥ 21.5	Pt = 6234.7 x Hn - 7535.7
	21.5 ≥ Hn ≥ 15.5	Pt = 7142.9 x Hn - 27042.5
	15.5 ≥ Hn ≥ 10.6	Pt = 8418.4 x Hn - 46811.2
Allowed overload (MABO)	Hn ≥ 24.0	Pt = 137906.0
	24.0 ≥ Hn ≥ 18.8	Pt = 6836.2 x Hn - 26163.3
	18.8 ≥ Hn ≥ 15.5	Pt = 7311.9 x Hn - 35083.7
	15.5 ≥ Hn ≥ 13.0	Pt = 9081.0 x Hn - 62490.5
	13.0 ≥ Hn ≥ 10.6	Pt = 8056.3 x Hn - 48662.9

Table 1. Power vs. head

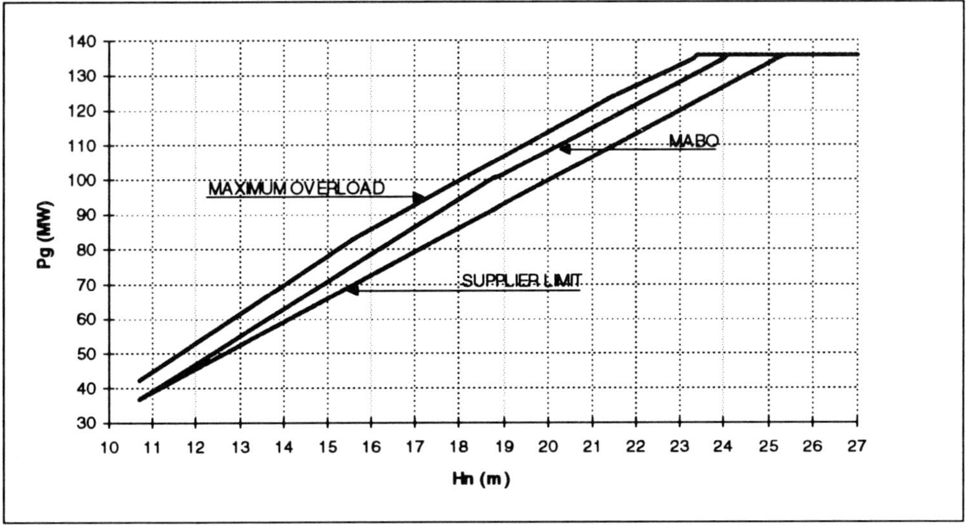

Figure 1. Power vs. head curves

The following power vs. head curves were considered:
- The supplier's power limit curve. The power limit for each head as guaranteed by the supplier.
- The maximum overload curve. The maximum power overload for each head determined through prototype tests.
- The allowed overload curve (MABO). This curve was determined as a result of these studies.

The maximum overload corresponds, for high heads, to the maximum runner blade tilt and for low heads, to the maximum guide vane opening.
The maximum overload curve and the supplier's limit curve were drawn on the hill chart of the Model Acceptance Tests of the Salto Grande Kaplan turbine (Fig.2) to know for each operating point on these curves, the values of Q_{11} and η_m, using the following method.
A series of points (p) was considered for both of the power curves defined by a power (P_t) and a head (H_n). For each point (p) the following expressions were calculated:

$$n_{11} = \frac{n_D}{\sqrt{H_n}} = \frac{75x8.5}{\sqrt{H_n}} \tag{1}$$

$$n'_{11} = n_{11} \sqrt{\frac{\eta_{mmax}}{\eta_{pmax}}} \tag{2}$$

where: η_{mmax} was obtained from the hill chart for each head

$$\eta_p = \eta_m + \frac{2}{3} \Delta_\eta \tag{3}$$

with
$$\Delta_\eta = (1-\eta_{mmax}) \left[1-\left(\frac{d}{D}\right)^{0.2}\right] \tag{4}$$

$$Q'_{11}\ \eta_p = \frac{P_t}{705.5\ H_n\ \sqrt{H_n}} \quad (5) \text{ from } \begin{cases} P_t = (KW)\dfrac{1000}{102}\ \gamma\ Q\ H_n\ \eta_p \\ Q = Q'_{11}\ D^2\ \sqrt{H_n} \\ \gamma = 0.996 \qquad (20^\circ C) \end{cases}$$

$$Q'_{11} = Q_{11}\ \sqrt{\frac{\eta_{mmax}}{\eta_{pmax}}} \tag{6}$$

$$Q_{11}\ \left(\eta_m + \frac{2}{3}\ \Delta_\eta\right) = \frac{P_t}{705.5\ H_n\ \sqrt{H_n}}\ \sqrt{\frac{\eta_{pmax}}{\eta_{mmax}}} \tag{7}$$

The values of Q_{11} and η_m corresponding to each point (p) were determinated on the hill chart with the n_{11} calculated for the head using the ecuation (7).
The supplier's power limit curve was drawn on the hill chart. For this reason, in this case we verified the described method and calculated the values of Q_{11} and η_m for the series of points of the mentioned curve.
Points A, B, α, β, γ, δ and Π of the maximum overload curve were determinated using the above mentioned method. These points were in the model tested range. Points A, B and α belong to the rated power line. (Figure 2).
Points α, β, γ and δ determined on the hill chart show a portion of the maximum runner blade tilt curve (+17°). The rest of the curve was drawn on the hill chart with a similar curvature as that for the blade tilt corresponding to +15°. This curve was used to determine points φ, Θ and μ because these points were not in the model tested range. From these points we knew the power (P_t), n_{11} and the runner blade tilt of +17°. With n_{11} corresponding to each set of points and the maximum runner blade tilt curve on the hill chart, we determined Q_{11}; furthermore with equations (7) and (3) η_m and η_p were determined respectively.
From the prototype tests it was determined that point μ corresponded to maximum guide vane opening (100%) and maximum runner blade tilt (+17°).
Points Ψ, Δ and Π belong to the maximum guide vane opening curve, however only point Π is in the model tested range. With points Π and μ, following a similar curvature as for 95% guide vane opening, the curve of maximum guide vane opening was drawn on the hill chart. Q_{11} and η_m of points Ψ and Δ were determined with the corresponding n_{11} and the

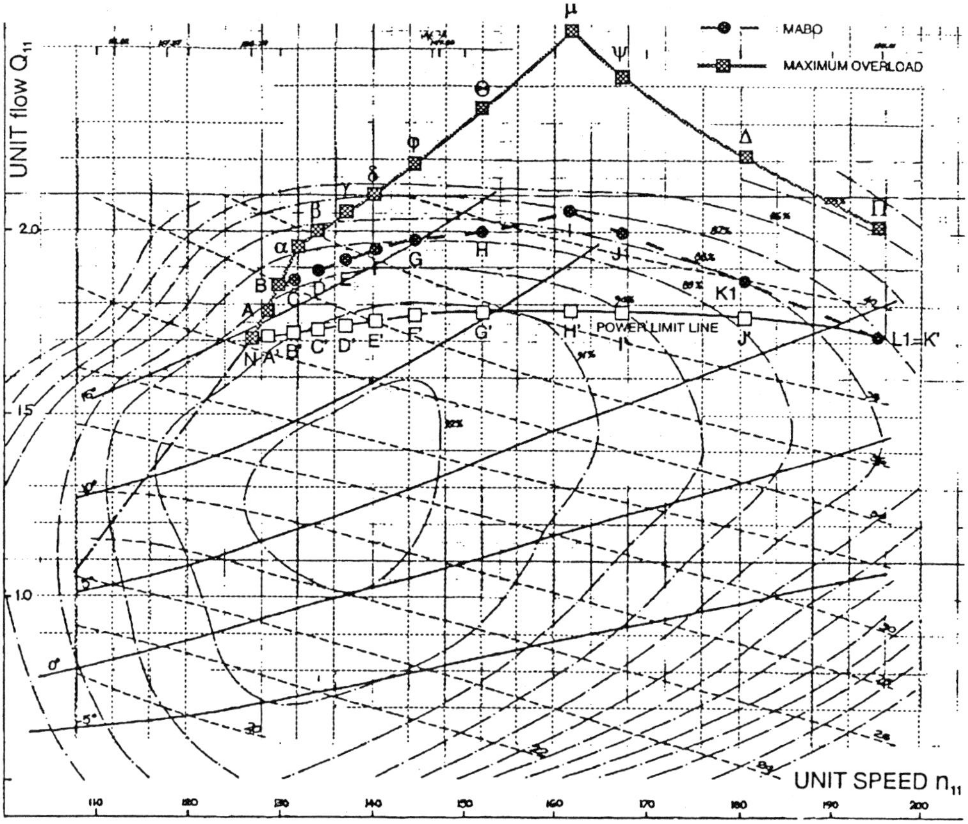

Figure 2. Hill chart of the Model Acceptance Test (1979).

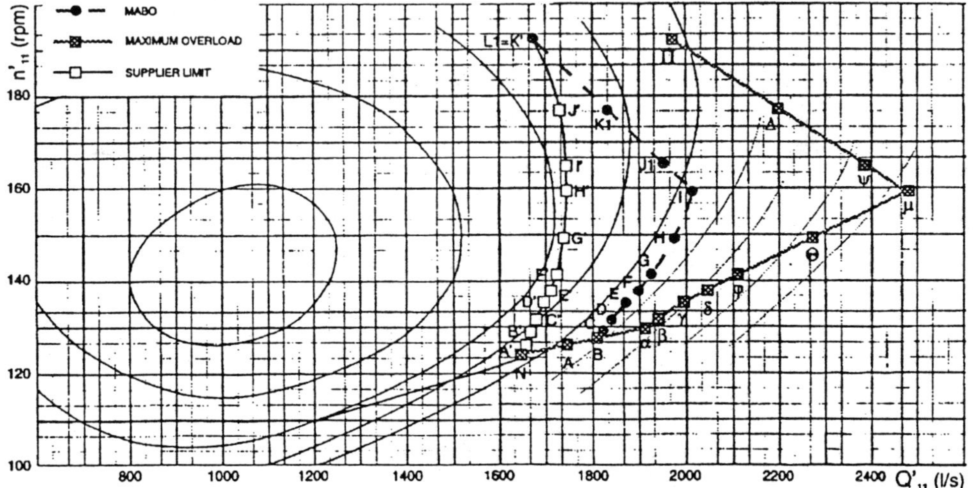

Figure 3. Supplier's Model pressure pulsations isomeric diagram.

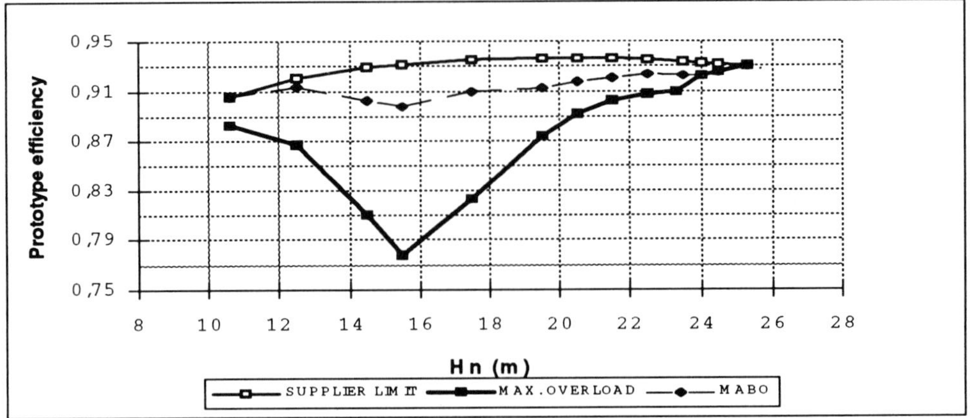

Figure 4.1

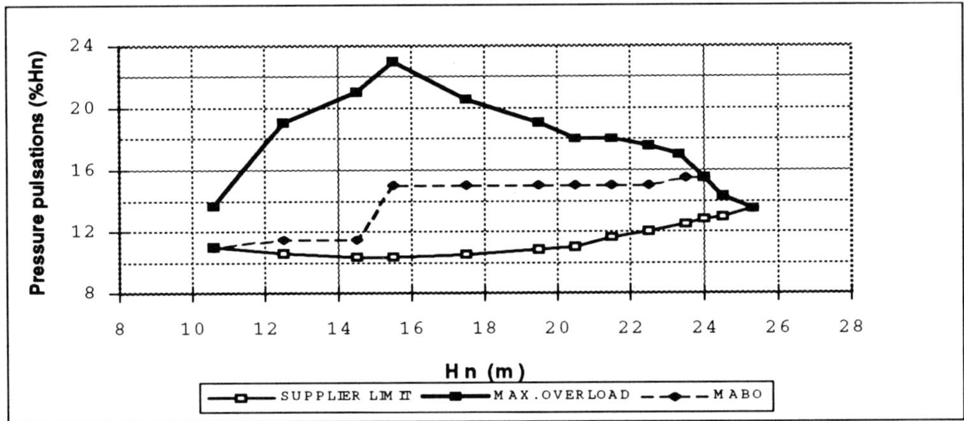

Figure 4.2

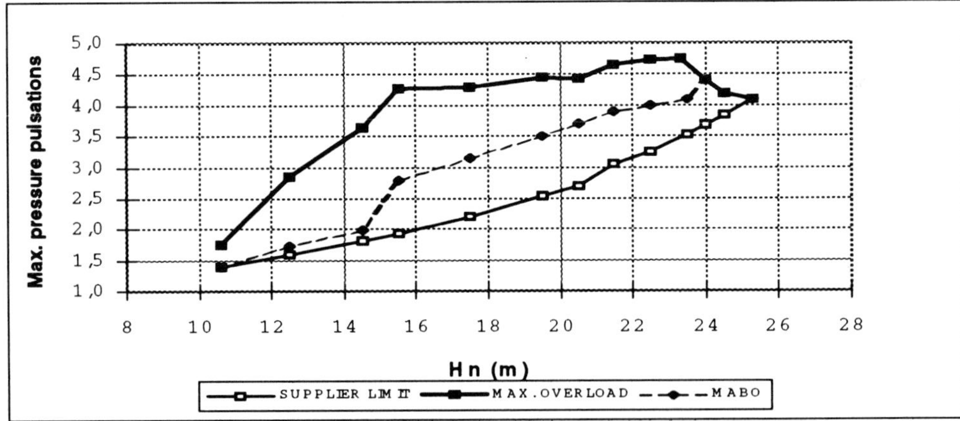

Figure 4.3

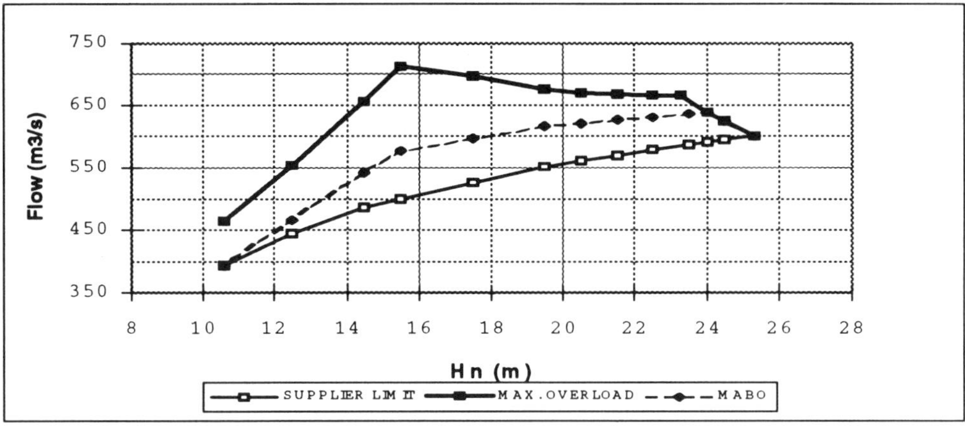

Figure 4.4

maximum guide vane opening curve drawn.
For each point, with the corresponding Q'_{11} and n'_{11} calculated from ecuations (6) and (2), using the supplier's Model pressure pulsation isomeric diagram (Figure 3) we determined the mean double amplitude pressure pulsations in the draft tube (below the runner) as a per cent of head ($2A/H_n$ %). The diagram shows the mean pressure pulsation curves down to 14 %; the curves corresponding to greater than 14 % pulsations were traced taking into consideration the data obtained from the prototype tests.
According to the supplier's Model Test Report the maximun amplitudes ($2A_{max}$) were 1.2 times the mean amplitudes, consequently this factor was used to calculate them.
The prototype flow was calculated using the following ecuation:

$$Q = Q'_{11} \; D^2 \; \sqrt{H_n}$$

Tables 2 and 3 show the calculated parameters and Figure 4 shows their variation with head.

Point p	Hn m	n11	Pg MW	Pt MW	$\sqrt{\frac{\eta_{mmax}}{\eta_{pmax}}}$	n'11	Q11 m³/s	η mod	η prot	Q'11 m³/s	Q m³/s	2A/Hn %	2A m	2Amax m
N	25.3	126.7	135.0	137.9	0.987	125.1	1.69	0.906	0.930	1.65	600	13.5	3.4	4.1
A'	24.5	128.8	129.8	132.4	0.987	127.1	1.71	0.907	0.931	1.66	594	13.0	3.2	3.8
A"	24.0	130.1	126.4	129.0	0.987	128.4	1.72	0.908	0.932	1.67	591	12.8	3.1	3.7
B'	23.5	131.5	123.1	125.6	0.987	129.8	1.72	0.909	0.933	1.68	587	12.5	2.9	3.5
C'	22.5	134.4	116.4	118.7	0.987	132.6	1.74	0.911	0.935	1.69	578	12.0	2.7	3.2
D'	21.5	137.5	109.6	111.9	0.987	135.7	1.75	0.912	0.936	1.70	569	11.6	2.5	3.1
E'	20.5	140.8	103.3	105.0	0.987	139.0	1.76	0.912	0.936	1.71	561	11.0	2.2	2.7
F'	19.5	144.4	96.2	98.2	0.987	142.5	1.78	0.912	0.936	1.73	551	10.8	2.1	2.5
G'	17.5	152.4	82.7	84.4	0.987	150.4	1.78	0.911	0.935	1.75	526	10.5	1.8	2.2
H'	15.5	161.9	69.2	70.6	0.985	159.5	1.79	0.904	0.931	1.76	500	10.4	1.6	1.9
I'	14.5	167.4	62.4	63.7	0.985	164.9	1.79	0.900	0.929	1.76	485	10.4	1.6	1.8
J'	12.5	180.3	48.9	49.9	0.982	177.1	1.77	0.887	0.920	1.74	445	10.6	1.3	1.6
K'=L1	10.6	195.8	36.0	36.8	0.981	192.1	1.70	0.871	0.906	1.67	393	11.0	1.2	1.4

Table 2. Supplier power limit curve.

Point p	Hn m	n11	Pg MW	Pt MW	$\sqrt{\frac{\eta_{mmax}}{\eta_{pmax}}}$	n'11	Q11 m³/s	η mod	η prot	Q'11 m³/s	Q m³/s	2A/Hn %	2A m	2Amax m
N	25.3	126.7	135.0	137.9	0.987	125.1	1.69	0.906	0.930	1.65	600	13.5	3.4	4.1
A	24.5	128.8	135.0	137.9	0.987	127.1	1.79	0.901	0.925	1.74	623	14.3	3.5	4.2
B	24.0	130.1	135.0	137.9	0.987	128.4	1.85	0.899	0.922	1.80	638	15.5	3.7	4.4
α	23.3	132.1	135.0	137.9	0.987	130.3	1.96	0.887	0.910	1.91	666	*17.0*	4.0	4.8
β	22.5	134.4	130.0	132.7	0.987	132.6	2.00	0.884	0.908	1.94	665	*17.5*	3.9	4.7
γ	21.5	137.5	124.0	126.5	0.987	135.7	2.05	0.879	0.903	1.99	667	*18.0*	3.9	4.6
δ	20.5	140.8	117.0	119.4	0.987	139.0	2.11	0.868	0.892	2.05	669	*18.0*	3.7	4.4
φ	19.5	144.4	110.0	112.2	0.986	142.3	2.18	0.850	0.874	2.12	675	*19.0*	3.7	4.4
Θ	17.5	152.4	96.0	98.0	0.985	150.1	2.34	0.798	0.823	2.31	697	*20.5*	3.6	4.3
μ	15.5	161.9	82.0	83.7	0.983	159.2	2.55	0.751	0.778	2.51	713	*23.0*	3.6	4.3
ψ	14.5	167.4	73.7	75.3	0.982	164.4	2.43	0.782	0.810	2.38	655	*21.0*	3.0	3.7
Δ	12.5	180.3	57.3	58.4	0.981	176.9	2.21	0.834	0.867	2.21	552	*19.0*	2.4	2.9
Π	10.6	195.8	41.6	42.4	0.980	192.0	2.02	0.848	0.883	1.98	465	13.7	1.5	1.7

Table 3. Maximum overload.

Point p	Hn m	n11	Pg MW	Pt MW	$\sqrt{\frac{\eta_{mmax}}{\eta_{pmax}}}$	n'11	Q11 m³/s	η mod	η prot	Q'11 m³/s	Q m³/s	2A/Hn %	2A m	2Amax m
N	25.3	126.7	135.0	137.9	0.987	125.1	1.69	0.906	0.930	1.65	600	13.5	3.4	4.1
A	24.5	128.8	135.0	137.9	0.987	127.1	1.79	0.901	0.925	1.74	623	14.3	3.5	4.2
B	24.0	130.1	135.0	137.9	0.987	128.4	1.85	0.899	0.922	1.80	638	*15.5*	3.7	4.4
C	23.5	131.5	131.7	134.4	0.987	129.8	1.86	0.898	0.922	1.81	635	*15.5*	3.6	4.1
D	22.5	134.4	125.0	127.6	0.987	132.6	1.88	0.899	0.923	1.84	629	*15.0*	3.4	4.0
E	21.5	137.5	118.3	120.7	0.987	135.7	1.92	0.896	0.920	1.87	625	*15.0*	3.2	3.9
F	20.5	140.8	111.7	113.9	0.987	139.0	1.95	0.893	0.917	1.90	620	*15.0*	3.1	3.7
G	19.5	144.4	105.0	107.1	0.987	142.5	1.98	0.889	0.913	1.93	616	*15.0*	2.9	3.5
H	17.5	152.4	91.0	92.9	0.986	150.2	2.00	0.886	0.910	1.97	596	*15.0*	2.6	3.2
I	15.5	161.9	76.7	78.3	0.985	159.5	2.06	0.871	0.898	2.02	576	*15.0*	2.3	2.8
J1	14.5	167.4	67.8	69.2	0.984	164.7	2.00	0.876	0.903	1.97	542	11.5	1.7	2.0
K1	12.5	180.3	51.0	52.0	0.982	177.0	1.86	0.881	0.914	1.83	466	11.5	1.4	1.7
L1	10.6	195.8	36.0	36.8	0.981	192.1	1.70	0.871	0.906	1.67	393	11.0	1.2	1.4

Table 4. Allowed overload (MABO).

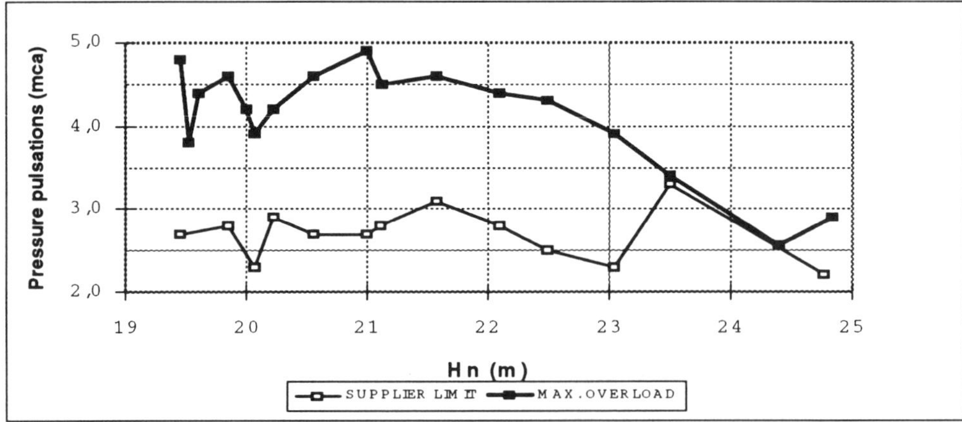

Figure 5. Prototype test.

3. Prototype tests

Prototype tests were carried out during 1994. The pressure pulsations below the runner were mesured and recorded each 0.5 m of head variation, on the supplier's power limit curve and at maximum overload conditions. Figure 5 shows the maximum pressure pulsations recorded for each head during these tests. Table 5 and Figure 6 show data from the prototype tests in comparisson with data from the model test.

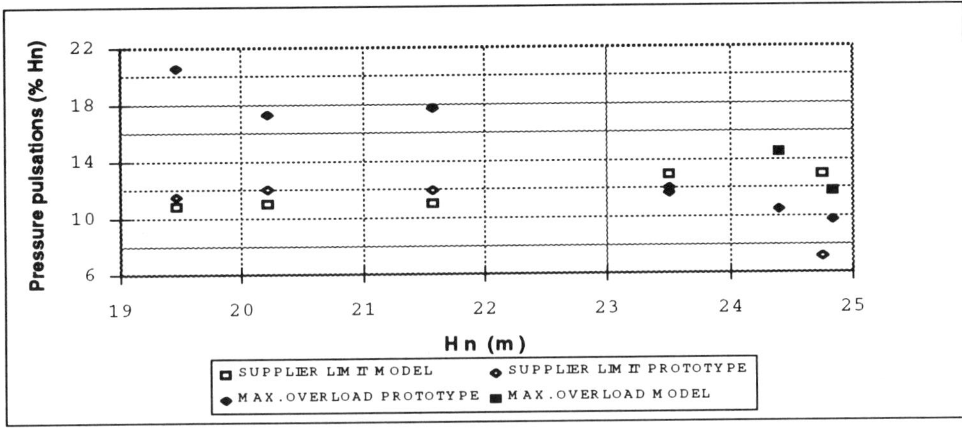

Figure 6

PROTOTYPE TEST				MODEL TEST												
Hn m	TWL m	Pg MW	2A m.c.a.	Test number	n11	n'11	Pt MW	Q11 m^3/s	η mod	η prot	Q'11 m^3/s	Q m^3/s	2A	2A/Hn prot %	2A/Hn mod %
19.46	13.4	95	2.7	1	144.5	142.6	97	1.77	0.912	0.936	1.73	551	2.3	11.5	10.8
19.46	13.4	111	4.8	2	144.5	142.5	113	2.18	0.850	0.874	2.14	682	4.0	20.5	
20.22	13.6	100	2.9	3	141.8	139.9	102	1.76	0.912	0.935	1.70	553	2.4	12.0	11.0
20.22	13.6	115	4.2	4	141.8	139.9	117	2.11	0.868	0.891	2.06	668	3.5	17.3	
21.58	12.8	105	3.1	5	137.2	135.4	107	1.66	0.915	0.939	1.61	541	2.6	11.9	11.0
21.58	12.8	125	4.6	6	137.2	135.4	128	2.05	0.879	0.903	2.00	671	3.8	17.7	
23.50	11.2	125	3.3	7	131.5	129.8	128	1.75	0.908	0.931	1.70	597	2.8	11.7	13.0
23.50	11.2	135	3.4	8	131.5	129.8	138	1.92	0.893	0.916	1.87	655	2.8	12.0	

Table 5

4. Allowed overload (MABO)

η_p under maximum overload, reaches unacceptable values for larger turbines (Figure 4.1). According to the suppliers' recomentations, η_p should not be less than 90% within the range of operation. For this reason, another overload curve was determined, taking into consideration the following conditions:

- η_p was not less than 90% within the operating range.
- Pressure pulsations (2A) did not surpass the maximum allowed by the supplier on the Model Test for rated conditions.
- Vibrations mesured on the prototype did not surpass the maximum allowed values by standards.

Table 4 shows the calculated parameter for points on the MABO curve and Figure 4 shows their variation with head. This curve was within the model tested range.

In the zone near the rated condition, pressure pulsations (2A) determined on the Model Test slightly surpassed the maximum values allowed by the supplier. Taking into consideration that the values obtained from the prototype tests on this zone were less than those obtained from the model test, we can confirm that the maximum allowed by the supplier was not surpassed when operating the prototype on the MABO curve.

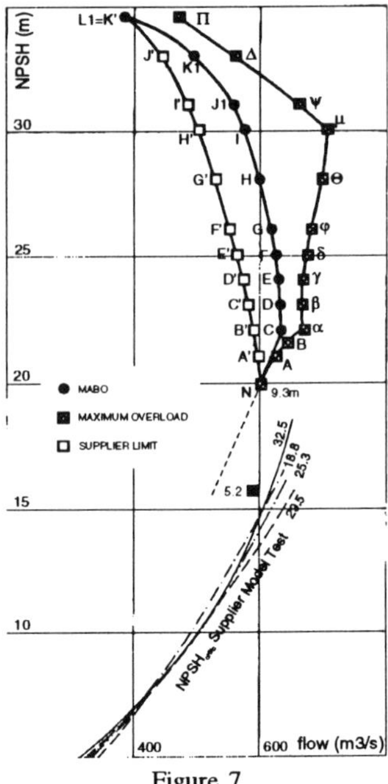

Figure 7

5. Cavitation

Figure 7 shows the calculated NPSH vs. flow, for points on supplier limit, MABO, maximum overload curves and $NPSH_{critic}=\sigma_{critic} H_n$ obtained from the Model Test by Eng. D. Algorta. Operating the prototype on the MABO curve is far from critical cavitation point.

6. Conclusion

These studies enabled an increase in power without modification to the hydrogenerator. The average overload operating hours are 30 % of the total year operating hours, thus an increase of the yearly generation of aprox. 3 % has been achieved, with the resulting economic benefit to the Company.

Others studies to increase the turbine output by either replacing the turbine runners or alternatively replacing only the runner blades are also taking place at this time.

Nomenclature and References

D	(m)	prototype throat diameter=8.5	d	(m)	model throat diameter=0.46
H_n	(m)	net nominal head=25.3	n	(rpm)	rotation speed=75
n_{11}	(rpm)	unit speed	P_t	(KW)	turbine power
P_g	(KW)	generator power	Q_{11}	(m^3/s)	unit flow
Q	(m^3/s)	prototype flow	NPSH	(m)	net pressure suction head
η_p		prototype efficiency	η_{pmax}		max. prototype efficiency
η_m		model efficiency	η_{mmax}		max. model efficiency
σ		cavitation number			

(1) Eluen, A. & Baccino, M.; "Experiences about relationships between cavitation erosion and operating conditions in Kaplan turbines at Salto Grande", Procs. 16th Symposium of the IAHR, section on Hydraulic machinery and cavitation, Brazil, sept. 1992.

FULL SIZED TESTS ON A FRENCH MAIN COOLANT PUMP UNDER TWO-PHASE FLOW

J.C. HUCHARD, C. BORE, E. DUEYMES
EDF-DER, 6 quai Watier
78400 Chatou, FRANCE

Abstract

The French Safety Authorities required EDF to demonstrate the ability of the new main coolant pump to withstand two-phase flow conditions without damage. Therefore three full sized tests, simulating a bleeding flow on the primary system, were performed on a laboratory test loop at real operating conditions (temperature = 290 °C, pressure = 155 b, flowrate = 7 m^3/s, electrical power 7 MW). The maximum value of the mean void fraction reached 75 %. The outcome of the tests is very positive : the mechanical behaviour of the main coolant pump is good, even at high void fraction. The maximum vibration levels were below the limits fixed by the manufacturer. Correlations between the mechanical behaviour of the pump and the pressure pulsation in the test loop have been found.

1. Introduction

In a nuclear power plant, the main coolant pumps function is to insure the circulation of the water between the reactor and the steam generators. They constitute essential components for the safety and reliability of nuclear power plants.

In case of loss of coolant accident (LOCA), in nuclear reactor, which may result from a break in the primary system, the coolant may undergo rapid depressurization, form vapor-liquid two-phase flow and the main coolant pumps could have to operate under those extremely hard conditions.

In 1993, the French Safety Authorities required Electricité de France (EDF) to demonstrate the ability of the new Pressurized Water Reactor (PWR) pumps to withstand mechanically those conditions.

Although appreciable testings have been carried out around the world on pumps operating under two-phase flow conditions, their dynamic behaviour has rarely been studied, especially for these type of pumps.

Therefore, three full sized tests have been performed on a laboratory test loop at real operating conditions. The purpose of this paper is to :

E. Cabrera et al. (eds.), Hydraulic Machinery and Cavitation, 1063–1072.

-describe the main characteristics of the main coolant pumps

-outline the main features of the test loop

-present the three tests and their results :

thermo-hydraulic behaviour of the pump

mechanical behaviour of the pump

hydro-acoustic caracteristics in the test loop

2. Description of the main coolant pump

The main coolant pump tested is from the new French standardized nuclear power plant called N4. It was designed by the French manufacturer Jeumont-Industrie. The main features of the pump are illustrated in figure 1 and may be summarized as follows :

Operating conditions : The pump operates at a temperature and suction pressure of 293 °C and 15,4 Mpa respectively. At the nominal point, the pump delivers 6,9 m3/s at a 106 m head for a power of 6700 kW.

Impeller design The pump has a mixed flow geometry and the impeller has four vanes.

Motor The pump rotates at 1500 rpm, driven by an asynchronous motor.

Bearings The impeller is overhung by a hydrostatic bearing pressurized from the pump discharge. In case of loss of this bearing, a hydrodynamic bearing is located in the upper part of the pump. The motor shaft's lateral and axial support is provided by two guide bearings.

Casing The casing is a single-suction, single-radial-discharge arrangement.

3. Description of the test loop

As owner and operator of PWR reactors, EDF maintains a facility to test at real conditions full sized PWR main coolant pumps. The test loop is a closed loop of 30 m^3 of capacity. The main characteristics of the loop may be summarized as follows (see also figures 2&3) :

Piping arrangement The suction and discharge header are horizontal (diameter : 910 mm, 750 mm near the pump). The total length of the loop is about 30 m. The casing is mounted on a concrete support. This is the only fixed point of the test loop.

Temperature control	Due to the power of the pump, there is no need to heat the water. The temperature is controlled by a water-cooled heat exchanger.
Pressure control	The pressure in the test loop is controlled by a pressurizer as in the primary circuit in PWR nuclear power plant.
Flow control	The flow rate control is assured by a butterfly vane. The maximal flow's range is from 0 m^3/s to 8 m^3/s.
Bleeding circuit	To realize two-phase flow test, a bleeding circuit was added. Water was taken away from the main circuit in order to depressurize the test loop and form vapor-liquid two-phase flow. The bleeding flow rate is controlled by a vane over the rang of 0 m^3/h to 100 m^3/h and is measured by a diaphragm (the vapor-liquid mixture is cooled with cold water before the diaphragm in order to have one-phase flow).
	To refill the test loop, positive displacement pumps were used, taking the water from the volume control tank at the temperature of 30 °C. The refilling flow rate was 7.2 m^3/h.

4. Instrumentation

In order to analyze the results, specific instrumentation was installed on the pump (see figure 1) and the test loop (see figure 2) :

Motor shaft	8 radial accelerometers.
	1 axial accelerometer.
Motor casing	4 radial accelerometers.
Test loop	26 accelerometers.
	10 pressure fluctuations sensors.
	10 strains gauges.

The void fraction was calculated by integrating the bleeding flow rate. It is only a mean void fraction : it is not representative of the void fraction at the suction or discharge pipe.

For the last test, gamma densitometer measurements were used to determine the local void fraction at the suction pipe and the discharge pipe. And short life radioactive tracers measurements of the liquid phase and the gas phase flow rates have been made.

5. Tests performed

Three two-phase flow tests have been performed on the test loop. The first one with a bleeding flow rate of 1 m^3/h (about 3 % of void fraction per hour). The maximal void fraction reached was 28 %, the two others with a bleeding flow rate of 13 m^3/h (about 50 % void fraction per hour). The maximal void fraction reached were 75 % and 70 % respectively.

The thermo-hydraulic behaviour and the mechanical behaviour of the pump during the three different tests were very similar.

6. Thermo-hydraulics behaviour of the pump

6.1. VOID FRACTION

Figure 4 represents the void fraction variation during the third test. On this figure, the mean void fraction is plotted with the suction and discharge void fraction (taken from the gamma densitometer). This graph brings out two points :

• As long as the suction void fraction is less than 30 %, the discharge void fraction is quasi-null. This happens during the bleeding phase of the test but also during the refilling phase. This phenomenon has already been observed and studied [1]. For low suction void fraction, as the steam-water mixture flows through the impeller channel, the pressure increases and condensation takes place. At the discharge pipe, the flow is almost in a one-phase flow state.

• For mean fraction void larger than 40 %, the suction void fraction is smaller than the discharge void fraction. This might be explained by the configuration of the test loop. The discharge pipe is located in the upper part of the test loop, while the suction pipe is located in the lower part and some water may have stayed in this part for high void fraction.

On figure 5, the spectral analysis between 0 Hz and 2 Hz of the discharge pump is shown over the entire duration of the third test. This graph reveals the presence of a peak around 0.4-0.6 Hz and its harmonics during the bleeding phase of the test and especially during the refilling phase. It also can be seen on the suction void fraction, on the shaft, motor casing and test loop accelerometers and the pressure fluctuation sensors (but at very low level). The fact that this frequency appears on all the probes seems to show a slug flow in the test loop.

6.2.ΔP CREATED BY THE PUMP

Figure 6 shows the Δp created by the pump as function of the mean void fraction. As may be seen, the Δp steady decreases with the void fraction. At the maximum of the mean void fraction (75 %), the value of Δp is about 70 kPa. This shows that, even at high void fraction, the pump does not seem to dewater.

Tests made on another pump [2], have shown that the curve Δp=f(void fraction) strongly depends on the water temperature. For high temperatures, as it is the case here, the Δp is less degrading with the void fraction as for low temperatures.

7. mechanical behaviour of the pump

7.1. MOTOR CASING

Figure 7 summarizes the motor casing mechanical behaviour during the most severe test (maximum value of void fraction reached : 75 %). As it can be seen, there are four vibration crises : two during the bleeding phase and two during the refilling phase. As may be noticed, the vibration crises appear for the same void fraction range during the bleeding phase and the refilling phase : they seem to be symmetric and the mechanical behaviour of the motor casing seems to depend strongly on the void fraction.

The coherence study shows good agreement in the range 0 - 10 Hz between the motor casing accelerometers and the test loop accelerometers and also between the motor casing accelerometers and the pressure fluctuation probes. This is not the case for the shaft and motor casing accelerometers. These results could tend to demonstrate that the vibration crises are mainly due to the fluid : the vapour-water mixture produces a large bandwidth excitation at low frequency. This induces response of the motor casing and test loop on their different low frequency modes.

When the mean void fraction reaches its maximum, the level of the motor casing vibrations are very low, even lower than in the normal operating conditions. This can be explained by the low excitation level : a very small amount of water is pumped at high void fraction.

7.2. SHAFT

On figure 7 is also drawn the shaft mechanical behaviour. At normal conditions, the vibration level is around 50 μm (peak-to-peak). During the two-phase flow tests, it increases regularly with the mean void fraction, up to 100 μm.

The spectral analysis of the shaft accelerometers shows, at high void fraction, the presence of a peak around $\omega/2$. For those high void fraction, the Δp created by the pump is very low (under 100 kPa) and therefore, the hydrostatic bearing is not well furnished. Some bearing instabilities occur. The presence of the peak around $\omega/2$ reveals these instabilities. This was well predicted by a calculation taking account the hydrostatic bearing loss.

After the tests, the condition of the key components was checked by dismantling and direct inspection. Particular attention was given to the hydrostatic journal and bearing. It has shown contact between the rotor and the stator parts but without significant damage.

7.3. INFLUENCE OF THE WATER TEMPERATURE ON VIBRATION

On figure 8 the maximal value of the motor casing vibrations level against the test loop water temperature is plotted. One may notice, the lower the temperature is, the larger the vibrations are. Such phenomenon was already seen (2). This is due to the two-phase flow mixture : at high temperature, the vapour-liquid mixture is more homogeneous than at low temperature and therefore the excitation level caused is lower. This would tend to confirm that the fluid is mainly responsible for the vibration crises (cf. 7.1).

8. Pressure fluctuations in the test loop

Figure 9 illustrates the pressure pulsation spectral variation during a two-phase test (on the abscissa axis frequency (range 0 - 30 Hz), on the co-ordinates axis, the time and the void fraction). The graph brings out two points :

• When the mean void fraction is under 30 %, the pulsation level is the highest for the two-phase flow test. This is caused by the fact that, at the discharge pipe, the flow stays in a one-phase state and pressure fluctuations are not damped by the presence of steam.

• As long as the mean void fraction is larger than 30 %, the level of the fluctuation is very low, comparatively to the normal operating conditions. The aspects of the spectrums are typical turbulence spectrums. No particular frequency appears.

9. Conclusion

Three full sized tests of the new French main coolant pump under two-phase flow conditions have been performed on a laboratory test loop. The outcome of the test is very positive :

• The maximum value of the mean void fraction reached 75 %.

• The tests have shown the ability of the main coolant pump to withstand extremely hard void fraction without damage. The level of vibrations were always below the limits fixed by the manufacturer.

• Even at high void fraction, the pump does not seem to dewater.

The tests have proved the reliability of the new main coolant pump. They will improve the understanding of safety matters and the corresponding simulation models.

References

[1] G.R. Noghrehkar, M. Kawaji, A.M.C. Chan, H. Nakamura, Y. Kukita, Investigation of centrifugal pump performance under two-phase flow conditions, Journal of Fluids Engineering, March 1995, Vol 117, pp. 129-137.

[2] G. Rzentkowsky, Analysis of two-phase flow instabilities in a closed loop, pump-piping system, ASME Pressure Vessels and Piping Conference, Honolulu, July 1995. Presentation without publication.

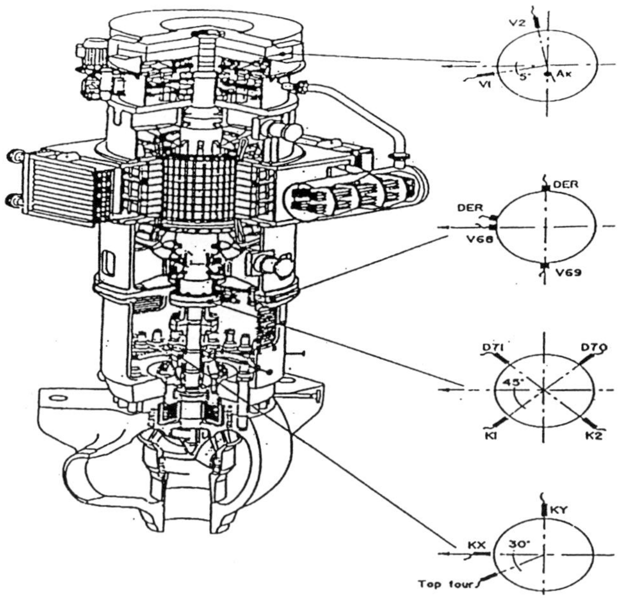

Figure 1 : View of the main coolant pump and its instrumentation

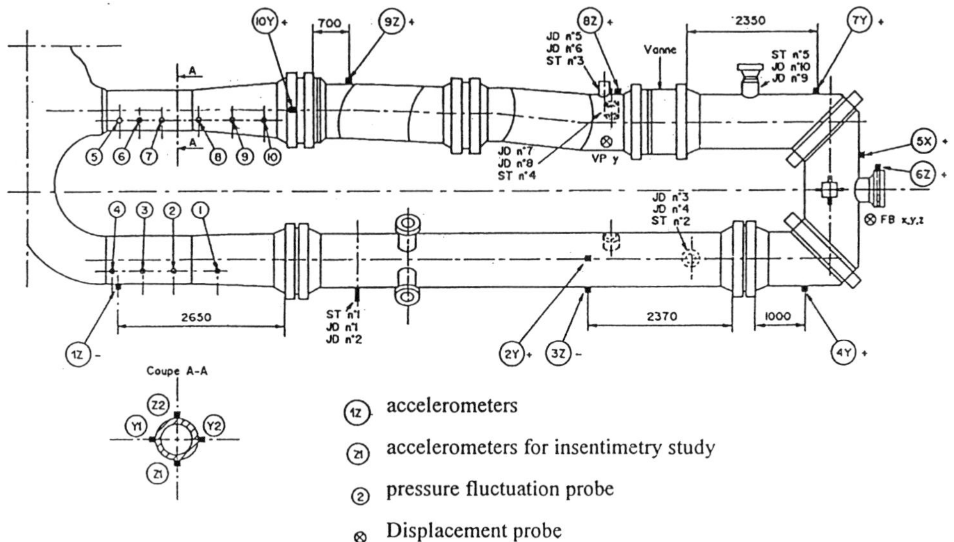

Figure 2 : View of the test loop and its instrumentation

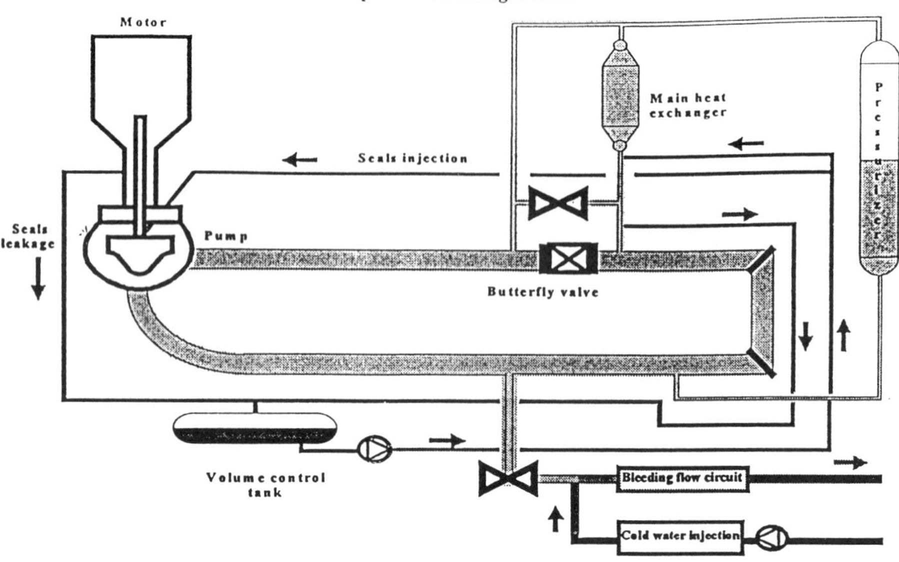

Figure 3 : Main coolant pump test loop : two phase flow configuration

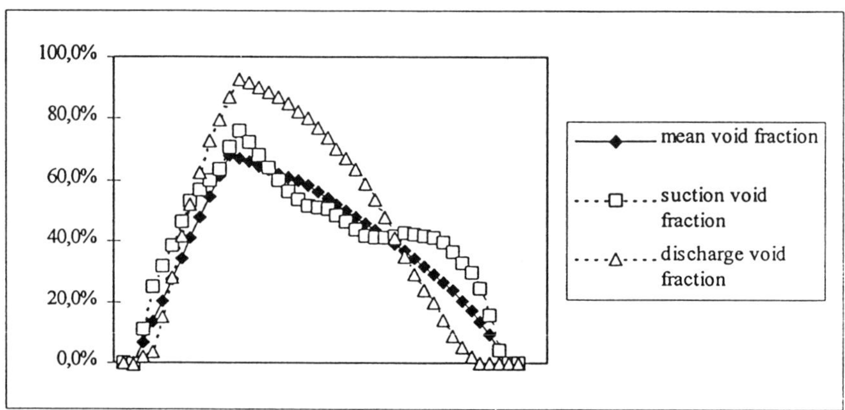

Figure 4 : Mean, suction and discharge void fraction during the third test

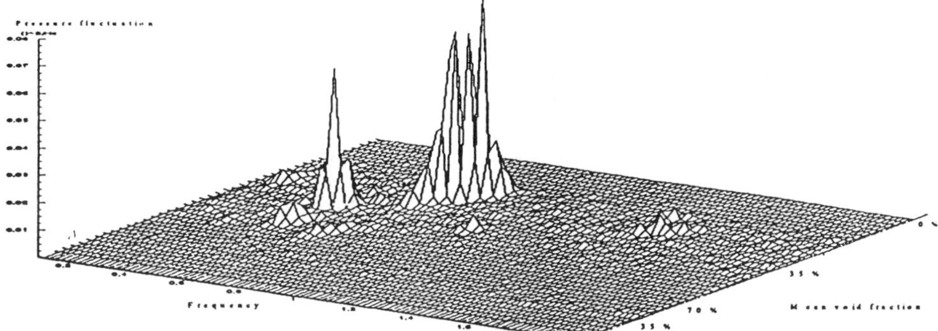

Figure 5 : Spectral analysis of the discharge void fraction

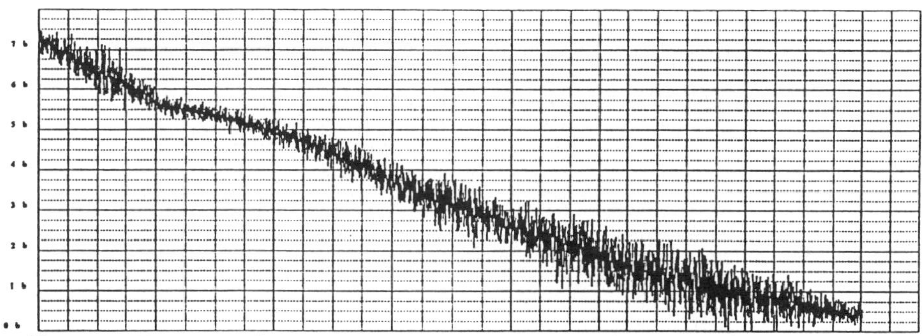

Figure 6 : Δp created by the pump versus the mean void fraction

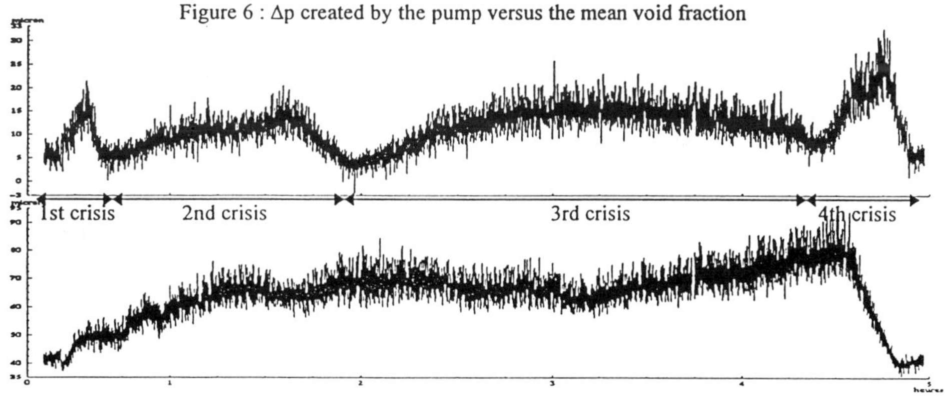

1st crisis 2nd crisis 3rd crisis 4th crisis

Figure 7 : Mechanical behavior of the motor casing (upper curve) and the shaft (lower curve)

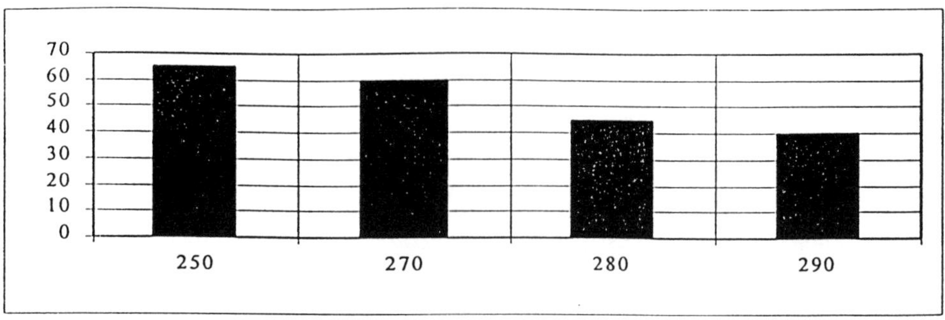

Figure 8 : Maximum value of motor casing vibration against the water temperature

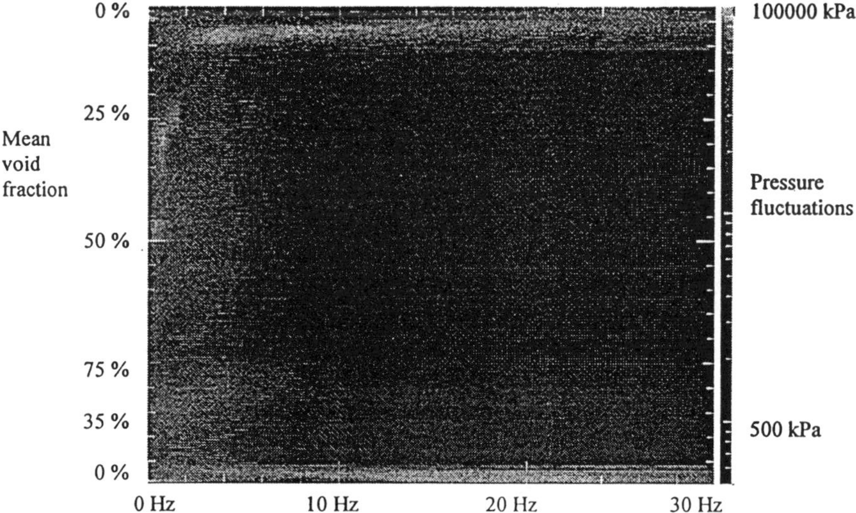

Figure 9 : Pressure fluctuation during the second two-phase flow test

COMPUTER SIMULATIONS OF DYNAMIC PERFORMANCE OF 400 MW ADJUSTABLE SPEED PUMPED STORAGE UNITS

EIZO KITA
The Kansai Electric Power Co. , Inc. , Osaka , Japan
HIROTO NAKAGAWA
The Kansai Electric Power Co. , Inc. , Osaka , Japan
TAKAO KUWABARA
Hitachi ,Ltd. , Tokyo , Japan
MASATAKA HARADA
Hitachi ,Ltd. , Tokyo , Japan

1. Abstract

Two units of the world largest 400 MW adjustable speed pumped storage units for Ohkawachi Power Station , The Kansai Electric Power Inc. , Japan were commissioned in December 1993 and June 1995 respectively .

They were designed to utilize the flywheel energy of the rotor to the maximum extent possible so that they can change in stepwise at least 32 MW in the generate mode and at least 80 MW in the pump mode within 0.2 second . The pump turbines have two kinds of local multivalued speed versus discharge characteristics , so-called S characteristics in the turbine operation zone and so-called hump characteristics in the pump operation zone , against which careful considerations are required to minimize their influences . The two pump turbines are connected to the same penstock and receive water hammer interferences from each other through the bifurcation .

Since the adjustable speed machines were considered too difficult to be adjusted by trial and error in the field , a complete set of simulation studies corresponding to the series of the commissioning tests were run on the computer and adjustment know-hows were established prior to the tests . As a result , the commissioning tests were completed very smoothly without any particular

E. Cabrera et al. (eds.), Hydraulic Machinery and Cavitation, 1073–1082.

readjustments . This paper is to present some of the computer simulations as well as brief explanation of the sophisticated computer simulation programs for the adjustable speed pumped storage units . Computer simulation studies focussed on the electrical equipment such as the generator motors and the cycloconverters were done separately from those focussed on the mechanical equipment . Hereinafter discussed are only the latter .

2. Configuration of Computer Simulation Program

The two adjustable speed pump turbines for the Ohkawachi power station , Kansai Electric Power Inc. are connected to the same penstock through a bifurcation , while , on the downstream side , they have their own tailraces respectively . If discharge of one pump turbine changes , water hammmers occur on both sides of the pump turbine , upstream and downstream , and the upstream side water hammers are propagated to the other pump turbine through the bifurcation . Due to the water hammer interferences , the second pump turbine changes its discharge and torque , which result in water hammers and speed fluctuations on the second pump turbine as well . The computer simulation program is capable of handling such complicated pipeline configuration and water hammer interferences . Fig. 1 shows four quadrant performance curves of unit speed versus unit discharge of the pump turbines . As shown in Fig. 1 , the so-called S characteristics appears in the turbine operation zone and the so-called hump characteristics appears in the pump operation zone , where multiple unit discharge ,Q1 ,values and multiple unit torque , T1 ,values are given against a given unit speed N1 . Therefore , for computer simulation cases where operation of the machine is to be studied in the special characteristics , " from Q1 to N1 " and " from Q1 to T1 " reading procedures were applied in order to increase accuracy .

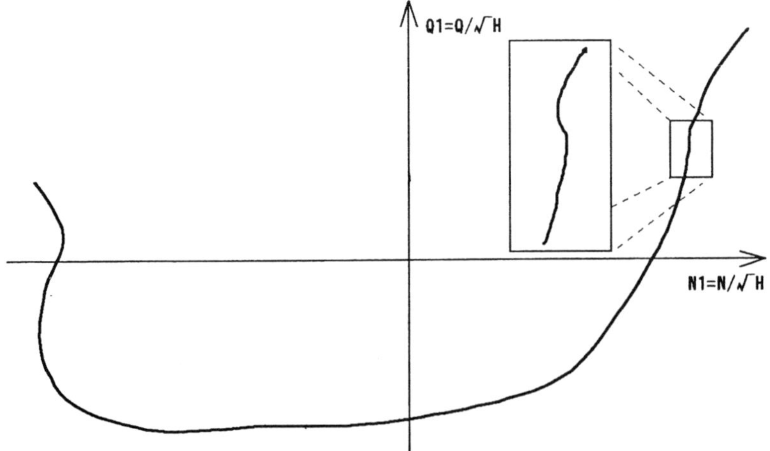

Figure1.Four Quadrant Performance Curves of Pump Turbines, Unit Speed N1 versus Unit Discharge Q1

Fig. 2 is a typical block diagram of the control systems for the adjustable speed pumped storage units in the generate mode which shall be incorporated in the simulation program . For the pump mode , similar block diagrams as cited in Fig. 5 of [1] are incorporated .

Firstly , the generate mode is discussed referring to Fig. 2 . Response speed of the turbine output could not be increased arbitrarily because of water hammer limitations . On the other hand , response speed of the generator output can be increased as desired substantially with no limitation , since it is controlled directly from the power command signal as shown in Fig. 2 . As a result , rotating speed of the machine starts to change according to the difference between the generator output and the turbine output . In parallel , the optimum speed command signal from the function generator for optimum speed is changed according to the change of the power command signal . Then the speed governor controls rotating speed of the machine to follow the new speed command signal . The speed governor may be considered as an output balance keeper as well since it controls the turbine output to follow the generator output until new balanced condition is achieved .

Secondly , the pump mode is discussed . The motor input power is controlled directly in response to the power command signal similarly to the generate mode . The wicket gates , however , are not available for control of the pump input in this case . Therefore , in the pump mode , the rotating speed is also required to be controlled on the motor side , more specifically , to be controlled by the cycloconverter . The wicket gates are assigned exclusively for the best efficiency control as cited in Fig. 2 of [1] .

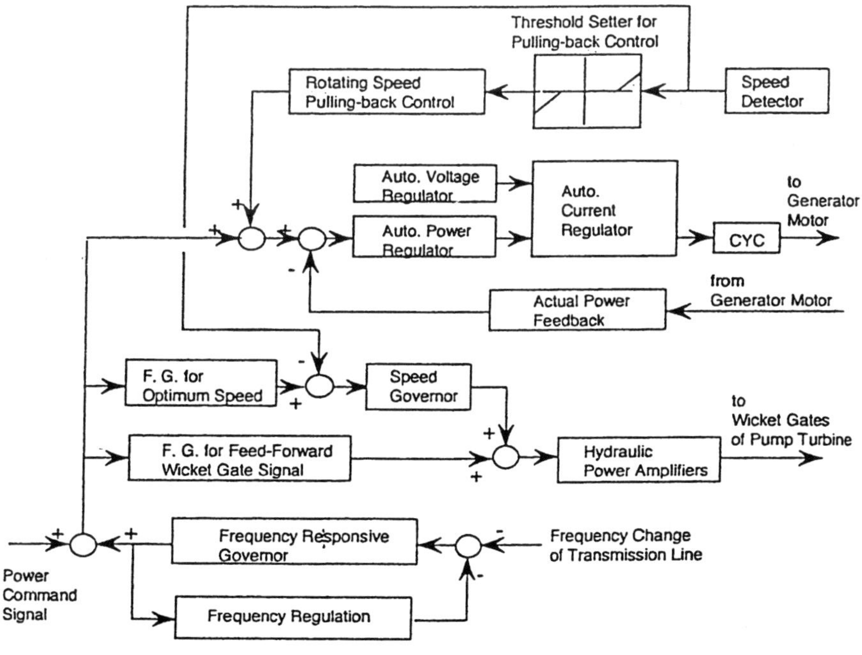

Figure2.Control Scheme of Adjustable Speed Machine in Generate Mode

Fig. 3 shows overall configuration of the computer simulation programs for the adjustable speed pumped storage units . The water hammers are calculated by the method of characteristics . For the main program , the Dynamic Digital Simulator developed by our research laboratory is used , which is a general use program based on the Runge-Kutta-Gill method . Generally , effects of dynamic load forces to the wicket gate servomotor can be neglected .

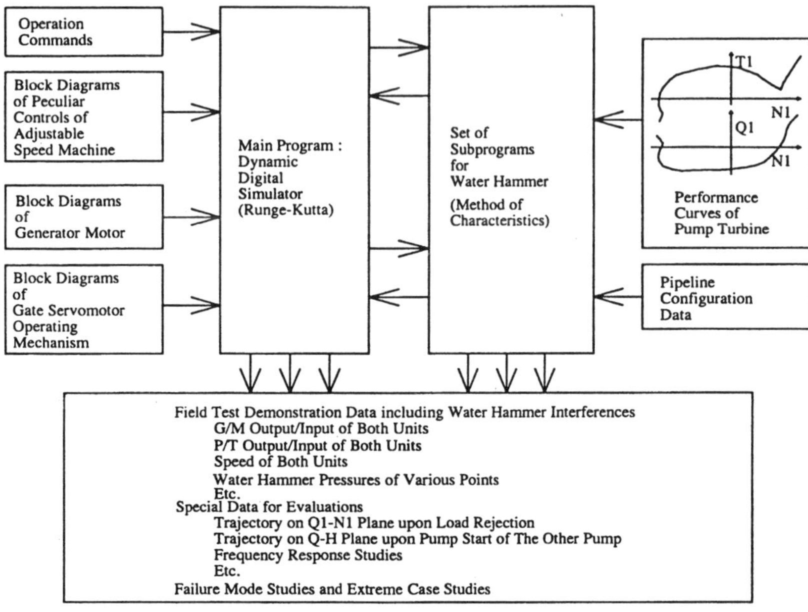

Figure3.Configuration of Computer Simulation Program for Adjustable Speed Pumped Storage Units

3. Tuning of Adjustable Speed Pumped Storage Units

3.1. DIRECT CONTROL OF ELECTRICAL POWER OF GENERATOR / MOTOR

In both of the generate mode and the pump mode , the generator motor power were adjusted to respond to the power command signal within 0.2 second so that the flywheel energy of the rotor can be utilized to the maximum possible level .

3.2. FREQUENCY RESPONSIVE GOVERNOR FOR GENERATE MODE

The frequency responsive governor consists of the PID type forward circuit and the feedback circuit for the speed regulation . The frequency responsive governor is quite different from the speed governor of Item 3.3 below in that PID gains of the latter are to be adjusted for required stability and response speed of the machine itself in consideration of the water starting time Tw and the mechanical starting time Tm , as in the case of the governor for the conventional synchronous unit , while the settings of the former is to be adjusted for required stability and response speed of the transmission line . Naturally , PID gains required for the former are higher than those for the latter .

3.3. SPEED GOVERNOR FOR GENERATE MODE

In the case of the adjustable speed machine , the synchronizing power from the transmission line is no longer available for speed control of the machine even if the unit is on the transmission line and the speed governor control loop is considered as an entirely isolated loop regardless of amount of load on the generator . Therefore , PID gains of the speed governor shall be adjusted to assure stability of the isolated loop by themselves considering the water starting time Tw and the mechanical starting time Tm . On the other hand , considerations shall also be paid to increase response speed of the speed governor as high as possible in order to minimize transient speed excursions .

3.4. FREQUENCY RESPONSIVE GOVERNOR FOR PUMP MODE

The settings of the frequency responsive governor for the pump mode are to be adjusted for required stability and response speed of the transmission line , similarly to the frequency responsive governor for the generate mode .

3.5. SPEED GOVERNOR FOR PUMP MODE

PID settings of the speed governor shall be adjusted to assure stability . Regarding response speed , situation is somewhat different from the case of the speed governor for the generate mode , because high response speed is not always good . In the pump mode , both of the motor input power and the speed are controlled by the cycloconverter at the same time and , therefore , some interferences between the two controls are unavoidable . If too much importance is given to increase the speed response , the motor power response tends to overshoot . Therefore , some compromise is required .

3.6. OPTIMUM SPEED SIGNAL GENERATOR FOR GENERATE MODE

This is to give the most desirable speed command to achieve the highest efficiency under given head and given power command signal as cited in Fig. 1 of [1] .

3.7. OPTIMUM SPEED AND GATE OPENING FUNCTION GENERATOR FOR PUMP

This is to give the most desirable speed command and the most desirable gate opening command to achieve the highest efficiency under given head and given power command signal as cited in Fig. 2 of [1] .

3.8. REVERSE FLOW PREVENTIVE CONTROLS

In case of the adjustable speed pumped storage unit , there is a need to expand input power adjustable range as much as possible for economical reasons . And it is not seldom that operation point of the machine is located close to the reverse flow region as shown in Fig. 4 during operation under the minimum power command . Therefore , provisions are required to prevent the machine from entering the reverse flow region in any transient conditions , especially under water hammer interferences from other machines sharing the same pipeline . In the case of the Ohkawachi power station , provisions are made either to raise the optimum speed command signal or to lower the optimum gate opening command signal temporarily when larger water hammer interferences are expected , for example , in such case that the other machine is going to be pump started .

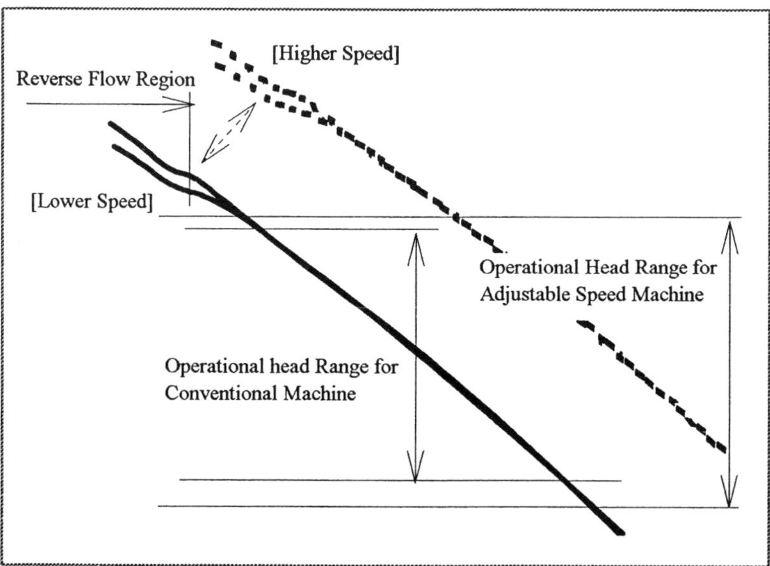

Figure4.Typical Total Dynamic Head Versus Discharge Characteristics of Pump Turbine

3.9. EXCITATION FREQUENCY GIVEN MODE

In this mode , the excitation frequency " fc " is given as desired so that the rotating speed (in frequency) " fr " of the machine is adjusted to " f - fc " in accordance with the given " fc " , where f is the frequency of the transmission line . If fluctuation of the line frequency is negligible , the machine in this mode can be considered as a synchronous machine with synchronous speed arbitrarily adjustable . In the case of Ohkawachi power station , this mode was used during the pump priming , the mechanical quick shutdown and such operation that stability of the speed is required rather than stability of the power .

4. Computer Simulations for Ohkawachi Power Station

4.1. ONE UNIT IN SLIP-PHASE CONTROL LOADED AND UNLOADED WHILE THE OTHER RUNNING IN EXCITATION FREQUENCY GIVEN MODE

Fig. 5 shows a typical computer simulation study of water hammer interferences between two adjustable speed machines , the unit 4 in the slip-phase control and the unit 3 in the excitation frequency given mode . The unit 4 is loaded as shown in the right column while the unit 3 is running in the excitation frequency given mode as shown in the left column . Fig. 6 shows oscillograph charts of actual machines for comparison .

It is here noted that the system power output of the unit 3 changes reflecting entirely water hammer interferences from the unit 4 . In other words , in the excitation frequency given mode , water hammer interferences are conveyed to the transmission line entirely . It means that the same comment applies to the conventional pump turbines .

In sharp contrast with this , the adjustable speed machine in the slip-phase control is perfect as described in Item 4.2. below .

4.2. ONE UNIT LOADED AND UNLOADED WHILE THE OTHER RUNNING , BOTH IN GENERATE MODE

Fig. 7 shows a typical computer simulation study of water hammer interferences between two adjustable speed machines , both in the slip-phase control . The unit 3 is loaded and unloaded as shown in the left column while the unit 4 is running carrying a constant generator output as shown in the right column . Fig. 8 shows oscillograph charts of actual machines for comparison .

It is here noted that the system power output of the unit 4 in the slip-phase control does not change at all even with water hammer interferences . The water hammer interferences appear only on the rotating speed " fr " but it can be easily compensated by counter adjustment of " fc " so that the machine can run continuously tied with the transmission line of the same frequency " f " .

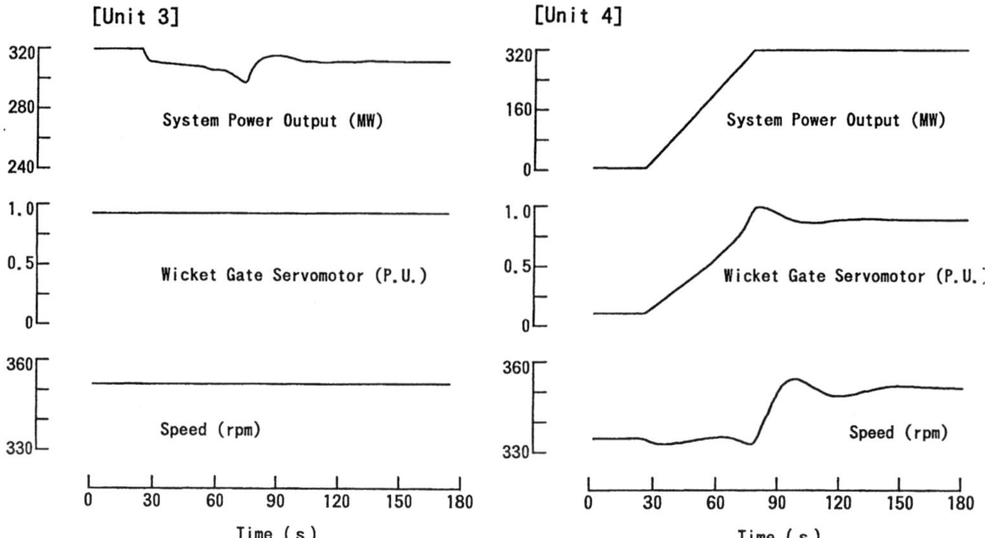

Figure5.Computer Simulation of Water Hammer Interferences between 2 Adjustable Speed Machines of Ohkawachi in Generate Mode,Unit 4 Being Loaded,Unit 3 Sitting in Constant Frequency Excitation Mode

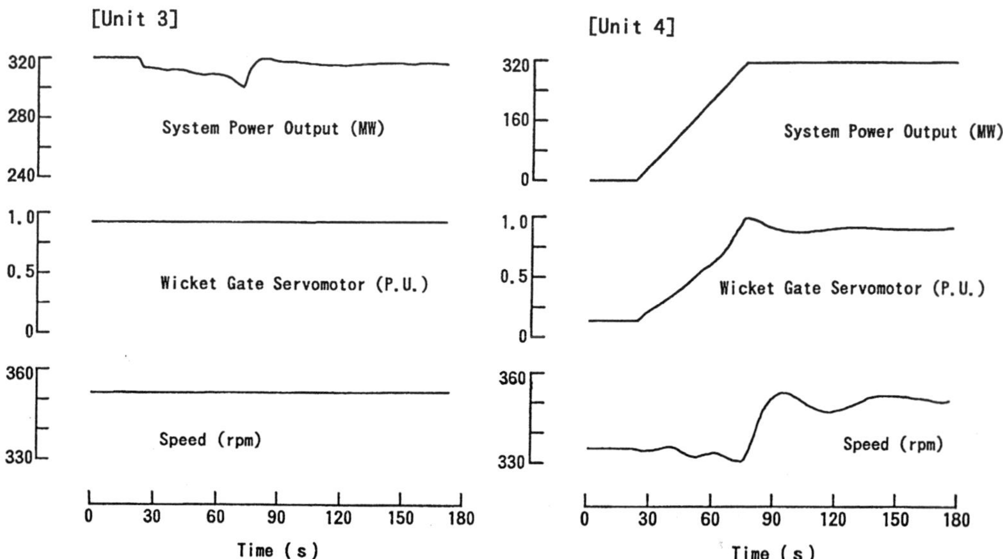

Figure6.Oscillograph Charts of 2 Adjustable Speed Machines of Ohkawachi for Direct Comparison with Fig.5

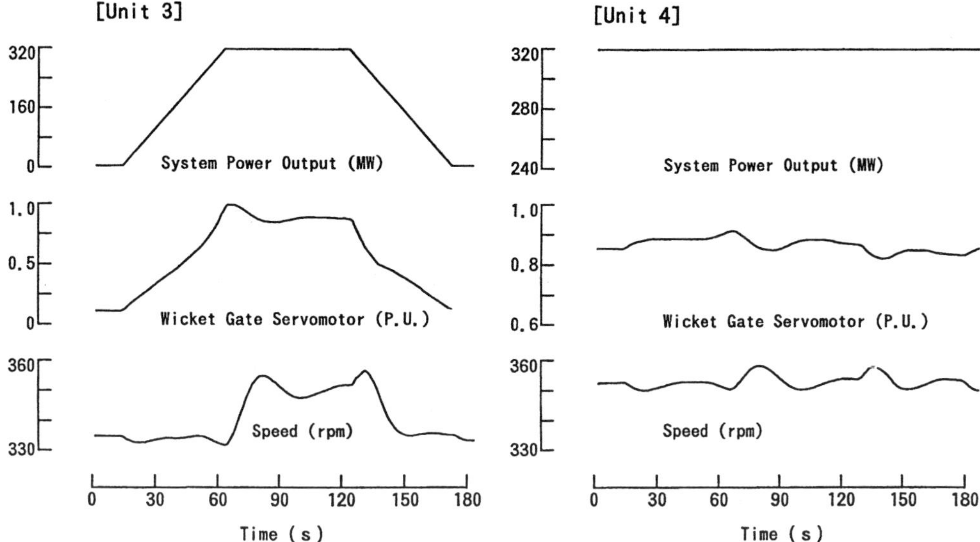

Figure7.Computer Simulation of Water Hammer Interferences between 2 Adjustable Speed Machines of Ohkawachi in Generate Mode,Unit 3 Being Loaded and Unloaded, Unit 4 Sitting with Same Load

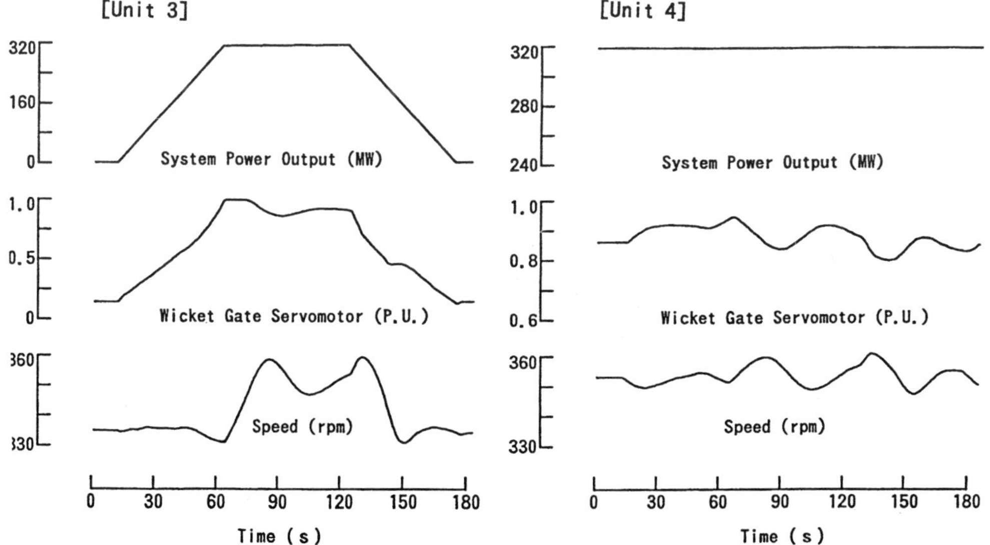

Figure8.Oscillograph Charts of 2 Adjustable Speed Machines of Ohkawachi for Direct Comparison with Fig.7

4.3. FREQUENCY RESPONSE OF ADJUSTABLE SPEED MACHINE IN PUMP MODE

A bode diagram is cited in Fig. 6 of [1] which shows pump mode performance of the adjustable speed machines for the Ohkawachi power station according to our computer simulations. Both the gain and the phase shift of the system power input (sum of the motor input and power consumption of the cycloconverter) do not start to drop up to almost 5 of angular frequency. On the other hand, the gain and the phase shift of the pump input start to drop from about 0.2 of angular frequency. It is here noted that the big response speed difference between the system power input and the pump input can be achieved only in the case where response of the system power input to the power command signal is set to be much faster than that of the rotating speed. If the system is adjusted oppositely, the flywheel energy of the rotating parts could no longer be utilized and the quick response advantage could no longer be enjoyed. The conventional fixed speed pump can be considered as an extreme example for the latter case, where response speed of the rotating speed is infinitely high while that of the motor input is zero. To summarize the above, the adjustable speed unit as adjusted to give much greater importance to the response speed of the power than the response speed of the rotating speed exhibits significant contribution potential to the transmission line in the pump mode as well.

5. Conclusion

The 400 MW adjustable speed pumped storage units for Ohkawachi power station were commissioned in December 1993 and June 1995 respectively. Prior to the commissioning tests, simulation studies were done on the computer to demonstrate every test from the starting to the stopping and adjustment know-hows were established at this stage to maximize advantages of the adjustable speed machines. In the field, each of the commissioning tests was done simply comparing the test results with the simulation studies and, after all, there was no particular " trial and error " adjustments required. Hereby presented are brief explanations of the simulation programs and some typical examples of the simulation studies as well as some principal adjustment know-hows for the adjustable speed units. It would now be concluded that the computer simulation studies can play so decisive rolls and are so important in the case of the adjustable speed machines that have never been realized in the case of the conventional pump turbines.

6. References

1. Takao Kuwabara et al. (1994) Performance of 400 MW Adjustable Speed Pumped Storage Unit for Ohkawachi Power Station, ASME 1994 Winter Meeting, Paper No. NE-Vol. 15, 1-10.

PERFORMANCE COMPARISON OF NUCLEAR REACTOR RECIRCULATION PUMPS TESTED UNDER LARGE REYNOLDS NUMBER DIFFERENCE

K.SAIKI, T.IIKURA, K.MATSUMOTO, H.KOMITA
M.KOBAYASHI, T.SAITO, H.TANAKA
Toshiba Corporation
Yokohama, Japan

ABSTRACT

The reactor coolant recirculation pump mounted on advanced boiling water reactor (ABWR) is a single-stage mixed flow pump. On 10 pumps, hydraulic performance tests were performed under largely different Reynolds number conditions. In this paper, pump characteristics tested under low Reynolds number conditions are strictly converted, with regard to the influence of the variations of Reynolds number in particular, and they were compared with the results of the tests under high Reynolds number conditions. Through these tests, the methods for converting performances into those under different Reynolds number conditions have been established for the hydraulic performances of mixed flow pumps mounted directly on a vessel without upstream and downstream pipes and the validity of the applied precise conversion method has been verified.

RESUME

La pompe de refoulement pour fluide de refroidissement de réacteur montée sur un réacteur à eau bouillante avancé (ABWR) est une pompe mono-étage à pignons. Des essais de performances hydrauliques ont été réalisés sur 10 pompes. avec des nombres de Reynolds différents. Dans cet essai, les différentes caractéristiques de pompes testés sous des petits nombres de Reynolds ont été strictement convertis, compte tenu de la variation du nombre de Reynolds en particulier, et comparées aux résultats de test obtenus avec des numéros de Reynolds élevés. Ces tests ont permis d'établir des méthodes de conversion des performances sous différents numéros de Reynolds pour les performances hydrauliques des pompes à pignons montées directement sur une cuve sans tuyaux en amont et en aval, et de vérifier la validité de la méthode de conversion précise appliquée.

E. Cabrera et al. (eds.), Hydraulic Machinery and Cavitation, 1083–1092.

1. Introduction

For the reactor coolant recirculation pump of ABWR, a reactor internal pump system that has no external recirculation piping is adopted to improve the safety, reliability, economy and to reduce the exposure rate. In the case of a 1350-MWe ABWR, 10 pumps are installed at uniform intervals on the circumference of the pump deck in the bottom of the down-comer, which consists of the reactor pressure vessel and the core shroud. The pump diffusers are mechanically connected to the nozzles on the pressure vessel bottom head, and the impeller shafts are connected to the lower motors through these nozzles (see Fig. 1). The motors are installed in the motor casings which are welded to the vessel bottom head. The reactor internal pumps suck in the coolant water from the down-comer and discharge the coolant water directly to the reactor core. The reactor output is controlled by regulating the circulation flow rate of coolant by adjusting pump rotating speed. Therefore, the measurement of the flow rate in the reactor core is important for the operation of a nuclear reactor. For this purpose, the pressure taps are provided on the suction and discharge ports of the pump. The differential pressure measured by these taps is converted into the flow rate in the reactor core using the Q-H characteristics obtained from the hydraulic performance test conducted beforehand. The tests are conducted on the pumps mounted on the testing vessel simulating the structures of the actual reactor. All the reactor internal pumps are tested under low temperature and low pressure first to confirm the basic characteristics, and then tested under high temperature and high pressure simulating the plant operating conditions.

Further, for the reduction of cost and manufacturing period of reactor internal pumps for future plants, it is desirable to rationalize the shop performance tests on similar pumps; especially, to eliminate or reduce the tests under high Reynolds number (high temperature and high pressure) condition. The results of the study presented in this paper provide basic data for the prediction of high Reynolds number performance based on low Reynolds number test results.

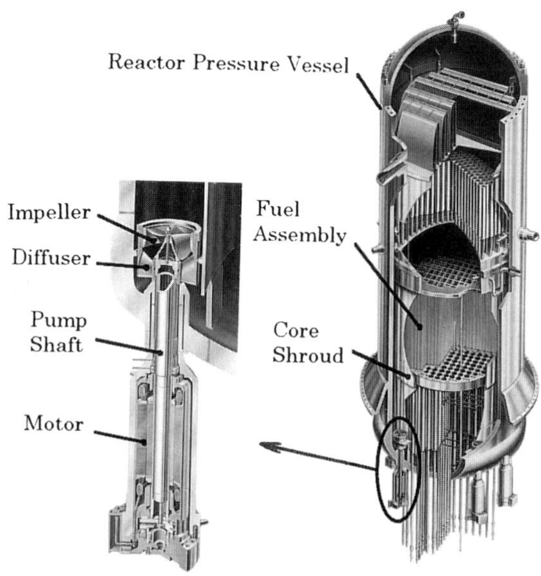

Fig .1 ABWR Reactor internal pump

2. Test Facility and Test Procedures

2.1 Test facility

The test equipment consisted of a testing vessel, flow-rate measuring loop and auxiliary systems for regulating temperature and pressure. Fig. 2 shows the layout of the test facility and the flow-rate measuring loop. The testing vessel had a height of 6.5 m and a diameter of 2.5 m. Further, the shroud wall, the pump deck, and the control rod guide tubes were installed in the testing vessel to simulate the partially reactor internals. The flow rate of circulating water was measured by a venturi tube installed on the flow-rate measuring loop and the system resistance was adjusted by a valve mounted on the loop. The differential pressure of the pump was measured by the pressure taps installed on the suction and discharge sides. The location of the taps was equivalent to that of the actual reactor. The pressure and temperature in the vessel were measured on the suction side of the pump. Further, the pump shaft power was calculated from the product of the motor input and the motor efficiency which had been measured in advance on each motor.

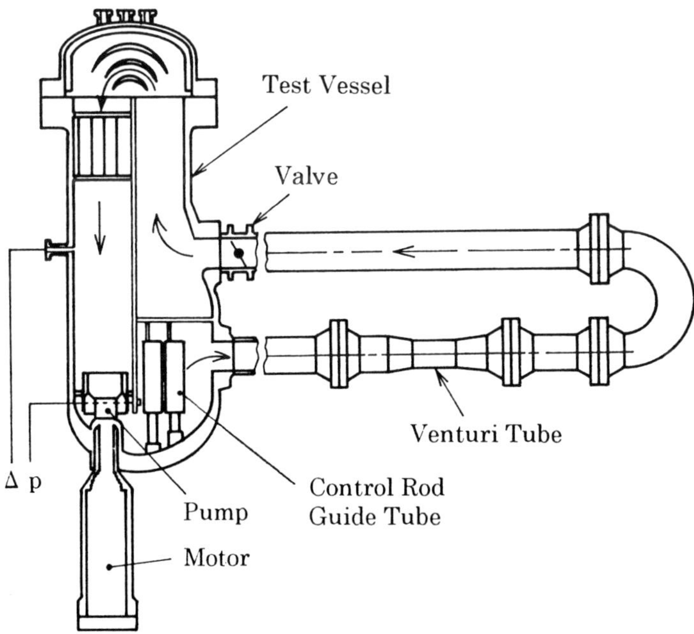

Fig .2 Arrangement of the test facility

2.2 Test procedures

In high Reynolds number test (Hot Test : Re=1.6×10^{8}), temperature and pressure conditions of the test loop were set to simulate the actual plant operating conditions approx. 280℃, approx. 8Mpa. The test speed of the pump was set to the rated rotating speed of 1450 rpm. Pump differential pressure and motor input power were measured while varying the flow rate by adjusting the opening of the valve. In low Reynolds number test (Cold Test : Re=2.8×10^{7}), temperature and pressure conditions were approx. 40℃ and approx. 1Mpa. The pump rotating speed was set to 1300 rpm to avoid overload of the motor due to high density of water at low temperature. Through these tests, the pump charactristic performances covering a very wide range of Reynolds number corresponding to 1:6 were obtained for 10 pumps of identical design and manufacture.

3. Test Results and Considerations

3.1 Hydraulic characteristics of the pump

Fig. 3 shows the comparisons of the measured head characteristics and the pump efficiency characteristics between the hot test (1450rpm) and the cold test (1300rpm) obtained by the representative pump.

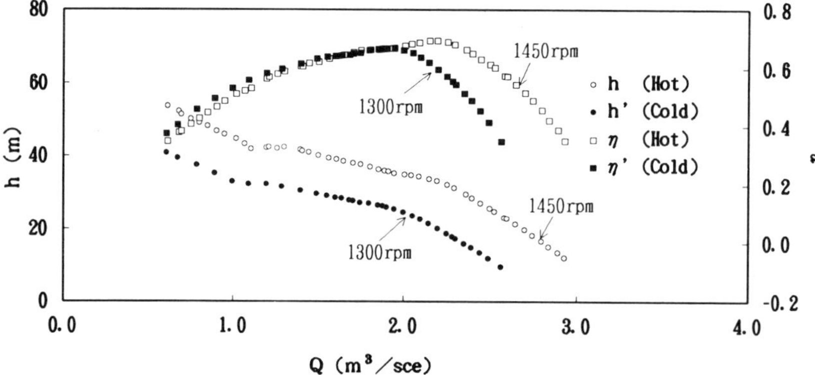

Fig .3 Performance comparison between the hot test and the cold test

The maximum pump efficiency is about 70% and seems to be low in comparison with ordinary mixed flow pumps. The reason is that the pump efficiency is calculated using the differential static pressure measured at upstream and downstream of the pump deck. Therefore, in this study, the pressure losses of the collision, reduction and expansion of the flow at both inlet and outlet of the pumps are regareded as internal losses of the pumps.

The pressure loss characteristics in association with the form of the passage were measured through a preliminary test prior to the pump performance tests. In the preliminary test, a dummy diffuser simulating the diffuser boss section and external cylinder was installed on the testing vessel and the circulating water was driven by an external pump. The relationship between the flow rate Q and the pump differential pressure was obtained and the pressure loss head with respect to flow rate Q was approximated by KQ^2 (K is the pressure loss coefficient: constant). The pump efficiency corrected by the pressure loss obtained in the preliminary test revealed 91% at the optimum point.

3.2 Conversion of performances

The pump characteristics obtained from the cold test were converted into those under hot test conditions based on various considerations including hydraulic similarity law.

(1) *Conversion of rotating speed and correction for thermal effects*
Conversion for the difference of rotating speed, changes in pump dimensions due to thermal expansion and changes in physical characteristic values of water were performed by converting the flow rate, head, and shaft power into non-dimensional values, without correction of Reynolds number effect.
Dimensionless pump characteristics obtained from the hot test and the cold test on the representative pump are shown in Fig. 4. As shown in this figure, hot test characteristics are higher than cold test characteristics in the head coefficient and the pump efficiency.

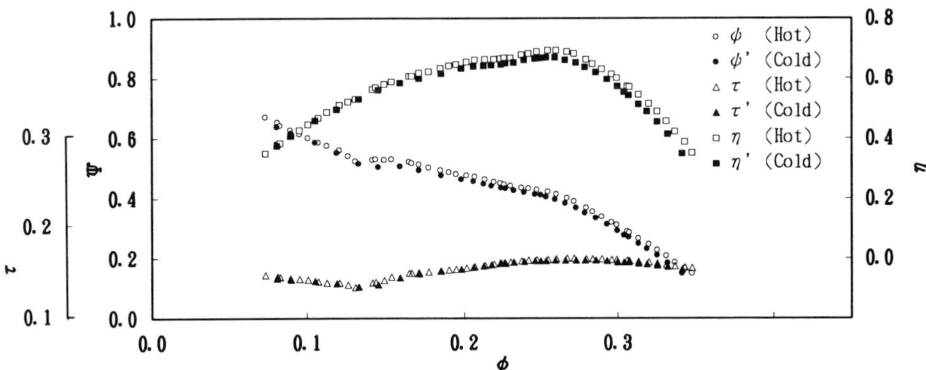

Fig.4 Normalized pump characteristics

Conversion taking into account the Reynolds number effect on leakage loss, power loss due to external hydraulic friction of impeller and specific energy loss due to hydraulic friction were made and the shift of performance was calculated in the following paragraphs.

(2) *Conversion of discharge coefficient* ϕ

Leakage flow ΔQ through the clearance between the impeller and the diffuser was evaluated and volumetic efficiency η_v was calculated.

$$\Delta Q=A_w u_w=C_w A_w(2g\Delta H_w)^{1/2} \qquad (1)$$

$$\eta_v=Q/(Q+\Delta Q) \qquad (2)$$

Here, the differential head across the seal ΔH_w can be calculated from the following equation.

$$\Delta H_w=H-u^2/(2g)=H-Q^2/(2A^2 g) \qquad (3)$$

Flow-rate coefficient at the seal section C_w is expressed as follows.

$$C_w=1/(\lambda_w L_w/2b_w+1+\zeta_w)^{0.5} \qquad (4)$$

Here, λ_w and ζ_w are expressed as follows.

$$\lambda_w=0.26Re_w^{-0.24}(1+0.191(R\omega_w/Re_w)^2)^{0.38} \qquad (5)$$

$$\zeta_w=0.5 \qquad (6)$$

Re_w and $R\omega_w$ represent axial-flow Reynolds number and rotational Reynolds number, respectively, and they are expressed as follows.

$$Re_w=b_w u_w/\nu=\Delta Q/(2\pi r_w \nu) \qquad (7)$$

$$R\omega_w=r_w \omega b_w/\nu \qquad (8)$$

Because the leakage flow ΔQ is unknown in the above relational expressions, iterating calculations were made on it. Then volumetric efficiency η_v' with respect to each data in cold test was obtained.

To convert η_v' to η_v corresponding to hot test, conversion or correction was made on rotating speed, flow rate, dimensions and dynamic viscosity so that they are adapted to the hot test conditions. Thus the volumetric efficiency η_v under the hot test conditions was estimated. As a result of these studies on volumetric efficiency η_v, conversion formula for discharge coefficient ϕ was established in the form of the following equation.

$$\phi=(\eta_v/\eta_v')\phi' \qquad (9)$$

(3) *Conversion of power coefficient* τ

The mechanical loss caused by the cylindrical friction of the impeller ring was evaluated and its conversion was made. Power efficiency of impeller

η_m can be calculated by the following expression.

$$\eta_m = 1 - \Delta M \omega L_w / P \qquad (10)$$

Here, ΔM(N·m/m) is the friction moment per unit length of the concentric rotating double cylindrical round tube, P(W) is the shaft input power and L_w(m) is the length of cylindrical section of the impeller ring
Then, the friction moment is expressed as follows.

$$\Delta M = C_M \cdot 2\pi \rho r_w^4 \omega^2 \qquad (11)$$

In the cold test, rotational Reynolds number was $R_{\omega w} = 5 \times 10^4$. When $R_{\omega w}$ is sufficiently large as observed in this test, it is known that C_M shows a constant value regardless of axial-flow Reynolds number Re_w. Therefore, its value accords with the value of C_M when there is no flow in the axial direction ($Re_w = 0$). C_M with no axial flow is expressed as follows.

$$C_M = 0.00759 R_{\omega w}^{-0.24} \qquad (12)$$

Using this equation, disk friction torque coefficient C_M' was calculated for each cold test data and the power efficiency of impeller η_m' was obtained. To obtain η_m corresponding to hot test condition, rotating speed, physical dimensions and dynamic viscosity were converted to meet the hot test conditions. Then, the disk friction torque coefficient C_M for the hot test conditions was estimated and the power efficiency of impeller corresponding to it was estimated using the following equation.

$$\eta_m = 1/[1 + \{(1/\eta_m') - 1\}(C_M/C_M')] \qquad (13)$$

By using η_m and η_m' obtained above, power coefficient τ_w was converted by using the following equation.

$$\tau = (\eta_m/\eta_m')\tau' \qquad (14)$$

(4) *Conversion of head coefficient* ϕ
Conversion of hydraulic efficiency η_h is made by Eq.(15) using loss distribution coefficient V and friction coefficient ratio Λ [1].

$$\eta_h = 1 - (1 - \eta_h')\{(1 - V) + V \cdot \Lambda\} \qquad (15)$$

Hydraulic efficiency under cold test condition η_h' is inversely obtained by Eq.(16) using η_v' and η_m' obtained in the preceding paragraphs.

$$\eta_h' = \eta'/(\eta_v' \cdot \eta_m') \qquad (16)$$

The loss distribution coefficient for mixed flow pumps at the optimum efficiency point can be calculated by the following equation from specific velocity Ns [1].

$$V_{opt}=1.4Ns^{-0.1}-0.07 \qquad (17)$$

The loss distribution coefficient at off-design flow rate can be obtained by the following equation using V_{opt} [1].

$$V=V_{opt}/[6.25\{(Q'/Q'_{opt})^{-1}-0.84\}^2+0.84] \qquad (18)$$

For the calculation of friction coefficient ratio Λ (= λ/λ') in Eq.(15), the following White's formula is applied because it suits the range of Reynolds number for the present test conditions.

$$\lambda^{-0.5}=1.8\log Re-1.4985 \qquad (19)$$

Since the measured pump head contains the pressure loss caused by the form of the vessel which not depend on the Reynolds number, there is a problem in applying Eq.(17) which gives a common value of V_{opt} for average mixed flow pumps.
Here in this study, both η_{opt} and η'_{opt} were measured, and $\eta_{h\,opt}$ and $\eta_h'{}_{opt}$ were calculated using η_v, η_v', η_m, η_m' at the optimum efficiency point obtained in the preceding paragraphs. Therefore, V_{opt} is inversely derived from them according to the following equation, which is inversely obtained from Eq.(15).

$$V_{opt}=(\eta_{h\,opt}-\eta_h'{}_{opt})/\{(1-\eta_h'{}_{opt})(1-\Lambda)\} \qquad (20)$$

Therefore, V_{opt} was calculated for each pump from the results of the cold and hot tests, and their average was obtained. From the result of these studies, V_{opt} was finally obtained as 0.36. By substituting this value into Eq.(18), loss distribution coefficient V at off-design flow rate was calculated. By substituting this value and η_h' obtained from Eq.(16) into Eq.(18), hydraulic efficiency η_h under the hot conditions was calculated, and conversion of head coefficient ψ was performed using the following equation.

$$\psi=(\eta_h/\eta_h')\psi' \qquad (21)$$

(5) *Results of the conversion of performances*
The result of the conversion of the cold test results to hot test condition according to the above procedure is plotted and compared with the hot test results in Fig. 5. The converted results show good agreement with the hot

test results and it is verified that the conversion procedure described above is valid and useful. Fig. 6 shows the breakdown of each internal efficiency and Fig. 7 the ratio of the internal efficiency from cold test to hot test.

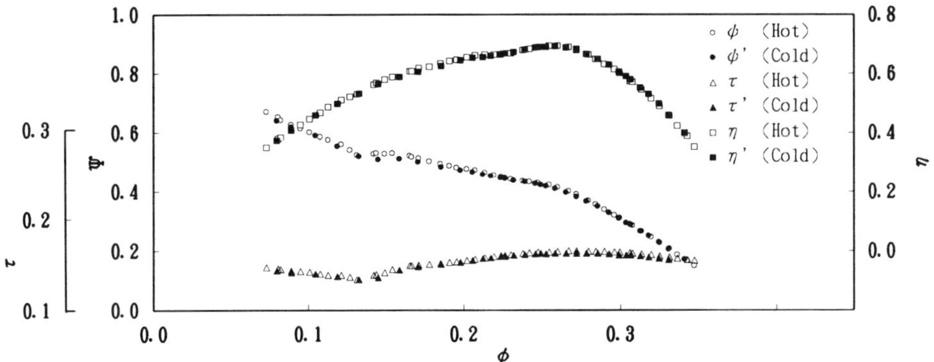

Fig .5 Comarison of converted characteristics

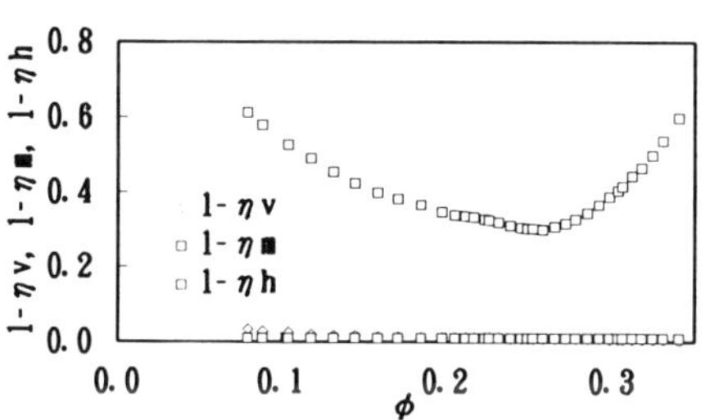

Fig .6 Breakdown of pump loss

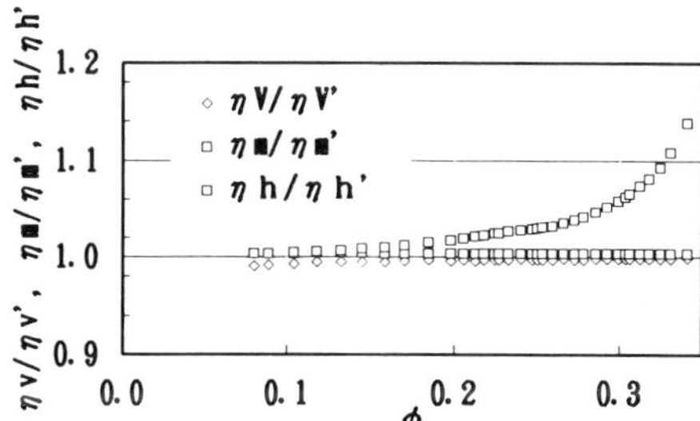

Fig .7 Variation of efficiency ratio

4. Conclusions

Tests under the conditions with Reynolds number difference by 6 times were performed on the mixed-flow pumps installed in a vessel having complex passage and following conclusions have been obtained from these tests.

(1) It has been confirmed that the performance tested under low Reynolds number condition can be converted into that under high Reynolds number condition with a high accuracy by using the conversion method described in this paper.

(2) Through this study, it has been confirmed from Figs. 6 and 7 that most portions of the Reynolds number effect appears on specific energy efficiency. The magnitude of the Reynolds number effect increases at larger flow rate.

(3) According to the analysis conducted on cold test results indicate that V_{opt}=0.36 for this type reactor internal pumps. This is compared with the value of V_{opt}=0.6 calculated from Eq.(17). The main reason of this disagreement is that the various hydraulic loss in external passage is included in pump efficiency as an internal loss.

NOTATIONS

A	:impeller cross-section area	b_w	: width of clearance between impeller and diffuser
C_M	: disk friction torque coefficient	C_w	: flow-rate coefficient
H	: total head	h	: pump measured head
K	: pressure loss coefficient	L_w	: impeller ring length
ΔM	: friction moment per unit length	Ns	: specific speed
P	: pump input power	Q	: discharge flow rate
ΔQ	: leakage flow rate	Re	: Reynolds number
R_ω	: rotational Reynolds number	r_w	: average radius of clearance
u	: discharge flow velocity	V	: loss distribution coefficient
ζ	: entrance loss coefficient	η	: total pump efficiency
η_m	: power efficiency of impeller	η_h	: hydraulic efficiency
η_v	: volumetric efficiency	Λ	: friction coefficient ratio
λ	: friction factor for pipe	τ	: power coefficient
ϕ	: discharge coefficient	ψ	: head coefficient
ω	: angular velocity		

Subscript
$_{opt}$:optimum point
$_w$:annulus between impeller and diffuser

Superscript
:cold test

Reference:

(1) JSME standard S-008, Performance conversion method for hydraulic turbines and pumps, (1989-10), JAPAN. Soc. Mech. Engrs.

PERFORMANCE OF CANDU HEAT TRANSPORT PUMPS UNDER TWO-PHASE FLOW CONDITIONS

H. Samarasekera, M.Sc., P.Eng.
Sulzer Bingham Pumps, Inc.
Burnaby, BC, Canada

and

A.N. Kumar Ph.D., P.Eng.
Atomic Energy of Canada Ltd.
Mississauga, ON, Canada

ABSTRACT

Heat Transport (HT) Pumps in a CANDU (**CAN**adian **D**euterium **U**ranium) reactor circulate heavy water coolant at high temperatures and pressures to remove fuel decay heat from the reactor core. Under a Loss-of-Coolant-Accident (LOCA), HT pumps are required to maintain their operational capability and structural integrity. Licensing requirements stipulate that HT pumps continue to operate under specific conditions, such as two-phase flow, for a specific duration (17 minutes). During this time period, decay heat has to be removed continuously from the reactor core before safe reactor shutdown is initiated.

This paper presents a summary of the simulated two-phase flow tests performed at the HT Pump supplier's test plant (Sulzer Bingham Pumps, Inc.) on a full-scale CANDU 6 HT pump-motor set. Results demonstrate that the HT pumps maintained their performance and structural integrity under the worst two-phase flow conditions attainable on the test loop for the specific duration of 17 minutes. The pump hydrodynamic bearing, lubricated by the pumped fluid, did not suffer any notable damage. Despite the lack of external seal injection during two-phase flow conditions, the mechanical seals showed no signs of damage.

E. Cabrera et al. (eds.), Hydraulic Machinery and Cavitation, 1093–1102.

1. Introduction

CANDU reactors have an excellent performance history and safety record due to the high quality of equipment in the reactor. The HT pumps are one such component which has contributed to this record.

CANDU 6 reactors have four HT pumps circulating heavy water (D_2O) coolant which removes decay heat from the reactor core during normal and certain accident conditions. A LOCA is one such condition under which HT pumps are required to operate. The LOCA may be caused by a pipe break or the failure of a valve in the main coolant system. HT pumps are required to operate and remove the decay heat from the reactor core for 17 minutes before safe reactor shutdown is initiated. It is a licencing requirement to demonstrate the HT pumps have this capability.

Past analytical methods used to demonstrate the operational integrity of HT pumps under LOCA conditions are not accurate nor reliable because they are based on erroneous assumptions. Details of current analytical methods used for qualification of pump-motor sets are discussed in the literature[2].

Some success has been achieved with tests using scaled down models of the pump-motor set. These models must accurately simulate pump specific speed and other hydraulic conditions of the full scale pump-motor set. Models could be costly, particularly if new mechanical and hydraulic designs are required. Because smaller forces are generated in a scaled down model, it is easier to control the test loop.

The most accurate data is obtained from tests using full scale models of the pump-motor set. Reliability of data is dependent on simulation of actual operating conditions and on accuracy of measurement. Results are also affected by test loop piping and geometry as well as the water level and volume above the pump impeller centre line. HT pump-motor sets for CANDU reactors were first tested under two-phase flow conditions in 1982[3]. This paper presents the results of two-phase flow tests performed on a slightly different design pump-motor set with a different test loop geometry.

2. Design of CANDU 6 Heat Transport Pump Motor Sets

HT pumps used for two-phase flow tests were designed and manufactured, to AECL specifications, by Sulzer Bingham Pumps, BC, Canada. The pumps comply with the ASME Code Section III Class 1 and Canadian Standards Association Code Z299.1 Quality Standards. HT pumps are vertical, single stage, single suction, double discharge, and double volute centrifugal pumps (Figure 1). Discharge nozzles are located diametrically opposite to each other and are connected to 90° long radius elbows. The stuffing box, located above the pump volute, contains the pump bearing, the mechanical seal package, and an auxiliary impeller for circulation of seal water. At the rated operating point, the HT pump delivers 2.23 m^3/s, with a head of 213 m, at a nominal speed of 1800 RPM .

Located above the pump impeller is a hydrodynamic, carbon-graphite bearing comprised of a carbon bushing in a stainless steel housing. The housing is free to rotate about a vertical axis. The bearing is lubricated by D_2O by means of four axial lubrication grooves, located 90° apart (Figure 2).

The mechanical seal package consists of three stages in series (Figure 2). Under normal operating conditions, each stage carries one third of the system pressure, but is capable of withstanding full system pressure. The seals are supplied with external gland injections at 345 kPa above pump suction pressure. This excludes any hot water entering the seal package from the pump. The external seal injection is connected directly to the internal recirculation loop, supplying water under controlled temperatures to the pump bearing and seals.. Under normal operating conditions, a small quantity of water is continuously lost from the recirculation loop and is made up by the external injection flow. When seal injection is lost during a LOCA, the pump bearing and seals can operate for at least thirty minutes using water from this loop. A tube-in-tube type heat exchanger, mounted on the driver stand, provides cooling to dissipate heat generated by the pump bearing and seals.

The HT pump is driven by a 6700 kW electric motor, directly coupled to the pump via a rigid, spacer-type coupling. Electric motors were supplied by General Electric Canada, Ontario, Canada. Located at the top and bottom ends of the motor, are eight shoe-tilting pad guide bearings to carry radial loads. An upward and downward thrust, eight shoe Kingsbury tilting pad bearing carries the axial thrust loads. Located on the motor rotor are two inertia packets providing an additional 1264 kg-m^2 of inertia. During the loss of Class IV power, this additional inertia provides the motor with a long run down time, enabling the pump to continuously remove decay heat from the reactor core.

The pump discharge lines, connected to the reactor header, are rigidly supported by structural steel hangers embedded in the building concrete (Figure 3). The motor is mounted on a driver stand which is bolted onto the lower end of the pump volute. The motor is also supported by two hanger rods, connected to spring supports, permitting thermal growth at the upper end of the pump-motor set. Three seismic supports, located at 120° intervals around the motor frame, restrict lateral movement of the motor frame during an earthquake. The stiffness of each lateral restraint is 1.25×10^6 kg/m.

3. The Test Loop

3.1 MECHANICAL SIMULATION IN THE TEST LOOP

To obtain reliable test data, it is critical to closely simulate Reactor Site conditions. Accurate simulation was not possible because testing was performed in an existing test loop and not in a test loop built specifically for LOCA testing.

The pump volute was identical to the volute supplied to the Reactor Site. The only exception was the orientation of the discharge elbows. The angle between the plane passing through the discharge elbows and the plane passing through the suction elbow was 35.5° as opposed to 90° at the Reactor site. The pump weight was supported by suction and discharge pipes connected to the loop piping. Lateral movements of the loop piping were constrained by supports. Because the pump-motor set used in testing was to be installed in the reactor, all internal components of the pump and the motor were fully simulated.

Mechanical and dynamic stiffness were verified analytically. Natural frequencies of the pump-motor set, the piping, and the support system of the test model were similar to field conditions.

3.2 HYDRAULIC SIMULATION

Simulation of hydraulic behaviour in the test loop requires identical acoustic resonances in both field and test systems. Because attainment of this state is not realistic, only "near field" hydraulic simulations were duplicated[4]. This ensured accurate results in terms of acoustic source strength assessment. Acoustic magnification effects, which are loop dependent, could not be compared or simulated.

To accurately simulate pump suction conditions, the head of water and volume above the impeller centre line should be similar to that found in the field. Because testing was performed on an existing test loop, this simulation was not possible.

4. LOCA Simulation

It is not practical to test HT pumps under all possible LOCA conditions. Therefore, the pump was tested for the 17 minutes under the worst LOCA condition (the point at which the highest vibrations in the pump-motor frame and shaft are produced).

When the system is depressurized during a LOCA, pump suction conditions go through two phases. During the first phase, suction pressure decreases from the normal value to the sum of vapour pressure and NPSH available to the pump. The Loop Average Void (LAV) fraction remains at 0%. Under these conditions, loop temperature and suction pressure influence the behaviour of the pumps.

During the second phase, two-phase flow conditions at the pump suction occur as liquid is removed from the loop. The LAV fraction range is 0% - 60%. Under these conditions, temperature and void fraction influence the behaviour of the pumps.

Under normal single-phase operating conditions, the temperature of the coolant is 266°C and the LAV fraction 0%. Immediately following a LOCA, there is a broad spectrum of temperatures (100°C - 266°C) and void fractions (0% - 100%) under which the worst LOCA conditions may occur. To reduce the number of tests, these temperature and void fraction ranges were divided into convenient intervals.

A matrix of temperature versus void fraction covering the entire LOCA operating range was created. Temperatures of 100°C, 150°C, 200°C, 220°C, and 266°C, and void fractions ranging from 0% to 100%, in 1% intervals, were chosen. Matrix Tests were conducted at established matrix points. Based on the results of the Matrix Tests, the worst LOCA condition was identified for the 17 minute test.

5. **Instrumentation and Data Acquisition**

A high speed, data-acquisition system was used to gather data during the tests. A total of 38 channels of data were monitored. Dynamic signals: vibrations, shaft run-outs, pressure pulsations, and air-borne noise, were gathered at 2000 Hz. The balance of the signals: pressures, temperatures, flows, horse power, and LAV fractions, were gathered at 1 Hz. Placement of the transducers and detectors is depicted in Figure 1.

6. **Test Procedure**

6.1 MATRIX TESTS

The pump was stabilized at 100°C, normal operating conditions of 2.23 m^3/s, and a suction pressure of 9522 kPa. The suction pressure was then slowly reduced to 101 kPa, which corresponds, approximately, to the vapour pressure of the liquid at 100°C. The pressure was reduced at a rate permitting stabilization of the loop. Thus far, no liquid had been extracted from the loop (0% LAV). This part of the test corresponds to the first phase described in Section 4.0.

At this point, the loop drain valve was opened and water slowly bled, permitting stabilization of the loop, until the discharge head of the pump was totally degraded. The LAV fraction increased from 0% to approximately 10%, corresponding to the second phase described in Section 4.0 Upon completion of the first part of the Matrix Test, the pump-motor set was turned off and data acquisition stopped. Note that during the entire test, loop temperature was maintained at 100°C.

The above procedure was repeated at 150°C, 200°C, 220°C, and 266°C.

Based on Matrix Test analysis, the worst LOCA condition was identified (Figure 4). At 200°C, at a LAV fraction of 25%, the pump shaft run-out peaked at 285 μm pk to pk and the pump-motor frame vibrations reached 5.3 mm/s pk. Under normal operating conditions, these values are 64 μm pk to pk and 1.3 mm/s pk, respectively.

6.2 17 MINUTE WORST-LOCA TEST

The pump was stabilized at 200°C under normal operating conditions of 2.23 m^3/s and 9522 kPa suction pressure. Suction pressure was then slowly reduced to approximately 1560 kPa. At this point, the drain valve was opened to bleed water from the loop. When the LAV fraction reached 25%, the drain valve was closed, and the seal injection to the pump stopped. This was the beginning of the worst-LOCA test. The pump was run under these conditions for 17 minutes and then turned off.

Following completion of the LOCA tests, the pump was restarted and a Hydraulic Performance Test conducted at normal operating conditions of 266°C. Results indicated that there was no deterioration of pump internals due to operation at two-phase flow conditions and high vibration levels.

7. Analysis and Discussion

7.1 LAV VERSUS SUCTION VOID FRACTION

The void fraction distribution at the pump suction does not remain stable or homogenous at reduced NPSH values of 90% NSPH $_{(req)}$ and lower. Suction flow distribution could change instantaneously from uniform bubbly flow to stratified, slug, or annular flow. Maintaining a stable suction void at values less than 90% NPSH $_{(req)}$ is impractical. The 90% NPSH $_{(req)}$ point occurs during the first phase of each matrix test, at a LAV fraction of 0 %. During the second phase, at a LAV fraction greater than 0%, the instability at the suction is more severe. For this reason, tests have been conducted only with LAV fraction measurements. If necessary, a thermohydraulic program, such as SOPHT, may be used to estimate suction void fractions based on LAV values.

7.2 EFFECTS OF LOOP GEOMETRY

Tests conducted in a different test loop[3], on a single discharge HT pump, indicated that the worst LOCA conditions occurred at a much lower temperature (100°C) and void fraction (approximately 10%). This indicates that test loop geometry and pump design play a major role in determining the worst LOCA condition. A critical difference in the two test loops was the head of liquid and volume above the impeller centre line, resulting in a different NPSH $_{(avl)}$ in each case.
Loop geometry plays a major role in determining the void fraction distribution, within a two-phase medium, approaching the impeller. At low LAV fractions, the entire volume of void could remain stagnant at the highest point in the loop. A loop with a simple piping arrangement is the best means of uniformly distributing the void by maintaining uniform fluid momentum from the pump discharge to suction..

Deviation from results of previous tests can also be explained by variances in the natural frequencies of the test systems. Meaningful results can be attained by maintaining the highest level of simulation, as described in Section 3.0.

7.3 VIBRATION LEVELS

Two peaks of shaft vibration levels are observed at each temperature (Figure 4). The first peak occurs during the first phase when the LAV fraction is 0%. The second peak occurs during the second phase when the loop is voided. Pressure pulsations behave in a similar manner, displaying two peaks at each temperature. This tendency is not evident in the pump and motor frame vibrations.

The following is a plausible explanation for the above observation. The first phase, with 0% LAV, is similar to a standard NPSH test. As the suction pressure is lowered and the resulting NPSH $_{(avl)}$ is less than required, cavitation at the impeller eye occurs. Local pressure at the impeller eye drops below the vapour pressure of the liquid, resulting in the formation of local voids. The void then enters the impeller, and due to repressurization, implodes, causing tremendous shock loads. This generates high pressure pulsations and vibrations in the pump-motor set. When suction pressure is further reduced, the number and volume of bubbles at the impeller eye increases. Some of these bubbles fail to collapse within the impeller. These bubbles then dampen and cushion shock loads generated by the collapsing bubbles. Thus, shaft vibration levels and pressure pulsations reach peak values, decreasing as suction pressure is further lowered.

As liquid is extracted from the loop during the second phase, voids that leave the impeller remain without collapsing, and could reenter the impeller. During this phase,

NPSH remains unchanged. The high levels of vibrations can be caused by acoustic and convective wave propagation, pressure drop oscillations, and thermal oscillations[5].

Results indicate that pump-shaft vibrations, pump-motor frame vibrations, and pressure pulsations are not totally synchronized. This is expected as system response is dependent upon the relationship between the exciting frequencies and resonance frequencies (both acoustic and structural) of the systems.

8. CONCLUSIONS

The HT pump-motor sets will maintain their operational capability and structural integrity under postulated " worst-LOCA conditions".

The mechanical seals of the pump and the hydrodynamic bearing will perform exceptionally well, without seal injection, under two-phase flow conditions.

The motor hanger supports and bearings can perform under the worst-LOCA conditions without sustaining any damage.

Licensing requirements for HT pumps for CANDU reactors have been satisfactorily met.

In a test set-up, the head and volume of liquid above the impeller centre line, plays a major role in the two-phase flow behaviour of the pump.

9. REFERENCES

1. Kumar, A.N. "CANDU Heat Transport Pumps - A Quality Product". Paper presented at the International Pump Specialists' Meeting on Pump Performance and Reliability, 1990. Nov. 26-28 held at Cologne, Germany. Organized by I.A.E.A. - O.E.C.D.

2. Kumar, A.N. "Current Methods in Qualification of CANDU Heat Transport Pumps For Operation Under Loss-Of-Coolant-Accident (LOCA) Conditions". Paper presented at CNA/CNS Conference, June 12-15, 1988, at Winnipeg, Canada.

3. Kumar, A.N. and Burchett, P.R. "Operational Capability of Heat Transport Pumps Under Two-Phase Flow Conditions". Paper published in "Tribology International", June 1989, Volume 22, No. 3, Butterworth & Co. (Publishers) Ltd., U.K.

4. Guelich, J.F., Bolleter, U., "Pressure Pulsations in Centrifugal Pumps", Trans. of ASME, Vol. 114, April 1992.

5. Rzentkowski, G., "Two-Phase Flow Tests of a Centrifugal Pump" Proceedings of the Symp. on Engineering Applications of Mechanics, June 1994, Montreal, pp. 525-537.

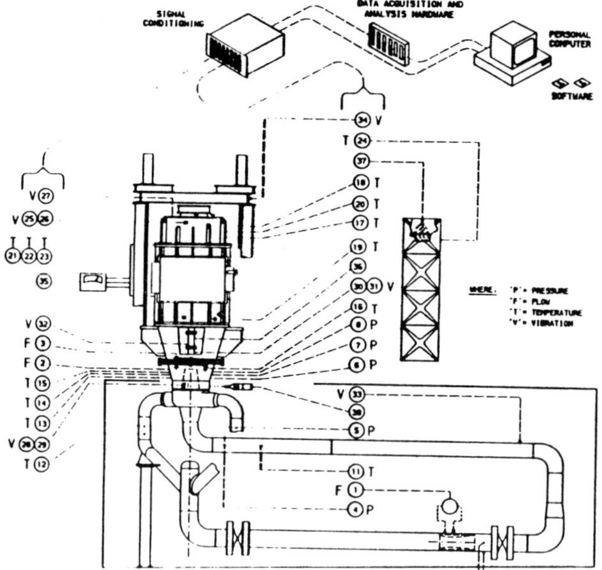

Figure 1

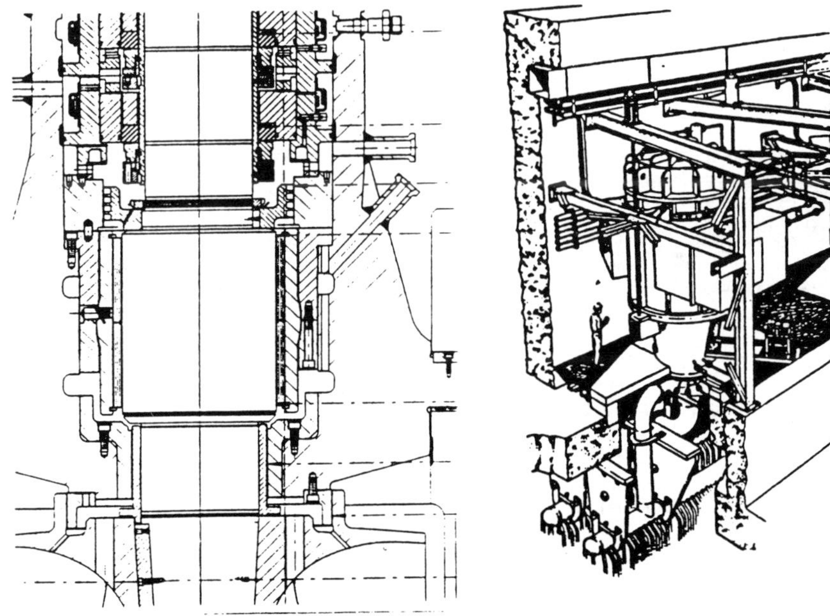

Figure 2

Figure 3

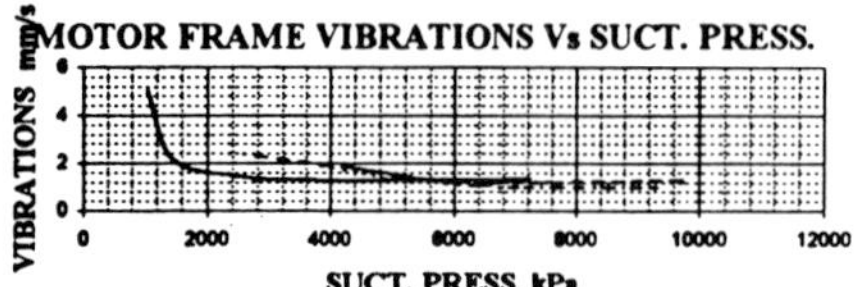
MOTOR FRAME VIBRATIONS Vs SUCT. PRESS.
VIBRATIONS mm/s
SUCT. PRESS. kPa

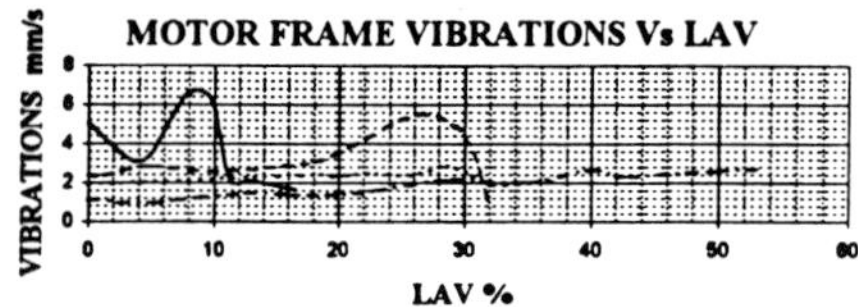
MOTOR FRAME VIBRATIONS Vs LAV
VIBRATIONS mm/s
LAV %

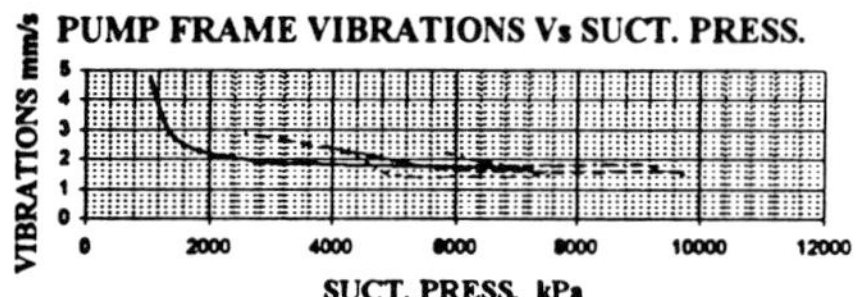
PUMP FRAME VIBRATIONS Vs SUCT. PRESS.
VIBRATIONS mm/s
SUCT. PRESS. kPa

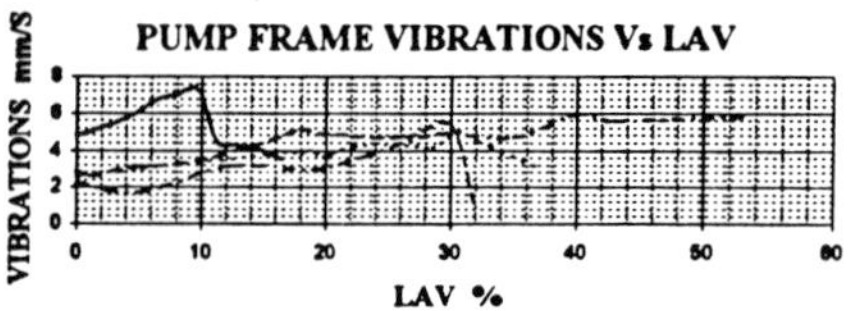
PUMP FRAME VIBRATIONS Vs LAV
VIBRATIONS mm/S
LAV %

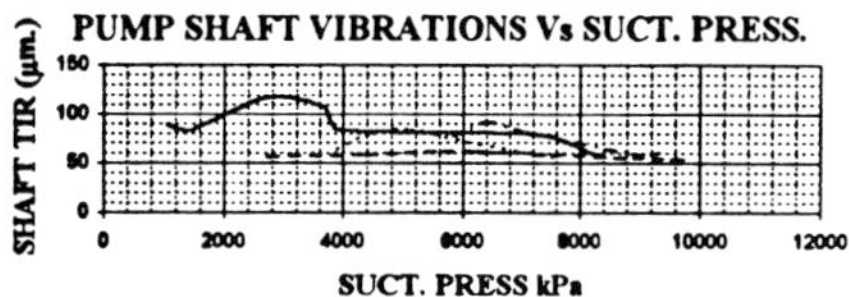
PUMP SHAFT VIBRATIONS Vs SUCT. PRESS.
SHAFT TIR (μm.)
SUCT. PRESS kPa

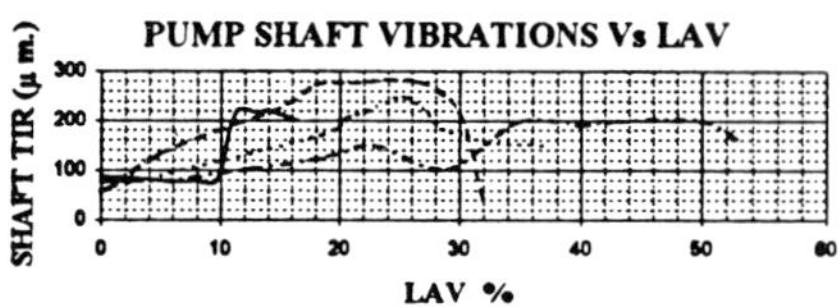
PUMP SHAFT VIBRATIONS Vs LAV
SHAFT TIR (μ m.)
LAV %

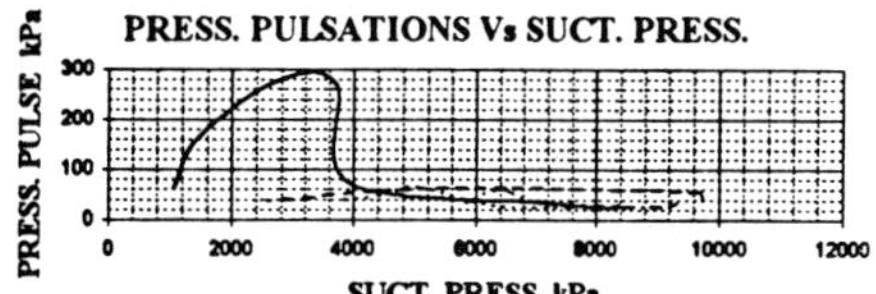
PRESS. PULSATIONS Vs SUCT. PRESS.
PRESS. PULSE kPa
SUCT. PRESS. kPa

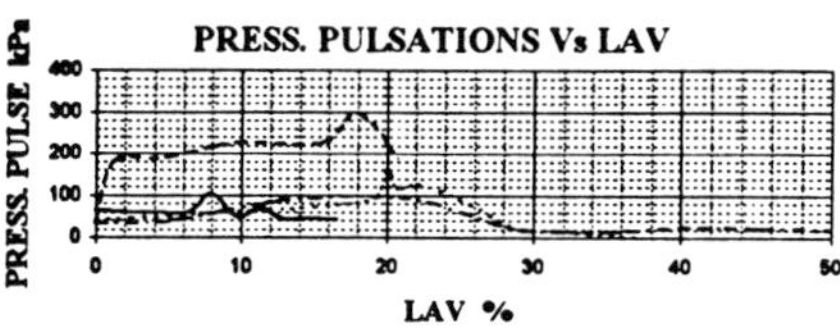
PRESS. PULSATIONS Vs LAV
PRESS. PULSE kPa
LAV %

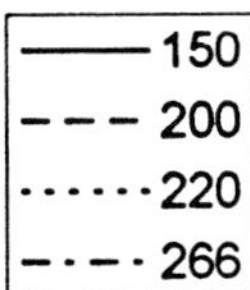
150
200
220
266

Figure 4

INTERDEPENDENCE OF DRAFT TUBE AND TAILWATER FLOW IN BULB TURBINE POWER PLANTS

CH. SCHNEIDER, W. KNAPP AND R. SCHILLING
Institute for Hydraulic Machinery and Plants
Technical University of Munich, Germany

1. Introduction

The advantages of bulb turbines result mainly from smaller dimensions and better efficiencies in comparison to the vertical-axis Kaplan type used earlier. Bulb turbines are characterized by high values of specific speed n_q which is closely related to the unit discharge Q_1'. Optimum results have been obtained with Q_1'-values that rise in proportion with the specific speed. As the specific velocity head h_{c1}/H at the runner exit increases with the square of the unit flow Q_1' an efficient deceleration of the flow in the draft tube is decisive for the performance of high specific speed turbines. Still the exit losses increase with the specific speed n_q and may reach values of 5% or more of the available head.

However, the further deceleration of the flow in the tailwater results in a certain amount of head recovery. As a consequence the corresponding depression of the water surface at the draft tube exit increases the available net head. The importance of draft tube and tailwater as flow decelerating elements therefore calls for a joint optimization which has to overcome the actual system boundary of mechanical and civil engineering at the draft tube exit. This paper presents some results which demonstrate the interdependence of draft tube and tailwater flow.

2. The low head test rig

A suitable low head test rig has been set up for the investigation of the subject mentioned above, fig. 1 gives an idea of the installation. An axial pump provides a flow of max. $2.0 m^3/s$ from a reservoir in the cellar of the laboratory building. A series of tanks is used to model the open flume

E. Cabrera et al. (eds.), Hydraulic Machinery and Cavitation, 1103–1112.

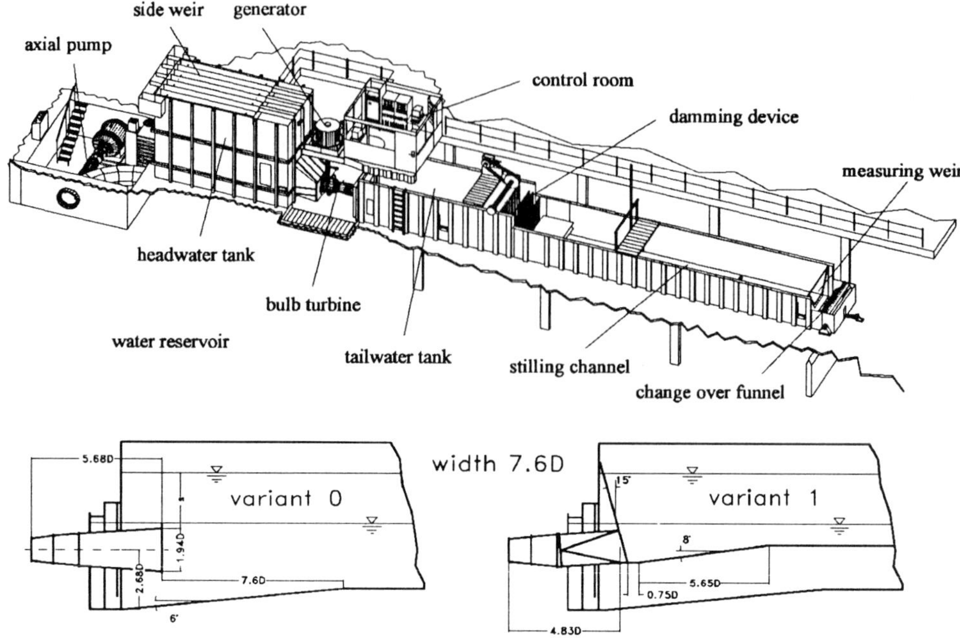

Fig. 1: The low head test rig of the Laboratory for Hydraulic Machinery and Plants, Technical University of Munich

upstream and downstream the actual power plant. The headwater level is stabilized by an adjustable side weir, the tailwater level can be varied by means of a damming device. Variable sheet elements allow to model the slope of the tailwater bottom and the power house wall. Discharge measurement is accomplished by a measuring weir at the end of the stilling channel where the water drops back into the reservoir. By means of a change over funnel the flow can be guided into a calibration basin for volumetric calibration.

The bulb turbine itself is equiped with a three bladed runner of $D = 320mm$ and $n_{q,opt} = 154\ min^{-1}$. A torque transducer incorporated into the turbine bulb is used to determine the hydraulic torque. The repeatability of a single efficiency reading near to the optimum point was found to be about $\pm 0.2\%$.

3. State of the art

Regarding the interface of draft tube and tailwater flow the non-uniformities of the velocity distribution at the draft tube exit are a major concern of the turbine manufacturers. The definition of the net head according to the IEC code considers the cross sectional mean velocity to determine the velocity head at the draft tube exit. Due to the non-uniformity of the velocity distribution the real velocity head as an integral value may be considerably higher:

$$h_{c2} = \frac{1}{2gQ}\int_A c_2^2 c_m dA = \alpha \cdot \bar{c}_2^2/2g = \alpha \cdot \bar{h}_{c2} = \alpha \cdot h_{c2,IEC} \quad (1)$$

Due to these factors $\alpha > 1$ the net head according to the definition is actually too high. Thus the turbine manufacturers try to reduce the non-uniformities as their extent decreases the turbine efficiency. A possible influence of tailwater parameters on the velocity distribution at the draft tube exit is still indistinct. Numerical investigations of elbow draft tubes, however, have shown that the velocity distribution even upstream the exit cross section may be influenced by the boundary condition in the tailwater, see [1].

The importance of the head recovery in the tailwater has been recognized repeatedly. In a first approximation the transition of flow from the draft tube exit to the tailwater can be regarded as an unsteady diffuser, known as Carnot diffuser. The law of conservation of momentum yields the pressure recovery of such diffusers:

$$\Delta p = \rho c_3(c_2 - c_3) \quad (2)$$

Based on this simple approach the head recovery in the open tailwater flume may be evaluated. The following equations indicate the dependency on the unit flow and the area ratio $AR_{tw} = A_{tw}/A_{dt}$, i.e. on the submersion of the draft tube exit s:

$$\frac{\Delta h_{tw}}{H} = \frac{16}{g\pi^2}\frac{1}{AR_{dt}^2}\frac{AR_{tw}-1}{AR_{tw}^2}Q_1'^2 \qquad c_p = \frac{2}{AR_{tw}^2}(AR_{tw}-1) \quad (3)$$

Comprehensive tests regarding the head recovery in the tailwater have shown a significant influence of the tailwater geometry, see [2]. Another investigation indicates a notable influence of the turbine operating point on the head recovery, see [3].

4. Preliminary tests

To assess the extent of the expected interdependence of tailwater and draft tube flow a first series of tests has been carried out with the given simple geometry shown in fig. 1 as variant 0. In these tests the influence of the draft tube submersion at constant head has been investigated, s. Fig. 2.

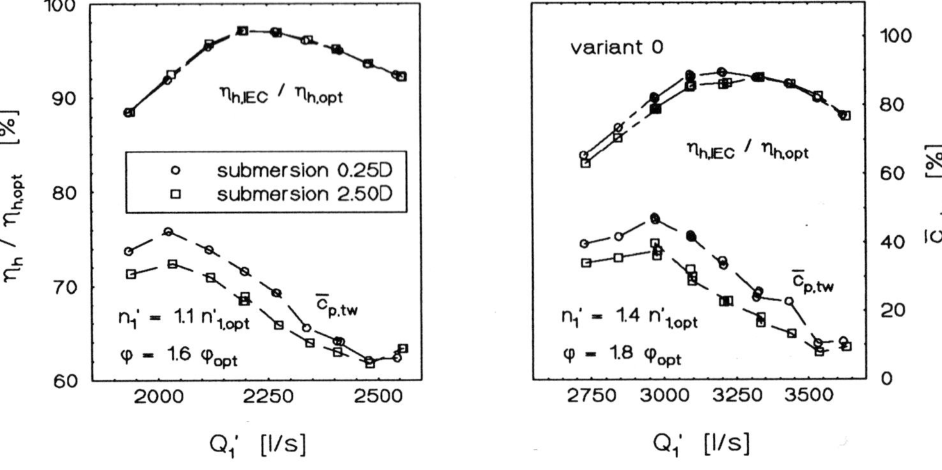

Fig. 2: Influence of draft tube submersion on turbine efficiency and tailwater performance

The left diagram corresponds to nearly optimum unit speed and a runner blade angle of $\varphi = 1.6\varphi_{opt}$. The efficiency curves for different rates of submersion coincide, indicating that the draft tube flow is not measureably affected by this parameter. The right diagram, on the contrary, shows the efficiency curves at $1.4n'_{1,opt}$ and full runner opening $\varphi = 1.8\varphi_{opt}$. Here the curves are clearly divergent at lower values of Q'_1, indicating an efficiency drop of ca. 1.5% for the higher value of submersion.

The tests included level readings of h_3 at a distance of 9.5D downstream the draft tube exit. The difference of h_3 to the static pressure head h_2 at the draft tube exit yields the head recovery of the tailwater. Fig. 2 also shows the values of $\bar{c}_{p,tw} = 2g\Delta h_{2-3}/\bar{c}_2^2$ which have been determined by referencing on the cross sectional mean velocity head. According to the Carnot relation the better values of $\bar{c}_{p,tw}$ have been found at low submersion. Furthermore at optimum and lower values of unit discharge the influence of the submersion results in a nearly constant difference of about $\Delta\bar{c}_{p,tw} \sim 9\%$. The evaluation based on the Carnot relation yields a difference of $\Delta c_{p,tw} \sim 6.9\%$ which coincides quite well.

Apart from the submersion the input swirl seems to be of significant importance for the head recovery. The increase of swirl at the circumference of the draft tube at lower values of Q'_1 causes a significant rise of the head recovery. The measured optimum values of nearly 40% resp. 50% exceed by far the values calculated by means of the Carnot relation of $c_{p,tw} = 12\%$ resp. $c_{p,tw} = 19\%$.

5. Systematic tests

Encouraged by the results of the preliminary tests a systematic investigation has been defined to study the

- influence of the draft tube submersion on the turbine efficiency and the
- influence of the turbine operating point on the tailwater performance.

In contrast to the preliminary tests a realistic geometry of draft tube and tailwater was to be applied for the systematic tests. A survey of current designs showed that there is a certain bandwidth of common draft tube geometries. A conservative layout confines itself to a moderate area ratio of about $AR_{dt} \sim 4$ at a rated length of $l/D \sim 5$ in order to ensure a reliable operation. On the other hand forced layouts have been developed to reduce the outlet losses. Such designs realize area ratios of up to $AR_{dt} \sim 6$ at comparable length. These designs inevitably lead to a higher susceptibility to stall. Fig. 3 shows two representative designs used as variants one and two for the systematic tests. The results obtained with the conservative draft tube variant one will be presented in the following. Fig. 1 shows the geometry of the chosen tailwater arrangement.

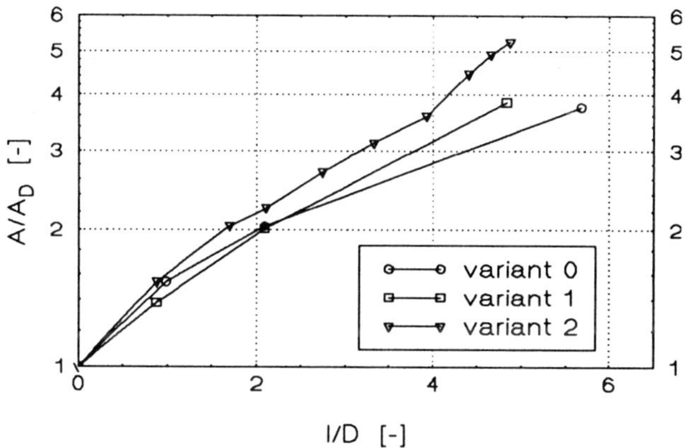

Fig. 3: Area development of the investigated draft tubes

5.1. PRESSURE RECOVERY IN THE DRAFT TUBE

Efficiency measurements at $(Q'_1, n'_1)_{opt}$, as expected, did not show any differences at low and high rate of submersion. Measurements at $(Q'_1, n'_1) > (Q'_1, n_1)_{opt}$ resulted in slightly lower optimum efficiencies at the high rate of submersion, a sample is shown in fig. 4. However, the observed differences of some tenth of a percent are actually below the threshold of significance. To verify the differences supplementary measurements of the pressure recovery in the draft tube have been carried out to determine the value of the draft tube pressure recovery factor $\bar{c}_{p,dt} = 2(\Delta p_{2-1})/\rho \bar{c}_1^2$; p_1 and p_2 have been measured by means of four pressure taps each. Fig. 4 shows the results obtained for the sample operating point. The values of $\bar{c}_{p,dt}$ also indicate a performance decrease at high submersion level. For a more detailed analysis the pressure development has been measured by means of nine pressure taps evenly distributed over the draft tube length. For every position the quality of the energy conversion downstream can be assessed by considering a local pressure recovery coefficient $\bar{c}_{p,i} = 2(p_9 - p_i)/\rho\bar{c}_i^2$, s. fig. 5. At high submersion the $c_{p,i}$-values were found to be about 3% worse at the far end of the draft tube. As the major portion of kinetic energy is converted already at the near end of the draft tube the total $\bar{c}_{p,dt}$-value thus is affected only by some tenth of a percent. Also noteworthy is the drop of $\bar{c}_{p,i}$ at the position 5 as a consequence of the slight bend in the draft tube axis.

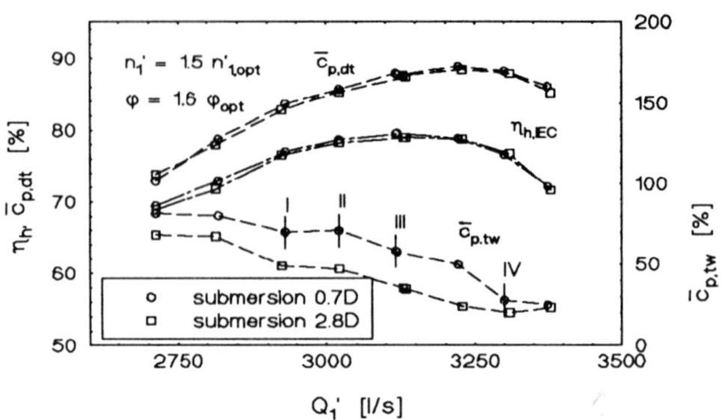

Fig. 4: Influence of submersion on turbine and draft tube performance

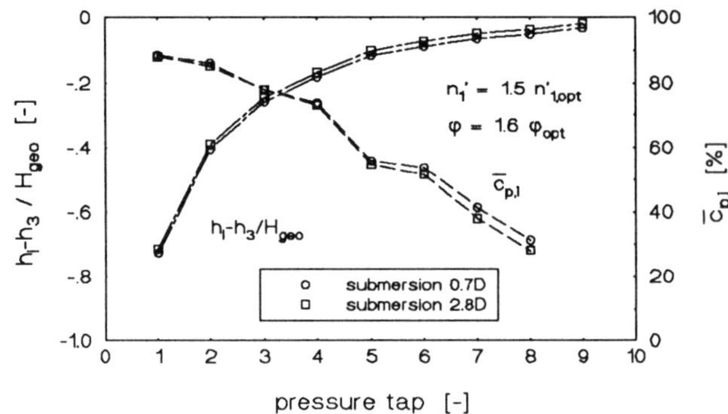

Fig. 5: Quality of energy conversion versus draft tube length, variant 1.

The observation of the flow near to the draft tube outlet shows a different mechanism of jet propagation into the tailwater, corresponding to the extent of submersion. At high submersion there is ample space for the expansion of the jet. Due to the large cross sectional area available for the backflow beside the jet the flow is rather quiet. At low submersion, on the contrary, there is a strong backflow which seems to constrain the emergent jet. Based on these observations a distortion of the flow dependent on different levels of submersion is imaginable even upstream the outlet cross section of the draft tube.

Finally a detailed investigation of the flow field at the draft tube exit has been carried out by means of a five hole probe. Velocity and static pressure

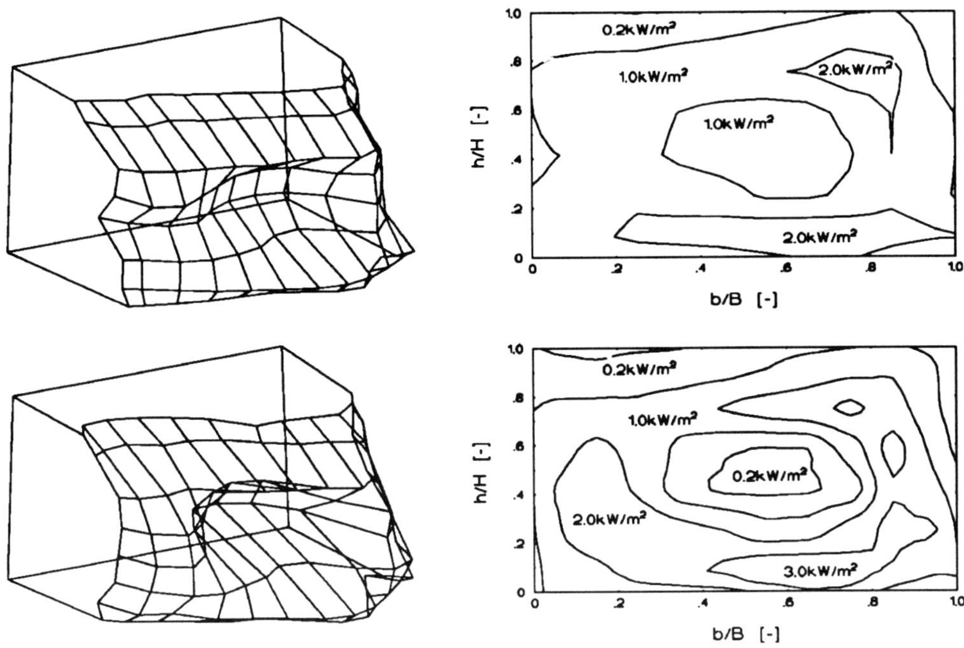

Fig. 6: Velocity distribution (left) and power density (right) at the draft tube outlet at $s = 0.7D$ (above) and $s = 2.8D$ (below), $n_1' = 1.5n_{1,opt}'$, $\varphi = 1.3\varphi_{opt}$

have been measured on a grid of 96 nodes $15mm$ downstream the actual outlet section. Fig. 6 shows on the left hand side the measured velocity distribution. Clearly visible is the higher rate of velocity at the perimeter of the outlet cross section caused by the swirl of the flow. The wake of the runner hub leads to a reduction of the velocity in the center of the outlet section. Also obvious is the intensification of these non-uniformities at the high rate of submersion. The contour plots on the right hand side of fig. 6 show the power density distribution $dP/dA = (\rho c^2/2 + p) \cdot c_m$ of the flow. The power density amplifies the velocity differences to the third power and therefore displays the observed flow distortion more clearly. The integration of the values measured on the grid yields the results shown in table 1.

The values of α indicate the increased non-uniformity of the flow at high submersion. To assess the accuracy of the measurements they have been checked for continuity. The total discharge obtained by integrating the measured values of c_m results in an error of -2.9% at $s = 2.8D$ resp. -0.2% at $s = 0.7D$.

TABLE 1. Integral evaluation of the grid measurement

		$s = 0.7D$	$s = 2.8D$
$\frac{1}{2g}(\frac{1}{A}\int_A c\,dA)^2$	$[mm]$	83.8	79.2
$\frac{1}{2g}\frac{1}{Q}\int_A c^2 c_m\,dA$	$[mm]$	105.2	109.6
α	$[-]$	1.23	1.38

5.2. HEAD RECOVERY IN THE TAILWATER

The tests included the determination of the $\bar{c}_{p,tw}$-values. The characteristics shown in fig. 4 have been found to be almost constant in the operation range $Q_1' \geq Q_{1,opt}'$ and $n_1' \geq n_{1,opt}'$. Similar to the preliminary tests the lower submersion level yielded the better results. The negative gradient at increasing Q_1' also has been found. However, with maximum values of ca. 80% the total level of $\bar{c}_{p,tw}$ has increased substantially.

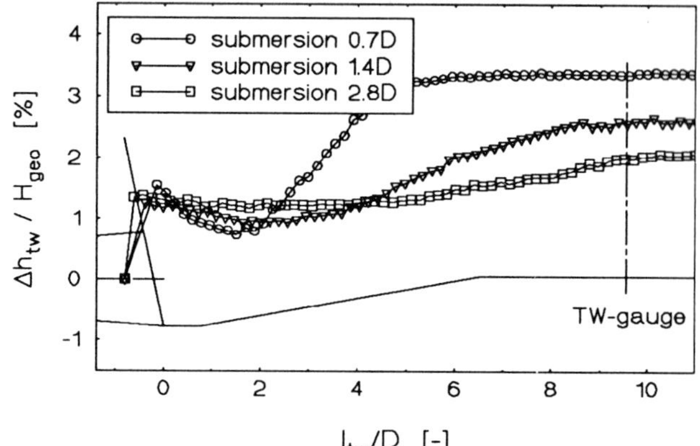

Fig. 7: Distribution of the head recovery versus flume length

To achieve a more detailed analysis of the head recovery in the tailwater the time averaged values of the water level have been measured on a grid of about 3500 nodes by means of magnetostrictive displacement sensors. The unsteady surface structures perceptible in a real time observation could not be resolved in the time averaged measurement and therefore are considered to be of stochastical nature. The predominant feature found is a steady rise of the water level. Fig. 7 shows the distribution of the level rise versus flume length at different rates of submersion. The single values have been obtained by averaging the values perpendicular to the flume axis. It is noticeable that the level difference between the pressure taps in front of the draft tube exit and the water level immediately behind the exit section is nearly constant. This fact is of particular interest for the practice of acceptance tests. The influence of the submersion on the head recovery therefore is obviously restricted to the flow deceleration in the actual tailwater. The other question of interest was the influence of swirl on the head recovery in the tailwater. As in the previous measurements the true velocity head in the outlet section has been unknown a series of grid measurements of the velocity distribution has been carried out at the four operating points indicated in figure 4.

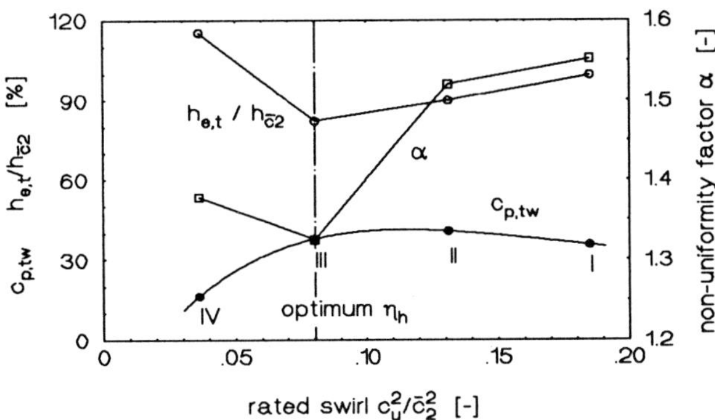

Fig. 8: Influence of swirl on overall exit loss and head recovery in the tailwater.

Fig. 8 shows the α-values versus rated swirl $\frac{1}{Q}\int_A c_u^2\, c_m\, dA/\bar{c}_m^2$. The corresponding correct values of $c_{p,tw}$ thus are considerably lower than the cross sectional mean values $\bar{c}_{p,tw}$. Neglecting the α-values therefore is the reason for the exaggerated values of $\bar{c}_{p,tw}$ shown in the fig. 4 and 2. However, the increase of swirl at lower Q_1'-values seems to be advantageous for the head recovery in the tailwater, the optimum of the real tailwater performance $c_{p,tw}$ is found at higher rates of swirl. Due to this fact the total exit loss head $h_{e,t}$ grows only slightly with increasing rates of swirl. $h_{e,t}$ considers the influence of non-uniformities as well as the head recovery in the tailwater. Thus a joint consideration of draft tube and tailwater may lead to a modified determination of the optimum amount of swirl to be supplied to the draft tube.

6. Conclusion

High amount of kinetic fluid energy is a characteristic feature of high specific speed turbines. Due to this fact their efficiency strongly depends on the quality of the flow decelaration, i.e. on the pressure recovery in the draft tube and on the head recovery in the tailwater. Joint considerations of these flow decelarating elements are still uncommon. As a consequence a specifically adapted test rig has been set up at the Laboratory for Hydraulic Machinery and Plants at the Technical University of Munich. Experimental results show an interdependence of the flow in the draft tube and in the tailwater:

- The flow in the rear part of the draft tube is affected by the flow in the tailwater at off-design points of operation. Low submersion renders the better results.

- The head recovery in the tailwater depends on the inlet boundary condition, i.e. the flow distribution at the draft tube exit. Swirl in the turbine's sense of rotation increases the head recovery.
- The geometry of the tailwater influences the extent of head recovery. Low area ratios of the tailwater and the draft tube cross section are favourable.

Nomenclature

A	$[m^2]$	area	h	$[m]$	level, pressure head
AR	$[-]$	area ratio	h_c	$[m]$	velocity head
b	$[m]$	width	H	$[m]$	head
B	$[m]$	draft tube exit width	l	$[m]$	length coordinate
c	$[m/s]$	velocity	n	$[min^{-1}]$	rotational speed
$\bar{c}$	$[m/s]$	cross sectional mean velocity	n_q	$[min^{-1}]$	specific speed
c_m	$[m/s]$	meridional velocity	p	$[Pa]$	pressure
c_p	$[-]$	pressure recovery coeff.	P	$[W]$	power
c_u	$[m/s]$	circumfer. velocity	Q	$[m^3/s]$	discharge
α	$[-]$	non-uniformity factor	ρ	$[kg/m^3]$	density
η	$[-]$	efficiency	φ	$[deg]$	runner blade angle

Indices

$_{dt}$	draft tube	$_{opt}$	at the optimum
$_{e,i}$	exit, pos. i	$_{tw}$	tailwater
$_{geo}$	geodetic	$_{1,2}$	at draft tube entrance, exit
$_{h}$	hydraulic	$_{3}$	at tailwater gauge

References

1. A. Ruprecht: *Finite Elemente zur Berechnung dreidimensionaler, turbulenter Strömungen in komplexen Geometrien*, Universität Stuttgart, Mitt. des Institutes für Hydr. Strömungsmaschinen, Nr. 3, 1989
2. J. Chevalier, H. Giraud: *Études théoriques et expérimentales de l'évolution de l'énergie dans les ouvrages aval des installations hydroélectriques de basse chute*, La Houille Blanche, Nr. 2-3, p.155-166, 1968
3. E. Kita, K. Kubota, Y. Kimoto: *Tailwater Level and Net Head in Low Head Power Plant*, Proceedings of the 3rd Japan-China Joint Conference on Fluid Machinery, Vol. 1, Osaka, 1990
4. Ch. Schneider, W. Knapp, R. Schilling: *Untersuchung der Wechselwirkung schnellläufiger Wasserturbinen mit dem Unterwasser*, Zwischenbericht zum BMBF-Forschungsvorhaben, March 1996

STUDY OF HYDRAULIC TRANSIENTS USING THE BOND GRAPHS METHOD

GERALDO LÚCIO TIAGO FILHO
Escola Federal de Engenharia de Itajubá (EFEI)
Instituto de Engenharia Mecânica (IEM)
Av. BPS, 1303 - 37500-000 Itajubá/MG - Brazil

1. Introduction

One of the latest modeling technique for dynamic systems studies came through the 1950's. Created by H.M. Payner (Speranza, 1992) who, according to Thoma (1975), felt the urge to create methods and criteria able to generalize the accomplishment of state equations representative of a system. That is the "Bond Graphs" technique. It is based on the power flow among the system components and provides a generalized approach to the system state equations, allowing simulation and interaction of different power domains like: hydraulical with mechanical, or mechanical with electrical, and others. The "bond graphs" techiniques became a powerful tool used in studies and modeling of dynamical systems as seen in works by Karnopp and Rosenberg (1983 and 1975).

Thus, in face of the potentialities presented in Tiago F. (1980) the bond graphs technique was used to model the behavior of a relief valve coupled to a system made up of a reservoir, a penstock, and a fast closure butterfly valve. The bond graphs technique was used both for modeling the relief valve control and the transient behavior of the pipeline.

The present work aims to show the use of the bond graphs technique in simple configuration penstock systems. The method applicability in this kind of study is checked through the study of some cases.

2. Application of Bond Graphs to Hydraulic Systems

Basically, the bond graphs technique uses four generalized variables, customized here for hydraulic systems: two state variables and two integral variables.

The two state variables are strain and flow. In hydraulics, strain and flow, when expressed in unit quantities, correspond to the unit load, h, and to the unit flow, q, respectively. With:

$$h = H/H_o \tag{1}$$

and

$$q = Q/Q_o \tag{2}$$

The state variables are also termed power variables due to the product between both having power dimension, that is:

E. Cabrera et al. (eds.), Hydraulic Machinery and Cavitation, 1113–1122.

$$pot = h \cdot q \tag{3}$$

The integral variables, or power ones, are defined by the time integrals of the state variables, that is:

The movement quantity, represented by the letter P_p is given by the strain integral that in hydraulics corresponds to the load h. Thus,

$$P_p = \int^t h.dt \tag{4}$$

and the displacement, given by the flow integral. In hydraulics the displacement corresponds to the volume, V, since the unit flow has the flow dimension. That is,

$$V = \int^t q\, dt \tag{5}$$

The relationship between the two distinct elements in a system to be modeled is represented by a half-arrow that indicates the connection between the elements besides permitting the indexing of the strain and flow variables, h and q, the direction of the power, indicated by the direction of the half-arrow and what is cause and what is effect in the intersection. The latter is indicated by the causal bar. It is a vertical bar inserted in one of the ends of the bond and that indicates the direction of the strain effort. In the opposite sense the flow direction is understood. Figure 1 shows a general idea about indications inserted in a bond between two elements.

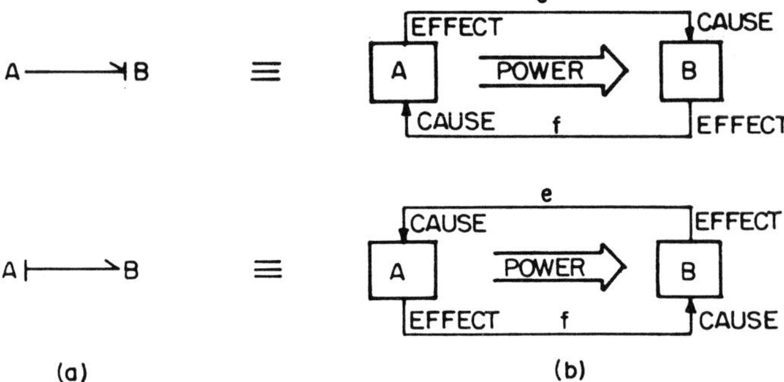

Figure 1. Elements part of a bond: a) bond graphs; b) block diagram.

The relationships between state variables, load h and flow q, and the integral variables, movement quantity P_p and volume V, define three other constitutive laws of which correspond to the inertia of the system, to its storage capacity and to the power dissipation. In hydraulics these elements correspond to:
Fluid inertance, represented by the letter "I", is associated to the inertia or to the kinetic power of the fluid. It relations the movement quantity, P_p, to the fluid flow.

$$q = \frac{1}{I} P_p \tag{6}$$

that is, according to the expression (4),

$$q = \frac{1}{I}\int^t h.dt \tag{7}$$

where:

$$h = \dot{P}_p \tag{8}$$

Fluid Capacitance, represented by the letter "C", is associated to the capacity of the system in storing the potential power. Concerning hydraulic systems, it is the element that takes into account the compressibility of the fluid and the deformability of the pipe. Its constitutive law relations the load h to the volume V.

$$h = \frac{1}{C} V$$

that is, taking into account the expression (5)

$$h = \frac{1}{C}\int^t q.dt$$

where

$$q = \dot{V} \tag{9}$$

The Fluid Resistance, represented by the letter "R", that corresponds to the power dissipation of the system, in hydraulics it is associated to load losses distributed in a pipeline or located in a valve or any other accessory. It relations the state variables: load, h and flow, q. When it comes to turbulent outflows the constitutive law is like:

$$h = R\, q|q| \tag{10}$$

The graphical representation of these elements and the respective block diagrams are given in Figure 2.

The interaction of the system with the environment, that is, with the outside the system to be modeled, two kinds of sources were conceived, namely:

Pressure source, S_e, that is the element that has the capacity of supplying a time-constant or time-varying pressure, irrespective the direction and the intensity of its flow, that is:

$$S_e = h(t) \qquad S_e \xrightarrow{h(t)}| \tag{11}$$

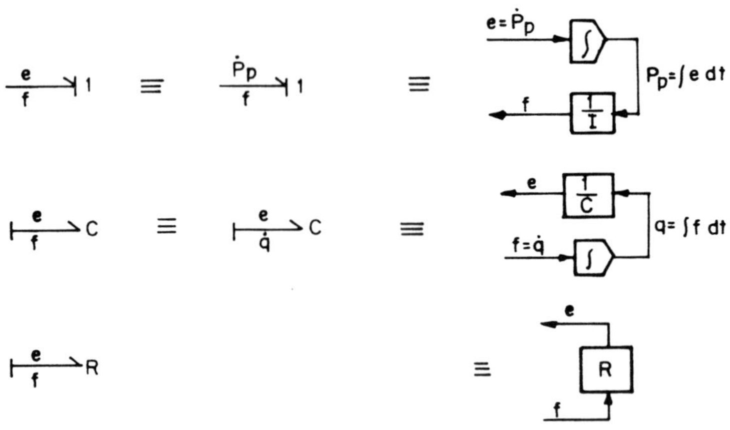

Figure 2. Graphical representation of the inertance, capacitance and fluid resistance elements

Flow source, S_f, just like pressure source, it supplies the system with a time-constant or -varying flow, irrespective the direction and the intensity of its pressure, that is:

$$S_f = q(t) \qquad S_f \vdash \xrightarrow[q(t)]{} \tag{12}$$

The system topology is obtained by the bond elements which have as a function the interlinking of the elements part of the system. They can be of two kinds: the common strain, represented by the number "0" and the common flow, represented by the number "1". Figure 3 shows the representation of these two elements. In both elements, the power transfer only happens between the other interlinked elements.
The constitutive laws of these elements are:

- common strain

$$h_1 = h_2 = h_3 \quad ; \quad q_3 = q_1 - q_2 \tag{13}$$

- common flow

$$q_1 = q_2 = q_3 \quad ; \quad h_3 = h_1 - h_2 \tag{14}$$

In both elements the summation of the powers is void, that is:

$$h_1 q_1 - h_2 q_2 - h_3 q_3 = 0 \tag{15}$$

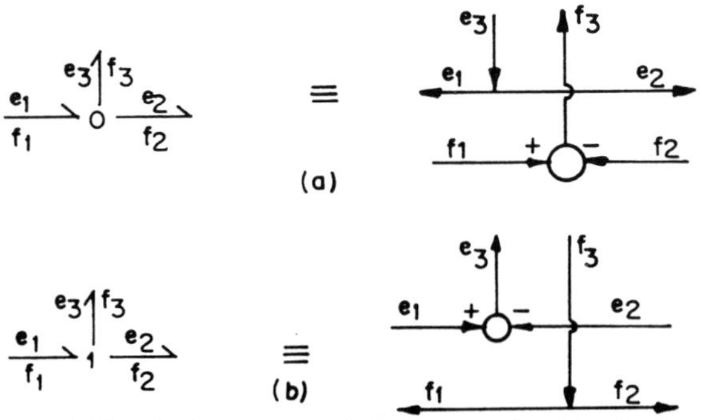

Figure 3. Bond elements and the respective block diagrams: a) common strain; b) common flow

Table 1 gives the physical schemes as well as the respective graphical representations and the constitutive laws of the elements used by the bond graphs technique to model a hydraulic system. The fluid inertance, I, and the fluid resistance, R, are associated to a bond "1" and the fluid capacitance "C" is associated to a bond "0".

TABLE 1. Active and passive elements of the bond graphs necessary to the modeling of a penstock

ELEMENT		PHISICAL MODEL	BOND GRAPH	CONSTITUTIVE LOW: DIMENSIONAL	CONSTITUTIVE LOW: UNITY
FLUID INERTANCE		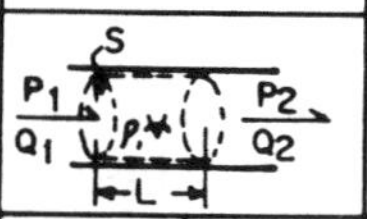 S, P_1, Q_1, ρ, $\forall$, P_2, Q_2, L	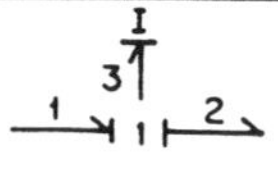 I, 3, 1, 1, 2	$I_3=\rho\frac{L}{S}$ $I=\int_0^L \frac{\rho}{S_x}\,dx$	$I=\frac{Q_o}{gS_oH_o}L$
FLUID CAPACITANCE	COMPRES-SIBILITY	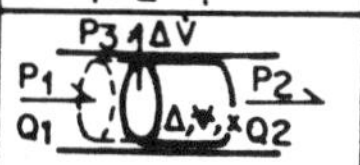 P_3, ΔV, P_1, Q_1, $\Delta\forall$, P_2, Q_2	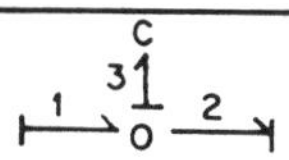 C, 3, 1, 0, 2	$C_3=\frac{\forall_o}{K}$	
FLUID CAPACITANCE		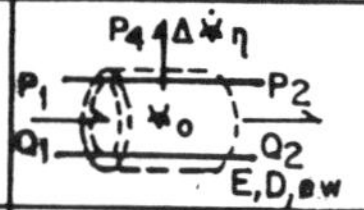 P_4, $\Delta\forall\eta$, P_1, Q_1, $\forall_o$, P_2, Q_2, E,D,ew	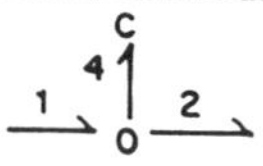 C, 4, 1, 0, 2	$C_4=\forall_o\frac{D_o}{E\,e_w}$	
FLUID CAPACITANCE	EQUIVALENT	C_3+C_4	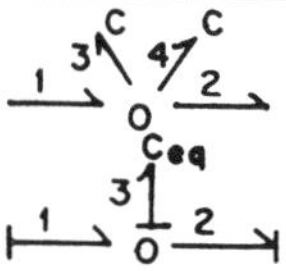 C, C, 3, 4, 1, 0, 2 Ceq, 3, 1, 0, 2	$C_{eq}=\forall_o\left(\frac{1}{K}+\frac{D_o}{E\,e_w}\right)$ $C_{eq}=\frac{Q_o}{\rho a^2 S}$	$C=\frac{1}{I}\left(\frac{L}{a}\right)$
FLUID CAPACITANCE	VOLUMETRIC	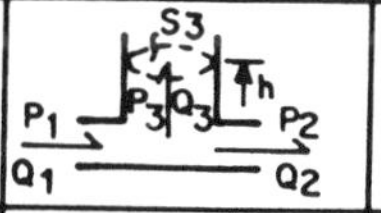 S_3, h, P_3, Q_3, P_1, Q_1, P_2, Q_2	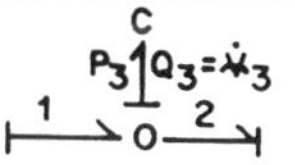 C, P_3, $Q_3=\dot{\forall}_3$, 1, 0, 2	$C_3=\frac{S_3}{\rho g}$	$C_v=S_b\frac{Y_{máx.}}{Q_o}$
FLUID RESISTANCE	DISTRIBUTED	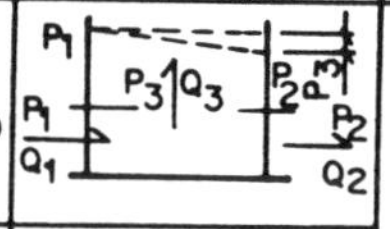 P_1, P_3, Q_3, P_2, Q_2, Q_1	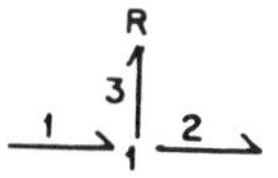 R, 3, 1, 1, 2	LAMINAR $R_3=128\frac{\mu L}{\pi D^4}$ TURBULENTY $R_3=fa\frac{L}{2aDS^2}$	$R=\bar{I}.F_T$ $F_T=2\frac{faQ_o}{\pi D_o^3}$
FLUID RESISTANCE	LOCATED	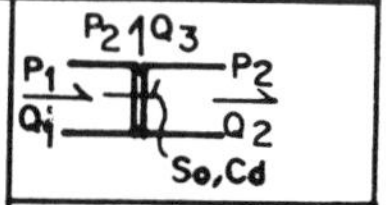 P_2, Q_3, P_1, Q_1, P_2, Q_2, So,Cd		$R_3=\frac{\rho}{2Cd^2 So^2}$	
FLUID RESISTANCE	LOCATED	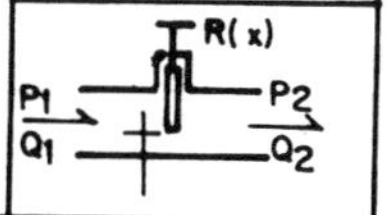 R(x), P_1, Q_1, P_2, Q_2		$R_3=\frac{\rho}{2Cd^2(x)S^2(x)}$	
SOURCE	PRESSURE	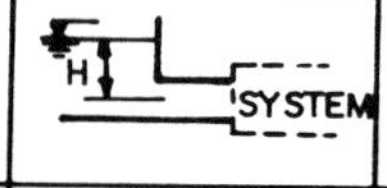 H, SYSTEM	Se, H(t)	Se = H(t)	$\overline{Se}=1$
SOURCE	FLOW	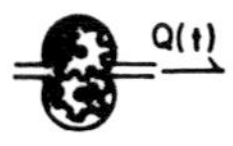 Q(t)	Sf, Q(t)	Sf = Q(t)	$\overline{Sf}=1$

3. Modeling a Penstock

Considering a finite stretch of a penstock, length L, diameter D, wall thickness e_w, made with a material with coefficient of elasticity E, in which a fluid of specific mass ρ flows, and kinematic viscosity μ and presents a compressibility coefficient K, according to Figure 1.a this element can be represented by the bond graph shown in Figure 4.b. where the fluid inertance I and the fluid resistance R are associated to a bond "1" and the fluid capacitance C, to a bond "0".
The same finite stretch of the penstock can be represented by the bond graphs (4.c), (4.d) and (4.e).

From them all, the configuration (4.e) has been shown as the most suitable for modeling of penstocks discretized in several stretches. In this configuration, where inertance I and the resistance R referring to the pipe stretch were divided and each one put upstream and downstream the capacitance, has the advantage of needing only pressure sources downstream and upstream the segment. This is concluded by the application of the causalities rules such as expected from the "bond graphs" technique, Rosemberg (1980), what will always provide an integrative causality to the passive elements of inertance I and capacitance C. This is an important factor in the bond

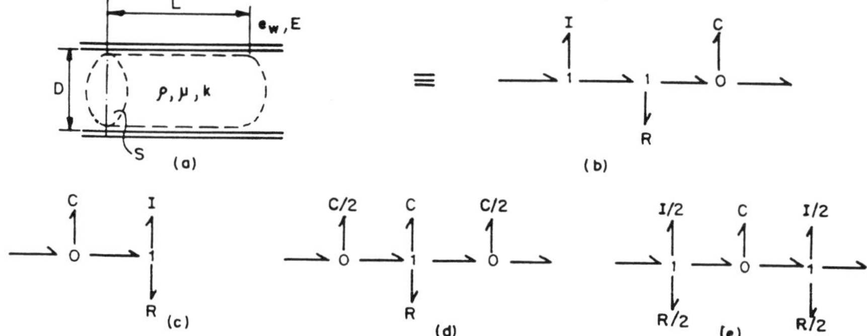

Figure 4. Finite stretch of a penstock: a) physical model; b,c,d,e) equivalent bond graphs

graphs method since each element I and C with integrative causality will correspond to a state equation, relative to the constitutive law of each element. The inertance I will have a pressure corresponding to the timeless derivative of the movement quantity; and by the capacitance will have the timeless derivative of the fluid volume contained in the one corresponding to the liquid flow in the control volume, that is:

$$h = \dot{P}_p \quad \text{and} \quad q = \dot{V} \tag{16}$$

By taking into account the constitutive procedures and of causalities predicted in the bond graphs technique, Rosemberg (1980), in the graph of Figure 5, one obtains the state equation system for the finite stretch of the penstock, that in matrix form is:

$$\begin{bmatrix} \dot{P}_{p2} \\ \dot{P}_{p4} \\ \dot{V}_3 \end{bmatrix} = \begin{bmatrix} \frac{-2R}{I} & 0 & -\frac{1}{C} \\ 0 & \frac{-2R}{I} & \frac{1}{C} \\ \frac{2}{I} & -\frac{2}{I} & 0 \end{bmatrix} \begin{bmatrix} P_{p2} \\ P_{p4} \\ q_3 \end{bmatrix} + \begin{bmatrix} 1 \\ -1 \\ 0 \end{bmatrix} \begin{bmatrix} h_1(t)\, h_5(t)\, 0 \end{bmatrix} \tag{17}$$

where $\dot{P}_{p2}$ corresponds to the load, h_2, in the first half of the pipe segment; $\dot{P}_{p4}$ corresponds to the load h_4, in the second half of the pipe segment; $\dot{V}_3$ to the load in the pipe segment and $h_1(t)$ and $h_5(t)$ to the loads in the ends of the pipe segment.

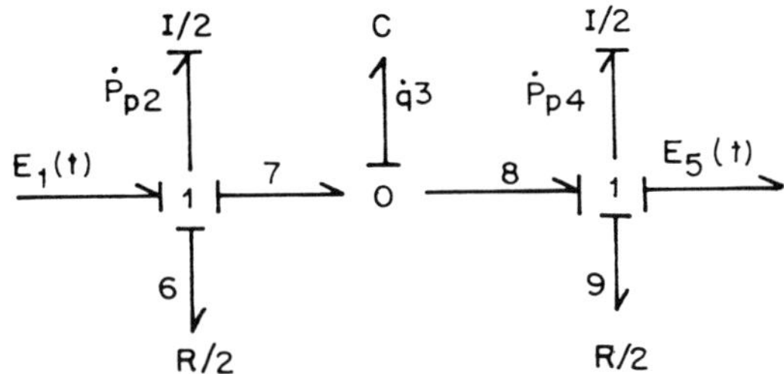

Figure 5. Bond graphs of a penstock stretch

The solution of the state equation system (17) supplies the load and flow variation in the penstock segment, besides permitting stability analysis of the system.

An adducting system formed by a reservoir, a penstock and a shutter device as shown in Figure 6.a can be represented by the bond graph of Figure 6.b, that outlined is just like in the graph of Figure 6.c.

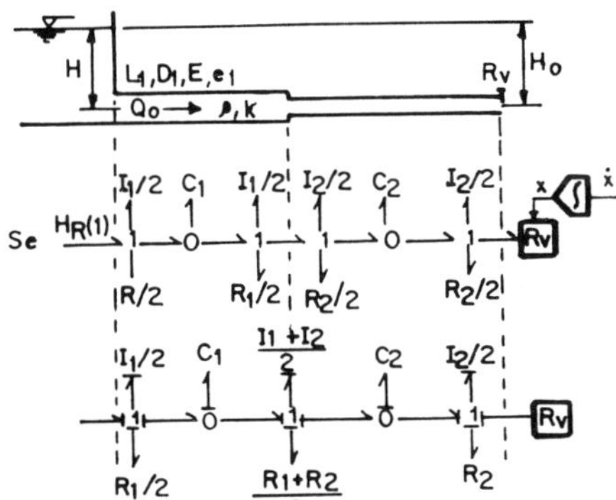

Figure 6. Bond graph of a system: reservoir - pipe - valve: a) physical model, b) extended bond graph, c) final bond graph

where R_v corresponds to the constitutive law relative to the valve operation law, where $\dot{X}$ is the speed with which this one is closed, X corresponding to its position and R_1 is the operation law that linearized by the Taylor series one, is given by the block diagram of Figure 7.

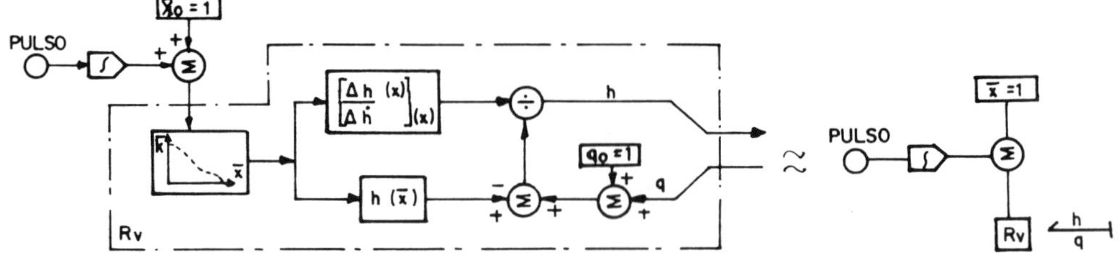

Figure 7. Operation law of a shutter: a) block diagram, b) bond graph

4. Simulation of Some Cases Through the Bond Graph and the Characteristics Method

The results obtained by Carvalho (1995) on the simulations of cases supplied by Streeter (1978), Chaudry (1979), Koele (1989) are presented, where the characteristics methods and the bond graph one were applied.

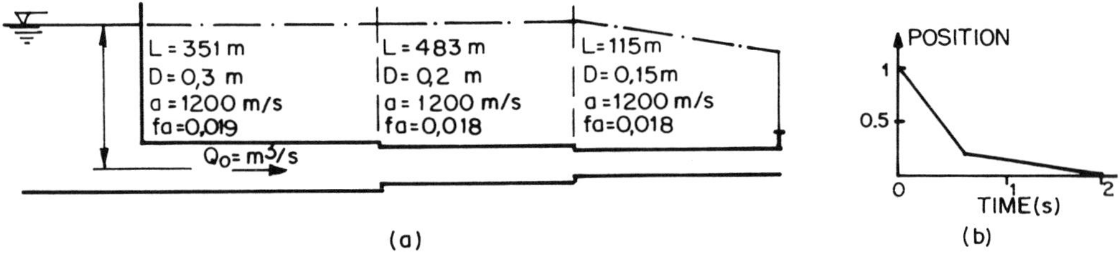

Figure 8. Example proposed by Streeter (1979): a) physical model; b) valve closing law

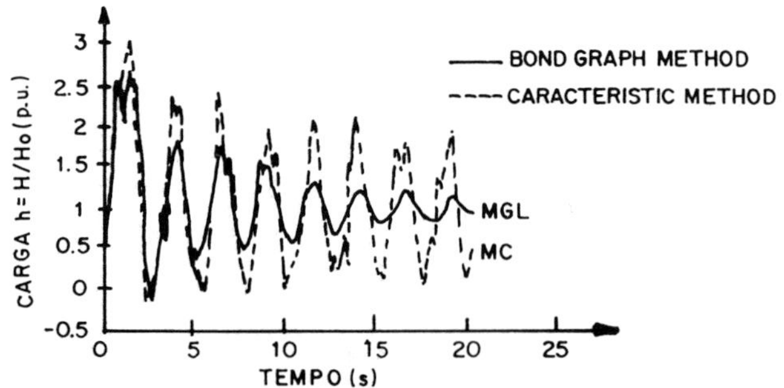

Figure 9. Results obtained on the simulation of the example of Streeter: a) characteristics method, b) bond graph

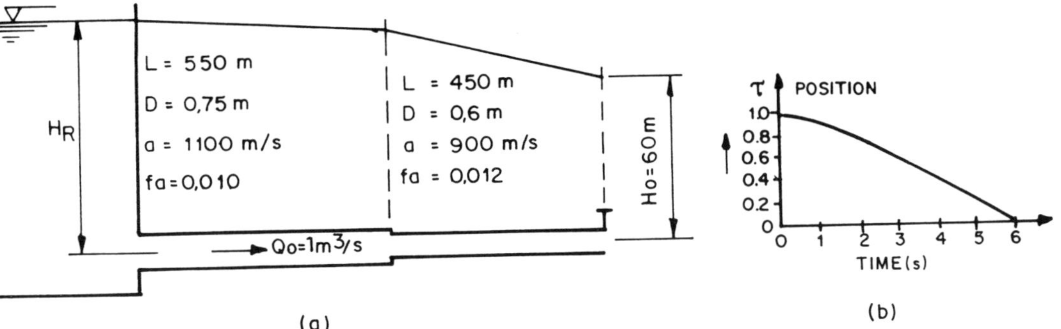

Figure 10. Example proposed by Chaudry (1978):
a) physical model, b) valve closing law

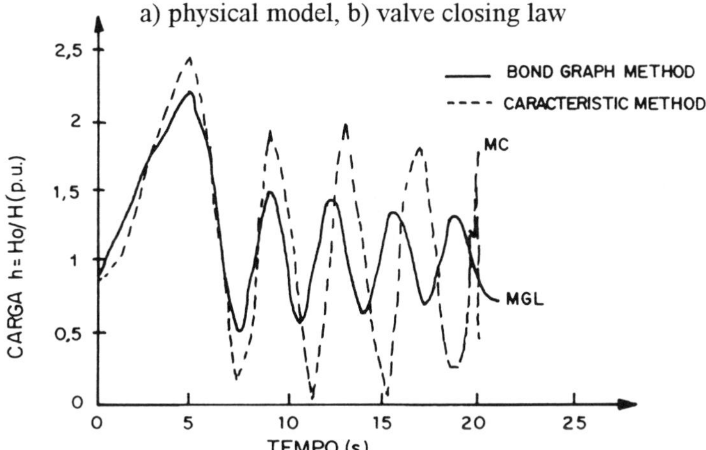

Figure 11. Results obtained in the simulation of the Chaudry's example:
a) characteristics method, b) bond graphs

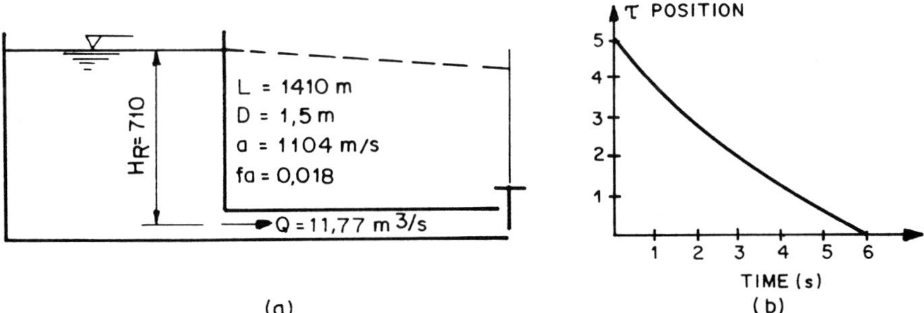

Figure 12. Example supplied by the Cubatão Hydro Plant: a) physical model,
b) valve closing law

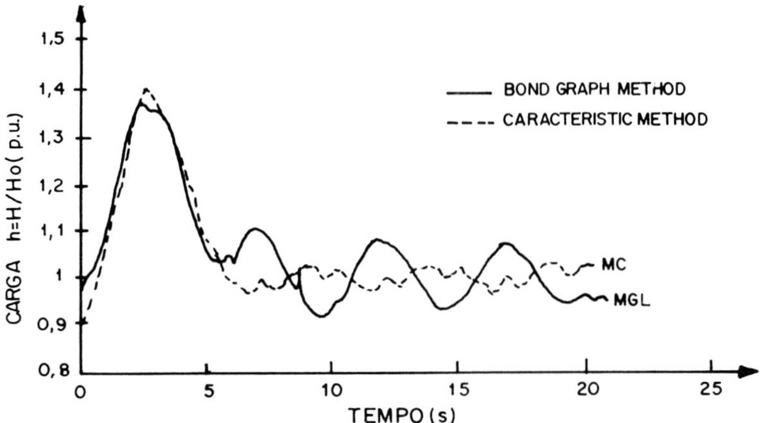

Figure 13. Results obtained in the simulation of the example of the Cubatão Hydro Plant: a) characteristics method, b) bond graph

By the results obtained in the cases simulated, the following table 2 can deployed.

TABLE 2. Results obtained in the simulation of the cases proposed

Case	Method	Maximum Pressure		Minimum Pressure	
		Value [ρV]	Instant of taking place (s)	Value in P.4	Instant of taking place (s)
Streeter	CM*	2.85	1.5	-0.04	2.5
	BG*	2.75	1.5	-0.4	2.4
Chaudry	CM	2.37	5.0	0	11.0
	BG	2.30	5.0	0.47	10.5
Cubatão plant	CM	1.46	2.5	0.95	10.0
	BG	1.45	2.6	0.85	10.0

*CM - Characteristics method
BG - Bond graph

Final Comments

With the results presented in hands, it is observed that besides not many deviations with respect to the maximum and minimum pressures occurred, in most cases, they took place in the same period of time. It is to be noted that the characteristics method presents smaller smoothening for the cases of Streeter (1970) and Chaudry (1978) while at Cubatão the bond graph method presented a smaller smoothening.

The fact that in the bond graph method the friction factor is taken as constant can be related to these differences.

Last but not least, in face of the results obtained in many other cases, made in Carvalho (1995), and not presented in this work, it can be concluded that the bond graph method, despite some constraints can be applied to the hydraulic transients study in simple cases of penstocks. Further studies are necessary so as to determine a discretization criterion and convergence conditions.

Bibliography

Carvalho, I.L (1995) Avaliação de Aplicabilidade do Método dos Gráficos de Ligações no Estudo de Escoamentos Transitórios em Condutos Forçados. *Dissertação de Mestrado*, EFEI, Itajubá/MG, Brazil.

Chaudrhy, M.H. (1970) Resonance in Pipe Systems, *Water Power*, no. 7/8, p. 241-245.

Streeter, V.L.; Wyle, E.B. (1978) *Fluid Transients*. McGraw-Hill, New York.

Tiago Filho, G.L. (1994) Aplicação do Método dos Gráficos de Ligações na Simulação de uma Válvula de Alívio, Anti-Golpe de Aríete, Auto-Operada. *Tese de Doutorado*, EPUSP, São Paulo, 277 p.

NUMERICAL FLOW ANALYSIS OF A KAPLAN TURBINE

DRAGICA JOŠT
Senior Research Associate
Turboinstitut, Ljubljana, Slovenia

ANDREJ LIPEJ
Senior Research Associate
Turboinstitut, Ljubljana, Slovenia

KAZIMIR OBERDANK
Senior Research Associate
Turboinstitut, Ljubljana, Slovenia

MATEJA JAMNIK
Research Associate
Turboinstitut, Ljubljana, Slovenia

BORIS VELENŠEK
Professor
Faculty of Mechanical Engineering, University of Ljubljana, Slovenia

1. Introduction

This paper details the numerical analysis of the flow in a Kaplan turbine. The purpose of the research is related to the refurbishment of two hydropower plants on the Drava River in Slovenia. This project was begun at the beginning of 1994 and is now nearly finished.

The flow in the spiral casing, tandem cascade, runner and draft tube was analysed by finite volume method using a TASCflow computer code, k - ε turbulence model was used. As it is still too time consuming to analyse all components of the turbine together, each part of the turbine was calculated separately. The computational domains overlap. Inlet conditions were obtained from a previous component analysis.

The results were experimentally validated with a high degree of agreement.

E. Cabrera et al. (eds.), Hydraulic Machinery and Cavitation, 1123–1132.

2. Spiral Casing

A semi-spiral casing is an integral part of a hydropower plant. Its task is to distribute the flow coming from the penstock into a swirling flow around the distributor. The spiral casing was analysed to predict the flow angle and velocity distribution at the inlet of the stay vanes.

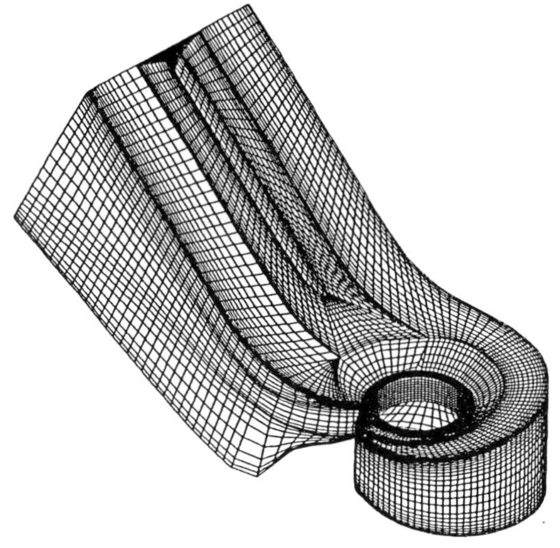

Figure 1. Computational grid of the spiral casing.

The spiral casing is a 215° curved channel with decreasing cross sectional area of trapezoidal shape. The computational domain (Fig. 1) includes the intake section with pier, the casing itself and the distributor housing. The grid is a composition of 9 subgrids with 19 interfaces. The grid consists of about 81000 nodes. There is an "O" type submesh around the pier.

Uniform flow distribution was assumed as the inlet condition. At the pinch symmetry boundary condition was applied. At the outlet portion of the domain pressure specified opening was prescribed. The CPU time required for the calculation was 13 hours on the supercomputer Convex 3860.

3. Stay Vanes and Wicket Gates

The flow in the tandem cascade was analysed in order to obtain the inlet conditions for runner analysis. The results were also used to estimate the value of the hydraulic torque acting on the guide vanes.

The tandem cascade consists of 12 stay vanes and 24 guide vanes. The domain of calculation was reduced to the region between two stay vanes and around two guide vanes (Fig. 2). Periodical boundary conditions were applied. At the inlet, the discharge and the flow angle were prescribed. The flow angle was obtained by means of a previous analysis of the flow in the spiral casing.

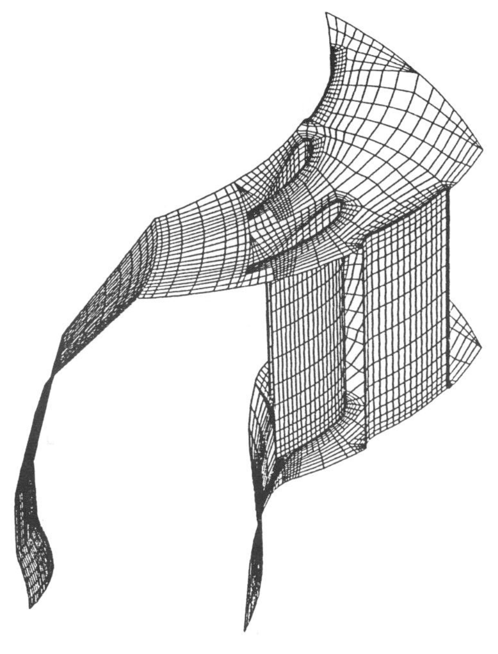

Figure 2.Tandem cascade, computational domain.

The grid consists of 24 subgrids. Around the stay and guide vanes, an "O" type grid was used. The whole domain consists of about 32,000 nodes. The analysis was done for several relative guide vane openings (A_0 =1.3, 1.6, 1.7, 1.95, 2.05, 2.2).

The CPU time for one calculation was between 7 and 9 hours on an IBM workstation. The calculation time depended on the number of iterations needed to obtain converged solution. The convergence was slower for small guide vane openings.

The inlet conditions for the runner analysis have to be axially symmetric, so the calculated velocity components were averaged in a circumferential direction.

4. Runner

In order to design a new and improved runner in a refurbishment project it is essential to have a thorough flow analysis of the old one.

First, a flow analysis through the old runner was done with various operating points treated.

Inlet velocity distributions for the flow calculation were obtained from the previous calculation of the flow between the guide vanes. Numerical grids for all calculations through the runner consist of 32000 nodes.

The results of the flow calculation are fully 3-D velocity vector field, streamlines, the pressure distribution on the suction and pressure side of the runner blades, pressure distribution on the crown and band, and turbulent kinetic energy and dissipation in the whole computational domain. From the obtained results the torque on the shaft, losses caused by friction forces and efficiency can be calculated.

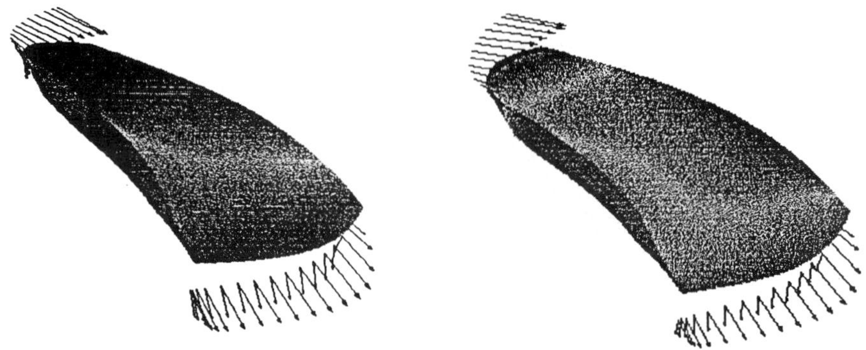

Figure 3. Relative and absolute velocity vectors for two operating points.

Special emphasis was placed on analysing the pressure distribution on the runner blades in order to predict the cavitation.

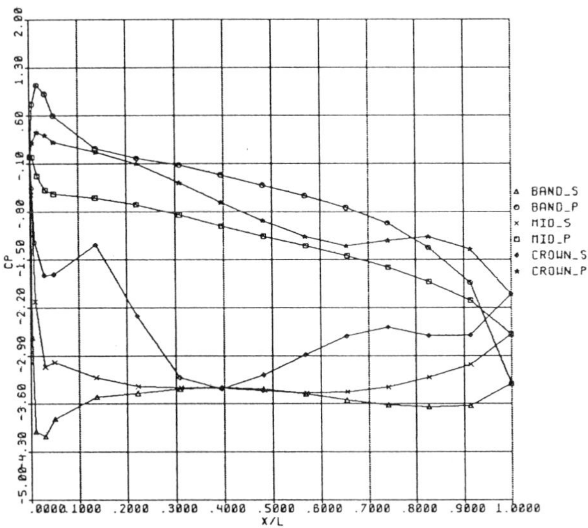

Figure 4. Pressure distribution.

After the complete analysis of the flow properties, and the energetic and cavitation characteristics of the old runner, the design and analysis of the new runner was begun. The inflow conditions for the flow analysis through the newly designed runners remained unchanged.

On the basis of all the known kinematics and geometric parameters, the new runner was designed. In order to obtain the resulting design, the new runner was modified and numerically analysed for different operating points. After each calculated geometry, the efficiency, torque on the shaft, velocity distribution at the inlet and outlet of the runner blades and the pressure

Figure 5. Geometry of designed runner.

distribution on the pressure and suction side was analysed. Special emphasis was placed on observing low pressure zones. On the basis of all the analysed features a new hydraulic design for the runner blades was proposed.

The whole procedure was repeated many times until the runner with desired energetic and cavitation characteristics was obtained .

5. Draft tube

A detailed numerical analysis of the flow inside the draft tubes of the two hydropower plants was carried out.

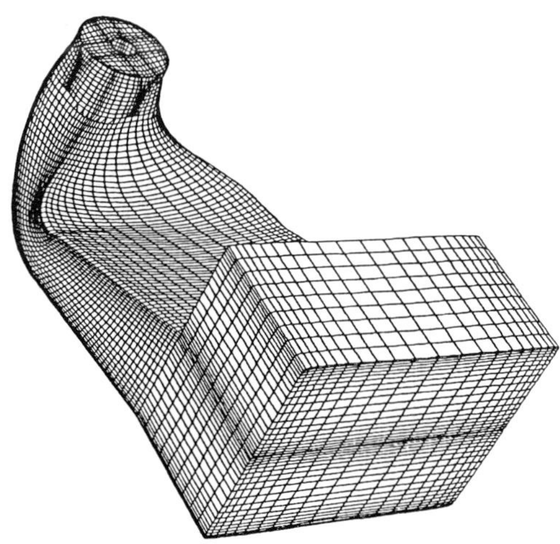

Figure 6. Draft tube grid with additional part at the outlet.

The numerical results of the flow in the draft tube is partly described in [1] and [2]. In [2] the results of the calculation for one operating point were presented. Subsequently, a detailed numerical analysis for more operating points was done. A special effort was made to research the influence of the piers on the draft tube characteristics. It was established that there is no general rule. Even for the same runner and for the same draft tube the influence of the piers is positive for some operating points and negative for others. So a numerical analysis should be done for many operating points.

The velocity components at the inlet were specified by interpolating calculated values from the previous runner analysis. Non-slip conditions were applied on solid walls. At the outlet, the so called inlet-outlet condition was applied. This condition enables the fluid to flow in or out of the domain. Constant pressure on one volume surface was prescribed. This boundary condition is not very robust, but it is the only one suitable to model the recirculation at the outlet. In cases where recirculation at the outlet occurred, the convergence was very slow. To improve the rate of convergence and still have the possibility of modelling the backflow at the outlet, an additional grid was attached at the outlet (Fig. 6). The height of this piece is the same as the height of the lower level of the water in the HPP. TASCflow is not able to model free surfaces, so symmetrical boundary condition was prescribed there. At the new outlet, the so called outlet condition was prescribed. In case of recirculation, this condition makes an artificial wall at the outlet. This boundary condition is more robust, and the convergence is more stable and faster.

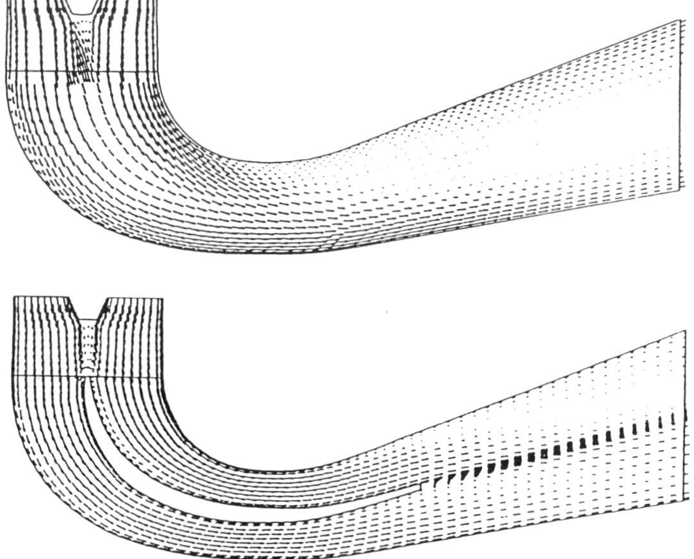

Figure 7. The influence of the horizontal pier on flow field.

The calculation was done on two grids, the first one consisting of about 31,000 nodes and the second one of about 67,000 nodes. The grid with the additional part at the outlet consists of about 40,000 nodes.

From the computed velocity and pressure distribution, the coefficient of pressure recovery was calculated. The coefficient of pressure recovery is defined as:

$$C_p = \frac{\int_{S_2} p v_t dS - \int_{S_1} p v_t dS}{\frac{\rho}{2} \int_{S_1} v^2 v_t dS} , \qquad (1)$$

where S_1 is the draft tube inlet section and S_2 is the draft tube outlet section. v_t is the transport velocity component, Q is the discharge, v is absolute velocity.

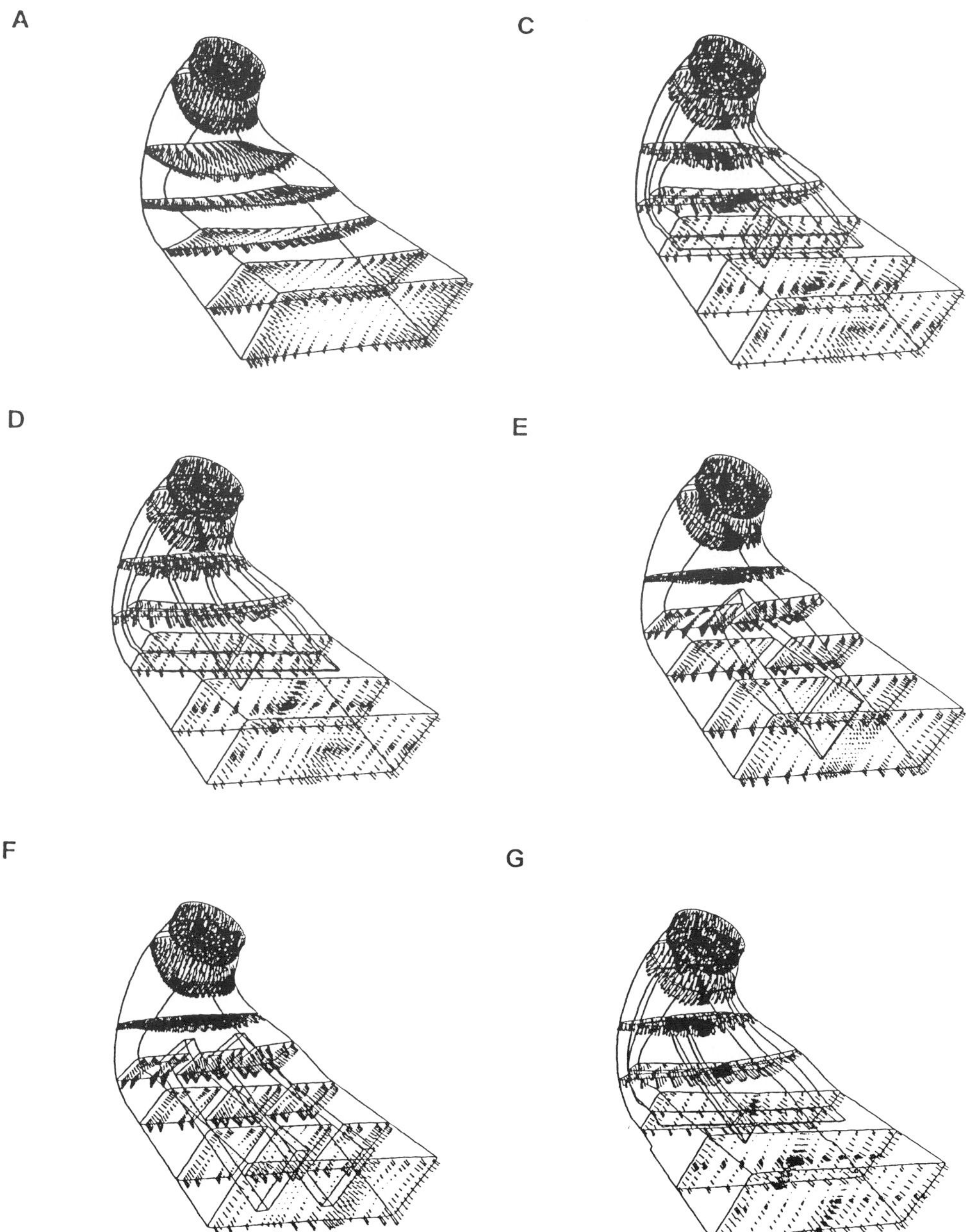

Figure 8. Velocity vector plot for the existing draft tube (A) and for the modified variants(C, D, E, F, G).

The existing geometry of the draft tube was analysed first. There was a large region of low velocity and recirculation. Most of the water flowed at the bottom of the draft tube. On the basis of the numerical results it was decided to put a horizontal pier inside the draft tube. The optimum length of the horizontal pier was determined by numerical analysis. Following this modification, the flow was more uniform and the region of recirculation was much smaller (Fig.7). The calculated coefficient of pressure recovery was much higher than for the draft tube with no piers. The model tests showed a higher efficiency for the turbine with the draft tube having a horizontal pier, than for the turbine with the draft tube with no piers, especially at full load.

It is difficult to install the horizontal pier inside the existing draft tube. It is easier if a vertical pier is also in the draft tube. Several variants of draft tubes were numerically analysed in order to find the optimal solution (Fig. 8):

A. existing draft tube with no piers,
B. draft tube with a horizontal pier only,
C. draft tube with a horizontal pier and short vertical pier,
D. draft tube with a horizontal pier and long vertical pier,
E. draft tube with a vertical pier,
F. draft tube with two vertical piers,
G. draft tube with vertical piers below the horizontal pier.

The results (calculated coefficient of pressure recovery C_p) for the best efficiency point are presented in Table 1. It was determined that the influence of the vertical piers is negative. The draft tubes E and F have a lower C_p than the draft tube with no piers. All the variants of the draft tube with the horizontal pier (B,C,D,G) have a higher C_p than the draft tube with no piers. The coefficient of pressure recovery is the highest for the draft tubes B and C. For the draft tubes D and G the C_p is a bit lower as for the draft tubes B and C, but it is still much higher than for the draft tube with no piers. For structural reasons it was decided that the prototype would have a draft tube with vertical piers below the horizontal pier (G).

Model tests for all variants of the draft tube except for the draft tube with two vertical piers were done. The agreement was very good except for the draft tube D. In this case the measured efficiency of the turbine was a bit lower that was expected on the basis of the numerical results.

The global solution is nearly the same for all grids. The values of the coefficient of pressure recovery obtained on the refined grids are a bit higher than on the coarse grids. The agreement between numerical and experimental results is the best for grids with the additional part at the outlet. These results are presented in Table 1.

$Cp_{relative} = \frac{C_p - C_{p0}}{C_{p0}} * 100\ \%$, C_{p0} is the coefficient of pressure recovery for the existing draft tube with no piers	$Cp_{relative}$
A. existing draft tube with no piers	0.0
B. draft tube with a horizontal pier only	6.9
C. draft tube with a horizontal pier and short vertical pier	6.9
D. draft tube with a horizontal pier and long vertical pier	6.0
E. draft tube with a vertical pier	-2.0
F. draft tube with two vertical piers	-9.9
G. draft tube with vertical piers below a horizontal pier	6.3

Table 1. The coefficient of pressure recovery for the best efficiency point

6. Flow kinematics measurements

Measurements were carried out on a model Kaplan turbine of HPP on Drava River at Turboinstitute. Velocities have been measured with a five hole cylindrical probe. The measurement section at the inlet of the draft tube with the horizontal and shorter vertical pier is shown on Fig 9. The traverse M_1 with 12 measuring points from the wall to the centre of the inlet section was chosen at a distance of 100mm under the runner hub. Measurements were performed in the one operating points at full load. As a result distributions of mean axial, tangential and radial velocity along the traverse have been obtained (Fig. 10). In the centre of the flow field a low velocity region influenced by the runner hub and the horizontal pier. is observed. In this region the tangential component of velocity has opposite direction of the runner rotation. Toward the outer wall tangential velocity is increasing, while the axial velocity is almost constant. Near the wall velocity measurements are less accurate because of the wall influence. The comparison between numerical and experimental results of velocity components at the outlet of the runner has been done in order to estimate the accuracy of boundary condition for draft tube analysis.

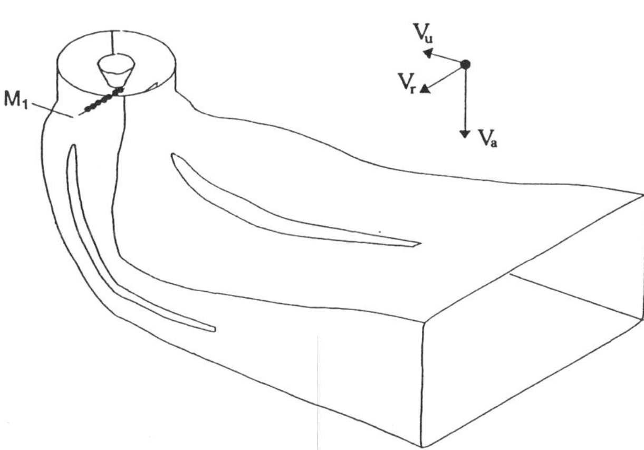

Figure 9. Schematic view of the draft tube with measuring traverse M_1 at the inlet of draft tube.

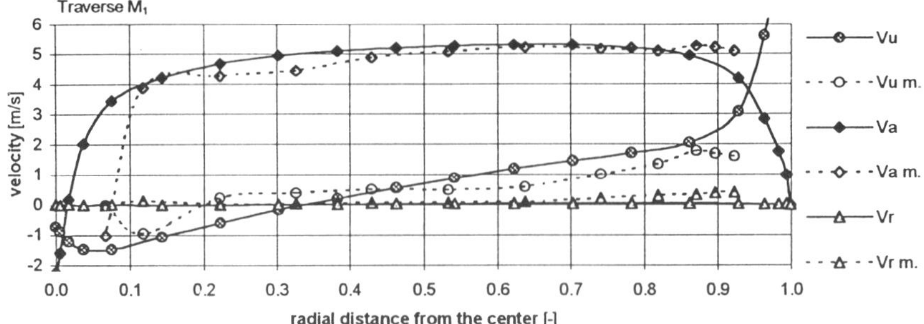

Figure 10. Comparison between measured (m.) and calculated velocity components along the traverse M_1.

7. Conclusion

A numerical analysis was found to be very useful in the refurbishment project on the Drava River. It was much easier and faster to modify the shape of the runner blades on a computer and analyse the flow than to make models and do the model tests. Additionally the geometry of the draft tube of the HPP on the Drava River was modified on the basis of the numerical results.

It can be concluded that the overall picture of velocity and pressure distribution is correct even on coarse grids, but to establish accurate enough characteristics (the torque on the shaft, the pressure recovery factor, the flow energy losses,...), the grids have to be refined. The accuracy of the numerical prediction would be increased by a simultaneous calculation of the flow in the rotating and nonrotating parts of a turbine.

Acknowledgement

The CFD analysis was realised with the financial assistance of the Ministry of Science and Technology of the Republic of Slovenia (project J2-7497).

8. References

1. Kercan, V., Djelic, V., Močnik, Z., Šturm, M. (1993) "Hydro power plants on the Drava River in Slovenia - Two approaches for refurbishment", International Water Power & Dam Construction;

2. Jošt, D., Velenšek, B. (1994) "Refurbishment of the Drava River HPP Chain, Numerical Prediction of the Draft Tube Performance", Proceedings of Modelling, Testing & Monitoring for Hydro Powerplants, Budapest, 1994;

3. TASCflow USERS MANUAL, Version 2.2, 1993, Waterloo, Ontario, Canada.

18–PATHS ACOUSTIC FLOWMETER AND TRANSDUCER PROTRUSION

J.–M. LÉVESQUE, J. NÉRON and C. M. TRAN, Engineers
Essais et Études techniques, Hydro–Québec
5655 de Marseille, Montréal, Canada, H1N 1J4
Telephone 514 251–7043 Telefax 514 251–7117

1. Introduction

Hydro–Québec's network includes several power plants of more than 1000 MW. Over the past 10 years, work was done in order to improve the accuracy of the turbine efficiency determination, which in turn, leads to a better management of the hydraulic resources and to a better income.

We practice and possess the following four methods: acoustic, pressure–time, current meter and thermodynamic. Often, two methods are applied on the same occasion to increase the accuracy and the confidence level of the measurement. The neutral zone of the efficiency curve, for which no bonus or penalty may be applied, is accordingly adjusted but is currently between ± 0.5 and ± 1 %.

In June 1994, our testing department obtained a grant from Hydro–Québec to apply simultaneously the four above methods at a power plant. The goals were to learn about the uncertainties of these methods and manners to improve them by comparing their feasibility and applicability. Laforge–1 power plant was chosen and the tests were done in December of 1994.

2. Description of the Laforge–1 power plant

Figure 1 shows a general view of the installation. The power plant has 6 units of 145 MW each with a discharge of 280 m^3/s under a 57 meter head. The contractual tests already included the acoustic and pressure–time methods. The grant permitted to elaborate more the acoustic and pressure–time methods and to add the current meter and thermodynamic ones. The present article will present only the results obtained from the application of the acoustic method with 36 transducers

E. Cabrera et al. (eds.), Hydraulic Machinery and Cavitation, 1133–1141.

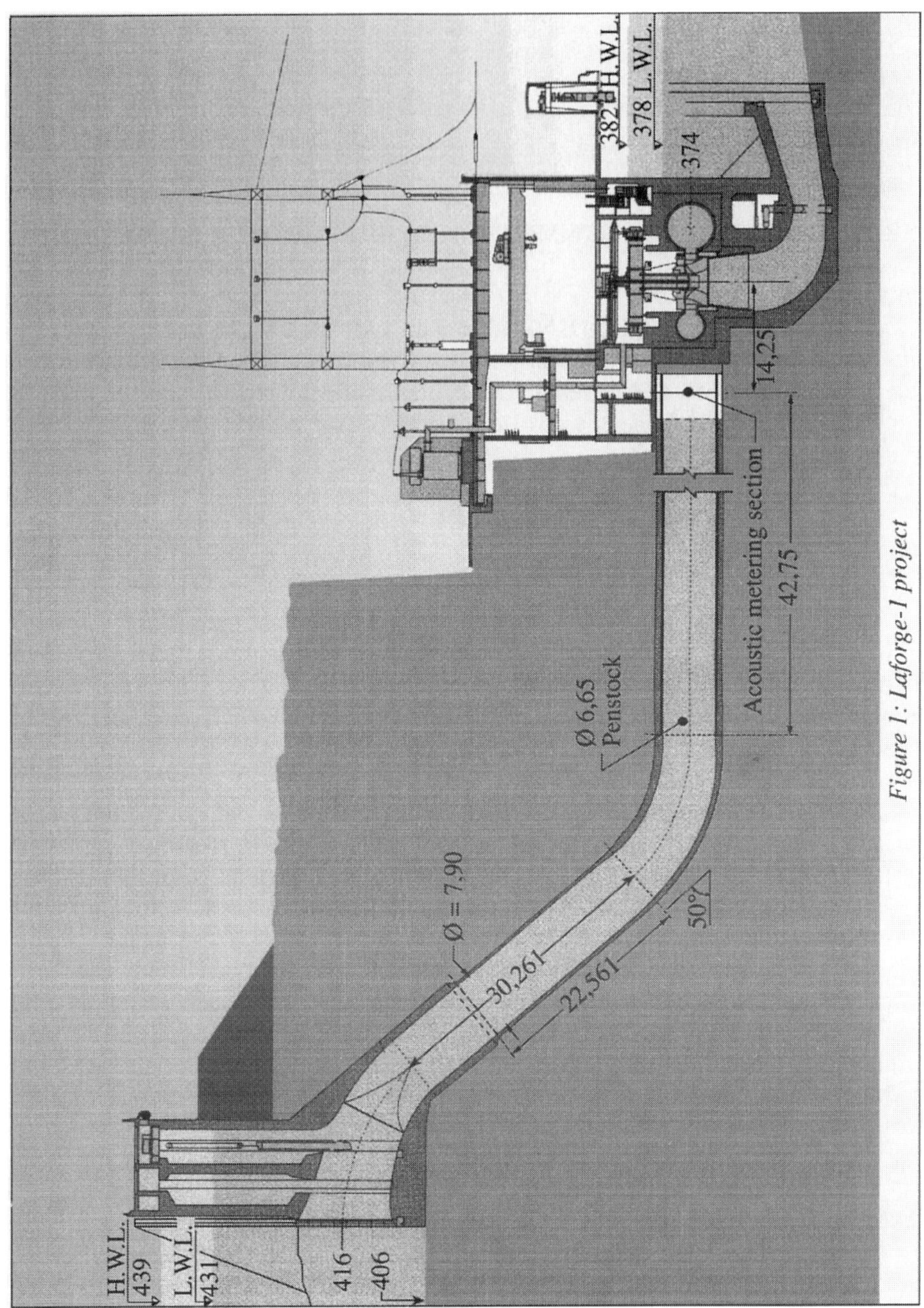

Figure 1: Laforge-1 project

3. Description of the acoustic method

The metering section was located just 1D upstream of the scroll case entrance and less than 7D downstream of a 50 degrees smooth shaped elbow. The 6.65 meter diameter penstock is steel lined but the upper conduit and the elbow are made of concrete.

The standard acoustic method consists of 8 cross–paths equipped of 16 transducers with fixed protrusion, over 4 levels. Five more levels were added for another 10 cross–paths of 20 transducers with adjustable protrusion. The protrusion of the transducers was adjusted on one level only for the tests of December 1994. More tests to identify the effect of the protrusion were performed in July 1995, during which the protrusion was adjusted on all the transducers on 4 of the 5 added levels.

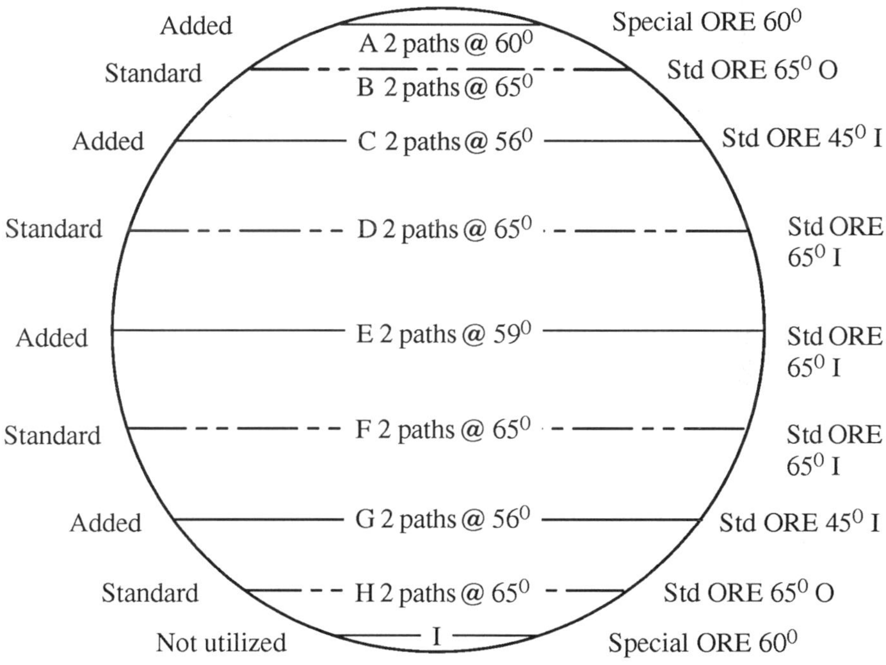

Figure 2: Arrangement of transducers and paths

Figure 2 describes the location of the standard 4 levels and the 5 added levels. One may note the different path angles of the flowmeters. Two flowmeters were used, one for the 8 cross–path flowmeter consisting of levels B, D, F and H, and the other one for the 10 cross–path flowmeter consisting of levels A, C, E and G. To avoid the use of a third flowmeter, level I velocities were not measured but assumed to be equal to those of level A.

Standard ORE 65^0 transducers were used for levels B, D, F and H with standard feed-through fittings, the inside part being terminated by a 62 mm diameter by 10 mm thick mount. The resulting protrusions, between the transducer centers and the penstock wall, were 25 mm for levels B and H and 14 mm for levels D and F.

The transducers used for levels A, C, E and G were mounted using pipe fittings in a manner to allow adjustment of the protrusion by steps of 12.7 mm up to a maximum of 55 mm. The minimum protrusions were 8, 5, 5 and 3 mm for levels A, C, G and E respectively.

Although necessary measurements were done, during the construction stage of the power plant, for the installation of the standard 8 cross–paths flowmeter, other measurements were also necessary for the installation of the new 10 cross–path flowmeter for the December 1994 test program, and the occasion was taken to repeat the previous measurements to ensure maximum quality control. All the measurements were checked and compared to make sure that the angle variations were compatible with the path length changes. For the July test, all the data was put on a floppy disk to allow rapid change without error. That turned out to be valuable with the protrusion changes on 16 transducers, following the unforeseen flowmeter operating changes.

The tests were conducted at two different discharges: a small discharge of about 48 m^3/s which is the speed–no–load discharge and a large discharge of about 240 m^3/s which is near the best turbine efficiency discharge. The small and large discharges give Reynold numbers of $7 \cdot 10^6$ and $3.2 \cdot 10^7$ respectively in the penstock.

When the testing started, both flowmeters were functioning together but we realized that interference occurred between the two, since measured discharges were obviously too small or large with many rejects. The flowmeters were then operated separately. Consequently, it took more time for each test run and the results were less accurate because the discharges were not simultaneously measured by the two flowmeters.

The testing included 12 hours of measuring time, the waiting time to stabilize the discharges, at least one hour each time for a complete change of protrusion of the transducers, and the time for 70 change–overs from one flowmeter to the other. For each run of 5 minutes, the discharge and the different velocities of each path, were measured at rate of one measurement per 2 seconds. For the two studied small and large discharges, measurements were done for protrusions of about 4, 17, 30, 42 and 55 mm, 7 runs with each flowmeter, for a grand total of nearly 140 test runs.

4. Results

The first results of prime interest are the velocity profiles presented in figure 3, combining the 8 and 10 cross–path measurements for small and large discharges. It can be observed that the flow velocity is slightly greater at the bottom of the penstock than at the top and this can be attributed to the upstream elbow.

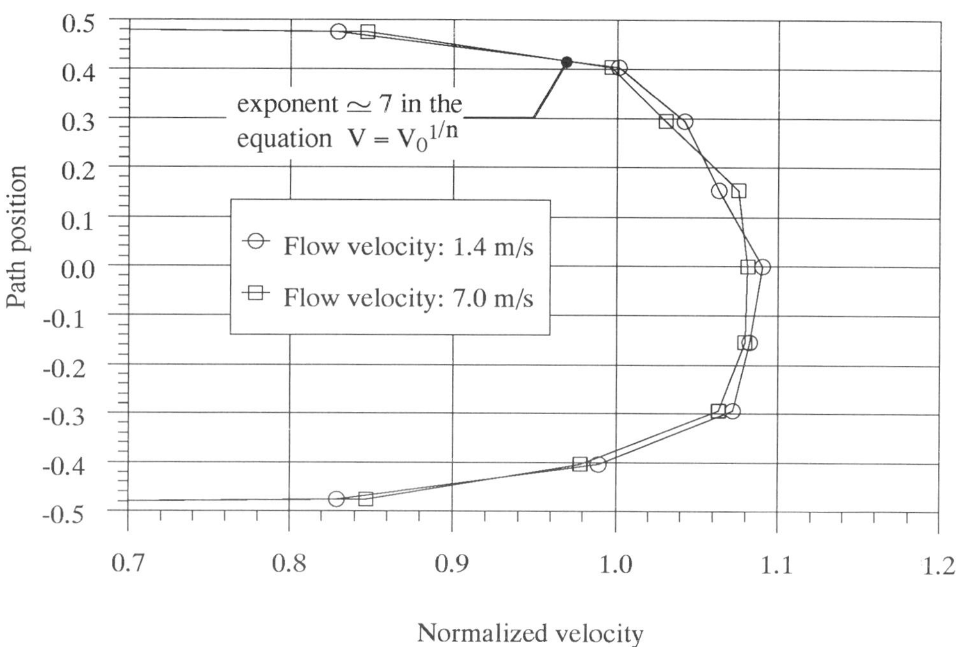

Figure 3: Normalized velocity profile

The main results, concerning the comparisons between the 8 and 10 cross–path flowmeters and protrusion effect, are presented in figure 5. However, these measured results must be completed by the influence of the protrusion on the path positions. It is the object of figure 4 to give the appropriate explanations.

When changing the protrusion P from zero to P_1 and P_2, the transducer is moved along the radial axis, therefore the transducer head also changes from path elevation C_0 to new elevation C_1 or C_2. This path shift created mean velocity changes that falsify the results. Two choices exist: to correct the weighing coefficient or to correct the mean velocity.

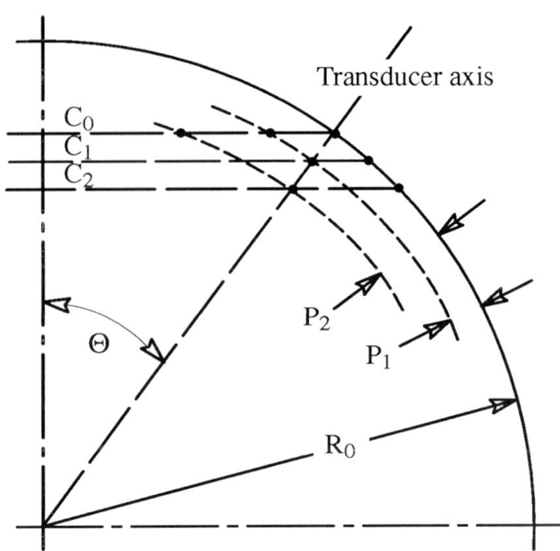

Figure 4: Path shift vs protrusion

When using a system composed of 36 cross–paths, each two cross–paths are positioned at an 18^0 angle compared to the adjacent ones. The weighting coefficient of each path can be calculated using orthogonal polynomials Chebyshev Polynomials of Second Kind according to the following equation given at page 889 of Handbook of Mathematical Functions of US National Bureau of Standards:

$$W = \pi/(n+1) \sin^2 \pi\, i/(n+1)$$

This equation assumes that all paths are at their exact locations. When changing the transducer protrusions by radial displacement, the radii change, modifying the equivalent spacing between the transducers and thus falsifying the weighting coefficient.

For small changes, it is possible to correct the weighting coefficient for radial protrusion. These corrections were done and validated by small steps numerical integrations on a model consisting of a velocity profile having an exponential form with an exponent of 0.15 (to simulate the inverse of $n = 7 \simeq 0.15$). It is also possible to make the corrections on the velocities, using the velocity profile of figure 3 and making adjustment on the mean velocities for small changes in path positions.

Both methods were applied and gave the same results as can be seen in figure 5 which shows the difference, of the ten cross–path discharge minus the eight cross–path discharge versus the protrusion. The solid lines are the corrected and final results for small and large discharges identified as flow velocities of 1.4 and 7.0 meters per second.

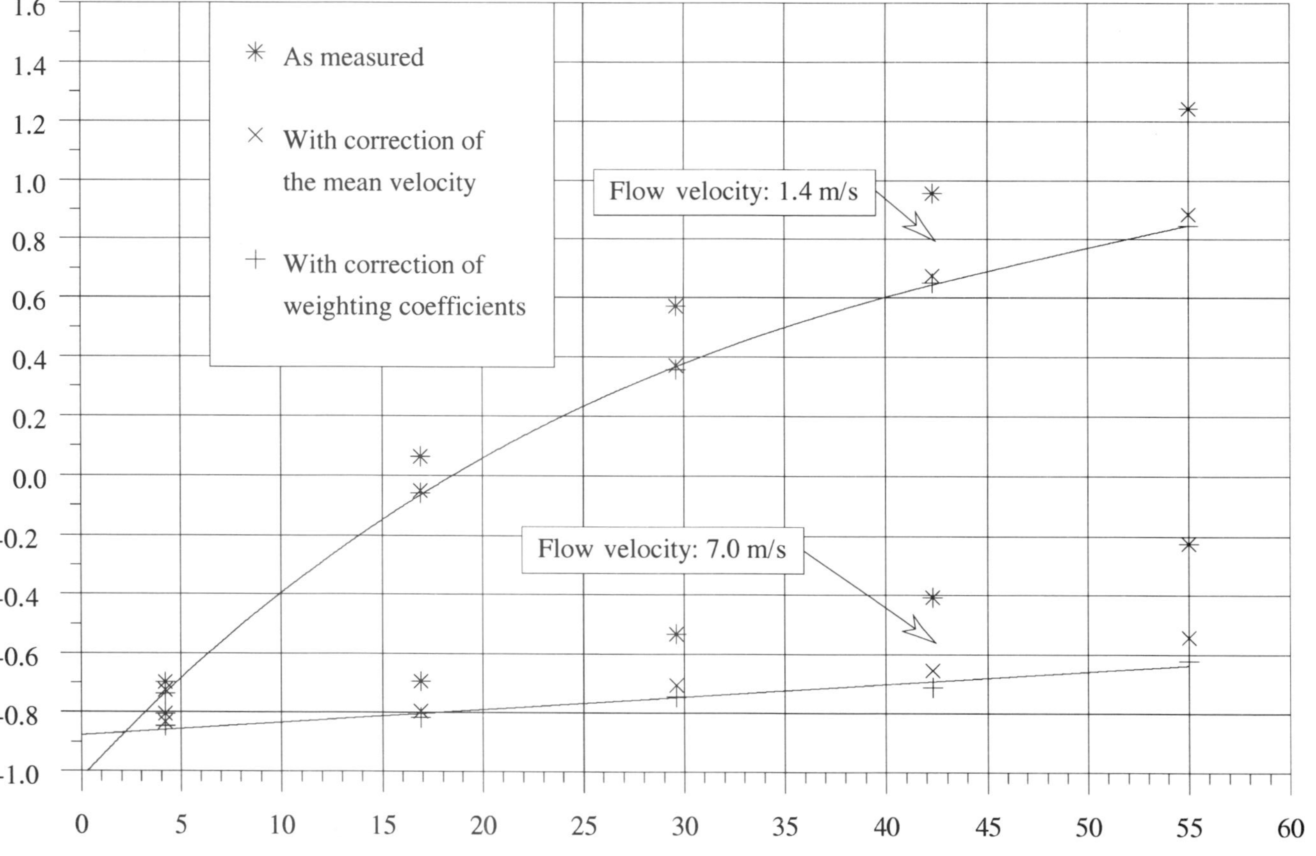

Figure 5: Ten cross-path discharge minus eight cross-path discharge in function of protrusion

From the corrected results, it is observed:

- For the same protrusion of 17 mm on all the transducers, the discharge of the 10 cross–path flowmeter minus the discharge of the 8 cross–path flowmeter amounts to –0.1 % for small discharge and –0.8 % for large discharge. We attribute these difference variations mainly to the velocity profile.
- The protrusion of 17 mm always yields an over–estimate of discharge: small ones by 0.9 % and large ones by only 0.1 %.
- Compared to the real discharge, the standard 8 cross–path flowmeter always measured higher discharge, 1 % at small flow velocity and 0.5 % at large flow velocity.
- The protrusion effect is very important for small velocity and minute for large velocity. The difference attributed to singularities of the velocity profile, between the 10 and the 8 cross–paths, seems to compensate the protrusion effect.
- If a unique –0.5 % correction is applied to the measured standard 8 cross–path discharges, the resulting discharges would be within 0.5 % of the true values. This correction of –0.5 % is the mean of the protrusion effect at small and large velocity.
- The mean correction factor of –0.5 % has an uncertainty of ±0.4 % and this important uncertainty is not due to tests inaccuracies but to the fact that the phenomenon is more complex than expected.

From the article entitled "Correction method to improve the accuracy of multi–path acoustic flowmeters" by Sugishita et al, IAHR, São Paulo 92, the protrusion correction was evaluated to –0.5 %, matching quite well with the present results.

Using the actual measurements by the current meters, the velocity profiles were calculated at locations corresponding to paths A, B, ..., and the discharge was computed for equivalent 10 and 8 cross–path flowmeters. These studies confirm the previous results with the acoustic method, particularly a computed discharge difference of –1 % between the 10 cross–path discharge and the 8 cross–path discharge.

Using the current meter measured velocity profile, the equivalent of a protrusion effect of 17 mm yields an over–estimation of the discharge by 0.35 %. However, it was not ascertained why the protrusion effect could change from the small to the large discharges since we did not observe any significant change in velocity profile. This last point is baffling and we were not able to understand the phenomenon completely in order to extrapolate the present results to other cases.

5. Conclusions

The PTC and IEC Test Codes recommend to locate the metering section more than 10D from an elbow followed by a 3D or longer straight section before the scroll case entrance. These distances were 7D and 1D respectively for our test at the Laforge–1 power plant. Nevertheless, we are convinced that the measured 8 cross–path discharge fell within 1 % of the true discharge. We thus believe that the uncertainties from 1 to 2 % as mentioned in PTC and IEC publications are pessimistic or the specified conditions of application are conservative for a 6.65 meter diameter penstock.

The extrapolation of these results for a 2 meter penstock would yield a discharge correction over 1 % for protrusion, and this precarious situation is certainly undesirable for all parties, including the equipment suppliers who must give efficiency guaranties, the owners who want to know the true machine efficiency and also the flowmeter suppliers who must sustain their instruments against criticism and skepticism and demonstrate their accuracy.

Since the protrusion effect was measured and appeared to vary with the discharge, our main recommendation is obviously to enforce the placing of the transducer face flush with the wall, so to avoid doubts and corrections.

STEP-UP IN REHABILITATION: NOT A MYTH, A SCIENCE

B. MAHÉ
Gec Alsthom
Tracy, Qc, Canada

V. DE HENAU
Gec Alsthom
Tracy, Qc, Canada

M. SABOURIN
Gec Alsthom
Tracy, Qc, Canada

Abstract:

New turbines, model tested, manufactured and site tested according to state of the art techniques, show consistently an efficiency step-up at least equal to IEC 995 prediction. Although some manufacturers have cast doubts about the capability to reproduce the step-up in rehabilitation, we provide evidence that it has been achieved on well conducted projects. We successively analyze major factors, on the model and on the prototype, that influence the step-up. We finally review some examples to illustrate the desired attitudes that will maximize all the benefits of the rehabilitation.

Résumé:

Pour les turbines neuves, lorsque essais modèle, fabrication et essais prototype sont bien menés, on démontre toujours une majoration au moins égale au CEI 995. Bien que certains manufacturiers doutent qu'on puisse faire aussi bien en réhabilitation, nous apportons la preuve que cette majoration se vérifie effectivement sur des projets bien gérés. Nous analysons successivement les principaux facteurs qui influencent la majoration, sur le modèle et le prototype. Nous faisons finalement à travers quelques exemples, une revue des comportements qui permettent de tirer un maximum de bénéfices de la réhabilitation.

Resumen:

Las turbinas nuevas, cuyos ensayos del modelo y del prototipo tanto como la fabricación han seguido las últimas técnicas , muestran un incremento de eficiencia al menos igual al CEI 995. A pesar de que ciertos fabricantes sospechan que no se puede conseguir lo mismo en rehabilitación, presentamos pruebas de que el incremento se obtiene effectivamente para proyectos bien manejados. Analizamos los factores que afectan este incremento, sean del modelo o del prototipo. Finalmente, con ejemplos, establecimos las conductas propicias a la rentabilidad máxima de la rehabilitación.

E. Cabrera et al. (eds.), Hydraulic Machinery and Cavitation, 1142–1151.

1. Introduction

The efficiency step-up for hydraulic turbines results from a number of phenomena, whose fundamental variable is the Reynolds number. On projects that combined quality model tests, sophisticated site tests and first grade manufacturing, we identified a step-up in excess of the value predicted by the IEC 995 formula, as shown by Couston [1]. The same conclusions were anticipated from former high precision site and model testing for Hydro-Québec [2,3]. With the availability of five axis machining and robotized techniques, consistent step-up is now expected from any new realization. The recent site tests at Laforge 1 and La Grande 1 confirm this expectation, according to the preliminary results that will be presented at Table 2, and by Lévesque at IGHEM [4].

Regarding the rehabilitations, some manufacturers and utilities still had doubts on the capability to achieve the step-up. Demers [5] made a pessimistic assessment of the situation, following a number of failures to achieve expected ratings and performances, after runner replacements somewhat simplistic.

Fortunately methods are improving: GEC Alsthom, in collaboration with B.C. Hydro, took all the necessary precautions to obtain a high step-up for the Kootenay rehabilitation and succeeded [6]. Based on efficiency only, the 575 MW power plant increases its annual generation by 131 Gwh for less than 18 Cdn$ / MWh.

The fact that certain manufacturers failed to achieve their guarantees does not mean that the step-up should be ignored. The solutions that tend to limit the rehabilitation benefits to the hydraulic profile redesign are far too limiting. Alleging that the step-up was uncertain, some laboratories and manufacturers proposed to measure this improvement through the sole comparison of the new model with a supposedly homologous representation of the existing turbine. This « model to model » comparison is most likely misleading and detrimental:

- Misleading since it demonstrates only a fraction of the potential for improvement, as the original manufacture is clearly far below the present standards and much too well represented by a perfectly periodic model, numerically idealized.
- Detrimental if it eliminates the engineers motivation to improve the unit from tip to toe: many aspects like surface refurbishment, clearance optimization, efficient air admission, being totally missed out.
- Detrimental also if manufacturers obligations are limited to minimum homology standards that do not correspond any more to the actual value of energy losses.

For utilities that intend to take full advantage of a rehabilitation, (full step-up), it is important to identify the points that will influence the efficiency prediction, from model testing to prototype refurbishment. This will be the purpose of section 2.
Then, there exists joint decisions from the different parties that will facilitate the search for the best economical solution. As perfect guide lines cannot be established at the bidding stage, many actions have to be confirmed during the engineering and testing phase. At section 3., we give examples of such actions resulting in substantial benefits for the project.

2. Factors that influence the efficiency prediction:

Too often, questions about precision, homology, roughness are raised only after everybody was disappointed by poor uprating results. Questions that come too late. The unit was shut-down for a long time ... Ample time for grinding, sand-blasting and painting that was wasted on gambling on the unrealistic ability of water to flow over sharp edges and barnacles and tolerate surface deviations. The following is a check list and reasons for a sound questioning on most aspects of the rehabilitation performance.

2.1. MODEL MANUFACTURING AND TESTING:

2.1.1. Model Homology and Existing Drawings:

Existing non modified hydraulic passages should be measured on at least one turbine: We recently discovered systematic thickness variations of half an inch (20%) on the Beauharnois stay-vanes (Hydro-Québec). Hydro-Ontario measured the same defects at Sir Adam Beck 2 combined with ¾ inch distortions. Kootenay Canal guide vanes were giving less opening than expected. It takes only two days and a few per cent of the model test cost to check the casing, the stay-vanes and the draft-tube of a large unit.

Modeling of these existing passages: We will call « IEC Code » the Committee Draft edition (December 1994) of the Publication 193 R. This publication provides tolerances. Their accumulation is certainly not desirable and some of them could be more stringent. Some details are left to the manufacturers professionalism. For an example the profile quality at the stay vane inlet edge may have considerable influence on the performance. For the Kootenay Canal rehabilitation, the stay-vane streamlining produced a two per cent (2%) increase on the turbine mean weighted efficiency. On the very sensitive inlet edge, machining scallops of 4.5 mm (within the IEC tolerance for the prototype) would not produce the full benefit of a smooth grinding. These details must be reproduced on models and numerical processes should be selected to manufacture the model stay vanes and guide vanes to the exact profile of the prototype.

Model new components must define precisely the theoretical geometry: Model periodicity, precision and smoothness are normally achieved by numerical machining, careful polishing and accurate and rigid assembly. Excellent results were achieved with carbon-epoxy molding of blades when supported by exceptional craftsmanship. However, this technique was limited to Francis runners and reduced test heads.

2.1.2. Model Clearances:

Upper runner seal for Francis turbines: In the past, most models [2,3] were tested without leakage from this upper seal and we consider that the step-up formula incorporates this non-similarity. With the IEC Code now stating that both prototype seal clearances shall not exceed the scaled model ones, a better step-up can be expected. Itaipu [1] tested with homologous seals has effectively demonstrated a high step-up.

Homologous runner chamber: At least, similar clearances between rotating and fixed surfaces should be respected at similar locations, rather than a strict homology, since model manufacture generally requires a thicker crown and band.

Blade tip clearance for Kaplan, Propellers and Diagonal turbines as well as lower seal clearance for Francis: The new gap clearance must first be established for the uprated prototype ahead of the model fabrication. Existing clearance may be optimized or increased according to the new runner design. It will take into consideration the local runner expansion at runaway, the concentricity and circularity of all components involved between the turbine bearing and the seal, the bearing clearance and its thermal behavior and finally the shaft deflection. For Francis turbines, sophisticated labyrinths might be difficult to reproduce on the model. Then, a second order correction can be applied using either experimental results such as those compiled by Viano [7], or a Navier-Stokes simulation such as the one shown on Figure 1.

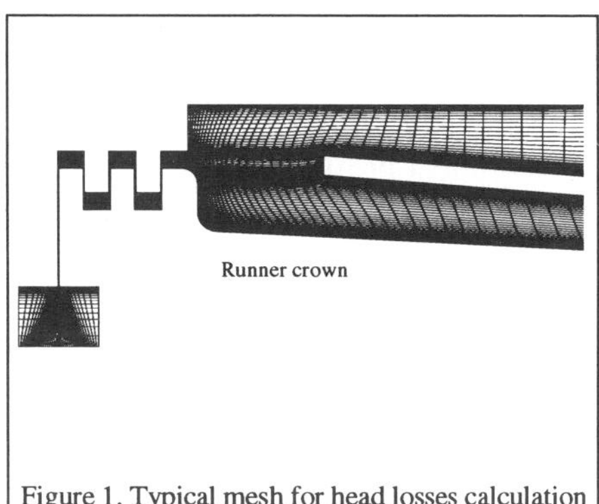

Figure 1. Typical mesh for head losses calculation in upper labyrinth.

Cavitation influence on seal losses: In some operating and testing conditions, the above problem could be complicated by a diphasic flow. Under certain testing conditions, the model efficiency will vary with the Thoma coefficient, even for sigma values above plant sigma. Gap cavitations of Kaplan turbines at the blade to hub and blade to discharge ring interfaces can be observed and correlated with this phenomenon. We must consequently admit that the desired homology between model and prototype gaps might not have existed if they were tested under different cavitating conditions. We give some examples on Figure 2, at paragraph *2.1.6. Test conditions..* In the future, based on current investigations, model testing conditions may change. Then, a correction between former testing results on which step-up formula were based, and the new ones, will have to be established. In any case, the cavitation similitude (test head, Thoma coefficient, nuclei and gas content) will not modify radically the step-up, it will just give a better prediction of prototype performance.

Wicket gate clearances: This influence has been extensively discussed by Brekke [8] and introduced in the IEC Code. The geometrical homology that is then essential for high head turbines can only improve the model to prototype comparison for others.

2.1.3. Model Rigidity:

Ideally, each model component should have a rigidity homologous to that of the prototype according to the test head over prototype head ratio. For most components, the distortions are smaller than the manufacturing tolerances and this principle may be ignored. For runner blades, different corrections are required according to the test head:

- *Model tested at prototype head:* Both the model and the prototype will have the same modulus of elasticity and the blade attachments should have the same rigidity.

The blade distortion in operation is added negatively to the theoretical design for both the model and prototype manufacture.

- *Model considered infinitely rigid:* This is the case of bronze models with test heads at least five times smaller than the prototype head. The prototype will be corrected negatively by the computed distortion in normal operation (highest weighting).
- *Model distortions non-negligible but non-homologous:* Based on a common theoretical profile, the model and prototype manufactures will require two different distortions calculated for the operating condition with highest weighting.

2.1.4. Model Roughness:

As explained in 2.2. *PROTOTYPE MANUFACTURING:*, it is profitable to upgrade the damaged prototype surfaces each time that roughness influences the losses. Concrete is generally in acceptable condition. Then, it is logical to realize model surfaces hydraulically smooth at the selected Reynolds number, as requested by the IEC Code.

2.1.5. Test Stand Measurement Precision

All the results that we will present were obtained on very reliable test stands including:
- A generator stator mounted in balance for torque measurement.
- Flowmeters whose calibration coefficients remain stable year after year.
- Double measurement on head and flow.
- A fully visible loop excluding valve leakage.

In Table 1., we compared the results obtained on two different test stand: TP3 or T4 at GEC ALSTHOM, IMHEF or CREMHYG in the case of Tanggari. The Reynolds number and Thoma coefficient were similar. All of them demonstrate the reliability of results obtained at these test stands as a base for performance prediction.

TABLE 1. Comparisons of model efficiencies as measured at GEC ALSTHOM and IMHEF: Differences measured at 100%, 115% & 85 % of the best load for design n_{11}

Project	Best	115%	85%	Mean
Berke	.14%	.05%	-.04%	.05%
Kootenay Canal	-.10%	.02%	.00%	-.03%
Koyna IV	.09%	.00%	.00%	.03%
Sir Adam Beck 2	-.13%	.11%	.09%	.02%
Alto Lindoso	-.09%	-.05%	-.05%	-.06%
La Grande 1	-.04%	.02%	-.40%	-.14%
Tanggari	.03%	.15%	.15%	.11%
Laforge 1	.16%	.00%	.30%	.15%
Mojave Siphon	.00%	-.03%	.07%	.03%

2.1.6. Model Test Conditions:

Test head, dissolved air and nuclei content all have variable influences on model efficiency at plant and/or high sigma, at least for maximum flow. Test head and nuclei content influence were investigated by Gindroz et al. [9] on Francis turbines.
Curves (A) and (B) of Figure 2 show the same influence for a Kaplan turbine tested at different heads and air contents (in two laboratories, corrected Reynolds). Both turbines are absolutely free of inlet or profile cavitation in these conditions.

As explained at *2.1.2. Seal Clearances* , flow perturbations sensitive to the local pressure do exist on both the model and the prototype. The similitude of the efficiency variations, however, is not granted. Prudently, the future IEC Code recommends to limit to 0.5% the model efficiency increase resulting from the test at plant sigma. The tendency of curve (A) to maintain a slope up to high sigma values was substantially amplified by a request for testing at high head (15 m) with a maximum air content (incipient sparkling). There was no scientific foundation for imposing such a test condition. The IEC Code admits (6.1.6.2) that appropriate test conditions have not yet been established. Curve (B) shows that the break curve for the same operating condition of this Kaplan turbine has only a small hump when tested according to our usual standards (6m net head, 2ppm oxygen content). Investigations will be conducted on a single test stand to clarify the shift existing between curves (A) and (B) at sigma plant and the air content influence on the curve stabilization.

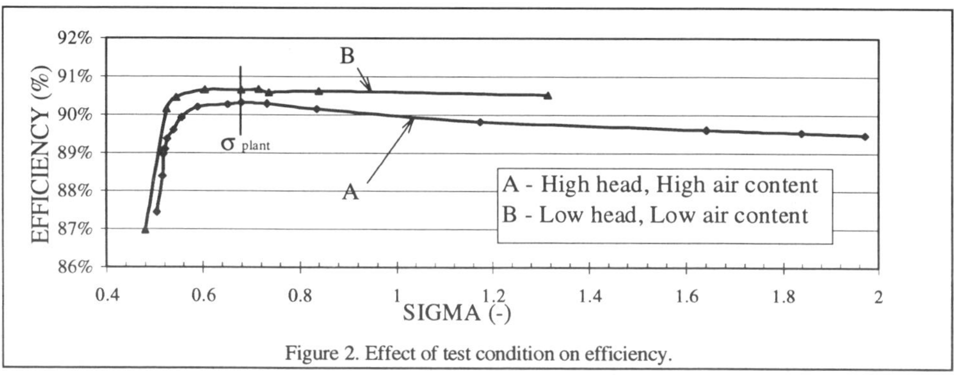

Figure 2. Effect of test condition on efficiency.

2.2. PROTOTYPE QUALITY AND SITE TEST PRECISION:

2.2.1. Site Test Precision:

The Table 2 presents a series of projects for which very high precision tests were realized by our clients, sometimes with considerable care and investments [3,4]. Models for Kootenay, Laforge 1 and La Grande 1 were tested twice, on our test stands and at IMHEF. Uncertainties were estimated by the chief of test. The step-up presented is the value obtained at peak efficiency.

TABLE 2. Model and Site Test Comparison of Medium and Low Head Turbines

Project name	Net Head (m)	Site Measur. Method	Site Test Uncertainty	Mod. Test Uncertainty	Step-up at peak	Δη IEC expected
La Grande 2	137	(1)	0.50 %	0.20 %	2.00 %	1.73 %
La Grande 4	117	(2)	0.55 %	0.20 %	1.70 %	1.69 %
Kootenay	79	(3)	0.50 %	0.15 %	1.87 %	1.71 %
Manic 5 R	142	(1)	0.50 %	0.15 %	1.68 %	1.82 %
Laforge 1	57	(4)	0.50 %	0.18 %	2.20 %	2.13 %
Itaipu	119	(4)	1.20 %	0.25 %	2.50 %	2.05 %
La Grande 1	28	(5)	0,70 %	0,17 %	3.10 %	2.24 %

Names in ***bold italic*** are rehabilitation projects. Legend for measurement methods on following page.

Site Measurement : (1) Thermodynamics, 2*3 Probes, LG-2: Flow mixing by intake gate, see [2].
(2) Ultrasonic, 4 Crossed plans, 16 Paths, see [3].
(3) Ultrasonic, Validation repeating EPRI tests, see [6], [10].
(4) Mean of very close results by Pressure Time and Ultrasonic methods. [4]
(5) Preliminary - Currentmeters: 561 effective measuring locations in 6 steps.

2.2.2. *Prototype Geometrical Precision:*

The need for geometrical precision in prototype turbine manufacture is essential to obtain performances homologous to those of the model plus the step-up. It is formalized in the IEC code under homology requirements. However, this code gives general guidelines on tolerances that may be too large in the case of projects where the efficiency has an economical importance.

The 5 axis machining of runner blades, permits, when coupled with very careful assembly and welding methods, to reduce the geometrical deviations to one order of magnitude below the IEC Code tolerances as presented by Lang, Thomas & Sabourin [6], [11] and summarized at Table 3.

TABLE 3. Typical homology deviation for the Kootenay Canal runners.

	Inlet	Outlet	IEC tolerances
Maximum Profile Deviation	0.89 mm	-0.88 mm	5.0 mm
Angular Deviation	0.1 deg.	0.09 deg.	1.0 to 1.5 deg.
Individual opening Deviations	-----------	1.2% max	3% to 5%
Mean opening deviations	-----------	0.4% max	-1% to 3%

As already discussed, the rehabilitation project of Kootenay produced the expected step-up and the precautions taken during the realization are responsible of that success.

Regarding the runner quality control, as for any quality program, it is subject to these main principles:

- Integration and adaptation of inspection to the manufacturing process. Any machining or inspection is fully numerical and the results are processed on line.
- Detection of problems at a step where corrections will be precise and economical.

The quality program for the runner identifies 3 steps:

- After machining of blades and before assembly: Inspection of the first blade by numerical theodolites, then, inspection of the rest by the NC probe.
- For Francis runners, inspection of a partial assembly with 4 blades (by theodolites).
- Also for Francis, inspection of outlet openings after assembly before welding and at the end of the fabrication process and heat treatment.

The location of inspections on the prototype is the same than on the model. Doing it that way helps to keep the numerical inspection process more transparent.

With this inspection sequence, any error in NC programming or in blade positioning would be detected in time with no incidence on quality.

2.2.3. *Prototype Roughness:*

In a recent study, Grenier and Vu [12] demonstrated that the surface roughness in hydraulic passages, and in particular the wicket gate channels, can have a significant influence on the efficiency of hydraulic turbines. This study supports the fact that precautions to smoothen the hydraulic passages, such as wicket gate refurbishment or stay vane and stay ring painting, are essential to obtain the full efficiency step-up between the model and the prototype.

GEC ALSTHOM is currently investigating the potential of a three-dimensional turbulent flow code to be used as a tool to evaluate the effect of prototype surface roughness on the efficiency step-up. It is hoped that such a tool can assist the manufacturer and the utilities in the decision process regarding the need to upgrade the hydraulic passages during a rehabilitation project. At the present time, the main difficulty associated with flow simulations for prototype turbines is the refinement of the numerical grid needed near solid walls. Figure 3 shows a typical refined grid that is used for the evaluation of the head loss in a model turbine distributor. Typically, the Reynolds number for a prototype is about 15 to 70 times greater than for the model, requiring a much finer grid near solid walls for the turbulence model to be valid. This condition is sometimes difficult to satisfy due to the geometry and present computer resources, even if studies are limited to sample zones.

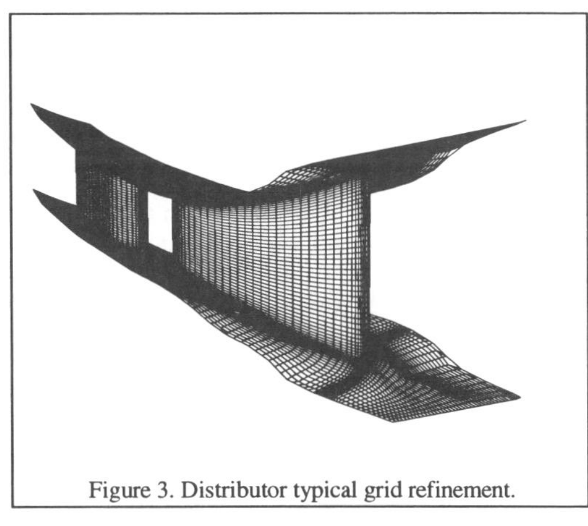

Figure 3. Distributor typical grid refinement.

3. Some of Extra Means to Maximize the Rehabilitation Benefits:

The same way it is important to identify which surfaces have to be hydraulically smooth, many actions can be taken all along a rehabilitation to ensure the owner of the maximum of benefits from its investment. We just mention three typical situations were a partnering approach is quite valuable.

3.1. SIMPLIFIED SITE WORKS WITHOUT COMPROMISING: SIR ADAM BECK 2

The upgrading of the Sir Adam Beck 2 Generating Station, located on the Niagara River in Ontario, Canada, consists of a combination of runner replacement along with stay vane modifications and minor changes to the guide vanes. After final model testing and within a general Research & Development program, a detailed three-dimensional turbulent analysis of the flow inside the distributor was performed to evaluate the contribution to the efficiency gain from each individual stay vane modification (De Henau and Markovich 1995)[13]. This study reveals that some of the stay vanes could be left unmodified without any detrimental effects on the efficiency of the turbine. In addition, a comparison of the estimated cost to streamline those stay vanes and the dollar value attached to the gain in efficiency ensuing from such a modification indicates that there are no benefits in this operation. GEC ALSTHOM therefore could recommend to Ontario Hydro, acting as the site contractor, to keep the original profile of the stay vanes in question provided, however, that some grinding and sand blasting be applied to remove all defects and make the profiles hydraulically smooth.

3.2. INSTABILITIES SOLVED EFFICIENTLY: KOOTENAY CANAL

The Kootenay Canal draft-tube is a not a masterpiece of hydraulics. In spite of a four percent global efficiency increase and respectable step-up, the Kootenay turbines are still one per cent behind the expected performance of a Francis of this specific speed, precisely because of the draft-tube. Knowing its instabilities, the original manufacturer had installed a quadripode that penalized efficiency. When commissioning the new runner, the air admission through the headcover proved useless against full load penstock resonance. Air admission through the shaft using a free air valve developed by GEC Alsthom was the effective solution that preserves the full step-up.

3.3. SEARCH FOR MAXIMUM IMPROVEMENTS: CHUTE DES PASSES

The Chute des Passes turbines have been operated by ALCAN (SECAL) in Québec, Canada since 1959. In view of an efficiency increase, SECAL first tested the existing prototypes and built a model that was manufactured and tested by IMHEF. Copying the eleven channels of a single piece casting was not feasible or at least not economical. A few blade surfaces only were analyzed . It was certainly equally difficult to copy the rough cavitation repairs. Initially, assembling pressure sides and suction sides of different blades even resulted in negative thicknesses. And finally, the Computer Aided Copying did so well that the model efficiency was equal to the prototype one. SECAL was renouncing to the step-up but GEC Alsthom, however, insisted to build its guarantees for both the model and the prototype based on IEC step-up. Today, model guarantees have been obtained but the development still goes on: Navier Stokes analyses show that something more can be done on guide vanes and joint research efforts are emerging to take advantage of this new development, with an even better guarantee for the upgraded prototype turbine.

4. Conclusion:

The time for the non-step-up rehabilitation is over. For certain projects like Kootenay Canal and Chute des Passes, the major economical improvement depends on the efficiency, not extra output. As the step-up is of the same order of magnitude as the model improvement itself, achieving this step-up is a major issue of the economical decision. Then, replacing a runner will be more than a design change. It will bring the perfection of modern manufacturing methods that also improve cavitation and stability. It will bring incentives to re-engineer any other component and refurbish them. Available coatings, if carefully applied, will last decades. And to make sure that these improvements come true, nothing will replace a full guarantee on the end product.

5. References.

[1] Couston M., Machado Fernandes Filho A.,Temes Ruiz Diaz P.P. (1994) "Hydraulic Performances of Itaipu Francis Turbines", XVII IAHR Symposium, Beijing, China.

[2] Coulson D., Henry P., Lévesque J-M., Miron J-G. (1986) "Performance of LG 2 Turbines", XIII IAHR Symposium, Montréal, Canada.

[3] Henry P., Leroux A., Lévesque J-M., Miron J-G. (1988) "Performance of the LG-4 Turbines", XIV IAHR Symposium, Trondheim, Norway.

[4] Lévesque J-M. (1996) "Principaux résultats à Laforge 1", International Group for Hydraulic Efficiency Measurements (IGHEM) Seminar, Montréal, Canada.

[5] Demers A., Désy N. (1990) "Experience with Hydraulic Performance for Major Hydro Modernization Projects", XV IAHR Symposium, Belgrade, Yugoslavia.

[6] Lang H., Thomas D.R. (1996) " Kootenay Canal Turbine Upgrade - Homology To Field Tests", ASME, Phoenix.

[7] Viano M. (1970) "Considérations et Essais sur les Labyrinthes Lisses des Machines Hydrauliques", SHF, La Houille Blanche / No 1-1970.

[8] Brekke H. (1988) "The Influence from the Guide Vanes Clearance Gap on Efficiency and Scale Effect for Francis Turbines", XIV IAHR Symposium, Trondheim, Norway.

[9] Gindroz B., Henry P., Avellan F. (1992) "Francis Cavitation Tests With Nuclei Injection: A New Test Procedure", XVI IAHR Symposium, São Paulo, Brazil.

[10] Mollicone, S. (1984) "EPRI - Comparaison des méthodes de mesure de débit turbiné - Résultats de Kootenay Canal", Hydro-Québec, Rapport Interne No. 15200.

[11] Thomas D.R., Lang H., Sabourin M. (1994) "Homology Assurance Program for Kootenay Canal Turbine Upgrade - A Case Study", ASME, Phoenix.

[12] Grenier, R., Vu, T.C. (1994) "Methods of predicting the effect of rough surfaces on the efficiency of hydraulic turbines", XVII IARH Symposium, Beijing, China.

[13] De Henau, V., Markovich, M.S. (1995) "Optimization of the Sir Adam Beck 2 Turbine Distributor Using Computational Fluid Dynamics", Uprating & Refurbishing Hydro Powerplants V Conference, Nice, France.

SUPER SYSTEM: A HYDROELECTRIC UNIT CONDITION MONITORING SYSTEM IN OPERATION AT HYDRO-QUEBEC

Y.MOSSOBA, Eng.
Production Maintenance Department, Hydro-Québec
C.P. 11006, Station A, 18F, Montréal, Qué. Canada, H3C 4T8

Abstract

The knowledge of the status and the dynamic behavior of a hydroelectric unit at any time is a major asset to optimize and to timely plan maintenance activities. That knowledge can become available through a user-friendly on-line monitoring system capable of processing and storing a large amount of data, retrievable in various meaningful configurations.

After eight years of research and development, Hydro-Québec started the implementation of a monitoring system at the end of 1993. This implementation is in progress reaching all Hydro-Québec's hydraulic units rated 100 MW and up and most of the smaller capacity units. The system is presently operational on more than (65) hydroelectric units in Brisay, LG1 (La Grande-1), LG2, LG3, LG4 , LA1 (LAFORGE-1) and Manic-5 plants.

Our objective in this paper is to show how machine condition monitoring can help solving maintenance problems. We will first describe what are the preventive maintenance goals that led to the decision of implementing such a system; how the design criteria has been chosen and what are the developments breakthroughs realized to reach these maintenance goals. Then, we will use recently acquired real life data to illustrate how data available from a monitoring system can help in the diagnostic of the machines' condition. Finally, we will discuss some problems encountered during implementation and their solutions.

1. Introduction

There are many preventive maintenance goals that warranted the need for the development and implementation of the SUPER (PERmanent SUrveillance) system. Besides the obvious goals of reducing forced outages and optimization of planned outages, a valuable goal to aim at, is the optimization of the unit's overhauling planning. That is, after say 30 years of operation, is the unit due for an overhaul, or could it wait safely for another three or five years? Other important goals are having

E. Cabrera et al. (eds.), Hydraulic Machinery and Cavitation, 1152–1161.

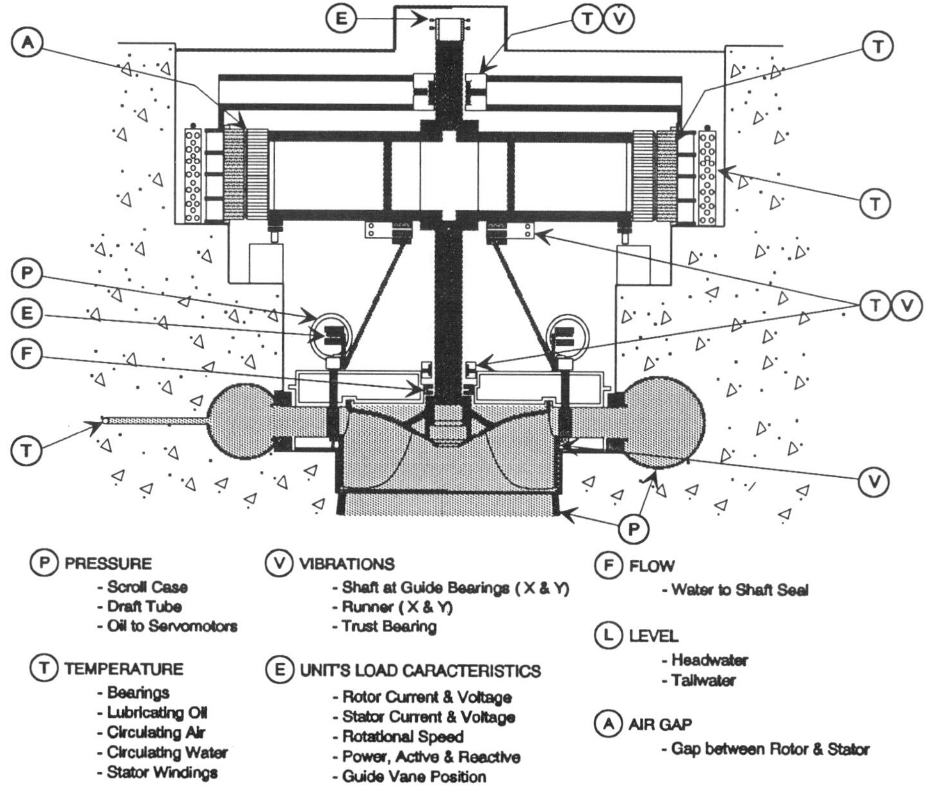

Figure 1 - Monitored Parameters

continuity and uniformity of data collection to counter balance employee turnover. Historically, maintenance crews would make visual and hands-on inspections and file reports on a systematic basis. While an analysis based on these reports provided assistance to the decision maker, such reports were not always consistent as different persons were preparing them. On-line monitoring now provides the essential information, graphically and numerically, allowing for accurate assessments of the hydroelectric unit condition upon which, users can make well-informed decisions. It assists a user to correct a problem in a convenient period of low demand for power or predicting the need of intervention before failure occurs. The monitoring system answers also to the goals of optimizing water utilization and operating at optimal levels to reduce cavitation. It facilitates as well warranty's management for newly commissioned units.

2. Choice of Design Criteria

There are basically three families of functions required from a monitoring system:

The first family of functions responds to the need of building-up a bank of information to identify each unit's signature at different operating loads. That is to be able to detect early on any developing problems. Accordingly the monitoring system must be an integrated high capacity computerized system designed to manage all of the data originating from the units measuring devices. It must handle the acquisition of these data in a regular basis, as well as the mathematical treatment to obtain representative values of each parameter in a condensed volume of meaningful data.

The second family of functions is to obtain data that is more detailed in time, in order to be able to diagnose rapidly a sudden problem that might lead to a forced outage. Hence, it must to be designed with the ability to "intelligently" process and store a larger amount of data when needed. This means, that it must be able to change the storage mode from normal to alarm or trip mode when either of these events occurs.

The third family of functions is the instrument's emulation. That is to allow the user to have on-line real-time testing capability in order to measure a phenomenon with a choice of parameters most suitable to the situation.

Also, the monitoring system has to be reliable and at optimum cost. For that reason, we had to develop software solutions to solve hardware problems whenever possible. That is we chose to reduce hardware content to the minimum. Software is not expensive to install nor to update, and it does not wear out, hence it is more reliable.

Another criterion is that data has to be labeled with the state of the unit; that is stable or transitory. That is because, when the load level is changing in a unit, the analog data can evolve with a time lag that could reach an hour time. In order to assure the quality of the data, the stability of the unit has to be determined. That should be done from the analog input data itself and not from the numerical data that characterizes the evolution of the unit.

3. Functional Description of the system

3.1. ARCHITECTURE OF THE SYSTEM

The main components of the system are as follows:

- A set of measuring devices like proximity probes, accelerometers, temperature and pressure probes. The monitored parameters are shown in figure 1.
- A Data Acquisition Unit for each hydroelectric unit, capable of monitoring and reading typically 64 analog and 64 digital inputs respectively.
- A Site Work Station consisting of a 486 IBM PC running under OS/2 platform.
- Remote Work stations that are linked to the Site Work Station through the existing communication network using TCP/IP protocol.

There are also Air Gap sensors that are part of a separate system but made running on the same PC as SUPER. The Air Gap system has been in use at Hydro-Québec since 1987 and will be fully integrated to SUPER in the very near future.

3.2. FUNCTIONS OFFERED BY THE SYSTEM

The functions of the system as detailed in [1], correspond to the preventive maintenance goals. For example, the *Analysis* function which contain data acquired every hour is useful for trending and solving long term problems. While the function *Alarm and Trip Events* provides data on all the parameters during the hour preceding and after an event occurring on any of the monitored parameters. That function is useful for the diagnostic of a forced outage. The main functions of the system are summarized in Figure 2 and briefly described as follows:

3.2.1. *Surveillance functions*

Plant summary status and Unit summary status. These two functions offer a color coded display [3] that shows whether the monitored parameters are normal, beyond the pre-set limits or insignificant. For each measured parameter, the system allows the user to set upper minor and major limits and where applicable lower minor and major limits.

Dials. This is an instrument emulating « dials » function that allows the user to visualize the data in real time. Data acquired through slow channels as temperatures for example are refreshed every second. Data like vibrations and pressure that are acquired through fast and semi-fast channel are refreshed every minute.

Journal of events. The user can use this tool to visualize the display of data graphically and numerically related to the alarm and trip events that are recorded chronologically as they occur. Upon choosing any event they can visualize a more detailed amount of data on any of all the monitored parameters.

Counters. This tool offers to the user a graphical display of specific numerical data such as the number of time and the duration that an equipment is on and off.

Turbine Generator Efficiency. This is a real-time display of the overall unit's efficiency as well as the power losses at different components of the unit.

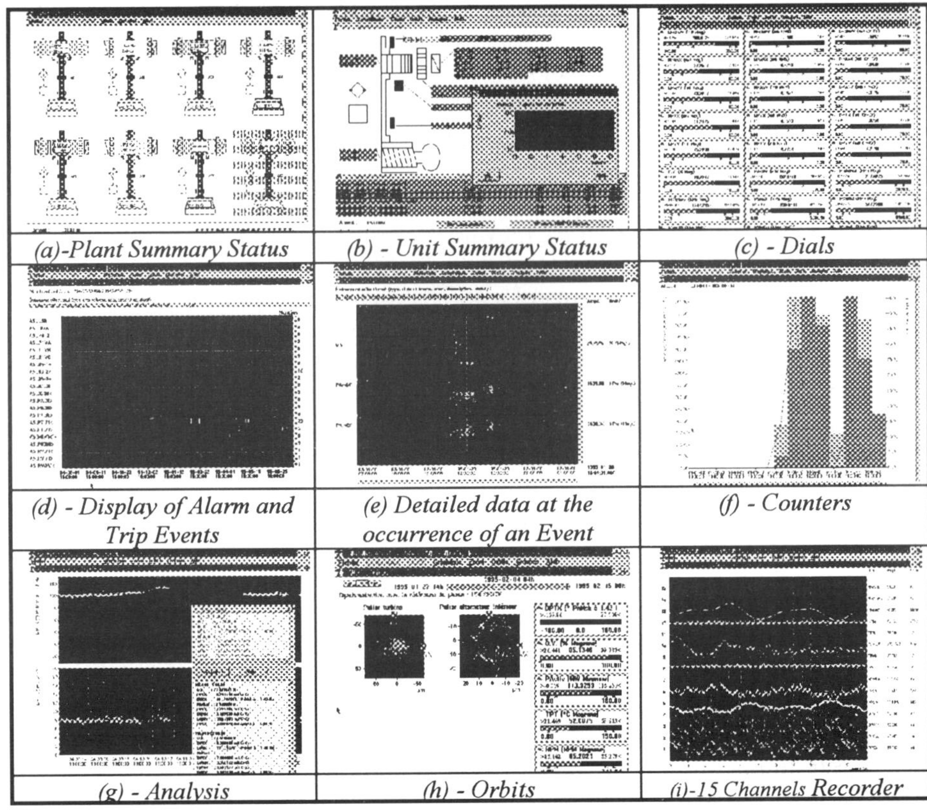

Figure 2 - Main Functions offered by the system

3.2.2. *Long term Analysis functions*

Analysis This function offers a graphical display of recorded data after choosing the parameters, the time span to consider as well as the time interval. The main purpose of this tool is to provide the user with a "signature" or known behavior of each hydroelectric unit on each of the monitored parameters.

Orbits. This tool displays the orbits of the shaft in each bearing individually as well as any other (5) parameters that the user wants to correlate with the orbits. It allows the users to visualize the relationships between the orbits at different bearings and the evolution of their stability with time. It indicates also the shaft's center of rotation.

3.2.3. *Testing functions*

15-Channel Recorder. This function allows the user to collect and save data at the frequencies and durations of the user's choice on (15) parameters at a time. The graphical display of collected data can be seen against time or against frequency and/or against one of the parameters in 2 or 3 dimensions.

Oscilloscope. As in a testing instrument, it allows the user to visualize graphically in a quasi-real time the behavior of the signals from two analogical channels in relation to time or in relation to each other.
Spectral analyzer. It allows the user to visualize graphically in a quasi-real time the behavior of the signals of a channel in relation to the frequency in the form of a spectral density. This tool offers an event trigger on any analog or numeric channel.

3.2.4. *Configuration function*

A user can choose his own configurations of data and save it; that feature is available with all the functions. This means that in the dials function for example, if a user wants to see all vibrations' data, he can group all the vibrations' parameters in a configuration file. In the same way he can group all the monitored data around the turbine's bearing like the radial displacement, axial acceleration, temperatures and oil pressure. In that way the user can make his own software *modules.* Accordingly different users can have their own different modules on the same unit.

3.3. DATA ACQUISITION AND STORAGE

There are three types of analog inputs: slow (e.g., temperature), semi-fast (e.g., pressure) and fast acquisition analog inputs (e.g., vibrations). Slow analog inputs are refreshed every second. The last sample at the hour is stored every hour in the normal acquisition mode. If any alarm or trip event is recognized by the system at any time through the numeric input channels, then the last sample at every minute is stored for typically 60 minutes before and after the event. Semi-fast and fast acquisition inputs are acquired at the rate of 100 samples per second and treated every minute during 16 machine's revolutions. Average, peak-to peak and RMS values are calculated. The last treatment on the hour is stored except when an alarm or trip event is declared, then the values of typically 60 minutes before and after the events are kept. The fast channels' inputs are also acquired at a rate varying from 640 to 1280 samples per second according to the unit's RPM, once every hour. A Fourier Fast Transform is performed and then stored hourly on 16 machine's revolutions.

4. Development Breakthroughs

We will emphasize here on the main development breakthroughs that were achieved to reflect the design criteria. To minimize hardware's costs, software solutions have been employed, for example trigger emulation for phase reference and skew error, avoiding the use of costly specialized acquisition hardware for mechanical vibrations of rotating machines. The phase reference is acquired in exactly the same manner as the other displacement measurements. Spectral data processing is performed on the phase reference channel followed by extraction of the phase of the spectral components of the machine rotation as cited in [2]. This phase serves as a reference for the phase of the spectral components extracted from the other dynamic channels. Furthermore, the skew error or *channel-to-channel sampling delay* due to the use of multiplexed acquisition is offset by correcting the phase of the spectral components. Also, mathematical tools are used to reduce the amount of data stored. For instance, a fast Fourier transform is performed on each dynamic channel. Then, for the first ten main lobes of the spectrum and for each channel, the magnitudes, frequencies, and phases are kept in the database together with peak to peak, RMS, and mean values. That means that a considerable reduction of data is obtained from more than a thousand values in the time domain to only 33 value for each dynamic channel.

5. Cases Studies

We will demonstrate, using the following cases how data available from a monitoring system can help in the diagnostic of the machines' condition as well as sometimes for retro-action design purposes.

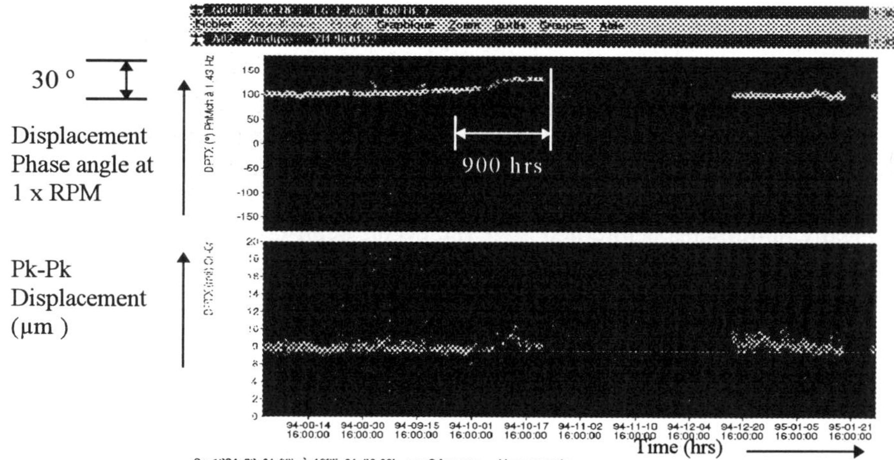

Figure 3 - Phase angle change during the detachement of the runner's cone

5.1. LOSS OF A RUNNER'S CONE AT LG1

This is the case of a new 114 MW hydroelectric unit that was commissioned in 1994 at Laforge 1 powerhouse. During a cavitation inspection after 3000 operating hours, the loss of its runner's cone has been noticed. After analyzing data retrieved from the monitoring system, we found that this loss was detectable, i.e., that a particular aspect of the vibrations behavior of the unit was changing during the detachment of the cone. Figure 3 shows the level of vibration peak-to-peak and the phase value in the 3 o'clock direction at 1X RPM. It can be seen that the cone's loss was not preceded by a higher level of vibrations amplitudes, but only by a change of phase of 30°. This slow variation in the phase angle at 1X RPM indicates a progressive eccentricity between the cone and the runner, hence a progressive detachment. The variation of the phase angle was noticeable at 900 operating hours before the complete detachment of the cone. The detachment occurred at 2000 hours from commissioning. Start-up tests did not show anything abnormal. This case demonstrates that this kind of problem could have been at least observed, should someone had looked at the monitoring system during the 900 hour's period. It pinpoints as well that monitoring has an important advantage over periodical tests for early on detection of developing anomalies. Finally, it also shows that the success of implementing a monitoring system depends on the users' experience and their ability to make a diagnostic from the available data.

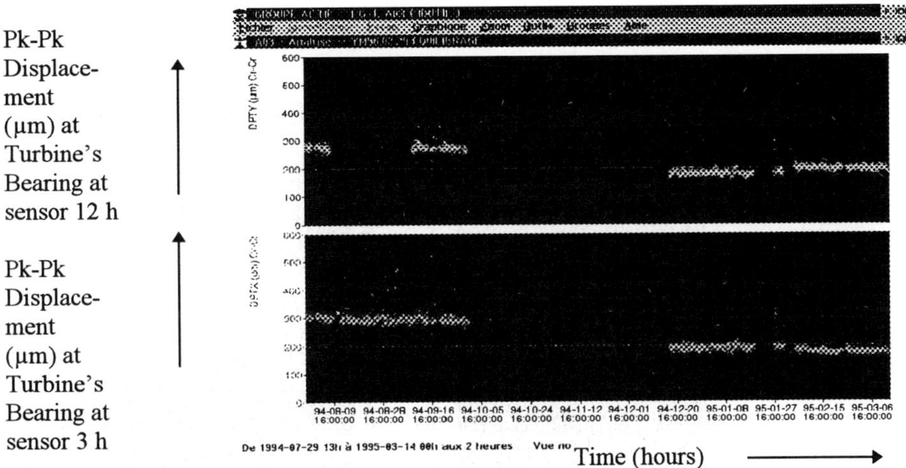

Figure 4 - Vibration data from the Analysis function before and after balancing

5.2. MECHANICAL BALANCING

This case shows displacement vibration at the turbine's bearing of unit #3 at LG1 plant. Here figure 4, indicates the peak-to peak levels at both X and Y bearing's axis against time. During the period of time shown at the left of the graph, the lower warning level of 200 μm was exceeded. Later, the 114 MW unit's helical runner has been mechanically balanced with a 250 Kg plate attached to the runner's cone. The user can verify after balancing, the behavior of the bearing vibrations, where peak-to-peak levels came down to below the 200 μm mark. The user can access to the same case using other configurations or other tools as well. Here the peak displacement at 1.43 Hz or 1xRPM has been around 110 μm and 60 μm before and after balancing.

5.3. THE IMPORTANCE OF THE SIGNATURE OF A UNIT

The notion of having a *signature* or known behavior is important as units of different plants, operates at different loads. There are units in a same plant that have different

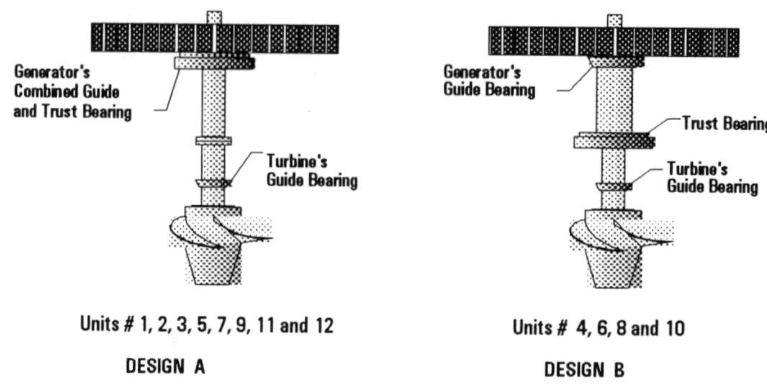

Figure 5 - Bearings arrangement at LG1

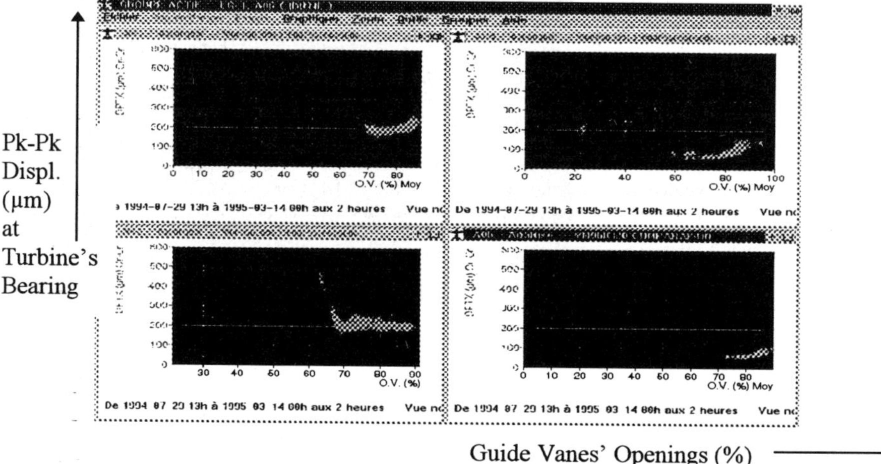

Figure 6 - Vibrations signature of two sets of units of different bearing configurations at LG1: Design A (left) vs. Design B (right)

designs; accordingly have different signatures. At LG1, there are units made by two different manufacturers. Units #1 and 2 and units #4 and 6 are made by manufacturers A and B respectively. Both sets of units have the same capacity; and among the differences between the two designs are the bearings configurations as shown in figure 5. Units of manufacturer A has the Trust bearing combined with the Generator's Guide bearing. Units of manufacturer B has an independent Trust bearing located between the stator and the runner. Figure 6 shows a comparison between the same (2) sets of designs at the turbine's guide bearing level. In that case, we see that the vibration's behavior of units #4 and #6 have been generally lower than that of units #1 and #2 particularly where guide vanes' openings are above 60%. We would like to mention that for the same period of time, both sets of units have acceptable levels of radial displacement vibration at the generator bearing. At the trust bearing, some of the axial vibration monitored by the accelerometer exceeds the lower warning level of 3 m/sec^2 from Units #1 and #2. However, the same data for Units #4 and #6 is satisfactory. By knowing the signature, when any parameter starts to change from its normal behavior at a particular load, the user can detect an upcoming problem and try to fix it before it becomes too late.

6. Implementation Problems and their solutions

Implementation has not been without encountering minor problems. We have experienced some noises in the signals coming from the vibrations and pressure probes in some plants. That happened, despite the systematic use of current transmitters (4-20 mA) to reduce electromagnetic noises on the signals while transmitting them on great distances as cited in [3]. To solve these problems we had to assure the elimination of ground loops and improve the shielding connection techniques. In some of these cases we had to add high frequency filters in the acquisition units. Mounting brackets were satisfactory as they were designed strong. On some pressure probes, we found out that noises have been reduced considerably by replacing the metal fittings by plastic ones.

7. Conclusions

Machine Condition Monitoring should be part of the maintenance activities in order to ensure the reliability of the hydroelectric units. The monitoring system has to offer on-line data in real time as well as stored data over a long period of time. Software development and mathematical treatment help reduce the amount of data to be stored and yet provide sufficient data to the user. In the very near future, Hydro-Québec is planning to fully integrate the Air Gap system to SUPER. Further on, more measured parameters will be added to SUPER like the Stator's bars vibrations and Cavitation. A Diagnostic Expert system is considered for development; but not before sufficient data is accumulated by on-line monitoring.

8. References

1. Mossoba Y. (June 1995); Presentation of the SUPER machine Condition Monitoring implemented at Hydro-Québec; COMADEM 95 (Condition Monitoring and Diagnostic Engineering Management) *Queen's University, Kingston, Ontario, Canada*, vol.**2**, pp. 689-696.
2. Léonard F., (June 1989); Analyseur spectral pour machines tournantes: Phase et coherance; CACAM'89, *Canadian congress of applied mechanics*, p.514.
3. Pastorel H., Léonard F., Watts A. (August 1994); Research on Monitoring Systems at Hydro-Québec; *Symposium on Hydro's Leading Edge Technology; Phoenix, Arizona, USA*

APPLICATION OF ROTOR RESPONSE ANALYSIS TO FAULT DETECTION IN HYDRO POWERPLANTS.

L. P. NASCIMENTO
UNESP/FEIS/DEM
Av. Brasil, 56
15385-000 Ilha Solteira - Brazil

E. EGUSQUIZA
UPC/ETSEIB
Av. Diagonal, 647
08028 Barcelona - Spain

Abstract

In this article some considerations obtained during the utilization of rotor response analysis tecniques in hydraulic powerplants are discussed. An applied research work was carried out in two hydraulic turbines analysing the rotor response both theoretically and experimentally. A developed mathematical model was used to simulate the rotordynamic behaviour of Francis and Kaplan turbines. The main dynamical effects that appear during the operation of the machines are discussed too. A series of measurements were carried out in the turbines using impact hammers to determine the modal behaviour of the units. The tests were carried out with the machine still and in operation. Some results and the comparison with the theory is presented in this article. The improved theoretical model was used for a sensitivity analysis of the different bearings to the main excitations that take place during the machine operation. From this analysis, the best measuring points for condition monitoring were determined.

1. Introduction.

The analysis of the rotor response can be useful in many ways. The detection and control of the rotor resonances is a useful diagnostic method [1,2]. As the natural frequencies depend on bearing parameters and shaft condition, a variation in the natural frequencies can be linked to a specific type of damage (like excessive bearing wear, for instance).

The mathematical model of the rotor can be used for rotordynamic simulation and sensitivity analysis. When building up a model that represents accurately the actual behaviour of the machine, several problems are normally found, and it is sometimes impossible to know exactly the physical details of such muchines, such as gaps in seals and mass distribution, especially in old machines [3]. For example, the bearing

E. Cabrera et al. (eds.), Hydraulic Machinery and Cavitation, 1162–1171.

coefficients can vary substantially depending on the machine type and mounting.

Normally, the experimental analysis for rotor response is carried out with an impact hammer when the machine is still. When the machine is in operation an impact hammer can be used, but special signal treatment techniques must be employed [4]. The natural frequencies vary depending on the operating load of the machine. This is due to a change in the stiffness of the bearings, seal effects, added mass, etc. Moreover, for large machines it is difficult to produced excition of the rotor with an impact hammer.

This article discusses the analysis of two different machines, one of Francis and another of Kaplan type. The Francis machine has a power of 15 MW, 375 rpm, with three guide bearings and a thrust bearing located in the upper part of the unit. The Kaplan machine has a power of 60 MW, 136.5 rpm, with two guide bearings (generator and turbine) and an thrust bearing located in the intermediate point of the shaft.

2. Theoretical analysis.

A theoretical simulation of the modal characteristics of the machines was carried out using a Finite Elements Method. In the program the shaft is represented by elastic beam elements, the guide bearings by single spring elements representing the stiffness of the fluid film, structure and foundation. The thrust bearing is represented by a torsional spring element and the rotors (generator and turbine) by disk elements. The mechanical properties of the elements were obtained from the drawings of the machine. Nevertheless, the bearing stiffnesses (film of oil + structures) were determined from the experimental results.

The digital simulation of the modal behaviour of the undamped systems is presented in figure 1. The four first vibration modes are shown. The main dynamic effects that occur when the machine is operating were considered in the mathematical model. These are: fluid-structure effects, seal effects in the turbine and gyroscopic and magnetic stiffness in the generator. Figure 2 shows a study of the influence of these effects in the Francis machine.

Figure 1 shows that there is a considerable difference between the modes of the two machines. Also, some modes induce high displacement in the turbine and in the generator. These modes, if excited, can produce an improper operation of the generator or friction in the turbine.

In Figure 2 (Francis group) it can be seen that the fluid-structure effect intervenes in the frequency of the first mode. The seal effect is important for stiffness greater than 5×10^7 N/m. The magnetic stiffness intervenes especially in the low frequencies and it is important for values (negative) greater than 1×10^8 N/m. The gyroscopic effect also must be considered, mainly in rotors with high moments of inertia. Evidently, these parameters are difficult to estimate, and they act on the system simultaneously. Therefore, they should be introduced in the models based into the experimental results.

3. Experimental analysis.

To determine the experimental modal behaviour of the machines, several blows were dealt to the shaft with an impact hammer. Responses to the impacts were obtained with accelerometers located at several points of the shaft and with proximity probes fixed on the structure of the guide bearings. The signals were stored in a tape recorder and later analyzed in laboratory with a multi-channel signal analyzer. Very often, the experimental tests are difficult to carry out. In general, access to the machines to fix the sensors is limited and very strong blows are necessary to energize the system. Better results were obtained from the smaller machine. Tests were carried out with the machine not working and with the machine in operation. Tables 1 and 2 present, in terms of the natural frequency, a comparison between the theoretical and experimental results.

The experimental vibration modes were obtained using a modal analysis package. The frequency response functions from the signal analyzer were used in this package and the modal parameters were determined. The results obtained from the experimental tests present a good approximation to the theoretical results. Figure 3 shows a comparison between these results for the Francis machine. It can be seen that some vibrating modes are quite similar. The experimental analysis shows that the modal damping ratio is greater in the first modes. Probably due to that it was not possible to determine the first mode. This analysis also indicates a high deformation of the shaft in the third mode.

4. Sensibility analysis.

A particular issue in vibratory systems is how to obtain the relationships between the parameter variations and their behaviour [5]. The parameters dictate a behaviour, but they can vary quickly with the operating time of the machine. In this context, sensibility analysis is of great interest in predicting the tendency of this behaviour depending on the variations of the parameters. The parameters vary due to a set of factors. Among them, a failure or excessive wear of the bearing represents changes in the stiffness and damping. An imbalance or a misalignment represents an additional force in the system. Operations with partial loads represent variations in the amplitude and frequency of the hydraulic forces. The water within the turbine acts as an additional added mass.

For the predictive maintenance and control of the state of the machine, the sensibility analysis can be used to determine the optimal position of sensors. For a cost effective condition monitoring and fault diagnosis the minimal number of the sensors have to be used. Sensors have to be located in the appropriate position of the machine so that they can detect the excitations generated by the machine and with sensitivity enough to detect the variation in these excitations produced by damage. This is important for signature and trend analysis.

The precision of the sensibility analysis is a function of the mathematical model. Therefore, it is necessary for the model to be adjusted sufficiently for the system to be represented dynamically. From the motion equation of the system, expressions of the frequency response sensibility due to parameter variations (mass, damp, stiffness and excitation force) can be obtained [6].

Figure 4 shows the response sensibility on the shaft due to the main excitations that take place in the turbine (Francis group). It can be seen that the sensibility in the rotation frequency (ff) and their harmonics are maximum in the turbine bearing. In the upper guide bearing the sensibility is relatively high too. Meanwhile, in the intermediate bearing the sensibility is much reduced. So, the points near to the turbine bearing are more sensitive to these excitations, consequently, more interesting for measurements, than the intermediate bearing. However, sensibility in the intermediate bearing increases notably for the higher excitation frequencies, as does that of blades passage (fb) and their harmonics.

In Figure 5 it can be seen that the sensibility due to the excitations produced in the generator is greater in the upper guide bearing. The sensibility in the intermediate guide bearing is low. On the other hand, the sensibility in this bearing increases at the highest frequencies. All this analysis indicates that, in general, the most adequate point to detect radial excitations in this machine is at the upper guide bearing. The second option is in the lower guide bearing.

In the Kaplan machine sensibility at the the lower guide bearing is considerably greater than at the others points of the shaft, for the excitations produced in the turbine and generator. Therefore, the lower bearing is the more interesting position to detect excitations.

The sensibility analysis due to the stiffness variation of the bearings indicate similar conclusions for the Francis and Kaplan turbines.

5 - Conclusion

The analysis of the modal rotor response can be a powerful technique for condition monitoring and fault detection in hydro powerplants. Analysis of rotor response can improve the typical vibration analysis normally used for troubleshooting. For this, experimental and theoretical work must be carried out. However, experimental tests are always difficult tasks to accomplish since the machines are large and difficult of access. On the other hand, to carry out the theoretical work, some difficulties can be found to obtain the accurate parameters of the mathematical model.

Experimental analysis indicates that modal rotor response of the machines in operation can be notably different of the response with the machine still. This is due to the dynamic effects that appear with the operation (added mass, seal, bearing stiffness variation, etc). Consequently, for simulation of the rotor response, these effects should be introduced into the mathematical model.

It is very important to compare the theoretical and experimental results. From these, it is possible to develop a most accurate model to simulate the behavior of the system, especially in those cases where many uncertainties in the values of the dynamic parameters exist.

Sensibility varies depending on the type of machine, number and position of the bearings, etc. Sensibility analysis has been effective in determining the appropriate position of the sensors for taking measurements. This analysis is very important for condition monitoring and fault detection.

6 - References

1. Egusquiza, E. and Nascimento, L. P. (1994) Fault detection in hydropower plants, *IMEKO Congress*, Turin, Italy, Vol. 2, 1231-1236.

2. Egusquiza, E. and Nascimento, L. P. (1994) Vibration diagnostic of a Francis turbine, *XVII IAHR Symposium*, Beijing, China.

3. Ohashi, H., editor (1991) Vibration and oscilation of hydraulic machinery, *Hydraulic Machinery Series Book*, University Press, Cambridge.

4. Nascimento L. P. and Egusquiza, E. (1995) Natural rotor frequencies of a Francis turbine. Experimental and theoretical approach, *VI Symposium on Dynamic Problems of Mechanics - DINAME/95*, Caxambú, Brazil, 128-131.

5. Zimoch, Z. (1987) Sensitivity analysis of vibrating systems, *Journal of Sound and Vibration,* Vol. 3, 447-458.

6. Nascimento, L. P. (1995) Análisis del comportamiento vibratorio para el mantenimiento predictivo de grupos hidráulicos, *Doctoral Thesis, Polytech. Univ. of Catalonia*, Barcelona, Spain.

7. Nascimento, L. P. and Egusquiza, E. (1995) Vibration in dynamic behaviour of a turbine after refurbishing, *Hydropower International Congress and Exhibition,* Barcelona, Spain, 687-698.

TABLE 1. Natural frequencies of the Francis machine

Machine Still				
Frequencies (Hz)	f_1	f_2	f_3	f_4
Experimental	21.5	35	47.5	75
Theoretical	21.57	34.63	47.05	73.54
Machine in Operation				
Experimental	—	—	44	70
Theoretical	25.45	34.02	47.32	71.26

TABLA 2. Natural frequencies of the Kaplan machine.

Machine Still				
Frequencies (Hz)	f_1	f_2	f_3	f_4
Experimental	7.25	10.75	23	38
Theoretical	7.15	10.7	23	39.4
Machine in Operation				
Experimental	—	9.5	23	36.75
Theoretical	6.71	10.45	24.78	36.86

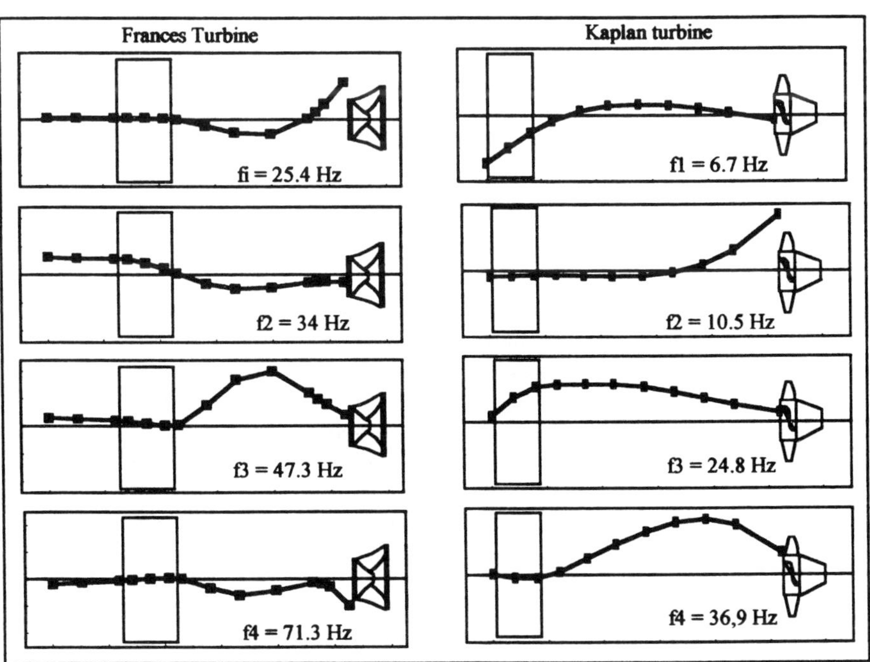

Figure 1. Theoretical vibration modes of the Francis and Kaplan machines.

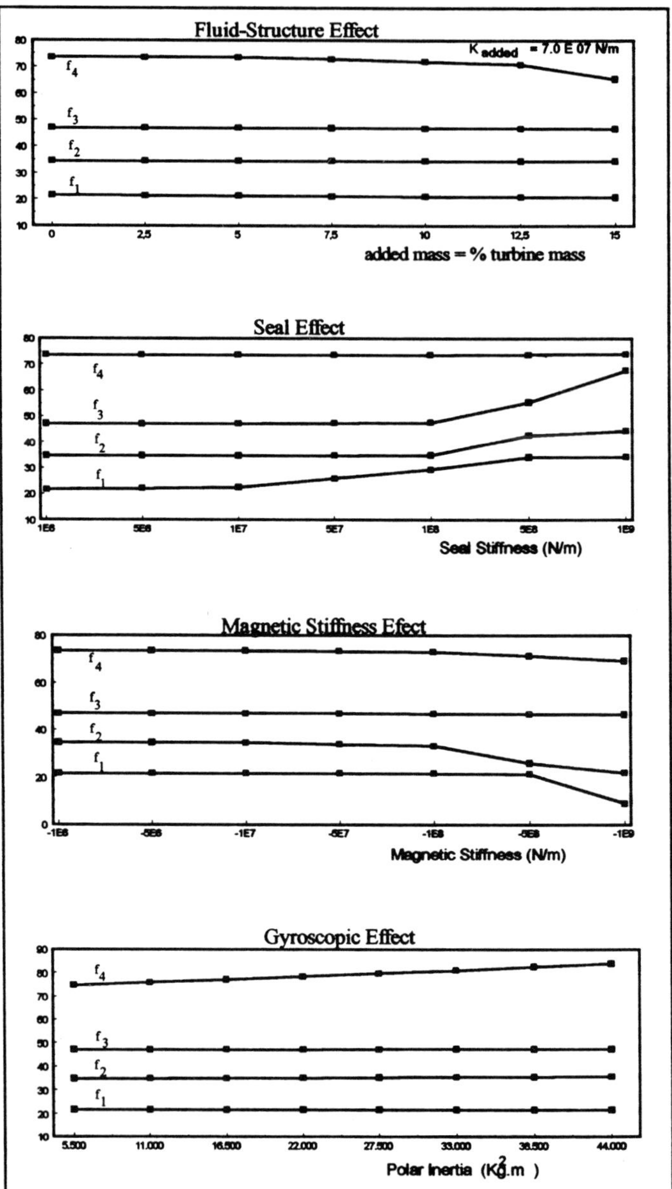

Figure 2. **Influence of the fluid-structure, seal, magnetic stiffiness and gyroscopic effects on the natural frequencies of the Francis machine.**

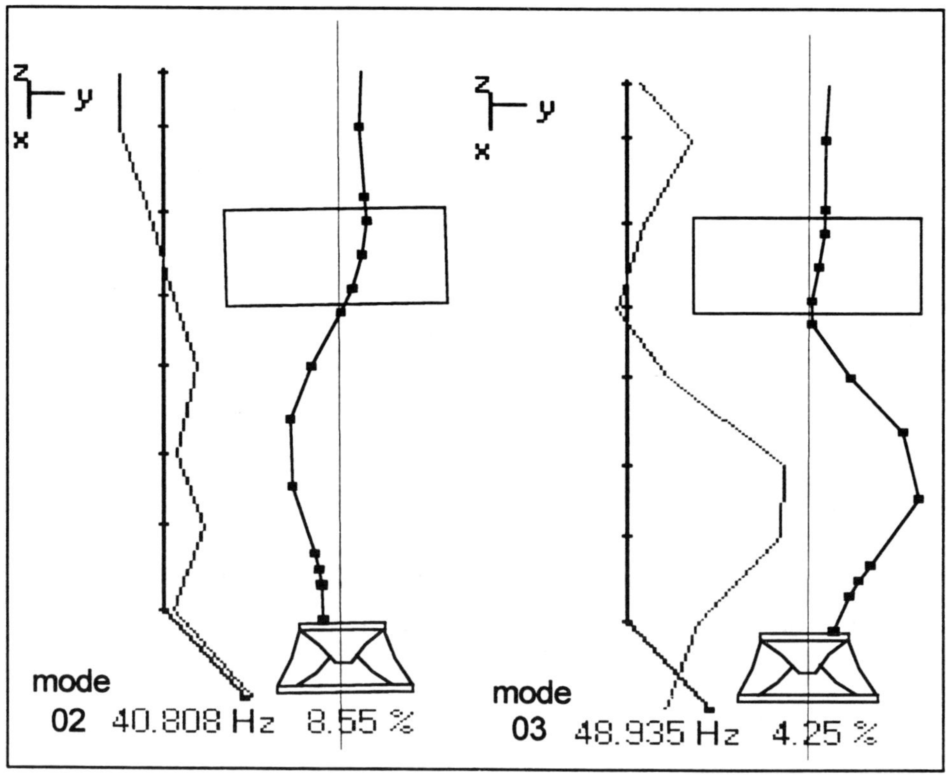

Figure 3. Comparison among theoretical and experimental vibration modes of the Francis machine.

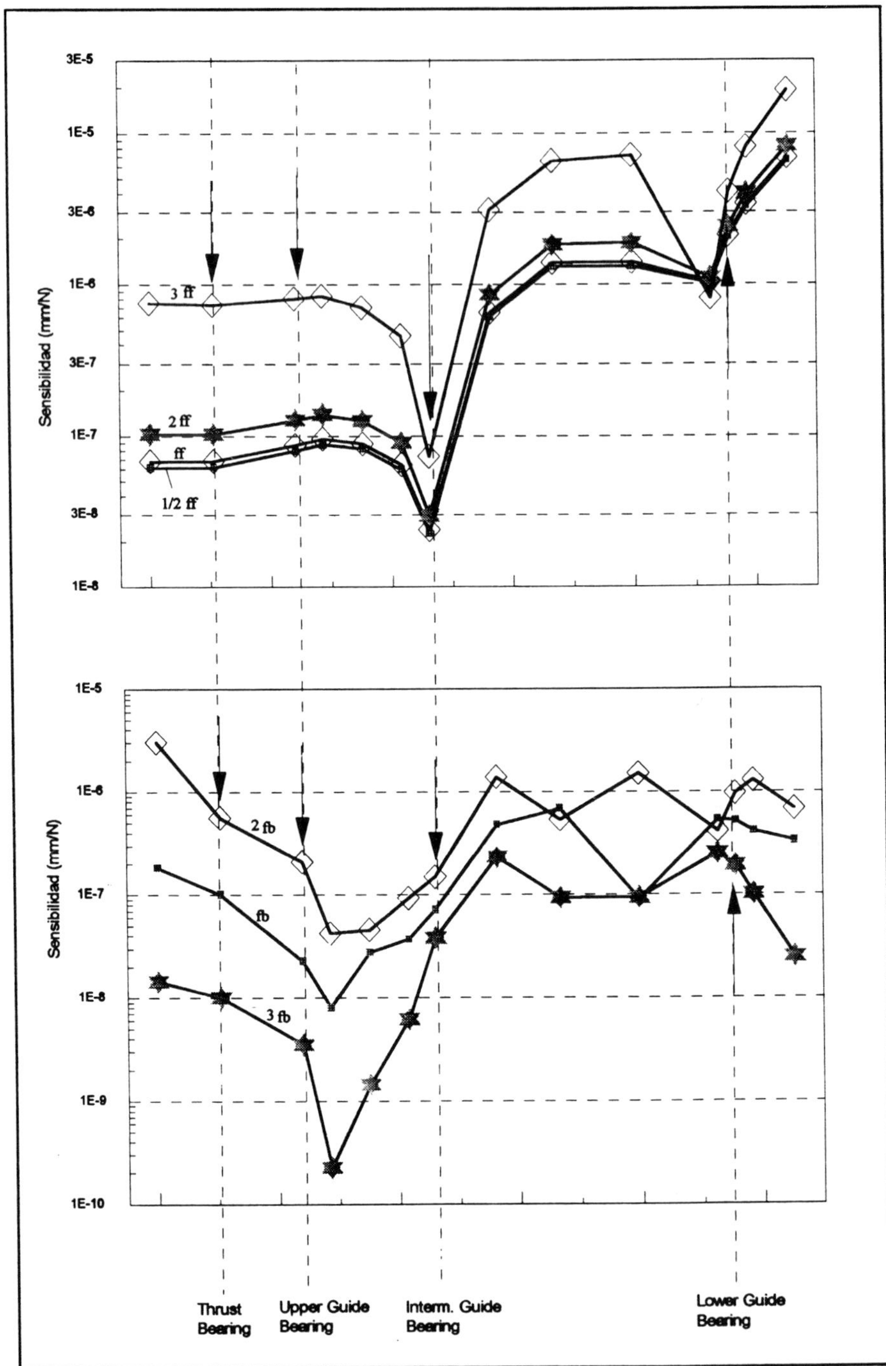

Figure 4. Sensibility of the frequency response due to main excitations on the Francis turbine.

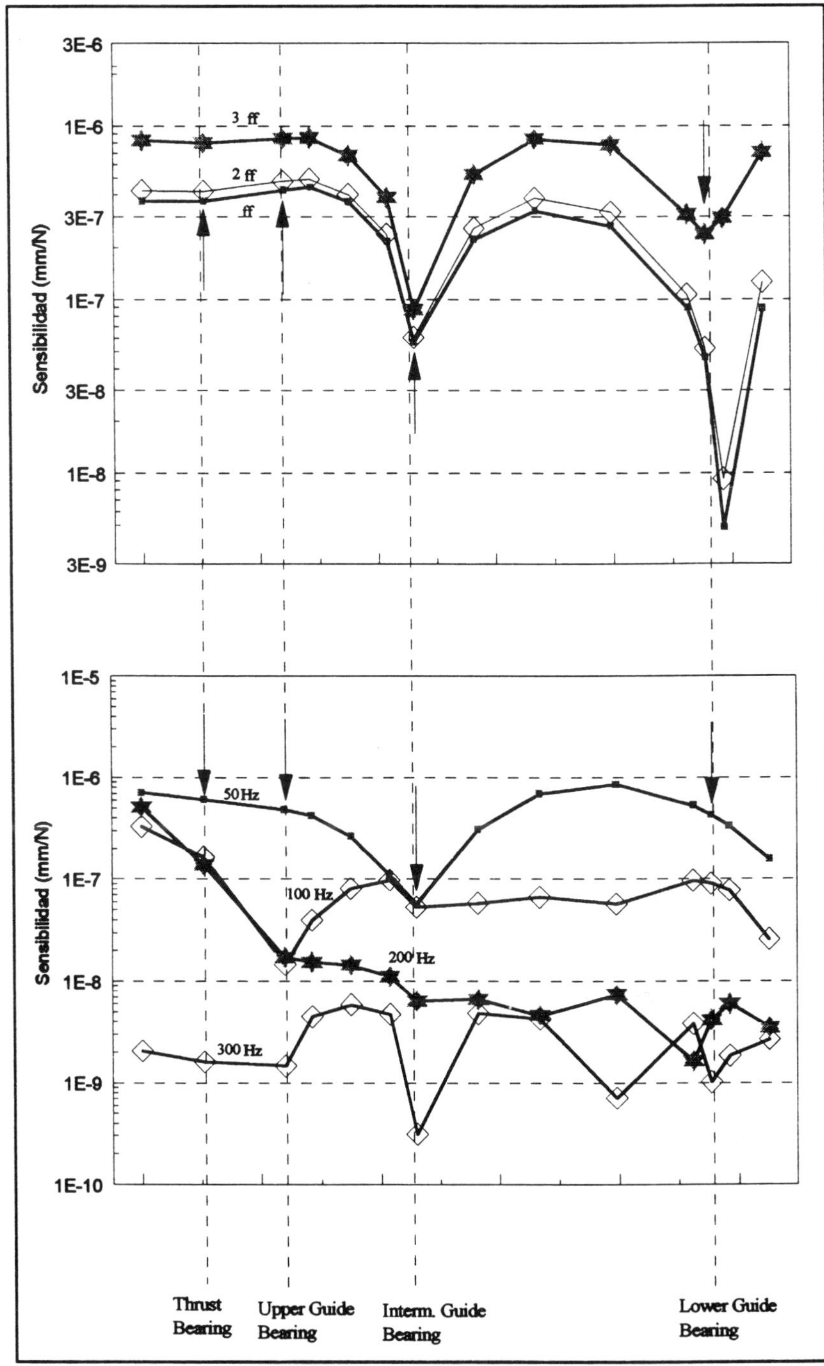

Figure 5. **Sensibility of the frequency response due to main excitations on the generator of the Francis machine.**

THE PLACEMENT EXPLORATION OF DIAGNOSTIC MEASURING POINTS ON OPERATING UNIT EQUIPMENT

XIAO-TING, LIU

The Electric Power Testing & Research Lmstitute of Central China

47 Xudong Rd Eastlake Wuhan P. R. C

1. Introduction

With the development of diagnostic technology in abroad, the research and application of failure diagnostic technology on operating unit equipment of water power station in China attract many departments have been rapidly developed in recent years, Its realization and expansion has very important siguificance and effect to the unit safe and economic operation , cost decrease, system simplification, operating level improvement of entire power network and science adminstration. The Ministry of Electrical Power Industry has envolled the study of diagnostic technology as key item of science and technology deveiopment plan for year 2000.

Following the appearence of large unit and large capacity of water turbine, the unit size is greater and greater, parameters higher and higher, equipment structure more and more difficult; of course problems on operating equipment are gradually remarkable; the research for the unit feature is significantly necessary and realigation of the diagnostic technoly is more and more important. In the research and application of diagnostic technology on operating equipment, it must solve five key problems in which the most important is the problem of measuring point selection and their placement. Focus on this problem, the article puts forward some points of view and gives a initial scheme of measnring poimt arrangement for water power station units.

2. The Key Points of FDI System and Its Application

2. 1 FUNCTIONS OF FDI SYSTEM

Unit operating equipment diagnostic system includes three functions, or failure detection, failure diagnosis and failure evaluation. The system which has failure detection and diagnosis functions is called FDI system[1]。 After select a reasonable diagnostic technologic line and a practical diagnostic type, the system can be used as

E. Cabrera et al. (eds.), Hydraulic Machinery and Cavitation, 1172–1181.

on line or off line supervision diagnosis, On the purpose of equipment diagnosis, it can fulfil the requirement of multi-function or single-function supervision diagnosis. On the control and adjustion, it can be conected to computer supervision system of power plant (station) and network or provincial power adjustion supervision system via communication interface for remote communication. Realize centralization of supervision, diagnosis, administration and adjustion.

To realize the functions of FDI system, it should Fulfiu following characteristics and requirements:

a. high response feature and sensitivity;

b. high diagnostic accuracy, none verror report and absent report;

c. short system supervition delay time;

d. strong diagnostic capability and usable extent;

e. strong anti-interference capability.

2. 2 KEY PROBLEMS IN THE APPLICATION OF FDI SYSTEM FUNCTIONS

Unit operating equipment diagnostic technology generally includes three aspects: the first is the get of equipment operating status signals; the second is the signal processing and analyze; the third is the evaluation and judgment. So if it wants to reach the diagnosis purpose should for unit operating equipment, the diagnostic system should solve following five problems:

a. selection and arrangement of measuring points;

b. selection and usage of supervision diagnosis primary detecting element (include sensor and tansducer);

c. usage of practical unit operationg equipment diagnosis supervision system and dynamic analysis system (include single- function system and multi- function system);

d. establishment of practical diagnostic technology archives, on knowledge library, data library and material library;

e. rules and standards suitable to the unit operating diagnostic technology.

3. Selection and Arrangement of Measuring Points

Because of the different type, structure and operating status of water turbine , the measuring points selection and arrangement is rather difficult. If one is familiam to the unit design, structure, operation regulation and the meaning of operation parameters, the selection and arrangement of unit diagnostic measuring point can reach the diagnosis purpose, Here we should point out that the equipment diagnostic measuring point is different not only with the former testing and research point but also with the unit computer supervision measuring point. The diagnostic measuring points have their special purpose though some of the measuring point arrangemeut are in the same position.

Here ofter mainly introduce the unit operating supervision diagnostic measuring point selection and arrangement taking the vertical-shaft water turbine for example.

3. 1 BASIC PARAMETERE OF MEASUREMEBT

Basic parameters of measurement is the basis of supervision diagnostic measring point selection and arrangement. There are lots of parameters expressing unit equipment operating status in which dynamic parameters express unit dynamic features and static parameters express unit position features (include eguipment location-movement) According to diagnostic aims and its features ,parameters can be classified as electrical, mechenical, hydraulic, mechanics, adjusting and thermal parameter, etc, Their name and quantity is ralative to detail diagnostic contents. Different diagnostic contents follow different measuring parameter and quantity. Generally, General diagnostic measuring parameters are lisfed in table 1.

TABLE 1 Basic parameters of diagnostic measurement

Parameter classification	parameter name and feature
Electricity	unit power (include active and reactive power); rated voltage and current; stator three phase voltage and current; rotor exciting voltage; frequency (include pole frequency and system frequency); generator rotational speed.
mechanism	unit location- movement : main distributing valve opening, servomotor (blade)stroke, gap at unit every individual positon; vibration throw value; threestep bearing throw (or two- step bearing) unit support element vibration and water- pass part vibration; oil- film thick ness of thrust bearing; etc.
hydraulic	unit water-pass part pressure and pressure pulsation: pressure pipe, scroll case, top lid (or support lid) and tailrace pipe, etc pressure and pressure pulsation, unit water head, flow capacity and flow speed, etc.
mechanics	unit big part(include rotating part, fixed part and water-pass part)stress; main shaft axial hydro- thrust and torgue, blade torque, bearing (include thrust bearing)load. etc.
thermal	unit bearing (include thrust bearing) shoe temperature, oil temperature; generator stator, rotor and winding temperature; generotor cool and heat air temperature and unit cooler temperature.
adjusting	static characteristic parameter: electric liquid transducer, adjusting system static charactenstics and adjusting quality parameter. dynamic characteristic parameter: dynamic adjusting quality parameter.

3. 2 SELECTION AND ARRANGEMENT OF MEASURING POINT

The reasonableness, accuracy and quantity of equipment supervision diagnostic measuring points are depend on equipment operating performance, structure characteristic and operating regularity.

3. 2. 1 *Optimization Principle*

Optimization principle is the most basic principle of measuring point selection and arrangement. Seeing from supervision diagnosis, minimization is important for unit safe operation and maintanance, but it is also nead considering feasibility and practical application. It is also very important to select the most delegable and capable reflecting operating equipment status measuring points. For example, the measuring parameters for sapervision and diagnosis of unit thrust bearing operating performance are oil film thickness, oil film pressure, bearing temperature(include shoe temperature and oil temperature)and thrust bearing load, etc. The measuring point arrangement is very complicated and the points are large quantity. If according to optimization principle to analyse thrust bearing' s operation performance and structure characterists, it is enough only to measure the bearing temperature and oil film thickness while these two types of measuring points are selected and arranged with logic diagnostic tenet. If necessary, add one pair of points to meaeure bearing load for expanding diagnostic analysis range. The position and quantity of thrust bearing temperature measuring point can be determined by the bearing shoe' s guantity and structure (it is not centain to arrange measuring point on every shoe), The oil thickness measuring points are determined by operating performance of hydro-turbine and structure of thrust bearing. The measuring points are fixed on the thrust bearing outlet edge of bearing supporting plate along x, -y direction respectively . The measuring point arrangement requirements of spring metal plastic shoe or two line bearing shoe (iuclude inner bearing shoe) are the same. The measuring points for stress supervision diagnosis of turbine scroll case and base ring, in general condition, are selected and arranged for 2- 3 points on the maximum hydro pressure NO. 1 meridian section inside scroll case, in which the first point should be placed near the connector of scuoll case and base ring and the other two points may be placed near the connector of base ring and fixed blade along x, -y direction(shown as Fig. 1).

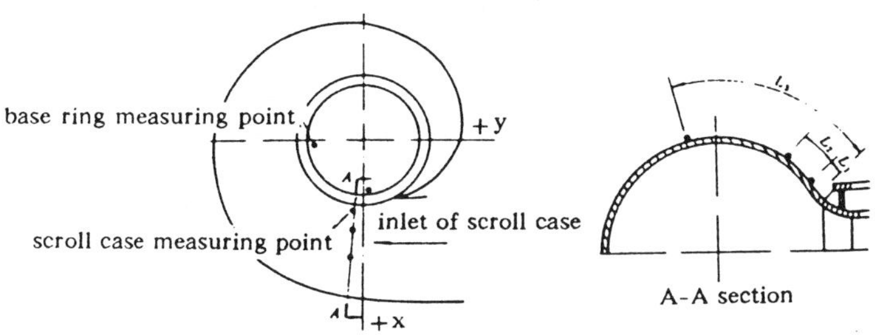

Figure 1. Arramgement of deform supervision measuring point of turbine scroll case and base ring

3. 2. 2 *Hydro Features of Water Turbine*

Hydro stability of water turbine effects unit' s safe operation in great extent. Measaring parameters are mainly the pressure pulsation factor of turbine water pass part scroll case , top lid (or suppor lid), tailrace pipe and in front and back of movable propeller water turbine blade, The dimension of these values are related with the selection and arrangement of measuring point. So, we should select correct position on water pass part where stands maximum pressure according to water turbine hydro characteistics. when selecting maximum measuring point , it is needed to consider the place on turbine water pass part where bave maximum water pressure pulsation factor and water flowing characteristics. For example, scroll case pressure measuring point (one point) is placed on the inlet section of scroll case. On turbine top lid (or support lid)·, place a point very near to main shaft line on place two points near main shaft line along x, -y direction for the study of unit hydro pulsation or vibration in vertical direction. The meesuring point of tailrace pipe pressure pulsation is placed generally on tailrace pipe inlet section in down stream side. Its detail position is related with the type of water turbine. For mixed flow turbine, the point is placed blow the inlet section of tailrace pripe hammer at (0. 4 ~ 0. 5) D_1, down stream side. For propeller water turbine, the point is placed blow the inlet section of tailrace pipe at (0. 7 ~ 0. 8) D_1, down stream side. [3] In addition because pressure pulsation measuring point of mixed flow turbine leakproof ring in related with the supervision and diagnosis of unit operating stability, place two points at upper and lower leakproof ring. For the supervision of hydro stability proplem excited by pressure water hammer when water turbine is in load change operating condition, place pressur and pressure pulsation measuring point on pressure steel pipe near a certain distance from scroll case inlet. The detail position and poirt quantity are related with the type, structure and size of the spiral steal case.

3. 2. 3 *Rotating Mechanical Feature*

The unit operating stability is related not only with unit' s hydro-features, electrical features, but also with unit unit' s mechanical features. Rotating mechanism have high speed and low speed two types of dynamic equipment. Water turbine belongs to low speed rotating equipment. So the measuring point arrangement of unit vibration, throw and air gap should be placed reasonably and accuratly considering the interferance of unit hydro and electrical factor. Unit stability vibration throw measuring points are placed on turbine frame (upper frame , lower frame and thrust supporting frame), turbine top lid (or supporting lid) and tuibine spiral steel pipe along unit vertical and horirontal direction, Throw measuring points are placed along x, -y direction according to unit three step bearings (or two step bearings). when necessary the peasuring points can be placed on commutator or slip ring. Generator stator core vibration measuring point can be placed on the outer ring of stator core along vertical, horizontal (radial) and tagent direction. [4] Figure 2 shows the measuring point position example for four segment stator. For supervising the flexible of stator silicon steel piece, stator measuring points in vertical direction had better place on rack of thread bolt.

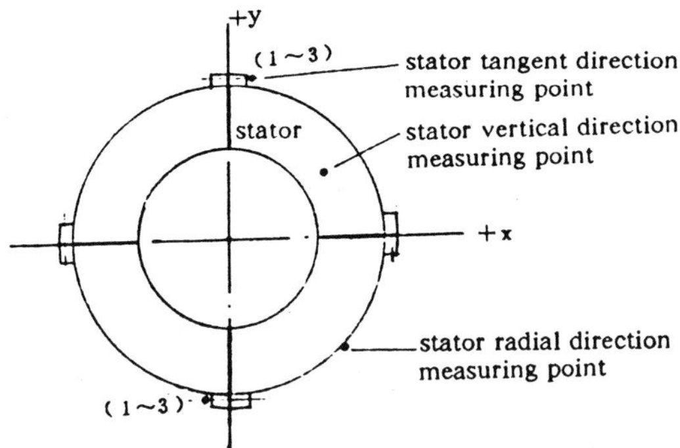

Figure 2. Measuring point arrangement of generator stator core vibration

Measuring ponits in tagent direction should be consided according upper ,middle and lower of stator combination crack. The supervision measuring points of genenator air gap should consider to be placed at stator and rotor respectively, That is to set two measuring points on top of rotor and stator respectively along90° direction or set 2~12 points at generator stator whose guantity and direction are based on unit stator and rotor' s dimention and structure. And also arrange a phase signal measuring point on main shaft of rotor for synchronize locating(shown as air gap measuring system of Vibro-meter, Switzerland). For example, the generator air gap superrision applied in Ge Heyan Water Power Station adopts 4 measuring points which locate at generator stator along ±x, ±y direction, shown as Fig 3.

3.2.4 *Equipment Structure Characteristics*

The structure change of unit operating equipment affects very much to the unit sofe and economic operation. If there are any abnormal or failure it will affects unit normal operation and decreases unit service life. So the measuring point should be arranged correctly accorting to the structure characteristic of operating equipment.

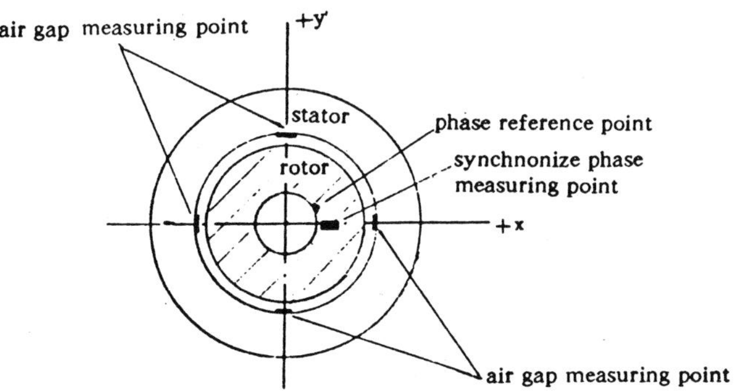

Figure 3. Generator air gap supervision measuring point arrangement

For mixed flow turbine wheel stress supervision, the measuring points may be arranged only on two positions at whecl upper crown outlet and lower ring inlet where the stress changes with the value of turbine output (as shown in Fig. 4b). For propeller water turbine, the measuring points should be put at the blade root and middle, outer section layer a certain distance from inlet edge (see Fig. 4a)[5]. For pelton turbine, the wheel stress measuring points should be put at the blade handle of

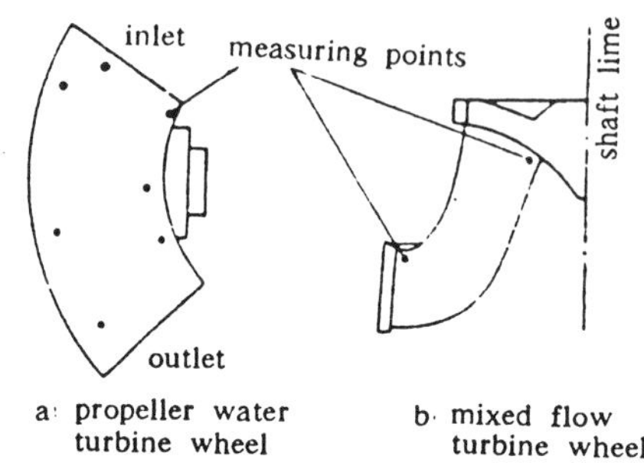

Figure 4. wheel stress supervision measuring point arrangement

bucket root. For supervision diagnosis of unit main shaft mechanics features (include main shaft thrust force and twisting torque), the measuring point position should be set at maximum stress standing place where differs according to unit shaft structure and dimension. For supervision diagnosis of thrust bearing standed force, two measuring points should be arranged along 90° direction on supporting part of bearing supporting ring respectively, shown as Fig 5.

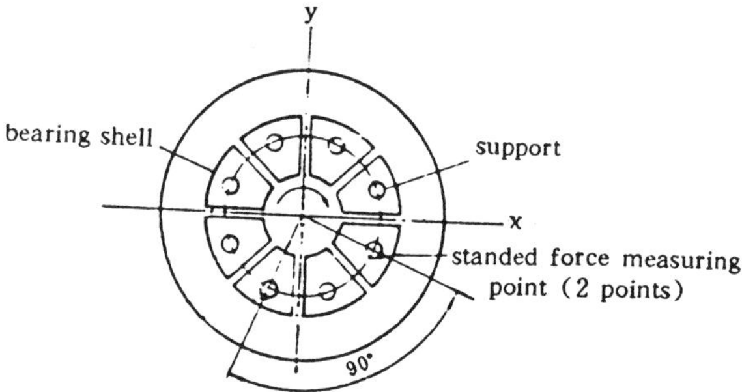

Figure 5. Thrust bearing standed force supervision measuring points arrangement

3. 2. 5 *Generator Electrical Features*

The electrical interference that will affect unit safe operation is also the important supervision diagnostic parameter. The measuring point selection and arrangement should analyze according to generator electrical feature and structure. For example, because of the phase to phase unbalance caused by generator negative current or three-phase load current unbalance, it is necessary to consider measuring point of phase voltage, phase current, negative current , etc . Arrange not only air gap measuring points but also exciting voltage and current points to supervise electromagnetic vibration caused by magnetic ploe magnetic boots uneven and magnetic pulling force uneven.

For temperature supervision measurement of generator stator core and winding, the measuring point arrangement should be determined according to stator core structure, ventilation system and cooling type optimally. wire band insulation supervision measurement diagnosis are mainly depend on partial discharge measuring point and wire band end point vibration now . Better measuring point is being investigation.

3. 2. 6 *Others.*

The other measuring points refer to unit operation supervision are unit head water level and down water level, turbine flow rate, scroll case pressure difference, unit liguit level (include bearing oil level and water level of water pressure and adjusting parameters[5]. They are almost the same as general testing measuring points and computer supervision points.

3. 3 MEASURING POINT ARRANGEMENT EXAMPLE AND QUESTIONS NEEDED TO BE CO NSIDERED

The contents of water turbine unit equipment supervision and diagnosis are very extensive. Among them, unit stability supervision is the important one . It is one of three tangets of unit operation performance (unit energy target, stability target and turbine cavitation target). Generally there are three aspects to expree unit stability: unit operating throw, unit operating vibration and turbine water pass part pressure andpressure pulsation (or unit hydrolic power characteristic). So, the arrangement of unit stability supervision measuring point should consider not only the difference of turbine type and structure, but also the measuring point selection of unit stability basic parameter (include measuring point quautity, position and direction). Figure 6 shows the example of two type turbine measuring point arrangement method and stability measuring point arrangement condition.

The following questions should be consided in measuring points application:

a. Convenience to unit safe operation, equipment safety and maintenance;

b. Reasonably select and use primary sensor element, In accordance with equipment supervision measuring point selection and arrangement, consider both its feasibility and its necessity;

c. prevent measuring accuracy decease due to installation technology or enviroment condition .

d. use measuring points that have applicated success fully find new points continuously and optimize them continuously.

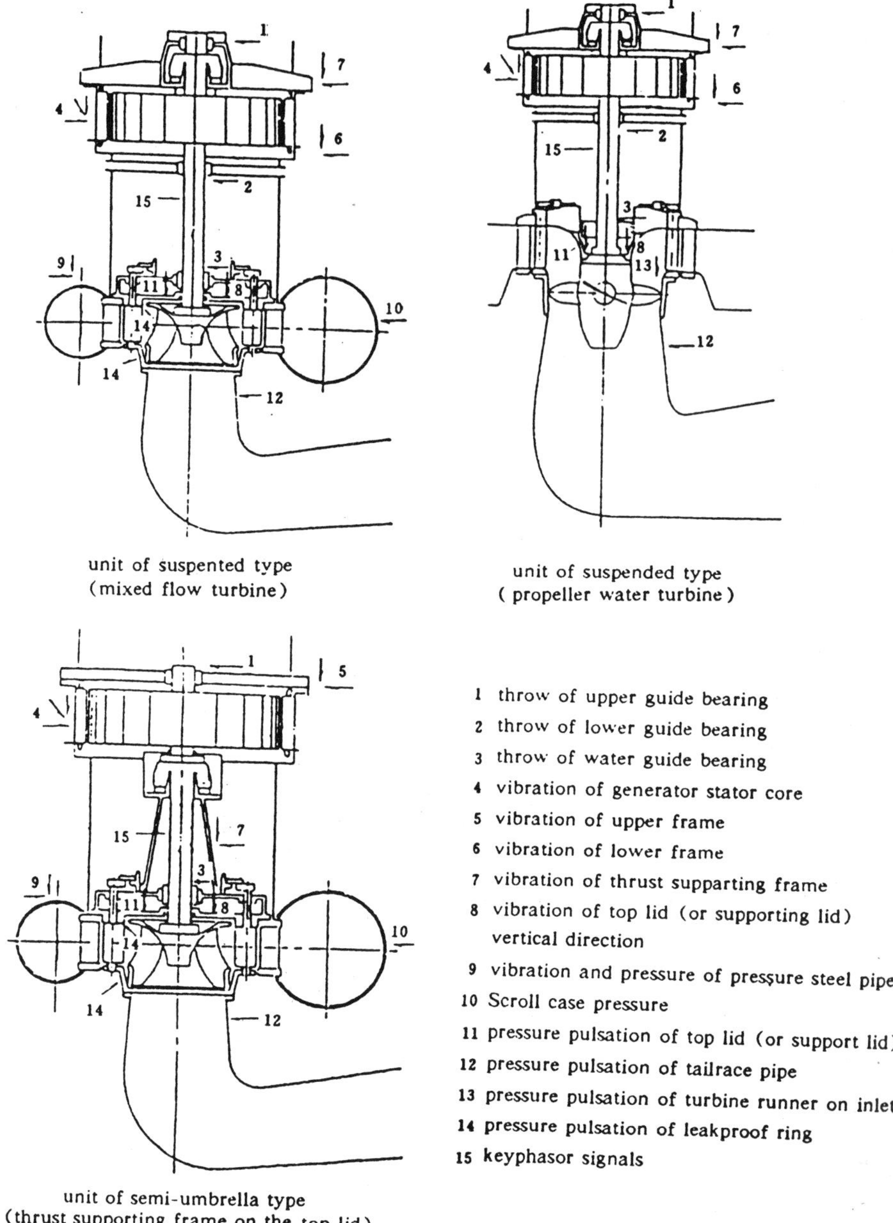

Figure 6. Stability measuring point arrangement example of different type water turbines

4. Conclusion

a. In function system realization of unit operating eguipment diagnostic technology, we should' notice the function characteristics and reguirments of FDI system and solve 5 key problems first in application.

b. The optimizing selection and reasonable arrangement are the most important question of unit operating equipment diagnostic technology data collection. It affects directly the reliability and accuracy of equipment status supervision measurement . while selecte and arrange measuring points , there should consider equipment operating performance and also make it conform to equipment structure characteristic and operating regularity.

c. The application of measuring points is related with the detail contents of unit operating equipment supervision diagnosis. Different supervision contents correspond different supervision parameters and different measurine point arrangement. A measuring point arrangement is applicable for both single- function diagnosis and multi-function diagnosis. In practical condition, there should pay attention to not only general questions of measuring point arrangement but also the failure speciality of diagnosed unit.

d. Because the testing research and practice is not enough, some measuring points need further discussion, For example, the supervision measurement of unit none rotating parts, namely, big parts rigidity and the stress of some parts, their point selection and arrangement have certain difficulty and need more practical application , test and comparision with desigh to optimize the most delegable typic measuring point. As to turbine cavitation supervision diagnosis , because the principle determination and the unstable process of cavitation affect the accuracy of the signals, the measuring points can not be arranged exactly. Some countries arrange measuring points and study the effects and endangers to unit operation caused by cavitation according to cavitation caused vibration change or tail pipe pressure pulsation. Some departments of china through finding the relation of cavitation and supersonic stress have produced cavitation detector , Its measuring points are placed at turbine tail pipe, about (0. 4～0. 5)D_1 under the tail pipe inlet section in down stream side[6], In general , the measuring point selection and measurement of cavitation need further study and practice.

5. Reference

[1][2] X. T. Liu. (1994) The study of Dignostic Technology Applying on Operating unit Equipment in Water Power Station, Proceedings of XV I IAHR Symposium, Beijing Vol. 3.

[3]Q. Z. Li, Z. M. Li, L. Z. Fu. (1993) The Measurement Method Suggestion of Circular Belt Pressure Pulsation Amplitude, National Hydraulic Equipment Symposium, Changdao

[4]X. T. Liu, (1988) # 1, # 2 Units Stator Vibration Analysis and Treatment in Huang Longtan Water Power Station, *Water Power Station Meehanical-electrical Technology* 1, 26-31

[5]X. T. Liu, W. F. Li, (1993) *Water Power Units Site Measurement Handbook*, Hydraulic and Electric Power Press, Beijing

[6]C. T. You, (1995) Water Turbine Cavitation Supersonic Detection Test and Application , *Water Power Station Mechanicl-electric Technology* 1 , 21-24